Reference Values for *Frequently* Assayed Clinical Chemistry Analytes* (continued)

Analyte	Specimen†	Conventional Units	Reference Interval‡	
			SI Units	
Gamma immunoglobulins	S	IgG 800–1200 mg/dL		
		IgA 70–312 mg/dL		
		IgM 50–280 mg/dL		
		IgD 0.5–2.8 mg/dL		
		IgE 0.01–0.06 mg/dL	IgE 0.1–0.6 mg/L	
Glucose (fasting)	P	Fasting 70–110 mg/dL	3.9–6.0 mmol/L	223
	CSF	40–70 mg/dL	2.2–3.9 mmol/L	
High-density lipoprotein (HDL) cholesterol	S	Male 29–60 mg/dL	0.75–1.6 mmol/L	325
		Female 38–75 mg/dL	1.0–1.94 mmol/L	
Iron	S	Male 65–170 µg/dL		284
		Female 50–170 µg/dL		
Lactate dehydrogenase (LD)	S	(L → P) 100–225 U/L		220
		(P → L) 80–280 U/L		
Lactate dehydrogenase isoenzymes (as percentage of total)	S	LD-1 14–26%		219
		LD-2 29–39%		
		LD-3 20–26%		
		LD-4 8–16%		
		LD-5 6–16%		
Lipase	S	0–1.0 U/mL		228
Magnesium (Mg^{2+})	S	1.2–2.1 mEq/L	0.63–1.0 mmol/L	267
	U	6.0–10.0 mEq/24 hr	3.00–5.00 mmol/24 hr	
Osmolality	S	275–295 mOsmol/kg		258
	U (24 hr)	300–900 mOsmol/kg		
Urine-serum ratio		1.0–3.0		
Phosphate	S	2.7–4.5 mg/dL	0.87–1.45 mmol/L	272
Potassium (K^+)	S	3.4–5.0 mEq/L	3.4–5.0 mmol/L	264
		Neonate 3.7–5.9 mEq/L	3.7–5.9 mmol/L	
	U (24 hr)	25–125 mEq/day	25–125 mmol/day	
Protein (total)	S	6.5–8.3 g/dL	65–83 g/L	185
	CSF	0.5% of plasma		
Sodium (Na^+)	S	135–145 mEq/L	135–145 mmol/L	261
	U (24 hr)	40–220 mEq/L	40–220 mmol/L	
	CSF	138–150 mEq/L	138–150 mmol/L	
Thyroid-stimulating hormone (TSH)	S	0.5–5.0 µU/mL		409
Thyroxine (T_4)	S	4.5–13 µg/dL	58–167 nmol/L	434
Triglycerides	S	67–157 mg/dL	0.11–2.15 mmol/L	323
Uric acid	S	Male 3.5–7.2 mg/dL	208–428 µmol/L	350
		Female 2.6–6.0 mg/dL	155–357 µmol/L	

* The list represents those analytes most frequently assayed in the clinical chemistry laboratory. For drug therapeutic ranges, refer to Appendix T, Summary Table of Pharmacokinetic Parameters.

† S = serum; P = plasma; U = urine; CSF = cerebrospinal fluid; WB = whole blood.

‡ Values may vary according to method and population; values for enzymes are at 37° C unless otherwise noted.

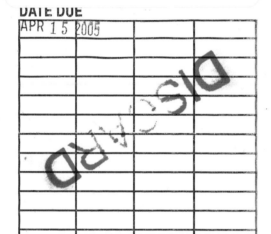

Clinical Chemistry

Clinical Chemistry

PRINCIPLES, PROCEDURES, CORRELATIONS

Fourth Edition

Michael L. Bishop, MS, MT(ASCP), CLS(NCA)

Training, Application, and Support Specialist
Sales Training Department
Organon Teknika Corporation
Durham, North Carolina

Janet L. Duben-Engelkirk, EdD, MT(ASCP), CLS(NCA)

Director, Allied Health Education and
Program in Clinical Laboratory Science
Scott & White Memorial Hospital,
Scott, Sherwood, and Brindley Foundation
Temple, Texas

Edward P. Fody, MD

Chief, Department of Pathology
Erlanger Medical Center
Chattanooga, Tennessee

38 Contributors

LIPPINCOTT WILLIAMS & WILKINS
A **Wolters Kluwer** Company

Philadelphia · Baltimore · New York · London
Buenos Aires · Hong Kong · Sydney · Tokyo

Acquisitions Editor: Lawrence McGrew
Editorial Assistant: Frank J. Musick
Senior Project Editor: Erika Kors
Senior Production Manager: Helen Ewan
Production Coordinator: Patricia McCloskey
Design Coordinator: Doug Smock
Cover Designer: Larry Didona
Indexer: Gaye F. Tarallo

4th Edition

9 8 7 6 5 4 3 2 1

Library of Congress Cataloging-in-Publication Data

Clinical chemistry : principles, procedures, correlations / [edited
  by] Michael L. Bishop, Janet L. Duben-Engelkirk, Edward P. Fody. — 4th
  ed.
      p.   cm.
   Includes bibliographical references and index.
   ISBN 0-7817-1776-0 (alk. paper)
   1. Clinical chemistry.   I. Bishop, Michael L.   II. Duben
-Engelkirk, Janet L.   III. Fody, Edward P.
   [DNLM: 1. Chemistry, Clinical.   QY 90 C6413 2000]
RB40.C576 2000
616.07'56—dc21
DNLM/DLC
for Library of Congress   99-36820
                                                   CIP

Care has been taken to confirm the accuracy of the information presented and to describe generally accepted practices. However, the authors, editors, and publisher are not responsible for errors or omissions or for any consequences from application of the information in this book and make no warranty, express or implied, with respect to the contents of the publication.

The authors, editors, and publisher have exerted every effort to ensure that drug selection and dosage set forth in this text are in accordance with current recommendations and practice at the time of publication. However, in view of ongoing research, changes in government regulations, and the constant flow of information relating to drug therapy and drug reactions, the reader is urged to check the package insert for each drug for any change in indications and dosage and for added warnings and precautions. This is particularly important when the recommended agent is a new or infrequently employed drug.

Some drugs and medical devices presented in this publication have Food and Drug Administration (FDA) clearance for limited use in restricted research settings. It is the responsibility of the health care provider to ascertain the FDA status of each drug or device planned for use in their clinical practice.

Contributors

Shauna C. Anderson, PhD
Microbiology Department
Brigham Young University
Provo, Utah
Chapter 18 Endocrinology

Mary Ruth Beckham, MEd, MT(ASCP)
Education Coordinator
Program in Clinical Laboratory Science
Scott & White Memorial Hospital
Scott, Sherwood, and Brindley Foundation
Temple, Texas
Chapter 2 Laboratory Safety and Regulations

Michael L. Bishop, MS, MT(ASCP), CLS(NCA)
Training, Application, and Support Specialist
Sales Training Department
Organon Teknika Corporation
Durham, North Carolina
Answers to Review Questions, Exercises, Practice Problems, and Case Studies
Appendices

Steven L. Bradley, MT(ASCP)
Regulations Compliance Manager
Department of Pathology
Scott & White Memorial Hospital
Temple, Texas
Chapter 7 Point-of-Care Testing

Anthony W. Butch, PhD, DABCC
Director, Clinical Chemistry
Pathology and Laboratory Medicine
UCLA Medical Center
Los Angeles, California
Chapter 27 Tumor Markers

Eileen Carreiro-Lewandowski, MS, CLS
Department of Medical Laboratory Science
University of Massachusetts, Dartmouth
North Dartmouth, Massachusetts
Chapter 1 Basic Principles and Practices of Clinical Chemistry

George S. Cembrowski, MD, PhD
Associate Professor
Department of Laboratory Medicine and Pathology
University of Alberta
Director, Medical Biochemistry
Regional Laboratory Services
Capital Health Authority
Edmonton, Alberta
Chapter 3 Quality Control and Statistics

Janet L. Duben-Engelkirk, EdD, MT(ASCP), CLS(NCA)
Director, Allied Health Education and
Program in Clinical Laboratory Science
Scott & White Memorial Hospital
Scott, Sherwood, and Brindley Foundation
Temple, Texas
Chapter 29 Clinical Chemistry and the Geriatric Patient
Glossary and Appendices

Sharon S. Ehrmeyer, PhD, MT(ASCP)
Pathology and Laboratory Medicine
University of Wisconsin
Madison, Wisconsin
Chapter 16 Blood Gases, pH, and Buffer Systems

Kevin D. Fallon, PhD
Director of Scientific Affairs
Instrumentation Laboratory
Lexington, Massachusetts
Chapter 16 Blood Gases, pH, and Buffer Systems

Edward P. Fody, MD
Chief, Department of Pathology
Erlanger Medical Center
Chattanooga, Tennessee
Chapter 17 Liver Function
Chapter 22 Pancreatic Function
Chapter 23 Gastrointestinal Function

Vicki S. Freeman, PhD, MT(ASCP)SC
Chair and Associate Professor
Department of Clinical Laboratory Science
School of Allied Health Science
The University of Texas Medical Branch at Galveston
Galveston, Texas
Chapter 10 Carbohydrates

H. Jesse Guiles, EdD
Clinical Laboratory Sciences
School of Health Related Professions
University of Medicine and Dentistry of New Jersey
Newark, New Jersey
Chapter 19 Thyroid Function

Tammy L. Hofer, BSC, MLT
Point-of-Care Technologist
Department of Laboratory Medicine
University of Alberta Hospital
Edmonton, Alberta
Chapter 3 Quality Control and Statistics

Susan A. Hurgunow, MS, MT(ASCP)SC
Clinical Sciences Education
MCP Hahnemann University
Philadelphia, Pennsylvania
Chapter 15 Trace Elements

Lynn R. Ingram, MS
Clinical Laboratory Sciences
University of Tennessee, Memphis
Memphis, Tennessee
Chapter 20 Cardiac Function

Carolyn Houghton Insall, MS, MT(ASCP), DLM
Director, Blood Bank
Scott & White Memorial Hospital
Scott, Sherwood, and Brindley Foundation
Temple, Texas
Chapter 2 Laboratory Safety and Regulations

Dennis W. Jay, PhD, DABCC
Technical Director
Proficiency Testing Service
American Association of Bioanalysts
Brownsville, Texas
Chapter 12 Nonprotein Nitrogen

Robin Gaynor Krefetz, MEd, MT(ASCP), CLS(NCA)
Associate Professor and CLT Department Chair
Community College of Philadelphia
Philadelphia, Pennsylvania
Chapter 9 Enzymes

Louann W. Lawrence, DrPH, CLSpH(NCA), MT(ASCP)SH
Associate Professor and Department Head
Clinical Laboratory Sciences
School of Allied Health Professions
Louisiana State University Medical Center
New Orleans, Louisiana
Chapter 13 Porphyrins and Hemoglobin

Barbara Lindsey
Clinical Laboratory Sciences
Virginia Commonwealth University
Richmond, Virginia
Chapter 8 Amino Acids and Proteins

Nicole A. Massoll, MD
Pathology and Laboratory Medicine
University of Arkansas for Medical Sciences
Little Rock, Arkansas
Chapter 27 Tumor Markers

Judith R. McNamara, MT
Lipid Research Laboratories,
Jean Mayer USDA Human Nutrition Research
Center on Aging at Tufts University, and
New England Medical Center
Boston, Massachusetts
Chapter 11 Lipids

Sharon M. Miller, PhC, CLS(NCA), MT(ASCP)
Professor and Associate Dean
College of Health and Human Sciences
Northern Illinois University
DeKalb, Illinois
Chapter 29 Clinical Chemistry and the Geriatric Patient

Alex A. Pappas, MD
Professor, Department of Pathology and Health Related Professions
Pathology and Laboratory Medicine
University of Arkansas for Medical Sciences
Little Rock, Arkansas
Chapter 27 Tumor Markers

Joan E. Polancic, MSEd, CLS(NCA), MT(ASCP)
Associate Professor
Clinical Laboratory Science Program
University of Illinois at Springfield
Springfield, Illinois
Chapter 14 Electrolytes

Larry Scheoff, MS, MT(ASCP)
Department of Pathology
University of Utah
Salt Lake City, Utah
Chapter 6 Principles of Clinical Chemistry Automation

Frank A. Sedor, PhD
Clinical Laboratories
Duke University Medical Center
Durham, North Carolina
Chapter 24 Body Fluid Analysis

Catherine Sheehan, MS, CLS(NCA), SI(ASCP)
Department of Medical Laboratory Science
University of Massachusetts, Dartmouth
North Dartmouth, Massachusetts
Chapter 5 Immunoassays and Nucleic Acid Probe Techniques

John E. Sherwin, PhD
Chief, Genetic Disease Laboratory
State of California
Berkeley, California
Chapter 30 Pediatric Clinical Chemistry

Carol J. Skarzynski, BA, SC(ASCP)
Clinical Instructor
Department of Pathology and Laboratory Medicine
Hartford Hospital School of Allied Health
Hartford, Connecticut
Chapter 21 Renal Function

Juan Sobenes, MD
Chair, Department of Pathology
Veterans Administration Hospital
Fresno, California
Chapter 30 Pediatric Clinical Chemistry

A. Michael Spiekerman, PhD
Department of Clinical Pathology
Scott & White Clinic and Memorial Hospital,
Texas A & M University, School of Medicine
Temple, Texas
Chapter 28 Vitamins and Nutritional Assessment

Anne M. Sullivan, MS
Department of Biomedical Technologies
University of Vermont
Burlington, Vermont
Chapter 3 Quality Control and Statistics

David P. Thorne, PhD, MT(ASCP)
Medical Technology Program
Michigan State University
East Lansing, Michigan
Chapter 25 Therapeutic Drug Monitoring
Chapter 26 Toxicology

G. Russell Warnick, MS, MBA
Pacific Biometrics Research Foundation
Issaquah, Washington
Chapter 11 Lipids

Alan H. B. Wu, PhD
Department of Pathology and Laboratory Medicine
Hartford Hospital
Hartford, Connecticut
Chapter 4 Analytical Techniques and Instrumentation
Chapter 21 Renal Function

Lily L. Wu, PhD
Research Associate Professor
Department of Pathology
Adjunct Associate Professor
Department of Internal Medicine
University of Utah
Salt Lake City, Utah
Chapter 11 Lipids

Foreword

It seems only a short time ago that I wrote the foreword to the third edition of this text. In looking back, however, I find that my word processor has changed twice since that time and that our laboratory has changed almost all of its analytical systems, installed a new specimen transport system, and is now changing to a new information system. Much has changed since the third edition and it is to the credit of the editors and contributors of this book that they are committed to keeping the educational contents current with the changes occurring in laboratory medicine.

Many new tests, methods, and measurement systems continue to be introduced in a health care market that is becoming more and more competitive. Laboratory tests seem to be getting easier to perform, but the testing processes often require more complex technology that is more difficult to understand. New tests are also evolving because of new research findings and new measurement processes that again often depend on complex technology. The one constant seems to be the increasing knowledge that is required to understand clinical laboratory science today.

This rapid change and evolution of laboratory testing makes it increasingly difficult to capture the knowledge that defines the current state of practice for clinical laboratory scientists. The task becomes more daunting with each passing year and also becomes more critical for laboratory professionals. Fortunately, the editors and contributors of this text have been willing to tackle this seemingly impossible task in order to support and advance the profession. I want to both thank and congratulate them!

The fourth edition of *Clinical Chemistry: Principles, Procedures, Correlations* continues its mission of addressing the formal educational needs of our students in clinical laboratory science, as well as the ongoing needs of professionals in the field. It facilitates the educational process by identifying the learning objectives, focusing on key concepts and ideas, and applying the theory through case studies. It covers the basics of laboratory testing, as well as the many special areas of clinical chemistry testing. And it is still possible to carry this text with you to class, to the laboratory, to the office, or home to study!

Having personally worked with some of the editors and contributors, I know they have high standards both in the laboratory and in the classroom. Their interests and backgrounds provide an excellent balance between the academic and the practical, ensuring that both we and our students are exposed to a well-developed base of knowledge that has been carefully refined by experience. They provide the "shifting and winnowing" that is necessary to separate the wheat from the chaff and to provide the real sustenance for our students and ourselves.

For the many students for whom this book is intended, let me offer some advice from my close friend and mentor, Hagar the Horrible. It seems his young Viking son was embarking on a voyage to the real world of work. Hagar's son asked him "How do I get to the top?" Hagar's advice was "You have to start at the bottom and work your way up!" After pondering this for a moment, his son asked "How do I get to the bottom?" Hagar replied, "You have to know somebody!" As students in clinical laboratory science, you should study and heed the advice offered in this book, but you should also search for mentors. The authors of this book, as well as the instructors in your courses and your teachers in the laboratory, are keys to getting started in your careers. You need to seek them out to profit from their learning and experiences. They are the professionals who know the state of the art, possess current knowledge of the field, and are dedicated to helping you.

James O. Westgard, PhD
Professor, Pathology and Laboratory Medicine
University of Wisconsin
Madison, Wisconsin

Preface

Clinical chemistry continues to be one of the most rapidly advancing areas of laboratory medicine. Since the publication of the first edition of this textbook in 1985, many changes have taken place. New technologies and analytical techniques have been introduced with a dramatic impact on the practice of clinical chemistry. In addition, the health care system is changing, there is an increased emphasis on improving the quality of patient care, individual patient outcomes, overall cost effectiveness, and total quality management. Point-of-care testing (POCT) is also at the forefront of health care practice and has brought forth both challenges and opportunities to clinical laboratorians. Now, more than ever, clinical laboratorians need to be concerned with disease correlations, interpretations, problem solving, quality assurance, and cost effectiveness; they need to know not only the *how* of tests, but also the *what*, *why* and *when*. The editors of *Clinical Chemistry: Principles, Procedures, Correlations* have designed the fourth edition to be an even more valuable resource to both students and practitioners.

Like the previous three editions, the fourth edition of *Clinical Chemistry: Principles, Procedures, Correlations* is a comprehensive, up-to-date, and easy to understand textbook for students at all levels. It is also intended to serve as a practically organized resource for both instructors and practitioners. The editors have tried to maintain the book's readability and further improve its content. Because clinical laboratorians utilize their interpretative and analytical skills in the daily practice of clinical chemistry, an effort has been made to maintain an appropriate balance between analytical principles, techniques, and the correlation of results with disease states.

In this fourth edition, the editors have made several significant changes in response to requests by our readers, students, instructors, and practitioners. Chapter outlines, objectives, key terms, and summaries have been updated and expanded. Every chapter now includes current, frequently encountered case studies, and practice questions or exercises. Answers to the case studies and assessment exercises and questions are included in the appendix of the textbook. The glossary of key terms has been expanded and an appendix on test correlations has been added. To provide a thorough, up-to-date study of clinical chemistry, chapters on Point-of-Care Testing and Cardiac Function were added. The Automated Techniques chapter has been expanded to include total laboratory automation. The Vitamins and Nutritional Assessment chapters have been combined for concise and improved coverage. The Immunochemistry chapter has been expanded to cover the most recent advances in immunoassays, flow cytometry, and molecular diagnostics. The basic principles of the analytical procedures discussed in the chapters reflect the most recent or commonly performed techniques in the clinical chemistry laboratory. Detailed procedures have been omitted because of the variation in equipment and commercial kits utilized in today's clinical laboratories. Instrument manuals and kit package inserts are the most reliable reference for detailed instructions on current analytical procedures. All chapter material has been updated, improved, and rearranged for better continuity and readability. As a new offering, an Instructor's Manual with teaching resources, teaching tips, additional references, and teaching aids is available from the publisher to assist in the use of this textbook.

Acknowledgments

A project as large as this requires the assistance and support of many individuals. The Editors wish to express their appreciation to the contributors of this fourth edition of *Clinical Chemistry: Principles, Procedures, Correlations*—the dedicated laboratory professionals and educators whom the editors have had the privilege of knowing and exchanging ideas with over the years. These individuals were selected because of their expertise in particular areas and their commitment to the education of clinical laboratorians. Many have spent their professional careers in the clinical laboratory, at the bench, teaching students, or consulting with clinicians. In these front-line positions, they have developed their perspective of what is important for the next generation of clinical laboratorians.

We extend appreciation to our students, colleagues, teachers, and mentors in the profession who have helped to shape our ideas about clinical chemistry practice and education. Also, we want to thank the many companies and professional organizations that have provided product information and photographs or have granted permission to reproduce diagrams and tables from their publications. Many National Committee for Clinical Laboratory Standards (NCCLS) documents have been important sources of information as well. These documents are referenced directly in the appropriate chapters.

The editors would like to acknowledge the contribution and effort of all individuals to previous editions. Their efforts provided the framework for many of the current chapters. Finally, we gratefully acknowledge the cooperation and assistance of the staff at Lippincott Williams & Wilkins Publishers, particularly Larry McGrew, Editor, Allied Health, Holly Chapman and Frank Musick, Editorial Assistants, Erika Kors, Project Editor, Pat McCloskey, Production Coordinator, and Doug Smock, Designer.

The editors are continually striving to improve future editions of this book. We again request and welcome our readers' comments, criticisms, and ideas for improvement.

Michael L. Bishop
Janet L. Duben-Engelkirk
Edward P. Fody

Contents

List of Case Studies

BASIC PRINCIPLES AND PRACTICE OF CLINICAL CHEMISTRY

CHAPTER 1

Basic Principles and Practice of Clinical Chemistry

Eileen Carreiro-Lewandowski

Units of Measure
Temperature
Reagent Preparation
Chemicals
Reference Materials
Solutions
Water Specifications
Clinical Laboratory Supplies
Glass and Plasticware
Desiccators and Desiccants

Balances
Basic Separation Techniques
Centrifugation
Filtration
Dialysis
Laboratory Mathematics and
   Calculations
Significant Figures
Logarithms
Concentration

Dilutions
Water of Hydration
Specimen Considerations
Types of Samples
Sample Processing
Sample Variables
Chain of Custody
Summary
Exercises
References

Objectives

Upon completion of this chapter, the clinical laboratorian should be able to:

- *Convert results from one unit format to another using the SI system.*
- *List and describe the types of thermometers used in the clinical laboratory.*
- *Identify the varying chemical grades used in reagent preparation and indicate their correct use.*
- *Define the following terms: primary standard, SRM, secondary standard.*
- *Describe each of the following terms that are associated with solutions and, when appropriate, provide the respective units: percent, molarity, normality, molality, saturation,*

colligative properties, redox potential, conductivity, specific gravity.

- *Define a buffer and give the formula for the pH and pK calculations.*
- *When given either the pK and pH or the pK and concentration of the weak acid and its conjugate base, use the Henderson-Hasselbalch equation to determine the missing variable.*
- *Describe the specifications for each type of laboratory water.*
- *When given an actual pipet or its description, classify the type of pipet.*
- *Describe two ways to calibrate a pipetting device.*
- *Define a desiccant and discuss its use in the clinical laboratory.*

(continued)

- *Correctly perform the laboratory mathematical calculations provided in this chapter.*
- *Identify and describe the types of samples used in clinical chemistry.*
- *Outline the general steps for processing blood samples.*
- *Identify the preanalytical, precollection, collection, and postcollection variables that can adversely affect laboratory results.*
- *List the proper drawing order for collection tubes.*
- *When given a collection stopper color, identify the additive or preservative if present.*

KEY TERMS

Analyte	Evacuated tube	Pipet
Anhydrous	Filtrate	pK
Arterial blood	Filtration	Primary standard
Beer's Law	Hemolysis	Redox potential
Buffer	Henderson-	Reduced
Buret	Hasselbalch	Secondary standard
Centrifugation	Hydrate	Serial dilution
Cerebral spinal fluid	Hygroscopic	Serum
(CSF)	Icterus	Significant figures
Character	Ionic strength	Solute
Colligative property	Mantissa	Solution
Conductivity	Molality	Solvent
Deionized water	Molarity	Specific gravity
Deliquescent	NCCLS	Standard
substances	Normality	Standard reference
Delta absorbance	One point	materials (SRMs)
Density	calibration	Systeme
Desiccant	Osmotic pressure	International (SI)
Desiccator	Oxidized	d'Unités
Dialysis	Oxidizing agent	Thermistor
Dilution	Paracentesis	Valence
Distilled water	Percent solution	Venipuncture
EDTA	pH	Whole blood
Equivalent weight	phlebotomy	

The primary goal of a clinical chemistry laboratory is to correctly perform analytical procedures that yield accurate and precise information to aid in patient diagnosis. To achieve reliable results, the clinical laboratorian must have the ability to use basic supplies and equipment correctly and have an understanding of fundamental concepts critical to any analytical procedure. This chapter includes the topics of units of measure, properties of a solution, classification of chemicals, reagents, glassware and plasticware, laboratory mathematics, and specimen collection and processing.

UNITS OF MEASURE[1]

Any meaningful result comprises two parts. The first component is a number and the second component is the units label. The number describes the numeric value, whereas the unit defines the physical quantity or dimension, such as mass, length, time, or volume. Although there are several systems of units that various scientific divisions have traditionally used, the *Systeme International (SI) d'Unités,* adopted internationally in 1960, is the only system used in many countries. The units of the system are referred to as *SI units.* The three types of SI units are basic, supplemental, and derived. The basic unit for length is the meter; for mass, the kilogram; and for the quantity of a substance, the mole. Only one basic unit is given for each physical quantity so that length, for example, would be expressed in terms of meters rather than any other non–SI unit such as an inch or foot. This convention avoids the use of several different terms to describe the same physical quantity. Prefixes can be added to indicate decimal fractions or multiples of the basic SI units. These prefixes are listed in Table 1-1. For example, 0.5 liters also could be expressed using the prefix *milli* (which is equivalent to 1/1000, 0.001 or 10^{-3}) as 500 milliliters.

In addition to the basic unit, the SI system includes derived units, which are related mathematically to the basic or supplemental unit. Supplemental units are units that have not been classified as either basic or derived. Some long-standing units, such as hour, minute, day, plane angles expressed as degrees, gram, and liter, are not listed as basic SI units but are frequently used "traditional" units.

Reporting of laboratory results is often expressed in terms of substance concentration (*eg,* moles) or the mass of a substance (*eg,* mg/dL, g/dL, g/L, mEq/L, and so on). These familiar and traditional units can cause confusion during interpretation. It has been recommended that analytes be reported using moles of solute per volume of solution (substance concentration) and that the liter be used as the reference volume. In tables listing laboratory reference values, traditional values as well as SI-recommended values are provided. Appendix E,

TABLE 1-1. Prefixes to be Used with SI Units

Factor	Prefix	Symbol
10^{-18}	atto	a
10^{-15}	femto	f
10^{-12}	pico	p
10^{-9}	nano	n
10^{-6}	micro	μ
10^{-3}	milli	m
10^{-2}	centi	c
10^{-1}	deci	d
10^{1}	deka	da
10^{2}	hecto	h
10^{3}	kilo	k
10^{4}	mega	M
10^{9}	giga	G
10^{12}	tera	T
10^{15}	peta	P
10^{18}	exa	E

Note: Prefixes are used to indicate a subunit or multiple of a basic SI unit.

Conversion of Traditional Units to SI Units for Common Clinical Chemistry Analytes, lists both units along with the conversion factor from traditional to SI units for common analytes.

TEMPERATURE[2]

The predominant practice is to make laboratory temperature readings using the Celsius (C) scale. However, Fahrenheit (F) and Kelvin (K) scales are also used. The SI designation for temperature is the Kelvin scale. Appendix D, *Basic Clinical Laboratory Conversions,* lists the various conversion formulas between each scale.

All analytical reactions occur at an optimal temperature. Some laboratory procedures, such as enzyme determinations, require precise temperature control, whereas others work well over a wide range of temperatures. Reactions that are very temperature dependent use some type of heating cell, heating block, or water/ice bath to provide the correct temperature environment. Temperatures of laboratory refrigerators often are critical and need periodic verification. Thermometers are either an integral part of an instrument or need to be placed in the device for temperature maintenance. The two major types of thermometers are the liquid (mercury)-in-glass and the electronic thermometer or thermistor probe.

In mercury-in-glass thermometers, the mercury column should be inspected to ensure that it is continuous and free from bubbles. If either condition is present, cooling the thermometer in ice may force the mercury toward the bulb. A circular "shake" of the thermometer also can help. Once the mercury column is intact, an ice point and desired bath/block temperature may be made and compared with a certified *Standard Reference Material (SRM)* thermometer. It is important that the thermometer be immersed to its appropriate depth and calibrated in that fashion.

Mercury-in-glass thermometers should be calibrated against a National Institute of Standards and Technology (NIST), U.S. Department of Commerce–certified thermometer. The NIST has an SRM thermometer with various calibration points (0°, 25°, 30°, and 37°C) for use with mercury-in-glass thermometers. Gallium, another SRM, has a known melting point and can be used for thermometer verification.

As automation advances and miniaturizes, the need for an accurate, fast-reading electronic thermometer (*thermistor*) has increased. The advantages of a thermistor over the more traditional mercury-in-glass thermometers are its size—particularly in use with flow-through cuvettes—and millisecond response time. A disadvantage can be the initial cost, although the use of a thermistor probe attached to an already owned volt–ohm meter (VOM) can be very cost efficient. Like the mercury-in-glass thermometers, the thermistor can be calibrated against an SRM thermometer or the gallium melting-point cell.[3,4] Once the thermistor is calibrated against the gallium cell, it can then be used as a reference for any type of thermometer.

REAGENT PREPARATION

In today's highly automated laboratory, there seems to be little need for reagent preparation by the laboratorian. Most instruments are sold by manufacturers who also make the reagents, usually in a readily available "kit" form (*ie,* all necessary reagents and respective storage containers are prepackaged as a unit). Because of heightened awareness of the hazards of certain chemicals and the existence of numerous regulatory agencies, clinical chemistry laboratories have readily eliminated their massive stocks of chemicals and have opted instead for the ease of using prepared reagents. Periodically, especially in hospital laboratories involved in research and development or specialized analyses, the laboratorian may still be faced with preparing various reagents. As a result of reagent deterioration, supply and demand, or the institution of cost-containment programs, the decision to prepare reagents in-house may be made. Therefore, a thorough knowledge of chemicals, standards, solutions, buffers, and water requirements is necessary.

Chemicals[5]

Chemicals exist in varying grades of purity: analytical reagent grade (AR); ultrapure, chemically pure (CP), United States Pharmacopeia (USP); National Formulary (NF); and technical or commercial grade. The American Chemical Society (ACS) has established specifications for analytical reagent-grade chemicals. Labels on these reagents either state the actual impurities for each chemical lot or list the maximum allowable impurities. The label should reveal the percentage of impurities present and have either the initials "AR" or "ACS" or the term "for laboratory use" clearly printed. Chemicals of this category are suitable for use in most analytical laboratory procedures. Ultrapure chemicals have been put through additional purification steps for use in specific procedures such as chromatography, atomic absorption, fluorometry, standardization, or other techniques that require extremely pure chemicals.

Because USP and NF grade chemicals are used to manufacture drugs, the limitations established for this group of chemicals are based only on the criterion of not being injurious to individuals. Chemicals in this group may be pure enough for use in most chemical procedures, but consider that their purity standards are not based on the needs of the laboratory and, therefore, they may or may not meet all assay requirements.

Reagent designations of CP or pure grade indicate that the impurity limitations are not stated and that preparation of these chemicals is not uniform. Melting point analysis is used quite often to ascertain the acceptable purity range. It is not recommended that clinical laboratories use these chemicals for reagent preparation unless further purification or a reagent blank is included. Technical- or commercial-grade reagents are used primarily in manufacturing and should never be used in the clinical laboratory.

Aside from the purity aspects of the chemicals, laws such as the Occupational Safety and Health Act (OSHA) require manufacturers to clearly indicate the lot number, plus the hazard and precautions needed for the safe use and storage of any chemical. A more detailed discussion on this topic may be found in Chapter 2, Laboratory Safety and Regulations.

Reference Materials[1,6,7,8]

Unlike other areas of chemistry, clinical chemistry is involved in the analysis of biochemical by-products found in *serum,* which makes purifying and providing their exact composition almost impossible. For this reason, traditionally defined standards do not necessarily exist for use in clinical chemistry.

Recall that a *primary standard* is a highly purified chemical that can be measured directly to produce a substance of *exact* known concentration. The ACS purity tolerances for primary standards are $100 \pm 0.02\%$. Because most biologic constituents are unavailable within these limitations, NIST–certified SRMs are used in lieu of ACS primary standards.

The NIST developed certified SRMs for use in clinical chemistry laboratories. These substances may not have the properties and purity equivalent of a primary standard, but each one has been characterized for certain chemical or physical properties and can be used in place of an ACS primary standard in clinical work. A human serum SRM (no. 909a) is available and is certified for calcium, chloride, cholesterol, creatinine, glucose, lithium, magnesium, potassium, sodium, urea, uric acid, and the trace metals cadmium, chromium, copper, iron, lead, and vanadium.[9]

A *secondary standard* is a substance of lower purity whose concentration is determined by comparison with a primary standard. The secondary standard depends not only on its composition, which cannot be determined directly, but also on the analytical reference method. Once again, because physiologic primary standards are generally unavailable, clinical chemists do not by definition have "true" secondary standards. Manufacturers of secondary standards will list the SRM or primary standard used for comparison. This information is often needed during laboratory accreditation processes.

Solutions

In clinical chemistry, substances found in biologic fluids (*ie,* serum, urine, spinal fluid, and so on) are measured. A substance that is dissolved in a liquid is called a *solute,* and in laboratory science, these biologic solutes are also known as *analytes.* The liquid in which the solute is dissolved—in this instance, a biologic fluid—is the *solvent.* Together they represent a *solution.* Any chemical or biologic solution has basic properties that describe it. These properties include concentration, saturation, the colligative properties, redox potential, conductivity, density, pH, and ionic strength.

Concentration

Concentration of an analyte in solution can be expressed in many ways. Routinely, concentration is expressed as percent solution, molarity, molality, or normality.

Percent solutions are equal to parts per hundred or the amount of solute per 100 total units of solution. Three expressions of percent solutions are weight per weight (w/w), volume per volume (v/v), and, most commonly, weight per volume (w/v). It is recommended that for v/v solutions, the units milliliters per liter (mL/L) be used instead of 70% (v/v).

Molarity is expressed as the number of moles per 1 L of solution. One mole of a substance equals its gram molecular weight (gmw). The SI representation for the traditional molar concentration is moles of solute per volume of solution, with the volume of the solution given in terms of liters. The SI expression for concentration should be represented as moles per liter (mol/L), millimoles per liter (mmol/L), micromoles per liter (μmol/L), and nanomoles per liter (nmol/L). The familiar concentration term of "molarity" has not been adopted by the SI as an expression of concentration.

Molality represents the amount of solute per 1 kg of solvent. Molality is sometimes confused with molarity, but it can be distinguished easily from molarity because molality is always expressed in terms of weight per weight or moles per kilogram and describes moles per 1000 g (1 kg) of solvent. The preferred expression for molality is moles per kilogram (mol/kg).

Normality is defined as the number of gram equivalent weights per 1 L of solution. An *equivalent weight* is equal to the molecular weight of a substance divided by its valence. The *valence* is the number of units that can combine with or replace 1 mole of hydrogen ions. Normality is no longer endorsed as an SI expression of concentration, and the expression milliequivalents per liter (mEq/L) for reporting sodium, potassium, and chloride concentration should be reported as millimoles per liter (mmol/L).

Saturation gives little specific information about the concentration of a solution. Temperature, as well as the presence of other ions, can influence the solubility constant for a given solution and thus affect the saturation. Terms routinely used in the clinical laboratory to describe the extent of saturation are dilute, concentrated, saturated, and supersaturated. A *dilute solution* is one in which there is relatively little solute. In contrast, a *concentrated solution* has a large quantity of solute in solution. A solution in which there is an excess of undissolved solute particles is called a *saturated solution.* As the name implies, a *supersaturated solution* has an even greater concentration of undissolved solute particles than does a saturated solution of the same substance. Because of the greater concentration of solute particles present, a supersaturated solution is thermodynamically unstable. Addition of a crystal of solute, or mechanical agitation, disturbs the supersaturated solution, resulting in crystallization of any excess material out of solution. An example of this is seen when measuring serum osmolality by freezing point depression.

Colligative Properties

The behavior of particles in solution demonstrates four repeatable properties based only on the relative number of each kind of molecule present. The properties of osmotic pressure, freezing point, boiling point, and vapor pressure are called *col-*

ligative properties. Vapor pressure is the pressure at which the liquid solvent is in equilibrium with the water vapor. *Freezing point* is the temperature at which the vapor pressures of the solid and liquid phases are the same. *Boiling point* is the temperature at which the vapor pressure of the solvent reaches one atmosphere.

Osmotic pressure is the pressure that allows solvent flow through a semipermeable membrane to establish equilibrium between compartments of different osmolality. The osmotic pressure of a dilute solution is proportional to the concentration of the molecules in solution. When a solute is dissolved in a solvent, these colligative properties change: the freezing point is lowered, the boiling point is raised, the vapor pressure is lowered, and the osmotic pressure is increased. In the clinical setting, freezing point and vapor pressure depression are measured as a function of osmolality.

Redox Potential

Redox potential, or *oxidation-reduction potential,* is a measure of the ability of a solution to accept or donate electrons. Substances that donate electrons are called *reducing agents* and those that accept electrons are considered *oxidizing agents.* The pneumonic—LEO (lose electrons = *oxidized*) "the lion says" GER (gain electrons = *reduced*)—may prove useful when trying to recall the relationship between reducing/oxidizing agents and redox potential.

Conductivity

Conductivity is a measure of how well electricity passes through a solution. A solution's conductivity quality depends principally on the number of respective charges of the ions present. *Resistivity,* the reciprocal of conductivity, is a measure of a substance's resistance to the passage of electrical current. The primary application of resistivity in the clinical laboratory is for assessing the purity of water.

Buffers

Buffers are weak acids or bases and their related salts, which, because of their dissociation characteristics, minimize any changes in hydrogen ion concentration. Hydrogen ion concentration is often expressed as pH. A lowercase "p" in front of certain letters or abbreviations means the "negative logarithm of" or "inverse log of" that substance. In keeping with this convention, the term *pH* represents the negative or inverse log of the hydrogen ion concentration. Mathematically, pH is expressed as

$$pH = \log(1/[H^+])$$

$$pH = -\log[H^+], \qquad \text{(Eq. 1–1)}$$

where $[H^+]$ is equal to the concentration of hydrogen ions in moles per liter.

The pH scale ranges from 0 to 14 and is a convenient way to express hydrogen ion concentration.

A buffer's capacity to minimize changes in pH is related to the dissociation characteristics of the weak acid or base in the presence of its respective salt. Unlike a strong acid or

base, which dissociates almost completely, the dissociation constant for a weak acid or base solution tends to be very small, meaning little dissociation occurs.

The ionization of acetic acid (CH_3COOH), a weak acid, can be illustrated as follows:

$$[HA] \leftrightarrow [A^-] + [H^+]$$

$$[CH_3COOH] \leftrightarrow [CH_3COO^-] + [H^+]$$

$$\text{(Eq. 1–2)}$$

where HA = weak acid, A^- = conjugate base, and H^+ = hydrogen ions.

Note that the dissociation constant, K_a, for a weak acid may be calculated using the following equation:

$$K_a = \frac{[A^-][H^+]}{[HA]} \qquad \text{(Eq. 1–3)}$$

Rearrangement of this equation reveals

$$[H^+] = K_a \times \frac{[HA]}{[A^+]} \qquad \text{(Eq. 1–4)}$$

Taking the log of each quantity and then multiplying by minus 1 (−1), the equation can be rewritten as

$$-\log[H^+] = -\log K_a \times -\log \frac{[HA]}{[A^+]} \qquad \text{(Eq. 1–5)}$$

By definition, lower case "p" means "negative log of"; therefore, $-\log [H^+]$ may be written as pH, and $-K_a$ may be written as pK_a. The equation now becomes

$$pH = pK_a - \log \frac{[HA]}{[A^+]} \qquad \text{(Eq. 1–6)}$$

Eliminating the minus sign in front of the log of the quantity $[HA] / [A^+]$ results in an equation known as the *Henderson-Hasselbalch* equation, which mathematically describes the dissociation characteristics of weak acids and bases and the effect on pH:

$$pH = pK_a + \log \frac{[A^+]}{[HA]} \qquad \text{(Eq. 1–7)}$$

When the ratio of $[A^+]$ to $[HA]$ is 1, the pH equals the pK and the buffer has its greatest buffering capacity. The dissociation constant K_a, and therefore the pK_a, remains the same for a given substance. Any changes in pH are then due only to the ratio of base/salt $[A^+]$ concentration to weak acid $[HA]$ concentration.

Another important aspect of buffers is their ionic strength, particularly in separation techniques. *Ionic strength* is the concentration or activity of ions in a solution or buffer. It is defined[6] as follows:

$$\mu = I = \frac{1}{2} \sum c_i z_i^2 \text{ or}$$

$$\frac{\sum\{(c_i) \times (z_i)^2\}}{2}, \qquad \text{(Eq. 1–8)}$$

where c_i is the concentration of the ion, z_i is the charge of the ion, and $\sum$ is the sum of the quantity $(c_i) \times (z_i)^2$ for each ion present. In mixtures of substances, the degree of dissociation must be considered. It is known that increasing ionic strength can promote compounds to dissociate into ions and increase the solubility of some salts, which can affect electrophoretic migration.

Water Specifications

The most frequently used reagent in the laboratory is water. Because tap water is unsuitable for laboratory applications, most procedures, including reagent and standard preparation, use water that has been substantially purified. Water solely purified by distillation results in *distilled water,* whereas water purified by ion exchange produces *deionized water.* Reverse osmosis, which pumps water across a semipermeable membrane, produces *RO water.* Laboratory requirements generally call for reagent grade water that, according to the *National Committee for Clinical Laboratory Standards (NCCLS),* belongs to one of three types (Type I, II, and III). It is recommended that water be classified in terms of type rather than the method of preparation. The NCCLS criteria for reagent grade water is discussed in NCCLS Document C3-A2, *Preparation and Testing of Reagent Water in the Clinical Laboratory.*[10]

Prefiltration can be used to remove particulate matter for municipal water supplies before any additional treatments. Filtration cartridges are composed of glass; cotton; activated charcoal, which removes organic materials and chlorine; and submicron filters (~0.2 mm), which remove any substances larger than the filter's pores, including bacteria. The manner in which these filters are used depends on the quality of the municipal water and the other purification methods used. For example, hard water (contains calcium, iron, and other dissolved elements) may require prefiltration with a glass or cotton filter rather than activated charcoal or submicron filters, which would quickly become clogged and expensive to operate. The submicron filter might be better used following distillation, deionization, or reverse osmosis treatment.

Distilled water has been purified to remove almost all organic materials using the technique of distillation like that found in organic chemistry laboratory distillation experiments. Water is boiled and vaporized. The vapor rises and enters into the coil of a condenser, which is a glass tube that contains within it a glass coil. Cool water surrounds this condensing coil, lowering the temperature of the water vapor. The water vapor returns to a liquid state, which is then collected. Many impurities do not rise in the water vapor but remain in the boiling apparatus. The water collected after condensation has less contamination. Because laboratories use thousands of liters of water per day, stills are used instead of small condensing apparatuses, but the principles are basically the same. Water may be distilled more than once. Each distillation cycle removes more impurities.

Deionized water has some or all ions removed, although organic material may still be present, so it is neither pure nor sterile. In general, deionized water is purified from previously treated water, such as prefiltered or distilled water. Deionized water is produced using either an anion or a cation exchange resin, followed by replacement of the removed particles with hydroxyl or hydrogen ions. The anticipated ions to be removed from the water will dictate the type of ion exchange resin to be used. One column cannot service all ions present in water. A combination of several resins will produce different grades of deionized water. A two-bed system uses an anion resin followed by a cation resin. The different resins may be in separate columns or in the same column.

Depending on the quality of the feed water, Type I water can be obtained by initially filtering it to remove particulate matter, followed by reverse osmosis, deionization, and a 0.2-mm filter. Type III water is acceptable for glassware washing but not for analysis or reagent preparation. Type II water is acceptable for most analytical requirements, including reagent, quality control, and standard preparation. Type I water is used for test methods requiring minimum interference, such as trace metal, iron, and enzyme analyses. Use with high-performance liquid chromatography may require less than a 0.2-mm final filtration step. Because Type I water should be used immediately, storage is discouraged because the resistivity changes, and Type II water should be stored in a manner that will reduce any chemical or bacterial contamination and for short periods.

Testing procedures to determine the quality of reagent-grade water include measurements of resistance; pH; colony counts (for assessing bacterial contamination) on selective and nonselective media for the detection of coliforms; chlorine; ammonia; nitrate or nitrite; iron; hardness; phosphate; sodium; silica; carbon dioxide; chemical oxygen demand (COD); and metal detection. The College of American Pathologists (CAP) recommends that a laboratory document culture growth, pH, and specific resistance on water used in reagent preparation. Resistance is measured because pure water, devoid of ions, conducts electricity poorly. The relationship of water purity to resistance is a linear one: as the purity increases, so does the resistance. This one measurement does not suffice for determination of true water purity, because a nonionic contaminant may be present that has little effect on resistance.

CLINICAL LABORATORY SUPPLIES

Many different supplies are required in today's medical laboratory, but there are several items that are common to all. These items include pipets, flasks, beakers, burets, desiccators, and filtering material. A brief discussion of the composition and general use of these supplies follows.

Glass and Plasticware

Until recently, laboratory supplies (pipets, flasks, beakers, burets) consisted of some type of glass and could be correctly termed *glassware.* As plastic material was refined and made available to manufacturers, plastic has increasingly been used to make laboratory utensils. Before the discussion of general laboratory supplies can begin, a brief summary of the types and uses of glass and plastic commonly seen in today's laboratories is given. (See Appendix O, *Characteristics of Types of Glass;* Appendix P, *Characteristics of Types of Plastic;* and Appendix Q, *Chemical Resistance of Types of Plastic*).

Glassware used in the clinical laboratory may fall into one of the following categories: high thermal, high silica, high alkali resistant, low actinic, or soda lime glass.[11,12] Whenever possible, routinely used clinical chemistry glassware should consist of high thermal borosilicate or aluminosilicate glass and meet Class A tolerances prescribed by the NIST.[13] Glassware that *does not* meet Type A specifications may have twice the tolerance range despite that its appearance may be identical to a piece of Type A glassware. The best source of information about specific uses, limitations, and accuracy specifications for glassware is the manufacturer.

Plasticware is beginning to replace glassware in the laboratory setting. Its unique high resistance to corrosion and breakage, as well as its varying flexibility, has made it most appealing. It is relatively inexpensive, allowing for most items to be completely disposable after each use. The major types of resins frequently used in the clinical chemistry laboratory are polystyrene, polyethylene, polypropylene, Tygon, Teflon, polycarbonate, and polyvinylchloride. Once again, the individual manufacturer is the best source of information concerning the proper use and limitations of any plastic materials.

In most laboratories, the glass or plastic directly associated with testing is almost always disposable. However, should the need arise, cleaning of glass or plastic may require special techniques. Immediately rinsing glass or plastic supplies after use, followed by washing with a powder or liquid detergent designed for cleaning laboratory supplies and several distilled water rinses, may be sufficient. Presoaking glassware in soapy water is highly recommended whenever immediate cleaning is impractical. Many laboratories use automatic dishwashers and dryers for cleaning. Detergents and temperature levels should be compatible with the material and the manufacturer's recommendations. To ensure that all detergent has been removed from the labware, multiple rinses with Type II water is recommended. Check the pH of the final rinse water and compare it with the initial pH of the prerinse water. Detergent-contaminated water will have an alkaline pH. Visual inspection should reveal spotless vessel walls. Any biologically contaminated labware should be disposed of according to the precautions followed by that laboratory.

Some determinations, such as enzymes, iron, and other heavy metals, require scrupulously clean glassware. Cleaning solutions that have been used successfully are acid dichromate and nitric acid. It is suggested that disposable glass and plastic be used wherever possible.

Dirty pipets should be placed immediately in a container of soapy water with the pipet tips up. The container should be long enough to allow the pipet tips to be covered with solution. Use of a specially designed pipet soaking jar and

washing/drying apparatus is recommended. For final water rinses, fresh Type I or II water should be provided for each rinse. If possible, set aside a pipet container for final rinses only. Cleaning brushes are available to fit almost any size glassware and are recommended for any articles that are washed routinely.

Plastic material is often easier to clean because of its nonwettable surface. A brush or harsh abrasive cleaner should not be used on plasticware. Acid rinses or washes are not required. The initial cleaning procedure described in Appendix R, *Cleaning Labware,* can be adapted for plasticware as well.

Pipets

Pipets are utensils made of glass or plastic that are used to transfer liquids, and they may be reusable or disposable. In many institutions, automatic pipetting devices have replaced the manual glass or plastic pipets. To minimize confusion, Table 1-2 lists the classification scheme further described here. Examples of pipets are found in Figure 1-1.

Pipets are designed to contain (TC) or to deliver (TD) a particular volume of liquid. Near the top of the pipet, most manufacturers stamp the initials *TC* or *TD* to alert the user as to the type of pipet. A TC pipet holds a particular volume but does not dispense that exact volume, whereas a TD pipet will dispense the volume indicated. When using either pipet, the tip of the pipe must be immersed in the liquid to be transferred to a level that will allow it to remain in solution after the volume of liquid has entered the pipet and without touching the vessel walls. The pipet is held upright, not at an angle (Fig. 1-2). A slight suction using a pipet bulb or similar device is applied to the opposite end until the liquid enters the pipet and the meniscus is brought above the desired graduation line (Fig. 1-3A); suction is then stopped. While the meniscus level is held in place, the pipet tip is raised slightly out of the solution and wiped of any adhering liquid with a

---

TABLE 1-2. Pipet Classification

   I. Design
     A. To contain (TC)
     B. To deliver (TD)
  II. Drainage characteristics
     A. Blow-out
     B. Self-draining
 III. Type
     A. Measuring or graduated
       1. Serologic
       2. Mohr
       3. Bacteriologic
       4. Ball, Kolmer, or Kahn
       5. Micropipet
     B. Transfer
       1. Volumetric
       2. Ostwald-Folin
       3. Pasteur pipets
       4. Automatic macro- or micropipets

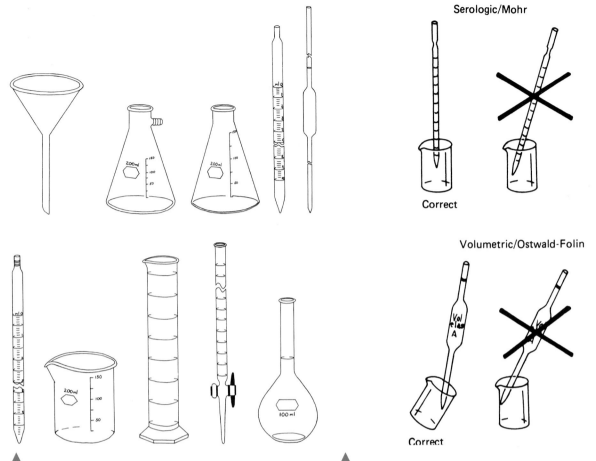

Figure 1-1. Laboratory glassware.

Figure 1-2. Correct and incorrect pipet positions.

laboratory tissue. The liquid is allowed to drain until the bottom of the meniscus touches the desired calibration mark (Fig. 1–3*B*). With the pipet held in a vertical position and the tip against the side of the receiving vessel, the pipet contents are allowed to drain into the vessel (*ie,* test tube, cuvet, flask, and so on). If the pipet has a continuous etched ring or two small continuous rings very close together located near the top of the pipet, it is called a *blow-out pipet*. This means that the last drop of liquid should be expelled into the receiving vessel. When these markings are absent from a pipet, it is *self-draining,* and the user allows the contents of the pipet to drain by gravity. The tip of the pipet should not be in contact with the accumulating fluid in the receiving vessel during drainage. With the exception of the Mohr pipet, the tip should remain in contact with the side of the vessel for several seconds after the liquid has drained. The pipet is then removed. Various examples of pipet bulbs are illustrated in Figure 1-4. Mouth pipetting is strictly *forbidden* because of the possibility of aspirating hazardous material.

Measuring or graduated pipets are capable of dispensing several different volumes. Because the graduation lines located on the pipet may vary, they should be indicated on the top of each pipet. For example, a 5-mL pipet can be used to

measure 5, 4, 3, 2, or 1 mL of liquid, with further graduations between each milliliter. The pipet is designated as 5 in ¹⁄₁₀ increments (Fig. 1–5), and could deliver any volume in tenths of a milliliter, up to 5 mL. Another pipet, such as a 1-mL pipet, may be designed to dispense 1 mL and have subdivisions of hundredths of a milliliter. The markings at the top of a measuring or graduated pipet indicate the volume(s) it is designed to dispense.

The subgroups of measuring or graduated pipets are Mohr, serologic, and micropipets. A *Mohr pipet* does not have grad-

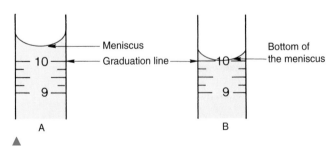

Figure 1-3. Pipetting technique. (**A**) Meniscus is brought above the desired graduation line. (**B**) Liquid is allowed to drain until the bottom of the meniscus touches the desired calibration mark.

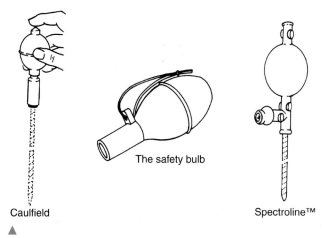

Caulfield

The safety bulb

Spectroline™

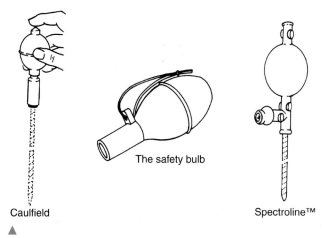

Figure 1-4. Types of pipet bulbs.

uations to the tip. It is a self-draining pipet, but the tip should not be allowed to touch the vessel while the pipet is draining and its full volume should be used to achieve proper accuracy. A *serologic pipet* has calibration marks to the tip and is generally a blow-out pipet. A *micropipet* is a pipet with a total holding volume of less than 1 mL: it may be designed as either a Mohr or serologic pipet.

The next major category of pipets consists of the *transfer pipets*. These pipets are designed to dispense one volume without further subdivisions. *The bulblike enlargement in the pipet stem easily distinguishes the Ostwald-Folin and volumetric subgroups.* Ostwald-Folin pipets are used with biologic fluids having a viscosity that is greater than that of water. They are blow-out pipets, indicated by two etched continuous rings at the top. The volumetric pipet is designed to dispense or transfer aqueous solutions and is always self-draining. This type of pipet usually has the greatest degree of accuracy and precision and should be used when diluting standards, calibrators, or quality-control material. *Pasteur pipets* do not have any calibration marks and are used to transfer solutions or biologic fluids without consideration of a specific volume. These pipets should not be used in any quantitative analytical techniques.

The *automatic pipet* is by far the most routinely used pipet in today's clinical chemistry laboratory. The major advantages of automatic and semiautomatic pipets are time savings, safety, ease of use, stability, increase in precision, and lack of

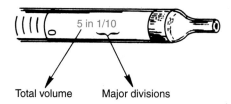

5 in 1/10

Total volume   Major divisions

Figure 1-5. Volume indication of a pipet.

required cleaning, because the contaminated portions of the pipet, such as the tips, are often disposable. Figure 1-6 illustrates many common automatic pipets. Many models have different volumes that may be selected and are termed *variable,* but only one volume at a time may be used. The range of available volumes is from 1 mL to 1 L. The widest volume range usually seen in a single pipet is 0 to 1 mL. Any pipet that has a pipetting capability of less than 1 mL is considered a *micropipet,* and a pipet that dispenses greater than 1 mL is called an *automatic macropipet.*

The term *automatic* implies that the mechanism that draws up the liquid and dispenses it is an integral part of the pipet itself. There are three general types of automatic pipets: air-displacement, positive-displacement, and dispenser pipets. An *air-displacement pipet* relies on a piston for suction creation to draw the sample into a disposable tip. The piston does not come in contact with the liquid. A *positive-displacement pipet* operates by moving the piston in the pipet tip or barrel, much like a hypodermic syringe. It does not require a different tip for each use. Because of carryover concerns, rinsing and blotting between samples may be required. *Dispensers* and *dilutor/dispensers* are automatic pipets that obtain the liquid from a common reservoir and dispense it repeatedly. The dispensing pipets may be bottle-top, motorized, or hand-held, or attached to a dilutor. The dilutor often combines sampling and dispensing functions. Figure 1-7 provides examples of different types of automatic pipetting devices. These pipets should be used according to the individual manufacturer's directions. Many automated pipets use a wash in between samples to eliminate carryover problems. However, to minimize carryover contamination with manual or semiautomated pipets, careful wiping of the tip may be necessary to remove any liquid that may have adhered to the outside of the tip before dispensing any liquid. Care should be taken to ensure that the orifice of the pipet tip is not blotted, thereby drawing sample from the tip. Another precaution in using manually operated semiautomatic pipets is to move the plunger in a continuous, slow manner.

Pipet tips are designed to be used with positive- and air-displacement pipets. The laboratorian needs to ensure that the pipet tip is seated snugly on the end of the pipet and is free from any deformity. Plastic tips used on air-displacement pipets are particularly likely to vary. Different brands can be used for one particular pipet but do not necessarily perform in an identical manner. Plastic burrs, which sometimes cannot be detected by the naked eye, can be present. A method using a 0.1%-solution of phenol red in distilled water has been used to compare the reproducibility of different brands of pipet tips.[14] When using this method, the pipet and the operator should remain the same so that variation is due solely to changes in the pipet tips.

Class A pipets, like all other Class A glassware, do not need to be recalibrated by the laboratory. Automatic pipetting devices, as well as non–Class A materials, do need recalibration.

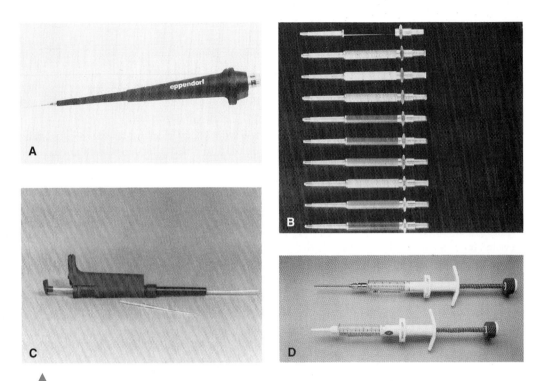

Figure 1-6. (**A**) Fixed volume ultra micro-digital air displacement pipetter with tip ejector. (**B**) Fixed-volume air-displacement pipet. (**C**) Digital electronic positive displacement pipetter. (**D**) Syringe pipetters.

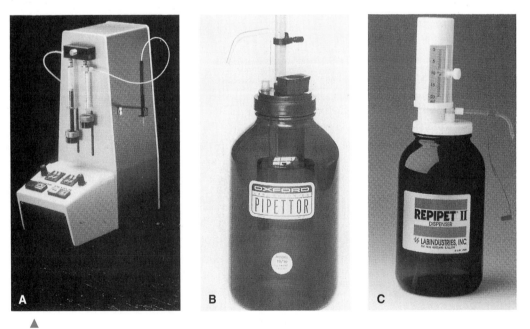

Figure 1-7. (**A**) Digital dilutor/dispenser. (**B**) Dispenser. (**C**) Dispenser.

A gravimetric method can be used to accomplish this task. It is done by delivering and weighing a solution of known specific gravity, such as water. It is suggested that a currently calibrated analytical balance and at least Class S weights be used.

Gravimetric Pipet Calibration[15]

MATERIALS

Pipet.

10–20 pipet tips, if needed.

Balance capable of accuracy and resolution to ±0.1% of dispensed volumetric weight.

Weighing vessel large enough to hold volume of liquid.

Type I water.

Thermometer and barometer.

PROCEDURE

1. Record the weight of the vessel. Record the temperature of the water. It is recommended that all materials be at room temperature. Obtain the barometric pressure.
2. Place a small volume (0.5 mL) of the water into the container. To prevent effects from evaporation, loosely covering each container using a substance such as Parafilm™ is desirable. Avoid handling of the containers.
3. Weigh each container plus water to the nearest 0.1 mg *or* set the balance to zero.
4. Using the pipet to be tested, draw up the specified amount. Carefully wipe the outside of tip. Care should be taken not to touch the end of the tip. This will cause liquid to be wicked out of the tip, introducing an inaccuracy as a result of technique.
5. Dispense the water into the weighed vessel. Touch the tip to the side.
6. Record the weight of the vessel.
7. Subtract the weight obtained in step 3 from that obtained in step 6. Record the result.
8. If plastic tips are used, change the tip between each dispensing. Repeat steps 1 to 6 for a minimum of nine additional times.
9. Obtain the average or mean of the weight of the water. Multiply the mean weight by the corresponding density of water at the given temperature and pressure. This may be obtained from the *Handbook of Chemistry and Physics*[16] or in Table 1-3[6] for a quick reference.
10. Determine the accuracy or the ability of the pipet to dispense the expected (selected or stated) volume according to the following formula:

$$\frac{\text{Mean volume}}{\text{Expected volume}} \times 100\% \qquad \textit{(Eq. 1–9)}$$

The manufacturer usually gives acceptable limitations for a particular pipet, but they should not be used if the value differs by more than 1.0% from the expected value.

Precision can be indicated as the percent coefficient of variation (%CV) or standard deviation (SD) for a series of repetitive pipetting steps. A discussion of %CV and SD can be found in Chapter 3 Quality Control and Statistics. The equations to calculate the SD and %CV are as follows:

$$SD = \sqrt{\frac{\Sigma(x - \bar{x})^2}{n - 1}}$$

$$\%CV = \frac{SD}{\bar{x}} \times 100 \qquad \textit{(Eq. 1–10)}$$

Required imprecision is usually ±1 SD. The %CV will vary with the expected volume of the pipet, but the smaller the %CV value, the greater is the precision. When *n* is large, the data are more statistically valid.

Although gravimetric validation is the most desirable, pipet calibration may also be accomplished by using photometric methods, particularly for automatic pipetting devices. When a spectrophotometer is used, the molar extinction coefficient of a compound, such as potassium dichromate, is obtained. After an aliquot of diluent is pipetted, the change in concentration will reflect the volume of the pipet. Another photometric technique to assess pipet accuracy uses the comparison of the absorbances of dilutions of potassium dichromate made using Class A volumetric glassware versus equivalent dilutions made with the pipetting device.

These calibration techniques are time-consuming and therefore impractical for use in daily checks. It is recommended that pipets be checked initially and subsequently three to four times per year. A quick daily check for many larger volume automatic pipetting devices uses volumetric flasks. For example, a bottle-top dispenser that routinely delivers 2.5 mL of a reagent may be checked by dispensing four aliquots of the reagent into a 10-mL Class A volumetric flask. The bottom of the meniscus should meet with the calibration line on the volumetric flask.

Burets

A *buret* looks much like a wide, long, graduated pipet with a stopcock at one end. A buret usually holds a total volume ranging from 25 mL to 100 mL of solution and is used to dispense a particular volume of liquid during a titration (Fig. 1-8).

TABLE 1-3. Water Density

°C	Density
20	0.9982
21	0.9980
22	0.9978
23	0.9976
24	0.9973
25	0.9971
26	0.9968
27	0.9965
28	0.9963
29	0.9960
30	0.9957

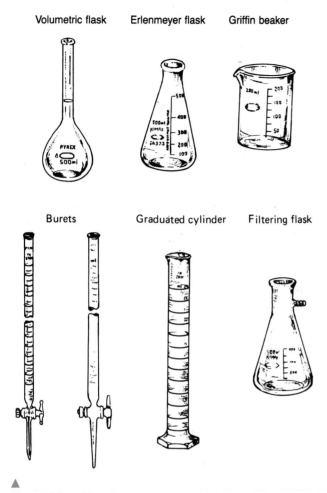

Volumetric flask   Erlenmeyer flask   Griffin beaker

Burets   Graduated cylinder   Filtering flask

▲

Figure 1-8. Examples of laboratory glassware.

Syringes

Syringes are used for transfer of very small volumes (less than 500 μL) in separation techniques such as chromatography or electrophoresis. The syringes are made of glass and have very fine barrels. The plunger is often made of a fine piece of wire. When syringes are used for injection of sample into a gas chromatographic system, tips are not used. However, in electrophoresis work, disposable Teflon tips are used. Expected inaccuracies of volumes less than 5 μL is 2%, whereas for greater volumes, the inaccuracy is approximately in the range of 1%.[17]

Laboratory Vessels

Flasks, beakers, and graduated cylinders are used to hold solutions. Volumetric and Erlenmeyer flasks are two types of containers in general use in the clinical laboratory.

A *volumetric flask,* like any volumetric utensil, is calibrated to hold one exact volume of liquid. The flask has a round, flat lower portion and a long, thin neck with the calibration line etched into the neck. Volumetric flasks are used to bring a given reagent to its final volume with the prescribed diluent, and should be of Class A quality. When bringing the bottom of the meniscus up to the calibration mark, a pipet

should be used when adding the final drops of diluent so that maximum control is maintained and the calibration line is not missed.

Erlenmeyer flasks and *Griffin beakers* are designed to hold different volumes rather than one exact amount. Because Erlenmeyer flasks and Griffin beakers are often used in reagent preparation, the flask size, chemical inertness, and thermal stability should be considered. The Erlenmeyer flask has a wide bottom that gradually evolves into a smaller, short neck. The Griffin beaker has a flat bottom, straight sides, and an opening as wide as the flat base with a small spout in the lip.

Graduated cylinders are long, cylindrical tubes usually held upright by an octagonal or circular base. The cylinder has calibration marks along its length and is used to measure volumes of liquids. Graduated cylinders do not have the accuracy of volumetric glassware. The sizes routinely used are 10, 25, 50, 100, 500, 1000, and 2000 mL.

All laboratory utensils should be Class A whenever possible to maximize accuracy and precision and thus decrease calibration time. Figure 1-8 illustrates representative laboratory glassware. Table 1-4 lists the Class A tolerances for some commonly used volumes.

Desiccators and Desiccants

Many compounds combine with water molecules to form loose chemical crystals. The compound and its associated water are called a *hydrate.* When the water of crystallization is removed from the compound, it is said to be *anhydrous.*

TABLE 1-4. Class A Tolerances	
Size (mL)	**Tolerances (mL)**
Burets	
5	±0.01
10	±0.02
25	±0.03
50	±0.05
100	±0.10
Pipets (Transfer)	
0.5–2	±0.006
3–5	±0.01
10	±0.02
15–25	±0.03
50	±0.05
Volumetric flasks	
1–10	±0.02
25	±0.03
50	±0.05
100	±0.08
200	±0.10
250	±0.12
500	±0.20
1000	±0.30
2000	±0.50

Some substances take up water on exposure to atmospheric conditions and are called *hygroscopic*. Materials that are very hygroscopic can remove moisture from the air as well as from other materials. These make excellent drying substances and are sometimes used as *desiccants* (drying agents) to keep other chemicals from becoming hydrated. Commonly used desiccants[6] are listed in Table 1-5, beginning with the most hygroscopic and ending with the least effective drying agent. If these compounds absorb enough water from the atmosphere to cause dissolution, they are called *deliquescent substances*.

Desiccants are most effective when placed in a closed chamber called a *desiccator* (Fig. 1-9). The desiccant is placed below the perforated platform inside the desiccator. Placing a fine film of grease on the rim of the cover seals a glass or plastic desiccator, and it is correctly opened or sealed by slowly sliding the lid sideways. The greased seal prevents the desiccator from being opened with a direct pull upward. Open a desiccator slowly and with caution because the air pressure inside the desiccator could be below atmospheric pressure. Desiccants containing indicators that signify desiccant exhaustion and that can be regenerated using heat are particularly helpful. Desiccants that produce dust should also be avoided. In the laboratory, desiccants are primarily used to prevent moisture absorption by chemicals, gases, and instrument components.

Balances

A properly operating balance is essential in producing high-quality reagents and standards. Balances may be classified according to their design, number of pans (single or double), whether they are mechanical or electronic, or based on their operating ranges, as determined by precision (readability ~2 µg) balances, analytical balances (readability ~0.001 g), or microbalances (readability ~0.1 µg; Fig. 1-10).

TABLE 1-5. Common Desiccants

Agent	Formula
Magnesium perchlorate (Dehydrite)	$Mg(ClO_4)$
Barium oxide	BaO
Alumina	Al_2O_3
Phosphorous pentoxide	P_4O_{10}
Lithium perchlorate	$LiClO_4$
Calcium chloride	$CaCl_2$
Calcium sulfate (Drierite)	$CaSO_4$
Silica gel	SiO_2
Ascarite	NaOH on asbestos

Analytical and electronic balances are the most popular balances used in today's clinical laboratory. Analytical balances are required for the preparation of any primary standards. The mechanical analytical balance is also known as a *substitution balance*. It has a single pan enclosed by sliding transparent doors, which minimize environmental influences on pan movement. The pan is attached to a series of calibrated weights, which are counterbalanced by a single weight at the opposite end of a knife-edge fulcrum. The operator adjusts the balance to the desired mass and places the material, contained within a tared weighing vessel, on the sample pan. An optical scale allows the operator to visualize the mass of the substance. The weight range for an analytical balance is from 0.01 mg up to 160 g on some models.

Electronic balances are single-pan balances that use an electromagnetic force to counterbalance the weighed sample's mass. Their measurements equal the accuracy and precision of any available mechanical balance, with the advantage of a very fast response time (less than 10 seconds).

It is recommended that NIST Class S weights are used for calibrating balances on a monthly basis. Balances should be

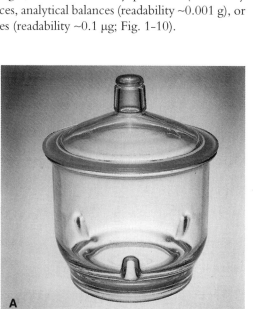
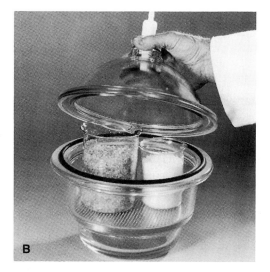

Figure 1-9. (**A**) Glass desiccator. (**B**) Glass desiccator with dry seal ring (no grease is necessary to seal cover to body).

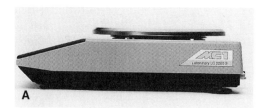

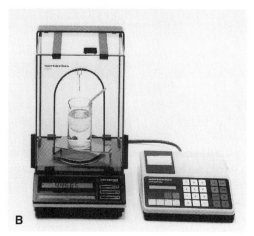

Figure 1-10. (**A**) Electronic toploading balance. (**B**) Electronic analytical balance with printer.

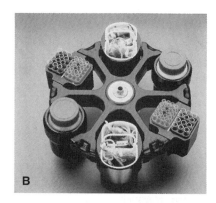

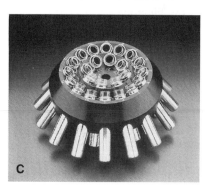

Figure 1-11. (**A**) Benchtop centrifuge. (**B**) Swinging-bucket rotor. (**C**) Fixed head rotor.

kept scrupulously clean and located in an area away from heavy traffic, large pieces of electrical equipment, and open windows. A slab of marble separated from its supporting surface by a flexible material is sometimes placed under balances to minimize any vibration interference that may occur. The level checkpoint should always be corrected before any weighings.

BASIC SEPARATION TECHNIQUES

Centrifugation

Centrifugation is a process whereby centrifugal force is used to separate solid matter from a liquid suspension. The *centrifuge* is an apparatus that carries out this action. A centrifuge consists of a head or rotor, carriers, or shields (Fig. 1-11) that are attached to the vertical shaft of a motor and enclosed in a metal covering. The centrifuge always has a lid and an on/off switch, but many models also include a brake or a built-in tachometer, which indicates speed, and can be refrigerated. Centrifugal force depends on three variables: mass, speed, and radius. The speed is expressed in revolutions per minute (rpm), and the centrifugal force generated is expressed in terms of relative centrifugal force (RCF) or gravities (g). The speed of the centrifuge is related to the RCF by the following equation:

$$RCF = 1.118 \times 10^{-5} \times r \times (rpm)^2, \quad (Eq. 1–11)$$

where 1.118×10^{-5} is a constant, determined from the angular velocity, and r is the radius in centimeters, measured from the center of the centrifuge axis to the bottom of the test-tube shield. The RCF value also may be obtained from a nomogram like that found in Appendix J, *Relative Centrifugal Force Nomograph*. Classification of centrifuges may be based on several criteria, including benchtop or floor model, refrigeration, rotor head (*eg*, fixed, hematocrit, swinging-bucket, or angled; Figure 1-11), or maximum speed attainable (*ie*, ultracentrifuge). Centrifuges are generally used to separate serum from a blood clot as the blood is being processed; to separate a supernatant from a precipitate during an analytical reaction; to separate two immiscible liquids; or to expel air.

Care of a centrifuge includes daily cleaning of any spills or debris, such as blood or glass, and checking to see that the centrifuge is properly balanced (Fig. 1-12) and free from any excessive vibrations. Balancing the centrifuge load is critical. Many newer centrifuges will automatically decrease their speed if the load is not evenly distributed, but more often, the centrifuge will shake and vibrate or make more noise than is usual. The centrifuge cover should remain closed until the centrifuge has come to a complete stop to avoid any aerosol contamination. It is recommended that the timer, brushes (if present), and speed be checked periodically. The brushes, which are graphite bars attached to a retainer spring, create an electrical contact in the motor. The specific manufacturer's service manual should be consulted for details on how to change brushes and on lubrication requirements. The speed of a centrifuge is easily checked using a tachometer or strobe light. The hole located in the lid of many centrifuges is designed for speed verification using these devices.

Filtration

Filtration was used in place of centrifugation for the separation of solids from liquids. In today's laboratory, paper filtration is no longer routinely used, and thus only the basics are discussed here. Filter material is made of paper, cellulose and its derivatives, polyester fibers, glass, and a variety of resin column materials. Traditionally, filter paper was folded in a manner that allowed it to fit into a funnel. In method A, round filter paper is folded like a fan (Fig. 1-13A), and in method B, the paper is folded into fourths (Fig. 1-13B).

Filter paper differs in terms of pore size and should be selected according to separation needs. Filter paper should not be used when using strong acids or bases. Once the filter

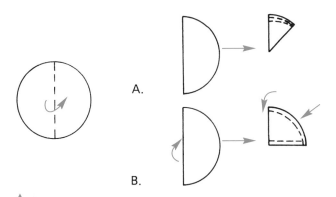

Figure 1-13. Methods of folding a filter paper. (**A**) In a fan. (**B**) In fourths.

paper is placed inside the funnel, the solution slowly drains through the filter paper within the funnel and into a receiving vessel. The liquid that passes through the filter paper is called the *filtrate*.

Dialysis

Dialysis is another method for separating macromolecules from a solvent. It was made popular when used in conjunction with the Technicon Autoanalyzer™ system. Basically, a solution is put into a bag or is contained on one side of a semipermeable membrane. Larger molecules are retained within the sack or on one side of the membrane, while smaller molecules and solvents diffuse out. This process is very slow. Use of columns that contain a gel material have replaced manual dialysis separation in most analytical procedures.

LABORATORY MATHEMATICS AND CALCULATIONS

Significant Figures

Significant figures are the minimum number of digits needed to express a particular value in scientific notation without loss of accuracy. The number 814.2 has four significant figures, because in scientific notation it is written as 8.142×10^2. The number 0.000641 has three significant figures, and the scientific notation expression for this value is 6.41×10^{-4}. The zeros are merely holding decimal places and are not needed to express the number in scientific notation.

Logarithms

The base 10 *logarithm* (log) of a positive number N greater than zero is equal to the exponent to which 10 must be raised to produce N. Therefore, it can be stated that N equals 10^x, and the log of N is equal to x. The number N is the antilogarithm (antilog) of x.

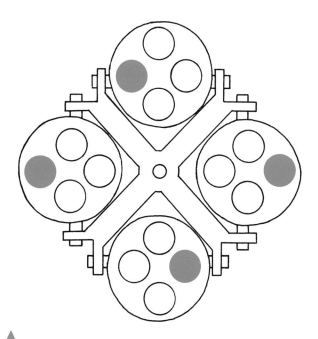

Figure 1-12. Properly balanced centrifuge. Color circles represent counter-balanced positions for sample tubes.

The logarithm of a number, which is written in a *decimal* format, consists of two parts: the character, or characteristic, and the mantissa. The *character* is the number to the left of the decimal point in the log and is derived from the exponent, and the *mantissa* is that portion of the logarithm to the right of the decimal point and is derived from the number itself. Although several approaches can be taken to determine the log, one approach is to write the number in scientific notation. The number 1424 expressed in scientific notation is 1.424×10^3, making the character a 3. The character also can be determined by adding the number of significant figures and subtracting 1 from the sum. The mantissa is derived from a log table or calculator having a log function for the remainder of the number. For 1.424, a calculator would give a mantissa of 0.1535, and a log table would reveal what is shown in Table 1-6. Some calculators with a log function do not require conversion to scientific notation.

When using a log table, the initial step is to determine N. N is obtained from the first two digits in the number. In this example, N is 14. Look under the N column in the log tables (Table 1-6) until the number 14 is located. The next digit in this example, 1.424, is 2. Continue along the row and read the numbers under 142, which, in this case, are 1523. The last part of the mantissa is obtained from the proportional parts section of the log table and is added to 1523. In this instance (1.424), the remaining digit is 4. The value under 4 in the table is 12. The mantissa becomes 1523 + 12 = 1535. The logarithm of 1.425×10^3 is equal to 3.1535. The character is 3, and the mantissa is 1535. Because there are only four significant figures in the original number, only four significant figures should be written in the log. The log now becomes 3.154. For numbers less than 1, the character usually has a bar over the top. The number 2.12×10^{-2} has a log of $\overline{2}.3263$, or $\overline{2}.33$ to satisfy the number of significant figures.

To determine the original number from a log value, the process is done in reverse. This process is termed the *antilogarithm*. Using the previous example, determine the antilogarithm of $\overline{2}.33$. The character $\overline{2}$, which has a bar over the top of it, indicates 10^{-2}. The mantissa, 0.3263, or 0.33, may lead to the number in two ways. The mantissa may be located in a logarithm table, and N may be determined from it. A second approach is to use an antilogarithm table. Table 1-7 illustrates both methods. The rounding of numbers does affect the antilogarithm. Most calculators have logarithm and antilogarithm functions. Consult the specific manufacturer's directions to become acquainted with the proper use of these functions.

Negative or Inverse Logarithms

The character of a log may be positive or negative, but the mantissa is always positive. In some circumstances, the laboratorian must deal with inverse or negative logs. Such is the case with pH. The pH of a solution is defined as minus the log of the hydrogen ion concentration. The following is a convenient formula used to determine the negative logarithm when dealing with pH:

$$pH = x - \log N \qquad (Eq.\ 1\text{--}12)$$

For example, if the hydrogen ion concentration of a solution is 5.4×10^{-6}, then $x = 6$. The logarithm of N (5.4) is equal to 0.7324, or 0.73. The pH becomes

$$pH = 6 - 0.73 = 5.27 \qquad (Eq.\ 1\text{--}13)$$

The same formula can be applied to obtain the hydrogen ion concentration of a solution when only the pH is given. Using a pH of 5.27, the equation becomes

$$5.27 = x - \log N \qquad (Eq.\ 1\text{--}14)$$

The *x* term is always the next largest whole number. For this example, the next largest whole number is 6. Substituting for *x* the equation becomes

$$5.27 = 6 - \log N \qquad (Eq.\ 1\text{--}15)$$

Multiply all the variables by −1 and solve the equation for the unknown quantity, that is, log N. The result is −0.73, which is the antilogarithm value of *N*, which is 5.37, or 5.4. The hydrogen ion concentration for a solution with a pH of 5.27 is 5.4×10^{-6}.

Concentration

A detailed description of each concentration term (ie, molarity, normality, and so on) may be found at the beginning of this chapter. The following discussion focuses on basic mathematical expressions needed to prepare reagents of a stated concentration.

Percent Solution

A percent solution is determined in the same manner regardless of whether weight/weight, volume/volume, or weight/volume units are used. *Percent* means "parts per 100," which is represented as percent (%) and is independent of the molecular weight of a substance.

TABLE 1-6. Portion of a Logarithm Table

| | | | | | | | | | | | Proportional Parts | | | | | | | |
N	0	1	2	3	4	5	6	7	8	9	1	2	3	4	5	6	7	8	9
14	1461	1492	1523	1553	1584	1614	1644	1673	1703	1732	3	6	9	12	15	18	21	24	27

TABLE 1-7. Antilogarithm Determination

A. Log Table

N	0	1	☐2	3
☐21	3222	3243	<u>3263</u>	3284

The number, as determined from the mantissa 3263, is 212.

B. Antilog Table

											Proportional Parts
N	0	1	2	3	4	5	☐6	7	8	9	12☐3456789
☐.32	2089	2094	2099	2104	2109	2113	<u>2118</u>	2123	2128	2133	011223344
☐.33	<u>2138</u>	2143									

$$.3263 = 2118 + 1 = 2119 = 2.12$$
$$.33 = 2138 = 2.14$$

Example 1-1: Weight/Weight (w/w)

To make up 100 g of a 5% aqueous solution of hydrochloric acid (using 12 M HCl), multiply the total amount by the percent expressed as a decimal. The calculation becomes

$$5\% = \frac{5}{100} = 0.050 \qquad \text{(Eq. 1–16)}$$

Therefore,

$$0.050 \times 100 = (5 \text{ g of HCl}) \qquad \text{(Eq. 1–17)}$$

Another way of arriving at the answer is to set up a ratio so that

$$\frac{5}{100} = \frac{x}{100}$$
$$x = 5 \qquad \text{(Eq. 1–18)}$$

Example 1-2: Weight/Volume (w/v)

The most frequently used term for a percent solution is weight per volume, which is often expressed as grams per 100 mL of diluent. To make up 1000 mL of a 10% (w/v) solution of NaOH, use the preceding approach. The calculations become

$$\underset{\text{(\% expressed as a decimal)}}{0.10} \times \underset{\text{(total amount)}}{1000} = 100 \text{ g}$$

or

$$\frac{10}{100} = \frac{x}{1000}$$
$$x = 100 \qquad \text{(Eq. 1–19)}$$

Therefore, add 100 g of NaOH to a 1000 mL volumetric Class A flask and dilute to the calibration mark with Type II water.

Example 1-3: Volume/Volume (v/v)

Make up 50 mL of a 2% (v/v) concentrated hydrochloric acid solution.

$$0.02 \times 50 = 1 \text{ mL}$$

or

$$\frac{2}{100} = \frac{x}{50}$$
$$x = 1 \qquad \text{(Eq. 1–20)}$$

Therefore, add 40 mL of water to a 50-mL Class A volumetric flask, add 1 mL of HCl, mix, and dilute up to the calibration mark with Type II water. Remember, always add acid to water!

Molarity

Molarity (M) is routinely expressed in units of moles per liter (mol/L) or sometimes millimoles per milliliter (mmol/mL). Remember that 1 mol of a substance is equal to the gmw of that substance. When trying to determine the amount of substance needed to yield a particular concentration, initially decide what final concentration *units* are needed. In the case of molarity, the final units will be moles per liter (mol/L) or millimoles per milliliter (mmol/mL). The second step is to consider the existing units and the relationship they have to the final desired units. Essentially, try to put as many units as possible into "like" terms so that the only units that are not identical are those wanted in the final answer.

Example 1-4

How many *grams* are needed to make 1 L of a 2 M solution of HCl?

Step 1: Which units are needed in the final answer? *Answer:* Grams per liter (g/L).

Step 2: Assess other mass/volume terms used in the problem. In this case, molarity is the concentration needed, so moles are needed for the calculation: How many grams are equal to 1 mol? The gmw of HCl, which can be determined from the periodic table, will be equal to 1 mol. For HCl, the gmw is 36.5, so the equation may be written as

$$\frac{36.5 \text{ g HCL}}{\text{mol}} \times \frac{2 \text{ mol}}{\text{L}} = \frac{73 \text{ g HCl}}{\text{L}} \qquad \text{(Eq. 1–21)}$$

Cancel out "like" units, and the final units should be grams per liter. In this example, 73 g HCl per liter is needed to make up a 2 M solution of HCl.

Example 1-5

A solution of NaOH is contained within a Class A 1-L volumetric flask filled to the calibration mark. The content label reads 24 g of NaOH. Determine the molarity.

Step 1: What *units* are ultimately needed? *Answer:* Moles per liter (mol/L).
Step 2: The units that exist are grams and 1 L. NaOH may be expressed as moles and grams. The gmw of NaOH is calculated to equal 40 g/mol.
Step 3: The equation becomes

$$\frac{24\ \text{g NaOH}}{\text{L}} \times \frac{1\ \text{mol}}{40\ \text{g NaOH}} = 0.6\ \frac{\text{mol}}{\text{L}} \qquad \text{(Eq. 1–22)}$$

By canceling out "like" units and performing the appropriate calculations, the final answer of 0.6 M or 0.6 mol/L is derived.

Example 1-6

Make up 250 mL of a 4.8 M solution of HCl.

Step 1: *Units* needed? *Answer:* Grams (g).
Step 2: Determine the gmw of HCl (36.5 g), which is needed to calculate the molarity.
Step 3: Set up the equation, cancel out "like" units, and perform the appropriate calculations:

$$\frac{36.5\ \text{g HCL}}{\text{mol}} \times \frac{4.8\ \text{mol HCL}}{\text{L}} \times \frac{250\ \text{mL} \times 1\ \text{L}}{1000\ \text{mL}} = 43.8\ \text{g HCl}$$

$$\text{(Eq. 1–23)}$$

In a 250-mL volumetric flask, add 200 mL of Type II water. Add 43.8 g of HCl and mix. Dilute up to the calibration mark with Type II water.

Although there are various methods to calculate laboratory mathematics problems, this technique of canceling "like" units can be used in most clinical chemistry situations, regardless of whether the problem requests molarity, normality, or exchanging one concentration term for another.

Normality

Normality (N) is expressed as the number of equivalent weights per liter (Eq/L) or milliequivalents per milliliter (mEq/mL). Equivalent weight is equal to the gmw divided by the valence V. Normality has often been used in acid–base calculations because an equivalent weight of a substance is also equal to its combining weight. Another advantage in using equivalent weight is that an equivalent weight of one substance is equal to the equivalent weight of any other chemical.

Example 1-7

Give the equivalent weight, in grams, for each substance listed below.

1. NaCl (gmw = 58 g, valence = 1)

$$58/1 = 58\ \text{g per equivalent weight} \qquad \text{(Eq. 1–24)}$$

2. HCl (gmw = 36, valence = 1)

$$36/1 = 36\ \text{g per equivalent weight} \qquad \text{(Eq. 1–25)}$$

3. H_2SO_4 (gmw = 98, valence = 2)

$$98/2 = 49\ \text{g per equivalent weight} \qquad \text{(Eq. 1–26)}$$

A. What is the normality of a 500-mL solution that contains 7 g of H_2SO_4? The approach used to calculate molarity could be used to solve this problem as well.

Step 1: Units needed? *Answer:* Equivalents per liter (Eq/L).
Step 2: Units you have? *Answer:* Milliliters and grams.
Step 3: Change grams into equivalent weight terms and set up the final equation. This equation is

$$\frac{7\ \text{g}\ H_2SO_4}{500\ \text{mL}} \times \frac{1\ \text{Eq}}{49\ \text{g}\ H_2SO_4} \times \frac{1000\ \text{mL}}{1\ \text{L}}$$

$$= 0.285\ \text{Eq/L} = 0.285\ N \qquad \text{(Eq. 1–27)}$$

Because 500 mL is equal to 0.5 L, the final equation could be written by substituting 0.5 L for 500 mL and thereby eliminating the need for including the 1000 mL/L conversion factor in the equation.

B. What is the normality of a 0.5 M solution of H_2SO_4? Continuing with the previous approach, the final equation is

$$\frac{0.5\ \text{mol}\ H_2SO_4}{\text{L}} \times \frac{98\ \text{g}\ H_2SO_4}{\text{mol}\ H_2SO_4} \times \frac{1\ \text{Eq}\ H_2SO_4}{49\ \text{g}\ H_2SO_4}$$

$$= 1\ \text{Eq/L} = 1\ N \qquad \text{(Eq. 1–28)}$$

When changing molarity into normality or vice versa, the following conversion formula may be applied:

$$M \times V = N \qquad \text{(Eq. 1–29)}$$

where V is the valence of the compound. Using this formula, Example 1–7B becomes

$$0.5\ M \times 2 = 1\ N \qquad \text{(Eq. 1–30)}$$

Example 1-8

What is the molarity of a 2.5 N solution of HCl? This problem may be solved in several ways. One way is to use the stepwise approach in which existing units are exchanged for units needed. The equation is

$$\frac{2.5\ \text{Eq HCl}}{\text{L}} \times \frac{36\ \text{g HCl}}{1\ \text{Eq}} \times \frac{1\ \text{mol HCl}}{36\ \text{g HCl}}$$

$$= 2.5\ \text{mol/L HCl} \qquad \text{(Eq. 1–31)}$$

The second approach is to use the normality to molarity conversion formula. The equation now becomes

$$M \times V = 2.5\ N$$

$$V = 1$$

$$M = \frac{2.5\ N}{1} = 2.5\ N \qquad \text{(Eq. 1–32)}$$

When the valence of a substance is 1, the molarity will be equal to the normality.

Specific Gravity

Density is expressed as mass per unit volume. The specific gravity is the ratio of the density of a material when compared to the density of water at a given temperature. The units for specific gravity are grams per milliliter. Specific gravity is often used with very concentrated materials, such as commercial acids like sulfuric and hydrochloric acids.

The density of a concentrated acid also can be expressed in terms of an assay or percent purity. The actual concentration is equal to the *specific gravity* multiplied by the *assay* or *percent purity value* (expressed as a decimal) stated on the label of the container. (See Appendix G, *Concentrations of Commonly Used Acids and Bases With Related Formulas,* for the specific gravity.)

Example 1-9

A. What is the actual weight of a supply of concentrated HCl whose label reads specific gravity 1.19 with an assay value of 37%?

$$1.19 \text{ g/mL} \times 0.37 = 0.44 \text{ g/mL of HCl} \quad \textit{(Eq. 1–33)}$$

B. What is the molarity of this stock solution? The final units desired are moles per liter (mol/L). The molarity of the solution is

$$\frac{0.44 \text{ g HCl}}{\text{mL}} \times \frac{1 \text{ mol HCl}}{3.5 \text{ g HCl}} \times \frac{1000 \text{ mL}}{\text{L}}$$

$$= 12.05 \text{ mol/L or } 12 \, M \quad \textit{(Eq. 1–34)}$$

Conversions

To convert one unit into another, the same approach of crossing out "like" units can be applied. In some instances, a chemistry laboratory may report a given analyte using two different concentration units. An example of this is calcium. The recommended SI unit for calcium is millimoles per liter. The better known and more traditional units are milligrams per deciliter (mg/dL).

Example 1-10

Convert 8.2 mg/dL calcium to millimoles per liter (mmol/L). The gmw of calcium is 40 g. So, if there are 40 g per mol, then it follows that there are 40 mg per mmol. The units wanted are mmol/L. The equation becomes

$$\frac{8.2 \text{ mg}}{\text{dL}} \times \frac{1 \text{ dL}}{100 \text{ mL}} \times \frac{1000 \text{ mL}}{\text{L}} \times \frac{1 \text{ mmol}}{40 \text{ mg}} = \frac{2.05 \text{ mmol}}{\text{L}}$$

(Eq. 1–35)

Once again, the systematic stepwise approach of deleting similar units can be used for this conversion problem.

A frequently encountered conversion is when you need a weaker concentration or different volume than the substance available, but the concentration *terms* are the same. The following formula is used:

$$V_1 \times C_1 = V_2 \times C_2 \quad \textit{(Eq. 1–36)}$$

This formula is useful only if the concentration and volume units between the substances are the same and if three out of the four variables are known.

Example 1-11

What volume is needed to make 500 mL of a 0.1 *M* solution of tris buffer from a solution of 2 *M* tris buffer?

$$V_1 \times 2 \, M = 0.1 \, M \times 500 \text{ mL}$$

$$(V_1)(2 \, M) = (0.1)(500) = (V_1)(2) = 50 = V_1 = 50/2 = 25$$

(Eq. 1–37)

It requires 25 mL of the 2 M solution to make up 500 mL of a 0.1-M solution. This problem differs from the other conversions in that it is actually a dilution of a stock solution. A more involved discussion of dilutions follows.

Dilutions

A *dilution* represents the ratio of concentrated or stock material to the total final volume of a solution and consists of the volume or weight of the concentrate plus the volume of the diluent, with the concentration units remaining the same. This ratio of concentrated or stock solution to the total solution volume equals the *dilution factor*. Because a dilution is made by adding concentrated stock to a diluent, it always represents a less concentrated substance. The relationship of the dilution factor to concentration is an inverse one; thus, the dilution factor increases as the concentration decreases. To determine the dilution factor required, simply take the amount needed and divide by the stock concentration, leaving it in a reduced-fraction form.

Example 1-12

What is the dilution factor needed to make a 100 mEq/L sodium solution from a 3000 mEq/L stock solution? The dilution factor becomes

$$\frac{100}{3000} = \frac{1}{30} \quad \textit{(Eq. 1–38)}$$

The dilution factor indicates that the ratio of stock material is 1 part stock made to a *total volume* of 30. To actually make this dilution, 1 mL of stock is added to 29 mL of diluent. Note that the dilution factor indicates the parts per total amount, but in making the dilution, the sum of the amount of the stock material plus the amount of the diluent must equal the total volume or dilution fraction denominator. The dilution factor may be written as a fraction or can be expressed as 1 : 30. Either format is correct. Any total volume can be used as long as the fraction reduces to give the dilution factor.

Example 1-13

If in the preceding example 150 mL of the 100 mEq/L sodium solution was required, the dilution ratio of stock to total volume must be maintained. Set up a ratio between the desired total volume and the dilution factor to determine the amount of stock needed. The equation becomes

$$\frac{1}{30} = \frac{x}{150}$$

$$x = 5 \quad \textit{(Eq. 1–39)}$$

Note that ⁵⁄₁₅₀ reduces to the dilution factor of ¹⁄₃₀. To make up this solution, 5 mL of stock is added to 145 mL of the appropriate diluent, making the ratio of *stock volume* to *diluent volume* equal to ⁵⁄₁₄₅. Recall that the dilution factor includes the total volume of *both* stock plus diluent.

Simple Dilutions

When making a *simple dilution,* the laboratorian must decide on the total volume desired and the amount of stock to be used.

Example 1-14

A $1:10$ (⅒) dilution of serum can be achieved by using any of the following approaches:

A. 100 μL of serum and 900 μL of saline.
B. 20 μL of serum and 180 μL of saline.
C. 1 mL of serum and 9 mL of saline.
D. 2 mL of serum and 18 mL of saline.

Note that the ratio of serum to diluent needed to make up each dilution satisfies the dilution factor of stock material to total volume.

The dilution factor is also used to determine the concentration of a dilution or stock material by multiplying the original concentration by the dilution factor.

Example 1-15

Determine the concentration of a 200 N mol/mL human chorionic gonadotropin (hCG) standard that was diluted ⅟₅₀. This value is obtained by multiplying the original concentration, 200 N mol/mL hCG, by the dilution factor, ⅟₅₀. The result is 4 μmol/mL hCG. Quite often, the concentration of the original material is needed.

When determining the original stock or undiluted concentration, multiply the concentration of the dilution by the dilution factor denominator.

Example 1-16

A $1:2$ dilution of serum with saline had a creatinine result of 8.6 mg/dL. Calculate the actual serum creatinine concentration.

Dilution factor: ½

Dilution result = 8.6 mg/dL

Because this result represents ½ of the concentration, the actual serum creatinine value is

$$2 \times 8.6 = 17.2 \text{ mg/dL} \qquad (Eq.\ 1\text{--}40)$$

Serial Dilutions

A *serial dilution* may be defined as multiple progressive dilutions ranging from more concentrated solutions to less concentrated solutions. Serial dilutions are extremely useful when the volume of concentrate or diluent is in short supply and needs to be minimized. The volume of patient serum available to the laboratory may be very small (*eg,* pediatric samples), and a serial dilution may be needed to ensure that sufficient sample is available. The serial dilution is initially made in the same manner as a simple dilution. Subsequent dilutions will then be made from each preceding dilution. When a serial dilution is made, certain criteria may need to be satisfied. The criteria vary from one situation to the next but usually include such considerations as the total volume desired, the amount of diluent or concentrate available, the dilution factor, and the support materials needed.

Example 1-17

A serum sample is to be diluted $1:2$, $1:4$, and finally, $1:8$. It is arbitrarily decided that the total volume for each dilution is to be 1 mL. Note that the least common denominator for these dilution factors is 2. Once the first dilution is made (½), a $1:2$ (least common denominator between 2 and 4) dilution of it will yield

$$½ \times ½ = ¼$$

(initial dilution factor)(next dilution factor)

$$= \text{(final dilution factor)} \qquad (Eq.\ 1\text{--}41)$$

By making a $1:2$ dilution of the first dilution, the second dilution factor of $1:4$ is satisfied. Making a $1:2$ dilution of the $1:4$ dilution will result in the next dilution ($1:8$). To establish the dilution factor needed for subsequent dilutions, it is helpful to solve the following equation for (x):

Stock/preceding concentration $\times$ (x)

$$= \text{final dilution factor needed} \qquad (Eq.\ 1\text{--}42)$$

To make up these dilutions, three test tubes are labeled $1:2$, $1:4$, and $1:8$, respectively. One mL of diluent is added to each test tube. To make the primary dilution of $1:2$, 1 mL of serum is added to test tube number 1. The solution is mixed, and 1 mL of the primary dilution is removed and added to test tube number 2. After mixing, this solution contains a $1:4$ dilution. Then 1 mL of the $1:4$ dilution from test tube number 2 is added to test tube number 3. After discarding 1 mL from test tube number 3, the resultant dilution in this test tube is $1:8$, and this satisfies all of the previously established criteria. Refer to Figure 1-14 for an illustration of this serial dilution.

Example 1-18

Another type of serial dilution combines several dilution factors that are not multiples of one another. In our previous example, $1:2$, $1:4$, and $1:8$ dilutions are all related to one another by a factor of 2. Consider the situation when $1:10$, $1:20$, $1:100$, and $1:200$ dilution factors are required. There are several approaches to solving this type of dilution problem. One method is to treat the $1:10$ and $1:20$ dilutions as one serial dilution problem, the $1:20$ and $1:100$ dilutions as a second serial dilution, and the $1:100$ and $1:200$ dilutions as the last serial dilution. Another approach is to consider what dilution factor of the concentrate is needed to yield the final dilution. In this example, the initial dilution is $1:10$, with subsequent dilutions of $1:20$, $1:100$, and $1:200$. The first dilution may be accomplished by adding 1 mL of concentrate to 9 mL of diluent. The total volume of solution is 10 mL. Our initial dilution factor has been satisfied. In

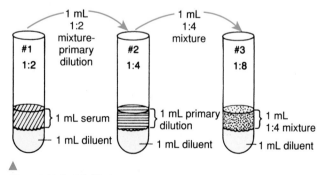

Figure 1-14. Serial dilution.

making the remaining dilutions, 2 mL of diluent is added to each test tube.

Initial/preceding dilution $\times$ (x) = dilution needed

Solve for x.

Using the dilution factors listed above and solving for (x), the equations become

$1:10 \times (x) = 1:20$,
where $(x) = 2$ (or 1 part stock to 1 part diluent)

$1:20 \times (x) = 1:100$,
where $(x) = 5$ (or 1 part stock to 4 parts diluent)

$1:100 \times (x) = 1:200$,
where $(x) = 2$ (or 1 part stock to 1 part diluent)   *(Eq. 1–43)*

In practice, the 1:10 dilution must be diluted by a factor of 2 to obtain a subsequent 1:20 dilution. Because the second tube already contains 2 mL of diluent, 2 mL of the 1:10 dilution should be added (1 part stock to 1 part diluent). In preparing the 1:100 dilution from this, a 1:5 dilution factor of the 1:20 mixture is required (1 part stock to 4 parts diluent). Because this tube already contains 2 mL, the volume of diluent in the tube is divided by its parts, which is 4; thus, 500 mL, or 0.500 mL, of stock should be added. The 1:200 dilution is prepared in the same manner using a 1:2 dilution factor (1 part stock to 1 part diluent) and adding 2 mL of the 1:100 to the 2 mL of diluent already in the tube.

Water of Hydration

Some compounds are available in a hydrated form. To obtain a correct weight for these chemicals, the attached water molecule(s) must be included.

Example 1-19

How much $CuSO_4 \cdot 5H_2O$ must be weighed to prepare 1 L of 0.5 M $CuSO_4$? When calculating the gmw of this substance, the water weight must be considered so that the gmw is 250 g rather than gmw of $CuSO_4$ alone (160 g). Therefore,

$$\frac{250 \text{ (g) } CuSO_4 \cdot 5H_2O}{\text{mol}} \times \frac{0.5 \text{ mol}}{1 \text{ (L)}} = 125 \text{ g/L} \quad \text{(Eq. 1–44)}$$

Cancel out "like" terms to obtain the result of 125 g/L. A reagent protocol often designates the use of an anhydrous form of a chemical; frequently, however, all that is available is a hydrated form.

Example 1-20

A procedure requires 0.9 g of $CuSO_4$. All that is available is $CuSO_4 \cdot 5H_2O$. What weight of $CuSO_4 \cdot 5H_2O$ is needed?

Calculate the percentage of $CuSO_4$ present in $CuSO_4 \cdot 5H_2O$. The percentage is

$$\frac{160}{250} = 0.64, \quad \text{or} \quad 64\% \quad \text{(Eq. 1–45)}$$

Therefore, 1 g of $CuSO_4 \cdot 5H_2O$ contains 0.64 g of $CuSO_4$ so the equation becomes:

$$\frac{0.9 \text{ g } CuSO_4 \text{ needed}}{0.64 \; CuSO_4 \text{ in } CuSO_4 \cdot 5H_2O}$$

$$= 1.41 \text{ g } CuSO_4 \cdot 5H_2O \text{ required} \quad \text{(Eq. 1–46)}$$

Graphing Beer's Law

The *BeerBouger/Lambert Law (Beer's Law)* mathematically establishes the relationship between concentration and absorbance in many photometric determinations. Beer's Law is expressed as

$$A = abc \quad \text{(Eq. 1–47)}$$

where A = absorbance; a = absorptivity constant for a particular compound at a given wavelength under specified conditions of temperature, pH, and so on; b = length of the light path; and c = concentration.

If a method follows Beer's Law, then absorbance is proportional to concentration as long as the length of the light path and the absorptivity of the absorbing species remains unaltered during the analysis. Even in automated systems, adherence to Beer's Law is often determined by checking the linearity of the test method over a wide concentration range. At the end of manual assays, the concentration results are obtained by using a Beer's Law graph, known as a standard graph. This graph is made by plotting absorbance versus the concentration of known standards (Fig. 1-15). Because most photometric assays set the initial absorbance to zero (0) using a reagent blank, the initial data points are 0,0. Graphs should be labeled properly and the concentration units must be given. The horizontal axis is referred to as the x-axis, whereas the vertical line is the y-axis. It is not important which variable (absorbance or concentration) is assigned to an individual axis, but it is very important that the values assigned to them are uniformly distributed along the axis. By convention in the clinical laboratory, concentration is usually plotted on the x-axis. On a standard graph, only the standard and the associated absorbances are plotted.

Once a standard graph has been established, it is permissible to run just one standard, or calibrator, as long as the system remains the same. *One point calculation* or *calibration* refers to the calculation of the comparison of the known standard/calibrator concentration and its corresponding absorbance to the absorbance of the unknown value according to the following ratio:

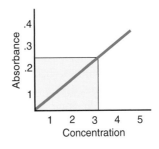

Unknown absorbance=.250
Concentration from graph=3.2

▲

Figure 1-15. Standard curve.

$$\frac{\text{Concentration of standard } (Cs)}{\text{Absorbance of standard } (As)}$$

$$= \frac{\text{Concentration of unknown } (Cu)}{\text{Absorbance of the unknown } (Au)} \quad (Eq.\ 1\text{–}48)$$

Solving for the concentration of the unknown, the equation becomes:

$$C_u = \frac{(A_u)(C_s)}{A_s} \quad (Eq.\ 1\text{–}49)$$

Example 1-21

The biuret protein assay is very stable and follows Beer's Law. Rather than make up a complete new standard graph, one standard (6 g/dL) was assayed. The absorbance of the standard was 0.400, and the absorbance of the unknown was 0.350. Determine the value of the unknown in g/dL.

$$Cu = \frac{(0.350)(6 \text{ g/dL})}{(0.400)} = 5.25 \text{ g/dL} \quad (Eq.\ 1\text{–}50)$$

This method of calculation is acceptable as long as everything, including the instrument and lot of reagents, remains the same. If anything in the system changes, a new standard graph should be done.

Enzyme Calculations

Another application of Beer's Law is the calculation of enzyme assay results. When calculating enzyme results, the *rate of absorbance change* is often monitored continuously during the reaction to give the difference in absorbance, known as the *delta absorbance,* or ΔA. Instead of using a standard graph or a one point calculation, the molar absorptivity of the product is used. If the absorptivity constant and absorbance, in this case ΔA, is given, Beer's Law can be used to calculate the enzyme concentration directly, as follows:

$$A = abc$$

$$C = \frac{A}{ab} \quad (Eq.\ 1\text{–}51)$$

When the absorptivity constant (*a*) is given in units of grams per liter (moles) through a 1-centimeter (cm) light path, the term *molar absorptivity* (ϵ) is used. Substitution of ϵ for *a*, and ΔA for *A* produces the following Beer's Law formula:

$$C = \frac{\Delta A}{\epsilon} \quad (Eq.\ 1\text{–}52)$$

Units used to report enzyme activity traditionally have included weight, time, and volume. In the early days of enzymology, method-specific units (*eg*, King-Armstrong, Caraway) were all different and very confusing. In 1961, the Enzyme Commission of the International Union of Biochemistry recommended using one unit, the international unit (IU), for reporting enzyme activity. The IU is the amount of enzyme that will catalyze 1 µmol/min. These units were often expressed as units per liter (U/L). The designations IU, U, or IU/L were adopted by many clinical laboratories to represent the IU. Although the reporting unit is the same, unless the conditions of the analysis are identical, use of the IU does not standardize the actual enzyme activity. The SI recommended unit is the *katal,* which is expressed as mol/L/second. Whichever unit is used, calculation of the activity using Beer's Law required inclusion of the dilution and, depending on the reporting unit, required conversion to the appropriate term (*eg*, µmol to mol, mL to L, minute to second, temperature factors, and so on). Beer's Law for the IU now becomes:

$$C = \frac{(\Delta A)10^{-6}\,(TV)}{(\epsilon)(b)(SV)} \quad (Eq.\ 1\text{–}53)$$

where TV = total volume of sample plus reagents in mL and SV = sample volume used in mL. The 10^{-6} converts moles to nmol for the IU. If another unit of activity is used, such as the katal, conversion into liters and seconds would be needed, but the conversion to and from micromoles would not be needed.

Example 1-22

The ΔA per minute for an enzyme reaction is 0.250. The product measured has a molar absorptivity of 12.2×10^3 at 425 nm at 30°C. The incubation and reaction temperature are also kept at 30°C. The assay calls for 1 mL of reagent and 0.050 mL of sample. Give the enzyme activity results in international units.

Applying Beer's Law and the necessary conversion information, the equation becomes:

$$c = \frac{(0.250)(10^{-6})(1.050 \text{ mL})}{(12.2 \times 10^3)(1)(0.050 \text{ mL})} = 430 \text{ U} \quad (Eq.\ 1\text{–}54)$$

Note: *b* is usually always given as 1 cm, and because it is a constant, may not be considered in the calculation.

SPECIMEN CONSIDERATIONS

Specimen collection, handling, and processing remains one of the *primary* areas of preanalytical error. Careful attention to each phase is necessary to ensure proper subsequent testing and reporting of meaningful results. All accreditation agencies require laboratories to clearly define and delineate the procedures used for proper collection, transport, and processing of patient samples and the steps used to minimize and detect any errors.

Types of Samples

Phlebotomy, or *venipuncture,* is the act of actually obtaining a blood sample from a vein using a needle attached to either a syringe or a stoppered *evacuated tube.* The most frequent site for venipuncture is the antecubital of the arm. A tourniquet made of pliable rubber tubing or a strip with Velcro™ at the end is wrapped around the arm, causing a cessation of blood flow and dilation of the veins, making them easier to detect. Sites adjacent to IV therapy should be avoided, but in the event both arms are involved in IV therapy and the IV cannot be discontinued for a short time, a site *below* the IV site should be sought. The initial sample drawn (5 mL) should be discarded because it is most likely contaminated with IV fluid and only subsequent sample tubes used for analytical purposes.

In addition to venipuncture, blood samples can be collected using a skin puncture technique that customarily involves the outer area of the bottom of the foot known as a "heel stick," or the fleshy part of the middle of the last phalanx of the third or fourth (ring) finger and is known as a "finger stick." A sharp lancet is used to pierce the skin and a capillary tube (ie, short, narrow glass tube) for sample collection.

Additional information regarding phlebotomy and skin puncture is available from several sources[18,19] plus many accredited training programs in phlebotomy and laboratory science include formal guidelines and procedures for the proper collection of blood samples. Unfortunately, other hospital personnel who may be collecting samples (ie, nurses, physicians, and so on) may not have the benefit of such training. This lack of training may result in an increase in the errors as the result of improper collection.

Analytical testing of blood involves the use of whole blood, serum, or plasma. *Whole blood,* as the name implies, uses both the liquid portion of the blood called plasma and the cellular components (red blood cells, white blood cells, and platelets). This requires blood collection into a vessel containing an anticoagulant. Complete mixing of the blood immediately following venipuncture is necessary to ensure the anticoagulant can adequately inhibit the blood's clotting factors. As whole blood sits, the cells begin to fall toward the bottom, leaving a clear yellow supernate on top. This clear yellow fluid is called plasma. Centrifuging the sample accelerates the process of separating the plasma and cells. Specimens should be centrifuged for approximately 10 minutes at an RCF of 1000 g to 2000 g. In a tube that is sterile but does not contain an anticoagulant, the blood's clotting factors are active to form a clot. The clot is encapsulated by the large protein fibrinogen. The remaining liquid is called serum rather than plasma. Most testing in the clinical chemistry laboratory is performed on serum. The major difference between plasma and serum is that serum does not contain fibrinogen (ie, there is less protein in serum than plasma) and some potassium is released from platelets (serum potassium is slightly higher in serum than in plasma). It is important that serum samples be allowed to completely clot (about 20 minutes) before being centrifuged.

Arterial blood samples are used to measure blood gases (partial pressures of oxygen and carbon dioxide) and pH. Syringes are used because evacuated tubes are inappropriate because of the pressure in an arterial blood vessel. The primary arterial sites are the radial, brachial, and femoral arteries. Arterial punctures are more difficult to perform because of the inherent arterial pressure, difficulty in stopping bleeding afterward, and the undesirable development of a hematoma, which cuts off the blood supply to the surrounding tissues. Because of the severity of the possible adverse consequences, physicians most often perform arterial punctures. NCCLS document H11–A2, 1992, provides further information on arterial punctures.[20]

In some instances, such as glucose, continued metabolism will occur if the serum or plasma remains in contact with the cells for any period. Evacuated tubes may incorporate within them plastic gel-like material that serves as a barrier between the cells and the plasma or serum and seals off these compartments from one another during centrifugation. Some of the gels can interfere with certain analytes, notably trace metals, and drugs such as the tricyclic antidepressants.

Any preservatives required before analytical testing can also be incorporated into the collection tubes. Usually, either the collection tube (ie, capillary tubes) or the stopper is color coded for easy identification of the presence of an anticoagulant or preservative (Table 1-8). Not all tests have the same sample requirements, so the laboratorian needs to ascertain the specific collection requirements for each test before sample collection. To prevent contamination, tubes should be drawn starting with blood culture tubes, then tubes without anticoagulants or preservatives (red stoppered), followed by tubes containing anticoagulants and preservatives (blue, then green), with the tubes containing *ethylenediaminetetraacetic acid (EDTA)* (lavender) and sodium fluoride/iodoacetate (gray) being drawn last.

Blood is not the only sample analyzed in the clinical chemistry laboratory. The next most common fluid for determination is urine. Most quantitative analyses of urine require a timed sample (usually 24 hours), and a complete sample (all urine in the specified time must be collected) can be difficult because many timed samples are collected by the patient in an outpatient situation. Creatinine analysis is often used to assess the completeness of a 24-hour urine sample because the output of creatinine is relatively free from interference and is very stable, with little change in output between individuals. On average, an adult excretes 1–2 g of creatinine per 24 hours. The volume of urine differs widely among individuals, but a container able to hold 4 L is adequate (average output is about 2 L).

Other body fluids analyzed by the clinical chemistry laboratory include *cerebral spinal fluid (CSF); paracentesis* fluids (pleural, pericardial, and peritoneal); and amniotic fluids. For all of these samples, the color and characteristics of the fluid *before* centrifugation should be noted. A laboratorian should also verify that the sample is designated for clinical chemistry analysis only *before* centrifugation, because a single fluid sample may be shared among several departments (ie, hematology or microbiology) and centrifugation could invalidate some tests in those areas.

CSF fluid is an ultrafiltrate of the plasma and will, ordinarily, reflect the values seen in the plasma. For glucose and protein analysis (total and specific proteins), it is recommended that a blood sample be analyzed concurrently with the analysis of those analytes in the CSF. This will assist in determining the clinical utility of the values obtained on the CSF sample. This is also true for lactate dehydrogenase and protein assays requested on paracentesis fluids. All fluid samples should be handled immediately without delay between sample procurement, transport, and analysis.

Amniotic fluid is used to assess fetal lung maturity (L/S ratio), congenital diseases, hemolytic diseases, genetic defects, and gestational age. The laboratorian should verify the specific handling of this fluid with the manufacturer of the testing procedure(s).

TABLE 1-8. Commonly Used Evacuated Tubes

Stopper Color	Additive	Action	Use
Red	None	Allows blood clotting resulting in serum	Most chemistry, blood bank, and immunology assays
Red/gray Red/black	Contains separator material	Allows blood clotting resulting in serum; material serves as a barrier between serum and cells	Most chemistry tests
Lavender	EDTA (Na_2 or K_2)	Anticoagulant; binds Ca^{++} resulting in whole blood or plasma	CEA, lead, CBC with differential, platelet and reticulocyte counts
Orange	Thrombin	Accelerates clot formation resulting in serum	STAT serum tests (less time needed for clot formation)
Blue	Na Citrate	Anticoagulant; binds Ca^{++} resulting in whole blood or plasma	Coagulation testing; factor assays, fibrinogen, PT, PTT, thrombin time
Gray	a) Na fluoride/K oxalate	Inhibits the glycolytic enzyme enolase and acts as an anticoagulant resulting in whole blood or plasma. Interferes with Na, K, and most BUN (urease) determinations	Glucose (OGTT)/lactate
	b) Iodoacetate	Inhibits the glycolytic enzyme glyceraldehyde-3-phosphate; results in serum; will not interfere with Na, K, or BUN (urease) assays	
Green	Heparin (Na, Li, or NH_4); interferes with many ions	Inhibits thrombin activation resulting in whole blood or plasma	Ammonia, carboxy/methemoglobin, lead

Sample Processing

Once a sample arrives in the laboratory, the samples are then "processed." In the clinical chemistry laboratory, this means proper matching of the blood collection tube(s) with the appropriate analyte request and patient identification labels. This is a particularly sensitive area of preanalytical error. Bar code readers have become popular as a means to detect errors at this point of the processing and to minimize clerical errors. In some facilities, samples are numbered or entered into work lists or a second identification system useful in the analytical phase. The laboratorian must also ascertain if the sample is acceptable for further processing. The criteria used depends on the test involved, but usually include volume considerations (ie, is there sufficient volume for testing needs); use of proper anticoagulants or preservatives; that timing is clearly indicated for timed testing; that specimen is intact; and that specimen has been transported properly (ie, cooled or on ice, within a reasonable period, protected from light, properly stoppered, and so on). Unless a whole blood analysis is being performed, the sample is then centrifuged as previously described and the serum or plasma should be separated from the cells.

Once processed, the laboratorian should note the presence of any serum or plasma characteristics such as *hemolysis, icterus,* or turbidity. Samples should be analyzed within 4 hours and to minimize effects of evaporation, they should be properly capped, away from areas of rapid airflow, light, and heat. If testing is to occur after that time, samples should be appropriately stored. For most, this means refrigeration at 4°C for 8 hours. Most analytes are quite stable at this temperature, with the exception of alkaline phosphatase (increases) and lac-

tate dehydrogenase (decreases as a result of temperature labile fractions four and five). Samples may be frozen at −20°C and stored for longer periods without deleterious effects on the results. Repeated freezing and thawing, like those that occur in so called "frost-free" freezers, should be avoided.

Sample Variables

Sample variables include physiologic considerations, proper patient preparation, and problems in collection, transportation, processing, and storage. Although laboratorians must include mechanisms to minimize the effect these factors have on testing, and to document each preanalytical incident, it is often very frustrating to try and control all the variables that largely depend on individuals outside of the laboratory. The best course of action is to critically assess or predict the weak areas or links, identify potential problems, and put a plan of action in place, which contains policies, procedures, or "checkpoints" throughout the sample's journey to the laboratorian actually performing the test. Good communication with all personnel involved helps to ensure that whatever plans are in place meet the needs of the laboratory and, ultimately, the patient and physician.

Physiologic variation refers to changes that occur within the body such as cyclic changes (diurnal or circadian variation) or those resulting from exercise; diet; stress; gender; age; underlying medical conditions (ie, fever, asthma, obesity); drugs; or posture (Table 1-9). Most samples are drawn on patients who are fasting (usually overnight for at least 8 hours), but for those tests that are most affected by diet or fasting, the length of time and what was consumed during

TABLE 1-9. Factors Affecting Common Serum Constituents

Constituent	Prolonged Tourniquet Application	Hemolysis	Posture Supine to Standing	Diurnal Variation	EDTA	Meals	Other
Ammonia	↑	↑				↑	Cigarette smoking; ↑ with exercise
TP/Albumin	↑	↑ (Alb ↓*)	↑				Lipemia
Bilirubin		↓					Light
Iron	↑	↑		√		↑	Menstruation
Glucose		√				↑	Caffeine, smoking
Phosphate		↑				↓	
Electrolytes							
Calcium			↑	↓		↑ (ionized)	Marked changes in albumin and pH must be considered in evaluation
Na		↓ (severe)					
K	↓	↑				↑	↑ with opening/closing of hand before collection
Chloride					↓		
Magnesium		↑			↓		Lipemia*
Enzymes							
ALT		↑	↑				↑ with strenuous exercise
AST		↑	↑				↑ with strenuous exercise
Amylase			↑	↓			Serum and only heparinized plasma
CK		↑ (moderate)*					Light and pH sensitive—keep covered to minimize CO$_2$ loss and in dark
LD		↑					Cold labile fractions
ALP		↑	↑	↓			
ACP		↑	↑				Lipemia; pH stabilized at 5.4 for proper storage
Lipids							
Cholesterol	↑	↑*	↑				
Triglyceride		↑*	↑				
Hormones							
Prolactin			√			↑ (protein)	Stress, ambulation, exercise = ↑
Insulin						↑	
ACTH				√			Adheres to glass; use polystyrene
Catecholamines		↑		√		↑	Stress; ↑ with intoxication, tobacco
Cortisol				√			Caffeine
Gastrin						↑	
GH				√		↑	↑ with exercise
PTH				√	√		Affected by Ca levels; short half-life
Renin			√		√		Sample should not be chilled; affected by licorice
Aldosterone					√		Na & K values must be considered
Glucagon						↑	
Thyroid			↑				
TSH				√			

* Possible interference in assay methods.

that time should be determined before sample procurement because "overnight" and "fasting" are relative terms. Patient preparation for timed samples or those requiring specific diet or other instructions must be well written and verbally explained to patients. Elderly patients often misunderstand or are overwhelmed by the directions given to them by physician office personnel. A laboratory information telephone number listed on these directions can often serve as an excellent reference for proper patient preparation.

Drugs can affect various analytes,[21] so it is important to ascertain what, if any, medications the patient is taking that may interfere with the test. Unfortunately, many laboratorians do not have either the time or access to this information, and the interest in this type of interference only arises when the physician questions a result. Some frequently encountered influences are smoking, which, through the action of nicotine, causes an increase in glucose, growth hormone, cortisol, cholesterol, triglycerides, and urea. High amounts or chronic consumption of alcohol causes hypoglycemia, increased triglycerides, and an increase in the enzyme gamma-glutamyltransferase. Intramuscular injections are known to increase the enzyme creatine kinase and the skeletal muscle fraction of lactate dehydrogenase. Opiates such as morphine or meperidine causes increases in liver and pancreatic enzymes, and oral contraceptives may affect many analytical results. Many drugs affect liver function tests. Diuretics can cause decreased potassium and hyponatremia. Thiazides can cause hyperglycemia and prerenal azotemia secondary to the decrease in blood volume.

Proper patient identification is the first step in sample collection. The importance of using the proper collection tube, avoiding prolonged tourniquet application, drawing tubes in the proper order, and proper labeling of tubes cannot be stressed enough. Prolonged tourniquet application causes a stasis of blood flow and an increase in hemoconcentration and anything bound to proteins or the cells. Having patients open and close their fist during phlebotomy is of no value but may cause an increase in potassium and should be avoided. Intravenous contamination should be considered if a very large increase in the substances is being infused such as glucose, potassium, sodium, and chloride with a decrease of other analytes such as urea and creatinine. In addition, the proper antiseptic must be used. For example, isopropyl alcohol "wipes" are used for cleaning and disinfecting the collection site. However, this is not the proper antiseptic for disinfecting the site when drawing blood alcohol levels.

Postcollection variations are related to those factors discussed previously under specimen processing. Clerical errors are the most frequently encountered, followed by inadequate separation of cells from serum, and improper storage.

Chain of Custody

In situations in which laboratory tests are likely linked to a crime or accident, they become forensic in nature. In these cases, documented specimen identification is required at each phase of the process. Each facility has its own forms, but the patient, and usually a witness, must identify the sample. It should have a tamper-proof seal. Any individual that comes in contact with the sample must document the receipt of the sample, its condition at the time of receipt, plus the date and time it was received. In some instances, a single witness verifies the entire process and co-signs as the sample moves along. However, any analytical test could be used as part of legal testimony. Therefore the laboratorian should give each sample, even without the documentation, the same attention given to a forensic sample.

SUMMARY

The primary goal of any clinical chemistry laboratory is to correctly perform analytical procedures that yield accurate and precise information to aid in patient diagnosis. To achieve accurate and reliable results that reflect the patient's physiologic status, the clinical laboratorian must be familiar with the use of basic supplies and equipment. He or she also must have a fundamental understanding of the concepts critical to clinical chemistry testing. This chapter provides the information necessary for the important practice of clinical chemistry including units of measure; temperature; reagents (chemicals, standards, solutions, water specifications); laboratory supplies (glass and plasticware, pipets, burets, balances, and desiccants); separation techniques; specimen collection, transport, and processing; and laboratory mathematics and calculations.

EXERCISES

1. Determine each of the following for a solution containing 64 g of NaCl made up to 500 mL with distilled water.
   a. Molarity
   b. Normality
   c. Percent (w/v)
   d. Dilution factor
2. Determine the concentration values for each of the following for a solution containing 10 mg of $CaCl_2$ made with 200 mL of distilled water.
   a. mg/dL
   b. Molarity
   c. Normality
3. You must make 1 L of 0.1 M acetic acid (CH_3COOH). All you have available is concentrated glacial acetic acid whose assay value is 98% and whose specific gravity is 1.05 g/mL. How many mL are needed to make this solution?
4. What is the hydrogen ion concentration of an acetate buffer whose pH is 4.24?
5. Use the Henderson-Hasselbalch equation to solve each of the following buffer problems:
   a. What is the ratio of salt to weak acid for a Veronal buffer with a pH of 8.6 and a pK_a of 7.43?

b. The pK_a for acetic acid is 4.76. What is the expected pH if the concentration of salt is 5 mmol/L and that of acetic acid is 10 mmol/L?

6. The hydrogen ion concentration of a solution is 0.0000937. What is the pH?

7. You are making up a standard curve for an albumin assay. The standard values needed are:

1.0 g/dL
2.0 g/dL
4.0 g/dL
6.0 g/dL
8.0 g/dL
10.0 g/dL

You have a 30 g/dL stock solution.

a. What is the dilution needed to obtain the standard values needed?

b. You need a total volume of 15 mL per standard. List the amount of stock plus the amount of diluent needed to make up each standard.

8. Perform the following conversions.

8×10^3 mg = _____ μg
40 μg = _____ mg
13 dL = _____ mL
200 μL = _____ mL
5 μL = _____ mL
50 mL = _____ L
4 cm = _____ mm

9. What volume of 14 N H_2SO_4 is needed to make 250 mL of 3.2 M H_2SO_4 solution?

10. A 24-hour urine has a total volume of 1200 mL. A 1:200 dilution of the urine specimen gives a creatinine result of 0.8 mg/dL.

a. What is the creatinine concentration for the undiluted urine specimen?

b. What is the concentration in terms of milligrams per milliliter?

c. What is the result in terms of grams per 24 hours?

11. How much $CuSO_4 \cdot 5H_2O$ is needed to make up 500 mL of 2.5 M $CuSO_4$?

12. An enzyme reaction yields a ΔA of 0.031 per 30 seconds. The assay calls for 1.5 mL of reagent and 50 μL of sample. The ϵ of the product is 5.3×10^3. Give the enzyme results in international units.

13. A blood sample is received in the laboratory. The test requests include electrolytes, CK, LD, and AST. After centrifuging, hemolysis is noted. Which tests, if any, are invalid?

14. A blood sample is received in the lab with a label having only a room number written on it. It is a private room, so only one person is assigned for that room. Is this sample acceptable?

15. A 3-hour glucose tolerance from a remote site is received. This tube uses sodium fluoride as glycolytic inhibitor. The physician's office calls to add a test for urea. Can you satisfy this request?

REFERENCES

1. National Committee for Clinical Laboratory Standards. Quantities and units: SI. Committee Report C11-CR. Villanova, PA: NCCLS, 1979.
2. National Committee for Clinical Laboratory Standards. Temperature calibration of water baths, instruments, and temperature sensors. NCCLS Document 12-A2. Villanova, PA: NCCLS, 1990.
3. Bowers GN, Inman SR. The gallium melting-point standard. Clin Chem 1977;23:733.
4. Bowie L, Esters F, Bolin J, et al. Development of an aqueous temperature-indicated technique and its application to clinical laboratory instruments. Clin Chem 1976;22:449.
5. Harris D. Quantitative chemical analysis. San Francisco: Freeman, 1982.
6. International Union of Pure and Allied Chemistry and International Federation of Clinical Chemistry. Quantities and units in clinical chemistry, Bulletin No. 20. Oxford: IUPAC, 1972.
7. National Committee for Clinical Laboratory Standards. Development of certified reference materials for the national reference system for the clinical laboratory. NCCLS Document NRSCLS-3A. Villanova, PA: NCCLS, 1991.
8. National Institute of Standards and Technology. Standard reference materials. Publication 260. Washington, DC: US Department of Commerce, 1991.
9. Mears TW, Young DS. Measurement and standard reference materials in clinical chemistry. Am J Clin Pathol 1968;50:411.
10. National Committee for Clinical Laboratory Standards. Preparation and testing of reagent water in the clinical laboratory. NCCLS Document C3-A2. Villanova, PA: NCCLS, 1991.
11. Glass engineering handbook. 2nd ed. New York: McGraw-Hill, 1958.
12. Morey GW. The properties of glass (American Chemical Society monograph). New York: Reinhold, 1954.
13. National Institute of Standards and Technology. The international system of units (SI). NIST Publication 330. Gaithersburg, MD: NIST, 1991.
14. Bio Rad Laboratories. Procedure for comparing precision of pipet tips. Product Information No. 81-0208, 1993.
15. National Committee for Clinical Laboratory Standards. Determining performance of volumetric equipment. NCCLS document 18-P. Villanova, PA, 1984.
16. Lide DR, ed. Handbook of chemistry and physics. 76th ed. Boca Raton, FL: CRC Press, 1995.
17. Kaplan L, Pesce A, eds. Clinical chemistry: theory, analysis, correlation. 3rd ed. St. Louis, MO: Mosby-Year Book, 1996.
18. National Committee for Clinical Laboratory Standards. Procedures for the collection of diagnostic blood specimens by skin puncture. Approved standard H4—A3. Villanova, PA: NCCLS, 1991.
19. National Committee for Clinical Laboratory Standards. Procedures for the collection of diagnostic blood specimens by venipuncture. Approved standard H3—A3. Villanova, PA: NCCLS, 1991.
20. National Committee for Clinical Laboratory Standards. Collection technique and patient preparation for arterial blood. Approved standard H11–A2. Villanova, PA: NCCLS, 1992.
21. Young, DS. Effects of drugs on clinical laboratory tests. Washington, DC: AACC, 1990.

CHAPTER 2

Laboratory Safety and Regulations

Carolyn Houghton, Mary Ruth Beckham

Objectives

Upon completion of this chapter, the clinical laboratorian should be able to:

- *Discuss safety awareness for clinical laboratory personnel.*

- *List the responsibilities of employer and employee in providing a safe workplace.*

- *Identify hazards related to handling chemicals, biologic specimens, and radiologic materials.*

- *Choose appropriate personal protective equipment when working in the clinical laboratory.*

- *Identify the classes of fires and the type of fire extinguishers to use for each.*

- *Describe steps used as precautionary measures when working with electrical equipment, cryogenic materials, and compressed gases, and avoiding mechanical hazards associated with laboratory equipment.*

- *Select the correct means for disposal of waste generated in the clinical laboratory.*

- *Outline the steps required in documentation of an accident in the workplace.*

KEY TERMS

Airborne pathogens
Biohazard
Bloodborne pathogens
Carcinogen
Chemical hygiene plan
Corrosive chemical
Cryogenic material
Fire tetrahedron
Hazard communication standard
Hazardous material
High Efficiency Particulate Air
   (HEPA) Filter
Laboratory standard

Mechanical hazard
Medical waste
Material safety data sheet
   (MSDS)
National Fire Protection
   Association (NFPA)
Occupational Safety and Health
   Act (OSHA)
Radioactive material
Reactive chemical
Standard precautions
Teratogen

LABORATORY SAFETY AND REGULATIONS

All clinical laboratory personnel, by the nature of the work they perform, are exposed daily to a variety of real or potential hazards: electric shock, toxic vapors, compressed gases, flammable liquids, radioactive material, corrosive substances, mechanical trauma, poisons, and the risks of handling biologic materials, to name a few. As G. C. Roach stated, "A cautious attitude and an awareness of hazards are essential to . . . safety and productivity in the clinical laboratory. Maintaining a continuous concern for safety is an additional challenge to the already demanding field of clinical laboratory sciences."*

Laboratory safety necessitates the effective control of all hazards that exist in the clinical laboratory at any time. Safety begins with the recognition of hazards and is achieved through the application of common sense, a safety-focused attitude, good personal behavior, good housekeeping in all laboratory work and storage areas, and, above all, the continual practice of good laboratory technique. In most cases, accidents can be traced directly to two primary causes: unsafe acts (not always recognized by personnel) and unsafe environmental conditions. This chapter discusses laboratory safety as it applies to the clinical laboratory.

Occupational Safety and Health Act (OSHA)

In 1970, Public Law 91-596, better known as the *Occupational Safety and Health Act (OSHA),* was enacted by Congress. The goal of this federal regulation was to provide all employees (clinical laboratory personnel included) with a safe work environment. Under this legislation, the Occupational Safety and Health Administration is authorized to conduct on-site inspections to determine whether an employer is complying with the mandatory standards. Safety is no longer only a moral obligation, but also a federal law.

Other Regulations and Guidelines

There are other federal regulations relating to laboratory safety, such as the Clean Water Act, the Resource Conservation and Recovery Act, and the Toxic Substances Control Act. In addition, clinical laboratories are required to comply with applicable local/state laws such as fire and building codes. OSHA standards that regulate safety in the laboratory include Bloodborne Pathogen Standard, Formaldehyde Standard, Laboratory Standard, Hazard Communication Standard, Respiratory Standard, Air Contaminants Standard, and Personal Protective Equipment Standard. Because laws, codes, and ordinances are updated frequently, current reference materials should be reviewed. Assistance can be obtained from local libraries and from federal, state, and local regulatory agencies.

Safety is also an important part of the requirements for accreditation and reaccreditation of health care institutions by voluntary accrediting bodies such as the Joint Commission on Accreditation of Health Care Organizations (JCAHO) and the Commission on Laboratory Accreditation of the College of American Pathologists. JCAHO publishes a yearly accreditation manual for hospitals and an *Accreditation Manual for Pathology and Clinical Laboratory Services,* which includes a detailed section on safety requirements.

*See JCAHO website—www.jcaho.org

SAFETY AWARENESS FOR CLINICAL LABORATORY PERSONNEL

Safety Responsibility

The employer and the employee share safety responsibility. The employer has the ultimate responsibility for safety and delegates authority for safe operations to supervisors. Safety management in the laboratory should start with a written safety policy. Laboratory supervisors, who reflect the attitudes of management toward safety, are essential members of the safety program.

Employer's Responsibilities

- Establish laboratory work methods and safety policies.
- Provide supervision and guidance to employees.
- Provide safety information, training, personal protective equipment, and medical surveillance to employees.
- Provide and maintain equipment and laboratory facilities that are adequate for the tasks required.

The employee also has a responsibility for his or her own safety and the safety of coworkers. Employee conduct in the laboratory is a vital factor in the achievement of a workplace without accidents or injuries.

Employee's Responsibilities

- Know and comply with the established laboratory work methods.
- Have a positive attitude toward supervisors, coworkers, facilities, and training.
- Give prompt notification of unsafe conditions to the immediate supervisor.
- Engage in the conduct of safe work practices and use of personal protective equipment.

Signage and Labeling

Appropriate signs to identify hazards are critical not only to alert laboratory personnel to potential hazards, but also to identify specific hazards that arise because of an emergency such as fire or explosion. The *National Fire Protection Association (NFPA)* has developed a standard hazards-identification system (diamond-shaped, color-coded symbol), which also has been adopted by many clinical laboratories. At a glance, emergency personnel can assess health hazards (blue quadrant), flammable hazards (red quadrant), reactivity/stability hazards (yellow quadrant), and other special information (white quadrant). In addition, each quadrant shows the magnitude of severity, graded from a low of 0 to a high of 4, of the hazards within the posted area. (Note the NFPA hazard-code symbol in Fig. 2-1).

Manufacturers of laboratory chemicals also provide precautionary labeling information for users. Indicated on the product label is information such as a statement of the hazard, precautionary measures, specific hazard class, first aid

Cat. **C4324-5GL**   Qty. **5 gallons (18.9L)**

Baxter
Scientific Products
S|P™ Methyl Alcohol

(Methanol - Anhydrous)
CH₃OH   FW 32.04

(7)
For Laboratory Use
Store at 68-86°F (20-30°C)

Flash Point: 11°C (52°F Closed Cup)
Maximum Limits and Specifications
Methyl Alcohol: 99.8% Minimum by volume
Residue after Evaporation: 0.001% Maximum
Water: 0.10% Maximum
CAS — (67-56-1);

Distributed by:
Baxter Healthcare Corporation
Scientific Products Division
McGaw Park, IL 60085-6787 USA
Rev. 11/87

Made in USA

Lot No. **KEAM** (8)

(2) **DANGER! FLAMMABLE**

DANGER! POISON

(1)

(4)
NFPA

When using, the following safety precautions are recommended:

SAFETY GLASSES & SHIELD

VENT HOOD

LAB COAT, APRON & GLOVES

FIRE EXTINGUISHER

(3)

(5)

FLAMMABLE • VAPOR HARMFUL • MAY BE FATAL OR CAUSE BLINDNESS IF SWALLOWED • CANNOT BE MADE NON-POISONOUS • HARMFUL IF INHALED • CAUSES IRRITATION • NON-PHOTOCHEMICALLY REACTIVE

Keep away from heat, sparks and flame. Keep container tightly closed and upright to prevent leakage. Avoid breathing vapor or spray mist. Avoid contact with eyes, skin and clothing. Use only with adequate ventilation. Wash thoroughly after handling. In case of fire, use water spray, alcohol foam, dry chemical or CO₂. In case of spillage, absorb and flush with large volumes of water immediately.

FIRST AID: In case of skin contact, flush with plenty of water; **for eyes,** flush with plenty of water for 15 minutes and get medical attention. **If swallowed,** if conscious, induce vomiting by giving two glasses of water and sticking finger down throat. Have patient lie down and keep warm. Cover eyes to exclude light. Never give anything by mouth to an unconscious person. **If inhaled,** remove to fresh air. If necessary, give oxygen or apply artificial respiration.

NOTE: It is unlawful to use this fluid in any food or drink or in any drug or cosmetic for internal or external use. Not for internal or external use on man or animal.

DOT Description: Methyl Alcohol, Flammable Liquid, UN1230

(6)

Figure 2.1. Sample chemical label: (**1**) statement of hazard; (**2**) hazards class; (**3**) safety precautions; (**4**) National Fire Protection Agency (NFPA) hazard code; (**5**) fire extinguisher type; (**6**) safety instructions; (**7**) formula weight; (**8**) lot number.
　　Color of the diamond in the NFPA label indicates hazard:
　　Red = flammable. Store in an area segregated for flammable reagents.
　　Blue = health hazard. Toxic if inhaled, ingested, or absorbed through the skin. Store in a secure area.
　　Yellow = reactive and oxidizing reagents. May react violently with air, water, or other substances. Store away from flammable and combustible materials.
　　White = corrosive. May harm skin, eyes, or mucous membranes. Store away from red-, blue-, and yellow-coded reagents.
　　Gray = presents no more than moderate hazard in any of categories. For general chemical storage.
　　Exception = reagent incompatible with other reagents of same color bar. Store separately.
　　Hazard code (**4**)—Following the NFPA usage, each diamond shows a red segment (flammability), a blue segment (health, ie, toxicity), and yellow (reactivity). Printed over each color-coded segment is a black number showing the degree of hazard involved. The fourth segment, as stipulated by the NFPA, is left blank. It is reserved for special warnings, such as radioactivity. The numerical ratings indicate degree of hazard: 4 = extreme hazard; 3 = severe hazard; 2 = moderate hazard; 1 = slight hazard; 0 = none according to present data.

instructions for internal/external contact, the storage code, the safety code, and personal protective gear and equipment needed. This information is in addition to specifications on the actual lot analysis of the chemical constituents and other product notes (see Fig. 2-1). All "in-house" prepared reagents and solutions should be labeled in a standard manner with the chemical identity, concentration, hazard warning, special handling, storage conditions, date prepared, expiration date (if applicable), and preparer's initials.

SAFETY EQUIPMENT

Safety equipment has been developed specifically for use in the clinical laboratory. The employer is required by law to have designated safety equipment available, but it is also the responsibility of the employee to comply with all safety rules and to use safety equipment.

All laboratories are required to have safety showers, eyewash stations, and fire extinguishers and to periodically test and inspect the equipment for proper operation. It is recommended that safety showers deliver 30–50 gal/min of water at 20–50 psi. Other items that must be available for personnel include fire blankets, spill kits, and first aid supplies.

Mechanical pipetting devices must be used for manipulating all types of liquids in the laboratory, including water. Mouth pipetting is strictly prohibited.

Hoods

Fume Hoods

Fume hoods are required to expel noxious and hazardous fumes from chemical reagents. Fume hoods should be visually inspected for blockages. A piece of tissue paper placed at the hood opening will indicate airflow direction. The hood should never be operated with the sash fully opened. Chemicals stored in hoods should not block airflow. Periodically ventilation should be evaluated by measuring the face velocity with a calibrated velocity meter. The velocity at the face of the hood (with the sash in normal operating position) must be 100 to 120 ft/min. Smoke testing is also recommended to locate dead or turbulent areas in the working space. Additional monitoring should be in accordance with the chemical hygiene plan of the facility.

Biosafety Hoods

Biohazard hoods remove particles that may be harmful to the employee who is working with infective biologic specimens. The Centers for Disease Control and Prevention (CDC) and the National Institutes of Health have described four levels of biosafety, which consist of combinations of laboratory practices and techniques, safety equipment, and laboratory facilities. The biosafety level of a laboratory is based on the operations performed, the routes of transmission of the infectious agents, and the laboratory function or activity. Accordingly, biohazard hoods are designed to offer various levels of protection, depending on the biosafety level of the specific laboratory (Table 2-1).

Chemical Storage Equipment

Safety equipment is available for the storage and handling of chemicals and compressed gases. Safety carriers should always be used to transport 500 mL bottles of acids, alkalis, or other solvents, and approved safety cans should be used for storing, dispensing, or disposing of flammables in volumes greater than 1 qt. Safety cabinets are required for the storage

of flammable liquids, and only specially designed explosion-proof refrigerators should be used to store flammable materials. Only that amount of chemical needed for the day should be available at the bench. Gas-cylinder supports or clamps must be used at all times, and large tanks should be transported using handcarts.

Personal Protection Equipment

The parts of the body most frequently subject to injury in the clinical laboratory are the eyes, skin, and respiratory and digestive tracts. Hence, the use of personal protective equipment is very important. Safety glasses, goggles, visors, or work shields protect the eyes and face from splashes and impact. Contact lenses do not offer eye protection. It is strongly recommended that they not be worn in the clinical chemistry laboratory. If any solution is accidentally splashed into the eye(s), thorough irrigation is required.

Gloves and rubberized sleeves protect the hands and arms when using caustic chemicals. Latex gloves are required for routine laboratory use; however, polyvinyl, nonlatex gloves are an acceptable alternative for people with latex allergies. Lab coats are to be full length, buttoned, and made of liquid-resistant material. Proper footwear is required; shoes constructed of porous materials, open-toed shoes, or sandals are considered hazardous.

Respirators are required for various procedures in the clinical laboratory. Whether used for biologic or chemical hazards, the correct type of respirator must be used for the specific hazard. Respirators with *High Efficiency Particulate Air (HEPA)* filters must be worn when engineering controls are not feasible, for example, when working directly

TABLE 2-1. Comparison of Biologic Safety Cabinets

	Cabinets			Applications		
Type	*Face Velocity (lfpm)*	*Airflow Pattern*		*Radionuclides/ Toxic Chemicals*	*Biosafety Level(s)*	*Product Protection*
Class I, open front	75	In at front; out rear and top through HEPA filter		NO	2,3	NO
Class II: Type A	75	70% recirculated through HEPA; exhaust through HEPA		NO	2,3	YES
Type B1	100	30% recirculated through HEPA; exhaust via HEPA and hard-ducted		YES (Low levels/ volatility)	2,3	YES
Type B2	100	No recirculation; total exhaust via HEPA and hard-ducted		YES	2,3	YES
Type B3	100	Same as IIA, but plena under negative pressure to room and exhaust air is ducted		YES	2,3	YES
Class III	NA	Supply air inlets and exhaust through 2 HEPA filters		YES	3,4	YES

* Glove panels may be added and will increase face velocity to 150 lfpm; gloves may be added with an inlet air pressure release that will allow work with chemicals/radionuclides.

Source: Centers for Disease Control and Prevention and the National Institutes of Health. Table 3, Comparison of Biological Safety Cabinets. Biosafety in Microbiological and Biomedical Laboratories. U.S. Government Printing Office, 1993. Reprinted with permission.

with tuberculosis (TB) patients or performing procedures that may aerosolize specimens of patients with suspected or confirmed cases of TB. Training, maintenance, and written protocol for use of respirators is required according to the respiratory protection standard.

Each employer must provide (at no charge) lab coats, gloves, or other protective equipment to all employees who may be exposed to biologic or chemical hazards. It is the employer's responsibility to clean and maintain all personal protective equipment. All contaminated personal protective equipment must be removed and properly disposed of before leaving the laboratory.

BIOLOGIC SAFETY

General Considerations

All blood samples and other body fluids should be collected, transported, handled, and processed using strict precautions. Gloves, gowns, and face protection must be used if splash or splattering is likely to occur.

Centrifugation of biologic specimens produce finely dispersed aerosols that are a high-risk source of infection. Ideally, specimens should remain "capped" during centrifugation. As an additional precaution, the use of a centrifuge with an internal shield is recommended.

Spills

Any blood, body fluid or other potentially infectious material spill must be cleaned up and the area or equipment disinfected immediately. Recommended cleanup includes the following:

- Wear appropriate protective equipment.
- Use mechanical devices to pick up broken glass or other sharp objects.
- Absorb the spill with paper towels, gauze pads, or tissue, and so on.
- Clean the spill site using a common aqueous detergent.
- Disinfect the spill site using approved disinfectant or 10% bleach using appropriate contact time.
- Rinse the spill site with water.
- Dispose of all materials in appropriate biohazard containers.

Bloodborne Pathogen Exposure Control Plan

In December 1991, OSHA issued the final rule for occupational exposure to *bloodborne pathogens*. To minimize employee exposure, each employer must have a written exposure control plan. The plan must be available to all employees whose reasonable anticipated duties may result in occupational exposure to blood or other potentially infectious materials. The exposure control plan must be discussed with all employees and be available to them while they are working.

The employee must be provided with adequate training of all techniques described in the exposure control plan at initial work assignment and annually thereafter. All necessary equipment and supplies must be readily available and inspected on a regular basis.

Clinical laboratory personnel are knowingly or unknowingly in frequent contact with potentially biohazardous materials. In recent years, new and serious occupational hazards to personnel have arisen, and this problem has been complicated because of the general lack of understanding of the epidemiology, mechanisms of transmission of the disease, or inactivation of the causative agent. Special precautions must be taken when handling all specimens because of the continual increase of infectious samples received in the laboratory. Therefore, in practice, specimens from patients with confirmed or suspected hepatitis, acquired immunodeficiency syndrome, Creutzfeldt-Jakob disease, or other potentially infectious diseases should be handled no differently than other routine specimens. Adopting a *standard precautions* policy, which considers blood and other body fluids from all patients as infective, is required.

Airborne Pathogens

Because of the recent resurgence of TB, in 1993 OSHA issued a statement that the agency would enforce the CDC *Guidelines for Preventing the Transmission of Tuberculosis in Health Care Facilities.* The purpose of the guidelines is to encourage early detection, isolation, and treatment of active cases. A TB exposure control program must be established and risks to laboratory workers must be assessed. Those workers in high-risk areas may be required to wear a respirator for protection. All health care workers at risk must be screened for TB infection.

Shipping

Clinical laboratories routinely ship regulated material. The U.S. Department of Transportation (DOT) has specific requirements for carrying regulated materials. There are two types of specimens. Infectious specimens are those known or reasonably believed to cause diseases in humans or animals. Diagnostic specimens are those tested as routine screening or for diagnosis. Each type of specimen has rules and packaging requirements. The DOT guidelines are found in *Code of Federal Regulations 49.*

CHEMICAL SAFETY

Hazard Communication

In the August 1987 issue of the *Federal Register,* OSHA published the new *Hazard Communication Standard* ("Right to Know Law"). The Right to Know Law was developed for employees who may be exposed to hazardous chemicals.

Employees must be informed of the health risks associated with those chemicals. The intent of the law is to ensure that the health hazards are evaluated for all chemicals that are produced and that this information is relayed to employees.

To comply with the regulation, clinical laboratories must:

- Plan and implement a written hazard communication program.
- Obtain *material safety data sheets* (MSDSs) for each hazardous compound present in the workplace and have the MSDSs readily accessible to employees.
- Educate all employees annually on how to interpret chemical labels, MSDSs, and health hazards of the chemicals, and how to work safely with the chemicals.

Material Safety Data Sheet (MSDS)

The MSDS is a major source of safety information for employees who may use *hazardous materials* in their occupations. Employers are responsible for obtaining from the chemical manufacturer or developing an MSDS for each hazardous agent used in the workplace. A standardized format is not mandatory, but all requirements listed in the law must be addressed. A summary of the MSDS information requirements includes the following:

- Product name and identification.
- Hazardous ingredients.
- Permissible Exposure Limit (PEL).
- Physical and chemical data.
- Health hazard data and carcinogenic potential.
- Primary routes of entry.
- Fire and explosion hazards.
- Reactivity data.
- Spill and disposal procedures.
- Personal protective equipment recommendations.
- Handling.
- Emergency and first aid procedures.
- Storage and transportation precautions.
- Chemical manufacturer's name, address, and phone number.
- Special information section.

The MSDS must be printed in English and provide the specific compound identity along with all common names. All information sections must be completed, and the date that the MSDS was printed must be indicated. Copies of the MSDS will be readily accessible to employees during all shifts.

Laboratory Standard

Occupational Exposure to Hazardous Chemicals in Laboratories, also known as the *laboratory standard,* was enacted in May 1990 to provide laboratories specific guidelines for handling hazardous chemicals. This OSHA standard requires each laboratory that uses hazardous chemicals to have a written *chemical hygiene plan.* This plan provides procedures and work practices for regulating and reducing exposure of laboratory personnel to hazardous chemicals. *Hazardous chemicals* are those that pose a risk of damage to a worker's lungs, skin, eyes, or mucous membranes following long- or short-term exposure. Procedures describing how to protect employees against *teratogens* (substances that affect cellular development in a fetus or embryo), carcinogens, and other toxic chemicals must be described in the plan. Training in use of hazardous chemicals to include recognition of signs and symptoms of exposure, location of MSDS, a chemical hygiene plan, and how to protect themselves against hazardous chemicals must be provided to all employees. A chemical hygiene officer must be designated for any laboratory using hazardous chemicals. The protocol must be reviewed annually and updated when regulations are modified or chemical inventory changes.

Toxic Effects From Hazardous Substances

Toxic substances have the potential of deleterious effects (local or systemic) by direct chemical action or interference with the function of body systems. They can cause acute or chronic effects related to the duration of exposure (*ie,* short-term or single contact versus long-term or prolonged repeated contact). Almost any substance, even the most harmless, can be toxic in excess. Moreover, some chemicals are toxic at very low concentrations. Exposure to toxic agents can be through direct contact, inhalation, or inoculation/injection.

In the clinical chemistry laboratory, personnel should be particularly aware of toxic vapors from chemical solvents, such as acetone, chloroform, methanol, or carbon tetrachloride, that do not give explicit sensory-irritation warnings, as do bromide, ammonia, and formaldehyde. Another source of poisonous vapors that is frequently disregarded is mercury. It is highly volatile and toxic and is rapidly absorbed through the skin and respiratory tract. Mercury spill kits should be available in areas where mercury thermometers are used. Many laboratories are phasing out the use of mercury and mercury-containing compounds. Laboratories should have a policy and method for legally disposing of mercury.

Storage and Handling of Chemicals

To avoid accidents when handling chemicals, it is important to develop respect for all chemicals and to have a complete knowledge of their properties. This is particularly important when transporting, dispensing, or using chemicals that, when in contact with certain other chemicals, could result in the formation of substances that are toxic, flammable, or explosive. For example, acetic acid is incompatible with other acids such as chromic and nitric, carbon tetrachloride is incompatible with sodium, and flammable liquids are incompatible with hydrogen peroxide and nitric acid.

Arrangements for the storage of chemicals will depend on the quantities of chemicals needed and the nature or type of chemicals. Proper storage is essential to prevent and control laboratory fires and accidents. Ideally, the storeroom should be organized so that each class of chemicals is isolated in an area that is not used for routine work. An up-to-date inventory should be kept that indicates location of chemicals, minimum/maximum quantities required, shelf life, and so on. Some chemicals deteriorate over time and become hazardous (eg, ether forms explosive peroxides). Storage should not be based solely on alphabetical order because incompatible chemicals may be stored next to each other and react chemically. They must be separated for storage as follows:

Flammable liquids	Flammable solids
Mineral acids	Organic acids
Caustics	Oxidizers
Perchloric acid	Water-reactive substances
Air-reactive substances	Others

Heat-reactive substances requiring refrigeration
Unstable substances (shock-sensitive explosives)
(Refer to Appendix H, *Examples of Incompatible Chemicals*.)

Flammable/Combustible Chemicals

Flammable and combustible liquids, which are used in numerous routine procedures, are among the most hazardous materials in the clinical chemistry laboratory because of possible fire or explosion. They are classified according to flash point, which is the temperature at which sufficient vapor is given off to form an ignitable mixture with air. A flammable liquid has a flash point below 37.8°C (100°F), and combustible liquids, by definition, have a flash point at or above 37.8°C (100°F). Some of the commonly used flammable and combustible solvents are acetone, benzene, ethanol, heptane, isopropanol, methanol, toluene, and xylene. It is important to remember that flammable chemicals also include certain gases and solids such as paraffin.

Corrosive Chemicals

Corrosive chemicals are injurious to the skin or eyes by direct contact or to the tissues of the respiratory and gastrointestinal tracts if inhaled or ingested. Typical examples include acids (acetic, sulfuric, nitric, and hydrochloric) and bases (ammonium hydroxide, potassium hydroxide, and sodium hydroxide).

Reactive Chemicals

Reactive chemicals are substances that, under certain conditions, can spontaneously explode or ignite or that evolve heat or flammable or explosive gases. Some strong acids or bases react with water to generate heat (exothermic reactions). Hydrogen is liberated if alkali metals (sodium or potassium) are mixed with water or acids, and spontaneous combustion also may occur. The mixture of oxidizing agents, such as peroxides, and reducing agents, such as hydrogen, generate heat and may be explosive.

Carcinogenic Chemicals

Carcinogenics are substances that have been determined to be cancer-causing agents. OSHA has issued lists of confirmed carcinogens and suspected carcinogens and detailed standards for the handling of these substances. Benzidine is a common example of a known carcinogen. If possible, a substitute chemical or different procedure should be used to avoid exposure to carcinogenic agents.

Chemical Spills

Strict attention to good laboratory technique can help prevent chemical spills. However, emergency procedures should be established to handle any accidents. If a spill occurs, the first step should be to assist/evacuate personnel, then confinement and cleanup of the spill can begin. There are several commercial spill kits available for neutralizing and absorbing spilled chemical solutions (Fig. 2-2). However, no single kit is suitable for all types of spills. Emergency procedures for spills also should include a reporting system.

RADIATION SAFETY

Environmental Protection

A radiation-safety policy should include environmental and personnel protection. All areas where *radioactive materials* are used or stored must be posted with caution signs, and traffic in these areas should be restricted to essential personnel only. Regular and systematic monitoring must be emphasized, and decontamination of laboratory equipment, glassware, and work areas should be scheduled as part of routine procedures.

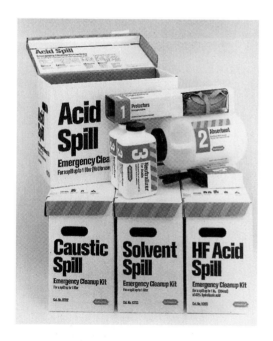

Figure 2-2. Spill cleanup kit.

Records must be maintained as to the quantity of radioactive material on hand as well as the quantity that is disposed.

Personal Protection

It is essential that only properly trained personnel work with radioisotopes and that users are monitored to ensure that the maximal permissible dose of radiation is not exceeded. Radiation monitors must be evaluated regularly to detect degree of exposure for the laboratory employee. Records must be maintained for the length of employment plus 30 years.

FIRE SAFETY

The Chemistry of Fire

Fire is basically a chemical reaction that involves the rapid oxidation of a combustible material or fuel, with the subsequent liberation of heat and light. In the clinical chemistry laboratory, all the elements essential for fire to begin are present—fuel, heat or ignition source, and oxygen (air). However, recent research suggests that a fourth factor is present. This factor has been classified as a reaction chain in which burning continues and even accelerates. It is caused by the breakdown and recombination of the molecules that compose the material burning with the oxygen in the atmosphere.

The fire triangle has been modified into a three-dimensional pyramid known as the *fire tetrahedron* (Fig. 2-3). This modification does not eliminate established procedures in dealing with a fire but does provide additional means by which fires may be prevented or extinguished. A fire will extinguish if any of the three basic elements (heat, air, or fuel) are removed.

Classification of Fires

Fires have been divided into four classes based on the nature of the combustible material and requirements for extinguishment:

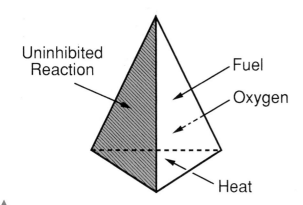

Figure 2-3. Fire tetrahedron.

Class A: ordinary combustible solid materials such as paper, wood, plastic, and fabric.

Class B: flammable liquids/gases and combustible petroleum products.

Class C: energized electrical equipment.

Class D: combustible/reactive metals such as magnesium, sodium, and potassium.

Types/Applications of Fire Extinguishers

Just as fires have been divided into classes, fire extinguishers are divided into classes that correspond to the type of fire to be extinguished. Be certain to choose the right type–using the wrong type of extinguisher may be dangerous. For example, do not use water on burning liquids or electrical equipment.

Pressurized-water extinguishers, as well as foam and multipurpose dry-chemical types, are used for Class A fires. Multipurpose dry-chemical and carbon dioxide extinguishers are used for Class B and C fires. Halogenated hydrocarbon extinguishers are recommended particularly for use with computer equipment. Class D fires present special problems, and extinguishment is left to trained firefighters using special dry-chemical extinguishers (Fig. 2-4).

Personnel should know the location and type of portable fire extinguisher near the work area and know how to use an extinguisher before a fire occurs. In the event of a fire, first evacuate all personnel, patients, and visitors who are in immediate danger and then activate the fire alarm, report the fire, and attempt to extinguish the fire, if possible. Personnel should work as a team to carry out emergency procedures. Fire drills must be conducted regularly and with appropriate documentation.

CONTROL OF OTHER HAZARDS

Electrical Hazards

Most individuals are aware of the potential hazards associated with the use of electrical appliances and equipment. Hazards of electrical energy can be direct and result in death, shock, or burns. Indirect hazards can result in fire or explosion. Therefore, there are many precautionary procedures to follow when operating or working around electrical equipment:

- Use only explosion-proof equipment in hazardous atmospheres.
- Be particularly careful when operating high-voltage equipment, such as electrophoresis apparatus.
- Use only properly grounded equipment (three-prong plug).
- Check for "frayed" electrical cords.
- Promptly report any malfunctions or equipment producing a "tingle" for repair.
- Do not work on "live" electrical equipment.

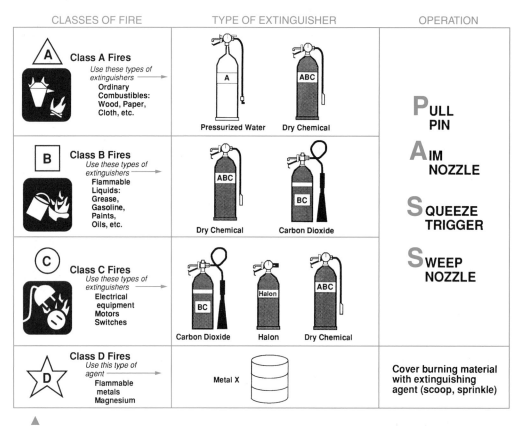

Figure 2-4. Proper use of fire extinguishers. (Adapted with permission from the Clinical and Laboratory Safety Department, The University of Texas Health Science Center at Houston.)

- Never operate electrical equipment with wet hands.
- Know the exact location of the electrical control panel for the electricity to your work area.
- Use only approved extension cords and do not overload circuits. (Some local regulations prohibit the use of any extension cord.)
- Have ground checks and other periodic preventive maintenance performed on equipment.

Compressed Gases Hazards

Compressed gases, which serve a number of functions in the laboratory, present a unique combination of hazards in the clinical laboratory: danger of fire, explosion, asphyxiation, or mechanical injuries. There are several general requirements for safely handling compressed gases:

- Know the gas that you will use.
- Store tanks in a vertical position.
- Keep cylinders secured at all times.
- Never store flammable liquids and compressed gases in the same area.
- Use the proper regulator for the type of gas in use.
- Do not attempt to control or shut off gas flow with the pressure relief regulator.
- Keep removable protection caps in place until the cylinder is in use.

- Make certain that acetylene tanks are properly piped (the gas is incompatible with copper tubing).
- Do not force a "frozen" or stuck cylinder valve.
- Use a hand truck to transport large tanks.
- Always check tanks on receipt and then periodically for any problems such as leaks.
- Make certain that the cylinder is properly labeled to identify the contents.
- Empty tanks should be marked "empty."

Cryogenic Materials Hazards

Liquid nitrogen is probably one of the most widely used cryogenic fluids (liquefied gases) in the laboratory. There are, however, several hazards associated with the use of any *cryogenic material:* fire or explosion, asphyxiation, pressure buildup, embrittlement of materials, and tissue damage similar to that of thermal burns.

Only containers constructed of materials designed to withstand ultralow temperatures should be used for cryogenic work. In addition to the use of eye-safety glasses, hand protection to guard against the hazards of touching supercooled surfaces is recommended. The gloves, of impermeable material, should fit loosely so that they can be taken off quickly if liquid spills on them. Also, to minimize violent boiling/frothing and splashing, specimens to be frozen should always be inserted into the coolant very slowly. Cryogenic fluids should be stored in well-insulated but loosely stoppered containers that minimize loss of fluid resulting from evaporation by "boil-off" and that prevent "plugging" and pressure buildup.

Mechanical Hazards

In addition to physical hazards such as fire and electric shock, laboratory personnel should be aware of the *mechanical hazards* of equipment such as centrifuges, autoclaves, and homogenizers.

Centrifuges, for example, must be balanced to distribute the load equally. The operator should never open the lid until the rotor has come to a complete stop. Safety locks on equipment should never be rendered inoperable.

Laboratory glassware itself is another potential hazard. Agents such as glass beads should be added to help eliminate bumping/boilover when liquids are heated. Tongs or gloves should be used to remove hot glassware from ovens, hot plates, or water baths. Glass pipettes should be handled with extra care, as should sharp instruments such as corkborers, needles, scalpel blades, and other tools.

DISPOSAL OF HAZARDOUS MATERIALS

The safe handling and disposal of chemicals and other materials requires a thorough knowledge of their properties and hazards. Generators of hazardous wastes have a moral and legal responsibility, as defined in applicable local, state, and federal regulations, to protect both the individual and the environment when disposing of wastes. There are four basic waste-disposal techniques: flushing down the drain to the sewer system, incineration, landfill burial, and recycling.

Chemical Waste

In some cases, it is permissible to flush water-soluble substances down the drain with copious quantities of water. However, strong acids or bases should be neutralized before disposal. Foul-smelling chemicals should never be disposed of down the drain. Possible reaction of chemicals in the drain and potential toxicity must be considered when deciding if a particular chemical can be dissolved or diluted and then flushed down the drain. For example, sodium azide, which is used as a preservative, forms explosive salts with metals such as the copper in pipes.

Other liquid wastes, including flammable solvents, must be collected in approved containers and segregated into compatible classes. If practical, solvents such as xylene and acetone may be filtered or redistilled for reuse. If recycling is not feasible, disposal arrangements should be made by specifically trained personnel. Flammable material also can be burned in specially designed incinerators with afterburners and scrubbers to remove toxic products of combustion.

Also, before disposal, hazardous substances such as carcinogens and peroxides, which are explosive, should be transformed to less hazardous forms whenever feasible. Solid chemical wastes that are unsuitable for incineration must be buried in a landfill. This practice, however, has created an environmental problem, and there is now a shortage of "safe" sites.

Radioactive Waste

The manner of use and disposal of isotopes is strictly regulated by the Nuclear Regulatory Commission (NRC) and depends on the type of waste (soluble or nonsoluble) and its level of radioactivity. The radiation safety officer should always be consulted about policies dealing with radioactive waste disposal. Many clinical laboratories transfer radioactive materials to a licensed receiver for disposal.

Biohazardous Waste

On November 2, 1988, President Reagan signed into law The Medical Waste Tracking Act of 1988. Its purpose was to (1) charge the Environmental Protection Agency with the responsibility to establish a program to track medical waste from generation to disposal, (2) define medical waste, (3) establish acceptable techniques for treatment and disposal, and (4) establish a department with jurisdiction to enforce the new laws. Several states have implemented the federal guidelines and incorporated additional requirements. Some entities covered by the rules are any health care-related facility including but not limited to ambulatory surgical centers; blood banks and blood drawing centers; clinics, including medical, dental, and veterinary; clinical, diagnostic, pathologic, or biomedical research laboratories; emergency medical services; hospitals; long-term care facilities; minor emergency centers; occupational health clinics and clinical laboratories; and professional offices of physicians and dentists.

Medical waste is defined as "special waste from health care facilities" and is further defined as solid waste that, if improperly treated or handled, "may transmit infectious diseases."* It comprises animal waste, bulk blood and blood products, microbiologic waste, pathologic waste, and sharps. The approved methods for treatment and disposition of medical waste are incineration, steam sterilization, burial, thermal inactivation, chemical disinfection, or encapsulation in a solid matrix.

Generators of medical waste must implement the following procedures:

1. Employers of health care workers must establish and implement an infectious waste program.
2. All biomedical waste should be placed into a bag that is marked with the biohazard symbol, and then placed into a leakproof container that is puncture resistant and equipped with a solid, tight-fitting lid. All containers must be clearly marked with the word *biohazard* or its symbol.

*OSHA Subpart Z 29CFR 1910.1000–.1450

3. All sharp instruments, such as needles, blades, and glass objects, should be placed into special puncture-resistant containers before placing them inside the bag and container. Needles should not be transported, recapped, bent, or broken by hand.

4. All biomedical waste must then be disposed of by one of the recommended procedures.

Potentially biohazardous material, such as blood or blood products and contaminated laboratory waste, cannot be discarded directly. Contaminated combustible waste can be incinerated. Contaminated noncombustible waste, such as glassware, should be autoclaved before being discarded. Special attention should be given to the discarding of syringes, needles, and broken glass that also could inflict accidental cuts or punctures. Appropriate containers should be used for discarding these sharp objects.

ACCIDENT DOCUMENTATION AND INVESTIGATION

Any accidents involving personal injuries, even minor ones, should be reported immediately to a supervisor. Under OSHA regulations, employers are required to maintain records of occupational injuries and illnesses for the length of employment plus 30 years. The record-keeping requirements include a first report of injury, an accident investigation report, and an annual summary that is recorded on an OSHA injury log (Form 200).

The first report of injury is used to notify the insurance company and the human resources or employee relations department that a workplace injury has occurred. The employee and the supervisor usually complete the report, which contains information on the employer and injured person, as well as the time and place, cause, and nature of the injury. The report is signed and dated; then it is forwarded to the institution's risk manager or insurance representative.

The investigation report should include information on the injured person; a description of what happened; the cause of the accident (environmental or personal); other contributing factors; witnesses; the nature of the injury; and actions to be taken to prevent a recurrence. This report should be signed and dated by the person who conducted the investigation.

Annually, a log and summary of occupational injuries and illnesses should be completed and forwarded to the U.S. Department of Labor, Bureau of Labor Statistics (OSHA Injury Log No. 200). The standardized form requests information similar to the first report of injury and the accident investigation report. Information about every occupational death, nonfatal occupational illness, and nonfatal occupational injury that involved loss of consciousness, restriction of work or motion, transfer to another job, or medical treatment (other than first aid) must be reported.

Because it is important to determine why and how an accident occurred, an accident investigation should be conducted. Most accidents can be traced to two underlying causes: environmental (unsafe conditions) or personal (unsafe acts). Environmental factors include inadequate safeguards, use of improper or defective equipment, hazards associated with the location, or poor housekeeping. Personal factors include improper laboratory attire, lack of skills or knowledge, specific physical or mental conditions, and attitude. The employee's positive motivation is very important in all aspects of safety promotion and accident prevention.

It is particularly important that if any individual sustains a needle puncture during blood collection or a cut during subsequent specimen processing or handling, the appropriate authority should be notified immediately. For a summary of recommendations for the protection of laboratory workers, refer to *Protection of Laboratory Workers from Instrument Biohazards and Infectious Disease Transmitted by Blood, Body Fluids and Tissue,* Approved Guideline M29-A (NCCLS).

SUMMARY

The cardinal safety rules of the clinical laboratory are to develop foresight and accident perception, use common sense, and develop and practice the following:

1. *Good personal behavior/habits.*
   Wear proper attire and protective clothing.
   Tie back long hair.
   Do not eat, drink, or smoke in the work area.
   Never mouth pipette.
   Wash hands frequently.
2. *Good housekeeping.*
   Keep work areas free of chemicals, dirty glassware, and so on.
   Store chemicals properly.
   Label reagents and solutions.
   Post warning signs.
3. *Good laboratory technique.*
   Do not operate new or unfamiliar equipment until you have received instruction and authorization.
   Read all labels and instructions carefully.
   Use the personal safety equipment that is provided.
   For the safe handling, use, and disposal of chemicals, learn their properties and hazards.
   Learn emergency procedures and become familiar with the location of fire exits, fire extinguishers, blankets, and so on.
   Be careful when transferring chemicals from container to container and always add acid to water slowly.

REVIEW QUESTIONS

1. Which of the following standards require that MSDSs are accessible to all employees who come in contact with a hazardous compound?

a. Bloodborne Pathogen Standard

b. Hazardous Communication Standard

c. CDC Regulations

d. Personal Protection Equipment Standard

2. Chemicals should be stored:

a. Alphabetically, for easy accessibility

b. Inside a safety cabinet with proper ventilation

c. According to their chemical properties and classification

d. Inside a fume hood, if toxic vapors can be released when opened

3. Proper personal protection equipment (PPE) in the chemistry laboratory for routine testing includes:

a. Respirators with HEPA filter

b. Gloves with rubberized sleeves

c. Safety glasses for individuals not wearing contact lenses

d. Impermeable lab coat with latex gloves

4. A fire caused by a flammable liquid should be extinguished by using which type of extinguisher?

a. Halogen

b. Class B

c. Pressurized water

d. Class C

5. Which of the following is the proper means of disposal for the type of waste?

a. Xylene into the sewer system

b. Microbiologic waste by steam sterilization

c. Mercury by burial

d. Radioactive waste by incineration

SUGGESTED READINGS

American Chemical Society, Committee on Chemical Safety, Smith GW, ed. Safety in academic chemistry laboratories. Washington, DC: American Chemical Society, 1985.

Allocca JA, Levenson HE. Electrical and electronic safety. Reston, VA: Reston Publishing Company, 1985.

Boyle MP. Hazardous chemical waste disposal management. Clin Lab Sci 1992;5:6.

Brown JW. Tuberculosis alert: an old killer returns. Med Lab Obs 1993;25:5.

Bryan RA. Recommendations for handling specimens from patients with confirmed or suspected Creutzfeldt-Jakob disease. Lab Med 1984;15:50.

Centers for Disease Control, National Institutes of Health. Biosafety in microbiological and biomedical laboratories. 3rd ed. Washington, DC: US Government Printing Office, 1993.

Chervinski D. Environmental awareness: it's time to close the loop on waste reduction in the health care industry. Adv Admin Lab 1994;3:4.

Furr AK. Handbook of laboratory safety. 3rd ed. Boca Raton, FL: CRC Press, 1990.

Gile TJ. An update on lab safety regulations. Med Lab Obs 1995;27:3.

Gile TJ. Hazard-communication program for clinical laboratories. Clin Lab 1988;1:2.

Hayes DD. Safety considerations in the physician office laboratory. Lab Med 1994;25:3.

Hazard communication. Federal Register 59:27, Feb 4, 1994.

Karcher, RE. Is your chemical hygiene plan OSHA proof? Med Lab Obs 1993;25:7.

Le Sueur CL. A three-pronged attack against AIDS infection in the lab. Med Lab Obs 1989;21:37.

Miller SM. Clinical safety: dangers and risk control. Clin Lab Sci 1992;5:6.

National Committee for Clinical Laboratory Standards. Clinical laboratory safety (approved guideline). Villanova, PA: NCCLS, 1996.

National Committee for Clinical Laboratory Standards. Clinical laboratory waste management (approved guideline). Villanova, PA: National Committee for Clinical Laboratory Standards, 1993.

National Committee for Clinical Laboratory Standards. Protection of laboratory workers from instrument biohazards and infectious disease transmitted by blood, body fluids and tissue (Approved Guideline M29-A). Villanova, PA: National Committee for Clinical Laboratory Standards, 1997.

National Institutes of Health, Radiation Safety Branch. Radiation: The National Institutes of Health safety guide. Washington, DC: US Government Printing Office, 1979.

National Regulatory Committee, Committee on Hazardous Substances in the Laboratory. Prudent practices for disposal of chemicals from laboratories. Washington, DC: National Academy Press, 1983.

National Regulatory Committee, Committee on Hazardous Substances in the Laboratory. Prudent practices for the handling of hazardous chemicals in laboratories. Washington, DC: National Academy Press, 1981.

National Regulatory Committee, Committee on the Hazardous Biological Substances in the Laboratory. Biosafety in the laboratory: prudent practices for the handling and disposal of infectious materials. Washington, DC: National Academy Press, 1989.

Occupational exposure to bloodborne pathogens; final rule. Federal Register 56:235, Dec 6, 1991.

Otto CH. Safety in health care: prevention of bloodborne diseases. Clin Lab Sci 1992;5:6.

Pipitone DA. Safe storage of laboratory chemicals. New York: Wiley, 1984.

Roach, GC. Laboratory Safety: Principle for personal practice. Washington, DC: American Society for Medical Technology, 1975.

Rose SL. Clinical laboratory safety. Philadelphia: Lippincott, 1984.

Rudmann SV, Jarus C, Ward KM, Arnold DM. Safety in the student laboratory: a national survey of university-based programs. Lab Med 1993;24:5.

Stern A, Ries H, Flynn D, et al. Fire safety in the laboratory: Part I. Lab Med 1993;24:5.

Stern A, Ries H, Flynn D, et al. Fire safety in the laboratory: Part II. Lab Med 1993; 24:6.

CHAPTER 3

Quality Control and Statistics

George S. Cembrowski, Anne M. Sullivan, Tammy L. Hofer

Objectives

Upon completion of this chapter, the clinical laboratorian should be able to:

- *Define the following terms: quality assurance, quality control, control, standard, accuracy, precision, descriptive statistics, inferential statistics, reference interval, random error, systematic error, dispersion, delta check, and confidence intervals.*

- *Calculate the following: sensitivity, specificity, efficiency, predictive value, mean, median, range, variance, and standard deviation.*

- *Evaluate laboratory data using the multirule system for quality control.*

- *Given laboratory data, graph the data and determine significant constant or proportional errors.*

- *Describe the pre-analytic and post-analytic phases of quality assurance.*

- *Given laboratory data, determine if there is a trend or a shift.*

- *Discuss the role of clinical laboratorians in point-of-care testing.*

- *Describe the important features/requirements of point-of-care analyzers.*

- *Discuss the processes involved in method selection and evaluation.*

- *Discuss proficiency-testing programs in the clinical laboratory.*

KEY TERMS

Accuracy	Histogram	Random error
Analytical variations	Inferential	Reference interval
CLIA	statistics	Reference
Control	Predictive value	method
Control rule	theory	Shift
Delta check	Precision	Standard
Descriptive statistics	Proficiency testing	Systematic error
Dispersion	Quality assurance	Trend
F-test	Quality control	t-test

STATISTICAL CONCEPTS

Statistics may be defined as the science of gathering, analyzing, interpreting, and presenting data. The volume of data generated by the clinical chemistry laboratory is enormous and must be summarized to be maximally useful to laboratorian and clinician. The introduction of a new test illustrates the extensive use of statistics in the laboratory. First, the laboratorian should introduce the test only after reviewing the data that document the usefulness of the test for diagnosing or monitoring a disease state. If several methods are available for performing the test, the laboratorian should study the published evaluations and select the most practical as well as the optimally accurate and precise method. During in-house evaluation of the method, precision and accuracy must be evaluated. If the method's performance is acceptable, reference-interval (normal range) data must be accumulated either to verify the manufacturer's recommended interval or else to set up a laboratory-specific reference interval. The clinician can then properly interpret patient data. Once the method is in use, accuracy and precision must be continually assessed to ensure reliable analyses. The following sections review some of the statistical concepts that must be understood by the laboratorian. For more comprehensive material, the reader is referred to several introductory textbooks.[1,2,3]

Descriptive Statistics

Descriptive statistics are used to summarize the important features of a group of data. Another type of statistics, *inferential statistics,* is used to compare the features of two or more groups of data. The descriptive statistics in this chapter are applied to groups of single observations as well as to groups of paired observations.

Descriptive Statistics of Groups of Single Observations

One of the most useful ways of summarizing groups of data is by graphing them. Figure 3-1 shows the results obtained from the repeated analysis of a patient plasma specimen for the analyte antithrombin III (ATT), which is a potent inhibitor of many of the activated clotting factors. The results are plotted as a frequency diagram, or *frequency histogram,* with the number or frequency of each result plotted on the *y*-axis and the value of the result plotted on the *x*-axis. The frequency histogram of the repeated measurements has a bell shape, with most of the results falling close to the center of the distribution. Repeated measurements of the same specimen are nonidentical and are due to diverse causes, including instrument, reagent, and operator variations. The laboratorian often classifies these variations as being *analytical variations*.

For most assays, the analytical variation is usually much lower than the variation of samples obtained from different individuals (the *inter-individual variations*). Table 3-1 shows the results of a reference-interval study for ATT. The blood of fasting, healthy, ambulatory subjects was drawn in a stan-

TABLE 3-1. Antithrombin III Values From a Reference Interval

Value	Frequency	Cumulative Frequency
88	1	1
92	1	2
93	2	4
95	1	5
96	1	6
97	3	9
98	1	10
99	1	11
100	3	14
101	4	18
102	4	22
103	3	25
104	2	27
105	4	31
106	4	35
107	5	40
108	3	43
109	2	45
110	9	54
111	4	58
112	4	62
113	4	66
114	7	73
115	7	80
116	7	87
117	4	91
118	7	98
120	4	102
121	1	103
122	4	107
124	1	108
125	2	110
126	3	113
127	1	114
129	1	115
133	2	117
138	1	118
140	1	119

Mean = 111.6; median = 112; mode = 110; s = 9.5 units.

dard manner, with the plasma analyzed for ATT. Frequency histograms of the ATT data are shown in Figure 3-2. The scale for the concentration is expressed in intervals of 2 units for Figure 3-2*A* and 5 units for Figure 3-2*B*. With a decrease in the number of intervals, the shape of the frequency histogram becomes more regular. Both histograms are almost symmetrical and bell-shaped. The bell shape of this distribution approximates the shape of a Gaussian distribution and allows analysis of the data by standard (parametric) statistical tests. Data that deviate greatly from the Gaussian distribution should be analyzed with distribution-free or nonparametric statistics. Small departures from Gaussian distributions do not seriously affect the results of parametric tests. The most common type of deviation from the Gaussian distribution in clinical laboratory observations is *skewness,* or the presence of in-

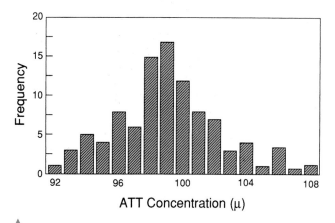

Figure 3-1. Frequency histogram of ATT results obtained from the repeated analysis of a single patient specimen (*x* = 100; *s* = 3 units).

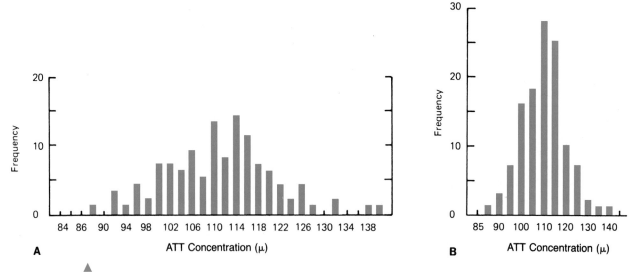

Figure 3-2. Frequency histograms of antithrombin III concentrations from a reference-interval study. The data in **A** have been grouped in intervals of 2 units. The data in **B** have been grouped by 5 units ($\overline{x} = 111.6$; $s = 9.5$ units).

creased numbers of observations in one of the tails of the distribution. (An example of a skewed distribution is shown in Fig. 3-9).

Another type of plot is the *cumulative-frequency histogram*. In this plot, the number of observations that are less than or equal to a certain observation are plotted against the value of that observation. Figure 3-3 shows a cumulative-frequency histogram for the normal-range ATT data. The cumulative

frequencies in Table 3-1 have been divided by the total number of observations (*n*) and then multiplied by 100 to obtain relative cumulative frequencies that range from 0% to 100%.

Groups of Gaussian observations can be described by statistics that summarize their location and dispersion. The most common statistical test that summarizes location is the *mean*, which is calculated by summing the observations and dividing by the number of the observations. If the observations are $x_1, x_2, x_3, \ldots, x_n$, then the mean, or $\overline{x}$, is

$$\overline{x} = \frac{x_1 + x_2 + x_3 + \cdots + x_n}{n}$$

$$= \frac{\Sigma x_i}{n} \qquad\qquad \text{(Eq. 3–1)}$$

Three other measures of location are commonly used: median, mode, and percentiles. The *median* is the value of the observation that divides the observations into two groups, each containing equal numbers of observations. The values in the one group are smaller than the median, and those in the other are larger than the median. If the observations are arranged in increasing order and there is an odd number of observations, the median is the middle observation. If there is an even number of observations, the median is the average of the two innermost observations.

The *mode* is the most frequent observation. For the ATT data, the mean is 111.6, the median is 112, and the mode is 110 units. For data with approximately a Gaussian distribution, the mean, mode, and median are approximately equal, and thus reporting of the mean is usually sufficient. Data that are significantly non-Gaussian should have the mode, median, and mean reported. The *percentile,* the only nonparametric statistic to be described in this chapter, is the value of an observation below which a certain proportion

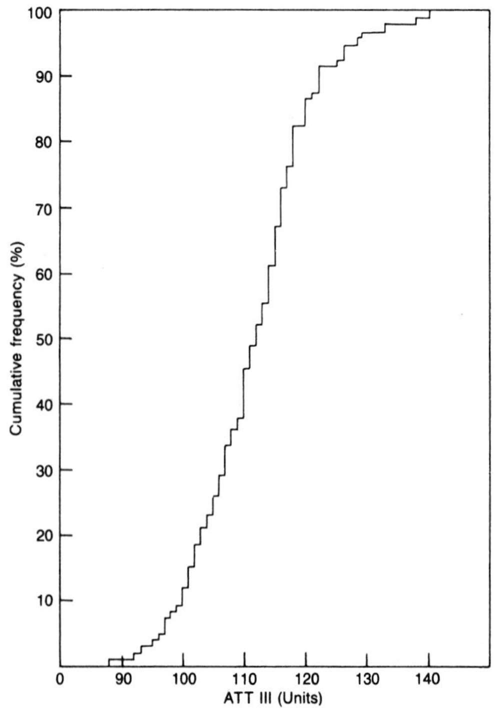

Figure 3-3. Cumulative frequency histogram for antithrombin III.

of the observations fall and which may be obtained from the cumulative-frequency histogram. The 5th percentile, or P_5, is the value below which 5% of the observations fall. For ATT, P_5 is 96. The 97.5 percentile, or $P_{97.5}$, is the value below which 97. 5% of the observations fall. For ATT, $P_{97.5}$ is 133. The median is, of course, P_{50}. *Dispersion,* or the spread of data around its location, is most simply estimated by the range, which is the difference between the largest and smallest observations. The most commonly used statistic for describing the dispersion of groups of single observations is the *standard deviation,* which is usually represented by the symbol s. The standard deviation of the observations $x_1, x_2, x_3, \ldots, x_n$ is

$$s = \sqrt{\frac{\Sigma(\overline{x} - x_i)^2}{n - 1}} \qquad (Eq.\ 3\text{--}2)$$

Figure 3-4 shows an idealized frequency histogram of glucose control values, which have a mean of 120 and a standard deviation of 5 mg/dL. For these data, as well as for other data with Gaussian distributions, approximately 68.2% of the observations will be between the limits of $\overline{x} - s$ and $\overline{x} + s$, 95.5% will be between $\overline{x} - 2s$, and $\overline{x} + 2s$, and 99.7% will be between $\overline{x} - 3s$ and $\overline{x} + 3s$. Additionally, 99% of the observations will be between $\overline{x} - 2.58s$ and $\overline{x} + 2.58s$. Limits can be constructed to include a specific proportion of the population. The usual limits for reference ranges include 95% of the population and correspond to $\overline{x} - 1.96s$ and $\overline{x} + 1.96s$. For ATT, the 95% or $\pm1.96s$ limits would be 92.5 to 130.6 or 92 to 131 units (U). The percentile also may be used to express dispersion. The percentile limits that would enclose 95% of the population would be $P_{2.5}$ to $P_{97.5}$, or 93 to 133 U.

In Equation 3-2 calculation of s requires that the mean be calculated first. There is an alternate equation (Eq. 3-3) that does not need prior calculation of the mean:

$$s = \sqrt{\frac{n\Sigma x_i^2 - (\Sigma x_i)^2}{n(n - 1)}} \qquad (Eq.\ 3\text{--}3)$$

This equation is frequently used in computer programs to minimize computation time. Another way of expressing s is in terms of the coefficient of variation (CV), which is ob-

tained by dividing s by the mean and multiplying by 100 to express it as a percentage:

$$CV\ (\%) = \frac{100s}{\overline{x}} \qquad (Eq.\ 3\text{--}4)$$

The CV, a unitless number, simplifies comparison of standard deviations of test results expressed in different units and concentrations. The CV of the glucose control data of Figure 3-4 is thus 100 times 5 mg/dL divided by 120 mg/dL, or 4.2%. The CV is used extensively in quality control summary reports.

The *mean absolute deviation* (MAD), also known as the *average deviation,* is another measure of dispersion of groups of single observations and is calculated using the following equation:

$$MAD = \frac{\Sigma|x_i - \overline{x}|}{n} \qquad (Eq.\ 3\text{--}5)$$

The *standard deviation of the mean,* also called the *standard error of the mean* (SEM), is calculated from the following equation, in which n represents the number of observations averaged to calculate the mean:

$$SEM = \frac{s}{\sqrt{n}} \qquad (Eq.\ 3\text{--}6)$$

The SEM may be used to calculate the statistical limits for the mean. The SEM can be interpreted as the average error encountered if the sample mean was used to estimate the population mean. The 95% limits for the mean $\overline{x}$ would be $\overline{x} \pm 1.96s\sqrt{n}$. The SEM decreases as sample size increases, and the mean of a large sample is likely to be closer to the true mean than the mean of a small sample.

Descriptive Statistics of Groups of Paired Observations

Perhaps the most informative step in the evaluation of a new analytical method is the comparison-of-methods experiment, in which patient specimens are measured by both the new method and the old, or comparative, method.[4] The data obtained from this comparison consist of two values for each patient specimen. Graphing is the simplest way to visualize and summarize the paired-method comparison data. By convention, the values obtained by the old (comparative) method are plotted on the *x*-axis and the values obtained by the new (test) method are plotted on the *y*-axis. Figure 3-5 shows a plot of potassium determinations performed using two different instruments, (1) the Hitachi 917 analyzing patient plasma specimens plotted on the *x*-axis and (2) a point-of-care analyzer, the ISTAT™, analyzing whole blood specimens plotted on the *y*-axis. There is a linear relationship between the two methods over the entire range of potassium values.

The agreement between the two methods may be estimated from the straight line that best fits the points. Whereas

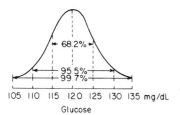

▲
Figure 3-4. Idealized Gaussian frequency histogram of glucose control values with a mean of 120 and standard deviation of 5 mg/dL. The percentages indicate the area under the curve bounded by the $\pm1, \pm2$, and ±3, limits.

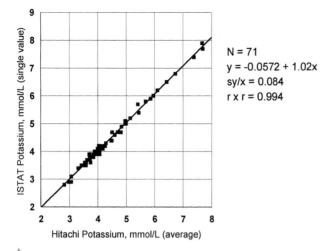

Figure 3-5. Central laboratory ISTAT versus central laboratory Hitachi Graphic presentation of a comparison-of-methods experiment. Potassium measured by the Hitachi 917 (plasma specimen) is compared with the potassium measured by the ISTAT (whole blood). Whole blood specimens were obtained from the emergency department, transported to the chemistry laboratory where the whole blood specimens were mixed with pairs of aliquots removed, one for ISTAT testing and the other separated into plasma for Hitachi analysis.

visual estimation may be used to draw the line, the use of a statistical technique, *linear regression analysis,* will result in an impartial choice of line and will provide the laboratorian with measures of location and dispersion for the line. The straight line through the data will have the equation of

$$y = mx + y_0 \qquad \text{(Eq. 3—7)}$$

The slope of the line will be *m;* the value of the *y* intercept (the value of *y* at *x* = 0) will be y_0. If there is perfect agreement between the two methods, each value measured by the test method will be identical to that measured with the comparative method. The equation of the line would be *y = x,* with *m* being 1 and y_0 being 0. Figure 3-6*A* shows perfect agreement between the test and comparative methods. Figure 3-6*B* shows the situation in which values from the test method are consistently higher than those of the comparative method. The best line through the data still has a slope of 1 but a *y* intercept of 5.0. Figure 3-6*C* shows the situation in which values from the test method are higher than values from the comparative method for nonzero concentrations; the slope is greater than 1 (1.1), but the *y* intercept is 0. For increasing concentrations, there are greater differences in the test values measured by the two methodologies. In Figure 3-6*D,* the *y* intercept is still 0 and the slope is less than 1 (0.9), showing that the test-method values are lower than the comparative-method values for all nonzero values.

Linear regression analysis usually provides unbiased estimates of the slope and *y* intercept. In linear regression, the line of best fit is one that minimizes the sum of the squares of the vertical distances of the observed points from the line.

For the points $(x_1, y_1), (x_2, y_2), \ldots, (x_i, y_i), \ldots, (x_n, y_n)$, the equation of the slope of the regression line is

$$m = \frac{\Sigma(x_i - \overline{x})(y_i - \overline{y})}{\Sigma(x_i - \overline{x})^2}$$

$$= \frac{n\Sigma x_i y_i - \Sigma x_i \Sigma y_i}{n\Sigma x_i^2 - (\Sigma x_i)^2} \qquad \text{(Eq. 3–8)}$$

The *y* intercept is calculated from *m* and the means of x_i and y_i:

$$y_0 = \frac{\Sigma y_i}{n} - m \frac{\Sigma x_i}{n}$$

$$= \overline{y} - m\overline{x} \qquad \text{(Eq. 3–9)}$$

Linear regression assumes that there is no measurement error in the comparative method and that the standard deviation of the regression line is due to random errors in the test method. The dispersion of the points about the regression line is referred to as the *standard deviation* of the regression line and is abbreviated as $s_{y/x}$. Another name for this dispersion is the *standard error* of the estimate. It is calculated using the following equation:

$$S_{y/x} = \sqrt{\frac{\Sigma(y_i - Y_i)^2}{n - 2}} \qquad \text{(Eq. 3–10)}$$

The method-comparison plots in Figure 3-6*E* and *F* show the influence of increased scatter of points about the regression line. In Figure 3-6*E* and *F,* the value of either 2 or 5, respectively, was alternately added to or subtracted from the values of *y* in Figure 3-6*A.* The slope and intercept did not change. Only $s_{y/x}$ increased to 2 or 5, respectively.

The *correlation coefficient r* is a measure of the strength of the relationship between the *y* and *x* variables. The correlation coefficient can have values from −1 to +1, with the sign indicating the direction of relationship between the two variables. A positive *r* indicates that both variables increase or decrease together, whereas a negative *r* indicates that as one variable increases, the other decreases. An *r* value of 0 indicates no relationship. The usual equation for the calculation of *r* is

$$r = \sqrt{\frac{n\Sigma x_i y_i - \Sigma x_i \Sigma y_i}{[n\Sigma x_i^2 - (\Sigma x_i)^2] \times [(n\Sigma y_i)^2 - (\Sigma y_i)^2]}} \qquad \text{(Eq. 3–11)}$$

Whereas many laboratorians equate high positive values of *r* (0.95 or higher) with excellent agreement between the test and comparative methods, most clinical chemistry comparisons should have correlation coefficients greater than 0.98. The absolute value of the correlation coefficient can be significantly increased by widening the range of samples being compared. The correlation coefficient does have a use, however. When *r* is less than 0.99, use of the regression formula results in an estimate of the slope that is too small and a *y* intercept that is too large. Waakers and associates have recommended that if *r* is less than 0.99, alternate regression

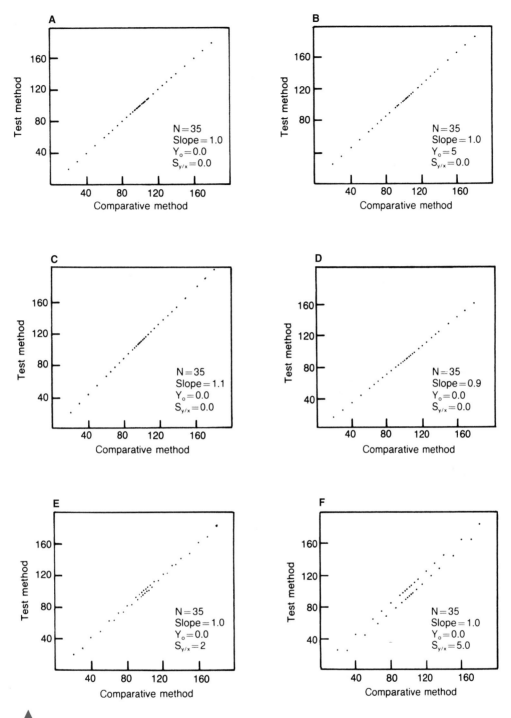

Figure 3-6. Comparison-of-methods experiments using simulated data. (**A**) shows no error; (**B**) shows constant error; (**C**) and (**D**) show proportional error; (**E**) and (**F**) show random error.

statistics should be used to derive more realistic estimates of the regression, slope, and y intercept.[5]

Error accounts for the difference between the test- and comparative-method results. Two kinds of error are measured in comparison-of-methods experiments: random and systematic. *Random error* is present in all measurements, is due to chance, and can be either positive or negative. The measure of dispersion $s_{y/x}$ provides an estimate of random error.

Systematic error is error that influences observations consistently in one direction. Unlike random error, systematic error should not be present in a method. The measures of location, the slope and y intercept, provide measures of the systematic error.

Because $s_{y/x}$ is an estimate of the standard deviation about the regression line, statistical limits can be calculated for any point on the line, just as for groups of single observations. The

95% limits of any y value on the regression line are $y \pm 1.96 s_{y/x}$; that is, 95% of the y values for any x value will fall within the calculated value of y. If $mx + y_0$ is substituted for y (Eq. 3-7), then the 95% limits are $mx + y_0 \pm 1.96 s_{y/x}$.

There are two types of systematic error: constant error and proportional error. *Constant systematic error* exists when there is a constant difference between the test method and the comparative method values regardless of the concentration. In Figure 3-6B, there is a constant difference of 5 between the test-method values and the comparative-method values. This constant difference, reflected in the y intercept, is called *constant systematic error*. *Proportional error* exists when the differences between the test method and the comparative method values are proportional to the analyte concentration. In Figures 3-6C and D, the difference between the test method and comparative method is proportional to the measured concentration. This difference, manifested by a slope different from unity, is therefore due to proportional systematic error.

Inferential Statistics

The inferential statistical tests described in this chapter are used to compare the means or standard deviations of two groups of data. The t-*test,* described by Gosset in 1908, is used to determine whether there is a statistically significant difference between the means of two groups of data. The F-*test* is used to determine whether there is a statistically significant difference between the standard deviations of two groups of data. Both tests have limited usefulness in method-evaluation studies.

For both the t- and F-tests, a statistic is calculated and then compared to critical values found in the t and F tables of statistics books. The critical values define the significance level or the probability that the differences are due to chance. By convention, the t- or F-test is said to be *statistically significant* if the probability of the difference occurring due to chance is less than 5%. If the probability of the difference occurring due to chance exceeds 5%, then the difference is usually said to be *not statistically significant.* The lower the probability, the more statistically significant is the difference: for example, a difference that occurs 1% of the time due to chance has greater significance than one that occurs 5% due to chance.

To apply the t-test, the t statistic is calculated and compared to a table of critical t values for selected significance levels and degrees of freedom. Values of t should be obtained from the t table if a small number of observations (30 or fewer) is averaged. If more than 30 observations are averaged, the critical values are almost independent of the number of observations and depend primarily on the significance level. The critical values listed in Table 3-2 may be used if more than 30 observations are averaged.

If the calculated t statistic exceeds the critical value, then a significant difference is said to exist. The larger the difference, the larger will be the t statistic and the lower will be the significance level or the probability that the difference is due to chance. The one-tailed critical values are used to test whether one mean of a group of numbers is either signifi-

TABLE 3-2. Critical t Values

Significance Level	Two-Tailed	One-Tailed
5% (0.05)	1.96	1.64
1% (0.01)	2.58	2.33

cantly greater or less than the other mean. The two-tailed critical values are used to test whether the means are significantly different.

In its simplest application, the t-test is used to determine whether the mean of a group of data ($\bar{x}$) is different from the true mean (abbreviated as M). The equation for the calculation of the t statistic is

$$t = \frac{|\bar{x} - \text{M}|}{s/\sqrt{n}}$$

Degrees of freedom $= n - 1$ *(Eq. 3–12)*

In this example, the t value is the absolute value of the arithmetic difference of the true mean and the mean of the group of data divided by the SEM. For example, if the mean of a group of glucose determinations on one control product were tested and shown to be different from the usual mean value, the **two-tailed** critical t values would be used. If the calculated t value were less than 1.96, then the difference between the means would not be considered significant at the 5% level. A t value between 1.96 and 2.58 indicates a statistically significant difference; such a difference would occur due to chance with a probability between 1% and 5%. A difference with a t value greater than 2.58 is more statistically significant and has less than a 1% probability of occurring due to chance.

The **one-tailed** critical values would be used to determine whether the mean of a group of data is either significantly larger or smaller than the true mean. For example, a physician may have several cholesterol values from a patient and may wish to determine whether these values are significantly greater than the upper limit of acceptability. After calculating the t value, the physician should use the table of one-tailed critical values.

In clinical chemistry, the t-test is sometimes applied to method-comparison data obtained by measuring patient specimens by both the test and the comparative methods. The measured values are averaged for each method, with the averages tested for statistically significant differences. If x_i and y_i are the values obtained by the comparative and test methods, respectively, the t value for a group of paired observations $(x_1, y_1), (x_2, y_2), \ldots, (x_i, y_i), \ldots, (x_n, y_n)$ is

$$t = \frac{\dfrac{\Sigma y_i}{n} - \dfrac{\Sigma x_i}{n}}{s_\text{d}/\sqrt{n}}$$

$$= \text{bias}/(s_\text{d}/\sqrt{n}) \qquad \textit{(Eq. 3–13)}$$

The numerator of the expression is the difference between the mean of the test method ($\Sigma y_i / n$) and the mean of the comparative method ($\Sigma x_i / n$). This difference between the means is called the *bias*. The symbol s_d stands for the standard deviation of the differences:

$$s_d = \sqrt{\frac{\Sigma[y_i - x_i - \text{bias}]^2}{n - 1}} \qquad \textit{(Eq. 3–14)}$$

Equation 3-13 shows that the *t* value is simply the bias, or difference of the means, divided by a standard error. Westgard and Hunt have shown that the interpretation of the *t*-test without regard for s_d and *n* may be misleading.[6,7] Statistically significant differences may exist between methods, but the size of the bias may not be clinically significant. The user is cautioned to determine whether the bias is clinically significant whenever the difference between the means is found to be statistically significant.[7] Also, the bias and s_d may be clinically large, but the resulting *t* value may be small. Thus all terms, bias, s_d, and *t* value, must be critically evaluated when interpreting the results of the *t*-test.

The second inferential statistical test, the *F*-test, has been used to compare the sizes of the standard deviations of two methods. To calculate the *F* statistic, the square of the larger standard deviation (s_L) is divided by the square of the smaller standard deviation (s_S):

$$F = \frac{(s_L)^2}{(s_S)^2} \qquad \textit{(Eq. 3–15)}$$

Like the *t*-test, the *F* statistic is then compared with critical *F* values in statistical tables that are tabulated by degrees of freedom and significance level. Because the *F*-test provides information about statistical significance but not clinical significance, Westgard and Hunt state that the *F* value should not be used as an indicator of acceptability of a test.[7] Rather, acceptability should depend on the size of random error.

REFERENCE INTERVALS (NORMAL RANGE)

Definition of Reference Interval

Physicians order laboratory tests for a variety of reasons. The most important of these are diagnosis of disease, screening for disease, and monitoring of levels of drugs and endogenous substances such as electrolytes. Other reasons for testing include determining prognosis, confirming a previously abnormal test, physician education, and medicolegal purposes. When a test is used for diagnosis, screening, or prognosis, the test result is usually compared with a reference interval (normal range) that is defined as the usual values for a healthy population. For example, if a patient appears to have signs and symptoms of hyperthyroidism, one of the follow-up tests ordered by the attending physician would be the serum thyroxine level. If the patient's thyroxine level exceeds the upper

reference limit, the physician may require further testing to determine the cause of the hyperthyroidism. When a test is used for monitoring, the test result is usually compared with values that were previously obtained from the same patient. For example, in patients with surgically removed colonic carcinomas, the presence of carcinoembryonic antigen (CEA) is often used to detect recurrence of the carcinoma. In these patients, each new CEA value is compared with previous CEA values. The acceptable range for the CEA values should be derived from the previous test values of each patient.[8]

The International Federation of Clinical Chemistry (IFCC) has recommended use of the term *reference interval* to denote the usual limits of laboratory data.[9] The presence of health is not implied in the definition, and thus reference intervals may be constructed for ill as well as healthy populations. The IFCC recommends[9] that the following five factors be specified when reference intervals are established: (1) makeup of the reference population with respect to age, sex, and genetic and socioeconomic factors; (2) the criteria used for including or excluding individuals from the reference sample group; (3) the physiologic and environmental conditions under which the reference population was studied and sampled, including time and date of collection, intake of food and drugs, posture, smoking, degree of obesity, and stage of menstrual cycle; (4) the specimen-collection procedure, including preparation of the individual; and (5) the analytical method used, including details of its precision and accuracy. The IFCC considers the terms *normal values* and *normal range* to be specific reference intervals that correspond to the health-associated (central 95%) reference interval. This chapter discusses the health-associated reference interval almost exclusively and thus uses the terms *normal values, normal range,* and *reference range* and *reference interval* interchangeably.

In the past, many hospital laboratories have either used the reference interval recommended by the instrument or test manufacturer or the values published in medical or laboratory textbooks. Because of the diversity of instrumentation, methodologies, reagents, and populations, it is important that moderate- to large-sized hospital laboratories determine their own reference intervals. Smaller laboratories will lack the resources to conduct such work; instead they should analyze far fewer specimens (at least 20) and verify the reference interval specified in the method's package insert, which is provided by the manufacturer.

The selection of subjects for the reference-interval study is very important. Many laboratory data depend on age and sex. For example, plasma testosterone level is low in both prepubertal boys and girls and increases during puberty, attaining higher levels in boys. If a physician obtains a testosterone level to rule out a testosterone-secreting tumor in a pubertal girl, the physician must be able to compare the girl's testosterone value with normal values for girls of her age. Similarly, alkaline phosphatase levels are elevated during growth (Fig. 3-7)[10] as well as in men and women older than 60 years. Ideally, the laboratory should have age and sex-stratified normal values for all populations tested. Thus,

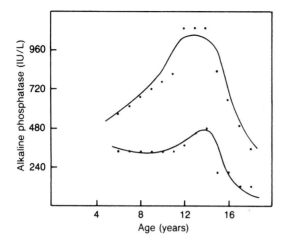

Figure 3-7. $P_{2.5} - P_{97.5}$ reference intervals for alkaline phosphatase in healthy boys determined by the Bowers-McComb method.[23]

if a laboratory tests many specimens from older adults, proper reference intervals should be provided for this population.

To derive reliable estimates of reference intervals, at least 120 individuals should be tested in each age and sex category. However, it is often necessary to carry out reference range studies using fewer individuals. Sampling of 120 males and females would be adequate for determining the reference interval of an analyte that does not vary substantially with age and sex (eg, sodium). An analyte such as creatine kinase, in which there are substantial differences between males and females, would require sampling of at least 120 males and 120 females. If reference intervals were desired for both males and females from birth to age 70 years, and if each age category equaled a 10-year interval, then the minimum number of individuals to be tested would be 100 subjects times 2 sexes times 7 age classifications, or 1400 subjects. Systematic testing of such a large number of subjects is almost always prohibitive. Winsten[11] has suggested that four age categories be used: newborns, prepubertal, adult population (postpubertal and premenopausal), and older adult population (males older than age 60 years and postmenopausal females). Although these divisions are not optimal, they reduce the number of categories that should be tested. The dependence of alkaline phosphatase on age (shown in Figure 3-7) indicates that, unless there is further stratification of alkaline phosphatase by age, the prepubertal reference interval may be limited in its usefulness.

Reference-interval studies on healthy children are limited because of the psychological and physical pain of phlebotomy. Many such studies are done on pediatric patients for whom serum and plasma samples are already available. Unfortunately, the diseases and treatments of these children can result in systematic shifts in their laboratory data. Use of their laboratory data can result in erroneous reference intervals. Because very few centers can undertake systematic reference-interval studies on healthy newborns and children, we recommend that published reference values be used and compared critically with the clinical laboratory's adult reference values. Soldin et al[12] and Meites[13] have compiled reference values for pediatric patients in comprehensive monographs. The earlier edition of Meites's monograph contains very useful information not included in the third edition and should not be discarded.

Collection of Data for Reference-Interval Studies

If reference-interval data are to be maximally useful, the reference population should consist of individuals in a state of good health. Hospital employees are the most readily available healthy individuals for reference-interval studies, but unfortunately, there is a sampling bias, with the population consisting primarily of premenopausal females. Great effort often must be expended to recruit enough male adults and postmenopausal females. Once the desired reference population is identified, a consent form must be drafted and presented to the hospital's human subject experimentation committee or institutional review board so that volunteers can be recruited for donation of blood or other specimens.

All potential donors should be interviewed to gather the following data: age, sex, health status, activity level, height and weight, alcohol consumption, drug usage (including oral contraceptives), smoking history, and stage of menstrual cycle. If a normal-range or health-associated reference interval is desired, individuals with acute or chronic disease should be excluded. The donors should be instructed about preparation before specimen collection. Fasting specimens are usually required, in which case the donor should be instructed not to eat after 10 p.m. and to drink only noncaloric decaffeinated fluids before blood drawing. All donors should be sampled in a similar fashion, with care taken by the phlebotomist not to apply the tourniquet for longer than 1 minute. It should be noted that many analytes (eg, iron, ACTH [adrenocorticotropic hormone], cortisol, and so on) exhibit significant diurnal variations. The time of sampling for these substances should be controlled.

Once the specimens are acquired, they must be labeled and handled in the same manner as the regular specimens. The instruments used to analyze these specimens should be in good running operation. Ideally, no more than 5 to 10 subjects should be sampled and analyzed daily. With this long-term analysis, the normal range will reflect the long-term state of analytical control. Analysis over a short period may introduce systematic differences or *shifts* in the reference range due to transient instrument or reagent differences.

Statistical Analysis of the Reference-Interval Data

Rigorous Definition of the Reference Interval

The analysis of the large amount of data derived from reference-interval studies used to be very laborious. Today this task is simplified by the ready availability of microcomputer-based spreadsheet or statistical programs. The test data are first entered into the computer along with donor demographic data, such as identifier code, sex, and age. Frequency histograms then are plotted for all the tests. Results for donors with outlying laboratory data should be forwarded to their physicians. If the reason for the outlying results is known (*eg,* elevated creatine phosphokinase levels after participation in a football game), the outlying results should be eliminated from further analysis.

Until approximately 1990, the standard practice was to plot the reference-interval data as cumulative-frequency histograms on probability paper and then derive reference intervals from the probability plot. Figure 3-2 shows the frequency histogram of ATT. Figure 3-3 shows the cumulative-frequency histogram of ATT using a linear scale for the *y*-axis. Figure 3-8 shows a cumulative-frequency histogram for ATT plotted on probability paper. Cumulative frequency probability plots make Gaussian distributions linear and enables drawing of the best straight line through the points. Most of the reference-interval data are thus used to determine the reference interval, and the effect of outliers is diminished. The outer 2.5% and 97.5% limits of the population are determined by selecting values for ATT on the straight line that correspond to cumulative frequencies of 2.5% and 97.5%, respectively. The reference-intervals derived from Figure 3-8 are 92.5 U to 129 U. If the distribution of the population is very smooth and

Gaussian, then the 95% limits also can be calculated directly from $\bar{x} \pm 1.96s$. In the case of ATT, the limits calculated in this manner are $115.6 \pm 1.96 \times 9.5$ U, or 92.5 U to 130.6 U. Alternately, the 95% limits can be determined from the 2.5 and 97.5 percentile limits, or 93 U to 133 U (see Figure 3-3).

If the data are not Gaussian, determination of the 95% limits is more difficult. Figure 3-9 shows a frequency histogram and Figure 3-10 presents a probability plot for gamma-glutamyl transpeptidase (GGT) in 118 females. The frequency histogram shows a skewing toward increased GGT values. The probability plot is nonlinear and indicates a non-Gaussian distribution. Percentiles do not depend on the shape of the distribution and may be used to determine the 2.5% and 97.5% limits for the population. The 2.5 and 97.5 percentile limits for GGT are 10 and 35 IU/L.

Since the early 1990s, there has been less enthusiasm for deriving reference intervals with probability plots and more for using percentile analysis. This change is due to at least three factors, including the fact that many sets of reference-interval data are not Gaussian, that the construction and interpretation of probability plots is complex, and that the recent National Committee for Clinical Laboratory Standards (NCCLS) document *Approved Guideline for How To Define and Determine Reference Intervals in the Clinical Laboratory C28-A,*[14] promotes the percentile approach.

Reference intervals are occasionally widened to include the lower limit of the analyte. For example, Figure 3-11 shows the frequency histogram and the probability plot for total bilirubin in 228 male and female subjects. The probability plot is nonlinear so the 2.5 and 97.5 percentile limits are used to define the reference interval: 0.3 to 1.4 mg/dL. Because the published lower limit of bilirubin is usually 0, and because there are few, if any, pathologic reasons for a bilirubin of 0, the reference interval derived from these data was set as 0 to 1.4 mg/dL.

Occasionally, the results of a new reference-interval study may be quite different from the old reference intervals, even though the same instrument and methodology were used. In such a situation, careful investigation must be undertaken to discover whether there are differences in the analytical method or in the reference population.

The distribution of patient data has been analyzed by various computational methods in an attempt to derive health-associated reference intervals. Martin and associates have recommended that reference intervals be derived from the numerical analysis of distributions of patient data.[15] Problems exist in this approach to reference-interval determination and account for its lack of acceptance.

A normal range or health-associated reference interval is adequate for most tests. There are a few tests, such as that for glycosylated hemoglobin (HbA_{1c}), for which an alternate reference interval is desirable. HbA_{1c} is a measure of an individual's average blood glucose level over the past 6 to 12 weeks; it usually is measured to determine and improve compliance. The frequency histogram in Figure 3-12 compares the HbA_{1c} values of normal subjects with those of patients with diabetes.

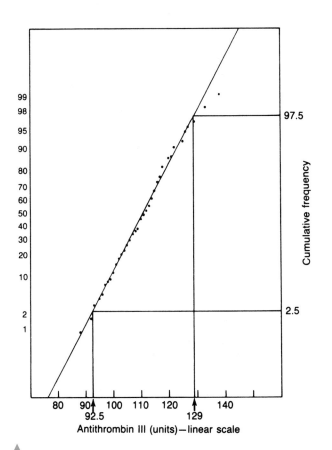

Figure 3-8. Probability plot for antithrombin III (linear scale).

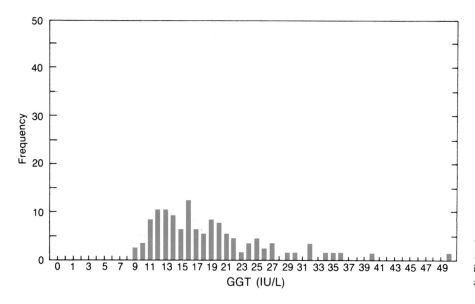

Figure 3-9. Frequency histogram of GGT in 118 females. The corresponding probability plots are shown in Figure 3-10.

There is little overlap between nondiabetic and diabetic patients. The upper range of normal has little significance to diabetic patients; many physicians use a somewhat arbitrary limit of less than 7% to designate good diabetic control. Probably the most useful reference interval to improve patient compliance is that based on the patient's previous values.

Validation of the Reference Interval

Many laboratories do not have adequate resources to perform statistically rigorous reference-interval studies. Even without the data entry and analysis steps, the recruitment and sampling of a minimum of 120 subjects is very demanding. Most manufacturers of clinical laboratory analyzers provide reference in-

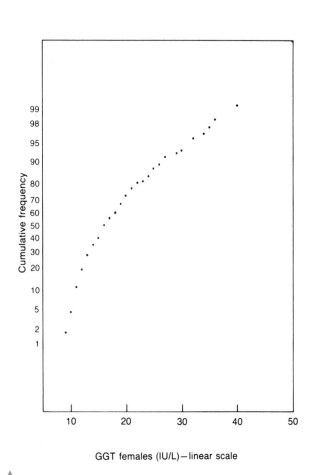

Figure 3-10. Probability plot for GGT: linear scale.

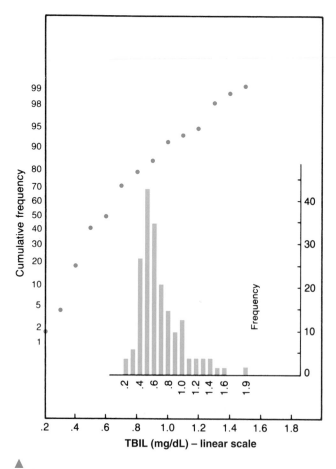

Figure 3-11. Frequency histogram of total bilirubin in 228 males and females (*inset*) and the corresponding probability plot (*linear scale*).

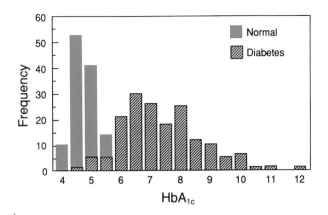

Figure 3-12. Comparison of frequency histograms of hemoglobin A_{1c} for subjects with normal fasting glucoses and patients with diabetes.

tervals for those analytes that are measured with their reagents. It is unnecessary for each laboratory to develop its own reference interval for a manufacturer's analyzer and reagents. Rather, a far smaller validation study is required and is described in the same NCCLS document *Approved Guideline for How To Define and Determine Reference Intervals in the Clinical Laboratory C28-A.*[14] Only 20 reference individuals need to be sampled for analysis on the test instrument if the laboratorian determines that the test instrument and test subject population are similar to those described in the manufacturer's package insert. The manufacturer's reported 95% reference limits may be considered valid if no more than 2 of the 20 tested subjects fall outside the original reported limits. If 3 or more test results are outside these limits, another 20 reference specimens should be obtained; if no more than 2 of these new results are outside the manufacturer's limits, then the manufacturer's limits can be used. If the second attempt at validation fails, the laboratorian should reexamine the analytical procedure and identify any differences between the laboratory's population and the population used by the manufacturer for the package insert information. If no differences are identified, the laboratory may need to do the full 120 individual reference-interval study.

DIAGNOSTIC EFFICACY

Until recently, many years might elapse between the introduction of a new diagnostic test and its rigorous clinical evaluation. Now, in as little as 1 year to 2 years, a new diagnostic test may gain sudden fame and then, just as quickly, disfavor. The analysis of prostatic acid phosphatase (PAP) by radioimmunoassay (RIA) is such an example. A 1977 *New England Journal of Medicine* article concluded that the RIA for PAP had the potential to detect well more than half the cases of prostate cancer that had not yet penetrated the prostatic capsule.[16] The same journal published an article 2½ years later that stated that "the most economical and most reliable probe for the detection [of prostate cancer] remains not a needle in the patient's arm, but the gloved finger of the physician, engaged in a thorough rectal examination of the prostate."[17] The controversy

associated with the PAP test probably caused the urologic community to be wary about other prostatic tumor markers. The subsequent introduction of a test for prostate-specific antigen (PSA), a sensitive but not specific test for prostate cancer, was not accompanied by the fanfare of the RIA test for PAP. The PSA test has turned out to be a more useful test in the management of prostate cancer.[18]

The material in this section allows the reader to analyze frequency histograms of patient test data and to calculate a test's diagnostic sensitivity and specificity and predictive value. The reader also will be able to compare the diagnostic efficacies of various laboratory tests and to determine the most useful test.

Predictive Value Theory

Whereas the terms *diagnostic sensitivity, specificity,* and *predictive value* were used by radiologists beginning in the early 1950s, it was not until the 1970s that laboratorians became familiar with their meanings. The diagnostic sensitivity should not be confused with analytical sensitivity. For a test that is used to diagnose a certain disease, the *diagnostic sensitivity* is the proportion of individuals with that disease who test positively with the test. Sensitivity is usually expressed as a percentage:

$$\text{Sensitivity (\%)} = \frac{100 \times \text{the number of diseased individuals with a positive test}}{\text{total number of diseased individuals tested}} \quad \textit{(Eq. 3–16)}$$

The specificity of a test is defined as the proportion of individuals without the disease who test negatively for the disease. The specificity (%) is defined as follows:

$$\text{Specificity (\%)} = \frac{100 \times \text{the number of individuals without the disease with a negative test}}{\text{total number of individuals tested without the disease}} \quad \textit{(Eq. 3–17)}$$

Ideally, sensitivity and specificity should each be 100%. The sensitivity and specificity of a test depend on the distribution of test results for the diseased and nondiseased individuals and also on the value of the test that defines the abnormal levels. Figure 3-13 shows frequency histograms of PSA of two patient populations. These two populations are subsets of a group of 4962 male volunteers 50 years old or older who were evaluated for the presence of prostate cancer by way of digital rectal examination of the prostate and a PSA measurement.[19] Of these volunteers, 770 had abnormal findings. The upper frequency histogram represents the subset of 578 patients with abnormal PSA values or abnormal digital rectal examinations who were biopsied for possible prostate cancer and found to be cancer free. The lower frequency histogram represents the subset of 192 patients with abnormal PSA values or abnormal digital rectal examinations who were biopsied for, and found to have, prostate cancer. It can be seen that, although patients with prostate

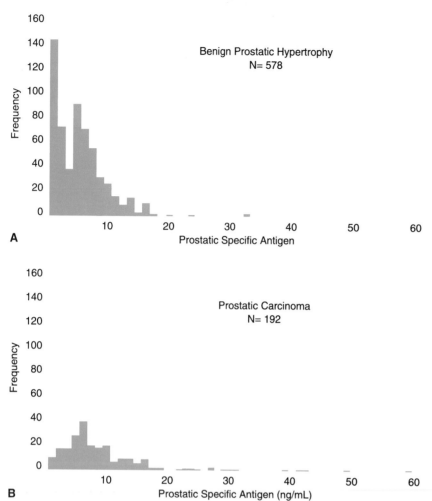

Figure 3-13. Frequency histograms of PSA results (Hybritech Tandem assay) of patients evaluated for possible prostate cancer. (**A**) The top plot shows the PSA of patients being biopsied without prostate cancer; (**B**) the bottom shows the PSA of patients with biopsy-positive prostate cancer. (Adapted with permission from Catalona WJ, Richie JP, deKernion JB, et al. Comparison of prostate specific antigen concentration versus prostate specific antigen density in the early detection of prostate cancer: Receiver operating characteristic curves. J Urol, 1994;152:2031.)

carcinoma tend to have higher values of PSA, there is a great deal of overlap between the two populations.

The sensitivity and specificity can be calculated for any value of PSA. If high values of PSA are used to indicate the presence of disease (*eg,* if values exceeding 35 ng/mL are used to indicate prostate cancer), then the specificity of the test will be 100% (all the patients without prostate cancer are classified as negative using the test). The sensitivity, however, is quite low (2.6%), with only 5 of the 192 patients with carcinoma having PAP values greater than 35 ng/mL. The sensitivity can be increased by decreasing the test value. Thus, if values of PSA in excess of 4.0 ng/mL (the upper reference limit) are used to diagnose carcinoma, the sensitivity increases to 79%. The specificity, however, decreases to 46%. There are very few laboratory tests with sensitivities and specificities close to 100%. The sensitivity and specificity of the MB fraction of creatine kinase for the diagnosis of myocardial infarction are approximately 95% each. Troponin, although having roughly the same sensitivity for diagnosing myocardial infarction, has a slightly higher specificity. The urine hCG (human chorionic gonadotropin) pregnancy tests used in the clinical laboratory have sensitivities and specificities approaching 100%. Many tests have sensitivities and specificities that are close to

50%. Galen and Gambino have stated that if the sum of the sensitivity and specificity of a test is approximately 100%, the test is no better than a coin toss.[20]

The sensitivity and specificity may be calculated from simple ratios. Patients with a disease who are correctly classified by a test to have the disease are called *true positives* (TP). Patients without the disease who are classified by the test not to have the disease are called *true negatives* (TN). Patients with the disease who are classified by the test as disease-free are called *false negatives* (FN). Patients without the disease who are incorrectly classified as having the disease are called *false positives* (FP). The sensitivity may be calculated from the formula 100TP/(TP + FN). Specificity may be calculated from 100TN/(TN + FP).

Three other ratios are important in evaluating diagnostic tests: predictive value of a positive test (PV+), predictive value of a negative test (PV−), and the efficiency. PV+ is that fraction of positive tests that are true positives: PV+ (%) = 100TP/(TP + FP). PV− is that fraction of negative tests that are true negatives: PV^a (%) = 100TN/(TN + FN). The efficiency of a test is the fraction of all test results that are either true positives or true negatives: efficiency (%) = 100(TP + TN)/(TP + TN + FP + FN). The calculations for sensitivity, specificity, and predictive values for PSA exceeding 4.0 ng/mL are shown in Table 3-3. If a patient has a negative test result, he has an 87% probability of not having cancer. The probability that a patient has cancer if he has a positive test is rather low—approximately 33%.

The predictive value of a test can be expressed as a function of sensitivity, specificity, and disease prevalence, or the proportion of individuals in the population who have the disease:

$$PV^+ = \frac{prevalence \times sensitivity}{(prevalence)(sensitivity) + (1 - prevalence)(1 - specificity)} \qquad \textit{(Eq. 3–18)}$$

The PV$^+$ for PSA for various prevalences and a cutoff of 4.0 ng/mL (sensitivity =, specificity = are shown in Table 3-4). It can be seen that, even in the situation in which disease has a high prevalence (0.2, or one-fifth of the population), the probability that the positive test truly indicates carcinoma is only 27%.

Because the selection of a cutoff level to define disease can be arbitrary, it is preferable to calculate and plot sensitivity and specificity for all values of a test. This allows comparison of sensitivities of two or more tests at defined specificities or comparison of specificities for certain sensitivities. Receiver operating characteristic (ROC) curves, which are plots of sensitivity (true-positive rate) versus 1− specificity (false-positive rate), have been used to compare different laboratory tests.[21] Figure 3-14 illustrates two different ROC curves. Curve A is an ROC curve for a test in which there is wide separation between the test values of the diseased and nondiseased patients. Curve B illustrates the ROC curve obtained from a test for which there is little separation between diseased and nondiseased patients. The lower left part of the curve, where the false-positive rate is close to 0 (100% specificity) and the true-positive rate is close to 0 (0% sensitivity), corresponds to extreme test values in the diseased population. The upper right part of the curve, in which both the true-positive rate and the false-positive rate are high, corresponds to typical test values in the nondiseased population. For intermediate test values, a good test should have a high sensitivity (high true-positive rate) and a high specificity

(low false-positive rate) and will form an ROC curve with its point close to the upper lefthand corner of the plot. Curve A corresponds to such a test. The test represented by curve B, in which the false-positive rate is equal to the true-positive rate, conveys no useful diagnostic information. Increasingly, clinical evaluations of diagnostic tests are being presented in the form of ROC curves. The NCCLS has published guidelines for clinically evaluating laboratory tests, including a comprehensive guide to ROC curves.[22]

Figure 3-15 shows the ROC curve of the PSA data of Figure 3-13. Also shown are the PSA concentrations at which the sensitivity and specificity were calculated. It can be seen that no PSA value offers both a high sensitivity and high specificity. Urologists are now recommending a more specific variant of the PSA test: the percentage of free PSA.[23]

METHOD SELECTION AND EVALUATION

Method Selection

Before a new test or methodology is introduced into the laboratory, both managerial and technical information must be compiled and carefully considered. The information should be collected from many different sources, including manufacturer and sales representatives, colleagues, scientific presentations, and the scientific literature. The managerial information should include instrument cost, throughput, sample volume, personnel requirements, cost per test, specimen types, instrument size, and power and environmental requirements. The technical information must include analytical sensitivity, analytical specificity, detection limit, linear range, interfering substances, and estimates of imprecision and inaccuracy.

The *analytical sensitivity* or *detection limit* refers to the smallest concentration that can be measured accurately. One professional group has defined the *detection limit* as equal to three times the standard deviation of the blank, or as located three standard deviations above the measured blank.[24] *Specificity* refers to a method's ability to measure only the analyte of interest. The *linear range* (sometimes called the *analytical* or *dynamic range*) is the concentration range over which the mea-

TABLE 3-3. Sample Calculation of Sensitivity, Specificity, PV$^+$, PV$^-$, and Efficiency Calculated for Values of PSA > 4.0 ng/mL		
	Number of Patients With Positive PSA (>4.0 ng/mL)	**Number of Patients With Negative PSA (4.0 ng/mL)**
Number of patients with prostate cancer	TP (151)	FN (41)
Number of patients without prostate cancer	FP (313)	TN (265)

Sensitivity = 100 × TP/(TP + FN) = 100 × 151/192 = 79%.
Specificity = 100 × TN/(TN + FP) = 100 × 265/578 = 46%.
PV$^+$ = 100 × TP/(TP + FP) = 100 × 154/464 = 33%.
PV$^-$ 100 × TN/(TN + FN) = 100 × 265/306 = 87%.

TABLE 3-4. Dependence of Predictive Value on Disease Prevalence (Sensitivity = 79%, Specificity = 46%)

Prevalence of Disease	PV+
0.001	0.1%
0.01	1.5%
0.10	14.0%
0.2	26.8%
0.5	59.4%

sured concentration is equal to the actual concentration without modification of the method. The wider the linear range, the less frequent will be specimen dilutions. Estimates of the inaccuracy of instruments may be obtained by studying the summaries of proficiency-testing programs that use fresh plasma or serum (misleading comparisons can be made from proficiency programs that analyze reconstituted lyophilized product). Estimates of intra-instrument imprecision are available from the group summary reports that are provided by vendors of quality-control products.

Method Evaluation

Once a method is brought in-house, the laboratorian must become proficient in using it. Then, in advance of the complete evaluation, a short initial evaluation should be carried out. This preliminary evaluation should include the analysis of a series of standards to determine the linear range, the replicate analysis (at least 8 measurements) of two to four patient

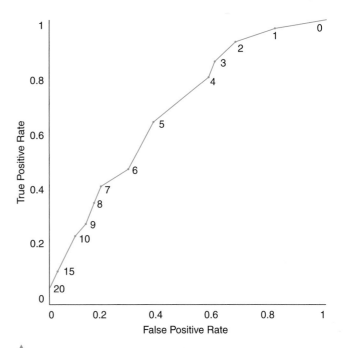

▲
Figure 3-15. ROC curve for PSA (Hybritech Tandem assay) for the diagnosis of prostate cancer. (Adapted with permission from Catalona WJ, Richie JP, deKernion JB, et al. Comparison of prostate specific antigen concentration versus prostate specific antigen density in the early detection of prostate cancer: Receiver operating characteristic curves. J Urol 1994;152:2031.)

samples to obtain estimates of short-term imprecision, and finally, preliminary interference and recovery studies. If any of these initial results falls short of the specifications published in the method's product information sheet (package insert), the method's manufacturer should be consulted. Without improvement in the method, more extensive evaluations would be pointless.[25] Figure 3-16 shows a simple data input form that can be used to simplify collection of method evaluation data.

Once the method evaluation data are collected, the imprecision and inaccuracy of a method are estimated and compared with the maximum allowable error, which usually is based on medical criteria. If the imprecision or inaccuracy exceeds this maximum allowable error, the method is judged as unacceptable and must be modified and reevaluated or rejected. *Imprecision,* the dispersion of repeated measurements about the mean, is due to the presence of random analytical error. Imprecision is estimated from studies in which aliquots of a specimen with a constant concentration of analyte are analyzed repetitively. *Inaccuracy,* the difference between a measured value and its true value, is due to the presence of systematic analytical error, which can be either constant or proportional. Inaccuracy can be estimated from three studies: recovery, interference, and a comparison-of-methods study.

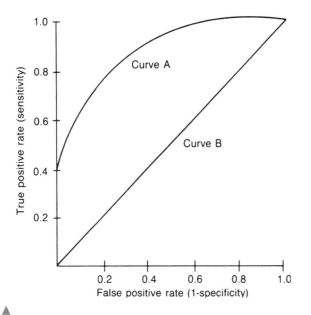

▲
Figure 3-14. ROC curves. Curve A corresponds to a test with wide separation between the diseased and nondiseased patients. Curve B corresponds to a test that provides no additional information.

Measurement of Imprecision

The first step in method evaluation is the *precision* study. This study estimates the random error associated with the

PATIENT COMPARISONS:

COMPARATIVE METHOD: _____

TEST METHOD: _____

MID RANGE				LO RANGE				HI RANGE		
	DATE	COMP	TEST		DATE	COMP	TEST		DATE	COMP	TEST
1				31				61			
2				32				62			
3				33				63			
4				34				64			
5				35				65			
6				36				66			
7				37				67			
8				38				68			
9				39				69			
10				40				70			
11				41				71			
12				42				72			
13				43				73			
14				44				74			
15				45				75			
16				46				76			
17				47				77			
18				48				78			
19				49				79			
20				50				80			
21				51				81			
22				52				82			
23				53				83			
24				54				84			
25				55				85			
26				56				86			
27				57				87			
28				58				88			
29				59				89			
30				60				90			

LINEARITY:

	A	B	C	D	E
TARGET					
1					
2					
3					
MEASURED					

PRECISION:

WITHIN RUN		LO	MID	HI
	1			
	2			
	3			
	4			
	5			
	6			
	7			
	8			
	9			
	10			
	$\overline{\overline{X}}$			
	SD			
	CV			

CONTROL PRODUCT: _____
I: _____
II: _____
III: _____

BETWEEN RUN		DATE	I	II	III
	1				
	2				
	3				
	4				
	5				
	6				
	7				
	8				
	9				
	10				
	11				
	12				
	13				
	14				
	15				
	16				
	17				
	18				
	19				
	20				
	$\overline{X}$				
	SD				
	CV				

◄ *Figure 3-16.* Data input form for method evaluation experiment. (Courtesy of Kristen Lambrecht.)

test method and points out any problems affecting reproducibility. It is recommended that this study be done over a 10- to 20-day period, incorporating one or two analytical runs per day.[26,27] An *analytical run* is defined as a group of patient specimens and control materials that are analyzed, evaluated, and reported together. Imprecision should be measured at more than one concentration with control materials spanning the clinically meaningful range of concentrations. For example, glucose should be studied in the hypoglycemic (~50 mg/dL), hyperglycemic (~150 mg/dL), and normoglycemic (~100 mg/dL) ranges.

Once the precision data are collected, the mean, standard deviation, and coefficient of variation are calculated. The random error or imprecision associated with the test procedure is indicated by the standard deviation and the coefficient of variation. The within-run imprecision is indicated by the standard deviation of the controls analyzed within one run. The total imprecision may be obtained from the standard deviation of control data with one data point accumulated per day. A statistical technique, analysis of variance, may be used to analyze all the available precision data to provide estimates of the within-run between-run, and total imprecision.[28]

Measurement of Inaccuracy

Once the short-term imprecision of the method is estimated and deemed adequate, the accuracy experiments[27] can begin. *Accuracy* can be estimated in three ways: recovery, interference, and patient-sample comparison studies. Recovery and interference studies should be performed and documented by the manufacturer. Because these two types of studies can be prohibitive both in terms of effort and materials, they are usually performed in larger clinical laboratories. Recovery experiments will show whether a method measures all the analyte or only part of it. In a recovery experiment, a test sample is prepared by adding a small aliquot of concentrated analyte in diluent to the patient sample. Another sample of the patient specimen is diluted by adding the same volume of diluent alone. Both diluted specimens are then analyzed by the test method. The amount recovered is the difference between the two measured values. Care should be taken to ensure that the original patient samples are diluted by no more than 10%; in this way, the solution matrix of the samples is minimally affected. The comparative method can also be used to measure the diluted specimens,[27] and thus assure appropriate specimen preparation. Table 3-5 shows a sample calculation of recovery, which is expressed as percentage recovered.

The interference experiment is used to measure systematic errors caused by substances other than the analyte. An interfering material can cause systematic errors in one of two ways. The material itself may react with the analytical reagents, or it may alter the reaction between the analyte and the analytical reagents. The interference experiment[29] is similar to the recovery experiment, except that the substance suspected of interference is added to the patient sample. A sample calculation of interference is shown in Table 3-6. The concentration of the potentially interfering material should be in the maximally elevated range. If an effect is observed, its concentration should be lowered to discover the concentration at which test results are first invalidated. Materials to be tested should be selected from literature reviews and recent method-specific references. Young[30] and others[31] have published extensive listings of the effects of drugs on laboratory tests. Common in-

terferences (eg, hemoglobin, lipids, bilirubin, anticoagulants, preservatives, and so on) also should be tested. Glick and Ryder have presented "interferographs"[32,33] for various chemistry instruments—these are graphs relating analyte concentration measured versus interferent concentration. They have demonstrated that tens of thousands of dollars of rework can be saved by the acquisition of instruments that minimize hemoglobin, triglyceride, and bilirubin interference.[34]

Comparison-of-Methods Experiment

Comparison-of-methods studies are done on the same patient samples using the method being evaluated (test method) and a comparative method. The quality of the comparative method will affect the interpretation of the experimental results. The best comparative method that can be used is the *reference method,* which is a method with negligible inaccuracy in comparison with its imprecision. Reference methods may be laborious and time-consuming (eg, the reference method for cholesterol). Because most laboratories are not staffed and equipped to perform reference methods, the results of the test method are usually compared with those of the method routinely in use. The routine method will have certain inaccuracies, both known and unknown, depending on how well the laboratory has studied them or how well the method is documented in the literature. If the test method will be replacing the comparative method, the differences between the two methods should be well characterized.

Westgard et al[27] and the NCCLS[35] recommend that at least 40 samples, and preferably 100 samples, be run by both methods. They should span the clinical range and should represent many different pathologic conditions. Duplicate analyses of each sample by each method are recommended,

TABLE 3-5. Example of a Recovery Study

Sample Preparation

Sample 1: 2.0 mL serum + 0.1 mL H$_2$O
Sample 2: 2.0 mL serum + 0.1 mL 20 mg/dL calcium standard
Sample 3: 2.0 mL serum + 0.1 mL 50 mg/dL calcium standard

		Concentration		
	Calcium Measured	*Added*	*Recovered*	*Recovery*
Sample 1	7.50 mg/dL			
Sample 2	8.35 mg/dL	0.95 mg/dL	0.85 mg/dL	89%
Sample 3	9.79 mg/dL	2.38 mg/dL	2.29 mg/dL	96%

Calculation of Recovery

$$\text{Concentration added} = \text{standard concentration} \times \frac{mL\ standard}{mL\ standard + mL\ serum}$$

$$\text{Concentration recovered} = \text{concentration (diluted test)} - \text{concentration (baseline)}$$

$$\text{Recovery} = \frac{\text{concentration recovered}}{\text{concentration added}} \times 100\%$$

TABLE 3-6. Example of an Interference Study

Sample Preparation

Sample 1: 1.0 mL serum + 0.1 mL H_2O
Sample 2: 1.0 mL serum + 0.1 mL of 10 mg/dL magnesium standard
Sample 3: 1.0 mL serum + 0.1 mL of 20 mg/dL magnesium standard

	Calcium Measured	Magnesium Added	Interference
Sample 1	9.80 mg/dL		
Sample 2	10.53 mg/dL	0.91 mg/dL	0.73 mg/dL
Sample 3	11.48 mg/dL	1.81 mg/dL	1.68 mg/dL

Calculation of Interference

$$Concentration\ added = standard\ concentration \times \frac{mL\ standard}{mL\ standard + mL\ serum}$$

$$Interference = concentration\ (diluted\ test) - concentration\ (baseline)$$

with the duplicate samples analyzed in different runs and in a different order of analysis on the two runs. Analysis by both methods should be performed on the same day, preferably within 4 hours.

If duplicates are not analyzed, the validity of the experimental results must be checked by comparing the test and comparative-method results immediately after analysis, identifying those samples with large differences, and, if necessary, repeating the analyses. If 40 specimens are compared, two to five patient specimens should be analyzed daily for a minimum of 8 days. If 100 specimens are compared, the comparison study should be carried out during the 20-day replication study.

A graph of the test-method data (plotted on the y-axis) versus the comparative-method data (plotted on the x-axis) helps to visualize the method comparison data (see Fig. 3-5). Data should be plotted on a daily basis and inspected for linearity and outliers. In this way, original samples can be available for reanalysis. Visual confirmation of linearity is usually adequate; in some cases, however, it may be necessary to evaluate linearity more quantitatively.[36] Many analysts have used the F-test, the paired t-test, and the correlation coefficient for the interpretation of the experimental data. The F-test is used to compare the magnitude of the imprecision of the test method with that of the imprecision of the comparative method. The paired t-test is used to compare the magnitude of the bias (difference between the means of the test and that of the comparative method) with that of the random error. Both the F-test and t-test indicate only whether a statistically significant difference exists between the two standard deviations or means, respectively. They provide no information about the magnitude of the existing error relative to the clinically allowable limits of error.[7]

Despite regular admonishments about the misuse of the correlation coefficient r, laboratorians continue to use it as an indicator of the acceptability of the test method. The value of r can be increased by increasing the range of the patient specimens. The main application of the correlation coefficient in method-evaluation studies should be in determining the type of regression analysis to be used. If the correlation coefficient is 0.99 or greater, the range of patient samples is adequate for the standard linear regression analysis described in the statistics section. If r is less than 0.99, then alternate regression analysis should be used.[5,37,38] Linear regression analysis is far more useful than the t-test and F-test for evaluating method-comparison data.[7] The constant systematic error can be estimated from the y intercept, proportional systematic error from the slope, and random error from the standard error of the estimate ($s_{y/x}$). If there is a nonlinear relationship between the test and comparative values, then linear regression can only be used over the linear range. Because outlying points are weighted more heavily in linear regression, it is important to ensure that these outliers are genuine and not the result of laboratory error.

Once all the estimates of the imprecision and inaccuracy are calculated, they are compared with predefined limits of medically allowable analytical error.[39] If the error is smaller than the allowable error, the performance is considered acceptable. If the error is larger than the allowable error, the errors must be reduced or the method rejected.

The allowable analytical error represents total error and includes components of both random and systematic error. Many approaches have been used to estimate medically allowable error including, including using multiples of the reference interval[40] using pathologist opinions[41] and physiologic variation.[42] Under the federal law, the Clinical Improvement Amendments of 1988 (CLIA-88), the Health Care Financing Administration (HCFA) has published allowable errors for a wide array of clinical laboratory tests.[43] Although the *CLIA* allowable error limits are used to specify the maximum error allowable in federal-mandated *proficiency testing,* these limits are now being used as guidelines to determine the acceptability of clinical chemistry analyzers.[44,45] Table 3-7 shows the CLIA limits for common clinical chemistry tests. Westgard et al[39] have recommended two different sets of criteria for the evaluation of error: confidence-interval criteria and single-value criteria. Because of the complexity of confidence-interval criteria, the reader is referred to the original description.[46] The single-value criteria are found in Table 3-8. In using the single-value criteria, estimates of the random error, proportional error, constant error, and systematic error are calculated and then compared to the allowable error. Several of the error estimates depend on the concentration of analyte measured and are usually calculated at critical concentrations of the analyte. In applying these performance criteria, all error criteria (random, proportional, constant, and systematic error) must be less than the allowable error for a method to be judged acceptable. Otherwise, the analytical method must be rejected or modified to reduce the error. As an ultimate criterion, Westgard et al[39] have suggested a total-error criterion, which combines random and systematic components of error, to estimate the magnitude of error that can be expected when a patient specimen is measured. The use of the single-value criteria is illustrated in Table 3-9, in which whole blood potassium measured by a point-of-care method are compared to central laboratory plasma potassium. The method-evaluation experiment for this comparison is plotted in Figure 3-5. The CLIA error limit for potassium is 0.5 mmol/L.

Currently, there is disagreement regarding the selection of the multiplier of the standard deviation in the random error calculation. Westgard et al originally suggested 1.96, which permits 5% of observations to be outside of the allowable error limits. Choice of just a slightly larger multiplier would allow considerable fewer outliers; for example, if the multiplier was 2.58, no more than 1% of the observations could be outside the allowable error limits. The CLIA regulations forced laboratorians to reevaluate the proportion of outliers that they would accept, which fell outside of the CLIA limits. Clinical laboratories are at high risk of failing proficiency testing if they use analyzers that produce 5% of

TABLE 3-8. Single-Value Criteria of Westgard et al[60]

Analytical Error	Criterion
Random error (RE)	$2.58s < E_A$
Proportional error (PE)	$\lvert (\text{Recovery} - 100) \times X_0/100) \rvert < E_A$
Constant error (CE)	$\lvert \text{Bias} \rvert < E_A$
Systematic error (SE)	$\lvert (y_0 + mX_c) - X_0 \rvert < E_A$
Total error (TE = RE + SE)	$1.96s + \lvert (y_0 + mX_c) - X_c \rvert < E_A$

E_A = medically allowable error.
X_c = critical concentration.

results that deviate by more than the CLIA limits. We are now using 2.58 and are recommending all laboratories use a multiplier of at least 2.58 to calculate random error. Although some authors have proposed multipliers as high as 4.0, most instruments would need to be extremely precise and have very small standard deviations for their random error be lower than the CLIA limit.[33]

Before a method can be put into routine use, the manufacturer's reference range must be validated or even reestablished, or the procedure must be written or updated and personnel should be trained to use the method. A statistical quality-control program should be implemented for the procedure using the accumulated replication data to establish quality-control limits. It may be necessary to adjust the control limits once the procedure is in routine service.

Because of constraints of personnel, time, and budget, a laboratory may be unable to perform comprehensive comparison-of-methods experiments on every new method to be introduced. An understanding of the principles of method evaluation and a familiarity with statistical tests will allow the laboratory supervisor to choose a method that would probably fit the laboratory's performance criteria.[47] A series of abbreviated experiments could then be undertaken to estimate imprecision and inaccuracy.[3]

The NCCLS has recently published guidelines for just such an abbreviated application. This protocol, the User Demonstration of Precision and Accuracy, can be used to demonstrate that a laboratory can obtain precision and accuracy performance consistent with manufacturer's claims. Because these precision and accuracy studies can be completed in 5 working days, it is very likely they will be used by many laboratories for setting up new methods.

TABLE 3-7. Performance Standards for Common Clinical Chemistry Analytes as Defined by the CLIA[43]

Calcium, total	Target ± 1.0 mg/dL
Chloride	Target ± 5%
Cholesterol, total	Target ± 10%
Cholesterol, HDL	Target ± 30%
Glucose	Greater of target ± 6 mg/dL or ± 10%
Potassium	Target ± 0.5 mmol/L
Sodium	Target ± 4 mmol/L
Total protein	Target ± 10%
Triglycerides	Target ± 25%
Urea nitrogen	Greater of target ± 2 mg/dL or ± 9%
Uric acid	Target ± 17%

QUALITY ASSURANCE AND QUALITY CONTROL

The *quality-control system* is the laboratory's system for recognizing and minimizing analytical errors.[48,49] Quality control is one component of the *quality-assurance system*, which has been defined as all systematic actions necessary to provide adequate

TABLE 3-9. Example of the Application of the Single-Value Criteria of Westgard et al[39] to Potassium Evaluation Data. Allowable error (E_A) for potassium, as defined in the CLIA Regulations is 0.5 mmol/L.

1. Random error (RE) = 1.96 s.
   Random error is estimated from analyzing a control product once daily for 20 days.

   $$X = 5.5 \text{ mmol/L}, s = 0.07 \text{ mmol/L}$$
   $$RE = 2.58 \times s$$
   $$= 2.58 \times 0.07 \text{ mmol/L}$$
   $$= 0.18$$

   Because RE < E_A, RE is acceptable.
2. Proportional error (PE) = | (Recovery − 100) × (X_C/100) |.
   Proportional error is estimated from a series of recovery experiments, after which the average recovery is calculated.

   $$\text{Average recovery for potassium} = 99\%$$
   $$PE = |(99 - 100) \times (5.5/100)|$$
   $$= 0.05 \text{ mmol/L}$$

3. Constant Error (CE) = bias.
   Constant error is estimated from the bias | Y − X |, derived from the comparison-of-methods experiment.

   $$CE = |Y - X|$$
   $$= |5.31 - 5.28|$$
   $$= 0.03 \text{ mmol/L}$$

   Because CE < E_A, CE is acceptable.
4. Systematic Error (SE) = |($Y_O + mX_C$) − X_C|.
   Systematic error is estimated from the above equation in which Y_O and m are derived from a comparison-of-methods experiments (Figure 3-5).

   $$Y_0 = -0.057, m = 1.02$$
   $$SE = |(-0.057 + 1.02 \times 5.5) - 5.5)|$$
   $$= |5.55 - 5.5|$$
   $$= 0.05$$

   Because SE E_A, SE is acceptable.
5. Total error (TE) = RE + SE.
   Total error is estimated from the sum of RE and SE as calculated above.

   $$TE = 0.18 + 0.05$$
   $$= 0.23 \text{ mmol/L}$$

   Because TE < E_A, the method is acceptable.

confidence that laboratory services will satisfy given medical needs for patient care.[50] The quality-assurance system encompasses pre-analytical, analytical, and post-analytical factors. The monographs *Laboratory Quality Management* by Cembrowski and Carey[51] and *Cost-Effective Quality Control* by Westgard and Barry[52] provide detailed information about quality-control and quality-assurance practices in the clinical chemistry laboratory.

There are many pre-analytical factors that can influence analytical results, including patient preparation, sample collection, sample handling, and storage. Young provides the most comprehensive reviews of pre-analytical factors on chemistry tests.[30,53] Pre-analytical factors are difficult to monitor and con-trol because most occur outside the laboratory. Health care professionals, especially physicians and nurses, should become more aware of the importance of patient preparation and how it can affect laboratory tests. Patient preparation for a test is critical. For example, nutritional status, a recent meal, alcohol, drugs,[30] smoking, exercise, stress, sleep, and posture all affect various laboratory tests. The laboratory must provide instructions, usually in the form of a procedure manual,[54] for proper patient preparation and specimen acquisition. This procedure manual should be found in all nursing units and thus be available to all medical personnel. Additionally, easy-to-understand patient handouts must be available for outpatients.

Specimen-collection procedures should follow specific guidelines such as those established by the NCCLS.[55,56,57,58] Blood-collection teams should be reminded periodically about the guidelines for duration of application of tourniquets and the types of specimen-collection tubes and anticoagulants to be used. Methods of specimen transport, separation, aliquoting, and storage are critical.[59,60] The length of time elapsed between drawing and separation of the serum or plasma from the cells can be a factor in analytical testing. For example, leukocytes and erythrocytes metabolize glucose and cause a steady decrease in glucose concentration in clotted, uncentrifuged blood. The centrifuging and aliquoting of samples to secondary containers may be critical.[57,60] Contamination of the specimen may occur at this point, rendering a less than optimal specimen for analytical testing. For example, secondary containers for specimens submitted for lead analysis must be scrupulously clean because of the ubiquitous presence of lead and the low levels of the substance that must be measured.

Specimen storage also may lead to errors in the reported results. Guidelines on storage requirements for specimens should be established for each analyte. Specimens may be affected by evaporation (*eg*, electrolytes) and exposure to light (*eg*, bilirubin), refrigeration (*eg*, lactate dehydrogenase [LD]), freezing, and so on. Clerical errors may occur at any step in the processing of specimens. Although the use of computers has simplified clerical tasks, one specimen may still be mistaken for another. Obviously, such mistakes should be minimized.

The laboratorian is better able to control the analytical factors, which primarily depend on instrumentation and reagents. A schedule of daily and monthly preventive maintenance for each piece of equipment is essential. Instrument function checks that are to be routinely performed should be detailed in the procedure manual and their performance should be documented. The NCCLS has developed standards for monitoring variables such as water quality,[61] calibration of analytical balances, calibration of volumetric glassware and pipettes, stability of electrical power,[62] and the temperature of thermostatically controlled instruments.[63] Reagents and kits should be dated when received and also when opened. New lots of reagents should be run in parallel with old reagent lots before being used for analysis.

If the reagents are to be used as standards or calibrators, the most highly purified chemicals, reagent grade or ACS (American Chemical Society) grade, should be used. Different types of standards are available. A *primary standard* is a stable, non-hygroscopic (does not absorb water), highly purified substance that can be dried, preferably at 104°C to 110°C without a change in composition. Primary standards can thus be dried and then weighed to prepare solutions of selected concentrations. When purchased, primary standards are supplied with a record of analysis for contaminating elements, which should not exceed 0.05% by weight. Some standards have been certified to be pure by various official bodies such as the National Bureau of Standards and the College of American Pathologists (CAP).

A *primary standard* is the most highly purified substance currently available that can be weighed analytically. A *secondary standard* is one whose concentration is usually determined by analysis by an acceptable reference method that is calibrated with a primary standard. Its concentration cannot be determined directly from the weight of solute and volume of solution. Calibrators are used in calibration processes to establish concentrations of patient specimens. Calibration materials should meet the identity, labeling, and performance requirements of NCCLS guideline C22 *Tentative Guideline for Calibration Materials in Clinical Chemistry.*[64] Whenever possible, calibrators should have their concentrations assigned through the use of either reference methods or other very specific methods. The comparative analytical response of the calibrator and the specimens provides the basis for calculating values for patient specimens. The same material should not serve as calibrator and control.

Laboratory error can be minimized if attention is paid to proper laboratory procedures and techniques. The quality-control system for individual test methodologies can focus on controlling the test-specific variables. The post-analytical factors consist of the recording and reporting of the patient data to the physician within the appropriate time interval. With automation and computer-generated patient reports, the incidence of errors in the post-analytical phase has decreased greatly.

Quality Control

The purpose of the quality-control system is to monitor analytical processes, detect analytical errors during analysis, and prevent the reporting of incorrect patient values. Analytical methods are usually monitored by measuring stable control materials and then comparing the measured values with their expected value. The laboratory's budget must reflect the importance of quality control. The monetary commitment is important to ensure an adequate system for monitoring and improving the laboratory's performance. Some large laboratories have a full-time technologist responsible for quality control. This individual is in charge of reviewing quality-control data and keeping the staff informed about the status of the analytical methods, acting as liaison among the laboratory director, the chief technologist, and the bench technologists.

The statistical system used to interpret the measured concentrations of the controls is called the *statistical quality-control system.* The principles of statistical quality control were established by Shewhart[65] early in this century. In 1950, Levey and Jennings[66] used these same basic principles when they introduced statistical quality control to the clinical laboratory. Since 1950, statistical quality-control systems in the laboratory have undergone many modifications.

Analytical error may be separated into random-error and systematic-error components (Fig. 3-17). *Random error* affects precision and is the basis for varying differences between repeated measurements. Increases in random error may be caused by factors such as variations in technique and temperature. *Systematic error* arises from factors that contribute a constant difference, either positive or negative, and directly affects the estimate of the mean. Increases in systematic error can be caused by poorly made standards or reagents, failing instrumentation, poorly written procedures, and so on.

Control materials should behave like real specimens, be available in sufficient quantity to last a minimum of 1 year, be stable over that period, be available in convenient vial volumes, and vary minimally in concentration and composition from vial to vial.[67] The control material should closely resemble the specimen it is simulating in both its physical and chemical characteristics. The control material should be tested in exactly the same manner as patient specimens. Control ma-

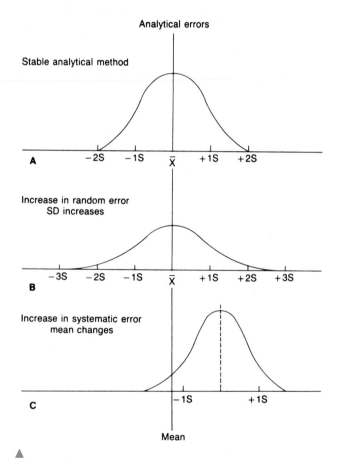

Figure 3-17. Schematic representation of distributions of control data. (**A**) represents the situation with no analytical error; (**B**) represents increased random error; (**C**) represents systematic shift.

terials should span the clinically important range of the analyte's concentrations. Levels of the control should be at appropriate decision levels; for example, for sodium, they might be at 130 and 150 mmol/L, levels that define hyponatremia and hypernatremia. For general chemistry quality control, two levels of control are typically used; for immunochemistry, three levels are often used. The manufacturer may assay its controls with various instruments and methodologies and then offer target ranges for its control products. These "assayed" controls are more expensive but can be used as external checks for accuracy.

Many commercially prepared control materials are lyophilized and require reconstitution before use. In reconstitution, care should be exercised to add and mix the proper amount of diluent. Incomplete mixing yields a partition of supernatant liquid and underlying sediment and will result in incorrect control values. Frequently, the reconstituted material will be more turbid (cloudy) than the actual patient specimen. Stabilized frozen controls do not require reconstitution but may behave differently from patient specimens in some analytical systems. It is important to carefully evaluate these stabilized controls with any new instrument system.

Although some chemistry laboratories used to prepare all their control materials,[68] there are serious shortcomings in their preparation and use. It is difficult to minimize the infectious disease risk of "homemade" materials. Additionally, compared with commercial controls, these control materials are more susceptible to deterioration and contamination. On occasion, it is necessary to prepare control pools for selected analytes, such as drugs. Proper procedures should be followed,[68] but the task is more manageable because far smaller quantities are required than for a laboratory-wide pool.

Most control materials are produced from human serum. With greater emphasis placed on cost containment, more laboratories are using bovine-based control materials, which are lower in price than the human-based materials. The stability of bovine-based control materials is similar to that of human-based materials. For most analytes, bovine material meets the necessary requirements for monitoring imprecision.[69] Because bovine proteins differ greatly from human proteins, bovine material is inappropriate for immunochemistry assays of specific proteins. Similarly, bovine material is inappropriate for some dye-binding procedures for albumin and certain bili-

rubin methods. Bovine-based material can be used as a control in electrophoresis, but its electrophoretic pattern differs from that of human control serum and resembles a polyclonal gammopathy.

General Operation of a Statistical Quality-Control System

The statistical quality-control system in the clinical laboratory is used to monitor the analytical variations that occur during testing. In certain instances, these variations may be systematic and are caused by procedural errors due to technique, instrumentation, or failures of reagents or other materials. In other instances, however, random variations may appear despite tightly controlled, well-calibrated analytical methods.

The statistical quality-control program can be thought of as a three-stage process:

1. Establishing allowable statistical limits of variation for each analytical method.
2. Using these limits as criteria for evaluating the quality-control data generated for each test.
3. Taking remedial action when indicated (ie, finding causes of errors, rectifying them, and reanalyzing patient and control data).

To establish statistical quality control on a new instrument or on new lots of control material, the different levels of control material must be analyzed for 10 to 20 days. The means ($\bar{x}$) and standard deviations (s) of these control data are then calculated. Because of the small number of observations and possible outliers in the data, these initial estimates may not be totally reliable and should be revised as more data become available. When changing between different lots of similar material, many laboratorians use the newly obtained mean as the target mean but retain the standard deviation used for the previous control product. As more data are obtained, all of the data should be averaged to derive the best estimates of the mean and standard deviation.[70]

Control values may be compared with statistical limits numerically or by display on a control chart. This chart is simply an extension of a Gaussian distribution curve (Fig. 3-18), with

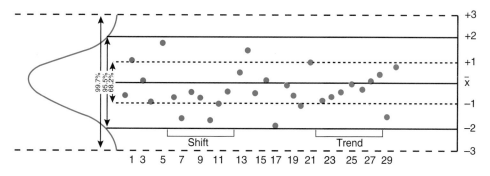

Figure 3-18. Control chart showing the relationship of control limits to the Gaussian distribution. Daily control values are graphed, and they show examples of a shift, an abrupt change in the analytical process, and a trend, a gradual change in the analytical process.

time expressed on the x-axis. The y-axis usually is scaled to provide a concentration made from $\bar{x} - 3s$ to $\bar{x} + 3s$. Horizontal lines corresponding to multiples of s are drawn around the x-axis. The $2s$ lines correspond to the 95.5% limits for the control. If the analytical process is in control, approximately 95% of the points will be within these limits and approximately 5% of the points will be outside these limits. The $3s$ limits correspond to approximately the 99.7% limits. If the process is in control, no more than 0.3% of the points will be outside the $3s$ limits. An analytical method is considered in control when there is symmetrical distribution of control values about the mean and there are few control values outside the $2s$ control limits. Historically, many laboratories have defined an analytical method as out of control if a control value fell outside its $2s$ limits. In laboratories that have used the $2s$ limits as warning limits and $3s$ limits as error limits, a control value between $2s$ and $3s$ would alert the technologist to a potential problem. A point outside the $3s$ limits would require corrective action.

Laboratories use different criteria for judging whether control results indicate out-of-control situations. Westgard and Groth have studied the error-detection capabilities of most of these criteria. They used the term *control rule* to indicate the criterion for judging whether the analytical process is out of control. To simplify comparison of the various control rules, Westgard and associates used abbreviations for the different control rules. Table 3-10 lists most of the frequently used control rules and their abbreviations.

The abbreviations have the form A_L, where A is a symbol for a statistic or is the number of control observations per analytical run and L is the control limit.[72] For example, a 1_{2s} rule violation indicates the situation in which one control observation is outside the $\bar{x} \pm 2s$ limits. An *analytical run* is defined as a set of control and patient specimens assayed, evaluated, and reported together.

Ideally, if a method is in control, none of the control rules should be violated and there should be no rejection of the analytical run. Unfortunately, some analytical runs will be rejected as out of control even when there is no additional analytical error. For example, when the 1_{2s} control rule is used with only one control being analyzed, per analytical run, then 5% of the runs will be outside the $2s$ limits when only the usual analytical variation is present. When more than one control is analyzed in an analytical run, and no additional error is present, the probability that at least one control is outside the $2s$ limits becomes higher. When two controls are used, there is approximately a 10% chance that at least one control will be outside the $2s$ limits. When four controls are used, there is a 17% chance. For this reason, many analysts usually do not investigate the analytical method if a single control exceeds the $2s$ limits when two or more controls are used. They merely re-assay the controls or the entire analytical run.

Unfortunately, this intuitive approach to quality control achieves an unknown level of quality. What is needed is a control system that will reliably signal the presence of significant analytical error but will not respond to small errors. Defining such a control system requires an understanding of the response of control rules to analytical error.

Response of Control Rules to Error

Westgard and associates[71] have studied the response of control rules, either singly or in groups, to the presence of error, either systematic or random. Different control procedures (groups of control rules) have distinct responses, depending on the control rules and number of control observations (n) used. Using computers, Westgard and Groth[73] simulated the analysis of control materials by instruments with varying levels of error. A large number of simulations was done at each error level, with the proportion of out-of-control situations tabulated and then plotted against the size of the error. The resulting graphs, depicting probability of rejection versus size of analytical error, are called *power functions*. Ideally, a control rule should have a 0% probability of detecting no error and a 100% probability of detecting significant error. Figure 3-19A shows a graph of a family of power functions for the detection of systematic error by the 1_{2s} control rule. The probability of rejection is plotted against the size of the systematic error. The size of the systematic error ranges from 0 to $5s$, where s is the standard deviation. The different lines correspond to different numbers of controls analyzed. When one control is used and there is no error, the probability of rejection is approximately 5%. The probability of rejection when no error is present is called the *probability of false rejection* (P_{fr}). The probability of

TABLE 3-10. Popular Control Rules	
1_{2s}	Use as a rejection or warning when one control observation exceeds the $\bar{x} \pm 2s$ control limits; usually used as a warning.
1_{3s}	Reject a run when one control observation exceeds the $\bar{x} \pm 3s$ control limits.
2_{2s}	Reject a run when two consecutive control observations are on the same side of the mean and exceed the $\bar{x} + 2s$ or $\bar{x} - 2s$ control limits.
4_{1s}	Reject a run when four consecutive control observations are on the same side of the mean and exceed either the $\bar{x} + 1s$ or $\bar{x} - 1s$ control limits.
$10_{\bar{x}}$	Reject a run when ten consecutive control observations are on the same side of the mean.
R_{4s}	Reject a run if the range or difference between the maximum and minimum control observation out of the last 4 to 6 control observations exceeds $4s$.
$\bar{x}_{0.01}$	Reject a run if the mean of the last N control observations exceeds the control limits that give a 1% frequency of false rejection ($P_{fr} = 0.01$).
R0.01	Reject a run if the range of the last N control observations exceeds the control limits that give a 1% frequency of false rejection (Pfr = 0.01).

s, standard deviation; $\bar{x}$, mean

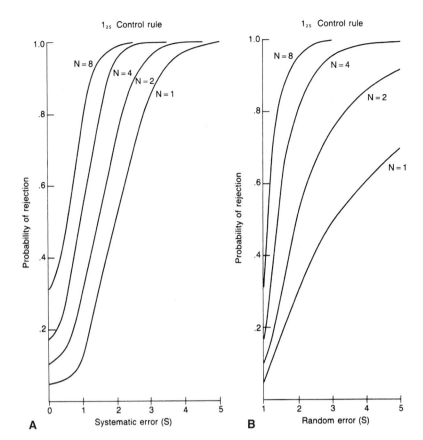

Figure 3-19. Power function curves for the 1_{2s} control rule for systematic error (**A**) and random error (**B**). (Provided by P. Douville)

error detection (P_{ed}) is the probability of rejecting an analytical run as out of control when an error exists. The P_{ed} can be determined for any size error by locating the size of the error on the *x*-axis and erecting a vertical line that intersects the power-function curve for the desired *n*. From this intersection, a horizontal line is drawn to the *y*-axis. The value of the probability on the *y*-axis is the P_{ed}. In Figure 3-19*A*, P_{ed} for the 1_{2s} control rule and a systematic error of $2s$ is 0.5 if one control is analyzed. The power-function graphs may be used to determine the effectiveness of various control procedures in detecting analytical errors. Two sets of power functions are necessary, one for systematic error (shift of the mean) and one for the random error (increase in imprecision). Figure 3-19*B* shows a family of power functions for the detection of random error by the 1_{2s} control rule. Because analytical processes have random error present, the *x*-axis originates at $1s$. In Figure 3-19*B*, P_{ed} for the 1_{2s} control rule and a doubling of the random error is 0.3 if one control is analyzed. The best control procedure is the one with the lowest P_{fr} and the highest P_{ed} for detecting analytical errors of a size that can compromise the quality of analytical results. Figures 3-20*A* and *B* show power functions for the 1_{3s} control rule for systematic and random error, respectively. Although the 1_{2s} control rule results in a high detection of moderate-sized systematic errors, the P_{fr} is unacceptably high for *2 or more controls*. On the other hand, the 1_{3s} control rule is less responsive to moderate increases in systematic error but has a low P_{fr}.

Westgard and associates have suggested a manual implementation of a combination of control rules with at least two control observations per analytical run.[70] In addition to using the 1_{2s} rule as a warning rule and the 1_{3s} rule for rejection, this system also can include the 2_{2s}, R_{4s}, 4_{1s}, and $10_{\bar{x}}$ rules. This combination of rules allows for improved detection of both random and systematic error. The counting rules (the 2_{2s}, 4_{1s}, and $10_{\bar{x}}$ rules) are effective in detecting systematic error. The R_{4s} and 1_{3s} rules are effective in detecting random error. The 1_{3s} rule also can detect systematic error. Examples of the violations of all of the above rules are shown in Figure 3-21.

To apply this combination of control rules, all new control results are evaluated to ensure that they are within their $\pm 2s$ limits. If they are, there is no further inspection. Otherwise, the other rules are applied in this order: 2_{2s}, 4_{1s}, $10_{\bar{x}}$, and R_{4s}. The 2_{2s} rule is invoked whenever two consecutive control observations on the same side of the mean exceed either the $\bar{x} + 2s$ or $\bar{x} - 2s$ control limits. This rule responds most often to systematic errors. The 2_{2s} rule is initially applied to the control observations within the most recent analytical run (across materials and within run). The rule can then be applied to the last two observations on the same control material but from consecutive runs (within materials and across runs), or it can be applied on the last two consecutive observations of the different control materials (across materials and across runs).

The 4_{1s} rule is violated when four consecutive control observations on the same side of the mean exceed the $\bar{x} + 1s$ or $\bar{x} - 1s$ control limits. It is most responsive to systematic errors. This rule is applied within and across materials and runs. The $10_{\bar{x}}$ rule is sensitive to systematic error and is violated when 10 consecutive control observations fall on one side of the mean. This rule is applied within materials and across runs as well as across materials and across runs.

The R_{4s} rule is violated when the range or difference between the highest and lowest control observations within the run exceeds 4_s. This rule is most responsive to random error or increased imprecision. The rule is invoked when the observation on one control material exceeds a $+2s$ limit and the observation on the other exceeds a $-2s$ limit. The two observations are outside $2s$ limits but in opposite directions, resulting in a range that exceeds $4s$. The range rule is intended for use within a single run with a maximum of four to six control observations, not across runs.

The combination of the 1_{2s}, 1_{3s}, 2_{2s}, 4_{1s}, $10_{\bar{x}}$, and R_{4s} control rules used in conjunction with a control chart has been called the *multirule Shewhart procedure*. The power functions for this multirule procedure are shown in Figure 3-22. This multirule procedure yields increased error detection over the use of the 1_{3s} rule alone. The inclusion of the 1_{3s} and R_{4s} rules improves the detection of random error, and the 2_{2s}, 4_{1s}, and $10_{\bar{x}}$ rules increase the detection of systematic error. This multirule procedure provides the laboratory with improved error detection and less false rejection of analytical runs. It is easily adapted to existing control procedures and has become extremely popular since its initial description in 1981.

Implementation of the multirule procedure involves the following:

1. Calculation of the means and standard deviations of the different concentrations of control materials.
2. Construction of control charts, with lines indicating the 0, ±1, 2, and 3 standard deviation limits.
3. Analysis of the different levels of control material in each analytical run, with plotting of the control data on the appropriate chart.
4. Accepting the analytical run if each observation falls within $2s$ limits.
5. Checking the 1_{3s}, 2_{2s}, R_{4s}, 4_{1s}, and $10_{\bar{x}}$ rules for violations if one of the control materials is outside its $2s$ limits. If none of the rules is violated, the run is accepted. If a violation has occurred, the run is rejected, with the most likely type of error determined (random versus systematic). Once the error condition is identified and corrected, the patient and control specimens are reanalyzed.

Figure 3-23 shows a two-level control chart that simplifies implementation of the multirule procedure. Application of the multirule procedure to a two-level control system is illustrated in Figure 3-24. Interpretation of the control data is summarized in Table 3-11. Note that the R_{4s} rule is not applied across runs. The astute reader will note that the multirule procedure will detect $10_{\bar{x}}$ or 4_{1s} rule violations only if

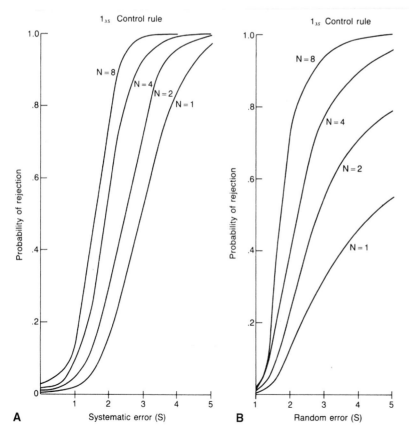

A

B

Figure 3-20. Power function curves for the 1_{3s} control rule for (**A**) systematic error and (**B**) random error. (Provided by P. Douville)

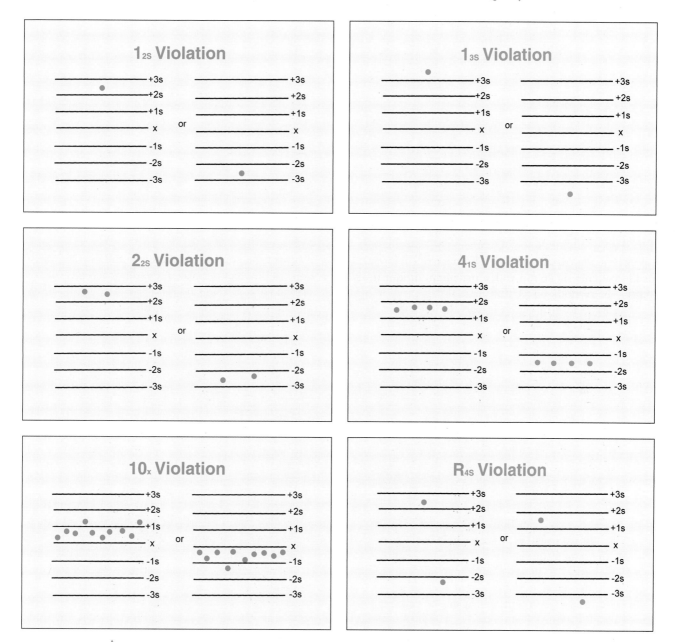

Figure 3-21. Examples of violations of the 6 rules comprising the Westgard multirule control procedure.

there is a 1_{2s} rule violation. On occasion, the $10_{\bar{x}}$ and 4_{1s} rules may be violated without the activation of the 1_{2s} warning rule. Such occurrences will be infrequent and would be indicative of small systematic errors that would eventually be detected by the multirule procedure.

The power functions in Figure 3-22 show that when the aforementioned multirule procedure is used with four control observations, it will detect moderate-sized errors (*eg,* a 2*s* shift or a doubling of the random error) with a probability of approximately 50%. Smaller errors cannot be reliably detected. More sensitive control procedures are required for assays in which a 2*s* shift or a doubling of the standard deviation will cause misclassifications of clinical status. For example, the standard deviation of many cal-

cium assays is 0.15 mg/dL for a normal range calcium. A positive shift of 0.30 mg/dL can easily place many normocalcemic patients into the hypercalcemic category. The sensitivity of the multirule procedure can be increased to detect smaller systematic errors by increasing the number of observations considered. Other control procedures can be used. The mean and range control procedure[74] has significantly greater power for the detection of error if each control material is analyzed in replicate in each analytical run. Practically, this procedure requires computer implementation, because calculations of the mean and range are required with each analytical run.

The *cumulative summation (cusum) technique* has been described for routine laboratory use[75]; it also requires computer

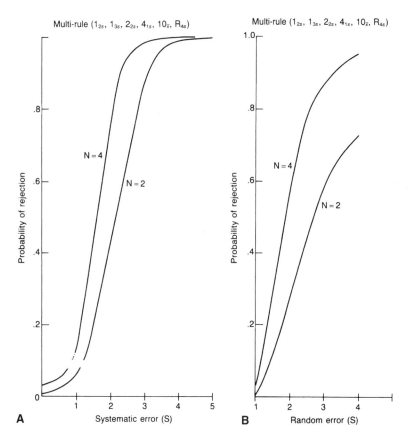

Multi-rule (1_{2s}, 1_{3s}, 2_{2s}, 4_{1s}, $10_{\bar{x}}$, R_{4s})

Multi-rule (1_{2s}, 1_{3s}, 2_{2s}, 4_{1s}, $10_{\bar{x}}$, R_{4s})

A Probability of rejection — Systematic error (S)

B Probability of rejection — Random error (S)

Figure 3-22. Power functions for the multirule control procedure (**A**) for systematic error and (**B**) random error. (Provided by P. Douville)

implementation. The cusum procedure is very responsive to systematic error and can be used with the 1_{3s} rule. Cusum is very sensitive to small, persistent shifts that commonly occur in the modern, low-calibration-frequency analyzer. There are other control procedures that use exponentially smoothed averages and standard deviations.[73,76] These procedures have been implemented on several commercially available laboratory information quality-control systems.

Many of today's analyzers are capable of very high accuracy and precision. The magnitude of the analytical standard deviation may be very small when compared with that of the medically allowable error (*eg*, the standard deviation of glucose on certain analyzers approaches 1–2 mg/dL, which is much smaller than the medically allowable error of 10 mg/dL). Only very large errors need be detected when these analyzers measure analytes such as glucose. The sensitivity of the control procedure must be reduced rather than increased. One way of doing this is by incorporating fewer observations, for example, using the 1_{3s} control rule; or even expanding the control limits, for example, using $\pm3.5s$ control limits (*ie,* the $1_{3.5s}$ control rule).[77] Koch et al showed that the application of optimized analyte-specific quality control practices reduced the frequency of falsely rejected runs, reduced quality control expenses and increased the efficiency of their high-volume chemistry analyzer, the Hitachi 737.[78] Their current analyte specific quality control program comprises the $1_{3.5s}$ rule for sodium, potassium, glucose, and blood urea nitrogen; the $1_{2.5s}$ rule

for albumin, chloride, and carbon dioxide; and the $1_{2.5s}$ rule for calcium, which are run in duplicate and averaged.

A variety of approaches can be used to derive these analyte-specific control rules. Westgard et al have constructed "selection grids" consisting of 3×3 tables that list optimal control procedures.[79] Figure 3-25 shows such a grid. Multirule procedures with optimal performance characteristics form each of the nine entries in the grid. A commercially available computer program[80] allows the laboratorian to specify an analyte, its imprecision, and the medically allowable error (usually the proficiency test limits as specified by CLIA-88). The program then proposes optimal quality control procedures for that analyte.

Most laboratories just use the same subset of the multirule control procedure for all of their analytes: the 1_{2s}, rule for screening; 1_{3s}, rule for detecting random error; and the 2_{2s}, rule for detecting systematic error. This three-rule combination is probably the most common control procedure used in North American clinical laboratories.

Use of Patient Data for Quality Control

Various algorithms have been proposed for manipulating patient data to determine whether processing or analytical errors have occurred. This section focuses on two different control procedures that use patient data: the average of patient data and delta checks. Other quality-control procedures that use patient data include the review of individual out-

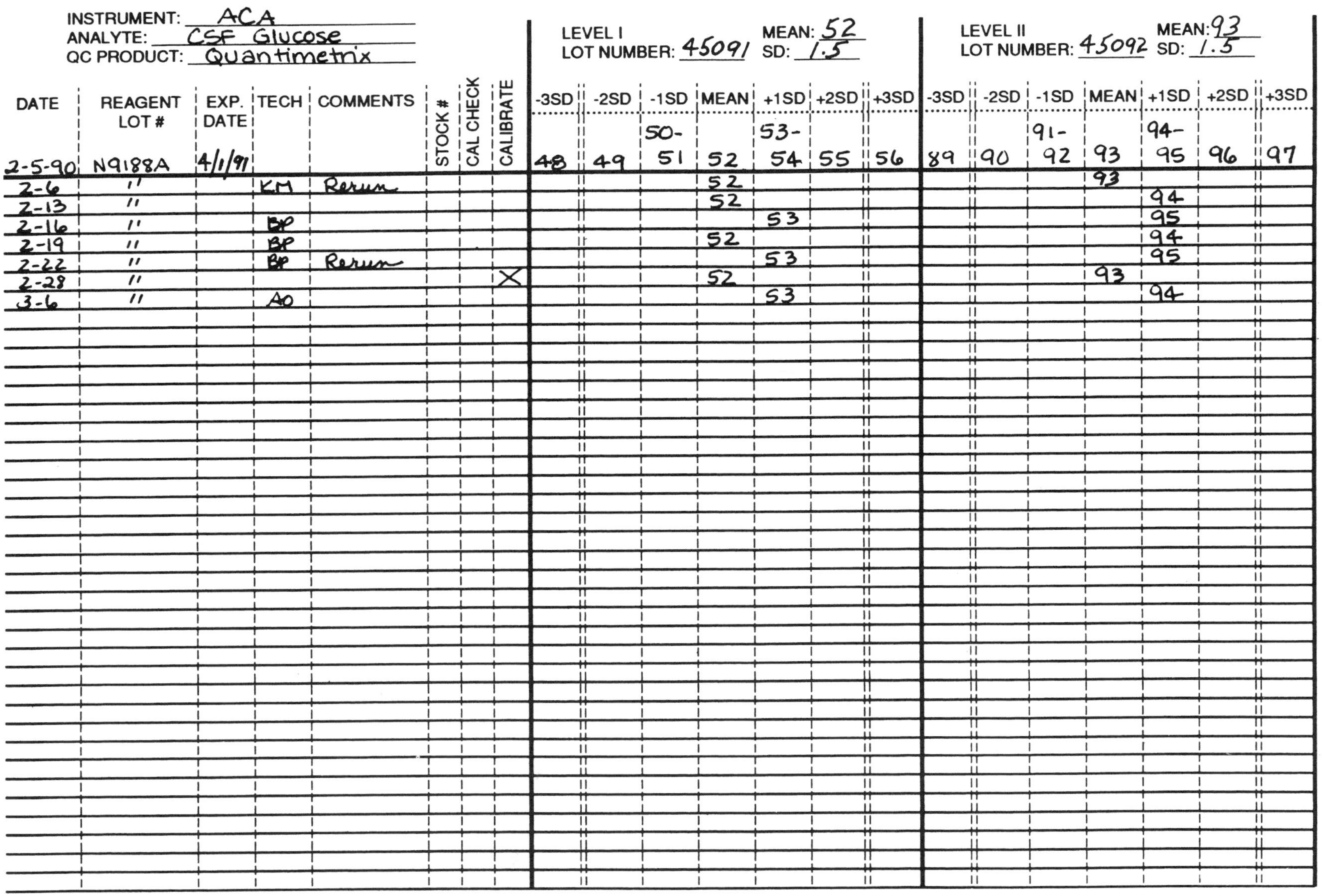

INSTRUMENT: __ACA__
ANALYTE: __CSF Glucose__
QC PRODUCT: __Quantimetrix__

LEVEL I LOT NUMBER: __45091__   MEAN: __52__   SD: __1.5__

LEVEL II LOT NUMBER: __45092__   MEAN: __93__   SD: __1.5__

DATE	REAGENT LOT #	EXP. DATE	TECH	COMMENTS	STOCK #	CAL CHECK	CALIBRATE	Level I (−3SD 48 / −2SD 49 / −1SD 50− 51 / MEAN 52 / +1SD 53− 54 / +2SD 55 / +3SD 56)	Level II (−3SD 89 / −2SD 90 / −1SD 91− 92 / MEAN 93 / +1SD 94− 95 / +2SD 96 / +3SD 97)
2-5-90	NQ188A	4/1/91					X		
2-6	"		KM	Rerun				52	93 94
2-13	"		BP					52 53	95
2-16	"		BP					52	94 95
2-19	"		BP	Rerun				52 53	95
2-22	"							52 53	93 94
2-28	"		AO						
3-6	"								

Figure 3-23. Two-level control chart for implementing multirule procedures.

67

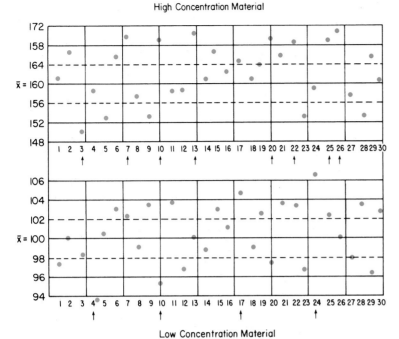

Figure 3-24. Control data to illustrate use of the multirule procedure. The arrows correspond to days in which there are violations of at least one rule.

lying results to identify gross clerical errors (sometimes called *limit checks*) and the routine analysis of duplicate specimens, as is frequently done in endocrinology assays. Some laboratories use duplicate analyses of another type: patient-sample comparisons. These comparisons require the regular analysis of split samples on instruments that measure the same analyte. Differences between instruments that exceed predetermined limits are investigated and corrected.

The use of the averages of patient data was described by Hoffman and Waid in 1965.[81] In their "average of normals" method, an error condition was signaled when the average of consecutive centrally distributed patient data was beyond the control limits established for the average of the patient data. The assumption underlying the average of normals is that the patient population is stable. Any shift would thus be

secondary to a systematic analytical error. Cembrowski, Chandler, and Westgard studied the use of average of patients with computer simulations and found that its error-detection capabilities depended on several factors.[82] The most important were the number of patient results averaged and the ratio of the standard deviation of the patient population (s_p) to the standard deviation of the analytical method (s_a). Other important factors included the limits for evaluating the mean (control limits), the limits for determining which patient data are averaged (truncation limits), and the magnitude of the population lying outside the truncation limits. Cembrowski et al stated that the technique could be used to supplement reference-sample quality control and recommended computer implementation. Douville, Cembrowski, and Strauss evaluated the averages of patient endocrine data

Control Material	Day	Rule Violation
High	Day 3	1_{2s} rule violation—warning—accept run
Low	Day 4	1_{3s} rule violation—reject run
High + low	Day 7	4_{1s} rule violation—reject run—across runs and materials
High + low	Day 10	R_{4s} rule violation—reject run—within run, across materials
High	Day 13	1_{2s} rule violation—warning—accept run
Low	Day 17	1_{2s} rule violation—warning—accept run
Low	Day 20	1_{2s} rule violation—warning—accept run
High	Day 22	$10_{\bar{x}}$ rule violation—reject run—within materials, across runs
Low	Day 24	1_{3s} rule violation—reject run
High	Day 25	1_{2s} rule violation—warning—accept run
High	Day 26	2_{2s} rule violation—reject run—within materials, across runs

TABLE 3-11. Interpretation of the Multirule Shewhart Control Charts of Figure 3-22 for High and Low Concentration Materials.

Westgard Multi-rule QC Selection Grid		PROCESS STABILITY (Frequency of Errors, f)		
		>10%	2% – 10%	<2%
PROCESS CAPABILITY (Magnitude of errors, SEc)	<2.0s	1_{3S} / 2_{2S} / R_{4S} / 4_{1S} / $\overline{12x}$ N = 6	1_{3S} / 2_{2S} / R_{4S} / 4_{1S} / $\overline{8x}$ N = 4	1_{3S} / 2_{2S} / R_{4S} / 4_{1S} N = 2
	2.0s to 3.0s	1_{3S} / 2_{2S} / R_{4S} / 4_{1S} / $\overline{8x}$ N = 4	1_{3S} / 2_{2S} / R_{4S} / 4_{1S} N = 2	1_{3S}/2_{2S}/R_{4S}/(4_{1S} W) N = 2
	>3.0s	1_{3S} / 2_{2S} / R_{4S} / 4_{1S} N = 2	1_{3S}/2_{2S}/R_{4S}/(4_{1S} W) N = 2	1_{3S}/(4_{1S} W) N = 2

◄ *Figure 3-25.* QC selection grid for Westgard multirule algorithm. (Reproduced with permission)

and demonstrated high error-detection capabilities for thyroid testing.[83]

Cembrowski et al investigated the use of the anion gap for quality control and found that the practice of reanalyzing single specimens with abnormally high or low anion gaps usually resulted in the needless repetition of analyzing patient specimens.[84] More often, these single specimens had genuinely abnormal anion gaps. An improved control procedure consists of averaging eight consecutive patient anion gaps and comparing the average with the control limits for the anion-gap average.[85]

The last algorithm to be discussed is the *delta check,* in which the most recent result of a patient is compared with the previously determined value. The differences between consecutive laboratory data (deltas) may be compared with limits that have been established by various investigators.[86,87] The differences are usually calculated in two ways: as numerical differences (current value minus last value) and as percentage differences (numerical differences times 100 divided by the current value). Wheeler and Sheiner evaluated the performance of several delta check methods and classified each delta check investigated as either a true or false positive.[88] They found that the percentage of true positives ranged from 5% to 29% and concluded that the delta check methods could detect errors otherwise overlooked but at the cost of investigating many false positives. In their tertiary-care hospital population, there were many false positives caused by large excursions in laboratory values secondary to disease or therapy.

External Quality Control

Proficiency-testing programs periodically provide samples of unknown concentrations of analytes to participating laboratories. Participation in these programs is mandated by the HCFA under the CLIA-88. The CAP and the American Association of Bioanalysts are probably the two largest providers of proficiency testing programs in the United States. They provide samples for all major qualitative and quantitative chemistry areas, including general chemistry, protein chemistry, urinalysis, toxicology, and endocrinology. Samples for quantitative analysis contain multiple analytes and are usually provided as lyophilized materials. Once a laboratory receives its samples, it must analyze and return its results within a specified time to a computer center for compilation and comparison with the results of other participating laboratories. The computer center establishes target values and ranges of acceptable results based on either the average of the participants' values or reference laboratories' values.

For the CAP proficiency program, the means and standard deviations of all results from peer (similar instrument and methodology) laboratories are computed. Then values beyond three standard deviations from the mean are discarded, and the mean and standard deviation are recomputed. A participant result is classified as acceptable if the difference between the result and the target answer (usually the peer mean) is less than the allowable error. CLIA-88 has defined allowable errors for a large number of regulated analytes[43]; some are shown in Table 3-7. These allowable errors are sometimes referred to as "fixed limits" and are expressed either in measurement units of the analyte (*eg,* ±0.5 mmol/L from the mean for potassium) or as percentages (*eg,* ±10% for total cholesterol).

For a far smaller number of analytes (*eg,* thyroid-stimulating hormone), CLIA-88 has statistically defined limits of acceptability. For these analytes, the participant result is acceptable if it falls within ±3 SDI of the group mean. The comparison with statistical limits requires calculation of the deviation of the survey results from the mean, which is expressed in numbers of standard deviation indices (SDI) above or below the mean. The SDI is the numerical difference between an individual laboratory's results and the mean, divided by the standard deviation. A deviation outside ±3 SDIs usually is considered unacceptable because such deviations will occur only 0.3% of the time as a result of chance.

Compared with statistical limits, fixed-limits criteria will more often demonstrate the need for replacement of outmoded or unreliable instrumentation. Ehrmeyer and Laessig have used computer simulations to evaluate the ability of many different proficiency-testing schemes to detect poorly performing laboratories. The performance of all of these schemes is imperfect, with some good laboratories being judged as poor and some poor laboratories escaping detection.[89,90]

Cembrowski, Hackney, and Carey have proposed a multirule system to evaluate the HCFA-mandated proficiency test result.[91,92,93] The multirule is illustrated in Figure 3-26. When significant deviations are detected in a set of five survey results (one or more observations exceeding ±3 SDIs, or the range of the observations exceeding 4 SDIs, or the mean of the 5 results exceeding ±1.5 SDIs), the laboratory records, including the internal quality-control results, should be reviewed. Mix-ups of proficiency specimens or of proficiency and clinical specimens should be ruled out. Whenever possible, an aliquot of the survey specimen should be saved and assayed again for the analytes yielding erroneous results. Results that still deviate significantly after retesting indicate a long-term bias. If the deviations are variable in magnitude and direction, there may be a problem with imprecision (random error). In the event that repeat analysis yields satisfactory results, the error probably represented a random error or transient bias encountered during the testing period.

External proficiency testing is also used to determine estimates of the state-of-the-art of interlaboratory performance. The CAP regularly publishes summaries of its interlaboratory comparisons in the *Archives of Pathology and Laboratory Medicine.*

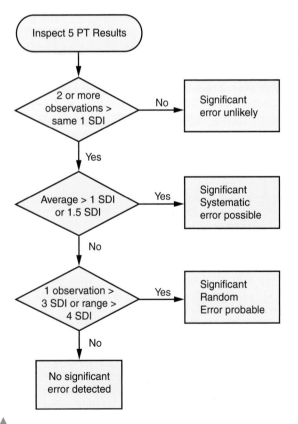

Figure 3-26. Screening PT Results. Flow chart to investigate groups of 5 proficiency results for significant systematic and random error.

Point-of-Care Testing: The Newest Challenge

Point-of-Care Testing (POCT) is defined as analytical testing performed outside the confines of the central laboratory, usually by nonlaboratorian personnel (nurses, respiratory therapists, and so on). Synonyms for POCT include near patient testing, decentralized testing, bedside testing, and alternate site testing. One of the most common examples of POCT is the use of portable whole blood glucose meters for the management of patients with diabetes.[94] Many different analytes can now be measured accurately and quickly near the patient, whether the patient is in a hospital, an ambulance, or even in an airplane. Part of the of success of POCT arises from the inability of the centralized clinical laboratory to respond to clinician demands for faster turnaround times (TATs).[95,96] The rapid provision of test results can increase the throughput of patients through bottleneck areas such as emergency departments and can even decrease a patient's length of stay and reduce the total cost of patient care.[97]

A successful POCT program requires careful planning, implementation, and ongoing evaluation of equipment, education, and quality control. The first step to implementing a POCT program is forming the POCT steering committee. Membership on this committee should include the director of laboratory medicine (chair holder); the POCT coordinator (usually a laboratory technologist); and clinician, nursing, and respiratory care representatives. If possible, information systems and finance personnel should also attend. Meeting attendance will vary depending on items to be covered on the agenda. This collaboration sets the stage for communication and the resolution of POCT problems. Before the POCT steering committee can recommend any POCT device, they must evaluate all available systems and then choose the one that best meets the needs of the requesting area. The quality requirements of POCT analyzers are summarized in Table 3-12. Many criteria must be considered before a POCT program can be approved, including the perceived need for POCT, potential for improvements in patient outcome, as well as testing frequency, method reliability, and training and staffing requirements. It is important to choose an analyzer that is relatively simple to operate and control. It is almost impossible to fit extensive POCT training programs clinicians' already hectic schedules. Data management capabilities must also be considered. The interfacing of POCT analyzers to the laboratory information system can provide institution-wide access to the patient data.

The overall cost of POCT depends on the volume of samples tested and the personnel performing the testing (nurse vs. technician).[98] When calculating cost per test at the point-of-care or the central laboratory, it is important to include the following: labor costs, instrumentation costs, supplies, training, and depreciation and indirect costs (*ie,* reporting costs or hospital overhead). In most cases, the central laboratory will be less expensive on a cost per test basis than alternate site test-

TABLE 3-12. Quality Requirements of Point-of-Care Analyzers

Speed (test[s] should be completed within minutes of sample introduction)

Accuracy and precision approaching that of the central laboratory analyzer

Small and portable equipment

Ability to analyze an "unprepared specimen" (eg, whole blood)

Low sample size

Flexible test menu

Broad dynamic range to minimize repeats, dilutions, and confirmatory tests

Ease of use by nonlaboratory personnel

Lock-out ability

    To prevent testing by unauthorized users

    When patient identification is not entered

    When quality control is not entered

Cost per test approaching that provided by the main laboratory

Low capital equipment outlay

Quantitative readout (no subjectivity on the part of the observer)

Automatic calibration

Automated quality control interpretation

Seamless interface with laboratory or hospital information system (communication by wireless system: infrared or radio frequency)

Low maintenance

Minimal troubleshooting requirements

High reliability with minimum downtime

Backup capability

Bar-code reading capabilities

Use of either no reagent or ready-for-use reagents

Minimal waste production

Minimal and recyclable disposables.

Source: Cembrowski GS, Kiechle FL. Point-of-care testing: Critical analysis and practical application. Adv Pathol Lab Med 1994;7:3–26.

ing. Although it may appear that it is less expensive to perform analysis in the central laboratory, one must consider the impact of TAT on patient throughput efficiency and length of stay. Selective use of POCT in critical care areas can result in long-term cost savings to the health care facility.

Toward Quality Patient Care

Much of this chapter has focused on the delivery of *accurate* laboratory test results. Accuracy in laboratory testing is only one quality characteristic that is required of the clinical chemistry laboratory.[51] Other equally important quality characteristics include effective test request forms; clear instructions for patient preparation and specimen handling; appropriate turnaround times for specimen processing, testing, and result reporting; appropriate reference ranges; and intelligible result reports. Most clinical laboratories provide accurate laboratory testing. We are just beginning to appreciate that most laboratory failures or mistakes occur in the pre-analytical or post-analytical realm.

For example, Ross and Boone reviewed 363 incidents that occurred in a large tertiary-care hospital in 1987.[99] Of the 336 medical records investigated, they found that pre-analytical and post-analytical mistakes accounted for 46% and 47% of the total incidents, respectively. Pre-analytical mistakes included missed or incorrectly interpreted laboratory orders, improper patient preparation, incorrect patient identification, wrong specimen container, and mislabeled or mishandled specimen. Post-analytical mistakes included delayed, unavailable, or incomplete results. Nonlaboratory personnel were responsible for 29% of the mistakes.

Most of these errors were interdepartmental; prevention of such errors requires a coordinated interdepartmental group approach. Committed representation of all relevant players is a prerequisite for success, whether they be physician, nurse, ward clerk, phlebotomist, or analyst. These individuals should form a *quality team* or *improvement team* and should meet regularly to characterize the problem and, eventually, to make recommendations for its solution. The group should be able to use various statistical and group techniques. For these *group processes* to succeed, there must be total commitment from management. Management must be trained in this "quality-improvement process"[100,101] and must support its growth throughout the institution. Such approaches are working extraordinarily well in Japanese industry and are now being successfully transferred to American businesses and even the hospital and clinic environment.[102,103] It is only through such quality-improvement efforts that we can significantly improve total patient care.

The development of portable, simple to use, clinical analyzers has made point-of-care or near patient testing a reality.

PRACTICE PROBLEMS

Problem 3-1: Calculation of Sensitivity and Specificity

Alpha-fetoprotein (AFP) levels are used by obstetricians to help diagnose neural tube defects (NTD) in early pregnancy. For the following data, calculate the sensitivity, specificity, and efficiency of AFP for detecting NTD, as well as the predictive value of a positive AFP.

Number of Pregnancies Interpretation of AFP Findings

Outcome of Pregnancy	Positive (NTD)	Negative (No NTD)	Total
NTD	5	3	8
No NTD	4	843	847
Total	9	846	855

Problem 3-2: A Management Decision in Quality Control

You are in charge of the clinical laboratory when a technologist presents you with the technologist's glucose worksheet. A 2_s rule violation has occurred across runs and within materials on the high-concentration material. You ask to see the patient data and the previous control data. They follow:

Glucose Worksheet January 8

| Samples | Results | January Glucose Control Values | | |
		Date	Low	High
Control—High	224	1/1	86	215
Patient	117	1/2	82	212
Patient	85	1/3	83	218
Patient	98	1/4	87	214
Patient	74	1/5	85	220
Patient	110	1/6	81	217
Control—Low	83	1/7	88	223
Patient	112	1/8	83	224
Patient	120			
Patient	97			
Patient	105			

1. Plot these control data.
2. What do you observe about these control data?
3. What might be a potential problem?
4. Should you report the patient data for today? Why or why not?

See Figure 3-27.

Problem 3-3: Interdepartmental Communication

You are having a problem with the medical intensive care unit (MICU) and the arterial blood gas specimens they submit to the laboratory. In the past 3 weeks, you have refused to perform blood gas analyses on six different MICU specimens because of small clots found in the specimens. The MICU staff is furious with the rejection policy, yet you believe the analyses will be incorrect if these specimens are used.

1. Outline where the problem lies.
2. What can be done to remedy this problem?
3. Why would your present quality-control system not detect this sort of error?

The following problems represent the steps in a method-evaluation study. An abbreviated data set is used to encourage hand calculations by the student. Perform the calculations for the following experimental data, which were obtained from a glucose study. The test method is a coupled glucose oxidase procedure. The comparative method is the hexokinase method currently in use.

Problem 3-4: Precision (Replication)

For the following precision data, calculate the mean, standard deviation, and coefficient of variation for each of the two control solutions A and B. These control solutions were chosen because their concentrations were close to medical decision levels (X_c) for glucose: 120 mg/dL for control solution A and 300 mg/dL for control solution B. Control solution A was analyzed daily, and the following values were obtained:

118, 120, 121, 119, 125, 118, 122, 116, 124, 123, 117, 117, 121, 120, 120, 119, 121, 123, 120, and 122 mg/dL

Control solution B was analyzed daily and gave the following results:

295, 308, 296, 298, 304, 294, 308, 310, 296, 300, 295, 303, 305, 300, 308, 297, 297, 305, 292, and 300 mg/dL

Problem 3-5: Recovery

For the recovery data below, calculate the percent recovery for each of the individual experiments and the average of all the recovery experiments. The experiments were performed by adding two levels of standard to each of five patient samples (A through E) with the following results:

Sample	0.9 mL Serum + 0.1 mL Water	0.9 mL Serum + 0.1 mL 500 mg/dL STD	0.9 mL Serum + 0.1 mL 1000 mg/dL STD
A	59	110	156
B	63	112	160
C	76	126	175
D	90	138	186
E	225	270	320

What do the results of this study indicate?

Problem 3-6: Interference

For the interference data that follows, calculate the concentration of ascorbic acid added, the interference for each individual sample, and the average interference for the group

HIGH   JANUARY

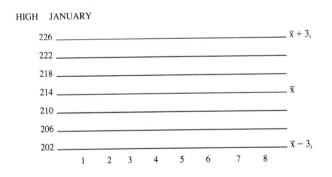

LOW

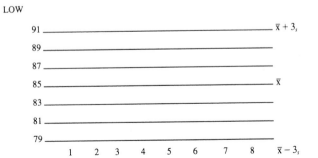

Figure 3-27. Blank control chart for Problem 3-2.

of patient samples. The experiments were performed by adding 0.1 mL of a 150-mg/dL ascorbic acid standard to 0.9 mL of five different patient samples (A through E). A similar dilution was prepared for each patient sample using water as the diluent. The results follow:

Sample	0.9 mL Serum + 0.1 mL Water	0.9 mL Serum + 0.1 mL 150 mg/dL STD
A	54	46
B	99	91
C	122	112
D	162	152
E	297	286

What do the results of this study indicate?

Problem 3-7: Linear Regression

The following method-comparison data were obtained (mg/dL):

Sample	By Hexokinase (mg/dL)	By Coupled Glucose Oxidase (mg/dL)
1	191	192
2	97	96
3	83	85
4	71	72
5	295	299
6	63	61
7	127	131
8	110	114
9	320	316
10	146	141

1. Graph these results. From inspection of the graph, determine whether there is significant constant or proportional error.
2. Calculate the linear regression statistics as follows:
   a. Set up a table with the following column headings:
   $$x_1, y_1, x^2i, y^2i, x_iy_i, Y_i$$
   b. Enter the x_i and y_i data into the table (remember that x is the comparative method and y the test method) and calculate $\Sigma x_i, \Sigma y_i, \Sigma x^2i, \Sigma y^2i$, and Σx_iy_i. Enter these data into the table.
   c. From the summations in (b), calculate $\bar{x}$ and $\bar{y}$. Use Equations 3–8 and 3–9 to calculate the slope m and the y intercept (y_0), respectively.
   d. Using the regression equation $Y = mx + y_0$, calculate y_i for each x_i, enter the Y_i values in the Y_i column, and then calculate:
   $$(y_i - Y_i), (y_i - Y_i)^2, \text{ and } \Sigma(y_i - Y_i)^2$$
   Enter these data into the table.
   e. From the summation in (d) and Equation 3-10, calculate $s_{y/x}$.
   f. From the summations in (b) and Equation 3-11, calculate r.
3. Report the following statistics for the comparison of methods experiment: $m, y_0, s_{y/x}$, and r.

Problem 3-8: Interpretation

Use the statistics calculated in Problem 3-7 to answer the following questions. Explain your answers by referring to the statistical value you used to arrive at your answer. Where appropriate, calculate errors at the two medical decision levels $X_{c1} = 120$ mg/dL and $X_{c2} = 300$ mg/dL.

1. What is the random error (RE) of the test method? What is the magnitude of the statistic that quantitates RE *between* the methods?
2. What is the constant error (CE) and the proportional error (PE)?
3. Calculate the systematic error (SE) at both X_c. What is the predominant nature of the SE (refer to item 2)?
4. What is the total error (TE = RE + SE) of the method? *Note:* Linear regression statistics should be used for error estimates only within the concentration range studied; we should have collected data up to 300 mg/dL in the comparison-of-methods experiment.
5. a. What is the statistic that quantitates random error between methods?
   b. What value did you calculate for this statistic?
6. Judge the acceptability of the performance of the test method. To reach your judgment, apply the following criteria:
   a. For the method to be accepted, all errors must be less than the allowable error (E_a) for a given X_c.
   b. The following are the E_a for glucose:
   $X_{c1} = 120$ mg/dL$E_{a1} = 10$ mg/dL
   $X_{c2} = 300$ mg/dL$E_{a2} = 25$ mg/dL

Problem 3-9

1. You are appointed as a clinical laboratorian to a point-of-care (POC) work team. Who else might serve on this team?
2. What needs to be in place before implementing a POC testing protocol to ensure accurate and precise patient results?
3. Once you decide to provide physicians and patients with POC testing, what are important features or requirements of POC analyzers?

REVIEW QUESTIONS

1. A Gaussian distribution is usually:
   a. Rectangular
   b. Bell-shaped
   c. Uniform
   d. Skewed
2. The following chloride (mmol/L) results were obtained using a point-of-care analyzer in the emergency department:

106	111	104	106	112	110
115	127	85	110	108	109
83	119	105	106	108	114
120	100	107	110	109	102

What is the mean?
a. 105
b. 108
c. 109
d. 107

What is the median?
e. 105
f. 108
g. 109
h. 107

3. The correlation coefficient should be used:
a. For determining method acceptability
b. For determining regression type used to derive slope and y intercept
c. Always expressed as R_2
d. To express method imprecision

4. On doing a correlation study, the test method and reference method generate equal data points. When plotted, the slope equals: _____ and the y intercept equals: _____.
a. 0.0, 1.0
b. 1.0, 1.0
c. 1.0, 0.0
d. 0.0, 0.0

5. Using the following ROC curve, which is the best test?
a. Test A
b. Test B
c. Test C
d. Test D

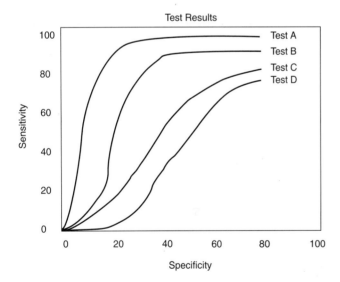

6. Interference studies typically use _____ as an interferent:
a. Hemolyzed red blood cells
b. Intralipids
c. Highly icteric specimens
d. All of the above

7. Which Westgard rule detects random error?
a. 1_{3s}
b. 4_{1s}
c. 2_{2s}
d. 10_0

8. Which of the following rule(s) probably detect small systematic error and should hardly be used?
a. R4s
b. 10_0, O_{R4}
c. 22s, 41s
d. 1_{3S}

9. Which rules can be used to evaluate sets of five proficiency testing results?
a. Mean > 1.0 SDI and 2/5 results > ± 1.0 SDI
b. 1 result > 2SDI
c. 5/5 results > ± 1.0 SDI and mean > 1.5 SDI
d. 2/5 results > ± 1.0 SDI and mean > 1.5 SDI

10. The primary reason for implementing POCT is?
a. Reduced testing cost
b. Enhanced outcomes of patient care
c. Lower central laboratory workload
d. Use nonlaboratorians as analysts

REFERENCES

1. Barnett RN. Clinical laboratory statistics. 2nd ed. Boston: Little, Brown, 1979.
2. Chatfield C. Statistics for technology: a course in applied statistics. 3rd ed. New York: Chapman & Hall, 1983.
3. Weisbrot IM. Statistics for the clinical laboratory. Philadelphia: Lippincott, 1985.
4. Westgard JO. Precision and accuracy: concepts and assessment by method evaluation testing. Crit Rev Clin Lab Sci 1981;13:28c.
5. Waakers PJM, Hellendoorn HBA, Op De Weegh GH, et al. Applications of statistics in clinical chemistry: a critical evaluation of regression lines. Clin Chem Acta 1975;64:173.
6. Westgard JO, deVos DJ, Hunt MR, et al. Concepts and practices in the evaluation of clinical chemistry methods: III. Statistics. Am J Med Tech 1978;44:552.
7. Westgard JO, Hunt MR. Use and interpretation of common statistical tests in method-comparison studies. Clin Chem 1973;19:49.
8. Harris EK, Cooil BK, Shakarji G, et al. On the use of statistical models of within person variation in long-term studies of healthy individuals. Clin Chem 1980;26:383.
9. Grasbeck R, Siest G, Wilding P, et al. Provisional recommendation on the theory of reference values: I. The concept of reference values. Clin Chem 1979;25:1506.
10. Fleisher GA, Eickelberg ES, Elveback LR. Alkaline phosphatase activity in the plasma of children and adolescents. Clin Chem 1977;23:469.
11. Winsten S. The ecology of normal values in clinical chemistry. Crit Rev Clin Lab Sci 1976;6:319.
12. Soldin WJ, Hicks JM, Gunter KC, et al. Pediatric reference ranges. 2nd ed. Washington, DC: American Association for Clinical Chemistry, 1997.
13. Meites S, ed. Pediatric clinical chemistry: reference (normal) values. Washington, DC: American Association of Clinical Chemistry, 1989.
14. National Committee for Clinical Laboratory Standards. Approved guideline for how to define and determine reference intervals in the clinical laboratory. Document C28-A. Villanova, PA: NCCLS, 1995.
15. Martin HF, Gudzinowicz BJ, Driscoll JL. An algorithm for the selection of proper group intervals for histograms representing clinical laboratory data. Am J Clin Pathol 1975;64:327.
16. Foti AG, Cooper JF, Herschman H, et al. Detection of prostatic cancer by solid-phase radioimmunoassay of serum prostatic acid phosphatase. N Engl J Med 1977;297:1357.
17. Guinan P, et al. The accuracy of the rectal examination in the diagnosis of prostatic carcinoma. N Engl J Med 1980;303:499.
18. Brawer MK, Lange PH. Prostate-specific antigen and premalignant change: implications for early detection. CA Cancer J Clin 1989;39:361.
19. Catalona WJ, et al. Comparison of prostate specific antigen concentration versus prostate specific antigen density in the early detection

of prostate cancer: receiver operating characteristic curves. J Urol 1994;152:203.

20. Galen RS, Gambino SR. Beyond normality: the predictive value and efficacy of medical diagnoses. New York: Wiley, 1975.
21. Robertson EA, Zweig MH, Van Steirteghem AC. Evaluating the clinical efficacy of laboratory tests. Am J Clin Pathol 1983;79:78.
22. National Committee for Clinical Laboratory Standards. Approved guideline for assessment of clinical sensitivity and specificity of laboratory tests using receiver operating characteristic (ROC) plots. Document GP10-A. Villanova, PA: NCCLS, 1995.
23. Catalona WJ, et al. Use of the percentage of free prostate-specific antigen to enhance differentiation of prostate cancer from benign prostatic disease. JAMA 1998;279:1542–1547.
24. American Chemical Society, Committee on Environmental Improvement, Subcommittee on Environmental Analytical Chemistry. Guidelines for data acquisition and data quality evaluation in environmental chemistry. Anal Chem 1980;52:2242.
25. Westgard JO, deVos DJ, Hunt MR, et al. Concepts and practices in the evaluation of clinical chemistry methods: I. Background and approach. Am J Med Tech 1978;44:290.
26. National Committee for Clinical Laboratory Standards. Approved guideline for precision performance of clinical chemistry devices. Document EP5-A. Villanova, PA: NCCLS, 1998.
27. Westgard JO, deVos DJ, Hunt MR, et al. Concepts and practices in the evaluation of clinical chemistry methods: II. Experimental procedures. Am J Med Tech 1978;44:420.
28. Krouwer JS, Rabinowitz R. How to improve estimates of imprecision. Clin Chem 1984;30:290.
29. National Committee for Clinical Laboratory Standards. Proposed guideline for interference testing in clinical chemistry. Document EP7-P. Villanova, PA: NCCLS, 1986.
30. Young DS. Effects of drugs on clinical laboratory tests. 4th Edition, American Association for Clinical Chemistry, 1995.
31. Siest G, Galteau MM. Drug effects on laboratory test results. Littleton, MA: PSG Publishing, 1988.
32. Glick MR, Ryder KW, Jackson SA. Graphical comparisons of interferences in clinical chemistry instrumentation. Clin Chem 1986;32:470.
33. Glick MR, Ryder KW. Analytical systems ranked by freedom from interferences. Clin Chem 1987;33:1453.
34. Ryder KW, Glick MR. Erroneous laboratory results from hemolyzed, icteric, and lipemic specimens. Clin Chem 1993;39: 175–176.
35. National Committee for Clinical Laboratory Standards. Approved guideline for method comparison and bias estimation using patient samples. Document EP9-A. Villanova, PA: NCCLS, 1995.
36. National Committee for Clinical Laboratory Standards. Proposed guideline for evaluation of linearity of quantitative analytical methods. Document EP6-P. Villanova, PA: NCCLS, 1986.
37. Cornbleet PJ, Gochman N. Incorrect least-squares regression coefficients in method-comparison analysis. Clin Chem 1979;25:432.
38. Feldman U, Schmeider B, Klinkers H. A multivariate approach for the biometric comparison of analytical methods in clinical chemistry. Clin Biochem 1981;19:121.
39. Westgard JO, deVos DJ, Hunt MR, et al. Concepts and practices in the evaluation of clinical chemistry methods: IV. Decisions of acceptability. Am J Med Tech 1978;44:727.
40. Tonks D. A study of the accuracy and precision of clinical chemistry determinations in 170 Canadian laboratories. Clin Chem 1963;9:217.
41. Barnett RN. Medical significance of laboratory results. Am J Clin Pathol 1968;50:671.
42. Fraser CG: Data on biological variation: essential prerequisites for introducing new procedures [editorial]? Clin Chem 1994;40:1671–1673.
43. US Department of Health and Human Services. Medicare, Medicaid, and CLIA programs. Regulations implementing the clinical laboratory improvement amendments of 1988 (CLIA) final rule. Federal Register 1992;57:7002.
44. Ehrmeyer SS, et al. Medicare/CLIA final rules for proficiency testing: minimum intra-laboratory performance characteristics (CV and bias) need to pass. Clin Chem 1990;36:1736.
45. Westgard JO, Seehafer JJ, Barry PL. European specifications for imprecision and inaccuracy compared with operating specifications that assure the quality required by US CLIA proficiency-testing criteria. Clin Chem 1994;40;1228.
46. Westgard JO, Carey RN, Wold S. Criteria for judging precision and accuracy in method development and evaluation. Clin Chem 1974; 20:825.

47. Westgard JO, deVos DJ, Hunt MR, et al. Concepts and practices in the evaluation of clinical chemistry methods: V. Applications. Am J Med Tech 1978;44:803.
48. National Committee for Clinical Laboratory Standards. User demonstration of performance for precision and accuracy: proposed guideline. Document EP15-P. Wayne, PA: NCCLS, 1998.
49. Buttner J, Borth R, Boutwell JH, et al. Provisional recommendation on quality control in clinical chemistry. Clin Chem 1976;22:532.
50. Elin RJ. Elements of cost management for quality assurance. Pathologist 1980;34:182.
51. Cembrowski GS, Carey RN. Laboratory quality management. Chicago: ASCP Press, 1989.
52. Westgard JO, Barry PD. Cost-effective quality control: managing the quality and productivity of analytical processes. Washington, DC: AACC Press, 1986.
53. Young DS. Effects of pre-analytical variables on clinical laboratory tests. 2nd ed. Washington, DC: American Association for Clinical Chemistry, 1997.
54. National Committee for Clinical Laboratory Standards. Approved guideline for clinical laboratory procedure manuals. 3rd ed. Document GP2-A3. Villanova, PA: NCCLS, 1996.
55. National Committee for Clinical Laboratory Standards. Approved standard for procedures for the collection of diagnostic blood specimens by venipuncture. 3rd ed. Document H3-A4. Villanova, PA: NCCLS, 1998.
56. National Committee for Clinical Laboratory Standards. Approved standard for procedures for the collection of diagnostic blood specimens by skin puncture. 3rd ed. Document H4-A3. Villanova, PA: NCCLS, 1991.
57. National Committee for Clinical Laboratory Standards. Approved standard for the percutaneous collection of arterial blood for laboratory analysis. 2nd ed. Document H11-A2. Villanova, PA: NCCLS, 1992.
58. National Committee for Clinical Laboratory Standards. Approved guideline for devices for collection of skin puncture specimens. 2nd ed. NCCLS Document H14-A2. Villanova, PA: NCCLS, 1990.
59. National Committee for Clinical Laboratory Standards. Approved standard for procedures for the handling and transport of domestic diagnostic specimens and etiologic agents. 3rd ed. Document H5-A3. Villanova, PA: NCCLS, 1985.
60. National Committee for Clinical Laboratory Standards. Approved guideline for procedures for the handling and processing of blood specimens. Document H18-A. Villanova, PA: NCCLS, 1990.
61. National Committee for Clinical Laboratory Standards. Approved guideline for preparation and testing of reagent water in the clinical laboratory. 3rd ed. Document C3-A3. Villanova, PA: NCCLS, 1991.
62. National Committee for Clinical Laboratory Standards. Approved standard for power requirements for clinical laboratory instruments and for laboratory power sources. Document I5-A. Villanova, PA: NCCLS, 1980.
63. National Committee for Clinical Laboratory Standards. Approved standard for temperature calibration of water baths, instruments, and temperature sensors. 2nd ed. Document I2-A2. Villanova, PA: NCCLS, 1990.
64. National Committee for Clinical Laboratory Standards. Tentative guideline for calibration materials in clinical chemistry. Document C22-T. Villanova, PA: NCCLS, 1982.
65. Shewhart WA. Economic control of quality of the manufactured product. New York: Van Nostrand, 1931.
66. Levey S, Jennings ER. The use of control charts in the clinical laboratories. Am J Clin Pathol 1950;20:1059.
67. National Committee for Clinical Laboratory Standards. Tentative guideline for control materials in clinical chemistry. Document C23-T. Villanova, PA: NCCLS, 1982.
68. Bowers GN, Burnett RW, McComb RB. Preparation and use of human serum control materials for monitoring precision in clinical chemistry. In: Selected methods for clinical chemistry. Vol 8. Washington, DC: American Association of Clinical Chemists, 1977;21.
69. Caputo MJ, et al. Bovine-based serum as quality control material: a comparative analytical study of bovine vs. human controls. Fullerton, CA: Hyland Diagnostics, 1981.
70. Westgard JO, Barry PL, Hunt MR, et al. A multirule Shewhart chart for quality control in clinical chemistry. Clin Chem 1981;27:493.
71. Westgard JO, Groth T. Power functions for statistical control rules. Clin Chem 1979;25:863.
72. Westgard JO, Groth T, Aronsson T, et al. Performance characteristics of rules for internal quality control: probabilities for false rejection and error detection. Clin Chem 1977;23:1857.

73. Westgard JO, Groth T. Design and evaluation of statistical control procedures: applications of a computer "quality control simulator" program. Clin Chem 1981;27:1536.

74. Hainline A Jr. Quality assurance: theoretical and practical aspects. In: Selected methods for the small clinical chemistry laboratory. Washington, DC: American Association of Clinical Chemistry, 1982;17.

75. Westgard JO, Groth T, Aronsson T, et al. Combined Shewhart-cusum control chart for improved quality control in clinical chemistry. Clin Chem 1977;23:1881.

76. Cembrowski GS, Westgard JO, Eggert AA, et al. Trend detection in control data: optimization and interpretation of Trigg's technique for trend analysis. Clin Chem 1975;21:139.

77. Cembrowski GS, Carey RN. Considerations for the implementation of clinically derived quality control procedures. Lab Med 1989; 20:400.

78. Koch DD, Oryall JJ, Quam EF, et al. Selection of medically useful quality control procedures for individual tests on a multi-test analytical system. Clin Chem 1990;36:230–233.

79. Westgard JO, Quam BS, Barry PL. QC selection grids for planning QC procedures. Clin Lab Sci 1990;3:271.

80. Westgard JO: Charts of operational process specifications ("OP-Specs charts") for assessing the precision, accuracy, and quality control needed to satisfy proficiency testing performance criteria. Clin Chem 1992;38:1226–33.

81. Hoffman RG, Waid ME. The "average of normals" method of quality control. Am J Clin Pathol 1965;43:134.

82. Cembrowski GS, Chandler EP, Westgard J. Assessment of "average of normals" quality control procedures and guidelines for implementation. Am J Clin Pathol 1984;81:492.

83. Douville P, Cembrowski GS, Strauss J. Evaluation of the average of patients, application to endocrine assays. Clin Chim Acta 1987;167:173.

84. Cembrowski GS, Westgard JO, Kurtycz DFI. Use of anion gap for the quality control of electrolyte analyzers. Am J Clin Pathol 1983;79:688.

85. Bockelman HW, et al. Quality control of electrolyte analyzers: evaluation of the anion gap average. Am J Clin Pathol 1984;81:219.

86. Ladenson JH. Patients as their own controls: use of the computer to identify "laboratory error." Clin Chem 1975;21:1648.

87. Sheiner LB, Wheeler LA, Moore JK. The performance of delta check methods. Clin Chem 1979;25:2034.

88. Wheeler LA, Sheiner LB. A clinical evaluation of various delta check methods. Clin Chem 1981;27:5.

89. Ehrmeyer SS, Laessig RH, Schell K. Use of alternate rules (other than the 1_{2s}) for evaluating interlaboratory performance data. Clin Chem 1988;34:250.

90. Ehrmeyer SS, Laessig RH. External proficiency testing. In: Cembrowski GS, Carey RN, eds. Laboratory quality management. Chicago: ASCP Press, 1989;227.

91. Cembrowski GS, Anderson PG, Crampton CA, et al. Pump up your PT IQ, Medical Laboratory Observer 1996;28(1):46–51.

92. Cembrowski GS, Crampton C, Byrd J, et al. Detection and classification of proficiency testing errors in HCFA regulated analytes. Application to ligand assays. J Clin Immunoassay 1995;17:210.

93. Cembrowski GS, Hackney JR, Carey N. The detection of problem analytes in a single proficiency test challenge in the absence of the Health Care Financing Administration rule violations. Arch Pathol Lab Med 1993;117:437.

94. Portable blood glucose monitors. Health Devices 1992;21:43–79.

95. Cembrowski GS, Kiechle FL. Point of care testing: critical analysis and practical application. Adv Pathol Lab Med 1994;7:3.

96. Goldsmith BM. New POCT guide establishes testing uniformity. MLO. Aug 1995;50–52.

97. Tsai WW, Nash DB, Seamonds B, et al. Point-of-care versus central laboratory testing: an economic analysis in an academic medical center. Clinical Therapeutics 1994;16:898–910.

98. Bailey TM, Topham TM, Wantz S, et al. Laboratory process improvement through point-of-care testing. J Quality Improvement 1997;23:363–380.

99. Ross JW, Boone DJ. Assessing the effect of mistakes in the total testing process on the quality of patient care. [Abstract] Presented at the 1989 Institute on Critical Issues in Health Laboratory Practice, sponsored by the Centers for Disease Control and the University of Minnesota, April 9–12 1989, Minneapolis, MN.

100. Harrington HJ. The improvement process. New York: McGraw-Hill, 1987.

101. Westgard JO, Barry PL. Total quality control: evolution of quality management systems. Lab Med 1989;20:377.

102. Berwick DM. Continuous improvement as an ideal in health care. N Engl J Med 1989;320:53.

103. Laffel G, Blumenthal D. The case for using industrial quality management science in health care organizations. JAMA 1989; 262:2869.

CHAPTER 4

Analytical Techniques and Instrumentation

Alan H. B. Wu

Objectives

Upon completion of this chapter, the clinical laboratorian should be able to:

- *Explain the general principles of each of the analytical methods.*

- *Discuss the limitations of each analytical technique.*

- *Compare and contrast the various analytical techniques.*

- *Discuss existing clinical applications for each analytical technique.*

- *Describe the operation and component parts of the following instruments: spectrophotometer, atomic absorption spectrometer, fluorometer, gas chromatograph, osmometer, ion-selective electrode, and pH electrode.*

- *Outline the quality assurance and preventive maintenance procedures involved with the following instruments: spectrophotometer, atomic absorption spectrometer, fluorometer, gas chromatograph, osmometer, ion-selective electrode, and pH electrode.*

KEY TERMS

Atomic absorption	Flame photometry	Liquid chromatography
Chemilumines-	Fluorometry	Point-of-care testing
cence	Gas chromatography	(POCT)
Electrochemistry	Ion-selective	Spectrophotometry
Electrophoresis	electrodes	

Analytical techniques and instrumentation provide the foundation for all measurements made in a modern clinical chemistry laboratory. The majority of techniques fall into one of four basic disciplines within the field of analytical chemistry: spectrometry (including spectrophotometry, atomic absorption, and mass spectrophotometry); luminescence (including fluorescence, chemiluminescence, and nephelometry); electroanalytical methods (including electrophoresis, potentiometry, and amperometry); and chromatography (including gas, liquid, and thin-layer). With the improvements in optics, electronics, and computerization, instrumentation has become miniaturized. This miniaturization has enabled the development of point-of-care testing (POCT) devices that produce results as accurate as large laboratory-based instrumentation.

SPECTROPHOTOMETRY AND PHOTOMETRY

The instruments that measure electromagnetic radiation have several concepts and components in common. Shared instrumental components are discussed in some detail in a later section. Photometric instruments measure light intensity without consideration of wavelength. Most instruments today either use filters (photometers), prisms, or gratings (spectrometers) to select (isolate) a narrow range of the incident wavelength. Radiant energy that passes through an object will be partially reflected, absorbed, and transmitted.

Electromagnetic radiation is described as photons of energy traveling in waves. The relationship between wavelength and energy E is described by Planck's formula:

$$E = h\nu, \qquad\text{(Eq. 4–1)}$$

where h = a constant (6.62×10^{-27} erg sec)
    ν = frequency

Because the frequency of a wave is inversely proportional to the wavelength, it follows that the energy of electromagnetic radiation is inversely proportional to wavelength. Figure 4-1A shows this relationship. Electromagnetic radiation includes a spectrum of energy from short-wavelength, highly energetic gamma and x-rays on the left in Figure 4-1B to long-wavelength radio frequencies on the right. Visible light falls in between, with the color violet at 400 nm and red at 700 nm wavelength being the approximate limits of the visible spectrum.

The instruments discussed in this section measure either absorption or emission of radiant energy to determine concentration of atoms or molecules. The two phenomena, absorption and emission, are closely related. For a ray of electromagnetic radiation to be absorbed, it must have the same frequency as a rotational or vibrational frequency in the atom or molecule that it strikes. Levels of energy that are absorbed move in discrete steps, and any particular type of molecule or atom will absorb only certain energies and not others. When energy is absorbed, valence electrons move to an orbital with a higher energy level. Following energy absorption, the excited electron will fall back to the ground state by emitting a discrete amount of energy in the form of a characteristic wavelength of radiant energy.

Absorption or emission of energy by atoms results in a line spectrum. Because of the relative complexity of molecules, they absorb or emit a bank of energy over a large region. Light emitted by incandescent solids (tungsten or deuterium) is in a continuum. The three types of spectra are shown in Figure 4-2.[1-3]

Beer's Law

The relationship between absorption of light by a solution and the concentration of that solution has been described by Beer and others. Beer's Law states that the concentration of a substance is directly proportional to the amount of light absorbed or inversely proportional to the logarithm of the transmitted light. Percent transmittance ($\%T$) and absorbance (A) are related photometric terms that are explained in this section.

Figure 4-3A shows a beam of monochromatic light entering a solution. Some of the light is absorbed. The remainder passes through, strikes a light detector, and is converted to an electric signal. *Percent transmittance* is the ratio of the radiant energy transmitted T divided by the radiant energy incident on the sample I. All light absorbed or blocked results in $0 \%T$. A level of $100\%\,T$ is obtained if no light is absorbed. In practice, the solvent without the constituent of interest is placed in the light path, as in Figure 4-3B. Most of the light is transmitted, but a small amount is absorbed by the solvent and cuvet or is reflected away from the detector. The electrical readout of the instrument is set arbitrarily at $100\%\,T$, while the light is passing through a "blank" or reference. The sample containing absorbing molecules to be measured is placed in the light path. The difference in amount of light transmitted by the blank and that transmitted by the sample is due only to the presence of the compound being measured. The $\%T$ measured by commercial spectrophotometers is the ratio of the sample transmitted beam divided by the blank transmitted beam.

Equal thicknesses of an absorbing material will absorb a constant fraction of the energy incident upon the layers. For

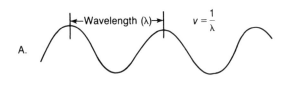

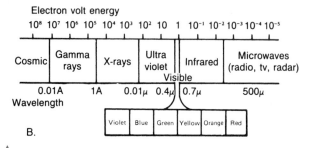

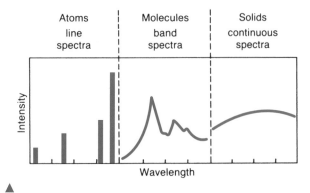

Figure 4-2. Characteristic absorption or emission spectra. (From Coiner D. Basic concepts in laboratory instrumentation. ASMT Education and Research Fund, Inc., 1975–1979.)

▲
Figure 4-1. Electromagnetic radiation—relationship of energy and wavelength.

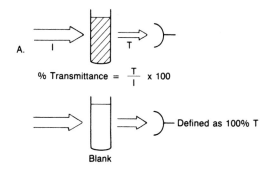

% Transmittance = $\frac{T}{I}$ × 100

Defined as 100% T

Blank

B.

%T = $\frac{\text{Sample beam signal}}{\text{Blank beam signal}}$ × 100

Sample

Figure 4-3. % transmittance defined.

example, in a tube containing layers of solution (Fig. 4-4A), the first layer transmits 70% of the light incident upon it. The second layer will, in turn, transmit 70% of the light incident upon it. Thus, 70% of 70% (49%) is transmitted by the second layer. The third layer transmits 70% of 49%, or 34% of the original light. Continuing on, successive layers transmit 24% and 17%, respectively. The % T values, when plotted on linear graph paper, yields the curve shown in Figure 4-4B. Considering each equal layer as many monomolecular layers, we can translate layers of material to concentration. If semilog graph paper is used to plot the same figures, a straight line is

obtained (Fig. 4-4C), indicating that, as concentration increases, % T decreases in a logarithmic manner.

Absorbance A is the amount of light absorbed. It cannot be measured directly by a spectrophotometer, but rather is mathematically derived from % T as follows:

$$\% \; T = \frac{I}{I_0} \times 100, \qquad (Eq. \; 4\text{--}2)$$

where I_0 = incident light
I = transmitted light

Absorbance is defined as

$$A = -\log(I/I_0) = \log(100\%) - \log \% \; T = 2 - \log \% \; T$$

$$(Eq. \; 4\text{--}3)$$

According to Beer's Law, absorbance is directly proportional to concentration:

$$A = \epsilon \times b \times c, \qquad (Eq. \; 4\text{--}4)$$

where ϵ = molar absorptivity, the fraction of a specific wavelength of light absorbed by a given type of molecule
b = length of light path through the solution
c = concentration of absorbing molecules

Absorptivity depends on molecular structure and the way in which the absorbing molecules react with different energies. For any particular molecular type, absorptivity changes as wavelength of radiation changes. The amount of light absorbed at a particular wavelength depends on molecular and ion types present, and may vary with concentration, pH, or temperature.

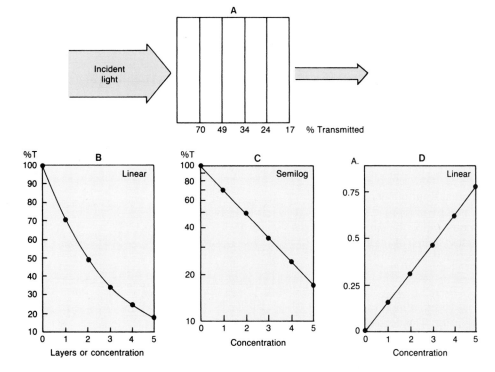

Figure 4-4. (**A**) % of original incident light transmitted by equal layers of light-absorbing solution; (**B**) % T versus concentration on linear graph paper; (**C**) % T versus concentration on semilog graph paper; (**D**) A versus concentration on linear graph paper.

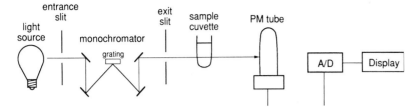

Figure 4-5. Single-beam spectrophotometer.

Because the path length and molar absorptivity are constant for a given wavelength,

$$A \propto C \qquad (Eq.\ 4\text{--}5)$$

Unknown concentrations are determined from a calibration curve that plots absorbance at a specific wavelength versus concentration for standards of known concentration. For calibration curves that are linear and have a zero y intercept, unknown concentrations can be determined from a single calibrator. Not all calibration curves result in straight lines. Deviations from linearity are typically observed at high absorbances. The stray light within an instrument will ultimately limit the maximum absorbance that a spectrophotometer can achieve. This is typically 2.0 absorbance units.

Spectrophotometric Instruments

A spectrophotometer is used to measure the light transmitted by a solution to determine the concentration of the light-absorbing substance in the solution. Figure 4-5 illustrates the basic components of a single-beam spectrophotometer, which are described in subsequent sections.

Components of a Spectrophotometer

Light Source

The most common source of light for work in the visible and near-infrared region is the incandescent tungsten or tungsten-iodide lamp. Only about 15% of the radiant energy emitted falls in the visible region, with most emitted as near-infrared.[1-3] Often, a heat-absorbing filter is inserted between the lamp and sample to absorb the infrared radiation.

The lamps most commonly used for ultraviolet work are the deuterium-discharge lamp or the mercury-arc lamp. Deuterium provides continuous emission down to 165 nm. Low-pressure mercury lamps emit a sharp-line spectrum, with both ultraviolet and visible lines. Medium and high-pressure mercury lamps emit a continuum from ultraviolet to the mid-visible region. The most important factors for a light source are range, spectral distribution within the range, the source of radiant production, stability of the radiant energy, and temperature.

Monochromators

Isolation of individual wavelengths of light is an important and necessary function of a monochromator. The degree of wavelength isolation is a function of the type of device used and the width of entrance and exit slits. The bandpass of a monochromator defines the range of wavelengths transmitted and is calculated as width at more than half the maximum transmittance (Fig. 4-6).

Numerous devices are used for obtaining monochromatic light. The least expensive are colored-glass filters. These filters usually pass a relatively wide band of radiant energy and have a low transmittance of the selected wavelength. They are not precise, but they are simple, inexpensive, and useful.

Interference filters produce monochromatic light based on the principle of constructive interference of waves. Two pieces of glass, each mirrored on one side, are separated by a transparent spacer that is precisely one-half the desired wavelength. Light waves enter one side of the filter and are reflected at the second surface. Wavelengths that are twice the space between the two glass surfaces will reflect back and forth, reinforcing others of the same wavelengths, and finally pass on through. Other wavelengths will cancel out because of phase differences (destructive interference). Because interference filters also transmit multiples of the desired wavelengths, they require accessory filters to eliminate these harmonic wavelengths. Interference filters can be constructed to pass a very narrow range of wavelengths with good efficiency.

The prism is another type of monochromator. A narrow beam of light focused on a prism is refracted as it enters the more dense glass. Short wavelengths are refracted more than long wavelengths, resulting in dispersion of white light into a continuous spectrum. The prism can be rotated, allowing only the desired wavelength to pass through an exit slit.

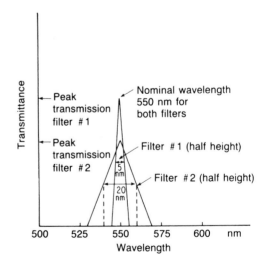

Figure 4-6. Spectral transmittance of two monochromators with band pass at half height of 5 nm and 20 nm.

Diffraction gratings are most commonly used as monochromators. This consists of many parallel grooves (15,000 or 30,000 per inch) etched onto a polished surface. Diffraction, the separation of light into component wavelengths, is based on the principle that wavelengths are bent as they pass a sharp corner. The degree of bending depends on the wavelength. As the wavelengths move past the corners, wave-fronts are formed. Those that are in phase reinforce one another, whereas those that are not in phase cancel out and disappear. This results in complete spectra. Gratings with very fine line rulings produce a widely dispersed spectrum. They produce linear spectra, called *orders,* in both directions from the entrance slit. Because the multiple spectra have a tendency to cause stray light problems, accessory filters are used.

Sample Cell

The next component of the basic spectrophotometer is the sample cell or cuvet, which may be round or square. The light path must be kept constant to have absorbance proportional to concentration. This is easily checked by preparing a colored solution to read midscale when using the wavelength of maximum absorption. Fill each cuvet to be tested, take readings, and save those that match within an acceptable tolerance (eg, $\pm 0.25\% \, T$). Because it is difficult to manufacture round tubes with uniform diameters, they should be etched to indicate the position for use. Cuvets are sold in matched sets. Square cuvets have plane-parallel optical surfaces and a constant light path. They have an advantage over round cuvets in that there is less error from the lens effect, orientation in the spectrophotometer, and refraction. Cuvets with scratches on their optical surface scatter light and should be discarded. Inexpensive glass cuvets can be used for applications in the visible range, but they absorb light in the ultraviolet region. Quartz cuvets must therefore be used for applications requiring ultraviolet radiation.

Photodetectors

The purpose of the detector is to convert the transmitted radiant energy into an equivalent amount of electrical energy. The least expensive of the devices is known as a *barrier-layer cell,* or *photocell.* The photocell is composed of a film of light-sensitive material, frequently selenium, on a plate of iron. Over the light-sensitive material is a thin, transparent layer of silver. When exposed to light, electrons in the light-sensitive material are excited and released to flow to the highly conductive silver. In comparison with the silver, a moderate resistance opposes the electron flow toward the iron, forming a hypothetical barrier to flow in that direction. Consequently, this cell generates its own electromotive force, which can be measured. The produced current is proportional to incident radiation. Photocells require no external voltage source but rely on internal electron transfer to produce a current in an external circuit. Because of their low internal resistance, the output of electrical energy is not easily amplified. Consequently, this type of detector is used mainly in filter photometers with a wide bandpass, producing a fairly high level of illumination so that there is no need to amplify the signal. The photocell is inexpensive and durable, but it is temperature-sensitive and is nonlinear at very low and very high levels of illumination.

Figure 4-7 shows a *phototube,* which is similar to a barrier-layer cell in that it has photosensitive material that gives off electrons when light energy strikes it. It differs in that an outside voltage is required for operation. Phototubes contain a negatively charged cathode and a positively charged anode enclosed in a glass case. The cathode is composed of a material such as rubidium or lithium that will act as a resistor in the dark but will emit electrons when exposed to light. The emitted electrons jump over to the positively charged anode, where they are collected and return through an external, measurable circuit. The cathode usually has a large surface area. Varying the cathode material changes the wavelength at which the phototube gives its highest response. The photocurrent is linear, with the intensity of the light striking the cathode as long as voltage between the cathode and anode remain constant. A vacuum within the tubes avoids scattering of the photoelectrons by collision with gas molecules.

The third major type of light detector is the *photomultiplier (PM) tube,* which detects and amplifies radiant energy. As shown in Figure 4-8, incident light strikes the coated cathode, emitting electrons. The electrons are attracted to a series of anodes, known as *dynodes,* each having a successively higher positive voltage. These dynodes are of a material that will give off many secondary electrons when hit by single electrons. Initial electron emission at the cathode thus triggers a multiple cascade of electrons within the PM tube itself. Because of this amplification, the PM tube is 200 times more sensitive than the phototube. PM tubes are used in instruments designed to be extremely sensitive to very low light levels and light flashes of very short duration. The accumulation of electrons striking the anode produces a current signal, measured in amperes, that is proportional to the initial intensity of the light. The analog signal is converted first to a voltage and then to a digital signal through the use of an analog-

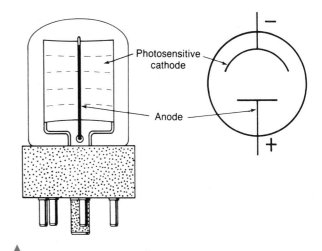

Figure 4-7. Phototube drawing and schematic.

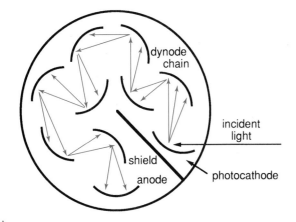

▲
Figure 4-8. Dynode chain in a photomultiplier.

to-digital (A/D) converter. Digital signals are processed electronically to produce absorbance readings.

In a *photodiode,* absorption of radiant energy by a reversed-biased *pn*-junction* diode produces a photocurrent that is proportional to the incident radiant power. Although photodiodes, because of the lack of internal amplification, are not as sensitive as PM tubes, their excellent linearity (6 to 7 decades of radiant power), speed, and small size make them useful in applications where light levels are adequate.[4] *Photodiode array* (PDA) detectors are available in integrated circuits containing 256 to 2048 photodiodes in a linear arrangement. A linear array is shown in Figure 4-9. Each photodiode responds to a specific wavelength and, as a result, a complete UV/Visible spectrum can be obtained in less than 1 second. Resolution is 1 nm to 2 nm and depends on the number of dis-

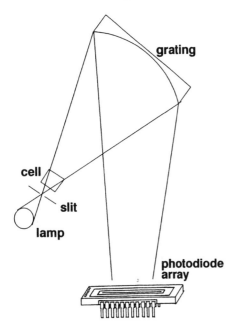

▲
Figure 4-9. Photodiode array spectrophotometer illustrating the placement of the sample cuvet before the monochromator.

pn-junction, positive–negative junction

crete elements. In spectrophotometers using PDA detectors, the grating is positioned *after* the sample cuvet and disperses the *transmitted* radiation onto the PDA detector (Fig. 4-9).

For single-beam spectrophotometers, the absorbance reading from the sample must be blanked using an appropriate reference solution that does not contain the compound of interest. Double-beam spectrophotometers permit automatic correction of sample and reference absorbance, as shown in Figure 4-10. Because the intensities of light sources vary as a function of wavelength, double-beam spectrophotometers are necessary when the absorption spectrum for a sample is to be obtained.

Spectrophotometer Quality Assurance

Performing at least the following checks should validate instrument function: wavelength accuracy, stray light, and linearity. *Wavelength accuracy* means that the wavelength indicated on the control dial is the actual wavelength of light passed by the monochromator. It is most commonly checked using standard absorbing solutions or filters with absorbance maxima of known wavelength. Didymium or holmium oxide in glass are stable and are frequently used as filters. The filter is placed in the light path, and the wavelength control is set at the wavelength at which maximal absorbance is expected. The wavelength control is then rotated in either direction to locate the actual wavelength that has maximal absorbance. If these two wavelengths do not match, the optics must be adjusted to calibrate the monochromator correctly.

Some instruments with narrow bandpass use a mercury-vapor lamp to verify wavelength accuracy. The mercury lamp is substituted for the usual light source, and the spectrum is scanned to locate mercury emission lines. The wavelength indicated on the control is compared with known mercury emission peaks to determine the accuracy of the wavelength indicator control.

Stray light refers to any wavelengths outside the band transmitted by the monochromator. The most common causes of stray light are reflection of light from scratches on optical surfaces or from dust particles anywhere in the light path, and higher-order spectra produced by diffraction gratings. The major effect is absorbance error, especially in the high-absorbance range. Stray light is detected by using cutoff filters, which eliminate all radiation at wavelengths beyond the one of interest. For example, to check for stray light in the near-ultraviolet region, insert a filter that does not transmit in the region of 200 nm to 400 nm. If the instrument reading is greater than 0% *T,* there is stray light present. Certain liquids, such as $NiSO_4$, $NaNO_2$, and acetone, absorb strongly at short wavelengths and can be used in the same way to detect stray light in the UV range.

Linearity is demonstrated when a change in concentration results in a straight-line calibration curve, as discussed under Beer's Law. Colored solutions may be carefully diluted and used to check linearity, using the wavelength of

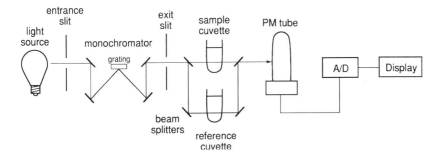

▲

Figure 4-10. Double-beam spectrophotometer.

maximal absorbance for that color. Sealed sets of different colors and concentrations are available commercially. They should be labeled with expected absorbance for a given bandpass instrument. Less than expected absorbance is an indication of stray light or of a bandpass that is wider than specified. Sets of neutral-density filters are also available commercially to check linearity over a range of wavelengths.

A routine system should be devised for each instrument to check and record each of the parameters. The probable cause of a problem and the maintenance required to eliminate it are generally described in the instrument's manual.

Atomic Absorption Spectrophotometer

The *atomic absorption* spectrophotometer is used to measure concentration by detecting absorption of electromagnetic radiation by atoms rather than by molecules. Figure 4-11 shows the basic components. The usual light source, known as a *hollow-cathode lamp,* consists of an evacuated gastight chamber containing an anode, a cylindrical cathode, and an inert gas such as helium or argon. When voltage is applied, the filler gas is ionized. Ions attracted to the cathode collide with the metal,

knock atoms off, and cause the metal atoms to be excited. When they return to the ground state, light energy is emitted that is characteristic of the metal in the cathode. A separate lamp generally is required for each metal (*eg,* a copper hollow-cathode lamp is used to measure Cu).

Electrode-less discharge lamps are a relatively new light source for atomic absorption spectrophotometers. A bulb is filled with argon and the element to be tested. A radio-frequency generator around the bulb supplies the energy to excite the element, causing a characteristic emission spectrum of the element.

The sample being analyzed must contain the reduced metal in the atomic vaporized state. This is commonly done by using the heat of a flame to break the chemical bonds and form free, unexcited atoms. The flame is the sample cell in this instrument, rather than a cuvet. There are various burner designs, but the most common one is known as a *premix long-path burner.* The sample, in solution, is aspirated as a spray into a chamber, where it is mixed with air and fuel. This mixture passes through baffles, where large drops fall and are drained off. Only fine droplets reach the flame. The burner is a long and narrow slit, to permit a longer path length for absorption of incident radiation to occur. Light from the hollow-

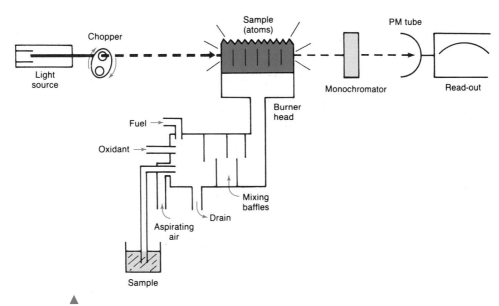

▲

Figure 4-11. Single-beam atomic absorption spectrophotometer—basic components.

cathode lamp passes through the sample of ground-state atoms in the flame. The amount of light absorbed is proportional to the concentration. When a ground-state atom absorbs light energy, an excited atom is produced. The excited atom then returns to the ground state, emitting light of the same energy as it absorbed. The flame sample thus contains a dynamic population of ground-state and excited atoms, both absorbing and emitting radiant energy. The emitted energy from the flame will go in all directions, and it will be a steady emission. Because the purpose of the instrument is to measure the amount of light absorbed, the light detector must be able to distinguish between the light beam emitted by the hollow-cathode lamp and that emitted by excited atoms in the flame. To do this, the hollow-cathode light beam is modulated by inserting a mechanical rotating chopper between the light and the flame or by pulsing the electric supply to the lamp. Because the light beam being absorbed enters the sample in pulses, the transmitted light also will be in pulses. There will be less light in the transmitted pulses because part of it will be absorbed. There are thus two light signals from the flame: an alternating signal from the hollow-cathode lamp and a direct signal from the flame emission. The measuring circuit is tuned to the modulated frequency. Interference from the constant flame emission is eliminated electronically by accepting only the pulsed signal from the hollow cathode.

The monochromator is used to isolate the desired emission line from other lamp emission lines. In addition, it serves to protect the photodetector from excessive light emanating from flame emissions. A PM tube is the usual light detector.

Flame-less atomic absorption requires an instrument modification that uses an electric furnace to break chemical bonds (electrothermal atomization). A tiny graphite cylinder holds the sample, either liquid or solid. An electric current passes through the cylinder walls, evaporates the solvent, ashes the sample, and finally heats the unit to incandescence to atomize the sample. This instrument, like the spectrophotometer, is used to determine the amount of light absorbed. Thus, again, Beer's Law is used for calculating concentration. A major problem is that background correction is considerably more necessary and critical for electrothermal techniques than for flame-based atomic absorption methods. Currently, the most common approach is to use a deuterium lamp as a secondary source and measure the difference between the two absorbance signals. However, there has also been extensive development of background correction techniques based on the Zeeman effect.[1] Atomic absorption spectrophotometry is very sensitive and precise. Its direct use, however, is limited to elements that exist in the atomic state.

Flame Photometry

The flame-emission photometer is used to measure light emitted by excited atoms. It is used primarily to determine concentration of Na^+, K^+, or Li^+. Figure 4-12 illustrates the basic components of a typical flame photometer. The sample solu-

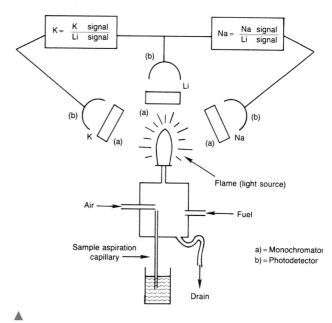

Figure 4-12. Flame-emission photometer—basic components.

tion is aspirated into the flame. The purpose of the flame is twofold. Chemical bonds are broken to produce atoms, and then atoms absorb energy from the flame and enter an excited electronic state. The excited atoms return to the ground state by emitting light energy that is characteristic for that atomic species. The emitted light, focused by lenses or mirrors, passes through monochromators that are selected to transmit radiant energy from the specific atoms. The monochromator for Na^+ is set at 589 nm, for K^+ at 767 nm, and for Li^+ at 671 nm. The emitted light then strikes a photodetector, which generates an electric signal proportional to the concentration of the atoms in the flame.

Fluctuation in the light source of the flame photometer is generally caused by changes in fuel or air pressure that affect flame temperature or rate of sample aspiration. Stability is achieved by using an internal-standard system. For sodium or potassium measurements, lithium can be added in equal amount to all standards and samples to act as an internal standard. The readout, as illustrated in Figure 4-12, is a ratio comparing the emission intensity for the internal standard with that of the element being analyzed. The same amount of lithium is present in the blank, sample, and standard. Any change in aspiration rate or flame temperature will affect the sodium, potassium, and lithium emissions proportionately. The ratio remains constant, thus compensating for possible fluctuations.

Several criteria are required for an internal standard. Concentration of the internal standard must be precisely the same in all samples and standards so that the reference beam is truly constant. The amount of energy required to excite the internal standard must be close to that required to excite the elements being measured. Emission lines must be far enough apart to be separated by available monochromators. Furthermore, the internal standard must not be normally found in the solution being analyzed. Cesium generally fulfills these requirements and is commonly used as an internal standard. Some instruments reverse the K/Cs ratio and use a K$^+$ solution as internal reference so that

$$Cs^+ \text{ concentration} = \frac{Cs^+ \text{ signal}}{K^+ \text{ signal}} \qquad (Eq.\ 4\text{--}6)$$

Older instruments use a three-filter photodetector with lithium as the internal standard. In this case, the readout of Na$^+$ or K$^+$ is the ratio using lithium emission as the reference signal.

Comparison of Flame Methods

Flame-emission photometry is commonly used to measure sodium, potassium, and lithium because these alkali metals are fairly easy to excite. A flame using propane as fuel has sufficient energy to raise the valence electrons of approximately 1% to 5% of the atoms present in the flame to a higher energy level. Although this is relatively inefficient, it is sensitive enough to measure the concentrations found in biologic specimens. Most other metals (*eg*, calcium) are less easily excited and are present in lower concentrations. In these cases, flame-emission photometry is not applicable.

Atomic absorption spectrophotometry is routinely used to measure concentration of trace metals that are not easily excited. It is generally more sensitive than flame emission, because the vast majority of atoms produced in the usual propane or air-acetylene flame remain in the ground state available for light absorption. It is accurate, precise, and very specific. One of the disadvantages, however, is the inability of the flame to dissociate samples into free atoms. For example, phosphate may interfere with calcium analysis by formation of calcium phosphate. This may be overcome by adding cations that compete with calcium for phosphate. Routinely, lanthanum or strontium is added to samples to form stable complexes with phosphate. Another possible problem is the ionization of atoms following dissociation by the flame. This can be decreased by reducing the flame temperature. Matrix interference is another source of error. This is due to the enhancement of light absorption by atoms in organic solvents or formation of solid droplets as the solvent evaporates in the flame. Such interference may be overcome by pretreatment of the sample by extraction.[5]

Recently, inductively coupled plasma has been used to increase sensitivity for atomic emission. The torch is an argon plasma maintained by the interaction of a radiofrequency field and an ionized argon gas, and is reported to have used temperatures between 5500°K and 8000°K. At these temperatures, complete atomization of elements is thought to occur. Use of inductively coupled plasma as a source is recommended for determinations involving refractory elements such as uranium, zirconium, and boron.

Fluorometry

As seen with the spectrophotometer, light entering a solution may pass mainly on through or may be absorbed partly or entirely, depending on the concentration and the wavelength entering that particular solution. Whenever absorption occurs, there is a transfer of energy to the medium. Each molecular type possesses a series of electronic energy levels and can pass from a lower energy level to a higher level only by absorbing an integral unit (quantum) of light that is equal in energy to the difference between the two energy states. There are additional energy levels owing to rotation or vibration of molecular parts. The excited state lasts about 10^{-5} seconds before the electron loses energy and returns to the ground state. Energy is lost by collision, heat loss, transfer to other molecules, and emission of radiant energy. Because the molecules are excited by absorption of radiant energy and lose energy by multiple interactions, the radiant energy emitted is less than the absorbed energy. The difference between the maximum wavelengths, excitation, and emitted fluorescence is called *Stokes shift*. Both excitation (absorption) and fluorescence (emission) energies are characteristic for a given molecular type. For example, Figure 4-13 shows the absorption and fluorescence spectra of quinine in 0.1 *N* sulfuric acid. The dashed line on the left shows the short-wavelength excitation energy that is maximally absorbed, whereas the solid line on the right is the longer-wavelength (less energy) fluorescent spectrum.

Basic Instrumentation

Filter fluorometers are used to measure the concentrations of solutions that contain fluorescing molecules. A basic instrument is shown in Figure 4-14. The source emits short-wavelength high-energy excitation light. A mechanical attenuator controls the light intensity. The wavelength that is best absorbed by the solution to be measured is selected by the primary filter placed between the radiation source and the sample. The fluorescing sample in the cuvet emits radiant energy in all directions. The detector, placed at right angles to the sample cell, and a secondary filter that passes the longer wavelengths of fluorescent light prevent incident light from striking the photodetector. The electrical output of the photodetector is proportional to the intensity of fluorescent energy. In spectrofluorometers, filters are replaced by prisms or grating monochromators.

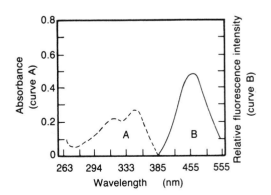

Figure 4-13. Absorption and fluorescence spectra of quinine in 0.1 N sulfuric acid. (From Coiner D. Basic concepts in laboratory instrumentation. ASMT Education and Research Fund, Inc., 1975–1979.)

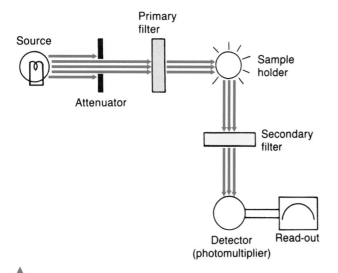

Figure 4-14. Basic filter fluorometer. (From Coiner D. Basic concepts in laboratory instrumentation. ASMT Education and Research Fund, Inc., 1975–1979.)

Gas–discharge lamps (mercury and xenon-arc) are the most frequently used sources of excitation radiant energy. Incandescent tungsten lamps are seldom used because little energy is released by them in the ultraviolet region. Mercury-vapor lamps are commonly used in filter fluorometers. Mercury emits a characteristic line spectrum. Resonance lines at 365 nm to 366 nm are commonly used. Energy at wavelengths other than the resonance lines is provided by coating the inner surface of the lamp with a material that absorbs the 254-nm mercury radiation and emits a broad band of longer wavelengths. Most spectrofluorometers use a high-pressure xenon lamp. Xenon has a good continuum, which is necessary for determining excitation spectra.

Monochromator fluorometers make use of grating, prisms, or filters for isolation of incident radiation. Light detectors are almost exclusively PM tubes because of their higher sensitivity to low light intensities. Double-beam instruments are used to compensate for instability due to electric-power fluctuation.

Fluorescence concentration measurements are related to molar absorptivity of the compound, intensity of the incident radiation, quantum efficiency of the energy emitted per quantum absorbed, and length of the light path. In dilute solutions with instrument parameters held constant, fluorescence is directly proportional to concentration. Generally, a linear response will be obtained until the concentration of the fluorescent species is so high that the sample begins to absorb significant amounts of excitation light. Figure 4-15 shows a curve demonstrating nonlinearity as concentration increases. The solution must absorb less than 5% of the exciting radiation for a linear response to occur.[6] As with all quantitative measurements, a standard curve must be prepared to demonstrate that the concentration used falls in a linear range.

In fluorescence polarization, the radiant energy is polarized in a single plane. When the sample (fluorophor) is excited, it will emit polarized light along the same plane as the incident light if the fluorophor is attached to a large molecule. In contrast, a small molecule will emit depolarized light, because it will rotate out of the plane of polarization during its excitation lifetime. This technique is widely used for the detection of therapeutic and abused drugs. In the procedure, the sample analyte is allowed to compete with a fluorophor-labeled analyte for a limited antibody to the analyte. The lower the concentration of the sample analyte, the higher the macromolecular antibody-analyte-fluorophor formed and the lower the depolarization of the radiant light.

Advantages and Disadvantages of Fluorometry

The two advantages of *fluorometry* over conventional spectrophotometry are specificity and sensitivity. Fluorometry increases specificity by selecting the optimal wavelength for both absorption and fluorescence, rather than just the absorption wavelength, as with spectrophotometry.

Fluorometry is about a thousand times more sensitive than most spectrophotometric methods.[6] One of the reasons is that, because emitted radiation is measured directly, it can be increased simply by increasing the intensity of the exciting radiant energy. In addition, fluorescence measures the amount of light intensity present over a zero background. In absorbance, on the other hand, the quantity of absorbed light is measured indirectly as the difference between the transmitted beams. At low concentrations, the small difference between 100% *T* and the transmitted beam is difficult to measure accurately and precisely, thereby limiting the sensitivity.

The biggest disadvantage is that fluorescence is very sensitive to environmental changes. Changes in pH affect availability of electrons, and temperature changes the probability of loss of energy by collision rather than fluorescence. Contaminating chemicals or a change of solvents may change the structure. Ultraviolet light used for excitation can cause

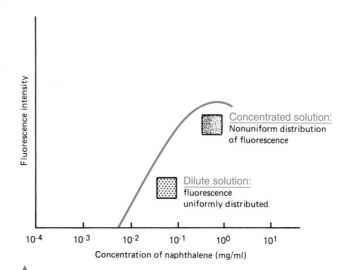

Figure 4-15. Dependence of fluorescence on the concentration of fluorophor. (From Guilbault GG. Practical fluorescence, theory, methods and techniques. New York: Marcel Dekker, 1973.)

photochemical changes. Any decrease in fluorescence resulting from any of these possibilities is known as *quenching*. Because so many factors may change the intensity or spectra of fluorescence, extreme care is mandatory in analytical technique and instrument maintenance.

Chemiluminescence

In *chemiluminescence* reactions, part of the chemical energy generated produces excited intermediates that decay to a ground state with the emission of photons.[7] The emitted radiation is measured with a PM tube, and the signal is related to analyte concentration. Chemiluminescence is different than fluorescence in that no excitation radiation is required and no monochromators are needed because the chemiluminescence arises from one species. Most important, chemiluminescence reactions are oxidation reactions of luminol, acridinium esters, and dioxetanes, and are characterized by a rapid increase in intensity of emitted light followed by a gradual decay. Usually, the signal is taken as the integral of the entire peak. Enhanced chemiluminescence techniques increase the chemiluminescence efficiency by including an enhancer system in the reaction of a chemiluminescent agent with an enzyme. The time course for the light intensity is much longer (60 minutes) than conventional chemiluminescent reactions, which last for about 30 seconds (Fig. 4-16).

Advantages of chemiluminescence assays include subpicomolar detection limits, speed (with flash-type reactions, light is measured for 10 seconds only), ease of use (most assays are one-step procedures), and simple instrumentation.[7] The main disadvantage is that impurities can cause background signal that degrades sensitivity and specificity.

Turbidity and Nephelometry

Turbidimetric measurements are made with a spectrophotometer to determine concentration of particulate matter in a sample. The amount of light blocked by a suspension of particles depends on concentration, but also on size. Because particles tend to aggregate and settle out of suspension, sample handling becomes critical. Instrument operation is the same as for any spectrophotometer.

Nephelometry is similar, except that light scattered by the small particles is measured at an angle to the beam incident on the cuvet. Figure 4-17 demonstrates two possible optical arrangements for a nephelometer. Light scattering depends on wavelength and size of particle. For macromolecules whose size is close to or larger than the wavelength of incident light, sensitivity is increased by measuring forward light scatter.[8] Instruments are available with detectors placed at various forward angles, as well as at 90° to the incident light. Monochromatic light is used to obtain uniform scatter and to minimize sample heating. Some instruments use lasers as a source of monochromatic light, but any monochromator may be used.

Measuring light scatter at an angle other than at 180° in turbidimetry minimizes error from colored solutions and increases sensitivity. Because both methods depend on particle size, some instruments quantitate initial change in light scatter rather than total scatter. Reagents must be free of any particles, and cuvets must have no scratches.

Applications of Lasers

Light amplification by stimulated emission of radiation (LASER) is based on the interaction of radiant energy and suitably excited atoms or molecules. The interaction leads to stimulated emission of radiation. The wavelength, direction of propagation, phase, and plane of polarization of the emitted light are the same as those of the incident radiation. Laser light has narrow spectral width, small cross-sectional area with low divergence, and is polarized and coherent. The radiant emission can be very powerful and can be either continuous or pulsating.

Laser light can serve as the source of incident energy in a spectrometer or nephelometer. Some lasers can produce bandwidths of a few kilohertz in both the visible and infrared regions, thereby making these applications about three to six orders more sensitive than conventional spectrometers.[9]

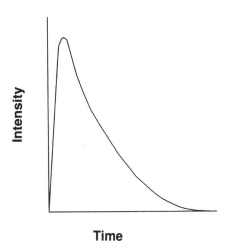

Figure 4-16. Representative intensity versus time curve for a transient chemiluminescence signal.

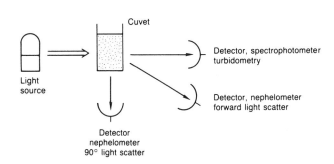

Figure 4-17. Nephelometer versus spectrophotometer—optical arrangements.

Laser spectrometry also can be used for the determination of structure and identification of samples, as well as for diagnosis. Quantitation of samples depends on the spectrometer used. An example of the clinical application of the laser is the Coulter counter, which is used for differential analysis of white blood cells.[10]

ELECTROCHEMISTRY

Many types of electrochemical analyses are used in the clinical laboratory. Examples are potentiometry, amperometry, coulometry, and polarography. The two basic electrochemical cells involved in these analyses are galvanic and electrolytic cells.

Galvanic and Electrolytic Cells

An electrochemical cell can be set up as shown in Figure 4-18. It consists of two half-cells and a salt bridge, which can be a piece of filter paper saturated with electrolytes. The electrodes can be immersed in a single large beaker containing a salt solution instead of the two shown. In such a setup, the solution serves as the salt bridge.

In a galvanic cell, as the electrodes are connected, there is spontaneous flow of electrons from the electrode with the lower electron affinity (oxidation; *eg,* silver), and these electrons pass through the external meter to the cathode (reduction), where OH^- ions are liberated. This reaction continues until one of the chemical components is depleted, at which point the cell is "dead" and cannot produce electrical energy to the external meter.

Current may be forced to flow through the "dead" cell only by applying an external electromotive force E. This is called an *electrolytic cell.* In short, a galvanic cell can be built from an electrolytic cell. When the external E is turned off, accumulated products at the electrodes will spontaneously produce current in the opposite direction of the electrolytic cell.

Half-Cells

It is impossible to measure the electrochemical activity of one half-cell. Two reactions must be coupled, and one reaction compared with the other. To rate half-cell reactions, a specific electrode reaction is arbitrarily assigned 0.00 V. Every other reaction coupled with this arbitrary zero reaction is either positive or negative, depending on the relative affinity for electrons. The electrode defined as 0.00 V is the standard hydrogen electrode: H_2 gas at 1 atmosphere (atm). The hydrogen gas in contact with H^+ in solution develops a potential. The hydrogen electrode coupled with a zinc half-cell is cathodic, with the reaction $2H^+ + 2e^- \rightarrow H_2$, because H_2 has a greater affinity than Zn for electrons. Cu, however, has a greater affinity than H_2 for electrons, and thus the anodic reaction $H_2 \rightarrow 2H^+ + 2e^-$ occurs when coupled to the Cu-electrode half-cell.

The potential generated by the hydrogen-gas electrode is used to rate the electrode potential of metals in 1 mol/L solution. Reduction potentials for some metals are shown in Table 4-1.[11] A hydrogen electrode is used to determine the accuracy of reference and indicator electrodes, the stability of standard solutions, and the potentials of liquid junctions.

Ion-Selective Electrodes (ISE)

Potentiometric methods of analysis involve the direct measurement of electrical potential due to the activity of free ions. ISEs are designed to be sensitive toward individual ions.

pH Electrodes

An ISE universally used in the clinical laboratory is the pH electrode. Figure 4-19 presents an illustration of the basic components of a pH meter.

Indicator Electrode

The pH electrode consists of a silver wire coated with AgCl, immersed into an internal solution of 0.1 mmol/L HCl, and placed into a tube containing a special glass membrane tip. This membrane is sensitive only to hydrogen ions. Glass membranes that are selectively sensitive to H^+ consist of spe-

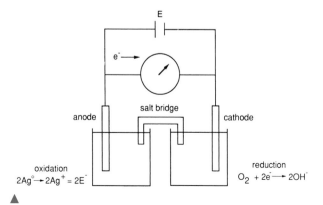

Figure 4-18. Electrochemical cell.

TABLE 4-1. Standard Reduction Potentials	
	Potential, V
$Zn^{2+} + 2e \leftrightarrow Z$	−0.7628
$Cr^{2+} + 2e \leftrightarrow Cr$	−0.557
$Ni^{2+} + 2e \leftrightarrow Ni$	−0.23
$2H^+ + 2e \leftrightarrow H_2$	0.000
$Cu^{2+} + 2e \leftrightarrow Cu$	0.3402
$Ag^+ + e \leftrightarrow Ag$	0.7996

(Data presented are examples from CRC Handbook of Chemistry and Physics, 61st ed., 1980-1981.)

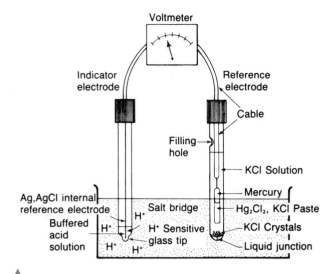

Figure 4-19. Necessary components of a pH meter.

cific quantities of lithium, cesium, lanthanum, barium, or aluminum oxides in silicate. When the pH electrode is placed into the test solution, movement of H^+ ions near the tip of the electrode produces a potential difference between the internal solution and the test solution, which is measured as pH and read by a voltmeter. The combination pH electrode also contains a built-in reference electrode, either Ag/AgCl or calomel (Hg/Hg_2Cl_2) immersed in a solution of saturated KCl.

The specially formulated glass continually dissolves from the surface. The present concept of the selective mechanism that causes formation of electromotive force at the glass surface is that an ion-exchange process is involved. Cationic exchange occurs only in the gel layer. There is no penetration of H^+ through the glass. Although the glass is constantly dissolving, the process is slow, and the glass tip generally lasts for several years. pH electrodes are highly selective for hydrogen ions, but other cations in high concentration do interfere, the most common of which is sodium. Electrode manufacturers should list the concentration of interfering cations that may cause error in pH determinations.

Reference Electrode

The reference electrode commonly used is the calomel electrode. Calomel, a paste of mercurous chloride and potassium chloride, is in direct contact with metallic mercury in an electrolyte solution of potassium chloride. As long as the electrolyte stays at a constant concentration and the temperature remains constant, a stable voltage is generated at the interface of the mercury and its salt. A cable connected to the mercury leads to the voltmeter. The filling hole is needed for adding potassium chloride solution. A tiny opening at the bottom is required for completion of electric contact between the reference and indicator electrodes. The liquid junction consists of a fiber or ceramic plug that allows a small flow of electrolyte filling solution.

Construction varies, but all reference electrodes must generate a stable electrical potential. In general, reference elec-

trodes consist of a metal and its salt in contact with a solution containing the same anion. Mercury/mercurous chloride, as in this example, is a frequently used reference electrode, but it has a disadvantage in that it is slow to reach a new stable voltage following temperature change and is unstable above 80°C.[1,2] Ag/AgCl is another common reference electrode. It can be used at high temperatures, up to 275°C, and the AgCl-coated Ag wire makes a more compact electrode than that of mercury. In measurements in which chloride contamination must be avoided, a mercury sulfate and potassium sulfate reference electrode may be used.

Liquid Junctions

Electrical connection between the indicator and reference electrodes is achieved by allowing a slow flow of electrolyte from the tip of the reference electrode. A junction potential is always set up at the boundary between two dissimilar solutions because of positive and negative ions diffusing across the boundary at unequal rates. The resultant junction potential may increase or decrease the potential of the reference electrode. Therefore, it is important that the junction potential be kept to a minimum reproducible value when the reference electrode is in solution.

KCl is a commonly used filling solution because K^+ and Cl^- ions have nearly the same mobilities. When KCl is used as the filling solution for Ag/AgCl electrodes, addition of AgCl is required to prevent dissolution of the AgCl salt. A way of producing a lower junction potential is to mix K^+, Na^+, NO_3^-, and Cl^- in appropriate ratios.

Readout Meter

Electromotive force produced by the reference and indicator electrodes is in the millivolt range. Zero potential for the cell indicates that each electrode half-cell is generating the same voltage, assuming there is no liquid junction potential. The isopotential is that potential at which a temperature change has no effect on the response of the electrical cell. Manufacturers generally achieve this by making midscale (pH 7.0) correspond to 0 V at all temperatures. They use an internal buffer whose pH changes due to temperature compensate for the changes in the internal and external reference electrodes.

Nernst Equation

The electromotive force generated because of H^+ at the glass tip is described by the Nernst equation, which is shown in a simplified form:

$$\epsilon = \Delta pH \times \frac{RT \ln 10}{F} = \Delta pH \times 0.059 \text{ V}$$

(Eq. 4–7)

where ϵ = the electromotive force of the cell
F = Faraday's constant (96,500 C/mol)
R = the molar gas constant
T = temperature, in Kelvin

As temperature increases, hydrogen ion activity increases and the potential generated increases. Most pH meters have a temperature-compensation knob that amplifies the millivolt response when the meter is on pH function. pH units on the meter scale are usually printed for room-temperature use. On the voltmeter, 59.16 is read as 1 pH unit change. The temperature compensation changes millivolt response to compensate for changes due to temperature from 54.2 at 0°C to 66.10 at 60°C. However, most pH meters are manufactured for greatest accuracy in the 10°C to 60°C range.

Calibration

The steps necessary to standardize a pH meter are fairly straightforward. First, balance the system with the electrodes in a buffer whose pH is 7.0. The balance or intercept control shifts the entire slope, as shown in Figure 4-20. Next, replace the buffer with one of a different pH. If the meter does not register the correct pH, amplification of the response changes the slope to match that predicted by the Nernst equation. If the instrument does not have a slope control, the temperature compensator performs the same function.

pH Combination Electrode

The most common pH electrode used in laboratories today has both the indicator and reference electrodes combined in one small probe. This is convenient when small samples are tested. It consists of an Ag/AgCl internal reference electrode sealed into a narrow glass cylinder with a pH-sensitive glass tip. The reference electrode is an Ag/AgCl wire wrapped around the indicator electrode. The outer glass envelope is filled with KCl and has a tiny pore near the tip of the liquid junction. The solution to be measured must cover the glass tip completely. Examples of other ISEs are shown in Figure 4-21. The reference electrode, electrometer, and calibration system described for pH measurements are applicable to all ISEs.

ISEs are of three major types: inert-metal electrodes in contact with a redox couple, metal electrodes that participate in a redox reaction, and membrane electrodes. The membrane can be of solid material (eg, glass), liquid (eg, ion-exchange electrodes), or special membrane (eg, compound electrodes) such as gas-sensing and enzyme electrodes.

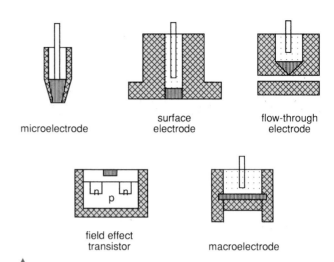

Figure 4-21. Other examples of ion-selective electrodes.

The standard hydrogen electrode is an example of an inert-metal electrode. The Ag/AgCl electrode is an example of the second type. The electrode process $AgCl + e^- \leftrightarrow Ag^+ + Cl^-$ produces an electrical potential proportional to Cl^- ion activity. When the chloride ion is held constant, the electrode is used as a reference electrode. The electrode in contact with varying Cl^- concentrations is used as an indicator electrode to measure chloride concentration.

The H^+ sensitive gel layer of the glass pH electrode is considered a membrane. A change in the glass formulation makes the membrane more sensitive to sodium ions than to hydrogen ions, creating a sodium ISE. Other solid-state membranes consist of either a single crystal or fine crystals immobilized in an inert matrix such as silicone rubber. Conduction depends on a vacancy defect mechanism, and the crystals are formulated to be selective for a particular size, shape, and change. Examples include F^- selective electrodes of LaF, Cl^- sensitive electrodes with AgCl crystals, and AgBr electrodes for the detection of Br^-.

The calcium ISE is a liquid-membrane electrode. An ion-selective carrier, such as dioctyphenyl phosphate dissolved in an inert water-insoluble solvent, diffuses through a porous membrane. Because the solvent is insoluble in water, the test sample cannot cross the membrane, but Ca^{2+} ions are exchanged. The Ag/AgCl internal reference in a filling solution of $CaCl_2$ is in contact with the carrier by means of the membrane.

Potassium-selective liquid membranes use the antibiotic valinomycin as the ion-selective carrier. Valinomycin membranes show great selectivity for K^+. Liquid-membrane electrodes are recharged every few months to replace the liquid ion exchanger and the porous membrane.

Gas-Sensing Electrodes

Gas electrodes are similar to pH glass electrodes but are designed to detect specific gases (eg, CO_2 and NH_3) in solutions and are usually separated from the solution by a thin, gas-

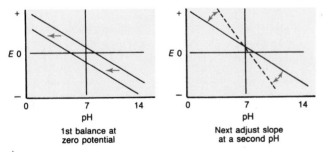

Figure 4-20. pH meter calibration. (From Willard HH, Merritt LL, Dean JA, Settle FA. Instrumental methods of analysis. Belmont, CA: Wadsworth, 1981.)

permeable hydrophobic membrane. Figure 4-22 shows a schematic illustration of the PCO_2 electrode. The membrane in contact with the solution is permeable only to CO_2, which diffuses into a thin film of sodium bicarbonate solution. The pH of the bicarbonate solution is changed as follows:

$$CO_2 + H_2O \leftrightarrow H_2CO_3 \leftrightarrow H^+ + HCO_3^- \quad \textit{(Eq. 4–8)}$$

The change in pH of the HCO_3^- is detected by a pH electrode. The PCO_2 electrode is widely used in clinical laboratories as a component of instruments for measuring serum electrolytes and blood gases.

In the NH_3 gas electrode, the bicarbonate solution is replaced by ammonium chloride solution, and the membrane is permeable only to NH_3 gas. As in the PCO_2 electrode, NH_3 changes the pH of NH_4Cl as follows:

$$NH_3 + H_2O \leftrightarrow NH_4^+ + OH^- \quad \textit{(Eq. 4–9)}$$

The amount of OH^- ions produced varies linearly with the log of the partial pressure of NH_3 in the sample.

Other gas-sensing electrodes function on the basis of amperometric principle, that is, measurement of the current flowing through an electrochemical cell at a constant applied electrical potential to the electrodes. Examples are the determination of PO_2, glucose, and peroxidase.

The chemical reactions of the PO_2 electrode (Clark electrode) are illustrated in Figure 4-18. It is an electrochemical cell with a platinum cathode and an Ag/AgCl anode. The electrical potential at the cathode is set to -0.65 V and will not conduct current without oxygen in the sample. The membrane is permeable to oxygen, which diffuses through to the platinum cathode. Current passes through the cell and is proportional to the PO_2 in the test sample.

Glucose determination is based on the reduction in PO_2 during glucose oxidase reaction with glucose and oxygen. Unlike the PCO_2 electrode, the peroxidase electrode has a polarized platinum anode and its potential is set to $+0.6$ V. Current flows through the system when peroxide is oxidized at the anode as follows:

$$H_2O_2 \rightarrow 2H^+ + 2e^- + O_2 \quad \textit{(Eq. 4–10)}$$

Enzyme Electrodes

The various ISEs may be covered by immobilized enzymes that can catalyze a specific chemical reaction. Selection of the ISE is determined by the reaction product of the immobilized enzyme. Examples include urease, which is used for the detection of urea, and glucose oxidase, which is used for glucose detection. A urea electrode must have an ISE that is selective for NH_4^+ or NH_3, whereas glucose oxidase is used in combination with a pH electrode.

Coulometric Chloridometers and Anodic Stripping Voltametry

Chloride ISEs have largely replaced coulometric titrations for determination of chloride in body fluids. Anodic stripping voltametry was widely used for analysis of lead and is best measured by electrothermal (graphite furnace) atomic absorption spectroscopy.

ELECTROPHORESIS

Electrophoresis is the migration of charged solutes or particles in an electrical field. *Iontophoresis* refers to the migration of small ions, whereas *zone electrophoresis* is the migration of charged macromolecules in a porous support medium such as paper, cellulose acetate, or agarose-gel film. An electrophoretogram is the result of zone electrophoresis and consists of sharply separated zones of a macromolecule. In a clinical laboratory, the macromolecules of interest are proteins in serum, urine, cerebrospinal fluid, other biologic body fluids, erythrocytes, and tissues.

Electrophoresis consists of five components: the driving force (electrical power), the support medium, the buffer, the sample, and the detecting system. A typical electrophoretic apparatus is illustrated in Figure 4-23.

Charged particles migrate toward the opposite charged electrode. The velocity of migration is controlled by the

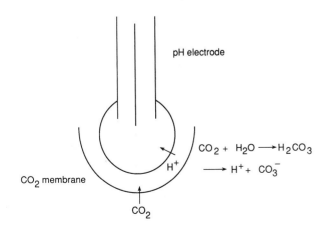

Figure 4-22. The PCO_2 electrode.

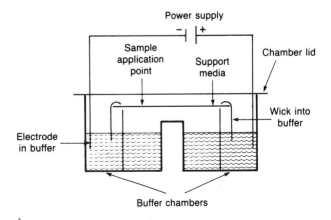

Figure 4-23. Electrophoresis apparatus—basic components.

net charge of the particle, the size and shape of the particle, the strength of the electric field, chemical and physical properties of the supporting medium, and the electrophoretic temperature. The rate of mobility[12] of the molecule (μ) is given by

$$\mu = \frac{Q}{k} \times r \times n, \qquad \text{(Eq. 4–11)}$$

where Q = net charge of particle
k = constant
r = ionic radius of the particle
n = viscosity of the buffer

From the equation, the rate of migration is directly proportional to the net charge of the particle and inversely proportional to its size and the viscosity of the buffer.

Procedure

The sample is soaked in hydrated support for approximately 5 minutes. The support is put into the electrophoresis chamber, which was previously filled with the buffer. Enough buffer must be added to the chamber to maintain contact with the support. Electrophoresis is carried out by applying a constant voltage or constant current for a specific length of time. The support is removed and placed in a fixative or is rapidly dried to prevent diffusion of the sample. This is followed by staining the zones with appropriate dye. The uptake of dye by the sample is proportional to sample concentration. After excess dye is washed away, the supporting medium may need to be placed in a clearing agent. Otherwise, it is completely dried.

Power Supply

Power supplies operating at either constant current or constant voltage are available commercially. In electrophoresis, when current flows through a medium that has resistance, heat is produced. This results in an increase in thermal agitation of the dissolved solute (ions), leading to a decrease in resistance and an increase in current. The latter increase leads to increases in the heat and evaporation of water from the buffer. This increases the ionic concentration of the buffer and subsequent further increases in the current. The migration rate can be kept constant by using a power supply with constant current. This is true because, as electrophoresis progresses, a decrease in resistance as a result of heat produced also decreases the voltage.

Buffers

Two properties of a buffer that affect the charge of ampholytes are its pH and its ionic strength. The ions carry the applied electric current and also allow the buffer to maintain constant pH during electrophoresis. An ampholyte is a molecule, such as protein, whose net charge can be either positive or negative. If the buffer is more acidic than the isoelectric point (pI) of the ampholyte, it binds H^+ ions, becomes positively charged, and migrates toward the cathode. If the

buffer is more basic than the pI, the ampholyte loses H^+ ions, becomes negatively charged, and migrates toward the anode. A particle without a net charge will not migrate, remaining at the point of application. During electrophoresis, ions cluster around a migrating particle. The higher the ionic concentration, the higher is the size of the ionic cloud and the lower is the mobility of the particle. Greater ionic strength produces sharper protein-band separation but leads to increased heat production. This may cause denaturation of heat-labile proteins. Consequently, for any electrophoretic system, the optimal buffer concentration should be determined. Generally, the most widely used buffers are made of monovalent ions because their ionic strength and molality are equal.

Support Materials

Cellulose Acetate

Paper electrophoresis is no longer widely used in clinical laboratories. It has been replaced by cellulose acetate or agarose gel. Cellulose is acetylated to form cellulose acetate by treating it with acetic anhydride. Cellulose acetate is produced commercially. It is a dry, brittle film composed of about 80% air space. When the film is soaked in buffer, the air spaces are filled with electrolyte, and the film becomes pliable. After electrophoresis and staining, cellulose acetate can be made transparent for densitometer quantitation. The dried transparent film can be stored for long periods. Cellulose acetate prepared to reduce electroendosmosis is available commercially. Cellulose acetate is also used in isoelectric focusing.

Agarose Gel

Agarose gel is another widely used supporting medium. It is used as a purified fraction of agar, is neutral, and therefore does not produce electroendosmosis. After electrophoresis and staining, it is detained (cleared), dried, and scanned with a densitometer. The dried gel can be stored indefinitely. Agarose-gel electrophoresis requires small amounts of sample (approximately 2 mL), it does not bind protein, and therefore, migration is not affected.

Polyacrylamide Gel

Polyacrylamide-gel electrophoresis involves separation of protein on the basis of charge and molecular size. Layers of gel with different pore sizes are used. The gel is prepared before electrophoresis in a tubular-shaped electrophoresis cell. At the bottom is the small-pore separation gel, followed by a large-pore spacer gel, and finally another large-pore gel containing the sample. Each layer of gel is allowed to form a gelatin before the next gel is poured over it. At the start of electrophoresis, the protein molecules move freely through the spacer gel to its boundary with the separation gel, which slows their movement. This allows for concentration of the sample before separation by the small-pore gel. Polyacrylamide-gel electrophoresis separates serum proteins into 20 or more fractions rather than the usual 5 fractions separated by cellulose acetate or agarose. It is widely used to study individual proteins (*eg,* isoenzymes).

Starch Gel

Starch-gel electrophoresis separates proteins on the basis of surface charge and molecular size, as does polyacrylamide gel. The procedure is not widely used because of technical difficulty in preparing the gel.

Treatment and Application of Sample

Serum contains a high concentration of protein, especially albumin, and, therefore, it is routine to dilute serum specimens with the buffer before electrophoresis. In contrast, urine and cerebrospinal fluid (CSF) are usually concentrated. Hemoglobin hemolysate is used without further concentration. Generally, preparation of a sample is done according to the suggestion of the manufacturer of the electrophoretic supplies.

Cellulose acetate and agarose-gel electrophoresis require approximately 2 mL to 5 mL of sample. These are the most commonly performed routine electrophoreses in clinical laboratories. Overloading of agarose gel with sample is not a frequent problem because most commercially manufactured plates come with a thin plastic template that has small slots through which samples are applied. After serum is allowed to diffuse into the gel for approximately 5 minutes, the template is blotted to remove excess serum before being removed from the gel surface. Sample is applied to cellulose acetate with a twin-wire applicator that is designed to transfer a small amount.

Detection and Quantitation

Separated protein fractions are stained to reveal their locations. Different stains come with different plates from different manufacturers. The simplest way to accomplish detection is visualization under UV light, whereas densitometry is the most common and reliable way for quantitation. Most densitometers will automatically integrate the area under a peak, and the result is printed as percentage of the total. A schematic illustration of a densitometer is shown in Figure 4-24.

Electroendosmosis

The movement of buffer ions and solvent relative to the fixed support is called *endosmosis* or *electroendosmosis*. Support media such as paper, cellulose acetate, and agar gel take on a negative charge from adsorption of hydroxyl ions. When current is applied to the electrophoresis system, the hydroxyl ions remain fixed, while the free positive ions move toward the cathode. The ions are highly hydrated, resulting in net cathodic movement of solvent. Molecules that are nearly neutral are swept toward the cathode with the solvent. Support media such as agarose and acrylamide gel are essentially neutral, thus eliminating electroendosmosis. The position of proteins in any electrophoresis separation depends on the nature of the protein, but also on all other technical variables.

Isoelectric Focusing

Isoelectric focusing is a modification of electrophoresis. An apparatus similar to that in Figure 4-23 is used. Charged proteins migrate through a support medium that has a continuous pH gradient. Individual proteins move in the electric field until they reach a pH equal to their isoelectric point, at which point they have no charge and cease to move.

Capillary Electrophoresis

In capillary electrophoresis (CE), separation is performed in narrow-bore fused silica capillaries (inner diameter 2575 mm). Usually, the capillaries are filled only with buffer, although gel media can also be used. A schematic of CE instrumentation is shown in Figure 4-25. Initially, the capillary is filled with buffer, then the sample is loaded and applying an electric field performs the separation. Detection can be made near the other end of the capillary directly through the capillary wall.[13]

A fundamental concept in CE is the *electro-osmotic flow* (EOF). EOF is the bulk flow of liquid toward the cathode upon application of electric field and it is superimposed on electrophoretic migration. EOF controls the amount of time solutes remain in the capillary. Cations migrate fastest because both EOF and electrophoretic attraction are toward the cathode, neutral molecules are all carried by the EOF but

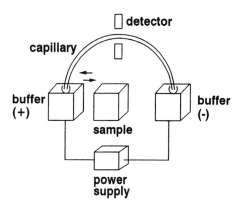

Figure 4-25. Schematic of capillary electrophoresis instrumentation. Sample is loaded onto the capillary by replacing the anode buffer reservoir with the sample reservoir. (From Heiger DN. High performance capillary electrophoresis. Waldbronn, Germany: Hewlett-Packard, 1992.)

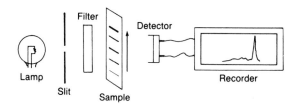

Figure 4-24. Densitometer—basic components.

are not separated from each other, and anions move slowest because, although they are carried to the cathode by the EOF, they are attracted to the anode and repelled by the cathode (Fig. 4-26).

UV-visible detection is widely used for monitoring separated analytes and is performed directly on the capillary, but sensitivity is poor because of the small dimensions of the capillary, resulting in a short path-length. Fluorescence, laser-induced fluorescence, and chemiluminescence detection can be used for higher sensitivity.

CE has been used for the separation, quantitation, and determination of molecular weights of proteins and peptides; for the analysis of polymerase chain reaction (PCR) products; and for the analysis of inorganic ions, organic acids, pharmaceuticals, optical isomers, and drugs of abuse in serum and urine.[14]

CHROMATOGRAPHY

Chromatography refers to a group of techniques used to separate complex mixtures on the basis of different physical interactions between the individual compounds and the stationary phase of the system. The basic components in any chromatographic technique are the mobile phase (gas or liquid), which carries the complex mixture (sample); the stationary phase (solid or liquid), through which the mobile phase flows; the column holding the stationary phase; and the separated components (eluate).

Modes of Separation

Adsorption

Adsorption chromatography, also known as *liquid-solid chromatography,* is based on the competition between the sample and the mobile phase for adsorptive sites on the solid stationary phase. There is an equilibrium of solute molecules being adsorbed to the solid surface and desorbed and dissolved in the mobile phase. Those molecules, which are most soluble in the mobile phase, move fastest, whereas those, which are least soluble, move slowest. Thus a mixture is typically separated into classes according to polar functional groups. The stationary phase can be either acidic polar (*eg,* silica gel), basic polar (*eg,*

alumina), or nonpolar (*eg,* charcoal). The mobile phase can be either a single solvent or a mixture of two or more solvents, depending on the analytes to be desorbed. Liquid-solid chromatography is not widely used in clinical laboratories because of technical problems with the preparation of a stationary phase that has homogeneous distribution of absorption sites.

Partition

Partition chromatography is also referred to as *liquid-liquid chromatography*. Separation of solute is based on relative solubility in an organic (nonpolar) solvent and an aqueous (polar) solvent. In its simplest form, partition (extraction) is performed in a separatory funnel. Molecules containing polar and nonpolar groups in an aqueous solution are added to an immiscible organic solvent. After vigorous shaking, the two phases are allowed to separate. Molecules that are polar remain in the aqueous solvent, whereas nonpolar molecules are extracted in the organic solvent. This results in the partitioning of the solute molecules into two separate phases.

The ratio of the concentration of the solute in the two liquids is known as the *partition coefficient:*

$$K = \frac{\text{solute in stationary phase}}{\text{solute in mobile phase}} \qquad (Eq.\ 4-12)$$

Modern partition chromatography uses pseudo-liquid stationary phases that are chemically bonded to the support or high-molecular-weight polymers that are insoluble in the mobile phase.[15] Partition systems are called *normal phase* when the mobile solvent is less polar than the stationary solvent and are termed *reverse phase* when the mobile solvent is more polar.

Partition chromatography is applicable to any substance that may be distributed between two liquid phases. Because ionic compounds are generally soluble only in water, partition chromatography works best with nonionic compounds.

Steric Exclusion

Steric exclusion, a variation of liquid-solid chromatography, is used to separate solute molecules on the basis of size and shape. The chromatographic column is packed with porous material, as shown in Figure 4-27. A sample containing different-sized molecules moves down the column dissolved in the mobile solvent. Small molecules enter the pores in the

Figure 4-26. Differential solute migration superimposed on electroosmotic flow in capillary zone electrophoresis. (From Heiger DN. High performance capillary electrophoresis. France: Hewlett-Packard, 1992.)

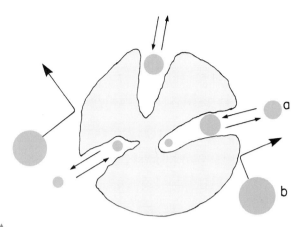

Figure 4-27. Pictorial concept of steric exclusion chromatography. Separation of sample components by their ability to permeate pore structure of column-packing material. Smaller molecules (**a**) permeating the interstitial pores; large excluded molecules (**b**). (From Parris NA. Instrumental liquid chromatography: A practical manual on high performance liquid chromatographic methods. New York: Elsevier, 1976.)

packing and are momentarily trapped. Large molecules are excluded from the small pores and so move quickly between the particles. Intermediate-sized molecules are partially restricted from entering the pores and therefore move through the column at a rate that is intermediate between those of the large and small molecules.

Early methods used hydrophilic beads of cross-linked dextran, polyacrylamide, or agarose, which formed a gel when soaked in water. This method was termed *gel filtration.* A similar separation process using hydrophobic gel beads of polystyrene with a nonaqueous mobile phase was called *gel permeation chromatography.* Current porous packing uses rigid inorganic materials such as silica or glass. The term *steric exclusion* includes all these variations. The size of the pores is controlled by the manufacturer, and packing materials can be purchased with different pore sizes, depending on the size of the molecules being separated.

Ion-Exchange Chromatography

In ion-exchange chromatography, solute mixtures are separated by virtue of the magnitude and charge of ionic species. The stationary phase is a resin, consisting of large polymers of substituted benzene, silicates, or cellulose derivatives, with charge functional groups. The resin is very insoluble in water, and the functional groups are immobilized as side chains on resin beads that are used to fill the chromatographic column. Figure 4-28*A* shows resin with sulfonate functional groups. H^+ ions are held loosely and are free to react. This is an example of a cation-exchange resin. When a cation such as Na^+ comes into contact with these functional groups, an equilibrium is formed, following the law of mass action. Because there are many sulfonate groups, Na^+ ions are effectively and completely removed from solution. The Na^+ ions that are concentrated on the resin column can be eluted from the resin by pouring acid through the column, driving the equilibrium to the left.

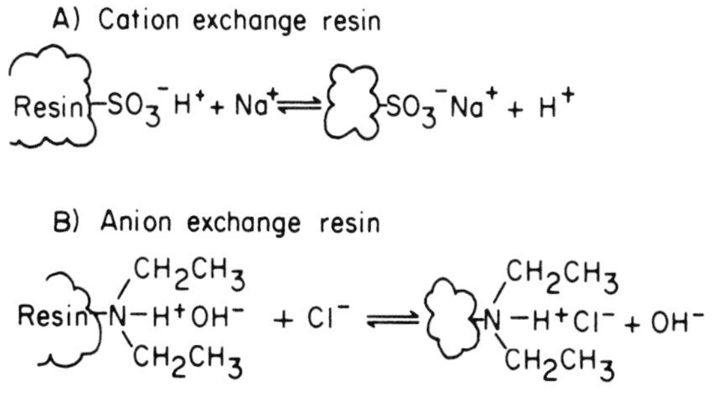

A) Cation exchange resin

$$Resin{-}SO_3^-H^+ + Na^+ \rightleftharpoons SO_3^-Na^+ + H^+$$

B) Anion exchange resin

$$Resin{-}N{-}H^+OH^- + Cl^- \rightleftharpoons N{-}H^+Cl^- + OH^-$$

Figure 4-28. Chemical equilibrium of ion-exchange resins. (**A**) Cation exchange resin. (**B**) Anion exchange resin.

Anion-exchange resins are made with exchangeable hydroxyl ions such as the diethylamine functional group illustrated in Figure 4-28*B*. They are used like cation-exchange resins, except that hydroxyl ions are exchanged for anions. The example shows Cl^- ions in sample solution exchanged for OH^- ions from the resin functional group. Anion and cation resins mixed together (mixed-bed resin) are used to deionize water. The displaced protons and hydroxyl ions combine to form water. Other ionic functional groups besides the illustrated examples are used for specific analytical applications. Ion-exchange chromatography is used to remove interfering substances from a solution, to concentrate dilute ion solutions, and to separate mixtures of charged molecules, such as amino acids. Changing pH and ionic concentration of the mobile phase allows separation of mixtures of organic and inorganic ions.

Chromatographic Procedures

Thin-Layer Chromatography (TLC)

TLC is a variant of column chromatography. A thin layer of *sorbent,* such as alumina, silica gel, cellulose, or cross-linked dextran, is uniformly coated on a glass or plastic plate. Each sample to be analyzed is applied as a spot near one edge of the plate, as shown in Figure 4-29. The mobile phase (solvent) is usually placed in a closed container until the atmosphere is saturated with solvent vapor. One edge of the plate

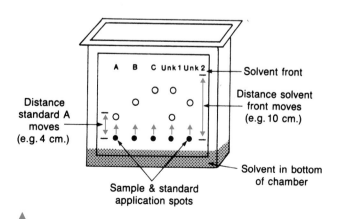

Figure 4-29. TLC plate in chromatographic chamber.

is placed in the solvent, as shown. The solvent migrates up the thin layer by capillary action, dissolving and carrying sample molecules. Separation can be achieved by any of the four processes previously described, depending on the sorbent (thin layer) and solvent chosen. After the solvent reaches a predetermined height, the plate is removed and dried. Sample components are identified by comparison with standards on the same plate. The distance a component migrates, compared with the distance the solvent front moves, is called the *retention factor* R_f:

$$R_f = \frac{\text{distance leading edge of component moves}}{\text{total distance solvent front moves}}$$

(Eq. 4–13)

Each sample-component R_f is compared with the R_f of standards. Using Figure 4-29 as an example, standard *A* has an R_f value of 0.4, standard *B* has an R_f value of 0.6, and standard *C* is 0.8. The first unknown contains *A* and *C,* because the R_f values are the same. This ratio is valid only for separations run under identical conditions. Because R_f values may overlap for some components, further identifying information is obtained by spraying different stains on the dried plate and comparing colors of the standards.

TLC is most commonly used as a semiquantitative screening test. Technique refinement has resulted in the development of semiautomated equipment and the ability to quantitate separated compounds. For example, sample applicators apply precise amounts of sample extracts in concise areas.

Use of plates prepared with uniform sorbent thickness, finer particles, and new solvent systems has resulted in the technique of high-performance thin-layer chromatography (HPTLC).[16] Absorbance of each developed spot is measured using a densitometer, and the concentration is calculated by comparison with a reference standard chromatographed under identical conditions.

High-Performance Liquid Chromatography (HPLC)

Modern liquid chromatography uses pressure for fast separations, controlled temperature, in-line detectors, and gradient elution techniques.[17,18] Figure 4-30 illustrates its basic components.

Pump

A pump forces the mobile phase through the column at a much greater velocity than is accomplished by gravity-flow columns. The pump can be pneumatic, syringe-type, reciprocating, or hydraulic amplifier. Pneumatic pumps are used for preoperative purposes, whereas hydraulic amplifier pumps are no longer commonly used. The most widely used pump today is the mechanical reciprocating pump, which is now used as a multihead pump with two or more reciprocating pistons. During pumping, the pistons operate out of phase (180° for two heads, 120° for three) to provide constant flow.

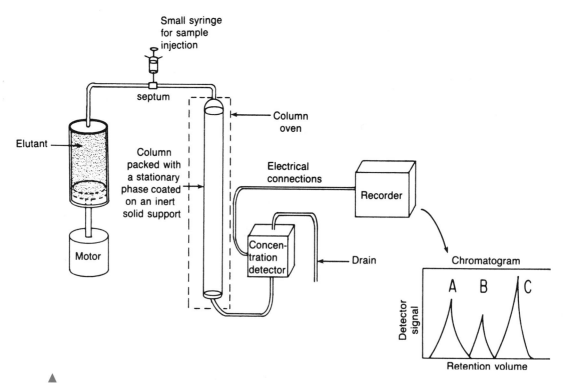

Figure 4-30. HPLC basic components. (From Bender GT. Chemical instrumentation: A laboratory manual based on clinical chemistry. Philadelphia: WB Saunders, 1972.)

Column

The stationary phase is packed into long stainless steel columns. Usually, HPLC is run at ambient temperatures, although columns can be put in an oven and heated to enhance the rate of partition. Fine, uniform column packing results in much less band broadening but requires pressure to force the mobile phase through. The packing also can be either pellicular (an inert core with a porous layer), inert and small particles, or macroporous particles. The most common material used for column packing is silica gel. It is very stable and can be used in different ways. It can be used as a solid packing in liquid-solid chromatography or coated with a solvent, which serves as the stationary phase (liquid-liquid). Owing to the short lifetime of coated particles, molecules of the mobile-phase liquid are now bonded to the surface of silica particles.

Reversed-phase HPLC is now very popular, and the stationary phase is nonpolar molecules (eg, octadecyl C-18 hydrocarbon) bonded to silica-gel particles. For this type of column packing, the mobile phase commonly used is acetonitrile, methanol, water, or any combination of solvents. A reversed-phase column can be used to separate ionic, non-ionic, and ionizable samples. A buffer is used to produce the desired ionic characteristics and pH for separation of the analyte. Column packings vary in size (3 mm to 20 mm), with the smaller particles used mostly for analytical separations and the larger ones for preparative separations.

Sample Injector

As shown in Figure 4-30, a small syringe can be used to introduce the sample into the path of the mobile phase that carries it into the column. The best and most widely used method, however, is the loop injector. The sample is introduced into a fixed-volume loop. When the loop is switched, the sample is placed into the path of the flowing mobile phase and is flushed onto the column.

Loop injectors have high reproducibility and are used at high pressures. Many HPLC instruments now have loop injectors that can be programmed for automatic injection of samples. When the sample size is less than the volume of the loop, the syringe containing the sample is often filled with the mobile phase to the volume of the loop before filling the loop. This prevents the possibility of air being forced through the column, because such a practice may reduce the lifetime of the column packing material.

Detectors

Modern HPLC detectors monitor the eluate as it leaves the column and, ideally, produce an electronic signal proportional to the concentration of each separated component. Spectrophotometers that detect absorbances of visible or ultraviolet light are most commonly used. PDA and other rapid scanning detectors are also used for spectral comparisons and compound identification and purity. These detectors have been used for drug analyses in urine. Obtaining an ultraviolet scan of a compound as it elutes from a column can provide important information as to its identity. Unknowns can be compared against library spectra in a similar manner as mass spectrometry. Unlike gas chromatography/mass spectrometry, which requires volatilization of targeted compounds, however, LC-PDA (liquid chromatography-photodiode array) enables direct injection of aqueous urine samples.

Fluorescence detectors are also used, because many biologic substances fluoresce strongly. The principles involved are the same as discussed in the section on spectrophotometric measurements. Another common HPLC detector is the amperometric or electrochemical detector. These devices measure current produced when the analyzer of interest is either oxidized or reduced at some fixed potential set between a pair of electrodes.

A mass spectrometer (MS) can also be used as a detector not only for the identification and quantitation of compounds, but also for structural information and molecular weight determination.[19] In an MS, the sample is first volatilized and then ionized to form charged molecular ions and fragments that are separated according to their mass-to-charge (m/z) ratio; the sample is then measured by a detector, which gives the intensity of the ion current for each species. Identification of molecules is based on the formation of characteristic fragments. Coupling a liquid chromatograph to a mass spectrometer is difficult because of the large amount of solvent in the eluate. Electrospray (ES) is a technique that allows ions to be transferred from solution to the gas phase.[20] The sample is passed through a metal capillary tip and converted, under the influence of a high electric field (10^6 V/m), to a fine mist of positively charged droplets, from which the solvent rapidly evaporates. The solute ions that remain are then transferred to a mass spectrometer to be analyzed.

Recorder

The recorder is used to record detector signal versus the time the mobile phase passed through the instrument starting from the time of sample injection. The graph is called a *chromatogram* (Fig. 4-31). The retention time is used to identify compounds when compared with standard retention times run under identical conditions. Peak area is proportional to concentration of the compounds that produced the peaks.

When the elution strength of the mobile phase is constant throughout the separation, it is called *isocratic elution*. For samples containing compounds of widely differing relative compositions, the choice of solvent is a compromise. Early eluting compounds may have retention times close to zero, thereby producing a poor separation (resolution), as shown in Figure 4-31A. Basic compounds often have low retention times because C-18 columns cannot tolerate high pH mobile phases. The addition of cation-pairing reagents to the mobile phase, for example, octane sulfonic acid, can result in better retention of negatively charged compounds onto the column.

The late-eluting compounds may have long retention times, producing broad bands resulting in decreased sensitivity. In some cases, certain components of a sample may have such a great affinity for the stationary phase that they

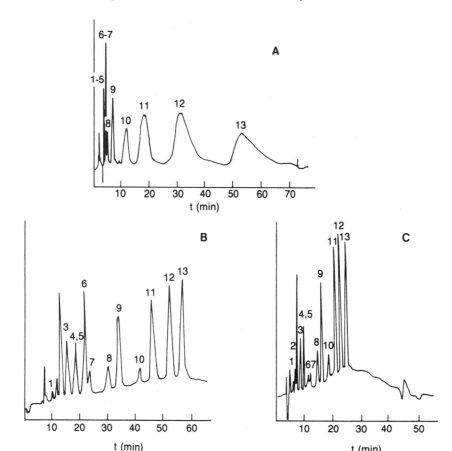

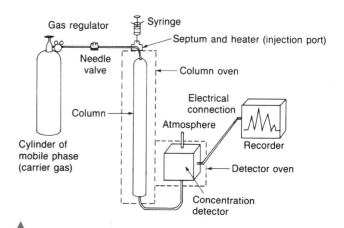

Figure 4-31. Chromatograms: (A) Isocratic ion-exchange separation-mobile phase contains 0.055 M NaNO₃. (B) Gradient elution-mobile phase gradient from 0.01 to 0.1 MNaNO₃ at 2%/min. (C) Gradient elution—5%/min. (From Horváth C. High performance liquid chromatography, advances and perspectives. New York: Academic Press, 1980.)

do not elute at all. Gradient elution is an HPLC technique that can be used to overcome this problem. The composition of the mobile phase is varied to provide a continual increase in the solvent strength of the mobile phase entering the column (Fig. 4–31B). The same gradient elution can be performed with a faster change in concentration of the mobile phase (Fig. 4–31C).

Gas Chromatography

Gas chromatography is used to separate mixtures of compounds that are volatile or can be made volatile.[21] Gas chromatography may be gas-solid chromatography (GSC), with a solid stationary phase, or gas–liquid chromatography (GLC), with a nonvolatile liquid stationary phase. GLC is commonly used in clinical laboratories. Figure 4-32 illustrates the basic components of a gas chromatographic system. The setup is almost the same as for HPLC, except that the mobile phase is a gas and samples are partitioned between a gaseous mobile phase and a liquid stationary phase. The carrier gas can be either nitrogen, helium, or argon. The selection of a carrier gas is determined by the detector used in the instrument. The instrument can be operated at a constant temperature or programmed to run at different temperatures if a sample has components with different volatilities. This is analogous to gradient elution described for HPLC.

The sample, which is injected through a septum, must be injected as a gas, or the temperature of the injection port must be above the boiling point of the components so that they vaporize upon injection. Sample vapor is swept through the column partially as a gas and partially dissolved in the liquid phase. Volatile compounds that are present mainly in the gas phase will have a low partition coefficient and will move quickly through the column. Compounds with higher boil-

Figure 4-32. GLC basic components. (From Bender GT. Chemical instrumentation: A laboratory manual based on clinical chemistry. Philadelphia: WB Saunders, 1972.)

ing points will move slowly through the column. The effluent passes through a detector that produces an electric signal proportional to the concentration of the volatile components. As in HPLC, the chromatogram is used both to identify the compounds by the retention time and to determine their concentration by the area under the peak.

Columns

GLC columns are available in a variety of coil configurations and sizes. They are generally made of glass or stainless steel. Packed columns are filled with inert particles such as diatomaceous earth or porous polymer or glass beads coated with a nonvolatile liquid (stationary) phase. These columns usually have a diameter of ⅛ to ¼ inch and a length of 3 ft to 12 feet. Capillary-wall coated open tubular columns have inside diameters in the range of 0.25 mm to 0.50 mm and are up to 60 m long. The liquid layer is coated on the walls of the column. A solid support coated with a liquid stationary phase may in turn be coated on column walls.

The liquid stationary phase must be nonvolatile at the temperatures used, must be thermally stable, and must not react chemically with the solutes to be separated. The stationary phase is termed *nonselective* when separation is primarily based on relative volatility of the compounds. Selective liquid phases are used to separate polar compounds based on relative polarity (as in liquid-liquid chromatography).

Detectors

There are many types of detectors, but only thermal conductivity (TC) and flame ionization detectors are explained here because they are the most stable (Fig. 4-33). TC detectors contain wires (filaments) that change electrical resistance with change in temperature. The filaments form opposite arms of a Wheatstone bridge and are heated electrically to raise their temperature. Helium, which has a high thermal conductivity, is usually the carrier gas. Carrier gas from the reference column flows steadily across one filament, cooling it slightly. Carrier gas and separated compounds from the sample column flow across the other filament. The sample components usually have a lower thermal conductivity so that the temperature and resistance of the sample filament increase. The change in resistance results in an unbalanced bridge circuit. The electrical change is amplified and fed to the recorder. The electrical change is proportional to the concentration of the analyte.

Flame ionization detectors are widely used in the clinical laboratory. They are more sensitive than TC detectors. The column effluent is fed into a small hydrogen flame burning in excess air or atmospheric oxygen. The flame jet and a collector electrode around the flame have opposite potentials. As the sample burns, ions form and move to the charged collector. Thus, a current proportional to the concentration of the ions is formed and fed to the recorder.

Definitive determinations of samples eluting from gas chromatographic columns are possible when a mass spectrometer is used as a detector.[20] In an electron-impact mass spectrometer, samples are bombarded by electrons to form charged molecular ions and fragments that are filtered in terms of their mass-

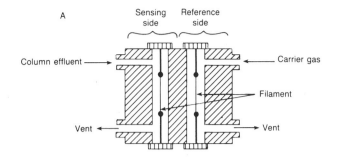

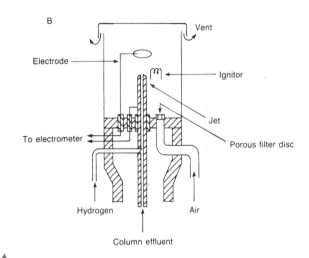

Figure 4-33. (**A**) Schematic diagram of a thermal conductivity detector. (**B**) Schematic diagram of a flame ionization detector. (From Tietz NW, ed. Fundamentals of clinical chemistry. Philadelphia: WB Saunders, 1987.)

to-charge (*m/z*) ratio and measured by an electron multiplier. The characteristic fragmentation patterns produced by these ions are used for their identification. GC/MS systems are widely used for measuring unknown drugs in urine toxicology confirmations. Figure 4-34 illustrates the mass spectrum of carboxytetrahydrocannabinol, a metabolite of marijuana. Tandem mass spectrometers (GC/MS/MS), obtained by the addition of a second mass spectrometer to a GC/MS system, can be used for greater selectivity and lower detection limits. The first mass spectrometer allows only ions of a specific *m/z* ratio to pass to the second mass spectrometer, where they are further fragmented and analyzed (see Fig. 4-34). Other gas chromatography detectors used for selective applications include electron capture and helium-ionization flame emission detectors.

Scintillation Counting

Radioimmunoassays were widely used to measure trace concentrations of hormones and drugs. Detection of radioactive signals required the use of scintillation counters. With the proliferation of non-isotopic immunoassays, the need to measure gamma rays has been greatly diminished. For a discussion of scintillation counting techniques, the reader is referred the any of numerous reference books.

GC / MS / MS

Tandem-in-Space

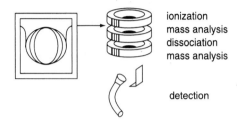

ionization   mass analysis   dissociation   mass analysis   detection

Tandem-in-Time

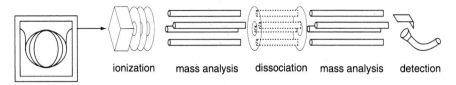

ionization
mass analysis
dissociation
mass analysis

detection

Figure 4-34. Triple quadruple mass spectrometer.

OSMOMETRY

An osmometer is used to measure the concentration of solute particles in a solution. The mathematical definition is

$$\text{Osmolality} = \phi \times n \times C, \qquad \textit{(Eq. 4–14)}$$

where ϕ = osmotic coefficient

    n = number of dissociable particles (ions) per molecule in the solution

    C = concentration in moles per kilogram of solvent

The osmotic coefficient is an experimentally derived factor to correct for the fact that some of the molecules, even in a highly dissociated compound, exist as molecules rather than as ions.

The four physical properties of a solution that change with variations in the number of dissolved particles in the solvent are osmotic pressure, vapor pressure, boiling point, and freezing point. Osmometers measure osmolality indirectly by measuring one of these colligative properties, which change proportionally with osmotic pressure. Osmometers in clinical use measure either freezing-point depression or vapor-pressure depression, and results are expressed in milliosmolal per kilogram (mOsm/kg) units.

Freezing-Point Osmometer

Figure 4-35 illustrates the basic components of a freezing-point osmometer. The sample in a small tube is lowered into a chamber with cold refrigerant circulating from a cooling unit. A thermistor is immersed in the sample. To measure temperature, a wire is used to gently stir the sample until it is cooled to several degrees below its freezing point. It is possible to cool water to as low as −40°C and still have liquid water, provided no crystals or particulate matter are present. This is referred to as a *supercooled solution.* Vigorous agitation when the sample is supercooled results in rapid freezing. Freezing also can be started by "seeding" a supercooled solution with crystals. When the supercooled solution starts to freeze because of the rapid stirring, a slush is formed, and the solution actually warms to its freezing-point temperature. The slush, an equilibrium of liquid and ice crystals, will stay at the freezing-point temperature until the sample freezes solid and drops below its freezing point.

Impurities in a solvent will lower the temperature at which freezing or melting occurs by reducing the bonding forces between solvent molecules so that the molecules break away from each other and exist as a fluid at a lower temperature. The decrease in the freezing-point temperature is proportional to the number of dissolved particles present.

The thermistor is a material that has less resistance when temperature increases. The readout uses a Wheatstone bridge

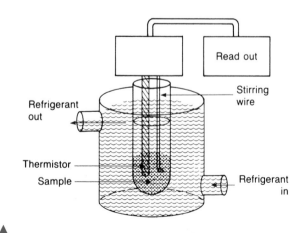

Figure 4-35. Freezing-point osmometer. (From Coiner D. Basic concepts in laboratory instrumentation. ASMT Education and Research Fund, Inc., 1975–1979.)

circuit that detects temperature change as proportional to change in thermistor resistance. Freezing-point depression is proportional to the number of solute particles. Standards of known concentration are used to calibrate the instruments in mOsm/kg.

Vapor-Pressure Osmometer

A vapor pressure osmometer indirectly measures osmolality by measuring the dew point temperature depression. In this method, the physical state and temperature of the sample are not changed by freezing-point depression. Advantages of vapor pressure osmometers over freezing-point depression osmometers are the small sample size requirement (6–8 μL) and the use of paper disks as sample holders. Most osmolality results in the clinical laboratory are determined by the freezing-point depression method.

ANALYTICAL TECHNIQUES FOR POCT

Point-of-care testing devices are now widely used for a variety of clinical applications, including physician offices, emergency departments, intensive care units, and even for self-testing. Because analyses can be done at patient side by primary care givers, the major attraction to POCT is the reduced turnaround times needed to deliver results. In some cases, total costs can be reduced if the devices can eliminate the need for laboratory-based instrumentation, or if increased turnaround times lead to shorter hospital lengths of stay. POCT relies on the same analytical techniques as laboratory-based instrumentation: spectrometry, electroanalytical techniques, and chromatography. As such, the same steps needed to perform an analysis from the central laboratory are needed for POCT, including instrument validation, periodic assay calibration, quality control testing, operator training, proficiency testing, and so on. Chapter 7, *Point-of-Care Testing,* provides an in-depth discussion of this technology. The analytical techniques used in these devices is given in this section.

The most commonly used POCT devices used at bedside, in physician offices, and at home are the fingerstick blood glucose monitors. The first generation devices use a photometric approach, whereby glucose produces hydrogen peroxide with glucose oxidase immobilized onto test strips. The H_2O_2 is coupled to peroxidase to produce a color whose intensity is measured as a function of concentration and measured using reflectance photometry. A schematic of this technique is shown in Figure 4-36. These strips are measured for glucose concentration without the need to wipe the blood off the strips themselves.

The strip technology in a POCT platform can also be used to measure proteins and enzymes such as cardiac markers. The separation of analytes from the matrix is accomplished by paper chromatography, whereby specific antibodies im-

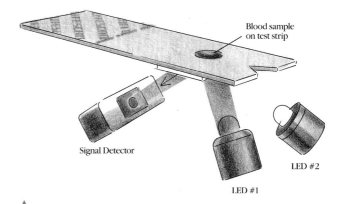

Figure 4-36. Dual-wavelength reflectance photometry used in a point-of-care glucose monitors. A microporous hydrophilic membrane is used as a reservoir of the sample, to filter out solid cellular material from the reservoir, and provide a smooth optical surface for reflectance measurements. (Figure courtesy of Lifescan Inc., from the chapter, "One Touch System Technology," *Challenges in Diabetes Management: Clinical Protocols for Professional Practice,* Health Education Technologies, New York, 1988.)

mobilized onto the chromatographic surface capture the target analyte as it passes through. For qualitative analysis, the detection is made by visual means. Reflectance meters similar to those used for glucose are also available for quantitative measurements.

The next generation of POCT devices make use of biosensors.[22] A biosensor couples a specific biodetector such as an enzyme, antibody, or nucleic acid probe to a transducer for the direct measurement of a target analyte without the need to separate it from the matrix (Fig. 4-37). The field has exploded in recent years with the development of micro silicon chip fabrication, because biosensors can be miniaturized and made available at low costs. An array of biosensors can be produced onto a single silicon wafer to produce a multipanel of results, such as an electrolyte profile. Commercial POCT devices use electrochemical (such as micro *ion selective electrodes*) and optical biosensors for the measurement of glucose, electrolytes, and arterial blood gases. With the immobilization of antibodies and specific DNA sequences, biosensor probes will soon be available for detection of hormones, drugs and drugs of abuse, and hard-to-culture bacteria and viruses such as *Chlamydia,* tuberculosis, or human immunodeficiency virus.[22]

SUMMARY

The techniques and general principles used in a clinical chemistry laboratory are identical to those used in other analytical testing laboratories. Clinical laboratories do have special needs that require analyzers to have high throughput and sample turnaround times. The current generation of chemistry analyzers operates under the "random access" mode, that is, any combination of tests can be performed on

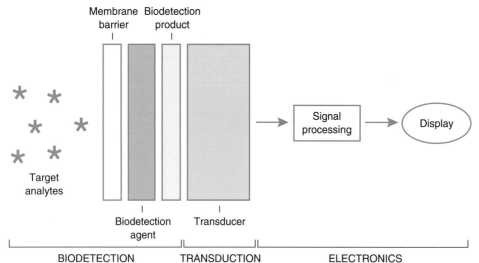

Figure 4-37. Schematic diagram of a biosensor. (From Rosen A. Biosensors: where do we go from here. MLO 1995;27(3):24).

a sample from an onboard menu of analytes. For general chemistry analytes, such as glucose or phosphorus, spectrophotometry is the most widely used technique. For electrolytes such as sodium and potassium, ion-specific electrodes are extensively used and have largely replaced flame photometers. Atomic absorption spectrophotometry is still used for metals such as zinc and copper, and remains the reference method for calcium and magnesium. However, colorimetric assays and ISEs are now used routinely for these latter two metals.

Electrophoresis is still an important technique in clinical laboratories, although the development of new assays have challenged its role. For example, lipoprotein electrophoresis largely has been replaced by direct measurement of high-density lipoprotein (HDL) cholesterol and calculation of low-density lipoprotein (LDL) cholesterol. The development of a specific assay for LDL cholesterol will further diminish the role of electrophoresis. With the development of specific immunoassays for creatine kinase isoenzyme form MB (CK-MB) and inhibition assays for lactate dehydrogenase form 1 (LD1), electrophoresis is not as popular today as before; however, an automated electrophoresis assay has recently been released for CK-MB isoforms. Electrophoresis will play a critical role in the area of molecular pathology for the identification of gene products and mutations. Capillary electrophoresis is an emerging technology that promises to have many clinical applications.

With the development of immunoassays, the role of liquid chromatography has shifted away from therapeutic drug monitoring toward toxicology. HPLCs with rapid scanning UV detectors are being used for comprehensive drug screens. GC/MS is the mainstay for confirmation analysis. The interface of LC to MS through electrospray atomization opens many new applications to the clinical laboratory, because this technique works well with aqueous materials and high molecular weight substances. However, HPLC/MS instrumentation is very expensive and will likely not be used in routine clinical laboratories in the near future.

EXERCISES

1. Which of the following is NOT necessary for obtaining the spectrum of a compound from 190–500 nm?
    a. Deuterium light source
    b. Double beam spectrophotometer
    c. Quartz cuvets
    d. Tungsten light source
    e. Photomultiplier
2. Stray light in a spectrophotometer places limits on:
    a. Sensitivity
    b. Upper range of linearity
    c. Photometric accuracy below 0.1 absorbance units
    d. Ability to measure in the UV range
    e. Use of a grating monochromator
3. Which of the following light sources is used in atomic absorption spectrophotometry?
    a. Hollow cathode lamp
    b. Xenon arc lamp
    c. Tungsten light
    d. Deuterium lamp
    e. Laser
4. Which of the following is true concerning fluorometry?
    a. Emission wavelengths are always set at lower wavelengths than excitation
    b. The detector is always placed at right angles to the excitation beam
    c. All compounds undergo fluorescence
    d. Fluorescence is an inherently more sensitive technique than absorption
    e. Fluorometers require special detectors
5. Which of the following techniques has the highest potential sensitivity?
    a. Chemiluminescence
    b. Fluorescence
    c. Turbidimetry
    d. Nephelometry
    e. Phosphorescence

6. Which electrochemical assay measures current at fixed potential?
   a. Anodic stripping voltametry
   b. Amperometry
   c. Coulometry
   d. Analysis with ion selective electrodes
   e. Electrophoresis
7. Which of the following refers to the movement of buffer ions and solvent relative to the fixed support?
   a. Isoelectric focusing
   b. Iontophoresis
   c. Zone electrophoresis
   d. Electroendosmosis
   e. Plasmapheresis
8. "Reverse-phase" liquid chromatography refers to:
   a. Polar mobile phase and nonpolar stationary phase
   b. Nonpolar mobile phase and polar stationary phase
   c. Distribution between two liquid phases
   d. Size is used to separate solutes instead of charge
   e. Charge is used to separate solutes instead of size
9. Which of the following is NOT an advantage of capillary electrophoresis?
   a. Very small sample size
   b. Rapid analysis
   c. Use of traditional detectors
   d. Multiple samples can be assayed simultaneously on one injection
   e. Cations, neutrals, and anions move in the same direction at different rates
10. Tandem mass spectrometers:
    a. Are two mass spectrometers placed in series with each other
    b. Are two mass spectrometers placed in parallel with each other
    c. Requires use of a gas chromatograph
    d. Requires use of an electrospray interface
    e. Does not require an ionization source
11. Which of the following is FALSE concerning the principles of POCT devices?
    a. They use principles that are identical to laboratory-based instrumentation.
    b. Biosensors have enabled miniaturization particularly amendable for POCT.
    c. Devices do not require quality control testing.
    d. Onboard microcomputers control instrument functions and data reduction.
    e. Whole blood analysis is the preferred specimen.

REFERENCES

1. Christian GD, et al. Instrumental analysis. 2nd ed. Boston: Allyn and Bacon, 1986.
2. Willard HH, et al. Instrumental methods of analysis. Belmont, CA: Wadsworth, 1981.
3. Coiner D. Basic concepts in laboratory instrumentation. ASMT Education and Research Fund, Inc, 1975–1979.
4. Ingle JD, Crouch SR. Spectrochemical analysis. NJ: Prentice-Hall, Upper Saddle River, 1988.
5. Holland JF, et al. Mass spectrometry on the chromatographic time scale: realistic expectations. Anal Chem 1983;55:997A.
6. Guilbault GG. Practical fluorescence, theory, methods and techniques. New York: Marcel Dekker, 1973.
7. Kricka LJ. Chemiluminescent and bioluminescent techniques. Clin Chem 1991;37:1472.
8. Wild D. The immunoassay handbook. London: Macmillan Press Ltd, 1994.
9. Svelto O, et al. Principles of lasers. 2nd ed. New York: Plenum Press, 1982.
10. Coulter hematology analyzer: multidimensional leukocyte differential analysis, Vol. 11 (1), 1989.
11. Weast RC. CRC handbook of chemistry and physics. 61st ed. 1980–1981. Cleveland, OH: CRC Press, 1981.
12. Burtis CA. Tietz textbook of clinical chemistry. 2nd ed. Philadelphia: WB Saunders, 1993.
13. Heiger, DN. High performance capillary electrophoresis: an introduction. 2nd ed. Waldbronn, Germany: Hewlett-Packard Co, 1992.
14. Monnig CA, Kennedy RT. Capillary electrophoresis. Anal Chem 1994;66:280R.
15. Parris NA. Instrumental liquid chromatography: a practical manual on high performance liquid chromatographic methods. New York: Elsevier, 1976.
16. Jurk H. Thin-layer chromatography. Reagents and detection methods. Vol. 1a, Weinheim, Germany: Verlagsgesellschaft, 1990.
17. Bender GT. Chemical instrumentation: a laboratory manual based on clinical chemistry. Philadelphia: WB Saunders, 1972.
18. Horváth C. High performance liquid chromatography, advances and perspectives. New York: Academic Press, 1980.
19. Constantin E, et al. A. Mass spectrometry. New York: Ellis Horwood, 1990.
20. Kebarle E, Liang T. From ions in solution to ions in the gas phase. Anal Chem 1993;65:972A.
21. Karasek FW, Clement RE. Basic gas chromatography: mass spectrometry. New York: Elsevier, 1988.
22. Rosen A. Biosensors: where do we go from here. MLO 1995;27 (3):24.

CHAPTER 5

Immunoassays and Nucleic Acid Probe Techniques

Catherine Sheehan

Objectives

Upon completion of this chapter, the clinical laboratorian should be able to:

- *State the principle of each of the following methods:*
  - *Double diffusion*
  - *Radial immunodiffusion*
  - *Immunoelectrophoresis*
  - *Immunofixation electrophoresis*
  - *Nephelometry*
  - *Turbidimetry*
  - *Competitive immunoassay*
  - *Noncompetitive immunoassay*
  - *Immunoblot*
  - *Direct immunocytochemistry*
  - *Indirect immunocytochemistry*
  - *Immunophenotyping by flow cytometry*
  - *Polymerase chain reaction*
  - *Southern blot*
  - *In situ hybridization*
  - *Restriction fragment length polymorphism*
- *Compare and contrast the general types of labels used in immunoassays.*
- *Given the outline of an immunoassay, classify it as homogeneous or heterogeneous, competitive or noncompetitive, and by its label.*
- *For competitive and noncompetitive immunoassays, explain how the concentration of the analyte in the test sample is related to the amount of bound labeled reagent.*
- *Describe the three methods used to separate free labeled reagent from bound labeled reagent.*
- *Describe the data reduction in the classic competitive radioimmunoassay.*
- *Compare and contrast EMIT™, DELFIA, MEIA, RIA, FPIA, ELISA, CEDIA, ICON®, and OIA® methodologies.*

- *Explain the principles of hybridization.*
- *State the role of reverse transcriptase, polymerase, and restriction endonuclease in nucleic acid probe assays.*

KEY TERMS

Affinity	Hybridization	Polyclonal
Amplicon	Immunoblot	Polymerase chain
Anneal	Immunocyto-	reaction (PCR)
Antibody	chemistry	Postzone
Antigen	Immunoelectro-	Prozone
Avidity	phoresis (IEP)	Radial immuno-
Competitive	Immunofixation	diffusion (RID)
immunoassay	electrophoresis	Restriction
Counterimmuno-	(IFE)	fragment length
electrophoresis	Immunohisto-	polymorphism
Cross reactivity	chemistry	Reverse
Direct	Immunophenotyping	transcriptase
immunofluores-	Indirect immuno-	polymerase chain
cence (DIF)	fluorescence (IIF)	reaction (RT-
DNA index (DI)	In situ hybridization	PCR)
Duplex	Ligase chain	Rocket technique
Epitope	reaction (LCR)	Self-sustained
Flow cytometry	Monoclonal	sequence
Hapten (Hp)	Nephelometry	replication (3SR)
Heterogeneous	Noncompetitive	Solid phase
immunoassay	immunoassay	Southern blot
Homogeneous	Northern blot	Tracer
immunoassay	Nucleic acid probe	Turbidimetry
		Western blot

This chapter introduces two generic analytic methods used in the clinical laboratory: one involves the binding of antibody to antigen and the other relies on the binding of a nucleic acid sequence to its target nucleic acid sequence. Common to both methods are the complementary nature of the reactants, which determines the specificity, and the detector system, which indicates the extent of the binding reaction and is related to the analytic sensitivity. In immunoassays, an antigen binds to an antibody. The antigen–antibody interactions may involve unlabeled reactants in less analytically sensitive techniques or a labeled reactant in more sensitive techniques. The design, label, and detection system combine to create a large number of different assays, which enable the measurement of a wide range of molecules.

In nucleic acid binding, typically a probe will bind to a complementary nucleic acid sequence. Thus, nucleic acid based methods are designed to detect changes at the DNA or RNA level rather than to detect a synthesized gene product, such as a protein detected in immunoassays. Again there are many assay designs that use probes. Immunoassays are considered separately from probe assays to more clearly explain each. This chapter reviews the concepts of binding, describes the nature of the reagents used, and discusses basic assay design of selected techniques used in the clinical laboratory; as such, it is intended to be an overview rather than an exhaustive review.

IMMUNOASSAYS

General Considerations

In an immunoassay, an *antibody* molecule recognizes and binds to an *antigen*. This binding is related to the concentration of each reactant, the specificity of the antibody for the antigen, the affinity and avidity for the pair, and the environmental conditions. Although this chapter focuses on immunoassays that use an antibody molecule as the binding reagent, other assays, such as receptor assays and competitive binding protein assays, use receptor proteins or transport proteins as the binding reagent, respectively. The same principles apply to these assays. An antibody molecule is an immunoglobulin with a functional domain known as the F(ab); this area of the immunoglobulin protein binds to a site on the antigen. An antigen is relatively large and complex and usually has multiple sites that can bind to antibodies with different specificities; each site on the antigen is referred to as an antigenic determinant or *epitope*. Some confusion exists in the terminology used: some immunologists refer to an immunogen as the molecule, which induces the biologic response and synthesis of antibody, and use antigen to refer to that which binds to antibody. However, all agree that the antigenic site to which an F(ab) can bind is the epitope.

The degree of binding is an important consideration in an immunoassay. The binding of an antibody to an antigen is directly related to the affinity and avidity of the antibody for the epitope as well as the concentration of the antibody and epitope. Under standard conditions, the *affinity* of an antibody is measured using a *hapten (Hp)* because the hapten is a low molecular weight antigen considered to have only one epitope. The affinity for the hapten is related to the likelihood to bind or to the degree of complementary nature of each. The reversible reaction is summarized in Equation 5-1:

$$\text{Hapten} + \text{Antibody} \rightleftharpoons \text{Hapten-Antibody Complex}$$

(Eq. 5–1)

The binding between a hapten and the antibody obeys the law of mass action and is expressed mathematically in Equation 5-2:

$$K_a = \frac{k_1}{k_2} = \frac{[\text{Hp} - \text{Ab}]}{[\text{Hp}][\text{Ab}]} \qquad \textit{(Eq. 5–2)}$$

K_a is the affinity or equilibrium constant and represents the reciprocal of the concentration of free hapten when 50% of the binding sites are occupied. The greater the affinity of the hapten for the antibody, the smaller the concentration of hapten needed to saturate 50% of the binding sites of the antibody. For example, if the affinity constant of a monoclonal antibody is 3×10^{11} L/mole, it means that a hapten concentration of 3×10^{-11} moles/L is needed to occupy half of the binding sites. Typically, the affinity constant of antibodies used in immunoassay procedures ranges from 10^9 to 10^{11} L/mole, whereas the affinity constant for transport proteins ranges from 10^7 to 10^8 L/mole and the affinity for receptors ranges from 10^8 to 10^{11} L/mole.

As with all chemical (molecular) reactions, the initial concentrations of the reactants and the products affect the extent of complex binding. In immunoassays, the reaction moves forward (to the right), as written in Equation 5-1, when the concentration of reactants (Ag and Ab) exceeds the concentration of the product (Ag-Ab complex) and when there is a favorable affinity constant.

The forces that unite an antigenic determinant and an antibody are noncovalent, reversible bonds that result from the cumulative effects of hydrophobic, hydrophilic, and hydrogen bonding and van der Waals forces. The most important factor that affects the cumulative strength of bonding is the goodness (or closeness) of fit between the antibody and the antigen. The strength of most of these interactive forces is inversely related to the distance between the interactive sites. The closer the antibody and antigen can physically approach one another, the greater the attractive forces.

After the antigen–antibody complex is formed, the likelihood of separation (which is inversely related to the tightness of bonding) is referred to as *avidity*. The avidity represents a value-added phenomenon in which the strength of binding of all antibody-epitope pairs exceeds the sum of single antibody-epitope binding. In general, the stronger the affinity and avidity, the greater the possibility of cross reactivity.

The specificity of an antibody is most often described by the antigen that induced the antibody production, the homologous antigen. Ideally, this antibody would react only with that antigen. However, an antibody can react with an antigen that is structurally similar to the homologous antigen; this is referred to as *cross reactivity*. Considering that an antigenic determinant can be five or six amino acids or one immunodominant sugar, it is not surprising that antigen similarity is common. The greater the similarity between the cross-reacting antigen and the homologous antigen, the stronger the bond with the antibody.[1] Reagent antibody production is achieved by polyclonal or monoclonal techniques. In *polyclonal* antibody production, the stimulating antigen is injected into an animal responsive to the antigen; the animal detects this foreign antigen and mounts an immune response to eliminate the antigen. If part of this immune response includes strong antibody production, then blood is collected and antibody is harvested, characterized, and purified to yield the commercial antiserum reagent. This polyclonal antibody reagent is a mixture of antibody specificities. Some antibodies react with the stimulating epitopes and some are endogenous to the host. Multiple antibodies directed against the multiple epitopes on the antigen are present and can crosslink the multivalent antigen.

In contrast, an immortal cell line produces *monoclonal* antibodies; each line produces one specific antibody. This method developed as an extension of the hybridoma work published by Kohler and Milstein in 1975.[2] The process begins by selecting cells with the qualities that will allow the synthesis of a homogeneous antibody. First, a host (commonly a mouse) is immunized with an antigen (the one to which an antibody is desired); later the sensitized lymphocytes of the spleen are harvested. Second, an immortal cell line (usually a nonsecretory mouse myeloma cell line that is hypoxanthine guanine phosphoribosyltransferase deficient) is required to ensure that continuous propagation in vitro is viable. These cells are then mixed in the presence of a fusion agent (such as polyethylene glycol) that promotes the fusion of two cells to form a hybridoma. In a selective growth medium, only the hybrid cells will survive. B cells have a limited natural life span in vitro and cannot survive, and the unfused myeloma cells cannot survive due to their enzyme deficiency. If the viable fused cells synthesize antibody, then the specificity and isotype of the antibody are evaluated. Monoclonal antibody reagent is commercially produced by growing the hybridoma in tissue culture or in compatible animals. An important feature about monoclonal antibody reagent is that the antibody is homogeneous (a single antibody not a mixture of antibodies). Therefore it recognizes only one epitope on a multivalent antigen and cannot crosslink a multivalent antigen.

Unlabeled Immunoassays

Immune Precipitation in Gel

In one of the simplest unlabeled immunoassays introduced into the clinical laboratory, unlabeled antibody was layered on top of unlabeled antigen (both in the fluid phase); during the incubation period, the antibody and antigen diffused and the presence of precipitation was recorded. The precipitation occurred because each antibody recognized an epitope and that multivalent antigens were crosslinked by multiple antibodies. When the antigen–antibody complex is of sufficient size, the interaction with water is limited so that the complex becomes insoluble and precipitates.

It has been observed that if the concentration of antigen is increased while the concentration of antibody remains constant, the amount of precipitate formed is related to the ratio of antibody to antigen. As shown in Figure 5-1, there is an optimal ratio of the concentration of antibody to the concentration of antigen that results in the maximal precipitation; this is the zone of equivalence. Outside the zone of equivalence, the amount of precipitate is diminished or absent because the ratio of antibody to antigen is out of proportion and the crosslinking of antigen is decreased. When the antibody concentration is in excess and the crosslinking is decreased, the assay is in *prozone*. Conversely, when the antigen concentration is in excess and the crosslinking is decreased, the assay is in *postzone*. Although originally described with precipitation reactions, this concept applies to other assays in which the ratio of antibody to antigen is critical.

Precipitation reactions in gel are commonly performed in the clinical laboratory today. Gel is dilute agarose (typically less than 1%) dissolved in an aqueous buffer and provides a semisolid medium through which soluble antigen and antibody can easily pass. Immune precipitation methods in gel can be classified as passive methods or those using electrophoresis and are summarized in Table 5-1. The simplest and least sensitive method is double diffusion (or the Ouchterlony technique).[3] Agarose is placed on a solid surface and allowed to solidify. Wells are cut into the agarose. A common template is six antibody wells, which surround a single antigen well in the center. Soluble antigen and soluble antibody are added to separate wells and diffusion occurs. The intensity and pattern of the precipitation band are interpreted. As shown in Figure 5-2, the precipitin band of an unknown sample is compared with the precipitin band of a sample known to contain the antibody. A pattern of identity confirms the presence of the antibody in the unknown sample.

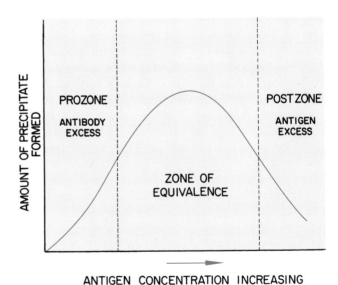

Figure 5-1. Precipitin curve showing the amount of precipitate versus antigen concentration. The concentration of antibody is constant.

TABLE 5-1. Immune Precipitation Methods

Gel
  Passive
    Double diffusion (Ouchterlony technique)
    Single diffusion (radial immunodiffusion)
  Electrophoresis
    Counterimmunoelectrophoresis
    Immunoelectrophoresis
    Immunofixation electrophoresis
    Rocket electrophoresis

Soluble phase
  Turbidimetry
  Nephelometry

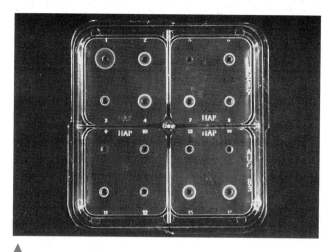

Figure 5-3. A radial immunodiffusion plate to detect haptoglobin. The diameter of the precipitin circle is related to the concentration of haptoglobin in the serum.

Patterns of partial identity and nonidentity are ambiguous. This technique is used to detect antibodies associated with autoimmune diseases such as the Sm antibody detected in systemic lupus erythematosus and the Scl-70 antibody in progressive systemic sclerosis.

The single-diffusion technique, *radial immunodiffusion (RID),* is an immune precipitation method used to quantitate protein (the antigen). In this method, monospecific antiserum is added to the liquefied agarose; then the agarose is poured into a plate and cooled. Wells are cut into the solidified agarose. Multiple standards, one or more quality control samples, and patient samples are added to the wells. The antigen diffuses from the well in all directions, binds to the soluble antibody in the agarose, and forms a complex seen as a concentric precipitin ring shown in Figure 5-3.

The diameter of the ring is related to the concentration of the antigen that diffused from the well. A standard curve is constructed to determine the concentration in the quality control and patient samples. The usable analytic range is between the lowest and highest standards. If the ring is greater than the highest standard, the sample should be diluted and retested. If the ring is less than the lowest standard, the sample should be run on a low-level plate. Two variations exist: the endpoint (or Mancini) method[4] and the kinetic (Fahey-McKelvey) method.[5] The endpoint method requires that all antigen diffuse from the well and the concentration of the antigen is related to the square of the diameter of the precipitin ring; the standard curve is plotted on linear graph paper and is the line of best fit. To ensure all antigen has diffused, the incubation time is 48 to 72 hours depending on the molecular weight of the antigen; for instance, the quantitation of IgG requires 48 hours and IgM requires 72 hours. In contrast, the kinetic method requires that all rings be measured at a fixed time of 18 hours; a sample with a greater concentration will diffuse at a faster rate and will be larger at a fixed time. Using semilogarithmic graph paper, the concentration of the antigen is plotted against the diameter of the precipitin ring; the line is drawn point to point. For those performing RID, the endpoint method is favored because of its stability and indifference to temperature variations.

Counterimmunoelectrophoresis is an immune precipitation method that uses an electrical field to cause the antigen and antibody to migrate toward each other. Two parallel line of wells are cut into agarose; antibody is placed in one line and antigen is placed in the other. Antibody will migrate to the cathode and the antigen to the anode; a precipitin line forms where they meet. This qualitative test is useful to detect bacterial antigens in cerebrospinal fluid and other fluids when a rapid laboratory response is needed.

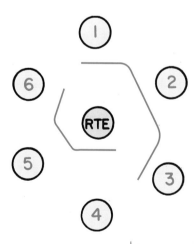

Figure 5-2. A schematic demonstrating the pattern of identity. The center well contains the antigen, rabbit thymus extract. Well 1 is filled with a serum known to contain Sm antibody. Test sera are in wells 2 and 3; the pattern of identity, the smooth continuous line between the 3 wells, confirms the presence of Sm antibody in the test sera. Well 4 is filled with a serum known to contain U1-RNP antibody. Test sera in wells 5 and 6 also contain U1-RNP antibody confirmed by the pattern of identity between the known serum and the test sera.

In the clinical laboratory, *immunoelectrophoresis (IEP)* and *immunofixation electrophoresis (IFE)* are two methods used to characterize monoclonal proteins in serum and urine. In 1964, Grabar and Burtin published methods for examining serum proteins using electrophoresis coupled with immunochemical reactions in agarose.[6] Serum proteins are electrophoretically separated and then reagent antibody is placed in a trough running parallel to the separated proteins. The antibody reagent and separated serum proteins diffuse; when the reagent antibody recognizes the serum protein and the reaction is in the zone of equivalence, a precipitin arc is seen as shown in Figure 5-4. The agarose plate is stained (typically with protein stain such as Amido black 10), destained, and dried to enhance the readability of precipitin arcs, especially weak arcs. The size, shape, density and location of the arcs aid in the interpretation of the protein. All interpretation is made by comparing the arcs of the patient sample with the arcs of the quality control sample, a normal human serum. Because IEP is used to evaluated a monoclonal protein, the heavy chain class and light chain type must be determined. To evaluate the most common monoclonal proteins, the following antisera are used: anti-humans whole serum (which contains a mixture of antibodies against the major serum proteins), anti-human IgG (gamma chain specific), anti-human IgM (mu chain specific), anti-human IgA (alpha chain specific), anti-human lambda (lambda chain specific), and anti-human kappa (kappa chain specific). The test turnaround time and the subtlety in interpretation have discouraged the use of IEP as the primary method to evaluate monoclonal antibodies.

IFE[7] has replaced IEP in many laboratories. A serum, urine, or cerebrospinal fluid sample is placed in all six lanes of an agarose gel and is electrophoresed to separate the proteins. Cellulose acetate (or some other porous material) is saturated with antibody reagent and then applied to one lane of the separated protein. If the antibody reagent recognizes the protein, then an insoluble complex is formed. After staining and drying the agarose film, interpretation is based on the migration and appearance of bands. As shown in Figure 5-5, the monoclonal protein will appear as a discrete band (with both a heavy chain monospecific antiserum and a light chain monospecific antiserum occurring at the same position). Polyclonal proteins will appear as a diffuse band. The concentration of patient sample may need to be adjusted to assure the reaction is in the zone of equivalence.

The last immune precipitation method in gel is the *rocket technique* (or Laurell technique or electroimmunoassay).[8,9] In this quantitative technique, reagent antibody is mixed with agarose; antigen is placed in the well and electrophoresed. As the antigen moves through the agarose, it reacts with the reagent antibody and forms a "rocket" with stronger precipitation along the edges. The height of the rocket is proportional to the concentration of antigen present; the concentration is determined based on a calibration curve. The narrow range of linearity may require the dilution or concentration of the unknown sample.

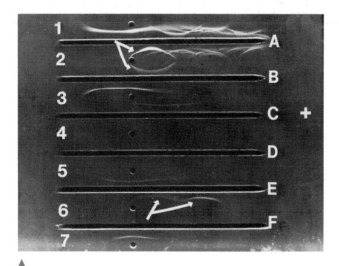

▲

Figure 5-4. Immunoelectrophoresis. Wells 1, 3, 5 and 7 contain normal human serum and Wells 2, 4, and 6 contain the test serum. Antiserum reagent is in the troughs:

A contains anti-human whole serum
B contains anti-human IgG
C contains anti-human IgA
D contains anti-human IgM
E contains anti-human kappa
F contains anti-human lambda

The arrows at the top point to an abnormal gamma chain which reacts with the anti-IgG reagent. A similar band is shown at the bottom with the anti-kappa reagent. There is also a pattern of identity with the anti-kappa reagent that shows free kappa chains in the test serum.

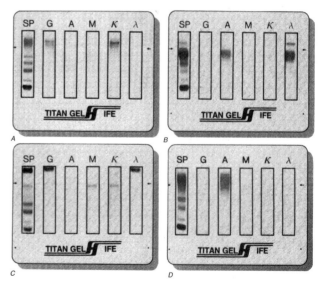

▲

Figure 5-5. Immunofixation electrophoresis. **A.** IgG kappa monoclonal immunoglobulin. **B.** IgA kappa monoclonal immunoglobulin with free kappa light chains. **C.** IgG lambda and IgM kappa biclonal immunoglobulins. **D.** Diffuse IgA heavy chain band without a corresponding light chain.

Detection of Fluid-Phase Antigen–Antibody Complexes

A different strategy to quantitate antigen–antibody complexes is to use an instrument to detect the soluble antigen–antibody complexes as they interact with light. When antigen and antibody combine, they form complexes that act as particles in suspension and thus can scatter light. The size of the particles determines the type of scatter that will dominate when the solution interacts with nearly monochromatic light.[10] When the particle (such as albumin or IgG) is relatively small compared to the wavelength of incident light, the particle will scatter light symmetrically, in both forward and backward directions. A minimum of scattered light is detectable at 90° from the incident light. Larger molecules and antigen–antibody complexes have diameters that approach the wavelength of incident light and scatter light with a greater intensity in the forward direction. The wavelength of light used is selected based on its ability to be scattered in a forward direction and the ability of the antigen–antibody complexes to absorb the wavelength of light.

Turbidimetry measures the light transmitted and *nephelometry* measures the light scattered. Turbidimeters are spectrophotometers or colorimeters and are designed to measure the light passing through a solution so the photodetector is placed at an angle of 180° from the incident light. If light absorbance is insignificant, turbidity can be expressed as the absorbance, which is directly related to the concentration of suspended particles and the pathlength. Nephelometers measure light at an angle other than 180° from the incident light; most measure forward light scattered at less than 90° because the sensitivity is increased (see Chapter 4, *Analytical Techniques and Instrumentation*). The relative concentration of the antibody reagent and antigen is critical to ensure that the size of the complex generated is best detected by the nephelometer or turbidimeter and that the immune reaction is not in postzone or prozone. Thus it may be important to test more than one concentration of patient sample, to monitor the presence of excess antibody, or to add additional antiserum and monitor the peak rate. Excess antibody would indicate that there is little antigen and that the reaction is underestimated.

For both turbidimetry and nephelometry, all reagents and sera must be free of particles that could scatter the light. Pretreatment of serum with polyethylene glycol, a nonionic, hydrophilic polymer, enhances the antigen–antibody interaction. The polymer is more hydrophilic than the antigen or antibody and thus attracts water from the antigen and antibody to the polyethylene glycol. This results in a faster rate and greater quantity of antigen–antibody complex formation.

Both methods can be performed in an endpoint or kinetic mode. In the endpoint mode, a measurement is taken at the beginning of the reaction (the background signal) and one is taken at a set time later in the reaction (plateau or endpoint signal); the concentration is determined using a calibration curve. In the kinetic mode, the rate of complex formation is continuously monitored and the peak rate is determined.

The peak rate is directly related to the concentration of the antigen, although this is not necessarily linear. Thus, a calibration curve is required to determine the concentration in the unknown samples.

Labeled Immunoassays

General Considerations

In all labeled immunoassays, a reagent (either the antigen or antibody) is labeled usually by attaching a particle, or molecule that will better detect lower concentrations of antigen–antibody complexes. Thus the label improves the analytic sensitivity. All assays have a binding reagent, which can bind to the antigen or ligand. If the binding reagent is an antibody, the assay is an immunoassay. If it is a receptor (*eg,* estrogen or progesterone receptors), the assay is a receptor assay. If the binding reagent is a transport protein (*eg,* thyroid binding globulin or transcortin), the assay may be called competitive protein binding assay. Today immunoassays are used almost exclusively with two notable exceptions: estrogen and progesterone receptor assays and the thyroid hormone binding ratio, which uses thyroid binding globulin.

Immunoassays may be described based on the label, which reactant is labeled, the relative concentration and source of the antibody, the method used to separate free from bound labeled reagent, the signal that is measured, and the method used to assign the concentration of analyte in the sample. Thus the design of immunoassays has many variables to consider and has led to diverse assays.

Labels

The simplest way to identify an assay is by the label used. Table 5-2 lists the labels commonly used and the methods used to detect the label.

Radioactive Labels. Atoms with unstable nuclei that spontaneously emit radiation are radioactive and referred to as radionuclides. The emission is known as radioactive decay and is independent of chemical or physical parameters, such as temperature, pressure, or concentration. Of the three forms of radiation, only beta and gamma are used in the clinical laboratory. In beta emission, the nucleus can emit negatively charged electrons or positively charged particles called positrons. The emitted electrons are also known as beta particles. Tritium (^{3}H) is the radionuclide commonly used in cellular immunology assays for diagnosis and research.

Gamma emission is electromagnetic radiation with very short wavelengths originating from unstable nuclei. As a radionuclide releases its energy and becomes more stable, it disintegrates or decays, releasing energy. A specific spectrum of energy levels is associated with each radionuclide. The standardized unit of radioactivity is the becquerel (Bq), which is equal to one disintegration per second. The traditional unit is the curie (Ci), which equals 3.7×10^{10} Bq; 1 μCi = 37 kBq. The half-life of the radionuclide is the time needed for 50% of the radionuclide to decay and to become more sta-

TABLE 5-2. Labels and Detection Methods

Immunoassay	Common Labels	Detection Methods
RIA	^{3}H	Liquid scintillation counter
	^{125}I	Gamma counter
EIA	Horseradish peroxidase	Photometer, fluorometer, luminometer
	Alkaline phosphatase	Photometer, fluorometer, luminometer
	β-D galactosidase	Fluorometer, luminometer
	Glucose-6-phosphate dehydrogenase	Photometer, luminometer
CLA	Isoluminol derivative	Luminometer
	Acridinium esters	Luminometer
FIA	Fluorescein	Fluorometer
	Europium	Fluorometer
	Phycobiliproteins	Fluorometer
	Rhodamine B	Fluorometer
	Umbelliferone	Fluorometer

RIA, radioimmunoassay; EIA, enzyme immunoassay; CLA, chemiluminescent assay; FIA, fluorescent immunoassay.

ble. The longer the half-life, the more slowly it decays, thereby increasing the length of time it can be measured. For those radioactive substances used in diagnostic tests, it is preferable that the emission have an appropriate energy level and that the long half-life be relatively long; ^{125}I satisfies these requirements and is the most commonly used gamma-emitting radionuclide in the clinical laboratory.

Gamma-emitting nuclides are detected using a crystal scintillation detector (also known as a gamma counter). The energy released during decay excites a fluor, such as thallium-activated sodium iodide. The excited fluor releases a photon of visible light, which is amplified and detected by a photo-multiplier tube; the amplified light energy is then translated into electrical energy. Detectable decay of the radionuclide is expressed as counts per minute (CPM).

In immunoassays, one reactant is radiolabeled. In competitive assays, the antigen is labeled and is called the *tracer*. The radiolabel must allow the tracer to be fully functional and to compete equally with the unlabeled antigen for the binding sites. When the detector antibody is radiolabeled, the antigen-combining site must remain biologically active and unhindered.

Enzyme Labels. Enzymes are commonly used to label the antigen/hapten or antibody.[11,12] Horseradish peroxidase (HRP), alkaline phosphatase (ALP), and glucose-6-phosphate dehydrogenase are used most often. Enzymes are biologic catalysts, which increase the rate of conversion of substrate to product and are not consumed by the reaction. As such, an enzyme can catalyze many substrate molecules, thus amplifying the amount of product generated. The enzyme activity may be monitored directly by measuring the product formed or by measuring the effect of the product on a coupled reaction. Depending on the substrate used, the product can be photometric, fluorometric, or chemiluminescent. For example, a typical photometric reaction using

HRP-labeled antibody (Ab-HRP) and the substrate (a peroxide) generate the product (oxygen). The oxygen can then oxidize a reduced chromogen (reduced orthophenylenediamine [OPD]) to produce a colored compound (oxidized OPD), which is measured using a photometer.

$$Ab - HRP + peroxide \rightarrow Ab = HRP + O_2$$

$$O_2 + reduced\ OPD \rightarrow oxidized\ OPD + H_2O \quad \textit{(Eq. 5–3)}$$

Fluorescent Labels. Fluorescent labels (fluorochromes or fluorophores) are compounds that absorb radiant energy of one wavelength and emit radiant energy of a longer wavelength in less than 10^{-4} seconds. Generally, the emitted light is detected at a 90° angle from the path of excitation light using a fluorometer or a modified spectrophotometer. The difference between the excitation wavelength and emission wavelength is Stokes shift and usually ranges between 20 nm and 80 nm for most fluorochromes. Some fluorescence immunoassays simply substitute a fluorescent label (such as fluorescein) for an enzyme label and quantitate the fluorescence.[13] Another approach, time-resolved fluorescence immunoassay, uses a highly efficient fluorescent label, such as a europium chelate,[14] which fluoresces approximately 1000 times more slowly than the natural background fluorescence and has a wide Stokes shift. The delay allows the fluorescent label to be detected with minimal interference from background fluorescence. The long Stokes shift facilitates measurement of emission radiation while excluding the excitation radiation. The resulting assay is highly sensitive and time-resolved, with minimized background fluorescence.

Luminescent Labels. Luminescent labels emit a photon of light as the result of a electrical, biochemical or chemical reaction.[15,16] Some organic compounds become excited when oxidized and will emit light as they revert to the ground state. Oxidants include hydrogen peroxide, hypochlorite, or oxygen. Sometimes a catalyst, such as peroxidase, alkaline phosphatase, or metal ions, is needed.

Luminol, the first chemiluminescent label used in immunoassays, is a cyclic diacylhydrazide that emits light energy under alkaline conditions in the presence of peroxide and peroxidase. Because peroxidase can serve as the catalyst, assays may use this enzyme as the label; the chemiluminogenic substrate, luminol, will produce light that is directly proportional to the amount of peroxidase present as shown in Equation 5-4:

$$\text{Luminol} + 2H_2O_2 + OH^- \xrightarrow{\text{Peroxidase}}$$
$$\text{3-aminophthalate} + \text{light (425 nm)} \qquad \textit{(Eq. 5-4)}$$

A popular chemiluminescent label, acridinium esters, is a triple-ringed organic molecule linked by an ester bond to an organic chain. In the presence of hydrogen peroxide and under alkaline conditions, the ester bond is broken and an unstable molecule (N-methylacridon) remains. Light is emitted as the unstable molecule reverts to its more stable ground state.

$$\text{Acridinium ester} + 2H_2O_2 + OH^- \rightarrow$$
$$\text{N-methylacridon} + CO_2 + H_2O + \text{light (430 nm)}$$

$$\textit{(Eq. 5-5)}$$

Alkaline phosphatase commonly conjugated to an antibody has been used in automated immunoassay analyzers to produce some of the most sensitive chemiluminescent assays. ALP catalyzes adamantyl 1,2-dioxetane aryl phosphate substrates (AMPPD) to release light at 477 nm. The detection limit approaches 1 zmol or approximately 602 enzyme molecules.[17,18]

Assay Design

Competitive Immunoassays. The earliest immunoassay was a *competitive immunoassay* in which the radiolabeled antigen (Ag*; also called the tracer) competed with unlabeled antigen (Ag) for a limited number of binding sites (Ab) (Fig. 5-6). The proportion of Ag and Ag* binding

with the Ab is related to the Ag and Ag* concentration and requires limited antibody in the reaction. In the competitive assay, the Ag* concentration is constant and limited. As the concentration of Ag increases, more will bind to the antibody, resulting in less binding of Ag*. These limited reagent assays were very sensitive because low concentrations of unlabeled antigen yielded a large measurable signal from the bound labeled antigen. If the competitive assay is designed to reach equilibrium, often the incubation times are long.

The antigen–antibody reaction can be accomplished in one step when labeled antigen (Ag*), unlabeled antigen (Ag), and reagent antibody (Ab) are simultaneously incubated together to yield bound labeled antigen (Ag*Ab), bound unlabeled antigen (AgAb), and free label (Ag*) as shown in Figure 5-6 and Equation 5-6:

$$\text{Ag* (fixed reagent)} + \text{Ag} + \text{Ab (limited reagent)} \rightarrow$$
$$\text{Ag*Ab} + \text{AgAb} + \text{Ag*} \qquad \textit{(Eq. 5-6)}$$

A generic heterogeneous, competitive simultaneous assay begins by pipetting the test sample (quality control, calibrator, or patient) into test tubes. Next, labeled antigen and antibody reagents are added. After incubation and separation of free (unbound) labeled antigen, the bound labeled antigen is measured.

Alternatively, the competitive assay may be accomplished in sequential steps. First, labeled antigen is incubated with the reagent antibody and then labeled antigen is added. After a longer incubation time and a separation step, the bound labeled antigen is measured. This approach increases the analytic sensitivity of the assay.

Consider the example in Table 5-3. A relatively small, yet constant number of Ab combining sites is available to combine with a relatively large, constant amount of Ag* (tracer) and calibrators with known antigen concentrations. Because the amount of tracer and antibody are constant, the only variable in the test system is the amount of unlabeled antigen. As the concentration of unlabeled antigen increases, the concentration (or percentage) of free tracer increases.

By using multiple calibrators, a dose response curve is established. As the concentration of unlabeled Ag increases, the concentration of tracer that binds to the Ab decreases. In the example presented in Table 5-3, if the amount of unlabeled antigen is zero, maximum tracer will combine with the antibody. When no unlabeled antigen is present, maximum binding by the tracer is possible; this is referred to as B_0, B_{max}, maximum binding, or the zero standard. When the amount of unlabeled antigen is the same as the tracer, each will bind equally to the antibody. As the concentration of antigen increases in a competitive assay, the amount of tracer that complexes with the binding reagent decreases. If the tracer is of low molecular weight, then often free tracer is measured. If the tracer is of high molecular weight, then the bound tracer is measured. The data may be plotted in one of three ways, as shown in Figure 5-7: bound/free versus the

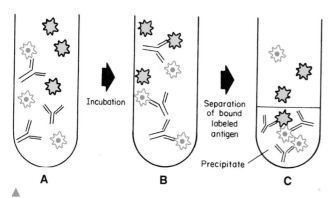

Figure 5-6. Competitive labeled immunoassay. During simultaneous incubation, labeled antigen ( ) and unlabeled antigen ( ) compete for the antibody binding sites ( ). The bound label in the precipitate is frequently measured.

TABLE 5-3. Competitive Binding Assay Example

| Ag | + | Ag* | + | Ab | → | AgAb | + | Ag*Ab | + | Ag* |

Concentration of Reactants			Concentration of Products		
Ag	Ag*	Ab	AgAb	Ag*Ab	Ag*
0	200	100	0	100	100
50	200	100	20	80	120
100	200	100	34	66	134
200	200	100	50	50	150
400	200	100	66	34	166

Sample calculations

Dose of [Ag]	%B	B/F
0	$\frac{100}{200} = 50$	$\frac{100}{100} = 1$
50	$\frac{80}{200} = 40$	$\frac{80}{120} = .67$
100	$\frac{66}{200} = 33$	$\frac{66}{134} = .49$
200	$\frac{50}{200} = 25$	$\frac{50}{150} = .33$
400	$\frac{34}{200} = 17$	$\frac{34}{166} = .20$

arithmetic dose of unlabeled antigen; percentage bound versus the log dose of unlabeled antigen; and logit Bound/B_0 versus the log dose of the unlabeled antigen.

The bound fraction can be expressed in several different formats. Bound/Free (B/F) is CPM of the bound fraction compared with the CPM of the free fraction. Percent bound (%B) is the CPM of the bound fraction compared with the CPM of maximum binding of the tracer (B_0) times 100. Logit B/B_0 transformation is the natural log of $(B/B_0)/(1 - B/B_0)$.

When using logit-log graph paper on which B/B_0 is plotted on the ordinate and the log dose of the unlabeled antigen is plotted on the abscissa, a straight line with a negative slope is produced. More often, microcomputers calculate the best straight line using linear regression; patient values may then be calculated by the computer using this relationship.

It is important to remember that the best type of curve fitting technique is determined by experiment and that there is no assurance that a logit-log plot of the data will always generate a straight line. Several different methods of data plotting should be tried when a new assay is introduced to determine which method is best. Then every time the assay is performed, a dose response curve should be prepared to check the performance of the assay. Remember that the relative error for all radioimmunoassay (RIA) dose response curves is minimal when B/B_0 is 0.5 and increases at both high and low concentrations of the plot. As shown in Figure 5-7 in the plot of B/B_0 versus log of the antigen con-

centration, a relatively large change in the concentration at either end of the curve produces little change in the B/B_0 value. Patient values derived from a B/B_0 value greater than 0.9 or less than 0.1 should be interpreted with caution. When the same data are displayed using the logit-log plot, it is easy to forget the error at either end of the straight line.

Noncompetitive Immunoassays. Sometimes known as immunometric assays, *noncompetitive immunoassays* use a labeled reagent antibody to detect the antigen. Excess labeled antibody is required to assure that the labeled antibody reagent does not limit the reaction. The concentration of the antigen is directly proportional to the bound labeled antibody as shown in Figure 5-8. The relationship is linear up to a limit and then is subject to the high-dose hook effect.

In the sandwich assay to detect antigen (also known as an antigen capture assay), immobilized unlabeled antibody captures the antigen. After washing to remove unreacted molecules, the labeled detector antibody is added. After another washing to remove free labeled detector antibody, the signal from the bound labeled antibody is proportional to the antigen captured. This format relies on the ability of the antibody reagent to react with a single epitope on the antigen. The specificity and quantity of monoclonal antibodies has allowed the rapid expansion of diverse assays. A schematic is shown in Figure 5-9.

Another type of noncompetitive assay is the sandwich assay to detect antibody. Here the immobilized antigen captures specific antibody. After washing, the labeled detector

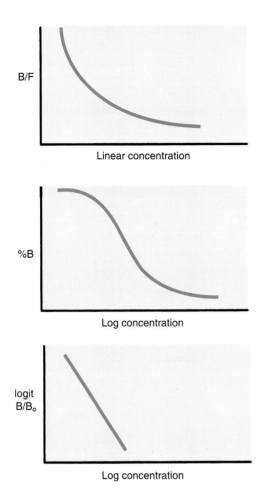

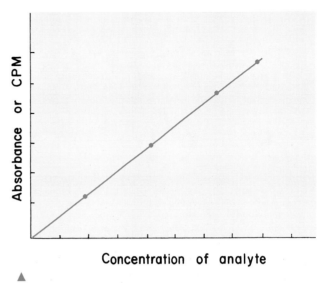

Figure 5-8. Dose response curve in a noncompetitive immunoassay.

Figure 5-7. Dose response curves in a competitive assay. B = bound labeled antigen. F = free labeled antigen. B_o = maximum binding. $\%B = B/B_o \times 100$.

antibody is added and binds to the captured antibody. The amount of bound labeled antibody is directly proportional to the amount of specific antibody present. A schematic is shown in Figure 5-10. This assay can be modified to determine the immunoglobulin class of the specific antibody present in serum. For example, if the detector antibody were labeled and monospecific (such as a rabbit anti-human IgM [μ-chain specific]), it would detect and quantitate only human IgM captured by the immobilized antigen.

Separation Techniques

All immunoassays require that free labeled reactant be distinguished from bound labeled reactant. In *heterogeneous assays*, physical separation is necessary and is achieved by adsorption, precipitation, or interaction with a solid phase as listed in Table 5-4. The better the separation of bound from free reactant, the more reliable the assay will be. This is in contrast to *homogeneous assays* in which the activity or expression of the label depends on whether the labeled reactant is free or bound. No physical separation step is needed in homogeneous assays.

Adsorption. Adsorption techniques use particles to trap the small antigens, labeled or unlabeled. Most commonly, a mixture of charcoal and crosslinked dextran is used. Charcoal is porous and readily combines with small molecules to remove them from solution; dextran prevents nonspecific protein binding to the charcoal. The size of the dextran influences the size of the molecule that can be adsorbed; the lower the molecular weight of dextran used, the smaller the molecular weight of free antigen that can be adsorbed. Other adsorbents include silica, ion exchange resin, and sephadex. After adsorption and centrifugation, the free labeled antigen is found in the precipitate.

Precipitation. Nonimmune precipitation occurs when the environment is altered affecting the solubility of protein. Compounds such as ammonium sulfate, sodium sulfate, polyethylene glycol, and ethanol precipitate protein nonspecifically; both free antibody and antigen–antibody complexes will precipitate. Ammonium sulfate and sodium sulfate "salt out" free globulins and antigen–antibody complexes. Ethanol

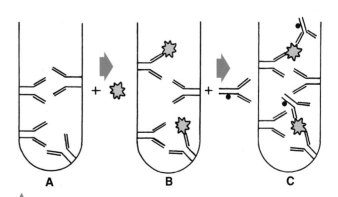

Figure 5-9. Two site noncompetitive sandwich assay to detect antigen. Immobilized antibody (⤙) captures the antigen (✿). Then labeled antibody (⤙) is added, binds to the captured antigen, and is detected.

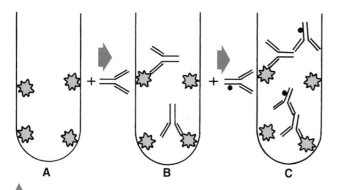

▲ *Figure 5-10.* Two-site noncompetitive sandwich assay to detect antibody. Immobilized antigen (❀) captures the antibody (⤜). Then labeled antibody (⤜) is added, binds to the captured antibody, and is detected.

denatures protein and antigen–antibody complexes, causing precipitation. Polyethylene glycol precipitates larger protein molecules with or without the antigen attached. Ideally, after centrifugation, all bound labeled antigen will be in the precipitate, leaving free labeled antigen in the supernatant.

Soluble antigen–antibody complexes can be precipitated by a second antibody that recognizes the primary antibody in the soluble complex. The result is a larger complex that becomes insoluble and will precipitate. Centrifugation is again used to aid in the separation. This immune precipitation method is also known as the double-antibody or second-antibody method. For example, in a growth hormone assay, the primary or antigen-specific antibody produced in a rabbit recognizes growth hormone. The second antibody, produced in a sheep or goat, would recognize rabbit antibody. Labeled antigen–antibody complexes, unlabeled antigen–antibody complexes, and free primary antibodies are precipitated by the second antibody. This separation method is more specific than nonimmune precipitation because only

the primary antibody is precipitated. A similar separation occurs when staphylococcal protein A (SPA) replaces the second antibody. SPA binds to human IgG causing precipitation.

Solid Phase. The use of a *solid phase* to immobilize reagent antibody or antigen provides a method to separate free from bound labeled reactant after washing. The solid-phase support is an inert surface to which reagent antigen or antibody is attached. The solid-phase support may be, but is not limited to, polystyrene surfaces, membranes, and magnetic beads. The immobilized antigen or antibody may be adsorbed or covalently bound to the solid-phase support; covalent linkage prevents spontaneous release of the immobilized antigen or antibody. Immunoassays using solid-phase separation are easier to perform and to automate and require less manipulation and time to perform than other immunoassays. However, a relatively large amount of reagent antibody or antigen is required to coat the solid-phase surface, and consistent coverage of the solid phase is difficult to achieve. Solid-phase assays are more expensive to produce and require greater technical skill to perform to minimize intra-assay and interassay variability. Insufficient washing is a common source of error.

Examples of Labeled Immunoassays

Particle enhanced turbidimetric inhibition immunoassay (PETINIA) is a homogeneous competitive immunoassay in which low molecular weight haptens bound to particles compete with unlabeled analyte for the specific antibody. The extent of particle agglutination is inversely proportional to the concentration of unlabeled analyte and is assessed by measuring the change in transmitted light in a DuPont (now Dade) Automated Clinical Analyzer (ACA).[19]

Enzyme linked immunosorbent assays (ELISA) is a very popular group of heterogeneous immunoassays that all have an enzyme label and use a solid phase as the separation tech-

TABLE 5-4. Characteristics of Separation Techniques		
Separation Technique	**Examples**	**Action**
Adsorption	Charcoal and dextran	Traps free labeled antigen
	Silica	Separation by centrifugation
	Ion exchange resin	
	Sephadex	
Precipitation		
Nonimmune	Ethanol	Denatures bound labeled antigen
	Ammonium sulfate	Separation by centrifugation
	Sodium sulfate	
	Polyethylene glycol	
Immune	Second antibody	Primary antibody is recognized
	Staph protein A	and forms an insoluble complex
Solid phase	Polystyrene	Separation by centrifugation
	Membranes	One reactant is adsorbed or covalently attached to the inert surface
	Magnetized particles	Separation by washing

nique. Four formats are available: a competitive assay using labeled antigen, a competitive assay using labeled antibody, a noncompetitive assay to detect antigen, and a noncompetitive assay to detect antibody.

One of the earliest homogeneous assays was an enzyme immunoassay named enzyme multiplied immunoassay technique (EMIT™) and is currently produced by Syva Corporation.[20] As shown in Figure 5-11, the reactants in most test systems include an enzyme-labeled antigen (commonly a low molecular weight analytes, such as a drug), an antibody directed against the antigen, the substrate, and test antigen. The enzyme is catalytically active when the labeled antigen is free (not bound to the antibody). It is thought that when the antibody combines with the labeled antigen, the antibody sterically hinders the enzyme. The conformational changes that occur during antigen–antibody interaction inhibit the enzyme activity. In this homogeneous assay, the unlabeled antigen in the sample competes with the labeled antigen for the antibody binding sites; as the concentration of unlabeled antigen increases, less enzyme-labeled antigen can bind to the antibody. Therefore, more labeled antigen is free, and the enzymatic activity is greater.

Cloned enzyme donor immunoassays (CEDIA) are competitive, homogeneous assays in which the genetically engineered label is β-galactosidase.[21] The enzyme is in two inactive pieces: the enzyme acceptor and the enzyme donor. When these two pieces bind together, the enzyme activity is restored. In the assay, the antigen labeled with the enzyme donor and unlabeled antigen in the sample compete for specific antibody binding sites. When the antibody binds to the labeled antigen, the enzyme acceptor cannot bind to the enzyme donor. Thus, the enzyme is not restored and the enzyme is inactive. More unlabeled antigen in the sample results in more enzyme activity.

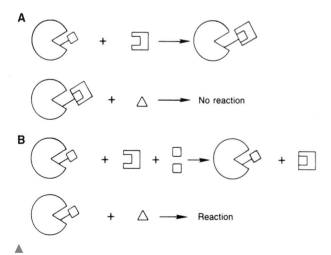

Figure 5-11. Enzyme multiplied immunoassay technique. *(A)* When enzyme labeled antigen is bound to the antibody, the enzyme activity is inhibited. *(B)* Free patient antigen binds to the antibody and prevents antibody binding to the labeled antigen. The substrate indicates the amount of free labeled antigen.

Microparticle capture enzyme immunoassay (MEIA) is an automated assay available on the Abbott IMx. The microparticles serve as the solid phase, and a glass fiber matrix separates the bound labeled reagent. Both competitive and noncompetitive assays are available. Although the label is an enzyme (alkaline phosphatase), the substrate (4-methylumbelliferyl phosphate) is fluorogenic.

Solid phase fluorescence immunoassays (SPFIA) are analogous to the ELISA methods except that the label fluoresces. Of particular note is FIAX® (BioWhittaker, Inc., Walkersville, MD). In this assay, fluid phase unlabeled antigen is captured by antibody on the solid phase; after washing, the detector antibody (with a fluorescent label attached) reacts with the solid-phase captured antigen.

Particle concentration fluorescence immunoassay (PCFIA) is a heterogeneous, competitive immunoassay in which particles are used to localize the reaction and concentrate the fluorescence. Labeled antigen and unlabeled antigen in the sample compete for antibody bound to polystyrene particles. The particles are trapped and the fluorescence is measured. The assay can also be designed so that labeled antibody and unlabeled antibody compete for antigen fixed onto particles.

Fluorescence excitation transfer immunoassay (FETI) is a competitive, homogeneous immunoassay using two fluorophores (such as fluorescein and rhodamine).[22] When the two labels are in close proximity, the emitted light from fluorescein will be absorbed by rhodamine. Thus the emission from fluorescein is quenched. Fluorescein labeled antigen and unlabeled antigen compete for rhodamine labeled antibody. More unlabeled antigen lessens the amount of fluorescein labeled antigen that binds; thus more fluorescence is present (less quenching).

Substrate level fluorescence immunoassay (SLFIA) is another competitive, homogeneous assay. This time the hapten is labeled with a substrate; when catalyzed by an appropriate enzyme, fluorescent product is generated. Substrate-labeled hapten and unlabeled hapten in the sample compete with antibody; the bound labeled hapten cannot be catalyzed by the enzyme.

Fluorescence polarization immunoassay (FPIA) is another assay that uses a fluorescent label.[23,24] This homogeneous immunoassay uses polarized light to excite the fluorescent label. Polarized light consists of parallel light waves oriented in one plane and is created when light passes through special filters. When polarized light is used to excite a fluorescent label, the emitted light could be polarized or depolarized. Small molecules, such as free fluorescent-labeled hapten, rotate rapidly and randomly, thus interrupting the polarized light. Larger molecules, such as those created when the fluorescent labeled hapten binds to an antibody, rotate more slowly and emit polarized light parallel to the excitation polarized light. The polarized light is measured at a 90° angle compared with the path of the excitation light. In a competitive FPIA, fluorescent-labeled hapten and unlabeled hapten in the sample compete for limited antibody sites. When no unlabeled hapten is present, the labeled hapten binds maximally to the antibody, creating large complexes that rotate slowly and emit a high level of polarized light. When hapten is present, it competes with the labeled hapten for the antibody sites; thus, as the hapten concentration increases, more labeled hapten is displaced and is free. The free labeled hapten rotates rapidly and emits less polarized light. The degree of labeled hapten displacement is inversely related to the amount of unlabeled hapten present.

Dissociation-enhanced lanthanide fluoroimmunoassay (DELFIA) is an automated system (Pharmacia) that measures time-delayed fluorescence from the label europium. The assay can be designed as a competitive, heterogeneous assay or a noncompetitive (sandwich), heterogeneous assay.[25]

The classic RIA is a heterogeneous, competitive assay with a tracer.[26] When bound tracer is measured, the signal from the label (counts per minute) is inversely related to the concentration of the unlabeled antigen in the sample.

Rapid Immunoassays

The sensitivity and specificity of automated labeled assays and the trend for decentralized laboratory testing have led to the development of assays that are easy to use, simple (many classified as waived or moderately complex related to the Clinical Laboratory Improvement Amendments of 1988), fast, site-neutral, and require no instrumentation. Those discussed here are representative of currently available commercial kits; however, this discussion is not intended to be exhaustive. Three categories of rapid immunoassays emerge: (1) latex particles for visualization of the reaction, (2) fluid flow and labeled reactant, and (3) changes in a physical or chemical property following antigen–antibody binding.

The earliest rapid tests were those in which a latex particle suspension was added to the sample; if the immunoreactive component attached to the particle recognized its counterpart in the sample, macroscopic agglutination occurred. Colored latex particles are now available to facilitate reading the reaction.

Self-contained devices that use the liquid-nature of the specimen have evolved. In flow-through systems, a capture reagent is immobilized onto a membrane, the solid phase. The porous nature of membranes increases the surface area to which the capture reagent can bind. The more capture reagent that binds to the membrane, the greater the potential assay sensitivity. After the capture reagent binds to the membrane, other binding sites are saturated with a nonreactive blocking chemical to reduce nonspecific binding by substances in the patient sample. In the assay, the sample containing the analyte is allowed to pass through the membrane and the analyte is bound to the capture reagent. Commonly the liquid is attracted through the membrane by an absorbent material. The analyte is detected by a labeled reactant, as well as the signal from the labeled reactant.

The next step in the development of self-contained single use devices was to incorporate internal controls. One scheme, the ImmunoConcentration™ Assay (ICON®, Hybritech, Inc.)[27,28] to detect human chorionic gonado-

tropin, creates three zones in which specifically treated particles are deposited. In the assay zone, particles are coated with reagent antibody specific for the assay; in the negative control zone, particles are coated with nonimmune antibody; and in the positive control zone, particles are coated with an immune complex specific for the assay. The patient sample (serum or urine) passes through the membrane, and the analyte is captured by the specific reagent antibody in the assay zone. Next, a labeled antibody passes through the membrane, which fixes to the specific immune complex formed in the assay zone or the positive control zone. Following color development, a positive reaction is noted when the assay and positive zones are colored.

A second homogeneous immunoassay involves the tangential flow of fluid across a membrane. The fluid dissolves and binds to the dried capture reagent; the complex flows to the detection area, where it is concentrated and viewed.

A third homogeneous immunoassay, enzyme immunochromatography, involves vertical flow of fluid along a membrane.[29] This is quantitative and does not require any instrumentation. A dry paper strip with immobilized antibody is immersed in a solution of unlabeled analyte and an enzyme-labeled analyte; the liquid migrates up the strip by capillary action. As the labeled and unlabeled analyte migrate, they compete and bind to the immobilized antibody. A finite amount of labeled and unlabeled analyte mixture is absorbed. The migration distance of the labeled analyte is visualized when the strip reacts with a substrate reagent and develops a colored reaction product. Comparing the migration distance of the sample with the calibrator allows the concentration of the unlabeled ligand to be assigned.

The next generation of rapid immunoassays involves change in physical or chemical properties after an antigen–antibody interaction occurs. One example is the Optical Immunoassay (OIA®).[30] A silicon wafer is used to support a thin film of optical coating; this is then topped with the capture antibody. Sample is applied directly to the device. If an antigen–antibody complex is formed, the thickness of the optical surface increases and this changes the optical path of light. The color changes from gold to purple. Some studies suggest that this method has better analytic sensitivity compared to immunoassays, which rely on the fluid flow.

Immunoblots

Most assays described thus far are designed to measure a single analyte. In some circumstances, it is beneficial to separate multiple antigens by electrophoresis so as to be able to simultaneously detect multiple serum antibodies. The *Western blot* is a transfer technique used to detect specific antibodies. As shown in Figure 5-12, multiple protein antigens (such as those associated with the human immunodeficiency virus [HIV]) are isolated, denatured, and separated by sodium dodecyl sulfate-polyacrylamide gel electrophoresis (SDS-PAGE). SDS denatures the protein and adds an overall negative charge proportional to the molecular weight of the protein. The PAGE of SDS-treated proteins thus allows the separa-

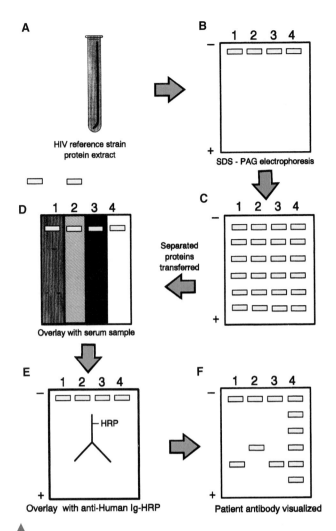

Figure 5-12. Immunoblot (Western blot) to detect antibodies to HIV antigens. (**A**) The HIV is disrupted and extracted to generate its antigens. (**B**) The HIV antigens are separated by sodium dodecyl sulfate polyacrylamide gel electrophoresis. (**C**) The separated antigens are visualized and then transferred to a membrane. (**D**) Each lane of the membrane is overlaid with a patient or control serum. (**E**) After incubation and washing, a labeled antibody reagent is overlaid onto all lanes. (**F**) The bound labeled antibody is detected.

Immunocytochemistry and Immunohistochemistry

When antibody reagents are used to detect antigens in cells or tissues, the methods are known as *immunocytochemistry* and *immunohistochemistry,* respectively. When the antigen is an integral part of the cell or tissue, this is direct testing. A second strategy, indirect testing, uses cells or tissue as a substrate (the source of antigen) to capture serum antibody; this complex is then detected using a labeled antibody reagent.

Fluorescent labels are most commonly used in immunocytochemistry and immunohistochemistry. When used to microscopically identify bacteria or constituents in tissue (such as immune complexes deposited in vivo), the method is called *direct immunofluorescence (DIF)* or the direct fluorescence assay (DFA). A specially configured fluorescence microscope is necessary. Appropriate wavelengths of light are selected by a filter monochromator to excite the fluorescent label; the fluorescent label emits light of a second wavelength that is selected for viewing by a second filter monochromator. Examples of DFA include detection of *Treponema pallidum* in lesional fluid or detection of immune complexes in glomerulonephritis associated with Goodpasture's syndrome and lupus nephritis.

When a fluorescent label is used in indirect testing, this is *indirect immunofluorescence (IIF)* or an indirect immunofluorescence assay (IFA). The substrate is placed on a microscopic slide, serum is overlaid and allowed to react with the antigen, and the bound antibody is detected by the labeled anti-human globulin reagent. The slide is viewed using a fluorescence microscope. The most common IFA performed in the clinical laboratory is to detect antibodies to nuclear antigens (ANA). Both the titer and pattern of fluorescence provide useful information to diagnose connective tissue diseases. Other autoantibodies and antibodies to infectious disease can be detected by IFA.

Immunophenotyping

An important and more recent advance in immunocytochemistry is the use of a flow cytometer to detect intracellular and cell surface antigens. This technique, *immunophenotyping,* is used to classify cell lineage and to identify the stage of cell maturation. In particular, immunophenotyping is useful to aid in the diagnosis of leukemias and lymphomas. Differentiating between acute myelogenous leukemia and acute lymphoblastic leukemia is difficult morphologically and requires additional information to identify the phenotypically expressed molecules. In lymphoid leukemias and lymphomas, identification of tumor cells as either T or B lymphocytes can be an important predictor of clinical outcome. Another application is to determine the CD4/CD8 ratio (the ratio of the number of helper T lymphocytes to cytotoxic T lymphocytes) and more recently the absolute number of CD4 positive cells. This is the standard method to diagnose infection and to initiate and monitor treatment though viral load quantitation is considered to be a better marker by many.

Immunophenotyping begins with a living cell suspension. The cells may come from the peripheral blood, bone mar-

tion of protein based on molecular weight. The separated proteins are then transferred to a new medium (*eg,* nylon, nitrocellulose, or polyvinylidene difluoride membrane). The separated proteins will be fixed onto the new medium and may be stained to assure separation. Each lane on the membrane is then incubated with patient or control sample. The antibody recognizes and binds to the antigen forming an insoluble complex. After washing, a labeled antibody reagent detects the complex. Depending on the label, it may yield a photometric, fluorescent, or chemiluminescent product, which appears as a band. The bands from the patient sample are compared to those from the control containing antibody that will react with known antigens. Molecular weight markers are also separated and stained to provide a guide for interpretation of molecular weight.

row, or a solid tissue. Leukocytes or mononuclear cells can be isolated using density gradient separation (centrifugation through Ficoll-Hypaque) or by red blood cell lysis. Tissues, such as lymph node and bone marrow, require mechanical removal of cells from the tissue to collect a cell suspension. Based on the patient history and type of specimen, a panel of fluorochrome-labeled monoclonal antibodies (MAbs) is used. Fluorochromes commonly used in immunophenotyping include fluorescein isothiocyanate, phycoerythrin, Quantum Red™, and tetramethyl rhodamine isothiocyanate. An aliquot of the cell suspension is incubated with one or more MAb, depending on the design of the flow cytometer. If the cell expresses the antigen, then the labeled MAb binds and the fluorescent label can be detected. *Flow cytometry* is based on cells transported under fluidic pressure passing one by one through a laser beam. The forward light scatter, side light scatter, and light emitted from fluorescent labels are detected by photomultiplier tubes. Forward light scatter is related to the size of the cell and side light scatter is related to the granularity of the cell. When these two parameters are used together in a scattergram (2-parameter histogram), the desired cell population can be electronically selected. This cell population is also evaluated for emission from the labeled MAb. For a single parameter, the frequency of cells versus the intensity of fluorescence (the channel number) is recorded and displayed as a single-parameter histogram. Alternatively, if two parameters are evaluated on the same cell, then a scattergram is generated that diagrams the expression of two antigens simultaneously. By using a panel of MAbs, the cell can be identified and the relative or absolute number of the cell can be determined.

DNA Analysis by Flow Cytometry

In *flow cytometry,* another use of the flow cytometer is to measure the nuclear deoxyribonucleic acid (DNA) content and proliferative capacity of malignant cells. This can help to distinguish between benign and malignant disease, to monitor disease progression, and to predict response to treatment. Resting normal cells are in G_0 and G_1 phase of the cell cycle, and the nuclear DNA content is 2 sets of 23 chromosomes, known as diploid. As a cell prepares to replicate, the DNA is synthesized (S phase) and the DNA content is aneuploid. Mitosis (M phase) and cell division follow. The ploidy status reflected in the *DNA index (DI)* compares the amount of measured DNA in tumor cells compared with that in normal cells as calculated in Equation 5-7.

$$DI = \frac{\text{peak channel number of aneuploid } G_0/G_1 \text{ peak}}{\text{peak channel number of diploid } G_0/G_1 \text{ peak}}$$

(Eq. 5–7)

If normal diploid cells were measured, then the DI is one. If the DI is not one, then the cells are aneuploid. If the DI is less than one, the cells are hypodiploid; conversely a DI greater than one is hyperploid. The percentage of cells in S phase is also determined and normally is less than 5%.

Fresh, alcohol-fixed cells or rehydrated cells removed from a paraffin block may be used. The cells are treated with a detergent to enable the stain to enter the nucleus. Stains, such as propidium iodide or ethidium bromide, intercalate the DNA and the fluorescence is measured using the flow cytometer. The DNA content is related to the fluorescence intensity (channel number). Measuring the DI and %S phase cells are prognostic indicators in breast, ovarian, bladder, and colorectal cancer.

NUCLEIC ACID PROBES

A rapidly expanding area is the study and use of nucleic acids. Nucleic acids store all genetic information and direct the synthesis of specific proteins. By evaluating nucleic acids, insight into cellular changes may be realized before specific protein products are detectable. Genetically based diseases, presence of infectious organisms, differences between individuals for forensic and transplantation purposes, and altered cell growth regulation are areas that have been investigated using nucleic acid hybridization.

Nucleic Acid Chemistry

DNA stores human genetic information and dictates the amino acid sequence of peptides and proteins. DNA is composed of two strands of nucleotides; each strand is a polymer of deoxyribose molecules linked by strong 3'5' phosphodiester bonds that join the 3' hydroxyl group of one sugar to the 5' phosphate group of a second sugar. A purine or pyrimidine base is also attached to each sugar. The two strands are arranged in a double helix with the bases pointing toward the center. The strands are antiparallel, so that the 3'5' end of one strand bonds with a strand in the 5' 3' direction. Because the phosphate esters are strong acids and are dissociated at neutral pH, the strand has a negative charge, which is proportional to its length. The purine bases (adenine [A] and thymine [T]) and the pyrimidine bases (cytosine [C] and guanine [G]) maintain the double helix by forming hydrogen bonds between base pairs as shown in Figure 5-13. Adenine pairs with thymine with two hydrogen bonds and cytosine pairs with guanine with three hydrogen bonds. The strands are complementary due to the fixed manner by which the base pairs bond.

Under physiologic conditions, the helical structure of double stranded DNA (dsDNA) is very stable due to the numerous, although weak, hydrogen bonds between base pairs and the hydrophobic interaction between the bases in the center of the helix. However, the weak bonds can be broken in vitro by changing environmental conditions; the strands are denatured and separate from each other. Once denatured, the negative charge of each strand causes the strands to repel each other. The two complementary strands can be reassociated or reannealed if the conditions change and favor this. The renaturation will follow the rules of base pairing so the original DNA molecule is recovered.

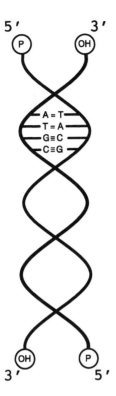

Figure 5-13. Representation of a DNA molecule.

Ribonucleic acid (RNA) is also present in human cells and is chemically similar to DNA. RNA differs from DNA in three ways: ribose replaces deoxyribose as the sugar; uracil replaces thymine as a purine base; and RNA is single stranded. DNA and RNA work together to synthesize proteins. Genomic dsDNA is enzymatically split into its two strands, one of which serves as the template for synthesis of complementary messenger RNA (mRNA). As mRNA is released from the template DNA, the DNA strands reanneal. The mRNA specifies the amino acid to be added to the peptide chain by transfer RNA, which transports the amino acid to the ribosome, where peptide chain elongates.

This discussion of protein synthesis highlights that physiologically DNA routinely is denatured, binds to RNA, and reanneals to reestablish the original DNA, always following base pair rules. These processes form the foundation of nucleic acid *hybridization* assays in which complementary strands of nucleic acid from unrelated sources bind together to form a hybrid or *duplex*.

Hybridization Techniques

A *nucleic acid probe* is a short strand of DNA or RNA that is well characterized and complementary for the base sequence on the test target. Probes may be fragments of genomic nucleic acids, cloned DNA (or RNA), or synthetic DNA. The genomic nucleic acids are isolated from purified organisms. Some probes are molecularly cloned in a bacterial host. First the sequence of DNA to be used as the probe must be isolated using bacterial restriction endonucleases to cut the

DNA at a specific base sequence. The desired base sequence (the probe) is inserted into a plasmid vector, circular dsDNA. The vector with the insert is incorporated into a host cell, such as *Escherichia coli,* where the vector replicates. The replicated desired base sequence is then isolated and purified. For short DNA segments, an oligonucleotide probe can be synthesized using an automated process; if the amino acid sequence of the protein is known, it is possible to determine the base sequence based on amino acid sequence.

In a hybridization reaction, the probe must be detected. The probe can be labeled directly with a radionuclide (such as ^{32}P), enzyme, or biotin. [32]P is detect by autoradiography when the radioactive label exposes x-ray film wherever the probe is located. If the probe is directly labeled with an enzyme, then an appropriate substrate must be added to generate a colorimetric, fluorescent, or chemiluminescent product. Biotin labeled probes can bind to avidin, which is complexed to an enzyme (such as ALP or HRP); the enzyme activity can then be detected. Alternatively, the biotinylated probe can be detected by a labeled avidin antibody reagent.

The probe techniques to be discussed are listed in Table 5-5. A classic method for DNA analysis attributed to EM Southern[31] is the *Southern blot.* In this method, DNA is extracted from a sample using a phenolic reagent and then enzymatically digested using restriction endonucleases to produce DNA fragments. These fragments are then separated by agarose gel electrophoresis. The separated DNA fragments are denatured and transferred to a solid support medium, most commonly nitrocellulose or a charged nylon membrane. The transfer occurs by way of capillary action of a salt solution, transferring the DNA to the membrane or using an electric current to transfer the DNA. Once the DNA is on the membrane, a labeled probe is added that binds to the complementary base sequence and appears as a band. In a similar method, the *Northern blot,* RNA is extracted.

TABLE 5-5. Probe Techniques

Unamplified
  Southern blot
  Northern blot
  In situ hybridization
  Restriction fragment length polymorphism (RFLP)

Target amplification
  Polymerase chain reaction (PCR)
  Reverse transcriptase polymerase chain reaction (RT-PCR)
  Self-sustained sequence replication (3SR)

Probe amplification
  Q-beta replicase
  Ligase chain reaction (LCR)

Signal amplification
  Multiple labels per probe
  Multiple probes per target
  Two-tiered probe system
  Branched DNA assay (bDNA)

The *polymerase chain reaction (PCR)* developed by KB Mullis of Cetus Company[32,33] is an amplified hybridization technique that enzymatically synthesizes millions of identical copies of the target DNA to increase the analytic sensitivity. The test reaction mixture includes the test DNA sample (lysed cells or tissue enzymatically digested with RNAase and proteinase and then extracted), oligonucleotide primers, thermostable DNA polymerase (such as *Taq* polymerase from *Thermus aquaticus*), and nucleotide triphosphates (ATP, GTP, CTP, and TTP) in a buffer. The process, shown in Figure 5-14, begins by heating the target DNA to denature it, separating the strands. Two oligonucleotide primers (probes) that recognize the edges of the target DNA are added and anneal to the target DNA. Thermostable DNA polymerase and nucleotide triphosphates extend the primer. The process of heat denaturation, cooling to allow the primers to anneal and heating again to extend the primers, is repeated many fold (15 to 30 times or more).

The amplified target DNA sequences, known as *amplicons,* can be analyzed by gel electrophoresis, Southern blot, or using directly labeled probes. When microbial RNA or mRNA is the target, the RNA must be enzymatically converted to DNA by reverse transcriptase; the product, complementary DNA (cDNA), can then be analyzed by PCR. This method is referred to a *reverse transcriptase polymerase chain reaction (RT-PCR).* Recently, an enzyme with dual functions (reverse transcriptase and DNA polymerase) has been described.

PCR is limited by its expense, need for special thermocyclers, potential aerosol contamination from one sample to another, nonspecific annealing, and degree of stringency. Stringency is related to the stability of the bonding between target DNA or RNA and the probe and is based on the degree of match and to the length of the probe. Stability of the duplex is strongly influenced by the temperature, pH, and ionic strength of the hybridization solution. Under low stringency conditions (low temperature or increased ionic strength), imperfect binding occurs.

Other techniques have evolved to overcome some of these shortcomings, to make methods more standardized for use in clinical laboratories, or to provide new proprietary approaches. Amplification of the target, probe, and signal have been described. The classic target amplification method, PCR, increases the number of target nucleic acids so that simple signal detection systems can be used. Another target amplification method is *self-sustained sequence replication(3SR),* which detects target RNA and involves continuous isothermic cycles of reverse transcription.[34,35] One primer (T7RNA polymerase) attaches to the RNA and reverse transcriptase extends the annealed primer. Ribonuclease H then degrades the RNA and allows the second primer to bind, followed by synthesis of the cDNA and RNA sequences. The cDNA serves as a template for the production multiple copies of antisense RNA, which are converted to cDNA. Thus the cycle is self-perpetuating.

The *ligase chain reaction (LCR)* is a probe amplification technique that uses two pairs of labeled probes that are complementary for two short target DNA sequences in close proximity.[36] After hybridization, the DNA ligase interprets the break between the ends as a nick and links the probe pairs.

Q-beta replicase system uses Q-beta replicase to synthesize additional copies of the MDV-1 sequence of the single-stranded Q-beta RNA.[37] After target DNA or target RNA is heated to denature it, a probe with target specific sequence and the MDV-1 sequence is added and hybridizes. Unbound probe is removed and the Q-beta replicase and excess ribonucleotide bases are added and amplification follows.

Signal amplification methods are designed to increase the signal strength by increasing the concentration of the label. A probe may have multiple labels attached or several short probes complementary for the target each could be labeled. A two-tiered probe system has been developed in which part of the primary probe attaches to the target DNA and part of the primary probe extends away from the target DNA. Labeled secondary probes are added that hybridize the portion of the primary probe, which is not bound to the target

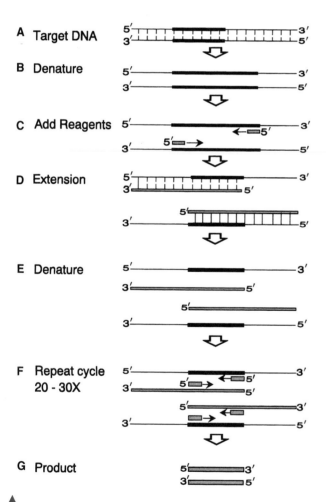

▲

Figure 5-14. Polymerase chain reaction. (**A**) The target DNA sequence is indicated by the bold line; (**B**) the double stranded DNA is denatured (separated) by heating; (**C**) reagents are added, and the primer binds to the target DNA sequence; (**D**) polymerase extends the primers; (**E–G**) heating, annealing of the primer and extension are repeated.

DNA. A very powerful signal amplification system uses multiple probes. Multiple primary probes are used to bind to the target DNA. One arm of a secondary probe can recognize one end of the primary probe. The other arms of the secondary probe are recognized by an enzyme labeled tertiary probe. The branched DNA (bDNA) assay is an example of the last method.[38]

In situ hybridization is performed on cells, tissue, or chromosomes that are fixed on a microscope slide.[39] After the DNA is heat denatured, a labeled probe is added and will hybridize the target sequence after the slide is cooled. Colorimetric or fluorescent products are generally used. A strength of this method is the morphologic context in which the localization of target DNA is viewed.

Restriction fragment length polymorphisms (RFLP) is a technique to evaluate differences in genomic DNA sequences.[40] This can be used to help establish identity or nonidentity in forensic or paternity testing or to identify a gene associated with a disease. Genomic DNA is extracted from a sample (*eg*, peripheral blood leukocytes) and is purified and quantitated. A restriction endonuclease, which cleaves DNA sequences at a specific site, is added. If there is a mutation or change in the DNA sequence, this may cause the length of the DNA fragment to be different than usual. Southern blotting can be used to identify the different lengths of the DNA fragments. A labeled specific probe could be used to identify a specific aberration. PCR can be used to amplify the target DNA sequence before RFLP analysis.

Nucleic Acid Probe Applications

Nucleic acid probes are used to detect infectious organisms; to detect gene rearrangements, chromosomal translocations, or chromosomal breakage; to detect changes in oncogenes and tumor suppressor factors; to aid in prenatal diagnosis of an inherited disease or carrier status; to identify polymorphic markers used to establish identity or nonidentity; and to aid in donor selection. Nucleic acid probes are useful in identifying microorganisms in a patient specimen or confirming an organism isolated in culture. Probes are currently available to confirm *Mycobacterium* species, *Legionella* species, *Salmonella*, diarrheogenic *Escherichia coli* strains, *Shigella*, and *Campylobacter* species. Probes also are available to identify fungi such as *Cryptococcus neoformans*, *Coccidioides immitis*, *Histoplasma capsulatum*, and *Blastomyces dermatitidis*. Direct identification of microorganisms in patient specimens include *Chlamydia trachomatis*, *Neisseria gonorrhea*, and herpes simplex virus. In addition, viral load testing for HIV and the hepatitis C virus are detected using probe technology.

Gene rearrangement studies by Southern blot are helpful to distinguish between T and B lymphocyte lineage.[41] Also the chromosome translocation associated with most follicular, non-Hodgkin's lymphomas and some diffuse large cell lymphomas have been detected and monitored. The Philadelphia chromosome in chronic myelogenous leukemia is associated with the translocation that results in the detection of the bcr/abl fusion gene.[42]

Prenatal diagnosis of genetic diseases such as sickle cell anemia, cystic fibrosis,[43] Huntington's chorea,[44] Duchenne type muscular dystrophy,[45] and von Willebrand's disease[46] have been made possible using probe technology. In addition, the carrier status in Duchenne type muscular dystrophy and von Willebrand's disease can be determined.

PCR has been used to detect major histocompatibility complex class I and class II polymorphism.[47] The increased accuracy of detecting differences in the genes rather than the gene product has been used to improve transplant compatibility.

SUMMARY

This chapter has introduced the foundations of immunoassays and nucleic acid probe techniques. All immunoassays use antibodies to detect target analytes. Unlabeled immunoassays in gel include methods to characterize monoclonal proteins and to identify specific antibodies to nuclear components. Unlabeled immunoassays in the soluble phase are typified by nephelometry and turbidimetry. Labeled immunoassays have increased analytic sensitivity. The variety of assay designs, improved and reliable specificity of monoclonal antibodies, and automation have made immunoassays increasingly popular in the clinical laboratory. Immunoassays are described in terms of the label used, the method to detect the label, the need for separation (homogeneous or heterogeneous), competitive or noncompetitive binding, and qualitative or quantitative measurement. Specialized techniques using antibodies include immunoblots, immunocytochemistry, immunohistochemistry, and flow cytometry.

Nucleic acid probes are used to detect DNA or RNA sequences in human, bacterial, or viral genomes. It is a diverse group of methods that are highly specific under stringent conditions. Unamplified techniques are generally used to identify qualitative changes. Amplified techniques that increase the amount of target nucleic acid, probe, or signal are used to increase analytic sensitivity.

The future of immunoassays and nucleic acid probe technology will be to do more tests simultaneously on a single sample in a single reaction vessel or on a single chip surface. The ability to detect different labels and to recognize capture molecules will allow the simultaneous detection of an exponential number of analytes or sequences. Thus multiplex testing will replace the test panel, which now consists of separately performed individual tests.

REVIEW QUESTIONS

1. The strength of binding between an antigen and antibody is related to the:
   a. Concentration of antigen and antibody
   b. Source of antibody production because monoclonal antibodies bind better
   c. Goodness of fit between the epitope and the F(ab)
   d. Specificity of the antibody

2. In monoclonal antibody production, the specificity of the antibody is determined by the:
   a. myeloma cell line
   b. sensitized B lymphocytes
   c. sensitized T lymphocytes
   d. selective growth medium

3. Which unlabeled immune precipitation method in gel is used to quantitate a serum protein?
   a. Double diffusion
   b. Radial immunodiffusion
   c. Counter immunoelectrophoresis
   d. Immunofixation electrophoresis

4. In immunofixation electrophoresis, discrete bands appear at the same electrophoretic location; one reacted with anti-human IgA (alpha chain specific) reagent and the other reacted with anti-human lambda reagent. This is best described as:
   a. An IgA λ monoclonal protein
   b. An IgA λ polyclonal protein
   c. IgA biclonal proteins
   d. Cross reactivity

5. In nephelometry, the antigen–antibody complex formation is enhanced in the presence of:
   a. High ionic strength saline solution
   b. Normal saline
   c. Polyethylene glycol
   d. Complement

6. In a classic competitive, heterogeneous RIA to measure T4, which statements are true and which are false?
   _____ a. The assay requires a separation step.
   _____ b. The tracer is T4 with a radiolabel.
   _____ c. The detector molecule is radiolabeled anti-T4.
   _____ d. As the concentration of analyte in the test sample increases, the concentration of bound labeled molecule decreases.

7. Which homogeneous immunoassay relies on inhibiting the activity of the enzyme label when bound to antibody reagent to eliminate separating free from bound labeled reagent?
   a. EMIT™
   b. CEDIA
   c. MEIA
   d. ELISA

8. Number the steps in a Western blot in the order in which it is performed. Step 1 is the first step in the assay.
   _____ a. Protein antigens are isolated and denatured.
   _____ b. Test serum is incubated with the proteins.
   _____ c. Proteins are transferred to a nitrocellulose membrane.
   _____ d. Proteins are separated by SDS-PAGE.
   _____ e. Labeled antiserum reagent that reacted is evaluated.

9. In flow cytometry, the side scatter is related to the:
   a. DNA content of the cell
   b. Granularity of the cell
   c. Size of the cell
   d. Number of cells in G_0 and G_1

10. The nucleic acid technique in which RNA is converted to cDNA, which is then amplified, is known as:
    a. PCR
    b. RT-PCR
    c. RFLP
    d. In situ hybridization

REFERENCES

1. Sheehan C. An overview of antigen–antibody interaction and its detection. In Sheehan C, ed. Clinical immunology: principles and laboratory diagnosis. 2nd ed. Philadelphia: Lippincott-Raven Publishers, 1997;109.
2. Kohler G, Milstein C. Continuous cultures of fused cells secreting antibody of predefined specificity. Nature 1975;256:495.
3. Ouchterlony O. Antigen–antibody reactions in gel. Acta Pathol Microbiol Scand 1949;26:507.
4. Mancini G, Carbonara AO, Heremans JF. Immunochemical quantitation of antigens by single radial immunodiffusion. Immunochemistry 1965;2:235.
5. Fahey JL, McKelvey EM. Quantitative determination of serum immunoglobulins in antibody-agar plates. J Immunol 1965;98:84.
6. Grabar P, Burtin P. Immunoelectrophoresis. Amsterdam, The Netherlands: Elsevier, 1964.
7. Alper CC, Johnson AM. Immunofixation electrophoresis: a technique for the study of protein polymorphism. Vox Sang 1969;17:445.
8. Laurell CB. Quantitative estimation of proteins by electrophoresis in agarose gel containing antibodies. Anal Biochem 1966;15:45.
9. Laurell CB. Electroimmunoassay. Scand J Clin Lab Invest 1972;29(suppl 124):21.
10. Kusnetz J, Mansberg HP: Optical considerations: nephelometry. In Ritchie RF, ed. Automated immunoanalysis. Part 1. New York: Marcel Dekker, 1978.
11. Engvall E, Perlmann P. Immunochemistry 1971;8:871.
12. Van Weemen BK, Schuurs AHWM. Immunoassay using antigen-enzyme conjugates. FEBS Lett 1971;15:232.
13. Nakamura RM, Bylund DJ. Fluorescence immunoassays. In Rose NR, deMarcario EC, Folds JD, Lane HC, Nakamura RM, eds. Manual of clinical laboratory immunology. 5th ed. Washington, DC: ASM Press 1997;39.
14. Diamandis EP, Evangelista A, Pollack A, et al. Time-resolved fluoroimmunoassays with europium chelates as labels. Am Clin Lab 1989; 8(8):26.
15. Kricka LJ. Chemiluminescent and bioluminescent techniques. Clin Chem 1991;37:1472.
16. Kricka LJ. Selected strategies for improving sensitivity and reliability of immunoassays. Clin Chem 1994;40:347.
17. Bronstein I, Juo RR, Voyta JC. Novel chemiluminescent adamantyl 1,2-dioxetane enzyme substrates. In Stanley PE, Kricka LJ, eds. Bioluminescence and chemiluminescence: current status. Chichester, England: Wiley, 1991;73.
18. Edwards B, Sparks A, Voyta JC, et al. In Campbell AK, Kricka LJ, Stanley PE, eds. Bioluminescence and chemiluminescence: fundamentals and applied aspects. Chichester, England: Wiley, 1994.
19. Litchfield WJ. Shell-core particles for the turbidimetric immunoassays. In Ngo TT, ed. Nonisotopic immunoassay. New York: Plenum Press, 1988.
20. Rubenstein KD, Schneider RB, Ullman EF. "Homogeneous" enzyme immunoassay, a new immunochemical technique. Biochem Biophys Res Comm 1972;47:846.
21. Henderson DR, Freidman SB, Harris JD, et al. CEDIA, a new homogeneous immunoassay system. Clin Chem 1986;32:1637.
22. Ullman EF, Schwartzberg M, Rubinstein KD. Fluorescent excitation transfer assay: a general method for determination of antigen. J Biol Chem 1976;251:4172.
23. Dandliker WB, Dandiker BJ, Levison SA, et al. Fluorescence methods for measuring reaction equilibria and kinetics. Methods Enzymol 1978;48:380.

24. Dandliker WB, Kelly RJ, Dandiker BJ, et al. Fluorescence polarization immunoassay: theory and experimental methods. Immunochemistry 1973;10:219.

25. Diamandis EP. Immunoassays with time-resolved fluorescence spectroscopy: principles and applications. Clin Biochem 1988;21:139.

26. Yalow RS, Berson SA. Immunoassay of endogenous plasma insulin in man. J Clin Invest 1960;39:1157.

27. Valkirs RF, Barton R. Immunoconcentration^c: a new format for solid phase immunoassays. Clin Chem 1985;31:1427.

28. Rubenstein AS, Hostler RD, White CC, et al. Particle entrapment: application to ICON^c immunoassay. Clin Chem 1986;32:1072.

29. Zuk RF, Ginsberg VK, Houts T, et al. Enzyme immunochromatography: a quantitative immunoassay requiring no instrumentation. Clin Chem 1985;31:1144.

30. Harbeck RJ, Teague J, Crossen GR, et al. Novel, rapid optical immunoassay technique for detection of group A streptococci from pharyngeal specimens: comparison with standard culture methods. J Clin Microbiol 1993;31:839.

31. Southern EM. Detection of specific sequences among DNA fragments separated by gel electrophoresis. J Mol Biol 1975;98:503.

32. Mullis KB, Faloona FA. Specific synthesis of DNA in vitro via a polymerase-catalyzed chain reaction. Methods Enzymol 1987;155:335.

33. Mullis KB. The unusual origin of the polymerase chain reaction. Sci Am 1990;262:56.

34. Guatelli JC, Whitfield KM, Kwoh DY, et al. Isothermal in vitro amplification of nucleic acids by a multienzyme reaction modeled after retroviral replication. Proc Natl Acad Sci USA 1990;87:1874.

35. Fahy E, Kwoh DY, Gingeras TR. Self-sustained sequence replication (3SR): an isothermal transcription-based amplification system alternative to PCR. PCR Methods and Applications 1991;1:25.

36. Barany F. Genetic disease detection and DNA amplification using cloned thermostable ligase. Proc Natl Acad Sci USA 1991;88:189.

37. Klinger JD, Pritchard CG. Amplified probe-based assay: possibilities and challenges in clinical microbiology. Clin Microbiol Newsletter 1990;12:133.

38. Wiedbrauk DL. Molecular methods for virus detection. Lab Med 1992;23:737.

39. Hankin RC. In situ hybridization: principles and applications. Lab Med 1992;23:764.

40. Bishop JE, Waldholz M. Genome. New York: Simon & Schuster, 1990.

41. Farkas DH. The Southern blot: application to the B- and T-cell gene rearrangement test. Lab Med 1992;23:723.

42. Ni H, Blajchman MA. Understanding the polymerase chain reaction. Transfusion Med Rev 1994;8:242.

43. Gasparini P, Novelli G, Savoia A, et al. First-trimester prenatal diagnosis of cystic fibrosis using the polymerase chain reaction: Report of eight cases. Prenat Diag 1989;9:349.

44. Thies U, Zuhlke C, Bockel B, et al. Prenatal diagnosis of Huntington's disease (HD): experience with six cases and PCR. Prenat Diag 1992;12:1055.

45. Clemens PR, Fenwick RG, Chamberlain JS, et al. Carrier detection and prenatal diagnosis in Duchenne and Becker muscular dystrophy families, using dinucleotide repeat polymorphism. Am J Hum Genet 1991;49:951.

46. Peake IR, Bowen D, Bignell P. Family studies ad prenatal diagnosis in severe von Willebrand disease by polymerase chain reaction amplification of variable number tandem repeat region of the von Willebrand factor gene. Blood 1990;76:555.

47. Erlich H, Tugawan T, Begovich AB, et al. HLA-DR, DQ, and DP typing using PCR amplification and immobilized probes. Eur J Immunogenet 1991;18:33.

CHAPTER 6

Principles of Clinical Chemistry Automation

Larry Schoeff

Objectives

Upon completion of this chapter, the clinical laboratorian should be able to:

- *Define the following terms: automation, continuous flow, discrete analysis, channel, flag, random access, dwell time, and throughput.*

- *Discuss the history of the development of automated analyzers in the clinical chemistry laboratory.*

- *List four driving forces behind the development of more and better automated analyzers.*

- *Name the three basic approaches to sample analysis used by automated analyzers.*

- *Explain the major steps in automated analysis.*

- *State an example of a commercially available discrete analyzer and a centrifugal analyzer.*

- *Compare the different approaches to automated analysis used by instrument manufacturers.*

- *Discriminate between an open versus a closed reagent system.*

- *Relate three considerations in the selection of an automated analyzer.*

- *Explain the concept of total laboratory automation.*

- *Differentiate the three phases of the laboratory testing process.*

- *Discuss future trends in automated analyses.*

KEY TERMS

Automation	Discrete analysis	Robotics
Bar code	Dry chemistry slide	Rotor
Centrifugal analysis	Flags	Total laboratory
Channel	Probe	automation
Continuous flow	Random access	

HISTORY OF AUTOMATED ANALYZERS

Since the introduction of the first automated analyzer by Technicon in 1957, automated instruments have proliferated from many manufacturers.[1] This first "Auto Analyzer" (AA)© was a continuous flow, single *channel,* sequential batch analyzer capable of approximately 40 tests per hour. The next generation of Technicon instruments to be developed was the Simultaneous Multiple Analyzer (or SMA) series. As the name implies, SMA-6© and SMA-12© were analyzers with multiple channels (for different tests) working synchronously to produce 6 or 12 test results simultaneously at the rate of 360 or 720 tests per hour. It was not until the mid-1960s that these continuous-flow analyzers had any significant competition in the marketplace.

In 1970, the first commercial centrifugal analyzer was introduced as a spinoff technology from NASA outerspace research. Dr. Norman Anderson developed a prototype in 1967 at the Oak Ridge National Laboratory as an alternative to continuous-flow technology, which had significant carryover problems and costly reagent waste. He wanted to perform analyses in parallel and also take advantage of advances in computer technology. The second generation of these instruments in 1975 was more successful, as a result of miniaturization of computers and advances in the polymer industry for high-grade optical plastic cuvets.

The next major development that revolutionized clinical chemistry instrumentation occurred in 1970 with the introduction of the DuPont (now Dade) Automated Clinical Analyzer (ACA©). It was the first noncontinuous flow, discrete analyzer, as well as the first instrument to have *random access* capabilities, whereby *STAT* specimens could be analyzed out of sequence from the batch as needed. Plastic test packs, positive patient identification, and infrequent calibration were among the unique features of the ACA©.

The last major milestones were the introduction of thin film analysis technology in 1976, and the production of the Ektachem (now Vitros) Analyzer© by Kodak (now Johnson & Johnson) in 1978. This instrument was the first to use microvolumes of sample and reagents on slides for dry chemistry analysis, and it was the first to incorporate computer technology extensively into its design and use.

Since 1980, several primarily discrete analyzers have been developed that incorporate such characteristics as ion selective electrodes (ISE), fiber optics, polychromatic analysis, continually more sophisticated computer hardware and software for data handling, and larger test menus. Some of the popular and more successful analyzers using these and other technologies since 1980 are the Beckman ASTRA© (now Synchron) analyzers extensively using ISEs; Dade Paramax©, using reagent tablet dispensing and primary tube sampling; the Hitachi analyzers by Boehringer Mannheim with reusable reaction disks and fixed diode arrays for spectral mapping; and the Chem 1© by Technicon (now Bayer), which uses encapsulated oil segments of sample and reagents in a single continuous-flow tube. Other automated systems that are commonly used in clinical chemistry are the Abbott Spectrum© and the Olympus and Roche series of analyzers.

Many of the manufacturers of these instrument systems have adopted the more successful features and technologies of other instruments, where possible, to make each generation of their product more competitive in the marketplace. The differences among the manufacturers' instruments, their operating principles, and their technologies are less distinct now than they were in the beginning years of laboratory automation.

DRIVING FORCES TOWARD MORE AUTOMATION

Since 1995, the pace of changes with current routine chemistry analyzers and the introduction of new ones have slowed down considerably, compared to the first half of the 1990s. Certainly analyzers are faster and easier to use, due to continual reengineering and electronic refinements. Methods are more precise, sensitive, and specific, although some of the same principles are found in today's instruments as in earlier models. Manufacturers have worked successfully toward automation with "walk away" capabilities and minimal operator intervention.[2] Manufacturers have also responded to the physicians' desire to bring laboratory testing to the patient. The introduction of small, portable, easy-to-operate benchtop analyzers in physician office laboratories (POL), as well as in surgical and critical care units that demand immediate lab results has resulted in a hugely successful domain of point-of-care (POC) analyzers. Another specialty area with a rapidly developing arsenal of analyzers is immunochemistry. Immunologic techniques for assaying drugs, specific proteins, tumor markers, hormones, and so on, have evolved

to an increased level of automation. Instruments that use techniques such as fluorescence polarization (FPIA), nephelometry, and chemiluminescence have become very popular in laboratories.

Other forces are also driving the market toward more focused automation. A higher volume of testing and faster turnaround time have resulted in fewer and more centralized core labs performing more comprehensive testing.[3] The use of laboratory panels or profiles has declined, with more diagnostically directed individual tests as dictated by recent policy changes from Medicare and Medicaid. Researchers have known for many years that chemistry panels only occasionally lead to new diagnoses in patients who appear healthy.[4] The expectation of quality results with higher accuracy and precision is ever present with the regulatory standards set by the Clinical Laboratory Improvement Amendment (CLIA), Joint Commission on Accreditation of Healthcare Organizations (JCAHO), College of American Pathologists (CAP), and so on. Intense competition among instrument manufacturers has driven automation into more sophisticated analyzers with creative technologies and unique features. Furthermore, escalating costs have spurred health care reform and, more specifically, managed care and capitation environments within which laboratories are forced to operate.

BASIC APPROACHES TO AUTOMATION

There are many advantages to automating procedures. One purpose is to increase the number of tests performed by one laboratorian in a given period. Labor is a very expensive commodity in laboratories. Through mechanization, the labor component devoted to any single test is minimized, and this effectively lowers the cost per test. A second purpose is to minimize the variation in results from one laboratorian to another. By reproducing the components in a procedure as identically as possible, the coefficient of variation is lowered and reproducibility is increased. Accuracy is then not dependent on the skill or workload of a particular operator on a particular day. This allows better comparison of results from day to day and week to week. Automation, however, cannot correct for deficiencies inherent in methodology. A third advantage is gained because automation eliminates the potential errors of manual analyses such as volumetric pipetting steps, calculation of results, and transcription of results. A fourth advantage accrues because instruments can use very small amounts of samples and reagents. This allows less blood to be drawn from each patient. In addition, the use of small amounts of reagents decreases the cost of consumables.

There are three basic approaches with instruments: continuous flow, centrifugal analysis, and discrete analysis. All three can use batch analysis (ie, large number of specimens in one run), but only discrete analyzers offer random access, or STAT capabilities.

In continuous flow, liquids (reagents, diluents, and samples) are pumped through a system of continuous tubing. Samples are introduced in a sequential manner, following each other through the same network. A series of air bubbles at regular intervals serve as separating and cleaning media. Continuous flow, therefore, resolves the major consideration of uniformity in performance of tests, because each sample follows the same reaction path. Continuous flow also assists the laboratory that needs to run many samples requiring the same procedure. The more sophisticated continuous-flow analyzers use parallel single channels to run multiple tests on each sample, for example, SMA©, and SMAC©. The major drawbacks that contributed to the eventual demise of traditional continuous-flow analyzers (ie, AA, SMA, and SMAC) in the marketplace were significant carryover problems and wasteful use of continuous flowing reagents. Technicon's answer to these problems was a noncontinuous-flow discrete analyzer (the RA1000©), using random-access fluid, which is a hydrofluorocarbon liquid to reduce surface tension between samples/reagents and their tubing, and thereby reduce carryover. Later, the Chem 1© was developed by Technicon to use Teflon tubing and Teflon oil, virtually eliminating carryover problems. The Chem 1© is a continuous-flow analyzer but only remotely comparable to the original continuous-flow principle.

Centrifugal analysis uses the force generated by centrifugation to transfer and then contain liquids in separate cuvets for measurement at the perimeter of a spinning *rotor*. Centrifugal analyzers are most capable of running multiple samples, one test at a time, in a batch. Batch analysis is their major advantage, because reactions in all cuvets are read virtually simultaneously, taking no longer to run a full rotor of about 30 samples than it would take to run a few. Laboratories with a high workload of individual tests for routine batch analysis may use these instruments. Again, each cuvet must be uniformly matched to each other to maintain quality handling of each sample. The Cobas-Bio© by Roche, with its xenon flash lamp and longitudinal cuvets,[5] and the IL Monarch©, with its fully integrated walk away design, are two of the more successful centrifugal analyzers.

Discrete analysis is the separation of each sample and accompanying reagents in a separate container. Discrete analyzers have the capability of running multiple tests one sample at a time or multiple samples one test at a time. They are the most popular and versatile analyzers. However, because each sample is in a separate reaction container, uniformity of quality must be maintained in each cuvet so that a particular sample's quality is not affected by the particular space that it occupies. The Dimension© (Dade), Vitros©, Paramax©, Synchron©, and Hitachi analyzers are all popular examples of discrete analyzers with random access capabilities.

STEPS IN AUTOMATED ANALYSIS

In clinical chemistry, *automation* is the mechanization of the steps in a procedure. Manufacturers design their instruments to mimic manual techniques. The major steps in a procedure may be listed as follows:

- Specimen preparation and identification
- Specimen measurement and delivery
- Reagent systems and delivery
- Chemical reaction phase
- Measurement phase
- Signal processing and data handling

In this section, each step of automated analysis is explained, and several different applications are discussed. Several instruments have been chosen because they have components that represent either common features used in chemistry instrumentation or a unique method of automating a step in a procedure. None of the representative instruments is completely described, but rather the important components are described in the text as examples.

Specimen Preparation and Identification

Preparation of the sample for the analysis has been and remains a manual process in most laboratories. The clotting time (if using serum), centrifugation, and the transferring of the sample to an analyzer cup (unless using primary tube sampling) cause delay and expense in the testing process. One alternative to manual preparation is to automate this process by using *robotics,* or front-end automation, to "handle" the specimen through these steps and load the specimen onto the analyzer. Another option is to bypass the specimen preparation altogether by using whole blood for analysis, for example, Abbott Vision. Robotics for specimen preparation has already become a reality in some clinical laboratories in the United States and other countries. More discussion about "pre-analytical" specimen processing, or front-end automation, appears later in this chapter.

The sample must be properly identified and its location in the analyzer must be monitored throughout the test. The simplest means of identifying a sample is by placing a manually labeled sample cup in a numbered analysis position on the analyzer, in accordance with a manually prepared worksheet or a computer-generated load list. The most sophisticated approach uses a bar code label affixed to the primary collection tube. This label contains patient demographics and also may include test requests.

The *bar code*–labeled tubes are then transferred to the loading zone of the analyzer, where the bar code is scanned and the information is stored in the computer's memory. The analyzer is then capable of monitoring all functions of identification, test orders and parameters, and sample position. Some analyzers may take test requests downloaded from the laboratory information system and run them when the appropriate sample is identified and ready to be pipetted.

Specimen Measurement and Delivery

Most instruments use either circular carousels or rectangular racks as specimen containers for holding disposable cups or primary sample tubes in the loading or pipetting zone of the analyzer. These cups or tubes hold standards, controls, and patient specimens to be pipetted into the reaction chambers of the analyzers. The slots in the trays or racks usually are numbered to aid in identification of the sample. The trays or racks move automatically in 1-cup steps at preselected speeds. The speed determines the number of specimens to be analyzed per hour. As a convenience, the instrument can determine the slot number containing the last sample and terminate the analysis after that sample. The instrument's computer holds the number of cups in memory and aspirates only in slots containing sample cups.

To facilitate the loading of samples onto an automated sampling device, the necessity for exact measurement by the laboratorian should be eliminated. The configuration of the sample cup should determine the minimum and maximum volumes required. Indentations or scorings molded into the plastic can indicate these volumes. The operator can then pour or pipet the samples into the containers easily and quickly. One of the most commonly used containers is the 2-mL sample cup (Fig. 6-1). These cups are available from many manufacturers and are relatively inexpensive.

On the Vitros© analyzer, sample cup trays are quadrants that hold 10 samples each in cups with conical bottoms. The four quadrants fit on a tray carrier (Fig. 6-2). Although the tray carrier accommodates only 40 samples, more trays of samples can be programmed and then loaded in place of completed trays while tests on other trays are in progress. A disposable sample tip is hand loaded adjacent to each sample cup on the tray.

In centrifugal analyzers, the loading of the samples and reagents is accomplished by pipetting the appropriate liquid into a rotor with 20 or more positions. Each position contains a sample compartment, a reagent compartment, and a cuvet located at the periphery of the rotor (Fig. 6-3).

The Paramax© allows sampling from primary collection tubes, or for limited samples there are micro-sample tubes. The tubes are placed in a circular tray that holds 96 speci-

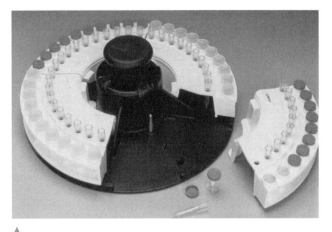

Figure 6-2. Vitros©. The four quadrant trays, each holding ten samples, fit on a tray carrier. (Photograph courtesy of Johnson & Johnson Co.)

mens at one time. Bar code labels for each sample, complete with patient name and identification number, are printed on demand by the operator (Fig. 6-4). This allows samples to be loaded in any order. The loading carousel dispenses the tubes to a transfer carousel, from which the sampling occurs, and then the samples are transferred to an unloading carousel (Fig. 6-5). This analyzer makes use of a continuous belt of flexible, disposable plastic cuvets carried through the analyzer's water bath on a main drive track. The cuvets are loaded onto the Paramax from a continuous spool. These cuvets index through the instrument at the rate of one every 5 seconds. The cuvets are cut into sections or groups as required.

Figure 6-1. 2-mL sample cup with conical bottom.

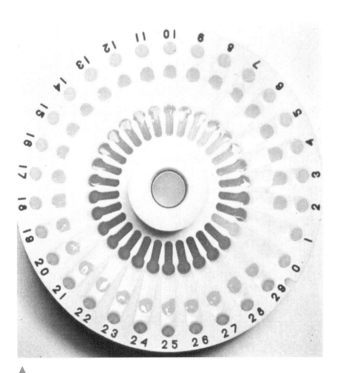

Figure 6-3. Centrifugal analyzer rotor.

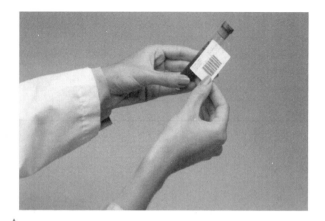

Figure 6-4. Paramax©. Sample collection tubes are identified with bar-code labels. (Photograph courtesy of Dade International.)

A problem with sample handling results from the exposure of the aliquot to the air, which can produce errors in analysis. Evaporation of the sample may be significant and may cause the concentration of the constituents being analyzed to rise in relation to the time of exposure. With instruments measuring electrolytes, the carbon dioxide present in the samples will be lost to the atmosphere, resulting in low carbon dioxide values. Manufacturers have devised a variety of mechanisms to minimize this effect, for example, lid covers for trays and individual caps that can be pierced.

The actual measurement of each aliquot for each test must be very accurate. This is generally done through aspiration of the sample into a *probe*. When the discrete instrument is in operation, the probe automatically dips into each sample cup and aspirates a portion of the liquid. After a preset, computer-controlled time interval, the probe quickly rises from the cup. Sampling probes on instruments using specific sampling cups are programmed or adjusted to reach a prescribed depth in those cups to maximize use of available

sample. Those analyzers capable of aspirating sample from primary collection tubes usually have a parallel liquid level-sensing probe that will control entry of the sampling probe to a minimal depth below the surface of the serum, allowing full aliquot aspiration while avoiding clogging of the probe with serum separator gel or clot (Fig. 6-6).

In continuous-flow analyzers, when the sample probe rises from the cup, air is aspirated for a specified time to produce a bubble in between sample and reagent plugs of liquid. Then the probe descends into a container where wash solution is drawn into the probe and through the system. The wash solution is usually deionized water, possibly with a surfactant added. Remembering that all samples follow the same reaction path, the necessity for the wash solution between samples becomes obvious. Immersion of the probe into the wash reservoir cleanses the outside, whereas aspiration of an aliquot of solution cleanses the lumen. The reservoir is continually replenished with an excess of fresh solution. The wash aliquot, plus the previously mentioned air bubble, maintains sample integrity and minimizes sample carryover.

Some pipettors use a disposable tip and an air-displacement syringe to measure and deliver reagent. When this is used, the pipettor may be reprogrammed to measure sample and reagent for batches of different tests comparatively easily. Besides eliminating the effort of priming the reagent delivery system with the new solution, no reagent is wasted or contaminated, because nothing but the pipet tip contacts it.

Figure 6-5. Paramax©. The loading carousel dispenses tubes to a transfer carousel, from which the sampling occurs, and then the samples are transferred to an unloading carousel. (Photograph courtesy of Dade International.)

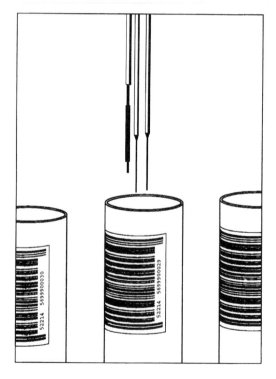

Figure 6-6. Dual sample probes of the Hitachi 736 analyzer. Note the liquid level sensor to the left of probes. (Photograph courtesy of Boehringer Mannheim Corp.)

The cleaning of the probe and tubing after each dispensing to minimize the carryover of one sample into the next is a concern for many instruments. In some systems, the reagent or diluent is also dispersed into the cuvet through the same tubing and probe. Deionized water may be dispensed into the cuvet after the sample to produce a specified dilution of the sample and also to rinse the dispensing system. In the Technicon (now Bayer) RA1000©, a random-access fluid, a fluorocarbon, is the separation medium. The fluid is a viscous, inert, immiscible, nonwetting substance that coats the delivery system. The coating on the sides of the delivery system prevents carryover due to the wetting of the surfaces and, forming a plug of the solution between samples, prevents carryover by diffusion. A small amount (10-mL) of this fluid is dispensed into the cuvet with the sample. Surface tension leaves a coating of the fluid in the dispensing system.

If a separate probe or tip is used for each sample and discarded after use, as in the Vitros©, the issue of carryover is a moot point. The Vitros© has a unique sample-dispensing system. A "proboscis" presses into a tip on the sample tray, picks it up, and moves over the specimen to aspirate the volume required for the tests programmed for that sample. The tip is then moved over to the slide-metering block. When a slide is in position to receive an aliquot, the proboscis is lowered so that a dispensed 10-mL drop touches the slide, where it is absorbed from the nonwettable tip. A stepper motor-driven piston controls aspiration and drop formation. The precision of dispensing is specified at ±5%.

In the centrifugal-analysis instrument, a loader is used for dispensing samples and reagents. Typically, it consists of two Hamilton syringes, a keyed turntable for holding the transfer disk or disposable rotor, a ring surrounding the disk to hold the sample cups, an autostop module, and a digital control panel. For each assay, the operator selects the appropriate volumes by pressing digital switches on the front panel of the loader. Sample and diluent, or a second reagent, are pipetted into one compartment on the disk, and the reagent and diluent are pipetted into a second compartment. The Hamilton syringes are driven by stepper motors for pipetting. After all the samples and reagents have been pipetted, the disk is removed from the loader and placed on the analyzer.

In several discrete systems, the probe is attached by means of nonwettable tubing to precision syringes. The syringes draw a specified amount of sample into the probe and tubing. Then the probe is positioned over a cuvet, into which the sample is dispensed. The Hitachi 736© uses two sample probes to simultaneously aspirate a double volume of sample in each probe immersed in one specimen container, and thereby deliver sample into four individual test channels, all in one operational step (Fig. 6-7). The loaded probes pass through a fine mist shower bath before delivery to wash off any sample residue adhering to the outer surface of the probes. After delivery the probes move to a rinse bath station for cleaning the inside and outside surfaces of the probes.

The ACA Star© filling station puts the appropriate amounts of sample and diluent into each analytical test pack. A filled sample cup is followed in the loading tray by the test packs for the tests required on the specimen. When the system is activated, a shuttle pushes the first sample cup left to the sampling position. While the specimen is moving left, a spring-loaded pack pusher pushes the first test pack onto the filling station rail below a decoder plate. The decoder senses the binary code on the top of the test pack. This code, when translated, provides the instrument with instructions determining the sample volume, type of diluent, and special handling characteristics. The pump flushes with diluent to be used for the particular method to purge the lines and needle before diluent intake. After the flush, the pump intakes diluent into the pump cylinder. The sample needle is positioned over the sample cup, dips down into the cup, and aspirates the proper volume of specimen. The needle rises out of the sample cup and moves right to a position over the test pack. The sample and diluent are injected into the test pack through a special pack-fill opening. After the pack is filled, the needle, tubing, and pump are flushed. The test pack is moved onto the transport system. The instrument also transfers the patient sample-identification number on the sample pack to the results-reporting system.

The Paramax© uses computer-controlled stepping motors to drive both the sampling and washout syringes. Every 5 seconds, the sampling probe enters a specimen container, withdraws the required volume, moves to the cuvet, and dispenses the aliquot with a volume of water to wash the probe. The washout volume is adjusted to yield the final reaction volume. If a procedure's range of linearity is exceeded, the system will retrieve the original sample tube, repeat the test

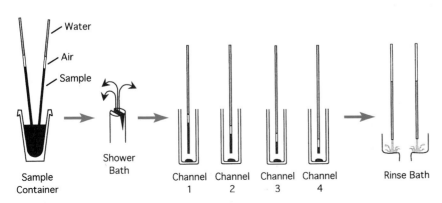

Figure 6-7. Sampling operation of the Hitachi 736 analyzer. (Courtesy of Boehringer Mannheim Corp.)

using one-fourth the original sample volume for the repeat test, and calculate a new result, taking the dilution into consideration.

Economy of sample size is a major consideration in developing automated procedures, but methodologies have limitations to maintain proper levels of sensitivity and specificity. The factors governing sample and reagent measurement are interdependent. Generally, if sample size is reduced, then reagent concentration must be increased to ensure sufficient color development for accurate photometric readings.

Reagent Systems and Delivery

Reagents may be classified as liquid or dry systems for use with automated analyzers. Liquid reagents may be purchased in bulk volume containers or in unit dose packaging as a convenience for *STAT* testing on some analyzers. Dry reagents are packaged in various forms. They may be bottled as lyophilized powder, which requires reconstitution with water or a buffer. Unless the manufacturer provides the diluent, water quality may be compromised by the laboratory. Other dry reagents may be in tablet form, such as those used by the ACA Star© and Paramax©. The ACA Star© crushes and dissolves reagent tablets in the plastic test pouch on board the instrument. The Paramax© uses an ultrasonic horn to break up and dissolve the tablet in a plastic cuvet filled with water. A third and unique type of dry reagent is the multilayered *dry chemistry slide* for the Vitros© analyzer. These slides have microscopically thin layers of dry reagents mounted on a plastic support. The slides are approximately the size of a postage stamp and not much thicker.

Reagent handling varies according to instrument capabilities and methodologies. Many test procedures use sensitive, short-lived working reagents, so contemporary analyzers use a variety of techniques to preserve them. One technique is to keep all reagents refrigerated until the moment of need, and then quickly preincubate them to reaction temperature or store them in a refrigerated compartment on the analyzer that feeds directly to the dispensing area. Another means of preservation is to provide reagents in a dried, tablet form and reconstitute them when the test is to be run. A third is to manufacture the reagent in two stable components that will be combined at the moment of reaction. If this approach is used, the first component also may be used as a diluent for the sample. The various manufacturers often use combinations of these reagent-handling techniques.

Reagents also must be dispensed and measured accurately. Many instruments use bulk reagents to decrease the preparation and changing of reagents. Instruments that do not use bulk reagents have unique reagent packaging.

In continuous-flow analyzers, reagents and diluents are supplied from bulk containers into which tubing is suspended. The inside diameter, or bore, of the tubing governs the amount of fluid that will be dispensed. A proportioning pump, along with a manifold, continuously and precisely introduces, proportions, and pumps liquids and air bubbles throughout the continuous-flow system.

To deliver reagents, many discrete analyzers use techniques similar to those used to measure and deliver the samples. A syringe or syringes driven by stepping motors pipet the reagents into reaction containers. Piston-driven pumps connected by tubing may also dispense reagents. Another technique for delivering reagents to reaction containers uses pressurized reagent bottles connected by tubing to dispensing valves. The computer controls the opening and closing of the valves. The fill volume of reagent into the reaction container is determined by the precise amount of time the valve remains open.

The Vitros© analyzers use slides to contain their entire reagent chemistry system. Multiple layers on the slide are backed by a clear polyester support. The coating itself is sandwiched in a plastic mount. There are three or more layers: (1) a spreading layer, which accepts the sample; (2) one or more central layers, which can alter the aliquot; and (3) an indicator layer, where the analyte of interest may be quantitated (Fig. 6-8). The number of layers varies depending on the assay to be performed. The color developed in the indicator layer varies with the concentration of the analyte in the sample. Physical or chemical reactions can occur in one layer, with the product of these reactions proceeding to another layer, where subsequent reactions can occur. Each layer may offer a unique environment and the possibility to carry out a reaction comparable to that offered in a chemistry assay, or it may promote an entirely different activity that does not occur in the liquid phase. The ability to create multiple reaction sites allows the possibility of manip-

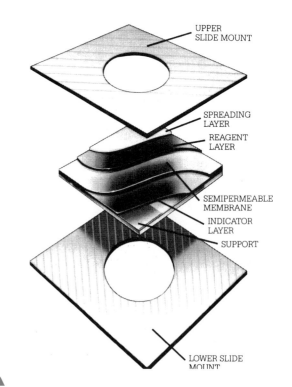

Figure 6-8. Vitros©. Slides with multiple layers contain the entire reagent chemistry system. (Courtesy of Johnson & Johnson Co.)

ulating and detecting compounds in ways not possible in solution chemistries. Interfering materials can be left behind or altered in upper layers.

All the reagents for the ACA© analyzers are contained in special test packs, which are compartmentalized plastic envelopes. A separate pack is used for each test performed on a specimen (Fig. 6-9). Compartments along the top of the envelope contain liquid or tablet reagents. The code name and a binary code depicting the test are printed on a plastic bar at the top of each test pack. All of the test packs require refrigerated storage.

The Paramax© has a reagent-dispensing carousel with positions for 32 disposable tablet dispensers, which contain a maximum of 300 tablets each. When activated by the instrument, the dispenser ejects a single tablet into a cuvet (Fig. 6-10). The reagent contained in each dispenser is identified to the instrument by an optical code, allowing dispensers to be loaded at random. The primary reagent for each test is contained in a tablet. Secondary reagents, which are liquid, are added after the tablet has been dissolved and the sample and diluent have been injected into the cuvet.

Chemical Reaction Phase

This phase consists of mixing, separation, incubation, and reaction time. In most discrete analyzers, the chemical reactants are held in individual moving containers that are either disposable or reusable. These reaction containers also function as the cuvets for optical analysis. If the cuvets are reusable, then wash stations are set up immediately after the read stations to clean and dry these containers (Fig. 6-11). This arrangement allows the analyzer to operate continuously without replacing cuvets. Examples of this type are the Hitachi and Synchron© analyzers. Alternatively, the reactants may be placed in a stationary reaction chamber (*eg,* ASTRA©)

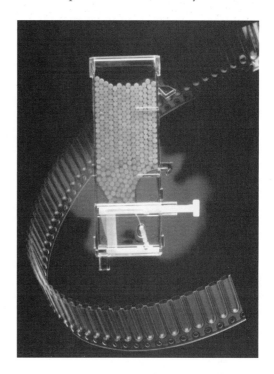

▲
Figure 6-10. Paramax©. The primary reagents are contained in a tablet. (Photograph courtesy of Dade International.)

in which a flow-through process of the reaction mixture occurs before and after the optical reading. In continuous-flow systems, flow-through cuvets are used and optical readings are taken during the flow of reactant fluids. The Chem 1© uses this approach with its single analytical pathway Teflon tube (Fig. 6-12).

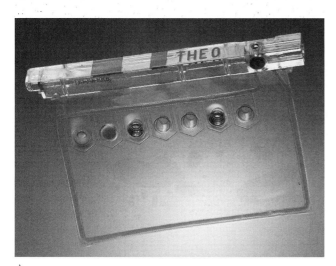

▲
Figure 6-9. ACA. A test pack with binary code on a bar at the top. (Photograph courtesy of Dade International.)

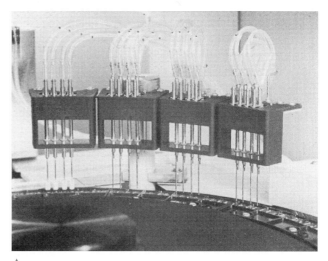

▲
Figure 6-11. Wash stations on Hitachi 736 analyzer perform the following: (1) aspirate reaction waste and dispense water; (2) aspirate and dispense rinse water; (3) aspirate rinse water and dispense water for measurement of cell blank; (4) aspirate cell blank water to dryness. (Photograph courtesy of Boehringer Mannheim Corp.)

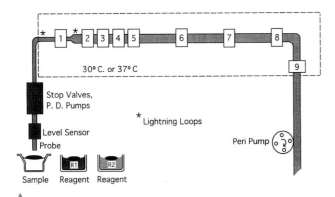

Figure 6-12. Technicon™ system overview. Chem 1 analytical pathway of single Teflon tube. (Courtesy of Bayer Co.)

Mixing

A vital component of each procedure is the adequate mixing of the reagents and sample. Instrument manufacturers go to great lengths to ensure complete mixing. Nonuniform mixtures can result in noise in continuous-flow analysis and in poor precision in discrete analysis.

Mixing is accomplished in continuous-flow analyzers (*eg*, the Chem 1©) through the use of coiled tubing. When the reagent and sample stream goes through coiled loops, the liquid rotates and tumbles in each loop. The differential rate of liquids falling through one another produces mixing in the coil.

The RA1000© uses a rapid start–stop action of the reaction tray. This causes a sloshing action against the walls of the cuvets, which mixes the components.

Centrifugal analyzers may use a start–stop sequence of rotation or bubbling of air through the sample and reagent to mix them while these solutions are moving from transfer disk to rotor. This process of transferring and mixing occurs in just a few seconds. The centrifugal force is responsible for the mixing as it pushes sample from its compartment, over a partition into a reagent-filled compartment, and finally into the cuvet space at the perimeter of the rotor.

In the Vitros© slide technology, the spreading layer provides a structure that permits a rapid and uniform spreading of the sample over the reagent layer(s) for even color development.

The ACA© analyzers have components specially designed for mixing: the breaker-mixers. Platens press selectively against and collapse the reagent compartments along the top of the test pack, releasing the reagents into the interior environment of the pack. Meanwhile, a lower platen presses against the bottom portion of the test-pack envelope to force the fluids up into the ruptured reagent compartments. Then, with a patting motion, the breaker-mixer thoroughly mixes the reagents with the fluids.

The Paramax© uses ultrasonic sound waves for 45 seconds to dissolve the reagent tablets in the deionized water in each cuvet. Ultrasound is also used to mix the sample with the previously prepared reagents and to mix, if necessary, the contents of the cuvets after a second reagent addition.

The Hitachi analyzers use stirring paddles that dip into the reaction container for a few seconds to stir sample and reagents, after which they return to a wash reservoir (Fig. 6-13). Other instruments, such as the ASTRA©, use magnetic stir bars lying in the bottom of the reaction container that, when activated, produce a whirling motion to mix. Still others may use forceful dispensing to accomplish mixing.

Separation

In chemical reactions, undesirable constituents that will interfere with an analysis may need to be separated from the sample before the other reagents are introduced into the system. Protein causes major interference in many analyses. One approach without separating protein is to use a very high reagent-to-sample ratio (the sample is highly diluted) so that any turbidity caused by precipitated protein is not sensed by the spectrophotometer. Another is to shorten reaction time to eliminate slower reacting interferents.

In the older continuous-flow systems, a dialyzer was the separation or filtering module. It performed the equivalent of the manual procedures of precipitation, centrifugation, and filtration, using a fine-pore cellophane membrane.

In the Vitros© slide technology, the spreading layer of the slide traps cells, crystals, and other small particulate matter but also retains large molecules such as protein. In essence, what passes through the spreading layer is a protein-free filtrate.

Many discrete analyzers have no automated methodology by which to separate interfering compounds from the reaction mixture. Therefore, methods have been chosen that have few interferences or that have known interferences that can be compensated for by the instrument (*eg*, using correction formulas).

The use of automated column chromatography in some methods gives the ACA© the ability to remove from the sample substances that might adversely influence the reaction. Three types of columns can be used: gel filtration, ion

Figure 6-13. Stirring paddles on Hitachi 736 analyzer. (Photograph courtesy of Boehringer Mannheim Corp.)

exchange, and protein-removal. The column is located at the top of the test pack, immediately under the plastic pack header. If the test pack contains a chromatographic column, the sample and a prescribed quantity of diluent are injected into the column fill site opposite the usual sample and diluent fill site. The sample is moved through the column by the pressure of the diluent and into the test pack for analysis.

Incubation

A heating bath in discrete or continuous-flow systems maintains the required temperature of the reaction mixture and provides the delay necessary to allow complete color development. The principal components of the heating bath are the heat-transfer medium (ie, water or air), the heating element, and the thermoregulator. A thermometer is located in the heating compartment of an analyzer and is monitored by the system's computer. On many discrete analyzer systems, the multi-cuvets float in a waterbath incubator maintained at a constant temperature of usually 37°C.

Slide technology incubates colorimetric slides at 37°C. There is a precondition station to bring the temperature of each slide close to 37°C before it enters the incubator. The incubator moves the slides at 12-second intervals in such a manner that each slide is at the incubator exit four times during the 5-minute incubation time. This feature is used for two point-rate methods and enables the first point reading to be taken part way through the incubation time. Potentiometric slides are held at 25°C. The slides are kept at this temperature for 3 minutes to ensure stability before reading.

Reaction Time

Before the optical reading by the spectrophotometer, the reaction time may depend on the rate of transport through the system to the "read" station, timed reagent additions with moving or stationary reaction chambers, or a combination of both processes.

An environment conducive to the completion of the reaction must be maintained for a sufficient length of time before spectrophotometric analysis of the product is made. Time is a definite limitation. To sustain the advantage of speedy multiple analyses, the instrument must produce results in as short a time as possible.

It is possible to monitor not only completion of a reaction but also the rate at which the reaction is proceeding. The instrument may delay the measurement for a predetermined length of time or may present the reaction mixtures for measurement at constant intervals of time. Use of rate reactions may have two advantages: the total analysis time is shortened and interfering chromogens that react slowly may be negated. Reaction rate is controlled by temperature; therefore, the reagent, timing, and spectrophotometric functions must be coordinated to work in harmony with the chosen temperature.

The ACA© has five delay stations located in the pack-processing area. At these points, the instrument performs no operations on the test packs. This interval allows enough re-

action time for endpoint or blank reactions to go to completion. The entire pack-processing area is maintained at 37°C through a closed environment with circulating fans.

The Paramax© has eight photometric stations located along the cuvet track. The addition of the sample initiates each reaction, which is monitored from 40 seconds to 10 minutes by the photometric stations for rate or endpoint reactions. The environment of the cuvets is maintained by a constant water bath in which the cuvets move.

Measurement Phase

After the reaction is completed, accommodation must be made to measure the products that are formed. Almost all available systems for measurement have been used, such as ultraviolet, fluorescent, and flame photometry; ion-specific electrodes; gamma counters; and luminometers. Still, the most common is visible and ultraviolet light spectrophotometry, although adaptations of traditional fluorescence measurement, such as fluorescence polarization, chemiluminescence, and bioluminescence, have become popular. The Abbott TDX©, for example, is a very popular instrument for drug analysis that uses fluorescence polarization to measure immunoassay reactions.

Analyzers that measure light require a monochromator to achieve the desired component wavelength. Traditionally, analyzers have used filters or filter wheels to separate light. The old Auto Analyzers used filters that were manually placed in position in the light path. Many instruments still use rotating filter wheels, which the computer controls to position the appropriate filter into the light path. However, newer and more sophisticated systems offer the higher resolution afforded by diffraction gratings to achieve light separation into its component colors. Many instruments now use such monochromators with either a mechanically rotating grating or a fixed grating that spreads its component wavelengths onto a fixed array of photo diodes, for example, Hitachi analyzers (Fig. 6-14). This latter grating arrangement, as well as rotating filter wheels, easily accommodates polychromatic light analysis, which offers improved sensitivity and specificity over monochromatic measurement. By recording optical readings at different wavelengths the instrument's computer can then use these data to correct for reaction mixture interferences that may occur at adjacent, as well as desired, wavelengths.

Many newer instruments use fiber optics as a medium to transport light signals from remote read stations back to a central monochromator detector box for analysis of these signals. The Paramax© has fiber optic cables, or "light pipes" as they are sometimes called, attached from multiple remote stations where the reaction mixtures reside, to a centralized filter wheel/detector unit that, in conjunction with the computer, sequences and analyzes a large volume of light signals from multiple reactions (Fig. 6-15).

The containers holding the reaction mixture also play a vital role in the measurement phase. The reagent volume

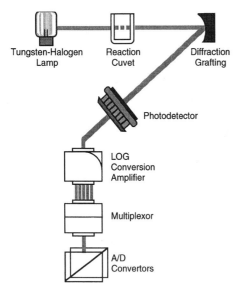

Photometer

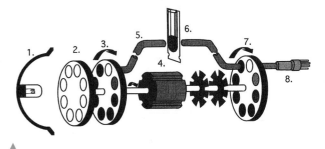

Figure 6-15. Paramax instrument photo optical system: (**1**) Source with reflector, (**2**) fixed focusing lens, (**3**) filter wheel, (**4**) double shafted motor, (**5**) fiber optic bundle, (**6**) cuvet, (**7**) filter wheel, (**8**) photomultiplier tube. (Courtesy of Dade International.)

Figure 6-14. Photometer for Hitachi© 736 analyzer. Fixed diffraction grating separates light into specific wavelengths and reflects them onto a fixed array of 11 specific photodetectors. Photometer has no moving parts. (Courtesy of Boehringer Mannheim Corp.)

and, therefore, sample size, speed of analysis, and sensitivity of measurement are some aspects influenced by the method of analysis.

A flow-through cuvet is used in continuous-flow analysis. The reagent stream under analysis flows continuously through the flow-cell tubing. The Chem 1© "captures" absorbance signals in between air bubbles of the flowing stream (Fig. 6-16). This means that no debubbling is required as in the older continuous-flow analyzers. As the stream flows through the flow-cell, a steady beam of light is focused through the stream. The amount of light that exits from the flow-cell is dictated primarily by the absorbance of light by the stream. The exiting light strikes a photodetector, which converts the light into electrical energy. Filters and light-focusing compo-

nents permit the desired light wavelength to reach the photodetector. The photometer continuously senses the sample photodetector output voltage and, as is the process in most analyzers, compares it with a reference output voltage. The electrical impulses are sent to a readout device, such as a printer or computer, for storage and retrieval.

In discrete analyzers, such as Hitachi or Synchron© or ACA© systems, the cuvet used for analysis is also the reaction vessel in which the entire procedure has occurred.

The ACA Star© photometer consists of a cuvet-forming device and a photometric system for making absorbance measurements. When compressed by the photometer jaws, the test pack forms a cuvet between quartz windows. A wetting solution is introduced between the quartz windows and the test-pack walls to achieve a good optical interface.

Centrifugal analysis measurement occurs while the rotor is rotating at a constant speed of approximately 1000 rpm. Consecutive readings are taken of the sample, the dark current (readings between cuvets), and the reference cuvet. Each cuvet passes through the light source every few milliseconds. After all the data points have been determined, centrifugation stops and the results are printed. The rotor is removed from the analyzer and discarded. For endpoint analyses, an initial absorbance is measured before the constituents have had time to react, usually a few seconds, and is considered a blank

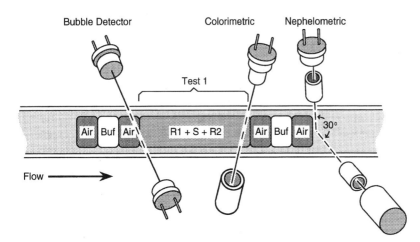

Figure 6-16. In-line reaction detector. Flow-through cuvet for Chem 1 Analyzer. (Courtesy of Bayer Co.)

measurement. After enough time has expired for the reaction to be completed, another absorbance reading is taken. For rate analyses, the initial absorbance is measured, and then a lag time is allowed (preset into the instrument for each analysis). For each assay, several data points are determined at a programmed data time interval. The instrument monitors the absorbance measurements at each data point and calculates a result.

Slide technology depends on reflectance spectrophotometry, as opposed to traditional transmittance photometry, to provide a quantitative result. The amount of chromogen in the indicator layer is read after light passes through the indicator layer, is reflected from the bottom of a pigment containing layer (usually the spreading layer), and is returned through the indicator layer to a light detector. For colorimetric determinations, the light source is a tungsten-halogen lamp. The beam focuses on a filter wheel holding up to eight interference filters, which are separated by a dark space. The beam is focused at a 45° angle to the bottom surface of the slide, and a silicon photodiode detects the portion of the beam that reflects down. Three readings are taken for the computer to derive reflectance density. The three recorded signals taken are (1) the filter wheel blocking the beam, (2) reflectance of a reference white surface with the programmed filter in the beam, and (3) reflectance of the slide with the selected filter in the beam (Fig. 6-17).

After a slide is read, it is shuttled back in the direction from which it came, where a trap door allows it to drop into a waste bin. If the reading was the first for a two-point rate test, the trap door remains closed, and the slide reenters the incubator.

The Ciba-Corning ACS:180©, a fully automated, random access immunoassay system (Fig. 6-18), uses chemiluminescence technology for reaction analysis. In chemiluminescence assays, quantitation of an analyte is based on emission of light resulting from a chemical reaction.[6] The principles of chemiluminescence immunoassays are similar to those of radioimmunoassay (RIA), except that an acridinium ester is used as the tracer and paramagnetic particles are used as the solid phase. Sample, tracer, and paramagnetic particle reagent are added and incubated in disposable plastic cuvets, depending on the assay protocol. After incubation, magnetic separation and washing of the particles is performed automatically. The cuvets are then transported into a light-sealed luminometer chamber, where appropriate reagents are added to initiate the chemiluminescent reaction. On injection of the reagents into the sample cuvet, the system luminometer generates the chemiluminescent signal. Luminometers are similar to gamma counters in that they use a photomultiplier tube detector. However, unlike gamma counters, luminometers do not require a crystal to convert gamma rays to light photons. Light photons from the sample are detected directly, converted to electrical pulses, and then converted to counts.

Signal Processing and Data Handling

Because most automated instruments print the results in reportable form, accurate calibration is essential to obtaining accurate information. There are many variables that may enter into the use of calibration standards. The matrices of the standards and unknowns may be different. Depending on the methodology, this may or may not present problems. If secondary standards are used to calibrate an instrument, the methods used to derive the standard's constituent values should be known. Standards containing more than one assayable value per vial may cause interference problems. Because there are no primary standards available for enzymes, either secondary standards or calibration factors based on the molar extinction coefficients of the products of the reactions may be used.

Many times, a laboratory will have more than one instrument capable of measuring a constituent. Unless there are different normal ranges published for each method, the

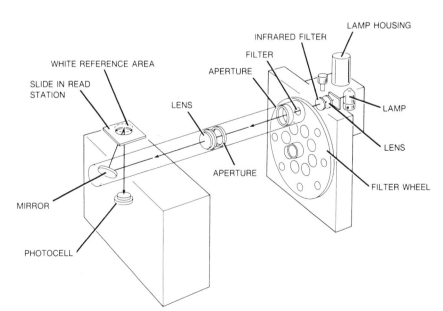

Figure 6-17. Components of the system for making colorimetric determinations in slide technology. (Courtesy of Johnson & Johnson Co.)

Figure 6-18. ACS:180© Automated Chemiluminescence System. (Photo courtesy of Ciba-Corning Diagnostics Corp.)

instruments should be calibrated so that the results are comparable. The advantage of calibrating an automated instrument is the long-term stability of the standard curve, which requires only monitoring with controls on a daily basis. Some analyzers use low- and high-concentration standards at the beginning of each run and then use the absorbances of the reactions produced by the standards to produce a standard curve electronically for each run. Other instruments are self-calibrating after analyzing standard solutions.

The original continuous-flow analyzers used six standards assayed at the beginning of each run to produce a calibration curve for that particular batch. Now, continuous-flow analyzers use a single-level calibrator to calibrate each run with water used to establish the baseline.

The centrifugal analyzer uses standards pipetted into designated cuvets in each run for endpoint analyses. After the delta absorbance for each sample has been obtained, the computer calculates the results by determining a constant for each standard. The constants are derived by dividing the concentration of the standard (pre-entered into the computer) by the delta absorbance and averaging the constants for all the standards to obtain a factor. The concentration of each control and unknown is determined by multiplying the delta absorbance of the unknown by the factor. If the concentration of an unknown exceeds the range of the standards, the result is printed with a *flag*. Enzyme activity is derived by a linear regression fit of the delta absorbance versus time. The slope of the produced line is multiplied by the enzyme factor (preentered) to calculate the activity.

Slide technology requires more sophisticated calculations to produce results. The calibration materials require a protein-based matrix because of the necessity for the calibrators to behave as serum when reacting with the various layers of the slides. Calibrator fluids are bovine serum based, and the concentration of each analyte is determined by reference methods. Endpoint tests require three calibrator fluids, blank-requiring tests need four calibrator fluids, and enzyme methods require three calibrators. Colorimetric tests use spline fits to produce the standardization. In enzyme analysis, a curve-fitting algorithm estimates the change in reflec-

tion density per unit time. This is converted to either absorbance or transmission-density change per unit time. Then, a quadratic equation converts the change in transmission density to volume activity (U/L) for each assay.

The ACA Star© retains the calibrations for each lot of a particular method until the laboratorian programs the instrument for recalibration. Instrument calibration is initiated or verified by assaying a minimum of three levels of primary standards or, in the case of enzymes, reference samples. The values obtained are compared with the known concentrations by using linear regression, with the x-axis representing the expected values and the y-axis representing the mean of the values obtained. The slope (scale factor) and y intercept (offset) are the adjustable parameters on the ACA©. On earlier models of this instrument, the parameters were determined and entered manually into the instrument computer by the operator, but in the more automated models, the calibration is performed by the instrument on notification by the operator.

After calibration has been performed and the chemical or electrical analysis of the specimen is either in progress or completed, the instrument's computer goes into the data acquisition and calculation mode. The process may involve signal averaging, which may entail hundreds of data pulses per second, as with a centrifugal analyzer, and blanking and corrections formulas for interferents that are programmed into the computer for calculation of results.

All advanced automated instruments have some method of reporting printed results with a link to sample identification. In sophisticated systems, the demographic-sample information is entered in the instrument's computer along with the tests required. Then the sample identification is printed with the test results. The Paramax© prints bar code labels for sample identification after the operator enters the patient information and tests requested into the computer terminal. Once the label is applied to the sample, the sample can be loaded on the analyzer. Microprocessors control the tests, reagents, and timing, while verifying the bar code for each sample. This is the link between the results reported and the specimen identification. Even the simplest of systems sequentially number the test results to provide a connection with the samples.

Because most instruments now have either a built-in or attached video monitor, the sophisticated software programs that come with the instrument can be displayed for readily accessible inspection of different aspects of the testing process. Computerized monitoring is available for such parameters as reaction and instrument linearity, quality control data with various options for statistical display and interpretation, short sample sensing with flags on the printout, abnormal patient results flagged, clot detection, reaction vessel or test chamber temperature, and reagent inventories. The printer can also display patients' results as well as various warnings previously mentioned. Most instrument manufacturers offer computer software for preventive maintenance schedules and algorithms for diagnostic troubleshooting. Some manufacturers also install phone modems on the analyzer for a direct communication link between the instrument and their service center for instant troubleshooting and diagnosis of problems.

SELECTION OF AUTOMATED ANALYZERS

Each manufacturer's approach to automation is unique. The instruments being evaluated should be rated according to previously identified needs. One laboratory may need a *STAT* analyzer, whereas another's need may be a batch analyzer for high-test volumes. When considering cost, the price of the instrument, and even more important, the total cost of consumables, are significant. The high capital cost of an instrument may actually be small when divided by the large number of samples to be processed. It is also important to calculate the total cost per test for each instrument that is considered. Moreover, a break-even analysis to study the relationship of fixed costs, variable costs, and profits would be helpful in the financial justification and economic impact on a laboratory. Of course, the mode of acquisition, that is, purchase, lease, rental, and so on, has to be factored into this analysis also. The variable cost of consumables will increase as more tests are performed or samples are analyzed. The ability to use reagents produced by more than one supplier (open vs. closed reagent systems) can provide a laboratory with the ability to customize testing and, possibly, save money. The labor component also should be evaluated. For instruments that have the CAP's workload recording units assigned, this provides an excellent basis for comparison, using time-motion studies. Unfortunately, the newest instruments on the market may not have been timed for labor, and thus this judgment may be more subjective than desired. With the large number of instruments available on the market, the goal is to find the right instrument for each situation.

Another major concern toward the selection of an instrument is its analytic capabilities. What is the instrument's performance characteristics for accuracy, precision, linearity, specificity, and sensitivity (which may be method dependent), calibration stability, and stability of reagents (both shelf life and onboard or reconstituted)? The best way to verify these performance characteristics of an analyzer before making a decision on an instrument is to see it in operation. Ideally, if a manufacturer will place an instrument in the prospective buyer's laboratory on a trial basis, then its analytic performance can be evaluated to the customer's satisfaction with studies to verify accuracy, precision, and linearity. At the same time, laboratory personnel can observe such design features as its test menus, true "walkaway" capability, "user friendliness," and the space that the instrument and its consumables occupy in their lab.

Clinical chemistry instrumentation provides speed and precision for assays that would otherwise be performed manually. The chosen methodologies and adherence to the requirements of the assay provide accuracy. No one may assume that the result produced is the correct value. Automated methods must be evaluated completely before being accepted as routine. It is important to understand how each instrument actually works.

TOTAL LABORATORY AUTOMATION[7]

The pressures of health care reform and managed care have caused increasing interest and improved productivity of the pre-analytical and post-analytical phases of laboratory testing. As for the analytical process itself, routine analyzers in clinical chemistry today have about all the mechanization they need. The next generation of automation is to replicate the Japanese practice of "black box" labs, in which the sample goes in at one end and the printed result comes out the other end.[8] Much effort has been expended during the past 5 years with the development of automated "front-end" feeding of the sample into the analytical "box" and computerized/automated management of the data that come out the back end of the box. There have been many developments in the three phases of the laboratory testing process, that is, pre-analytical (sample processing), analytical (chemical analyses), and post-analytical (data management) as they merge closer into an integrated *total laboratory automation* (TLA) system.

Pre-Analytical Phase (Sample Processing)

The sample handling protocol currently available on all major chemistry analyzers is to use the original specimen collection tube (primary tube sampling) of any size (after plasma or serum separation) as the sample cup on the analyzer and to use bar code readers also on the analyzer to identify the specimen. An automated process is gradually replacing this manual handling and presentation of the sample to the analyzer. Increasing efficiency while decreasing costs has been a major impetus for laboratories to start integrating some aspect of total laboratory automation into their operations. Conceptually TLA refers to automated devices and robots integrated with existing analyzers to perform all phases of laboratory testing. Most of the attention to date has been devoted to development of the front-end systems that can identify and label specimens, centrifuge the specimen and prepare aliquots, and sort and deliver samples to the analyzer or to storage.[9] "Back-end" systems may include removal of the specimens from the analyzer to storage, retrieval from storage for retesting, realiquoting, or disposal, as well as comprehensive management of the data from the analyzer for interfacing with the laboratory's information system (LIS).

Ever since Dr. Sasaki and his colleagues installed the first fully automated clinical laboratory in the world at Koshi Medical School in Japan,[10] the concept has gradually but steadily become a reality in the United States. The University of Nebraska and the University of Virginia have been the pioneers for development of TLA systems. In 1992 at the

University of Nebraska, a prototype of a laboratory automation platform was developed, the key components being a conveyance system, bar coded specimens, and a computer software package to control specimen movement and tracking, and coordination of robots with the instruments as work cells.[11] Some of the first automated laboratories in the United States have reported their experiences with front-end automation with a wealth of information for others interested in the technology.[12,13] The first hospital laboratory to have an automated system installed was the University of Virginia Hospital in Charlottesville in 1995. Their Medical Automation Research Center cooperated with Johnson & Johnson and Coulter Corporation to use a Vitros 950 attached to a Coulter/IDS "U" lane for direct sampling from a specimen conveyor without using intervening robotics.[14] The only commercially available turnkey system is the Hitachi Clinical Laboratory Automation System (CLAS), which is marketed by Boehringer-Mannheim Diagnostics. It couples the Hitachi line of analyzers to a conveyor belt system to provide a completely operational system with all interfaces.[15] Some laboratories have taken a modular phased-end approach with devices for only certain automated functions. Ciba-Corning Clinical Laboratories has installed Coulter/IDS robotic systems in several regional laboratories.[15]

Robotics and front-end automation are changing the face of the clinical laboratory.[16] Several instrument manufacturers are currently working on or are already marketing interfacing front-end conveyor devices along with software for their own chemistry analyzers. Johnson & Johnson Clinical Diagnostics introduced their Vitros 950 AT (Automation Technology) system in 1995 with an open architecture design to allow laboratories to select from many front-end automation systems rather than being locked into a proprietary interface. A Lab-Track interface is now available on the Dimension RxL from Dade International Chemistry Systems that is compatible with major laboratory automation vendors and allows for direct sampling from a track system. Beckman Instruments offers an open interface design called Accel-Net Laboratory Automation Network, for use with their Synchron CX7 Delta analyzers. Accel-Net can also be used in combination with other TLA conveyor systems for multitasking functions. Also, the technology now exists for micro-centrifugal separators to be integrated into clinical chemistry analyzers.[17] The bottom line is that robotics and front-end automation are here to stay. As more and more clinical laboratories reengineer for total laboratory automation, they are building core labs containing all of their automated analyzers as the necessary first step to more easily link the different instruments into one TLA system.[18]

Analytical Phase (Chemical Analyses)

There have been changes and improvements that are now common to many general chemistry analyzers. They include ever smaller micro-sampling and reagent dispensing with multiple additions possible from randomly replaced reagents; expanded onboard and total test menus, especially drugs and hormones; accelerated reaction times with chemistries for faster throughput and lower dwell time; higher resolution optics with grating monochromators and diode arrays for polychromatic analysis; improved flow-through electrodes; enhanced user-friendly interactive software for quality control, maintenance, and diagnostics; integrated modems for on-line troubleshooting; LIS interfacing data management systems; reduced frequencies of calibration and controls; automated modes for calibration, dilution, rerun, and maintenance; as well as ergonomic and physical design improvements for operator ease, serviceability, and maintenance reduction. According to recent CAP survey data, the five most popular general chemistry analyzers are Dimension (Dade International); Vitros© (Johnson & Johnson Clinical Diagnostics); Hitachi systems (Boehringer Mannheim Diagnostics); Paramax© (Dade International); and the Synchron systems (Beckman Instruments).[19] Features and specifications of these five systems are summarized in Table 6-1.

Post-Analytical Phase (Data Management)

Although most of the attention in recent years in total laboratory automation concept has been devoted to front-end systems for sample handling, several manufacturers have been developing and enhancing back-end handling of data. Bidirectional communication between the analyzer(s) and the host computer or LIS has become an absolutely essential link to request tests and enter patient demographics, automatically transfer this customized information to the analyzer(s), as well as post the results in the patient's record. Evaluation and management of the data between analysis and posting has become more sophisticated and automated with the integration of work station managers into the entire communication system.[20] Most of the data management devices are personal computer–based modules with manufacturers' proprietary software that interfaces with one or more of their analyzers and the host LIS. They offer automated management of quality control data with storage and evaluation of quality control results against the lab's predefined quality control perimeters with multiple plotting, displaying, and reporting capabilities. Review and editing of patient results before verification and transmission to the host is enhanced by user-defined perimeters for reportable range limits, panic value limits, delta checks, and quality control comparisons for clinical change, repeat testing, and algorithm analysis. Reagent inventory and quality control, along with monitoring of instrument functions, are also managed by the workstation's software. Most LIS vendors have interfacing software available for all the major chemistry analyzers.

Dade has developed the Data Fusion System Integrator for information management from up to seven linked Dimension or ACA analyzers, including those at remote locations. It allows the operator to program which tests are sent to specific analyzers and provides a true query mode with any LIS. The Ektanet workstation manager from Johnson &

TABLE 6-1. Comparison of Operational Features of Clinical Chemistry Analyzers

	Dimension RxL©	Hitachi 911©	Paramax Rx©	Synchron CX7D©	Vitros 950 IRC©
Throughput (tests/h)	740–920	720	720	900	900
Dwell Time	60 s	5 min	2 min	52 s	4 min
Methods available	57	70	56	80	43
Methods onboard	48	41	48	33	48
User-defined channels	10	41	10	24	none
Reagents	Liquid	Liquid	Tablets	Liquid	Slides
Reagents added "on the fly"	Yes	No	Yes (with pause)	Yes	After sampling
Onboard reagent stability	30 days	8 h–30 days	90 days	3–30 days	1–5 weeks
Sample volume	3–35 µL	2–35 µL	2–50 µL	3–69 µL	10 µL
Primary tube sampling	Yes	Yes	Yes	Yes	Yes
Closed container sampling	Yes	No	Yes	No	No
Calibration stability	3 mo	Auto-calibrate within run	15–90 days	8 h–42 days	3–6 mo
QC frequency	2 levels/24 h	2 levels/8 h	2 levels/24 h	2 levels/8 h	2 levels/24 h
Auto dilution	Yes	Yes	Yes	Preset	No
Auto rerun	Yes	Yes	Yes	Enzymes only	No
Cuvets	Disposable; sealed	Reusable	Disposable; sealed	Reusable	None
Optics	Halogen lamp with filter wheel	Halogen lamp with diffraction grating; fixed diode array (12)	Halogen lamp with filter wheels	Xenon lamp with diffraction grating; fixed diode array (10)	Halogen lamp with filter wheel; reflectance
Robotic interface	Lab-track	CLAS	No	Accel-Net	950-AT
Data management system	Data-fusion	No	ADAPT	Quantum IV	Ektanet

Johnson is the information hub for up to 5 Vitros systems. It uses broadcast downloading for specimen and test assignment to any analyzer; and patient means and histograms are included in the quality control software to enhance interpretation of results. Beckman's Quantum IV operating system software is integrated directly into the Synchron's computer. In addition to routine management functions, it has built-in calibration verification and linearity capabilities that conform to CLIA and CAP guidelines for results reporting. The Paramax Rx's management console uses ADAPT software to manage patient and quality control data, calibration, and user-defined testing.

FUTURE TRENDS IN AUTOMATION

Clinical chemistry automation will continue to evolve at a rapid pace from 2000 to 2010 as it has in the 1990s. With most of the same forces driving the automation market through 2005, as those discussed in this chapter, analyzers will continue to perform more cost effectively and efficiently. More integration and miniaturization of components and systems will persist to accommodate more sophisticated portable analyzers for the very successful point-of-care testing market.

More new tests for expanded menus will be developed with a mixture of measurement techniques used on the ana-

lyzers to include more immunoassays and PCR-based assays. Spectral-mapping, or multiple wavelength monitoring, with high-resolution photometers in analyzers will be routine for all specimens and tests as more instruments are designed with the monochromator device in the light path after the cuvet, not before. Spectral mapping capabilities will allow simultaneous analysis of multiple chemistry analytes in the same reaction vessel. This will have a tremendous impact on throughput and turnaround time of test results. Mass spectrometry and capillary electrophoresis will be used more extensively in clinical laboratories for identification and quantitation of elements and compounds in extremely small concentrations. In the coming years, more system and workflow integration will occur with robotics and data management for more inclusive total laboratory automation.[21] To accomplish this, more companies will form alliances to place their instrumentation products into the laboratories. The incorporation of artificial intelligence into analytical systems will evolve, using both experts systems and neural networks.[22,23] This will greatly advance the technologies of robotics, digital processing of data, computer-assisted diagnosis, and data integration with electronic patient records.

Finally, technological advances in chip technology and biosensors[24, 25] will accelerate the development of noninvasive, in vivo testing. Transcutaneous monitoring is already available with some blood gases. "True" or dynamic values from in vivo monitoring of constituents in blood and other body fluids will revolutionize laboratory medicine as we know it today. It all sounds futuristic, but then so did the first Auto Analyzer 50 years ago.

SUMMARY

Since the introduction of the first automated analyzer by Technicon, automated instruments have proliferated in the clinical chemistry laboratory. The driving forces behind the development of more and better automated analyzers included the increased volume of testing, faster turnaround time, and escalating costs. Automated analyzers use three basic approaches to sample analysis: continuous flow, centrifugal analysis, and discrete analysis. Manufacturers have designed their instruments to mimic the steps in a manual procedure to include specimen preparation and identification, specimen measurement and delivery, reagent systems and delivery, chemical reaction phase, measurement phase, and signal processing and data handling. Each manufacturer's approach to automation is unique.

When selecting an automated analyzer for the clinical chemistry laboratory, several factors need to be considered: needs, cost, mode of acquisition, consumables, analytic capabilities, space, and user friendliness. Clinical chemistry automation will continue to evolve. TLA now provides integration of all three phases of laboratory testing, linking the "front-end" sample processing and "back-end" data management with the analytical phase. Future trends will undoubtedly include more computer power, more portability, increased use of robotics, spectral mapping, continuous improvement in computer software, and the development of artificially intelligent computer systems. New technologies, such as polymerase chain reactions and noninvasive in vivo testing, will also have a large impact.

REVIEW QUESTIONS

1. Which of the following is not a driving force for more automation?
   a. High volume of testing
   b. Fast turnaround time
   c. Expectation of high-quality, accurate results
   d. Increase in the use of chemistry "panels"
2. Which of the following approaches of automated analyzers uses a rotor to mix reagents?
   a. Centrifugal analysis
   b. Continuous flow
   c. Discrete analysis
   d. Dry chemistry slide analysis
3. Which of the following types of analyzers offers random access capabilities?
   a. Continuous-flow analyzers
   b. Centrifugal analyzers

c. Discrete analyzers
d. None of the above

4. All of the following are primary considerations in the selection of an automated chemistry analyzer EXCEPT:
   a. Cost of consumables
   b. Total instrument cost
   c. Labor component
   d. How reagents are added or mixed
5. An example of a centrifugal analyzer would be the:
   a. Cobas-Bio
   b. Paramax
   c. Synchron
   d. Chem 1
6. *Dwell time* refers to:
   a. Number of tests an instrument can handle in a specified time
   b. Ability of an instrument to perform a defined workload in a specified time
   c. Time between initiation of a test and the completion of the analysis
   d. None of the above
7. The first commercial centrifugal analyzer was introduced in what year?
   a. 1957
   b. 1967
   c. 1970
   d. 1976
8. All of the following are advantages to automation EXCEPT:
   a. Increase number of tests performed
   b. Minimize labor component
   c. Correction for deficiencies inherent in methodologies
   d. Use of small amounts of samples and reagents in comparison to manual procedures
9. Which of the following steps in automation generally remains a manual process in most laboratories?
   a. Preparation of the sample
   b. Specimen measurement and delivery
   c. Reagent delivery
   d. Chemical reaction phase
10. Which of the following chemistry analyzers uses "slides" to contain the entire reagent system?
    a. Vitros© analyzers
    b. ACA© analyzers
    c. Paramax© analyzers
    d. None of the above

REFERENCES

1. Hodnett J. Automated analyzers have surpassed the test of time. Adv Med Lab 1994; 8.
2. Schoeff L, et al. Principles of laboratory instruments. St Louis, MO: Mosby–Year Book, 1993.
3. Boyce N. Why hospitals are moving to core labs. Clin Lab News 1996;22(11): 1–2.
4. Boyce N. Why labs should discourage routine testing. Clin Lab News 1996;22(12): 1,9.

5. Eisenwiener H, Keller M. Absorbance measurement in cuvets lying longitudinal to the light beam. Clin Chem 1979;25(1):9:117.

6. Dudley R.F., Chemiluminescence immunoassay: an alternative to RIA. Lab Med 1990;21(4):216.

7. Schoeff L. Clinical instrumentation (general chemistry analyzers). Anal Chem 1997; 69(12):200R–203R.

8. Ringel M. Automation is everywhere. Med Lab Observer 1996; Dec: 28:38–43.

9. Felder R. Front-end automation. In: Kost G, ed. Handbook of clinical laboratory automation and robotics. New York: Wiley, 1995.

10. Sasaki M. A fully automated clinical laboratory. Lab Inform Mgmt 1993; 21:159–168.

11. Markin R, Sasaki M. A laboratory automation platform: the next robotic step. Med Lab Observer 1992; 24:24–29.

12. Bauer S, Teplitz C. Total laboratory automation: a view of the 21st century. Med Lab Observer 1995; 27:22–25.

13. Bauer S. Total laboratory automation: system design. Teplitz C. Med Lab Observer 1995; 27:44–50.

14. Felder R. Cost justifying laboratory automation. Clin Lab News 1996; 22(4):10–11, 17.

15. Felder R. Laboratory automation: strategies and possibilities. Clin Lab News 1996; 22(3):10–11.

16. Boyd J, Felder R, Savory J. Robotics and the changing face of the clinical laboratory. Clin Chem 1996; 42(12):1901–1910.

17. Richardson P, Molloy J, Ravenhall R, et al. High speed centrifugal separator for rapid online sample clarification in biotechnology. Biotechnol 1996; 49(1–3):111–118.

18. Zenie F. Re-engineering the laboratory. Autom Chem 1996; 18(4): 135–141.

19. CAP surveys, proficiency testing program. Northfield, IL: College of American Pathologists, 1997.

20. Saboe T. Managing laboratory automation. J Autom Chem 1995; 17 (3):83–88.

21. Brzezicki L. Workflow integration: does it make sense for your lab? Adv for Lab 1996; 23: 57–62.

22. Place J, Truchaud A, Ozawa K, et al. Use of artificial intelligence in analytical systems for the clinical laboratory. J Autom Chem 1995; 17(1): 1–15.

23. Boyce N. Neural networks in the lab: new hope or just hype? Clin Lab News 1997; 23(1): 2–3.

24. Boyce N. Tiny "lab chips" with huge potential? Clin Lab News 1996; 22(12): 21.

25. Rosen S. Biosensors: where do we go from here? Med Lab Observer 1995; 27: 24–29.

CHAPTER 7

Point-of-Care Testing

Steve Bradley

Objectives

Upon completion of this chapter, the clinical laboratorian should be able to:

- *Describe point-of-care testing (POCT).*

- *Explain and outline how POCT is used in patient management and distinguish between the concept of perceived versus actual advantages.*

- *Provide an outline indicating the need for continual interaction between laboratory and nursing departments in establishing and developing an efficient point-of-care system.*

- *List regulatory requirements in monitoring the quality of POCT.*

KEY TERMS

Point-of-care testing (POCT)
Clinical Laboratory
  Improvement Amendment
  (CLIA)

Quality improvement
Waived tests

POCT continues to grow as a facet of clinical laboratory science. Although POCT may be the responsibility of the clinical laboratory, testing may be predominately performed by other medical professionals or by the patient. This chapter focuses on the who, what, why, and when of POCT, including implementation and monitoring and cost and regulatory issues. The analytical principles used by POCT instruments is discussed in Chapter 4, *Analytical Techniques and Instrumentation*.

POINT-OF-CARE TESTING (POCT)

Point-of-care testing (POCT) is analytical testing of patient specimens at or near the location of patient care. It could be at the bedside, on the nursing unit, in a utility room, or any other location outside the main laboratory. It has been referred to as near-patient testing, ancillary testing, bedside testing, alternative site testing, and decentralized testing, among others. Basically, if the testing is not done in the main clinical laboratory, and it is a mobile system, it is called POCT.

Most POCT is classified as waived by government standards. This listing is reviewed and updated on a frequent basis, with tests continually being added (Table 7-1). *Waived tests* use systems that are designed to be fault tolerant to some extent. If the operator does not perform the test properly, the incorrect results should not pose significant risk to the patient. In a laboratory environment, this is unacceptable. Clinical laboratories provide the best quality analytical results on a consistent basis. The concept of accepting error is contradictory to a clinical laboratorian's education and goals. The prime reason for the development of waived tests and POCT is to provide the patient a quick and efficient means of performing and monitoring clinical results. The patient or physician uses these results to appropriately regulate or adjust medication or activities.

USE AND APPLICATIONS FOR POCT

Rationale

The primary reason for POCT in a health care environment is to obtain laboratory results at a decision point during patient care. It does not help the clinician to know the patient

TABLE 7-1. Current CLIA Waived Tests*

Bladder tumor associated antigen	Hematocrit
Body fluid pH	Hemoglobin
Cholesterol	Infectious mononucleosis antibodies
Erythrocyte sedimentation rate, nonautomated	Microalbumin
Ethanol	Nicotine and/or metabolites
Fecal occult blood	Ovulation test by visual color comparison
	Prothrombin time
Fructosamine	Spun microhematocrit
Gastric occult blood	Streptococcus, group A
Glucose (whole blood)	Triglyceride
Glycosylated hemoglobin (Hgb AlC)	
HDL cholesterol	Urine catalase
	Urine creatinine
	Urine dipstick or tablet analytes, nonautomated
Helicobacter pylori	Urine human chronic gonadotropin visual color comparison tests
Helicobacter pylori antibodies	

List as of June 4, 1999.

TABLE 7-2. Major Differences Between Clinical Laboratories and POCT

Aspect	Clinical Laboratory Testing	POCT
Turnaround time	Slow	Fast
Specimen identification	Complex	Simple
Specimen processing	Usually required	None
Calibration	Frequent	Infrequent
Reagent	Preparation required	Ready to use
Consumables	Few	Many
Analyzers	Complex	Simple
Operator competency	High	Variable
Cost per test	Low	High

had a life-threatening result 20 minutes to 30 minutes after they should have taken appropriate action. The time to obtain that result is when a decision needs to be made to provide immediate quality care to the patient. In most cases, these situations will be in focused areas within a hospital or clinic. Urgent care situations exist primarily with emergency medical services, emergency departments, intensive care units, and the operating rooms. These areas are not exclusive. Patients can have episodes where immediate laboratory results would be beneficial in any aspect of care.

The primary emphasis is on immediate results. Using conventional systems used throughout hospitals and clinics, there are numerous inherent delays. Laboratories are continually striving to reduce turnaround time—the time it takes to get a sample collected, processed, analyzed, and the results delivered to the clinician. Technological advancements in instrumentation have almost eliminated analyzer delays, with direct tube micro-sampling, bar code scanners, and host query. Primary consideration in extended turnaround time should focus on preanalytical delays. Preanalytical delays are measured from the time the order is documented to the time the specimen can be analyzed. The clinician's concept of preanalytical begins when the clinician first thinks of a need for a test, which at times compounds the perception of preanalytical delays. POCT essentially eliminates the preanalytical delays that are inherent in clinical laboratory systems. The testing begins when the clinician orally requests the test. Patient identification is uncomplicated. Sample collection is simplified with usually no processing required. The specimen is analyzed and the results are made available within minutes of the request. Table 7-2 summarizes these and other differing aspects between POCT and testing in large clinical laboratories. The clinical labora-

tory will never be able to succeed in providing this speed in providing results. That is why it is essential that a collaborative effort be made among laboratorians and primary care providers to ensure that POCT systems used in direct patient care are the best possible.

Staff Requirements

Anyone can perform POCT when using primarily waived tests. Remember that waived tests have been designed to provide results in situations in which the complications of an incorrect result are minimal. What is done if a result is inaccurate? Could the complications be severe? How is an inaccurate result recognized? What happens when the clinician performs a test that is more complex? POCT is more involved and complex than most laboratorians or medical professionals comprehend.

Waived tests can, and do, give inaccurate results. In a home setting, if a result is unusual, the patient may not even take notice, thinking the device is not working properly. In a hospital setting, nursing staff may not comprehend that instruments can and do give inaccurate results. Most inaccuracies are a result of improper or poor technique. This is usually not acknowledged, and without appropriate follow-up, complications could occur. For example, using inaccurate whole blood glucose results as a guide, inadequate insulin doses may be administered. This would cause unanticipated fluctuations in the patient's blood glucose concentration. Lack of adequate control causes delays in improvement, increased potential for complications, and a longer stay in the hospital.

Many people perform POCT. Physicians, nurses, aides, phlebotomists, technologists, technicians, and therapists perform laboratory testing. Each facility is different. Deciding factors include availability of personnel, staffing, regulations from certifying agencies, state health guidelines, and need. Basically, anyone can perform POCT with proper training and documentation of competency. Regular review of skills is required by all certifying agencies. This helps ensure that the results obtained are accurate and the operators are skilled, or competent, in the testing they perform.

Factors Affecting Performance

There are numerous factors involved in determining when and where POCT should be performed. If left uncontrolled, POCT would soon become unmanageable. Without a coordinated effort and monitoring, it would soon become an issue of physician convenience. This would lead to virtually any test performed anywhere in the facility. On the other hand, too much control of when and where, or an extremely inflexible system, would compromise patient care when some latitude and flexibility are needed. This basically defeats the purpose and design of POCT. Obviously, the ideal system—if one could exist—would include checks and balances throughout the system, evaluating need, quality, efficiency, cost, ease of use, competency, and patient benefit. Not all of these can be monitored realistically, and each situation will have a different focus.

The first step in determining when and where POCT should be performed is need. Is the test really needed where the request originated and would the test improve the patient's condition, or outcome, during that episode of care? Ideally, there should be a committee to evaluate such requests; the committee would comprise clinicians, pathologists, clinical scientists, and nursing and support personnel. This group would determine if the test could be beneficial in the proposed care setting, evaluate all aspects of the test as previously indicated, and approve or deny requests for evaluation based on the findings.

Initially, the adequacy of the current system should be evaluated. Why can't the laboratory provide the service? Why is there a perception that the result is not provided in a timely manner? Could the laboratory improve services? Could a bedside system provide improved patient outcome? Is a system available? Incorporating clinicians in the group is critical. Their understanding of the requester's perception is crucial to an efficient system. Pathologists will be able to provide consultative support regarding regulations, laboratory resources, and availability of POCT systems. Clinical laboratorians could clarify technical aspects regarding quality, reliability, accuracy, cost, and efficiency of POCT systems, and would be necessary in the evaluation of any systems. Nursing and support personnel would be needed to provide feedback on functionality within the nursing and patient care environment. Efficiency of a POCT system will be primarily an evaluation of faster results. Without improved result turnaround time, there is no need to investigate POCT.

Cost aspects can be simple or complex. The simplest perspective would involve a comparison of reagents and supplies (consumables) between the POCT system and the main laboratory. This should include items such as reagents, controls, sampling devices, containers, and calibrators. Additional cost comparisons could include labor. It is usually more costly to have a nurse rather than a technician perform a task for 10 minutes. Hidden costs should also be investigated. Maintenance agreements should be specifically outlined regarding responsibilities and exchange policies and procedures. The most complex evaluations of costs have involved comparison of a patient's length of stay. This type of evaluation is extremely complex. The goal is to determine if the POCT has decreased the patient's length of stay. This type of study cannot realistically determine if POCT is solely responsible for decreased length of stay. To do so, it would be necessary to eliminate influence from all other aspects of patient care.

Ease of use and competency are other factors requiring investigation. Most POCT systems are listed as waived tests. Nonwaived testing is more complex and requires a more detailed understanding of the test, procedures, quality control, and clinical applications. It also lends itself to a more detailed and complex training system. The more complex a system used, the more imperative the need for a comprehensive training system.

Another fairly abstract aspect of evaluation of a POCT system is that of patient benefit. The primary question when dealing with this situation is, Will the clinician be better prepared to care for the patient with the added information provided by the POCT result? This question is difficult because it depends on many of the conditions previously discussed. Are the results accurate and do they appropriately indicate the patient's condition? If it is a complex system, are the operators competent if they perform the test only once a month? Has the training been sufficient to provide the operators the information they need to operate independently? Patient benefit may focus on any one or a combination of these conditions. It should never be allowed to revolve around the primary care provider's impression of patient benefit. If this were to be allowed, we would be in the situation of uncontrolled and uncoordinated testing.

IMPLEMENTATION AND MONITORING OF POCT

Evaluation of POCT Systems

Evaluation of a test is expensive and time consuming. It is frustrating for clinicians as well because they may be unfamiliar with the requirements. They feel that if the system is available and on the market, it must be all right to use. This is not always accurate. Evaluation of a system requires analysis of many samples with calculations for correlation (statistical accuracy), establishment of reference ranges, and calculation of control ranges. The more complex the system, the more detailed the evaluation. Waived tests are the least complex to perform and evaluate, and have minimal requirements. Basic requirements include training, proficiency testing, data management, and inspections. Dealing with moderately complex tests is more involved, with additional requirements of personnel qualifications, quality assurance, and more strict inspection guidelines.

Correlation of data obtained from the POCT system with the main laboratory is vital. If patient care is to be applied consistently, results must match as closely as possible between the two areas. If there are discrepancies, patient care may not be applied consistently, compromising the patient's condition. Correlation lends itself well to numeric data. Analysis should include (but is not limited to) mean, standard deviation, coefficient of variation, and correlation coefficient. Additional parameters may include median, mode, normality parameters (variance, skewness, kurtosis), and statistical deviations (such as student's t, F-test, and z tests).

Qualitative data must be analyzed in a different manner. Sensitivity, specificity, and accuracy are the primary parameters when dealing with qualitative data. Patients' diagnosis is needed to determine whether the test result is a true or false positive/negative.

Establishment of reference ranges for patients and controls is vital in evaluation in POCT systems. The results obtained must make sense, and if clinicians do not know where they should be with a result, they will not know the patient's status. Reference ranges will be necessary in evaluating a patient's status. Clinicians rely on the laboratory to provide this information. They may be unaware that it is information that must be calculated for each test. Using published ranges does not negate the need for data collection. Published ranges must be confirmed, indicating that they are appropriate for the regional population. Additional concern lies in regulatory requirements that state there must be periodic confirmation of the accuracy of the reference ranges. As in any laboratory test, quality control ranges are also important. To consistently monitor a system and check for mechanical or procedural failures, quality control material must be assayed at least once with each run of patient specimens.

Evaluation of a POCT system is incomplete until a subgroup of the personnel who will be responsible for the test has been trained. It is the subgroup's responsibility to determine how effective the system will be in the targeted patient care setting. It would be difficult to ever have a successful program if the operators do not agree with the situation, regardless of how well the test performs. With input from the proposed operators, it will be a more thorough evaluation. Their input will contribute key information regarding operations. If the system cannot be used properly, for whatever reason, it is best to determine this during the evaluation phase, rather than find out after expenditure of extensive time, effort, and capital.

Physical Facilities (Space and Connectivity)

Most POCT systems are designed to be mobile. The only considerations when dealing with space requirements would be for centralized organization of materials and supplies. With more complex and involved systems, there may be a need for counter space. Placement will need to be investigated. Dealing with biohazardous materials will require an isolated area to help prevent personnel not actively involved in the testing process from being exposed to possible contaminants. Another consideration is data management. The more advanced systems become, the more the need for network connectivity. It may be necessary to select an area with easy access to a network or computer connections. If the device is not designed with this type of data management, space will be required to set up an alternate manual data management system.

Some facilities may have small laboratories set up in areas with specific needs (*eg,* emergency department, intensive care units, operating rooms). These areas still qualify as POCT when the testing is limited to the specific needs of that area. Staffing and management are not criteria used to define POCT. Most situations require considerations of more than one area in managing the sites. Numerous groups must agree on accessibility and menu selection. Staffing may be the responsibility of the laboratories, or it may be arranged with another group. Alternate resources may include therapists, nurses, or aides. The main emphasis when using non–laboratory-trained personnel, is proper training and documentation of competency.

Training and Documentation

An efficient and effective training system is imperative. It is essential that laboratory personnel be actively involved in all training of POCT, regardless of who will be performing the tests. Laboratorians must also understand that the most difficult aspect to the training will be explaining the need for quality control and quality assurance. Nonlaboratory personnel generally do not comprehend the necessity, or the ramifications, of improper monitoring of a laboratory testing system. In most cases, they will view these as unnecessary and time consuming. Some may perceive that they are not laboratory staff, they did not go to school for laboratory testing, they do not have enough time to do it, and it is not important anyway. Many have the attitude that if the POCT device gives them a number it is all right—"if it didn't work we wouldn't be using it." "Attitude" is the most difficult hurdle for laboratory personnel who are training nonlaboratory staff. It is necessary for trainers to clearly and comprehensively grasp all situations related to the test. They must know the clinical significance of the test as well as what situations it would be best used for, and be able to convey complications from improperly performed tests or inaccurate results.

Outlines, which should be used whenever possible, will ensure that all appropriate information is presented and eliminate inconsistencies among trainers. In many cases, the vendor provides consultants to perform the training seminars to help alleviate some of the stress when bringing a system online. Once a majority of the personnel have been trained, the laboratory can assume the primary responsibility of training the remaining staff, and any other new employees as they are hired. The vendor should provide outlines and training materials. Clinicians should work with the vendor's representative regarding any training concerns. It is in their best interest to address any issues regarding their system.

Documentation of training should include forms that acknowledge an understanding of the principles of operation and an understanding of the clinical significance of the test(s), and that outline aspects of testing including patient identification, specimen collection, instrument setup, quality control, maintenance, and basic troubleshooting. Copies of procedure manuals should be available for review as necessary. Emphasis will vary, depending on previous experience with the personnel being trained. Some may emphasize instrument setup; others, maintenance; and some, troubleshooting. Each facility will need to fine-tune their system to ensure that all testing personnel understand the system and its functions. They must also understand the complications that could arise from improper use or failure to acknowledge when the test is invalid.

Quality Control

One of the main difficulties in the establishment of POCT instilling the concept of quality control. Most nonlaboratorians will perceive quality control as a waste of their time and do not want to deal with it.[1] With continual staffing reductions, time is an extremely valuable asset. It will be crucial to educate everyone in the concept of quality control as an asset, rather than a hindering, time-consuming process.

Quality control has two primary functions. First, it is a means of evaluating the technical skills of the person performing the test(s). If the person is unfamiliar with the system or has a limited understanding of the operations, the person's quality control may reflect any deficiencies he or she has while performing the test.[2] The second goal of quality control is to monitor and detect problems with the equipment or reagents. If a person's skills are not questioned, and the quality control indicates problems, it can be presumed to be with the equipment or supplies used in determining the results.

The concept of electronic quality control has been discussed and has been received with mixed feelings. Proponents have stated that electronic quality control is a more accurate means of monitoring systems functions because it tests the extreme operating ranges of the device. It is independent of the operator and, therefore, provides more consistent results and allows for more accurate longitudinal monitoring of analyzers. Opponents have stated that the entire operating range of a device is not critical. Monitoring should be performed in a manner consistent with that used with patient specimens. Quality control should focus on physiologic ranges rather than extremes and the laboratorian determines if a technique is adequate when electronic quality controls remove that aspect of testing.

Monitoring and Review of Records/Logs

A number of agencies involved in health care require regular review of POCT logs to ensure that accurate records are maintained for patient and quality control test results. The easiest way to review logs is to keep in mind an audit trail. The audit trail should include the patient's name, identification number, test and result, units, operator identification, date and time of testing, lot numbers of reagents and controls, expiration dates, and any necessary corrective action. It is important to monitor whether quality control was performed according to specified protocol (required with each run of specimens). This is identical to the required information maintained for each test performed within the clinical laboratory.

A regular review of logs will always find problems. A means of providing feedback and corrective action needs to be developed and continually enhanced so that the most effective bidirectional communication occurs. Feedback involves providing information to personnel, usually nursing managers or directors, who are responsible for ensuring accuracy in their areas. The information must include sufficient information and clarification so that nursing managers may appropriately council their employees. Tracking the problems may assist the nursing managers in determining which personnel may be having difficulties that would require further education, or in reinforcing education previously provided. If tracking shows that problems are not isolated to specific personnel, then investigation can be initiated to examine problems with the overall system. In any situation, if the logs are not reviewed regularly, problems will go unnoticed and uncorrected. Unrecognized problems compromise patient care and health.

Systems that can be integrated into a networked computer system are becoming more prevalent in the POCT environment. The primary focus with these systems has been on the higher volume tests, such as whole blood glucose testing. Although integrating the system into a computer-oriented environment does not eliminate the problems associated with accurate record-keeping, it does allow for enhanced automated monitoring of the records. Automated monitoring provides more timely feedback to the nursing managers so that corrective action will be most effective. It is easier to correct improper behavior when the activity has recently occurred, rather than attempting to correct situations that occurred during a previous week or month. Integration into a network also allows for more accurate patient records by providing a means of interfacing with laboratory or hospital information systems. Such interfacing allows for more complete records when compared to manual tracking and documentation into the patient's chart. It also allows for all health care providers to have access to the results when they need them. If results are manually tracked, they may not be recorded in the chart, which would require a physician to know where the results are recorded and review an additional log to monitor the patient's status. Network interfacing also also eliminates the potential for duplicate testing when consulting health care providers are unaware of previous test results.

Quality Improvement and Proficiency Testing

As with any system, there will be problems. *Quality improvement* is a system that acknowledges these problems and establishes procedures for correction and mechanisms to prevent reoccurrence. When dealing with POCT, it may be necessary to have multidisciplinary involvement to appropriately deal with situations. The laboratory is responsible for monitoring testing situations for compliance and detecting errors or potential pitfalls in the testing process. The multidisciplinary group is then responsible for determining how to correct the situation and establishing procedures to ensure that the same failure does not replicate in other areas. Less complicated systems would allow for direct interaction between the laboratory and the testing department. In any situation, the key is to detect and correct the problems early, thereby reducing the potential for harm to the patient.

One established method for detecting technical or analytical errors is to use proficiency testing materials. These specimens may be purchased from an agency that works with government departments to monitor the effectiveness of the test results. Blind samples are regularly sent to be tested. These samples are to be tested in a manner consistent with patient testing. Results are recorded and returned to the agency for analysis. Reports are generated and sent for review. The agency will also provide results to government agencies responsible for laboratory accreditation. Corrective action is required for any result exceeding defined limits. Integration of proficiency testing is not difficult with POCT; however, it can be time consuming in terms of distribution of the specimen(s). A coordinated effort between the laboratory and all areas performing proficiency testing is essential. Once protocols are established, the system should function similarly to those systems used within the laboratory. This effort will instill an exchange of information and ideas between the laboratory and areas performing testing, thus gradually leading to improvement in the overall processes.

Proficiency testing systems may be fairly unique for each institution. Regulatory agencies have stated that it is more effective to monitor sites consistently, establishing consistent background and monitoring. This may not always be practical in larger settings. It may be more appropriate to vary the locations to detect problems that normally would be neglected if monitoring were performed repetitively by site. Once a system has been initiated, the problems encountered will compel fine-tuning of the system into an efficient process.

COST ISSUES

Budget (Instruments/Equipment)/ Cost Containment

As with the laboratory, ancillary sites are under cost restraints. Laboratory testing is one of the largest revenue-producing activities in the health care environment. This revenue provides incentive for areas to want to perform testing. This incentive, however, may be detrimental to the facility overall. In contrast to testing in large volumes on large analyzers in the main laboratory, POCT equipment is usually much more expensive to operate. It may appear on paper that one area has increased reimbursements due to laboratory testing, but the overall result is a deficit for the facility because of the increased costs per test. As an example, batch testing for a specific test in the main clinical laboratory costs approximately $2.50 per test as opposed to $7.99 per test at a point-of-care site. This includes controls and standards within the laboratory, but only one patient at the POCT site. The POCT site would need to run controls, which would increase the cost to $15.99. When focusing on a single test, including any necessary standards and controls, the main laboratory cost would be $16.07. The cost for a single test at the POCT site would be $23.97.

The more elaborate the POCT in a facility, the more costly it will become. As testing systems are simplified and easier to operate, they will become more prevalent in the patient care setting. Instrument-based systems may need to be justified to show the potential financial benefit to the facility. Unlike large analyzers in the laboratory, the instruments used at POCT sites are less expensive to obtain, but are more costly per reportable test. This alone may make the POCT prohibitive for various tests in some institutions. It may still be necessary to work these systems into the budget, accounting for costs of equipment, reagents, quality controls, supplies, and any connectivity requirements. The more elaborate and enhanced the capabilities, the more elaborate the considerations when dealing with budget concerns. Storage requirements for reagents and supplies will need to be anticipated. If reagents must be refrigerated, sufficient refrigerators will be needed to handle the supply.

Investigation of costs and overall impact to the facility is important. In some situations, cost will not be the deciding factor. If patient care can be significantly improved, then the added cost is justified. If the test will be performed primarily for physician convenience, then it would not be advisable to pursue POCT. Cost containment is critical to the survival of an institution. If the same procedure could be performed at two or more locations, it would be advisable to perform that test where it is least costly. Fortunately an increasing number of health care providers are becoming more aware of this limitation. There are some who have a more limited focus regarding these issues and will pursue POCT regardless of the costs involved. Under these circumstances, physician education will be an essential component when dealing with budget and cost issues related to POCT.

Reagents and Storage

Manufacturers label their reagents with various information, including any storage requirements. These may include restrictions on lighting, humidity, or temperature. Regulatory agencies require compliance with these requirements. Such compliance will require monitoring of the storage conditions. If requirements are exceeded, corrective action will need to be taken and documented. Physical storage space is an aspect rarely considered when investigating POCT. If reagents need to be stored in a refrigerator, is there sufficient space available at the ancillary site? If not, what other alternatives are available? If the requirements are for room temperature storage, are the storage locations monitored for compliance? Additional documentation is needed to provide evidence of compliance. Temperature logs, room temperature, or refrigerator need to be maintained and reviewed.

Supplies for testing need to be stored in an area readily accessible to personnel who will be performing the test. Are sufficient and safe storage facilities available? Anticipation of testing volumes and needed storage may affect the decision. If supplies cannot be adequately stored, and under proper conditions, the testing process will be compromised and could potentially affect the results. If there will be unstable storage conditions, the results will be inaccurate and could potentially lead to an inaccurate diagnosis and possibly compromise patient care.

REGULATORY ISSUES

Federal, state, and private regulatory issues abound in the laboratory. Federal requirements are usually the most broad and provide basic guidelines when dealing with laboratory operations. They outline issues such as certifications and complexity; proficiency testing enrollment, successful participation, and reinstatement after failure to participate successfully; patient test management; quality control; personnel standards; quality assurance; and inspections. State agencies may impose more stringent guidelines if they have determined such guidelines are necessary. Private agencies may add additional requirement.

The most frequently cited federal regulation, with which most laboratories must comply, is the CLIA regulation (CLIA-88). Health Care Financing Administration (HCFA) is the governing agency responsible for administration of the CLIA guidelines. All laboratories must obtain a "CLIA license" to operate a laboratory. The management of those personnel licensed in each facility will vary. Some institutions may opt for all laboratory testing to be managed under one license. Other institutions may deem it necessary to have multiple licenses. The only exceptions to CLIA certification are those states that have received exempt status. Laboratories in those states must obtain a state license and must comply with the requirements that are, at times, more stringent than the CLIA guidelines.

POCT is not excluded from these regulations. The most common types of POCT are waived tests and physician-performed microscopy procedures. CLIA waived tests are those designed primarily with respect to patient self-testing. Physician performed microscopy procedures may include urinalysis sediment, KOH, wet mount, and gram stains. There is a duality in the regulations when working in an institutional setting. Waived tests are basic, and when performed at home have almost no requirements. When waived tests are performed in an institutional setting, all aspects of laboratory operations apply. Proficiency testing, patient test management, quality control, quality assurance, and personnel standards must be considered. There has been, and will continue to be, confusion regarding this duality. The primary reason is the differing degrees of requirements among the differing agencies with which a laboratory deals. If the POCT sites and the main laboratory operate under one license, then the POCT sites are required to comply with all requirements imposed on the laboratory. If they operate separately, they must still comply with the basic federal requirements in operating a laboratory. Laboratories are better prepared in dealing with these regulations and requirements. It is beneficial to the laboratory to assume the lead role in the education and training of those personnel less prepared in these regulations. Laboratorians are better prepared in patient test management, quality control, and quality assurance issues related to analytical testing of patient specimens.

Regulatory requirements in the health care industry have been increasing. Accountability will be distributed in any number of ways. A recent requirement, effective in 1996, has been documentation of medical necessity for all laboratory tests.[1] This has resulted in a decrease in the availability of paneled tests. Each test must have comments associated with it by the ordering physician, or it jeopardizes the reimbursement for the entire panel. The rational is that, if one test was not needed, were all the tests needed, and why should the laboratory be reimbursed for performing unnecessary tests?

SUMMARY

POCT is an extremely diverse and dynamic aspect of laboratory testing. There are essential and basic requirements of testing that are identical to those in the main laboratory. Application and administration of the testing in multiple patient care settings is a challenging and increasingly controversial topic. Some institutions have been expanding and enhancing their POCT systems, whereas others have been eliminating POCT in some settings.[4]

There are varying viewpoints on the value of testing outside the main laboratory. Primary controversies involve the aspects of patient-focused care versus single standard of care. POCT has advanced dramatically in the past few years. As these systems are developed, there will be an increase in the divergence in the perception of the value of POCT. It is the laboratory's responsibility to take the lead role in dealing with these changes. The laboratory has adapted to dealing with changes in technology and regulatory guidelines. These changes are now expanding outside the laboratory to the patient's bedside. The laboratory should accept the responsibility to assist and lead others in this aspect of the health care environment, which will provide the best care possible for the patient at the least cost to the patient.

REVIEW QUESTIONS

1. POCT tests classified as waived by government standards include all of the following except:
   a. Fructosamine
   b. Cholesterol
   c. Triglycerides
   d. Ionized calcium

2. Which of the following comparisons between clinical laboratory testing and POCT does not coincide:

   Aspect Laboratory Testing POCT
   a. Turnaround time: Slow—Rapid
   b. Calibration: Frequent—Infrequent
   c. Consumables: Few—Many
   d. Cost per test: High—Low

3. Factors used in determining whether POCT should be used include all of the following EXCEPT:
   a. Need
   b. Efficiency
   c. Benefit to the laboratory
   d. Cost

4. In the evaluation of a POCT system, data obtained from the POCT system and the main laboratory must be correlated. Which of the following analyses should be used when dealing with qualitative data?
   a. Mean
   b. Sensitivity
   c. Standard deviation
   d. Mode

5. POCT includes which of the following:
   a. Near-patient testing
   b. Decentralized testing
   c. Ancillary testing
   d. All of the above

6. One of the primary difficulties in the implementation of POCT in a health care facility is:

   a. Instilling the importance of quality control in nonlaboratorians performing POCT
   b. Instruction on how to perform the test
   c. Documentation of training for those performing POCT
   d. None of the above

7. The federal regulation most cited and the one with which most laboratories must comply when implementing a POCT system is:
   a. National Accrediting Agency for Clinical Laboratory Science
   b. Clinical Laboratory Improvement Amendment
   c. National Committee for Clinical Laboratory Standards
   d. American Society for Clinical Laboratory Science

8. Which of the following statements most appropriately describes proficiency testing as it applies to POCT?
   a. It allows for the detection of technical or analytical errors
   b. It ensures that the laboratory quality assurance policies are being followed
   c. It ensures that specimens are being properly collected and processed
   d. None of the above

REFERENCES

1. Travers E. The effects of federal regulations and managed care. Adv Admin of the Lab 1998;7(3):8–10.
2. Westgard, J. Taking care with point-of-care quality control. Clin Lab News 1997;23(8):4.
3. US Department of Health and Human Services. Medicare, Medicaid and CLIA programs: regulations implementing the Clinical Laboratory Improvement Amendments of 1988 (CLIA). Final rule. Federal Register 1992;57:7002–7186.
4. Horton G. Quality control: How to check costs and quality of point-of-care testing. Med Lab Observer 1997;29(9):31–37.

PART II

CLINICAL CORRELATIONS AND ANALYTICAL PROCEDURES

CHAPTER 8

Amino Acids and Proteins

Barbara J. Lindsey

Objectives

Upon completion of this chapter, the clinical laboratorian should be able to:

- *Describe the structures and properties of amino acids and proteins.*
- *Differentiate between conjugated and simple proteins.*
- *Outline protein synthesis and catabolism.*
- *Discuss the general characteristics of the aminoacidopathies, including the metabolic defect in each and the procedure used for detection.*
- *Briefly discuss the function and clinical significance of the following proteins:*
  - *prealbumin*
  - *albumin*
  - *α_1-antitrypsin*
  - *α_1-fetoprotein*
  - *haptoglobin*
  - *ceruloplasmin*
  - *transferrin*
  - *fibrinogen*
  - *C-reactive protein*
  - *immunoglobulins*
  - *troponin*

- *Discuss at least five general causes of abnormal serum protein concentrations.*
- *List the reference intervals for total protein and albumin and discuss any nonpathologic factors that influence the levels.*
- *Describe and compare methodologies used in the analysis of total protein and albumin, and include the structural characteristics or chemical properties that are relevant to each measurement.*
- *Briefly explain the principle and clinical usage of separation of proteins by:*
  - *Electrophoresis*
  - *High-resolution electrophoresis*
  - *Isoelectric focusing*
- *Given a densitometric scan of a serum protein electrophoresis using the routine method (five zones), recognize and name the fractions, interpret any abnormality in the pattern, and associate these patterns with common disease states.*
- *Differentiate the types of proteinuria on the basis of etiology and type of protein found in the urine and describe the principle of the methods used for both qualitative and*

(continued)

quantitative determination and identification of urine proteins.

- *Describe the diseases associated with alterations in cerebrospinal fluid proteins.*

KEY TERMS

Albumin	Hyperproteinemia	Peptide bond
Amino acid	Hypoproteinemia	Phenylketonuria
Aminoacidopathies	Isoelectric point (pI)	(PKU)
Amphoteric	Monoclonal	Proteinuria
Conjugated protein	immunoglobulins	Quaternary
Denaturation	Nitrogen balance	structure
Globulins	Opsonization	Simple protein

In 1839, the Dutch chemist G.J. Mulder was investigating the properties of a substance found in milk and egg whites that coagulated when heated. The Swedish scientist J. J. Berzelius suggested to Mulder that these substances should be called *proteins* from the Greek word *proteis,* meaning "first rank of importance," because he suspected that they might be the most important of all biologic substances. Mulder also thought that all protein substances were the same because they contained carbon, nitrogen, and sulfur. This was found to be untrue when Dumas discovered slightly varying nitrogen content in protein obtained from different sources.[1]

Today we know that proteins are macromolecules composed of polymers of covalently linked amino acids and that they are involved in every cellular process. This chapter discusses the general properties of amino acids and proteins, abnormalities related to each, and methods of analysis.

AMINO ACIDS

Basic Structure

α-amino acids are small biomolecules containing at least one amino group ($-NH_2$) and one carboxyl group ($-COOH$) bonded to the α-carbon. They differ from one another by the chemical composition of their R groups (side chains). The general structure of an α-amino acid is depicted in Figure 8-1. There are 20 different amino acids that are used as building blocks for protein. The R groups found on these α-amino acids are shown in Table 8-1.

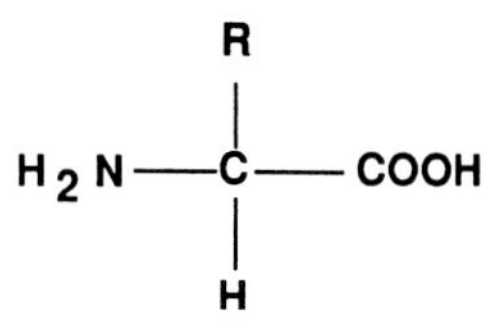

Figure 8-1. General structure of an α-amino acid.

TABLE 8-1. Amino Acids Required in the Synthesis of Proteins

Amino Acid	R
Glycine	$-H$
Alanine	$-CH_3$
Valine*	$-CH-CH_3$ (with $-CH_3$)
Leucine*	$-CH_2-CH-CH_3$ (with CH_3)
Isoleucine*	$-CH-CH_2-CH_3$ (with CH_3)
Cysteine	$-CH_2-SH$
Methionine*	$-CH_2-CH_2-S-CH_3$
Tryptophan*	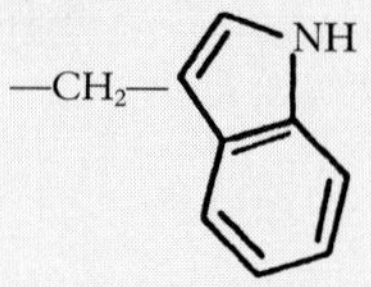 $-CH_2-$ (indole ring, NH)
Phenylalanine*	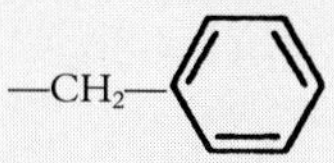 $-CH_2-$ (benzene ring)
Asparagine	$-CH_2-\overset{O}{\overset{\|}{C}}-NH_2$
Glutamine	$-CH_2-CH_2-\overset{O}{\overset{\|}{C}}-NH_2$
Serine	$-CH_2$ (with OH)
Threonine*	$-CH-CH_3$ (with OH)
Tyrosine	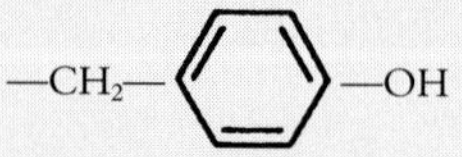 $-CH_2-$ (benzene ring) $-OH$
Lysine*	$-CH_2-CH_2-CH_2-CH_2-NH_2$
Arginine	$-CH_2-CH_2-CH_2-N-\overset{NH_2}{\overset{\|}{C}}-NH_2$ (with H)
Histidine*	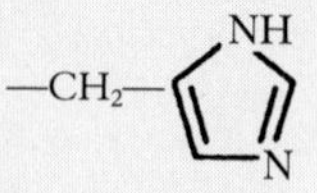 $-CH_2-$ (imidazole ring, NH, N)
Aspartate	$-CH_2-COOH$
Glutamate	$-CH_2-CH_2-COOH$
Proline†	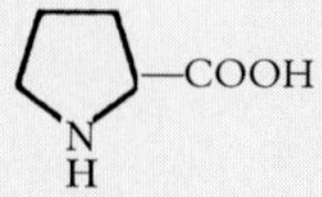 (pyrrolidine ring) $-COOH$

The R group is the group attached to the alpha carbon.
* Nutritionally essential.
† Exception to alpha attachment of R group is proline.

Metabolism

About half of the 20 amino acids needed by humans cannot be synthesized at a rapid enough rate to support growth. These nine nutritionally essential amino acids must be supplied by the diet in the form of proteins. Under normal circumstances, proteolytic enzymes, such as pepsin and trypsin, completely digest dietary proteins into their constituent amino acids. Amino acids are then rapidly absorbed from the intestine into the portal blood and subsequently become part of the body pool of amino acids. Other contributors to the amino acid pool are newly synthesized amino acids and those released by the normal breakdown of body proteins.

The amino acid pool is drawn on primarily for the synthesis of body proteins, including plasma, intracellular, and structural proteins. Amino acids are also used for the synthesis of nonprotein nitrogen-containing compounds such as purines, pyrimidines, porphyrins, creatine, histamine, thyroxine, epinephrine, and the coenzyme NAD. In addition, protein provides 12% to 20% of the total daily body energy requirement. The amino group is removed from amino acids by either deamination or transamination. The resultant ketoacid can enter into a common metabolic pathway with carbohydrates and fats. Those amino acids that generate precursors of glucose, for example, pyruvate or a citric acid cycle intermediate, are referred to as *glucogenic*. Examples include alanine, which can be deaminated to pyruvate; arginine, which is converted to α-ketoglutarate; and aspartate, which is converted to oxaloacetate. Amino acids that are degraded to acetyl CoA or acetoacetyl CoA, such as leucine or lysine, are termed *ketogenic* because they give rise to ketone bodies. Some amino acids can be both ketogenic and glucogenic because some of their carbon atoms emerge in ketone precursors and others appear in potential precursors of glucose. Figure 8-2 shows the points of entry of the carbon atoms of amino acids into the metabolic pathways of carbohydrates and fats. The ammonium ion that is produced during deamination of the amino acids is converted into urea by the urea cycle in the liver.

Aminoacidopathies

Aminoacidopathies are rare inherited disorders of amino acid metabolism. The abnormalities exist in either the activity of a specific enzyme in the metabolic pathway or in the membrane transport system for amino acids. More than 100 diseases have been identified that result from inborn errors of amino acid metabolism.

Phenylketonuria

Phenylketonuria (PKU) is inherited as an autosomal recessive trait and occurs in approximately 1 in 10,000 births. Although there are several variants of this disease, the biochemical defect in the classic form of PKU is an almost total absence of activity of the enzyme phenylalanine hydroxylase (also called phenylalanine-4-mono-oxygenase), which catalyzes the conversion of phenylalanine to tyrosine (Fig. 8-3). In the absence of the enzyme, phenylalanine accumulates and is metabolized by an alternate degradative pathway. The catabolites include

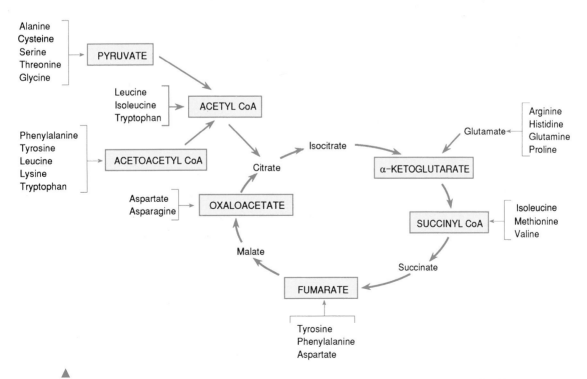

Figure 8-2. Conversion of the carbon skeletons of amino acids into pyruvate, acetyl CoA, or TCA cycle derivatives for further catabolism.

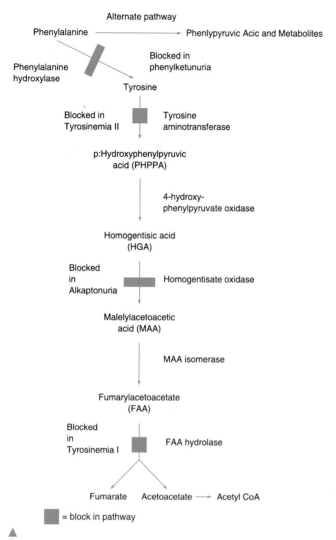

Figure 8-3. Metabolism of phenylalanine and tyrosine.

phenylpyruvic acid, which is the product of deamination of phenylalanine; phenyllactic acid, which is the reduction product of phenylpyruvic acid; phenylacetic acid, which is produced by decarboxylation and oxidation of phenylpyruvic acid; and phenylacetyl glutamine, which is the glutamine conjugate of phenylacetic acid. Although phenylpyruvic acid is the primary metabolite, all these compounds are seen in both the blood and the urine of a phenylketonuric patient, giving the urine a characteristic musty odor.

In infants and children with this inherited defect, retarded mental development occurs as a result of the toxic effects on the brain of phenylpyruvate or one of its metabolic by-products. The deterioration of brain function begins in the second or third week of life. Brain damage can be avoided if the disease is detected at birth and the infant is maintained on a diet containing very low levels of phenylalanine. Until recently, once the child reached age 5 or 6 years, the diet was terminated. This was based on the belief that, at that age, the brain was no longer vulnerable to damage from hyperphenylaninemia. However, it has been reported that there is a slight reduction in IQ after discontinuation of the diet.

Also, offspring of women with PKU that were untreated during pregnancy were almost always microcephalic and mentally retarded. The fetal effects of maternal PKU are preventable if the mother is maintained on a phenylalanine-restricted diet from before conception through term.[2] For both these reasons, it is now recommended that the PKU patient continue dietary treatment indefinitely.

There are cases of hyperphenylalaninemia, however, that are not responsive to dietary treatment. The defect in these cases is a deficiency in the enzymes needed for the regeneration and synthesis of tetrahydrobiopterin (BH_4). BH_4 is a cofactor required for the enzymatic hydroxylation of the aromatic amino acids phenylalanine, tyrosine, and tryptophan. A deficiency of BH_4 results in elevated blood levels of phenylalanine and deficient production of neurotransmitters from tyrosine and tryptophan. Examination of urinary proteins is helpful in diagnosis. Although cofactor defects account for only 1% to 3% of all cases of elevated phenylalanine levels, they must be identified so that appropriate treatment can be initiated. Patients must be given the active cofactor along with the neurotransmitter precursors L-DOPA and 5-OH tryptophan.[3]

Today, all states have enacted legislation to enforce screening programs so that therapeutic measures can be taken early to prevent the resulting disability and mortality associated with hyperphenylaninemia. The screening procedure used routinely is the Guthrie bacterial inhibition assay. In this test, spores of the organism *Bacillus subtilis* are incorporated into an agar plate that contains β_2-thienylalanine, a metabolic antagonist to *B. subtilis* growth. A filter paper disk impregnated with blood from the infant is placed on the agar. If the blood level of phenylalanine exceeds a range of 2 mg/dL to 4 mg/dL, the phenylalanine counteracts the antagonist, and bacterial growth occurs. To avoid false-negative results, the infant must be at least 24 hours of age to ensure adequate time for enzyme and amino acid levels to develop. Furthermore, the sample should be taken before administration of antibiotics or transfusion of blood or blood products. Premature infants can show false-positive results due to the immaturity of the liver's enzyme systems.

A more recent approach to the screening for PKU involves a microfluorometric assay for the direct measurement of phenylalanine in dried blood filter discs. This method yields quantitative rather than the semiquantitative results of the Guthrie test, is more adaptable to automation, and is not affected by the presence of antibiotics. The procedure is based on the fluorescence of a complex formed of phenylalanine-ninhydrin-copper in the presence of a dipeptide (ie, L-leucyl-L-alanine).[4] The test requires pretreatment of the filter paper specimen with trichloroacetic acid (TCA). The extract is then reacted in a microtiter plate with a mixture of ninhydrin, succinate, and leucylalanine in the presence of copper tartrate. The fluorescence of the complex is measured using excitation/emission wavelengths of 360 nm and 530 nm, respectively.[5]

Any positive results on the screening tests must be verified. The reference method for quantitative serum phenylalanine is HPLC; however, both fluorometric[4] and enzymatic[6] methods are available. The normal limits for serum phenylalanine levels for full-term, normal weight newborns range from 1.2 mg/dL to 3.4 mg/dL (70 μmol/L to 200 μmol/L).

Urine testing for phenylpyruvic acid, although not considered an initial screening procedure, can be used both for diagnosis in questionable cases and for monitoring of dietary therapy. The test, which may be performed by tube or reagent strip test (Phenistix, Miles Diagnostics, Elkhart, IN), involves the reaction of ferric chloride with phenylpyruvic acid in urine to produce a green color.

Prenatal diagnosis and detection of carrier status in families with PKU is now available using deoxyribonucleic acid (DNA) analysis. Molecularly, PKU results from multiple independent mutations at the phenylalanine hydroxylase (PAH) locus. Analysis, using cloned human PAH cDNA as a probe, has revealed the presence of numerous restriction fragment length polymorphisms in the PAH gene. Because the restriction fragment patterns and the disease state have been proven to be tightly linked in PKU families, these polymorphisms can be used for diagnosis.[7]

Tyrosinemia and Related Disorders

A range of familial metabolic disorders of tyrosine catabolism are characterized by excretion of tyrosine and tyrosine catabolites in urine. Normally, the major path of tyrosine metabolism involves the removal of an amine group by tyrosine amino transferase forming p-hydroxyphenylpyruvic acid (PHPPA), which is oxidized to homogentisic acid (HGA). HGA is further metabolized in a series of reactions to fumarate and acetoacetate, as shown in Figure 8-3. The defect in inherited tyrosine abnormalities is either a deficiency in tyrosine aminotransferase, resulting in *tyrosinemia II*, or, more commonly, a deficiency in fumarylacetoacetate FAA hydrolase, resulting in *tyrosinemia I*. FAA hydrolase cleaves fumarylacetoacetic acid to fumaric and acetoacetic acids. The absence of these enzymes results in abnormally high levels of tyrosine and, in some cases, increases in PHPPA and methionine. The elevated tyrosine leads to liver damage, which may be fatal in infancy, or to cirrhosis and liver cancer later in life. The incidence of tyrosinemia I is approximately 1 in 100,000 births.

Alkaptonuria

This disorder is of considerable historical interest in that it was one of the original "inborn errors of metabolism" described. At the turn of the 20th century, Archibald Garrod, a pediatrician, recognized that the syndrome, which had been called alkaptonuria, showed a pattern of familial inheritance. He proposed that alkaptonuria and certain other disorders were due to genetic defects, each of which resulted in the lack of activity of a particular metabolic enzyme.[8] Forty-five years later, it was confirmed that the biochemical defect in alkaptonuria is a lack of homogentisate oxidase in the ty-

rosine catabolic pathway (see Fig. 8-3). This disorder occurs in about 1 in 250,000 births. A predominant clinical manifestation of alkaptonuria is the darkening of urine upon standing exposed to the atmosphere. The phenomenon is due to an accumulation in the urine of HGA, which oxidizes to produce a dark polymer. Alkaptonuric patients have no immediate problems, but late in the disease, the high level of HGA gradually accumulates in connective tissue, causing generalized pigmentation of these tissues (ochronosis) and an arthritis-like degeneration.

Maple Syrup Urine Disease

As the name implies, the most striking feature of this hereditary disease is the characteristic maple syrup or burnt sugar odor of the urine, breath, and skin. *Maple syrup urine disease (MSUD)* results from an absence or greatly reduced activity of the enzyme, branched-chain keto acid decarboxylase, thereby blocking the normal metabolism of the three essential branched-chain amino acids leucine, isoleucine, and valine. Specifically, this enzyme is responsible for catalyzing the oxidative decarboxylation of all three branched-chain α-ketoacids to CO_2 and their corresponding acyl-CoA thioesters (Fig. 8-4). The result of this enzyme defect is an accumulation of the branched-chain amino acids and their corresponding ketoacids in the blood, urine, and cerebrospinal fluid (CSF).

Typically, infants with this inherited abnormality appear normal at birth but, by age 4 days to 7 days, develop lethargy, vomiting, and signs of failure to thrive. Central nervous system (CNS) symptoms follow, including muscle rigidity, and stupor and respiratory irregularities. If left untreated, the disease causes severe mental retardation, convulsions, acidosis, and hypoglycemia. In the classic form of the disease, death usually occurs during the 1st year; however, intermediate forms have been reported. In these less severe variants, the activity of the decarboxylase is approximately 25% of normal. Although this still results in a persistent elevation of the branched-chain amino acids, the levels frequently can be controlled by limiting dietary protein intake.[9]

Although the incidence of MSUD is low—occurring in only 1 in 216,000 births—early diagnosis is important to begin dietary treatment. Therefore, testing for MSUD is included in many of the metabolic screens required by law in certain states. A modified Guthrie test is commonly used for this neonatal screening. The metabolic inhibitor to *B. subtilis,* included in the growth media, is 4-azaleucine. In a positive test for MSUD, an elevated level of leucine from a filter paper disc impregnated with the infant's blood will overcome the inhibitor, and bacterial growth occurs.[10] Alternately, a microfluorometric assay for branched chain amino acids, using leucine dehydrogenase (EC 1.4.1.9), can be used for mass screening. In the procedure, the filter paper specimen is treated with a solvent mixture of methanol and acetone to denature the hemoglobin. Leucine dehydrogenase is added to an aliquot of this sample extract. The fluorescence of the NADH produced in the subsequent reaction is measured at 450 nm using an excitation wavelength of 360 nm.[5]

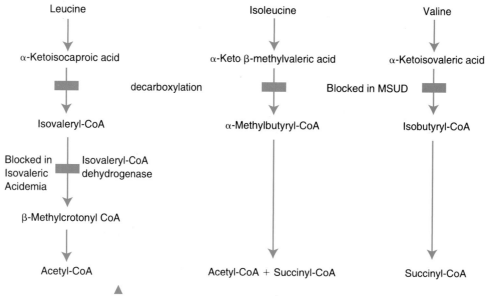

Figure 8-4. Metabolism of the branched-chain amino acids.

A confirmed diagnosis is based on finding increased plasma and urinary levels of the three branched-chain amino acids and their ketoacids, with leucine being in highest concentration. A leucine level above 4 mg/dL is indicative of MSUD. The presence of alloisoleucine, an unusual metabolite of isoleucine, is characteristic. MSUD can be diagnosed prenatally by measuring the decarboxylase enzyme concentration in cells cultured from amniotic fluid.

Isovaleric Acidemia

Isovaleric acidemia results from a deficiency of the enzyme isovaleryl-CoA dehydrogenase in the degradative pathway of leucine (see Figure 8-4). The resultant elevation of the glycine conjugate of isovaleric acid, isovalerylglycine, produces a characteristic "sweaty feet" odor. The abnormal organic acid levels can be identified by chromatography.

Homocystinuria

Homocysteine is an intermediate amino acid in the synthesis of cysteine from methionine. The synthesis of cysteine from methionine is illustrated in Figure 8-5. The usual cause of the hereditary disease, homocystinuria, is an impaired activity of the enzyme cystathionine β-synthase, which results in elevated plasma and urine levels of the precursors homocysteine and methionine. Newborns show no abnormalities, but physical defects develop gradually with age. Associated clinical findings in late childhood include thrombosis due to toxicity of homocysteine to the vascular endothelium, osteoporosis, dislocated lenses in the eye due to the lack of cysteine synthesis essential for collagen formation, and frequently, mental retardation.

The enzyme cystathionine β-synthase requires vitamin B$_6$ (pyridoxine) as its cofactor. Slightly different genetic defects lead to two forms of the disease: a vitamin B$_6$ responsive form, in which treatment consists of therapeutic doses of vitamin B$_6$, and a vitamin B$_6$–unresponsive form, in which the treatment is a diet low in methionine and high in cystine.

The incidence of homocystinuria is approximately 1 in 200,000 births. Neonatal screening can be accomplished with a Guthrie test using L–methionine sulfoximine as the metabolic inhibitor. Increased plasma methionine levels from affected infants will result in bacterial growth. A level of methionine greater than 2 mg/dL using an HPLC procedure confirms positive results on the screening test. Elevations in urinary homocystine can be detected by the cyanide-nitroprusside spot test. Cystine and homocystine are reduced by sodium cyanide to their free thiol forms, cysteine and homocysteine, which can then react with sodium nitroprusside to produce a red–purple color. Because cystine also produces a positive result, the presence of homocystine must be confirmed with a silver–nitroprusside test. Silver nitrate will reduce homocystine but not cystine, thus allowing only homocystine to react with the nitroprusside and produce a reddish color. Cystine remains in the oxidized form, which does not react with sodium nitroprusside.

Cystinuria

This inherited aminoacidopathy is caused by a defect in the amino acid transport system rather than a metabolic enzyme deficiency. Normally, amino acids are freely filtered by the glomerulus and then actively reabsorbed in the proximal renal tubules. In cystinuria, there is a 20- to 30-fold increase in the urinary excretion of cystine due to a genetic defect in the renal resorptive mechanism. The transport mechanism is not specific for cystine. Excretion of the other diamino acids, lysine, arginine, and ornithine, is also markedly elevated due to deficient resorption.

CASE STUDY 8-1[89]

A 13-month-old boy was admitted to a small rural hospital. He had been in a normal state of health until 10 days prior, at which time he had developed an upper respiratory tract infection. He experienced increasing problems with his balance and became lethargic. These symptoms prompted his mother to seek medical attention at the hospital where pertinent laboratory results at admission showed a serum glucose level of 23 mg/dL and moderate ketones in the urine and serum. An intravenous solution of 5% dextrose was initiated to correct the low glucose level. Because the clinical picture resembled Reye's syndrome, the child was transferred to a medical center hospital for a definitive diagnosis. Laboratory results on admission to the medical center hospital appear in Case Study Table 8-1.1.

Questions

1. Which laboratory result can be useful in ruling out the diagnosis of Reye's syndrome?
2. What correlations can be made using the blood pH, PCO_2 serum glucose, and serum and urine ketone findings.
3. What other laboratory tests should be performed to verify a defect in protein metabolism?

CASE STUDY TABLE 8-1.1. Admission Laboratory Results

Hematology		Urinalysis	
Hct: 37%		Specific gravity	1.022
WBC: 128×10^9/L		Protein	Trace
Bands	28%	Acetone	3+
Segmented	46%	Blood	1+
Lymphocytes	21%		
Monocytes	5%		

Chemistry (Reference Range)

Glucose	133 mg/dL (65–105)	Alk phos	129 U/L (20–70)
BUN	21 mg/dL (7–18)	AST	154 U/L (10–30)
Na	136 mmol/L (136–145)	ALT	133 U/L (8–20)
K	4.3 mmol/L (3.6–5.1)	CK	36 U/L (25–90)
TCO₂	10 mmol/L (23–29)	LD	119 U/L (45–90)
Cl	112 mmol/L (98–106)	Ketone	Moderate
NH₃	48 μmol/L (40–80)		

Arterial blood gases
pH: 7.17
PCO_2: 23 mm Hg
PO_2: 90 mm Hg

Of the four, cystine is the amino acid that causes complications of the disease. Because cystine is relatively insoluble, when it reaches these high levels in the urine, it tends to precipitate in the kidney tubules and form urinary calculi. The formation of cystine calculi can be minimized by a high fluid intake and alkalinizing the urine, which makes cystine relatively more soluble. If this does not succeed, treatment with regular doses of penicillamine can be initiated.[11]

Cystinuria can be diagnosed by testing the urine for cystine using cyanide-nitroprusside, which produces a red-purple color on reaction with sulfhydryl groups. False-positive results due to homocystine must be ruled out.

Amino Acid Analysis

Blood samples for amino acid analysis should be drawn after at least a 6- to 8-hour fast to avoid the effect of absorbed amino acids originating from dietary proteins. The sample is collected in heparin, and the plasma is promptly removed from the cells, taking care not to aspirate the layer of platelets and leukocytes. If this step is not carried out with caution, the plasma becomes contaminated with platelet or leukocyte amino acids, in which the contents of aspartic acid and glu-

tamic acid, for example, are about 100-fold higher than those in plasma. Hemolysis should be avoided for the same reason. Deproteinization should be carried out within 30 minutes of sample collection, and analysis should be performed immediately or the sample should be stored at −20°C to −40°C.

Urinary amino acid analysis can be performed on a random specimen for screening purposes, but for quantitation, a 24-hour urine preserved with thymol or organic solvents is required. Amniotic fluid also may be analyzed.

For preliminary screening, the method of choice is thin-layer chromatography. The application of either one- or two-dimensional separations depends on the purpose of the analysis. If one is searching for a particular category of amino acids, such as branched-chain amino acids, or even a single amino acid, usually one-dimensional separations are sufficient. Separation conditions can be selected in such a way as to offer good resolution of the amino acid in question.

For more general screening, two-dimensional maps are essential. In two-dimensional chromatography, the amino acids are allowed to migrate along one solvent front, and then the chromatogram is rotated 90° and a second solvent migration occurs. A variety of solvents has been used, including butanol–acetic acid–water and ethanol–ammonia–water

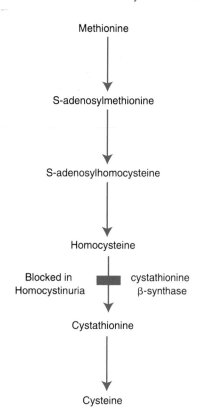

Methionine

↓

S-adenosylmethionine

↓

S-adenosylhomocysteine

↓

Homocysteine

Blocked in   cystathionine
Homocystinuria   β-synthase

↓

Cystathionine

↓

Cysteine

▲
Figure 8-5. Cysteine synthesis.

mixtures. The chromatogram is visualized by staining with ninhydrin, which gives a blue color with most amino acids.

When the possibility of an amino acid disorder has been indicated in the preliminary steps, amino acids can be separated and quantitated by cation-exchange chromatography using a gradient buffer elution.[12] The other choice is to use an HPLC reversed-phase system equipped with fluorescence detection.[13]

PROTEINS

General Characteristics

Proteins are an essential class of compounds composing 50% to 70% of the cell's dry weight. Proteins are found in all cells of the body, as well as in all fluids, secretions, and excretions.

Molecular Size

Biologically active proteins are macromolecules that range in molecular weight from approximately 6000 for insulin to several million for some structural proteins.

Structure

All proteins comprise covalently linked polymers of amino acids. The amino acids are linked in a head-to-tail fashion; in other words, the carboxyl group of one amino acid combines with the amino group of another amino acid (Figure 8-6). During this reaction, a water molecule is removed and the bond that is created is called a *peptide bond.* The amino acid that has the amino group free is the *N-terminal end,* and the amino acid that has the carboxyl group free is the *C-terminal end.* When two amino acids are joined, the molecule is called a *dipeptide;* three amino acids are called a *tripeptide;* and four together is a *tetrapeptide.* As the chain increases further, it is called a *polypeptide.* In human serum, proteins average about 100 to 150 amino acids in the polypeptide chains.

The conformation (shape) of a protein is determined by interaction between a polypeptide and its aqueous environment in which the polypeptide attains a stable three-dimensional structure. There are four aspects of the structure of a protein that relate to its conformation. The number and kinds of amino acids, as well as their sequence in the polypeptide chain, constitute the primary structure of a protein. The covalent peptide linkage is the only type of bonding involved at this level. The primary structure is crucial for the function and molecular characteristics of the protein. Any change in the amino acid composition can significantly alter the protein. For example, when the amino acid valine is substituted for glutamic acid in the β chain of hemoglobin A, hemoglobin S is formed, which results in sickle-cell anemia.

The secondary structure is the winding of the polypeptide chain. The usual pattern that is formed by the globular proteins (most serum proteins) is an helix requiring 3.6 amino acids to make one turn in the coil. A β-pleated sheet or a stable but irregular pattern also can be seen. The secondary structure is maintained by hydrogen bonds between the NH and CO groups of the peptide bonds either within the same chain or between different chains within the same molecule. Figure 8-7 schematically pictures the secondary structure.

$$NH_2—C—COOH \quad + \quad NH_2—C—COOH \quad \xrightarrow{-H_2O} \quad NH_2—C—C—N—C—COOH$$

**Peptide
Bond**

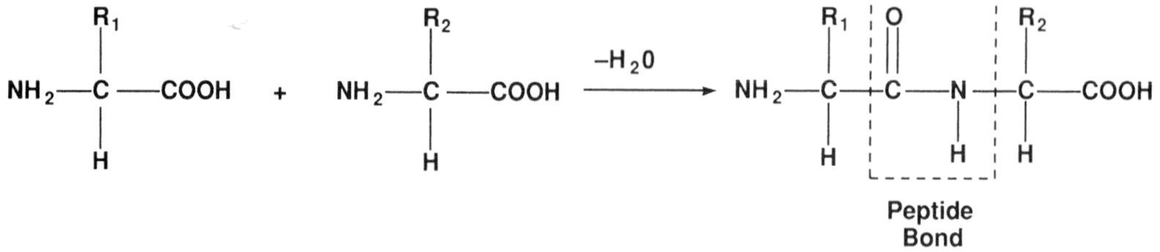

▲
Figure 8-6. Formation of a dipeptide.

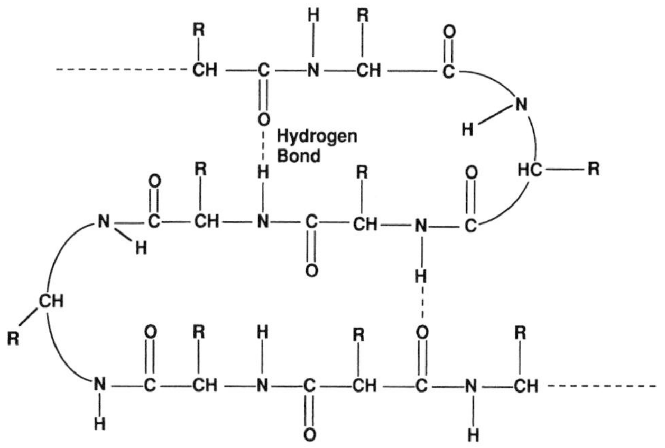

Figure 8-7. Secondary structure of proteins.

The next level of structure, the tertiary structure, refers to the way in which the twisted chain folds back on itself to form the three-dimensional structure. The specific convolutions a polypeptide undergoes are determined by the interaction of the R groups in the molecule. The reactions of the R groups include disulfide linkages, electrostatic attractions, hydrogen bonds, hydrophobic interactions, and van der Waals forces. The tertiary structure is responsible for many of the physical and chemical properties of the protein.

In addition to these first three levels that can exist in molecules comprising a single polypeptide chain, some proteins display a fourth level of organization called the quaternary structure. *Quaternary structure* is the arrangement of two or more polypeptide chains to form a functional protein molecule. *Albumin,* which comprises a single polypeptide chain, would thus have no quaternary structure. On the other hand, hemoglobin comprises four globin chains, lactate dehydrogenase consists of five peptide chains, and creatine kinase has two chains joined. The polypeptide chains are united by noncovalent attractions such as hydrogen bonds and electrostatic interactions.

When the secondary, tertiary, or quaternary structure of a protein is disturbed, the protein may lose its functional and molecular characteristics. This loss of its native or original character is called *denaturation.* Denaturation can be caused by heat, hydrolysis by strong acid or alkali, enzymatic action, exposure to urea or other substances, or exposure to ultraviolet light.

Nitrogen Content

Proteins comprise the elements carbon, oxygen, hydrogen, nitrogen, and sulfur. It is the nitrogen content that sets proteins apart from pure carbohydrates and lipids, which contain no nitrogen atoms. The nitrogen content of serum protein varies somewhat, but the average is approximately 16%. This characteristic is used in one method of total protein measurement.

Charge and Isoelectric Point

Because of their amino acid composition, proteins can bear positive and negative charges (*ie,* they are *amphoteric*). The reactive acid or basic groups that are not involved in the peptide linkage can exist in different charged forms, depending on the pH of the surrounding environment. Aspartic acid and lysine are examples of amino acids that have free carboxyl or amino groups, respectively (see Table 8-1). Figure 8-8 shows

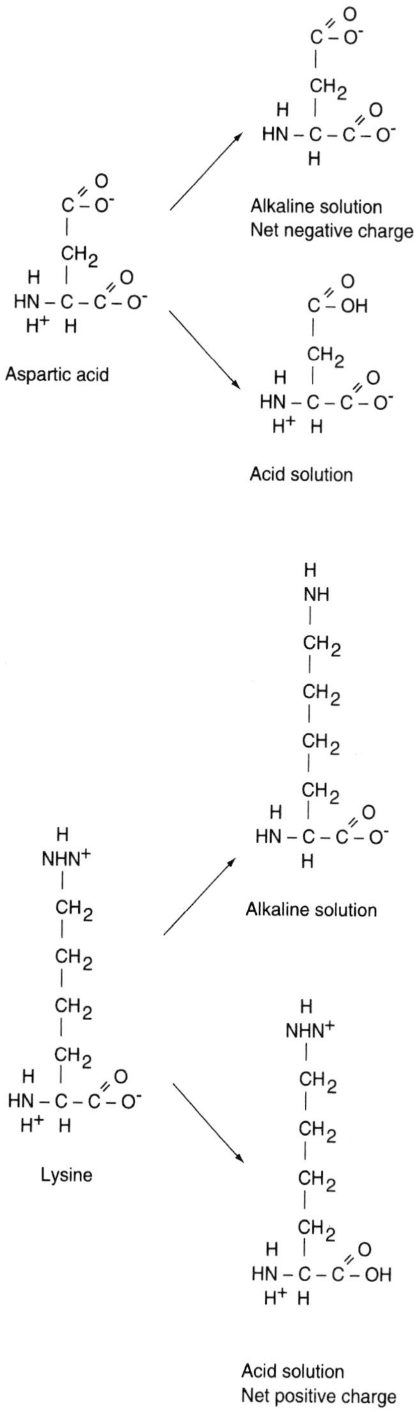

Figure 8-8. Charged states of amino acids.

these two amino acids and the charges they would have in solutions that are alkaline or acid.

At a pH of 2.98, L-aspartic acid has no net charge (equal or no ionization of the amino and carboxyl groups), whereas in an alkaline pH (9.47) aspartic acid has a net negative charge. On the other hand, L-lysine has no net charge at a pH of 9.47 but, in an acid environment, the gain in protons results in a net positive charge. The pH at which an amino acid or protein has no net charge is known as its *isoelectric point (pI)*. In other words, for a protein comprising several amino acids, the pI is the point at which the number of positively charged groups equals the number of negatively charged groups. If a protein is placed in a solution that has a pH greater than the pI, the protein will be negatively charged, whereas at a pH less than the pI, the protein will be positively charged. Proteins differ in the number and type of constituent amino acids; they also differ in their pI values. For example, albumin has a pI of 4.9. The pIs of some serum proteins are given in Table 8-2. Owing to the different pI values, proteins will carry different net charges at any given pH. This difference in magnitude of the charge is the basis for several procedures for separating and quantitating proteins such as electrophoresis. The procedure is discussed later in this chapter.

Solubility

Proteins in aqueous solution swell and enclose water. Natelson and Natelson[1] have advised that, for this reason, it is necessary when reconstituting lyophilized serum to mix the reconstituted vial gently for at least 30 minutes to complete the swelling process. Protein solutions are *colloidal emulsoids* or *micelles* because they are charged and because each molecule of protein has an envelope of water around it.

The solubility of proteins is promoted by a high dielectric character of the solvent and a high concentration of free water. Thus, low concentrations of salt (0.1 M) have been used to *salt in* globulins into solution. When the dielectric nature or the amount of free water is decreased, the charges on the protein promote aggregation. In a high salt concentration (~2 M), ionic salts compete with protein for water and thus decrease the amount of water available for hydration of the protein. The resultant precipitation of the globulins is called *salting-out*. By using a range of salt concentrations, a wide array of specific proteins can be separated from a mixture. Albumin remains in solution even in high salt concentrations because it has a higher dipole and holds water tighter and, because of its smaller size in relation to the globulins, it does not require as much water to keep it hydrated.

Water-miscible, neutral organic solvents also can be used to separate proteins based on their solubility. These solvents have a dielectric constant less than water, which suppresses the ionization of the R groups on the protein surface and thereby decreases the solubility. The use of alcohol concentrations to separate serum proteins into fractions bearing Roman numeral designations was developed and used by E. J. Cohn[14] during World War II, and human albumin, fraction V (Cohn), continues to be commercially available.

Immunogenicity

Because of their molecular mass, their content of tyrosine, and their specificity by species, proteins can be very effective antigens. When injected into another species (*ie,* rabbit, goat, or chicken), human serum proteins elicit the formation of antibodies specific for each of the proteins present in the serum. When it became possible to obtain these antibodies for each of the serum proteins, methods using antigen–antibody reactions that are very specific were introduced into the clinical laboratory. Some of the methods commonly used are described in Chapter 5, *Immunoassays and Nucleic Acid Probe Techniques.*

Synthesis

Most of the plasma proteins are synthesized in the liver and secreted by the hepatocyte into the circulation. The immunoglobulins are exceptions because they are synthesized in plasma cells.

The amino acids of a polypeptide chain are placed in a sequence that is determined by a corresponding sequence of bases (guanine, cytosine, adenine, and thymine) in the DNA that constitutes the appropriate gene. The double-stranded DNA unfolds in the nucleus, and one strand is used as a template for the formation of a complementary strand of messenger RNA (mRNA). The mRNA, now carrying the genetic code from the DNA, moves to the cytoplasm, where it attaches to ribosomes. A sequence of three bases (*codon*) on the mRNA is required to specify the amino acid to be transcribed. The code on the mRNA also contains initiation and termination codons for the peptide chain.

The next step in protein synthesis is getting the amino acid to the ribosomes. First, the amino acid is activated in a reaction that requires energy and a specific enzyme for each amino acid. This activated amino acid complex is then attached to another kind of RNA, transfer RNA (tRNA), with the subsequent release of the activating enzyme and adenosine monophosphate (AMP). The tRNA is a short chain of RNA that occurs free in the cytoplasm. Each amino acid has a specific tRNA that contains three bases that correspond to the three bases in the mRNA. The tRNA carries its particular amino acid to the ribosome and attaches to the mRNA in accordance with the matching codon. In this manner, the amino acids are aligned in sequence. As each new tRNA brings in the next amino acid, the preceding amino acid is transferred onto the amino group of the new amino acid, and enzymes located in the ribosomes form a peptide bond. The tRNA is released into the cytoplasm, where it can pick up another amino acid, and the cycle repeats itself. When the terminal codon is reached, the peptide chain is detached and the ribosome and mRNA dissociate. Figure 8-9 illustrates protein synthesis. Intracellular proteins are generally synthesized on free ribosomes, whereas proteins made by the liver for secretion are made on ribosomes attached to the rough endoplasmic reticulum. Protein synthesis occurs at the rate of approximately 2 to 6 peptide bonds per

TABLE 8-2. Characteristics of Selected Plasma Proteins

	Reference Value (Adult, g/L)	Molecular Mass (D)	Isoelectric Point, pI	Electrophoretic Mobility, pH 8.6, I = 0.1	Comments
Prealbumin	0.1–0.4	55,000	4.7	7.6	Indicator of nutrition; binds thyroid hormones and retinol-binding protein
Albumin	35–55	66,300	4.9	5.9	Binds bilirubin, steroids, fatty acids; major contributor to oncotic pressure
α_1-*Globulins*					
α_1-antitrypsin	2–4	53,000	4.0	5.4	Acute-phase reactant; protease inhibitor
α_1-fetoprotein	1×10^5	76,000	2.7	6.1	Principle fetal protein
α_1-acid glycoprotein (orosomucoid)	0.55–1.4	44,000		5.2	Acute-phase reactant
α_1-lipoprotein (HDL)	2.5–3.9	200,000		4.4–5.4	Transports lipids
α_1-antichymotrypsin	0.3–0.6	68,000		Inter α	Inhibits serine proteinases (*ie*, chymotrypsin)
Inter-α-trypsin inhibitor	0.2–0.7	160,000		Inter α	Inhibits proteinases (*ie*, trypsin)
Gc-globulin	0.2–0.55	59,000		Inter α	Binds Vitamin D and actin
α_2-*Globulins*					
Haptoglobins					
Type 1-1	1.0–2.2	100,000		4.5	Acute-phase reactant; binds hemoglobin
Type 2-1	1.6–3.0	200,000	4.1	3.5–4.0	Binds hemoglobin
Type 2-2	1.2–1.6	400,000		3.5–4.0	Binds hemoglobin
Ceruloplasmin	0.15–0.60	134,000	4.4	4.6	Peroxidase activity; contains copper
α_2-Macroglobulin	1.5–4.2	725,000	5.4	4.2	Inhibits thrombin, trypsin, pepsin
β-*Globulins*					
Pre β-lipoproteins (VLDL)	1.5–2.3	250,000		3.4–4.3	Transports lipids (primarily triglyceride)
Transferrin	2.04–3.60	76,000	5.9	3.1	Transports iron
Hemopexin	0.5–1.0	57,000–80,000		3.1	Binds heme
β-Liproteins (LDL)	2.5–4.4	3,000,000		3.1	Transports lipids (primarily cholesterol)
β_2-Microglobulin (B2M)	0.001–0.002	11,800		β^2	Component of human leucocyte antigen (HLA) molecules class I
C_4 complement	0.20–0.65	206,000		0.8–1.4	Immune response
C_3 complement	0.55–1.80	180,000		0.8–1.4	Immune response
C_1 q complement	0.15	400,000			Immune response
Fibrinogen	2.0–4.5	341,000	5.8	2.1	Precursor of fibrin clot
C-reactive protein (CRP)	0.01	118,000	6.2		Acute-phase reactant; motivates phagocytosis in inflammatory disease
γ-*Globulins*					
Immunoglobulin G	8.0–12.0	150,000	5.8–7.3	0.5–2.6	Antibodies
Immunoglobulin A	0.7–3.12	180,000		2.1	Antibodies (in secretions)
Immunoglobulin M	0.5–2.80	900,000		2.1	Antibodies (early response)
Immunoglobulin D	0.005–0.2	170,000		1.9	Antibodies
Immunoglobulin E	6×10^4	190,000		2.3	Antibodies (reagins, allergy)

second. Synthesis is controlled at two steps: selection of genes for transcription to mRNA and selection of mRNA for translation into proteins. Certain hormones are involved in controlling protein synthesis. Thyroxine, growth hormone, insulin, and testosterone promote synthesis, whereas glucagon and cortisol have a catabolic effect.

Proteins that are excreted after synthesis are thought to have a short sequence of amino acids, a secretory piece that attracts the protein to the Golgi apparatus for the consequent secretion by exocytosis. The secretory piece is removed during the secretion process. Secreted proteins thus usually exist as precursor proteins within the cells from which they arise.

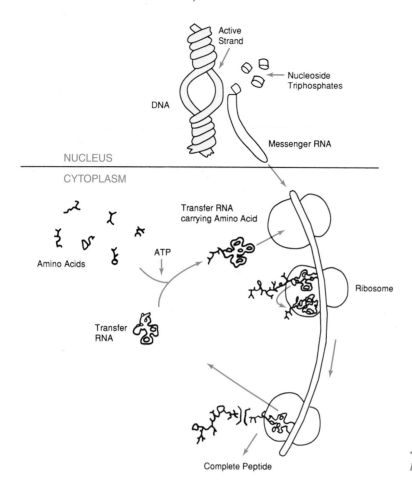

Figure 8-9. Schematic summary of protein synthesis.

Catabolism and Nitrogen Balance

Most proteins in the body are constantly being repetitively synthesized and then degraded. It has been demonstrated that over a wide range of rates of synthesis, protein catabolism and synthesis are equal. Normally, this turnover totals about 125 g to 220 g of protein each day. The rate of protein turnover, however, varies widely for individual proteins. For example, the plasma proteins and most intracellular proteins are rapidly degraded, having half-lives of hours or days, whereas some of the structural proteins such as collagen are metabolically stable and have half-lives of years.[15] The disintegration of protein occurs in the digestive tract, in the kidneys, and, particularly, in the liver. During catabolism, proteins are hydrolyzed to their constituent amino acids. The amino acids are deaminated, producing ammonia and ketoacids. The ammonia is converted to urea by the hepatocytes and is excreted in the urine. The ketoacids are oxidized by means of the citric acid cycle and converted to glucose or fat.

Ordinarily, a balance exists between protein anabolism (synthesis) and catabolism. In nutritional terms, this is called the *nitrogen balance*. When protein catabolism exceeds protein anabolism, the nitrogen excreted exceeds that ingested. This is "negative" nitrogen balance, and it occurs in conditions where there is excessive tissue destruction, such as burns, wasting diseases, continual high fevers, or starvation.

The converse, "positive" nitrogen balance, is seen when anabolism is greater than catabolism. A positive balance is found during growth, pregnancy, and repair processes.

Classification

Proteins are generally classified into two major groups based on composition: simple and conjugated.

Simple Proteins

Simple proteins contain peptide chains, which, on hydrolysis, yield only amino acids. Simple proteins may be globular or fibrous in shape. Globular proteins are relatively symmetrical, with compactly folded and coiled polypeptide chains. Albumin is an example of a globular protein. Fibrous proteins are more elongated and asymmetrical and have a higher viscosity. Collagen and troponin are examples of fibrous proteins.

Conjugated Proteins

Conjugated proteins comprise a protein (apoprotein) and a nonprotein moiety (prosthetic group). The prosthetic group may be lipid, carbohydrate, porphyrins, metals, and so on. These groups impart certain characteristics to the proteins. Most of the names given to these conjugated proteins are self-descriptive. The *metalloproteins* have a metal ion attached to the protein, either directly, as in ferritin (which contains iron) and ceruloplasmin (which contains copper), or as complex metals (metal plus another prosthetic group), such as hemoglobin and flavoproteins. The metal (iron) in hemoglobin is first attached to protoporphyrin, which is then attached to the globin chains; in flavoprotein, the metal is first attached to FMN or FAD.

When lipids such as cholesterol and triglyceride are linked to proteins, the molecules are called *lipoproteins*. When carbohydrates are joined to proteins, several terms may be used to describe the results. Generally, those molecules with 10% to 40% carbohydrate are called *glycoproteins*.[1] Examples of glycoproteins are haptoglobin and α_1-antitrypsin. When the percentage of carbohydrate linked to protein is higher, the proteins are often called *mucoproteins* or *proteoglycans* when heparan, keratan, or chondroitin sulfate are also present. An example of a mucoprotein is mucin, a compound that lubricates organ linings. *Nucleoproteins* are those proteins that are combined with nucleic acids (DNA or RNA). An example is chromatin.

General Function of Proteins

The broad array of molecules that make up proteins may function in general ways or, as individual molecules, may have some unique function. Plasma proteins and tissue proteins share the same amino acid pool, and thus alterations in one group will eventually affect the other. This is why the plasma proteins are important in tissue nutrition. The plasma proteins can be hydrolyzed to amino acids, which can be used for the production of energy by means of the citric acid cycle.

Another general function of plasma proteins is the distribution of water among the compartments of the body. Starling's law attributes a colloid osmotic force to plasma proteins, which, because of their size, cannot cross the capillary membranes. This osmotic force results in the absorption of water from the tissue into the venous end of the capillaries. When the concentration of plasma proteins is markedly decreased, the concomitant decrease in the plasma colloidal osmotic (oncotic) pressure results in increased levels of interstitial fluid and edema. This is often seen in renal disease when proteinuria results in a decreased plasma protein concentration and swelling of the face, hands, and feet occurs.

The amphoteric nature of proteins provides the mechanism for their participation as buffers within the plasma and interstitial tissues. The ionizable R groups can either bind or release excess hydrogen ions as needed.

Many plasma proteins function as specific transporters of metabolic substances. Examples include thyroxine-binding globulin, which carries thyroxine; haptoglobin, which binds free hemoglobin; and albumin, which transports free fatty acids, unconjugated bilirubin, calcium, sulfa drugs, and many other endogenous and exogenous compounds. In addition to transporting the molecules to the location where they can be used, proteins, through the binding process, serve to maintain the bound substance in a soluble state, provide a

storage site for excess substance (transcobalamin II), and prevent the loss of small molecular mass compounds through the kidney (*ie,* transferrin in conserving iron).

Several proteins are glycoproteins. A major function of glycoproteins is in distinguishing which cells are native and which are foreign to the body. These are notably evident as histocompatibility antigens and erythrocyte blood groups. Antigens can stimulate the synthesis of antibodies, which are also proteins. Antibodies and components of the complement system help protect the body against infection.

Many cellular proteins act as receptors for hormones. The receptor binds to its specific hormone and allows the hormonal message to be transmitted to the cell. In addition, certain hormones (*eg,* growth hormone and adrenocorticotropic hormone [ACTH]) are themselves proteins.

Proteins also serve a structural role. Collagen is the most abundant protein in mammals, constituting a quarter of their total weight. Collagen is the major fibrous element of skin, bone, tendon, cartilage, blood vessels, and teeth. Elastin and proteoglycans are two other connective-tissue proteins.

Another important biologic property of some proteins (enzymes) is their ability to catalyze biochemical reactions. Other proteins (clotting factors) aid in the maintenance of hemostasis. The vast array of functions attributed to various proteins is summarized in Table 8-3.

Plasma Proteins

Although the proteins in all fluid compartments play a major physiologic function, the plasma proteins are the most frequently analyzed. More than 500 plasma proteins have been identified. The properties of a few selected plasma proteins are discussed in the following section.

Prealbumin (Transthyretin)

Prealbumin is so named because it migrates ahead of albumin in the customary electrophoresis of serum or plasma proteins. The molecular characteristics of prealbumin can be seen in Table 8-2. Rarely is it observed as a distinct band on routine cellulose acetate electrophoretic patterns of serum, although it can be exhibited by high-resolution electrophoresis (HRE) or

---

TABLE 8-3. Functions of Proteins

Tissue nutrition
Maintenance of water distribution between cells and tissues, interstitial compartments, and the vascular system of the body
Participation as buffers to maintain pH
Transportation of metabolic substances
Part of defense system (antibodies)
Hormones and receptors
Connective-tissue structure
Biocatalysts (enzymes)
Participation in the hemostasis and coagulation of blood

immunoelectrophoresis. It is rich in tryptophan and contains 0.5% carbohydrate. Prealbumin combines with thyroxine and triiodothyronine to serve as the transport mechanism for these thyroid hormones. Prealbumin also binds with retinol-binding protein to form a complex that transports retinol (vitamin A). Prealbumin is decreased in hepatic damage, acute phase inflammatory response, and tissue necrosis. A low prealbumin level is also a sensitive marker of poor protein nutritional status. When dietary intake of protein-calories is deficient, hepatic synthesis of proteins is reduced and catabolism increases. This results in a decrease in the level of the proteins originating in the liver including prealbumin, albumin and β-globulins such as transferrin and complement C3.[16] Prealbumin, because of its short half-life of approximately 2 days, will decrease more rapidly than the other proteins. Prealbumin is increased in patients receiving steroids, in alcoholism, and in chronic renal failure.

Albumin

Albumin is the protein present in highest concentration in the serum (see Table 8-2). It is synthesized in the liver. The earliest method for its determination involved the salting-out of the globulins with sodium sulfate, leaving the albumin in solution. The albumin was then determined by the Kjeldahl method and, later, by the biuret color development. The method commonly used today is one involving dye binding and the shift in color when a dye is bound by albumin. When more information about proteins is needed, an electrophoretic pattern is obtained, and the albumin is calculated as a percentage of the total protein (usually approximately 60%). At birth, the reference value for serum albumin averages 39 g/L. The concentration falls to 28.4 g/L at about 9 months and then begins to increase slowly until adult values of 35 g/L to 55 g/L are reached.[17] Serum albumin level after age 60 years averages 38.3 g/L.[18]

Albumin has two well-known functions. One is the contribution albumin makes to the colloid osmotic pressure of the intravascular fluid. Due to its high concentration, albumin contributes nearly 80% of this pressure, which maintains the appropriate fluid in the tissues. The other prime function is its propensity to bind various substances in the blood. For example, albumin binds bilirubin, salicylic acid, fatty acids, calcium and magnesium ions, cortisol, and some drugs. This characteristic is also exhibited with certain dyes, providing a method for the quantitation of albumin.

Decreased concentrations of serum albumin may be caused by the following:

- An inadequate source of amino acids, which is seen in malnutrition and muscle-wasting diseases.
- Liver disease resulting in the inability of hepatocytes to synthesize albumin. The increase in globulins that occurs in early cirrhosis, however, will balance the loss in albumin to give a total protein concentration within acceptable limits. The decline in serum albumin is insignificant in infectious hepatitis.

- Gastrointestinal loss as interstitial fluid leaks out in inflammation and disease of the intestinal mucosa.
- Loss in the urine in renal disease. Albumin is normally excreted in very small amounts. This excretion is increased when the glomerulus no longer functions to restrict the passage of proteins from the blood.

Abnormalities in serum albumin are also exhibited by the absence of albumin (*analbuminemia*), or the presence of albumin that has unusual molecular characteristics—this abnormality is called *bisalbuminemia* and is demonstrated by the presence of two albumin bands instead of the single band usually seen by electrophoresis. Analbuminemia is an abnormality of genetic origin resulting from an autosomal recessive trait. These conditions are rare.

Increased serum albumin levels are seen in dehydration. This increase is a relative increase, because the albumin is contacted in a reduced volume of serum. Administering fluids to treat the dehydration will decrease serum albumin levels back to normal.

Globulins

The *globulin* group of proteins consists of α_1, α_2, β, and γ fractions. Each fraction consists of a number of different proteins with different functions. The following subsections describe selected examples of the globulins.

α_1-**Antitrypsin.** α_1-antitrypsin is an acute-phase reactant. Its main function is to neutralize trypsin-like enzymes (*ie*, elastase) that can cause hydrolytic damage to structural protein. α_1-antitrypsin is a major component (approximately 90%) of the fraction of serum proteins that migrates electrophoretically immediately following albumin. Molecular characteristics are seen in Table 8-2.

A deficiency of α_1-antitrypsin is associated with severe, degenerative, emphysematous pulmonary disease. The lung disease is attributed to the unchecked proteolytic activity of proteases from leukocytes in the lung during periods of inflammation. Juvenile hepatic cirrhosis is also a correlative disease in α_1-antitrypsin deficiency. The protein is synthesized but not released from the hepatocyte.

Several phenotypes of α_1-antitrypsin deficiency have been identified. The most common phenotype is *MM* (allele *Pi*[M]) and is associated with normal antitrypsin activity. Other alleles are *Pi*[S], *Pi*[Z], *Pi*[F], and *Pi*[−] (null). The homozygous phenotype *ZZ* individual is in serious jeopardy of liver and lung disease from a deficiency of α_1-antitrypsin. Individuals with the *MZ* or *MS* phenotype are usually not affected but should be counseled about having offspring who may be *ZZ*, *Z*[−], and in danger. The *Pi*[Z] allele occurs in 1 in 1500 Caucasians. Factors other than α_1-antitrypsin are also involved in disease, because some people who have abnormal phenotypes and low concentrations of proteins do not develop overt disease. Increased levels of α_1-antitrypsin are seen in inflammatory reactions, pregnancy, and contraceptive use.

The discovery of abnormal α_1-antitrypsin levels is most often made by the lack of an α_1-globulin band on protein electrophoresis. This discovery is followed with one of the quantitative methods. A widely used method is radial immunodiffusion. Immunonephelometric assays by automated instrumentation are also available. Phenotyping can be accomplished by immunofixation.

α_1-Fetoprotein. α_1-fetoprotein (AFP) is synthesized initially by the fetal yolk sac and then by the parenchymal cells of the liver.[2] In 1956, it was first discovered, in fetal serum, to have an electrophoretic mobility between that of albumin and α_1-globulin. It peaks in the fetus at about 13 weeks' gestation (3 mg/mL) and recedes at 34 weeks' gestation.[19] At birth, it recedes rapidly to adult concentrations, which are normally very low (see Table 8-2). The methods commonly used for AFP determinations are radioimmunoassay and enzyme labeled immunoassay.

The function of AFP is not well established. It has been proposed that the protein protects the fetus from immunolytic attack by its mother.[19] AFP is detectable in the maternal blood up to month 7 or 8 of pregnancy because it is transmitted across the placenta. Thus, measurement of the level of AFP in maternal serum is a screening test for any fetal conditions in which there is increased passage of fetal proteins into the amniotic fluid. Conditions associated with an elevated AFP level include spina bifida and neural tube defects, atresia of the gastrointestinal tract, and fetal distress in general.[20] Its use in determining neural tube defects before term is an important reason for its assay. It is also increased in ataxia-telangiectasia, in tyrosinosis, and in hemolytic disease of the newborn. It is interesting that maternal serum AFP is also increased in the presence of twins. Low levels of maternal AFP indicate a three-fold to four-fold increased risk for Down's syndrome.[21]

The normal time for screening is between 15 to 20 weeks' gestational age. The maternal AFP increases gradually during this period; thus, interpretation requires accurate dating of the pregnancy. Maternal AFP levels are also affected by maternal weight, which reflects blood volume (inverse relationship), race (10% higher in African-Americans), and diabetes (lowered value). Thus, test results need to be adjusted for these variables. It has been found that reporting AFP in multiples of the median (MoM) leads to good correlation with infant risk. MoM are calculated by dividing the patient's AFP value by the median reference value for that gestational age. Most screening programs use 2.0 MoM as the upper limit and 0.5 MoM as the lower limit for maternal serum AFP.

Serum levels of AFP can also be used as a tumor marker. Very high concentrations of AFP are found in many cases of hepatocellular carcinoma (approximately 80%) and certain gonadal tumors in adults[22] (see Chapter 27, *Tumor Markers*).

α_1-Acid Glycoprotein (Orosomucoid). α_1-acid glycoprotein comprises five carbohydrate units attached to a polypeptide chain. Owing to its low pI (2.7), it is negatively charged even in acid solutions, a fact that led to its name. Other molecular characteristics are seen in Table 8-2. Sequences of haptoglobin, immunoglobulins, and α_1-acid glycoprotein show considerable homology. This has led to speculation that these proteins share a precursor in the evolutionary past. α_1-acid glycoprotein is implicated in the formation of certain membranes and fibers in association with

---

CASE STUDY 8-2

Immediately following the birth of a baby girl, the attending physician requested a protein electrophoretic examination of the mother's serum. This was done on a sample that was obtained upon the mother's admission to the hospital the previous day. An electrophoretic examination was also performed on the cord-blood specimen. Laboratory reports were as follows:

CASE STUDY TABLE 8-2.1 Electrophoresis (Values g/dL)

	Adult Reference Values	Mother's Serum	Cord Blood
Albumin	3.5–5.0	4.2	3.3
α_1-globulins	0.1–0.4	0.3	0.0
α_2-globulins	0.3–0.8	1.2	0.4
β-globulins	0.6–1.1	1.3	0.7
γ-globulins	0.5–1.7	1.3	1.0

The appearance of the mother's electrophoretic pattern was within that expected for a healthy person. The electrophoretic pattern of the cord-blood serum resembled the one shown in Figure 8-13C.

Questions

1. What protein fraction(s) is/are abnormal in the mother's serum and the cord-blood serum?
2. An abnormality in this/these fraction(s) is/are most often associated with what disease?
3. What other test(s) may be done to confirm this abnormality?

collagen and may inactivate basic hormones such as progesterone.

The analytical methods used most commonly for the determination of orosomucoid are radial immunodiffusion and nephelometry. Immunofixation has been used to study inherited variants.[23]

Increased concentration of this protein is the major cause of an increased glycoprotein level in the serum during inflammation. It is also increased in pregnancy, cancer, pneumonia, rheumatoid arthritis, and other conditions associated with cell proliferation.

α_1-Antichymotrypsin. α_1-antichymotrypsin (α_1-ACT) is a serine proteinase with cathepsin G, pancreatic elastase, mast cell chymase, and chymotrypsin as target enzymes.[24] α_1-ACT carries four oligosaccharide side chains. It migrates between the α_1 and α_2 zones on high-resolution serum protein electrophoresis. Elevations are seen in inflammation, and it appears that the complex formed between plasma α_1-ACT and its target enzymes plays a major role in signaling for acute phase protein synthesis in response to injury. Hereditary deficiency of α_1-ACT is associated with asthma and liver disease.

Inter-α-Trypsin Inhibitor. Inter-α-trypsin (ITI) inhibitor comprises at least three distinct polypeptide subunits: two heavy chains designated H_1 and H_2 and a light chain designated L or *bikunin*. The light chain is responsible for the inhibition of the proteases trypsin, plasmin, and chymotrypsin.[25] On high-resolution serum electrophoresis, ITI migrates in the zone between the α_1 and α_2 fractions. Elevations are seen in inflammatory disorders.

Gc-globulin (Group-Specific Component; Vitamin D–Binding Protein). Gc-globulin is another protein that migrates in the α_1–α_2 interzone. This protein exhibits a high binding affinity for Vitamin D compounds and actin, the major constituent of the thin filaments of muscle. Due to genetic polymorphism, several phenotypes of Gc-globulin exist. Elevations of Gc-globulin are seen in the third trimester of pregnancy and in patients taking estrogen oral contraceptives. Severe liver disease and protein-losing syndromes are associated with low levels.

Haptoglobin. Haptoglobin, an α_2 glycoprotein, is synthesized in the hepatocytes and, to a very small extent, in cells of the reticuloendothelial system. Haptoglobin comprises two kinds of polypeptide chains: two α chains and one β chain. There are three possible α chains and only one form of β chain. On starch-gel electrophoresis, haptoglobin exhibits three types of patterns, which serve to illustrate the polymorphism in the α chains. Homozygous haptoglobin 1-1 gives 1 band. The peptide chains form polymers with each other and with haptoglobin 1 chains to provide the other two electrophoretic patterns, which have been designated as Hp 1-2 and Hp 2-2 phenotypes. Other characteristics are seen in Table 8-2.

Haptoglobin increases from a mean concentration of 0.02 g/L at birth to adult levels within the first year of life. As old age is approached, haptoglobin levels increase, with a more marked increase being seen in males. Radial immunodiffusion and immunonephelometric methods have been used for the quantitative determination of haptoglobin.

The function of haptoglobin is to bind free hemoglobin by its α chain. Abnormal hemoglobin such as Bart's and hemoglobin H have no α chains and cannot be bound. The reticuloendothelial cells remove the haptoglobin-hemoglobin complex from circulation within minutes of its formation. Thus haptoglobin prevents the loss of hemoglobin and its constituent iron into the urine.

Serum haptoglobin concentration is increased in inflammatory conditions. It is one of the proteins used to evaluate the rheumatic diseases. Increases are also seen in conditions such as burns and nephrotic syndrome when large amounts of fluid and lower-molecular-weight proteins have been lost. Determination of a decreased free haptoglobin level (or haptoglobin-binding capacity) has been used to evaluate the degree of intravascular hemolysis that has occurred in transfusion reactions or hemolytic disease of the newborn. Mechanical breakdown of red cells during athletic trauma may result in a temporary lowering of haptoglobin levels.

Ceruloplasmin. Ceruloplasmin is a copper-containing α_2-glycoprotein that has enzymatic activities (*ie*, copper oxidase, histaminase, and ferrous oxidase). It is synthesized in the liver, where six to eight atoms of copper, half as cuprous (Cu^+) and half as cupric (Cu^{2+}) ions, are attached to an apoceruloplasmin. Ninety percent or more of the total serum copper is found in ceruloplasmin. Molecular characteristics are given in Table 8-2.

The early analytical method of ceruloplasmin determination was based on its copper oxidase activity. Most assays today use immunochemical methods, including radial immunodiffusion and nephelometry.[26]

Low concentrations of ceruloplasmin at birth gradually increase to adult levels and slowly continue to rise with age thereafter. Adult females have higher concentrations than do males, and pregnancy, inflammatory processes, malignancies, oral estrogen, and contraceptives cause an increased serum concentration.

Certain diseases or disorders are associated with low serum concentrations. In Wilson's disease (hepatolenticular degeneration), an autosomal recessive inherited disease, the levels are typically low (0.1 g/L). Total serum copper is decreased, but the direct reacting fraction is elevated and the urinary excretion of copper is increased. The copper is deposited in the skin, liver, and brain, resulting in degenerative cirrhosis and neurologic damage. Copper also deposits in the cornea, producing the characteristic Kayser-Fleischer rings. Low ceruloplasmin is also seen in malnutrition, malabsorption, severe liver disease, nephrotic syndrome, and Menkes' kinky-hair syndrome, in which a decreased absorption of copper results in a decrease in ceruloplasmin.

α₂-Macroglobulin. α₂-macroglobulin, a dimeric, large protein (see Table 8-2), is synthesized by hepatocytes. It is found principally in the intravascular spaces because its movement is restricted due to its size. However, much lower concentrations of α₂-macroglobulin can be found in other body fluids, such as CSF. On binding with and inhibiting proteases, it is removed by the reticuloendothelial tissues. The analytical methods that have been used for the satisfactory assay of this protein are radial immunodiffusion and immunonephelometry.

This protein reaches a maximum serum concentration at the age of 2 years to 4 years[27] and then decreases to about one-third of that at about age 45 years. Later in life, a moderate increase is seen. This change with age is more pronounced in males than in females.[28] There is a distinct difference in the reference values for males and females, adult females having higher values than males.

α₂-macroglobulin inhibits proteases such as trypsin, pepsin, and plasmin. It also contributes more than one-fourth of the thrombin inhibition normally present in the blood. Very little is known of its correlation with disease or disorders, except in the case of renal disease.

In nephrosis, the levels of serum α₂-macroglobulin may increase as much as 10 times, because its large size aids in its retention. The protein is also increased in diabetes and liver disease. Use of contraceptive medications and pregnancy increase the serum levels by 20%.

Transferrin (Siderophilin). Transferrin, a glycoprotein, is synthesized primarily by the liver. Two molecules of ferric iron can bind to each molecule of transferrin. When the plasma is saturated with ferric iron, it assumes a pale pink color from the saturated molecules, whereas the ferrous iron complex is colorless.[1] Normally, only about 33% of the iron binding sites on transferrin are occupied. Transferrin is the major component of the β-globulin fraction and appears as a distinct band on high-resolution serum protein electrophoresis. Genetic variation of transferrin has been demonstrated by electrophoresis on polyacrylamide gel. Other characteristics of transferrin are given in Table 8-2.

The analytical methods used for the quantitation of transferrin are immunodiffusion and immunonephelometry. Both are considered to give precise and accurate results.

The major functions of transferrin are the transport of iron and the prevention of loss of iron through the kidney. Its binding of iron prevents iron deposition in the tissues during temporary increases in absorbed iron or free iron. Transferrin transports iron to its storage sites, where it is incorporated into another protein, apoferritin, to form ferritin. Transferrin also carries iron to cells such as bone marrow that synthesize hemoglobin and other iron-containing compounds.

The most common form of anemia is iron deficiency anemia, a hypochromic, microcytic anemia. In this type of anemia, transferrin in serum is normal or increased. A decreased transferrin level generally reflects an overall decrease in synthesis of protein, such as seen in liver disease or malnutrition, or it may be seen in protein-losing disorders such as nephrotic syndrome. A deficiency of plasma transferrin may result in the accumulation of iron in apoferritin or in histiocytes, or it may precipitate in tissues as hemosiderin. Patients with hereditary transferrin deficiencies have been shown to have marked hypochromic anemia. An increase of iron bound to transferrin is found in a hereditary disorder of iron metabolism, hemochromatosis, in which excess iron is deposited in the tissues, especially the liver and the pancreas. This disorder is associated with bronze skin, cirrhosis, diabetes mellitus, and low plasma transferrin levels.

Hemopexin. The parenchymal cells of the liver synthesize hemopexin, which migrates electrophoretically in the β-globulin region. Other characteristics are seen in Table 8-2. Hemopexin can be determined by radial immunodiffusion. The function of hemopexin is to remove circulating heme. When free heme (ferroprotoporphyrin IX) is formed during the breakdown of hemoglobin, myoglobin, or catalase, it binds to hemopexin in a 1:1 ratio. The heme-hemopexin complex is carried to the liver, where the complex is destroyed. Hemopexin also removes ferriheme and porphyrins.

The level of hemopexin is very low at birth but reaches adult values within the first year of life. Pregnant mothers have increased plasma hemopexin levels. Increased concentrations are also found in diabetes mellitus, Duchenne type muscular dystrophy, and some malignancies, especially melanomas. In hemolytic disorders, serum hemopexin concentrations decrease. Administration of diphenylhydantoin also results in decreased concentrations. A more complete list of conditions that cause increased and decreased levels of hemopexin may be found in a thorough review by Natelson and Natelson.[1]

Lipoproteins. Lipoproteins are complexes of proteins and lipids whose function is to transport cholesterol, triglycerides, and phospholipids in the blood. Lipoproteins are subclassified according to the apoprotein and specific lipid content. On high-resolution serum protein electrophoresis, high-density lipoproteins (HDL) migrate between the albumin and α₁-globulin zone; very low-density lipoproteins (VLDL) migrate at the beginning of the β-globulin fraction (pre-β); and the low-density lipoproteins (LDL) appear as a separate band in the β-globulin region. For a more detailed discussion of structure and methods of analysis, refer to Chapter 11, *Lipids and Lipoproteins.*

β₂-Microglobulin. β₂-microglobulin (B2) is the light chain component of the major histocompatibility complex (HLA). This protein is found on the surface of most nucleated cells and is present in high concentrations on lymphocytes. Due to its small size (MW 11,800), B2 is filtered by the renal glomerulus but most (>99%) is reabsorbed and catabolized in the proximal tubules. Elevated serum levels are the result of impaired clearance by the kidney or overproduction of the protein, as occurs in a number of inflammatory diseases such as rheumatoid arthritis and systemic lupus erythematosus. In patients with human immunodeficiency virus, a high B2 level in the absence of renal failure indicates a large lymphocyte turnover rate, which suggests the virus is killing lymphocytes. B2 may sometimes be seen on HRE, but because of its low concentration, it is usually measured by immunoassay.

Complement. *Complement* is a collective term for several proteins that participate in the immune reaction and serve as a link to the inflammatory response. Molecular characteristics of selected complement components are listed in Table 8-2. These proteins circulate in the blood as nonfunctional precursors. In the classic pathway, activation of these proteins begins when the first complement factor, C1q, binds to an antigen–antibody complex. The binding occurs at the Fc, or constant, part of the IgG or IgM molecule. Each complement protein (C2 through C9) is then activated sequentially and can bind to the membrane of the cell to which the antigen–antibody complex is bound. The final result is lysis of the cell. In addition, complement is able to enlist the participation of other humoral and cellular effector systems in the process of inflammation. An alternate pathway (properdin pathway) for complement activation exists in which the early components are bypassed and the process begins with C3. This pathway is triggered by different substances (does not require the presence of an antibody), but the lytic attack on membranes is the same (sequence C5–C9). Analytical methods have included a measurement of the titer of complement using complement activity in a hemolytic system and by immunochemical methods such as radial immunodiffusion and nephelometry.

Complement is increased in inflammatory states and decreased in malnutrition, lupus erythematosus, and disseminated intravascular coagulopathies. Inherited deficiencies of individual complement proteins also have been described. In most cases, the deficiencies are associated with recurrent infections.

Fibrinogen. Fibrinogen is one of the largest proteins in the blood plasma. It is synthesized in the liver, and because it has considerable carbohydrate content, it is classified as a glycoprotein. Other molecular characteristics are given in Table 8-2. On plasma electrophoresis, fibrinogen is seen as a distinct band between the β- and γ-globulins. The function of fibrinogen is to form a fibrin clot when activated by thrombin. Thus, fibrinogen is virtually all removed in the clotting process and is not seen in serum.

Fibrinogen customarily has been determined as clottable protein by a method to determine total protein (Biuret, Kjeldahl) performed on the clot that is formed when plasma is reacted with thrombin. The clot is washed, dried, dissolved, and analyzed. Fibrin split products (degradation products of fibrinogen and fibrin) are determined by immunoassay methods such as radial immunodiffusion, nephelometry,[29] and radioimmunoassay.

Fibrinogen is one of the *acute-phase reactants,* a term that refers to proteins that are markedly increased in plasma during the acute phase of the inflammatory process. Fibrinogen levels also rise with pregnancy and the use of birth control pills. Decreased values generally reflect extensive coagulation during which the fibrinogen is consumed.

C-Reactive Protein. C-reactive protein (CRP) appears in the blood of patients with diverse inflammatory diseases but is undetectable in healthy individuals. It is synthesized in the liver. Other characteristics are found in Table 8-2. CRP was so named because it precipitates with the C substance, a polysaccharide of pneumococci. However, it was found that CRP rises sharply whenever there is tissue necrosis, whether the damage originates from a pneumococcal infection or some other source. This led to the discovery that CRP recognizes and binds to molecular groups found on a wide variety of bacteria and fungi. CRP bound to bacteria promotes the binding of complement, which facilitates their uptake by phagocytes. This process of protein coating to enhance phagocytosis is known as *opsonization.*[30]

CRP is generally measured by its capacity to precipitate C substance or more commonly by immunologic methods including latex slide precipitation test, nephelometry, and enzyme immunoassay (EIA). CRP is elevated in acute rheumatic fever, bacterial infections, myocardial infarcts, rheumatoid arthritis, carcinomatosis, gout, and viral infections.

Immunoglobulins (Igs). There are five major groups of immunoglobulins in the serum: IgA, IgG, IgM, IgD, and IgE. They are synthesized in plasma cells. Their synthesis is stimulated by an immune response to foreign particles and microorganisms. The immunoglobulins are not synthesized to any extent by the neonate. IgG crosses the placenta, and the IgG present in the newborn's serum is that synthesized by the mother. IgM does not cross the placenta but rather is synthesized by the neonate. The concentration of IgM initially is 0.21 g/L, but this increases rapidly to adult levels by about age 6 months.[27] IgA is virtually lacking at birth (0.003 g/L), increases slowly to reach adult values at puberty, and continues to increase during the lifetime. IgD and IgE levels are undetectable at birth by customary methods and increase slowly until adulthood. IgA is generally higher in males than in females; IgM and IgG levels are somewhat higher in females. IgE levels vary with the allergic condition of the individual.

The immunoglobulins comprise two long polypeptide chains (heavy, or H, chains) and two short polypeptides (light, or L, chains), joined by disulfide bonds. An individual is capable of producing 1 million different immunoglobulin molecules. The differences among these molecules are found in a region of the molecule called the *variable region.* This variable region is located on the end of the molecule that contains both the light and heavy chains and is the site at which the immunoglobulin (antibody) combines with the antigen. In this way, there are many different antibodies that are relatively specific for corresponding antigens.

The differences in the heavy chains (H) are called *idiotypes* and are designated IgG, IgA, IgM, IgD, and IgE. The heavy chains are called γ, α, μ, σ, and ε, respectively. The light chains (L) for all the immunoglobulin classes are of two kinds, either κ or λ. Each immunoglobulin or antibody molecule has two identical H chains and two identical L chains. For example, IgG has two γ type H chains and two identical L chains (either κ or λ).

When a foreign substance (antigen) is injected into an animal (*eg,* rabbit or goat), an antibody that will react with that antigen is synthesized. This reaction is relatively specific; that is, the antibodies will react specifically and selectively with the antigen used to raise them. Proteins and polysaccharides are strong antigens. Antibodies can be raised in rabbits and other animals to the human immunoglobulins as well as the other serum proteins. These rabbit antihuman immunoglobulin antibodies are used to detect and to quantitatively assay IgG, IgA, IgM, IgD, and IgE. The immunoglobulins have been determined using radial immunodiffusion and radio-immunoassay. Fluorescent immunoassay techniques and immunonephelometric assays also have been used. The automated immunonephelometric method is available for IgG, IgA, and IgM.

A molecule of immunoglobulin derived from the proliferation of one plasma cell (clone) is called a *monoclonal immunoglobulin* or *paraprotein*. A marked increase in such a monoclonal Ig is found in the serum of patients who have plasma cell malignancy (myeloma). Monoclonal increases are seen on electrophoretic patterns as spikes. Although these immunoglobulins are typically in the β or γ fractions, on occasion one may appear in the α_2 area. The monoclonal immunoglobulin is typed by the immunofixation method to determine it as IgG, IgA, or IgM, and identify κ or λ light chains. Increases in IgD or IgE, or heavy chain disease should be considered if there is no reaction with the typical immunofixation electrophoresis antisera.

IgG is increased in liver disease, infections, and collagen disease. A decrease in IgG is associated with an increased susceptibility to infections and monoclonal gammopathy of one of the other idiotypes.

IgA is the idiotypic immunoglobulin present in the respiratory and gastrointestinal mucosa. IgA in fluids other than serum has an additional secretory peptide, called a *J piece.* This piece enables the IgA to appear in secretions. Polyclonal increases in the serum IgA (without the J piece) are found in liver disease, infections, and autoimmune diseases. Decreased serum concentrations are found in depressed protein synthesis, ataxia-telangiectasia, and hereditary immunodeficiency disorders.

IgM is the first antibody that appears in response to antigenic stimulation. IgM also is the type of antibody that acts as anti-A and anti-B antibody to red cell antigens, rheumatoid factors, and heterophile antibodies. Increased IgM concentration is found in toxoplasmosis, cytomegalovirus, rubella, herpes, syphilis, and various bacterial and fungal diseases. A monoclonal increase is seen in Waldenström's macroglobulinemia. This increase is seen as a spike in the

vicinity of the late β zone of a protein electrophoretic pattern. Decreases are seen in protein-losing conditions and immunodeficiency disorders.

IgD is the immunoglobulin with the fourth highest concentration in normal serum. Its concentration is increased in infections, liver disease, and connective-tissue disorders. IgD multiple myeloma also has been reported with a monoclonal spike in the late β zone of the electrophoretic pattern.

IgE is the idiotypic immunoglobulin associated with allergic and anaphylactic reactions. Polyclonal increases are seen in allergies, including asthma and hay fever. Monoclonal increases are seen in IgE myeloma, a rare disease.

Miscellaneous Proteins

Myoglobin

Myoglobin is a heme protein found in striated skeletal and cardiac muscles. It accounts for approximately 2% of the total muscle protein. A minor portion of the myoglobin found in cells is structurally bound but most of it is dissolved in the cytoplasm. It comprises one polypeptide chain containing 153 amino acids, coupled to a heme group. In size, myoglobin, with a molecular weight of 17,800 daltons, is slightly larger than one-fourth of a hemoglobin molecule. It can reversibly bind oxygen in a manner similar to the hemoglobin molecule, but myoglobin requires a very low oxygen tension to release the bound oxygen. The serum baseline level varies with physical activity and muscle mass and is in the range of 30 ng/mL to 90 ng/mL (mg/dL) for adult males. Females typically show myoglobin concentrations of less than 50 ng/mL.

When striated muscle is damaged, myoglobin is released, elevating the blood levels. In an acute myocardial infarction (AMI), this increase is seen within 1 hour to 3 hours of onset and reaches peak concentration in 5 hours to 12 hours. For the diagnosis of AMI, serum myoglobin should be measured serially. If a repeated myoglobin level doubles within 1 hour to 2 hours after the initial value, it highly diagnostic of an AMI[31] The degree of elevation indicates the size of the infarct. Because myoglobin is a small molecule, the kidney freely filters it, and blood levels return to normal in 18 hours to 30 hours after the AMI. Therefore, an increase in myoglobin in the circulation is an early indicator of myocardial infarction and is useful in determining patients who would benefit from thrombolytic therapy. Due to the speed of appearance and clearance of myoglobin, it also is a useful marker for monitoring the success or failure of reperfusion. Figure 8-10 shows the relative level versus time of onset after an AMI for some of the current cardiac markers. Although the diagnostic sensitivity of myoglobin elevations following an AMI has been reported to be between 75% and 100%, myoglobin is not cardiac specific. Elevations are also seen in conditions such as progressive muscular dystrophy and crushing injury in which skeletal muscle is damaged. As myoglobin is eliminated from the circulation through the kidneys, renal failure can also elevate the level of serum myoglobin. Table 8-4 lists some of the causes of a myoglobin elevation.

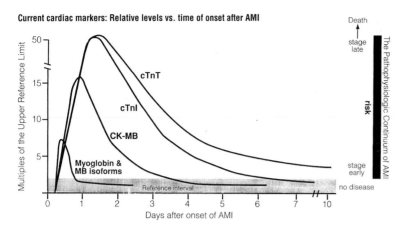

Current cardiac markers: Relative levels vs. time of onset after AMI

Figure 8-10. Current cardiac markers: relative level vs. time of onset after AMI. Source: Alan Wu, PhD, and Robert Jesse, MD, PhD, developed for a wall chart sponsored by Behring Diagnostics, Inc. (San Jose, Calif)

Latex agglutination, enzyme-linked immunosorbent assay (ELISA), immunonephelometry, and fluoroimmunoassays for myoglobin have been developed. A qualitative spot test using immunochromatography is also available.

Troponin

Troponin is a complex of three proteins that bind to the thin filaments of striated muscle (cardiac and skeletal) but are not present in smooth muscle. The complex consists of troponin T (TnT), troponin I (TnI), and troponin C (TnC). Together, they function to regulate muscle contraction. The muscle contraction cycle begins with a release of calcium in response to nerve impulses. Troponin C (MW 18,000) binds the calcium, causing a conformational change in the troponin-tropomyosin complex. Tropomyosin is a rod-shaped protein that stretches the entire length of the actin backbone (actin is the major constituent of thin filaments). This movement allows the head of the myosin molecule, which forms the thick filaments of muscle, to interact with actin. The binding of myosin and actin accelerates the myosin ATPase activity, and the result is muscle contraction. With the hydrolysis of ATP, the myosin head returns to its original

position and the cycle can begin again. Troponin I (MW 24,000) regulates this striated muscle contraction by preventing the binding of the myosin head to actin and thus inhibiting myosin ATPase activity. TnI also serves to bind the actin filament to TnC. The function of troponin T (MW 37,000) is to bind to tropomyosin and position the troponin complex along the actin filament.[32] The structure of the muscle thin filament is shown in Figure 8-11.

Three genes code for TnT: one each in cardiac muscle and fast- and slow-skeletal muscle. TnI, also encoded by three genes, has similar isoforms, whereas TnC, encoded by two genes, has only a cardiac and slow-skeletal muscle form. The isoforms have different amino acid structures and are biochemically distinct and therefore can be differentiated from one another. Of particular interest are the cardiac isoforms of troponin T (cTnT) and troponin I (cTnI). Cardiac troponin T levels in serum begin to rise within 3 hours to 4 hours following the onset of myocardial damage, peak in 10 hours to 24 hours and remain elevated for 10 days to 14 days following an AMI (see Figure 8-10). Because cardiac TnT is specific for heart muscle, and even small amounts of cardiac necrosis cause release of discernible amounts of cTnT into the serum, measurement of this protein is a valuable aid in the diagnosis of an AMI. Cardiac TnT is also useful in monitoring the

TABLE 8-4. Causes of Myoglobin Elevations

Acute myocardial infarction
Angina without infarction
Rhabdomyolysis
Multiple fractures; muscle trauma
Renal failure
Myopathies
Vigorous exercise
Intramuscular injections
Open heart surgery
Tonic–clonic seizures
Electric shock
Arterial thrombosis
Certain toxins

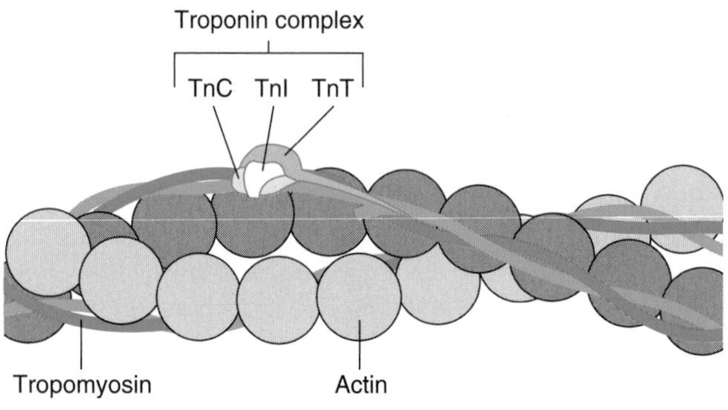

Figure 8-11. Schematic of a muscle thin filament.

effectiveness of thrombolytic therapy in myocardial infarction patients. The ratio of peak cardiac TnT concentration on day 1 to cardiac TnT concentration at day 4 discriminates between patients with successful (ratio greater than 1) and failed (ratio less than or equal to 1) reperfusion.[33] Another use of cTnT is in the risk assessment of patients with acute myocardial ischemia.[34,35] The prognosis of these patients is quite variable. The duration, frequency, and timing of ischemic symptoms can be used to determine the severity of unstable angina but are not predictive of events such as infarction, cardiogenic shock, heart failure, ventricular arrhythmia, or even death. However, elevated cTnT values have been shown to be associated with an increased incidence of adverse outcomes, and the higher the cTnT value, the greater the risk. The prognostic value of cTnT is independent of age, hypertension, number of antianginal drugs, and electrocardiographic changes. The identification of patients at risk of developing severe complications allows for institution of long-term antithrombotic protection. In a recent study, ischemic patients with cTnT levels ≥ 0.1 ng/mL were treated with low-molecular-weight heparin. The incidence of death or AMI in these patients occurred at a rate only one-half that of the placebo group (7.4% vs. 14.2%; $p < 0.01$).[36] One disadvantage of cTnT that has been reported is the potential for false-positive elevations in renal failure patients.[37]

Cardiac troponin I is also highly specific for myocardial tissue.[38,39] Because cTnI, like cTnT, does not normally circulate in the blood and it is 13 times more abundant in the myocardium than CK-MB on a weight basis, cTnI is a very sensitive indicator of even minor amount of cardiac necrosis. Following an AMI, cTnI levels begin to rise in 3 hours to 6 hours, reach peak concentration in 14 hours to 20 hours, and return to normal in 5 days to 10 days as shown in Figure 8-10. The relative increase in cTnI is greater than CK-MB or myoglobin following thrombolytic therapy reperfusion studies. Elevated cardiac troponin I is also associated with increased risk of mortality and morbidity in patients with ischemic heart disease. In evaluating cTnT and cTnI, both were found to offer comparable information. Some studies have reported that cTnT is more sensitive in unstable angina patients, whereas cTnI levels appear to have greater cardiac specificity and are not elevated in renal failure patients, but these claims remain an open and controversial issue.

Although single measurements are useful for risk stratification, serial measurements are needed for accurate diagnosis of AMI. Sequential cardiac troponin determinations on samples drawn at 3- to 8-hour intervals over a period of 48 hours following an AMI will demonstrate the classic rise and fall seen with other cardiac markers. Cardiac troponins can be measured on serum or heparinized plasma by ELISA or immunoenzymometric assays using two monoclonal antibodies directed against different epitopes on the protein. The reference interval for cTnT is <0.1 ng/mL (mg/L). The cutoff concentration for cTnI immunoassays varies from 0.1 ng/mL up to 3.1 ng/mL (mg/L). This difference can at least partly be explained by different specificities of the monoclonal antibodies used in the assays. Data have indicated that the largest portion of troponin I released into the bloodstream following myocardial tissue damage is in the form of a complex with cTnC, whereas only a small part is in the free form.[40] The ratio of total to free cTnI varies during the period the troponin circulates

---

CASE STUDY 8-3

A 76-year-old female was admitted to the hospital with gangrene of her right toe. She was disoriented and had difficulty finding the right words to express herself. On evaluation, it was revealed she lived alone and was responsible for her own cooking. A daughter who lived in the area said she was a poor eater even with much encouragement. An ECG, performed on admission, showed possible ectopic rhythm with occasional premature supraventricular contractions. The cardiologist suspected a possible inferior myocardial infarction of undetermined age. Lab results are as follows:

CASE STUDY TABLE 8-3.1 Laboratory Results

Day 1

CK-total	187 U/L	(40–325)
CK-MB Mass	6 g/L	(<8)
Troponin I	16.3 g/L	(0–2)
Prealbumin	15 mg/dL	(17–42)
Albumin	2.7 g/dL	(3.7–4.9)

Repeat (5 hours later)

CK-total	180 U/L
CK-MB Mass	5.4 g/L
Troponin I	17.5 g/L

Day 2

CK-total	177 U/L
CK-MB Mass	4.5 g/L
Troponin I	13.7 g/L
Myoglobin	>500 (g/L <76)

Questions

1. In this patient, what is the clinical value of the Troponin I measurements?
2. What is a possible explanation for the elevated myoglobin?
3. What condition is indicated by the low pre-albumin value?

in the bloodstream and is different in samples from different patients. Cardiac TnC, when complexed with cTnI, causes structural and chemical changes in cTnI that can mask certain epitopes and diminish the interaction of cTnI with certain monoclonal antibodies. In other assays, the pairs of antibodies used recognize epitopes that are not perturbed or sterically shielded by other troponin complexes. These antibodies will react equally with free and complexed cTnI, resulting in higher cutoff values.

There are also rapid immunochromatographic dry-strip assays available.[41,42] Immobilized antibodies bind troponin and produce a purplish band in the test window. The intensity and speed at which the color forms are related to the concentration of the specific troponin.

Total Protein Abnormalities

Measurement of total plasma protein content provides general information reflecting disease states in many organ systems.

Hypoproteinemia

A total protein level less than the reference interval—*hypoproteinemia*—occurs in any condition where a negative nitrogen balance exists. One cause of a low level of proteins in the plasma is excessive loss. Plasma proteins can be lost by excretion in the urine in renal disease (ie, nephrotic syndrome), leakage into the gastrointestinal tract in inflammation of the digestive system, and the loss of blood in open wounds, internal bleeding, or extensive burns. Another circumstance producing hypoproteinemia is decreased intake either because of deficiency of protein in the diet (malnutrition) or through intestinal malabsorption due to structural damage (ie, sprue). Without adequate dietary intake of proteins, there is a deficiency of certain essential amino acids and protein synthesis is impaired. A decrease in serum proteins due to decreased synthesis is also seen in liver disease (site of all nonimmune protein synthesis) or in the inherited immunodeficiency disorders, in which antibody production is diminished. Additionally, hypoproteinemia may result from accelerated catabolism of proteins, such as occurs in burns, trauma, or other injuries.

Hyperproteinemia

An increase in total plasma proteins—*hyperproteinemia*—is not seen as commonly as hypoproteinemia. One condition in which an elevation of all the protein fractions is observed is dehydration. When excess water is lost from the vascular system, the proteins, because of their size, remain within the blood vessels. Although the absolute quantity of proteins remains unchanged, the concentration is elevated due to a decreased volume of solvent water. Dehydration results from a variety of conditions, including vomiting, diarrhea, excessive sweating, diabetic acidosis, and hypoaldosteronism. In addition to dehydration, hyperproteinemia may be due to excessive production primarily of the γ-globulins.

Some disorders are characterized by the appearance of a monoclonal protein or paraprotein in the serum and often in the urine as well. This protein is an intact immunoglobulin molecule, or occasionally, κ or λ light chains only. The most common disorder is multiple myeloma, in which the neoplastic plasma cells proliferate in the bone marrow. The paraprotein in this case is usually IgG, IgA, or κ or λ light chains. IgD and IgE paraproteins occur very rarely. Paraproteins in multiple myeloma may reach a serum concentration of several g/dL.

Not all paraproteins are associated with multiple myeloma. IgM paraprotein is often found in patients with Waldenström's macroglobulinemia, a more benign condition. A wide variety of disorders including chronic inflammatory states, collagen vascular disorders, and other neoplasms may be associated with paraproteins. Polyclonal increases in immunoglobulins, which would be represented by increases in both κ or λ chains, are seen in the serum and urine in a variety of chronic diseases.

Methods of Analysis

Total Nitrogen

A total nitrogen determination measures all chemically bound nitrogen in the sample. The method can be applied to a variety of biologic samples, including plasma and urine.[43] In plasma, both the total protein and nonprotein nitrogenous compounds such as urea and creatinine are measured. The analysis of total nitrogen level is useful in assessing nitrogen balance. Monitoring the nitrogen nutritional status is particularly important in patients receiving total parenteral nutrition, such as individuals with neurologic injuries who are sustained on intravenous fluids for an extended period.

The method for total nitrogen analysis uses chemiluminescence. The sample, in the presence of oxygen, is heated to a very high temperature ($1100 \pm 20°C$). Any chemically bound nitrogen is oxidized to nitric oxide. The nitric oxide is then mixed with ozone (O_3) to form an excited nitrogen dioxide molecule (NO_2^*). When this molecule decays to the ground state, it emits a photon of light. The amount of light emitted is proportional to the concentration of nitrogen in the sample. This chemiluminescence signal is compared with that of a standard for quantitation.[44] There are now available specialized instruments that perform the measurement automatically (Antek Instruments, Houston, Texas).

Total Proteins

The specimen most often used to determine the total protein is serum rather than plasma. The specimen need not be collected while the patient is fasting, although interferences in some of the determination methods occur with the presence of lipemia. Hemolysis will falsely elevate the total protein result because of the release of RBC proteins into the serum. Clear serum samples, tightly stoppered, are stable for a week or longer at room temperature, for a month at $2°–4°C$, and for at least 2 months at $-20°C$.[45]

The reference interval for serum total protein is 6.5 g/dL to 8.3 g/dL (65 to 83 g/L) for ambulatory adults. In the recumbent position, the serum total protein concentration is 6.0 g/dL to 7.8 g/dL (60 g/L to 78 g/L). This lower normal range is due to shifts in water distribution in the extracellular compartments. The total protein concentration is lower at birth, reaching adult levels by age 3 years. There is a slight decrease with age. In pregnancy, lower total protein levels are also seen. Methods for the determination of total protein are described below and summarized in Table 8-5.

Kjeldahl. The classical method for quantitation of total protein is the Kjeldahl method. Because of its precision and accuracy, it is used as a standard by which other methods are compared. In this method, the nitrogen is determined, and an average of 16% nitrogen mass in protein is assumed to calculate the protein concentration.

The serum proteins are precipitated with an organic acid such as TCA or tungstic acid. The nonprotein nitrogen is removed with the supernatant. The protein pellet is digested in H_2SO_4 with heat (340°–360°C) and a catalyst, such as cupric sulfate, to speed the reaction. Potassium sulfate is also introduced to increase the boiling point to improve the efficiency of digestion. The H_2SO_4 oxidizes the C, H, and S in protein to CO_2, CO, H_2O, and SO_2. The nitrogen in the protein is converted to ammonium bisulfite (NH_4HSO_4), which is then measured by adding alkali and distilling the ammonia into a standard boric acid solution. The ammonium borate ($NH_4H_2BO_3$) formed is then titrated with a standard solution of HCL to determine the amount of nitrogen in the original protein solution.

This method is not used in the clinical laboratory because it is time consuming and too tedious for routine use. The nitrogen content of each individual protein may differ from the 16% assumed in the Kjeldahl calculation described. The actual nitrogen content of serum proteins varies from 15.1% to 16.8%. Thus, if one uses a protein standard (calibrated with the Kjeldahl) that differs in composition from the serum specimen to be analyzed, an error is introduced because the percentage of nitrogen will not be the same. It is also necessary to assume that no proteins of significant concentration in the unknown specimen are lost in the precipitation step.

Despite these assumptions, the Kjeldahl method is still considered by some to be the reference method for proteins.

Refractometry. Refractometry is useful when a rapid method that requires a very small volume of serum is needed. The velocity of light is changed as it passes the boundary between two transparent layers (ie, air and water), causing the light to be bent (refracted). When a solute is added to the water, the refractive index at 20°C of 1.330 for pure water is increased by an amount proportional to the concentration of the solute in solution. This proportionality holds fairly well over a twofold to three-fold increase in concentration (ie, from 520 g/dL). Because the majority of the solids dissolved in serum are protein, the refractive index reflects the concentration of protein. However, in addition to protein, serum contains several nonprotein solids, such as electrolytes, urea, and glucose that contribute to the refractive index of serum. Thus the built-in scale in the refractometer has to be calibrated with a serum of a known protein concentration that also has the nonprotein constituents present. An assumption is made that the test samples contain these other solutes in nearly the same concentration as in the calibrating serum. Error is introduced when these substances are increased or when the serum is pigmented (from bilirubin), lipemic, or hemolyzed. The refractive index is also temperature dependent, and some refractometers incorporate a built-in temperature correction.

The total protein is commonly measured with a handheld refractometer. A drop of serum is placed by capillary action between a coverglass and the prism. The refractometer is held so that light is refracted through the serum layer. The refracted rays cause part of the field of view to be light, producing a point at which there is a sharp line between light and dark. The number of grams per liter at this line on the internal scale is read. The temperature is corrected in the TS meter (American Optical Corp, Scientific Instruments Division, Buffalo, NY 14215) by a liquid crystal system.

The measurement of total protein by refractometry is very easy and fast. The accuracy is acceptable,[46] with a reported agreement of ±3% with the biuret method,[47] but it is subject to false-positive interferences.

TABLE 8-5. Total Protein Methods

Method	Principle	Comment
Kjeldahl	Digestion of protein; measurement of nitrogen content	Reference method; assume average nitrogen content of 16%
Refractometry	Measurement of refractive index due to solutes in serum	Rapid and simple; assume nonprotein solids are present in same concentration as in the calibrating serum
Biuret	Formation of violet-colored chelate between Cu^{2+} ions and peptide bonds	Routine method; requires at least two peptide bonds and an alkaline medium
Dye-binding	Protein binds to dye and causes a spectral shift in the absorbance maximum of the dye	Research use

Biuret. The most widely used method, and the one recommended by the International Federation of Clinical Chemistry expert panel for the determination of total protein, is the biuret procedure. In this reaction, cupric ions (Cu^{2+}) complex with the groups involved in the peptide bond. In an alkaline medium and in the presence of at least two peptide bonds, a violet-colored chelate is formed. The reagent also contains sodium potassium tartrate to complex cupric ions to prevent their precipitation in the alkaline solution, and potassium iodide, which acts as an antioxidant. The absorbance of the colored chelate formed is measured at 540 nm.[48] When small peptides react, the color of the chelate produced has a different shade than that seen with larger peptides. The color varies from a pink to a reddish violet. However, there is no discernible difference in the reaction given by the proteins normally seen in plasma. Thus, over a wide range of concentrations, the color that is formed is proportional to the number of peptide bonds present[49] and reflects the total protein level. However, in the presence of abnormally small proteins, such as those seen in multiple myeloma, the concentration of the protein would be underestimated due to the lighter shade of color produced. If lipemic sera must be analyzed, a method for overcoming this problem is available.[50]

In addition to the NHCO group that occurs in the peptide bond, cupric ions will react with any compound that has two or more of the following groups: $NHCH_2$ and NHCS. The method was given its name because a substance called biuret ($NH_2CONHCONH_2$) reacted with cupric ions in the same manner. There must be a minimum of two of the reactive groups; therefore, amino acids and dipeptides will not react.

Dye Binding. The dye-binding methods are based on the ability of most proteins in serum to bind dyes, although the affinity with which they bind may vary. The dyes bromphenol blue, Ponceau S, amido black 10B, lissamine green, and Coomassie brilliant blue have been used to stain protein bands after electrophoresis. Bradford[51] has described a dye-binding method for the determination of total protein using Coomassie blue 250. The binding of Coomassie brilliant blue 250 to protein causes a shift in the absorbance maximum of the dye from 465 nm to 595 nm. The increase in absorbance at 595 is used to determine the protein concentration.

Bradford showed the dye-binding responses of five proteins to be similar; however, others have reported that for several other proteins, this is not the case.[52,53] This drawback has prompted a recommendation for caution when applying this test to the complex mixture of protein that one finds in serum.

Ultraviolet Absorption. Serum proteins also have been estimated by the use of ultraviolet spectrophotometry. Proteins absorb light at 280 nm and at 210 nm. The absorptivity (absorbance of a 1% solution in a 1-cm light path) at 280 nm is related to the absorbance of tyrosine, tryptophan, and phenylalanine amino acids in the protein. Human albumin has only one tryptophan residue in the molecule and has an absorptivity of 5.31 compared with that of fibrinogen, which has 55 tryptophan residues and an absorptivity of 15.1.

The absorbance of proteins at 210 nm is due to the absorbance of the peptide bond at that wavelength. The wavelength at which maximal absorbance occurs depends to a small degree on the conformation of the protein. These methods have rarely been used in clinical laboratories but are used routinely in research laboratories to monitor eluates of protein separations from columns. To use these methods, the assumptions must remain that the composition of the unknown serum specimen is near that of the calibrating solution.

Fractionation, Identification, and Quantitation of Specific Proteins

In the assay of total serum proteins, useful diagnostic information can be obtained by determining the albumin fraction and the globulins. A reversal or significant change in the ratio of albumin and total globulin was first noticed in diseases of the kidney and liver. To determine the albumin–globulin (A/G) ratio, it is common to determine total protein and albumin. Globulins are calculated by subtracting the albumin from the total protein (the total protein − albumin = globulins).

When an abnormality is found in the total protein or albumin, an electrophoretic analysis is usually made. Serum proteins are separable into five or more fractions by the customary electrophoretic methods. If an abnormality is seen on the electrophoretic pattern, an analysis of the individual proteins within the area of abnormality is made.

The methods for measuring protein fractions are described in the following text. Table 8-6 lists the types of analyses used in quantitating albumin levels.

Salt Fractionation. Fractionation of proteins has been accomplished by various procedures using precipitation. Globulins can be separated from albumin by salting-out procedures using sodium salts.[54] Salts, by decreasing the water available for hydration of hydrophilic groups, will cause precipitation of the globulins. Several different concentrations (26% to 28% w/v) of different salts (ie, Na_2SO_4, Na_2SO_3, and so on) have been used. The albumin that remains in solution in the supernatant can then be measured by any of the routine total protein methods. The salting-out procedure is not used to separate the albumin fraction in most laboratories today because direct methods that react specifically with albumin in a mixture of proteins are available.

Dye Binding. The most widely used methods for determining albumin are the dye-binding procedures. The pH of the solution is adjusted so that albumin is positively charged. Then, by electrostatic forces, the albumin is attracted to and binds to an anionic dye. When bound to albumin, the dye has a different absorption maximum than the free dye. The amount of albumin can be quantitated by measurement of the absorbance of the albumin-dye complex. A variety of dyes have been used, including methyl orange, 2–4′-hydroxy-azobenzene)-benzoic acid (HABA), bromocresol green (BCG), and bromcresol purple (BCP). Methyl orange is nonspecific for albumin; β-lipoproteins and some α_1- and

TABLE 8-6. Albumin Methods		
Method	**Principle**	**Comment**
Salt precipitation	Globulins are precipitated in high salt concentrations; albumin in supernatant is quantitated by biuret reaction	Labor intensive
Dye binding		
Methyl orange	Albumin binds to dye; causes shift in absorption maximum	Nonspecific for albumin
HABA [2(4′ - hydroxyazobenzene) - benzoic acid]	Albumin binds to dye; causes shift in absorption maximum	Many interferences (salicylates, bilirubin)
BCG (bromcresol green)	Albumin binds to dye; causes shift in absorption maximum	Sensitive; overestimates low albumin levels; most commonly used dye
BCP (bromcresol purple)	Albumin binds to dye; causes shift in absorption maximum	Specific, sensitive, precise
Electrophoresis	Proteins separated based on electric charge	Accurate; gives overview of relative changes in different protein fractions

α_2-globulins also will bind to this dye. HABA, although more specific for albumin, has a low sensitivity. In addition, several compounds, such as salicylates, penicillin, conjugated bilirubin, and sulfonamides, interfere with the binding of albumin to the dye. BCG is not affected by interfering substances such as bilirubin and salicylates; however, hemoglobin can bind to the dye. For every 100 mg/dL of hemoglobin, the albumin is increased by 0.1 g/dL.[55] The measurement of albumin by BCG has also been reported to overestimate low albumin values. This was seen particularly in patients when the low albumin level was accompanied by an elevated α-globulin fraction, such as occurs in nephrotic syndrome or endstage renal disease.[56,57] It was found that the α-globulins such as ceruloplasmin and α_1-acid glycoprotein[58] would react with BCG, giving a color whose intensity is approximately one-third that of the reaction seen with albumin.[59] This reaction of the α-globulins contributed significantly to the absorbance of the test only after incubation times exceeded 5 minutes. Thus, the specificity of the reaction for albumin can be improved by taking absorbance readings within a standardized short interval after mixing.[58] The times used have varied from 0.5 second to 30 seconds after mixing.

Bromcresol purple (BCP) is an alternate dye that may be used for albumin determinations.[60] BCP binds specifically to albumin, is not subject to most interferences, is precise, and exhibits excellent correlation with immunodiffusion reference methods. Linear regression analysis with radial immunodiffusion (RID) gave an equation of y = 0.95 + 0.93.[60] Analysis of albumin by the BCP method, however, is not without its disadvantages. In patients with renal insufficiency, the BCP method underestimates the serum albumin.[61] It appears the serum of these patients contain either a substance tightly bound to albumin or a structurally altered albumin that affects the binding of BCP. Similarly, BCP binding to albumin is impaired in the presence of covalently bound bilirubin. BCG binding is unaffected in these situations. Today, both the BCG and BCP methods are commonly used to quantitate albumin.

Determination of Total Globulins. Another approach to fractionation of proteins is the measurement of total globulins. Albumin can then be calculated by subtraction of the globulin from total protein. The total globulin level in serum is determined by a direct colorimetric method using glyoxylic acid.[62] Glyoxylic acid, in the presence of Cu^{2+} and in an acid medium (acetic acid and H_2SO_4), condenses with tryptophan found in globulins to produce a purple color. Albumin has approximately 0.2% tryptophan, compared with 2% to 3% for the serum globulins. When calibrated using a serum of known albumin and globulin concentrations, the total globulins can be determined. The measurement of globulins based on their tryptophan content has never come into common use because of the ease and simplicity of the dye-binding methods for albumin.

Electrophoresis. Electrophoresis separates proteins on the basis of their electric charge densities. The charge characteristics of proteins were discussed earlier in this chapter. Proteins, when placed in an electric current, will move according to their net charge, which is determined by the pH of a surrounding buffer. At a pH greater than the pI, the protein is negatively charged, and vice versa. The direction of the movement depends on whether the charge is positive or negative; cations (positive net charge) migrate to the cathode (negative terminal), whereas anions (negative net charge) migrate to the anode (positive terminal). The speed of the migration depends to a large extent on the degree of ionization of the protein at the pH of the buffer. This can be estimated from the difference between the pI of the protein and the pH of the buffer. The more the pH of the buffer differs from the pI, the greater is the magnitude of the net charge of that protein and the faster it will move in the electric field. In addition to the net electric charge, the velocity of the movement also depends on the electric field strength, size, and shape of the molecule; temperature; and the characteristics of the buffer (*ie*, pH, qualitative composition, and ionic strength). The specific electrophoretic mobility μ of a protein can be calculated by:

$$\mu = \frac{(s/t)}{F}, \qquad \text{(Eq. 8–1)}$$

where s = distance traveled in cm
$\quad t$ = time of migration in seconds
$\quad F$ = field strength in V cm^{-1}

Tiselius developed electrophoresis using an aqueous medium. This is known as *moving-boundary* or *free electrophoresis*. Later, in the clinical laboratory, paper was used. The term given to the use of a solid medium is *zone electrophoresis*. Paper has been largely replaced by cellulose acetate or agarose gel as the support media used today.

Serum Protein Electrophoresis

In the standard method for serum protein electrophoresis (SPE), serum samples are applied close the cathode end of a support medium strip that is saturated with an alkaline buffer (pH 8.6). The support strip is connected to two electrodes and a current is passed through the strip to separate the proteins. All major serum proteins carry a net negative charge at pH 8.6 and will migrate toward the anode. Using the standard methods, the serum proteins arrange themselves into five bands: albumin travels farthest to the anode followed by α_1-globulins, α_2-globulins, β-globulins, and γ-globulins, in that order. The width of the band of proteins in a fraction depends on the number of proteins with slightly different molecular characteristics that are present in that fraction. Homogenous protein gives a narrow band.

After separation, the protein fractions are fixed by immersing the support strip in an acid solution (*eg*, acetic acid) to denature the proteins and immobilize them on the support medium. The proteins are then stained. A variety of dyes have been used, including Ponceau S, Amido black, or Coomassie blue. The protein appears as bands on the support medium. Typical cellulose acetate electrophoretic patterns are shown in Figure 8-12*B*, whereas Figure 8-12*A* shows the patterns obtained using agarose gel as the support medium.

Visual inspection of the membrane is made, or the cleared transparent strip is placed in a scanning densitometer. Reflectance measurements also may be made on the uncleared membranes; however, scanning densitometry is used more commonly. The pattern on the membrane is made to move past a slit through which light is transmitted to a phototube to record the absorbance of the dye that is bound to each of the fractions. Usually this absorbance is recorded on a strip-chart recorder to obtain a pattern of the fractions (Fig. 8-13).

Many scanning densitometers compute the area under the absorbance curve for each band and the percentage of total dye that appears in each fraction. The concentration is then calculated as a percentage of the total protein that was

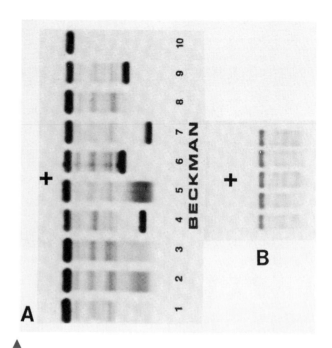

▲

Figure 8-12. Serum protein electrophoretic patterns on agarose and cellulose acetate. (**A**) Agarose gel—note the monoclonal γ-globulin; (**B**) cellulose acetate. (Courtesy of Department of Laboratory Medicine, The University of Texas M.D. Anderson Hospital, Drs. Liu, Fritsche, and Trujillo, and Ms. McClure, Supervisor.)

determined by one of the protein methods, such as the biuret procedure.

The computation also may be made by cutting out the small bands from the membrane and eluting the dye from each band in 0.1 M NaOH. The absorbances are added to obtain the total absorbance, and the percentage of the total absorbance found in each fraction is then calculated.

A reference serum control should be run with each electrophoretic run (see Fig. 8-13*A*), and the results should be monitored to maintain 95% confidence limits for the fractions. Reference values for each fraction are as follows: albumin, 53% to 65% of the total protein (3.55.0 g/dL); α_1-globulin, 2.5% to 5% (0.10.3 g/dL); α_2-globulin, 7% to 13% (0.61.0 g/dL); β-globulin, 8% to 14% (0.71.1 g/dL); and γ-globulin, 12% to 22% (0.81.6 g/dL).

Inadvertent use of plasma will result in a narrow band in the β_2-globulin region because of the presence of fibrinogen. The presence of free hemoglobin will cause a blip in the pattern in the late α_2 or early β zone, and the presence of hemoglobin-haptoglobin complexes will cause a small blip in the α_2 zone.

Often the information obtained by quantitation of each fraction is approximately equal to that obtained by visual inspection. The great advantage of electrophoresis compared with the quantitation of specific proteins is the overview it provides. The electrophoretic pattern can give information about the relative increases and decreases within the protein population, as well as information about the homogeneity of a fraction.

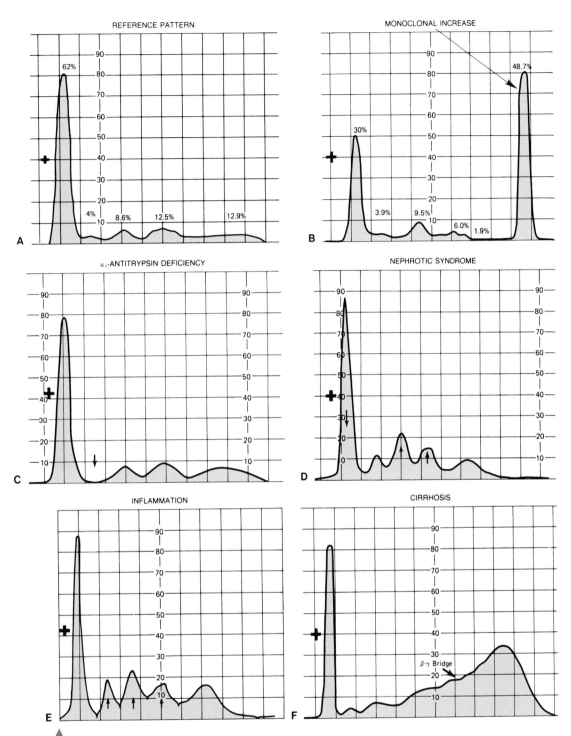

Figure 8-13. Selected densitometric patterns of protein electrophoresis. Albumin is at the anodal (+) end followed by α_1–, α_2, β–, and γ–globulin fractions. Arrows indicate decrease or increase in fractions. **(A)** Reference pattern (agarose). **(B)** Monoclonal increase in γ area (agarose). **(C)** α_1–Antitrypsin deficiency (cellulose acetate). **(D)** Nephrotic syndrome (cellulose acetate). **(E)** Inflammation (cellulose acetate). **(F)** Cirrhosis (cellulose acetate). (*A* and *B* are courtesy of Drs. Liu, Fritsche, and Jose Trujillo, Director, and Ms. McClure of the Department of Laboratory Medicine, The University of Texas M.D. Anderson Hospital. Others are courtesy of Dr. Wu of the Hermann Hospital Laboratory/The University of Texas Medical School)

Probably the most significant finding from an electrophoretic pattern is monoclonal immunoglobulin disease. The densitometric scan will show a sharp peak if the increase in immunoglobulins is due to a monoclonal increase (see Fig. 8-13B). A spike in the γ, β, or sometimes α₂ region signals the need for examination of the immunoglobulins and observation for clinical signs of myelomatosis. Likewise, a deficiency of the predominant immunoglobulin, IgG, is seen as a much paler stain in the γ area. Another significant finding is a decrease in α₁-antitrypsin (see Fig. 8-13C).

In nephrotic syndrome, the patient loses serum albumin and low-molecular-weight proteins in the urine. Some IgG is also lost. At the same time, an increase occurs in α₂-macroglobulin, β-lipoprotein, complement components, and haptoglobin.[63] These two events lead to a dramatic decrease in the relative amount of albumin and a marked increase in the relative amounts of α₂-globulin and β-globulin fractions (see Fig. 8-13D).

An inflammatory pattern indicating an inflammatory condition is seen when there is a decrease in albumin and an increase in the α₁-globulins (α₁-acid glycoprotein, α₁-antitrypsin), α₂-globulins (ceruloplasmin and haptoglobin), and β-globulin band (C-reactive protein; see Figure 8-13E). This type of pattern is also called an *acute-phase reactant pattern* and is seen in trauma, burns, infarction, malignancy, and liver disease. Acute-phase reactants phase are so named because they are found to be increased in the serum within days following trauma or exposure to inflammatory agents. Fibrinogen, haptoglobin, ceruloplasmin, and serum amyloid A increase several-fold, whereas CRP and α₂-macro-fetoprotein are increased several hundred-fold. Interleukin-1, a protein factor from leukocytes, is recognized as an important mediator of the synthesis of acute-phase proteins by hepatocytes. These acute-phase reactants are thought to play some role in immunoregulatory mechanisms. Chronic infections also produce a decrease in the albumin, but the globulin increase is found in the γ fraction as well as the α₁, α₂, and β fractions.

The electrophoretic pattern of serum proteins in liver disease shows the decrease in serum albumin concentration and the increase in γ-globulin. The pattern in cirrhosis of the liver is rather characteristic, with the abnormalities cited previously, but in addition there are some fast moving γ-globulins that prevent resolution of the β- and γ-globulins bands. This is known as the β–γ bridge of cirrhosis[64] (see Figure 8-13F).

In infectious hepatitis, the γ-globulin fraction rises with increasing hepatocellular damage. In obstructive jaundice, there is an increase in the α₂- and β-globulins. Also noted in obstructive jaundice is an increased concentration of lipoproteins, which is an indicator of its biliary origin. This is especially the case when there is little or no decrease of the serum albumin.[65]

High-Resolution Protein Electrophoresis

Standard SPE separates the protein into five distinct zones, which comprise many individual proteins. By modifying the electrophoretic parameters, these fractions may be further resolved into as many as 12 zones. This modification, known as *high-resolution electrophoresis (HRE)*, is accomplished by use of a higher voltage coupled with a cooling system in the electrophoretic apparatus and a more concentrated buffer. The support medium used most commonly is agarose gel. To obtain the HRE patterns, samples are applied on the agarose gel, electrophoresed in a chamber cooled by a gel block, stained, and then visually inspected. Each zone is compared to the same zone on a reference pattern for color density, appearance, migration rates, and appearance of abnormal bands or regions of density. A normal serum HRE pattern is shown in Figure 8-14. In addition, the patterns

---

CASE STUDY 8-4

A 32-year-old woman developed progressive fatigue and, later, edema in her ankles and in other dependent regions of her body when she lay down for prolonged periods. Although she produced near normal volumes of urine, dipstick urinalysis revealed 4-plus protein. Her serum albumin was below the reference range and her serum cholesterol was markedly elevated. Urine protein loss was in the range of 10 g/24 hours to 15 g/24 hours. Renal biopsy showed extensive glomerular involvement.

Serum protein electrophoresis showed reduced amounts of albumin (2.27 g/dL), α₁-globulins, and γ-globulins. The α₂ fraction was markedly elevated (34.4% of the total) as was the band of β-lipoprotein. The total serum protein was 4.7 g/dL. The patient became a candidate for renal transplantation. (*Case 8-4 courtesy of Dr. R. McPherson, Chairman, Clinical Pathology, Medical College of Virginia Hospitals, Virginia Commonwealth University*)

Questions:

1. What disease state is the most likely explanation for the patient's symptoms and laboratory results?
2. In this condition, why are the α₂- and β-globulin fractions elevated?
3. Why is the patient edematous?

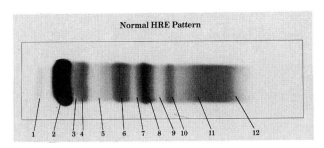

Normal HRE Pattern

Zones	Serum Proteins Found in Zones
1. PREALBUMIN ZONE	-Prealbumin
2. ALBUMIN ZONE	-Albumin
3. ALBUMIN-ALPHA-1 INTERZONE	-Alpha-lipoprotein (Alpha-fetoprotein)
4. ALPHA-1 ZONE	-Alpha 1-antitrypsin, Alpha-1-acid glycoprotein
5. ALPHA-1-ALPHA-2 INTERZONE	-Gc-globulin, Inter-alpha-trypsin inhibitor, Alpha-1-antichymotrypsin
6. ALPHA-2 ZONE	-Alpha-2-macroglobulin, Haptoglobin
7. ALPHA-2-BETA-1 INTERZONE	-Cold insoluble globulin, (Hemoglobin)
8. BETA-1 ZONE	-Transferrin
9. BETA-1-BETA-2 INTERZONE	-Beta-lipoprotein
10. BETA-2 ZONE	-C3
11. GAMMA-1 ZONE	-IgA, (Fibrinogen), IgM (Monoclonal Ig's, light chains)
12. GAMMA-2 ZONE	-IgG, (C-reactive protein) (Monoclonal Ig's, light chains)

Proteins listed in ( ) are normally found in too low a concentration to be visible in a normal pattern.

Figure 8-14. High-resolution electrophoretic pattern of serum. (Courtesy of Helena Laboratories, Beaumont, TX.)

may be scanned with a densitometer to obtain semiquantitative estimates of the protein found in each zone. HRE is particularly useful in detecting small monoclonal bands and in differentiating unusual bands or prominent increases of normal bands that can masquerade as a monoclonal gammopathy. (Case Study 8-5) For instance, in patients with nephrotic syndrome, an increased α_2-macroglobulin band in the α_2 region may be confused with a migrating monoclonal protein such as an IgA monoclonal protein gammopathy.[66]

Isoelectric Focusing

Isoelectric focusing (IEF) is zone electrophoresis that separates proteins on the basis of their p*I*. IEF uses constant power and polyacrylamide or agarose gel, which contains a pH gradient. The pH gradient is established by the incorporation of small polyanions and polycations (ampholytes) in the gel. The varying pIs of the polyions cause them, in the presence of an electric field, to seek their place in the gradient and to remain there. The pH gradient may range from 3.5 to 10.

When a protein is electrophoresed in the gel, it will migrate to a place on the gel where the pH is that of its pI. The protein becomes focused there because, if it should diffuse in either direction, it leaves its pI and gains a net charge. When this occurs, the electric current once again carries it back to its point of no charge, or its pI.

The clinical applications of IEF have included phenotyping of α_1-antitrypsin deficiencies,[67] determination of genetic variants of enzymes and hemoglobins,[68] detection of paraproteins in serum[69] and oligoclonal bands in CSF,[70] and isoenzyme determinations.

Immunochemical Methods

Specific proteins may be identified by immunochemical assays in which the reaction of the protein (antigen) and its antibody is measured. Methods using various modifications of this principle include RID, immunoelectrophoresis (IEP), immunofixation electrophoresis, electroimmunodiffusion, immunoturbidimetry, and immunonephelometry. These techniques are discussed in Chapter 5, *Immunoassays and Nucleic Acid Probe Techniques.*

Proteins in Other Body Fluids

The kinds of body fluids being studied for their protein content have increased. This is due in part to the increased sensitivity of the test methods that are now available. Proteins in 16 body fluids are included in a contemporary publication by Ritzmann and Killingsworth.[71] This section includes a discussion of the two fluids whose protein contents are studied most often: urine and CSF.

Urinary Proteins

The majority of proteins found in the urine arise from the blood; however, urinary proteins can also originate from the kidney and urinary tract, and from extraneous sources such as the vagina and prostate. The proteins in the blood appear in the urine because they have passed through the renal glomerulus and have not been reabsorbed by the renal tubules. The qualitative tests for proteinuria are most commonly performed using a reagent test strip. These methods are based on the change in the response of an indicator dye in the presence of protein, known as protein error of indicators. A more detailed discussion of these strips may be found in the package inserts of the products and in the literature.[72]

Quantitative examination of urine protein requires considerable attention to the volume or time of the urine collection, because the concentration may vary with time and volume. Most quantitative assays are performed on urine specimens of 12 hours or 24 hours. The 24–hour timing allows for circadian rhythmic changes in excretion at certain times of day. The patient should void, completely emptying the bladder, and discard this urine. Urine is collected from that time for the next 24 hours. At the end of the 24-hour period, the bladder is completely emptied and that urine included in the sample. The volume of the timed specimen is measured accurately and recorded. The results are reported

CASE STUDY 8-5

A 45-year-old man was undergoing continuing evaluation of possible recurrence of a plasmacytoma that had originally presented with a compression fracture of a vertebra. He had been treated with local radiation and chemotherapy. His serum protein electrophoresis showed normal amounts of albumin, α_1, α_2, and β fractions. The γ fraction demonstrated a slight monoclonal band in the fast γ region (close to β). Protein electrophoresis of concentrated urine showed a single monoclonal band that migrated slightly less than the serum band. *(Case 8-5 courtesy of Dr. R. McPherson, Chairman,*

Clinical Pathology, Medical College of Virginia Hospitals, Virginia Commonwealth University)

Questions

1. Does the presence of the monoclonal band in the serum indicate the recurrence of the patient's tumor?
2. What further information is obtained from a urine protein electrophoresis?
3. What other test is needed to confirm the type of urinary protein?

generally in terms of weight of protein per 24 hours by calculating the amount of protein present in the total volume of urine collected during that time.

Several methods for the determination of total protein in urine and other body fluids have been proposed. These include the measurement of turbidity when urinary proteins are mixed with an anionic organic acid such as sulfosalicylic acid, TCA, or benzethonium chloride.[73] These methods are sensitive, but the reagent does not react equally with each protein fraction. This is particularly true of sulfosalicylic acid, which produces four times more turbidity with albumin than it does with γ-globulin.[74]

Methods considered to give more accurate results consist of precipitation of the urine proteins, dissolution of the protein precipitate, and color formation with biuret reagent. Another chemical procedure for urinary protein uses the Folin-Ciocalteau reagent, which is a phosphotungsto-molybdic acid solution, frequently called phenol reagent because it oxidizes phenolic compounds. The reagent changes color from yellow to blue during the reaction with tyrosine, tryptophan, and histidine residues in proteins. This method is about

10 times more sensitive than the biuret method. Lowry and associates[75] increased the sensitivity of the Folin-Ciocalteau reaction by incorporating a biuret reaction as the initial step. After the binding of the Cu^{2+} to the peptide bonds, the Folin-Ciocalteau reagent is added. As the Cu^{2+} protein complex is oxidized, the reagent is reduced, forming the chromogens tungsten blue and molybdenum blue. This increased the sensitivity to 100 times greater than that of the biuret method alone. Another modification uses a pyrogallol red-molybdate complex that reacts with protein to produce a blue-purple complex.[76] This procedure is easily automated.

Dye-binding methods also have been used to determine the total protein content of body fluids. Methods using dyes such as Coomassie blue[77] and Ponceau S[78] are available in the literature. Table 8-7 summarizes the various methods for measurement of urinary total protein.[79]

The reference values or intervals for urinary proteins are highly method dependent,[80] ranging from 100 mg to 250 mg per 24 hours. Because of their ease of use, speed, and sensitivity, the techniques that are used most frequently today are turbidimetric procedures.

TABLE 8-7. Urine Protein Methods

Method	Principle	Comment
Turbidimetric methods (sulfosalicylic acid, trichloroacetic acid, or benzethonium chloride)	Proteins are precipitated as fine particles, turbidity is measured spectrophotometrically	Rapid, easy to use; unequal sensitivity for individual proteins
Biuret	Proteins are concentrated by precipitation, redissolved in alkali, then reacted with Cu^{2+}; Cu^{2+} form colored complex with peptide bonds	Accurate
Folin-Lowry	Initial biuret reaction; oxidation of tyrosine, tryptophan, and histidine residues by Folin phenol reagent (mixture of phosphotungstic and phosphomolybdic acids): measurement of resultant blue color	Very sensitive
Dye-binding (Coomassie blue, Ponceau S)	Protein binds to dye, causes shift in absorption maximum	Limited linearity; unequal sensitivity for individual proteins

Physiologic Significance of Proteinuria. *Proteinuria* in renal disease may be classified as resulting from either glomerular or tubular dysfunction. Glomerular proteinuria is a consequence of loss of glomerular membrane integrity, which normally keeps proteins from passing through to the urine because of their large molecular weight. In early selective glomerular proteinuria, the proteins responsible for the increase are typically albumin (>80%) and transferrin. As the glomerular lesion becomes more severe, the membrane becomes nonselective, and proteins of all sizes, including immunoglobulins, pass into the urine. Damage to the glomerular membrane occurs in diseases such as diabetes, amyloidosis, dysglobulinemia, and collagen disorders. The presence in the blood of toxic agents, such as mercury or heroin, may also result in loss of glomerular integrity. An early indicator of glomerular dysfunction is the presence of microalbuminuria. The term microalbuminuria is used to describe albumin concentrations in the urine that are greater than normal but not detectable with common urine dipstick assays. Recent studies[81] of patients with diabetes mellitus have showed that microalbuminuria precedes the nephropathy associated with diabetes, particularly in type I (insulin-dependent diabetes mellitus) diabetics. Progression from microalbuminuria to clinical nephropathy can be delayed with intensive therapy to normalize blood glucose and blood pressure. Therefore, it has been recommended that all diabetics be tested annually for microalbuminuria. The normal albumin excretion rate is <20 μg/min or <30 mg/day. Microalbuminuria can be measured by radioimmunoassay, radial immunodiffusion, immunonephelometry, and enzyme immunoassay.

Tubular proteinuria is a consequence of the renal tubules' either being unable to perform their usual function of reabsorption because of dysfunction, or because the amount of protein appearing in the tubular fluid exceeds the absorptive capability of a normal functioning tubule. In tubular proteinuria, small protein molecules that normally pass through the glomerulus and are reabsorbed, such as β_2-microglobulin (MW 11,800), retinol-binding protein (MW 21,000), and α_1-microglobulin (MW 30,000), appear in the urine.

Overflow proteinuria is an overabundance of proteins from the serum that appear in such high concentrations in the glomerular filtrate that the tubules are overwhelmed in their capacity to absorb them. This type of proteinuria is typified by the increased concentration of immunoglobulin light chains in multiple myeloma (Bence Jones protein). The determination of specific proteins yields much more information than total urinary protein, particularly in the study of tubular proteinuria.

To determine the type of proteins that are being excreted, most methods require initial concentration of the urine. This is accomplished by precipitation, ultrafiltration, dialysis, or gel filtration techniques. The urine can then be electrophoresed by a method similar to that described for serum. Quantitative measurement of specific urine proteins can be made by use of immunochemical methods, such as radioimmunoassay, radial immunodiffusion, electroimmunoassay, and immunonephelometry.[82]

A urinary protein not found in the serum is Tamm-Horsfall protein, a mucoprotein produced in the renal tubules. Its concentration in urine is 40.0 mg/day. An IgA (secretory) is also produced in the kidney, and 1.1 mg/day is excreted.

Cerebrospinal Fluid Proteins

CSF is formed in the choroid plexus of the ventricles of the brain by ultrafiltration of the blood plasma. Protein measurement is one of the tests that is usually requested, in addition to glucose level measurement and differential cell count, culture, and sensitivity. The accepted reference interval for patients between the ages of 10 years to 40 years is 15 mg/dL to 45 mg/dL of CSF protein.

The total CSF protein may be determined by several of the more sensitive chemical or optical methods referred to earlier in the discussion on urinary proteins. The most frequently used procedures are turbidimetric using TCA, sulfosalicylic acid with sodium sulfate,[83] or benzethonium chloride.[84] Also available are dye-binding methods (*ie,* Coomassie brilliant blue), a kinetic biuret reaction, and the Lowry method using a Folin phenol reagent.

Physiologic Significance of CSF Protein Analysis. Abnormally increased total CSF proteins may be found in conditions where there is an increased permeability of the capillary endothelial barrier through which ultrafiltration occurs. Examples of such conditions include bacterial, viral, and fungal meningitis; traumatic tap; multiple sclerosis; obstruction; neoplasm; disk herniation; and cerebral infarction. The degree of permeability can be evaluated by measuring the CSF albumin and comparing it with the serum albumin. Albumin is usually used as the reference protein for permeability because it is not synthesized to any degree in the CNS. The reference value for the CSF albumin–serum albumin ratio is less than 2.7 to 7.3.[85] A value greater than this indicates that the increase in the CSF albumin came from serum due to a damaged blood-brain barrier. Low CSF protein values are found in hyperthyroidism and when fluid is leaking from the CNS.

Although total protein levels in the CSF are informative, diagnosis of specific disorders often requires measurement of individual protein fractions. The pattern of types of proteins present can be seen by electrophoresis of CSF that has been concentrated about 100-fold. This may be performed on cellulose acetate or agarose gel. The pattern obtained normally from adult lumbar CSF shows prealbumin, a prominent albumin band, α_1-globulin composed predominantly of α_1-antitrypsin, an insignificant α_2 band, a β_1 band composed principally of transferrin, and a CSF-specific transferrin that is deficient in carbohydrate in the β_2 zone.[86]

Electrophoretic patterns of CSF from patients who have multiple sclerosis have multiple, distinct bands in the globulin zone.[87] This is called oligoclonal banding. More than 90% of patients with multiple sclerosis have oligoclonal bands, although the bands also have been found in inflammatory and infectious neurologic disease. These bands cannot be seen on routine cellulose acetate electrophoresis but require a high-resolution technique in which agarose is usually used.

Another finding in demyelinating diseases, such as multiple sclerosis, is the production of IgG within the CNS. However, the increase in IgG may result from either intrathecal synthesis or increased permeability of the blood-brain barrier. To identify the source of the elevated CSF IgG levels, the IgG-albumin index can be calculated as follows:

$$\text{CSF IgG Index} = \frac{\text{CSF IgG (mg/dL)} \times \text{serum albumin (g/dL)}}{\text{serum IgG (g/dL)} \times \text{CSF albumin (mg/dL)}}$$

(Eq. 8–2)

The CSF albumin concentration corrects for increased permeability. The reference range for the index is 0.25 to 0.80.[86] The index is elevated when there is increased CNS IgG production such as occurs in multiple sclerosis. IgG production is also increased in some bacterial infections and CNS inflammatory diseases. The IgG index is decreased when the integrity of the blood-brain barrier is compromised, as in some forms of meningitis and tumors.

Myelin basic proteins present in the CSF are also assayed because these proteins can provide an index of active demyelination. Myelin basic proteins are constituents of myelin, the sheath that surrounds many of the CNS axons. In very active demyelination, concentrations of myelin basic proteins of 17 ng/mL to 100 ng/mL were found by radioimmunoassay. In slow demyelination, values of 6 ng/mL to 16 ng/mL occurred, and in remission, the values were less than 4 ng/mL.[88]

SUMMARY

Amino acids are the building blocks of proteins. Inherited abnormalities in amino acid metabolism results in a variety of conditions most of which are associated with mental retardation. The synthesis of most proteins occurs in the liver, with the exception of the immunoglobulins, which are produced in the plasma cells. The linear sequence of amino acids in the protein, which composes the primary structure, is defined by the genetic code in cellular DNA. Proteins may be "simple," comprising only of amino acids, or "conjugated," in which the peptide chain is attached to a nonprotein moiety. With the large number of proteins in the plasma (more than 500 identified), the functions are varied. For example, albumin is primarily responsible for the colloid osmotic pressure; haptoglobin and transferrin are transport proteins, binding free hemoglobin and iron, respectively; immunoglobulins and components of the complement system help protect the body against infection; and other proteins such as fibrinogen aid in the maintenance of hemostasis. Low levels of total protein may be due to excessive loss, decreased synthesis, or accelerated catabolism, whereas elevated levels are associated with dehydration or excessive production.

The most widely used method for measuring total serum protein levels is the biuret reaction in which cupric ions complex with two or more peptide bonds. To determine the albumin fraction, dye-binding techniques are usually used with either BCG or BCP. The dye, when bound to albumin, is a different color than the free dye. Further fractionation of the serum proteins can be accomplished by electrophoresis, which uses the charge characteristics of the protein for separation. Routine SPE arranges the proteins in five bands: albumin travels farthest to the anode followed by α_1-globulins, α_2-globulins, β-globulins, and γ-globulins. HRE separates protein into 12 zones.

Protein can also be measured in other body fluids. Either glomerular damage or tubular dysfunction will result in elevated urinary proteins. Abnormally increased CSF protein is found in conditions where there is an increased permeability of the capillary endothelial barrier.

REVIEW QUESTIONS

1. The Guthrie screening test for increased serum phenylalanine is based on:
   a. A fluorometric method in which phenylalanine is reacted with ninhydrin
   b. The microbiologic procedure in which phenylalanine counteracts the effects of a metabolic antagonist on the growth of *B. subtilis*
   c. The thin layer chromatography method in which identification is made by comparison of the unknown Rf value to Rf values of the known
   d. A dipstick procedure in which phenylalanine is converted to phenylpyruvic acid and is then reacted with ferric chloride

2. The three dimensional spatial configuration of a single polypeptide chain as determined by disulfide linkages, hydrogen bonds, electrostatic attractions, and van der Waals forces is referred to as the:
   a. Primary structure
   b. Secondary structure
   c. Tertiary structure
   d. Quaternary structure

3. The plasma protein mainly responsible for maintaining colloidal osmotic pressure in vivo is:
   a. Hemoglobin
   b. Fibrinogen
   c. α_2-macroglobulin
   d. Albumin

4. Intra-individual differences in total serum protein concentrations attributable to erect versus recumbent postures are:
   a. 0–0.5 g/dL (virtually no difference)
   b. Approximately 0.5g/dL
   c. Approximately 2g/dL
   d. Approximately 5g/dL

5. Poor protein-caloric nutritional status is associated with:
   a. A low level of γ-globulins
   b. An elevated haptoglobin concentration
   c. A decreased level of prealbumin
   d. An increased level of α_1-fetoprotein

6. In which of the following conditions would a normal level of myoglobin be expected?
   a. Multiple myeloma
   b. Acute myocardial infarction
   c. Renal failure
   d. Crushing trauma received in a car accident

7. A total protein as assayed on serum from a patient with multiple myeloma. The result using the Biuret method was lower than the result given by the Kjeldahl method. The discrepancy was due to the fact that:
   a. The Kjeldahl method measures all nitrogen containing compounds including proteins and nonprotein nitrogens, and thus overestimates the protein level
   b. Because the two methods are based on different principles, a comparable result is not expected
   c. In the Biuret method, abnormally small proteins produce a complex that has a different shade of color than normal proteins, thus underestimating the protein level
   d. Hemolysis falsely elevates the Kjeldahl method but has no effect on the protein level determined by the Biuret method

8. The protein electrophoretic pattern of plasma, as compared to serum, reveals a:
   a. Broad increase in the γ-globulins
   b. Fibrinogen peak with the α_2-globulins
   c. A decreased albumin peak
   d. Fibrinogen peak between the β- and γ-globulins

9. The following pattern of serum protein electrophoresis is obtained:

   Albumin: decrease or normal
   α_1- and α_2-globulins: increased
   β- and γ-globulins: normal

   This pattern is characteristic of which of the following conditions?
   a. Cirrhosis
   b. Acute inflammation (primary response)
   c. Nephrotic syndrome
   d. Gammopathy

10. One advantage of high-resolution agarose electrophoresis over lower current electrophoresis is:
    a. High-resolution procedures detect monoclonal and oligoclonal bands at lower concentrations
    b. A smaller sample volume is required
    c. Results are obtained more rapidly
    d. More sample can be applied to the support medium

REFERENCES

1. Natelson S, Natelson E. Principles of applied clinical chemistry plasma proteins in nutrition and transport. Vol. 3. New York: Plenum, 1980.
2. Levy HL. Phenylketonuria–1986. Pediatr Rev 1986;7:269.
3. Matalon R, Michals K. Phenylketonuria: screening, treatment and maternal PKU. Clin Biochem 1991;24:337.
4. McCaman MW, Robins E. Fluorimetric method for the determination of phenylalanine in serum. J Lab Clin Med 1962;59: 885.
5. Yamaguchi A, Mizushima Y, Fukushi M, et al. Microassay system for newborn screening for phenylketonuria, maple syrup urine disease, homocystinuria, histidinemia and galactosemia with use of a fluorometric microplate reader. Screening 1992;1:49.
6. Shen RS, Abell CW. Phenylketonuria: a new method for the simultaneous determination of plasma phenylalanine and tyrosine. Science 1977;197:665.
7. Lidsky A, Guttler F, Woo S. Prenatal diagnosis of classic phenylketonuria by DNA analysis. Lancet 1985;1:549.
8. Armstrong F. Biochemistry. 3rd ed. New York: Oxford University Press, 1989.
9. Berry HK. Neonatal screening: III. The spectrum of metabolic disorders. Diagn Med 1984;7:39.
10. Feld R. Maple syrup urine disease. Clin Chem News Aug 1986;8.
11. Stephens AD. Cystinuria and its treatment: 25 years experience at St. Bartholomew's Hospital. J Inherit Metab Dis 1989;12:197.
12. Dickinson JC, Rosenblum H, Hamilton PB. Ion exchange chromatography of the free amino acids in the plasma of the newborn infant. Pediatrics 1965;36:2.
13. Deyl Z, Hyanek J, Horokova M. Profiling of amino acids in body fluids and tissues by means of liquid chromatography. J Chromatogr 1986;379:177.
14. Cohn EJ, Strong LE, Hughes WL Jr., et al. Preparation and properties of serum and plasma proteins: IV. A system for the separation into fractions of the protein and lipoprotein components of biological tissues and fluids. J Am Chem Soc 1946;68:459.
15. Champe P, Harvey R. Lippincott's illustrated reviews: biochemistry. Philadelphia: Lippincott, 1987.
16. Olusi SO, McFarlane H, Osunkoya BO, Adesina H. Specific protein assays in protein-calorie malnutrition. Clin Chem Acta 1975;62:107.
17. Hitzig WH, Joller PW. Developmental aspects of plasma proteins. In: Ritzmann SE, Killingsworth LM, eds. Body fluids, amino acids, and tumor markers: diagnostic and clinical aspects. 3rd ed. New York: Alan Liss, 1983;1.
18. Haferkamp O, SchlettweinGsell D, Schwick HG, et al. Serum protein in an aging population with particular reference to evaluation of immune globulins and antibodies. Gerontologia 1966;12:30.
19. Keyser JW. Human plasma proteins. New York: Wiley, 1979;280.
20. Seppala M, Ruoslahti E. Alpha-fetoprotein in maternal serum: a new marker for detection of fetal distress and intrauterine death. Am J Obstet Gynecol 1973;115:48.
21. Merkatz I, Nitowsky H, Macri J, et al. An association between low maternal serum fetoprotein and fetal chromosomal abnormalities. Am J Obstet Gynecol 1984;148:886.
22. Nichols WS, Nakamura RM. Biologic serum markers associated with human neoplasia. In: Ritzmann SE, Killingsworth LM, eds. Proteins in body fluids, amino acids, and tumor markers: diagnostic and clinical aspects. 3rd ed. New York: Alan Liss, 1983;378.
23. Johnson AM, Schmid K, Alper CA. Inheritance of human$_1$-acid glycoprotein variants. J Clin Invest 1969;48:2293.
24. Rubin H. The biology and biochemistry of antichymotrypsin and its potential role as a therapeutic agent. Biol Chem Hoppe Seyler 1992; 373(7):497.
25. Balduyck M, Mizo J. Inter-alpha-trypsin inhibitor and its plasma and urine derivatives. Ann Biol Clin 1991;49(5):273.
26. Pesce MA, Bodourian SH. Nephelometric measurement of ceruloplasmin with a centrifugal analyzer. Clin Chem 1982;28:516.
27. Geiger J, Hoffman P. Quantitative immunologische bestimimmung von 16 verschiedenen serum proteinen bei 260 normalen, 015 jahre alten kindern. Z Kinderheilk 1970;109:22.
28. Lyngbye J, Kroll J. Quantitative immunoelectrophoresis of proteins in serum from a normal population: season, age and sex-related variations. Clin Chem 1971;17:495.
29. Kang EP, Fitzpatrick MJ, O'Neill SP. Immunoturbidimetric assay of human fibrinogen using a centrifugal analyzer. [Abstract] Clin Chem 1982;28:1622.
30. Roitt I, Brostoff J, Male D. Immunology. St. Louis, MO: Mosby, 1985;1.
31. Wong SS. Strategic utilization of cardiac markers for the diagnosis of acute myocardial infarction. Ann Clin Lab Sci 1996;26(4):301.
32. Rawn JD. Biochemistry. Neil Patterson, 1989:1081.
33. Mair J, Dienstl F, Puschendorf B. Cardiac troponin T in the diagnosis of myocardial injury. Crit Rev Clin Lab Sci 1992;29(1):31.
34. Christenson RH. Cardiac troponin T in the risk assessment of acute coronary syndromes. Am Clin Lab June 1997:18.

35. Ohman EM, et al. Cardiac troponin T levels for risk stratification in acute myocardial ischemia. N Engl J Med 1996;335(18):1333.
36. Lindahl B, Venge P, Wallentin L. Troponin T identifies patients with unstable coronary artery disease who benefit from long-term antithrombotic protection. J Am Coll Cardiol 1997;29:43.
37. Li D, Keffer J, Corry K, et al. Nonspecific elevation of troponin T levels in patients with chronic renal failure. Clin Biochem 1995;28(4):474.
38. Adams JE, et al. Diagnosis of perioperative myocardial infarction with measurement of cardiac troponin I. N Engl J Med 1994;330(10):670.
39. Antman EM, et al. Cardiac-specific troponin I levels to predict the risk of mortality in patients with acute coronary syndromes. N Engl J Med 1996;335(18):1342.
40. Katrukha AG, et al. Troponin I is released in bloodstream of patients with acute myocardial infarction not in free form but as complex. Clin Chem 1997;43:1379.
41. Defilippi C, Parmar R. A rapid bedside troponin T assay to speed triage. Clin Chem News Oct 1997:6.
42. Hamm CW, Goldmann BU, Heeschen C, Kreymann G, Berger J, Meinertz T. Emergency room triage of patients with acute chest pains by means of rapid testing for cardiac troponin T or troponin I. N Engl J Med 1997;337(23):1648.
43. Ward MWN, Owens CWI, Rennie MJ. Nitrogen estimation in biological samples by use of chemiluminescence. Clin Chem 1980; 26:1336.
44. Gorimar TS, Hernandez HA, Polczynsk MW. Rapid protein determination using pyro-chemiluminescence. Am Clin Prod Rev Nov 1984.
45. Peters T Jr, Biamonte ET, Doumas BT. Protein (total) in serum, urine, cerebrospinal fluid, albumin in serum. In: Faulkner WR, Meites S, eds. Selected methods in clinical chemistry. Vol 9. Washington, DC: American Association for Clinical Chemistry, 1982;317.
46. Lines JG, Raines DN. Refractometric determination of serum concentration: 2. Comparison with biuret and Kjeldahl determination. Ann Clin Biochem 1970;7:6.
47. Ross JN, Fraser MD. Analytical chemistry precision: state of the art for fourteen analytes. Am J Clin Pathol 1977;68:130.
48. Doumas BT, Bayse D, Borner K, et al. A candidate reference method for determination of total protein in serum: I. Development and validation. Clin Chem 1981;27:1642.
49. Strickland RD, Freeman ML, Gurule FT. Copper binding by proteins in alkaline solution. Anal Chem 1961;33:545.
50. Chromy V, Fischer J. Photometric determination of total protein in lipemic serum. Clin Chem 1977;23:754.
51. Bradford MM. A rapid and sensitive method for the quantitation of microgram quantities of protein utilizing the principle of protein-dye binding. Anal Biochem 1976;72:248.
52. Pierce J, Svelter C. An evaluation of the Comassie blue G250 dye binding method for quantitative protein determination. Anal Biochem 1977;81:478.
53. Vankley J, Hales S. Assay for protein by dye binding. Anal Biochem 1977;81:485.
54. Majoor CLH. The possibility of detecting individual proteins in blood serum by differentiation of solubility curves in concentrated sodium sulfate solutions. J Biol Chem 1947;169:583.
55. Doumas BT, Watson WA, Biggs HG. Albumin standards and the measurement of serum albumin with bromcresol green. Clin Chem Acta 1971;31:87.
56. Speicher CE, Widish JR, Gaudot FJ, et al. An evaluation of the overestimation of serum albumin by bromcresol green. Am J Clin Pathol 1978;69:347.
57. Webster D, Bignell AHC, Attwood EC. An assessment of the suitability of bromcresol green for the determination of serum albumin. Clin Chem Acta 1974;53:101.
58. Gustafsson JEC. Improved specificity of serum albumin determination and estimation of "acute phase reactants" by use of the bromocresol green reaction. Clin Chem 1976;22:616.
59. Webster D. A study of the interaction of bromcresol green with isolated serum globulin fractions. Clin Chem Acta 1974;53:109.
60. Parviainen M, Harmoinen A, Jokela H. Serum albumin assay with bromcresol purple dye. Scand J Clin Lab Invest 1985;45(6):561.
61. Maguire G, Price C. Bromcresol purple method for serum albumin gives falsely low values in patients with renal insufficiency. Clin Chem Acta 1986;155(1):83.
62. Goldenberg J, Drewes PA. Direct photometric determination of globulin in serum. Clin Chem 1971;17:358.
63. Kawai T. Clinical aspects of the plasma proteins. Philadelphia: Lippincott, 1973.
64. Sunderman FW. Studies of the serum proteins: VI. Recent advances in clinical interpretation of electrophoretic fractions. Am J Clin Pathol 1964;42:1.
65. Sherlock S. Diseases of the liver and biliary system. 4th ed. Oxford, England: Blackwell, 1971;38.
66. Keren D. High resolution electrophoresis aids detection of gammopathies. Clin Chem News 1989; 14.
67. Johnson AM. Isoelectric focusing. In: Ritzmann SE, Killingsworth LM, eds. Proteins in body fluids, amino acids, and tumor markers: diagnostic and clinical aspects. New York: Alan Liss, 1983;101.
68. Vesterberg O. Isoelectric focusing of proteins. In: Colowick EJ and Kaplan LA, eds. Methods in enzymology. Vol 22. New York: Acapdemic Press, 1971;389.
69. Sinclair D, Kumeraratre DS, Forrester JB, et al. The application of isoelectric focusing to routine screening for paraproteinemia. J Immunol Methods 1983;64:147.
70. Roos RP, Lichter M. Silver staining of cerebrospinal fluid IgG in isoelectric focusing gels. J Neurosci Methods 1983;8:375.
71. Ritzmann SE, Killingsworth LM. Protein abnormalities, proteins in body fluids, amino acids, and tumor markers: diagnostic and clinical aspects. 3rd ed. New York: Alan Liss, 1983.
72. Michael BS. The routine examination of urine. In: Ross DL, Neely RA, eds. Textbook of urinalysis of body fluids. New York: Appleton-Century-Crofts, 1983;67.
73. Iwata J, Nishikaze O. New microturbidimetric method for determination of protein in cerebrospinal fluid and urine. Clin Chem 1979; 25:1317.
74. Schriever H, Gambino SR. Protein turbidity produced by trichloroacetic acid and sulfosalicylic acid at varying temperatures and varying ratios of albumin and globulin. Am J Clin Pathol 1965;44:667.
75. Lowry OH, Rosenbrough NH, Farr AL, et al. Protein measurement with the Folin-Phenol reagent. J Biol Chem 1951;193:265.
76. Fujita Y, Mori I, Kitano S. Color reaction between pyrogallol red-molybdenum (VI) complex and protein. Bunseki Kajaku 1983; 32:E379.
77. Heick HMC, Begin HN, Acharya C, et al. Automated determination of urine and cerebrospinal fluid proteins with Coomassie brilliant blue and the Abbott ABA-100. Clin Biochem 1980;13:81.
78. Pesce MA, Strande CS. A new micromethod for the determination of protein in cerebrospinal fluid and urine. Clin Chem 1973;19:1265.
79. McElderry LA, Tarbit IF, Cassells-Smith AJ. Six methods for urinary protein compared. Clin Chem 1982;28:356.
80. Buffone GJ, Ellis D. Urinary proteins. In: Ritzmann SE, Killingsworth LM, eds. Proteins in body fluids, amino acids and tumor markers: diagnostic and clinical aspects. New York: Alan Liss, 1983;117.
81. Cembrowski G. Testing for microalbuminuria: promises and pitfalls. Lab Med 1990;21 (8):491.
82. Christenson RH, Silverman LM. The determination of urinary protein adapted to the Beckman automated ICS system. [Abstract] Clin Chem 1982;23:1620.
83. Meulmans O. Determination of total protein in spinal fluid with sulfosalicylic acid and trichloroacetic acid. Clin Chem Acta 1960;5:757.
84. Flachaire E, Damour O, Bienvenu J, et al. Assessment of the benzethonium chloride method for routine determination of protein in cerebrospinal fluid and urine. Clin Chem 1983;29:343.
85. Tibbling G, Link H, Ohman S. Principles of albumin and IgG analyses in neurological disorders: I. Establishment of reference values. Scand J Clin Lab Invest 1977;37:385.
86. Killingsworth LM. Cerebrospinal fluid proteins. In: Ritzmann SE, Killingsworth LM, eds. Proteins in body fluids, amino acids, and tumor markers: diagnostic and clinical aspects. New York: Alan Liss, 1983;147.
87. Laterre EC, Callewaert A, Heremans JF, et al. Electrophoretic morphology of gamma globulins in cerebrospinal fluid of multiple sclerosis and other disease of the nervous system. Neurology 1970;20:982.
88. Cohen SR, Herndon RM, McKhann GM. RIA of myelin basic proteins in spinal fluid. N Engl J Med 1976;23:754.
89. Campbell P. Case studies. J Med Tech 1985;2:9.

CHAPTER 9

Enzymes

Robin Gaynor Krefetz

Objectives

Upon completion of this chapter, the clinical laboratorian should be able to:

- *Define enzyme, including physical composition and structure.*
- *Classify enzymes according to the International Union of Biochemistry (IUB).*
- *Discuss the different factors affecting the rate of an enzymatic reaction.*
- *Explain enzyme kinetics including zero-order and first-order kinetics.*
- *Discuss which enzymes are useful in the diagnosis of cardiac disorders, hepatic disorders, bone disorders, muscle disorders, malignancies, and acute pancreatitis.*
- *Discuss the tissue sources, diagnostic significance, and assays, including sources of error, for the following enzymes: CK, LD, AST, ALT, ALP, ACP, GGT, amylase, lipase, cholinesterase, and G-6-PD.*
- *Evaluate patient serum enzyme levels in relation to disease states.*
- *Explain why the measurement of serum enzyme levels is clinically useful.*

KEY TERMS

Activation energy	Enzyme	Hydrolase
Activators	Enzyme-substrate	International unit
Apoenzyme	(ES) complex	(IU)
Coenzyme	First-order kinetics	Isoenzyme
Cofactor	Holoenzyme	Kinetic assay

LD flipped pattern	Oxidoreductase	Zero-order kinetics
Michaelis-Menten	Transferase	Zymogen
constant		

*E*nzymes are specific biologic proteins that catalyze biochemical reactions without altering the equilibrium point of the reaction or being consumed or undergoing changes in composition. Other reactants are converted to products. The catalyzed reactions are frequently specific and essential to physiologic functions, such as the hydration of carbon dioxide, nerve conduction, muscle contraction, nutrient degradation, and energy use.

Found in all body tissues, enzymes frequently appear in the serum following cellular injury or, sometimes, in smaller amounts, from degraded cells or storage areas. Some enzymes, such as those that facilitate coagulation, are specific to plasma and, therefore, are present in significant concentrations in plasma. Hence, plasma or serum enzyme levels are often useful in the diagnosis of particular diseases or physiologic abnormalities. This chapter discusses the general properties and principles of enzymes, aspects relating to the clinical diagnostic significance of specific physiologic enzymes, and assay methods for those enzymes.

GENERAL PROPERTIES AND DEFINITIONS

Enzymes catalyze many specific physiologic reactions. These reactions are facilitated by the structure of the enzymes themselves and by several other factors. As a protein, each enzyme comprises a specific amino acid sequence (*primary structure*) with the resultant polypeptide chains twisting (*secondary structure*), which then folds (*tertiary structure*), resulting in structural cavities. If an enzyme contains more than one polypeptide unit, the *quaternary structure* refers to the spatial relationships between the subunits. Each enzyme contains an *active site*, often a water-free cavity, where the substance on which the enzyme acts (the *substrate*) interacts with particular charged amino acid residues. An *allosteric site*, which is a cavity other than the active site, may bind regulator molecules and thereby be significant to the basic enzyme structure.

Even though a particular enzyme maintains the same catalytic function throughout the body, that enzyme may exist in different forms within the same individual. The different forms, which may originate from genetic or nongenetic causes, may be differentiated from each other based on certain physical properties, such as electrophoretic mobility, solubility, or resistance to inactivation. The term *isoenzyme* is generally used when discussing such enzymes, but the International Union of Biochemistry (IUB) suggests restricting this term to multiple forms of genetic origin.

In addition to the basic enzyme structure, a nonprotein molecule, called a *cofactor*, may be necessary for enzyme activity. Inorganic cofactors, such as chloride or magnesium ions, are called *activators*. A *coenzyme* is an organic cofactor, such as nicotinamide adenine dinucleotide (NAD). When bound tightly to the enzyme, the coenzyme is called a *prosthetic group*. The enzyme portion (*apoenzyme*), with its respective coenzyme, forms a complete and active system, a *holoenzyme*.

$$\text{Holoenzyme} = \text{Apoenzyme} + \text{Coenzyme} \quad \textit{(Eq. 9–1)}$$

Some enzymes, mostly digestive enzymes, are originally secreted from the organ of production in a structurally inactive form, called a *proenzyme* or *zymogen*. Other enzymes later alter the structure of the proenzyme to make active sites available by hydrolyzing specific amino acid residues. This mechanism prevents digestive enzymes from digesting their place of synthesis.

ENZYME CLASSIFICATION AND NOMENCLATURE

Enzyme nomenclature has historically been a source of confusion. Some enzymes have been named arbitrarily (ptyalin, trypsin), whereas others have been named by attaching the suffix *-ase* to the name of the substrate on which the enzyme acted (*eg*, esterase, urease) or to the type of reaction catalyzed (*eg*, aminotransferase, oxidase). Some enzymes have been identified by different names in different parts of the world. The continual discovery of more enzymes has complicated enzyme identification by name.

To standardize enzyme nomenclature, the Enzyme Commission (EC) of the IUB adopted a classification system in 1961 and revised the standards in 1972 and 1978. The IUB system assigns a *systematic name* to each enzyme, defining the substrate acted on, the reaction catalyzed, and, possibly, the name of any coenzyme involved in the reaction. Because many systematic names are lengthy, a more usable, trivial, *recommended name* is also assigned by the IUB system.[1]

In addition to naming enzymes, the IUB system identifies each enzyme by an EC numerical code containing four digits separated by decimal points.[1] The first digit places the enzyme in one of the following six classes:

1. Oxidoreductases: Catalyze an oxidation–reduction reaction between two substrates.
2. Transferases: Catalyze the transfer of a group other than hydrogen from one substrate to another.
3. Hydrolases: Catalyze hydrolysis of a variety of bonds.
4. Lyases: Catalyze removal of groups from substrates without hydrolysis. The product contains double bonds.
5. Isomerases: Catalyze the interconversion of geometric, optical, or positional isomers.
6. Ligases: Catalyze the joining of two substrate molecules, coupled with breaking of the pyrophosphate bond in adenosine triphosphate (ATP) or a similar compound.

The second and third digits of the EC code number represent the subclass and sub-class of the enzyme, respectively, divisions that are made according to criteria specific to the enzymes in the class. The final number is a serial number that is specific to each enzyme in a sub-subclass. Table 9-1 provides the EC code numbers, as well as the systematic and recommended names, for enzymes frequently quantitated in the clinical laboratory.

Table 9-1 also lists common and standard abbreviations for commonly analyzed enzymes. Without IUB recommendation, capital letters have been used as a convenience to identify enzymes. The common abbreviations, sometimes developed from previously accepted names for the enzymes, were used until the standard abbreviations listed in the table were developed.[2,3] These standard abbreviations are being used in the United States and are used later in this chapter to indicate specific enzymes.

ENZYME KINETICS

Catalytic Mechanism of Enzymes

A chemical reaction may occur spontaneously if the free energy or available kinetic energy is higher for the reactants than for the products. The reaction then proceeds toward the lower energy if a sufficient number of the reactant molecules possess enough excess energy to break their chemical bonds and collide to form new bonds. The excess energy, called *activation energy*, is the energy required to raise all molecules in 1 mole of a compound at a certain temperature to the transition state at the peak of the energy barrier. At the transition state, each molecule is equally likely either to participate in product formation or to remain an unreacted molecule. Reactants possessing enough energy to overcome the energy barrier participate in product formation.

TABLE 9-1. Classification of Frequently Quantitated Enzymes

Class	Recommended Name	Common Abbreviation	Standard Abbreviation	EC Code No.	Systematic Name
Oxidoreductases	Lactate dehydrogenase	LDH	LD	1.1.1.27	L-lactate: NAD⁺ oxidoreductase
	Glucose-6-phosphate dehydrogenase	G-6-PDH	G6PD	1.1.1.49	D-glucose-6-phosphate: NADP⁺ 1-oxidoreductase
Transferases	Aspartate aminotransferase	GOT (glutamate oxaloacetate transaminase)	AST	2.6.1.1	L-Aspartate: 2-oxaloglutarate aminotransferase
	Alanine aminotransferase	GPT (glutamate transaminase)	ALT	2.6.1.2	L-Alanine: 2-oxaloglutarate aminotransferase
	Creatine kinase	CPK (creatine phosphokinase)	CK	2.7.3.2	ATP: creatine N-phospho-transferase
	Gamma glutamyltransferase	GGTP	GGT	2.3.2.2	(5-glutamyl) peptide: amino acid-5-glutamyltransferase
Hydrolases	Alkaline phosphatase	ALP	ALP	3.1.3.1	Orthophosphoric monoester phosphohydrolase (alkaline optimum)
	Acid phosphatase	ACP	ACP	3.1.3.2	Orthophosphoric monoester phosphohydrolase (acid optimum)
	α-Amylase	AMY	AMS	3.2.1.1	1,4-D-glucan gluconhydrolase
	Triacylglycerol lipase		LPS	3.1.1.3	Triacylglycerol acylhydrolase
	Cholinesterase	CHS	CHS	3.1.1.8	Acylcholine acylhydrolase

Adapted from Competence Assurance, ASMT, Enzymology, An Educational Program, RMI Corporation, 1980, with permission.

One way to provide more energy for a reaction is to increase the temperature and thus increase intermolecular collisions, but this does not normally occur physiologically. Enzymes catalyze physiologic reactions by lowering the activation energy level that the reactants (*substrates*) must reach for the reaction to occur (Figure 9-1). The reaction may then occur more readily to a state of equilibrium in which there is no net forward or reverse reaction, even though the equilibrium constant of the reaction is not altered. The extent to which the reaction progresses depends on the number of substrate molecules that pass the energy barrier.

The general relationship among enzyme, substrate, and product may be represented as follows:

$$E + S \rightleftharpoons ES \rightleftharpoons E + P, \qquad \textit{(Eq. 9-2)}$$

where E = enzyme
    S = substrate
  ES = enzyme-substrate complex
    P = product

The ES complex is a physical binding of a substrate to the active site of an enzyme. The structural arrangement of amino acid residues within the enzyme makes the three-dimensional active site available. The transition state for the ES complex has a lower energy of activation than has the transition state of S alone so that the reaction proceeds after the complex is formed. An actual reaction may involve two substrates and two products.

Different enzymes are specific to substrates to different extents or in different respects. Some enzymes exhibit *absolute specificity,* meaning that the enzyme combines with only one substrate and catalyzes only the one corresponding reaction. Other enzymes are *group-specific* because they combine with all substrates containing a particular chemical group,

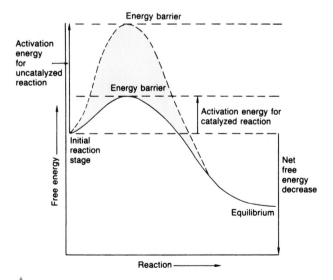

Figure 9-1. Energy versus progression of reaction, indicating the energy barrier that the substrate must surpass to react with and without enzyme catalysis. The enzyme considerably reduces the free energy needed to activate the reaction.

such as a phosphate ester. Still other enzymes are specific to chemical bonds and thereby exhibit *bond specificity.*

Stereoisometric specificity refers to enzymes that combine with only one optical isomer of a certain compound.

Factors That Influence Enzymatic Reactions

Substrate Concentration

The rate at which an enzymatic reaction proceeds and whether the forward or reverse reaction occurs depend on several reaction conditions. One major influence on enzymatic reactions is substrate concentration. In 1913, Michaelis and Menten hypothesized the role of substrate concentration in formation of the *enzyme-substrate (ES) complex.* According to the hypothesis, which is represented in Figure 9-2, the substrate readily binds to free enzyme at a low substrate concentration. With the amount of enzyme exceeding the amount of substrate, the reaction rate steadily increases as more substrate is added. The reaction is following *first-order kinetics,* because the **reaction rate is directly proportional to substrate concentration.** Eventually, however, the substrate concentration is high enough to saturate all available enzyme, and the reaction velocity reaches its maximum. When product is formed, the resultant free enzyme immediately combines with excess free substrate. The reaction is in *zero-order kinetics,* and **the reaction rate depends on enzyme concentration only.**

From their theory, Michaelis and Menten derived the *Michaelis-Menten constant (Km),* which is a constant for a specific enzyme and substrate under defined reaction conditions and is an expression of the relationship between the velocity of an enzymatic reaction and substrate concentration.

The assumptions are made that equilibrium among E, S, ES, and P is established rapidly and that the E + P → ES reaction is negligible. The rate-limiting step is the formation of product and enzyme from the ES complex. Then, maximum velocity is fixed, and the reaction rate is a function of only the enzyme concentration. As designated in Figure 9-2, K_m is specifically the substrate concentration at which the enzyme yields half the possible maximum velocity. Hence, K_m indicates the amount of substrate needed for a particular enzymatic reaction.

The Michaelis-Menten hypothesis of the relationship between reaction velocity and substrate concentration can be represented mathematically as follows:

$$V = \frac{V_{max}\,[S]}{K_m + [S]}, \qquad (Eq.\ 9\text{--}3)$$

where V = measured velocity of reaction
  V_{max} = maximum velocity
  $[S]$ = substrate concentration
  K_m = Michaelis-Menten constant of enzyme for specific substrate

Theoretically, V_{max} and then K_m could be determined from the plot in Figure 9-2. However, V_{max} is difficult to determine from the hyperbolic plot and often is not actually achieved in enzymatic reactions, because enzymes may not function optimally in the presence of excessive substrate. A more accurate determination of K_{min} may be made through a Lineweaver-Burk plot, a double-reciprocal plot of the Michaelis-Menten constant, which yields a straight line (Figure 9-3). The reciprocal is taken of both the substrate concentration and the velocity of an enzymatic reaction. The equation becomes

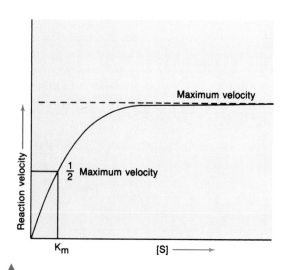

Figure 9-2. Michaelis-Menten curve of velocity versus substrate concentration for enzymatic reaction. K_m is the substrate concentration at which the reaction velocity is half of the maximum level.

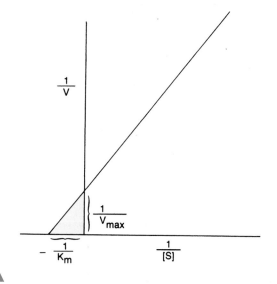

Figure 9-3. Lineweaver-Burk transformation of Michaelis-Menten curve. V_{max} is the reciprocal of the γ intercept of the straight line. K_m is the negative reciprocal of the x intercept of the same line.

$$\frac{1}{V} = \frac{K_m}{V_{max}} \frac{1}{[S]} + \frac{1}{V_{max}} \qquad \textit{(Eq. 9–4)}$$

As indicated in Figure 9-3, both V_{max} and K_m may conveniently be determined from the plot.

Enzyme Concentration

Because enzymes catalyze physiologic reactions, the enzyme concentration affects the rate of the catalyzed reaction. As long as the substrate concentration exceeds the enzyme concentration, **the velocity of the reaction is proportional to the enzyme concentration.** The higher the enzyme level, the faster the reaction will proceed, because more enzyme is present to bind with the substrate.

pH

Enzymes are proteins that carry net molecular charges. Extreme pH levels may denature an enzyme or influence its ionic state, resulting in structural changes or a change in the charge on an amino acid residue in the active site. Hence, each enzyme operates within a specific pH range and maximally at a specific pH. Most physiologic enzymatic reactions occur in the pH range of 7 to 8, but some enzymes are active in wider pH ranges than others. In the laboratory, the pH for a reaction is carefully controlled at the optimal pH by means of appropriate buffer solutions.

Temperature

Increasing temperature usually increases the rate of a chemical reaction by increasing the movement of molecules, the rate at which intermolecular collisions occur, and the energy available for the reaction. Such is the case with enzymatic reactions, until the temperature is high enough to denature the protein composition of the enzyme. The Q_{10} value, the increase in reaction rate for every 10°C increase in temperature, is approximately 2 for enzymes, indicating that this increase in temperature usually doubles the reaction rate.

Each enzyme functions optimally at a particular temperature, which is influenced by other reaction variables, especially the total time for the reaction. The optimal temperature is usually close to that of the physiologic environment of the enzyme. However, some denaturation may occur at the human physiologic temperature of 37°C. The rate of denaturation increases as the temperature increases and is usually significant at 40°C to 50°C.

Because low temperatures render enzymes reversibly inactive, many serum or plasma specimens for enzyme quantitations are refrigerated or frozen to prevent activity loss until analysis. Storage procedures may vary from enzyme to enzyme because of individual stability characteristics. However, repeated freezing and thawing tends to denature protein and should be avoided.

Because of their temperature sensitivity, enzymes should be analyzed under strictly controlled temperature conditions. Incubation temperatures should be accurate within ±0.1°C. Laboratories usually attempt to establish an analysis temperature for routine enzyme quantitations of 25°C, 30°C, or 37°C. Attempts to establish a universal temperature for enzyme analysis have been futile, and thus reference ranges for enzyme levels may vary significantly from laboratory to laboratory. In the United States, however, 37°C is most commonly used.

Cofactors

Cofactors are nonprotein entities that must bind to particular enzymes before a reaction occurs. Common *activators,* inorganic cofactors, are metallic (Ca^{2+}, Fe^{2+}, Mg^{2+}, Mn^2 Zn^{2+}, and K^+) and nonmetallic (Br^- and Cl^-). The activator may be essential for the reaction or may only enhance the reaction rate in proportion with concentration to the point at which the excess activator begins to inhibit the reaction. Activators function by alternating the spatial configuration of the enzyme for proper substrate binding, linking substrate to the enzyme or a coenzyme, or undergoing oxidation or reduction.

Some common *coenzymes,* organic cofactors, are nucleotide phosphates and vitamins. Coenzymes serve as second substrates for enzymatic reactions, and when bound tightly to the enzyme, they are called *prosthetic groups.* For example, NAD as a cofactor may be reduced to nicotinamide adenine dinucleotide phosphate (NADP) in a reaction in which the primary substrate is oxidized. Hence, increasing coenzyme concentration will increase the velocity of an enzymatic reaction in a manner synonymous with increasing substrate concentration. When quantitating an enzyme that requires a particular cofactor, that cofactor should always be provided in excess so that the extent of the reaction does not depend on the concentration of the cofactor.

Inhibitors

Enzymatic reactions may not progress normally if a particular substance, an *inhibitor,* interferes with the reaction. *Competitive inhibitors* physically bind to the active site of an enzyme and thereby compete with the substrate for the active site. With a substrate concentration significantly higher than the concentration of the inhibitor, the inhibition is reversible, because the substrate is more likely than the inhibitor to bind the active site, and the enzyme has not been destroyed.

A *noncompetitive inhibitor* binds an enzyme at a place other than the active site and may be reversible in the respect that some naturally present metabolic substances combine reversibly with certain enzymes. Noncompetitive inhibition also may be *irreversible* if the inhibitor destroys part of the enzyme involved in catalytic activity. Because the inhibitor binds the enzyme independently from the substrate, increasing substrate concentration does not reverse the inhibition.

In still another kind of inhibition, *uncompetitive inhibition,* the inhibitor binds to the ES complex so that increasing substrate concentration results in more ES complexes to which the inhibitor binds and thereby increases the inhibition. The enzyme-substrate-inhibitor complex does not yield product.

Each of the three kinds of inhibition is unique with respect to effects on the V_{max} and K_m of enzymatic reactions

(Fig. 9-4). In competitive inhibition, the effect of the inhibitor can be counteracted by adding excess substrate to bind the enzyme. The amount of the inhibitor is then negligible by comparison, and the reaction will proceed at a slower rate but to the same maximum velocity as an uninhibited reaction. The K_m is a constant for each enzyme and cannot be altered. However, because the amount of substrate needed to achieve a particular velocity is higher in the presence of a competing inhibitor, the K_m appears to increase when exhibiting the effect of the inhibitor.

The substrate and inhibitor, commonly a metallic ion, may bind an enzyme simultaneously in noncompetitive inhibition. The inhibitor may inactivate either an ES complex or just the enzyme by causing structural changes in the enzyme. Even if the inhibitor binds reversibly and does not inactivate the enzyme, the presence of the inhibitor when it is bound to the enzyme slows the rate of the reaction. Thus, for noncompetitive inhibition, the maximum reaction velocity cannot be achieved. Increasing substrate levels has no influence on the binding of a noncompetitive inhibitor, so the K_m is unchanged.

Because uncompetitive inhibition requires the formation of an ES complex, increasing substrate concentration increases inhibition. Thus, maximum velocity equal to that of an uninhibited reaction cannot be achieved, and the K_m appears to be decreased.

Measurement of Enzyme Activity

Because enzymes are usually present in very small quantities in biologic fluids and are often difficult to isolate from similar compounds, the most convenient quantitations of enzymes measure catalytic activity and relate this activity to concentration. A given method might photometrically measure an increase in product concentration, a decrease in substrate concentration, a decrease in coenzyme concentration, or an increase in the concentration of an altered coenzyme.

If the amount of substrate and any coenzyme is in excess in an enzymatic reaction, the amount of substrate or coenzyme used or product or altered coenzyme formed will depend only on the amount of enzyme present to catalyze the reaction. Hence, enzyme concentrations are always performed in zero-order kinetics, with the substrate in sufficient excess to ensure that not more than 20% of the available substrate is converted to product. Any coenzymes also must be in excess. NADH is a coenzyme frequently measured in the laboratory. NADH absorbs light at 340 nm, whereas NAD does not, and a change in absorbance at 340 nm is easily measured.

In specific laboratory methodologies, substances other than substrate or coenzyme are necessary and must be present in excess. NAD or NADH is often convenient as a reagent for a *coupled-enzyme* assay when neither NAD nor NADH is a coenzyme for the reaction. In other coupled-enzyme assays, more than one enzyme is added in excess as a reagent, and multiple reactions are catalyzed. After the enzyme under analysis catalyzes its specific reaction, a product of that reaction becomes the substrate on which an intermediate *auxiliary enzyme* acts. A product of the intermediate reaction becomes the substrate for the final reaction, which is catalyzed by an *indicator enzyme* and commonly involves the conversion of NAD to NADH or vice versa.

When performing an enzyme quantitation in zero-order kinetics, inhibitors must be lacking, and other variables that may influence the rate of the reaction must be carefully con-

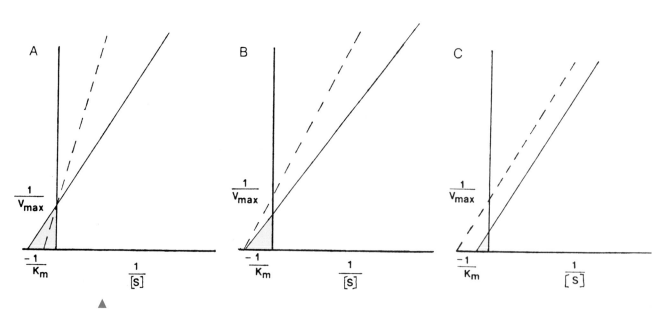

Figure 9-4. Normal Lineweaver-Burk plot (solid line) compared with each type of enzyme inhibition (dotted line). (**A**) Competitive inhibition V_{max} unaltered; K_m appears increased. (**B**) Noncompetitive inhibition V_{max} decreased; K_m unchanged. (**C**) Uncompetitive inhibition V_{max} decreased; K_m appears decreased.

trolled. A constant pH should be maintained by means of an appropriate buffer solution. The temperature should be constant within ±0.1°C throughout the assay at a temperature at which the enzyme is active (usually 25°C, 30°C, or 37°C).

During the progress of the reaction, the period for the analysis also must be carefully selected. When the enzyme is initially introduced to the reactants and the excess substrate is steadily combining with available enzyme, the reaction rate rises. After the enzyme is saturated, the rates of product formation, release of enzyme, and recombination with more substrate proceed linearly. After a time, usually 6 minutes to 8 minutes after reaction initiation, the reaction rate decreases as the substrate is depleted, the reverse reaction is occurring appreciably, and the product begins to inhibit the reaction. Hence, enzyme quantitations must be performed during the linear phase of the reaction.

One of two general methods may be used to measure the extent of an enzymatic reaction: (1) fixed-time and (2) continuous-monitoring or kinetic assay. For the *fixed-time method,* the reactants are combined, the reaction proceeds for a designated time, the reaction is stopped (usually by inactivating the enzyme with a weak acid), and a measurement is made of the amount of reaction that has occurred. The reaction is assumed to be linear over the reaction time, and so the larger the reaction, the more enzyme is present.

In *continuous-monitoring or kinetic assays,* multiple measurements, usually of absorbance change, are made during the reaction, either at specific time intervals, usually every 30 seconds or 60 seconds, or continuously by a continuous-recording spectrophotometer. Such assays are advantageous over fixed-time methods because the linearity of the reaction may be more adequately verified. If absorbance is measured at intervals, several data points are necessary to increase the accuracy of linearity assessment. Continuous measurements are preferred because any deviation from linearity is readily observable.

The most common cause of deviation from linearity occurs when the enzyme is so elevated that all substrate is used early in the reaction time. For the remainder of the reaction, the rate change is minimal, and the implication is that the coenzyme concentration is very low. With continuous monitoring, the laboratorian may observe a sudden decrease in the reaction rate (deviation from zero-order kinetics) of a particular determination and may repeat the determination using less patient sample. The decrease in the amount of patient sample operates as a dilution, and the answer obtained may be multiplied by the dilution factor to obtain the final answer. The sample itself is not diluted so that the diluent cannot interfere with the reaction. (Sample dilution with saline may be necessary to minimize negative effects in analysis caused by hemolysis or lipemia.)

Ensuring the accuracy of enzyme measurements has long been a concern of laboratorians. More recently, the Clinical Laboratory Improvement Amendments of 1988 (CLIA-88) has set guidelines for quality control and proficiency testing for all laboratories. Problems with quality control materials

for enzyme testing have been problematic. Differences between clinical specimens and control sera include species of origin of the enzyme, integrity of the molecular species, isoenzyme forms, matrix of the solution, addition of preservatives, and lypholization processes. Many studies have been conducted to ensure accurate enzyme measurements and good quality quality control materials.[4]

Immunoassay methodologies are also available for some enzymes. Because enzymes are proteins, they usually cause antibody production when injected into animals to which they are foreign. The antisera formed are very specific for the enzyme and can be used in conjunction with the sensitivity of immunoassay to quantitate an enzyme or isoenzyme that is present in serum in small amounts. Also, because the method is independent of the catalytic activity of the enzyme, inactive enzyme protein may be measured.

Calculation of Enzyme Activity

Because enzymes are quantitated relative to their activity rather than direct concentration, the units used to report enzyme levels are *activity units.* The definition for the activity unit must consider variables that may alter results (pH, temperature, substrate, and so on). Historically, specific method developers frequently established their own units, often named after themselves (ie, Bodansky and King units), for reporting results. To standardize the system of reporting quantitative results, the CE defined the *international unit (U)* as the amount of enzyme that will catalyze the reaction of 1 μmole of substrate per minute under specified conditions of temperature, pH, substrates, and activators. Because the specified conditions may vary from laboratory to laboratory, reference values are still often laboratory specific. Enzyme concentration is usually expressed in units per liter (U/L).

Another unit recommended for expressing an enzyme activity unit is the Katal (mol/s). The mole is the unit for substrate concentration, and the unit of time is the second. Enzyme concentration is then expressed as Katals per liter (Kat/L). $1.0 \text{ U} = 60 \text{ μKat.}$

When enzymes are quantitated by measuring the increase or decrease of NADH at 340 nm, the molar absorptivity (6.22×10^3) of NADH is used to calculate enzyme activity. Because an NADH solution with a concentration of 1.0 μmol/mL has an absorbance of 6.22, a 0.001 absorbance decrease represents a concentration change of 0.001/6.22 (1.6×10^{-4}) μmol/mL.

Enzymes as Reagents

Enzymes may be used as reagents to measure many nonenzymatic constituents in serum. For example, glucose, cholesterol, and uric acid are frequently quantitated by means of enzymatic reactions. Such methods measure the "true" level of the analyte due to the specificity of the enzyme. Enzymes are also used as reagents for methods of quantitating analytes that are substrates for corresponding enzyme quantitations. One such example, lactate dehydrogenase, may be a reagent when lactic acid levels are evaluated. For such methods, the enzyme is added in excess in a quantity sufficient to provide a complete reaction in a short period.

Immobilized enzymes are chemically bonded to adsorbents such as agarose or some types of cellulose by azide groups, diazo, and triazine. The enzymes act as recoverable reagents. When substrate is passed through the preparation, the product is retrieved and analyzed, and the enzyme is present and free to react with more substrate. Immobilized enzymes are convenient for continuous-flow or batch analyses and are more stable than enzymes in a solution. Enzymes are also used as reagents in enzyme-linked immunosorbent assay (ELISA), such as those used to measure human immunodeficiency virus antibodies in serum.

ENZYMES OF CLINICAL SIGNIFICANCE

Table 9-2 lists the commonly analyzed enzymes, including their systematic names and clinical significance. Each enzyme is discussed in this chapter with respect to tissue sources, diagnostic significance, assay methods, sources of error, and reference range.

Creatine Kinase

Creatine kinase (CK) is an enzyme that is generally associated with ATP regeneration in contractile or transport systems. Its predominate physiologic function occurs in muscle cells, where it is involved in the storage of high-energy creatine phosphate. Every contraction cycle of muscle results in creatine phosphate use, with the production of ATP. This

results in relatively constant levels of muscle ATP. The reversible reaction catalyzed by CK is shown in Equation 9-5.

$$\text{Creatine} + \text{ATP} \overset{\text{CK}}{\rightleftharpoons} \text{Creatine phosphate} + \text{ADP} \quad \textit{(Eq. 9–5)}$$

Tissue Sources

CK is widely distributed in tissues, with highest activities found in skeletal muscle, heart muscle, and brain tissue. Other tissue sources in which CK is present in much smaller quantities include the bladder, placenta, gastrointestinal tract, thyroid, uterus, kidney, lung, prostate, spleen, liver, and pancreas.

Diagnostic Significance

Because of the high concentrations of CK in muscle tissue, CK levels are frequently elevated in disorders of cardiac and skeletal muscle. The CK level is considered a very sensitive indicator of acute myocardial infarction (AMI) and muscular dystrophy, particularly the Duchenne type. Very striking elevations of CK occur in Duchenne type muscular dystrophy, with values reaching 50 to 100 times the upper limit of normal (ULN). Although total CK levels are sensitive indicators of these disorders, they are not entirely specific indicators inasmuch as CK elevation is found in various other abnormalities of cardiac and skeletal muscle.

Elevated CK levels also are seen occasionally in central nervous system disorders such as cerebral vascular accident, seizures, nerve degeneration, and central nervous system shock. Damage to the blood-brain barrier must occur to allow enzyme release to the peripheral circulation.

Other pathophysiologic conditions in which elevated CK levels occur are hypothyroidism, malignant hyperpyrexia, and Reye's syndrome. Table 9-3 lists the major disorders associated with abnormal levels of CK.

Because enzyme elevation is found in numerous disorders, the separation of total CK into its various isoenzyme fractions is considered a more specific indicator of various disorders than are total levels. Typically, the clinical relevance of CK activity depends more on isoenzyme fractionation than on total levels.

CK occurs as a dimer consisting of two subunits, which can be separated readily into three distinct molecular forms. The three isoenzymes have been designated as CK-BB (brain type), CK-MB (hybrid type), and CK-MM (muscle type). On electrophoretic separation, CK-BB will migrate fastest toward the anode and hence is called *CK-1*. CK-BB is followed by CK-MB (*CK-2*), and finally by CK-MM (*CK-3*), exhibiting the slowest mobility (Fig. 9-5). Table 9-3 indicates the tissue localization of the isoenzymes and the major conditions associated with elevated levels.

In the sera of healthy people, the major isoenzyme is the MM form. Values for the MB isoenzyme range from undetectable to trace (<6% of total CK). It also appears that CK-BB is present in small quantities in the sera of healthy people. However, the presence of CK-BB in serum depends on the method of detection. Most techniques cannot detect CK-BB in normal serum.

TABLE 9-2. Major Enzymes of Clinical Significance

Enzyme	Clinical Significance
Acid phosphatase (ACP)	Prostatic carcinoma
Alanine aminotransferase (ALT)	Hepatic disorders
Alkaline phosphatase (ALP)	Hepatic disorders
	Bone disorders
Amylase (AMS)	Acute pancreatitis
Aspartate aminotransferase (AST)	Myocardial infarction
	Hepatic disorders
	Skeletal muscle disorders
Creatine kinase (CK)	Myocardial infarction
	Skeletal muscle disorders
γ-Glutamyltransferase (GGT)	Hepatic disorders
Glucose-6-phosphate dehydrogenase (G–6–PD)	Drug-induced hemolytic anemia
Lactate dehydrogenase (LD)	Myocardial infarction
	Hepatic disorders
	Carcinomas
Lipase (LPS)	Acute pancreatitis
Pseudocholinesterase (PChE)	Organophosphate poisoning
	Genetic variants

TABLE 9-3. Creatine Kinase Isoenzymes—Tissue Localization and Sources of Elevation

Isoenzyme	Tissue	Condition
CK-MM	Heart	Myocardial infarction
	Skeletal muscle	Skeletal muscle disorders
		Muscular dystrophy
		Polymyositis
		Hypothyroidism
		Malignant hyperthermia
		Physical activity
		Intramuscular injections
CK-MB	Heart	Myocardial infarction
	Skeletal muscle	Myocardial injury
		Ischemia
		Angina
		Inflammatory heart
		disease
		Cardiac surgery
		Duchenne type
		muscular dystrophy
		Polymyositis
		Malignant hyperthermia
		Reye's syndrome
		Rocky Mountain
		spotted fever
		Carbon monoxide
		poisoning
CK-BB	Brain	Central nervous
		system shock
	Bladder	Anoxic encephalopathy
	Lung	Cerebrovascular
		accident
	Prostate	Seizures
	Uterus	Placental or
		uterine trauma
	Colon	Carcinoma
	Stomach	Reye's syndrome
	Thyroid	Carbon monoxide
		poisoning
		Malignant hyperthermia
		Acute and chronic
		renal failure

CK-MM is the major isoenzyme fraction found in striated muscle and normal serum. Skeletal muscle contains almost entirely CK-MM with a small amount of CK-MB. The majority of CK activity in heart muscle is also attributed to CK-MM, with approximately 20% due to CK-MB.[5] Normal serum consists of approximately 94% to 100% CK-MM. Injury to

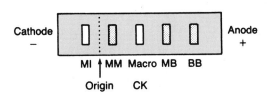

Figure 9-5. Electrophoretic migration pattern of normal and atypical CK isoenzymes.

both cardiac and skeletal muscle accounts for the majority of cases of CK-MM elevations (see Table 9-3). Hypothyroidism results in CK-MM elevations because of the involvement of muscle tissue (increased membrane permeability), the effect of thyroid hormone on enzyme activity, and, possibly, the slower clearance of CK as a result of slower metabolism.

Mild to strenuous activity may contribute to elevated CK levels, as may intramuscular injections of certain drugs. In physical activity, the extent of elevation is variable. However, the degree of exercise in relation to the exercise capacity of the individual is the most important factor in determining the degree of elevation.[6] Physically well-conditioned patients show lesser degrees of elevation than do less conditioned patients. Levels may be elevated for as long as 48 hours following exercise.

CK elevations seen following intramuscular injections are generally less than 5 × ULN and are usually not apparent after 48 hours, although elevations may persist for 1 week. The predominant isoenzyme is CK-MM.

The quantity of CK-BB in the tissues listed in Table 9-3 is usually small. The small quantity, coupled with its relatively short half-life (1–5 h), results in CK-BB activities that are generally low and transient and not usually measurable when tissue damage occurs. Highest concentrations are found in the central nervous system, the gastrointestinal tract, and the uterus during pregnancy.

Even though brain tissue has high concentrations of CK, serum rarely contains CK-BB of brain origin. Because of its molecular size (80,000), its passage across the blood-brain barrier is hindered. However, when extensive damage to the brain has occurred, significant amounts of CK-BB can sometimes be detected in the serum.

Recently, it has been observed that CK-BB may be significantly elevated in patients with carcinoma of various organs. It has been found in association with untreated prostatic carcinoma and other adenocarcinomas. These findings indicate that CK-BB may be useful as a tumor-associated marker.[7]

The most common causes of CK-BB elevations are central nervous system damage, tumors, childbirth, and the presence of macro-CK, an enzyme-immunoglobulin complex. In the majority of these cases, the CK-BB level is greater than 5 U/L, usually in the range of 10 U/L to 50 U/L. Other conditions listed in Table 9-3 usually show CK-BB activity below 10 U/L.[8]

The value of CK isoenzyme separation can be found principally in detection of myocardial damage. Cardiac tissue contains significant quantities of CK-MB, approximately 20% of all CK-MB. Whereas CK-MB is found in small quantities in other tissues, myocardium is essentially the only tissue from which CK-MB enters the serum in significant quantities. *Demonstration of elevated levels of CK-MB, greater than or equal to 6% of the total CK, is considered the most specific indicator of myocardial damage, particularly AMI.* Following myocardial infarction, the CK-MB levels begin to rise within 4 hours to 8 hours, peak at 12 hours to 24 hours, and return to normal levels within 48 hours to 72 hours. This time frame must be considered when interpreting CK-MB levels.

CASE STUDY 9-1

A 46-year-old overweight Caucasian man comes to see his family doctor complaining of "indigestion" of 5 days' duration. He has also had bouts of sweating, malaise, and headaches. His blood pressure is 148/105 and his family history includes a diabetic father dying at age 62 years of AMI secondary to diabetes mellitus. An electrocardiogram revealed changes from one done 6 months earlier. The following are the results of the patient's blood work:

CK	129 U/L	(30–60)
CK-MB	4%	(<6%)
LD	280 U/L	(100–225)
LD Isoenzymes	LD-1 > LD-2	
AST	35 U/L	(5–30)

Questions

1. Can a diagnosis of AMI be ruled out in this patient?
2. What further cardiac markers should be run on this patient?
3. Should this patient be admitted to the hospital?

CK-MB activity has been observed in other cardiac disorders (see Table 9-3). Therefore, increased quantities are not entirely specific for AMI but probably reflect some degree of ischemic heart damage. The specificity of CK-MB levels in the diagnosis of AMI can be increased if interpreted in conjunction with lactate dehydrogenase (LD) isoenzymes and if measured sequentially over a 48-hour period to detect the typical rise and fall of enzyme activity seen in AMI (Fig. 9-6).

Exchange of the subunits of CK isoenzymes resulting in transformation into another isoenzyme form has been observed both in vitro and in vivo. The clinical significance of this finding relates to the fact that CK-KB can be transformed into CK-MB by the following reaction:

$$CK - BB + CK - MM \rightleftharpoons CK - MB + CK - BB$$

$$+ CK - MM \qquad (Eq. 9–6)$$

This transformation could account for unexplained CK-MB levels in patients with lung cancer, acute cerebral disorders, and other disorders in which CK-MB activity would not normally be expected. It is also possible that this transformation of subunits results in the formation of CK-BB in sera with high CK-MB activity.

The MB isoenzyme also has been detected in the sera of patients with noncardiac disorders. CK-MB levels found in these conditions probably represent leakage from skeletal muscle, although in Duchenne type muscular dystrophy, there may be some cardiac involvement as well. CK-MB levels in Reye's syndrome also may reflect myocardial damage.

Despite the findings of CK-MB levels in disorders other than myocardial infarction, its presence still remains a highly significant indicator of AMI.[9] The typical time course of CK-MB elevation following AMI is not found in other conditions.

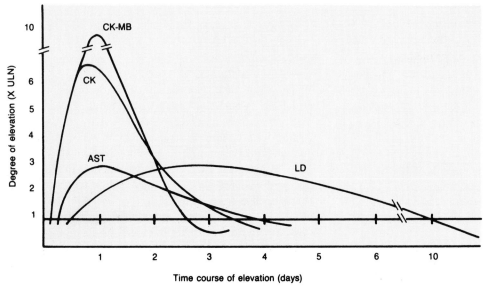

◄

Figure 9-6. Time activity curves of enzymes in myocardial infarction for AST, CK, CK-MB, and LD. CK, specifically the MB fraction, increases initially, followed by AST and LD. LD is elevated the longest. All usually return to normal within 10 days.

In recent years, there have been numerous reports describing the appearance of unusual CK isoenzyme bands displaying electrophoretic properties that differ from the three major isoenzyme fractions (see Figure 9-5).[10–14] These atypical forms are generally of two types and are referred to as macro-CK and mitochondrial CK.

Macro-CK appears to migrate to a position midway between that of CK-MM and CK-MB. This type of macro-CK largely comprises CK-BB complexed with immunoglobulin. In most instances, the associated immunoglobulin is IgG, although a complex with IgA also has been described. The term *macro-CK* has been used to describe complexes of lipoproteins with CK-MM as well.

The incidence of macro-CK in sera ranges from 0.8% to 1.6%. Currently, no specific disorder has been associated with its presence, although it appears to be age- and sex-related, occurring most frequently in women older than age 50 years.

Mitochondrial CK (CK-Mi) is bound to the exterior surface of the inner mitochondrial membranes of muscle, brain, and liver. It migrates to a point cathodal to CK-MM and exists as a dimeric molecule of two identical subunits. It occurs in serum in both the dimeric state and in the form of oligomeric aggregates of high molecular weight (350,000). CK-Mi is not present in normal serum, and it is typically not present following myocardial infarct. The incidence of CK-Mi ranges from 0.8% to 1.7%. Extensive tissue damage must occur, causing breakdown of the mitochondrion and cell wall, for it to be detected in serum. Its presence does not correlate with any specific disease state but appears to be an indicator of severe illness. It has been detected in cases of malignant tumor and cardiac abnormalities.

In view of the lack of a definite correlation between these atypical CK forms and a specific disease state, it appears that their significance relates primarily to the methods used for detecting CK-MB. In certain analytic procedures, these atypical forms may be measured as CK-MB, resulting in erroneously high CK-MB levels.

Methods that have been used for measurement of CK isoenzymes include electrophoresis, ion-exchange chromatography, radioimmunoassay (RIA), and immunoinhibition methods. Electrophoresis is the reference method. The electrophoretic properties of the CK isoenzymes are shown in Figure 9-5. Generally, the technique consists of performing electrophoresis on the sample, measuring the reaction using an overlay technique, and then visualizing the bands under ultraviolet light. With electrophoresis, the atypical bands can be separated, allowing their detection apart from the three major bands. Often a strongly fluorescent band appears, which migrates in close proximity to the CK-BB form. The exact nature of this fluorescence is unknown, but it has been attributed to the binding of fluorescent drugs or bilirubin by albumin.

In addition to visualizing atypical CK bands, other advantages of electrophoresis methods include detecting an unsatisfactory separation and allowing visualization of aden-

ylate kinase (AK). AK is an enzyme released from erythrocytes in hemolyzed samples and appearing as a band cathodal to CK-MM. AK may interfere with chemical or immunoinhibition methods, causing a falsely elevated CK or CK-MB value.

Ion-exchange chromatography has the potential for being more sensitive and precise than electrophoretic procedures performed with good technique. On an unsatisfactory column, however, CK-MM may merge into CK-MB, and CK-BB may be eluted with CK-MB. Also, macro-CK will elute with CK-MB.

Antibodies against both the M and B subunits have been used to determine CK-MB activity. Anti-M inhibits all M activity but not B activity. CK activity is measured before and after inhibition. Activity remaining after M inhibition is due to the B subunit of both MB and BB activity. The residual activity after inhibition is multiplied by 2 to account for MB activity (50% inhibited). The major disadvantage of this method is that it detects BB activity, which, although not normally detectable, will cause falsely elevated MB results when BB is present. In addition, the atypical forms of CK-Mi and macro-CK are not inhibited by anti-M antibodies and also may cause erroneous results for MB activity.

Recently, a double-antibody immunoinhibition technique has become available. This technique allows differentiation of MB activity due to adenylate kinase and the atypical isoenzymes, resulting in a more specific analytical procedure for CK-MB.[15]

RIA methods detect CK-MB reliably with no cross-reactivity. RIA methods measure the concentration of enzyme protein rather than enzymatic activity and can, therefore, detect enzymatically inactive CK-MB. This leads to the possibility of permitting detection of infarction earlier than other methods. RIA methods using a double-antibody system also have been described. The technique is time consuming and may be unsuitable for routine use, but the sensitivity is greater, which may be helpful in borderline cases.[11]

Physicians sometimes want to know a patient's CK-MB level rather quickly and do not feel they can wait the several hours needed to perform an electrophoresis or RIA procedure.

A new rapid assay for the subforms of CK-MB has been tested to help diagnose AMI in the first 6 hours after chest pain. The assay is performed on plasma samples for MB_2 and MB_1, activity using rapid high voltage electrophoresis on an automated analyzer. The assay takes approximately 25 minutes and allows emergency room physicians to better decide which patients need the costly care in the cardiac care unit of the hospital. This test may prove to be extremely helpful in the cost-containment horizon of health care reform.[16]

Assay

As indicated by Equation 9-5, CK catalyzes both forward and reverse reactions involving phosphorylation of creatine or ADP. Typically, for analysis of CK activity, this reaction is coupled with other enzyme systems, and a change in absorbance at 340 nm is determined. The forward reaction is coupled with the pyruvate kinase-lactate dehydrogenase-NADH system and proceeds according to Equation 9-7:

$$Creatine + ATP \overset{CK}{\rightleftharpoons} creatine\ phosphate + ADP$$

$$ADP + phosphoenolpyruvate \overset{PK}{\rightleftharpoons} pyruvate + ATP$$

$$Pyruvate + NADH + H^+ \overset{LD}{\rightleftharpoons} lactate + NAD^+$$

(Eq. 9–7)

The reverse reaction is coupled with the hexokinase-glucose-6-phosphate dehydrogenase-NADP system, as indicated in Equation 9-8:

$$Creatine\ phosphate + ADP \overset{CK}{\rightleftharpoons} creatine + ATP$$

$$ATP + glucose \overset{HK}{\rightleftharpoons} ADP + glucose\text{-}6\text{-}phosphate$$

$$Glucose\text{-}6\text{-}phosphate + NADP^+ \overset{G\text{-}6\text{-}PD}{\rightleftharpoons}$$

$$6\text{-}phosphogluconate + NADPH$$

(Eq. 9–8)

The reverse reaction proposed by Oliver and modified by Rosalki is the most commonly performed method in the clinical laboratory.[11] The reaction proceeds two to six times faster than the forward reaction, depending on the assay conditions, and there is less interference from side reactions. The optimal pH for the reverse reaction is 6.8, whereas that for the forward reaction is 9.0.

CK activity in serum is very unstable, being rapidly inactivated because of oxidation of sulfhydryl groups. Inactivation can be partially reversed by the addition of sulfhydryl compounds to the assay reagent. Compounds such as *N*-acetylcysteine, mercaptoethanol, thioglycerol, and dithiothreitol are among those used.

Sources of Error

Hemolysis of serum samples may be a source of elevated CK activity. Erythrocytes are virtually devoid of CK; however, they are rich in AK activity. AK reacts with ADP to produce ATP, which is then available to participate in the assay reaction, causing falsely elevated CK levels. This interference can occur with hemolysis of greater than 320 mg/L of hemoglobin, which releases sufficient AK to exhaust the AK inhibitors in the reagent. Trace hemolysis causes very little, if any, CK elevation. Serum should be stored in the dark, because CK is inactivated by daylight. Activity can be restored after storage in the dark at 4°C for 7 days or at −20°C for 1 month when the assay is conducted using a sulfhydryl activator.[17] Because of the effect of muscular activity on CK levels, it should be noted that physically well-trained people tend to have elevated baseline levels, whereas patients who are bedridden for prolonged periods may have decreased CK activity.

Reference Range

See Table 9-4. The higher values in males are attributed to increased muscle mass. Note that the reference ranges listed

TABLE 9-4. Reference Ranges for CK	
Total CK	
Male	15–160 U/L (37°C)
Female	15–130 U/L (37°C)
CK-MB	<6% of total CK

for all enzymes are subject to variation depending on the method used and the assay conditions.

Lactate Dehydrogenase

LD is an enzyme that catalyzes the interconversion of lactic and pyruvic acids. It is a hydrogen-transfer enzyme that uses the coenzyme NAD^+ according to Equation 9-9:

$$\begin{array}{ccc} CH_3 & & CH_3 \\ | & & | \\ HC-OH + NAD^+ \overset{LD}{\rightleftharpoons} & C=O + NADH + H^+ \\ | & & | \\ COOH & & COOH \\ Lactate & & Pyruvate \end{array}$$

(Eq. 9–9)

Tissue Source

LD is widely distributed in the body. Very high activities are found in the heart, liver, skeletal muscle, kidney, and erythrocytes. It is present in lesser amounts in the lung, smooth muscle, and brain.

Diagnostic Significance

Because of its widespread activity in numerous body tissues, LD is elevated in a variety of disorders. Increased levels are found in cardiac, hepatic, skeletal muscle, and renal diseases, as well as in several hematologic and neoplastic disorders. The highest levels of LD are seen in pernicious anemia and hemolytic disorders. Intramedullary destruction of erythroblasts causes elevations due to the high concentration of LD in erythrocytes. Liver disorders, such as viral hepatitis and cirrhosis, show slight elevations of two to three times ULN. AMI and pulmonary infarct also show slight elevations of approximately the same degree (2–3 × ULN). In AMI, LD levels begin to rise within 12 hours to 24 hours, reach peak levels within 48 hours to 72 hours, and may remain elevated for 10 days. Skeletal muscle disorders and some leukemias contribute to increased LD levels. Marked elevations can be observed in most patients with acute lymphoblastic leukemia in particular (see *Enzymes in Malignancy* later in this chapter).

Because of the many conditions that contribute to its increased activity, an elevated total LD value is a rather nonspecific finding. LD assays, therefore, assume more clinical significance when separated into isoenzyme fractions. The enzyme can be separated into five major fractions, each com-

TABLE 9-5. Lactate Dehydrogenase Isoenzymes—Tissue Localization and Sources of Elevation

Isoenzyme	Tissue	Disorder
LD-1 (HHHH) and LD-2 (HHHM)	Heart Red blood cells Renal cortex	Myocardial infarct Hemolytic anemia Megaloblastic anemia Acute renal infarct Hemolyzed specimen
LD-3 (HHMM)	Lung Lymphocytes Spleen Pancreas	Pulmonary embolism Extensive pulmonary pneumonia Lymphocytosis Acute pancreatitis Carcinoma
LD-4 (HMMM) and LD-5 (MMMM)	Liver Skeletal muscle	Hepatic injury or inflammation Skeletal muscle injury

prising four subunits. It has a molecular weight of 128,000 daltons. Each isoenzyme comprises four polypeptide chains with a molecular weight of 32,000 daltons each. Two different polypeptide chains, designated H (heart) and M (muscle), combine in five arrangements to yield the five major isoenzyme fractions.

Table 9-5 indicates the tissue localization of the LD isoenzymes and the major disorders associated with elevated levels. LD-1 migrates most quickly toward the anode, followed in sequence by the other fractions, with LD-5 migrating the slowest.

In the sera of healthy individuals, the major isoenzyme fraction is LD-2, followed by LD-1, LD-3, LD-4, and LD-5 (see Table 9-6 for the isoenzyme ranges). LD-1 and LD-2 are present to approximately the same extent in the tissues listed in Table 9-5. However, cardiac tissue and red blood cells contain a higher concentration of LD-1. Therefore, in conditions involving cardiac necrosis (AMI) and intravascular hemolysis, the serum levels of LD-1 will increase to a point at which they are present in greater concentration than is LD-2, resulting in a condition known as the *LD flipped pattern* (LD-1 > LD-2).[18] This flipped pattern is of greatest clinical significance when used with CK isoenzyme analysis in the diagnosis of AMI. LD-1/LD-2 ratios greater than 1 also may be observed in hemolyzed serum samples.[19] Elevations

of LD-3 occur most frequently with pulmonary involvement and are also observed in patients with various carcinomas. The LD-4 and LD-5 isoenzymes are found primarily in liver and skeletal muscle tissue, with LD-5 being the predominant fraction in these tissues. LD-5 levels have greatest clinical significance in the detection of hepatic disorders, particularly intrahepatic disorders. Disorders of skeletal muscle will reveal elevated LD-5 levels, as depicted in the muscular dystrophies.

A sixth LD isoenzyme has been identified, which migrates cathodic to LD-5.[20–22] LD-6 is neither an artifact nor an immunoglobulin complex but a distinct isoenzyme fraction. In the studies reporting its existence, it has been present in patients with arteriosclerotic cardiovascular failure. It is believed that its appearance signifies a grave prognosis and impending death. LD-5 is elevated concurrently with the appearance of LD-6, probably representing hepatic congestion due to cardiovascular disease. It is suggested, therefore, that LD-6 may reflect liver injury secondary to severe circulatory insufficiency.

LD has been shown to complex with immunoglobulins and to reveal atypical bands on electrophoresis. LD complexed with IgA and IgG usually migrates between LD-3 and LD-4. This macromolecular complex is not associated with any specific clinical abnormality.

Analysis of LD isoenzymes can be accomplished by electrophoretic or immunoinhibition methods or by differences in substrate affinity. The electrophoretic procedure is the most widely used method. After electrophoretic separation, the isoenzymes can be detected either fluorometrically or colorimetrically. LD can use other substrates in addition to lactate, such as α-hydroxybutyrate. The H subunits have a greater affinity for α-hydroxybutyrate than to the M subunits. This has led to the use of this substrate in an attempt to measure the LD-1 activity, which consists entirely of H subunits. The chemical assay, known as the measurement of α-hydroxybutyrate dehydrogenase activity (α-HBD), is outlined in Equation 9-10.

TABLE 9-6. LD Isoenzymes as a Percentage of Total LD[19]

Isoenzyme	%
LD-1	14–26
LD-2	29–39
LD-3	20–26
LD-4	8–16
LD-5	6–16

$$\begin{array}{ccc}
CH_3 & & CH_3 \\
| & & | \\
CH_2 + NADH + H^+ \overset{\alpha\text{-HBD}}{\rightleftharpoons} & & CH_2 + NAD^+ \\
| & & | \\
HC{=}O & & HC{-}OH \\
| & & | \\
COOH & & COOH \\
\alpha\text{-ketobutyrate} & & \alpha\text{-hydroxybutyrate}
\end{array}$$

(Eq. 9–10)

α-HBD is not a separate and distinct enzyme but is considered to represent the LD-1 activity of total LD. However, α-HBD activity is not entirely specific for the LD-1 fraction, because LD-2, LD-3, and LD-4 also contain varying amounts of the H subunit. HBD activity is increased in those conditions in which the LD-1 and LD-2 fractions are increased. In AMI, α-HBD activity is very similar to the time course of total LD activity and can be useful in the diagnosis when other LD isoenzyme procedures are not readily available.

Immunoinhibition methods that selectively measure LD-1 activity have been described for LD isoenzymes. One such method involves the use of antibodies that react with all M subunit-containing isoenzymes. LD-1 activity can then be quantitated following removal of the M subunit-inhibited fractions.[23] Several reagent manufacturers are releasing similar procedures to quantitate LD-1 levels quickly.

One recent study has supported eliminating the use of LD and its isoenzymes when evaluating patients in the emergency room to rule out AMI. The researchers found that the use of these two enzymes rarely gives additional information to support the diagnosis of AMI.[24] Eliminating the LD and LD-isoenzyme assay from a cardiac panel saves time and money. It is best used in evaluating patients who delay coming for treatment for chest pain for several days.

Assay

LD catalyzes the interconversion of lactic and pyruvic acids using the coenzyme NAD^+. The reaction sequence is outlined in Equation 9-11:

$$\text{Lactate} + NAD^+ \overset{LD}{\rightleftharpoons} \text{pyruvate} + NADH + H^+$$

(Eq. 9–11)

The reaction can proceed in either the forward, lactate (L), or reverse, pyruvate (P), direction. Both reactions have been used in clinical assays. The rate of the reverse reaction is approximately three times faster, allowing smaller sample volumes and shorter reaction times. However, the reverse reaction is more susceptible to substrate exhaustion and loss of linearity. The optimal pH for the forward reaction is 8.3 to 8.9, whereas that for the reverse reaction is 7.1 to 7.4.

Sources of Error

Erythrocytes contain an LD concentration approximately 100 to 150 times that found in serum. Therefore, any degree of hemolysis should render a sample unacceptable for analy-

TABLE 9-7. Reference Ranges for LD	
Forward (L → P)	100–225 U/L (37°C)
Reverse (P → L)	80–280 U/L (37°C)

sis. LD activity is unstable in serum regardless of the temperature at which it is stored. If the sample cannot be analyzed immediately, it should be stored at 25°C and analyzed within 48 hours. LD-5 is the most labile isoenzyme. Loss of activity occurs more quickly at 4°C than at 25°C. Serum samples for LD isoenzyme analysis should be stored at 25°C and analyzed within 24 hours of collection.

Reference Range
See Table 9-7.

Aspartate Aminotransferase

Aspartate aminotransferase (AST) is an enzyme belonging to the class of *transferases*. It is commonly referred to as a *transaminase* and is involved in the transfer of an amino group between aspartate and α-keto acids. The older terminology, *glutamic-oxaloacetic transaminase (GOT),* is also still used. Pyridoxal phosphate functions as a coenzyme. The reaction proceeds according to equation 9-12:

$$\begin{array}{ccccccc}
COOH & COOH & & COOH & & COOH \\
| & | & & | & & | \\
CH_2 & + \; CH_2 & \overset{AST}{\rightleftharpoons} & CH_2 & + & CH_2 \\
| & | & & | & & | \\
HC{-}NH_2 & CH_2 & & C{=}O & & CH_2 \\
| & | & & | & & | \\
COOH & C{=}O & & COOH & & HC{-}NH_2 \\
& | & & & & | \\
& COOH & & & & COOH \\
\text{Aspartate} & \alpha\text{-keto-} & & \text{Oxalo-} & & \text{Glutamate} \\
& \text{glutarate} & & \text{acetate} & &
\end{array}$$

(Eq. 9–12)

The transamination reaction is important in intermediary metabolism because of its function in the synthesis and degradation of amino acids. The ketoacids formed by the reaction are ultimately oxidized by the tricarboxylic acid cycle to provide a source of energy.

Tissue Sources

AST is widely distributed in human tissues. The highest concentrations are found in cardiac tissue, liver, and skeletal muscle, with smaller amounts found in the kidney, pancreas, and erythrocytes.

Diagnostic Significance

The clinical use of AST is confined mainly to the evaluation of myocardial infarct, hepatocellular disorders, and skeletal muscle involvement. In AMI, AST levels begin to rise within 6 hours to 8 hours, peak at 24 hours, and generally return to normal within 5 days. However, due to the wide tissue distribution, AST levels in the diagnosis of AMI are not as useful as CK and LD isoenzyme analysis.

AST elevations are frequently seen in pulmonary embolism. Following congestive heart failure, AST levels also may be increased, probably reflecting liver involvement due to inadequate blood supply to that organ. AST levels are highest in acute hepatocellular disorders. In viral hepatitis, levels may reach 100 times ULN. In cirrhosis, only moderate levels—approximately four times ULN—are detected (see Chapter 17, *Liver Function*). Skeletal muscle disorders, such as the muscular dystrophies, and inflammatory conditions also cause increases in AST levels ($4\text{–}8 \times$ ULN).

AST exists as two isoenzyme fractions located in the cell cytoplasm and mitochondria. The cytoplasmic isoenzyme is the predominant form occurring in serum. In disorders producing cellular necrosis, such as liver cirrhosis, the mitochondrial form may be markedly increased. Isoenzyme analysis of AST is not routinely performed in the clinical laboratory.

Assay

Assay methods for AST are generally based on the principle of the Karmen method, which incorporates a coupled enzymatic reaction using malate dehydrogenase (MD) as the indicator reaction and monitors the change in absorbance at 340 nm continuously as NADH is oxidized to NAD^+ (see Eq. 9-13). The pH optimum is 7.3 to 7.8.

$$\text{Aspartate} + \alpha\text{-ketoglutarate} \underset{}{\overset{\text{AST}}{\rightleftharpoons}}$$
$$\text{oxaloacetate} + \text{glutamate}$$

$$\text{Oxaloacetate} + \text{NADH} + \text{H}^+ \overset{\text{MD}}{\rightleftharpoons} \text{malate} + \text{NAD}^+$$

(Eq. 9–13)

Sources of Error

Hemolysis should be avoided because it can increase AST activity as much as tenfold. AST activity is stable in serum for 3 days to 4 days at refrigerated temperatures.

Reference Range

AST, 5–U/L (37°C).

Alanine Aminotransferase

Alanine aminotransferase (ALT) is a transferase with enzymatic activity similar to that of AST. Specifically, it catalyzes the transfer of an amino group from alanine to α-ketoglutarate with the formation of glutamate and pyruvate. The older terminology still in use is *glutamic-pyruvic transaminase (GPT)*. Equation 9-14 indicates the transferase reaction. Pyridoxal phosphate acts as the coenzyme.

(Eq. 9–14)

Tissue Sources

ALT is distributed in many tissues and is found in comparatively high concentrations in the liver. It is considered to be the more liver-specific enzyme of the transferases.

Diagnostic Significance

Clinical applications of ALT assays are confined mainly to evaluation of hepatic disorders. Higher elevations are found in hepatocellular disorders than in extra- or intrahepatic obstructive disorders. In acute inflammatory conditions of the liver, ALT elevations will frequently be higher than those of AST and will tend to remain elevated longer due to the longer half-life of ALT in serum.

Cardiac tissue contains a small amount of ALT activity, but the serum level usually remains normal in AMI unless subsequent liver damage has occurred. Assay of ALT levels will often be requested along with that of AST to help determine the source of an elevated AST level and to detect liver involvement concurrent with myocardial injury.

Assay

The typical assay procedure for ALT consists of a coupled enzymatic reaction using LD as the indicator enzyme, which catalyzes the reduction of pyruvate to lactate with the simultaneous oxidation of NADH. The change in absorbance at 340 nm measured continuously is directly proportional to ALT activity. The reaction proceeds according to Equation 9-15. The pH optimum is 7.3 to 7.8.

$$\text{Alanine} + \alpha\text{-ketoglutarate} \overset{\text{ALT}}{\rightleftharpoons} \text{pyruvate} + \text{glutamate}$$

$$\text{Pyruvate} + \text{NADH} + \text{H}^+ \overset{\text{LD}}{\rightleftharpoons} \text{lactate} + \text{NAD}^+$$

(Eq. 9–15)

Sources of Error

ALT is stable for 3 days to 4 days at 4°C. It is relatively unaffected by hemolysis.

Reference Range

ALT, 6–37 U/L (37°C).

Alkaline Phosphatase

Alkaline phosphatase (ALP) belongs to a group of enzymes that catalyze the hydrolysis of a wide variety of phosphomonoesters at an alkaline pH. Consequently, ALP is a nonspecific enzyme capable of reacting with many different substrates. Specifically, ALP functions to liberate inorganic phosphate from an organic phosphate ester with the concomitant production of an alcohol. The reaction proceeds according to Equation 9-16:

$$R-\overset{\overset{\textstyle O}{\|}}{\underset{\underset{\textstyle O^-}{|}}{P}}-O^- + H_2O \underset{\text{pH 9–10}}{\overset{\text{ALP}}{\rightleftharpoons}} R-OH + HO-\overset{\overset{\textstyle O}{\|}}{\underset{\underset{\textstyle O^-}{|}}{P}}-O^-$$

Phosphomonoester   Alcohol   Phosphate ion

(Eq. 9–16)

The optimal pH for the reaction is 9.0 to 10.0, but the optimum pH varies with the substrate used. The enzyme requires Mg^+ as an activator.

Tissue Sources

ALP activity is present in most human tissues. The highest concentrations are found in the intestines, liver, bone, spleen, placenta, and kidney. In the liver, the enzyme is located on both sinusoidal and bile canalicular membranes, whereas in bone, activity is confined to the osteoblasts, those cells involved in production of bone matrix. The specific location of the enzyme within these tissues accounts for the more predominant elevations in certain disorders.

Diagnostic Significance

Elevations of ALP are of most diagnostic significance in the evaluation of hepatobiliary and bone disorders. In hepatobiliary disorders, elevations are seen more predominantly in obstructive conditions than in hepatocellular disorders. In bone disorders, elevations are seen when there is involvement of the osteoblasts.

In biliary tract obstruction, ALP levels range from three to ten times ULN. Increases are due primarily to increased synthesis of the enzyme induced by cholestasis. In contrast, hepatocellular disorders such as hepatitis and cirrhosis show only slight increases, usually less than three times ULN. Owing to the degree of overlap of ALP elevations that occurs in the various liver disorders, a single elevated ALP level is difficult to interpret. It assumes more diagnostic significance when evaluated along with other tests of hepatic function (see Chapter 17, *Liver Function*).

Elevated ALP levels may be seen in a variety of bone disorders. Perhaps the highest elevations of ALP activity occur in Paget's disease (osteitis deformans). Other bone disorders include osteomalacia, rickets, hyperparathyroidism, and osteogenic sarcoma. In addition, increased levels are seen in the case of healing bone fractures and during periods of physiologic bone growth (see *Enzymes in Bone Disorders* later in this chapter).

In normal pregnancy, increased ALP activity, averaging approximately one and one-half times ULN, can be detected between the weeks 16 and 20 of pregnancy. ALP activity increases and persists until the onset of labor. Activity then returns to normal within 3 days to 6 days.[25] Elevations also may be seen in complications of pregnancy such as hypertension, preeclampsia, and eclampsia, as well as in threatened abortion.

ALP levels are significantly decreased in the inherited condition of hypophosphatasia. Subnormal activity is due to the absence of the bone isoenzyme and results in inadequate bone calcification.

ALP exists as a number of isoenzymes, which have been studied by a variety of techniques. The major isoenzymes that are found in the serum and that have been most extensively studied are those derived from the liver, bone, intestine, and placenta.[26]

Electrophoresis is considered the most useful single technique for ALP isoenzyme analysis. However, because there may still be some degree of overlap between the fractions, electrophoresis in combination with another separation technique may provide the most reliable information. A direct

---

CASE STUDY 9-2

An 11-year-old boy with an increased appetite complains to his mother of muscle pain in his legs. The boy is starting to show signs of early puberty such as facial hair, voice deepening, and darkening of the hair on his legs and arms. The boy is taken to the pediatrician who finds little wrong with the boy.

The pediatrician orders the following blood work:

CK	60 U/L	(30–160)
Alkaline phosphatase	120 U/L	(30–90)
Glucose	85 mg/dL	(65–110)
Amylase	75 U/L	(60–180)

Questions

1. Based on the laboratory results, what is the likely diagnosis?
2. Why is the alkaline phosphatase elevated?
3. Should any further tests be run at this time?

immunochemical method for the measurement of bone-related ALP is now available; this has made ALP electrophoresis unnecessary in most cases.

The liver fraction migrates the fastest, followed by bone, placental, and intestinal fractions. Because of the similarity between the liver and bone phosphatases, there often is not a clear separation between them. Quantitation by a densitometer is sometimes difficult because of the overlap between the two peaks. The liver isoenzyme can actually be divided into two fractions, the major liver band and a smaller fraction called *fast liver,* or α₁ liver, which migrates anodal to the major band and corresponds to the α₁ fraction of protein electrophoresis. When total ALP levels are increased, it is the major liver fraction that is most frequently elevated. Many hepatobiliary conditions cause elevations of this fraction, usually early in the course of the disease. The fast-liver fraction has been reported in metastatic carcinoma of the liver, as well as in other hepatobiliary diseases. Its presence is regarded as a valuable indicator of obstructive liver disease.[23] However, it is even present occasionally in the absence of any detectable disease state.[25]

The bone isoenzyme increases due to osteoblastic activity and is normally elevated in children during periods of growth and in adults older than age 50 years. In these cases, an elevated ALP level may be difficult to interpret.[27]

The presence of intestinal ALP isoenzyme in serum depends on the blood group and secretor status of the individual. Individuals who have B or O blood group and are secretors are more likely to have this fraction. Apparently, intestinal ALP is bound by erythrocytes of group A. Furthermore, in these individuals, increases in intestinal ALP occur after consumption of a fatty meal. Intestinal ALP may increase in several disorders, such as diseases of the digestive tract and cirrhosis. Increased levels are also found in patients undergoing chronic hemodialysis.

Difference in heat stability is the basis of a second approach used to identify the isoenzyme source of an elevated ALP. Typically, ALP activity is measured before and after heating the serum at 56°C for 10 minutes. If the residual activity after heating is less than 20% of the total activity before heating, then the ALP elevation is assumed to be due to bone phosphatase. If greater than 20% of the activity remains, the elevation is probably due to liver phosphatase. These results are based on the finding that placental ALP is the most heat stable of the four major fractions, followed by intestinal, liver, and bone fractions in decreasing order of heat stability. Placental ALP will resist heat denaturation at 65°C for 30 minutes.

Heat inactivation is an imprecise method for differentiation, because inactivation depends on many factors, such as correct temperature control, timing, and analytic methods sensitive enough to detect small amounts of residual ALP activity. In addition, there is some degree of overlap between heat inactivation of liver and bone fractions in both liver and bone diseases.

A third approach to identification of ALP isoenzymes is based on selective chemical inhibition. Phenylalanine is one of several inhibitors that have been used. Phenylalanine inhibits intestinal and placental ALP to a much greater extent than liver and bone ALP. With the use of phenylalanine, however, it is impossible to differentiate placental from intestinal ALP or liver from bone ALP.

In addition to the four major ALP isoenzyme fractions, there exist some abnormal fractions that are associated with neoplasms. The most frequently seen of these are the Regan and the Nagao isoenzymes. They have been referred to as *carcinoplacental alkaline phosphatases* because of their similarities to the placental isoenzyme. The frequency of occurrence ranges from 3% to 15% in cancer patients. The Regan isoenzyme has been characterized as an example of an ectopic production of an enzyme by malignant tissue. It has been detected in various carcinomas, such as lung, breast, ovarian, and colon, with highest incidences in ovarian and gynecologic cancers. Because of its low incidence in cancer patients, diagnosis of malignancy is rarely based on its presence. It is, however, useful in monitoring the effects of therapy because it will disappear on successful treatment.

The Regan isoenzyme migrates to the same position as does the bone fraction and is the most heat stable of all ALP isoenzymes, resisting denaturation at 65°C for 30 minutes. Its activity is inhibited by phenylalanine.

The Nagao isoenzyme may be considered a variant of the Regan isoenzyme. Its electrophoretic, heat-stability, and phenylalanine-inhibition properties are identical to those of the Regan fraction. However, Nagao also can be inhibited by L-leucine. Its presence has been detected in metastatic carcinoma of pleural surfaces and in adenocarcinoma of the pancreas and bile duct.

Assay

Because of the relative nonspecificity of ALP with regard to substrates, a variety of methodologies for its analysis have been proposed and are still in use today. The major differences between these relate to the concentration and types of substrate and buffer used and the pH of the reaction. A continuous-monitoring technique based on a method devised by Bowers and McComb allows calculation of ALP activity based on the molar absorptivity of *p*-nitrophenol.[20] The reaction proceeds according to Equation 9–17:

p-nitrophenyl-phosphate → *p*-nitro-phenol + Phosphate ion

(Eq. 9–17)

p-nitrophenylphosphate (colorless) is hydrolyzed to *p*-nitrophenol (yellow), and the increase in absorbance at 405 nm, which is directly proportional to ALP activity, is measured.

Sources of Error

Hemolysis may cause slight elevations, because ALP is approximately six times more concentrated in erythrocytes than in serum. ALP assays should be run as soon as possible after collection. Activity in serum increases approximately 3% to 10% on standing at 25°C or 4°C for several hours. Diet may induce elevations in ALP activity of blood group B and O individuals who are secretors. Values may be 25% higher following ingestion of a high-fat meal.

Reference Range

ALP, 30–90 U/L (30°C).

Acid Phosphatase

Acid phosphatase (ACP) belongs to the same group of phosphatase enzymes as ALP and is a *hydrolase* that catalyzes the same type of reactions. The major difference between ACP and ALP is the pH of the reaction. ACP functions at an optimal pH of approximately 5.0. Equation 9-18 outlines the reaction sequence:

$$R\overset{\overset{\displaystyle O}{\|}}{\underset{\underset{\displaystyle O^-}{|}}{P}}\!\!-\!\!O^- + H_2O \underset{pH\ 5}{\overset{ACP}{\rightleftharpoons}} R\!\!-\!\!OH + HO\overset{\overset{\displaystyle O}{\|}}{\underset{\underset{\displaystyle O^-}{|}}{P}}\!\!-\!\!O^-$$

Phosphomonoester   Alcohol   Phosphate ion

(Eq. 9–18)

Tissue Sources

ACP activity is found in the prostate, bone, liver, spleen, kidney, erythrocytes, and platelets. The prostate is the richest source, with many times the activity found in other tissues.

Diagnostic Significance

The major diagnostic significance of ACP measurement is its use as an aid in the detection of prostatic carcinoma, particularly metastatic carcinoma of the prostate. Prostatic cancer is the third most prevalent fatal male cancer in the United States. The incidence is more common in men older than age 50 years and increases with age.

Prostatic carcinoma is commonly classified into four stages. In stages A and B, the malignant tumor is confined to the prostate. Stage C tumors have spread beyond the prostate but are confined to the pelvic area, whereas stage D tumors have metastasized to various regions of the body. Total ACP determinations are relatively insensitive techniques and can detect elevated ACP levels due to prostatic carcinoma in the majority of cases only when the tumor has metastasized.

Because of the importance of detecting prostatic carcinoma before it has metastasized, when it may still be curable, attempts have been made to increase both the sensitivity and specificity of ACP measurements, with an emphasis on detection of the prostatic contribution only. Assay methods for this purpose include the use of differential substrates, specific chemical inhibitors, and immunologic techniques. One of the most specific substrates for the prostatic portion is thymolphthalein monophosphate. Other tissue sources of ACP can react, but to a much lesser degree, with this substrate.

Chemical-inhibition methods used to differentiate the prostatic portion most frequently use tartrate as the inhibitor. The prostatic fraction is inhibited by tartrate. Serum and substrate are incubated both with and without the addition of L-tartrate. ACP activity remaining after inhibition with L-tartrate is subtracted from total ACP activity determined without inhibition, and the difference represents the prostatic portion:

Total ACP − ACP after tartrate inhibition

$$= \text{prostatic ACP} \qquad \textit{(Eq. 9–19)}$$

The reaction is not entirely specific for prostatic ACP, but other tissue sources are largely uninhibited.

Neither of these methods of ACP determinations is sensitive to prostatic carcinoma that has not metastasized. Values are usually normal in the majority of cases and, in fact, may be elevated only in as many as 50% of cases of prostatic carcinoma that has metastasized.

A technique with much improved sensitivity over conventional ACP assays is the immunologic approach using antibodies that are specific for the prostatic portion. Immunochemical techniques, however, are not of value as screening tests for prostatic carcinoma.

Because of improved sensitivity and specificity of immunochemical assays, these methods are also of value in monitoring the course of the disease following therapy. After successful surgical removal or radiotherapy, the elevated enzyme levels usually return to the normal range.

In addition to using prostatic ACP to monitor prostatic carcinoma, there is an increasing trend to monitor prostatic-specific antigen (PSA; see Chapter 27, *Tumor Markers*). PSA is more likely than ACP to be elevated at each stage of prostatic carcinoma, even though a normal PSA level may be found in stage D tumors. PSA is particularly useful to monitor the success of treatment. PSA is controversial as a screening test for prostatic malignancy, because PSA elevation may occur in conditions other than prostatic carcinoma.[28–30]

Other prostatic conditions in which elevations of ACP activity have been reported include hyperplasia of the prostate and prostatic surgery. There are conflicting reports of elevations following rectal examination and prostate massage. Some studies have reported ACP elevations, whereas others have indicated no detectable change. When elevations are found, levels usually return to normal within 24 hours.[31]

ACP assays have proven useful in forensic clinical chemistry, particularly in the investigation of rape. Vaginal washings are examined for seminal fluid–ACP activity, which can persist for up to 4 days.[32] Elevated activity is presumptive evidence of rape in such cases.

Serum ACP activity may frequently be elevated in bone diseases. Activity has been shown to be associated with the osteoclasts.[33] Elevations have been noted in Paget's disease, breast cancer with bone metastases, and Gaucher's disease, in which there is an infiltration of bone marrow and other tissues by Gaucher cells rich in ACP activity. Because of ACP activity in platelets, elevations are seen when damage to platelets occurs, as in thrombocytopenia due to excessive platelet destruction.

Assay

Assay procedures for total ACP use the same techniques as in ALP assays but are performed at an acid pH (see Eq. 9-20):

$$p\text{-nitrophenylphosphate} \underset{\text{pH 5}}{\overset{\text{ACP}}{\rightleftharpoons}}$$

$$p\text{-nitrophenol} + \text{phosphate ion} \qquad \textit{(Eq. 9–20)}$$

The reaction products are colorless at the acid pH of the reaction, but the addition of alkali stops the reaction and transforms the products into chromogens, which can be measured spectrophotometrically.

Some substrate specificities and chemical inhibitors for prostatic ACP measurements have been discussed previously. Thymolphthalein monophosphate is the substrate of choice for quantitative endpoint reactions. For continuous monitoring methods, α-naphthyl phosphate is preferred.

Immunochemical techniques for prostatic ACP use several approaches, including RIA, counter immunoelectrophoresis, and immunoprecipitation. Also, an immunoenzymatic assay (Tandem E) includes incubation with an antibody to prostatic ACP followed by washing and incubation with p-NPP. The p-NP formed, measured photometrically, is proportional to the prostatic ACP in the sample.

Sources of Error

Serum should be separated from the red cells as soon as the blood has clotted to prevent leakage of erythrocyte and platelet ACP. Serum activity decreases within 1 hour to 2 hours if the sample is left at room temperature without the addition of a preservative. Decreased activity is due to a loss of carbon dioxide from the serum, with a resultant increase in pH. If not assayed immediately, serum should be frozen or acidified to a pH below 6.5. With acidification, ACP is stable for 2 days at room temperature. Hemolysis should be avoided because of contamination from erythrocyte ACP.

RIA procedures for measurement of prostatic ACP require nonacidified serum samples. Activity is stable for 2 days at 4°C.

TABLE 9-8. Reference Ranges for ACP

Total ACP	
Males	2.5–11.7 U/L
Females	0.3–9.2 U/L
Prostatic ACP	
Males	0.2–5.0 U/L
Females	0.0–0.8 U/L

Reference Range

See Table 9-8.

γ-Glutamyltransferase

γ-glutamyltransferase (GGT) is an enzyme involved in the transfer of the γ-glutamyl residue from γ-glutamyl peptides to amino acids, H_2O, and other small peptides. In most biologic systems, glutathione serves as the γ-glutamyl donor. Equation 9-21 outlines the reaction sequence:

$$\text{Glutathione} + \text{amino acid} \overset{\text{GGT}}{\longrightarrow}$$

$$\text{glutamyl} - \text{peptide} + \text{L-cysteinylglycine}$$

$$\textit{(Eq. 9–21)}$$

The specific physiologic function of GGT has not been clearly established, but it has been suggested that GGT is involved in peptide and protein synthesis, regulation of tissue glutathione levels, and the transport of amino acids across cell membranes.[34]

Tissue Sources

GGT activity is found primarily in tissues of the kidney, brain, prostate, pancreas, and liver. Clinical applications of assay, however, are confined mainly to evaluation of liver and biliary system disorders.

Diagnostic Significance

In the liver, GGT is located in the canaliculi of the hepatic cells and particularly in the epithelial cells lining the biliary ductules. Because of these locations, GGT is elevated in virtually all hepatobiliary disorders, making it one of the most sensitive of enzyme assays in these conditions (see Chapter 17, *Liver Function*). Higher elevations are generally observed in biliary tract obstruction.

Within the hepatic parenchyma, GGT exists to a large extent in the smooth endoplasmic reticulum and is, therefore, subject to hepatic microsomal induction. Thus, GGT levels will be increased in patients receiving enzyme-inducing drugs such as warfarin, phenobarbital, and phenytoin. Enzyme elevations may reach levels four times ULN.

Because of the effects of alcohol on GGT activity, GGT assays are considered sensitive indicators of alcoholism, particularly occult alcoholism, when other indications are lacking. Generally, enzyme elevations in alcoholics and heavy drinkers range from two to three times ULN, although higher levels have been observed. GGT assays are also useful in monitoring the effects of abstention from alcohol and are used as such by alcohol treatment centers. Levels usually return to normal within 2 weeks to 3 weeks after cessation but can rise again if alcohol consumption is resumed. In view of the susceptibility to enzyme induction, any interpretation of GGT levels must be done with consideration of the consequent effects of drugs and alcohol.

GGT levels are elevated in other conditions such as acute pancreatitis, diabetes mellitus, and myocardial infarction. The source of elevation in pancreatitis and diabetes is probably the pancreas, but the source of GGT in myocardial infarction is unknown. GGT assays are of limited value in the diagnosis of these conditions and are not routinely requested.

GGT activity is useful in differentiating the source of an elevated ALP level, because GGT levels are normal in skeletal disorders and during pregnancy. It is particularly useful in evaluating hepatobiliary involvement in adolescents, because ALP activity will invariably be elevated owing to bone growth.

Assay

The substrate most widely accepted for use in the analysis of GGT is γ-glutamyl-*p*-nitroanilide. The γ-glutamyl residue is transferred to glycylglycine, releasing *p*-nitroaniline, a chromogenic product with a strong absorbance at 405 nm to 420 nm. The reaction, which can be used as a continuous-monitoring or a fixed-point method, is outlined in Equation 9-22:

TABLE 9-9. Reference Ranges for GGT

Male	6–45 U/L
Female	5–30 U/L

Sources of Error

GGT activity is very stable, with no loss of activity for 1 week at 4°C. Hemolysis does not interfere with GGT levels, because the enzyme is lacking in erythrocytes.

Reference Range

See Table 9-9. Values are lower in females, presumably because of suppression of enzyme activity due to estrogenic or progestational hormones.

Amylase

Amylase (AMS) is an enzyme belonging to the class of hydrolases. It catalyzes the breakdown of starch and glycogen. Starch consists of both amylose and amylopectin. Amylose is a long, unbranched chain of glucose molecules, linked by α, 1–4 glycosidic bonds, whereas amylopectin is a branched-chain polysaccharide with α, 1–6 linkages at the branch points. The structure of glycogen is similar to that of amylopectin but is more highly branched. α-AMS attacks only the α, 1–4 glycosidic bonds to produce degradation products consisting of glucose, maltose, and intermediate chains, called *dextrins,* which contain α, 1–6 branching linkages. Cellulose and other structural polysaccharides consisting of linkages are not attacked by α-AMS. AMS is thus an important enzyme in the physiologic digestion of starches. The reaction proceeds according to Equation 9-23:

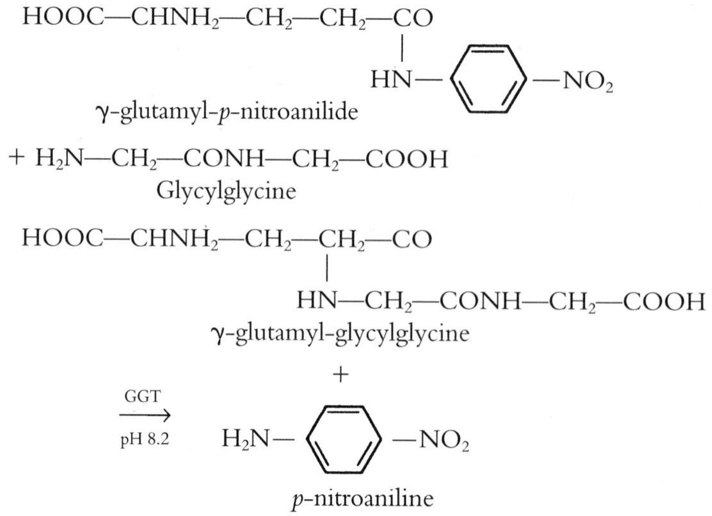

HOOC—CHNH$_2$—CH$_2$—CH$_2$—CO
|
HN—⟨ ⟩—NO$_2$

γ-glutamyl-*p*-nitroanilide

+ H$_2$N—CH$_2$—CONH—CH$_2$—COOH
Glycylglycine

HOOC—CHNH$_2$—CH$_2$—CH$_2$—CO
|
HN—CH$_2$—CONH—CH$_2$—COOH
γ-glutamyl-glycylglycine

+

$\xrightarrow[\text{pH 8.2}]{\text{GGT}}$ H$_2$N—⟨ ⟩—NO$_2$

p-nitroaniline

(Eq. 9–22)

(Eq. 9–23)

AMS requires calcium and chloride ions for its activation.

Tissue Sources

The acinar cells of the pancreas and the salivary glands are the major tissue sources of serum AMS. Lesser concentrations are found in skeletal muscle, small intestine, and fallopian tubes. AMS is the smallest in size of the enzymes, with a molecular weight of 50,000 to 55,000. Because of its small size, it is readily filtered by the renal glomerulus and also appears in the urine.

Digestion of starches begins in the mouth with the hydrolytic action of salivary AMS. Salivary AMS activity, however, is of short duration because, on swallowing, it is inactivated by the acidity of the gastric contents. Pancreatic AMS then performs the major digestive action of starches once the polysaccharides reach the intestine.

Diagnostic Significance

The diagnostic significance of serum and urine AMS measurements is in the diagnosis of acute pancreatitis.[35] Disorders of tissues other than the pancreas can also produce elevations in AMS levels. Thus, an elevated AMS level is a nonspecific finding. However, the degree of elevation of AMS is helpful, to some extent, in the differential diagnosis of acute pancreatitis. In addition, other laboratory tests, such as measurements of urinary AMS levels, AMS clearance studies, AMS isoenzyme studies, and measurements of serum lipase (LPS) levels, when used in conjunction with serum AMS measurement, increase the specificity of AMS measurements in the diagnosis of acute pancreatitis (see *Enzymes in Acute Pancreatitis* later in this chapter).

In acute pancreatitis, serum AMS levels begin to rise 2 hours to 12 hours after onset of an attack, peak at 24 hours, and return to normal levels within 3 days to 5 days. Values generally range from 250 Somogyi units/dL to 1000 Somogyi units/dL (2.55 × ULN). Values can reach much higher levels.

Other disorders causing an elevated serum AMS level include salivary gland lesions such as mumps and parotitis and other intraabdominal diseases such as perforated peptic ulcer, intestinal obstruction, cholecystitis, ruptured ectopic pregnancy, mesenteric infarction, and acute appendicitis. In addition, elevations have been reported in renal insufficiency and diabetic ketoacidosis. Serum AMS levels in intraabdominal conditions other than acute pancreatitis are usually less than 500 Somogyi units/dL.

An apparently asymptomatic condition of hyperamylasemia has been noted in approximately 1% to 2% of the population. This condition, called *macroamylasemia,* results when the AMS molecule combines with immunoglobulins to form a complex that is too large to be filtered across the glomerulus. Serum AMS levels increase because of the reduction in normal renal clearance of the enzyme and, consequently, the urinary excretion of AMS is abnormally low. The diagnostic significance of macroamylasemia lies in the need to differentiate it from other causes of hyperamylasemia.

Much interest has been focused recently on the possible diagnostic use of AMS isoenzyme measurements.[35,36] Serum AMS is a mixture of a number of isoenzymes that can be separated by differences in physical properties, most notably electrophoresis, although chromatography and isoelectric focusing also have been applied. In normal human serum, two major bands and as many as four minor bands may be seen. The bands are designated as P-type and S-type isoamylase. P isoamylases are derived from pancreatic tissue, whereas S isoamylase is derived from salivary gland tissue, as well as the fallopian tube and lung. The isoenzymes of salivary origin migrate most quickly (S1, S2, S3), whereas those of pancreatic origin are the slower ones (P1, P2, P3). In normal human serum, the isoamylases migrate in regions corresponding to the β- to γ-globulin regions of protein electrophoresis. The most commonly observed fractions are P2, S1, and S2. P1 is rare, occurring in less than 2% of patients studied.

In acute pancreatitis, there is typically an increase in P-type activity, with P3 being the most predominant isoenzyme. However, P3 also has been detected in cases of renal failure and, therefore, is not entirely specific for acute pancreatitis. S-type isoamylase represents approximately two-thirds of AMS activity of normal serum, whereas P-type predominates in normal urine. Currently, AMS isoenzyme analysis is not readily adaptable as a routine laboratory procedure. However, with further refinement, it would appear to hold considerable promise as an aid in the diagnosis of pancreatic disease.

Recent studies have shown a marked increase in serum AMS levels in patients with acquired immunodeficiency syndrome (AIDS). These levels ranged from 282 U/L to 2340 U/L (reference range, 92–278). Demographics of these AIDS patients indicate that African-Americans and Hispanics as well as intravenous drug abusers are more likely to have elevated serum AMS levels.[36]

Assay

AMS can be assayed by a variety of different methods, which are summarized in Table 9-10. The four main approaches are categorized as amyloclast, saccharogenic, chromogenic, and continuous-monitoring.

In the amyloclastic method, AMS is allowed to act on a starch substrate to which iodine has been attached. As AMS hydrolyzes the starch molecule into smaller units, the iodine is released, and a decrease in the initial dark-blue color intensity of the starch-iodine complex occurs. The decrease in color is proportional to the AMS concentration.

The saccharogenic method uses a starch substrate that is hydrolyzed by the action of AMS to its constituent carbohydrate molecules that have reducing properties. The amount

TABLE 9-10. Amylase Methodologies	
Amyloclastic	Measures the disappearance of starch substrate
Saccharogenic	Measures the appearance of the product
Chromogenic	Measures the increasing color from production of product coupled with a chromogenic dye
Continuous-monitoring	Coupling of several enzyme systems to monitor amylase activity

of reducing sugars is then measured where the concentration is proportional to AMS activity. The saccharogenic method is the classic reference method for determining AMS activity and is reported in Somogyi units. Somogyi units are an expression of the number of milligrams of glucose released in 30 minutes at 37°C under specific assay conditions.

Chromogenic methods use a starch substrate to which a chromogenic dye has been attached, forming an insoluble dye-substrate complex. As AMS hydrolyzes the starch substrate, smaller dye-substrate fragments are produced, and these are water soluble. The increase in color intensity of the soluble dye-substrate solution is proportional to AMS activity.

Recently, coupled-enzyme systems have been used to determine AMS activity by a continuous-monitoring technique in which the change in absorbance of NAD^+ at 340 nm is measured. Equation 9-24 is an example of a continuous-monitoring method. The optimal pH for AMS activity is 6.9.

$$Maltopentose \xrightarrow{AMS} maltotriose + maltose$$

$$Maltotriose + maltose \xrightarrow{\alpha\text{-glucosidase}} 5\text{-glucose}$$

$$5\text{-glucose} + 5\ ATP \xrightarrow{Hexokinase}$$

$$5\text{-glucose-6-phosphate} + 5\ ADP$$

$$5\text{-glucose-6-phosphate} + 5\ NAD^+ \xrightarrow{G\text{-}6\text{-}PD}$$

$$5,6\text{-phosphogluconolactone} + 5\ NADH$$

(Eq. 9–24)

Sources of Error

AMS in serum and urine is quite stable. Little loss of activity occurs at room temperature for 1 week or at 4°C for 2 months. Because plasma triglycerides suppress or inhibit serum AMS activity, AMS values may be normal in acute pancreatitis with hyperlipemia.

The administration of morphine and other opiates for the relief of pain before sampling of blood will lead to falsely elevated serum AMS levels. The drugs presumably cause constriction of Oddi's sphincter and pancreatic ducts, with consequent elevation of inarticulate pressure causing regurgitation of AMS into the serum.

Reference Range

Because of the variety of AMS procedures currently in use, activity is expressed according to each procedure. There is no uniform expression of AMS activity, although Somogyi

units are frequently used (Table 9-11). The approximate conversion factor between Somogyi units and international units is 1.85.

Lipase

LPS is an enzyme that hydrolyzes the ester linkages of fats to produce alcohols and fatty acids. Specifically, LPS catalyzes the partial hydrolysis of dietary triglycerides in the intestine to the 2-monoglyceride intermediate, with the production of long-chain fatty acids. The reaction proceeds according to Equation 9-25:

$$
\begin{array}{l}
CH_2-O-C(=O)-R_1 \\
CH-O-C(=O)-R_2 + 2H_2O \underset{LPS}{\rightleftharpoons} \\
CH_2-O-C(=O)-R_3 \\
\text{Triacylglycerol}
\end{array}
\quad
\begin{array}{l}
CH_2OH \\
CH-O-C(=O)-R_2 + 2\ \text{fatty acids} \\
CH_2OH \\
\text{2-monoglyceride}
\end{array}
$$

(Eq. 9–25)

The enzymatic activity of pancreatic LPS is specific for the fatty acid residues at positions 1 and 3 of the triglyceride molecule.

Tissue Sources

LPS concentration is found primarily in the pancreas, although it is also present in the stomach and small intestine.

Diagnostic Significance

Clinical assays of serum LPS measurements are confined almost exclusively to the diagnosis of acute pancreatitis. In this respect, it is similar to AMS measurements but is con-

TABLE 9-11. Reference Ranges for AMS	
Serum	60–180 SU*/dL
	95–290 U/L
Urine	35–400 U/h

Somogyi units.

sidered more specific for pancreatic disorders than is AMS measurement, although somewhat less sensitive. Elevated LPS levels also may be found in other intraabdominal conditions but with less frequency than elevations of serum AMS. Elevations have been reported in cases of penetrating duodenal ulcers and perforated peptic ulcers, intestinal obstruction, and acute cholecystitis. In contrast to AMS levels, LPS levels are normal in conditions of salivary gland involvement. Thus, LPS levels are useful in differentiating a serum AMS elevation due to pancreatic versus salivary involvement.

A new latex test for serum LPS levels has been reported as a possible future test to help evaluate acute abdominal pain. This may also be useful as a near patient testing procedure to be used as a screening test in the emergency department.[37]

Assay

Procedures used to measure LPS activity include estimation of liberated fatty acids and turbidimetric methods. The reaction is outlined in Equation 9-26:

$$\text{Triglyceride} + 2\ H_2O \underset{\text{pH 8.6–9.0}}{\overset{\text{LPS}}{\rightleftharpoons}}$$
$$\text{2-monoglyceride} + 2\ \text{fatty acids} \qquad \textit{(Eq. 9–26)}$$

Early methods were hampered by long incubation times, but numerous modifications have produced current methods with moderately short incubation times. The classic method of Cherry and Crandall used an olive oil substrate and measured the liberated fatty acids by titration after a 24-hour incubation. Modifications of the Cherry-Crandall method have been complicated by the lack of stable and uniform substrates. However, triolein is one substrate that now is used as a more pure form of triglyceride.

Turbidimetric methods are simpler and more rapid than titrimetric assays. Fats in solution create a cloudy emulsion. As the fats are hydrolyzed by LPS, the particles disperse, and the rate of clearing can be measured as an estimation of LPS activity.

Sources of Error

LPS is very stable in serum, with negligible loss in activity at room temperature for 1 week or for 3 weeks at 4°C. Hemolysis should be avoided, because hemoglobin inhibits the activity of serum LPS, causing falsely low values.

Reference Range

LPS, 0–1.0 U/mL.

Glucose-6-Phosphate Dehydrogenase

Glucose-6-phosphate dehydrogenase (G-6-PD) is an *oxidoreductase* that catalyzes the oxidation of glucose-6-phosphate to 6-phosphogluconate or the corresponding lactone. The reaction is important as the first step in the pentose-phosphate shunt of glucose metabolism with the ultimate production of NADPH. The reaction is outlined in Equation 9-27:

Glucose-6-phosphate

6-phosphogluconate

$$\textit{(Eq. 9–27)}$$

Tissue Sources

Sources of G-6-PD include the adrenal cortex, spleen, thymus, lymph nodes, lactating mammary gland, and erythrocytes. Very little activity is found in normal serum.

Diagnostic Significance

Most of the interest of G-6-PD focuses on its role in the erythrocyte. Here, it functions to maintain NADPH in its reduced form. An adequate concentration of NADPH is required to regenerate sulfhydryl-containing proteins, such as glutathione, from the oxidized to the reduced state. Glutathione in the reduced form, in turn, protects hemoglobin from oxidation by agents that may be present in the cell. A deficiency of G-6-PD consequently results in an inadequate supply of NADPH and, ultimately, in the inability to maintain reduced glutathione levels. When erythrocytes are exposed to oxidizing agents, hemolysis occurs because of oxidation of hemoglobin and damage of the cell membrane.

G-6-PD deficiency is an inherited sex-linked trait. The disorder can result in several different clinical manifestations, one of which is drug-induced hemolytic anemia. When exposed to an oxidant drug such as primaquine, an antimalarial drug, affected individuals experience a hemolytic episode. The severity of the hemolysis is related to the drug concentration. G-6-PD deficiency is most common in African-Americans but has been reported in virtually every ethnic group.

Increased levels of G6PD in the serum have been reported in myocardial infarction and megaloblastic anemias. No elevations are seen in hepatic disorders. G-6-PD levels, however, are not routinely performed as diagnostic aids in these conditions.

Assay

The assay procedure for G-6-PD activity is outlined in Equation 9-28:

$$\text{Glucose-6-phosphate} + NADP^+ \underset{}{\overset{\text{G-6-PD}}{\rightleftharpoons}}$$

$$6\text{-phosphogluconate} + NADPH + H^+$$

(Eq. 9–28)

A red cell hemolysate is used to assay for deficiency of the enzyme, whereas serum is used for evaluation of enzyme elevations.

Reference Range

See Table 9-12.

PATHOPHYSIOLOGIC CONDITIONS

The measurement of any *one* serum enzyme activity gives little information to the clinician on the condition of an organ or organ system. Traditionally, chemistry panels including several enzyme tests have been performed to rule out organ malfunction or monitor treatment. Panels may vary from laboratory to laboratory but trends are discussed as follows that are used to diagnose and monitor a variety of disorders. Changes in managed care and payment for chemistry panel testing has affected how many laboratories package their organ panels. More specifically, Medicare and insurance companies have limited what tests can be included on panels. The following is a general discussion of what enzyme tests may be useful in diagnosing different pathophysiologic conditions.

Enzymes in Cardiac Disorders

AMI occurs when there is an abrupt reduction in blood flow to a region of myocardial tissue, usually caused by atherosclerosis of the coronary arteries. Rupture of an atherosclerotic plaque is a common precipitating event in many patients with AMI. Because of the cessation of blood flow, the myocardium becomes ischemic and undergoes necrosis and, ultimately, its cellular contents are released into the circulation, reflecting the myocardial damage.

Increased activity of serum enzymes has been used in diagnostic testing for AMI for many years. Enzymes reach their peak levels of serum activity at different times following a myocardial infarct. Figure 9-6 shows the relationship of these enzymes to the occurrence of the infarct. The total enzyme levels that are used most routinely in the diagnosis of AMI are CK, AST, and LD. Of these, CK elevations are the first to appear, followed by AST and LD. Table 9-13 summarizes the time course of enzyme activity in AMI.

Despite the delayed rise of LD levels relative to the occurrence of the infarct, the main advantage to performing total LD levels is to aid in the detection of those infarctions that have occurred more than 5 days before enzyme analysis. The probability of CK or AST levels remaining elevated beyond 5 days is very small unless reinfarction has occurred.

The degree of enzyme elevation is roughly proportional to the size of the infarct. CK levels greater than eight times the ULN or AST levels greater than five times ULN are associated with large infarcts with an increased risk of mortality. However, lower values do not necessarily guarantee a more favorable prognosis.

Total serum enzyme levels are sensitive indices of AMI, but they are not specific for AMI, because elevations also occur in other cardiac conditions and following injury to noncardiac tissues (see Table 9-2). The development of assay methods for the isoenzymes of CK and LD, however, has produced diagnostic measures that are considered both highly sensitive and specific for AMI when interpreted according to specific criteria. The following four isoenzyme patterns represent the major criteria used for interpreting CK and LD isoenzyme results that are obtained during the 48-hour period following onset of symptoms.

TABLE 9-12. Reference Ranges for G-6-PD

Serum	0–0.18 U/L
Erythrocytes	0.117–0.143 U/10^9 cells

TABLE 9-13. Time Course of Enzyme Activity in Myocardial Infarction

Enzyme	Onset of Elevation (h)	Peak Activity (h)	Degree of Elevation (× ULN)	Duration of Elevation (days)
CK	4–8	12–24	5–10	3–4
CK-MB	4–8	24–38	5–15	2–3
AST	8–12	24	2–3	5
LD#	12–24	72	2–3	10
LD-1 > LD-2	12–24			5

Pattern 1: CK-MB > 6% of Total CK; LD-1 > LD-2

Pattern 1 is considered to provide the most reliable diagnostic criteria for detecting AMI. Such a finding confirms that AMI has occurred. Typically, the CK-MB band begins rising 4 hours to 8 hours after infarct and reaches peak levels 12 hours to 38 hours after infarct. Peak values range from 11% to 31% of total CK. CK-MB levels usually return to normal within 2 days to 3 days. The LD flipped pattern occurs 12 hours to 24 hours after infarct. It can persist for 5 days, being present even after CK-MB has returned to normal. This isoenzyme pattern is found in 80% to 85% of patients with AMI.

Pattern 2: CK-MB > 6% of Total CK; LD-2 > LD-1

The presence of CK-MB in quantities greater than 6% of total CK with a normal LD isoenzyme pattern reflects some degree of myocardial damage but not necessarily AMI. It does not provide biochemical confirmation of AMI. Approximately 15% to 20% of suspected AMI patients present with this pattern. However, this pattern may also occur in patients who have not sustained myocardial infarction.

Pattern 3: CK-MB < 6% of Total CK

CK-MB values that are less than 6% of total CK indicate that no myocardial infarction has occurred, regardless of the LD isoenzyme data. The lack of elevated CK-MB levels reveals the high degree of sensitivity and specificity of this isoenzyme in ruling out AMI.

Pattern 4: CK-MB < 6% of Total CK; LD-1 > LD-2

Pattern 4 can be found in a variety of noncardiac disorders. It is most frequently seen in intravascular hemolysis, renal cortex infarct, and megaloblastic anemia, and has been reported in hemolyzed serum samples. The lack of significant elevations of CK-MB however, rules out myocardial damage, even though the LD pattern is flipped.

Other conditions resembling AMI can usually be differentiated by the use of combined isoenzyme data. CK-MB has been noted in cases of angina pectoris and coronary insufficiency without evidence of infarction. However, in these cases, LD isoenzyme analysis does not reveal the flipped pattern.

Pulmonary embolism causes clinical symptoms similar to those of AMI. Isoenzyme analysis does not show abnormal CK-MB levels, because lung tissue contains the CK-BB isoenzyme. Total CK levels are usually normal because of the low activity of CK in lung tissue and the instability of CK-BB in blood. In addition, pulmonary embolism does not produce a flipped LD pattern unless the embolism is associated with a hemorrhagic pulmonary infarct. The typical LD isoenzyme pattern in pulmonary embolism is the appearance of elevated LD-2 and LD-3 fractions.

Congestive heart failure produces enzyme abnormalities suggestive of liver involvement due to inadequate blood perfusion. Increased levels of AST, ALT, LD, and LD-5 are seen, rather than the typical cardiac enzyme profile.

Commonly performed diagnostic procedures, such as cardiac catheterization and coronary arteriography, do not usually produce elevated CK-MB levels unless AMI also occurs. However, patients undergoing cardiac surgery, such as coronary bypass surgery, usually have increased CK-MB levels.

It is recommended that blood specimens for enzyme and isoenzyme analysis in the diagnosis of AMI be collected initially and every 12 hours over a 48-hour period until the diagnostic changes are observed. If the CK-MB fraction appears significantly elevated, LD isoenzymes should be determined for at least 48 hours, unless the flipped pattern appears earlier. It is further recommended that, in the event that total CK levels are not elevated, CK isoenzyme analysis be performed whenever the total CK levels fall within the upper one-third of the normal range of CK reference values. Elevated CK-MB levels may occur in the presence of normal total CK levels. If the laboratory is capable of performing these isoenzyme studies, routine assay of AST levels is unnecessary because it provides no further diagnostic information.

α-HBD assays are occasionally requested in the diagnosis of AMI. Because α-HBD represents primarily the LD-1 isoenzyme fraction, it may provide some useful information when other methods of LD isoenzyme evaluation are unavailable. α-HBD follows the same general time course as total LD levels. α-HBD analysis is unnecessary if LD isoenzyme fractionation is available.

Research has recently shown the existence of three subforms of the isoenzymes of CK. These *isoforms* have similar enzymatic activity but different isoelectric points. Those most useful in early identification of patients with AMI are CK-MB^2 and CK-MM^3. These are truly tissue CK isoforms and are released into circulation after tissue necrosis. Measurement of these isoforms and calculation of the MM^3/MM^1 and MB^2/MB^1 ratios may be useful in early diagnosis of myocardial damage. These elevations may be seen in 2 hours to 4 hours after onset of chest pain. This earlier diagnosis may allow the use of thromboembolytic drugs to minimize myocardial damage.[38] More rapid test kits are becoming available to measure these isoforms of CK with low turnaround times.

Troponins are structural proteins found in myocardial tissue. Although not enzymes, the cardiac isoforms of troponin—troponin-T (c-TnT) and troponin-1 (c-Tn1)—have been found to be useful as serum markers for the diagnosis of AMI. Troponin-1 is the most useful for a cardiac muscle injury because it is not elevated in any other condition. Levels will be elevated within 3 hours to 12 hours after AMI with a peak value found 24 hours after onset of chest pain.[39] Troponin-T levels have also been used to identify angina patients that are at increased risk of having an AMI or other cardiac event in the near future. Using both CK-MB and troponin-T can help identify those patients that should be watched closely over the next 6 months.[40] Rapid test kits as well as reagents for common analyzers are available for troponin-T and troponin-1.[41]

Enzymes in Hepatic Disorders

Enzyme assays perform a vital function as laboratory aids in the diagnosis of hepatic disorders. Specific enzyme profiles can be adapted to increase the specificity of enzyme assays in the differential diagnosis of liver disease. Those enzymes of most clinical significance in the diagnosis of hepatocellular disorders include AST, ALT, LD, and LD-5. Generally, in inflammatory hepatocellular disorders, ALT activity is higher than AST activity, whereas in liver cell necrosis AST shows greater elevations.

The enzymes most predominantly elevated in biliary tract obstruction include ALP and GGT. GGT is considered the most sensitive of all enzyme assays for all types of liver disorders, with highest elevations observed in obstructive disorders. Despite this presumptive classification of enzymes, there is a considerable overlap of enzyme elevations in hepatic disorders. However, the degree of elevation, as well as the combined use of enzyme assays and nonenzymatic liver function tests, is helpful, to some extent, in differential diagnosis. Refer to Chapter 17, *Liver Function,* for a more detailed discussion.

Enzymes in Skeletal Muscle Disorders

Serum enzyme levels play an important role in the diagnosis and assessment of primary muscle disorders. The enzymes most commonly assayed are CK, AST, LD, and aldolase. CK is considered the most sensitive enzyme index of primary muscle involvement and is the enzyme of choice in the evaluation of muscle disorders. Muscle conditions in which enzyme elevations are seen include myopathies, muscle trauma, physical exertion, intramuscular injections, malignant hyperthermia, and muscle hypoxia. In contrast to the enzyme elevations seen in these primary muscle disorders, muscle weakness due to neurogenic muscular disorders, such as multiple sclerosis, myasthenia gravis, and poliomyelitis, is characterized by enzyme levels that are normal or only slightly elevated.

Perhaps the highest elevations of CK activity are seen in the muscular dystrophies, particularly the most severe sex-linked form of Duchenne type muscular dystrophy. CK elevations occur early in the course of the disease, usually in the 1st year of life, and reach values 50 to 100 times ULN. CK elevations are attributed to leakage of intracellular enzymes due to genetically induced alterations in cell membrane function. As the disease progresses, enzyme levels decrease, reflecting the inability of diseased muscle fibers to function normally. In the terminal stages of the disease, enzyme activity may be within normal limits. Other forms of the muscular dystrophies, such as limb-girdle and facioscapulohumeral muscular dystrophies, reveal slight to moderate elevations of CK that do not tend to decrease as the disease

progresses. In addition to abnormal CK-MM levels, approximately 90% of Duchenne type muscular dystrophy patients also will have elevated CK-MB levels. These levels are attributed to both cardiac involvement and abnormal muscle composition resembling fetal muscle with CK-MB production.

LD-5 is the predominant LD isoenzyme present in skeletal muscle. However, in the muscular dystrophies, there is an increase in levels of the LD-1 and LD-2 forms. Levels of the other skeletal muscle enzymes are abnormal in more than 90% of affected muscular dystrophy patients but show lesser elevations than does CK.

Polymyositis, considered a connective-tissue disorder characterized by inflammatory and degenerative muscle changes, also reveals elevated enzyme levels. Serum CK and aldolase levels are elevated in approximately two-thirds of patients. Because CK levels in childhood polymyositis approximate the levels observed in Duchenne type muscular dystrophy, it may be difficult to differentiate the two diseases on the basis of enzyme levels. However, patients with polymyositis may respond to steroid therapy with a reduction in enzyme levels, which does not occur in muscular dystrophy.

Muscle trauma such as that occurring from surgery or intramuscular injections will cause serum elevations of muscle enzymes. CK elevations following surgical procedures may be similar to those occurring in AMI. However, the predominant isoenzyme is CK-MM. All muscle enzymes may be elevated following muscular exercise, although CK is most frequently and predominantly elevated.

Malignant hyperthermia is a serious complication of general anesthesia that occurs following the use of halothane and succinylcholine. It occurs usually in genetically susceptible people and is characterized by impaired respiration, muscle spasm, shock, and increased body temperature. CK elevations can rise to extremely high elevations (300 × ULN) in the early postanesthetic period. The CK isoenzyme pattern is abnormal, with the appearance of all three isoenzyme forms. The majority of affected individuals show evidence of subclinical myopathy. The presence of CK enzyme elevations are also used to detect other affected relatives, the majority showing elevations less than two times ULN.

Enzymes in Bone Disorders

The major enzyme that is increased in disorders of the skeletal system is ALP, particularly the bone isoenzyme, which is associated with osteoblastic activity. In addition, ACP is elevated in certain bone disorders.

Paget's disease of bone (osteitis deformans) is one of the more common chronic skeletal diseases. It is characterized by an excessive resorption of bone caused by osteoclastic activity. Bone resorption is followed by an osteoblastic phase in which new bone is formed and deposited in a random fashion, causing an irregular pattern of bone deposition. ALP activity is elevated because of the stimulation of the osteoblasts and can reach extreme elevations ranging from 10 to

100 times ULN. The incidence of occurrence is about 3% in individuals older than age 40 years and increases with advancing age.

Bone disorders resulting from vitamin D deficiency are known as *rickets* in children and *osteomalacia* in adults. The two diseases are not synonymous, however, owing to differences in growing bone in children and calcified bone in adults. Deposition of uncalcified bone matrix occurs in both conditions. Both conditions are characterized by an increase in osteoblastic activity with resultant elevations in serum ALP activity.

Hyperparathyroidism results from excessive secretion of parathyroid hormone. Because of the effects of parathyroid hormone on calcium and phosphorous metabolism, there is usually a consequent abnormality in bone metabolism. However, bone involvement is primarily due to abnormal bone resorption with little osteoblastic activity. Therefore, the majority of patients have normal serum ALP levels, with elevations seen only when there is significant bone involvement, in approximately 30% of cases.

Osteogenic sarcoma is one of the most common of primary bone tumors and occurs most frequently in the 10- to 20-year-old age group. ALP values are variable, but considerably higher values are observed in patients with osteoblastic lesions than in those with lytic lesions. Patients whose bone disorder is primarily confined to osteoclastic activity, such as those with multiple myeloma, usually have normal ALP values.

During the course of healing of bone fractures, ALP elevations of approximately two times ULN may be seen. Values return to normal within 2 months.

ALP levels can be expected to be increased during periods of physiologic bone growth. In children younger than age 10 years, the average ALP level is almost four times that of adult activity. Peak activity occurs in girls between the ages of 11 and 12 years and in boys, between the ages of 13 and 14 years. Elevations are more pronounced in boys than in girls and can reach levels 50% higher than those of the younger age groups. Adult levels are usually reached in girls by age 15 years, whereas boys still have values one and one-half times adult levels at age 18 years.[27]

Elevations in ACP activity in bone disorders appear to be associated more with osteoblastic activity. Therefore, those disorders characterized by excessive bone resorption will more likely reveal increased ACP levels. Additionally, metastases to bone will show elevations of both ACP and ALP.

Enzymes in Malignancy

Several of the enzymes and isoenzymes of clinical interest have some value in the diagnosis and management of various malignancies and, therefore, serve as tumor markers. A discussion of specific enzymes that may indicate particular malignancies is included in Chapter 27, *Tumor Markers*. Sensitive and specific immunologic methods are usually used for the serum quantitation.

The use of enzymes to detect, monitor, and follow treatment for malignancies is a rapidly changing field. Researchers are continually searching for serum levels of cellular enzymes that will be elevated during specific malignancies.[42] Specificity is one of the main concerns—there are not many enzymes that are elevated in the serum of patients only with certain malignancies. Many unusual procedures for serum levels of enzymes, such as glutamate dehydrogenase (GD), 5′ nucleotidase (5′ NT), ribonuclease, and polyhexose isomerase, are not performed in most clinical laboratories. Their limited diagnostic use and low test volume result in their being sent to a reference laboratory for analysis.

Enzymes in Acute Pancreatitis

Acute pancreatitis should be considered in the differential diagnosis of any patient presenting with persistent, severe abdominal pain. Acute pancreatitis occurs when inactive proteolytic and lipolytic enzymes of the pancreas become activated and initiate autodigestion of pancreatic tissue and blood vessels. The severity of the condition ranges from pancreatic edema, with a mortality of 5% to 10%, to hemorrhagic pancreatic necrosis, with a mortality of 50% to 80%. Acute pancreatitis is most frequently associated with alcoholism and biliary tract disease, although approximately 20% of cases are of an unknown etiology.[35]

The laboratory diagnosis of acute pancreatitis includes evaluation of serum and urine AMS levels and serum LPS levels and calculation of the AMS-creatinine clearance ratio (C_{am}/C_{cr}). Measurement of serum AMS isoenzymes may become an important diagnostic tool in the future.

Serum AMS levels generally become elevated within 2 hours to 12 hours of an attack, reach peak levels at 24 hours, and return to normal within 3 days to 5 days. There is a lack of good correlation between the degree of serum AMS elevation and the severity of the pancreatitis. In fact, normal levels have been observed in cases of severe, fulminating pancreatitis. Generally, AMS levels range between 250 Somogyi units/dL and 1000 Somogyi units/dL. Higher elevations tend to have a greater specificity for acute pancreatitis. Values greater than 500 Somogyi units/dL are highly suggestive of acute pancreatitis. Other causes of acute abdominal pain, such as acute appendicitis and intestinal obstruction, usually show serum AMS levels less than 500 Somogyi units/dL. Considering that AMS is readily filtered by the glomerulus, urine AMS is an expected normal finding, with elevations being detected in acute pancreatitis.

Urine levels may rise to relatively higher values than those of serum AMS and usually remain elevated longer, for 7 days to 10 days. Urine determinations are generally performed on a timed urine collection. If renal function is impaired, levels may be falsely low.

An attempt to enhance the specificity of AMS measurement for the diagnosis of acute pancreatitis is calculation of the C_{am}/C_{cr} ratio. The ratio, although not entirely specific for acute pancreatitis, has some bearing on the differential

CASE STUDY 9-3

A 56-year-old African-American woman presents to the emergency room with abdominal pain, weakness, and loss of appetite. She had not traveled in recent months. She has not been well for several days.

Questions

1. What procedures should be done to rule out appendicitis?
2. What enzyme tests will be useful in diagnosing this patient?
3. What treatment is indicated?

---

diagnosis. In acute pancreatitis, the renal clearance of AMS is generally much greater than that of creatinine, causing an elevated ratio. The mechanism accounting for this increase in renal clearance is, in part, attributed to a disturbance in tubular reabsorption of the enzyme. The ratio is derived from the creatinine clearance formula, and its calculation requires both AMS and creatinine values obtained from simultaneously collected samples of serum and urine. The calculation is according to the following formula:

$$C_{am}/C_{cr} \% = \frac{[AMS]\ urine}{[AMS]\ serum} \times \frac{[creatinine]\ serum}{[creatinine]\ urine} \times 100$$

(Eq. 9–29)

Because the ratio is independent of urine volume, a random urine sample is adequate. Comparison of AMS clearance to creatinine clearance is corrected for abnormal AMS clearance because of changes in the glomerular filtration rate, although an elevated ratio may be found in severe renal insufficiency.[42]

Normally, the ratio ranges from 2% to 5%. In acute pancreatitis, the ratio is greater than 5%, frequently ranging from 7% to 15%. A normal ratio is not commonly seen early in the course of acute pancreatitis. An elevated ratio also may occur in diabetic ketoacidosis, in extensive burns, and in duodenal perforation, thus decreasing the specificity of the test for acute pancreatitis. In acute pancreatitis, the elevated C_{am}/C_{cr} ratio returns to normal after serum and urine AMS levels become normal.

Another advantage of calculating the ratio is its ability to differentiate macroamylasemia from other causes of hyperamylasemia. Because of the large complex of macroamylase, its renal clearance is decreased, resulting in a low ratio. In macroamylasemia the C_{am}/C_{cr} ratio is usually less than 2%.

In conjunction with other laboratory measures, serum LPS measurements serve as a useful measure of acute pancreatitis. Serum LPS analysis has not achieved a degree of popularity equal to that of serum AMS primarily because of the lengthy procedures used in the past to measure LPS. LPS activity in acute pancreatitis generally parallels that of serum AMS. However, peak levels probably occur a bit later, at 12 hours to 24 hours, and levels may remain elevated longer than those of AMS. The combined sensitivity of both serum AMS and LPS lends greater confidence to exclusion of a diagnosis of acute pancreatitis when neither enzyme is elevated.

Serum AMS and LPS levels as well as the LPS/AMS ratio may be useful in determining the etiology and severity of acute pancreatitis. A recent study showed significant differences in the LPS/AMS ratios in patients with biliary, alcoholic, and nonbiliary–nonalcoholic (NBNA) acute pancreatitis. Correlations were also seen with imaging techniques such as ultrasound and computerized tomography scan.[43]

Measurement of AMS levels in pleural or peritoneal fluid also may be requested. In acute pancreatitis, pleural fluid invariably has an elevated AMS level, so that a normal pleural fluid AMS rules out acute pancreatitis.[35] However, elevations can also occur in primary or metastatic carcinoma of the lung and esophageal perforation.

SUMMARY

Enzymes are proteins that are found in all body tissues and catalyze biochemical reactions. The catalyzed reactions are specific and essential to physiologic well-being. In injury or in many disease states, certain enzymes are released from their normal locations and appear in increased amounts in the general circulation. Hence, an understanding of biologically significant enzymes and the reactions they catalyze is useful in the diagnosis and treatment of certain disease states.

This chapter has reviewed the general properties of enzymes and their classification and catalytic mechanisms. Several factors that affect the rate of enzymatic reactions and general methods for measuring enzyme activity were also discussed.

Many enzymes are clinically important. Quantitating serum levels of certain enzymes or isoenzymes can assist in the diagnosis and prognosis of hepatic disorders, skeletal muscle disorders, bone disorders, cardiac disorders, malignancy, or acute pancreatitis. This chapter has discussed the tissue sources, diagnostic significance, preferred assay methodologies, and reference ranges of several enzymes of major clinical significance.

Methodologies for identifying and quantitating enzymes are continuously becoming more sensitive and specific. As the methodologies improve, the importance of evaluating enzymes to assist in diagnosis and treatment of disease will continue to increase.

REVIEW QUESTIONS

1. When a reaction is performed in zero-order kinetics:
   a. The substrate concentration is very low
   b. The rate of reaction is directly proportional to the substrate concentration
   c. The rate of the reaction is independent of the substrate concentration
   d. The enzyme level is always high
2. Activation energy is:
   a. The energy needed for an enzyme reaction to stop
   b. Increased by enzymes
   c. Very high in catalyzed reactions
   d. Decreased by enzymes
3. Enzyme reaction rates are increased by increasing temperatures until they reach the point of denaturation at:
   a. 40°–60°C
   b. 25°–35°C
   c. 100°C
   d. 37°C
4. An example of using enzymes as reagents in the clinical laboratory is:
   a. The diacetyl monoxime blood urea nitrogen (BUN) method
   b. The hexokinase glucose method
   c. The alkaline picrate creatinine method
   d. The biuret total protein method
5. Serum levels of total CK are elevated after an acute myocardial infarction and in which of the following conditions?
   a. Diabetes mellitus
   b. Chronic renal failure
   c. Duchenne type muscular dystrophy
   d. Early pregnancy
6. The isoenzymes LD-4 and LD-5 are elevated in:
   a. Pulmonary embolism
   b. Liver disease
   c. Renal disease
   d. Myocardial infarction
7. Which of the following liver enzymes is the best to use to evaluate liver function?
   a. AST
   b. GGT
   c. ALT
   d. LD
8. The enzyme test that is used to evaluate metastatic cancer of the prostate is:
   a. CK
   b. AST
   c. LD
   d. ACP
9. The saccharogenic method for amylase determinations measures:
   a. The amount of product produced
   b. The amount of substrate consumed
   c. The amount of iodine present
   d. The amount of starch present
10. Elevation of tissue enzymes in serum may be used to detect:
    a. Presence of toxins
    b. Infectious diseases
    c. Tissue necrosis or damage
    d. Diabetes mellitus

REFERENCES

1. Enzyme nomenclature 1978, recommendations of the Nomenclature Committee of the International Union of Biochemistry on the Nomenclature and Classification of Enzymes. New York: Academic Press, 1979.
2. Baron DN, Moss DW, Walker PG, et al. Abbreviations for names of enzymes of diagnostic importance. J Clin Pathol 1971;24:656.
3. Baron DN, Moss DW, Walker PG, et al. Revised list of abbreviations for names of enzymes of diagnostic importance. J Clin Pathol 1975;28:592.
4. Rej R. Accurate enzyme measurements. Arch Pathol Lab Med 1998;117:352.
5. Galen RS. The enzyme diagnosis of myocardial infarction. Human Pathol 1975;6:141.
6. Wilkinson JH. The principles and practice of diagnostic enzymology. Chicago: Year Book Medical Publishers, 1976.
7. Silverman LM, Dermer GB, Zweig MH, et al. Creatine kinase BB: a new tumor-associated marker. Clin Chem 1979;25:1432.
8. Lang H, ed. Creatine kinase isoenzymes: pathophysiology and clinical application. Berlin, Germany: Springer-Verlag, 1981.
9. Irvin RG, Cobb FR, Roe CR. Acute myocardial infarction and MB creatine phosphokinase: relationship between onset of symptoms of infarction and appearance and disappearance of enzyme. Arch Intern Med 1980;140:329.
10. Bark CJ. Mitochondrial creatine kinase—a poor prognostic sign. J Am Med Assoc 1980;243:2058.
11. Batsakis J, Savory J, eds. Creatine kinase. Crit Rev Clin Lab Sci 1982;16:291.
12. Lang H, Wurzburg U. Creatine kinase, an enzyme of many forms. Clin Chem 1982;28:1439.
13. Lott JA. Electrophoretic CK and LD isoenzyme assays in myocardial infarction. Lab Management, Feb 1983;23.
14. Pesce MA. The CK isoenzymes: findings and their meaning. Lab Management, Oct 1982;25.
15. Roche Diagnostics. Isomune-CK package insert. Nutley, NJ: Hoffman-La Roche, Inc, 1979.
16. Bruns DE, Emerson JC, Intemann S, et al. Lactate dehydrogenase isoenzyme 1: changes during the first day after acute myocardial infarction. Clin Chem 1981;27:1821.
17. Faulkner WR, Meites S, eds. Selected methods for the small clinical chemistry laboratory, selected methods of clinical chemistry. Vol 9. Washington, DC: American Association for Clinical Chemistry, 1982.
18. Lott JA, Stang JM. Serum enzymes and isoenzymes in the diagnosis and differential diagnosis of myocardial ischemia and necrosis. Clim Chem 1980;26:1241.
19. Leung FY, Henderson AR. Influence of hemolysis on the serum lactate dehydrogenase1/lactate dehydrogenase-2 ratio as determined by an accurate thin-layer agarose electrophoresis procedure. Clin Chem 1981;27:1708.
20. Bhagavan NV, Darm JR, Scottolini AG. A sixth lactate dehydrogenase isoenzyme (LD-6) and its significance. Arch Pathol Lab Med 1982;106:521.
21. Cabello B, Lubin J, Rywlin AM, et al. Significance of a sixth lactate dehydrogenase isoenzyme (LDH6). Am J Clin Pathol 1980;73:253.
22. Goldberg DM, Werner M, eds. LD-6, A sign of impending death from heart failure, selected topics in clinical enzymology. New York: Walter de Gruyter, 1983.

23. Roche Diagnostics. Isomune-LD package insert. Nutley, NJ: Hoffman-La Roche, Inc, 1979.

24. Randall D, Jones D. Eliminating unnecessary lactate dehydrogenase testing. Arch Intern Med 1997;157:1441.

25. Posen S, Doherty E. The measurement of serum alkaline phosphatase in clinical medicine. Adv Clin Chem 1981;22:165.

26. Warren BM. The isoenzymes of alkaline phosphatase. Beaumont, TX: Helena Laboratories, 1981.

27. Fleisher GA, Eickelberg ES, Elveback LR. Alkaline phosphatase activity in the plasma of children and adolescents. Clin Chem 1977;23:469.

28. Gittes RF. Carcinoma of the prostate. N Engl J Med 1991;324:236.

29. Gittes RF. Prostate specific antigen. N Engl J Med 1987;318:954.

30. Oesterling JE. Prostate specific antigen: a critical assessment of the most useful tumor marker for adenocarcinoma of the prostate. J Urol 1991;145:907.

31. Griffiths JC. The laboratory diagnosis of prostatic adenocarcinoma. Crit Rev Clin Lab Sci 1983;19:187.

32. Lantz RK, Berg MJ. From clinic to court: acid phosphatase testing. Diagn Med, Mar/Apr 1981;55.

33. Yam LT. Clinical significance of the human acid phosphatases, a review. Am J Med 1974;56:604.

34. Rosalki SB. Gamma-glutamyl transpeptidase. Adv Clin Chem 1975;17:53.

35. Salt WB II, Schenker S. Amylase: its clinical significance. A review of the literature. Medicine 1976;4:269.

36. Murthy U, et al. Hyperamylasemia in patients with acquired immunodeficiency syndrome. Am J Gastroenterol 1992;87(3):332.

37. Benner RA, et al. Lipase latex test for acute abdominal pain. Ital J Gastroenterol Feb 1994;24(2).

38. Chesebro M. Using serum markers in the early diagnosis of myocardial infarction. Am Fam Physician 1997;55:2267.

40. Pettijohn T, et al. Usefulness of positive troponin-T and negative creatine kinase levels in identifying high-risk patients with unstable angina. Am J Cardio 1997;80:510.

41. Christiansen R. Cardiac troponin I measurement with the ACCESS immunoassay system: analytical and clinical performance characteristics. Clin Chem 1998;44:52.

42. Yoshito B, Murphy GP. Antigen marker assays in prostate cancer. Lab Management, Apr 1983;19.

43. Pezzillie R, et al. Serum amylase and lipase concentrations and lipase/amylase ratio in assessment of etiology and severity of acute pancreatitis. Dig Dis Sci 1993;38:1265.

SUGGESTED READINGS

Cornish-Bowden A. Fundamentals of enzyme kinetics. London: Butterworth, 1979.

Dixon M. Enzymes. New York: Academic Press, 1979.

Foster RL. The nature of enzymology. New York: Wiley, 1980.

Holley JW. Enzymology: basic text. Houston: CASMT, sponsored by BMC, 1980.

Pincus MR, Zimmerman HJ, Henry JB. Clinical enzymology. In: JB Henry, ed. Clinical diagnosis and management by laboratory methods. Philadelphia: WB Saunders, 1991;250.

Remaley AL, Wilding P. Macroenzymes: biochemical characterization, clinical significance, and laboratory detection. Clin Chem 1989;35(2):2261.

Roger GP. Fundamentals of enzymology. New York: Wiley, 1982.

Segel IH. Enzyme kinetics. New York: Wiley, 1975.

Wilkinson JH. Isoenzymes. 2nd ed. Philadelphia: Lippincott, 1970.

Wilkinson JH. The principles and practice of diagnostic enzymology. Chicago: Year Book Medical Publishers, 1976.

CHAPTER 10

Carbohydrates

Vicki S. Freeman

Objectives

Upon completion of this chapter, the clinical laboratorian will be able to:

- *Classify carbohydrates into their respective groups.*
- *Discuss the metabolism of carbohydrates in the body.*
- *Explain the mode of action of hormones in carbohydrate metabolism.*
- *Differentiate the types of diabetes by clinical symptoms and laboratory findings according to the American Diabetes Association (ADA)*
- *Explain the clinical significance of the three ketone bodies.*
- *Relate expected laboratory results and clinical symptoms to the following metabolic complications of diabetes:*
  - *Ketoacidosis*
  - *Hyperosmolar coma*
- *Distinguish between reactive and spontaneous hypoglycemia.*
- *Describe the principle, specimen of choice, and the advantages and disadvantages of the glucose analysis methods.*
- *Describe the use of hemoglobin A_1C in the long-term monitoring of diabetes.*
- *Discuss the methods of analysis for ketone bodies.*

KEY TERMS

Carbohydrates	Gluconeogenesis	Hypoglycemic
Diabetes mellitus	Glycogen	Insulin
Disaccharide	Glycogenolysis	Ketone
Embden-Myerhof	Glycolysis	Microalbuminuria
pathway	Haworth projection	Monosaccharide
Fisher projection	HbA_{1c}	Polysaccharide
Glucagon	Hyperglycemic	Triose

Organisms rely on the oxidation of complex organic compounds to obtain energy. Three general types of such compounds are carbohydrates, amino acids, and lipids. Although all three are used as a source of energy, carbohydrates are the primary source for brain, erythrocytes, and retinal cells in humans. Carbohydrates are the major food source and energy supply of the body and are stored primarily as liver and muscle glycogen.

GENERAL DESCRIPTION OF CARBOHYDRATES

Carbohydrates are compounds containing C, H, and O. The general formula for a carbohydrate is $C_x(H_2O)_y$. All carbohydrates contain C=O and −OH functional groups. There are some derivatives from this basic formula, because carbohydrate derivatives can be formed by the addition of other chemical groups such as phosphates, sulfates, and amines.

The classification of carbohydrates is based on four different properties: (1) the size of the base carbon chain, (2) the location of the CO function group, (3) the number of sugar units, and (4) the stereochemistry of the compound.

Classification of Carbohydrates

Carbohydrates can be grouped into generic classifications based on the number of carbons in the molecule. For example, *trioses* contain three carbons, tetroses contain four, pentoses contain five, and hexoses contain six. In actual practice, the smallest carbohydrate is glyceraldehyde, a three-carbon compound.

Carbohydrates are hydrates of aldehyde or ketone derivatives based on the location of the CO functional group (Fig. 10-1). The two forms of carbohydrates are aldose and ketose (Fig. 10-2). The aldose form has an aldehyde as its functional group, whereas the ketose form has a ketone as the functional group. The carbon in the functional group is called the anomeric carbon.

Several models are used to represent carbohydrates. The *Fisher projection* of a carbohydrate has the aldehyde or ketone at the top of the drawing. The carbons are numbered starting at the aldehyde or ketone end. The compound can be represented as a straight chain or might be linked to show a representation of the cyclic, hemiacetal form (Fig. 10-3). The *Haworth projection* represents the compound in the cyclic form that is more representative of the actual structure. This structure is formed when the functional group (ketone or aldehyde) reacts with an alcohol group on the same sugar to form a ring called the hemiacetal ring (Fig. 10-4).

Stereoisomers

The central carbons of a carbohydrate are asymmetric—four different groups attached are attached to the carbon atoms. This allows for various spatial arrangements of carbohydrate molecules called *stereoisomers*. Stereoisomers have the same order and types of bonds, but different spatial arrangements and different properties. For each asymmetric carbon, there are 2^n possible isomers; therefore, there are 2^1, or two, forms of glyceraldehyde. These isomers are mirror images. Because an aldohexose contains four asymmetric carbons, there are 2^4 or 16 possible isomers, two of which are paired stereoisomers named D-glucose and L-glucose. D and L are terms used to describe some of the possible optical isomers of glucose and other compounds that exist as stereoisomers. They are mirror images that cannot be overlapped. The D-configuration has the −OH on the lowest asymmetric carbon on the right; the

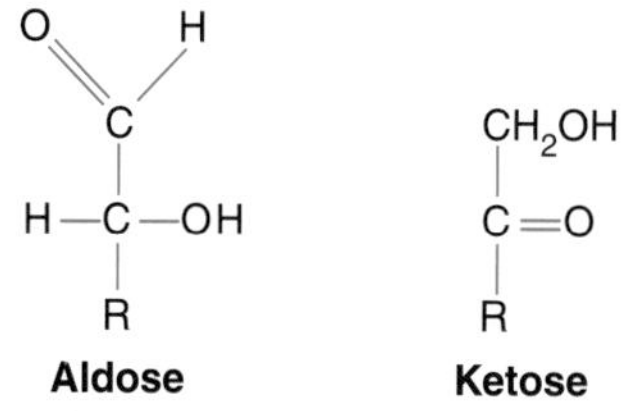

Figure 10-2. Two forms of carbohydrates.

L-form has the −OH on the left. In Figure 10-5, D-glucose is represented in the Fisher projection with the hydroxy group on carbon number five positioned on the right. L-glucose has the hydroxy group of carbon number five positioned on the left. Most sugars in humans are in the D-form.

Monosaccharides, Disaccharides, and Polysaccharides

Another classification of carbohydrates is based on number of sugar units in the chain: monosaccharides, disaccharides, oligosaccharides, and polysaccharides. This chaining of sugars relies on the formation of glycoside bonds that are bridges of oxygen atoms. When two carbohydrate molecules join, a water molecule is produced. When they split, one molecule of water is used to form the individual compounds. This reaction is called *hydrolysis*. The glycoside linkages of carbohydrate can involve any number of carbons; however, certain carbons are favored, depending on the carbohydrate.

Monosaccharides are simple sugars that cannot be hydrolyzed to a simpler form. These sugars can contain three, four, five, and six or more carbon atoms (known as trioses, tetroses, pentoses, and hexoses, respectively). The most common include glucose, fructose, and galactose.

Disaccharides are formed on the interaction of groups between two monosaccharides with the production of a molecule of water. On hydrolysis, disaccharides will be split into two monosaccharides by disaccharide enzymes (*eg*, lactase) located on the microvilli of the intestine. These monosaccharides are then actively absorbed. The most common di-

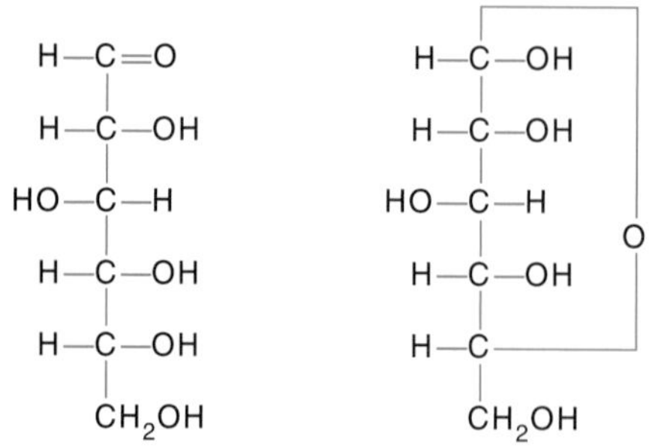

Figure 10-1. Aldehyde and ketone structure.

Figure 10-3. Fisher projection of glucose. The figure on the left is the open chain Fisher projections, and the figure on the right is a cyclic Fisher projection.

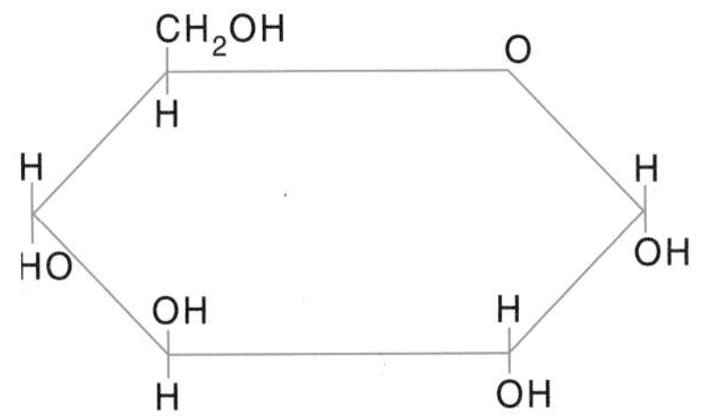

Figure 10-4. Haworth projection of glucose.

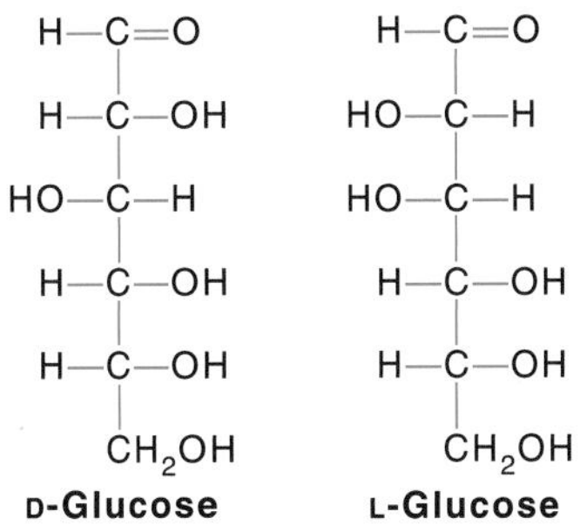

Figure 10-5. Stereoisomers of glucose.

saccharides are maltose (comprising 2-β-D-glucose molecules in a 1→ 4 linkage), lactose, and sucrose.

Oligosaccharides are the chaining of two to ten sugar units, whereas *polysaccharides* are formed by the linkage of many monosaccharide units. On hydrolysis, polysaccharides will yield more than ten monosaccharides. Amylase hydrolyzes starch to disaccharides in the duodenum. The most common polysaccharides are starch (glucose molecules) and glycogen (Fig. 10-6).

Chemical Properties of Carbohydrates

Some carbohydrates are reducing substances; these carbohydrates can oxidize or reduce other compounds. To be a reducing substance, the carbohydrate must contain a ketone or aldehyde group. Examples of reducing substances include glucose, maltose, fructose, lactose, and galactose. This property was used in many laboratory methods in the past in the determination of carbohydrates.

Carbohydrates can form glycosidic bonds with other carbohydrates and with noncarbohydrates. The aldose or ketone group on the carbohydrate forms an oxygen bond. If the bond forms with one of the other carbons on the carbohydrate other than the anomeric carbon, the anomeric carbon (functional group) is unaltered and the resulting compound remains a reducing substance.

If the bond is formed with the anomeric carbon on the other carbohydrate, the resulting compound is *no longer* a reducing substance. Nonreducing carbohydrates *do not* have an active ketone or aldehyde group. They *will not* oxidize or reduce other compounds. The most common nonreducing sugar is sucrose—table sugar (Fig. 10-7).

All monosaccharides and many disaccharides are reducing agents. This is because a free aldehyde or ketone (the open chain form) can be oxidized under the proper conditions. As a disaccharide, one of the aldehydes or ketones is usually alpha to the glycosidic linkage. The other aldehyde

or ketone is still free to function as a reducing agent. However, when both aldehydes or ketones are alpha to the glycosidic linkage, such as in sucrose, then the disaccharide is incapable of undergoing mutarotation and is therefore incapable of functioning as a reducing agent. Both maltose and lactose are reducing agents, whereas sucrose is not.

Glucose Metabolism

Glucose is a primary source of energy for humans. The nervous system, including the brain, totally depends on glucose from the surrounding extracellular fluid (ECF) for energy. Nervous tissue cannot concentrate carbohydrates nor can it store them; therefore, it is critical to maintain a steady supply of glucose to the tissues. For this reason, the concentration of glucose in the ECF must be maintained in a narrow range. When the concentration falls below a certain level, the nervous tissues lose their primary energy source and are incapable of maintaining normal function.

Fate of Glucose

Most of our ingested carbohydrates are polymers such as starch and *glycogen*. Salivary amylase and pancreatic amylase are responsible for the digestion of these nonabsorbable polymers to dextrins and disaccharides, which are further hydrolyzed to monosaccharides by maltase, an enzyme released by the intestinal mucosa. Sucrase and lactase are two other important gut-derived enzymes that hydrolyze sucrose to glucose and fructose, and lactose to glucose and galactose.

Once disaccharides are converted into monosaccharides, they are absorbed by the gut and transported to the liver by the hepatic portal venous blood supply. Glucose is the only carbohydrate to be either directly used for energy or stored as glycogen. Galactose and fructose must be converted to

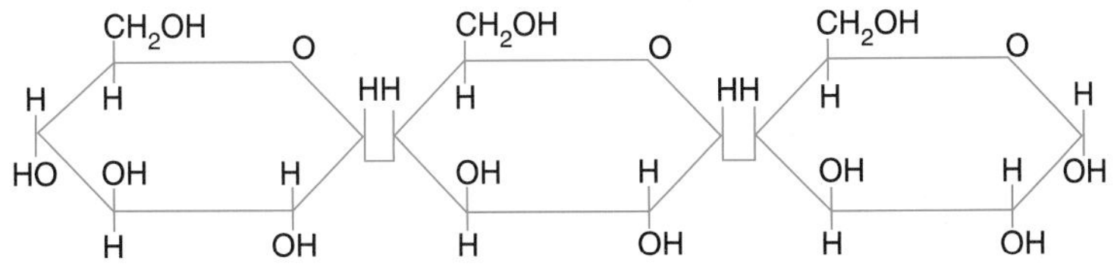

Figure 10-6. Linkage of monosaccharides.

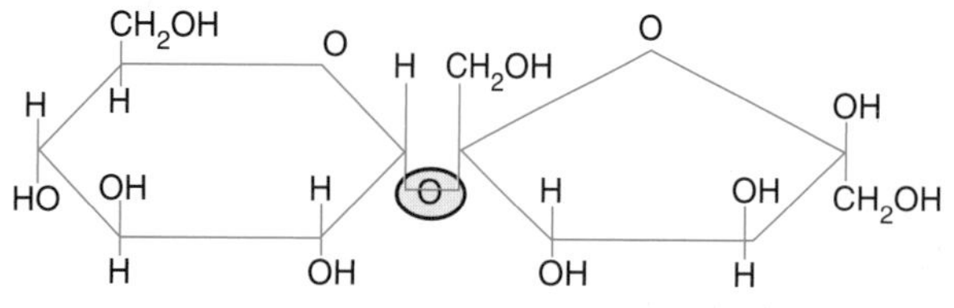

CH₂OH ... [Haworth structure]

Figure 10-7. Haworth projection of sucrose.

glucose before they can be used. After glucose enters the cell, it is quickly shunted into one of three possible metabolic pathways, depending on the availability of substrates or the nutritional status of the cell. The ultimate goal of the cell is to convert glucose to carbon dioxide and water. During this process, the cell obtains the high-energy molecule adenosine triphosphate (ATP) from inorganic phosphate and adenosine diphosphate (ADP). The cell requires oxygen for the final steps in the electron transport chain (ETC). Nicotinamide adenine dinucleotide (NAD) in its reduced form (NADH) will act as an intermediate to couple glucose oxidation to the ETC in the mitochondria where much of the ATP is gained.

The first step for all three pathways requires glucose to be converted to glucose-6-phosphate using the high-energy molecule ATP. This reaction is catalyzed by the enzyme hexokinase (Fig. 10-8). Glucose-6-phosphate can enter the *Embden-Myerhof pathway* or *the hexose monophosphate pathway,* or can be converted to glycogen (Fig. 10-8). The first two pathways are important for the generation of energy from glucose; the conversion to glycogen pathway is important for the storage of glucose.

In the Embden-Myerhof pathway, glucose is broken down into two three-carbon molecules of pyruvic acid that can enter the *tricarboxylic acid cycle* (TCA cycle) on conver-

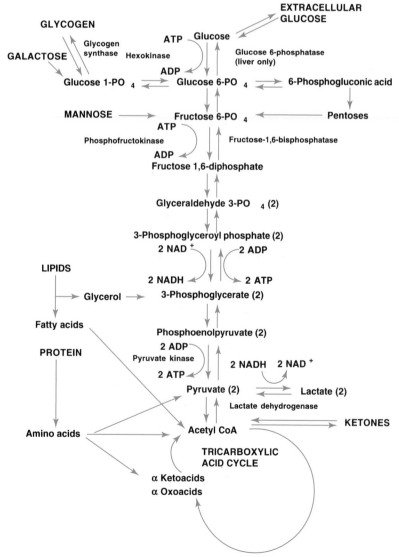

Figure 10-8 The Embden-Myerhof pathway for anaerobic glycolysis.

sion to acetyl coenzyme A (acetyl CoA). This pathway requires oxygen and is called the *aerobic pathway* (Fig. 10-8). Other substrates have the opportunity to enter the pathway at several points. Glycerol released from the hydrolysis of triglycerides can enter at 3-phosphoglycerate, and fatty acids and ketones and some amino acids are converted or catabolized to acetyl CoA, which is part of the TCA cycle. Other amino acids enter the pathway as pyruvate or as deaminated α-ketoacids and oxoacids. The conversion of amino acids by the liver, and other specialized tissues such as the kidney, to substrates that can be converted to glucose is called *gluconeogenesis*. Gluconeogenesis also encompasses the conversion of glycerol, lactate, and pyruvate to glucose.

Anaerobic *glycolysis* is important for tissues such as muscle, which often have important energy requirements without an adequate oxygen supply. These tissues can derive ATP from glucose in an oxygen-deficient environment by converting pyruvic acid into lactic acid. The lactic acid diffuses from the muscle cell, enters the systemic circulation, and is then taken up and used by the liver (Fig. 10-8). For anaerobic glycolysis to occur, 2 moles of ATP must be consumed for each mole of glucose; however, 4 moles of ATP are directly produced, resulting in a net gain of 2 moles of ATP. Further gains of ATP result from the introduction of pyruvate into the TCA cycle and NADH into the ETC.

The second energy pathway is the *hexose monophosphate shunt* (HMP shunt), which is actually a detour of glucose-6-phosphate from the glycolytic pathway to become 6-phosphogluconic acid. This oxidized product permits the formation of ribose-5-phosphate and nicotinamide dinucleotide phosphate in its reduced form (NADPH). NADPH is important to erythrocytes that lack mitochondria and are therefore incapable of the TCA cycle. The reducing power of NADPH is required for the protection of the cell from oxidative and free radical damage. Without NADPH, the lipid bilayer membrane of the cell and critical enzymes would eventually be destroyed, resulting in cell death. The HMP shunt also permits pentoses, such as ribose, to enter the glycolytic pathway.

When the cell's energy requirements are being met, glucose can be stored as glycogen. This third pathway, which is called *glycogenesis,* is relatively straightforward. Glucose-6-phosphate is converted to glucose-1-phosphate, which is then converted to uridine diphosphoglucose and then to glycogen by glycogen synthase. Several tissues are capable of the synthesis of glycogen, especially the liver and muscles. Hepatocytes are capable of releasing glucose from glycogen or other sources to maintain the blood glucose concentration. This is because the liver synthesizes the enzyme glucose-6-phosphatase. Without this enzyme, glucose is trapped in the glycolytic pathway. Muscle cells do not synthesize glucose-6-phosphatase, and therefore they are incapable of dephosphorylating glucose. Once glucose enters a muscle cell, it remains as glycogen unless it is catabolized. *Glycogenolysis* is the process by which glycogen is converted back to glucose-6-phosphate for entry into the glycolytic pathway. Table 10-1

TABLE 10-1. Pathways in Glucose Metabolism

Glycolysis	Metabolism of glucose molecule to pyruvate or lactate for production of energy
Gluconeogenesis	Formation of glucose-6-phosphate from noncarbohydrate sources
Glycogenolysis	Breakdown of glycogen to glucose for use as energy
Glycogenesis	Conversion of glucose to glycogen for storage
Lipogenesis	Conversion of carbohydrates to fatty acids
Lipolysis	Decomposition of fat

outlines the major energy pathways involved either directly or indirectly with glucose metabolism.

Overall, dietary glucose and other carbohydrates can either be used by the liver and other cells for energy or can be stored as glycogen for later use. When the supply of glucose is low, the liver will use glycogen and other substrates to elevate the blood glucose concentration. These substrates include glycerol from triglycerides, lactic acid from skin and muscles, and amino acids. If the lipolysis of triglycerides is unregulated, it results in the formation of ketone bodies, which the brain can use as a source of energy through the TCA cycle. The synthesis of glucose from amino acids is gluconeogenesis. This process is used in conjunction with the formation of ketone bodies when glycogen stores are depleted—conditions normally associated with starvation. The principle pathway for glucose oxidation is through the Embden-Myerhof pathway. NADPH can be synthesized through the HMP shunt, which is a side pathway from the anaerobic glycolytic pathway (Fig. 10-8).

Regulation of Carbohydrate Metabolism

The liver, pancreas, and other endocrine glands are all involved in the controlling the blood glucose concentrations within a narrow range. During a brief fast, glucose is supplied to the ECF from the liver through glycogenolysis. When the fasting period is longer than 1 day, glucose is synthesized from other sources through gluconeogenesis. Control of blood glucose is under two major hormones: insulin and glucagon, both produced by the pancreas. Their actions oppose each other. Other hormones and neuroendocrine substances also exert some control over blood glucose concentrations, permitting the body to respond to increased demands for glucose or to survive prolonged fasts. It also permits the conservation of energy as lipids when excess substrates are ingested.

Insulin is the primary hormone responsible for the entry of glucose into the cell. It is synthesized by the β-cells of islets of Langerhans in the pancreas. When these cells detect an increase in body glucose, they release insulin. The release of insulin causes an increased movement of glucose into the cells and increased glucose metabolism. Insulin is normally released when glucose levels are high and is *not* released when glucose levels are decreased. It decreases plasma glucose levels by increasing the transport entry of glucose in muscle and adipose tissue by way of nonspecific receptors. It also regulates glucose by increasing glycogenesis, lipogenesis, and glycolysis and inhibiting glycogenolysis. Insulin is the only hormone that decreases glucose levels and can be referred to as a *hypoglycemic* agent (Table 10-2).

Glucagon is the primary hormone responsible for increasing glucose levels. It is synthesized by the α-cells of islets of Langerhans in the pancreas and is released during stress and fasting states. When these cells detect a decrease in body glucose, they release glucagon. Glucagon acts by increasing plasma glucose levels by glycogenolysis in the liver and an increase in gluconeogenesis. It can be referred to as a *hyperglycemic* agent (Table 10-2).

Two hormones produced by the adrenal gland affect carbohydrate metabolism. *Epinephrine,* produced by the adrenal medulla, increases plasma glucose by inhibiting insulin secretion, increasing glycogenolysis and promoting lipolysis. Epinephrine is released during times of stress "fight or flight." *Glucocorticoids,* primarily cortisol, are released from the adrenal cortex on stimulation by adenocorticotropic hormone (ACTH). Cortisol increases plasma glucose by decreasing intestinal entry into the cell and increasing gluconeogenesis, liver glycogen, and lipolysis.

Two anterior pituitary hormones, growth hormone and ACTH, both promote increased plasma glucose. *Growth hormone* increases plasma glucose by decreasing the entry of glucose into the cells and by increasing glycolysis. Its release from the pituitary is stimulated by decreased glucose levels and is inhibited by increased glucose. Decreased levels of cortisol stimulate the anterior pituitary to release ACTH. ACTH, in turn, stimulates the adrenal cortex to release cortisol and increases plasma glucose levels by converting liver glycogen to glucose and by promoting gluconeogenesis.

Two other hormones affect glucose levels: thyroxine and somatostatin. The thyroid gland is stimulated by the production of thyroid-stimulating hormone (TSH) to release *thyroxine* that increases plasma glucose levels by increasing glycogenolysis, gluconeogenesis and intestinal absorption of glucose. *Somatostatin,* produced by the D cells of the islets of Langerhans of the pancreas, increases plasma glucose levels

by the inhibition of insulin, glucagon, growth hormone, and other endocrine hormones.

HYPERGLYCEMIA

Hyperglycemia is an increase in plasma glucose levels. In healthy patients, during a hyperglycemia state, insulin is secreted by the β-cells of the pancreatic islets of Langerhans. Insulin enhances membrane permeability to cells in the liver, muscle, and adipose tissue. It also alters the glucose metabolic pathways. Hyperglycemia, or increased plasma glucose levels, is caused by an imbalance of hormones.

Diabetes Mellitus

Diabetes mellitus is actually a group of metabolic diseases characterized by hyperglycemia resulting from defects in insulin secretion, insulin action, or both. In 1979, the National Diabetes Data Group developed a classification and diagnosis scheme for diabetes mellitus.[1] This scheme included dividing diabetes into two broad categories: type 1, insulin-dependent diabetes mellitus (IDDM), and type 2, non–insulin-dependent diabetes mellitus (NIDDM).

Established in 1995, the International Expert Committee on the Diagnosis and Classification of Diabetes Mellitus, working under the sponsorship of the ADA, was given the task of updating the 1979 classification system. The changes they proposed included eliminating the older terms of IDDM and NIDDM. The categories of type 1 and type 2 were retained with the adoption of Arabic numerals instead of Roman numerals (Table 10-3).[2]

Therefore, the ADA/World Health Organization (WHO) guidelines recommend the following categories of diabetes:

- Type 1 diabetes
- Type 2 diabetes
- Other specific types of diabetes
- Gestational diabetes mellitus (GDM)

Type 1 diabetes is characterized by inappropriate hyperglycemia primarily due to pancreatic islet β–cell destruction and proneness to ketoacidosis. Type 2 diabetes, in contrast, includes hyperglycemia cases that result from insulin resistance with an insulin secretory defect. An intermediate stage in which the fasting glucose in increased above-normal limits but not to the level of diabetes has been named *impaired fasting glucose.* The term *impaired glucose tolerance* to indicate glucose tolerance values above normal, but below diabetes levels, was retained. Also, the term *gestational diabetes mellitus* was retained for women who develop glucose intolerance during pregnancy.

Type 1 diabetes mellitus is a result of cellular-mediated autoimmune destruction of the β-cells of the pancreas, causing an absolute deficiency of insulin secretion. Upper limit of 110 mg/dL on the fasting plasma glucose is designated as the upper limit of normal blood glucose. Type 1 constitutes

---

TABLE 10-2. The Action of Hormones

The Action of Insulin
 Increases glycogenesis and glycolysis
 Glucose → Glycogen → Pyruvate → Acetyl CoA
 Increases lipogenesis
 Decreases glycogenolysis

The Action of Glucagon
 Increases glycogenolysis
 Glycogen → Glucose
 Increases gluconeogenesis
 Fatty acids → Acetyl CoA → ketone
 Proteins → Amino acids

TABLE 10-3. Classification of Diabetes Mellitus

Pathogenesis

Type 1
- β-cell destruction
- Absolute insulin deficiency
- Autoantibodies
- Islet cell autoantibodies
- Insulin autoantibodies
- Glutamic acid decarboxylase autoantibodies
- Tyrosine phosphatase IA-2 and IA2B autoantibodies

Type 2
- Insulin resistance with an insulin secretory defect
- Relative insulin deficiency

Other
- Associated with secondary conditions
- Genetic defects of β-cell function
- Pancreatic disease
- Endocrine diseases
- Drug or chemical induced
- Insulin receptor abnormalities
- Other genetic syndromes

Gestational
- Glucose intolerance during pregnancy
- Due to metabolic and hormonal changes

TABLE 10-4. Laboratory Findings in Hyperglycemia

Increased glucose in plasma and urine
Increased urine specific gravity
Increased serum and urine osmolality
Ketones in serum and urine (ketonemia and ketonuria)
Decreased blood and urine pH (acidosis)
Electrolyte imbalance

only 10% to 20% of all diabetes and commonly occurs in childhood and adolescence. This disease is usually initiated by an environmental factor or infection (usually a virus) in individuals with a genetic predisposition and causes the immune destruction of the β-cells of the pancreas and, therefore, a decreased production of insulin. Characteristics of Type 1 diabetes include abrupt onset, insulin dependence, and ketosis tendency. This diabetic type is genetically related. One or more of the following markers are found in 85% to 90% of individuals with fasting hyperglycemia: islet cell autoantibodies, insulin autoantibodies, glutamic acid decarboxylase autoantibodies, and tyrosine phosphatase IA-2 and IA2B autoantibodies.

Signs and symptoms include polydipsia (excessive thirst); polyphagia (increased food intake); polyuria (excessive urine production); rapid weight loss; hyperventilation; mental confusion; and possible loss of consciousness (due to increased glucose to brain). Complications include microvascular problems such as nephropathy, neuropathy, and retinopathy. Increased heart disease is also found in diabetic patients. Table 10-4 lists the laboratory findings in hyperglycemia. *Idiopathic type 1 diabetes* is a form of type 1 diabetes that has no known etiology, is strongly inherited, and does not have β-cell autoimmunity. Individuals with this form of diabetes have episodic requirements for insulin replacement.

Type 2 diabetes mellitus is characterized by hyperglycemia due to an individual's resistance to insulin with an insulin secretory defect. This resistance results in a relative, not an absolute, insulin deficiency. Type 2 constitutes the majority of

the diabetes cases. Most patients in this type are obese or have an increased percentage of body fat distribution in the abdominal region. This type of diabetes is associated with a strong genetic predisposition with patients at an increased risk with an increase in age, obesity, and lack of physical exercise, and often goes undiagnosed for many years. Characteristics usually include an adult onset of the disease and milder symptoms than in type 1, with ketoacidosis very seldom occurring. However, these patients are more likely to go into a hyperosmolar coma. These patients are at an increased risk of developing macrovascular and microvascular complications.

Other specific types of diabetes are associated with certain conditions (secondary) including genetic defects of β-cell function or insulin action, pancreatic disease, diseases of endocrine origin, drug or chemical induced insulin receptor abnormalities, and certain genetic syndromes. The characteristics and prognosis of this form of diabetes depends on the primary disorder. Maturity-onset diabetes of youth (MODY) is a very rare form of diabetes that is inherited in an autosomal dominant fashion.[3]

Gestational diabetes mellitus (GDM) is "any degree of glucose intolerance with onset or first recognition during pregnancy."[4] Causes of GDM include metabolic and hormonal changes. Patients with GDM frequently return to normal postpartum. However, this disease is associated with increased perinatal complications and an increased risk for development of diabetes in later years. Infants born to diabetic mothers are at increased risk for respiratory distress syndrome, hypocalcemia, and hyperbilirubinemia. In the neonate of a diabetic mother, fetal insulin secretion is stimulated. However, when the infant is born and the umbilical cord is severed, the infant's oversupply of glucose is abruptly terminated, causing severe hypoglycemia.

Pathophysiology of Diabetes Mellitus

In both type 1 and type 2 diabetes, the individual will be hyperglycemic, which can be severe. Glucosuria can also occur after the renal tubular transporter system for glucose becomes saturated. This happens when the glucose concentration of plasma exceeds roughly 180 mg/dL in an individual with normal renal function and urine output. As hepatic glucose overproduction continues, the plasma glucose concentration reaches a plateau around 300 mg/dL to 500 mg/dL

(17 mmol/L to 28 mmol/L). Provided renal output is maintained, glucose excretion will match the overproduction causing the plateau.

The individual with type 1 has a higher tendency to produce ketones. Type 2 patients seldom generate ketones, but instead have a greater tendency to develop hyperosmolar nonketotic states. The difference in glucagon and insulin concentrations in these two groups appears to be responsible for the generation of ketones through increased β-oxidation. In type 1, there is an absence of insulin with an excess of glucagon. This permits gluconeogenesis and lipolysis to occur. In type 2, insulin is present as is (sometimes) hyperinsulinemia; therefore, glucagon is attenuated. Fatty acid oxidation is inhibited in type 2. This causes fatty acids to be incorporated into triglycerides for release as very low-density lipoproteins.

The laboratory findings of a diabetic patient with ketoacidosis tend to reflect dehydration, electrolyte disturbances, and acidosis. Acetoacetate, β-hydroxybutyrate, and acetone are produced from the oxidation of fatty acids. The two former ketone bodies contribute to the acidosis. Lactate, fatty acids, and other organic acids can also contribute to a lesser degree. Bicarbonate and total carbon dioxide are usually decreased due to Kussmaul-Kien respiration (deep respirations). This is a compensatory mechanism to blow off carbon dioxide and remove hydrogen ions in the process. The anion gap in this acidosis can exceed 16 mmol/L. Serum osmolality is high due to hyperglycemia; sodium concentrations tend to be lower due in part to losses (polyuria) and in part to a shift of water from cells because of the hyperglycemia. Care must be taken not to falsely underestimate the sodium value because of hypertriglyceridemia. Grossly elevated triglycerides will displace plasma volume and give the appearance of decreased electrolytes when flame photometry or prediluted ion-specific electrodes are used for sodium determinations. Hyperkalemia is almost always present due to

the displacement of potassium from cells in acidosis. This is somewhat misleading because the patient's total body potassium is usually decreased.

More typical of the untreated type 2 patient is the nonketotic hyperosmolar state. The individual presenting with this syndrome has an overproduction of glucose; however, there appears to be an imbalance between production and elimination in urine. Often, this state is precipitated by heart disease, stroke, or pancreatitis. Glucose concentrations exceed 300 mg/dL to 500 mg/dL (17 mmol/L to 28 mmol/L) and severe dehydration is present. The severe dehydration contributes to the inability to excrete glucose in the urine. Mortality is high with this condition. Ketones are not observed, because the severe hyperosmolar state inhibits the ability of glucagon to stimulate lipolysis. The laboratory findings of nonketotic hyperosmolar coma include plasma glucose values exceeding 1000 mg/dL (55 mmol/L), normal or elevated plasma sodium and potassium, slightly decreased bicarbonate, elevated blood urea nitrogen (BUN) and creatinine, and an elevated osmolality. The gross elevation in glucose and osmolality, the elevation in BUN, and the absence of ketones distinguish this condition from diabetic ketoacidosis.

Other forms of impaired glucose metabolism that do not meet the criteria for diabetes mellitus include impaired fasting glucose and impaired glucose tolerance. These forms are discussed in the following section.

Criteria for the Diagnosis of Diabetes Mellitus

The diagnostic criteria for diabetes mellitus was modified by the Expert Committee to allow for earlier detection of the disease. According to the new ADA recommendations, all adults older than age 45 years should have a measurement of fasting blood glucose every 3 years unless the individual is otherwise

---

CASE STUDY 10-1

An 18-year-old male high school student who had a 4-year history of diabetes mellitus was brought to the emergency room because of excessive drowsiness, vomiting, and diarrhea. His diabetes has been well controlled with 40 units of NPH insulin daily until several days ago when he developed excessive thirst and polyuria. For the past 3 days, he has also had headaches, myalgia, and a low-grade fever. Diarrhea and vomiting began 1 day ago.

Plasma glucose	600 mg/dL
BUN	48 mg/dL
Creatinine	2.0 mg/dL
Serum ketones	4+

Questions

1. What is the probable diagnosis of this patient based on the data presented?
2. What laboratory test(s) should be performed to follow this patient and aid in adjusting insulin levels?
3. Why are the urine ketones positive?
4. What methods are used to quantitate urine ketones? Which ketone(s) do they detect?

Urinalysis		**Chemistry test results**	
Specific gravity	1.012	Sodium	126 mEq/L
pH	5.0	Potassium	6.1 mEq/L
Glucose	4+	Chloride	87 mEq/L
Ketone	Large	Bicarbonate	6 mEq/L

CASE STUDY 10-2

A 58-year-old patient comes into the emergency room in a coma. The following laboratory work is drawn and reported:

Arterial blood gases		Chem Profile	
pH	7.05	Glucose	600 mg/dL
PCO_2	35 mm Hg	BUN	20 mg/dL
PO_2	75 mm Hg	Creatinine	1.0 mg/dL

Electrolytes			
Total CO_2	6 mEq/L	Total bilirubin	1.0 mg/dL
Sodium	133 meq/L	Alkaline phosphatase	78 U/L
Potassium	4.0 meq/L	Serum glutamate pyruvate transaminase	20 U/L
Chloride	100 meq/L		
Anion gap	27	Serum osmolality	350 mOsm/kg

Questions

1. What is the probable diagnosis of this patient?
2. The doctor also requested a calculated osmolality. What would this value be?
3. Why is the difference between the calculated and measured osmolality significant?
4. How does the anion gap confirm your suspicions?
5. What other tests could be performed to confirm this?
6. What is the patient's acid–base status?

diagnosed with diabetes. Testing should be carried out at an earlier age or more frequently in individuals who display

- Obesity [120% of desirable body weight or body mass index (BMI) of 27 kg/M²]
- Family history of diabetes in a first-degree relative
- Membership in a high-risk minority population (*eg,* African-American, Hispanic-American, Native American, or Asian-American)
- History of GDM or delivering a baby >9 lb (>4.1 kg)
- Hypertension (>140/90)
- Low high-density lipoprotein (HDL) cholesterol concentrations (*eg,* <35 mg/dL)
- Elevated triglyceride concentrations (*eg,* >250 mg/dL)
- A history of impaired fasting glucose/impaired glucose tolerance

The criteria listed in Tables 10-5, 10-6, and 10-7 suggest three methods of diagnosis, each of which must be confirmed on a subsequent day by any one of the three methods. These methods are (1) symptoms of diabetes plus a random plasma glucose level of ≥200 mg/dL, (2) a fasting plasma glucose of ≥126 mg/dL, or (3) an oral glucose tolerance test (OGTT) with a 2-hour postload (75-g glucose load) level ≥200 mg/dL.

The preferred test for diagnosing diabetes is the measurement of the fasting plasma glucose level.

An intermediate group that did not meet the criteria of diabetes mellitus but had glucose levels above normal was defined by two methods. First, those patients with fasting glucose levels ≥110 mg/dL but <126 mg/dL was called the impaired fasting glucose group. Another set of patients who had 2-hour OGTT levels of ≥140 mg/dL but <200 mg/dL was defined as impaired glucose tolerance.

The diagnostic criteria for gestational diabetes follows the guidelines established by the American College of Obstetrics and Gynecology. Only high-risk patients should be screened for GDM. The criteria for women at high risk include any of the following: age older than 25 years, overweight, family history of diabetes, or ethnicity as an African-American, Hispanic-American, or Native American. The screening tests include the measurement of plasma glucose at 1-hour postload (50-g glucose load). If the value is ≥140 mg/dL (7.8 mmol/L), then the need to perform a 3-hour OGTT using a 100-g glucose load is indicated. GDM is diagnosed when any two of the following four values are met or exceeded: fasting >105 mg/dL, 1 h >190 mg/dL, 2 h ≥165 mg/dL or 3 h ≥145 mg/dL.

TABLE 10-5. Diagnostic Criteria for Diabetes Mellitus

1 Random plasma glucose ≥200 mg/dL (11.1 mmol/L) + symptoms of diabetes
2 Fasting plasma glucose ≥126 mg/dL (7.0 mmol/L)
3 2 hours plasma glucose ≥200 mg/dL (11.1 mmol/L) during an oral glucose tolerance test

Any of the three criteria must be confirmed on a subsequent day by any of the three methods.

TABLE 10-6. Categories of Fasting Plasma Glucose

Normal fasting glucose	FPG <110 mg/dL
Impaired fasting glucose	FPG ≥110 mg/dL and 126 mg/dL
Provisional diabetes diagnosis	FPG ≥126 mg/dL

FPG, fasting plasma glucose.

TABLE 10-7. Categories of Oral Glucose Tolerance

Normal glucose tolerance	2-h PG <140 mg/dL
Impaired glucose tolerance	2-h PG ≥140 mg/dL and <200 mg/dL
Provisional diabetes diagnosis	2-h PG ≥200 mg/dL

PG, plasma glucose.

HYPOGLYCEMIA

Hypoglycemia involves decreased plasma glucose levels and can have many causes—some are transient and relatively insignificant; others can be life-threatening (Table 10-8). The plasma glucose concentration at which glucagon and other glycemic factors are released is between 65 mg/dL to 70 mg/dL (3.6 mmol/L to 3.9 mmol/L), and at about 50 mg/dL to 55 mg/dL (2.8 mmol/L to 3.0 mmol/L), observable symptoms of hypoglycemia appear. The warning signs and symptoms of hypoglycemia are all related to the central nervous

CASE STUDY 10-3

A 14-year-old male student was seen by his physician. His chief complaints were fatigue, weight loss, and increases in appetite, thirst, and frequency of urination. For the past 3 weeks to 4 weeks, he had been excessively thirsty and had to urinate every few hours. He began to get up three to four times a night to urinate. Patient has a family history of diabetes mellitus.

Laboratory data

Urinalysis	Specific gravity	1.040
	Glucose	4+
	Ketones	Moderate
Fasting plasma glucose	160 mg/dL	

Questions

1. Based on the preceding information, can this patient be diagnosed with diabetes?
2. What further tests might be performed to confirm the diagnosis?
3. According the American Diabetes Association, what criteria is required for the diagnosis of diabetes?
4. Assuming this patient is diabetic, which type would he be diagnosed?

CASE STUDY 10-4

A 28-year-old woman delivered a 9.5 lb infant. The infant was above the 95th percentile for weight and length. The mother's history was incomplete; she claimed to have had no medical care through her pregnancy. Shortly after birth, the infant became lethargic and flaccid. A whole blood glucose and ionized calcium were performed in the nursery with the following results:

Whole blood glucose	25 mg/dL
Ionized calcium	4.9 mg/dL

Plasma glucose was drawn and analyzed in the main laboratory to confirm the whole blood findings.

Plasma glucose	33 mg/dL

An intravenous glucose solution was started and whole blood glucose was measured hourly.

Questions

1. Give the possible explanation for the infant's large birth weight and size.
2. If the mother was a gestational diabetic, why was her baby hypoglycemic?
3. Why was there a discrepancy between the whole blood glucose concentration and the plasma glucose concentration?
4. If the mother had been monitored during pregnancy, what laboratory tests should have been performed and what criteria would have indicated that she had gestational diabetes?

Laboratory tests were performed on a 50-year-old lean Caucasian female during an annual physical examination. She has no family history of type II diabetes or any history of an elevated glucose level during pregnancy.

Laboratory Results

Fasting blood glucose (FBG)	90 mg/dL
Cholesterol	140 mg/dL
HDL	40 mg/dL
Triglycerides	90 mg/dL

1. What is the probable diagnosis of this patient?
2. Describe the proper follow-up for this patient?
3. What is the preferred screening test for diabetes in nonpregnant adults?
4. What are the risk factors that would indicate a potential of this patient's developing diabetes?

system. The release of epinephrine into systemic circulation and norepinephrine at nerve endings of specific neurons act in unison with glucagon to increase plasma glucose. Glucagon is released from the islet cells of the pancreas and inhibits insulin. Epinephrine is released from the adrenal gland and increases glucose metabolism and inhibits insulin. In addition, cortisol and growth hormone are released and increase glucose metabolism.

Hypoglycemia can be classified as postabsorptive (fasting) and postprandial (reactive) hypoglycemia. In postabsorptive hypoglycemia, an individual has a loss of glycemic control during in a fasting state. The healthy individual can rely on gluconeogenesis to maintain the extracellular glucose concentration. However, an individual with fasting hypoglycemia, for one of a variety of reasons, is unable to maintain a stable plasma glucose level. Fasting hypoglycemia is a serious form of the disease, usually with some underlying metabolic condition. Postabsorptive hypoglycemia characteristics include a nonsuppressible insulin-like activity in which the glu-cose level drops below normal fasting levels. However, spontaneous recovery of glucose level *does not occur;* glucose must be given to the patient to relieve symptoms. Postabsorptive hypoglycemia may be due to β-islet cell insulinomas; other insulin-producing tumors (hepatic, adrenal, gastrointestinal); may be ethanol induced (glycogen depleted) or drug induced (hypoglycemic agents); or may be due to severe liver disease or enzymatic defects. Symptoms of the disease are increased hunger, sweating, nausea and vomiting, dizziness, nervousness and shaking, blurring of speech and sight, and mental confusion. Laboratory findings include decreased plasma glucose levels during hypoglycemic episode and extremely elevated insulin levels in β-cell tumors (insulinoma).

Postprandial (reactive) hypoglycemia, on the other hand, is usually not serious. The characteristics of postprandial hypoglycemia include an excessive release of insulin that results in glucose levels dropping below normal fasting levels. There is usually a spontaneous recovery of glucose level as the result of insulin levels returning to normal.

TABLE 10-8. Causes of Hypoglycemia

Postabsorptive (Fasting)	
Drug-induced	Hypoglycemic agents: reduce substrates or inhibit gluco-neogenesis
Ethanol-induced	Glycogen depleted
Organ failure	Diminished metabolic capacity, hepatic disease
Insulinoma	Beta islet cell tumors, other insulin-producing tumors
Neonatal	Transient hyperinsulinism, impaired gluconeogenesis
Congenital/enzyme	Glucose-6-phosphatase deficiency (von Gierke's disease)
Postprandial (Reactive)	
Alimentary (gastrointestinal surgery)	Accelerated absorption of glucose
Functional	Occurs in fasting states

Genetic Defects in Carbohydrate Metabolism

Glycogen storage diseases are result of the deficiency of a specific enzyme that causes an alternation of glycogen metabolism. The most common congenital form of glycogen storage disease is glucose-6-phosphatase deficiency type 1, which is also called von Gierke disease, an autosomal recessive disease. This disease is characterized by a severe hypoglycemia that coincides with metabolic acidosis, ketonemia, and elevated lactate and alanine. Hypoglycemia occurs because glycogen cannot be converted back to glucose by way of hepatic glycogenolysis. A buildup of glycogen is found in the liver, causing hepatomegaly. The patients usually have severe hypoglycemia, hyperlipidemia, uricemia, and growth retardation. A liver biopsy will show a positive glycogen stain. Although the glycogen accumulation is irreversible, the disease can be kept under control by avoiding the development of hypoglycemia. Liver transplantation corrects the hypoglycemic condition. Other enzyme defects or deficiencies that cause hypoglycemia include glycogen synthase, fructose-1, 6-bisphosphatase, phosphoenolpyruvate carboxykinase, and pyruvate carboxylase. Glycogen debrancher enzyme deficiency does not cause hypoglycemia but does cause hepatomegaly.

Galactosemia, a cause of failure to thrive syndrome in infants, is a congenital deficiency of one of three enzymes involved in galactose metabolism, resulting in increased levels of galactose in plasma. The most common is enzyme deficiency is galactose-1-phosphate uridyl transferase. Galactosemia is accompanied by diarrhea and vomiting, and occurs because of the inhibition of glycogenolysis. Galactose must be removed from the diet to prevent the development of irreversible complications. If left untreated, the patient will develop mental retardation and cataracts. The disorder can be identified by measuring erythrocyte galactose-1-phosphate uridyl transferase activity. Laboratory findings include hypoglycemia, hyperbilirubinemia, and galactose accumulation in the blood, tissues, and urine following milk ingestion. Another enzyme deficiency, fructose-1-phosphate aldolase deficiency, causes nausea and hypoglycemia after fructose ingestion.

Specific inborn errors of amino acid metabolism and long chain fatty acid oxidation are also responsible for hypoglycemia. There are also alimentary and idiopathic hypoglycemias in the postprandial category. Alimentary hypoglycemia appears to be caused by an increase in the release of insulin in response to rapid absorption of nutrients after a meal or the rapid secretion of insulin-releasing gastric factors. Idiopathic postprandial hypoglycemia is a controversial diagnosis that may be overused.[5]

ROLE OF LABORATORY IN DIFFERENTIAL DIAGNOSIS AND MANAGEMENT OF PATIENTS WITH GLUCOSE METABOLIC ALTERATIONS

The demonstration of hyperglycemia or hypoglycemia under specific conditions is used to diagnose diabetes mellitus and hypoglycemic conditions. Other laboratory tests have been developed to identify insulinomas and to monitor glycemic control and the development of renal complications.

Methods of Glucose Measurement

Glucose can be measured from serum, plasma, or whole blood. Today, most glucose measurements are performed on serum or plasma. The glucose concentration in whole blood is approximately 15% lower than the glucose concentration in serum or plasma. Serum or plasma must be refrigerated and separated from the cells within 1 hour to prevent substantial losses of glucose by the cellular fraction, particularly if the white blood cell count is elevated. Sodium fluoride ions (gray top tubes) are often used as an anticoagulant and preservative of whole blood, particularly if the analysis is delayed. Fluoride inhibits glycolytic enzymes. Fasting blood glucose (FBG) should be obtained after an approximately 10-hour fast (not >16 hours). Cerebrospinal fluid and urine can also be analyzed. Urine glucose measurement is not used in diabetes diagnosis; however, some patients use this measurement for monitoring purposes.

The ability of glucose to function as a reducing agent has been useful in the detection and quantitation of carbohydrates in body fluids. Glucose and other carbohydrates are capable of converting cupric ions in alkaline solution to cuprous ions. The solution loses its deep blue color and a red precipitate of cuprous oxide forms. Benedict's and Fehling's reagents, which contain an alkaline solution of cupric ions stabilized by citrate or tartrate, respectively, have been used to detect reducing agents in urine and other body fluids. Another chemical characteristic that used to be exploited to quantitate carbohydrates in the past is the ability of these molecules to form Schiff bases with aromatic amines. O-Toluidine in a hot acidic solution will yield a colored compound with an absorbance maxima at 630 nm. Galactose, an aldohexose, and mannose, an aldopentose, will also react with O-toluidine and produce a colored compound that can interfere with the reaction. The Schiff base reaction with O-toluidine is of historical interest only and has been replaced by more specific enzymatic methods, which are discussed in the following section.

The most used methods of glucose analysis use the enzymes glucose oxidase or hexokinase (Table 10-9). Glucose oxidase is the most specific enzyme reacting with only β-D-glucose. Glucose oxidase converts β-D-glucose to gluconic acid. Mutarotase may be added to the reaction to facilitate the conversion of α-D-glucose to β-D-glucose. Oxygen

---

TABLE 10-9. Methods of Glucose Measurement

Glucose oxidase

$$Glucose + O_2 + H_2O \xrightarrow{glucose\ oxidase} Gluconic\ acid + H_2O_2$$

$$H_2O_2 + reduced\ chromogen \xrightarrow{peroxidase} Oxidized\ chromogen + H_2O$$

Hexokinase

$$Glucose + ATP \xrightarrow{Hexokinase} Glucose\text{-}6\text{-}PO_4 + ADP$$

$$Glucose\text{-}6\text{-}PO_4 + NADP \xrightarrow{G6PD} NADPH + H^+ + 6\text{-}phosphogluconate$$

Clinitest

$$R\ reducing\ substance + Cu^{+2} \longrightarrow Cu^{+1}O$$

is consumed and hydrogen peroxide is produced. The reaction can be monitored polarographically either by measuring the rate of disappearance of oxygen using an oxygen electrode or by consuming hydrogen peroxide in a side reaction. Horseradish peroxidase is used to catalyze the second reaction, and the hydrogen peroxide is used to oxidize a dye compound. Two commonly used chromogens are 3-methyl-2-benzothiazolinone hydrazone and N,N-dimethylaniline. The shift in absorbance can be monitored spectrophotometrically and is proportional to the amount of glucose present in the specimen. This coupled reaction is known as the *Trinder reaction*. However, the peroxidase coupling reaction used in the glucose oxidase method is subject to positive and negative interference. Increased levels of uric acid, bilirubin, and ascorbic acid can cause falsely decreased values as a result of these substances' being oxidized by peroxidase, which then prevents the oxidation and detection of the chromogen. Strong oxidizing substances such as bleach can cause falsely increased values. An oxygen consumption electrode can be used to perform the direct measurement of oxygen by the polarographic technique, which avoids this interference. Oxygen depletion is measured and is proportional to the amount of glucose present. Polarographic glucose analyzers measure the rate of oxygen consumption, because glucose is oxidized under first-order conditions using glucose oxidase reagent. The H_2O_2 formed must be eliminated in a side reaction to prevent the reaction from reversing. Molybdate can be used to catalyze the oxidation of iodide to iodine by H_2O_2 or catalase can be used to catalyze oxidation of ethanol by H_2O_2, forming acetaldehyde and H_2O.

The hexokinase method is considered more accurate than glucose oxidase methods because the coupling reaction using glucose-6-phosphate dehydrogenase is highly specific; therefore, it has less interference than the coupled glucose oxidase procedure. Hexokinase in the presence of ATP converts glucose to glucose-6-phosphate. Glucose-6-phosphate and the cofactor $NADP^+$ are converted to 6-phosphogluconate and NADPH by glucose-6-phosphate dehydrogenase. NADPH has a strong absorbance maxima at 340 nm, and the rate of appearance of NADPH can be monitored spectrophotometrically and is proportional to the amount of glucose present in the sample. This method is generally accepted as the reference method. This method is not affected by ascorbic acid or uric acid; gross hemolysis and extremely elevated bilirubin may cause a false decrease in results. The hexokinase method may be performed on serum or plasma collected using heparin, ethylenediaminetetra-acetic acid (EDTA), fluoride, oxalate, or citrate. The method can also be used for urine, cerebrospinal fluid, and serous fluids.

Nonspecific methods of measuring glucose are still used in the urinalysis section of the laboratory primarily to detect reducing substances other than glucose. The method below is the Benedict's modification, also called the Clinitest reaction.

Self-Monitoring of Blood Glucose (SMBG)

The ADA has recommended that individuals with diabetes should monitor their blood glucose levels in an effort to maintain levels as close to normal as possible. For Type 1 diabetes, the recommendations are three to four times per day; for type 2, the optimal frequency is unknown. It is important that patients be taught how to use control solutions and calibrators to ensure the accuracy of their results.[6] Urine glucose testing should be replaced by SMBG, but urine ketone testing will remain for type 1 and gestation diabetes.

Glucose Tolerance and 2-Hour Postprandial Tests

New guidelines for the performance and interpretation of the *2-hour postprandial test* were set by the Expert Committee. A variation of this test is to use a standardized load of glucose. A solution containing 75 g of glucose is administered, and a specimen for plasma glucose measurement is drawn 2 hours later. Under the new criteria, a patient drinks a standardized (75-g) glucose load and a glucose measurement is taken 2 hours later. If that level is ≥200 mg/dL, and is confirmed on a subsequent day by either an increased random or fasting glucose level, the patient is diagnosed with diabetes (see earlier discussion).

The *oral glucose tolerance test (OGTT)* is not recommended for routine use under the new ADA guidelines. This procedure is inconvenient to patients and is not being used by physicians for diagnosing diabetes. However, if the OGTT is used, WHO recommends the criteria listed in Table 10-7. It is important that proper patient preparation be given before this test is performed. The patient should be ambulatory and on a normal-to-high carbohydrate intake for 3 days before the test. The patient should be fasting at least 10 hours and not more than 16 hours, the test should be performed in the morning because of the hormonal diurnal effect on glucose. Just before tolerance and while the test is in progress, patients should refrain from exercise, eating, drinking (except that the patient may drink water), and smoking. Factors that affect the tolerance results include medications such as large doses of salicylates, diuretics, anticonvulsants, oral contraceptives, and corticosteroids. Also gastrointestinal problems including malabsorption problems, gastrointestinal surgery, and vomiting and endocrine dysfunctions can affect the OGTT results. The new guidelines recommend that only the fasting and the 2-hour sample be measured, except when the patient is pregnant. The adult dose of glucola is 75 g. Children receive 1.75 g/kg of glucose to a maximum dose of 75 g.

Glycosylated Hemoglobin

The aim of diabetic management is to maintain the blood glucose concentration within or near the nondiabetic range with a minimal number of fluctuations. Serum or plasma glucose concentrations can be measured by laboratories in addition to patient self-monitoring of whole blood glucose concentrations. Long-term blood glucose regulation can be followed by measurement of glycosylated hemoglobins.

Many proteins are known to react with carbohydrates at the peptide N-terminus forming glycosylated peptides.

Glucose can rapidly react with hemoglobin to form a labile aldimine (Schiff base, Figure 10-9.[7] This equilibrium product can then undergo an Amadori rearrangement to form glycosylated hemoglobin (ketoamine). The ketoamine product is stable and cannot revert back to hemoglobin and glucose. HbA_{1c} is the largest subfraction of normal HbA in both diabetic and nondiabetic subjects. This fraction is formed by the reaction of the β-chain of HbA with glucose.

Glycosylated hemoglobin (GHb) is the term used to describe the formation of a hemoglobin compound formed when glucose (a reducing sugar) reacts with the amino group of hemoglobin (a protein). The glucose molecule attaches nonenzymatically to the hemoglobin molecule in a ketoamine structure to form a ketoamine. The rate of formation is directly proportional to the plasma glucose concentrations. Because the average red blood cell lives approximately 120 days, the glycosylated hemoglobin level at any one time reflects the average blood glucose level over the previous 2 months to 3 months. Therefore, measuring the glycosylated hemoglobin provides the clinician with a time-averaged picture of the patient's blood glucose concentration over the past 3 months. HbA_{1c}, the most commonly detected glycosylated hemoglobin, is a glucose molecule attached to one or both N-terminus valines of the β-polypeptide chains of normal adult hemoglobin.[8] Hemoglobin A_{1c} is a reliable method of monitoring long-term diabetic control rather than random plasma glucose (FBS). Normal values range from 4.5 to 8.0. Remember that two factors determine the glycosylated hemoglobin levels: the average glucose concentration and the red blood cell life span. If the red blood cell life span is decreased because of another disease state such as hemoglobinopathies, the hemoglobin will have less time to become glycosylated and the glycosylated hemoglobin level will be lower.

The specimen requirement for hemoglobin A_{1c} measurement is an EDTA whole blood sample. Before analysis, a hemolysate must be prepared. The methods of measurement are grouped into two major categories: (1) based on charge differences between GHb and non-GHb (cation-exchange chromatography, electrophoresis, and isoelectric focusing) and (2) structural characteristics of glycogroups on hemoglobin (affinity chromatography and immunoassay).[9] There is no consensus on the reference method and no single standard available to be used in the assays. Because of this, Hemoglobin A_{1c} values vary with the method and laboratory performing them (Table 10-10).

The preferred method of measurement is using affinity chromatography. In this method, the glycosylated hemoglobin attaches to the boronate group of the resin and is then is selectively eluted from the resin bed by using a buffer. This method is not temperature dependent and is not affected by hemoglobin F, S, or C. Another method of measurement uses cation exchange chromatography in which the negatively charged hemoglobins attach to the positively charged resin bed. The glycosylated hemoglobin is selectively eluted from the resin bed by using a buffer of specific pH in which the glycohemoglobins are the most negatively charged and elute first from the column. However, this method is highly temperature dependent and is affected by hemoglobinopathies. The presence of hemoglobin F yields false increased levels and the presence of hemoglobins S and C yields false decreased levels. High-performance liquid chromatography and electrophoresis methods are also used to separate the various forms of hemoglobin. With high-performance liquid chromatography, all forms of glycosylated hemoglobin, A_{1a}, A_{1b}, A_{1c}, can be separated.

Ketones

Ketone bodies are products of incomplete fat metabolism (B oxidation). The three ketone bodies are acetone (2%), acetoacetic acid (20%), and β-hydroxybutyric acid (78%). Ketones are produced in cases of carbohydrate deprivation or decreased carbohydrate use such as diabetes mellitus, starvation/fasting, prolonged vomiting, and glycogen storage disease. The term *ketonemia* refers to the accumulation of ketones in blood, and the term *ketonuria* refers to accumulation of ketones in urine (Fig. 10-10).

The specimen requirement is *fresh* serum or urine; the sample should be tightly stoppered and analyzed immediately. No method used for determination of ketones reacts with all three ketone bodies. The historical (Gerhardt's test) that used ferric chloride reacted with acetoacetic acid to produce a red color. The procedure had many interfering substances including salicylates. A more common method using sodium nitroprusside ($NaFe[CN]_5NO$) reacts with acetoacetic acid in an alkaline pH to form a purple color. If the

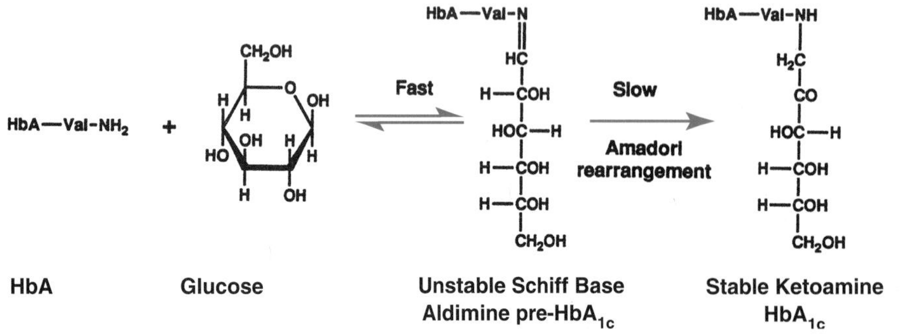

Figure 10-9 The nonenzymatic glycosylation of HbA to HbA_{1c} (From Higgis PJ, Bunn HF. Kinetic analysis of the nonenzymatic glycosylation of hemoglobin. J Biol Chem 1981; 256:5204–5208.)

TABLE 10-10. Methods of Glycated Hemoglobin Measurement

Methods Based on Structural Differences

Immunoassays	Polyclonal or monoclonal antibodies toward the glycated N-terminal group of the β-chain of Hgb	
Affinity chromatography	Separates based on chemical structure using borate to bind glycosylated proteins	Not temperature dependent Not affected by other hemoglobins

Methods Based on Charge Differences

Ion-exchange chromatography	Positive charged resin bed	Highly temperature dependent Affected by hemoglobinopathies
Electrophoresis	Separation is based on differences in charge	Hgb F values >7% interfere with analysis
Isoelectric focusing	Type of electrophoresis using isoelectric point to separate	Pre-Hgb A_{1c} interferes
High-performance liquid chromatography	Form of ion-exchange chromatography	Separates of all forms of glycosylated Hgb: A_{1a}, A_{1b}, A_{1c}

Hgb, hemoglobin.

reagent contains glycerin, then acetone is also detected. This method is used with the urine reagent strip test and Acetest tablets. A newer enzymatic method adapted to some automated instruments uses the enzyme β-hydroxybutrate dehydrogenase to detect either β-hydroxybutyric acid or acetoacetic acid, depending on the pH of the solution. A pH of 7.0 causes the reaction to proceed to the right (decreasing absorbance), whereas a pH of 8.5 to 9.5 causes the reaction to proceed to left (increasing absorbance; Table 10-11).

Microalbuminuria

Diabetes mellitus causes progressive changes to the kidneys and ultimately results in diabetic renal nephropathy. This complication progresses over years and may be delayed by aggressive glycemic control. An early sign that nephropathy is occurring is an increase in urinary albumin. Microalbumin measurements are useful to assist in diagnosis at an early stage and before the development of proteinuria. Microalbumin concentrations are between 20 mg/day to 300 mg/day. Proteinuria is typically greater than 0.5 g/day.[10]

Islet Autoantibody, Insulin, and C-Peptide Testing

Islet autoantibody testing should only be used to help clarify the etiology of diabetes (*eg,* autoimmune versus nonautoimmune) in patients with newly diagnosed diabetes. Insulin measurements are not required for the diagnosis of diabetes mellitus. However, in certain hypoglycemic states, it is important to know the concentration of insulin in relation to the plasma glucose concentration. To investigate an insulinoma, the patient is required to fast under controlled conditions. Males and females have different metabolic patterns in prolonged fasts. The healthy male will maintain plasma glucose of 55 g/dL to 60 mg/dL (3.1 mmol/L to 3.3 mmol/L) for several days. Healthy females will produce ketones more readily and permit plasma glucose to decrease to approximately 35 mg/dL (1.9 mmol/L) after around 36 hours. A decreased glucose coincident with a nonsuppressed insulin of greater than 8 units/mL indicates insulinoma. An insulinoma can also be diagnosed if serial measurements are taken and the glucose concentration decreases faster than the insulin

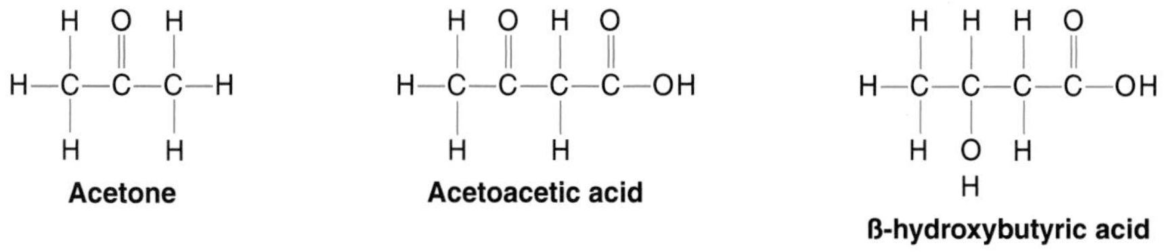

Acetone   **Acetoacetic acid**

ß-hydroxybutyric acid   *Figure 10-10.* The three ketone bodies.

TABLE 10-11. Methods of Ketone Measurement

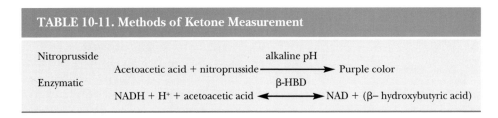

Nitroprusside		alkaline pH	
	Acetoacetic acid + nitroprusside $\longrightarrow$		Purple color
Enzymatic		β-HBD	
	NADH + H⁺ + acetoacetic acid $\longleftrightarrow$		NAD + (β– hydroxybutyric acid)

concentration. In the male, insulinoma may be ruled out if hypoglycemia is not observed after a 3-day fast.

Proinsulin assays may be used in the future because proinsulin represents less than 20% of the total circulating insulin in a healthy individual. In contrast, in an individual with an insulinoma, 30% to 90% of the total circulating insulin is in the proinsulin form. This assay, which uses monoclonal antibodies, may prove useful and convenient, allowing the clinician to avoid hospitalizing the patient for the fasting period and sparing the patient the discomfort and cost of the hospital stay.

C-peptide assays are of little diagnostic use, except in instances in which inappropriate insulin administration must be distinguished from endogenous insulin production. Hypoglycemia in a diabetic with access to insulin may be due to excessive use or due to endogenous insulin. Exogenous insulin has been purified and C-peptide has been removed. If insulin concentrations are greater than the C-peptide concentration, the plasma insulin is from exogenous administration.

SUMMARY

Carbohydrates have the general formula $C_x(H_2O)_n$. Glucose is a six-carbon aldohexose. There are 32 different possible isomers of aldohexoses with the glucose chemical formula. Glucose and other sugars can exist in the open chain or ring form. The open chain form permits the free carbonyl to reduce Benedict's or Fehling's reagents. β-D-glucose is a primary source of energy for humans. Energy in the form of ATP can be obtained from glucose through the anaerobic pathway. Additional energy is then obtained from the product pyruvate as it passes through the TCA cycle. The nervous system relies solely on glucose for energy in normal circumstances. Therefore, it is important to maintain the glucose concentration within a narrow range.

Insulin, which is produced in the β-cells of the pancreas, is responsible for the uptake of glucose into cells and the reduction of plasma glucose postprandially. Insulin also promotes glycogenolysis and triglyceride synthesis. Glucagon, which is produced in the β-cells of the pancreas, opposes the action of insulin. Both glucagon and epinephrine increase plasma glucose by activating gluconeogenesis and glycogenolysis in the liver. Gluconeogenesis is the formation of glucose from lactate, amino acids, pyruvate, and glycerol.

Diabetes mellitus can be classified as type 1 and type 2. The development of type 1 appears to be partly related to an individual's human leukocyte antigen (HLA) genotype.

Type 1 may also have an environmental component, which is thought to trigger an immune reaction that leads to an autoimmune response, which causes β-cell destruction. Untreated hyperglycemia in diabetes usually is no greater than 500 mg/dL (28 mmol/L) when healthy renal function is present. Ketoacidosis is more common in type 1, osmolality is increased, plasma potassium is increased, and plasma sodium is slightly decreased. Bicarbonate is decreased in response to the acidosis.

Type 2 is also thought to have a genetic factor. Type 2 individuals have no β-cell destruction and may have decreased, normal, or increased insulin concentrations—but they are insulin resistant at the tissues. In type 2, there is a greater tendency toward hyperosmolar nonketotic coma. Type 2 is characterized by a glucose concentration of greater than 600 mg/dL (33 mmol/L) and an absence of ketones. The BUN and osmolality are increased and urine output is decreased. GDM may be related to type 2. The three definitive tests for diabetes are (1) symptoms of diabetes plus a random plasma glucose level of ≥200 mg/dL, (2) a fasting plasma glucose of ≥126 mg/dL, or (3) an OGTT with a 2-hour postload (75-glucose load) level ≥200 mg/dL. Any of the three criteria must be confirmed on a subsequent day by any of the three methods.

Hypoglycemia can occur as one of two types: reactive (postprandial) or (fasting) postabsorptive. Common causes of fasting hypoglycemia include insulin excess, oral hypoglycemic drugs, and prolonged alcohol ingestion. Neonatal, congenital, and ketotic hypoglycemia occur in children. Congenital forms of hypoglycemia include von Gierke disease. Galactosemia is another relatively common congenital variety of hypoglycemia.

REVIEW QUESTIONS

1. Which of the following hormones promotes gluconeogenesis?
   a. Epinephrine
   b. Hydrocortisone
   c. Insulin
   d. Thyroxine
2. Glucose oxidase oxidizes glucose to gluconic acid and:
   a. H_2O_2
   b. CO_2
   c. HCO_3
   d. H_2O

3. Hexokinase catalyzes formation of   from glucose and ATP:
   a. Acetyl CoA
   b. Fructose 6-phosphate
   c. Glucose 6-phosphate
   d. Lactose

4. What is the preferred specimen for glucose analysis?
   a. EDTA plasma
   b. Fluoride oxalate plasma
   c. Heparinized plasma
   d. Serum

5. Hyperglycemic factor produced by the pancreas is:
   a. Follicle-stimulating hormone (FSH)
   b. Glucagon
   c. Insulin
   d. Luteinizing hormone (LH)

6. Polarographic methods of glucose assay are based on which principle?
   a. Nonenzymatic oxidation of glucose
   b. Rate of oxygen depletion measured
   c. Chemiluminescence caused by formation of ATP
   d. Change in electrical potential as glucose is oxidized

7. Select the enzyme that is most specific for β-D-glucose:
   a. Glucose oxidase
   b. Glucose-6-phosphate dehydrogenase
   c. Hexokinase
   d. Phosphohexisomerase

8. Select the coupling enzyme used in the hexokinase method for glucose:
   a. Glucose dehydrogenase
   b. Glucose-6-phosphatase
   c. Glucose-6-phosphate dehydrogenase
   d. Peroxidase

9. All of the following are characteristic of von Gierke disease EXCEPT:
   a. Hypoglycemia
   b. Hypolipidemia
   c. Increased plasma lactate
   d. Subnormal response to epinephrine

10. Following the 1997 ADA guidelines, the times of measurement for plasma glucose levels during an OGTT in nonpregnant patients are:
    a. Fasting and 2 hours
    b. Fasting and 60 minutes
    c. 30, 60, 90, and 120 minutes
    d. Fasting, 30, 60, 90, and 120 minutes

REFERENCES

1. National Diabetes Data Group. Classification and diagnosis of diabetes mellitus and other categories of glucose tolerance. Diabetes 1979;28:1039–1057.
2. Expert Committee on the Diagnosis and Classification of Diabetes Mellitus. Report of the Expert Committee on the Diagnosis and Classification of Diabetes Mellitus. Diabetes Care 1997;20(7):5–19.
3. Malchoff CD. Diagnosis and classification of diabetes mellitus. Conn Med 1991;55(11):625.
4. Expert Committee on the Diagnosis and Classification of Diabetes Mellitus. Report of the Expert Committee on the Diagnosis and Classification of Diabetes Mellitus. Diabetes Care 1997;20(7):5–19.
5. Cryer PE. Glucose homeostasis and hypoglycemia. In: Wilson JD, Foster DW, eds. Williams textbook of endocrinology. Philadelphia: WB Saunders, 1992.
6. American Diabetes Association. Tests of glycemia in diabetes. Diabetes Care 1998;21(Suppl. 1):S69–S71.
7. Higgis PJ, Bunn HF. Kinetic analysis of the nonenzymatic glycosylation of hemoglobin. J Biol Chem 1981;256:5204–5208.
8. Eckfeldt JH, Bruns DE. Another step towards standardization of methods for measuring hemoglobin A_{1c}. Clin Chem 1997;43(10):1811–1813.
9. Goldstein DE, et al. Tests of glycemia in diabetes (technical rev). Diabetes Care 1995;18(6):896–909.
10. Stehouwer CDA, Donker AJM. Clinical usefulness of measurement of urinary albumin excretion in diabetes mellitus. Neth J Med 1993;42:175.

CHAPTER 11

Lipids and Lipoproteins

Judith R. McNamara, G. Russell Warnick, Lily L. Wu

Objectives

Upon completion of this chapter, the clinical laboratorian should be able to:

- *Explain lipid/lipoprotein physiology and metabolism.*

- *Define the following terms: lipoprotein, exogenous, endogenous, chylomicrons, fatty acids, phospholipids, triglycerides, cholesterol, VLDL, LDL, HDL, and Lp(a).*

- *Describe the clinical tests used to assess lipids and lipoproteins, including the principles and procedures.*

- *Given clinical data, evaluate the patient's lipid or lipoprotein status.*

- *Identify the reference ranges for the major lipids discussed.*

- *Discuss the interaction in the body between the lipids and lipoproteins and various hormones.*

- *Relate the clinical significance of lipid and lipoprotein values in the assessment of coronary heart disease.*

- *Discuss the incidence and types of lipid and lipoprotein abnormalities.*

KEY TERMS

Arteriosclerosis	Fatty acids	Lipoprotein
Cholesterol	Friedewald	Lp (a)
Chylomicrons	calculation	Phospholipids
Dyslipidemias	Glycolipid	Triglycerides
Endogenous	HDL	VLDL
Exogenous	LDL	

Lipoproteins constitute the body's "petroleum industry." Like the great tankers that travel the world's oceans transporting petroleum for fuel needs, the large chylomicrons carry dietary triglycerides throughout the circulatory system to cells, finally docking at the liver to deposit the chylomicron remnants. The *very low-density lipoproteins (VLDL)* are like tanker trucks, carrying triglycerides assembled in the liver out to cells for energy needs or storage as fat. The *low-density lipoproteins (LDL),* rich in cholesterol, are the almost empty tankers that deliver cholesterol to the peripheral cells after the triglycerides have been off-loaded. The *high-density lipoproteins (HDL)* are the cleanup crew, gathering up extra cholesterol for transport back to the liver. Thus, cholesterol, which in excess contributes to heart disease, is used by the body for such useful functions as the maintenance of cell membranes and as a precursor for hormone synthesis, and facilitates triglyceride transport in serving the fuel needs of the body.

The lipids and lipoproteins, which are central to the metabolism of the body, have become increasingly important in clinical practice, primarily because of their association with coronary heart disease (CHD). Many national and international epidemiologic studies have demonstrated that, especially in affluent countries with high fat consumption, there is a clear association with the development of atherosclerosis. Decades of basic research have contributed to knowledge about the nature of the lipoproteins and their lipid and protein constituents, as well as their role in the pathogenesis of the atherosclerotic process. International efforts to reduce the impact of CHD on public health have focused attention on improving the reliability and convenience of the lipid and lipoprotein assays. Expert panels have developed guidelines for the detection and treatment of high cholesterol, as well as laboratory performance goals and detailed recommendations for reliable measurement of the lipid and lipoprotein analytes.[1–5] The lipids and lipoproteins are reviewed here primarily in the context of clinical and laboratory guidelines from the National Cholesterol Education Program (NCEP).

LIPID CHEMISTRY

The major *lipids* of the body—triglycerides, cholesterol, phospholipids, and glycolipids—play a variety of biologic roles. They serve as a primary source of fuel and are important components of cell membranes and many cell structures, they provide stability to the cell membrane and allow for transmembrane transport, and are transported through the blood stream in the form of lipoproteins.

Lipid and Lipoprotein Constituents

Fatty Acids
Fatty acids are the major constituents of triglycerides and phospholipids. There are short-chain (4–6 carbon atoms), medium-chain (8–12 carbon atoms), and long-chain (>12 carbon atoms) fatty acids. Many dietary fatty acids are straight-chain compounds with even numbers of carbon atoms (4–24 carbon atoms). Depending on the number of double-bonds in the molecule, the fatty acids may be saturated (no double-bonds), monounsaturated (one double-bond), or polyunsaturated (two or more double-bonds). The double-bonds of unsaturated fatty acids are usually arranged in *cis* form, which causes the fatty acids to bend. These bends increase the space that the fatty acids require and, as a result, increase the fluidity. Double-bonds occurring in the *trans* form do not bend, and have properties more closely resembling saturated fatty acids.

Triglycerides
The *triglyceride* molecule comprises one molecule of glycerol with three fatty acid molecules attached (usually three different fatty acids including both saturated and unsaturated molecules). Triglycerides containing saturated fatty acids without bends pack together closely and tend to be solids at room temperature. Unsaturated fatty acids do not pack together closely and tend to be liquids at room temperature. Most triglycerides from plants (*eg,* corn, sunflower seed, and safflower oil) comprise primarily polyunsaturated fatty acids, oriented in the *cis* formation, and are in liquid form even at 4°C. Triglycerides from animal sources, which contain largely saturated fats, are typically solid at room temperature. Vegetable oils that have been hydrogenated to solidify them for use as butter and lard replacements have had the *cis* double-bonds changed to *trans* double-bonds.

The source of triglycerides in the body can be either *exogenous* (dietary) or *endogenous*(synthesized in liver and other tissues). Triglyceride molecules allow the body to compactly store long carbon chains (fatty acids) for energy that can be used during fasting states between meals. The high-energy triglyceride molecules, which constitute 95% of fats stored in tissues, are transported in plasma mostly in the form of large triglyceride-rich lipoproteins called chylomicrons and VLDL. When the triglycerides are metabolized, their fatty acids are released to the cells and converted into energy. The glycerol of the triglyceride is recycled into additional triglyc-

erides. Triglycerides also provide insulation to vital organs in the form of fat deposits in adipose cells. The breakdown of triglycerides is facilitated by hormone-sensitive lipase, lipoprotein lipase (LPL), epinephrine, and cortisol.[6] The LPL molecules are attached to heparan sulfate stalks in the capillaries. As the triglyceride-rich lipoproteins (chylomicrons and VLDL) are carried through the circulation, the triglycerides are hydrolyzed as they come in contact with LPL.[7] The supply of free fatty acids available to the cells to be used in the tricarboxylic acid cycle for energy production depends on LPL interaction with chylomicrons and VLDL. Hormone-sensitive lipase acts inside adipose (fat) cells to release free fatty acids from triglyceride stores for energy when dietary sources are unavailable or are insufficient for the body's energy needs. Epinephrine and cortisol promote triglyceride breakdown when the cells need energy and glucose stores have been depleted.

Cholesterol
Cholesterol is an unsaturated steroid alcohol of high molecular weight, consisting of a perhydrocyclopentanthroline ring and a side chain of eight carbon atoms (Fig. 11-1). In its esterified form, it contains one fatty acid molecule. Cholesterol is found almost exclusively in animals. Virtually all cells and body fluids contain some cholesterol. Cholesterol is used for the manufacture and repair of cell membranes, for synthesis of bile acids and vitamin D, and is the precursor of five major classes of steroid hormones: progestins, glucocorticoids, mineralocorticoids, androgens, and estrogens. As with triglycerides, there are both exogenous (dietary) and endogenous (primarily hepatic) sources of cholesterol. Diet contributes 100 mg to 700 mg cholesterol per day (<300 mg/day is recommended for most adults; Table 11-1). Another 500 mg to 1000 mg is produced in the liver and other tissues, and an additional 600 mg to 1000 mg of biliary cholesterol is secreted into the intestine daily, about 50% of which is reabsorbed into the blood. In the body, about 70% of cholesterol is located in stationary pools located in the skin, adipose tissue, and muscle cells; the remaining 30% or so forms a mobile pool that is transported in the form of lipoproteins and circulates through the liver. In the blood circulation, two thirds of the cholesterol is esterified, and one third is in free form.

Phospholipids
Phospholipid, glycolipid, and cholesterol are the three major types of membrane lipids. *Phospholipids* are amphipathic, which means they contain polar hydrophilic (water-loving) head groups and nonpolar hydrophobic (water-hating) fatty acid side chains.[8] With hydrophilic and hydrophobic groups, they act as detergents and are particularly suited to serve as major constituents of biologic interfaces, such as cell membranes and the outer shells of lipoprotein particles. Most phospholipids are formed by the conjugation of two fatty acids and a phosphorylated glycerol. The two fatty acids normally are 14 to 24 carbon atoms long, with one being saturated and one

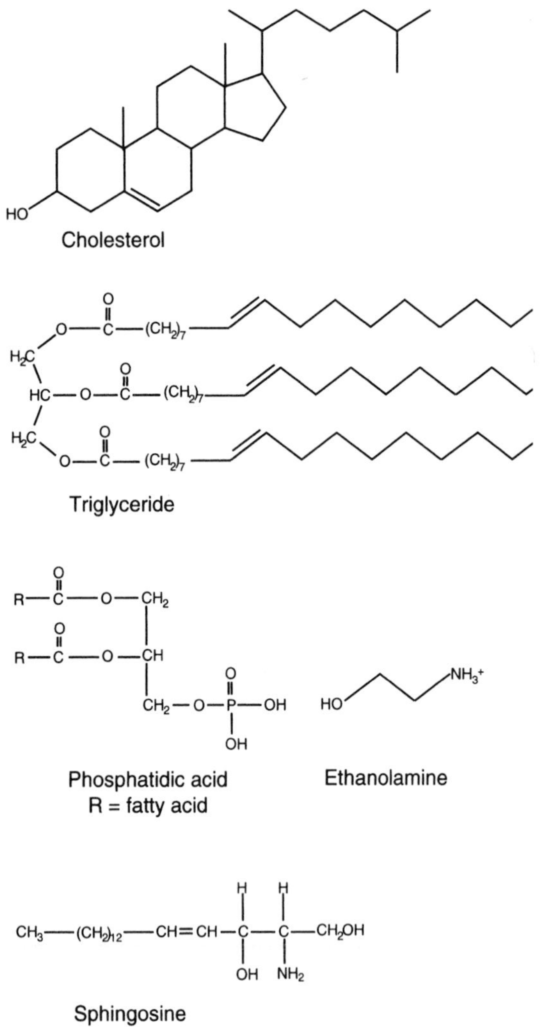

Figure 11-1. Lipid structures.

unsaturated. The phosphorus group can be complexed to choline to form phosphatidylcholine (lecithin), or to ethanolamine, inositol, or serine to form cephalins.

Different fatty acid residues result in several different lecithins and cephalins. Sphingomyelin is the only phospholipid in membranes that is not derived from glycerol but rather from an amino alcohol called *sphingosine* by affixing a fatty acid to its amino group and a phosphoryl choline to its hydroxy group.

Glycolipids

Glycolipids are sugar containing lipids that consist of a sphingosine molecule that has a fatty acid attached to its amino group and a sugar linked to the primary alcohol group. The whole molecule is often referred to as a *ceramide*. The simplest glycolipids are galactosylceramide and glucosylceramide. More complex glycolipids, such as gangliosides, may contain branched chains with as many as seven sugar residues. Gangliosides are the major lipids of cell membranes of the brain and central nervous system, and are synthesized from ceramides. Membrane glycosphingolipids also play an important role in cell recognition and blood typing.

Prostaglandins

Prostaglandins (PG) are long-chain polyunsaturated fatty acids, called *eicosanoids,* which contain 20 carbon atoms, including a five-carbon (cyclopentane) ring, and affect a wide variety of physiologic processes. PG are synthesized in almost all tissues from arachidonate and other polyunsaturated fatty acids. The major classes of PG are designated PGA through PGI, with a subscript number to indicate the number of carbon to carbon double-bonds outside the ring. Other PG-related compounds that are also derived from arachidonate include thromboxanes, with a six-member ether ring, and leukotrienes. PG, thromboxanes, and leukotrienes are short lived and can alter the activities of the cells in which they are synthesized as well as those of adjoining cells.

Apolipoproteins[9]

Apolipoproteins (apo) are the protein moieties associated with the plasma lipoproteins (Table 11-2). Apos are structural elements in the amphipathic shell of lipoprotein particles and help to keep the lipids in solution during circulation through the blood stream. Apos also regulate plasma lipid metabolism by activating and inhibiting enzymes that are involved in the process. They interact with specific cell-surface receptors and direct the lipids to the correct target organs and tissues in the body.

TABLE 11-1. Dietary Guidelines Developed by the American Heart Association and Recommended by the Adult Treatment Panel II of the NCEP (as Compared With the Average American Diet)

Dietary Nutrient	Step I Diet	Step II Diet	Average American Diet
Total fat (% of calories)	≤30%	≤30%	36%
Saturated	<10%	<7%	15%
Monounsaturated	<15%	<15%	15%
Polyunsaturated	<10%	<10%	6%
Cholesterol	<300 mg/day	<200 mg/day	>400 mg/day

TABLE 11-2. Characteristics of Human Apolipoproteins and Their Variants

Apolipoproteins	Functions	Major Source	Normal Plasma Concentration (mg/dL)	Relative Mass (Mr)	Isoelectric Point
Apo A-I	Major structural protein in HDL Activates LCAT Ligand for HDL binding	Liver and intestine	100–200	27,000	5.3–5.4
Apo A-II	Structural protein in HDL Activates LCAT Enhances hepatic triglyceride lipase activity	Liver	20–50	17,400 (dimer)	5.0
Apo A-IV	Component of intestinal lipoproteins	Intestine	10–20	38,000	—
Apo B-100	Major structural protein in VLDL and LDL Ligand for the LDL receptor	Liver	70–125	5.4×10^5	—
Apo B-48	Primarily structural protein in chylomicrons	Intestine	<5	2.6×10^5	—
Apo C-I	Activates lipoprotein lipase	Liver	5–8	6630	6.5
Apo C-II	Activates lipoprotein lipase Activates LCAT	Liver	3–7	8835	4.8
Apo C-III	Inhibits lipoprotein lipase Inhibits receptor recognition of apo E	Liver	10–12	9960	4.5–4.9
Apo E2,3,4	Binds to LDL-receptor and remnant-receptor	Liver	3–15	34,145	5.4–6.1
Apo (a)	Structural protein for Lp(a) May inhibit plasminogen binding	Liver	<30	$3–7 \times 10^5$	—

LCAT, lecithin-cholesterol acyltransferase.
*There are many minor apolipoproteins, such as apo D, apo J, apo H, apo F, and apo G.

Apo A–I is the major apo of HDL. Apo B, which is responsible for the binding of LDL to LDL receptors, is a large protein with a molecular weight of approximately 500 kDa and is the functional protein for transporting cholesterol to cells. More than 95% of the protein in LDL is apo B. Apo B is synthesized in two forms: apo B-100 in the liver and apo B-48 in the intestine. Apo B-48 consists of the first 48% of the amino acid sequence found in apo B-100. Apo B-100 is found in VLDL, IDL, and LDL; apo B-48 is found in chylomicrons. Apo E, which promotes binding of lipoproteins (LDL, VLDL, and apo E-HDL) to the LDL receptor and a specific chylomicron remnant receptor, is also associated both with transport of cholesterol ester in plasma and with the redistribution of cholesterol in tissues.[10] There are three major isoforms of apo E: apo E2, E3, and E4. Individuals have any one (homozygous) or a combination of any two (heterozygous) isoforms. Apos E3 and E4 both bind well to the LDL receptor, whereas E2 is defective in binding. The apo E isoform pattern influences circulating cholesterol levels. For example, individuals with apo E4 (4/2, 4/3, 4/4) average about 10% higher plasma cholesterol levels than ones having apo E3 (3/3).[11] The apo E2/2 phenotype is present in Type III hyperlipoproteinemia.[12] Apo (a) is a highly glycosylated apolipoprotein with approximately 70% structural homology with plasminogen. There are several isoforms of apo (a) ranging in molecular weight from 300 kDa to 700 kDa. The size heterogeneity is under genetic control and is due to the variability in the length of the polypeptide chain.

Lipoproteins[13–16]

Lipoproteins (Fig. 11-2) may be classified by either their density or their electrophoretic mobility. The four major classes of lipoproteins separated by ultracentrifugation according to their differences in density are chylomicrons, VLDL, LDL, and HDL. These lipoproteins also differ in their chemical composition, size, and potential atherogenicity (Table 11-3). Because of the need to transport water-insoluble lipids through the blood stream to various tissues, cholesterol and triglyceride are packaged in spherical lipoproteins. Each lipoprotein contains a nonpolar, hydrophobic lipid core consisting primarily of cholesteryl esters and triglycerides. This core is surrounded by a water-soluble surface comprising apolipoproteins, phospholipids, and free (or unesterified) cholesterol. The polar surface components confer the solubility, which makes possible the transport of the highly insoluble cholesterol esters and triglycerides in the blood circulation. Within each of the major lipoprotein classes, there are subpopulations with subtle variations in size and relative composition.

Chylomicrons

Chylomicrons are the largest of the lipoprotein particles with diameters ranging from 80 nm to 1200 nm and with density <0.95 g/mL. They are the major carriers of exogenous triglycerides.

Chylomicrons comprise 90% to 95% (by weight) triglyceride, 2% to 6% phospholipid, 2% to 4% cholesteryl ester, 1% free cholesterol, and 1% to 2% apolipoprotein. Apo C

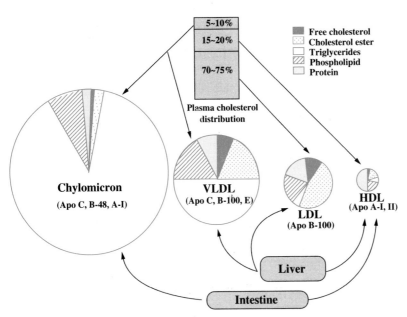

Figure 11-2. Classification of lipoproteins.

and apo B-48 are the most abundant apolipoproteins in chylomicrons, but small amounts of others, including apo E and apos A-I, II, and IV, can also be found. Following absorption of dietary lipids through the microvilli of the intestine, chylomicrons are responsible for transporting dietary triglycerides and some cholesterol to the rest of the body. The clearance time from the formation of chylomicrons after a meal and the removal of remnants by the liver is about 6 hours. Normally, chylomicrons are not found in 12- to 14-hour fasting blood specimens. A creamy layer rising to the top of a fasting serum specimen that has been cooled overnight indicates the presence of chylomicrons and signals a defect in their clearance.

Very Low-Density Lipoproteins
VLDL, like chylomicrons, are also rich in triglycerides. They are the major carriers of endogenous (liver synthesized) triglycerides and have a diameter ranging from 40 nm to 80

nm, with a density range of 0.95 g/mL to 1.006 g/mL. VLDL comprise 50% to 65% (by weight) triglyceride, 8% to 14% cholesteryl ester, 12% to 16% phospholipid, 4% to 7% free cholesterol, and 5% to 10% apolipoprotein. The protein composition includes 40% to 50% apo C, 30% to 40% apo B-100, and 10% to 15% apo E. Excess dietary intake of carbohydrate enhances hepatic synthesis of triglycerides, which, in turn, increases VLDL production. Elevations in VLDL are evidenced by increased serum triglyceride concentrations, and significant elevations can produce turbid serum because VLDL and chylomicrons are large enough particles to scatter light.

Low-Density Lipoproteins
LDL contain 50% cholesterol by weight and are the most cholesterol rich of the lipoproteins. They are synthesized in the liver and are responsible for transporting cholesterol from the liver to the peripheral tissues. LDL have diameters ranging

TABLE 11-3. Characteristics of Human Lipoproteins

Lipoprotein	Density (g/mL)	Electrophoretic Mobility	Source	Apolipoproteins Major & Minor (Italic)	Molecular Weight (10⁶)	Diameter (nm)
Chylomicron	~0.93	Origin	Intestine	CI, II, III; B-48 *A-I, II, IV; E*	50–1,000	80–1,200
VLDL	0.93–1.006	Pre-β	Liver	B-100 *CI, II, III; E*	10–80	40–80
IDL	1.006–1.019	Slow pre-β	Catabolism of VLDL	B-100 *CIII, E*	5–15	30–40
LDL	1.019–1.063	β	Catabolism of IDL	B-100 *C*	~3	18–30
HDL	1.063–1.21	α	Liver, intestine	A-I, II *CI, II, III; E, D*	0.36–0.20	5–12
Lipoprotein (a) [Lp(a)]	1.050–1.100 (sinking pre-β)	Pre-β	Liver	B-100 & apo(a)	~5	25–35

from 18 nm to 30 nm, with a density range of 1.019 g/mL to 1.063 g/mL, and contain 35% to 45% cholesteryl esters, 6% to 15% free cholesterol, 22% to 26% phospholipid, 4% triglyceride, and 22% to 26% apolipoprotein in the form of apo B-100. LDL are the most atherogenic lipoproteins, and high serum levels are regarded as a major CHD risk factor.[17–19]

LDL particles of different size and composition have been separated into as many as eight subclasses by density ultracentrifugation or by gradient electrophoresis gels.[20,21] LDL size is inversely related to serum triglyceride levels.[22] Smaller, more dense LDL are associated with a higher risk of CHD. Very small LDL, however, are not associated with a higher risk of CHD but are observed in severe elevations of triglycerides and a disease called pancreatitis, which is discussed later in the chapter.

High-Density Lipoproteins

HDL are the smallest lipoproteins with diameters ranging from 5 nm to 12 nm and density from 1.063 g/mL to 1.21 g/mL. HDL particles are synthesized by both the liver and intestine, and consist of 25% to 30% phospholipid, 15% to 20% cholesteryl esters, 5% free cholesterol, 3% triglyceride, and 45% to 59% apolipoprotein. The major apolipoproteins are apo A-I (70%) and apo A-II (10%–23%); minor proteins include apo C and apo E. HDL typically carry 20% to 35% of total plasma cholesterol, but unlike LDL, which carry cholesterol to the tissues, HDL take excess cholesterol from the tissues and return it to the liver (reverse transport). Based on density differences, there are two major groups of HDL subclasses: HDL_2 and HDL_3. HDL_2 are larger in size and richer in lipid than HDL_3 and may be the more efficient vehicles for the transfer of cholesterol from peripheral tissue to the liver.[23] In addition to the two major classes of HDL, there are as many as 14 subclasses that have been identified by various methods.[24–26]

Lipoprotein (a) [Lp(a)][27–29]

Lp(a) are LDL like lipoprotein particles with one additional molecule of apo (a) linked to apo B-100 by a disulfide bond. Lp(a) are heterogeneous both in size and density due to differing numbers of peptide sequences called *kringles,* which are named for their resemblance to Danish pastries. Lp(a) have a density of approximately 1.050 g/mL to 1.13 g/mL and migrate at the pre-beta position on plasma lipoprotein electrophoresis. The concentration of Lp(a) is inversely related to the size of the isoform.

The presence of elevated serum levels of Lp(a) is thought to be an independent risk factor for the development of premature CHD, myocardial infarction, and cerebrovascular disease. Because the kringle domains of Lp(a) have a high level of homology with plasminogen, which is an element of the blood coagulation cascade, Lp(a) may compete with plasminogen for binding sites. This interference with plasminogen binding may then retard the rate of clot lysis, thereby increasing the likelihood of myocardial events. Plasma levels of Lp(a) vary widely among individuals in a population, but the levels tend to remain relatively constant within the individual. Elevated plasma levels are found in approximately 10% to 15% of Caucasians, and levels average two times higher in African-Americans. Plasma levels of Lp(a) are predominately genetically determined, but a number of other factors, such as age, race, hormone status, and liver and renal function, moderate the values.

LIPOPROTEIN PHYSIOLOGY AND METABOLISM

Lipid Absorption[30,31]

Lipids constitute an important part of the diet. An average person ingests, absorbs, resynthesizes, and transports about 60 g to 130 g of fat daily in the body, mostly in the form of triglycerides. Because fats are insoluble in an aqueous system, special mechanisms are required for their absorption by the intestine. In the intestinal lumen, the process of digestion converts lipid into more polar compounds with amphophilic properties. Thus, triglycerides are transformed into monoglycerides, diglycerides, and free fatty acids; cholesterol esters are transformed into free cholesterol; and phospholipids are transformed into their lyso-derivatives. These compounds and bile salts form mixed micelles, in which the amphipathic lipids orient themselves with the hydrophobic regions on the inside of the micelles and the polar groups exposed to the aqueous environment. Lipid absorption occurs when the micellar solutions of lipids come in contact with the microvillus membranes of the mucosal cells. The smaller free fatty acids with ten or fewer carbon atoms can pass directly into the portal circulation and be carried by albumin to the liver. The longer-chain fatty acids, monoglycerides, and diglycerides can be absorbed into the intestinal mucosal cells from bile acid micelles. Inside the intestinal cells, the long-chain free fatty acids are immediately re-esterified to form triglycerides and cholesteryl esters. The triglycerides and cholesteryl esters are packaged into the core of chylomicrons for delivery to the blood circulation. Approximately 90% to 95% of triglycerides in the diet are absorbed from the intestinal lumen, whereas only about 50% of the cholesterol is absorbed.

Lipid Synthesis

Triglycerides[8]

Fatty acids are stored in cells as triglycerides. In the liver, adipose tissue, and many other organs and tissues, the synthesis of triglycerides begins with the activation of fatty acids to acyl-CoA ester. This process requires the input of energy from ATP. There are different routes involved in the formation of triglycerides. Two acyl-CoA molecules may either condense with glycerol-3-phosphate to form phosphatidate (diacylglycerol-3-phosphate) or be added to dihydroxyacetone phosphate through an esterification process. The final stage involves removal of the phosphoryl group followed by acylation to form the triglyceride.

Cholesterol[32]

The liver is the major site of cholesterol synthesis, although cholesterol is also produced in many other organs and tissues. Cholesterol is formed from acetyl-CoA. Three molecules of acetyl-CoA are condensed to produce 3-hydroxy-3-methylglutaryl coenzyme A (HMG-CoA), which, in turn, is converted to mevalonic acid through the action of HMG-CoA reductase. Mevalonic acid is converted into squalene after a series of condensation and rearrangement steps. Squalene then cyclizes to form lanosterol, which is further modified to yield cholesterol.

Intracellular cholesterol concentration and the activity of HMG-CoA reductase regulate the synthesis of cholesterol in the liver, which is the rate-limiting enzyme in cholesterol biosynthesis. Consequently, an important class of drugs targets this enzyme as a means of lowering plasma cholesterol levels. In the liver, some of the cholesterol is degraded to bile acids, primarily cholic acid and chenodeoxycholic acid. Formation of bile acids is essential for removing cholesterol from the body.

Lipoprotein Metabolism

The mechanisms by which lipids are used, transported, and removed from the body are complex. Several specialized receptors, enzymes (Table 11-4), transfer proteins, and transport pathways are involved, with each playing an important role or roles. Several defects in cholesterol transport and removal have been shown to lead to the development and progression of atherosclerosis.[31,33]

Lipoproteins transport lipids in three separate but interacting pathways through the body: the exogenous, endogenous, and reverse cholesterol transport pathways. The exogenous pathway transports dietary lipids, through the formation of chylomicrons in the small intestine, to the liver; the endogenous pathway is responsible for transporting hepatic lipids by way of VLDL and LDL to the peripheral tissues; and the reverse cholesterol pathway uses HDL to transport cholesterol from the peripheral tissues back to the liver for excretion or reuse.

Exogenous Pathway

Lipids of dietary origin, mainly dietary triglycerides, absorbed in the small intestine are packaged into chylomicrons (Fig. 11-3). The newly synthesized chylomicrons, collected in the intestinal lymphatics, are released into the blood stream by way of the thoracic duct. After being released into the blood circulation, chylomicrons travel to the capillaries of skeletal muscle, heart, and adipose tissues. On the surface of capillary endothelial cells, they are cleaved by the membrane-bound LPL, and free fatty acids and glycerol are released from the core triglycerides. Some of the free fatty acids are used as an immediate energy source but most are taken up by the tissue and re-esterified into triglycerides for storage. During this metabolic process, phospholipids, free cholesterol, and apos A-I, II, and IV are transferred from chylomicrons to HDL in exchange for apo C and apo E from HDL and VLDL. These processes result in the transformation of chylomicrons into cholesteryl ester–rich chylomicron remnants. The chylomicron remnants are rapidly taken up by the liver through interaction of apo E on the remnant particles with remnant receptors on the surface of liver cells. Once in the liver, lysosomal enzymes break down the remnants to release free fatty acids, free cholesterol, and amino acids. Some of the cholesterol is converted to bile acids. The bile acids and free cholesterol are excreted into the bile and hence into the intestine. However, as previously mentioned, approximately 50% of the excreted cholesterol in the intestine is reabsorbed and returns to the liver; only about 50% of the total excreted cholesterol ends up in the stool as fecal neutral steroids. On the other hand, about 97% to 98% of the bile acids are reabsorbed in the lower intestine, with the remainder excreted in stool as fecal acidic steroids.

Endogenous Pathway

Most triglycerides in the liver that are destined to be incorporated in VLDL are derived from the diet after recirculation from adipose tissue (Fig. 11-4). Only a relatively small fraction appear to be synthesized *de novo* from dietary carbohydrate. Cholesterol used in VLDL formation is derived

TABLE 11-4. Enzymes of Plasma Lipoprotein Metabolism

Enzyme	Major Tissue Source	Substrates	Function	Location	Cofactors
LPL	Adipose tissue (adipocytes) Muscle	Triglycerides and phospholipids of chylomicrons and VLDL	Hydrolyzes triglycerides	Muscle, adipose capillary	Apo CII Apo CIII inhibits
Hepatic lipase	Liver (hepatocytes)	Triglycerides and phospholipids of VLDL, LDL, and HDL	Hydrolyzes triglycerides	Liver	Apo C-II
LCAT	Liver	Cholesterol and phosphatidylcholine of HDL	Esterifies free cholesterol	Plasma	Apo A-I, C-I

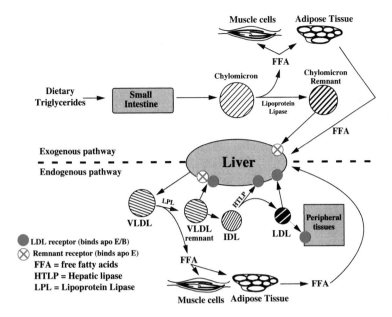

Figure 11-3. Overall exogenous pathway of triglyceride metabolism.

from both dietary cholesterol and that synthesized directly in the liver.

VLDL particles, once secreted into the blood stream, undergo a similar degradation process as chylomicrons. Through the action of LPL and apo C-II, VLDL lose their core lipids, mainly triglycerides, and also lose some of the apolipoproteins and phospholipids from the surface.[16] During this process, VLDL become VLDL remnants, which can then be converted to intermediate-density lipoproteins (IDL) and then further processed to LDL. However, only about 50% of VLDL are converted to LDL; the other 50% are taken up by either the remnant receptor or the LDL receptor of the liver through interaction with apo B or apo E.

LDL are the major cholesterol-carrying lipoproteins and function primarily to transport cholesterol to peripheral cells. LDL bind to high-affinity receptors of plasma membranes and, through internalization, deliver cholesterol to peripheral tissues. Endocytosis of an LDL particle into the cell is followed by its degradation into component parts by cellular lysozymes and hydrolases. The triglycerides are saponified into free fatty acids and glycerol to be used by the cell for energy or to be re-esterified for later use. The apolipoproteins are degraded to their constituent amino acids for use by the cell. Free cholesterol derived from degraded LDL can be used in cell membranes for hormone synthesis or can be re-esterified for storage. In cells of the adrenal and other steroid-synthesizing tissues, cholesterol is used to synthesize the steroids cortisol, testosterone, estrogen, and androgen. Regulation of intracellular cholesterol concentration is controlled by a metabolite of free cholesterol, which turns off cellular cholesterol synthesis and then down-regulates LDL receptors as intracellular cholesterol concentration rises to a certain level. This mechanism allows some control of the routing of cholesterol around the body. Abnormalities in the LDL receptor mechanism can result in elevations of LDL in the circulation and thus lead to hypercholesterolemia and premature atherosclerosis.[34,35]

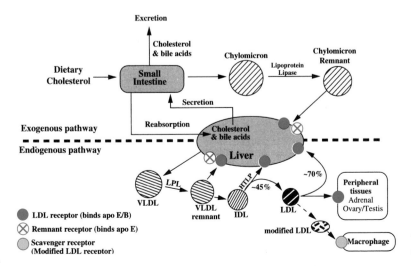

Figure 11-4. Overall pathway of cholesterol metabolism.

Reverse Cholesterol Transport Pathway

A major function of lipoproteins is the transport of cholesterol from the liver through the blood vessels to peripheral tissues. However, when supply exceeds need, cholesterol accumulates in tissues. Because cholesterol can only be excreted from the body through the liver as bile acid or free cholesterol, excess cholesterol must be transported back to the liver to be removed from peripheral tissues. The major role of HDL is in the reverse cholesterol transport pathway,[36] a process which removes cholesterol from peripheral tissues (Fig. 11-5). In this process, nascent HDL produced in the liver and intestine absorb free cholesterol from peripheral cells (even from macrophages)[37] and convert it to cholesteryl ester for storage inside the HDL core during transport. Both lecithin-cholesterol acyltransferase (LCAT) and apos A-I and D are essential for the esterification of cholesterol. There are several routes by which HDL can deliver esterified cholesterol to the liver[38]: direct uptake of HDL by the liver can be mediated by specific receptors capable of binding apo A; apo E–containing HDL may enter the liver through the binding of apo E to the LDL receptor or the remnant receptor on the liver; and, alternatively, with the help of cholesteryl ester transfer protein (CETP), the cholesteryl ester in HDL may be indirectly delivered to the liver by transferring it to triglyceride-rich lipoproteins, such as chylomicrons and VLDL, in exchange for triglycerides.

Lipoprotein Receptors

Perhaps the most important mechanism involving lipoprotein metabolism is the interaction between apolipoproteins on the lipoprotein surfaces and the receptors on various cell surfaces. This critical interaction is necessary not only for delivering lipoproteins to the cells, but also for efficient removal of potentially atherogenic lipoproteins from the blood and peripheral tissues. Lipoprotein receptors are plasma membrane proteins that are capable of binding with high affinity to circulating lipoprotein particles through their interaction with apolipoproteins. The following paragraphs discuss three of the major receptors.

LDL Receptors[33,34,39]

LDL receptors recognize both apo E and apo B-100 and mediate cellular binding, uptake, and degradation of LDL and the other apo B-100 containing lipoproteins (VLDL and IDL). They are widely expressed and, therefore, play an important role in cellular and systemic cholesterol homeostasis.[40]

The receptors are synthesized inside the cell and then migrate to the surface of the cell membrane. Once the receptor binds to LDL, they migrate together into coated pits. The *coated pits* are invaginations in the cell membrane, and act like protective harbors. Once an LDL particle binds to a receptor and migrates to the coated pit, it is drawn into the cell for degradation. Receptors that migrate to areas of the cell membrane outside the coated pits are unable to bind and internalize LDL. Synthesis of LDL receptors is inhibited by high intracellular cholesterol levels, thus allowing the amount of cholesterol entering the cell to be controlled. Conversely, insufficient or defective receptors will stimulate intracellular cholesterol synthesis. In patients who are heterozygous for a disease called familial hypercholesterolemia (approximately 1 in 500 persons in many populations), half of the LDL receptors are defective, causing insufficient LDL to be taken up by the cells. Consequently, LDL cholesterol removal rate from the circulation is about 40%, and the circulating LDL cholesterol level approximately doubles. Homozygotes without any normally functioning LDL receptors have LDL elevations that are even more severe.

Remnant Receptors[41,42]

Remnant receptors recognize apo E and are the major receptors for the clearance of chylomicron remnants and β-VLDL from blood circulation. They also bind apo E–containing HDL.

Scavenger Receptors[43]

Scavenger receptors can be found on the surfaces of macrophages and on a subset of other cells, such as muscle cells. These receptors mediate the removal of modified LDL, including oxidized LDL and β-VLDL (collective name for chylomicron remnants and VLDL remnants) from blood circulation. The scavenger receptors are biochemically and genetically distinct from native LDL receptors. Unlike LDL receptors, scavenger receptor expression is not regulated by intracellular cholesterol concentration. Macrophages can continuously take up cholesterol from modified LDL through scavenger receptors, resulting in cholesterol accumulation and the formation of foam cells, which is the hallmark of early atherosclerotic lesions.

Effect of Hormones

Insulin

One of insulin's major functions is the inactivation of hormone-sensitive lipase. In adipose tissues, the result is suppression of the release of free fatty acids, causing a reduction

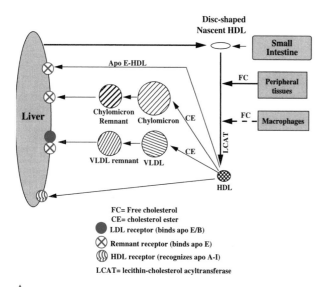

Figure 11-5. Reverse cholesterol transport pathway.

of free fatty acid delivery to the liver. However, insulin and glucose also promote esterification of free fatty acids into triglycerides. Thus, in the liver, insulin action tends to regulate VLDL synthesis. Insulin is also required for expression of LPL activity. This enzyme is essential for triglyceride-rich lipoprotein clearance. Genetic deficiency of the enzyme leads to chylomicronemia.[44] In insulin-dependent diabetes, severe insulin deficiency can lead to fatty liver and marked hypertriglyceridemia, as free fatty acids are released unchecked from adipose tissue, transported to the liver, re-esterified into triglyceride, and transported out as VLDL. The loss of LPL activity leads to defects in the clearance of chylomicrons and VLDL, greatly exacerbating the hypertriglyceridemia frequently found in this condition.

Insulin deficiency or insulin resistance (a diminution of the biologic response to a given concentration of insulin),[45,46] which is associated with diabetes, often leads to elevation of plasma triglycerides, reduction of HDL cholesterol concentrations, and production of atherogenic, small, dense LDL. These unfavorable changes in lipid metabolism can increase the risk for CHD in non–insulin-dependent diabetic patients.[46]

Growth Hormone[47]

Growth hormone can increase the production of VLDL by increasing the production of apo E and apo B-48 and by stimulating lipolysis in adipose tissues and triglyceride synthesis in the liver. It also stimulates the synthesis of LDL receptors and enhances the clearance of lipoproteins by the liver. In patients with growth hormone deficiency, administration of growth hormone results in a normalization of dyslipidemia.

Sex Hormones[48,49]

Estrogens appear to raise VLDL levels by accelerating the production rate of large, triglyceride-rich VLDL. On the other hand, an increase in the number of LDL receptors by estrogen results in a lowering of LDL levels, and an increase in apo A-I production results in an increase in HDL levels.[50] Progesterone balances these effects by lowering VLDL and triglyceride levels by increasing their rate of catabolism.

Thyroid Hormone

Thyroid hormone exerts multiple effects on the metabolism of fatty acids and glycerol. Thyroid hormone stimulates both LPL and hepatic triglyceride lipase synthesis, and also increases LDL receptor production. As a result, patients with hypothyroidism will exhibit elevated levels of plasma LDL-cholesterol.[51]

LIPID AND LIPOPROTEIN ANALYSES

Lipid Measurement

Accuracy is especially important in the analysis of the lipids and lipoproteins. Because these analytes are important indicators of CHD risk and require results from large popu-

lation studies to assign the levels associated with that risk, decision cut-points cannot be easily established by an individual laboratory or manufacturer as is done for many other diagnostic analytes. Decision cut-points have been set by NCEP expert panels based on population distributions and coronary disease risk relationships established in large long-term epidemiologic studies. Standardization of the lipid analytical methods in the participating research laboratories involved in these epidemiologic studies made results comparable among laboratories and over time. For other research laboratories to generate comparable results and for routine clinical laboratories to obtain reliable classification of patients using the national decision cut-points, the methods must be standardized to the same accuracy base used for the national studies, a process that will be described subsequently.

Cholesterol Measurement

The lipid work-up traditionally has begun with measurement of total serum cholesterol. The lipoproteins are also generally quantitated based on their cholesterol content. The early analytical methods[52] used strong acids (eg, sulfuric and acetic) and chemicals (eg, acetic anhydride or ferric chloride), which produce a measurable color with cholesterol. Because the strong acid reactions are relatively nonspecific, partial or full extraction by organic solvents was used to improve specificity. The current reference method for cholesterol uses hexane extraction after hydrolysis with alcoholic KOH followed by reaction with Liebermann-Burchard color reagent, which comprises sulfuric and acetic acids and acetic anhydride.[53,54] The multistep manual method is complicated but gives good agreement with the "gold standard" method developed and applied at the U.S. National Institute for Standards and Technology, the so-called Definitive Method using isotope dilution mass spectrometry.[55]

In recent years, virtually all routine clinical laboratories have progressed to using enzymic reagents. Enzymes, selected for specificity to the analyte of interest, provide reasonably accurate quantitation without the necessity for extraction or other pretreatment. The reagents are mild compared to earlier strong acid reagents and suitable for use in modern automated microprocessor-controlled robotic instruments. Cholesterol, and often triglycerides and HDL cholesterol, are included in routine test panels on automated batch and discrete chemistry analyzers. The lipoproteins HDL and LDL are generally quantified based on their cholesterol content.

Although several enzymic reaction sequences have been described, only one sequence is in common use for measuring cholesterol.[5,56,57] About two thirds of circulating cholesterol occurs as cholesteryl ester, so the enzyme cholesteryl ester hydrolase is used to cleave the fatty acid residue and thus convert cholesteryl esters to unesterified or free cholesterol. The free cholesterol is reacted by cholesterol oxidase, producing hydrogen peroxide, which is the substrate for a common enzymic color reaction using horseradish peroxidase to couple two colorless chemicals into a colored compound.

The intensity of the color is proportional to the amount of cholesterol in the samples and can be measured by a spectrophotometer at a wavelength of about 500 nm. The enzymes and reagents have improved over the years so most commercial reagents, when calibrated appropriately, can be expected to give reliable results. This reaction sequence is generally used on serum without an extraction step but can be subject to interference. For example, vitamin C and bilirubin are reducing agents that can interfere with the peroxidase catalyzed color reaction.[58]

Triglycerides

Measurement of triglycerides in conjunction with cholesterol is useful in detecting certain genetic and other types of metabolic disorders. However, in many cases the triglyceride measurement is made primarily for estimation of LDL cholesterol by the Friedewald equation (discussed subsequently). Several enzymic reaction sequences are available for triglyceride measurement, all having in common the use of lipases to cleave fatty acids from glycerol.[59] The freed glycerol participates in any one of several enzymic sequences. One of the more common earlier reactions, ending in a product measured in the ultraviolet (UV) region, used glycerol kinase and pyruvate kinase, culminating in the conversion of NADH to NAD$^+$ with an associated decrease in absorbency.[60] This reaction is quite susceptible to interference and side reactions. The UV endpoint is also less convenient for modern analyzers, so this and other UV sequences are gradually being replaced by a second sequence involving glycerol kinase and glycerol-phosphate oxidase, which feeds into the same peroxidase color reaction described for cholesterol.[61]

The enzymic triglyceride reaction sequences also react with endogenous free glycerol, which is universally present in serum and constitutes a significant source of interference.[62,63] In most specimens, the endogenous free glycerol contributes to a 10 mg/dL to 20 mg/dL overestimation of triglycerides. A few specimens, about 20%, will have higher glycerol, with levels increased in certain conditions such as diabetes and liver disease or from glycerol-based medications. Most research laboratories incorporate some type of correction for endogenous free glycerol, but this practice is uncommon in clinical laboratories. The most common correction, designated double cuvet blanking, results from the performance of a second parallel measurement using reagent without lipase to quantitate only the free glycerol blank. This measurement is subtracted from the total glycerol measurement of the complete reaction sequence to determine a net or blank-corrected triglyceride result.[64] Another approach, designated the single cuvet blank, begins with the lipase-free reagent. After a brief incubation, a blank reading is taken, which measures only the concentration of free glycerol. The lipase is then added as a second separate reagent and after additional incubation, a total reading is taken that, corrected for the blank by the instrument, represents net triglycerides.[65] Because these blank corrections add to the cost of the measurement and are perhaps unjustified by the accuracy

requirements in routine practice, a convenient and easily implemented alternative—designated calibration blanking—involves adjustment of the calibrator set points to compensate for the average free glycerol content of specimens. This approach is usually reasonably accurate because the free glycerol levels are relatively low and consistent in most specimens. Thus, only a few specimens will be undercorrected but will still be more accurate than uncorrected.

The triglyceride reference method involves alkaline hydrolysis, solvent extraction, and a color reaction with chromotropic acid.[66] The assay is tedious, poorly characterized, and not used except in the standardization laboratory at the Centers for Disease Control and Prevention (CDC). A more suitable reference method with solvent extraction followed by an optimized enzymic assay is in development. Accuracy in the triglyceride measurement is not as critical as it is for cholesterol, because the cut-points are widely spaced and because the physiologic variation is so large, with the coefficients of variation (CVs) for biologic variation in the 25% to 30% range, that analytical variation is relatively insignificant.

Phospholipids

Quantitative measurement of phospholipids is rare in routine clinical practice. At one time, there was a suggestion that HDL be quantified in terms of phospholipids because it is the predominant lipid constituent, but this practice has never bee adopted.[67] Phospholipids are sometimes measured in research (*eg,* in studies of dietary influences). An enzymic reaction sequence using phospholipase D, choline-oxidase, and horseradish peroxidase measures the choline-containing phospholipids lecithin, lysolecithin, and sphingomyelin, which account for at least 95% of total phospholipids in serum.[67,68] Commercial kit methods with this enzymic sequence can be purchased. Before the availability of enzymic reagents, the common quantitative method involved extraction and acid digestion with analysis of the total lipid-bound phosphorus.[69]

Certain choline-containing phospholipids within this class, or their ratios, have been determined in the clinical laboratory. For example, fetal lung maturity has been evaluated from characteristic patterns of phospholipids in amniotic fluid. In this instance, phospholipids may be recovered by solvent extraction, applied to a silica gel plate for separation by thin-layer chromatography, and quantified after visualization by treating with iodine vapor. The ratio between lecithin and sphingomyelin is predictive of the stage of development of the fetal lung.

Apolipoprotein Measurement

There are a number of apolipoproteins that are incorporated into the lipoproteins (see Table 11-3). Three apolipoproteins in particular have been of interest for clinical diagnostic purposes. Apo B, the major protein of LDL and VLDL, is an indicator of combined LDL and VLDL concentration and can be measured directly in serum by immunoassay methods.[70] This characteristic has provided some researchers with a rationale for replacing the alternative separation and measurement of LDL cholesterol.[71] Apo A-I, as the major protein of HDL, could be measured directly in serum in place of separation and analysis of HDL cholesterol, and Lp(a) could be measured as an independent indicator of CHD risk. These apolipoproteins are commonly measured in research laboratories and in some clinical laboratories supporting cardiovascular practices in addition to the conventional lipoprotein measurements. The apolipoproteins are useful in patient management by skilled practitioners, but they have not yet been accepted or recommended by any organization, such as NCEP, for use in routine practice. Prospective studies have not consistently shown them to be independent CHD risk factors, after adjusting for LDL and HDL cholesterol.

The apolipoproteins are measured by immunoassays of various types with several commercial kit methods available. Most common in routine laboratories are turbidimetric assays for chemistry analyzers or nephelometric assays for dedicated nephelometers. Especially for apo B and Lp(a), these light-scattering assays are often subject to interference from the larger triglyceride-rich lipoproteins, chylomicrons, and VLDL. Enzyme-linked immunosorbent assay (ELISA), radioimmunodetection (RID), and radioimmunoassay (RIA) methods are also available but, especially for the RID and RIA, are becoming less common. Antibodies used in the immunoassays may be polyclonal or monoclonal. International efforts to develop reference materials and standardization programs for the assays are in progress. Because Lp(a) is genetically heterogeneous and the levels and CHD risk correlate with the isoform size, qualitative assessment of isoform distribution may also be important.[29]

Lipoprotein Methods

Several methods have been used for the separation and quantitation of the serum lipoproteins. Lipoprotein separations take advantage of physical properties such as density, size, charge, and apolipoprotein content. The range in density observed among the lipoprotein classes is a function of the relative lipid and protein content and enables fractionation by ultracentrifugation. Electrophoretic separations take advantage of differences in charge and size. Chemical precipitation methods, which are most common in clinical laboratories, depend on particle size, charge, and differences in the apolipoprotein content. Antibodies specific to apolipoproteins can be used to bind and separate lipoprotein classes. Chromatographic methods take advantage of size differences in molecular sieving methods or composition in affinity methods using, for example, heparin sepharose.

Many ultracentrifugation methods have been used in the research laboratory, but ultracentrifugation is uncommon in the clinical laboratory.[72] The most common approach, called *preparative ultracentrifugation,* uses sequential density adjustments of serum to fractionate major and minor lipoprotein classes.[73] Density gradient methods[74–78]—either nonequilibrium techniques, in which separations are based on the rate of flotation, or equilibrium techniques in which separations run to completion—permit fractionation of several or all classes in a single run. The available methods use different types of ultracentrifuge rotors: swinging bucket, fixed angle, vertical, and zonal. Newer methods have trended toward smaller scale separations in small rotors using tabletop ultracentrifuges.[79,80] Ultracentrifugation, even though tedious, expensive, and technically demanding, remains a workhorse for separation of lipoproteins both for quantitative purposes and for preparative isolations. Ultracentrifugation is also used in the reference methods for lipoprotein quantitation, which is appropriate because the lipoproteins are classically defined in terms of hydrated density.

Electrophoretic methods[81,82] allow separation and quantitation of major lipoprotein classes and provide a visual display useful in detecting unusual or variant patterns. Agarose gel has been the most common medium for separation of intact lipoproteins, providing a clear background and convenient use.[83–87] Electrophoretic methods, in general, have been considered useful for qualitative analysis but less than desirable for lipoprotein quantitation because of poor precision and large systematic biases compared to other methods.[88] Evaluations of one of the newer commercial automated electrophoretic systems, however, demonstrate that electrophoresis can be precise and accurate.[89]

Electrophoresis in polyacrylamide gels is used for separation of lipoprotein classes,[90] of subclasses, and of the apolipoproteins. Of particular recent interest have been methods that fractionate LDL and HDL subclasses to characterize the more atherogenic,[21,24,26,91] heavier, lipid depleted, and smaller fractions versus the larger, lighter subclasses.

Separation methods involving chemical precipitation, usually with polyanions such as heparin and divalent cations such as manganese, can be used to separate any of the lipoprotein classes. Because of their convenience, such methods have become common in clinical laboratories in the separation of HDL.[92–94] Apo B in VLDL and LDL is rich in positively charged amino acids, which preferentially form complexes with the polyanions; the divalent cations neutralize the charged groups on the lipoproteins making them aggregate and become insoluble. The insoluble lipoproteins aggregate and precipitate, leaving HDL in the solution. By varying the concentration of polyanion and divalent cation any of the lipoprotein classes can be separated with reasonable specificity.

Separation of lipoproteins by immunochemical means using antibodies that are specific to epitopes on the apolipoproteins has potential for both research and routine use.[95,96] Antibodies can be immobilized on a solid support, such as a column matrix or latex beads, to facilitate separations. The apo B–containing lipoproteins as a group can be bound by antibodies to apo B. Selectivity within the apo B–containing lipoproteins, such as removing VLDL while retaining LDL, can be obtained by using antibodies to the minor apolipoproteins. HDL can be selectively removed using antibodies to apo A-I, the major protein of HDL. The use of latex immobilized antibodies is the basis for a recently introduced commercial method for direct quantitation of LDL cholesterol.[97] Immobilized monoclonal antibodies have also been used to separate a fraction of remnant lipoproteins designated RLP (remnant-like particles).[98–100]

HDL

The measurement of HDL cholesterol has assumed greater importance in the latest NCEP treatment guidelines. Previously, HDL cholesterol was measured as a risk factor but otherwise was not considered in treatment decisions. Following recommendations of a National Institutes of Health–sponsored consensus panel,[101] the NCEP guidelines now include HDL cholesterol measurement with total cholesterol in the first medical work-up. Because the risk associated with HDL cholesterol is expressed over a relatively small concentration range, accuracy in the measurement is especially important.

For routine diagnostic purposes, HDL has been separated almost exclusively by chemical precipitation, and for many years HDL cholesterol measurement has been a two-step procedure. The precipitation reagent is added to serum or plasma to aggregate non-HDL lipoproteins, which are sedimented by centrifugation. Early methods used centrifugation forces of approximately 1500 × gravity, requiring lengthy centrifugation times of 10 minutes to 30 minutes. Newer methods use high-speed centrifuges with forces of 10,000 to 15,000 × gravity, decreasing centrifugation times to 3 minutes to 5 minutes. The HDL is then quantified as the cholesterol in the supernate with analysis commonly by one of the enzymic assays modified for the lower HDL cholesterol range.

The earliest common precipitation method used heparin in combination with manganese to precipitate the apo B–containing lipoproteins.[102–104] Because manganese produced interference with enzymic assays, alternative reagents were developed.[105] Sodium phosphotungstate[106,107] with magnesium became common in routine use but, because of its sensitivity to reaction conditions and greater variability, is being replaced by dextran sulfate (a synthetic heparin) with magnesium.[108] The earliest dextran sulfate methods used material of 500 kDa, but 50 kDa, considered more specific, is now becoming more common.[94] Polyethylene glycol also precipitates lipoproteins, but 100-fold higher concentrations of reagent are required than for polyanions. The consequent larger dilutions and highly viscous reagents, which are difficult to pipet precisely, make this reagent less common.[109–111] Numerous commercial versions of these precipitation reagents are available. In the past, the various methods have sometimes given quite different results because there has been a lack of standardization, but recently a protocol has become available to reagent manufacturers that should reduce the amount of variability in the future.

A significant problem with HDL precipitation methods is interference from elevated triglyceride levels, which prevents sedimentation of the precipitate.[112] When the triglyceride-rich VLDL and chylomicrons are present, the low density of the aggregated lipoproteins may prevent them from sedimenting or may even cause floating during centrifugation. This incomplete sedimentation, indicated by cloudiness, turbidity, or particulate matter floating in the supernate, results in overestimation of HDL cholesterol. High-speed centrifugation will reduce the proportion of turbid supernates. Predilution of the specimen also promotes clearing but may lead to errors in the cholesterol analysis. Turbid supernates may also be cleared by ultrafiltration, a method that works well but is tedious and inefficient.

New, or so-called homogeneous, methods streamline and automate the preliminary HDL separation step. These methods use specific polymers, detergents, and even modified enzymes to suppress the cholesterol reaction in lipoproteins other than HDL.[113] This permits homogeneous assessment, with no preanalytical steps, on the chemistry analyzer. Reagent is added to "block" non-HDL lipoproteins, and then the remaining (HDL) cholesterol is measured, using enzymic reagent. These homogeneous assays appear to be highly precise and reasonably accurate.[114]

The accepted reference method for HDL cholesterol has been a three-step procedure developed at the CDC. This method involves ultracentrifugation to remove VLDL, heparin manganese precipitation from the 1.006 g/mL infranate to remove LDL, and analysis of supernatant cholesterol by the Abell-Kendall assay.[54] Because this method is tedious and expensive, a simpler direct precipitation method has been validated by the CDC Network Laboratory Group as a designated comparison method, it uses direct dextran sulfate (50 kDa) precipitation with an Abell-Kendall cholesterol analysis.[115]

LDL

LDL cholesterol, as the proven atherogenic lipoprotein, is the primary basis for treatment decisions in the NCEP clinical guidelines.[116,117] The common research method for accurate LDL cholesterol quantitation and the basis for the reference method is designated *beta-quantification,* where beta refers to the electrophoretic term for LDL. The beta-quantification technique involves a combination of ultracentrifugation and chemical precipitation.[104,118] Ultracentrifugation of serum at the native density of 1.006 g/L is used to float VLDL and any chylomicrons for separation. The fractions are recovered by slicing the tube between the fractions and pipetting. Ultracentrifugation is preferred for VLDL separation because other methods, such as precipitation, are not as specific for VLDL and are subject to interference from chylomicrons. Ultracentrifugation is a robust but tedious method that can give reliable results provided the technique is meticulous.

In a separate step, chemical precipitation is used to separate HDL from either the whole serum or the infranate obtained from ultracentrifugation. Compared to ultracentrifugation, the precipitation step is efficient and convenient, as well as relatively robust for HDL separation. Cholesterol is quantitated in serum, in the 1.006 g/mL infranatant, and in the HDL supernate by enzymic or other assay methods. LDL cholesterol is calculated as the difference between cholesterol measured in the infranate and in the HDL fraction. VLDL cholesterol is usually calculated as the difference between that in whole serum and the amount in the infranate fraction. The ultracentrifugation step makes beta-quantification tedious and inaccessible for routine diagnostic purposes.

A simpler technique for LDL cholesterol quantitation that is common in both routine and research laboratories and that bypasses ultracentrifugation is the so called *"Friedewald calculation"* or *derived beta-quantification.*[119] HDL is separated by precipitation, its cholesterol is assayed, and cholesterol and triglycerides are measured in the serum. VLDL cholesterol is estimated as triglycerides divided by 5 (when using mg/dL units), an approximation that works reasonably well in most normolipemic specimens. The presence of elevated triglycerides—400 mg/dL is the accepted limit, chylomicrons, and a β-VLDL characteristic of the rare type III hyperlipoproteinemia preclude this estimation. The estimated VLDL cholesterol and measured HDL cholesterol are subtracted from total serum cholesterol to estimate or derive LDL cholesterol.

The Friedewald estimation has been used almost universally in estimating LDL cholesterol in routine clinical practice. Investigations in lipid specialty laboratories have suggested the method is reliable for patient classification, provided the underlying measurements are made with appropriate accuracy and precision.[120,121] There is considerable concern about the reliability in routine laboratories, however, because the error in estimated LDL cholesterol includes the cumulative error in each of the underlying measurements used in the calculation of cholesterol, triglycerides, and HDL cholesterol. The NCEP laboratory expert panel reviewed performance data and concluded that the level of analytical performance required to derive LDL cholesterol accurately enough to meet clinical needs was beyond the capability of most routine laboratories. To meet the requisite NCEP precision goal of 4% CV for LDL cholesterol (Table 11-5), a laboratory would be required to achieve half the NCEP goals for each of the underlying measurements. The NCEP panel concluded that alternative methods are needed for routine diagnostic use, preferably methods that directly separate LDL for cholesterol quantitation.[3]

In response to the NCEP request, several direct LDL cholesterol methods have recently been developed or refined for general use.[88] One commercial method that is compatible as a pretreatment step with a variety of analytical systems uses immunochemical separation.[97] A mixture of antibodies specific to epitopes on the apolipoproteins of VLDL and HDL is immobilized on latex beads. Specimen is added to the beads in a microfiltration device, and is mixed and subjected to centrifugation. VLDL and HDL are retained by the filter while LDL passes through. Cholesterol in the LDL filtrate is assayed by enzymic reagent using an au-

TABLE 11-5. NCEP Analytical Performance Goals			
Cholesterol	**Precision CV** 3%	**Bias** ±3%	**Total Error** ±8.9%
HDL			
1993			
≥42 mg/dL	6%	±10%	±21.8%
<42 mg/dL	SD ≤ 2.5 mg/dL		
1998			
≥42 mg/dL	4%	±5%	±12.8%
<42 mg/dL	SD < 1.7 mg/dL		
LDL	4%	±4%	±11.8%
Triglycerides	5%	±5%	±14.8%

tomated chemistry analyzer. This convenient method has been found to correlate with beta-quantification on both fasting and nonfasting specimens.[97] Similarly to the homogeneous assays for HDL cholesterol, fully automated methods are now being reported for LDL cholesterol measurement.[122] Besides achieving full automation of the challenging LDL cholesterol separation, these assays have the potential to improve precision with acceptable accuracy. Separating LDL from other lipoproteins in this manner is more challenging than for HDL, and more experience will be required, especially with specimens of unusual composition, to judge the adequacy of the homogeneous separations.[123]

Fatty Acids

Analysis of fatty acids is used in the research laboratory for studies of dietary factors,[124] but is used less commonly in the routine laboratory for diagnosis of rare genetic conditions. Fatty acids are commonly analyzed by gas-liquid chromatography.[6] The fatty acids are extracted, undergo alkaline hydrolysis, and then are converted to methyl esters of diazomethane. Application of the sample to a gas chromatograph allows separation of fatty acids. A reference standard typically contains laurate, myristate, palmitate, palmitoleate, phytanate, stearate, oleate, linoleate, arachidate, and arachidonate.[125]

Stool Fat

Fat in stool or urine can have significant diagnostic implications. An adult with a normal diet will not have more than 6 g per day of fat in the feces. Higher values can indicate the presence of malabsorption in children and pancreatic insufficiency in adults. The common method in clinical laboratories requires a 72-hour collection of feces, followed by an analysis of lipid content and calculation of lipid as a percentage of the entire mass.[6]

Compact Analyzers

A major area of development in recent years has focused on compact analysis systems for use in point-of-care testing at the patient's bedside, in the physician's office, in wellness

CASE STUDY 11-1

A 52-year-old male went to his doctor for a physical. The patient was 24 pounds overweight and had been a district manager for an automobile insurance company for the last 10 years. He had missed his last two appointments to the physician because of business. The urinalysis dipstick was not remarkable. His blood pressure was elevated. The blood chemistry results listed in Case Study Table 11-1.1 were obtained.

CASE STUDY TABLE 11-1.1. Laboratory Results

Analyte	Patient Value	Reference Range
Na^+	151	135–143 mEq/L
K^+	4.5	3.0–5.0 mEq/L
Cl^-	106	98–103 mEq/L
CO_2 content	13	22–27 mmol/L
Total protein	5.7	6.5–8.0 g/dL
Albumin	1.6	3.5–5.0 g/dL
Calcium	7.9	9.0–10.5 mg/dL
Cholesterol	210	140–200 mg/dL
Uric acid	6.2	3.5–7.9 mg/dL

Creatinine	2.5	0.5–1.2 mg/dL
BUN	95	7–25 mg/dL
Glucose	88	75–105 mg/dL
Total bilirubin	1.2	0.2–1.0 mg/dL
Alkaline phosphatase	27	7–59 IU/L
Lactate dehydrogenase	202	90–190 IU/L
Aspartate transaminase	39	8–40 IU/L
Amylase	152	76–375 IU/L

Questions

1. Given the abnormal tests, what additional information would you like to have?
2. If this patient had triglycerides of 100 mg/dL (1.1 mmol/L) and an HDL cholesterol of 23 mg/dL (0.6 mmol/L), what would be his calculated LDL-cholesterol value?
3. If, on the other hand, his triglycerides were 476 mg/dL (5.4 mmol/L), with an HDL cholesterol of 23 mg/dL (0.6 mmol/L), what would be his calculated LDL-cholesterol value?

centers, and even in the home.[126] First-generation systems, introduced in the 1980s, were relatively large and measured cholesterol and triglycerides as well as other common analytes, usually separately and sequentially. HDL separations involving offline pretreatment steps were subsequently developed. Second-generation systems became smaller and more sophisticated, offering separation of HDL and analysis of cholesterol and triglycerides simultaneously from fingerstick blood in an integrated system. A third generation system measures cholesterol, triglycerides, HDL cholesterol, and glucose simultaneously from a fingerstick sample. Noninstrumented systems with a thermometer-like reading are available for total cholesterol and are in development for HDL cholesterol. These new technologies offer the capability of measuring lipids and lipoproteins reliably outside the conventional laboratory.

Standardization

Precision

Precision is a prerequisite for accuracy; a method may have no overall systematic error or bias, but if it is imprecise it will still be inaccurate on individual measurements. With the shift to modern automated analyzers, analytical variation, in general, has become less of a concern than biologic and other sources of preanalytical variation. Cholesterol levels are affected by many factors that can be categorized into biologic, clinical, and sampling sources.[127,128] Changes in lifestyle that affect usual diet, exercise, weight, and smoking patterns can

result in fluctuations in the observed cholesterol and triglyceride values and the distribution of the lipoproteins. Similarly, the presence of clinical conditions, various diseases, or the medications used in their treatment, affect the circulating lipoproteins. Conditions during blood collection, such as fasting status, posture, the choice of anticoagulant in the collection tube, and storage conditions, can alter the measurements. Typical observed biologic variation for more than 1 year for total cholesterol averages approximately 6.1% CV. Thus, in the average patient, measurements made over the course of a year would fall 66% of the time within ±6.1% of the mean cholesterol concentration and 95% of the time within twice this range. Some patients may exhibit substantially more biologic variation. Thus, preanalytical variation generally is relatively large in relation to the usual analytical variation, which is typically less than 3% CV, and must be considered in interpreting cholesterol results. Some factors, such as posture and blood collection, can be standardized to minimize the variation. The NCEP guidelines[117] recommend making decisions based on the average of two or three measurements to factor out the effect of both preanalytical and analytical sources. The use of stepped cut-points in the work-up also reduces the practical effect of variation.

Accuracy

It is essential for accuracy that a method be calibrated or traceable to its respective reference method. In the case of cholesterol, the reference system is quite advanced and complete, having served as a model for standardization of laboratory analytes.[5,115] The Definitive Method at the National Institute for Standards and Technology provides the accuracy target but is too expensive and complicated for frequent use.[55] The Reference Method developed and applied at the CDC, and calibrated by an approved primary reference standard to the Definitive Method, provides a transferable, practical reference link.[54] The Reference Method is now accessible in a network of standardized laboratories, which compose the Cholesterol Reference Method Laboratory Network. This network was established in the United States and some European and Asian countries to extend standardization to manufacturers and clinical laboratories.[115] The network was established primarily to provide direct accuracy comparisons using fresh native serum specimens, which are necessary for reliable accuracy transfer because of analyte—matrix interaction problems on processed reference materials.[129–131]

Matrix Interactions

In the early stages of cholesterol standardization, which were directed toward diagnostic manufacturers and routine laboratories, commercial lyophilized or freeze-dried materials were used. These materials, made in large quantities, often with spiking or artificial addition of analytes, were assayed by the Definitive Method or Reference Method and distributed widely for accuracy transfer. Subsequently, biases were observed in enzymic assays on fresh patient specimens. Even though such manufactured reference materials are convenient, stable, and amenable to shipment at ambient temperatures, the manufacturing process, especially spiking and lyophilization, altered the measurement properties in enzymic assays. These enzymic assays, sensitive to the nature of the analyte and specimen matrix, were altered such that results were not representative of results on patient specimens. To achieve reliable feedback on accuracy and transfer of the accuracy base, direct comparisons with the Reference Method on actual patient specimens were determined to be necessary.[115]

CDC Cholesterol Reference Method Laboratory Network

In response, the CDC cholesterol network program was organized. The network offers reliable accuracy comparisons and has developed a formal certification program whereby laboratories and manufacturers can document traceability to the National Reference System for Cholesterol.[115] Clinical laboratories performing total and HDL cholesterol analyses can select a commercial method that has been certified by the network. Certification does not ensure all aspects of quality in a reagent system but primarily ensures that the accuracy is traceable to the National Reference System for Cholesterol within accepted limits and that the precision is acceptable. Certification is most efficient through manufacturers, but individual laboratories desiring to confirm the performance of their system by completing a certification protocol can contact the CDC for information about the network.

Analytical Performance Goals

The NCEP laboratory panels[2–5] have established requisite analytical performance goals based on clinical needs for routine measurements (see Table 11-5). For analysis of total cholesterol, the performance goal for total error is 8.9%. That is, the overall error should be such that each individual cholesterol measurement falls within ±8.9% of the Reference Method value. Actually, because the goals are based on 95% certainty, 95 of 100 measurements should fall within the total error limit. One can assay a specimen many times and calculate the mean to determine the usual value or the central tendency. The scatter or random variation around the mean is described by the standard deviational interval of plus and minus one standard deviation around the mean and includes, by definition, two thirds of the observations. In the laboratory, because the scatter or imprecision is often proportional to the concentration, random variation is usually specified in relative terms as CV, the coefficient of variation or relative standard deviation, which equals the standard deviation divided by the mean. Overall accuracy or systematic error is described as bias, the difference between the mean and the true value. Bias is primarily a function of the method's calibration and may vary by concentration. Of greatest concern in this context is bias at the NCEP decision cut-points. The bias and CV targets presented in Table 11-5 are representative of performance that will meet the NCEP goals for total error.

Quality Control

Achieving acceptable analytical performance requires the use of reliable quality control materials, which should preferably closely emulate actual patient specimens. Currently, the most suitable are quality control pools that are prepared from freshly collected patient serum, aliquoted into securely sealed vials, quick frozen, and stored at –70°C. Pools of fresh frozen serum are essential for reliability when monitoring accuracy in lipoprotein separation and analysis, and are preferable for monitoring cholesterol and other lipid measurements. Because commercial pools often undergo matrix alterations that change their analysis characteristics, results may not represent results on patients. At least two pools should be analyzed, preferably with levels at or near decision points for each analyte.

Serum, usually collected in serum separator vacuum tubes with clotting enhancers, has been the fluid of choice for lipoprotein measurement in the routine clinical laboratory. Ethylenediaminetetraacetic acid (EDTA) plasma has traditionally been the choice in lipid research laboratories, especially for lipoprotein separations, because the anticoagulant is thought to enhance stability. EDTA does have potential disadvantages that discourage routine use. Microclots, which form during storage, plug the sampling probes on the modern chemistry analyzers. EDTA osmotically draws water from red cells, diluting the plasma constituents, and the dilution effect can vary depending on such factors as fill volume, the analyte being measured, and the extent of mixing.

Because the NCEP cut-points are based on serum values, cholesterol measurements made on EDTA plasma require correction by the factor of 1.03.

LIPID AND LIPOPROTEIN DISTRIBUTION IN THE POPULATION

In the general adult population, serum lipoprotein concentrations differ between men and women, the result, primarily, of differences in sex hormone levels.[132,133] Women generally have higher levels of HDL cholesterol and lower levels of total cholesterol and triglycerides than men due to higher estrogen levels.[134] The difference in total cholesterol values disappears after menopause as estrogen decreases.[48,135] Men and women both, however, show a tendency toward increases in concentrations of total cholesterol, LDL cholesterol, and triglycerides with age.[13,21,133] HDL cholesterol concentrations generally remain stable after the onset of puberty, and do not drop in women with the onset of menopause. General adult reference ranges are shown in Table 11-6.

Circulating levels of total cholesterol, LDL cholesterol, and triglycerides in young children are generally much lower than those seen in adults.[133,136,137] In addition, concentrations do not differ significantly between boys and girls. HDL cholesterol levels for both boys and girls are comparable to those of adult women. At the onset of puberty, however, HDL cholesterol concentrations in boys fall to adult male levels, a drop of approximately 20%, whereas those of girls remain the same. It is the lower concentration of HDL cholesterol in men, combined with their higher LDL cholesterol and triglyceride concentrations, that account for much of the observed association with increased risk of premature heart disease.

Epidemiologic studies have shown that the incidence of heart disease is strongly associated with serum cholesterol concentration,[19,138] and comparisons performed in various societies have shown that those individuals who traditionally eat less animal fat and more grains, fruits, and vegetables, such as many Asian populations, have lower levels of LDL cholesterol and lower rates of heart disease than societies that ingest more fat, particularly animal fat, in their diet.[139,140] These differences can be attributed to both genetic and dietary inputs. The importance of dietary factors was shown clearly in a study that compared the dietary patterns and heart disease rates in Japanese men living in Japan, Hawaii, and California.[18] In this study, as dietary intake became more westernized, with increased consumption of fat and cholesterol, the LDL cholesterol concentrations increased significantly, as did the rates of heart disease, so that Japanese men living in California had much higher rates of heart disease than Japanese men living in Japan; those in Hawaii were intermediate. Within societies in which diet tends to be more homogeneous, LDL cholesterol levels become somewhat less discriminatory as a risk factor, and HDL cholesterol levels become more important because of the ability of HDL to remove excess cholesterol from the circulation.[141]

A few years ago, as part of a campaign to alert the American population to the risk factors associated with heart disease and to try to reduce the rate of its morbidity and mortality, NCEP was formed. NCEP has used panels of experts, including the Adult Treatment Panel, the Children and Adolescents Treatment Panel, and the Laboratory Standardization Panel, to produce recommendations within the scope of each panel's activities.[1–5,116,117,136,142,143]

In 1988, the first Adult Treatment Panel (ATP I) developed a list of heart disease risk factors, which were reported in 1988.[116] These guidelines were revised and further refined by ATP II in 1993.[117,142] The current list of risk factors is shown in Table 11-7. The ATP II also has determined that all adults be screened for concentrations of both total and HDL cholesterol, and has developed guidelines for the diagnosis and follow-up treatment of individuals with abnormal levels (Table 11-8). The Children and Adolescents Treatment Panel has developed similar criteria for the pediatric population.[136]

The NCEP Laboratory Standardization Panel[1,5] and its successor, the Lipoprotein Measurement Working Group,[2–4] have set laboratory guidelines for the precision and accuracy associated with measurement of total cholesterol, triglycerides, and lipoprotein cholesterol (HDL and LDL cholesterol), as previously discussed (see Table 11-5).

Obviously, the best way to reduce the prevalence of heart disease in the general population is through prevention. Learning and practicing good dietary patterns early in life, maintaining these patterns throughout life,[144] and refraining from smoking[145,146] and controlling blood pressure are important means for reducing the incidence of CHD and stroke. Screening programs to measure total and HDL cholesterol, with follow-up testing for triglycerides and LDL cholesterol where indicated, is a method of identifying individuals who may have levels that put them at risk, so that they can receive treatment to reduce the level of risk. Treatment of other diseases that may affect lipoproteins, such as diabetes mellitus, hypothyroidism, and renal disease, is also important.

A prudent diet, low in fat and cholesterol, with a caloric intake adjusted to meet and maintain ideal body weight, along with regular exercise, can reduce the risk of heart disease, stroke, diabetes, and cancer.[147–150] Dietary intake of fat and cholesterol have been shown to have a synergistic effect, that is, dietary cholesterol is more efficiently absorbed when

TABLE 11-6. Adult Reference Ranges for Lipids

Analyte	Reference Range
Total cholesterol	140–200 mg/dL
HDL cholesterol	29–75 mg/dL
LDL cholesterol	57–130 mg/dL
Triglyceride	67–157 mg/dL

TABLE 11-7. Positive Risk Factors Associated With Coronary Heart Disease, as Determined by the NCEP Adult Treatment Panel

- Age: ≥45 y for men; ≥55 y or premature menopause for women
- Family history of premature CHD
- Current cigarette smoking
- Hypertension (BP ≥140/90 mm Hg or taking antihypertensive medication)
- LDL cholesterol concentration ≥160 mg/dL (≥4.1 mmol/L), with <2 risk factors
- LDL cholesterol concentration 130–159 mg/dL (3.4–4.1 mmol/L), with ≥2 risk factors
- HDL cholesterol concentration <35 mg/dL (<0.9 mmol/L)
- Diabetes mellitus

Negative Risk Factor Associated With Coronary Heart Disease

- HDL cholesterol concentration ≥60 mg/dL (≥1.6 mmol/L)

TABLE 11-8. Treatment Guidelines Established by the NCEP Adult Treatment Panel

Initial Testing (Can Be Nonfasting)

Risk Category	Action
TC <200 mg/dL (5.2 mmol/L) & HDLC ≥35 mg/dL (0.9 mmol/L)	Repeat within 5 y Provide risk reduction information
TC <200 mg/dL (5.2 mmol/L) & HDLC <35 mg/dL (0.9 mmol/L)	Perform lipoprotein analysis (see below)
TC 200–239 mg/dL (5.2–6.2 mmol/L), HDLC ≥35 mg/dL (0.9 mmol/L) & <2 risk factors	Repeat in 1–2 y Provide risk reduction information
TC 200–239 mg/dL (5.2–6.2 mmol/L), HDLC <35 mg/dL (0.9 mmol/L) or ≥2 risk factors	Perform lipoprotein analysis (see below)
TC ≥240 mg/dL (6.2 mmol/L)	Perform lipoprotein analysis (see below)

Follow-up Testing (Lipoprotein Analysis After 12-h Fast)

Risk Category	Action
LDLC <130 mg/dL (3.4 mmol/L)	Repeat TC and HDLC within 5 y Provide risk reduction information
LDLC 130–159 mg/dL (3.4–4.1 mmol/L) & <2 risk factors	Provide Step 1 diet and physical activity information, and reevaluate in 1 y
LDLC 130–159 mg/dL (3.4–4.1 mmol/L) & ≥2 risk factors	Do clinical evaluation, including family history Start dietary therapy (see below)
LDLC ≥160 mg/dL (4.1 mmol/L)	Do clinical evaluation, including family history Start dietary therapy (see below)

Treatment Decisions

Risk Category	Action Level	Goal
Dietary therapy		
No CHD; <2 risk factors	≥160 mg/dL (4.1 mmol/L)	<160 mg/dL (4.1 mmol/L)
No CHD; ≥2 risk factors	≥130 mg/dL (3.4 mmol/L)	<130 mg/dL (3.4 mmol/L)
CHD	>100 mg/dL (2.6 mmol/L)	100 mg/dL (2.6 mmol/L)
Drug therapy		
No CHD; <2 risk factors	≥190 mg/dL (4.9 mmol/L)	<160 mg/dL (4.1 mmol/L)
No CHD; ≥2 risk factors	≥160 mg/dL (4.1 mmol/L)	<130 mg/dL (3.4 mmol/L)
CHD	≥130 mg/dL (3.4 mmol/L)	≥100 mg/dL (2.6 mmol/L)

in the presence of fat,[151] and saturated fat has been found to be more atherogenic than unsaturated fat.[124,144,152] The American Heart Association has recommended dietary guidelines for the intake of fat and cholesterol for most adult Americans. These guidelines are shown in Table 11-1.

DISEASE PREVENTION, DIAGNOSIS, AND TREATMENT

Diseases associated with abnormal lipid concentrations are referred to as *dyslipidemia*. They can be caused directly by genetic abnormalities or through environmental/lifestyle imbalances, or they can develop secondarily, as a consequence of other diseases.[153–155] Dyslipidemia are generally defined by clinical characteristics of the patients and the results of blood tests and do not necessarily define the specific defect associated with the abnormality. Many dyslipidemia, however, regardless of etiology, are associated with CHD, or arteriosclerosis.

Arteriosclerosis

In the United States and many other developed countries *arteriosclerosis* is the single leading cause of death and disability. The mortality rate has decreased in the United States in the past few years, partly due to advances in diagnosis and treatment, but also due to changes in lifestyle in the American population, resulting from increased awareness of the relationship between cholesterol and heart disease. This increased awareness has resulted in an overall decrease in the average serum cholesterol concentration and in a lower prevalence of heart disease, but it still exceeds all other causes of death combined. As many women as men develop arteriosclerosis, but women develop it 10 years later than men, on average.

The relationship between heart disease and lipid abnormalities stems from the deposition of lipids, mainly in the form of esterified cholesterol, in the walls of the arteries. This lipid deposition starts with thin layers called fatty streaks. In studies examining blood vessels at autopsy, fatty streaks have been seen in almost everyone older than age 15 years, regardless of cause of death.[156,157] Under certain conditions, the fatty streaks develop over time into plaques, which partially block (occlude) blood flow. When the plaque formation develops in arteries of the arms or legs, it is called peripheral vascular disease (PVD); when it develops in the heart, it is referred to as coronary artery disease (CAD); and when it develops in the vessels of the brain, it is called cerebrovascular disease (CVD). CAD is associated with angina and myocardial infarction, and CVD is associated with stroke. Many genetic and acquired abnormalities may also lead to deposits of lipid in the liver and kidney, resulting in impaired function of these vital organs. Lipid deposits in skin are called *xanthomas* and are a clue to genetic abnormalities. They are nodules that disfigure and can indicate further lipid complications.

Plaque formation is a process of cell injury followed by infiltration and cell proliferation to repair the site. As blood travels through blood vessels, small injuries occur that signal macrophages and platelets to heal the break. LDL bring cholesterol to the site so that new cell membranes can be formed and macrophages can repair the area. LDL that have been modified by oxidative processes and chemical alterations can be taken up by the macrophages, producing foam cells.[158–160] These foam cells accumulate beneath the endothelial layer of the arterial wall. Future injury leads to more deposits, and eventually a plaque is formed. Continual injury and repair will lead to additional narrowing of the opening, or lumen, causing the blood to pass through under greater and greater pressure.

Deposits in the vessel walls are frequently associated with increased serum concentrations of LDL cholesterol or decreased HDL cholesterol.[138,161] Lowering the LDL cholesterol concentration is an important step in preventing and treating CHD.[101,133,162–170] It has been estimated that for every 1% decrease in LDL cholesterol concentration there is a 2% decrease in a person's of risk of developing arteriosclerosis.[166,171] For patients with established heart disease, recent studies have shown that aggressive treatment to reduce LDL cholesterol levels below 100 mg/dL (2.6 mmol/L) is effective in the stabilization and sometimes regression of plaques.[172–176] In some individuals, high levels of blood cholesterol or triglycerides are caused by genetic abnormalities whereby either too much is synthesized or too little is removed.[155,177–180] High levels of cholesterol or triglycerides in most people, however, are due to increased consumption of foods rich in fat and cholesterol, smoking, and lack of exercise, or to other disorders or disease states that have effects on lipid metabolism, such as diabetes, hypertension, hypothyroidism, obesity, other hormonal imbalances, liver and kidney diseases, and alcoholism. Although low levels of HDL cholesterol have been associated with increased risk of heart disease,[138,177] so far, due to a lack of well-tolerated drug therapies to raise HDL cholesterol levels, there are not enough studies to document whether raising HDL cholesterol decreases risk. Such studies are currently being completed and should provide valuable information in this regard.[181,182] There are, however, studies emerging that suggest such a relationship.[165,183]

Laboratory analyses can aid in the diagnosis of arteriosclerosis. Accurate determinations of total, HDL, and LDL cholesterol levels can indicate the need for diet or diet and drug therapy. As shown in Table 11-8, individuals on a lowfat diet who continue to have LDL cholesterol levels ≥190 mg/dL (≥4.9 mmol/L) on repeated measurement will benefit from drug intervention. If they have ≥2 CAD risk factors and continue to have LDL cholesterol levels of ≥160 mg/dL (≥4.1 mmol/L), they also would benefit from drug therapy. And if they have already been diagnosed with heart disease, drug therapy should be considered when the LDL cholesterol level is ≥130 mg/dL (≥3.4 mmol/L). The average of at least two measurements, taken 1 week to 8 weeks apart, should be used to determine treatment.[116,117]

CASE STUDY 11-2

A 30-year-old male was brought to the ER with chest pain after a softball game. He was placed in the coronary care unit when his ECG showed erratic waves in the ST region. A family history revealed that his father died of a heart attack when he was 45 years old. The patient had always been athletic in high school and college, so he had not concerned himself with a routine physical. The laboratory tests listed in Case Study Table 11-2.1 were run.

CASE STUDY TABLE 11–2.1. Laboratory Results

Analyte	Patient Values	Reference Range
Sodium	139	135–143 mEq/L
Potassium	4.1	3.0–5.0 mEq/L
Chloride	101	98–103 mEq/L
CO$_2$ content	29	22–27 mmol/L
Total protein	6.9	6.5–8.0 g/dL
Albumin	3.2	3.5–5.0 g/dL
Calcium	9.3	9.0–10.5 mg/dL
Cholesterol	278	140–200 mg/dL
Uric acid	5.9	3.5–7.9 mg/dL
Creatinine	1.1	0.5–1.2 mg/dL
BUN	20	7–25 mg/dL
Glucose	97	75–105 mg/dL
Total bilirubin	0.8	0.2–1.0 mg/dL
Alkaline phosphatase	20	7–59 IU/L
Lactate dehydrogenase	175	90–190 IU/L
Aspartate transaminase	35	8–40 IU/L
Amylase	98	76–375 IU/L

Questions

1. Given the symptoms and the family history, what additional tests should be recommended?
2. If his follow-up total cholesterol remains in the same range after he has been released from the hospital, and his triglycerides and HDL cholesterol are within the normal range, what course of treatment should be recommended?

Classic drug treatments with bile-acid sequestrants, such as cholestyramine, work by sequestering cholesterol in the gut to keep it from being absorbed, and are still considered to be the only safe drugs for use in young children because they are not absorbed.[166,171] They have uncomfortable side effects, however, such as bloating and constipation. Niacin is a very potent drug for reducing LDL cholesterol and raising HDL cholesterol; however, it can be hepatotoxic and can aggravate glucose intolerance and hyperuricemia.[162] The newest class of drugs includes the HMG-CoA reductase inhibitors lovastatin, simvastatin, pravastatin, fluvastatin, atorvastatin, and cerevastatin. These drugs are effective in reducing LDL cholesterol from 20% to 40%[167–170,173,174,183] and are generally well tolerated. They block intracellular cholesterol synthesis by inhibiting the enzyme HMG-CoA reductase. The major safety issues are hepatotoxic effects and the lack of data on possible long-term side effects. Patient monitoring in clinical trials has shown that <2% of patients have sustained increases in liver enzymes. A big drawback to these drugs is their cost. Other drugs being considered are probucol, which prevents lipid oxidation and macrophage uptake, and fibric acid derivatives, such as clofibrate, gemfibrozil, fenofibrate, and etiofibrate, which reduce triglyceride and VLDL cholesterol levels and increase HDL cholesterol.

Hyperlipoproteinemias

Lipoproteins are complex transport vehicles for moving cholesterol, cholesteryl esters, and triglycerides in the blood. Disease states associated with abnormal serum lipids are generally caused by malfunctions in the synthesis, transport, or catabolism of the lipoproteins.[180,184] Dyslipidemia can be subdivided into two major categories: *hyper*lipoproteinemias, which are diseases associated with elevated lipoprotein levels, and *hypo*lipoproteinemias, which are associated with decreased lipoprotein levels.[42] The hyperlipoproteinemias can be subdivided into hypercholesterolemia, hypertriglyceridemia, and combined hyperlipidemia, in which there are elevations of both cholesterol and triglycerides.

Hypercholesterolemia

Hypercholesterolemia is the lipid abnormality most closely linked to heart disease.[138] One form of the disease, which is associated with genetic abnormalities that predispose affected individuals to elevated cholesterol levels, is called familial hypercholesterolemia (FH). Homozygotes for FH are fortunately rare (1:1,000,000 in the population) but can have total cholesterol concentrations as high as 800 mg/dL to 1000 mg/dL (20–26 mmol/L), and patients frequently have their first heart attack while still in their teenage years.[185] Heterozygotes for the disease are seen much more frequently, because the disease is caused by an autosomal codominant disorder. Patients tend to have total cholesterol concentrations in the range of 300 mg/dL to 600 mg/dL (8–15 mmol/L) and, if not treated, will become symptomatic for heart disease in their 20s to 50s. Approximately 5% of patients younger than age 50 years with CAD are FH heterozygotes. Other symptoms associated with FH include tendinous and tuberous xanthomas, which are cholesterol deposits under the skin, and arcus, which indicates cholesterol deposition in the vessels of the eye.[180,184]

In both homozygotes and heterozygotes, the cholesterol elevation is primarily associated with an increase in LDL cholesterol. These individuals synthesize intracellular cholesterol normally, but lack, or are deficient in, active LDL receptors. Consequently, cholesterol derived through absorption and incorporated into LDL builds up in the circulation because there are no receptors to bind the LDL and transfer the cholesterol into the cells. The cells, on the other hand, which require cholesterol for use in cell membrane and hormone production, synthesize cholesterol intracellularly at an increased rate to compensate for the lack of cholesterol from the receptor mediated mechanism.

In FH heterozygotes, and other forms of hypercholesterolemia in which there is insufficient LDL receptor activity, reduction in the rate of internal cholesterol synthesis, through inhibition of HMG-CoA reductase activity by use of HMG-CoA reductase inhibitors (statins), stimulates the production of additional receptors, thereby increasing cell internalization of cholesterol from LDL, which in turn lowers the serum levels. These drugs have been associated with 20% to 40% decreases in serum cholesterol concentrations in these patients. Homozygotes, however, cannot benefit from this type of therapy, because they have no functional receptors to stimulate. Homozygotes rely primarily on a technique called *LDL pheresis,* in which blood is periodically drawn from the patient, processed to remove LDL, and returned to the patient in a similar fashion to the dialysis treatment used for kidney patients.

Most individuals with elevated LDL cholesterol levels do not have FH, but they are all at increased risk for premature CHD[186] and should be maintained on a lowfat, low-cholesterol diet, with the caloric intake adjusted to attain or maintain ideal body weight.[187–189] Regular physical activity should also be incorporated. Drug therapy is added when necessary (see Table 11-8).

Hypertriglyceridemia

The NCEP Adult Treatment Panel has identified borderline high triglycerides as levels of 200 mg/dL to 400 mg/dL (2.3–4.5 mmol/L), high as 400 mg/dL to 1000 mg/dL (4.5–11.3 mmol/L), and very high as >1000 mg/dL (11.3 mmol/L). *Hypertriglyceridemia* can derive from a genetic abnormality, and is then called familial hypertriglyceridemia, or from secondary causes, such as hormonal abnormities associated with the pancreas, adrenal glands, and pituitary, or from diabetes mellitus or nephrosis.[190] Diabetes mellitus, with its lack of insulin, leads to increased shunting of glucose into the pentose pathway, causing increased fatty acid synthesis. Nephrosis depresses the removal of large molecular weight constituents like triglycerides, causing increased serum levels. Hypertriglyceridemia is generally due to an imbalance between synthesis and clearance of VLDL in the circulation.[14,191,192] In most studies, hypertriglyceridemia has not been statistically implicated as an independent risk factor for CHD, but many CHD patients have moderately elevated triglycerides in conjunction with decreased HDL cholesterol levels.[193] It is difficult to separate the risk associated with increased triglycerides from that of decreased HDL cholesterol because the two are linked and serum concentrations are usually inversely related.

Triglycerides are influenced by a number of hormones, such as pancreatic insulin and glucagon, pituitary growth hormone, adrenocorticotropic hormone (ACTH) and thyrotropin, and adrenal medulla epinephrine and norepinephrine from the nervous system. Epinephrine and norepinephrine influence serum triglyceride levels by triggering production of hormone sensitive lipase, which is located in adipose tissue.[194] Other body processes that trigger hormone sensitive lipase activity are cell growth (growth hormone), adrenal stimulation (ACTH), thyroid stimulation (thyrotropin), and fasting (glucagon). Each of these processes,

CASE STUDY 11-3

A 43-year-old white man was diagnosed with hyperlipidemia at age 13 years when his father died of a myocardial infarction at age 34 years. The man's grandfather had died at age 43 years, also of a myocardial infarction. Currently, the man is active and asymptomatic with regard to CHD. He is taking 40 mg of Mevacor (lovastatin), two times per day (maximum dose). He has previously taken niacin, but could not tolerate it because of flushing and gastrointestinal distress, nor could he tolerate Questran (cholestyramine resin). His physical exam is remarkable for bilateral Achilles tendon thickening/xanthomas and a right carotid bruit.

Laboratory Results

Triglycerides, 91 mg/dL
Total cholesterol, 269 mg/dL
HDL cholesterol, 47 mg/dL
LDL cholesterol, 204 mg/dL
Aspartate aminotransferase (AST), 34 U/L
Alanine aminotransferase (ALT), 36 U/L
Alkaline phosphatase (ACP), 53 U/L
Electrolytes and fasting glucose, normal

1. What is his diagnosis?
2. Does he need further workup?
3. What other laboratory tests should be done?
4. Does he need further drug treatment, and, if so, what?

through its action on hormone sensitive lipase, will result in an increase in serum triglyceride values.

Severe hypertriglyceridemia (>1000 mg/dL or 11 mmol/L), although it does not put individuals at risk for CHD, is a potentially life-threatening abnormality because it can cause *pancreatitis* (inflammation of the pancreas). It is therefore imperative that these patients be diagnosed and treated with triglyceride-lowering medication and that they be monitored closely. Severe hypertriglyceridemia is generally caused by a deficiency of LPL or by a deficiency in apolipoprotein C-II, which is a necessary cofactor for LPL activity.[7] Normally, LPL hydrolyzes triglycerides carried in chylomicrons and VLDL to provide cells with free fatty acids for energy from exogenous and endogenous triglyceride sources. A deficiency in LPL or apo C-II activity keeps chylomicrons from being cleared, and serum triglycerides remain extremely elevated, even when the patient has fasted for 12 hours to 14 hours.

Treatment of hypertriglyceridemia consists of diet, fish oil, triglyceride-lowering drugs (primarily fibric acid derivatives) in cases of severe hypertriglyceridemia or where an accompanying low HDL cholesterol value indicates risk of CHD, and sometimes oils with specific fatty acid compositions.[195,196] It is possible that only certain subspecies of chylomicrons and VLDL are atherogenic. Chylomicron remnants and VLDL remnants, for example, are thought to be particularly atherogenic.[197–199] New methodology to isolate triglyceride RLP now makes it possible to study the potential atherogenicity of these particles.[98–100,200]

Combined Hyperlipoproteinemia

Combined hyperlipoproteinemia is generally defined as the presence of elevated levels of both serum total cholesterol and triglycerides. Individuals presenting with this syndrome are considered to be at increased risk for CHD. In the genetically derived form, called familial combined hyperlipoproteinemia (FCH), some individuals of an affected kindred may have only elevated cholesterol; others, only elevated triglycerides; and others, elevations of both.

Another rare genetic form of combined hyperlipoproteinemia is called familial dysbetalipoproteinemia, or Type III hyperlipoproteinemia. The name Type III hyperlipoproteinemia is a holdover from a former lipoprotein typing system developed by Fredrickson et al,[41] but which is otherwise generally no longer used. The disease stems from an accumulation of cholesterol-rich VLDL and chylomicron remnants due to defective catabolism of those particles. The disease is also associated with the presence of a relatively rare form of apo E, called apo E2/2. Individuals with Type III will frequently have total cholesterol values of 200 mg/dL to 300 mg/dL (5–8 mmol/L) and triglycerides of 300 mg/dL to 600 mg/dL (3–7 mmol/L). To distinguish them from other combined hyperlipoproteinemics it is first necessary to isolate the VLDL fraction of their serum by ultracentrifugation. A ratio derived from the cholesterol concentration in VLDL to total serum triglycerides will be >0.30 in the presence of Type III hyperlipoproteinemia. If the VLDL fraction is subjected to agarose electrophoresis, the particles will migrate in the beta region, rather than in the normal pre-beta region. Definitive diagnosis requires a determination of apo E isoforms by isoelectric focusing or deoxyribonucleic acid (DNA) typing, resulting in either apo E2/2 homozygosity or, very rarely, apo E deficiency. Treatment, however, does not rely on a diagnosis, because these patients are treated with standard therapy of diet, niacin, gemfibrozil, and HMG-CoA reductase inhibitors, as the clinical results dictate. Because of the cholesterol-enriched composition of

---

CASE STUDY 11-4

A 60-year-old female came to her physician because she was having urination troubles. Her previous history included hypertension and episodes of edema. The physician ordered various laboratory tests on blood drawn in his office. The results are shown in Case Study Table 11-4.1.

CASE STUDY TABLE 11–4.1. Laboratory Results

Analyte	Patient Values	Reference Range
Na$^+$	149	135–143 mEq/L
K$^+$	4.5	3.0–5.0 mEq/L
Cl$^-$	120	98–103 mEq/L
CO$_2$ content	12	22–27 mmol/L
Total protein	5.7	6.5–8.0 g/dL
Albumin	2.3	3.5–5.0 g/dL
Calcium	7.6	9.0–10.5 mg/dL

Cholesterol	201	140–200 mg/dL
Uric acid	15.4	3.5–7.9 mg/dL
Creatinine	4.5	0.5–1.2 mg/dL
BUN	87	7–25 mg/dL
Glucose	88	75–105 mg/dL
Total bilirubin	1.3	0.2–1.0 mg/dL
Triglycerides	327	65–157 mg/dL
Lactate dehydrogenase	200	90–190 IU/L
Aspartate transaminase	45	8–40 IU/L
Amylase	380	76–375 IU/L

Questions

1. What are the abnormal results in this case?
2. Why do you think the triglycerides are abnormal?
3. What is the primary disease exhibited by this patient's laboratory data?

CASE STUDY 11-5

A 49-year-old woman was referred for a lipid evaluation by her dermatologist after developing a papular rash over her trunk and arms. The rash consisted of multiple red raised lesions with yellow centers. She had no previous history of such a rash, and had no family history of lipid disorders or CHD. She is postmenopausal, on standard estrogen replacement therapy, and otherwise healthy.

Laboratory Results

Serum, grossly lipemic
Triglycerides, 6200 mg/dL

Total cholesterol, 458 mg/dL
Fasting glucose, 160 mg/dL
Liver function tests and electrolytes, normal

1. What is the cause of her rash and what is the rash?
2. Is her oral estrogen contributing? Is her glucose contributing?
3. What treatments are warranted and what is her most acute risk?

these particles, use of the Friedewald equation[119] to calculate LDL cholesterol levels will result in an underestimation of VLDL cholesterol, and therefore, an overestimation of LDL cholesterol, as compared to beta-quantification.[97,201]

Lp(a) Elevation

Elevations in the serum concentration of Lp(a) are currently thought to confer increased risk of CHD[28,202–205] and CVD.[27,206] Higher levels of Lp(a) have been seen in patients with CHD than in normal control subjects, although prospective studies have not yet conclusively determined the positive association.[207,208] Lp(a) are LDL that contain an extra apolipoprotein, called apo (a); the size and serum concentrations of Lp(a) are genetically determined.[29] Because apo (a) has a high degree of homology with plasminogen, a coagulation factor,[209] the general hypothesis involves a supposed competition between plasminogen and apo(a) for fibrin binding sites.[210–212] Under this hypothesis, if the apo(a) competes successfully, blocking plasminogen, clots that form along the arterial wall will not be dissolved. Most LDL-lowering drugs have no effect on Lp(a) concentration, even when LDL cholesterol becomes markedly reduced. The two drugs that have been shown to have some effect are niacin and replacement estrogen in postmenopausal women. Until prospective studies confirm Lp(a) atherogenicity, however, treatment with niacin is not advised except in conjunction with other dyslipidemic conditions in which niacin is also indicated.

Hypolipoproteinemia

Hypolipoproteinemia are abnormalities marked by decreased lipoprotein concentrations. They fall into two major categories: hypo*alpha*lipoproteinemia and hypo*beta*lipoproteinemia. Hypobetalipoproteinemia is associated with isolated low levels of LDL cholesterol (*ie,* without other accompanying lipoprotein disorders), but because it is not generally associated with CHD, it is not discussed further.

Hypoalphalipoproteinemia

Hypoalphalipoproteinemia indicates an isolated decrease in circulating HDL, usually defined as an HDL cholesterol concentration <35 mg/dL (0.9 mmol/L) without the presence of hypertriglyceridemia. The use of the term *alpha* denotes the region in which HDL migrate on agarose electrophoresis. There are several defects, often genetically determined, that are associated with hypoalphalipoproteinemia.[213–218] Virtually all of these defects are associated with increased risk of premature CHD. In the defect called *Tangier disease,* HDL cholesterol concentrations can be as low as 1 mg/dL to 2 mg/dL (0.03–0.05 mmol/L) in homozygotes accompanied by total cholesterol concentrations of 50 mg/dL to 80 mg/dL (1.3–2.1 mmol/L).

Treatment of individuals with isolated decreases of HDL cholesterol is limited. Niacin is somewhat effective but has side effects, such as hepatotoxicity, that sometimes precludes its use, although newer timed-release preparations may ameliorate those effects; estrogen replacement in postmenopausal women is also effective.[50,219,220]

Acute, transitory hypoalphalipoproteinemia can be seen in cases of severe physiologic stress, such as acute infections (primarily viral), other acute illnesses, surgical procedures, and so forth.[221] HDL cholesterol, as well as total cholesterol concentrations, can be markedly reduced under these conditions but will return to normal levels as recovery proceeds. For this reason, lipoprotein concentrations drawn during hospitalization or with a known disease state should be reassessed in the healthy, nonhospitalized state before intervention is considered.

SUMMARY

Lipoproteins are complex particles that interact with many other metabolic pathways of the body. Their homeostasis can be affected by imbalances of other pathways, such as hormone imbalances and diabetes, and, likewise, abnormal

CASE STUDY 11-6

Three patients are seen in clinic:

- The first patient is a 40-year-old man with hypertension, and who also smokes, but has not been previously diagnosed with CHD. His father developed CHD at age 53 years. He is fasting, and the results of his lipids include a total cholesterol concentration of 210 mg/dL, triglycerides of 150 mg/dL, and an HDL cholesterol value of 45 mg/dL. He has a fasting glucose level of 98 mg/dL.

- The second patient is a 60-year-old woman with no family history of CHD, who is normotensive, and does not smoke, with a total cholesterol concentration of 220 mg/dL, triglycerides of 85 mg/dL, and an HDL cholesterol value of 80 mg/dL. Her fasting glucose level is 85 mg/dL.

- The third patient is a 49-year-old man with no personal or family history of CHD, and who is not hypertensive and does not smoke. His fasting total cholesterol level is 260 mg/dL, his triglycerides are 505 mg/dL, his HDL cholesterol is 25 mg/dL, and his glucose level is 134 mg/dL.

For each of the three patients seen in clinic,

1. What is the LDL cholesterol level, as calculated by using the Friedewald calculation?
2. Which, if any, of the patients should have their LDL cholesterol measured, rather than calculated? Why?
3. How many known CHD risk factors does each have?
4. Based on what is known, are these patients recommended for lipid therapy (diet or drug), and, if so, on what basis?

---

lipoprotein metabolism can disrupt the body's system and cause disease, as seen in pancreatitis and coronary disease. Some aspects of lipoprotein balance are genetically determined (*ie,* gender, enzyme, or receptor defects), whereas others can be controlled through diet, decreased smoking, exercise, and the monitoring of blood pressure. Careful measurements are required for proper diagnosis of disease, and education is important to help prevent disease.

REFERENCES

1. National Cholesterol Education Program Laboratory Standardization Panel. Current status of blood cholesterol measurements in clinical laboratories in the United States. Clin Chem 1988;34:193.
2. Warnick GR, Wood PD JW, for the National Cholesterol Education Program Working Group on Lipoprotein Measurement. National Cholesterol Education Program recommendations for measurement of high-density lipoprotein cholesterol: executive summary. Clin Chem 1995;41:1427–1433.
3. Bachorik PS, Ross JW, for the National Cholesterol Education Program Working Group on Lipoprotein Measurement. National Cholesterol Education Program recommendations for measurement of low-density lipoprotein cholesterol: executive summary. Clin Chem 1995;41:1414–1420.
4. Stein EA, Myers GL, for the National Cholesterol Education Program Working Group on Lipoprotein Measurement. National Cholesterol Education Program recommendations for triglyceride measurement: executive summary. Clin Chem 1995;41:1421–1426.
5. Report from the Laboratory Standardization Panel of the National Cholesterol Education Program. Recommendations for improving cholesterol measurement. NIH publication no. 902964. Bethesda, MD: National Institutes of Health, 1990.
6. Tietz N. Fundamentals of clinical chemistry. 3rd ed. Philadelphia: WB Saunders, 1987.
7. Jackson RL, McLean LR, Ponce E, et al. Mechanism of action on LPL and hepatic triglyceride lipase. In: Malmendier CL, Alaupovic P, eds. Advances in experimental medicine and biology. Vol 210: Lipoproteins and atherosclerosis. New York: Plenum Press, 1987;73.
8. Lehninger AL. Biochemistry: the molecular basis of cell structure and function. 2nd ed. New York: Worth, 1975.
9. Mahley WR, Innerarity TL, Rall SC Jr, et al. Plasma lipoproteins: apolipoprotein structure and function. J Lipid Res 1984;25:1277.
10. Wu LH, Wu JT, Hopkins PN. Apolipoprotein E. Laboratory determinations and clinical significance. In: Rifai N, Warnick GR, eds. Handbook of lipoprotein testing. Washington, DC: AACC Press, 1997;329.
11. Davignon J, Gregg RE, Sing CF. Apolipoprotein E polymorphism and atherosclerosis. Arteriosclerosis 1988;8:1.
12. Hopkins PN, Wu LL, Schumacher MC, et al. Type III dyslipidemia in patients heterozygous for familial hypercholesterolemia and apolipoprotein E2. Evidence for a gene–gene interaction. Arterioscler Thromb 1991;11:1137.
13. Cohn JS, McNamara JR, Cohn SD, et al. Postprandial plasma lipoprotein changes in human subjects of different ages. J Lipid Res 1988;29:469.
14. Cohn JS, McNamara JR, Krasinski SD, et al. Role of triglyceride-rich lipoproteins from the liver and intestine in the etiology of postprandial peaks in plasma triglyceride concentration. Metabolism 1989;38:484.
15. Rifai N. Lipoproteins and apolipoproteins composition, metabolism and association with coronary heart disease. Arch Pathol Lab Med 1986;110:694.
16. Stein T, Stein O. Catabolism of VLDL and removal of cholesterol from intact cells. In: Schettler G, Strange E, Wissler RW, eds. Atherosclerosis: is it reversible? New York: Springer-Verlag, 1978.
17. Assmann G, Funke H, Schmitz G. Low-density lipoproteins and hypercholesterolemia. Drug Res 1989;39:996.
18. Kato H, Tillotson J, Nichaman MZ, et al. Epidemiologic studies of coronary heart disease and stroke in Japanese men living in Japan, Hawaii, and California. Am J Epidemiol 1973;97:372.
19. Keys A, ed. Coronary heart disease in seven countries. Circulation 1970;41(suppl I):1.
20. Austin MA, Breslow JL, Hennekens CH, et al. Low-density lipoprotein subclass patterns and risk of myocardial infarction. JAMA 1988;260:1917.
21. McNamara JR, Campos H, Ordovas JM, et al. Effect of gender, age, and lipid status on low-density lipoprotein subfraction distribution: results of the Framingham Offspring Study. Arteriosclerosis 1987;7:483.
22. McNamara JR, Jenner JL, Li Z, et al. Change in low-density lipoprotein particle size is associated with change in plasma triglyceride concentration. Arteriosclerosis 1992;12:1284.

23. Patsch JR, Prasad S, Gotto AM Jr, et al. Postprandial lipemia: a key for the conversion of HDL 2 into HDL 3 by hepatic lipase. J Clin Invest 1984;74:2017.

24. Blanche PJ, Gong EL, Forte TM, et al. Characterization of human high-density lipoproteins by gradient gel electrophoresis. Biochem Biophys Acta 1981;665:408.

25. Cheung MC, Albers JJ. Characterization of lipoprotein particles isolated by immuno-affinity chromatography. Particles containing A-I and A-II and particles containing A-I but no A-II. J Bio Chem 1984;259:12201.

26. Li Z, McNamara JR, Ordovas JM, et al. Analysis of high-density lipoproteins by a modified gradient gel electrophoresis method. J Lipid Res 1994;35:1698.

27. Armstrong VW, Cremer P, Eberle E, et al. The association between serum Lp(a) concentrations and angiographically assessed coronary atherosclerosis: dependence on serum LDL levels. Atherosclerosis 1986;62:249.

28. Hoefler G, Harnoncourt F, Faschke E, et al. Lipoprotein Lp(a): a risk factor for myocardial infarction. Arteriosclerosis 1988;8:398.

29. Marcovina SM, Koschinsky ML. Lipoprotein (a): structure, measurement, and clinical significance. In: Rifai N, Warnick GR, eds. Handbook of lipoprotein testing. Washington, DC: AACC Press, 1997;283.

30. Kuksis A, ed. Fat absorption. Vol 1. Boca Raton, FL: CRC Press, 1986.

31. Stein EA, Gary LM. Lipids lipoproteins and apolipoproteins. In: Tietz N, ed. Textbook of clinical chemistry. 2nd ed. Philadelphia: WB Saunders, 1994;1002.

32. Russell DW. Cholesterol biosynthesis and metabolism. Cardiovasc Drug Ther 1992;6:103.

33. Brown MS, Goldstein JL. Lipoprotein receptors, cholesterol metabolism, and atherosclerosis. Arch Pathol Lab Med 1975;99:181.

34. Brown MS, Goldstein JL. A receptor mediated pathway for cholesterol homeostasis. Science 1986;232:34.

35. Goldstein JL, Brown MS. Familial hypercholesterolemia. In: Scriver CR, Beaudet AL, Sly WS, et al, eds. The metabolic basis of inherited disease. 6th ed. New York: McGraw-Hill, 1989.

36. Small DM. The HDL system: a short review of structure and metabolism. Atherosclerosis Rev 1987;16:1.

37. Schmitz G, Robene H, Assmann G. Role of the high density lipoprotein receptor cycle in macrophage cholesterol metabolism. Atherosclerosis Rev 1987;16:95.

38. Johnson WJ, Mahlberg FH, Rothblat GH, et al. Cholesterol transport between cells and high-density lipoproteins. Biochem Biophys Acta 1991;1085:273.

39. Bilheimer DW, Grundy SM. Role of LDL receptor in lipoprotein metabolism. In: Malmendier CL, Alaupovic P, eds. Advances in experimental medicine and biology. Vol 210. Lipoproteins and atherosclerosis. New York: Plenum Press, 1987;123.

40. Kasaniemei YA, Grungy SM. Significance of low density lipoprotein production in the regulation of plasma cholesterol level in man. J Clin Invest 1982;70:13.

41. Fredrickson DS, Levy RI, Lees RS. Fat transport in lipoproteins: an integrated approach to mechanisms and disorders. N Engl J Med 1967;276:34, 94, 148, 215, 273.

42. Schaefer EJ, Levy RI. Pathogenesis and management of lipoprotein disorders. N Engl J Med 1985;312:1300.

43. Goldstein JL, Brown MS. Lipoprotein metabolism in the macrophage: implications for cholesterol deposition in atherosclerosis. Ann Rev Biochem 1983;52:223.

44. Chait A, Robertson HT, Brunzell JD. Chylomicronemia syndrome in diabetes mellitus. Diabetes Care 1981;4:343.

45. Bjorntorp. Fatty acid, hyperinsulinemia, and insulin resistance: which come first. Curr Opin Lipidol 1994;5:166.

46. Frayn KN. Insulin resistance and lipid metabolism. Curr Opin Lipidol 1993;4:197.

47. Angelin B, Rudling M. Growth hormone and hepatic lipoprotein metabolism. Curr Opin Lipidol 1994;5:160.

48. Furman RH, et al. Effects of androgens and estrogens on serum lipids and the composition and concentration of serum lipoproteins in normolipemic and hyperlipidemic states. Prog Biochem Pharmacol 1967;2:215.

49. Sacks FM, Walsh BW. Sex hormones and lipoprotein metabolism. Curr Opin Lipidol 1994;5:236.

50. Granfone A, Campos H, McNamara JR, et al. Effects of estrogen replacement on plasma lipoproteins and apolipoproteins in postmenopausal dyslipidemic women. Metabolism 1992;41:1193.

51. Heimberg M, Olubadewo JO, Wilcox HG. Plasma lipoproteins and regulation of hepatic metabolism of fatty acids in altered thyroid states. Endocr Rev 1985;6:590.

52. Zak B. Cholesterol methodologies: a review. Clin Chem 1977;23:1201.

53. Abell LL, Levy BB, Brody BB, Kendall FC. A simplified method for the estimation of total cholesterol in serum and demonstration of its specificity. J Bio Chem 1952;195:357.

54. Duncan IW, Mather A, Cooper GR. The procedure for the proposed cholesterol reference method. Atlanta, GA: Division of Environmental Health Laboratory Sciences, Center for Environmental Health, Centers for Disease Control.

55. Cohen A, Hertz HS, Mandel J, et al. Total serum cholesterol by isotope dilution/mass spectrometry: a candidate definitive method. Clin Chem 1980;26:854.

56. Allain CC, Poon LS, Chan CS, et al. Enzymatic determination of total serum cholesterol. Clin Chem 1974;20:470.

57. Richmond W. Preparation and properties of a cholesterol oxidase from *Nocardia sp.* and its application to the enzymatic assay of total cholesterol in serum. Clin Chem 1973;19:1350.

58. McGowan MW, Artiss JD, Zak B. Spectrophotometric study on minimizing bilirubin interference in an enzyme reagent mediated cholesterol reaction. Microchem J 1983;27:564.

59. Klotzsch SG, McNamara JR. Triglyceride measurements: a review of methods and interferences. Clin Chem 1990;36:1605.

60. Bucolo G, Yabut J, Chang TY. Mechanized enzymatic determination of triglycerides in serum. Clin Chem 1975;21:420.

61. McGowan M, Artiss J, Strandbergh DR, et al. A peroxidase-coupled method for the colorimetric determination of serum triglycerides. Clin Chem 1983;29:538.

62. Cole TG. Glycerol blanking in triglyceride assays: is it necessary? Clin Chem 1990;36:1267.

63. Stinshoff K, Weisshaar D, Staehler F, et al. Relation between concentrations of free glycerol and triglycerides in human sera. Clin Chem 1977;23:1029.

64. Warnick GR. Enzymatic methods for quantification of lipoprotein lipids. In: Albers JJ, Segrest JP, eds. Methods in enzymology. Vol 129. Orlando, FL: Academic Press, 1986;101.

65. Sullivan DR, Kruijswijk Z, West CE, et al. Determination of serum triglycerides by an accurate enzymatic method not affected by free glycerol. Clin Chem 1985;31:1227.

66. Lofland HB Jr. A semiautomated procedure for the determination of triglycerides in serum. Anal Biochem 1964;9:393.

67. Takayama M, Itoh S, Nagasaki T, et al. A new enzymatic method for determination of serum choline-containing phospholipids. Clin Chim Acta 1977;79:93.

68. McGowan, MW, Artiss JD, Zak B. A procedure for the determination of high-density lipoprotein choline-containing phospholipids. J Clin Chem Clin Biochem 1982;20:807.

69. Bartlett GR. Phosphorus assay in column chromatography. J Biol Chem 1959;234:466.

70. Labeur C, Shepherd J, Rosseneu M. Immunological assays of apolipoproteins in plasma: methods and instrumentation. Clin Chem 1990;36:591.

71. Sniderman AD. Is it time to measure apolipoprotein B? Arteriosclerosis 1990;10:665.

72. Hatch FT, Lees RS. Practical methods for plasma lipoprotein analysis. In: Pachetti R, ed. Advances in lipid research. Orlando, FL: Academic Press, 1968;6:1.

73. Havel RJ, Eder HA, Bragdon JH. The distribution and chemical composition of ultracentrifugally separated lipoproteins in human serum. J Clin Invest 1955;34:1345.

74. Chapman MJ, Goldstein S, Lagrange D, et al. A density gradient ultracentrifugal procedure for the isolation of the major lipoprotein classes from human serum. J Lipid Res 1981;22:339.

75. Kulkarni KR, Garber DW, Schmidt CF, et al. Analysis of cholesterol in all lipoprotein classes by single vertical ultracentrifugation of fingerstick blood and controlled-dispersion flow analysis. Clin Chem 1992;38:1898.

76. Patsch JR, Patsch W. Zonal ultracentrifugation. In: Albers JJ, Segrest JP, eds. Methods of enzymology. Orlando, FL: Academic Press, 1986;129:3.

77. Redgrave TG, Roberts DCK, West CE. Separation of plasma lipoproteins by density-gradient ultracentrifugation. Anal Biochem 1975;65:42.

78. Rosseneu M, Van Diervliet JP, Bury J, et al. Isolation and characterization of lipoprotein profiles in newborns by density gradient ultracentrifugation. Pediatr Res 1983;17(10):788.

79. Brousseau T, Clavey V, Bard JM, et al. Sequential ultracentrifugation micromethod for separation of serum lipoproteins and assays of lipids, apolipoproteins, and lipoprotein particles. Clin Chem 1993;39:960.

80. Wu LL, Warnick GR, Wu JT, et al. A rapid micro-scale procedure for determination of the total lipid profile. Clin Chem 1989; 35(7):1486.

81. Lewis LA, Opplt JJ. CRC Handbook of electrophoresis. Vol 1. Boca Raton, FL: CRC Press, Inc, 1980.

82. Lewis LA, Opplt JJ. CRC Handbook of electrophoresis. Vol 2. Boca Raton, FL: CRC Press, Inc, 1980.

83. Conlon D, Blankstein LA, Pasakarnis PA, et al. Quantitative determination of high-density lipoprotein cholesterol by agarose gel electrophoresis updated. Clin Chem 1979;24:227.

84. Lindgren FT, Silvers A, Jutaglr R, et al. A comparison of simplified methods for lipoprotein quantitation using the analytic ultracentrifuge as a standard. Lipids 1977;12:278.

85. Noble RP. Electrophoretic separation of plasma lipoproteins in agarose gel. J Lipid Res 1968;9:963.

86. Papadopoulos NM. Hyperlipoproteinemia phenotype determination by agarose gel electrophoresis updated. Clin Chem 1978;24:227.

87. Warnick GR, Nguyen T, Bergelin RO, et al. Lipoprotein quantification: an electrophoretic method compared with the Lipid Research Clinics method. Clin Chem 1982;28:2116.

88. Rifai N, Warnick GR, McNamara JR, et al. Measurement of low-density-lipoprotein cholesterol in serum: a status report. Clin Chem 1992;38:150.

89. Warnick GR, Leary ET, Goetsch J. Electrophoretic quantification of LDL-cholesterol using the Helena REP. [Abstract] Clin Chem 1993;39:1122.

90. Muniz N. Measurement of plasma lipoproteins by electrophoresis on polyacrylamide gel. Clin Chem 1977;23:1826.

91. Krauss RM, Lindgren FT, Ray RM. Interrelationships among subgroups of serum lipoproteins in normal human subjects. Clin Chem Acta 1980;104:275.

92. Burstein M, Legmann P. Lipoprotein precipitation. In: Clarkson TB, Kritchevsky D, Pollak OJ, eds. Monographs on atherosclerosis. Vol 2. New York: Karger, 1982;1.

93. Levin SJ. High-density lipoprotein cholesterol: review of methods. The American Society of Clinical Pathologists. Check Sample, Core Chemistry, no. PTS 892(PTS36), 1989;5(2).

94. Warnick GR, Benderson J, Albers JJ, et al. Dextran sulfate-Mg²⁺ precipitation procedure for quantitation of high-density-lipoprotein cholesterol. Clin Chem 1982;28:1379.

95. Kerscher L, Schiefer S, Draeger B, et al. Precipitation methods for the determination of LDL-cholesterol. Clin Biochem 1985;18:118.

96. Schumaker VN, Robinson MT, Curtiss LK, et al. Anti-apoprotein B monoclonal antibodies detect human low density lipoprotein polymorphism. J Biol Chem 1984;259:6423.

97. McNamara JR, Cole TG, Contois JH, et al. Immunoseparation method for measuring low-density lipoprotein cholesterol directly from serum evaluated. Clin Chem 1995;41:232.

98. Campos E, Nakajima K, Tanaka A, et al. Properties of an apolipoprotein E-enriched fraction of triglyceride-rich lipoproteins isolated from human blood plasma with a monoclonal antibody to apolipoprotein B-100. J Lipid Res 1992;33:369.

99. Nakajima K, Saito T, Tamura A, et al. Cholesterol in remnantlike lipoproteins in human serum using monoclonal anti apo B-100 and anti apo A-I immunoaffinity mixed gels. Clin Chim Acta 1993;223:53.

100. McNamara JR, Shah PK, Nakajima K, et al. Remnant lipoprotein cholesterol and triglyceride: reference ranges from the Framingham Heart Study. Clin Chem 1998;44:1224.

101. National Institutes of Health Consensus Conference. Triglyceride, HDL cholesterol and coronary heart disease. JAMA 1993;269:505.

102. Burstein M, Samaille J. Sur un dosage rapide du cholesterol lie aux α- et aux β-lipoproteines du serum. Clin Chim Acta 1960;5:609.

103. Fredrickson DS, Levy RI, Lindgren FT. A comparison of heritable abnormal lipoprotein patterns as defined by two different techniques. J Clin Invest 1968;47:2446.

104. Manual of laboratory operations, lipid research clinics program, lipid and lipoprotein analysis. Washington, DC: US Dept of Health and Human Services, National Institutes of Health, revised 1983.

105. Steele BW, Koehler DF, Azar MM, et al. Enzymatic determinations of cholesterol in high-density-lipoprotein fractions prepared by a precipitation technique. Clin Chem 1976;22:98.

106. Burstein M, Scholnick HR. Lipoprotein–polyanion–metal interactions. Adv Lipid Res 1973;11:68.

107. Lopes-Virella MF, Stone P, Ellis S, et al. Cholesterol determination in high-density lipoproteins separated by three different methods. Clin Chem 1977;23:882.

108. Warnick GR, Cheung MC, Albers JJ. Comparison of current methods for high-density lipoprotein cholesterol quantitation. Clin Chem 1979;25:596.

109. Allen JK, Hensley WJ, Nichols AV, et al. An enzymic and centrifugal method for estimating high-density lipoprotein cholesterol. Clin Chem 1979;25:325.

110. Demacker PNM, Vos-Janssen HE, Hijmans AGM, et al. Measurement of high-density lipoprotein cholesterol in serum: comparison of six isolation methods combined with enzymic cholesterol analysis. Clin Chem 1980;26:1780.

111. Viikari J. Precipitation of plasma lipoproteins by PEG6000 and its evaluation with electrophoresis and ultracentrifugation. Scand J Clin Lab Invest 1976;36:265.

112. Warnick GR, Albers JJ, Bachorik PS, et al. Multi-laboratory evaluation of an ultrafiltration procedure for high-density lipoprotein cholesterol quantification in turbid heparin-manganese supernates. J Lipid Res 1981;22:1015.

113. Sugiuchi H, Uji Y, Okabe H, et al. Direct measurement of high-density lipoprotein cholesterol in serum with polyethylene glycol-modified enzymes and sulfated a-cyclodextrin. Clin Chem 1995; 41:717.

114. Harris N, Galpchian V, Rifai N. Three routine methods for measuring high-density lipoprotein cholesterol compared with the reference method. Clin Chem 1996;42:738.

115. Myers GL, Cooper GR, Henderson LO, et al. Standardization of lipid and lipoprotein measurements. In: Rifai N, Warnick GR, eds. Handbook of lipoprotein testing. Washington, DC: AACC Press, 1997;223.

116. Expert Panel. Report of the National Cholesterol Education Program Expert Panel on detection, evaluation, and treatment of high blood cholesterol in adults. Arch Intern Med 1988;148:36.

117. Expert Panel. Summary of the second report of the National Cholesterol Education Program (NCEP) Expert Panel on detection, evaluation, and treatment of high blood cholesterol in adults (Adult Treatment Panel II). JAMA 1993;269:3015–3023.

118. Bachorik PS. Measurement of low-density lipoprotein cholesterol. In: Rifai N, Warnick GR, eds. Handbook of lipoprotein testing. Washington, DC: AACC Press, 1997;145.

119. Friedewald WT, Levy RI, Fredrickson DS. Estimation of the concentration of low-density lipoprotein cholesterol in plasma, without use of the preparative ultracentrifuge. Clin Chem 1972;18:499.

120. McNamara JR, Cohn JS, Wilson PWF, et al. Calculated values for low-density lipoprotein cholesterol in the assessment of lipid abnormalities and coronary disease risk. Clin Chem 1990;36:36.

121. Warnick GR, Knopp RH, Fitzpatrick V, et al. Estimating low-density lipoprotein cholesterol by the Friedewald equation is adequate for classifying patients on the basis of nationally recommended cutpoints. Clin Chem 1990;36:15.

122. Sugiuchi H, Irie T, Uji Y, et al. Homogeneous assay for measuring low-density lipoprotein cholesterol in serum with triblock copolymer and a-cyclodextrin sulfate. Clin Chem 1998;44:522.

123. Rifai N, Iannotti E, DeAngelis K, et al. Analytical and clinical performance of a homogeneous enzymatic LDL-cholesterol assay compared with the ultracentrifugation-dextran sulfate-Mg²⁺ method. Clin Chem 1998;44:1242.

124. Lichtenstein AH, Ausman LM, Carrasco W, et al. Hydrogenation impairs the hypolipidemic effect of corn oil in humans: hydrogenation, trans fatty acids, and plasma lipids. Arterioscler Thromb 1993; 13:154.

125. LePage G, Roy C. Direct transesterification of all classes of lipids in a one-step reaction. J Lipid Res 1986;27:114.

126. Warnick GR. Compact analysis systems for cholesterol, triglycerides and high-density lipoprotein cholesterol. Curr Opin Lipidol 1991;2:343.

127. Cooper GR, Myers GL, Smith SJ, et al. Blood lipid measurements. Variations and practical utility. JAMA 1992;267:1652.
128. Cooper GR, Smith SJ, Myers GL, et al. Estimating and minimizing effects of biologic sources of variation by relative range when measuring the mean of serum lipids and lipoproteins. Clin Chem 1994; 40:227.
129. Eckfeldt JH, Copeland KR. Accuracy verification and identification of matrix effects. The College of American Pathologist's protocol. Arch Pathol Lab Med 1993;117:381.
130. Greenberg N, Li ZM, Bower GN. National Reference System for Cholesterol (NRS-CHOL): problems with transfer of accuracy with matrix materials. Clin Chem 1988;24:1230.
131. Kroll MH, Chesler R, Elin RJ. Effect of lyophilization on results of five enzymatic methods for cholesterol. Clin Chem 1989;35:1523.
132. Heiss G, Tamir I, Davis CE, et al. Lipoprotein cholesterol distributions in selected North American populations: the lipid research clinics program prevalence study. Circulation 1980;61:302.
133. Lipid research clinics population studies data book. Vol 1. The prevalence study. NIH pub. no. 801527. Washington, DC: Dept of Health and Human Services, Public Health Service, 1980.
134. Schaefer EJ, Lichtenstein AH, Lamon-Fava S, et al. Lipoproteins, nutrition, aging, and atherosclerosis. Am J Clin Nutr 1995;61(suppl):726S.
135. Campos H, McNamara JR, Wilson PWF, et al. Differences in low-density lipoprotein subfractions and apolipoproteins in premenopausal and postmenopausal women. J Clin Endocrinol Metab 1988;67:30.
136. National Cholesterol Education Program (NCEP). Highlights of the report of the Expert Panel on blood cholesterol levels in children and adolescents. Pediatrics 1992;89:495.
137. Garcia RE, Moodie DS. Routine cholesterol surveillance in childhood. Pediatrics 1989;84:751.
138. Castelli WP, Garrison RJ, Wilson PWF, et al. Incidence of coronary heart disease and lipoprotein cholesterol levels: the Framingham Study. JAMA 1986;256:2835.
139. Stamler J, Stamler R, Shekelle RB. Regional differences in prevalence, incidence, and mortality from atherosclerotic coronary heart disease. In: deHaas JH, Hemker HC Snellen HA, eds. Ischaemic heart disease. Leiden, The Netherlands: Leiden University Press, 1970;84.
140. Thom TJ, Epstein FH, Feldman JJ, et al. Total mortality and mortality from heart disease, cancer, and stroke from 1950 to 1987 in 27 countries. NIH pub. no. 923088. Washington, DC: 1992.
141. Kannel WB. High-density lipoproteins: Epidemiologic profile and risks of coronary artery disease. Am J Cardiol 1983;52:9B.
142. Expert Panel. National Cholesterol Education Program. Second report of the Expert Panel on detection, evaluation, and treatment of high blood cholesterol in adults. (Adult Treatment Panel II). Circulation 1994;89:1329–1433.
143. American Academy of Pediatrics. National Cholesterol Education Program: Report of the Expert Panel on blood cholesterol levels in children and adolescents. Pediatrics 1992;89(3 Pt. 2):525–584.
144. Schaefer EJ, Lichtenstein AH, Lamon-Fava S, et al. Lipoproteins, nutrition, aging, and atherosclerosis. Am J Clin Nutr [Review] 1995;61(3 suppl):726S.
145. Hjermann I, Velve Byre K, Holme I, et al. Effect of diet and smoking intervention on the incidence of coronary heart disease. Report from the Oslo Study Group of a randomized trial in healthy men. Lancet 1981;2:1303.
146. Holme I, Hjermann I, Helgelend A, et al. The Oslo Study: diet and antismoking advice. Additional results from a 5-year primary preventive trial in middle-aged men. Prev Med 1985;14:279.
147. Hegsted DM, Ausman LM, Johnson JA, et al. Dietary fat and serum lipids: an evaluation of the experimental data. Am J Clin Nutr 1993;57:875.
148. Ornish D, Brown SE, Scherwitz LW, et al. Can lifestyle changes reverse coronary heart disease? The lifestyle heart trial. Lancet 1990;336:129.
149. Schuler G, Hambrecht R, Schlierf G, et al. Regular physical exercise and low fat diet. Effects on progression of coronary artery disease. Circulation 1992;86:1.
150. Shekelle RB, Shryock AM, Paul O, et al. Diet, serum cholesterol, and death from coronary heart disease. The Western Electric Study. N Engl J Med 1981;304:65.
151. Connor WE, Stone DB, Hodges RE. The interrelated effects of dietary cholesterol and fat upon human serum lipid levels. J Clin Invest 1964;43:1691.
152. Nicolosi RJ, Stucchi AF, Kowala MC, et al. Effect of dietary fat saturation and cholesterol on LDL composition and metabolism. In vivo studies of receptor and nonreceptor-mediated catabolism of LDL in cebus monkeys. Arteriosclerosis 1990;10:119.
153. Coleman MP, Key TJ, Wang DY, et al. A prospective study of obesity, lipids, apolipoproteins, and ischaemic heart disease in women. Atherosclerosis 1992;92:177.
154. Genest JJ Jr, Martin-Munley SS, McNamara JR, et al. Familial lipoprotein disorders in patients with premature coronary artery disease. Circulation 1992;85:2025.
155. Genest JJ Jr, Ordovas JM, McNamara JR, et al. DNA polymorphisms of the apolipoprotein B gene in patients with premature coronary artery disease. Atherosclerosis 1990;82:7.
156. Stary HC. Evolution and progression of atherosclerotic lesions in coronary arteries of children and young adults. Arteriosclerosis 1989; 9(suppl 1):I19.
157. Stary HC. Macrophages, macrophage foam cells, and eccentric intimal thickening in the coronary arteries of young children. Atherosclerosis 1987;64:91.
158. Brown MS, Goldstein JL. The LDL receptor concept: clinical and therapeutic implications. In: Stokes J III, Mancini M, eds. Hypercholesterolemia: clinical and therapeutic implications. Vol 18. Atherosclerosis reviews. New York: Raven Press, 1988;85.
159. Jialal I, Chait A. Differences in the metabolism of oxidatively modified low density lipoprotein and acetylated low density lipoprotein by human endothelial cells: inhibition of cholesterol esterification by oxidatively modified low density lipoprotein. J Lipid Res 1989; 30:1561.
160. Steinberg D, Parthasarathy S, Carew TE, et al. Beyond cholesterol. Modifications of low-density lipoprotein that increase its atherogenicity. New Engl J Med 1989;320:915.
161. Genest JJ, McNamara JR, Salem DN, et al. Prevalence of risk factors in men with premature coronary artery disease. Am J Cardiol 1991;67:1185.
162. Lipid research clinics program: the Lipid Research Clinic Coronary Primary Prevention Trial results. II. The relationship of reduction in incidence of coronary heart disease to cholesterol lowering. JAMA 1984;251:365.
163. National Institutes of Health Consensus Conference. Lowering blood cholesterol to prevent heart disease. JAMA 1985;253:2080.
164. Canner PL, Berge KG, Wenger NK, et al. Fifteen-year mortality in Coronary Drug Project patients: long-term benefit with niacin. J Am Coll Cardiol 1986;8:1245.
165. Frick MH, Elo O, Haapa K, et al. Helsinki heart study: Primary-prevention trial with gemfibrozil in middle-aged men with dyslipidemia. Safety of treatment, changes in risk factors, and incidence of coronary heart disease. N Engl J Med 1987;317:1237.
166. Carlson LA, Rosenhamer G. Reduction of mortality in the Stockholm Ischaemic Heart Disease Secondary Prevention Study by combined treatment with clofibrate and nicotinic acid. Acta Med Scand 1988;223:405.
167. Pedersen TR, et al. Randomized trial of cholesterol lowering in 4444 patients with coronary heart disease: the Scandinavian Simvastatin Survival Study. Lancet 1994;344:1383.
168. Shepherd J, et al. Prevention of coronary heart disease with pravastatin in men with hypercholesterolemia: the West of Scotland Coronary Prevention Study. N Engl J Med 1995;333:1301.
169. Jukema JW, et al. Effects of lipid lowering by pravastatin on progression and regression of coronary artery disease in symptomatic men with normal to moderately elevated serum cholesterol levels. The Regression Growth Evaluation Statin Study (REGRESS). Circulation 1995;91:2528.
170. Sacks FM, Pfeiffer MA, Moye LA, et al. The effect of pravastatin on coronary events after myocardial infarction in patients with average cholesterol levels. N Engl J Med 1996;335:1001.
171. Lipid Research Clinics program: the Lipid Research Clinics Coronary Primary Prevention Trial. I. Reduction in incidence of coronary heart disease. JAMA 1984;251:351.
172. Alderman E, Haskell WL, Fain JM, et al. Beneficial angiographic and clinical response to multifactor modification in the Stanford Coronary Risk Intervention Project (SCRIP). Circulation 1991;84(2):140.
173. Blankenhorn DH, Azen SP, Kramsch DM, et al. Coronary angiographic changes with lovastatin therapy. The monitored atherosclerosis regression study (MARS). Ann Intern Med 1993;119:969.

174. Brown BG, Albers JJ, Fisher LD, et al. Regression of coronary artery disease as a result of intensive lipid-lowering therapy in men with high levels of apolipoprotein B. N Engl J Med 1990;323:1289.

175. Brown BG, Zhao XQ, Sacco DE, et al. Lipid lowering and plaque regression. New insights into prevention of plaque disruption and clinical events in coronary disease. Circulation 1993;87:1781.

176. Kane JP, Malloy MJ, Ports TA, et al. Regression of coronary atherosclerosis during treatment of familial hypercholesterolemia with combined drug regimens. JAMA 1990;264:3007.

177. Schaefer EJ, Levy RI. The pathogenesis and management of lipoprotein disorders. N Engl J Med 1985;312:1300.

178. Schaefer EJ, McNamara JR, Genest J Jr, et al. Genetics and abnormalities in metabolism of lipoproteins. Clin Chem 1988;34:B9.

179. Welty FK, Lichtenstein AH, Barrett PHR, et al. Decreased production and increased catabolism of apolipoprotein B-100 in apolipoprotein B67/B-100 heterozygotes. Arterioscler Thromb Vasc Biol 1997;17:881.

180. Schaefer EJ, McNamara JR. Overview of the diagnosis and treatment of lipid disorders. In: Rifai N, Warnick GR, eds. Handbook of lipoprotein testing. Washington, DC: AACC Press, 1997;25.

181. Rubins HB, Robins SJ, Iwane MK, et al. Rationale and design of the Department of Veterans Affairs High-Density Lipoprotein Cholesterol Intervention Trial (HIT) for secondary prevention of coronary artery disease in men with low high-density lipoprotein cholesterol and desirable low-density lipoprotein cholesterol. Am J Cardiol 1993;71:45.

182. Behar S, Graff E, Reicher-Reiss H, et al. Low total cholesterol is associated with high total mortality in patients with coronary heart disease. The Bezafibrate Infarction Prevention (BIP) Study Group. Eur Heart J 1997;18:52.

183. Downs JR, Clearfield M, Weis S, et al. Primary prevention of acute coronary events with lovastatin in men and women with average cholesterol levels. Results of AFCAPS/TexCAPS. JAMA 1998;279:1615.

184. Schaefer EJ. Hyperlipoproteinemia. In: Rakel RE, ed. Conn's current therapy. Philadelphia: WB Saunders, 1991;515.

185. Sprecher DL, Schaefer EJ, Kent KM, et al. Cardiovascular features of homozygous familial hypercholesterolemia: analysis of 16 patients. Am J Cardiol 1984;54:20.

186. Stone NJ, Levy RI, Fredrickson DS, et al. Coronary artery disease in 116 kindred with familial type II hyperlipoproteinemia. Circulation 1974;49:476.

187. Grundy SM, Denke MA. Dietary influences on serum lipids and lipoproteins. J Lipid Res 1990;31:1149.

188. Hunninghake DB, Stein EA, Dujovne CA, et al. The efficacy of intensive dietary therapy alone or combined with lovastatin in outpatients with hypercholesterolemia. N Engl J Med 1993;328:1213.

189. Kromhout D, de Lezenne Coulander C. Diet, prevalence and 10-year mortality from coronary heart disease in 871 middle-aged men. The Zutphen Study. Am J Epidemiol 1984;119:733.

190. Avogaro P, Cazzolato G. Changes in the composition and physico-chemical characteristics of serum lipoproteins during ethanol-induced lipemia in alcoholic subjects. Metabolism 1975;24:1231.

191. Chait A, Albers JJ, Brunzell JD. Very low density lipoprotein overproduction in genetic forms of hypertriglyceridemia. Eur J Clin Invest 1980;10:17.

192. Cohn JS, McNamara JR, Schaefer EJ. Lipoprotein cholesterol concentrations in the plasma of human subjects as measured in the fed and fasted states. Clin Chem 1988;34:2456.

193. Schaefer EJ, McNamara JR, Genest J Jr, et al. Clinical significance of hypertriglyceridemia. Semin Thromb Hemost 1988;14:142.

194. Schonfeld G. Lipoproteins in atherogenesis. Artery 1975;5:305.

195. Harris WS, Connor WE, Alam N, et al. Reduction of postprandial triglyceridemia in humans by dietary n-3 fatty acids. J Lipid Res 1988;29:1451.

196. Mascioli EA, Lopes S, Randall S, et al. Serum fatty acid profiles after intravenous medium chain triglyceride administration. Lipids 1989;24:793.

197. Zilversmit DB. A proposal linking atherogenesis to the interaction of endothelial lipoprotein lipase with triglyceride-rich lipoproteins. Circ Res 1973;33:633.

198. Tatami R, Mabuchi H, Ueda K, et al. Intermediate density lipoprotein and cholesterol-rich very low-density lipoprotein in angiographically determined coronary artery disease. Circulation 1981;64:1174.

199. Zilversmit DB. Atherogenic nature of triglycerides, postprandial lipemia, and triglyceride-rich remnant lipoproteins. Clin Chem 1995;41:153.

200. Devaraj S, Vega G, Lange R, et al. Remnant-like particle cholesterol levels in patients with dysbetalipoproteinemia or coronary artery disease. Am J Med 1998;104:445.

201. Jialal I, Hirany SV, Devaraj S, et al. Comparison of an immunoseparation method for direct measurement of LDL cholesterol with beta-quantification (ultracentrifugation). Am J Clin Path 1995;104:76.

202. Dahlen GH, Guyton JR, Attar M, et al. Association of levels of lipoprotein Lp (a), plasma lipids, and other lipoproteins with coronary artery disease documented by angiography. Circulation 1986;74:758.

203. Genest J Jr, Jenner JL, McNamara JR, et al. Prevalence of lipoprotein (a) [Lp (a)] excess in coronary artery disease. Am J Cardiol 1991;67:1039.

204. Schaefer EJ, Lamon-Fava S, Jenner JL, et al. Lipoprotein(a) levels and risk of coronary heart disease in men. The Lipid Research Clinics Coronary Primary Prevention Trial. JAMA 1994;271:999.

205. Schwartzman RA, Cox ID, Poloniecki J, et al. Elevated plasma lipoprotein(a) is associated with coronary artery disease in patients with chronic stable angina pectoris. J Am Coll Cardiol 1998;31:1260.

206. Dahlen GH, Stenlund H. Lp (a) lipoprotein is a major risk factor for cardiovascular disease: pathogenic mechanisms and clinical significance. Clin Genet 1997;52:272.

207. Ridker PM, Hennekens CH, Stampfer MJ. A prospective study of lipoprotein(a) and the risk of myocardial infarction. JAMA 1993;270:2195.

208. Berg K, Dahlen G, Christophersen B, et al. Lp (a) lipoprotein level predicts survival and major coronary events in the Scandinavian Simvastatin Survival Study. Clin Genet 1997;52:254.

209. McLean JW, Tomlinson JE, Kuang WJ, et al. cDNA sequence of human apolipoprotein(a) is homologous to plasminogen. Nature 1987;330:132.

210. Hajjar KA, Gavish D, Breslow JL, et al. Lipoprotein(a) modulation of endothelial cell surface fibrinolysis and its potential role in atherosclerosis. Nature 1989;339:303.

211. Loscalzo J, Weinfield M, Fless GM, et al. Lipoprotein(a), fibrin binding, and plasminogen activation. Arteriosclerosis 1990;10:240.

212. Miles LA, Fless GM, Levin EG, et al. A potential basis for the thrombotic risks associated with lipoprotein(a). Nature 1989;339:301.

213. Genest JJ Jr, Bard JM, Fruchart JC, et al. Familial hypoalphalipoproteinemia in premature coronary artery disease. Arterioscler Thromb 1993;13:1728.

214. Schaefer EJ, Blum CB, Zech LA, et al. Metabolism of high density apolipoproteins in Tangier disease. New Engl J Med 1978;299:905.

215. Schaefer EJ, Heaton WH, Wetzel MG, et al. Plasma apolipoprotein A-I absence associated with a marked reduction of high density lipoproteins and premature coronary artery disease. Arteriosclerosis 1982;2:16.

216. Schaefer EJ, Ordovas JM, Law SW, et al. Familial apolipoprotein A-I and C-III deficiency, variant II. J Lipid Res 1985;26:1089.

217. Third JL, Montag J, Flynn M, et al. Primary and familial hypoalphalipoproteinemia. Metabolism 1984;33:136.

218. Vergani C, Bettale G. Familial hypo-alpha-lipoproteinemia. Clin Chim Acta 1981;114:45.

219. Barrett Connor E, Bush TL. Estrogen and coronary heart disease in women. JAMA 1991;265:1861.

220. Stampfer MJ, Colditz GA. Estrogen replacement therapy and coronary heart disease: a quantitative assessment of the epidemiologic evidence. Prev Med 1991;20:47.

221. Genest JJ, Corbett H, McNamara JR, et al. Effect of hospitalization on high-density lipoprotein cholesterol in patients undergoing elective coronary angiography. Am J Cardiol 1988;61:998.

CHAPTER 12

Nonprotein Nitrogen

Dennis W. Jay

Objectives

Upon completion of this chapter, the clinical laboratorian should be able to:

- *List the nonprotein nitrogen substances in the blood and recognize their relative concentration.*

- *Recognize the chemical structures of the following nonprotein nitrogen substances: urea, creatinine, uric acid, and ammonia.*

- *Describe the biosynthesis and excretion of urea, uric acid, creatine, creatinine, and ammonia.*

- *Relate the level of protein in the diet, protein metabolism, renal blood flow, and renal function to the plasma urea level.*

- *Discuss the blood urea nitrogen (BUN)/creatinine ratio and its use in distinguishing between prerenal and renal causes of an elevated blood urea.*

- *Recognize the major clinical conditions associated with increased and decreased levels of urea, creatine, and creatinine in the plasma.*

- *Describe specimen requirements and storage conditions for determinations of urea, uric acid, creatine, creatinine, and ammonia.*

- *Recognize the reference intervals for plasma urea, creatinine, uric acid, and ammonia.*

- *Relate plasma levels of creatinine to diet, muscle mass, and turnover and renal function.*

- *Discuss commonly used methods for the determination of creatine, creatinine, urea, uric acid, and ammonia in plasma and urine.*

- *Discuss sources of error in the Jaffe method for creatinine and methods that have been used to minimize these problems.*

- *Recognize the impact of specific methods for creatinine on the estimation of the glomerular filtration rate (GFR) with creatinine clearance.*

- *Recognize reference intervals for plasma and urine creatinine and the effect of age and sex on these values.*

- *Relate plasma levels of uric acid to diet, purine metabolism, and renal disease.*

- *Relate the solubility of uric acid to the pathologic consequences of increased plasma uric acid.*

- *Recognize reference intervals for plasma uric acid and the effect of age and sex on these values.*

- *Discuss clinical conditions and toxic effects related to increased plasma level of ammonia.*

- *Describe specimen collection and handling requirements for measurement of ammonia in plasma.*

- *Given patient values for urea, creatinine, uric acid, and supporting clinical history, suggest possible clinical conditions associated with the results.*

KEY TERMS

Allantoin	Clearance	Creatinine clearance
Ammonia	Coupled enzymatic	Distal tubule
Azotemia	method	Encephalopathy
BUN/creatinine	Creatine	GFR (glomerular
ratio	Creatinine	filtration rate)

(continued)

Glomerulus
Gout
Hyperuricemia
Hypouricemia
Kinetic method of
   measurement

NPN (nonprotein
   nitrogen)
Postrenal
Prerenal
Protein-free filtrate
Proximal tubule

Reabsorption
Secretion
Urea
Uric acid
Uremia or uremic
   syndrome

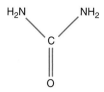

▲
Figure 12-1. Structure of urea.

The determination of nonprotein nitrogenous substances in the blood has traditionally been used to monitor renal function. The term *nonprotein nitrogen (NPN)* originated in the early days of clinical chemistry when analytical methodology required removal of protein from the sample before analysis. The concentration of nitrogen-containing compounds in this *protein-free filtrate* was quantitated spectrophotometrically by converting nitrogen to ammonia and subsequent reaction with Nessler's reagent (HgI_2/KI), forming a yellow color. This technique is technically difficult, but is still regarded as the most accurate method for the determination of total NPN concentration. However, much more valid clinical information is obtained by analyzing a patient's sample for the individual components of the NPN fraction.

The NPN fraction comprises about 15 compounds of clinical interest (Table 12-1). The majority of these compounds arise from the catabolism of proteins and nucleic acids.

UREA

Biochemistry

Urea constitutes nearly half the NPN substances in the blood (Fig. 12-1). It is synthesized in the liver from CO_2 and the ammonia arising from the deamination of amino acids by means of the ornithine or Krebs-Henseleit cycle. Urea constitutes the major excretory product of protein metabolism. Following synthesis in the liver, urea is carried in the blood to the kidney, where it is readily filtered from the plasma by the *glomerulus*. Most of the urea in the glomerular filtrate is excreted in the urine, although up to 40% is *reabsorbed* by passive diffusion during passage of the filtrate through the renal tubules. The amount reabsorbed depends on urine flow rate and level of hydration. Small amounts of urea (<10% of the total) are excreted through the gastrointestinal (GI) tract

TABLE 12-1. Major Components of the NPN Fraction

Compound	Approximate Plasma Concentration (% of Total NPN)
Urea	45
Amino acids	20
Uric acid	20
Creatinine	5
Creatine	1–2
Ammonia	0.2

and skin. The level of urea in the plasma is governed by renal function and perfusion, the protein content of the diet, and the amount of protein catabolism. The term *blood urea nitrogen (BUN)* is used extensively when referring to urea measurement because historical assays for urea were based on nitrogen measurement.

Disease Correlations

An elevated level of urea in the blood is called *azotemia.* Very high levels of plasma urea accompanied by renal failure is called *uremia* or the *uremic syndrome,* which is eventually fatal if not treated by dialysis. Conditions causing elevations of plasma urea are classified according to cause into three main categories: prerenal, renal, and postrenal.

Prerenal hyperuricemia is caused by reduced renal blood flow. Reduction in blood flow delivers less urea to the kidney and therefore less urea is filtered. Causative factors include congestive heart failure, shock, hemorrhage, dehydration, and any other factors resulting in a marked decrease in blood volume. The level of protein metabolism also causes prerenal changes in blood urea concentration. A high-protein diet, or increased protein catabolism such as occurs in fever, major illness, stress, corticosteroid therapy, and gastrointestinal hemorrhage, may increase urea levels. Levels will be decreased during periods of low protein intake or increased protein synthesis, such as late pregnancy and infancy.

Decreased renal function causes an increase in plasma urea concentration due to compromised urea excretion. Renal causes of an elevated urea include acute and chronic renal failure, glomerular nephritis, tubular necrosis, and other intrinsic renal disease (see Chapter 21, *Renal Function*).

Postrenal elevations of urea are due to obstruction to the urine flow anywhere in the urinary tract by renal stones, tumors of the bladder or prostate, or severe infection.

The major causes of decreased plasma urea levels include decreased protein intake and severe liver disease. The conditions affecting plasma urea levels are summarized in Table 12-2.

Differentiation of abnormal urea levels is aided by calculation of the *BUN/creatinine ratio,* which is normally 10:1 to 20:1. Prerenal conditions tend to elevate plasma urea, whereas plasma creatinine remains normal, causing a high BUN/creatinine ratio. A high BUN/creatinine ratio with an elevated creatinine is usually seen in postrenal conditions. A low BUN/creatinine ratio is seen in low protein intake, acute tubular necrosis, and severe liver disease.

TABLE 12-2. Causes of Abnormal Plasma Urea Concentration

Increased Levels

Prerenal	Congestive heart failure
	Shock, hemorrhage
	Dehydration
	Increased protein catabolism
	Corticosteroid therapy
Renal	Acute and chronic renal failure
	Glomerular nephritis
Postrenal	Urinary tract obstruction

Decreased Levels

	Decreased protein intake
	Severe liver disease
	Severe vomiting and diarrhea
	Pregnancy

Analytical Methods

Measurements of urea were originally done on a protein-free filtrate of whole blood and were based on measuring the amount of nitrogen. Current analytical methods have retained this custom and urea is reported in terms of nitrogen concentration rather than urea concentration. Urea's nitro-

gen concentration is converted to urea concentration by multiplying by 2.14, as shown in Equation 12-1.

$$\frac{1 \text{ mg urea } N}{dL} \times \frac{1 \text{ mmol } N}{14 \text{ mg } N} \times \frac{1 \text{ mmol urea}}{2 \text{ mmol } N}$$

$$\times \frac{60 \text{ mg urea}}{1 \text{ mmol urea}} = \frac{2.14 \text{ mg urea}}{dL} \qquad (Eq.\ 12\text{–}1)$$

Urea nitrogen concentration in milligrams per deciliter may be converted to millimoles per liter of urea by multiplying by 0.36.

Two analytical approaches have been used to assay for urea. The oldest and most often used involves the hydrolysis of urea by the enzyme urease (urea aminohydrolase EC 3.5.1.5) and quantitation of NH_4^+ by a variety of techniques. Early colorimetric methods were based on using Nessler's reagent or the Berthelot reaction to detect NH_4^+.[1] The most common clinically used kinetic method couples the urease reaction with L-glutamate dehydrogenase (EC 1.4.1.3) and monitors the rate of disappearance of NADH at 340 nm (Fig. 12-2).[2] A modification of this method using bacterial urease has been described for measuring urea in urine specimens.[3]

An alternative enzymatic method quantitates ammonium ion produced from urease hydrolysis by a series of coupled reactions that produce H_2O_2. The H_2O_2 is then quantitated by

CASE STUDY 12-1

A 65-year-old man was first admitted for treatment of chronic obstructive lung disease, renal insufficiency, and marked cardiomegaly. Pertinent laboratory data on admission (5/31) is shown in Case Study Table 12-1.1.

Because of severe respiratory distress, the patient was transferred to the intensive care unit, placed on a respirator, and given diuretics and intravenous (IV) fluids to promote diuresis. This treatment brought about a marked improvement in both cardiac output and renal function, as shown by laboratory results several days later (6/3). After 2 additional days on a respirator with IV therapy, the patient's renal function had returned to normal, and at discharge his laboratory results were normal (6/7).

The patient was readmitted 6 months later because of the increasing inability of his family to arouse him. On admission, he was shown to have a tremendously enlarged heart with severe pulmonary disease, heart failure, and probable renal failure. Laboratory studies on admission were as shown in Case Study Table 12-1.2. Numerous attempts were made to improve the patient's cardiac and pulmonary function, all to no avail, and the patient died 4 days later.

CASE STUDY TABLE 12-1.1. Laboratory Results—First Admission

Test	5/31	6/3	6/7
BUN, mg/dL	45	24	11
Creatinine, mg/dL	1.8	1.3	0.9
BUN/Creatinine	25	18.5	12.2
pH	7.22	7.5	
PCO₂, mm Hg	74.4	48.7	
PO₂, mm Hg	32.8	57.6	
O₂ sat, %	51.3	91.0	

CASE STUDY TABLE 12-1.2. Laboratory Results—Second Admission

BUN, mg/dL	90
Creatinine, mg/dL	3.9
Uric acid, mg/dL	12.0
BUN/Creatinine	23
pH	7.351
PCO₂, mm Hg	59.9
PO₂, mm Hg	34.6
O₂ sat, %	63.7

Question

1. What is the most likely cause of the patient's elevated BUN? What data support your conclusion?

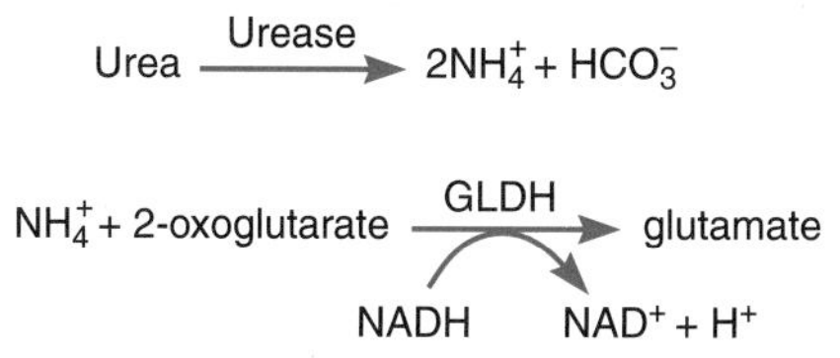

$$Urea \xrightarrow{\text{Urease}} 2NH_4^+ + HCO_3^-$$

$$NH_4^+ + \text{2-oxoglutarate} \xrightarrow{\text{GLDH}} glutamate$$
$$NADH \qquad NAD^+ + H^+$$

GLDH = glutamate dehydrogenase (EC 1.4.1.3).

▲

Figure 12-2. Enzymatic assay for urea.

the formation of a quinone-imine dye in the presence of phenol and 4-aminophenazone.[4] Ammonium from the urease reaction can also be measured by the color change associated with a pH indicator. This approach has been incorporated into instruments using a multilayer film format,[5,6] reagent strips,[7] or liquid reagents.[8]

An electrode can be used to measure the rate of increase in conductivity as urea is converted to ammonium ion.[9] Because the rate of change in conductivity is measured, ammonia contamination is not a problem as it is in other methods. Potentiometric methods using an ammonia ion-selective electrode have also been developed.[10]

Urea may be measured by direct condensation with diacetyl monoxime in the presence of strong acid and an oxidizing agent to form a yellow diazine derivative. Thiosemicarbazide is added to the reaction mixture to stabilize the color.[11] The major advantage of this method is that it measures urea directly and ammonia does not interfere.

Other direct approaches include reaction of urea to form colored compounds with *O*-phthalaldehyde and naphthylethylenediamine[12] or with salicylate, nitroferricyanide, and hypochlorite.[13] Isotope-dilution mass spectrometry has been proposed as a definitive method for urea.[14] The current reference method is the coupled urease/GLDH enzymatic method, which was described earlier. When used as a reference method, it is recommended that the assay be done on a protein-free filtrate.[15] Analytical methods are summarized in Table 12-3.

Specimen Requirements and Interfering Substances

Urea concentration may be measured in serum, plasma, or urine. When collecting plasma, ammonium ions and high concentrations of sodium fluoride or sodium citrate must be avoided. Fluoride or citrate will both inhibit urease. Although the overall protein content of the diet will influence the level of urea, the effect of a single protein-containing meal is minimal and a fasting sample is usually not required. A nonhemolyzed sample is recommended. Urea is quite susceptible to bacterial decomposition, so samples (especially urine) that cannot be analyzed within a few hours should be refrigerated. A few crystals of thymol may be added to urine samples as a preservative. Methods for serum or plasma may

require modification for use with urine because of high urea concentration and the presence of endogenous ammonia.

Reference Interval[16]

Urea nitrogen, 7 mg/dL to 18 mg/dL (2.5–6.4 mmol/L urea).

CREATININE/CREATINE

Biochemistry

Creatine is synthesized mainly in the liver from arginine, glycine, and methionine. It is then transported to other tissues, such as muscle, where it is converted to phosphocreatine, which serves as a high-energy source. Creatine phosphate loses phosphoric acid and creatine loses water to form *creatinine,* which is released into the plasma. The structures of these compounds are shown in Figure 12-3.

Creatinine is excreted into the circulation at a relatively constant rate that has been shown to be proportional to the individual's muscle mass. It is removed from the circulation by glomerular filtration and excreted in the urine. Additional amounts of creatinine are *secreted* by the *proximal tubule.* Small amounts may also be reabsorbed by the renal tubules, especially at low flow rates.[17]

Plasma levels of creatinine are related to the relative muscle mass, the rate of creatine turnover, and renal function. It has been accepted for many years that the plasma level of creatinine is relatively unaffected by diet. However, some reports have indicated that the protein content of the diet may indeed affect the plasma level of creatinine if it affects the individual's muscle mass.[18] Because of the observed constancy of endogenous production, the measurement of creatinine excretion has been used as a measure of the completeness of 24-hour urine collections in a given individual, although this method may be unreliable.[19]

Disease Correlations

Creatinine

Elevated levels of creatinine are mainly associated with abnormal renal function, especially as it relates to glomerular function. The *glomerular filtration rate (GFR)* is the volume of plasma filtered (V) by the glomerulus per unit of time (t).

$$GFR = \frac{V}{t} \qquad \textit{(Eq. 12–2)}$$

Assuming a substance, S, can be measured and is freely filtered at the glomerulus and neither secreted or reabsorbed by the tubules, the volume of plasma filtered would be equal to the mass of S filtered (M_S) divided by its plasma concentration (P_S).

$$V = \frac{M_S}{P_S} \qquad \textit{(Eq. 12–3)}$$

TABLE 12-3. Summary of Analytical Methods—Urea

Urease Methods

All methods use similar first step	Urea + H_2O $\xrightarrow{\text{Urease}}$ $2NH_4^+ + CO_3^{-2}$	
GLDH *coupled enzymatic*	$NH_4^+ + 2\text{-oxoglutarate} + NADH \xrightarrow{\text{GLDH}} NAD^+ + \text{glutamate} + H_2O$	Most popular on common automated instruments; best done kinetically; candidate reference method
Conductimetric	Conversion of unionized urea to NH_4^+, HCO_3^-; results in increased conductivity rate	Specific and rapid
Nesslerization	$NH_4^+ + 2HgI_2 + 4KI \rightarrow NH_2Hg_2I_3$ (yellow-orange colloid) $+ H_3O^+ + 4KI + NaI$	Nonspecific, historical interest only
Berthelot	$NH_4^+ + NaOCl + \text{phenol} \xrightarrow{\text{Nitroprusside}} \text{indophenol}$	Nonspecific; very sensitive to interference from endogenous NH_3
Indicator dye	$NH_4^+ + \text{pH indicator dye} \longrightarrow \text{color change}$	Used in multilayer film reagents, dry reagent strips, and automated systems
H_2O_2 *coupled*	$NH_4^+ + 2\text{-glutamate} + 2 ATP \xrightarrow{\text{Glutamine synthetase}} 2 ADP + 2 \text{L-glutamine}$ $ADP + 2 \text{ phosphoenolpyruvate} \xrightarrow{\text{Pyruvate kinase}} 2 \text{ pyruvate} + 2 ATP$ $\text{Pyruvate} + 2H_2PO_4^- + 2O_2 + 2H_2O \xrightarrow{\text{Pyruvate oxidase}} 2 \text{ acetyl phosphate} + CO_2 + 2H_2O_2$ $H_2O_2 + \text{phenol} + 4 \text{ aminophenazone} \xrightarrow{\text{Peroxidase}} \text{quinone-monoimine dye}$	Not commercially available

Direct Methods

Diacetyl monoxime	Diacetyl monoxime + urea $\xrightarrow{\text{Str acid}}$ diazine (yellow)	Nonspecific; uses toxic reagents
O-Phthaldehyde	Urea + *O*-phthalaldehyde $\xrightarrow{H^+}$ isoindoline Isoindoline + N-(I-naphthyl) ethylenediame $\xrightarrow{H^+}$ colored product	Used on some automated instruments; no NH_3 interference, positive interference from sulfa drugs

The mass of S filtered is equal to the product of its urine concentration (U_S) and the urine volume (V_U).

$$M_S = U_S V_U \qquad \text{(Eq. 12–4)}$$

These conditions allow calculation of GFR knowing the urine and plasma concentrations of S, the volume of urine collected, and the time over which the sample was collected.

$$GFR = \frac{U_S V_U}{P_S t} \qquad \text{(Eq. 12–5)}$$

The *clearance* of a substance is the volume of plasma from which that substance is cleared per unit time. The formula for *creatinine clearance (CrCl)* is given as follows, where U_{Cr} is urine creatinine concentration and P_{Cr} is plasma creatinine concentration.

$$CrCl = \frac{U_{Cr} V_U}{P_{Cr} t} \qquad \text{(Eq. 12–6)}$$

Creatinine clearance is usually reported in units of mL/min and can be corrected for body surface area (see Chapter 21, *Renal Function*). Because about 10% of creatinine is secreted by the tubules, creatinine clearance overestimates GFR, but it provides a good approximation.[20, 21]

From Equation 12-6, it can be seen that plasma creatinine is inversely proportional to clearance. Thus, when plasma creatinine is elevated, GFR is decreased, indicating renal damage. Unfortunately, plasma creatinine is a relatively insensitive monitor and may not be measurably increased until renal function has deteriorated more than 50%.[22] The observed relationship between plasma creatinine and the GFR and the observation that plasma creatinine levels are rela-

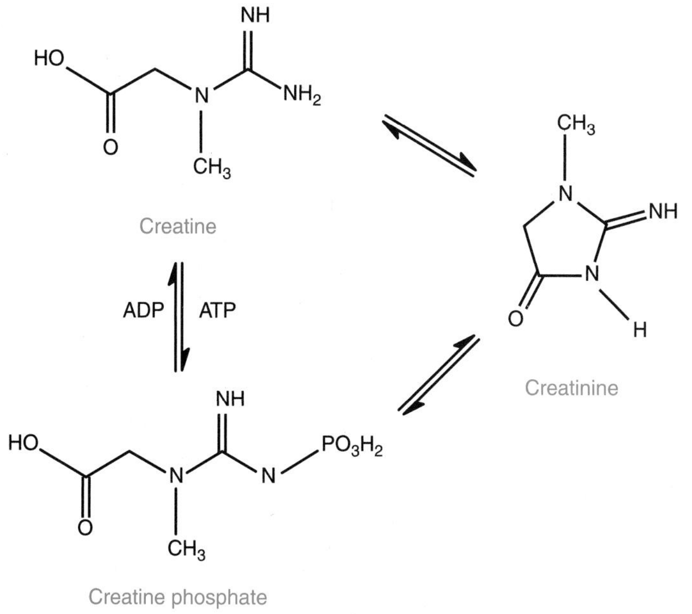

Figure 12-3. The structures of creatine, creatine phosphate, and creatinine.

tively constant and unaffected by diet should make it an excellent analyte for the assessment of renal function. However, because of the difficulties encountered in analyzing the small amount of creatinine normally present (see discussion under *Analytical Methods*), several authors have suggested that this use of plasma creatinine does not provide needed sensitivity in the detection of mild renal dysfunction and it has been suggested that it is better to use other methods for monitoring GFR.[23,24] Despite these problems, plasma creatinine is the most commonly used monitor of renal function.

Creatine

In muscle disease such as muscular dystrophy, poliomyelitis, hyperthyroidism, and trauma, both plasma creatine and urinary creatinine are often elevated. Plasma creatinine levels are usually normal in these patients. Because analytical methods for creatine are not readily available in most clinical laboratories, its measurement has been replaced by the measurement of creatine kinase levels for the diagnosis of muscle disease.[25] Plasma creatine levels are not elevated in renal disease.

Analytical Methods for Creatinine

The methods most frequently used to measure creatinine are based on the Jaffe reaction first described in 1886.[26] In this reaction, creatinine reacts with picric acid in alkaline solution to form a red-orange chromogen, which is a tautomer (convertible structures) of creatinine and picric acid.[19] This reaction was first adopted for the measurement of blood creatinine by Folin and Wu in 1919.[27] It has subsequently been shown to be nonspecific and subject to positive interference by a large number of compounds, such as ascorbate, pyruvate, acetone, and glucose.[19,28,29] The most accurate results are obtained when creatinine in a protein-free filtrate is adsorbed

onto Fuller's earth or Lloyd's reagent and then eluted and reacted with alkaline picrate.[28] Because this method is time-consuming and not readily automated, it is not used routinely.

Two approaches have been used to increase the specificity of the assay methods for creatinine: a *kinetic* Jaffee method and the use of various enzymes. In the kinetic Jaffee method, serum is mixed with alkaline picrate and the rate of change in absorbance is measured between two time points.[30–35] Although this method eliminates some of the nonspecific reactants, it is still subject to interference by α-keto acids[29,36] and cephalosporins.[37] Bilirubin and hemoglobin may cause a negative bias, probably due to their destruction in the strong base used.[38] The kinetic Jaffee method is popular despite these problems because it is inexpensive, rapid, and easy to do.

In an effort to eliminate the nonspecificity of the Jaffee reaction, several *coupled enzymatic methods* have been developed.[32,35,39–43] Of these, the method using creatininase (creatinine aminohydrolase [EC 3.5.2.10]), creatinase (creatine amidinohydrolase [EC 3.5.3.3]), sarcosine oxidase (EC 1.5.3.1), and peroxidase (EC 1.11.1.7) seems to show the most promise for an accurate and precise assay (Table 12-4).[38,40,44]

Another approach to the analysis of creatinine has been the measurement of the color formed when creatinine reacts with 3, 5-dinitrobenzenoic acid or its derivatives.[19,45] This method has been successfully adapted to a reagent strip method. However, the color produced is less stable than that of the Jaffee chromogen.[12]

Several high-performance liquid chromatography (HPLC) methods have been developed.[46–48] The most popular of these uses pretreatment of the sample with trichloroacetic acid to remove protein, cation-exchange chromatography, and ultraviolet (UV) detection of the creatinine.[47] This HPLC method has been recommended as a reference method to replace the Fuller's earth/Jaffe method.

CASE STUDY 12-2

An 80-year-old woman was admitted with a diagnosis of hypertension, congestive heart failure, anemia, possible diabetes, and chronic renal failure. She was treated with diuretics and IV fluids and released 4 days later. Her laboratory results are shown in Case Study Table 12-2.1.

CASE STUDY TABLE 12-2.1. Laboratory Results

	First Admission		Second Admission	
	2/11	2/15	7/26	7/28
BUN, mg/dL	58	*	*	61.0
Creatinine, mg/dL	6.2	6.2	6.4	6.0
Uric acid, mg/dL	10.0	*	*	9.2
BUN-creatinine	9.4	*	*	10.1
Glucose, mg/dL	86	80	113	*

Not done.

Five months later, she was readmitted for treatment of repeated bouts of dyspnea. She was placed on a special diet and drugs to control her hypertension and was discharged. It was felt that she had not been taking her medication as prescribed, and she was counseled regarding the importance of regular doses.

Questions

1. What is the most probable cause of the patient's elevated BUN? What data support your conclusion?
2. Note that this patient's admitting diagnosis is "possible diabetes." If this patient had truly been a diabetic with an elevated blood glucose and a positive acetone, what effect would this have had on the measured values for creatinine? Explain.

The development and use of more accurate methods for plasma creatinine that tend to give lower values will have a significant effect on results obtained for creatinine clearance, because results for urinary creatinine are not subject to as many interferences as plasma levels. Use of more specific methods for plasma will thus result in apparently higher clearance rates. Such results may mask mild renal dysfunction using current reference intervals.[24] Analytical methods for creatinine are summarized in Table 12-4.

Specimen Requirements and Interfering Substances

Creatinine can be measured in serum, plasma, or urine. Hemolyzed and icteric samples should be avoided especially when using a Jaffe method. Lipemic samples may give erroneous results for some methods. Fasting is not required, although heavy protein ingestion may transiently elevate serum levels. Urine should be refrigerated after collection or frozen if storage longer than 4 days is required.[16]

Sources of Error

When creatinine is measured by the Jaffe reaction, glucose, uric acid, α-keto acids and ascorbate may increase results, especially at temperatures above 30°C. This interference is markedly decreased when the reaction is measured kinetically. Depending on the concentration of reactants and measuring time, some interference from α-keto acids may still exist with kinetic Jaffe methods.[38] These substances do not

interfere in the enzymatic methods previously described. Bilirubin causes a negative bias in both Jaffe and enzymatic methods.[38]

Patients on cephalosporin antibiotics may show falsely elevated results when the Jaffe reaction is used.[37] Other drugs also have been shown to increase the results, particularly dopamine, which affects both enzymatic and Jaffe methods.[38,49] Ascorbate will interfere in any of the enzymatic methods that use peroxidase.[38] Lidocaine has been shown to cause a positive bias in some enzymatic methods.[41,50]

Reference Interval[16]

See Table 12-5. Reference intervals vary with assay type, age, and sex, and may decrease in the elderly population.[16]

Analytical Methods for Creatine

The traditional method for creatine analysis is based on analyzing the sample for creatinine before and after heating in acid solution using an endpoint Jaffe method. Heating converts creatine to creatinine and the difference between the two samples is the creatine concentration. Unfortunately, heating may result in the formation of additional nonspecific chromogens and the precision of this method is poor. In recent years, several enzymatic methods have been developed including the use of the creatininase assay previously described.[40] The initial enzyme is omitted and creatine kinase (EC 2.7.3.2), pyruvate kinase (EC 2.1.7.40), and lactate dehydrogenase (EC 1.1.1.27) are coupled.[51–53] At least one of these methods has been adapted for automated systems.

TABLE 12-4. Summary of Analytical Methods—Creatinine

Methods Based on Jaffe Reaction

$$Creatinine + picrate \xrightarrow{OH^-} Janovski\ complex\ (red)$$

Jaffe-Fuller's earth	Creatinine in protein-free filtrate adsorbed onto Fuller's earth; released with alkaline picrate to yield colored Janovski	Previously considered reference method
Jaffe without adsorbent	Creatinine in protein-free filtrate reacts with alkaline picrate to form colored complex	Subject to positive bias from glucose, uric acid, glutathione, ascorbic acid, α-keto acids, and cephalosporins
Jaffe-kinetic	Jaffe reaction done directly on serum; rate of color formation determined early in reaction	Subject to positive bias from α-keto acids and cephalosporins; requires automated equipment for precision

Enzymatic Methods

Creatininase-CK

$$Creatinine + H_2O \xrightarrow{Creatinine\ aminohydrolase} creatine$$
$$Creatine + ATP \xrightarrow{CK} creatine\ phosphate + ADP$$
$$ADP + phosphoenolpyruvate \xrightarrow{PK} ATP + pyruvate$$
$$Pyruvate + NADH \xrightarrow{LD} lactate + NAD^+$$

Requires large sample of pre-incubation; not widely used

Creatininase-H_2O_2

$$Creatinine + H_2O \xrightarrow{Creatininase} creatine$$
$$Creatine + H_2O \xrightarrow{Creatinase} sarcosine + urea$$
$$Sarcosine + H_2O + O_2 \xrightarrow{Sarcosine\ oxidase} glycine + formaldehyde + H_2O_2$$
$$H_2O_2 + phenol\ derivative + 4\ aminophenazone \xrightarrow{POD} benzoquinone\text{-}immine\ dye$$

Potential to replace Jaffee; some positive bias due to lidocaine; no interference from acetoacetate or cephalosporins

Other Methods

3,5-Dinitrobenzoic acid (DNBA)	$Creatinine + DNBA \xrightarrow{OH^-} purple\ color$	Used on reagent strips; colored product unstable
HPLC	Reversed-phase or cation-exchange chromatography	Highly specific; requires pretreatment step; proposed reference method

URIC ACID

Biochemistry

In the higher primates, such as humans and apes, *uric acid* is the final breakdown product of purine metabolism. Most other mammals have the ability to catabolize purines one step further to *allantoin,* a much more water-soluble end product. This reaction sequence is shown in Figure 12-4.

TABLE 12-5. Reference Ranges for Creatinine, mg/dL (mmol/L)

Population	Jaffe	Enzymatic
Children	0.3–0.7 (27–62)	0–0.6 (0–53)
Adult male	0.9–1.3 (80–115)	0.6–1.1 (53–97)
Adult female	0.6–1.1 (53–97)	0.5–0.8 (44–71)

Purines—such as adenosine and guanine—from the breakdown of ingested nucleic acids or from tissue destruction are converted into uric acid mainly in the liver. Uric acid is transported by the plasma from the liver to the kidney, where it is filtered by the glomerulus. *Reabsorption* of 98% to 100% of the uric acid in the glomerular filtrate occurs in the proximal tubules. Small amounts of uric acid are then secreted by the *distal tubules* into the urine. Triglycerides, ketone bodies, and lactic acid have been shown to compete with urate for excretory sites in the renal tubules. This route accounts for about 70% of the daily uric acid excretion. The remainder is excreted into the GI tract and degraded by bacterial enzymes.

Nearly all of the uric acid in plasma is present as monosodium urate. At the pH of plasma, urate in this form is relatively insoluble, and, at levels above 6.4 mg/dL, the plasma is saturated. As a result, urate crystals may form and precipitate in the tissues. In the urine at pH levels <5.75, uric acid is the predominant form and frequently appears as uric acid crystals.

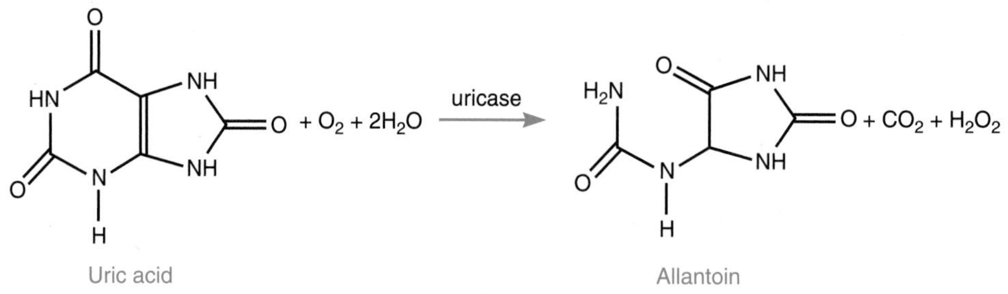

Figure 12-4. Conversion of uric acid to allantoin.

Disease Correlations

Three major disease states are associated with elevated levels of uric acid in the plasma: gout, increased nuclear catabolism, and renal disease. *Gout* is a disease found primarily in males and usually is first diagnosed between the ages of 30 and 50 years. Patients have pain and inflammation of the joints caused by precipitation of sodium urates. In 25% to 30% of these patients, hyperuricemia is due to overproduction of uric acid, although other common causes are drugs and alcohol. Plasma uric acid in these patients is usually above 6.0 mg/dL. Patients with gout are highly susceptible to the development of renal calculi, although not all patients with elevated serum urate levels develop these complications. In women, urate levels rise after menopause and some postmenopausal women may develop hyperuricemia and gout.

Another common cause of elevated plasma uric acid levels is increased breakdown of cell nuclei, as occurs in patients on chemotherapy for such proliferative diseases as leukemia, lymphoma, multiple myeloma, or polycythemia. Monitoring uric acid levels in these patients is important to avoid nephrotoxicity. Allopurinol, which inhibits xanthine oxidase, an enzyme in the uric acid synthesis pathway, is used to treat these patients. Patients with hemolytic and megaloblastic anemia may also exhibit elevated levels of uric acid.

Chronic renal disease also causes elevated levels of uric acid because filtration and secretion are hampered. However, uric acid is not useful as an indicator of renal function because so many other factors affect its plasma level (Table 12-6).

High levels of uric acid are also found secondary to glycogen storage disease and other congenital enzyme deficiencies. Production of excess metabolites, such as lactate and triglycerides, compete with urate for excretion in these diseases.[1]

Lesch-Nyhan syndrome is an X-linked genetic disorder (seen only in males) and is caused by a complete deficiency of hypoxanthine-guanine phosphoribosyltransferase (HGPRT), a major enzyme in the biosynthesis of purines. This disorder causes increased levels of uric acid in plasma and urine. This extremely rare disease is characterized by neurologic symptoms, mental retardation, and self-mutilation.

Hyperuricemia is also a common feature of toxemia of pregnancy and lactic acidosis presumably due to competition for binding sites in the renal tubules.[1] Elevated levels also may be found after ingestion of a diet rich in purines (*eg*, liver, kidney, sweetbreads, shellfish, and so forth) or a marked decrease in total dietary intake, resulting in increased tissue breakdown.

Decreased levels of uric acid are much less common than increased levels and are usually secondary to severe liver disease or defective tubular reabsorption as in Fanconi's syndrome.[16] *Hypouricemia* can also be caused by chemotherapy with 6-mercaptopurine or azathioprine as well as overtreatment with allopurinol.[1]

Analytical Methods

As shown in Figure 12-4, uric acid is readily oxidized to allantoin and thus can function as a reducing agent in many chemical reactions. This property formed the basis for nearly all the early analytical methods for uric acid. The most popular method of this type is the Caraway method, which is based on the oxidation of uric acid in a protein-free filtrate, with subsequent reduction of phosphotungstic acid to tungsten blue.[54] The Caraway method uses sodium carbonate to

TABLE 12-6. Causes of Increased Plasma Uric Acid

Increased dietary intake
Increased urate production
  Gout
  Treatment of myeloproliferative disease with cytotoxic drugs
Decreased excretion
  Lactic acidosis
  Toxemia of pregnancy
  Glycogen storage disease type I (glucose-6-phosphatase deficiency)
  Drug therapy
  Poisons: lead, alcohol
  Renal disease
Catabolic pathway enzyme defects
  Lesch-Nyhan syndrome

CASE STUDY 12-3

A 3-year-old girl was admitted with a diagnosis of acute lymphocytic leukemia. Her admitting laboratory data are shown in Case Study Table 12-3.1. After admission, she was treated with packed cells, 2 units of platelets, IV fluids, and allopurinol. On the 2nd hospital day, chemotherapy was begun, using IV vincristine and prednisone and intrathecal injections of methotrexate, prednisone, and cytosine arabinoside. She was discharged for home care 5 days later. She was continued on prednisone and allopurinol at home. She received additional chemotherapy 1 month later (11/1) and again on 11/14. On 12/6, she was readmitted because she had painful sores in her mouth and was unable to eat.

CASE STUDY TABLE 12-3.1. Laboratory Results

	10/1	10/2	10/3	10/4	11/14	12/6	6/20
BUN, mg/dL	12.0	*	*	15	4.0	2.0	*
Creatinine, mg/dL	0.7	*	*	1.0	0.7	*	0.7
Uric Acid, mg/dL	12.0	9.2	4.0	1.9	2.3	*	3.1
WBC mm³	56300				3700	2800	3700

*Not done.

Questions

1. How would you explain the marked elevations of uric acid on admission?
2. What two factors are responsible for the normal levels of uric acid seen in subsequent admissions?
3. What is the most likely cause of the abnormally low level of BUN seen on 12/6? What other laboratory result would be useful to confirm your suspicions?

provide the alkaline pH necessary for color development. The blue color produced can be intensified by the addition of cyanide to the reaction mixture, but adequate color can also be obtained by proper pH adjustment.

An alternative method that can be performed on serum is the reduction of Fe^{3+} to Fe^{2+} and subsequent reaction with a color-producing chelating agent.[55] The reduction methods are relatively nonspecific and attempts have been made to find more specific alternatives. Several methods using uricase (urate oxidase, EC 1.7.3.3) have been described.[56–59] This enzyme catalyzes the oxidation of uric acid to allantoin with subsequent production of H_2O_2.

The simplest of these methods is based on the fact that uric acid has a UV absorbance peak at 293 nm, whereas allantoin does not. The difference in absorbance before and after incubation with uricase is proportional to the uric acid concentration.[58,59] This method has been proposed as a candidate reference method by the American Association for Clinical Chemistry.[56] Proteins can cause high background absorbance in this method, reducing sensitivity. Negative interferences can occur with xanthine and hemoglobin.[49]

A more common approach uses a coupled enzymatic reaction measuring the amount of peroxide formed in the primary reaction. In these methods, either peroxidase or catalase is used to couple peroxide to a color-producing reaction. In peroxidase methods, a compound such as 4-aminoantipyrine is oxidized to yield a chromophore absorbing in the visible region.[57,60] The sequence of reactions using catalase and alcohol dehydrogenase is shown in Table 12-7.[61] Both methods are subject to interference from reducing agents such as bilirubin and ascorbic acid, which destroy peroxide. Commercial reagent preparations often include potassium ferricyanide and ascorbate acid oxidase to minimize these interferences.

HPLC methods using several types of columns also have been developed.[62,63] These methods show good specificity and sensitivity but may require pretreatment of the sample to remove protein. The National Institute of Standards and Technology has proposed a definitive method for uric acid based on isotope dilution/mass spectrometry.[64] Analytical methods are summarized in Table 12-7.

Specimen Requirements and Interfering Substances

Uric acid may be determined in serum, urine, or heparinized plasma. Serum should be removed from cells as quickly as possible to prevent dilution by intracellular contents. Diet may affect uric acid levels on a long-term basis, but a recent meal has no significant effect and fasting is unnecessary. Gross lipemia should be avoided. Increased levels of bilirubin may give decreased results in peroxidase methods. Significant hemolysis, which releases glutathione, also may give low values; a number of drugs, such as thiazides and salicylates, have been shown to cause elevated values for uric acid.[49]

Uric acid is stable in serum or plasma once separated from the cells. Serum samples may be stored refrigerated for 3 days to 5 days.[16]

Reference Interval[16]

Values are slightly lower in children and premenopausal females (Table 12-8).

TABLE 12-7. Summary of Analytical Methods—Uric Acid

Oxidation *Phosphotungstic acid*	Uric acid + phosphotungstate $\xrightarrow{Na_2CO_2/OH^-}$ CO_2 + allantoin + tungsten blue	Nonspecific; requires protein removal
Iron	Uric acid + Fe^{3+} $\xrightarrow{ligand}$ Fe^{2+}-chromophore	Can use serum; reducing agents interfere
Uricase	All based on same initial reaction with subsequent determination of H_2O_2 Uric acid + O_2 $\xrightarrow{Uricase}$ allantoin + CO_2 + H_2O_2	Enzyme highly specific
Spectrophotometric	Measurement of decrease in absorbance at 293 nm due to destruction of uric acid	Basis for candidate reference method; xanthine and hemoglobin interfere
Coupled enzymatic (I)		Most commonly automated clinical method; negative bias with reducing agents
Coupled enzymatic (II)		Readily automated, popular; reducing agents are interferents
HPLC	Ion exchange and reverse-phase columns	Very specific; usually require sample extraction
Isotope dilution/MS	Sample diluted with known amount of isotope; ratio of two isotopes detected by mass spectrometer	Proposed National Institutes of Standards definitive method

AMMONIA

Biochemistry

Ammonia arises from the deamination of amino acids, which occurs mainly through the action of digestive and bacterial enzymes on proteins in the intestinal tract. Ammonia is also released from metabolic reactions that occur in skeletal muscle during exercise. It is used by the parenchymal cells of the liver in the production of urea, as described earlier. In severe liver disease in which there is significant collateral circulation, or when parenchymal liver cell function is severely impaired, ammonia is not removed from the circulation and blood levels rise. Unlike some other NPN substances, plasma levels of ammonia are not dependent on renal function but on liver function and thus are not useful in the study of kidney disease.

Markedly elevated levels of ammonia are neurotoxic and are often associated with *encephalopathy*. It is thought that this toxicity may be due in part to increased extracellular glutamate concentration and subsequent depletion of adenosine triphosphate (ATP) in the brain.[65] Because of its important role in monitoring the progress of several severe clinical conditions and sample instability, this assay should be available in the laboratories of all tertiary care facilities.

Disease Correlations

The two major clinical conditions in which blood ammonia levels offer useful information are hepatic failure and Reye's syndrome. Severe liver disease represents the most common cause of disturbed ammonia metabolism, and the monitoring of blood ammonia is often used to determine the prognosis of these patients. Unfortunately, the correlation between the degree of hepatic encephalopathy and plasma NH_3 is not always consistent, and some patients may have encephalopathy with normal plasma ammonia levels.

Reye's syndrome is an often fatal disease that occurs most commonly in children. It is frequently preceded by a viral infection. It is now known that this is an acute metabolic disorder of the liver, and autopsy findings show severe fatty infiltration of that organ. Blood ammonia levels have shown good correlation with both the severity of the disease and prognosis. Fitzgerald and associates have shown that, when plasma ammonia levels are less than five times normal, the survival rate in Reye's syndrome is 100%.[66]

Blood ammonia levels are also elevated in the inherited deficiency of any of the urea-cycle enzymes except argininosuccinase. Patients with these disorders are mentally retarded and suffer from behavior problems. They often exhibit protein intolerance and elevated levels of glutamine and ammonia.[67]

Because of the importance of portal circulation in the removal of ammonia from the circulation, portal-systemic

TABLE 12-8. Reference Ranges for Uric Acid (Uricase Methods)

Population	mg/dL	mmol/L
Children	2.0–5.5	0.12–0.33
Adult males	3.5–7.2	0.21–0.43
Adult females	2.6–6.0	0.16–0.36

shunting of blood also will lead to elevated ammonia levels. The patency of a portocaval shunt may be evaluated by measuring the plasma ammonia before and after a standard dose of NH_4^+ salts.[68]

Analytical Methods

The accurate measurement of ammonia in plasma is complicated by its low concentration, sample instability, and the ubiquitous nature of NH_3 contamination. Two distinct approaches have been used for the measurement of plasma ammonia. One is a two-step approach in which ammonia is first isolated from the sample and then assayed. The second involves direct measurement of ammonia by an enzymatic method or ion-selective electrode.

One of the first analytical methods for ammonia, developed by Conway in 1935, involved use of a microdiffusion chamber.[69] In this method, ammonia is liberated from the sample by alkali, absorbed into acid, and then quantitated by titration or colorimetry. This method is laborious and is seldom used today.

A second, more successful approach for isolating NH_3 is the use of a cation-exchange resin (Dowex 50) followed by elution of the NH_3 with NaCl and quantitation by the Berthelot reaction.[70] Because it requires an isolation step, it is time-consuming and not readily automated.

A more specific and convenient approach is a coupled enzymatic assay illustrated in Table 12-9.[71-73] Under appropriate reaction conditions, the decrease in absorbance at 340 nm is proportional to the ammonia concentration. NADPH is the preferred coenzyme because it is used specifically by glutamate dehydrogenase and is not consumed by side reactions with endogenous substrates such as pyruvate. ADP is added to the reaction mixture to stabilize GLDH. This method is used on many automated systems and is available in kit form from numerous manufacturers.

Another approach is the use of an ammonia electrode.[74] This electrode measures the changes in pH of a solution of ammonium chloride as ammonia diffuses across a semipermeable membrane. Because the electrode membrane used is not very stable, this method has not gained much popularity. Analytical methods for ammonia are summarized in Table 12-9.

Specimen Requirements and Interfering Substances

Proper specimen handling is of the utmost importance in plasma ammonia assays. Whole blood ammonia levels rise rapidly after collection because of amino acid deamination.

Venous blood should be obtained without trauma and placed on ice immediately. Ethylenediaminetetraacetic acid (EDTA) is the preferred anticoagulant. Heparin is usually quoted as an acceptable anticoagulant, but one study has reported false decreases in some samples.[75] The tube type and anticoagulant should be evaluated for ammonia interference before use. Samples should be centrifuged at 0° to 4°C within 20 minutes of collection and the plasma removed. Some authors have reported that separated plasma samples are stable for up to 3½ h in an ice bath.[73] However, for best results, it is recommended that the assay be completed as soon as possible to prevent false increases due to in vitro deamination. Frozen plasma is reportedly stable for several days at −20°C. Hemolysis should be avoided because red cells contain two to three times as much ammonia as plasma.

In healthy patients, blood ammonia levels do not appear to be significantly affected by eating, but they are affected by cigarette smoking. It is recommended that patients not smoke for several hours before the sample is drawn.[67]

Sources of Error

The two biggest pitfalls in accurate ammonia analysis are sample handling and ammonia contamination. Ammonia levels rise rapidly in shed blood, and these samples must be handled as described previously. It has been shown that elimination of sources of ammonia contamination in the laboratory can significantly increase the accuracy of ammonia assay results.[76] Sources of contamination include tobacco smoke, urine, and the presence of ammonia in detergents, glassware, reagents, and water. Although enzymatic methods are affected less than resin methods, ammonia contamination is still a potential problem in all assay methods for ammonia.

Serum-based control material cannot be used for quality control because the ammonia content is unstable. Frozen aliquots of human serum albumin containing known amounts

---

TABLE 12-9. Summary of Analytical Methods—Ammonia

Conway microdiffusion	NH_3 released from alkaline solution determined by back titration with acid	Time-consuming; poor accuracy and precision; no longer used
Ion-exchange	NH_3 absorbed onto Dowex 50 resin, eluted, and quantitated with Berthelot reaction	Time-consuming manual method; good accuracy
Coupled enzymatic	$NH_4^+ + $ 2-oxoglutarate $+$ NADPH $\xrightarrow{\text{GLDH}}$ glutamate $+$ $NADP^+ + H_2O$	Most common on automated instruments; precise and accurate
Ion-selective electrode	Diffusion of NH_3 through selective membrane into NH_4Cl causing pH change, which is measured potentiometrically	Good accuracy and precision; some problem with membrane stability

of ammonium chloride or ammonium sulfate may be used. Solutions containing known amounts of ammonium sulfate are commercially available.

A number of substances have been shown to influence the ammonia level in vivo.[16,49] Ammonium salts, asparaginase, barbiturates, ethanol, hyperalimentation, narcotic analgesics, and diuretics all cause increased levels of ammonia in plasma. Levodopa, kanamycin, diphenhydramine, neomycin, *Lactobacillus acidophilus,* and lactulose all cause decreased values.

Reference Interval[16]

Ammonia, 19 mg/dL to 60 mg/dL (1135 mmol/L).

The values obtained will vary somewhat with the method used. Enzymatic methods tend to give slightly lower values than do ion-exchange methods. Higher values are usually seen in newborns.

SUMMARY

Clinically important NPN compounds found in the plasma include urea, creatinine, creatine, uric acid, amino acids, and ammonia. Urea, the major excretory product of protein metabolism, composes the major portion of the total NPN compounds. Its plasma level is related to the protein content of the diet, renal blood flow, level of protein catabolism, and renal function. Clinical conditions causing elevated urea levels are classified according to cause as prerenal, renal, and postrenal. The BUN/creatinine ratio can be used to distinguish among these conditions. Urea is usually measured by a coupled enzymatic method that quantitates the amount of ammonium produced by urease hydrolysis of the urea in the sample.

Creatinine is formed as creatine and phosphocreatine in muscle spontaneously lose water or phosphoric acid. In a given individual, it is excreted into the plasma at a constant rate correlating to muscle mass. GFR is estimated by calculating creatinine clearance, which requires measurement of plasma and urine creatinine. Plasma creatinine is inversely related to GFR and, although imperfect, it is most commonly used to monitor renal filtration function. Measurement of creatinine in plasma and urine has traditionally been done using the Jaffe reaction. This reaction is relatively nonspecific and subject to interference from many commonly occurring substances such as glucose, protein, and α-keto acids. Specificity of this reaction can be improved by using a kinetic rather than an endpoint method; however, the kinetic Jaffe method is still subject to interference from keto acids, bilirubin, and some drugs. A specific enzymatic method has been developed but has not been widely accepted. Use of a more specific method for measurement of creatinine may require readjustment of reference intervals for both plasma creatinine and creatinine clearance. Without such adjustment, mild levels of renal dysfunction may be overlooked.

Uric acid is the breakdown product of the purines from nucleic acid catabolism. It is filtered by the glomerulus, reabsorbed in the proximal tubules, and secreted by the distal tubules into the urine. Additional uric acid is excreted into the GI tract and degraded by bacterial enzymes. Uric acid is relatively insoluble in plasma, and elevated levels of urate can be deposited in the joints and tissues, causing painful inflammation. Elevations in plasma urate are seen in gout, increased nuclear breakdown, and renal disease. Increased uric acid may also be seen secondary to conditions that result in excess production of metabolites, such as lactate and triglycerides, which compete with urate for secretion in the distal tubule. Uric acid is usually quantitated by a coupled enzymatic method in which uric acid is oxidized by uricase to allantoin and peroxide. The peroxide produced is quantitated with a variety of dyes. Increased levels of bilirubin and other substances that destroy peroxide in the sample can cause falsely decreased results.

Ammonia is present in the plasma in low concentrations. It arises from the deamination of amino acids resulting from the breakdown of protein, and is normally rapidly removed from the circulation and converted to urea in the liver. Elevated levels of ammonia are neurotoxic. Elevations of plasma ammonia are usually associated with hepatic failure or Reye's syndrome. Ammonia is commonly measured by a coupled enzymatic method using glutamate dehydrogenase and measuring change in absorbance due to conversion of NADPH to NADP. The major source of error in ammonia determinations is improper sample handling. The ammonia level of freshly drawn blood rises rapidly on standing. The sample should be placed on ice and the plasma separated and analyzed as rapidly as possible. The presence of ammonia contamination and cigarette smoke may also affect the results.

REVIEW QUESTIONS

1. Which of the following is not an NPN substance:
   a. Urea
   b. Ammonia
   c. Creatinine
   d. Troponin T
2. Which NPN fraction constitutes nearly half of the NPN substances in the blood?
   a. Urea
   b. Creatine
   c. Ammonia
   d. Uric acid
3. Prerenal hyperuricemia is caused by:
   a. Congestive heart failure
   b. Chronic renal failure
   c. Renal tumors
   d. Glomerular nephritis
4. A high BUN/creatinine ratio with an elevated creatinine is usually seen in:
   a. Liver disease
   b. Low protein intake
   c. Tubular necrosis
   d. Postrenal conditions

5. Ammonia levels are usually measured to evaluate:
   a. Renal failure
   b. Acid-base status
   c. Hepatic coma
   d. Glomerular filtration

6. A technologist obtains a BUN value of 61 mg/dL and serum creatinine value of 3.1 mg/dL on a patient. These results indicate:
   a. Renal failure
   b. Kidney failure
   c. Gout
   d. Prerenal failure

7. In the Jaffee reaction, an red-orange chromogen is formed when creatinine reacts with:
   a. Picric acid
   b. Naphthylethylenediamine
   c. Diacetyl monoxime
   d. Nitroferricyanide

8. Substances known to increase results when measuring creatinine by the Jaffe reaction include all of the following EXCEPT:
   a. Glucose
   b. Ascorbate
   c. α-keto acids
   d. Bilirubin

9. A BUN of 9 mg/dL is obtained by a technologist. What is the urea concentration?
   a. 18.3 mg/dL
   b. 19.3 mg/dL
   c. 10.3 mg/dL
   d. 9.3 mg/dL

10. Uric acid is the final breakdown product of:
    a. Urea metabolism
    b. Purine metabolism
    c. Glucose metabolism
    d. Bilirubin metabolism

REFERENCES

1. Whelton A, Watson AJ, Rock RC. Nitrogen metabolites and renal function. In: Burtis CA, Ashwood ER, eds. Tietz textbook of clinical chemistry. 2nd ed. Philadelphia: WB Saunders, 1994;1513.
2. Tiffany TO, Jansen JM, Burtis CA, et al. Enzymatic kinetic rate and endpoint analysis of substrate by use of GEMSAEC fast analyzer. Clin Chem 1972;18:829.
3. Scott P, Maguire GA. A kinetic assay for urea in undiluted urine specimens. Clin Chem 1990;36:1830.
4. Lespinas F, Dupuy G, Revol F, et al. Enzymic urea assay: a new colorimetric method based on hydrogen peroxide measurement. Clin Chem 1989;35:654.
5. Spayd RW, Bruschi B, Burdick BA, et al. Multilayer-film elements for clinical analysis: applications to representative chemical determinations. Clin Chem 1978;24:1343.
6. Ohkubo A, Kamei S, Yamanaka M, et al. Multilayer-film analysis for urea nitrogen in blood, serum, or plasma. Clin Chem 1984;30:1222.
7. Akai T, Naka K, Yoshikawa C, et al. Salivary urea nitrogen as an index to renal function: a test-strip method. Clin Chem 1983;29:1825.
8. Orsonneau J, Massoubre C, Cabanes M, et al. Simple and sensitive determination of urea in serum and urine. Clin Chem 1992;38:619.
9. Paulson G, Ray R, Sternberg J. A rate sensing approach to urea measurement. Clin Chem 1971;17:644.
10. Georges J. Determination of ammonia and urea in urine and of urea in blood by use of an ammonia-selective electrode. Clin Chem 1979; 25:1888.
11. Marsh WH, Fingerhut B, Miller H. Automated and manual direct methods for determination of blood urea. Clin Chem 1965;11:624.
12. Pesce AJ, Kaplan LA. Methods in clinical chemistry. St. Louis, MO: CV Mosby, 1987.
13. Tabacco A, Meiattini F, Moda E, et al. Simplified enzymic/colorimetric serum urea nitrogen determination. Clin Chem 1979; 25:336.
14. Welch MJ, Cohen A, Hertz HS, et al. Determination of serum urea by isotope dilution mass spectrometry as a candidate definitive method. Anal Chem 1984;56:713.
15. Sampson EJ, Baird MA, Burtis CA, et al. A coupled-enzyme equilibrium method for measuring urea in serum: optimization and evaluation of the AACC Study Group on urea candidate reference method. Clin Chem 1980;26:816.
16. Tietz N. Clinical guide to laboratory tests. Philadelphia: WB Saunders, 1995.
17. Chesley LC. Renal excretion at low urine volumes and the mechanism of oliguria. J Clin Invest 1935;17:591.
18. Bleiler RE, Schedl HP. Creatinine excretion: variability and relationships to diet and body size. J Lab Clin Med 1972;59:945.
19. Narayanan S, Appelton H. Creatinine: a review. Clin Chem 1980;26: 1119.
20. Morgan DB, Dillon S, Payne RB. The assessment of glomerular function: creatinine clearance or plasma creatinine? Postgrad Med 1978;54: 302.
21. Shemesh O, Golbetz H, Kriss JP, et al. Limitations of creatinine as a filtration marker in glomerulopathic patients. Kidney Int 1985;28:830.
22. Renkin EM, Robinson RR. Glomerular filtration. N Engl J Med 1974;290:785.
23. Perrone RD, Madias NE, Levey AS. Serum creatinine as an index of renal function: new insight into old concepts. Clin Chem 1992;38:1933.
24. vanLente F, Suit P. Assessment of renal functions by serum creatinine and creatinine clearance: glomerular filtration rate estimated by four procedures. Clin Chem 1989;35:2326.
25. Thompson H, Rechnitz GA. Ion electrode based enzymated analysis of creatinine. Anal Chem 1974;46:246.
26. Jaffe M. Uber den niederschlag welchen pikrinsaure in normalen harn erzurgt und uber eine neue reaction des kreatinins. Z Physiol Chem 1886;10:391.
27. Folin O, Wu H. System of blood analysis. J Biol Chem 1919;31:81.
28. Haeckel R, Godsden RH, Sherwin JE, et al. Assay of creatinine in serum with use of Fuller's earth to remove interferents. In: Cooper GR, ed. Selected methods of clinical chemistry. Vol 10. Washington, DC: American Association for Clinical Chemistry, 1983;225.
29. Watkins PJ. The effects of ketone bodies on the determination of creatinine. Clin Chem Acta 1967;18:191.
30. Conway EJ. Apparatus for the microdetermination of certain volatile substances: the blood ammonia with observations on normal human blood. Biochem J 1935;29:2755.
31. Faulkner WR, Meites S, eds. Selected methods of clinical chemistry. Vol 9. Washington, DC: American Association for Clinical Chemistry, 1982;357.
32. Jaynes PK, Feld RD, Johnson GF. An enzymatic reaction rate assay for serum creatinine with a centrifugal analyzer. Clin Chem 1982;28:114.
33. Larsen K. Creatinine assay by a reaction kinetic principle. Clin Chem Acta 1972;41:209.
34. Lustgarten JA, Wenk RA. Simple, rapid kinetic method for serum creatinine measurement. Clin Chem 1972;18:1419.
35. Moss G, Bondar RJL, Buzelli D. Kinetic enzymatic method for determining serum creatinine. Clin Chem 1975;21:1422.
36. Soldin SJ, Henderson L, Hill GJ. The effect of bilirubin and ketones on reaction rate methods for the measurement of creatinine. Clin Biochem 1978;11:82.
37. Swain RR, Briggs SL. Positive interference with the Jaffe reaction by cephalosporin antibiotics. Clin Chem 1977;23:1340.
38. Weber JA, van Zanten AP. Interferences in current methods for measurements of creatinine. Clin Chem 1991;37:695.
39. Bissell MG, Ward E, Sanghavi P, et al. Multilaboratory evaluation of the new single slide enzymatic creatine method for the Kodak Ektachem analyzer. [Abstract] Clin Chem 1987;33:951.
40. Crocker H, Shephard MDS, White GH. Evaluation of an enzymatic method for determining creatinine in plasma. J Clin Pathol 1988;41:576.

41. Gosney K, Adachi-Kirkland J, Schiller HS. Evaluation of lidocaine interference in the Kodak Ektachem 700 analyzer single-slide method for creatinine. Clin Chem 1987;33:2311.

42. Lanser A, Blijenberg BG, Jeijnse B. Evaluation of a new diagnostic kit for the enzymatic determination of creatinine. J Clin Chem Biochem 1979;17:633.

43. Tanganelli E, Prencipe L, Bassi D, et al. Enzymatic assay of creatinine in serum and urine with creatinine iminohydrolase and glutamate dehydrogenase. Clin Chem 1982;28:1461.

44. Gruder WG, Hoffman GE, Hubbuch A, et al. Multicenter evaluation of an enzymatic method for creatinine using a sensitive color reagent. J Clin Chem Biochem 1986;24:889.

45. Langley WD, Evans M. The determination of creatinine with sodium 3,5-dinitrobenzoate. J Biol Chem 1936;115:333.

46. Brown ND, Sing HC, Neeley WE, et al. Determination of "true" serum creatinine by high-performance liquid chromatography combined with a continuous-flow microanalyzer. Clin Chem 1977;23:128.

47. Rosano TG, Ambrose RT, Wu AHB, et al. Candidate reference method for determining creatinine in serum: method development and interlaboratory validation. Clin Chem 1990;36:1951.

48. Soldin SJ, Hill GJ. Micromethod for determination of creatinine in biological fluids by high-performance liquid chromatography. Clin Chem 1978;24:747.

49. Young DS, Pestaner LC, Gibberman V. Effects of drugs on clinical laboratory tests. 3rd ed. Washington, DC: American Association for Clinical Chemistry Press, 1990.

50. Sena SF, Syed D, Romeo R, et al. Lidocaine metabolite and creatinine measurements in the Ektachem 700: steps to minimize its impact on patient care. Clin Chem 1988;34:2144.

51. Beyer, C. Creatine measurement in serum and urine with an automated enzymatic method. Clin Chem 1993;39:1613.

52. Marymount JH, Smith JN, Klotsch S. A simple method for urine creatine. Am J Clin Pathol 1968;49:289.

53. Oversteegen HJ, DeBoer J, Bakker AJ, et al. A simpler enzymatic determination of creatine in serum with a commercial creatinine kit. [Letter] Clin Chem 1984;33:720.

54. Caraway WT. Standard methods of clinical chemistry. Vol 4. New York: Academic Press, 1963;239.

55. Morin LG. Determination of serum urate by direct acid Fe^{3+} reduction or by absorbance change (at 293 nm) on oxidation of urate with alkaline ferricyanide. Clin Chem 1974;20:51.

56. Duncan PH, Gochman N, Cooper T, et al. A candidate reference method for uric acid in serum: I. optimization and evaluation. Clin Chem 1982;28:284.

57. Kabasakalian P, Kalliney S, Wescott A. Determination of uric acid in serum, with the use of uricase and a tribromophenol-aminoantipyrine chromogen. Clin Chem 1973;19:522.

58. Liddle L, Seegmillu JE, Loster L. The enzymatic spectrophotometric method for the determination of uric acid. J Lab Clin Med 1959;54:903.

59. Trivedi R, Berta E, Rebar L, et al. New ultraviolet method for assay of uric acid in serum or plasma. Clin Chem 1978;24:562.

60. Gochman N, Schmitz JM. Automated determination of uric acid with use of a uricase-peroxidase system. Clin Chem 1971;17:1154.

61. Haeckel R. The use of aldehyde dehydrogenase to determine H_2O_2 producing reactions: the determination of uric acid concentration. J Clin Chem Clin Biochem 1976;14:101.

62. Pachla LA, Kissinger PT. Measurement of serum uric acid by liquid chromatography. Clin Chem 1979;25:1847.

63. Tanaka M, Hama M. Improved rapid assay of uric acid in serum by liquid chromatography. Clin Chem 1988;34:2567.

64. Ellerbe P, Cohen A, Welch MJ, et al. Determination of serum uric acid by isotope dilution mass spectrometry as a new candidate reference method. Anal Chem 1990;62:2173.

65. Felipo V, Kosenko E, Minana MD, et al. Molecular mechanism of acute ammonia toxicity and of its prevention by L-carnitine. Adv Exp Med Biol 1994;368:65.

66. Fitzgerald JF, Clark JH, Angelides AG, et al. The prognostic significance of peak ammonia levels in Reye's syndrome. Pediatrics 1982;70:997.

67. Glasgow M. Clinical application of blood ammonia determinations. Lab Med 1981;12:151.

68. Conn H. Studies of the source and significance of blood ammonia: IV. early ammonia peaks after ingestion of ammonium salts. Yale J Biol Med 1972;45:543.

69. Cook JGH. Factors influencing the assay of creatinine. Ann Clin Biochem 1975;12:219.

70. Kingsley GR, Tager HS. Ion-exchange method for the determination of plasma ammonia nitrogen with the Berthelot reaction. In: MacDonald RP, ed. Standard methods of clinical chemistry. Vol 6. New York: Academic Press, 1970;115.

71. Ishihard A, Kurahasi K, Uihard H. Enzymatic determination of ammonia in blood and plasma. Clin Chem Acta 1972;41:255.

72. Kristen E, Gerez C, Kristen R. Eine enzymatische microbesimmung du ammoniasks geeignet fur extrakte tierischer bewebe und flussigkeiten. Biochem J 1963;337:312.

73. Mondzac A, Erlich GE, Seegmiller JE. An enzymatic determination of ammonia in biological fluids. J Med Clin Lab 1965;66:526.

74. Attili AF, Autizi D, Capocaccia L. Rapid determination of plasma ammonia, using an ion specific electrode. Biochem Med 1975;14:109.

75. Dowart W, Saner M. Heparinized plasma is an unacceptable specimen for ammonia determination. Clin Chem 1992;38:161.

76. Howanitz JH, Howanitz PJ, Skrodzki CA, et al. Influences of specimen processing and storage conditions on results for plasma ammonia. Clin Chem 1984;30:906.

CHAPTER 13

Porphyrins and Hemoglobin

Louann W. Lawrence

Objectives

Upon completion of this chapter, the clinical laboratorian should be able to:

- *Describe the chemical nature and structure of porphyrins and hemoglobin.*

- *Relate the role of porphyrins in the body.*

- *Outline the biochemical pathway of porphyrin and heme synthesis.*

- *Discuss the clinical significance of the porphyrias.*

- *Correlate the porphyrin disease states with clinical laboratory data.*

- *Compare and contrast the porphyrias with regard to enzyme deficiency and clinical symptoms.*

- *Explain the principles of the basic qualitative and quantitative porphyrin tests, to include PBG, ALA, uroporphyrin, coproporphyrin, and protoporphyrin.*

- *Relate the degradation of hemoglobin.*

- *Discuss the clinical significance and laboratory data associated with the hemoglobinopathies and thalassemias.*

- *Identify the tests used in the diagnosis of the hemoglobinopathies and thalassemias.*

- *Discuss the structure and clinical significance of myoglobin in the body.*

KEY TERMS

Cytochrome	Porphyria	Porphyrinuria
Hemoglobinopathy	Porphyrin	Pyrrole
Myoglobin	Porphyrinogens	Thalassemia

Because of chemical similarities, porphyrins, hemoglobin, and myoglobin are discussed together in this chapter. These compounds all contain the porphyrin ring, which comprises four pyrrole groups bonded by methene bridges (Fig. 13-1). Porphyrins are able to chelate metals to form the functional groups that participate in oxidative metabolism.

PORPHYRINS

Role in the Body

Porphyrins are chemical intermediates in the synthesis of hemoglobin, myoglobin, and other respiratory pigments called *cytochromes*. They also form part of the peroxidase and catalase enzymes, which contribute to the efficiency of internal respiration. Iron is chelated within porphyrins to form heme. Heme is then incorporated into proteins to become biologically functional hemoproteins. Porphyrins are analyzed in clinical chemistry to aid in the diagnosis of a group of disorders called the *porphyrias,* which result from disturbances in heme synthesis. The presence of excess amounts of these intermediate compounds in urine, feces, or blood indicates a metabolic block in heme synthesis.

Chemistry of Porphyrins

The porphyrins found in nature are all compounds in which side chains are substituted for the eight hydrogen atoms found in the four *pyrrole* rings that make up porphyrin (see Fig. 13-1). Because of the wide variety of substitutions, many porphyrins have been described in nature. The pigment chlorophyll is a magnesium porphyrin and is essential for plants to use light energy to synthesize carbohydrates. Four

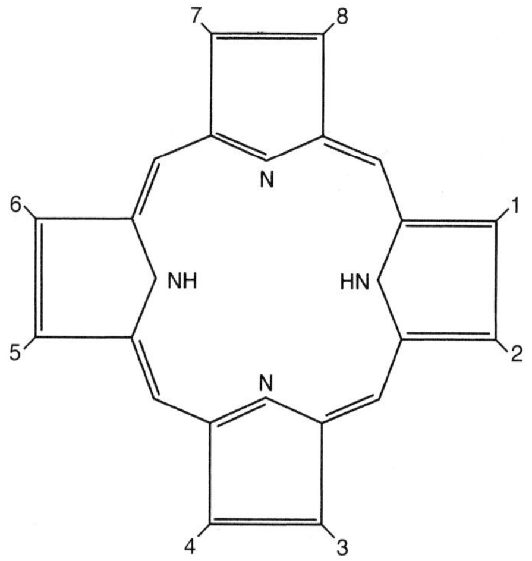

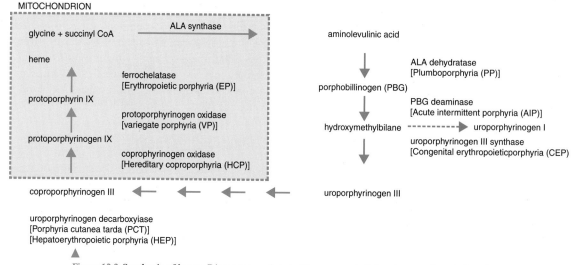

Figure 13-1. Basic structure of porphyrins.

The reduced forms of porphyrins are termed *porphyrinogens,* the functional form of the compound that must be used in heme synthesis. Porphyrinogens are highly unstable, colorless, and do not fluoresce, which makes them more difficult to analyze. In the presence of light, oxygen, or oxidizing agents, porphyrinogens are readily oxidized to their corresponding porphyrin form. Therefore, the porphyrin form is routinely analyzed in clinical laboratories.

Porphyrin Synthesis

All types of cells contain hemoproteins and can synthesize heme, but the main sites are the cells of the bone marrow and liver. The series of irreversible reactions is outlined in Figure 13-2. Some steps occur in the mitochondria of the cell and some steps in the cytoplasm. The transport of substrates across the mitochondrion membrane is a complex process and a potential point for interruptions in heme synthesis.

Control of the rate of heme synthesis in the cells in the liver is achieved largely through regulation of the enzyme δ-aminolevulinic acid (ALA) synthase. The main mechanism is repression of synthesis of new enzyme. A negative feedback mechanism exists whereby increases in the pool of hepatic heme diminish the production of ALA synthase. Conversely, ALA synthase production is increased with a depletion of heme. The size of the regulatory heme pool may be affected by the requirement for hemoproteins in the liver. Drugs and other compounds appear to induce ALA synthase production by several different mechanisms, but all result in a depletion of the regulatory heme pool. Therefore, the rate of heme synthesis is flexible and can change rapidly in response to a wide variety of external stimuli. In bone marrow erythrocytes, other enzymes in the pathway and the rate of cellular iron uptake seem to control the rate of heme synthesis.[1]

basic isomers may exist for every porphyrin compound, but only type I and type III occur in nature. The difference between types I and III isomers is in the arrangement of side chains. Only type III isomers form heme, but in some disorders, the functionless type I isomers may be present in excess in the tissues. Porphyrins are stable compounds, red-violet to red-brown in color, that fluoresce red when excited by light near 400 nm. Only three porphyrin compounds are clinically significant in humans: protoporphyrin (PROTO), uroporphyrin (URO), and coproporphyrin (COPRO). Their presence in excess in biologic fluids is a clinical sign of abnormal heme synthesis. The three compounds have different solubility properties and different degrees of ionization determined by the addition of various carboxyl groups to the basic porphyrin structure. This allows for separate assays of each. URO is excreted in urine, PROTO in the feces, and COPRO in either, depending on the rate of formation of the urine and its pH.

Figure 13-2. Synthesis of heme. Diseases associated with enzyme deficiencies are given in brackets.

Clinical Significance and Disease Correlation

The porphyrias are inherited or acquired enzyme deficiencies, which result in overproduction of heme precursors in the bone marrow (erythropoietic porphyrias) or the liver (hepatic porphyrias). Disease states corresponding to enzyme deficiencies have been identified in every step of heme synthesis except for ALA synthase. Some patients demonstrate an enzyme deficiency but do not show clinical or biochemical manifestations of porphyria, indicating that other factors, such as demand for increased heme biosynthesis, are also important in causing disease expression. An excess of the early precursors in the pathway of heme synthesis (ALA, porphobilinogen, or both) causes neuropsychiatric symptoms including abdominal pain, vomiting, constipation, tachycardia, hypertension, psychiatric symptoms, fever, leukocytosis, and parasthesias. Porphyrias in this category include plumboporphyria (PP) and acute intermittent porphyria (AIP). Excesses of the later intermediates (UROs, COPROs, and PROTOs) may cause cutaneous symptoms including photosensitivity, blisters, excess facial hair, and hyperpigmentation. Porphyria cutanea tarda (PCT), hepatoerythropoietic porphyria (HEP), erythropoietic porphyria (EP), and congenital erythropoietic porphyria (CEP) are associated with cutaneous symptoms. Porphyrin-induced photosensitivity manifests itself either by increased fragility of light-exposed skin, as in PCT, or by burning of light-exposed skin, as in EP. The photosensitizing effects of the porphyrins are attributable to their absorption of light. It is currently believed that light energy excites electrons of porphyrins to an elevated energy level, forming triplet-state porphyrins, which, in turn, react with molecular oxygen, forming compounds that damage tissues through several mechanisms.[2] There may also be excesses of both early and late intermediates, causing neurocutaneous symptoms. Hereditary coproporphyria (HCP) and variegate porphyria (VP) fall into this category. All of the porphyrias are inherited as autosomal dominant traits producing about a 50% reduction in enzyme level, except for PP and CEP, which are autosomal recessive and exhibit much lower levels of the affected enzyme.[3]

The diagnosis of porphyrias is made by a combination of history and physical and laboratory findings. The cutaneous porphyrias are easier to diagnose because photosensitivity is usually the presenting symptom. Laboratory diagnosis, if necessary, is made by analysis of the appropriate sample for intermediates in heme synthesis (Table 13-1). The differentiation of neurologic porphyrias from other disorders is more difficult based on history and physical examination and must be verified by laboratory findings.

Inherited PP (ALA dehydratase deficiency) is extremely rare, with only six cases having been reported.[3] The pattern of inheritance is autosomal recessive and affected patients are homozygous for the deficiency. Urinary ALA is markedly elevated with normal porphobilinogen (PBG) excretion. The much more common cause of low ALA dehydratase activity is lead poisoning, which must be ruled out before diagnosis of an inherited disorder.

AIP results from a deficiency of the enzyme PBG deaminase and usually appears in the third decade of life. Although the true prevalence is unknown, a reasonable estimate of gene frequency is 5 to 10 per 100,000 in the United States.[3] The enzyme deficiency is inherited as autosomal dominant, but only about 10% of the patients who inherit the enzyme deficiency suffer attacks of the disease, so other etiologic factors are involved. The most common precipitating causes of disease are drugs, especially barbiturates and sulfonamides, but a wide variety of drugs are potentially hazardous. This disease is characterized by multiple neurologic symptoms with colicky stomach pain and sometimes fever and vomiting. The characteristic laboratory findings are a marked elevation of ALA and PBG in the urine, but these test results may be normal between attacks. A patient's urine with clinically manifest AIP may turn red or black after exposure to light and air due to nonenzymatic polymerization of PBG to porphyrin or porphobilin. Measurement of levels of PBG deaminase in erythrocytes may be performed in reference laboratories to confirm the diagnosis.[4]

CEP, a deficiency of uroprophyrinogen III cosynthase, is one of the rarest porphyrias and usually appears shortly after birth. It is inherited as autosomal recessive and fewer than

TABLE 13-1. Metabolites Found in Excess in the Porphyrias				
Porphyria	**Urine**	**Feces**	**Erythrocytes**	**Symptoms**
PP	ALA	Normal	Normal	Neuropsychiatric
AIP	ALA, PBG	Normal	Normal	Neuropsychiatric
CEP	URO, COPRO	COPRO	Normal	Cutaneous
PCT	URO, COPRO	ISOCOPRO	Normal	Cutaneous
HEP	URO, COPRO	ISOCOPRO	Normal	Cutaneous
HCP	ALA,* PBG,* COPRO	COPRO	Normal	Neurocutaneous
VP	ALA,* PBG,* COPRO	PROTO > COPRO	Normal	Neurocutaneous
EP	Normal	PROTO	PROTO	Cutaneous

*During acute attacks.
PBG, porphobilinogen.

200 cases have been reported.[3] It is also known as *Günther's disease.* The teeth will fluoresce red under ultraviolet light and exhibit a red or brownish discoloration under normal light due to porphyrin deposits in the dentin. Urine porphyrins are markedly elevated, URO more than COPRO. The major fraction is the type I isomer. The urine is often red because of the presence of URO and COPRO. Fecal COPRO may also be elevated. Photosensitivity is a major clinical problem, resulting in lesions that may become infected, leaving the patient scarred, or with multiple occurrences, that may lead to mutilation of the ears, nose, or digits. Abnormal growth of hair is often seen in exposed areas. It is thought that the disfigurements of this disorder and the tendency to avoid daylight (hence only coming outside at night) led to the legend of the werewolf.[4] Patients may also develop a hemolytic anemia and splenomegaly with hemolysis serving as a stimulus for increased porphyrin production in the bone marrow.[3]

PCT, the heterozygous form of uroporphyrinogen decarboxylase deficiency, is the most common of the porphyrias. It usually presents in adulthood with cutaneous blistering and fragility in light-exposed areas, typically the hands, along with some abnormal growth of hair. Liver biopsy specimens from these patients show fluorescence, hemosiderosis, fatty infiltration, and variable degrees of necrosis and fibrosis. PCT is often acquired secondary to liver damage due to alcoholism, estrogen therapy, or hexachlorobenzene (insecticide) poisoning. Patients with healthy livers are able to synthesize heme despite the decreased levels of the enzyme and are asymptomatic. In addition, the disease has been identified in young women who use oral contraceptives and patients on chronic hemodialysis. In patients on chronic hemodialysis, the cause is probably related to iron overload due to multiple blood transfusions. It is well established that excess iron in the liver plays a role in the pathogenesis of PCT. The disease goes into remission if iron stores are decreased. However, the exact role of iron is still unknown. Both alcohol and estrogen are capable of causing modest increases in the activity of ALA synthase, but it is unknown if this is the primary cause or a secondary cause due to

inhibition of heme synthesis.[5] In the laboratory, PCT is characterized by increased levels of urinary URO and to a lesser extent COPRO. Fecal isocoproporphyrins (ISOCOPROs) are also increased.[6,7] Fractionation of urine porphyrins by high-performance liquid chromatography may be helpful in confirming the diagnosis.[4] Other laboratory findings that may also be seen in this disease are elevated serum iron, ferritin, and liver enzymes.

HEP is a rare form of porphyria resulting from a homozygous deficiency of uroporphyrinogen decarboxylase. Enzyme levels are approximately 5% to 10% of normal. It has been described in only 20 patients.[3] Photosensitivity begins in childhood. Patients are severely affected and develop excess facial hair and scarring of the hands and face. The severity of photosensitivity improves somewhat with age, but hepatic disease follows. Urine and fecal porphyrin levels are similar to those found in PCT.[3]

HCP, a deficiency of coproporphyrinogen oxidase, is a fairly mild condition, although attacks have been precipitated by exposure to certain drugs. Some carriers of the abnormal gene have normal excretion of porphyrins and porphyrin precursors. HCP is less common than AIP, but the true prevalence is unknown. The hallmark of HCP is markedly increased excretion of COPRO III in the feces and, to a lesser extent, in the urine. During acute attacks, urinary excretion of ALA and PBG is also increased. Clinical manifestations of HCP may be neurologic, as for AIP, or cutaneous, with lesions resembling those in PCT.

VP is prevalent in South Africa and can be traced to a single couple that immigrated from The Netherlands in 1688. The cause is a deficiency of protoporphyrinogen oxidase activity. Clinical manifestations include acute attacks of neurologic dysfunction, such as those in AIP, photodermatitis, as in HCP and PCT, or both. Hallmark laboratory findings are increased levels of COPRO and PROTO in the feces, with levels of PROTO exceeding those of COPRO. However, concentrations of porphyrins are higher in a duodenal aspirate of bile, which is now considered the preferred specimen

CASE STUDY 13-1

A 58-year-old male with a history of alcoholism complained of increased skin fragility and formation of skin lesions on his hands, forehead, neck, and ears on exposure to the sun. Also noted on physical exam was hyperpigmentation and hypertrichosis. Laboratory findings showed an increase in urinary uroporphyrin and a slight increase in coproporphyrin with normal levels of ALA and porphobilinogen. Isocoproporphyrin was elevated in the feces. Serum ferritin was increased as well as serum transaminase. Erythrocyte ZPP and FEP levels were normal.

Questions

1. What is the most probable disorder?
2. What confirmatory test should be done?
3. What is the most probable precipitating factor?
4. What are other causes of acquired cases of this type of porphyria?
5. How is this case differentiated from the homozygous deficiency of this enzyme?
6. How is this type of porphyria differentiated from other porphyrias causing cutaneous symptoms?

for analysis.[3,4] During acute attacks, urinary ALA and PBG excretion is increased, but in asymptomatic subjects, the excretion is often normal.

EP, the second most common porphyria, results from a deficiency of the enzyme, ferrochelatase. The major clinical symptom is photosensitivity, which is usually present from infancy. Patients complain of burning, itching, or pain in the skin on exposure to sunlight. Some patients also have severe liver disease. The diagnosis of EP is made by demonstrating increased levels of PROTO in erythrocytes, plasma, and stool, along with normal urinary porphyrins. Clinical expression of the disease is highly variable. Some individuals have no clinical manifestations of the disease but have increased levels of erythrocyte PROTO. They are considered to be clinically unaffected carriers of the gene defect.

Treatment in the inherited porphyrias is aimed at modifying the biochemical abnormalities causing clinical symptoms. The cutaneous symptoms are treated by avoiding sunlight, using sun-blocking agents and oral β-carotene, which acts as a singlet oxygen trap preventing skin damage. Reduction of the heme load can be accomplished by phlebotomy or by giving desferrioxamine to chelate iron. Intravenous hematin may be used to counteract acute attacks of neurologic dysfunction. Hematin, an enzyme inhibitor, limits synthesis of porphyrins in cells in the bone marrow. Cessation of precipitating factors, such as ingestion of alcohol or estrogens, may be useful in PCT.[8,9]

Secondary porphyrias or *porphyrinurias* is the term given to acquired conditions in which a mild to moderate increase in excretion of urinary porphyrins is seen. In this case, the disorders are not the result of an inherited biochemical defect in heme synthesis but are due to another disorder, toxin, or drug interfering with heme synthesis. Symptoms may be very similar to the inherited porphyrias in some cases. Various anemias, liver diseases, and toxins, such as lead and alcohol, fit into this category. Lead is known to inhibit both the activity of PBG synthase and the incorporation of iron into heme.[10] Secondary porphyrias can be distinguished from true porphyrias by measuring levels of urinary ALA and PBG. In

secondary porphyria, ALA levels are increased in the urine, whereas PBG excretion usually remains normal. Lead poisoning also classically exhibits increased COPRO in the urine and erythrocyte zinc protoporphyrin (ZPP), as well as increased ALA. However, determination of blood lead is the most sensitive method to detect lead poisoning.[1,5]

Methods of Analyzing Porphyrins

There are individual enzyme assays available for each of the defective enzymes that cause porphyria. However, most are still limited to use in specialized laboratories and are not discussed here. Screening tests can be performed easily and may be beneficial in emergency situations, but care should be taken in interpretation because false-negatives and false-positives occur.[10,11] Quantitative assays should follow all screening tests. Quantitative assays of the three porphyrins (URO, PROTO, and COPRO) and two porphyrin precursors (ALA and PBG) will serve to classify most porphyrias.

Tests for Urinary PBG and ALA

The two most common screening tests for urinary PBG are the Watson-Schwartz and the Hoesch tests.[10,12,13] Both tests are based on the principle of PBG forming a red-orange color when mixed with Ehrlich's reagent (acidic *p*-dimethylaminobenzaldehyde). In the Watson-Schwartz test, an extraction with chloroform or butanol is performed to differentiate PBG from interfering substances such as urobilinogen or indole. If a cherry-red color remains in the aqueous phase after addition of chloroform, this indicates a positive test for PBG. The reagent in the Hoesch test does not react with urobilinogen; therefore, it is sometimes used to confirm the results of the Watson-Schwartz test. Other substances that may be present in urine may form a variety of colors with Ehrlich's reagent, making the color interpretation ambiguous in either procedure. Both tests, whether positive or negative, should be followed with a quantitative test.

Porphobilinogen and ALA are determined quantitatively by isolating the compounds on ion-exchange columns.[10,14]

CASE STUDY 13-2

A young nurse from South Africa became emotionally disturbed and appeared to be hysterical a few days after a laparotomy for "intestinal obstruction." Before her operation, she had, for more than 1 week, taken barbiturate capsules to help her sleep. When first seen, she complained of severe abdominal and muscle pain and general weakness, her tendon reflexes were absent, and she was vomiting and constipated. Her urine was dark in color on standing and gave a brilliant pink-fluorescence when viewed in ultraviolet light.

Within 24 hours, she was totally paralyzed, and within 2 days she died.

Questions

1. What possible condition could this young woman have had, and why did it manifest itself at this time?
2. Would any members of her family have a similar disease?
3. What enzyme defect did she have?
4. What other confirmatory tests, if any, could be done?

A fresh sample of urine or a 24-hour sample of urine is obtained. For PBG analysis, the sample is added to an anion exchange column, where PBG is adsorbed and most interfering substances are removed by repeated water washes. PBG is then eluted with acetic acid. Ehrlich's reagent is added to the eluate and the resulting red color is measured spectrophotometrically. ALA is determined in a similar procedure except using a cation exchange column. Eluted ALA is reacted with Ehrlich's reagent after first being condensed with acetylacetone to form a pyrrole. The readings must be taken at 5 minutes to 15 minutes when the color has reached its maximum. This is considered the method of choice for determining PBG and ALA in most laboratories due to the relative simplicity and low cost.[4] In fact, it has been proposed that this method replace the use of screening tests because of the analytical problems previously discussed.[15,16]

Tests for Porphyrins

The screening test for porphyrins is based on their chemical properties, which produce a characteristic red fluorescence when viewed under ultraviolet light.[17,18] The porphyrins are extracted into an acidified organic solvent and then are extracted into aqueous acid. The test may be performed on urine, feces, or blood. The preferred specimen is a 24-hour urine, whole blood collected in any anticoagulant, or about 1 g for fecal analysis. The test should be performed using glass equipment, because synthetic materials may cause quenching of fluorescence. Urine is mixed with acetic acid–ethyl acetate reagent and allowed to separate. The top layer is examined with an ultraviolet lamp for the characteristic orange-red fluorescence. The top layer is removed and mixed with HCl to remove any interfering substances and reexamined under fluorescent light. Fluorescence in the lower (acidic) layer confirms the presence of porphyrins. The procedure is similar for analysis of blood or feces. Interpretation of results is subjective, and false-negative and false-positive results may occur. It is recommended that qualitative screening tests be replaced by a variety of quantitative tests, which are currently available.[10,16,19]

Most quantitative procedures for porphyrins are based on solvent extraction and measurement by fluorometric or chromatographic methods.[19] High-performance liquid chromatography is currently the method of choice for identification and quantitation of porphyrins.[10,20] Fractionation using this method reveals peaks corresponding to the quantity of various porphyrin types (URO, COPRO, ISOCOPRO, and so forth) present in the sample. High-performance thin-layer chromatography has also been used.[21] Measurement of porphyrin in blood also may be determined by analysis of ZPP, which is trapped in the heme-binding site in the hemoglobin molecule for the lifetime of the red blood cell. Because zinc competes with iron for insertion into PROTO, ZPP is also an indirect measure of iron availability in the developing erythrocyte. ZPP level is elevated in iron deficiency anemia, anemia of chronic disease and in lead poisoning.[4] A rapid screening method for determination of ZPP is by measuring the fluorescence of whole blood and washed erythrocytes using a hematofluorometer. A more definitive method of assaying blood porphyrins is by using traditional extraction techniques to measure free erythrocyte protoporphyrin (FEP). Details of these procedures have been published elsewhere.[10]

Molecular diagnostic techniques are becoming useful in the diagnosis of porphyrias. Most of the genes that encode the enzymes of heme synthesis have been identified and mutations, which cause various porphyrias, have been discovered. The use of these techniques to aid in the diagnosis of porphyrias has certain advantages over traditional biochemical assays. The interpretation of the traditional tests is complicated by the fact that analytes being measured may be normal except during an acute attack and the amount of porphyrin excreted in the various disorders is highly variable, causing significant overlap between affected patients and normal. By testing for the disease-causing mutation, these problems of biologic variability and disease activity can be overcome. However, perhaps the most important application of molecular testing is its use in detecting asymptomatic gene carriers, which are not easily identified by standard laboratory testing.[22]

HEMOGLOBIN

Role in the Body

Hemoglobin has many important functions in the body. Its major role is oxygen transport to the tissues and CO_2 transport back to the lungs. The hemoglobin molecule is designed to take up oxygen in areas of high oxygen tension and release oxygen in areas of low oxygen tension. Hemoglobin is carried to all tissues of the body by erythrocytes. Hemoglobin is also one of the major buffering systems of the body.

Structure of Hemoglobin

Hemoglobin is a large, complex protein molecule with a molecular weight of approximately 64,000. It is roughly spherical in shape and comprises two major parts: heme, which makes up 3% of the molecule, and globin proteins, which make up the remaining 97%. The heme portion comprises a porphyrin ring with iron chelated in the center. The iron atom is the site of reversible oxygen attachment. The protein portion comprises two pairs of globin chains that are twisted together so those heme groups are exposed on the exterior of the molecule (Fig. 13-3). The complete hemoglobin molecule contains four heme groups attached to each of four globin chains and may carry up to four molecules of oxygen. Each globin chain contains 141 or more amino acids.

The structure of each chain is four-fold. The primary structure consists of the individual amino acids and their sequences. Their sequences vary and are the basis of chain nomenclature: alpha, beta, delta, and gamma. The secondary structure is the three-dimensional arrangement of the amino acids making up the polypeptide chain. Regions of amino

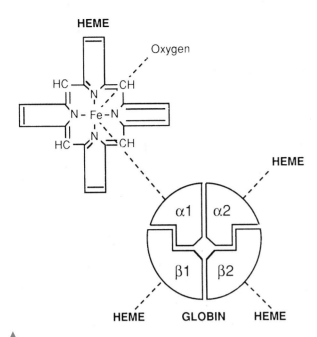

Figure 13-3. Hemoglobin A: Structure of the hemoglobin molecule. (From Linman JW. Hematology. New York: Macmillan, 1975. Reprinted with permission.)

acids may form helixes or a pleated structure. The tertiary structure is a larger fold superimposed on the helical or pleated forms. It represents the position taken by each chain or subunit in three-dimensional space. The quaternary structure represents the relationship of the four subunits to one another, particularly at the points of contact. Mutations at particular points of contact result in altered specific functional properties of the molecule, such as its oxygen affinity.

The majority of hemoglobin in normal adults is designated as hemoglobin A, or A_1, which contains two alpha and two beta chains (see Fig. 13-3). Hemoglobin A_2, which comprises two alpha and two delta chains, makes up less than 3% of normal adult hemoglobin. The remainder is composed of hemoglobin F, which contains two alpha and two gamma chains. Hemoglobin F is the main hemoglobin during fetal life and is about 60% of normal hemoglobin at birth. There is a gradual switch from production of delta chains to beta chains, and by about age 9 months, hemoglobin F usually constitutes less than 1% of total hemoglobin. Hemoglobin F has a greater affinity for oxygen than hemoglobin A; therefore, it is a more efficient oxygen carrier for the fetus. Hemoglobin F is more resistant to alkali than hemoglobin A, and this is the basis of one laboratory test to differentiate these two types of hemoglobin.

Two other hemoglobin chains, designated zeta and epsilon, are present only in embryonic life. Production of these chains stops by week 8 of gestation, and gamma-chain production takes over. The three embryonic hemoglobins are identified as Gower I, two zeta chains, and two epsilon chains; Gower II, two alpha chains, and two epsilon chains; and Portland I, two zeta chains, and two gamma chains.[23]

Genetic control of hemoglobin synthesis occurs in two areas: control of structure and control of rate and quantity of production. Defects in structure produce a group of diseases called the *hemoglobinopathies*. Defects in rate and quantity of production lead to disorders called the *thalassemias*. Structurally, each globin chain has its own genetic locus; therefore, it is the individual chains, not the whole hemoglobin molecule, that are under genetic control. The genes for the globin chains can be divided into two major groups: the α-genes, which is on chromosome 16, and the non-α-genes, which are on chromosome 11. In most persons, the α-gene locus is duplicated—there are two alpha-chain genes per haploid set of chromosomes. The α-gene and hence its polypeptide chains are identical in hemoglobins A, A_2, and F. The non-α-genes for the beta, delta, and gamma chains are sufficiently close in genetic terms to be subjected to nonhomologous crossover, with the resulting production of fused or hybrid globin chains such as hemoglobin Lepore (delta-beta globin chain) and Kenya (gamma-beta globin chain).

Based on the genetics of the globin chain production, the structural abnormalities can be divided into four groups:

1. Amino acid substitutions (*eg*, hemoglobins S, C, D, E, O, G, and so forth)
2. Amino acid deletion—deletions of three or multiples of three nucleotides in deoxyribonucleic acid (DNA; *eg*, hemoglobin Gun Hill)
3. Elongated globin chains resulting from chain termination, frame shift, or other mutations (*eg*, hemoglobin Constant Spring)
4. Fused or hybrid chains resulting from nonhomologous crossover (*eg*, hemoglobins Lepore and Kenya)

The amino acid substitutions are the most common abnormalities, with several hundred described so far. Approximately two thirds of the hemoglobinopathies have an affected beta chain. They may be clinically silent, or they may cause severe damage, as with hemoglobin S. Amino acid substitutions, for the majority of defects, result from a single-base nucleotide substitution in DNA.

Absent or diminished synthesis of one of the polypeptide chains of human hemoglobin characterizes the thalassemias, a heterogeneous group of inherited disorders. In α-thalassemia, alpha-globin chain synthesis is absent or reduced, and in β-thalassemia, beta-globin chain synthesis is absent ($β^0$-thal) or partially reduced ($β^1$-thal).

Synthesis and Degradation of Hemoglobin

Hemoglobin synthesis occurs in the immature red blood cells in the bone marrow: 65% in the nucleated cells and 35% in reticulocytes. Normal synthesis depends on adequate iron supply as well as normal synthesis of heme and protein synthesis to form the globin portion. Heme is synthesized in the mitochondria of the cells. Iron is transported to the developing red blood cells by transferrin, a plasma protein. Iron traverses the cell membrane and the mitochondria, where it is inserted into the PROTO ring to form heme. Protein synthesis of the globin chains occurs in the cytoplasmic polyribosomes. Heme leaves the mitochondria and is joined to the globin chains in the cytoplasm in the final step.

Two possible pathways degrade hemoglobin. The normal pathway is called extravascular because it occurs outside of the circulatory system in the reticuloendothelial, or mononuclear phagocyte, system. Within the splenic phagocytic cells, or macrophages, hemoglobin loses its iron to transferrin, its alpha carbon is expired as CO, the globin chains return to the amino acid pool, and the rest of the molecule is converted to bilirubin, which undergoes further metabolism. Normally, 90% of all hemoglobin is degraded in this manner (Fig. 13-4).

Normally, less than 10% hemoglobin is released directly into the blood stream and dissociated into alpha and beta dimers. Greater amounts are released during hemolytic episodes. The dimers are bound to haptoglobin, which prevents renal excretion of plasma hemoglobin and stabilizes the heme–globin bond. This complex is then removed from the circulation by the liver and processed in a fashion similar to extravascular degradation. If the amount of circulating haptoglobin is decreased, as during a hemolytic episode, the unbound dimers go through the kidneys, reabsorb, and are converted to hemosiderin. If greater than 5 g/day is broken down, free hemoglobin and methemoglobin will appear in the urine.

Hemoglobin that is not entirely bound by haptoglobin, or processed by the kidneys, is oxidized to methemoglobin. Heme groups are released and taken up by the protein hemopexin. The heme–hemopexin complex is cleared by the liver and catabolized. Then, heme groups present in excess of the binding capacity of hemopexin complex combine with albumin to form methemalbumin and are held by this protein until additional hemopexin becomes available for shuttle to the liver (Fig. 13-5). Laboratory measurement of any of these hemoglobin degradation products can help to determine increased red blood cell destruction, such as in a hemolytic anemia.

Clinical Significance and Disease Correlation

Hemoglobin Qualitative Defects: The Hemoglobinopathies

Hemoglobin S. The amino acid defect in hemoglobin S is at the sixth position on the beta chain, where glutamic acid is substituted by valine, giving the hemoglobin a less negative charge than hemoglobin A.

Individuals have either sickle cell trait (HbAS, the heterozygous state) or sickle cell disease (HbSS, the homozygous state). Black Africans and African-Americans have the highest incidence: 1 in 500 infants have sickle cell anemia and 8% to 10% carry the HbAS trait. It is also found in Mediterranean countries such as Greece, Italy, and Israel, as well as in Saudi Arabia and India.

Because of the high mortality and morbidity associated with homozygous expression of the gene, the frequency of the mutant gene would be expected to decline in the gene pool. However, a phenomenon known as balanced polymorphism exists, which indicates that the heterozygous state (HbAS) has a selective advantage over either of the homozygous states (HbAA or HbSS). It appears that the heterozygous condition offers protection from parasites, particularly *Plasmodium falciparum*, especially in children. When infected with *P. falciparum*, children with sickle cell trait have a lower parasite count, the infection is shorter in duration, and the inci-

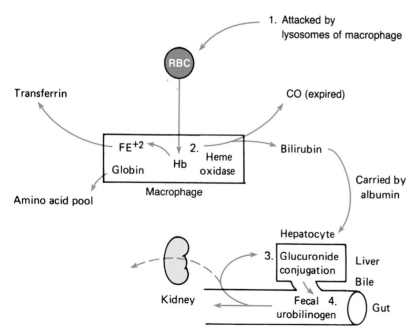

Figure 13-4. Extravascular degradation of hemoglobin.

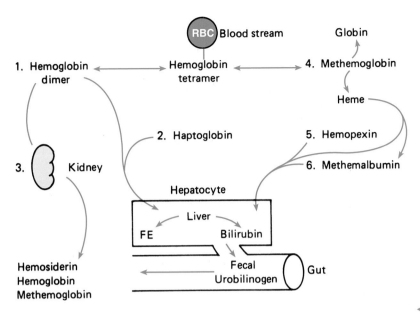

Intravascular Breakdown of Hemoglobin < 10%

Figure 13-5. Intravascular breakdown of hemoglobin.

dence of death is low. It is thought that the infected red blood cells are preferentially sickled and thereby efficiently destroyed by phagocytic cells.[24]

When hemoglobin S is deoxygenated in vitro under near-physiologic conditions, it becomes relatively insoluble as compared with hemoglobin A, and aggregates into long, rigid polymers called *tactoids*. These cells appear as sickle- or crescent-shaped forms on stained blood films. Sickled cells may return to their original shape when oxygenated; however, after several sickling episodes, irreversible membrane damage occurs and cells are phagocytized by macrophages in the spleen, liver, or bone marrow, causing anemia. The severity of the hemolytic process is directly related to the number of damaged cells in circulation. The rigid sickled cells are unable to deform and circulate through small capillaries resulting in blockage. Tissue hypoxia results, causing

extreme pain and leading to tissue death. Infarctions in the spleen are common, causing excessive necrosis and scarring, leading to a nonfunctional spleen in most adults with sickle cell anemia. This is referred to as autosplenectomy. The amount of sickling is related to the amount of hemoglobin S in the cells. The reported inhibitory effect of hemoglobins A and F is due to a dilutional effect. There is also less tendency for hemoglobin F to copolymerize with hemoglobin S than with hemoglobin A. This is considered responsible for the observed protective effect of elevated hemoglobin F levels in individuals with sickle cell anemia.

Laboratory findings in the homozygous disease include a normocytic, normochromic anemia, increased reticulocyte count, and variation in size and shape of red blood cells with target cells and sickle cells present. Polychromatophilia and nucleated red blood cells are common. The heterozygous

CASE STUDY 13-3

A 32-year-old African-American woman came to the obstetrics and gynecology clinic of her local hospital because she was feeling a little weak. A complete blood count (CBC) showed a hemoglobin of 9.9 g/dL with an MCV of 87 fL. The physician ordered a hemoglobin electrophoresis as a follow-up. The cellulose-acetate pattern showed a peak of 58% at the hemoglobin A position, a peak of 35% at the hemoglobin S position, and a peak of 5% at the A_2 position. Further studies indicated a positive dithionite solubility test result and a hemoglobin F value of 1%.

Questions

1. What is the best possible diagnosis for this woman?
2. Does this condition require further follow-up and treatment?
3. What implications does this disease have for her unborn child?
4. Are the values for hemoglobins A_2 and F normal for this condition?

disease is clinically asymptomatic and usually has a normal blood film. The dithionite solubility test will be positive in both homozygous and heterozygous forms, but should always be confirmed with hemoglobin electrophoresis. On cellulose acetate electrophoresis at an alkaline pH, hemoglobin S moves in a position between hemoglobin A and A_2. Of total hemoglobin, 85% to 100% will be hemoglobin S in the homozygous state and usually less than 50% in the heterozygous state. Other hemoglobins that migrate in the same position as hemoglobin S are hemoglobins D and G, but both would be negative with the dithionite solubility test. Electrophoresis on citrate agar at an acid pH is necessary to separate these hemoglobins from hemoglobin S (see Fig. 13-7).

Hemoglobin C. The glutamic acid in the sixth position of the beta chain is replaced by lysine, resulting in a net positive charge. Hemoglobin C is found in West Africa in the vicinity of North Ghana in 17% to 28% of the population and in 2% to 3% of African-Americans.

The heterozygous form, hemoglobin AC, is asymptomatic. The homozygous form usually causes a mild, well-compensated anemia characterized by abdominal pain and splenomegaly. The most prominent laboratory feature is the presence of target cells. There is a tendency to form large, oblong, hexagonal crystalloid structures within the red cell. These structures are most prominent in patients who have undergone splenectomy.

A differential diagnosis is obtained by cellulose acetate electrophoresis. Hemoglobin C moves with hemoglobin A_2 and is negative with the dithionite solubility test. In the heterozygous form, hemoglobin C falls in the range of 35% to 48%. Hemoglobins E, O, and C_{Harlem} migrate with hemoglobin C. These hemoglobin variants can be readily distinguished from hemoglobin C by citrate agar electrophoresis at an acid pH.

Hemoglobin SC. Hemoglobin SC disease is the most common mixed *hemoglobinopathy*. One beta gene codes for beta-S chains and the other beta gene codes for beta-C chains, thus leaving no normal beta chains to produce hemoglobin A. Clinically, this disease is less severe than homozygous sickle cell anemia but has similar clinical symptoms. The blood film characteristically shows many target cells and occasional abnormal shapes resembling both the sickle cell, the hexagonal hemoglobin C crystal, and a combination of the two. The dithionite solubility test is positive, and electrophoresis on cellulose acetate shows about equal amounts of hemoglobin S and hemoglobin C.

Hemoglobin E. Hemoglobin E is an amino acid substitution of lysine for glutamic acid in the 26th position of the beta chain, resulting in a net positive charge. Hemoglobin E is somewhat unstable when subjected to oxidizing agents.

It is found in Asia and is estimated to occur in about 20 million individuals, 80% of whom live in Southeast Asia.

In the homozygous form, there is a mild anemia with microcytosis and target cells. In the heterozygous form, the patient is asymptomatic. The differential diagnosis is obtained by electrophoresis. On cellulose acetate, hemoglobin E moves with A_2, C, and O. It is present in the heterozygous form in amounts varying from 30% to 45%, which is somewhat lower than the percentage for hemoglobin C. This is probably due to the somewhat unstable nature of hemoglobin E. On citrate agar, hemoglobin E migrates with A. It is more common to find this defect in association with both α- and β-thalassemia. E-β-thalassemia is a more severe disorder, with moderate anemia and splenomegaly.

Hemoglobin D. The letter *D* is given to any hemoglobin variant with an electrophoretic mobility on cellulose acetate similar to that of hemoglobin S but that has a negative dithionite solubility test. Hemoglobin $D_{Los\ Angeles}$ and its identical variant, Hemoglobin D_{Punjab}, are the most common, with glycine substituted for glutamic acid at the 121st position of the beta chain. Hemoglobin D_{Punjab} is found in Northwest India but occasionally can be seen in English, Portuguese, and French individuals because of the close historical connection of these countries with East India. Hemoglobin $D_{Los\ Angeles}$ is found in 0.02% of African-Americans.[23]

The homozygous state is rare. There is no anemia or splenomegaly and only a slight anisocytosis. The oxygen affinity is higher than in normal blood. Heterozygous individuals are asymptomatic. Differential diagnosis is accomplished with electrophoresis. On cellulose acetate, hemoglobin D migrates with hemoglobin S in proportions of 35% to 50%. On citrate agar, hemoglobin D migrates with A.

Hemoglobin Quantitative Defects: The Thalassemias

The thalassemias are a group of diseases in which a defect in the rate of synthesis of one or more of the hemoglobin chains occurs, but the chains are structurally normal. Gene deletions or point mutations are the cause for decreased or absent chain synthesis.[25] The two most common types are α-thalassemia, resulting from defective production of alpha chains, and β-thalassemia, resulting from a defect in production in beta chains. Defects in production of the delta and gamma chains have been described, but these are not involved in production of hemoglobin A and therefore are not clinically significant. Rarely, combinations of gene deletions, such as delta and beta, may lead to clinical disease. Any form of unbalanced production of globin chains causes the erythrocytes to be small, hypochromic, and sometimes deformed. Intracellular accumulation of unmatched chains in the developing erythrocytes causes precipitation of the proteins, which leads to cell destruction in the bone marrow. Although erythropoiesis is occurring, it is ineffective because mature cells do not reach the peripheral blood to carry oxygen.

Thalassemia is inherited as an autosomal dominant disorder with heterogeneous expression of the disease. It is one of the most common hereditary disorders and is distributed

CASE STUDY 13-4*

A 54-year-old African-American woman was admitted to the hospital with the chief complaint of left hip pain and lethargy. She had a long history of multiple emergency room visits for hip pain requiring medication. She had previously been found to have a positive solubility test for hemoglobin S, but denied a history of sickle cell disease. There was family history of sickle cell trait. She had had a mastectomy for breast cancer 10 years before. Admission laboratory values were as follows:

Hemoglobin	5.3 g/dL
Hematocrit	17%
MCV	82 fL
MCHC	31 g/dL
WBC	12,000/μL
Platelet count	53,000/μL
Differential	Normal
Reticulocyte count	6.4% (corrected 2.4%)
RBC morphology	Target cells, spherocytes, schistocytes, basophilic stippling, bizarre forms, including elongated, block-shaped, and more densely stained cells

A hemoglobin electrophoresis was ordered. Patterns from the cellulose acetate and citrate agar electrophoresis are shown in Case Study Figure 13-4.1. Chest x-ray showed a right lower lobe infiltrate with pulmonary vascular congestion and an enlarged spleen. Fluid aspirated from the nasogastric tube was positive for blood.

The patient was given medication for an aspiration pneumonia, gastrointestinal bleeding, and congestive heart failure. She was given packed RBCs, fresh-frozen plasma, and platelets, but her condition continued to worsen. Six hours later, laboratory tests confirmed disseminated intravascular coagulation (DIC). A bone marrow biopsy was performed and revealed extensive necrosis of marrow elements. Three hours later, the patient died of cardiac arrest.

Questions

1. What hemoglobinopathy is indicated by the hemoglobin electrophoresis patterns?
2. What clinical feature of this disease differs from the typical picture in sickle cell anemia?
3. What other hemoglobins interact with hemoglobin S, and how can these be differentiated from hemoglobin C?
4. Was this patient's death due to the hemoglobinopathy? Is it unusual for hemoglobin SC to be life-shortening?

* *Case study data provided by Margaret Uthman, MD, Assistant Professor of Pathology, University of Texas Medical School at Houston.*

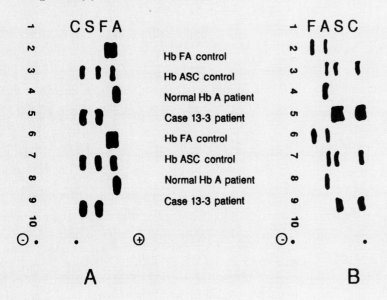

Case Study Figure 13-4.1. Hemoglobin electrophoretic patterns for Case Study 13-3. (*A*) Cellulose acetate at pH 8.4. (*B*) Citrate agar at pH 6.2. (Courtesy of Margaret Uthman, MD, University of Texas Medical School at Houston.)

worldwide. The prevalence of the thalassemia gene has been attributed to the protection it offers against falciparum malaria. The heterozygous state produces a disorder called *thalassemia minor,* which is clinically asymptomatic and resembles iron deficiency. The homozygous state, *thalassemia major,* is either lethal before birth or in childhood.

α-Thalassemias. α-thalassemia occurs with high frequency in Asian populations but is also seen in the Black African, African-American, Indian, and Middle Eastern populations. There are four principal clinical types of different severity known to occur in the population, and these four types can be explained, respectively, by deletions of 4, 3, 2, or 1 of the alpha-globin gene loci (Fig. 13-6). The type of α-thalassemia found in Black Africans and African-Americans is also associated with deletion of the alpha-globin genes but in a different pattern than is found in the Asian population (see Fig. 13-6). The four clinical types of α-thalassemia in order of most deletions to least deletions are the following:

1. Hydrops fetalis is the most clinically severe form of α-thalassemia because of the total absence of alpha-chain synthesis. Hemoglobin Bart's, which is a tetramer of gamma chains, is the main hemoglobin found in the red cells of affected infants. Hemoglobin Bart's has an extremely high O_2 affinity and allows almost no oxygen transport to the tissues. These infants are either stillborn or die of hypoxia shortly after birth.

2. Hemoglobin H disease has alpha chain synthesis at about one third the amount of beta-chain synthesis. As a result, beta chains accumulate and form tetramers, which are called hemoglobin H. Beta-chain precipitates (hemoglobin H inclusions) alter the shape and ability of the cell to deform, thus significantly shortening the life span of the cells. These individuals have a moderate hemolytic anemia, with 5% to 30% hemoglobin H, 1% hemoglobin A_2, and the remainder hemoglobin A. Hemoglobin H inclusions can be seen in the red cells with a supravital stain. Cord blood contains 10% to 20% hemoglobin Bart's.

3. α-thalassemia trait (or α-thalassemia minor) results from two gene deletions, either on the same or on different chromosomes. The deletion on two separate chromosomes is more common in Black Africans and African-Americans (see Fig. 13-6). These individuals have a mild microcytic, hypochromic anemia. Occasionally, excess beta chains may form hemoglobin H inclusions. Cord blood contains 2% to 10% hemoglobin Bart's, but after age 3 months, electrophoresis is normal.

4. Silent carriers are missing only one alpha gene, and the remaining genes direct production of sufficient alpha chains for normal hemoglobin production. This state can only be detected by expression of 1% to 2% of hemoglobin Bart's in a neonate. After age 3 months, it can only be detected by globin-gene analysis. It has been estimated that the frequency of this genetic disorder may be as high as 25% in the African-American population.[23]

β-Thalassemias. In contrast to α-thalassemia, gene deletions usually do not cause β-thalassemia. One of several types of mutations in the beta gene result in failure to produce normal amounts of beta-globin chains. In general, they may involve either faulty transcription, processing, or translation of the messenger ribonucleic acid (RNA) from the gene directing beta-chain production.[26] The β-thalassemias are classically divided into homozygous disease, called thalassemia major or Cooley's anemia, and heterozygous disease, called thalassemia minor. However, the clinical expression of the disease is very heterogeneous, depending on the type of genetic defect and involvement with other gene loci. The disease may be broadly divided into two major subtypes according to genetic expression: β^1, in which beta chains are produced in reduced amounts, and β^0, which is complete absence of beta chains.

β^1-thalassemia is the most common type. There is some synthesis of betaglobin chains but in significantly reduced amounts (5%–30%) of normal. The biochemical defect shows a quantitative deficiency of beta-globin mRNA. The hemoglobin electrophoresis pattern and hemoglobin F and A_2 quantitation show about 2% to 8% hemoglobin A_2, an elevated but varying amount of hemoglobin F, and the remainder hemoglobin A. The mean corpuscular volume (MCV) is very low, with a severe anemia, reticulocytes, nucleated red blood cells, basophilic stippling, target cells, extreme poikilocytosis, and anisocytosis.

β^0-thalassemia accounts for 10% of homozygous β-thalassemia, with a total absence of beta-chain synthesis but intact synthesis of gamma chains. In the homozygous form, there is 1% to 6% hemoglobin A_2 and 95% hemoglobin F. The hemoglobin concentration is very low, with a severe anemia. The heterozygous form appears clinically the same as homozygous β^1-thalassemia.

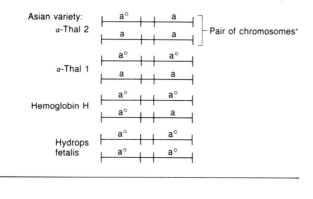

Alpha thalassemias

a- is normal alpha chain (normal gene loci)
a° is deleted alpha chain (deleted alpha gene loci)
+ note that a loci is duplicated on each chromosome

Figure 13-6. Deletions of α-globin gene loci in α-thalassemia.

Homozygous β-thalassemia—β-thalassemia major—either β^1 or β^0, is a crippling disease of childhood. This is unlike α-thalassemia, in which the child either dies shortly after birth or leads a normal life. The hypochromic, microcytic anemia is due to both the defect of functional hemoglobin tetramer synthesis and the premature destruction of red blood cells, both intramedullary and extramedullary, due to the increased presence of alpha chains. The bone marrow compensates by expanding enormously in size, sometimes causing structural bone abnormalities. Treatment of severe forms of the disease include regular transfusion therapy, iron chelation drugs to remove excess iron, and folic acid supplements.[25] Bone marrow transplantation has been successful if an HLA-identical donor is available.[27]

Heterozygous β-thalassemia—thalassemia minor—may be caused by inheritance of one thalassemia gene, either β^1 or β^0. The other gene directing beta-chain production is normal, and red blood cell survival is not shortened. About 1% of African-Americans are affected. Clinically, the condition is usually asymptomatic but may sometimes cause a mild microcytic anemia. The hematologic laboratory values resemble those of iron-deficiency anemia, and it is important to distinguish the two because quite different treatments are required. The red blood cell count in thalassemia minor is usually higher than would be expected with the accompanying hemoglobin concentration and a few target cells or basophilic stippling may be seen on a stained smear. The red cell distribution width (RDW) parameter on automated instruments, which is a quantitative measure of red blood cell variation in size, may be helpful to distinguish the two disorders. It is typically normal in thalassemia and increased in iron deficiency due to the heterogeneity of the red blood cells. Hemoglobin electrophoresis of thalassemia minor characteristically shows an increase in hemoglobin A_2. Quantitation of hemoglobin A_2 by column chromatography usually reveals values between 3.5% and 7%.

δβ-thalassemia is a rare type characterized by total absence of both beta-chain synthesis of hemoglobin A and delta-chain synthesis of hemoglobin A_2. Homozygous patients have 100% hemoglobin F. The heterozygous individuals have 93% hemoglobin A, 2% to 3% hemoglobin A_2, and 3% to 10% hemoglobin F.

Patients are anemic and show a thalassemic phenotype because the gamma-chain synthesis of hemoglobin F is not equal to alpha-chain synthesis. There is approximately one third as much gamma chain produced as alpha chain. Hemoglobin F is heterogeneously distributed among the erythrocytes, as revealed by the acid elution stain procedure.

Hereditary persistence of fetal hemoglobin (HPFH) is genetically and hematologically heterogeneous. In African

CASE STUDY 13-5

A 5-year-old Caucasian male was seen by a physician for an upper respiratory tract infection and splenomegaly was noted. A CBC was ordered and, subsequently, a hemoglobin electrophoresis. The following were the results:

		Reference Range
Hemoglobin	8.5 g/dL	11.7–15.7 g/dL
Hematocrit	27%	35–47%
RBC	4.3×10^{12}/L	$3.8–5.2 \times 10^{12}$/L
MCV	62.3 fL	80–100 fL
MCHC	32.0%	32–26%
RDW	18.5%	11.5–14.5%
Platelets	538×10^9/L	$150–440 \times 10^9$/L
WBC	10.7×10^9/L	$3.5–11.0 \times 10^9$/L
Reticulocyte count	5.6%	0.5–1.5%

WBC differential: Normal except for the presence of 1 nucleated RBC/100 WBC

RBC morphology: Moderate anisocytosis, moderate microcytosis, slight polychromasia, slight target cells, slight schistocytes

Hemoglobin electrophoresis: Hemoglobin C   89%
(Cellulose acetate)   Hemoglobin F   11%

A hemoglobin electrophoresis by citrate agar method confirmed the presence of hemoglobins C and F. Hemoglobin F was determined to be 7.5% by the alkali denaturation method. Hemoglobin A_2 could not be quantitated due to the presence of hemoglobin C. The patient's father had been told previously that he was slightly anemic due to a blood disorder called thalassemia. The patient's mother and four older siblings were healthy and unaware of the presence of any abnormal hemoglobin.

Questions

1. What combination of disorders did the patient most probably inherit?
2. Why were the mother and other siblings unaware of the presence of an abnormality?
3. Why was the patient unable to produce any hemoglobin A?
4. What caused the discrepancy in values for hemoglobin F from electrophoresis and the alkali denaturation test?
5. Why was it unusual to find hemoglobin C in a Caucasian family?

Blacks and African-Americans, there is a total absence of beta- as well as delta-chain synthesis because of deletions in chromosome 11. Gamma-chain synthesis is present in the adult at a high level, and in contrast to synthesis of hemoglobin F in β-thalassemia or δβ-thalassemia, it is uniformly distributed. In heterozygotes, there is no imbalance of globin-chain synthesis. There is 17% to 33% hemoglobin F. The patients are clinically normal. In homozygotes, there is 100% hemoglobin F, with no synthesis of hemoglobin A or A_2. There are no significant hematologic abnormalities, other than erythrocytosis, and these patients are also asymptomatic. In the Greek type, hemoglobin F production is slightly lower, 10% to 20%, and hemoglobin A_2 production is slightly higher at 1% to 3%. There is hypochromia, microcytosis, anisocytosis, and poikilocytosis. In the Swiss type, hemoglobin F ranges between 1% and 7% and is unevenly distributed (heterogeneously) among the erythrocytes.

Methodology

Most hemoglobinopathies and thalassemias can be diagnosed by use of the solubility test, cellulose acetate electrophoresis, and citrate agar electrophoresis. Thalassemias may require quantitation of hemoglobin A_2 or F by more definitive methods. The more complicated cases may require more specialized procedures such as globin chain electrophoresis,[28,29] anion exchange high-performance liquid chromatography,[30,31] or DNA technology testing.[28,32,33]

Dithionite Solubility Test [34,35] (Screening Test for Sickling Hemoglobins)

The dithionite solubility test is based on the principle that sickling hemoglobin, in the deoxygenated state, is relatively insoluble, and it forms a precipitate when placed in a high-

molarity phosphate buffer solution. The precipitate appears because the deoxygenated hemoglobin molecules form tactoids that refract and deflect light rays, thereby producing a turbid solution. A small amount of packed red blood cells is placed in a buffered solution of sodium dithionite with saponin to lyse the red blood cells. After standing at room temperature for 5 minutes, the tube containing the solution is placed approximately 1 inch in front of a heavy, black-lined index card. If there is no sickling hemoglobin present, the lines on the card will be easily read. If sickling hemoglobin is present, the lines will be very indistinct or impossible to read. The test is reported as positive or negative for sickling hemoglobin. A positive and a negative control should be run with each test batch.

Outdated reagents and reagents not at room temperature interfere with the test. False-negative tests may be due to anemia, recent transfusions or may occur in infants younger than age 6 months because of high concentrations of hemoglobin F. If whole blood is used instead of packed red blood cells, false-positive results may occur due to erythrocytosis, hyperglobulinemia, extreme leukocytosis, or hyperlipidemia, and false-negative results may occur in anemia. This test does not differentiate between the homozygous and heterozygous presence of hemoglobin S, and other rare sickling variants, such as hemoglobin C_{Harlem}, may give a positive test. This test is commonly used as a screening test for sickling hemoglobin, but it may also be used as a confirmatory test for sickling hemoglobin after initial evaluation with cellulose acetate electrophoresis.

Cellulose Acetate Hemoglobin Electrophoresis [36–38]

A fresh hemolysate made from a packed red blood cell sample is applied to a cellulose acetate plate using a buffer of alkaline pH (8.4–8.6) and electrophoresis is performed. After

---

CASE STUDY 13-6

A 22-year-old Caucasian female of Italian heritage had been told that she was slightly anemic and had been treated with iron periodically throughout her life. She was a student in a clinical laboratory science program and had a CBC performed as a part of a hematology class.

Laboratory values were as follows:

		Reference Range
Hemoglobin	11.0 g/dL	11.7–15.7 g/dL
Hematocrit	34%	35%–47%
RBC	5.8×10^{12}/L	3.8–5.2×10^{12}/L
MCV	59.4 fL	80–100 fL
MCHC	31.9%	32%–36%
RDW	14.2%	11.5%–14.5%

From these values, the hematology instructor suspected an inherited disorder instead of iron deficiency and suggested that the student contact her doctor for further testing. A hemoglobin electrophoresis revealed slightly increased amounts of hemoglobin F and A_2, which were subsequently quantitated to reveal hemoglobin F of 3.2% and hemoglobin A_2 of 4.2%. Iron studies were normal.

Questions

1. What is the most probable disorder?
2. What CBC values caused the instructor to suggest further testing?
3. Why is it important for this disorder to be correctly diagnosed?

electrophoresis, the membrane is stained and cleared. Interpretation is made by comparing the hemoglobin's migration with that of a control. A rough estimate of proportions of different hemoglobins may be made using a densitometer.

The order of electrophoretic mobility, from slowest to fastest, is hemoglobins C, S, F, and A (Fig. 13-7). (There are several mnemonic devices for remembering the migration pattern. The most sensible is *Accelerated, Fast, Slow,* and *Crawl.* The most fun is *A Fat Santa Claus!*) Any hemoglobin that migrates beyond hemoglobin A is termed a *fast hemoglobin.* Hemoglobin Bart's and hemoglobin H both migrate here. Any abnormal hemoglobin or any hemoglobin present in increased amounts, such as hemoglobin A_2 and F, should be confirmed. If there is an abnormal hemoglobin, confirm with a sickling test and citrate agar electrophoresis. If an increased amount of A_2 or F occurs, then quantify the amount (Fig. 13-8).

Citrate Agar Electrophoresis

Citrate agar electrophoresis is performed at an acid pH (6.0–6.2) after an abnormal hemoglobin is found on cellulose-acetate electrophoresis.[39,40] In this method, an important factor in determining the mobility of hemoglobin is solubility. Hemoglobin F, with the fastest cathodal mobility, is also the most soluble, probably because it is most resistant to denaturation at pH 6.0. Adult hemoglobins whose solubility is similar to that of hemoglobin A, such as G, D, E, O, I, and so forth, move with hemoglobin A. The relatively insoluble hemoglobin S moves behind hemoglobin A, and the even more insoluble hemoglobin C moves behind hemoglobin S (see Fig. 13-7).

Hemoglobin A_2 Quantitation

Hemoglobin A_2 can be estimated by hemoglobin electrophoresis; however, this yields only a rough estimate. Quantitation must be made by microcolumn chromatography[41–44] or high-performance liquid chromatography,[45] both of which use an ion-exchange resin.

Acid Elution Stain for Hemoglobin F

Erythrocytes containing an increased amount of hemoglobin F can be distinguished from normal adult cells by the acid-elution technique. This method may be helpful in the diagnosis of hereditary persistence of fetal hemoglobin or to detect fetal cells in maternal circulation during problem pregnancies. Adult hemoglobin, hemoglobin A, is eluted from the erythrocytes by incubation in an acid buffer. Hemoglobin F remains behind and is stained with eosin.[46] Adult cells are negative and have no staining, because they contain no hemoglobin F. Cord blood cells appear as smooth, homogeneous cells with a scarlet-red color and a distinct halo. Very old fetal cells and incompletely laked adult cells stain with a bluish cast. Lymphocyte nuclei stain pale red or pink and should not be mistaken for the fetal red cells.

Hemoglobin F Quantitation

Fetal hemoglobin may be quantitated based on the principle that it is resistant to alkali denaturation in 1.25 mol/L NaOH for 2 minutes. Denatured hemoglobin A is precipitated out with ammonium sulfate and removed by filtration. The optical density of the clear supernatant solution is read at 540 nm, and the percentage of fetal hemoglobin is calculated against the optical density of the total hemoglobin solutions.[47–49] High-performance liquid chromatography on an ion exchange resin also appears promising for analysis of fetal hemoglobin.[45]

The average adult has less than 1.5% fetal hemoglobin. However, elevated levels may be found in a number of inherited and acquired diseases. The hereditary persistence of fetal hemoglobin should be suspected in individuals who possess 10% or more of fetal hemoglobin with no other apparent clinical abnormalities.

DNA Technology

The definitive diagnosis of some hemoglobinopathies and thalassemias that involve combinations of genetic defects may require the analysis of DNA. With the increased use and efficiency of the polymerase chain reaction technique, the DNA sequence of interest may be easily analyzed from whole blood or spots of dried blood on filter paper. With currently available automated sequencing methods, the time required to perform this type of analysis is not significantly greater than standard methodology. Disadvantages of these methods are the higher costs and lack of availability in most routine laboratories. Advantages are that it provides defini-

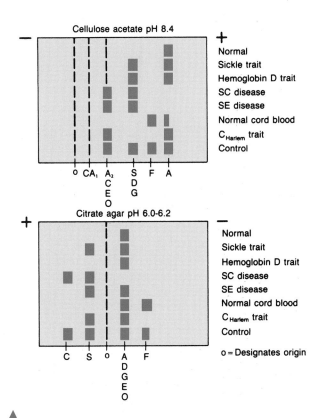

Figure 13-7. Comparison of various hemoglobin samples on cellulose acetate and citrate agar.

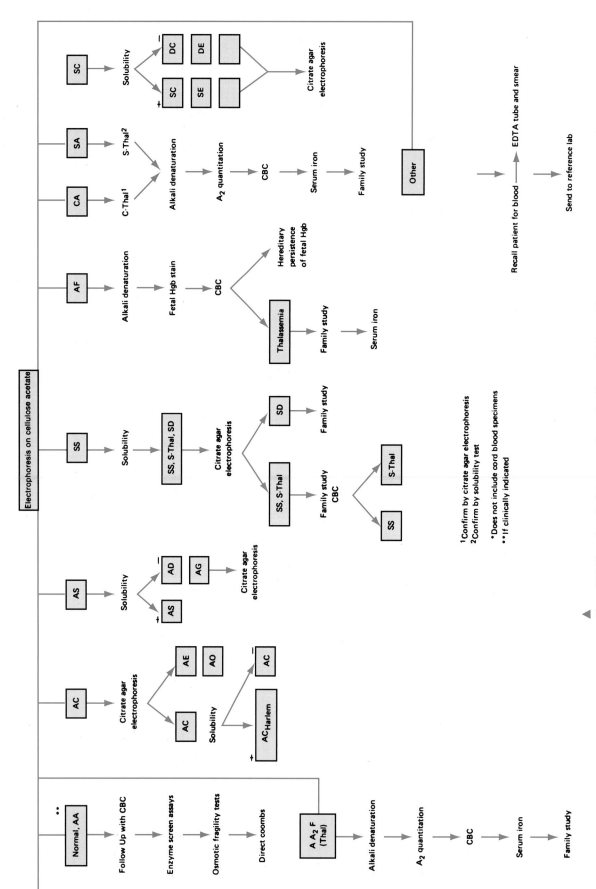

Figure 13-8. Flow sheet for laboratory diagnosis of hemoglobinopathies.

tive information on the genotype of individuals tested. In some cases, direct detection of the molecular lesions is possible. Specific techniques are discussed elsewhere.[28,32,33,50]

A special strength of the DNA technology is in the prenatal diagnosis of thalassemia major. Because the globin genes are represented in all tissues, including those in which they are not active, prenatal diagnosis of thalassemic states may be made by sampling tissues that are relatively easy to obtain, such as chorionic villi or amniotic fluid cells, rather than fetal blood, which is obtained with much greater difficulty and at a much greater risk to the fetus.[32]

DNA technology also has been used in the prenatal diagnosis of sickle cell anemia. Fetal cells obtained by amniocentesis or chorionic villus sampling may be analyzed using similar techniques as in the thalassemias. Hemoglobin electrophoresis on hemolysate of fetal blood cells also can be used in cases in which DNA probe technology is unavailable or when rapid results are needed owing to a patient's advanced gestational age.[32,33]

MYOGLOBIN

Structure and Role in the Body

Myoglobin is a heme protein found only in skeletal and cardiac muscle in humans. It can reversibly bind oxygen in a manner similar to the hemoglobin molecule, but myoglobin is unable to release oxygen, except under very low oxygen tension. Myoglobin is a simple heme protein containing one polypeptide chain and one heme group per molecule. The polypeptide chain contains 153 amino acids, making it slightly larger than one chain in the hemoglobin molecule. Therefore, its size is slightly larger than one fourth that of a hemoglobin molecule, with a molecular weight of approximately 17,000. The iron atom in the center of the heme group is the site of reversible oxygen binding, identical to the hemoglobin molecule. In the body, myoglobin acts as an oxygen carrier in the cytoplasm of the muscle cell. Transport of oxygen from the muscle cell membrane to the mitochondria is its main role.

Clinical Significance

Damage to muscles often results in elevated levels of serum and urine myoglobin (Table 13-2). Renal clearance is very rapid, and myoglobinemia following a single injury tends to be transient. In crush injuries, the myoglobin filtered through the kidneys may be sufficient to cause renal occlusion and death.

Measurement of myoglobin has been advocated to assist in the diagnosis of myocardial infarction. With the recent advent of thrombolytic therapy, the need for rapid and precise diagnosis of myocardial infarction has increased greatly. The damaged heart muscle cells release myoglobin within the first few hours of onset of myocardial infarction, and peak values are reached sooner than peak values of the enzyme creatine kinase. Due to the small molecular weight of myoglobin, it is able to diffuse more quickly through injured cell membranes.

TABLE 13-2. Causes of Elevations of Myoglobin
Acute myocardial infarction
Angina without infarction
Rhabdomyolysis
Multiple fractures; muscle trauma
Renal failure
Myopathies
Vigorous exercise
Intramuscular injections
Open heart surgery
Tonic–clonic seizures
Electric shock
Arterial thrombosis
Certain toxins

Therefore, an increase of myoglobin in the circulation provides an early indicator of myocardial infarction.[51] However, because myoglobin from cardiac muscle may be immunochemically identical to myoglobin liberated from skeletal muscle, false-positive results may occur from accompanying minor injury to skeletal muscle.[52] Despite the poor specificity of myoglobin for myocardium, skeletal muscle damage can be ruled out in many cases. Negative myoglobin results within the first few hours after chest pain can be used to rule out myocardial infarction. In cases in which thrombolytic therapy is used in the treatment of myocardial infarction, myoglobin levels may be used as an indicator of successful reperfusion of the occluded artery.[53] Myoglobin also has been investigated to aid in the diagnosis and differentiation of the different types of hereditary progressive muscular dystrophy.[54] Myoglobin is discussed in greater detail in Chapter 8 *Amino Acids and Proteins*.

Methodology

There are several methods for measurement and identification of myoglobin. Radioimmunoassay is the most specific method at present and can detect as little as 0.5 ng/mL of myoglobin. Latex agglutination, enzyme-linked immunosorbent assay (ELISA), immunonephelometry, and fluoroimmunoassays for myoglobin have been developed. These have the advantage of being less time-consuming and easier to perform than radioimmunoassay procedures, making them especially useful as emergency screening tests in the diagnosis of myocardial infarction.[52,55,56]

SUMMARY

Porphyrins are the intermediates in the multistep synthesis of heme, an iron chelating group that binds oxygen. The synthesis of heme is started in the mitochondrion by combination of glycine and succinyl CoA to form ALA. PBG is formed from ALA in the cytoplasm. ALA and PBG are precursors to the formation of porphyrins. Uroporphyrinogen and coproporphyrinogen are formed in the cytoplasm and protoporphyrinogen is synthesized in the mitochondrion. PROTO is then formed, which incorporates iron to form heme. ALA synthase is the rate-controlling enzyme of heme synthesis and is regulated by the amount of heme. Enzyme deficiencies can occur at almost every step of heme synthesis, resulting in a group of inherited disorders called the porphyrias. A buildup of intermediates occurs, which can cause cutaneous symptoms, neuropsychiatric symptoms, or both. Blocks in the early steps of heme synthesis tend to produce cutaneous symptoms, whereas buildups of later intermediates cause neuropsychiatric symptoms. Most porphyrias can be differentiated by laboratory analysis of ALA, PBG, URO, COPRO, and PROTO.

Hemoglobin is synthesized in immature erythroid cells in the bone marrow and functions in carrying oxygen to the tissues. Hemoglobin comprises two pairs of globin proteins, each containing a heme molecule capable of carrying oxygen. There are six types of globin chains: alpha, beta, gamma, delta, epsilon, and zeta, with epsilon and zeta being present only in the embryo. In the healthy adult, three types of hemoglobin are present: A ($\alpha_2\beta_2$), A_2 ($\alpha_2\delta_2$), and F ($\alpha_2\lambda_2$). Hemoglobinopathies result when defects are present in the structure of hemoglobin. These defects are caused by amino acid substitutions or deletions, globin chain elongation, and globin chain fusion, or hybridization. The most common hemoglobinopathy is hemoglobin S, which can be inherited as heterozygous (AS, sickle cell trait) or homozygous (SS, sickle cell disease). The deoxygenated form of hemoglobin S is relatively insoluble in vitro and can be detected by the dithionite solubility test. Heterozygotes are asymptomatic, whereas homozygotes may have anemia and compromised microcirculation. This and other hemoglobinopathies can be detected by a combination of alkaline and acid electrophoresis. Thalassemias are inherited disorders in the rate of synthesis of one or more globin chains, usually due to gene deletions or point mutations. Defects in the rate of alpha-globin synthesis are termed α-thalassemias, and β-thalassemia denotes defective beta-globin production. The heterozygous state is termed thalassemia minor and the homozygous state is called thalassemia major. These disorders are common in Asian, Black African, African-American, Indian, and Middle Eastern populations and cause anemias of varying degrees.

Myoglobin is a heme-containing protein present in skeletal and cardiac muscle. Its presence in serum or urine indicates skeletal or cardiac muscle damage. Because of its relatively small molecular weight, it is exuded into the plasma within a few hours after muscle damage. In this respect, it is useful as a marker of myocardial infarction or as a marker of reperfusion after thrombolytic therapy.

REVIEW QUESTIONS

1. The main purpose of porphyrins in the body is to:
   a. Carry oxygen to the tissues
   b. Transport iron
   c. Combine with free hemoglobin
   d. Contribute to the synthesis of heme

2. The two main sites in the body for accumulation of excess porphyrins are:
   a. Liver and bone marrow
   b. Heart and lung
   c. Muscle and blood
   d. Liver and spleen

3. The two main classes of porphyrias according to symptoms are:
   a. Erythropoietic and hepatic
   b. Neurologic and cutaneous
   c. Congenital and acquired
   d. Hematologic and muscular

4. Porphobilinogen is most commonly quantitated in the urine by:
   a. The Watson-Schwartz method
   b. Thin-layer chromatography
   c. Ion-exchange column
   d. Electrophoresis

5. Extremely high levels of ALA and PBG in the urine with normal porphyrin levels in the feces and blood most likely indicates:
   a. Acute intermittent porphyria (AIP)
   b. Erythropoietic porphyria (EP)
   c. Hereditary coproporphyria (HCP)
   d. Porphyria cutanea tarda (PCT)

6. Inherited disorders in which a genetic defect causes abnormalities in rate and quantity of synthesis of structurally normal polypeptide chains of the hemoglobin molecule are called:
   a. Hemoglobinopathies
   b. Porphyrias
   c. Molecular dyscrasias
   d. Thalassemias

7. A *decrease* in which of the following indicates increased intravascular hemolysis:
   a. Methemoglobin
   b. Methemalbumin
   c. Haptoglobin
   d. Hemopexin

8. Which of the following abnormal hemoglobins is found frequently in individuals from Southeast Asia and migrates with hemoglobin A_2 on cellulose acetate electrophoresis:
   a. Hemoglobin D
   b. Hemoglobin E
   c. Hemoglobin C
   d. Hemoglobin Lepore

9. Which type of α-thalassemia results from deletion of three genes and produces a moderate hemolytic anemia:
   a. Hemoglobin H disease
   b. Hemoglobin Bart's
   c. Hydrops fetalis
   d. Thalassemia trait

10. The most effective way to quantitate hemoglobin A_2 is by:
    a. Densitometry
    b. Citrate agar electrophoresis
    c. Alkali denaturation test
    d. Column chromatography

REFERENCES

1. Schreiber WE. Iron, porphyrin, and bilirubin metabolism. In: Kaplan LA, Pesce AJ, eds. Clinical chemistry—theory, analysis, correlation. 3rd ed. St. Louis, MO: CV Mosby, 1996;696.
2. Paslin DA. The porphyrias. Int J Dermatol 1992;31:527.
3. Kappas A, Sassa S, Galbraith RA, et al. The porphyrias. In: Scriver CR, Beaudet AL, Sly WS, et al, eds. The metabolic and molecular bases of inherited disease. 7th ed. New York: McGraw-Hill, 1995;2103.
4. Nuttall KL. Porphyrins and disorders of porphyrin metabolism. In: Burtis CA, Ashwood ER, eds. Tietz fundamentals of clinical chemistry. 4th ed. Philadelphia: WB Saunders, 1996;731.
5. Bloomer JR, Bonkovsky HL. The porphyrias. Dis Mon 1989;35:1.
6. Elder GH. Differentiation of porphyria cutanea tarda symptomatica from other types of porphyria by measurement of isocoproporphyrin in feces. J Clin Pathol 1975;28:601.
7. Tefferi A, Solberg LA, Ellefson RD. Porphyrias: clinical evaluation and interpretation of laboratory tests. Mayo Clin Proc 1994;69:289.
8. Elder GH. The cutaneous porphyrias. Semin Dermatol 1990;9:63.
9. Kiechle FL. The porphyrias. In: Bick RL, ed. Hematology—clinical and laboratory practice. St. Louis, MO: CV Mosby, 1993;553.
10. Nuttall KL. Porphyrins and disorders of porphyrin metabolism. In: Burtis CA, Ashwood ER, eds. Tietz textbook of clinical chemistry. 2nd ed. Philadelphia: WB Saunders, 1994;2073.
11. Buttery JE, Chamberlain BR, Beng CG. A sensitive method of screening for urinary porphobilinogen. Clin Chem 1989;35:2311.
12. Watson CJ, Schwartz S. A simple test for urinary porphobilinogen. Proc Soc Exp Biol Med 1941;47:393.
13. Lamon JM, With TK, Redeker AG. The Hoesch test: bedside screening for urinary porphobilinogen in patients with suspected porphyria. Clin Chem 1974;20:1438.
14. Mauzerall D, Granick S. The occurrence and determination of D-aminolevulinic acid and porphobilinogen in urine. J Biol Chem 1956;219:435.
15. Pinkham JM, Southall DM. Unusual case of suspected acute porphyria. Clin Chem 1995;41:323.
16. McCarroll NA. Diseases of metabolism (porphyrias). Anal Chem 1995;67:425R.
17. Cripps DJ, Peters HA. Fluorescing erythrocytes and porphyrin screening test in urine, stool and blood. Arch Dermatol 1967;96:712.
18. Haining RG, Hulse T, Labbe RF. Rapid porphyrin screening of urine, stool and blood. Clin Chem 1969;15:460.
19. Elder GH, Smith SG, Smyth SJ. Laboratory investigation of the porphyrias. Ann Clin Biochem 1990;27:395.
20. Lim CK, Li F, Peters TJ. High-performance liquid chromatography of porphyrins. J Chromatogr 1988;429:123.
21. Lai C-K, Lam CW, Chan YW. High-performance thin-layer chromatography of free porphyrins for diagnosis of porphyria. Clin Chem 1994;40:2026.
22. Schreiber WE. A molecular view of the neurologic porphyrias. Clin Lab Med 1997;17:73.
23. McKenzie SM. Textbook of hematology. 2nd ed. Philadelphia: Lea & Febiger, 1996;44,159,172.
24. Beutler E. The sickle cell diseases and related disorders. In: Beutler E, Lichtman MA, Coller BS, et al, eds. Williams hematology. 5th ed. New York: McGraw-Hill, 1995;616.
25. Dumars KW, et al. Practical guide to the diagnosis of thalassemia. Am J Med Gen 1996;62:29.
26. Weatherall DJ. The thalassemias. In: Beutler E, Lichtman MA, Coller BS, et al, eds. Williams hematology. 5th ed. New York: McGraw-Hill, 1995;581.
27. Lucarelli G, et al. Marrow transplantation for patients with thalassemia: results in class 3 patients. Blood 1996;87:2082.
28. Reddy PL, Bowie LJ. Sequence-based diagnosis of hemoglobinopathies in the clinical laboratory. Clin Lab Med 1997;17:85.
29. Schneider RG. Differentiation of electrophoretically similar hemoglobins—such as S, D, G, and P; or A₂, C, E, and O—by electrophoresis of the globin chains. Clin Chem 1974;20:1111.
30. Rao VB, Bannerjee MK, Shewale LH, et al. Detection of haemoglobin variants for the diagnosis of beta thalassemia and other hemoglo-

binopathies using anion exchange high performance liquid chromatography. Ind J Med Res 1996;104:365.
31. Eastman JW, Wong R, Liao CL, et al. Automated HPLC screening of newborns for sickle cell anemia and other hemoglobinopathies. Clin Chem 1996;42:704.
32. Old J. Haemoglobinopathies. Prenatal Diag 1996;16:1181.
33. Embry SH. Advances in the prenatal and molecular diagnosis of the hemoglobinopathies and thalassemias. Hemoglobin 1995;19:237.
34. Nalbandian RM, et al. Dithionite tube test—a rapid, inexpensive technique for the detection of hemoglobin. Clin Chem 1971;17:1028.
35. National Committee for Clinical Laboratory Standards. Solubility test to confirm the presence of sickling hemoglobins. 2nd ed; approved standard; document H10-A2. Villanova, PA: National Committee for Clinical Laboratory Standards, 1995.
36. Briere RO, Golias T, Gatsakis JG. Rapid qualitative and quantitative hemoglobin fractionation. Am J Clin Pathol 1965;44:695.
37. Graham JL, Grunbaum BW. A rapid method for microelectrophoresis and quantitation of hemoglobins on cellulose acetate. Am J Clin Pathol 1963;39:567.
38. National Committee for Clinical Laboratory Standards. Detection of abnormal hemoglobin using cellulose acetate electrophoresis. 2nd ed; approved standard; document H8-A2. Villanova, PA: National Committee for Clinical Laboratory Standards, 1994.
39. Milner PF, Gooden H. Rapid citrate-agar electrophoresis in routine screening for hemoglobinopathies using a simple hemolysate. Am J Clin Pathol 1975;64:58.
40. National Committee for Clinical Laboratory Standards. Citrate agar electrophoresis for confirming the identification of variant hemoglobins. Tentative guideline; document H23-T. Villanova, PA: National Committee for Clinical Laboratory Standards, 1988.
41. Efremov GD, Huisman THJ, Bowman K, et al. Microchromatography of hemoglobin: II. A rapid method for the determination of hemoglobin A₂. J Lab Clin Med 1974;83:657.
42. Huisman THJ, Schroeder WA, Brodie AN, et al. Microchromatography of hemoglobins: II. A simplified procedure for the determination of hemoglobin A₂. J Lab Clin Med 1975;86:700.
43. Muller CJ, Pik C. A simple and rapid method for the quantitative determination of hemoglobin-A₂. Clin Chem Acta 1962;7:92.
44. National Committee for Clinical Laboratory Standards. Chromatographic (microcolumn) determination of hemoglobin A₂. Approved standard; document H9-A. Villanova, PA: National Committee for Clinical Laboratory Standards, 1989.
45. Tan GB, et al. Evaluation of high performance liquid chromatography for routine estimation of haemoglobins A₂ and F. J Clin Pathol 1993;46:852.
46. Clayton E, Foster BE, Clayton EP. New stain for fetal erythrocytes in peripheral blood smears. Obstet Gynecol 1970;35:642.
47. Betke K, Marti HW, Shlicht I. Estimation of small percentage of fetal haemoglobin. Nature 1959;184:1877.
48. Huisman THJ. Normal and abnormal hemoglobins. Adv Clin Chem 1963;6:231.
49. National Committee for Clinical Laboratory Standards. Quantitative measurement of fetal hemoglobin using the alkali denaturation method. Approved guideline; document H13-A. Villanova, PA: National Committee for Clinical Laboratory Standards, 1989.
50. Sutcharitchan P, Embury SH. Advances in molecular diagnosis of inherited hemoglobin disorders. Curr Opin Hematol 1996;3:131.
51. Henderson AR. An overview and ranking of biochemical markers of cardiac disease. Clin Lab Med 1997;17:625.
52. Castaldo AM, et al. Plasma myoglobin in the early diagnosis of acute myocardial infarction. Eur J Clin Chem Clin Biochem 1994;32:349.
53. Zabel M, et al. Analysis of creatine kinase, CK-MB, myoglobin, and troponin T time-activity curves for early assessment of coronary artery reperfusion after intravenous thrombolysis. Circulation 1993; 87:1542.
54. Poche H, et al. Hereditary progressive muscular dystrophies: serum myoglobin pattern in patients with different types of muscular dystrophies. Clin Physiol Biochem 1989;7:40
55. Vaidya HC. Myoglobin. Lab Med 1992;23:306.
56. Tucker JF, et al. Value of serial myoglobin levels in the early diagnosis of patients admitted for acute myocardial infarction. Ann Emerg Med 1994;24:704.

CHAPTER 14

Electrolytes

Joan E. Polancic

Objectives

Upon completion of this chapter, the clinical laboratorian should be able to:

- *Define the following terms: electrolyte, osmolality, anion gap, anion, cation.*
- *Discuss the physiology of each electrolyte described in the chapter.*
- *State the clinical significance of each of the electrolytes mentioned in the chapter.*
- *Calculate osmolality, osmolal gap, and an anion gap and discuss the clinical usefulness of each.*
- *Discuss the analytical techniques used to assess electrolyte concentrations.*
- *Given patient data, correlate the information with disease state.*
- *Identify the reference ranges for sodium, potassium, chloride, bicarbonate, magnesium, and calcium.*
- *State the specimen of choice for the major electrolytes.*
- *Discuss the role of the kidney in electrolyte excretion and conservation in a healthy individual.*
- *Discuss the usefulness of urine electrolyte results: sodium, potassium, calcium, and osmolality.*

KEY TERMS

Active transport	Hyperkalemia	Hypovolemia
Anion	Hypermagnesemia	Intracellular fluid
Anion gap	Hypernatremia	(ICF)
Cation	Hyperphosphatemia	Osmolal gap
Diffusion	Hypocalcemia	Osmolality
Electrolyte	Hypochloremia	Osmolarity
Extracellular fluid	Hypokalemia	Osmometer
(ECF)	Hypomagnesemia	Polydipsia
Hypercalcemia	Hyponatremia	Tetany
Hyperchloremia	Hypophosphatemia	

*E*lectrolytes are ions capable of carrying an electric charge. They are classified as anions or cations based on the type of charge they carry. These names were determined years ago based on how the ion migrates in an electric field. *Anions* have a negative charge and move toward the anode, whereas *cations* migrate in the direction of the cathode because of their positive charge.

The numerous processes in which electrolytes are an essential component are volume and osmotic regulation (Na, Cl, K); myocardial rhythm and contractility (K, Mg, Ca); cofactors in enzyme activation (Mg, Ca, Zn, and so forth); regulation of adenosine triphosphatase (ATPase) ion pumps (Mg); acid–base balance (HCO_3, K, Cl); blood coagulation (Ca, Mg); neuromuscular excitability (K, Ca, Mg); and the production and use of ATP from glucose (Mg, PO_4, and so forth). Because many of the functions listed here require electrolyte concentrations to be held within narrow ranges, the body has complex systems for monitoring and maintaining the concentrations of electrolytes.

This chapter explores both the metabolic physiology and regulation of each electrolyte and relates these factors to the clinical significance of electrolyte measurements. In addition, methodologies used in determining concentrations of the individual analytes are discussed.

WATER

The average water content of the human body varies from 40% to 75% of the total body weight, with values declining with age and especially with obesity. Women have lower average water content than men due to a higher fat content. Water is the solvent for all processes in the human body. It transports nutrients to cells, determines cell volume by its transport into and out of cells, removes waste products by way of urine, and acts as the body's coolant by way of sweating. Water is located in both intracellular and extracellular compartments. The *intracellular fluid (ICF)* is fluid inside the cells and accounts for about ⅔ of total body water. *Extracellular fluid (ECF)* accounts for the other ⅓ of total body water and can be subdivided in the *intravascular extracellular fluid (plasma)* and the *interstitial cell fluid* that surrounds the cells in the tissues. Normal plasma is about 93% water, with the remaining volume occupied by lipids and proteins. The concentrations of ions within cells and in plasma are maintained both by energy-consuming active transport processes and by diffusion or passive transport processes.

Active transport is a mechanism that requires energy to move ions across cellular membranes. For example, maintaining a high intracellular concentration of potassium and a high extracellular (plasma) concentration of sodium requires use of energy from ATP in ATPase-dependent ion pumps. *Diffusion* is the passive movement of ions across a membrane. It depends on the size and charge of the ion being transported and on the nature of the membrane through which it is passing. The rate of diffusion of various ions also may be altered by physiologic and hormonal processes.

By maintaining the concentration of proteins and electrolytes in a controlled yet somewhat flexible environment, the distribution of water in these compartments also can be controlled. Because most biologic membranes are freely permeable to water, but not to ions or proteins, the concentration of ions and proteins on one side of the membrane or another will influence the flow of water across a membrane (an osmoregulator). In addition to the osmotic effects of sodium, other ions, proteins, and blood pressure influence the flow of water across a membrane.

Osmolality

Osmolality is a physical property of a solution, which is based on the concentration of solutes (expressed as millimoles) per kilogram of solvent (w/w). Osmolality is related to several changes in the properties of a solution relative to pure water, such as freezing point depression and vapor pressure decrease. These colligative properties (see Chapter 1) are the basis for routine measurements of osmolality in the laboratory. The term *osmolarity* is still occasionally used, with results reported in milliosmoles per liter (w/v), but it is inaccurate in cases of hyperlipidemia or hyperproteinemia, for urine specimens, or in the presence of certain osmotically active substances, such as alcohol or mannitol. Both the sensation of thirst and antidiuretic hormone (ADH) secretion are stimulated by the hypothalamus in response to an increased osmolality of blood. The natural response to the thirst sensation is to consume more fluids, thus increasing the water content of the ECF, diluting out the elevated solute (sodium) levels, and decreasing the osmolality of the plasma. Thirst, therefore, is important in mediating fluid intake. The other means of controlling osmolality is by secretion of ADH (vasopressin). This hormone is secreted by the posterior pituitary gland and acts on the cells of the collecting ducts in the kidneys to increase water reabsorption. As

water is conserved, osmolality decreases, turning off ADH secretion.[1]

Clinical Significance of Osmolality

Osmolality in plasma is important because it is the parameter to which the hypothalamus responds. The regulation of osmolality also affects the sodium concentration in plasma, largely because sodium and its associated anions account for approximately 90% of the osmotic activity in plasma. Another important process affecting the sodium concentration in blood is the regulation of blood volume. As discussed later, although osmolality and volume are regulated by separate mechanisms (except for ADH and thirst), they are related because osmolality (sodium) is regulated by changes in water balance, whereas volume is regulated by changes in sodium balance.[1,2]

To maintain a normal plasma osmolality (~275–295 mOsm/kg of plasma H_2O), osmoreceptors in the hypothalamus respond quickly to small changes in osmolality: A 1% to 2% increase in osmolality causes a four-fold increase in the circulating concentration of ADH, and a 1% to 2% decrease in osmolality shuts off ADH production. ADH acts by increasing the reabsorption of water in the cortical and medullary collecting tubules. ADH has a half-life in the circulation of only 15 minutes to 20 minutes.

Renal water regulation by ADH and thirst each play important roles in regulating plasma osmolality. Renal water excretion is more important in controlling water excess, whereas thirst is more important in preventing water deficit or dehydration. Consider what happens in several conditions.

Water Load

As excess intake of water (*eg*, in polydipsia) begins to lower plasma osmolality, both ADH and thirst are suppressed. In the absence of ADH, water is not reabsorbed, causing a large volume of dilute urine to be excreted, as much as 10 L to 20 L of water daily, well above any normal intake of water. Therefore, hypo-osmolality and hyponatremia usually occurs only in patients with impaired renal excretion of water.[1]

Water Deficit

As a deficit of water begins to increase plasma osmolality, both ADH secretion and thirst are activated. Although ADH contributes by minimizing renal water loss, thirst is the major defense against hyperosmolality and hypernatremia. Although hypernatremia rarely occurs in a person with a normal thirst mechanism and access to water, it becomes a concern in infants, unconscious patients, or anyone who is unable to either drink or ask for water. Osmotic stimulation of thirst progressively diminishes in people who are older than age 60 years. Particularly in the older patient with illness and diminished mental status, dehydration becomes increasingly likely. As an example of the effectiveness of thirst in preventing dehydration, a patient with diabetes insipidus (no ADH) may excrete 10 L of urine per day. However, because thirst persists, water intake matches output and plasma sodium remains normal.[1]

Regulation of Blood Volume

Adequate blood volume is essential to maintain blood pressure and ensure good perfusion to all tissues and organs. Regulation of both sodium and water are interrelated in controlling blood volume. The renin-angiotensin-aldosterone system responds primarily to a decreased blood volume. Renin is secreted near the renal glomeruli in response to decreases renal blood flow (decreased blood volume or blood pressure). Renin converts angiotensinogen to angiotensin I, which then becomes angiotensin II. Angiotensin II causes both vasoconstriction, which quickly increases blood pressure, and secretion of aldosterone, which increases retention of sodium and the water that accompanies the sodium. The effects of blood volume and osmolality on sodium and water metabolism are shown in Figure 14-1. Changes in blood volume (actually pressure) are initially detected by a series of stretch receptors located in areas such as the cardiopulmonary circulation, carotid sinus, aortic arch, and glomerular arterioles. These receptors then activate a series of responses (effectors) that restore volume by appropriately varying vascular resistance, cardiac output, and renal sodium and water retention.[1]

Four other factors affect blood volume: (1) atrial natriuretic peptide (ANP), released from the myocardial atria in response to volume expansion, promotes sodium excretion in the kidney; (2) volume receptors independent of osmolality stimulate the release of ADH, which conserves water by renal reabsorption; (3) glomerular filtration rate (GFR) increases with volume expansion and decreases with volume depletion; (4) all other things equal, an increased plasma sodium will increase urinary sodium excretion, and vice

versa. The normal reabsorption of 98% to 99% of filtered sodium by the tubules conserves nearly all of the 150 L of glomerular filtrate produced daily. A 1% to 2% reduction in tubular reabsorption of sodium can increase water loss by several liters per day.

Urine osmolality values may vary widely depending on water intake and the circumstances of collection. However, it is generally decreased in diabetes insipidus (inadequate ADH) and *polydipsia* (excessive H_2O intake due to chronic thirst), and increased in conditions such as the syndrome of inappropriate ADH secretion (SIADH) and hypovolemia (although urinary Na is usually decreased).

Determination of Osmolality

Specimen

Osmolality may be measured in serum or urine. The use of plasma is not recommended because osmotically active substances may be introduced into the specimen from the anticoagulant.

Discussion

The methods for determining osmolality are based on properties of a solution that are related to the number of molecules of solute per kilogram of solvent (colligative properties), such as changes in freezing point and vapor pressure. An increase in osmolality decreases the freezing-point temperature and the vapor pressure. Measurement of freezing-point depression and vapor pressure decrease (actually the "dewpoint") are the two most frequently used methods of analysis.[3] There are several different osmometers now on the laboratory market. For detailed information on theory and methodology, consult Chapter 4, *Analytical Techniques and Instrumentation,* or the operator's manual of the instrument in question.

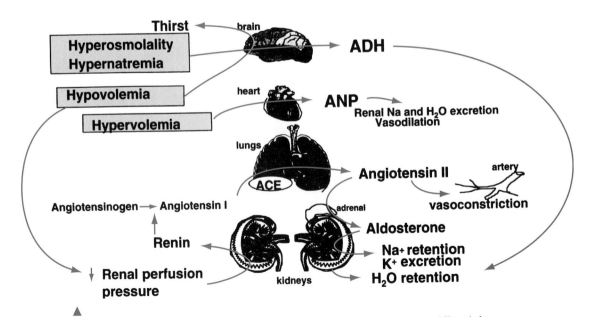

Figure 14-1. Responses to changes in blood osmolality and blood volume. ADH, antidiuretic hormone; ANP, atrial natriuretic peptide. The primary stimuli are shown in boxes (*eg,* hypovolemia).

Samples must be free of particulate matter to obtain accurate results. Serum and urine samples that are turbid should be centrifuged before analysis to remove any extraneous particles. If reusable sample cups are used, they should be thoroughly cleaned and dried between each use to prevent contamination.

Osmometers that operate by freezing-point depression are standardized using sodium chloride reference solutions. After calibration, the appropriate amount of sample is pipetted into the required cuvet or sample cup and placed in the analyzer. The sample is then supercooled to –7°C and seeded to initiate the freezing process. Once temperature equilibrium has been reached, the freezing point is measured, with results for serum and urine osmolality reported as milliosmoles per kilogram.

Calculation of osmolality has some usefulness either as an estimate of the true osmolality or to determine the *osmolal gap,* which is the difference between the measured osmolality and the calculated osmolality. The osmolal gap indirectly indicates the presence of osmotically active substances other than sodium, urea, or glucose, such as ethanol, methanol, ethylene glycol, lactate, or β-hydroxybutyrate.

Two formulas are presented here, each having theoretical advantages and disadvantages. Both are adequate for the purpose previously described. For more discussion, the reader may consult other references.[4]

$$2\,Na + \frac{glucose(mg/dL)}{20} + \frac{BUN(mg/dL)}{3}$$

(Eq. 14–1)

$$1.86\,Na + \frac{glucose}{18} + \frac{BUN}{2.8} + 9$$

Reference Ranges[5]
See Table 14-1.

THE ELECTROLYTES

Sodium

Sodium is the most abundant cation in the ECF, representing 90% of all extracellular cations, and largely determines the osmolality of the plasma. A normal plasma osmolality is approximately 295 mmol/L, with 270 mmol/L being the result of sodium and associated anions.

TABLE 14-1. Reference Ranges for Osmolality

Serum	275–295 mOsmol/kg
Urine (24-hour)	300–900 mOsmol/kg
Urine:serum ratio	1.0–3.0
Osmolal gap	<15

Sodium concentration in the ECF is much larger than inside cells. Because a small amount of sodium can diffuse through the cell membrane, the two sides would eventually reach equilibrium. To prevent equilibrium from occurring, active transport systems such as ATPase ion pumps are present in all cells. Potassium, as discussed in the section *Potassium,* is the major intracellular cation. Like sodium, potassium would eventually diffuse across the cell membrane until equilibrium is reached. The Na-K ATPase ion pump moves three sodium ions out of the cell in exchange for two potassium ions moving into the cell as ATP is converted to ADP. Because water follows electrolytes across cell membranes, the continual removal of sodium from the cell prevents osmotic rupture of the cell by also drawing water from the cell.

Regulation

The plasma sodium concentration depends greatly on the intake and excretion of water and, to a somewhat lesser degree, the renal regulation of Na. Three processes are of primary importance: (1) the intake of water in response to thirst, as stimulated or suppressed by plasma osmolality; (2) the excretion of water, largely affected by ADH release in response to changes in either blood volume or osmolality; and (3) the blood volume status, which affects sodium excretion through aldosterone, angiotensin II, and ANP. The kidneys have the ability to conserve or excrete large amounts of sodium, depending on the sodium content of the ECF and the blood volume. Normally, 60% to 75% of filtered sodium is reabsorbed in the proximal tubule; electroneutrality is maintained by either Cl reabsorption or H ion secretion. Some sodium is also reabsorbed in the loop and distal tubule and (under the control of aldosterone) is exchanged for K in the connecting segment and cortical collecting tubule. The regulation of osmolality and volume has been summarized in Figure 14-1.

Clinical Applications

Hyponatremia. Because extracellular sodium levels are influenced by total blood volume, the pathogenesis of *hyponatremia* is perhaps most easily understood by categorizing the causes of hyponatremia with the associated blood volume status, as shown in Table 14-2. Hypovolemic hyponatremia results from sodium loss in excess of water loss. Decreased volume status (hypovolemia) may be assessed by poor skin turgor (fullness), decreased jugular venous pressure, and dry mucous membranes.[6] Urinary sodium levels can assist in differentiating the source of fluid loss. If urine sodium is more than 20 mmol/day, renal loss of sodium and water is occurring.[6] There are several common causes of this condition that may be grouped into three main categories: renal loss, extrarenal loss, or cellular shift. The use of thiazide diuretics (but not loop diuretics) induces sodium and K loss without interfering with ADH-mediated water retention. Potassium depletion, which favors intracellular loss of K to the blood, may also cause sodium loss, and cellular loss of K promotes

TABLE 14-2. Hyponatremia Related to Blood Volume Status and Urine Sodium

Hypovolemia Urine Na >20 mmol/day (Renal Loss)
  Diuretics
  Potassium depletion
  Aldosterone deficiency
  Ketonuria
  Salt-losing nephropathy

Hypovolemia Urine Na <20 mmol/day (Extrarenal Loss or Cellular Shift)
  Vomiting
  Diarrhea
  Excess fluid loss as with burns, excess sweating, or trauma
  Potassium depletion

Normovolemia
  SIADH
  Pseudohyponatremia
  Excess water intake
  Adrenal insufficiency
  Reset osmostat

Hypervolemia Urine Na >20 mmol/day
  Acute or chronic renal failure

Hypervolemia Urine Na <20 mmol/day
  Nephrotic syndrome
  Hepatic cirrhosis
  Congestive heart failure

sodium movement into the cell with an associated decrease in plasma sodium and volume. Aldosterone deficiency increases renal loss of sodium and water, with sodium loss in excess of water loss. In diabetes mellitus, sodium loss occurs with ketonuria. Furthermore, salt-losing nephropathy may infrequently develop in renal tubular and interstitial diseases, such as medullary cystic and polycystic kidney disease, usually as renal insufficiency becomes severe [serum creatinine >610 mmol/L (>8 mg/dL)]. Hypovolemic hyponatremia accompanied by a urinary sodium of less than 20 mmol/day is caused by extrarenal loss of hypotonic fluid as with prolonged vomiting, diarrhea, sweating, burns, or trauma.[6] Thirst is stimulated by the hypovolemia, resulting in replacement by relatively more hypotonic fluid.

Normovolemic hyponatremia typically indicates a problem with water balance and can have the following six causes: (1) Syndrome of inappropriate ADH secretion (SIADH) in which there is a defect in ADH regulation. The most common causes are malignancies, pulmonary disease, and central nervous system (CNS) disorders.[6] Infections and trauma are also associated with SIADH. Up to 35% of hospitalized patients with acquired immunodeficiency syndrome have SIADH caused by *Pneumocystis carinii* pneumonia, CNS infection, or malignancy.[6] SIADH initiates mild hypervolemia, which then leads to the release of ANP, which causes excretion of sodium and water by inhibiting renal Na/K ATPase.[7] Urine sodium is usually more than 20 mmol/day.[6]

(2) Artifactual hyponatremia can occur in cases of severe hyperlipidemia or hyperproteinemia. Methods that dilute plasma before sodium analysis by ion-selective electrode (ISE) or flame-emission spectrophotometry (FES) give erroneously low sodium results on these samples because they measure mmol/L sodium per liter of plasma. "Direct" methods by ISEs, which do not dilute plasma or whole blood, give accurate sodium results because they detect the sodium concentration only in the plasma water. (3) Severe hyperglycemia induces water movement into plasma to normalize plasma osmolality, which results in hyponatremia. (4) Chronic excess intake of water, as with polydipsia (chronic thirst), can eventually lead to hyponatremia that is usually mild but occasionally severe. (5) Adrenal insufficiency can decrease cortisol and aldosterone, which contributes to the development of hyponatremia. Because cortisol usually inhibits ADH release, a deficiency of cortisol promotes ADH release and water retention. Although the initial phase of this process results in hypovolemia, the ADH-induced water retention typically restores volume status to normal.[1] (6) In pregnancy, the hypothalamic setpoint for osmolality is offset such that the plasma sodium concentration is regulated approximately 5 mmol/L lower than normal. This may be initiated by vasodilation, which leads to an ADH and possibly an aldosterone response to apparent hypovolemia.

Hypervolemic hyponatremia is nearly always a problem of water overload, which usually causes edema. Urine sodium levels can assist in differentiating the cause. A urine sodium of more than 20 mmol/day is associated with acute or chronic renal failure. Urine sodium of less than 20 mmol/day is associated with nephrotic syndrome, cirrhosis, or congestive heart failure.[6] The usual therapy is water restriction. Sodium should not be given, because it could increase the severity of edema. For example, congestive heart failure or hepatic cirrhosis increases venous backpressure in the circulation, which promotes movement of fluid from the blood to the interstitium, causing edema. Volume receptors sense *hypovolemia,* which leads to secretion of ADH and eventual hypervolemia and hyponatremia. Also, in advanced renal disease, the inability to excrete water promotes hypervolemia whenever fluid intake is excessive.

Symptoms of hyponatremia. Symptoms depend on the serum level. Between 125 mmol/L and 130 mmol/L, symptoms are primarily gastrointestinal (GI). Below 125 mmol/L more severe neuropsychiatric symptoms are seen. In general, symptoms include nausea and vomiting, muscular weakness, headache, lethargy, and ataxia. More severe symptoms include seizures, coma, and respiratory depression.[6] Serum electrolytes are monitored as treatment to return sodium levels to normal occurs.

Treatment of hyponatremia. Treatment is directed at correction of the condition that caused either water loss or sodium loss in excess of water loss. Appropriate management

A 48-year-old female was admitted to the hospital following 3 days of severe vomiting. Before this episode, she was reportedly well. Physical findings revealed decreased skin turgor and dry mucous membranes. Admission studies results were as follows:

Plasma

Na^+	130 mmol/L
K^+	5.0 mmol/L
Cl^-	77 mmol/L
HCO_3^-	9 mmol/L

Urine

Na^+	8 mmol/day
Ketones	trace

Questions

1. What is the cause for each of the abnormal plasma electrolyte results?
2. What is the significance of the urine sodium result?

---

of fluid administration is critical. Fluid administration and monitoring is required while the underlying cause of the hyponatremia is treated.

Hypernatremia. *Hypernatremia* (increased serum sodium concentration) results from excess loss of water relative to sodium loss, decreased water intake, or increased sodium intake. Hypernatremia is less commonly seen in hospitalized patients than hyponatremia.[6]

Loss of hypotonic fluid may occur either by the kidney or through profuse sweating, diarrhea, or severe burns. The measurement of urine osmolality is necessary to evaluate the cause of hypernatremia. With renal loss of water, the urine osmolality is low or normal. With extrarenal fluid losses, the urine osmolality is increased. Interpretation of the urine osmolality in hypernatremia is shown in Table 14-3.

Water loss through the skin and by breathing (insensible loss) accounts for about 1 L of water loss per day in adults. Any condition that increases water loss, such as fever, burns, diarrhea, or exposure to heat, will increase the likelihood of developing hypernatremia. Very commonly, hypernatremia occurs in adults with altered mental status and in infants, both of

whom may be thirsty but who are unable to ask for or obtain water. People who cannot fully concentrate their urine, such as neonates, young children, elderly persons, and some patients with renal insufficiency, may show a relatively lower urine osmolality.

Hypernatremia may result from loss of water in diabetes insipidus, either because the kidney cannot respond to ADH (nephrogenic diabetes insipidus) or ADH secretion is impaired (central diabetes insipidus). Diabetes insipidus is characterized by copious production of dilute urine (3–20 L/day). Because people with diabetes insipidus drink large volumes of water, hypernatremia usually does not occur in diabetes insipidus unless the thirst mechanism is also impaired. Partial defects of either ADH release or the response to ADH may also occur. In such cases, urine is concentrated to a lesser extent than appropriate to correct the hypernatremia. Excess water loss may also occur in renal tubular disease, such as acute tubular necrosis, in which the tubules become unable to fully concentrate the urine.

Chronic hypernatremia in an alert patient is indicative of hypothalamic disease, usually with a defect in the osmoreceptors rather than from a true resetting of the osmostat. A reset osmostat may occur in primary hyperaldosteronism, in which excess aldosterone induces mild hypervolemia, which retards ADH release, shifting plasma sodium upward by ~3 mmol/L to 5 mmol/L.[1]

Hypernatremia may be from excess ingestion of salt or administration of hypertonic solutions of sodium, such as sodium bicarbonate or hypertonic dialysis solutions. Neonates are especially susceptible to hypernatremia from this cause. In these cases, ADH response is appropriate, resulting in urine osmolality more than 800 mOsm/kg (see Table 14-3).

Symptoms of hypernatremia. Symptoms most commonly involve the CNS due to the hyperosmolar state. These symptoms include altered mental status, lethargy, irritability, restlessness, seizures, muscle twitching, hyperreflexes, fever, nausea or vomiting, difficult respiration, and increased thirst.

TABLE 14-3. Hypernatremia (150 mmol/L) Related to Urine Osmolality

Urine Osmolality <300 mOsmol/kg
  Diabetes insipidus (impaired secretion of ADH or kidneys cannot respond to ADH)

Urine Osmolality 300–700 mOsmol/kg
  Partial defect in ADH release or response to ADH
  Osmotic diuresis

Urine osmolality >700 mOsmol/kg
  Loss of thirst
  Insensible loss of water (breathing, skin)
  GI loss of hypotonic fluid
  Excess intake of sodium

Serum sodium of more than 160 mmol/L is associated with a mortality rate of 60% to 75%.[6]

Treatment of hypernatremia. Treatment is directed at correction of the condition that caused either water depletion or sodium retention. Because too rapid a correction of serious hypernatremia (>60 mmol/L) can induce cerebral edema and death, hypernatremia must be corrected gradually: the maximal rate should be 0.5 mmol/L per hour.[1]

Determination of Sodium

Specimen. Serum, plasma, and urine are all acceptable for sodium measurements. When plasma is used, lithium heparin, ammonium heparin, and lithium oxalate are suitable anticoagulants. Hemolysis does not cause a significant change in serum or plasma values due to decreased levels of intracellular sodium. However, with marked hemolysis, levels may be decreased due to a dilutional effect.

Whole blood samples may be used with some analyzers. Consult the instrument operator's manual for acceptability. The specimen of choice in urine sodium analyses is a 24-hour collection. Sweat is also suitable for analysis. Sweat collection and analysis is discussed in Chapter 24, *Body Fluid Analysis.*

Methods. Through the years, sodium has been measured in a variety of ways, including chemical methods, FES, atomic absorption spectrophotometry (AAS), and ISE. Chemical methods are outdated because of their large sample volume requirements and lack of precision. Of the methods now readily available, ISE is by far the most routinely used in clinical laboratories.

ISEs use a semipermeable membrane to develop a potential produced by having different ion concentrations on either side of the membrane. In this type of system, two electrodes are used. One electrode has a constant potential, making it the reference electrode. The difference in potential between the reference and measuring electrodes can be used to calculate the "concentration" of the ion in solution. However, it is the activity of the ion, not the concentration, that is being measured (see Chapter 4, *Analytical Techniques and Instrumentation*).

Most analyzers use a glass ion-exchange membrane in its ISE system for sodium measurement (Fig. 14-2). There are two types of ISE measurement, based on sample preparation: direct and indirect. Direct measurement provides an undiluted sample to interact with the ISE membrane. With the indirect method, a diluted sample is used for measurement. There is no significant difference in results, except when samples are hyperlipidemic or hyperproteinemic. Excess lipids or proteins displace plasma water, which leads to a falsely decreased measurement of ionic activity in mmol/L per liter of plasma, whereas the direct method measures in plasma water only. In these cases, direct ISEs are more accurate.

One source of error with ISEs is protein buildup on the membrane through continuous use. The protein-coated membranes cause poor selectivity, which results in poor reproducibility of results.

The Vitros analyzers (Johnson & Johnson Company) use a single-use direct ISE system. Each disposable slide contains a reference and measuring electrode (Fig. 14-3). A drop of sample fluid and a drop of reference fluid are simultaneously applied to the slide, and the potential difference between the two is measured with a voltmeter.[8]

In FES, a sample is aspirated into a flame, producing atoms in an excited state, which are capable of emitting light of a specific wavelength, depending on the element of interest. The intensity of the emitted light can then be correlated to the quantity of ion present in the sample.[9]

Reference Ranges[5]
See Table 14-4.

Potassium

Potassium is the major intracellular cation in the body, with a concentration 20 times greater inside the cells than outside. Many cellular functions require that the body maintain a low ECF concentration of K^+. As a result, only 2% of the body's total potassium circulates in the plasma. Functions of potassium in the body include regulation of neuromuscular excitability, contraction of the heart, ICF volume, and hydrogen ion concentration.[1]

The potassium ion concentration has a major effect on the contraction of skeletal and cardiac muscles. An elevated plasma potassium decreases the resting membrane potential (RMP) of the cell (the RMP is closer to zero), which decreases the net difference between the cell's resting potential and threshold (action) potential. A lower than normal difference increases cell excitability, leading to muscle weakness. Severe hyperkalemia can ultimately cause a lack of muscle excitability (due to a higher RMP than action potential), which may lead to paralysis or a fatal cardiac arrhythmia.[1] Hypokalemia decreases cell excitability by increasing the RMP, often resulting in an arrhythmia or

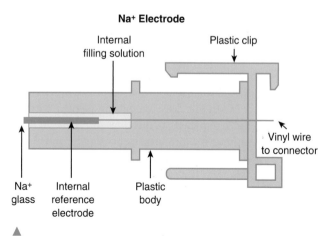

Na⁺ Electrode

Figure 14-2. Diagram of sodium ISE with glass capillary membrane. (Courtesy of Nova Biomedical, Waltham, MA.)

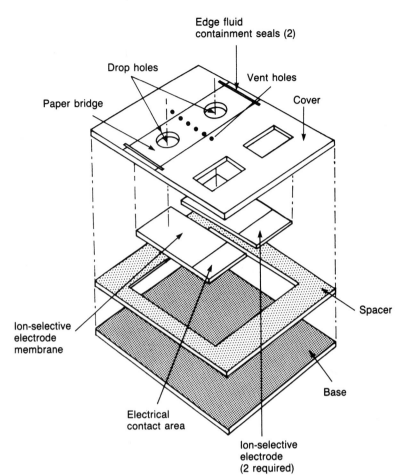

Edge fluid
containment seals (2)

Drop holes

Vent holes

Paper bridge

Cover

Ion-selective
electrode
membrane

Spacer

Base

Electrical
contact area

Ion-selective
electrode
(2 required)

◄

Figure 14-3. Schematic diagram of the ISE system for the potentiometric slide on the Ektachem. (Courtesy of Johnson & Johnson, Co., Rochester, NY.)

paralysis.[1] The heart may cease to contract in extreme cases of either hyperkalemia or hypokalemia.

The potassium concentration also affects the hydrogen ion concentration in the blood. For example, in hypokalemia (low serum potassium), as potassium ions are lost from the body, sodium and hydrogen ions move into the cell. The hydrogen ion concentration is therefore decreased in the ECF, resulting in alkalosis.

Regulation

The kidneys are important in the regulation of potassium balance. Initially, the proximal tubules reabsorb nearly all the potassium. Then, under the influence of aldosterone, additional potassium is secreted into the urine in exchange for sodium in both the distal tubules and the collecting ducts. Thus, the distal nephron is the principal determinant of urinary potassium excretion. Most individuals consume far more potassium than is needed; the excess is excreted in the urine but may accumulate to toxic levels if renal failure occurs.

TABLE 14-4. Reference Ranges for Sodium

Serum, plasma	136–145 mmol/L
Urine (24-hour)	40–220 mmol/day, varies with diet
CSF	136–150 mmol/L

Potassium uptake from the ECF into the cells is important in normalizing an acute rise in plasma K concentration due to an increased K intake. Excess plasma K rapidly enters cells to normalize plasma K. As the cellular K then gradually returns to the plasma, it is removed by urinary excretion. Note that chronic loss of cellular K may result in cellular depletion before there is an appreciable change in the plasma K concentration, because excess K is normally excreted in the urine.

There are three factors that influence the distribution of potassium between cells and ECF: (1) potassium loss frequently occurs whenever the Na/K ATPase pump is inhibited by conditions such as hypoxia, hypomagnesemia, or digoxin overdose; (2) insulin promotes acute entry of K ions into skeletal muscle and liver by increasing Na/K ATPase activity; and (3) catecholamines, such as epinephrine ($\beta2$-stimulator), promote cellular entry of K, whereas propranolol (β-blocker) impairs cellular entry of K. Dietary deficiency or excess is rarely a primary cause of hypokalemia or hyperkalemia. However, with a preexisting condition, dietary deficiency (or excess) can enhance the degree of hypokalemia (or hyperkalemia).

Exercise. Potassium is released from cells during exercise, which may increase plasma K by 0.3 mmol/L to 1.2 mmol/L with mild to moderate exercise and by as much as 2 mmol/L to 3 mmol/L with exhaustive exercise. These changes are usually reversed after several minutes of rest. Forearm exercise during venipuncture can cause erroneously high plasma K concentrations.[2]

Hyperosmolality. Hyperosmolality causes diffusion of water out of cells, carrying K ions with the water, which leads to gradual depletion of potassium.

Cellular Breakdown. Cellular breakdown releases K into the ECF. Examples are severe trauma, tumor lysis syndrome, and massive blood transfusions.

Clinical Applications

Hypokalemia. *Hypokalemia* is a plasma potassium concentration below the lower limit of the reference range. Hypokalemia can occur with GI or urinary loss of potassium or with increased cellular uptake of potassium. Common causes of hypokalemia are shown in Table 14-5. Of these, therapy with thiazide-type diuretics is by far the most common.[10] GI loss occurs when GI fluid is lost through vomiting, diarrhea, gastric suction, or discharge from an intestinal fistula. Increased K loss in the stool also occurs with some tumors, malabsorption, cancer therapy (chemotherapy or radiation therapy), and large doses of laxatives.

Renal loss of potassium can result from kidney disorders such as potassium-losing nephritis and renal tubular acidosis (RTA). In RTA, as tubular excretion of H+ decreases, K excretion increases. Because aldosterone promotes Na retention and K loss, hyperaldosteronism can lead to hypokalemia and metabolic alkalosis.[1] Hypomagnesemia can lead to hypokalemia by promoting urinary loss of potassium. Magnesium deficiency also diminishes the activity of Na/K ATPase and enhances the secretion of aldosterone. Effective treatment re-

quires supplementation with both Mg and K.[1] Renal K loss also occurs with acute myelogenous leukemia, acute myelomonocytic leukemia, and acute lymphocytic leukemia.[10] Although reduced dietary intake of K rarely causes hypokalemia in healthy persons, decreased intake may intensify hypokalemia caused by use of diuretics, for example.

Both alkalemia and insulin increase the cellular uptake of potassium. Because alkalemia promotes intracellular loss of H+ to minimize elevation of intracellular pH, both K and sodium enter cells to preserve electroneutrality. Plasma K decreases by about 0.4 mmol/L per 0.1 unit rise in pH.[1] Insulin promotes the entry of K into skeletal muscle and liver cells. Because insulin therapy can sometimes uncover an underlying hypokalemic state, plasma K should be monitored carefully whenever insulin is administered to susceptible patients.[1] A rare cause of hypokalemia is associated with a blood sample from a leukemic patient with a markedly elevated white blood cell count. The K present in the sample is taken up by the white cells if the sample is left at room temperature for several hours.[10]

Symptoms of hypokalemia. Symptoms often become apparent as plasma potassium decreases below 3 mmol/L. General symptoms are weakness, fatigue, and constipation. Hypokalemia can lead to muscle weakness or paralysis, which can interfere with breathing. The dangers of hypokalemia are of concern in all patients, but especially so in those with cardiovascular disorders because of an increased risk of arrhythmias, which may cause sudden death in certain patients with cardiac disorders. Mild hypokalemia (3.0–3.4 mmol/L) is usually asymptomatic.

Treatment of hypokalemia. Treatment includes oral or intravenous (IV) replacement of potassium. In some cases, chronic mild hypokalemia may be corrected simply by including food with high potassium content, such as dried fruits, nuts, bran cereals, bananas, and orange juice, in the diet. Plasma electrolytes are monitored as treatment to return K levels to normal occurs.

Hyperkalemia. The most common causes of hyperkalemia are shown in Table 14-6. Patients with *hyperkalemia* often have an underlying disorder, such as renal insufficiency, diabetes mellitus, or metabolic acidosis, that contributes to hyperkalemia.[2] For example, during administration of KCl, a person with renal insufficiency is far more likely to develop hyperkalemia than is a person with normal renal function. The most common cause of hyperkalemia in hospitalized patients is due to therapeutic K administration. The risk is greatest with IV K replacement.[10]

In healthy persons, an acute oral load of potassium will increase plasma K briefly, because most of the absorbed K rapidly moves intracellularly. Normal cellular processes gradually release this excess K back into the plasma, where it is normally removed by renal excretion. Impairment of urinary K excretion is almost always associated with chronic hyperkalemia.[1]

TABLE 14-5. Causes of Hypokalemia

GI Loss
   Vomiting
   Diarrhea
   Gastric suction
   Intestinal tumor
   Malabsorption
   Cancer therapy—chemotherapy, radiation therapy
   Large doses of laxatives

Renal Loss
   Diuretics—thiazides, mineralcorticoids
   Nephritis
   Renal tubular acidosis (RTA)
   Hyperaldosteronism
   Cushing's syndrome
   Hypomagnesemia
   Acute leukemia

Cellular Shift
   Alkalosis
   Insulin overdose

Decreased Intake

<table>
<tr><td colspan="1">TABLE 14-6. Causes of Hyperkalemia</td></tr>
</table>

Decreased Renal Excretion
  Acute or chronic renal failure (GFR <20 mL/min.)
  Hypoaldosteronism
  Addison's disease
  Diuretics

Cellular Shift
  Acidosis
  Muscle/cellular injury
  Chemotherapy
  Leukemia
  Hemolysis

Increased Intake
  Oral or IV potassium replacement therapy

Artifactual
  Sample hemolysis
  Thrombocytosis
  Prolonged tourniquet use or excessive fist clenching

If a shift of K from cells into plasma occurs too rapidly to be removed by renal excretion, acute hyperkalemia develops. In diabetes mellitus, insulin deficiency promotes cellular loss of K. Hyperglycemia also contributes by producing a hyperosmolar plasma that pulls water and K from cells, promoting further loss of K into the plasma.[1]

In metabolic acidosis, as excess H^+ moves intracellularly to be buffered, K leaves the cell to maintain electroneutrality. Plasma K increases by 0.2 mmol/L to 1.7 mmol/L for each 0.1 unit reduction of pH.[1] Because cellular K often becomes depleted in cases of acidosis with hyperkalemia (including diabetic ketoacidosis), treatment with agents such as insulin and bicarbonate can cause a rapid intracellular movement of K, producing severe hypokalemia.

A variety of drugs may cause hyperkalemia, especially in patients with either renal insufficiency or diabetes mellitus. These drugs include captopril (inhibits angiotensin converting enzyme), nonsteroidal anti-inflammatory agents (inhibit aldosterone), spironolactone (K-sparing diuretic), digoxin (inhibits Na-K pump), cyclosporine (inhibits renal response to aldosterone), and heparin therapy (inhibits aldosterone secretion).

Hyperkalemia may result when potassium is released into the ECF during enhanced tissue breakdown or catabolism, especially if renal insufficiency is present. Increased cellular breakdown may be caused by trauma, administration of cytotoxic agents, massive hemolysis, tumor lysis syndrome, and blood transfusions. In banked blood, K is gradually released from erythrocytes during storage, often causing elevated K concentration in plasma supernatant.

Patients on cardiac bypass may develop mild elevations in plasma potassium during warming after surgery, because warming causes cellular release of potassium. Hypothermia causes movement of potassium into cells.

Symptoms of hyperkalemia. Hyperkalemia can cause muscle weakness, tingling, numbness, or mental confusion by altering neuromuscular conduction.[7] Muscle weakness does not usually develop until plasma potassium reaches 8 mmol/L.[1]

Hyperkalemia disturbs cardiac conduction, which can lead to cardiac arrhythmias and possible cardiac arrest. Plasma potassium concentrations of 6 mmol/L to 7 mmol/L may alter the electrocardiogram, and concentrations more than 10 mmol/L may cause fatal cardiac arrest.[1]

Treatment of hyperkalemia. To offset the effect of potassium, which lowers the resting potential of myocardial cells, calcium may be given to reduce the threshold potential of myocardial cells. Therefore, calcium provides immediate but short-lived protection to the myocardium against the effects of hyperkalemia. Substances that acutely shift potassium back into cells, such as sodium bicarbonate, glucose, or insulin, may also be administered. Patients treated with these agents must be monitored carefully to prevent hypokalemia as K moves back into cells.

Collection of Samples

Proper collection and handling of samples for K analysis is extremely important, because there are many causes of artifactual hyperkalemia. First, the coagulation process releases K from platelets, so that serum K may be 0.1 mmol/L to 0.5 mmol/L higher than plasma K concentrations.[4] If the patient's platelet count is elevated (thrombocytosis), serum potassium may be further elevated. Second, if a tourniquet is left on the arm too long during blood collection, or if patients excessively clench their fists or otherwise exercise their forearms before venipuncture, cells may release potassium into the plasma. The first situation may be avoided by using a heparinized tube to prevent clotting of the specimen, and the second, by using proper care in the drawing of blood. Third, because storing blood on ice promotes the release of potassium from cells,[11] whole blood samples for potassium determinations should be stored at room temperature (never iced) and analyzed promptly or centrifuged to remove the cells. Fourth, if hemolysis occurs after the blood is drawn, potassium may be falsely elevated.

Determination of Potassium

Specimen. Serum, plasma, and urine may be acceptable for analysis. Hemolysis must be avoided because of the high K^+ content of erythrocytes. Heparin is the anticoagulant of choice. Whereas serum and plasma generally give similar potassium levels, serum reference intervals tend to be slightly higher. Markedly elevated platelet counts may result in the release of potassium during clotting from rupture of these cells, causing a spurious hyperkalemia. In this case, plasma is preferred. Whole blood samples may be used with some analyzers. Consult the instrument operator's manual for acceptability. Urine specimens should be collected over a 24-hour period to eliminate the influence of diurnal variation.

Methods. As with sodium, the current method of choice is ISE. For ISE measurements, a valinomycin membrane is used to selectively bind K^+, causing an impedance change that can be correlated to K^+ concentration. KCl is the inner electrolyte solution. FES may still be used in some labs; sodium, K, and Cl are measured simultaneously.

Reference Ranges[5]
See Table 14-7.

Chloride

Chloride (Cl^-) is the major extracellular anion. Its precise function in the body is not well understood; however, it is involved in maintaining osmolality, blood volume, and electric neutrality. In most processes, chloride ions shift secondarily to a movement of sodium$^+$ or bicarbonate ions.

Chloride ingested in the diet is almost completely absorbed by the intestinal tract. Chloride ions are then filtered out by the glomerulus and passively reabsorbed, in conjunction with sodium, by the proximal tubules. Excess chloride is excreted in the urine and sweat. Excessive sweating stimulates aldosterone secretion, which acts on the sweat glands to conserve sodium and chloride.

Chloride maintains electric neutrality in two ways. First, sodium$^+$ is reabsorbed along with Cl^- in the proximal tubules. In effect, Cl^- acts as the rate-limiting component in that Na^+ reabsorption is limited by the amount of Cl^- available. Electroneutrality is also maintained by chloride through the *chloride shift*. In this process, carbon dioxide (CO_2) generated by cellular metabolism within the tissue diffuses out into both the plasma and the red cell. In the red cell, CO_2 forms carbonic

TABLE 14-7. Reference Ranges for Potassium

Plasma, serum	3.4–5.0 mmol/L
Urine (24-hour)	25–125 mmol/day

acid (H_2CO_3), which splits into H^+ and HCO_3^- (bicarbonate). Deoxyhemoglobin buffers H^+, whereas the HCO_3^- diffuses out into the plasma and Cl diffuses into the red cell to maintain the electric balance of the cell (Fig. 14-4).

Clinical Applications
Chloride disorders are often due to the same causes that disturb Na levels, because Cl passively follows Na. There are a few exceptions. *Hyperchloremia* may occur when there is an excess loss of bicarbonate ion due to GI losses, RTA, or metabolic acidosis. *Hypochloremia* may occur with excessive loss of chloride from prolonged vomiting, diabetic ketoacidosis, aldosterone deficiency, or salt-losing renal diseases such as pyelonephritis. A low serum level of chloride may also be encountered in conditions associated with high serum bicarbonate concentrations, such as compensated respiratory acidosis or metabolic alkalosis.

Determination of Chloride
Specimen. Serum or plasma may be used, with lithium heparin being the anticoagulant of choice. Hemolysis does not cause a significant change in serum or plasma values due to decreased levels of intracellular chloride. However, with marked hemolysis, levels may be decreased due to a dilutional effect.

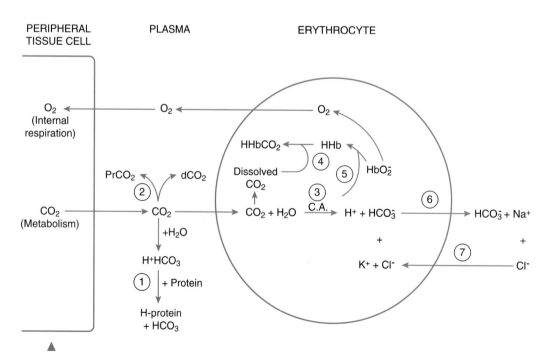

Figure 14-4. Chloride shift mechanism. See text for details. (From Burtis & Ashwood, Tietz Textbook of clinical chemistry, 2nd edition. Reprinted with permission. Philadelphia: WB Saunders, 1994.)

Whole blood samples may be used with some analyzers. Consult the instrument operator's manual for acceptability. The specimen of choice in urine chloride analyses is a 24-hour collection because of the large diurnal variation. Sweat is also suitable for analysis. Sweat collection and analysis is discussed in Chapter 24, *Body Fluid Analysis*.

Methods. There are several methodologies available for measuring chloride, including ISEs, amperometric-coulometric titration, mercurimetric titration, and colorimetry. The most commonly used is ISE. For ISE measurement, an ion-exchange membrane is used to selectively bind Cl ions.

Amperometric-coulometric titration is a method using coulometric generation of silver ions (Ag^+), which combine with Cl^- to quantitate the Cl ion concentration.

$$Ag^{+2} + 2Cl^- \rightarrow AgCl_2 \qquad (Eq.\ 14\text{--}2)$$

When all patient Cl^- ions are bound to Ag^+ ions, excess or free Ag^+ ions are used to indicate the endpoint of the reaction. As Ag^+ ions accumulate, the coulometric generator and timer are turned off. The elapsed time is used to calculate the concentration of Cl ions in the sample. The Cotlove Chloridometer (Buchler Instruments, Inc.) uses this principle in chloride analysis.

Mercurimetric titration is also a common method for chloride analysis, particularly the Schales-Schales method. This technique involves titrating Cl^- with a solution of mercury (Hg^{2+}), forming soluble but nonionized mercuric chloride.

$$2Cl^- + Hg^{+2} \rightarrow HgCl_2 \qquad (Eq.\ 14\text{--}3)$$

The endpoint is reached when excess Hg^{2+} forms a complex with an indicator, such as diphenylcarbazone, producing a violet-blue color. Both this method and amperometric-coulometric titration suffer from halogen interference. Other ions, such as Br^-, I^-, CN^-, CNS^-, and $-SH$, will react, creating positive errors.

There is also a colorimetric method for chloride that uses mercuric thiocyanate and ferric nitrate to form a reddish-colored complex with a peak at 480 nm.

Reference Ranges[5]
See Table 14-8.

Bicarbonate

Bicarbonate is the second most abundant anion in the ECF. Total CO_2 comprises the bicarbonate ion (HCO_3^-), carbonic acid (H_2CO_3), and dissolved CO_2, with bicarbonate accounting for more than 90% of the total CO_2 at physiologic pH. Because HCO_3^- composes the largest fraction of total CO_2, total CO_2 measurement is indicative of HCO_3^- measurement.

TABLE 14-8. Reference Ranges for Chloride	
Plasma, serum	98–107 mmol/L
Urine (24-hour)	110–250 mmol/day, varies with diet

Bicarbonate is the major component of the buffering system in the blood. Carbonic anhydrase in red blood cells converts CO_2 and H_2O to carbonic acid, which dissociates into H^+ and HCO_3^-.

$$CO_2 + H_2O \xleftrightarrow{\text{CA}} H_2CO_3 \xleftrightarrow{\text{CA}} H^+ + HCO_3^-$$

(Eq. 14–4)

CA, carbonic anhydrase

Bicarbonate diffuses out of the cell in exchange for chloride to maintain ionic charge neutrality within the cell (chloride shift; see Fig. 14-4). This process converts potentially toxic CO_2 in the plasma to an effective buffer: bicarbonate. Bicarbonate buffers excess hydrogen ion by combining with acid, then eventually dissociating into H_2O and CO_2 in the lungs where the acidic gas CO_2 is eliminated.

Regulation
In the kidneys, most (85%) of the bicarbonate ion is reabsorbed by the proximal tubules, with 15% being reabsorbed by the distal tubules. Because tubules are only slightly permeable to bicarbonate, it is usually reabsorbed as CO_2. This happens as bicarbonate, after filtering into the tubules, combines with hydrogen ions to form carbonic acid, which then dissociates into H_2O and CO_2. The CO_2 readily diffuses back into the ECF. Normally, nearly all the bicarbonate ions are reabsorbed from the tubules, with little lost in the urine. When bicarbonate ions are filtered in excess of hydrogen ions available, almost all excess HCO_3^- flows into the urine.

In alkalosis, with a relative increase in bicarbonate ion compared to CO_2, the kidneys increase excretion of HCO_3^- into the urine, carrying along a cation such as sodium. This loss of HCO_3^- from the body helps correct pH.

Among the responses of the body to acidosis is an increased excretion of H^+ into the urine. In addition, HCO_3 reabsorption is virtually complete, with 90% of the filtered bicarbonate reabsorbed in the proximal tubule, and the remainder in the distal tubule.[1]

Clinical Applications
Acid-base imbalances cause changes in bicarbonate and CO_2 levels. A decreased bicarbonate may occur from metabolic acidosis as bicarbonate combines with H^+ to produce CO_2, which is exhaled by the lungs. The typical response to metabolic acidosis is compensation by hyperventilation, which lowers PCO_2. Elevated total CO_2 concentrations occur in metabolic alkalosis as bicarbonate is retained, often with an increased PCO_2 due to compensation by hypoventilation. Typical causes of metabolic alkalosis include severe vomiting, hypokalemia, and excessive alkali intake.

Determination of Carbon Dioxide
Specimen. This chapter deals specifically with venous serum or plasma determinations. For discussion of arterial and whole blood PCO_2 measurements, refer to Chapter 16, *Blood Gases, pH, and Buffer Systems*.

Serum or lithium heparin plasma is suitable for analysis. Although specimens should be anaerobic for the highest accuracy, many current analyzers (excluding blood gas analyzers) do not permit anaerobic sample handling. In most instances, the sample is capped until the serum or plasma is separated and the sample is analyzed immediately. If the sample is left uncapped before analysis, CO_2 escapes. Levels can decrease by 6 mmol/L per hour.[4]

Carbon dioxide measurements may be obtained in several ways, but the actual portion of the total CO_2 being measured may vary with the method used. Two common methods are ISE and a colorimetric method.

One type of ISE for measuring total CO_2 (Beckman Instruments, Inc.) uses an acid reagent to convert all the forms of CO_2 to CO_2 gas. The gas then diffuses through a membrane into an alkaline reagent. The pH change of this reagent measured by a pH electrode is proportional to the amount of CO_2 gas present (the same type of electrode used to measure pCO_2). A second, reference electrode is also used, and the difference in electric impulses from these two electrodes is used to calculate the total CO_2 of the sample.

The Vitros (Johnson & Johnson, Co.) slide method uses an alkaline buffer layer to maintain CO_2, HCO_3^-, and CO_3^{2-} in equilibrium. A selective membrane is used for CO_2 measurement.[8]

The Hitachi (Boehringer Mannheim Corp.) uses an enzyme method that results in oxidation of NADH for measurement.[12] The main enzyme used is phosphoenolpyruvate (PEP) carboxylase, which catalyzes the formation of oxaloacetate with patient HCO_3^-.

$$\text{Phosphoenolpyruvate} + HCO_3^- \xrightarrow{\text{PEP carboxylase}}$$
$$\text{Oxaloacetate} + H_2PO_4^- \qquad (Eq.\ 14–5)$$

This is coupled to the following reaction, in which NADH is consumed due to the action of malate dehydrogenase (MDH).

$$\text{Oxaloacetate} + NADH + H^+ \xrightarrow{\text{MDH}} \text{Malate} + NAD^+$$
$$(Eq.\ 14–6)$$

The rate of change in absorbance of NADH is proportional to the concentration of HCO_3^-. Oxamate is added to inhibit interference from pyruvate.

Reference Ranges[5]

Carbon dioxide, venous 22 mmol/L to 29 mmol/L (plasma, serum).

Magnesium

Magnesium Physiology

Magnesium is the fourth most abundant cation in the body and second most abundant intracellular ion. The average human body (70 kg) contains 1 mole (24 g) of magnesium. Approximately 53% of body magnesium is found in bone, 46% in muscle and other organs and soft tissue, and less than 1% is present in serum and red blood cells.[13] Of the Mg present in serum, about one third is bound to protein, primarily albumin. Of the remaining two thirds, 61% exists in the free or ionized state and about 5% is complexed with other ions, such as phosphate and citrate. Similar to calcium, it is the free ion that is physiologically active in the body.[14]

The role of magnesium in the body is widespread. It is an essential cofactor of more than 300 enzymes, including those important in glycolysis, transcellular ion transport, neuromuscular transmission, synthesis of carbohydrates, proteins, lipids, and nucleic acids, and release of and response to certain hormones.

The clinical usefulness of serum magnesium levels has increased greatly in the past 10 years as more information about the analyte has been discovered. The most significant findings are the relationship between abnormal serum magnesium levels and cardiovascular, metabolic, and neuromuscular disorders. Although serum levels may not reflect total body stores of Mg, serum levels are useful in determining acute changes in the ion.

Regulation

Rich sources of Mg in the diet are raw nuts, dry cereals, and "hard" drinking water. Other sources include vegetables, meats, fish, and fruits.[13] Processed foods, which are an ever-increasing part of the average U.S. diet, have low levels of magnesium that may cause an inadequate intake. This in turn may increase the likelihood of a Mg deficiency. The small intestine may absorb from 20% to 65% of the dietary magnesium, depending on the need and intake.

The overall regulation of body magnesium is controlled largely by the kidney, which can reabsorb magnesium in deficiency states or readily excrete excess magnesium in overload states. Of the nonprotein–bound Mg that gets filtered by the glomerulus, 25% to 30% is reabsorbed by the proximal convoluted tubule (PCT), unlike Na, in which 60% to 75% is absorbed in the PCT. The Henle's loop is the major renal regulatory site, where 50% to 60% of filtered Mg is reabsorbed in the ascending limb. In addition, 2% to 5% is reabsorbed in the distal convoluted tubule.[15] The renal threshold for magnesium is approximately 0.60 mmol/L to 0.85 mmol/L (~1.46–2.07 mg/dL). Because this is close to the normal serum concentration, slight excesses of magnesium in serum are rapidly excreted by the kidneys.[2] Normally, only about 6% of filtered Mg is excreted in the urine per day.[13]

The regulation of magnesium appears to be related to that of calcium and sodium. Parathyroid hormone (PTH) increases the renal reabsorption of magnesium and enhances the absorption of magnesium in the intestine. However, for equivalent decreases in ionized calcium and ultrafilterable magnesium in blood, changes in ionized calcium have a far greater effect on PTH secretion.[16] Aldosterone and thyroxine apparently have the opposite effect of PTH in the kidney, increasing the renal excretion of magnesium.[14]

Clinical Applications

Hypomagnesemia. *Hypomagnesemia* is most frequently observed in hospitalized individuals in intensive care units or those receiving diuretic therapy or digitalis therapy. These patients most likely have an overall tissue depletion of magnesium as a result of severe illness or loss, which leads to the low serum levels. Hypomagnesemia is a rare occurrence in nonhospitalized individuals.[14]

The causes of hypomagnesemia are many, but can be grouped into general categories as seen in Table 14-9. Reduced intake is the least likely to cause severe deficiencies in the U.S. A magnesium-deficient diet as a result of starvation, chronic alcoholism, or Mg-deficient IV therapy can cause a loss of the ion.

A variety of GI disorders may cause decreased absorption by the intestine, which can result in an excess loss of magnesium by way of the feces. Malabsorption syndromes, intestinal resection or bypass surgery, nasogastric suction, pancreatitis, prolonged vomiting, diarrhea, or laxative use may

TABLE 14-9. Causes of Hypomagnesemia

Reduced Intake
  Poor diet/starvation
  Prolonged magnesium-deficient IV therapy
  Chronic alcoholism

Decreased Absorption
  Malabsorption syndromes
  Surgical resection of small intestine
  Nasogastric suction
  Pancreatitis
  Vomiting
  Diarrhea
  Laxative abuse
  Neonatal
  Primary
  Congenital

Increased Excretion—Renal
  Tubular disorder
  Glomerulonephritis
  Pyelonephritis

Increased Excretion—Endocrine
  Hyperparathyroidism
  Hyperaldosteronism
  Hyperthyroidism
  Hypercalcemia
  Diabetic ketoacidosis

Increased Excretion—Drug Induced
  Diuretics
  Antibiotics
  Cyclosporin
  Digitalis

Miscellaneous
  Excess lactation
  Pregnancy

(From Polancic JE. Magnesium: metabolism, clinical importance, and analysis. CLS 1991;4(2). Adapted with ASCLS permission.)

lead to a magnesium deficiency. Neonatal hypomagnesemia has been reported as a result of various surgical procedures in some neonates. A primary deficiency has also been reported in infants due to a selective malabsorption of the ion. A chronic congenital hypomagnesemia has also been reported secondary to an autosomal recessive disorder.[14]

Mg loss due to increased excretion by way of the urine can occur as a result of various renal disorders, various endocrine disorders, or due to the effects on the kidney by certain drugs. Renal tubular disorders and other select renal disorders may result in excess amounts of magnesium being lost through the urine because of decreased tubular reabsorption.

Several endocrine disorders can cause a loss of magnesium. Hyperparathyroidism and hypercalcemia may cause increased renal excretion of magnesium due to an excess of calcium ions. Excess serum sodium levels due to hyperaldosteronism may also cause increased renal excretion of magnesium. A pseudohypomagnesemia may also be the result of hyperaldosteronism due to increased water reabsorption. Hyperthyroidism may result in an increased renal excretion of magnesium and may also cause an intracellular shift of the ion. In diabetics, excess urinary loss of Mg is associated with glycosuria. Hypomagnesemia can aggravate the neuromuscular and vascular complications commonly found in this disease. Some studies have shown a relationship between Mg deficiency and insulin resistance; however, Mg is not thought to play a role in the pathophysiology of diabetes mellitus. The American Diabetic Association has made a statement regarding dietary intake of magnesium and measurement of serum Mg in diabetic patients.[17]

Several drugs, including diuretics, gentamicin, cisplatin, and cyclosporine, increase renal loss of magnesium and frequently result in hypomagnesemia. The loop diuretics, such as furosemide, are especially effective in increasing renal loss of Mg. Thiazide diuretics require a longer period of use to cause hypomagnesemia. Cisplatin has a nephrotoxic effect that inhibits the ability of the renal tubule to conserve magnesium. Cyclosporine, an immunosuppressant, severely inhibits the renal tubular reabsorption of magnesium and has many side effects, including nephrotoxicity, hypertension, hepatotoxicity, and neurologic symptoms such as seizures and tremors. Cardiac glycosides, such as digoxin and digitalis, can interfere with Mg reabsorption. The resulting hypomagnesemia is a significant finding because the decreased level of Mg can amplify the symptoms of digitalis toxicity.[14]

Excess lactation has been associated with hypomagnesemia due to increased use and loss through milk production. Mild deficiencies have been reported in pregnancy, which may cause a hyperexcitable uterus, anxiety, and insomnia.

Symptoms of hypomagnesemia. A hypomagnesemic patient may be asymptomatic until serum levels fall below 0.5 mmol/L.[14] A variety of symptoms can occur. The most frequent involve cardiovascular, neuromuscular, psychiatric, and metabolic abnormalities (Table 14-10). The cardiovascular and neuromuscular symptoms result primarily from the

TABLE 14-10. Symptoms of Hypomagnesemia

Cardiovascular	*Psychiatric*
Arrhythmias	Depression
Hypertension	Agitation
Digitalis toxicity	Psychosis
Neuromuscular	*Metabolic*
Weakness	Hypokalemia
Cramps	Hypocalcemia
Ataxia	Hypophosphatemia
Tremor	Hyponatremia
Seizure	
Tetany	
Paralysis	
Coma	

(From Polancic JE. Magnesium: metabolism, clinical importance, and analysis. CLS 1991; 4(2). Adapted with ASCLS permission.)

ATPase enzyme's requirement for Mg. Mg loss leads to decreased intracellular K levels because of a faulty Na-K pump (ATPase). This change in cellular RMP causes increased excitability and may lead to cardiac arrhythmias. This condition may also lead to digitalis toxicity.

Muscle contraction also requires magnesium and ATPase for normal calcium uptake following contraction. Normal nerve and muscle cell stimulation requires magnesium to assist with the regulation of acetylcholine, a potent neurotransmitter. Thus, hypomagnesemia can cause a variety of symptoms from weakness to tremors, tetany, paralysis, or coma. The CNS can also be affected, resulting in psychiatric disorders that range from very subtle changes to depression or psychosis.

Metabolic disorders are associated with hypomagnesemia. Studies have indicated that approximately 40% of hospital-ized patients with hypokalemia are also hypomagnesemic.[15] In addition, 20% to 30% of patients with hyponatremia, hypocalcemia, or hypophosphatemia are also hypomagnesemic.[15] Mg deficiency can impair PTH release and target tissue response, resulting in hypocalcemia. Replenishing any of these deficient ions alone, often does not remedy the disorder unless magnesium therapy is provided. Mg therapy alone may sometimes restore both ion levels to normal. Serum levels of the ions must be monitored during treatment.

Treatment of hypomagnesemia. The preferred form of treatment is by oral intake using Mg-lactate, Mg-oxide, Mg-Cl, or an antacid that contains Mg. In severely ill patients, a $MgSO_4$ solution is given parenterally. Before initiation of therapy, renal function must be evaluated to avoid inducing hypermagnesemia during treatment.[15]

Hypermagnesemia. *Hypermagnesemia* is observed less frequently than hypomagnesemia.[14] Causes for elevated serum magnesium levels are summarized in Table 14-11. The most common cause for hypermagnesemia is renal failure (GFR <30 mL/min). The most severe elevations are usually due to the combined effects of decreased renal function and increased intake of commonly prescribed magnesium-containing medications, such as antacids, enemas, or cathartics. Nursing home patients are at greatest risk for this occurrence.[14]

Hypermagnesemia has been associated with several endocrine disorders. Thyroxine and growth hormone cause a decrease in tubular reabsorption of Mg. A deficiency of either hormone may cause a moderate elevation in serum Mg. Adrenal insufficiency may cause a mild elevation due to decreased renal excretion of Mg.[14]

$MgSO_4$ may be used therapeutically with preeclampsia, cardiac arrhythmia, or myocardial infarction. Mg is a vasodilator, and can decrease uterine hyperactivity in eclampsic

---

CASE STUDY 14-2

A 60-year-old male entered the emergency room after two days of "not feeling so well." History revealed a previous myocardial infarction 5 years ago, after which he was prescribed digoxin. Two years ago, he was prescribed a diuretic after periodic bouts of edema. An electrocardiogram at time of admission indicated a cardiac arrhythmia. Admitting lab results are shown in Case Study Table 14-2.1.

CASE STUDY TABLE 14-2.1. Laboratory Results

Venous Blood

Digoxin	1.4 ng/mL, therapeutic 0.5–2.2
	(1.8 nmol/L, therapeutic 0.6–2.8)
Na^+	137 mmol/L
K^+	2.5 mmol/L
Cl^-	100 mmol/L
HCO^-	25 mmol/L
Mg^{+2}	0.4 mmol/L
Ion Ca^{+2}	1.0 mmol/L

Questions

1. Because the digoxin level is within the therapeutic range, what may be the cause for the arrhythmia?
2. What is the most likely cause for the hypomagnesemia?
3. What is the most likely cause for the decreased potassium and ionized calcium levels?
4. What type of treatment would be helpful?

TABLE 14-11. Causes of Hypermagnesemia

Decreased Excretion
Acute or chronic renal failure
Hypothyroidism
Hypoaldosteronism
Hypopituitarism (↓GH)

Increased Intake
Antacids
Enemas
Cathartics
Therapeutic—eclampsia, cardiac arrhythmia

Miscellaneous
Dehydration
Bone carcinoma
Bone metastases

(From Polancic JE. Magnesium: metabolism, clinical importance, and analysis. CLS 1991; 4(2). Adapted with ASCLS permission.)

TABLE 14-12. Symptoms of Hypermagnesemia

Cardiovascular
Hypotension
Bradycardia
Heart block

Dermatologic
Flushing
Warm skin

GI
Nausea
Vomiting

Neurologic
Lethargy
Coma

Neuromuscular
Decreased reflexes
Dysarthria
Respiratory depression
Paralysis

Metabolic
Hypocalcemia

Hemostatic
Decreased thrombin generation
Decreased platelet adhesion

(From Polancic JE. Magnesium: metabolism, clinical importance, and analysis. 1991;4(2). Adapted with ASCLS permission.)

states and increase uterine blood flow. This form of therapy can lead to maternal hypermagnesemia, as well as neonatal hypermagnesemia due to the immature kidney of the newborn. Premature infants are at greater risk to develop actual symptoms.[14] Studies have shown that IV Mg therapy in myocardial infarction patients may reduce early mortality.[13]

Dehydration can cause a pseudohypermagnesemia, which can be corrected with rehydration. Mild serum magnesium elevations can occur in individuals with multiple myeloma or bone carcinoma or metastases due to increased bone loss.

Symptoms of hypermagnesemia.
Symptoms due to hypermagnesemia typically do not occur until the serum level exceeds 1.5 mmol/L.[14] The most frequent symptoms involve cardiovascular, dermatologic, GI, neurologic, neuromuscular, metabolic, and hemostatic abnormalities, as listed in Table 14-12. Mild to moderate symptoms, such as hypotension, bradycardia, skin flushing, increased skin temperature, nausea, vomiting, and lethargy may occur when serum levels are 1.5 mmol/L to 2.5 mmol/L.[14] Life-threatening symptoms, such as electrocardiogram changes, heart block, asystole, sedation, coma, respiratory depression or arrest, and paralysis can occur when serum levels reach 5.0 mmol/L.[14]

Elevated Mg levels may inhibit PTH release and target tissue response. This may lead to hypocalcemia and hypercalciuria.[14] Normal hemostasis is a calcium-dependent process, which may be inhibited due to competition between increased levels of magnesium and calcium ions. Thrombin generation and platelet adhesion are two processes in which interference may occur.[14]

Treatment of hypermagnesemia.
Treatment of Mg excess associated with increased intake is to discontinue the source of Mg. Severe symptomatic hypermagnesemia requires immediate supportive therapy for cardiac, neuromuscular, respiratory, or neurologic abnormalities. Patients with renal failure require hemodialysis. Patients with normal renal function may be treated with a diuretic and IV fluid.

Determination of Magnesium
Specimen. Nonhemolyzed serum or lithium heparin plasma may be analyzed. Because the Mg^{2+} concentration inside erythrocytes is 10 times greater than that in the ECF, hemolysis should be avoided, and the serum should be separated from the cells as soon as possible. Oxalate, citrate, and ethylenediaminetetraacetic acid (EDTA) anticoagulants are unacceptable because they will bind with magnesium. A 24-hour urine is preferred for analysis because of a diurnal variation in excretion. The urine must be acidified with HCl to avoid precipitation.

Methods. The three most common methods for measuring total serum Mg are colorimetric: calmagite, formazen dye, and methylthymol blue. In the calmagite method (Hitachi, Synchron), Mg binds with calmagite to form a reddish-violet complex that may be read at 532 nm. In the formazen dye method (Vitros), Mg binds with the dye to form a colored complex that may be read at 660 nm. In the methylthymol blue method (Dimension), Mg binds with the chromogen to form a colored complex. Most methods use a calcium shelter to prohibit interference from this divalent cation.

As with calcium, measurement of the free or ionized fraction is a better indicator of the physiologically active ion. Nova Biomedical (Waltham, MA) recently introduced an ISE for clinical use (Fig. 14-5). Presently, ISE measurement is not commonly used. The reference method for measuring magnesium is AAS.

Although the measurement of total magnesium concentrations in serum remains the usual diagnostic test for detection of magnesium abnormalities, it has limitations. First, because

An 84-year-old nursing home resident was seen in the emergency room with the following symptoms: nausea, vomiting, decreased respiration, hypotension, and low pulse rate (46). Physical exam showed the skin was warm to the touch and flushed. Admission lab data is found in Case Study Table 14-3.1.

CASE STUDY TABLE 14-3.1. Laboratory Results

	Result	Reference Range
Serum		
Total protein	5.6 g/dL	6.0–8.0 g/dL
Albumin	3.0 g/dL	3.5–5.0 g/dL
AST	48 U/L	7–45 U/L
Total calcium	8.2 g/dL	8.6–10.0 g/dL
BUN	60 mg/dL	5–20 mg/dL
Creatinine	6.6 mg/dL	0.7–1.5 mg/dL
Magnesium	4.0 mmol/L	0.63–1.0 mmol/L
Plasma		
Na^+	131 mmol/L	136–145 mmol/L
K^+	5.3 mmol/L	3.4–5.0 mmol/L
Cl^-	95 mmol/L	98–107 mmol/L

Questions

1. What is the most likely cause for the patient's symptoms?
2. What is the most likely cause for the hypermagnesemia?
3. What could be the cause for the hypocalcemia?

approximately 25% of magnesium is protein bound, total magnesium may not reflect the physiologically active free ionized magnesium. Second, because magnesium is primarily an intracellular ion, serum concentrations will not necessarily reflect the status of intracellular magnesium. Even when tissue and cellular magnesium is depleted by as much as 20%, serum magnesium concentrations may remain normal.

A magnesium load or retention test may detect body depletion of magnesium in patients with a normal concentration of magnesium in serum. However, a magnesium load test requires more than 48 hours to complete. After collection of a baseline 24-hour urine, 30 mmol (729 mg) of Mg in a 5% dextrose solution is administered intravenously, as another 24-hour urine is collected. Whereas individuals with adequate body stores of magnesium will excrete 60% to 80% of the magnesium load within 24 hours, magnesium-deficient patients will excrete 50% or less.[18]

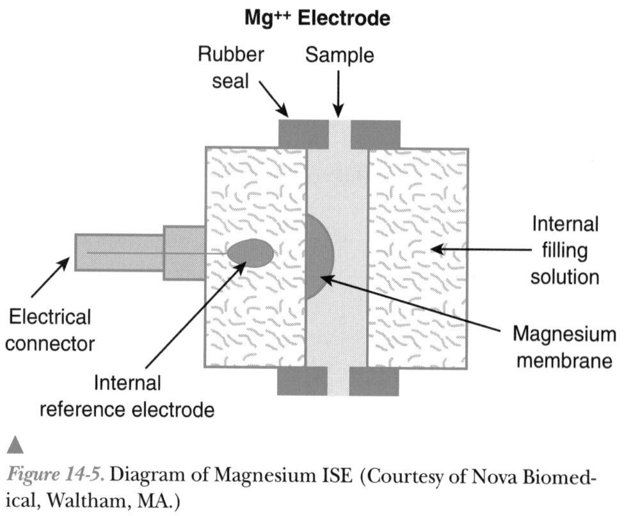

Reference Ranges[5]

See Table 14-13.

Calcium

Calcium Physiology

In 1883, Ringer showed that calcium was essential for myocardial contraction.[19] While attempting to study how bound and free forms of calcium affected frog heart contraction, McLean and Hastings showed that the ionized calcium concentration was proportional to the amplitude of frog heart contraction, whereas protein-bound and citrate-bound calcium had no effect.[20] From this observation, they developed the first assay for ionized calcium using isolated frog hearts. Although the method had poor precision by today's standards, the investigators were able to show that blood-ionized calcium was both closely regulated and had a mean concentration in humans of about 1.18 mmol/L. More recent studies have shown that, because decreased ionized calcium impairs myocardial function, it is important to maintain ionized calcium at a near normal concentration during surgery and in critically ill patients.[21]

Decreased ionized calcium concentrations in blood can cause neuromuscular irritability, which may become clinically apparent as irregular muscle spasms, called *tetany*. Studies have shown that the rate of fall in ionized calcium initiates tetany as much as the absolute concentration of ionized calcium.[22]

Figure 14-5. Diagram of Magnesium ISE (Courtesy of Nova Biomedical, Waltham, MA.)

TABLE 14-13. Reference Range for Magnesium

Serum, plasma	0.63–1.0 mmol/L (1.2–2.1 mEq/L)

Regulation

Three hormones are known to regulate serum calcium by altering their secretion rate in response to changes in ionized calcium. These hormones are PTH, vitamin D, and calcitonin. The actions of these hormones are shown in Figure 14-6.

PTH secretion into blood is stimulated by a decrease in ionized calcium and, conversely, PTH secretion is stopped by an increase in ionized calcium. PTH exerts three major effects on both bone and kidney. In the bone, PTH activates a process known as *bone resorption* in which activated osteoclasts break down bone and subsequently release calcium into the ECF. In the kidneys, PTH conserves calcium by increasing tubular reabsorption of calcium ions. PTH also stimulates renal production of active vitamin D.

Vitamin D_3, a cholecalciferol, is obtained either in the diet or from exposure of skin to sunlight. Vitamin D_3 is then converted in the liver to 25-hydroxycholecalciferol (25-OH-D_3), still an inactive form of vitamin D. In the kidney 25-OH-D_3 is specifically hydroxylated to form 1,25-dihydroxycholecalciferol [1,25-$(OH)_2$-D_3], the biologically active form. This active form of vitamin D increases calcium absorption in the intestine and enhances the effect of PTH on bone resorption.

Calcitonin, which originates in the medullary cells of the thyroid gland, is secreted when the concentration of calcium in blood increases. Calcitonin exerts its calcium-lowering effect by inhibiting the actions of both PTH and vitamin D. Although calcitonin is apparently not secreted during normal regulation of the ionized calcium concentration in blood, it is secreted in response to a hypercalcemic stimulus.

Distribution

More than 99% of calcium in the body is part of bone. The remaining 1% is mostly in the blood and other ECF. Very little is in the cytosol of most cells. In fact, the concentration of ionized calcium in blood is 5,000 to 10,000 times higher than in the cytosol of cardiac or smooth-muscle cells. Maintenance of this large gradient is vital to maintain the essential rapid inward flux of calcium ions.

Calcium in blood is distributed among several forms. About 45% circulates as free calcium ions, 40% is bound to protein, mostly albumin, and 15% is bound to anions such as bicarbonate, citrate, phosphate, and lactate. Clearly, this distribution can change in disease. It is especially noteworthy that concentrations of citrate, bicarbonate, lactate, phosphate, and albumin can change dramatically during surgery or critical care. This is the principal reason why ionized calcium cannot be reliably calculated from total calcium measurements, especially in acutely ill individuals.

Clinical Applications

Tables 14-14 and 14-15 summarize causes of hypocalcemic and hypercalcemic disorders. Although both total calcium and ionized calcium measurements are available in many laboratories, ionized calcium is usually a more sensitive and specific marker for calcium disorders.

Hypocalcemia. When PTH is not present, as with *primary hypoparathyroidism,* serum calcium levels are not properly regulated. Bone tends to "hang on" to its storage pool and the kidney increases excretion of calcium. Because PTH is also required for normal Vitamin D metabolism, the lack of

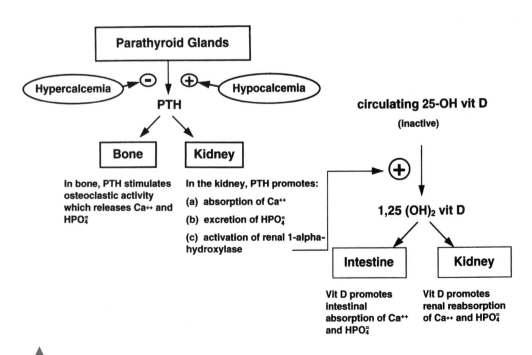

Figure 14-6. Hormonal response to hypercalcemia and hypocalcemia. PTH, parathyroid hormone; 25-OH vit D, 25 hydroxy vitamin D; 1,25(OH)$_2$ vit D, dihydroxy vitamin D.

TABLE 14-14. Causes of Hypocalcemia

Primary hypoparathyroidism—glandular aplasia, destruction, or removal
Hypomagnesemia
Hypermagnesemia
Hypoalbuminemia (total calcium only, ionized not affected by)—chronic liver disease, nephrotic syndrome, malnutrition
Acute pancreatitis
Vitamin D deficiency
Renal disease
Rhabdomyolysis
Pseudohypoparathyroidism

Vitamin D's effects also leads to a decreased level of calcium. Parathyroid gland aplasia, destruction, or removal are obvious reasons for primary hypoparathyroidism.

As hypomagnesemia in hospitalized patients has become more frequent, chronic hypomagnesemia has also become recognized as a frequent cause of *hypocalcemia*. Hypomagnesemia may cause hypocalcemia by three mechanisms: (1) it inhibits the glandular secretion of PTH across the parathyroid gland membrane, (2) it impairs PTH action at its receptor site on bone, and (3) it causes vitamin D resistance.[2] Elevated Mg levels may inhibit PTH release and target tissue response, perhaps leading to hypocalcemia and hypercalciuria.[14]

When total calcium is the only result reported, hypocalcemia can appear with hypoalbuminemia. Common causes are associated with chronic liver disease, nephrotic syndrome, and malnutrition. In general, for each 1 g/dL decrease in serum albumin, there is a 0.2 mmol/L (0.8 mg/dL) decrease in total calcium levels.[23]

About one half of the patients with acute pancreatitis develop hypocalcemia. The most consistent cause appears to be due to increased intestinal binding of calcium as increased intestinal lipase activity occurs.[23] Vitamin D deficiency and malabsorption can cause decreased absorption, which leads to increased PTH production or secondary hyperparathyroidism.

Patients with renal disease caused by glomerular failure often have altered concentrations of calcium, phosphate, albumin, magnesium, and hydrogen ion (pH). Very frequently in chronic renal disease, secondary hyperparathyroidism develops as the body tries to compensate for hypocalcemia caused either by hyperphosphatemia (phosphate binds and lowers ionized calcium) or altered vitamin D metabolism. Monitoring and controlling ionized calcium concentrations may avoid problems due to hypocalcemia, such as osteodystrophy, unstable cardiac output or blood pressure, or problems arising from hypercalcemia, such as renal stones and other calcifications. Rhabdomyolysis, as with major crush injury and muscle damage, may cause hypocalcemia due to increased phosphate release from cells, which bind to calcium ions.[23]

Pseudohypoparathyroidism is a rare hereditary disorder in which PTH target tissue response is decreased. PTH production responds normally to loss of calcium; however, without normal response (decreased cAMP [adenosine 3':5'-cyclic phosphate] production), calcium is lost in the urine or remains in the bone storage pool. Patients often have common physical features that include short stature, obesity, shortened metacarpals and metatarsals, and abnormal calcification.

Surgery and intensive care. Because appropriate calcium concentrations promote good cardiac output and maintain adequate blood pressure, the maintenance of a normal ionized calcium concentration in blood is beneficial to patients in either surgery or intensive care. Controlling calcium concentrations may be critical in open-heart surgery when the heart is restarted and during liver transplantation because large volumes of citrated blood are given.

Because these patients may receive large amounts of citrate, bicarbonate, calcium salts, or fluids, the greatest discrepancies between total calcium and ionized calcium concentrations may be seen during major surgical operations. Consequently, ionized calcium measurements are the only calcium measurement of greatest clinical value.

Hypocalcemia occurs commonly in critically ill patients, that is, those with sepsis, thermal burns, renal failure, or cardiopulmonary insufficiency. These patients frequently have abnormalities of acid-base regulation and losses of protein and albumin, which are best suited to monitoring calcium status by ionized calcium measurements. Normalization of ionized calcium may have beneficial effects on cardiac output and blood pressure.

Neonatal monitoring. Typically, blood-ionized calcium concentrations in neonates are high at birth and then rapidly decline by 10% to 20% after 1 day to 3 days. After about 1 week, ionized calcium concentrations in the neonate stabilize at levels slightly higher than in adults.[24]

The concentration of ionized calcium may decrease rapidly in the early neonatal period, because the infant may lose calcium rapidly and not readily reabsorb it. Several possible etiologies have been suggested: abnormal PTH and vitamin D metabolism, hypercholesterolemia, hyperphosphatemia, and hypomagnesemia.

TABLE 14-15. Causes of Hypercalcemia

Primary hyperparathyroidism—adenoma or glandular hyperplasia
Hyperthyroidism
Benign familial hypocalciuria
Malignancy
Multiple myeloma
Increased vitamin D
Thiazide diuretics
Prolonged immobilization

Symptoms of hypocalcemia. Neuromuscular irritability and cardiac irregularities are the primary groups of symptoms that occur with hypocalcemia. Neuromuscular symptoms include parasethesia, muscle cramps, tetany, and seizures. Cardiac symptoms may include arrhythmia or heart block. Symptoms usually occur with severe hypocalcemia, in which total calcium levels are below 1.88 mmol/L (7.5 mg/dL).[23]

Treatment of hypocalcemia. Oral or parenteral calcium therapy may occur, depending on the severity of the decreased level and the cause. Vitamin D may sometimes be administered in addition to oral calcium to increase absorption. If hypomagnesemia is a concurrent disorder, magnesium therapy should also be provided.

Hypercalcemia. Primary hyperparathyroidism is the main cause of *hypercalcemia.*[23] *Hyperparathyroidism,* or excess secretion of PTH, may show obvious clinical signs or may be asymptomatic. The patient population seen most frequently with primary hyperparathyroidism is elderly females.[23] Although either total or ionized calcium measurements are elevated in serious cases, ionized calcium is more frequently elevated in subtle or asymptomatic hyperparathyroidism. In general, ionized calcium measurements are elevated in 90% to 95% of cases of hyperparathyroidism, whereas total calcium is elevated in 80% to 85% of cases.

The second leading cause of hypercalcemia is associated with various types of malignancy, with hypercalcemia sometimes being the sole biochemical marker for disease.[23] Many tumors produce PTH-related peptide (PTH-rP), which binds to normal PTH receptors and causes increased calcium levels. Assays to measure PTH-rP are available, because this abnormal protein is not detected by most PTH assays.

Due to the proximity of the parathyroid gland to the thyroid gland, hyperthyroidism can sometimes cause hyperparathyroidism. A rare benign familial hypocalciuria has also been reported. Thiazide diuretics increase calcium reabsorption, leading to hypercalcemia. Prolonged immobilization may cause increased bone resorption. Hypercalcemia associated with immobilization is further compounded by renal insufficiency.

Symptoms of hypercalcemia. A mild hypercalcemia (2.62–3.00 mmol/L, 10.5–12 mg/dL) is often asymptomatic.[23] Moderate or severe calcium elevations include neurologic, GI, and renal symptoms. Neurologic symptoms may include mild drowsiness or weakness, depression, lethargy, and coma. GI symptoms may include constipation, nausea, vomiting, anorexia, and peptic ulcer disease. Hypercalcemia may cause renal symptoms of nephrolithiasis and nephrocalcinosis. Hypercalciuria can result in nephrogenic diabetes insipidus, which causes polyuria that results in hypovolemia, which further aggravates the hypercalcemia.[23] Hypercalcemia can also cause symptoms of digitalis toxicity.

Treatment of hypercalcemia. Treatment of hypercalcemia depends on the level of hypercalcemia and the cause. Often people with primary hyperparathyroidism are asymptomatic. These individuals are closely monitored for any rise in serum calcium and presence of symptoms. Estrogen deficiency in postmenopausal women has been implicated in primary hyperparathyroidism in elderly females.[23] Often, estrogen re-

CASE STUDY 14-4

Consider the following sets of laboratory results from adult patients:

- Primary hyperparathyroidism
- Malignancy
- Hypomagnesemic hypocalcemia

Questions

1. Which set of laboratory results (Case A, B, or C) is most likely associated with each of the following diagnoses:

Case Study Table 14–4.1. Laboratory Results

	Reference Ranges				
Case	*Ion Ca^{+2}* *1.16–1.32* *mmol/L*	*Total Mg^{+2}* *0.63–1.0* *mmol/L*	*PO$_4^-$* *0.87–1.45* *mmol/L*	*Hematocrit* *35–45%*	*Intact* *Parathyroid Hormone* *13–64 ng/L*
A	1.44	0.90	0.85	42	100
B	1.08	0.50	0.90	40	25
C	1.70	0.98	1.43	30	12

placement therapy reduces calcium levels. Parathyroidectomy may be necessary in some hyperparathyroidic patients. Patients with moderate to severe hypercalcemia are treated to reduce calcium levels. Salt and water intake is encouraged to increase calcium excretion and avoid dehydration, which can compound the hypercalcemia. Thiazide diuretics should be discontinued. Biphosphanates (a derivative of pyrophosphate) are the main class of drug used to lower calcium levels by its binding action to bone, which prevents bone resorption.[23]

Determination of Calcium

Specimen. The preferred specimen for total calcium determinations is either serum or lithium heparin plasma collected without venous stasis. Because anticoagulants such as EDTA or oxalate bind calcium very tightly and interfere with its measurement, they are unacceptable for use.

The proper collection of samples for ionized calcium measurements requires greater care. Because loss of CO_2 will increase pH, samples must be collected anaerobically. Although heparinized whole blood is the preferred sample, serum from sealed evacuated blood-collection tubes may be used if clotting and centrifugation are done quickly (<30 minutes) and at room temperature. No liquid heparin products should be used. Most heparin anticoagulants (sodium, lithium) partially bind to calcium and lower ionized calcium concentrations. A heparin concentration of 25 IU/mL, for example, decreases ionized calcium by about 3%. Dry heparin products are available either titrated with small amounts of Ca or Zn ions, or have small amounts of heparin dispersed in an inert "puff" that essentially eliminates the interference by heparin.

For analysis of calcium in urine, an accurately timed urine collection is preferred. The urine should be acidified with 6 mol/L HCl, with approximately 1 mL of the acid added for each 100 mL of urine.

Methods. The two commonly used methods for total calcium analysis use either orthocresolphthalein complexone (OCPC) or arsenzo III dye to form a complex with calcium. The OCPC method uses 8-hydroxyquinoline to prevent magnesium interference. OCPC is used on Hitachi (Boehringer Mannheim, Inc.) and Dimension (Dade) analyzers. The arsenzo III method can be found on the Vitros (Johnson & Johnson, Co.) and Synchron (Beckman) systems. AAS remains the reference method for total calcium, although is rarely used in the clinical setting.

Current commercial analyzers that measure ionized calcium use ISEs for this measurement. These systems may use membranes impregnated with special molecules that selectively, but reversibly, bind calcium ions. As calcium ions bind to these membranes, an electric potential develops across the membrane that is proportional to the ionized calcium concentration. A diagram of one such electrode is shown in Figure 14-7.

Reference Ranges[5]

For total calcium, the reference range varies slightly with age. In general, calcium concentrations are higher up through adolescence while bone growth is most active. Ionized calcium concentrations can change rapidly from day 1 to day 3 of life. Following this, they stabilize at relatively high levels, with a gradual decline through adolescence; see Table 14-16.

Phosphate

Phosphate Physiology

Compounds of phosphate are everywhere in living cells and participate in many of the most important biochemical processes. The genetic materials deoxyribonucleic acid (DNA)

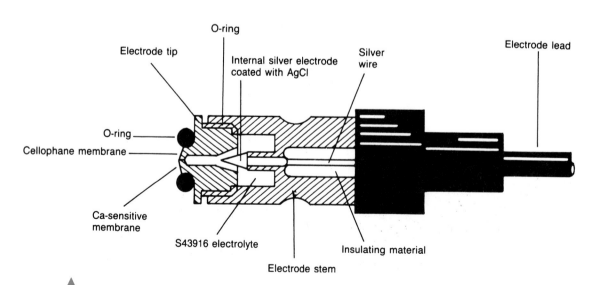

Figure 14-7. Diagram of ionized calcium electrode for the ICA ionized calcium analyzer. (Courtesy of Radiometer America, Westlake, OH.)

TABLE 14-16. Reference Ranges for Calcium

Total calcium (serum, plasma)
　Child: 2.20–2.70 mmol/L (8.8–10.8 mg/dL)
　Adult: 2.15–2.50 mmol/L (8.6–10.0 mg/dL)
Ionized calcium (serum)
　Neonate: 1.20–1.48 mmol/L (4.8–5.9 mg/dL)
　Child: 1.20–1.38 mmol/L (4.8–5.5 mg/dL)
　Adult: 1.16–1.32 mmol/L (4.6–5.3 mg/dL)
Urine (24-hour): 2.50–7.50 mmol/day (100–300 mg/day), varies with diet

and ribonucleic acid (RNA) are complex phosphodiesters. Most coenzymes are esters of phosphoric or pyrophosphoric acid. The most important reservoirs of biochemical energy are ATP, creatine phosphate, and phosphoenolpyruvate. Phosphate deficiency can lead to ATP depletion, which is ultimately responsible for many of the clinical symptoms observed in hypophosphatemia.

Alterations in the concentration of 2,3-bisphosphoglycerate (2, 3-BPG) in the red blood cells affect the affinity of hemoglobin for oxygen, with an increase facilitating the release of oxygen in the tissues and a decrease making oxygen bound to hemoglobin less available. By affecting the formation of 2,3-BPG, the concentration of inorganic phosphate indirectly affects the release of oxygen from hemoglobin.

Understanding the cause of an altered phosphate concentration in the blood is often difficult because transcellular shifts of phosphate are a major cause of hypophosphatemia in blood. That is, an increased shift of phosphate into cells can deplete phosphate in the blood. Once phosphate is taken up by the cell, it remains there to be used in the synthesis of phosphorylated compounds. As these phosphate compounds are metabolized, inorganic phosphate slowly leaks out of the cell into the blood, where it is regulated principally by the kidney.

Regulation

Phosphate in blood may be absorbed in the intestine from dietary sources, released from cells into blood, and lost from bone. In healthy individuals, all these processes are relatively constant and easily regulated by renal excretion or reabsorption of phosphate.

Disturbances to any of these processes can alter phosphate concentrations in the blood; however, the loss of regulation by the kidneys will have the most profound effect. Although other factors such as vitamin D, calcitonin, growth hormone, and acid–base status can affect renal regulation of phosphate, the most important factor is PTH, which overall lowers blood concentrations by increasing renal excretion.

Vitamin D acts to increase phosphate in the blood. Vitamin D increases both phosphate absorption in the intestine and phosphate reabsorption in the kidney.

Growth hormone, which helps regulate skeletal growth, can affect circulating concentrations of phosphate. In cases of excessive secretion or administration of growth hormone, phosphate concentrations in the blood may increase because of decreased renal excretion of phosphate.

Distribution

Although the concentration of all phosphate compounds in blood is about 12 mg/dL (3.9 mmol/L), most of that is organic phosphate and only about 3 mg/dL to 4 mg/dL is inorganic phosphate. Phosphate is the predominant intracellular anion, with intracellular concentrations varying, depending on the type of cell. About 80% of the total body pool of phosphate is contained in bone.

Clinical Applications

Hypophosphatemia. *Hypophosphatemia* is very common in hospitalized patients. Although most such cases are moderate and seldom cause problems, severe hypophosphatemia (<1.0 g/dL or 0.3 mmol/L) requires monitoring and possible replacement therapy. One study[25] found that the four most frequent causes of severe hypophosphatemia in hospitalized patients were as follows:

1. Infusion of dextrose solution. The flow of glucose into cells is often accompanied by an influx of phosphate. This cellular uptake of phosphate can be substantially enhanced by either insulin administration or respiratory alkalosis.
2. Nutritional recovery syndrome. This syndrome occurs as people who are severely malnourished are re-fed. Severe hypophosphatemia is often seen, and death sometimes occurs.
3. Use of antacids that bind phosphate. The divalent cations in antacids, such as aluminum hydroxide, magnesium hydroxide, or aluminum carbonate, will bind phosphate and prevent its absorption in the gut. Hypophosphatemia usually requires chronic excessive use of antacids.
4. Alcohol withdrawal. Because alcohol itself has little effect on phosphate excretion, the cause of hypophosphatemia in this condition is complex. Five factors may partially contribute: (a) poor diet, (b) use of antacids, (c) vomiting, (d) alcohol-induced magnesium deficiency, and (e) ketoacidosis.

In addition, hypophosphatemia can also be caused by increased renal excretion, as with hyperparathyroidism, and decreased intestinal absorption.

Hyperphosphatemia. An increased intake of phosphate or increased release of cellular phosphate may cause *hyperphosphatemia*. The most common cause for a decrease is renal phosphate excretion due to renal failure. Because they may not yet have developed mature PTH and vitamin D metabolism, neonates are especially susceptible to hyperphosphatemia caused by increased intake, such as from cow's milk or laxatives. Increased breakdown of cells can sometimes lead to hyperphosphatemia, as with severe infections, intensive exercise, neoplastic disorders, or intravascular hemolysis. Because immature lymphoblasts have about four times the phosphate content of mature lymphocytes, patients with lymphoblastic leukemia are especially susceptible to hyperphosphatemia.

Determination of Inorganic Phosphorus

Specimen. Serum or lithium heparin plasma is acceptable for analysis. Oxalate, citrate, or EDTA anticoagulants should not be used because they interfere with the analytical method. Hemolysis should be avoided because of the higher concentrations inside the red cells. Circulating phosphate levels are subject to circadian rhythm, with highest levels in late morning and lowest in the evening. Urine analysis for phosphate requires a 24-hour sample collection because of significant diurnal variations.

Methods. Most of the current methods for phosphorous determination involve the formation of an ammonium phosphomolybdate complex. This complex can be measured at 340 nm (Hitachi, Dimension, Synchron CX) or can be reduced to form molybdenum blue and read at 660 nm (Vitros).

Reference Ranges

Phosphate values vary with the age of the patient. Divided into age groups, the ranges are shown in Table 14-17.

TABLE 14-17. Reference Ranges for Phosphate

Serum, plasma	
Neonate:	1.45–2.91 mmol/L (4.5–9.0 mg/dL)
Child:	1.45–1.78 mmol/L (4.5–5.5 mg/dL)
Adult:	0.87–1.45 mmol/L (2.7–4.5 mg/dL)
Urine (24-hour)	13–42 mmol/day (0.4–1.3 g/day)

Lactate

Lactate Biochemistry and Physiology

Lactate is a by-product of an emergency mechanism that produces a small amount of ATP when oxygen delivery is severely diminished.[26] Pyruvate is the normal end product of glucose metabolism (glycolysis). The conversion of pyruvate to lactate is activated when a deficiency of oxygen leads to an accumulation of excess NADH (Fig. 14-8). Normally, sufficient oxygen maintains a favorably high ratio of NAD to NADH. Under these conditions, pyruvate is converted to acetyl coenzyme A (CoA), which enters the citric acid cycle and produces 38 moles of ATP for each mole of glucose oxidized. However, under hypoxic conditions acetyl-CoA formation does not occur and NADH accumulates, favoring the conversion of pyruvate to lactate through anaerobic metabolism. As a result, only 2 moles ATP are produced for each mole of glucose metabolized to lactate, with the excess lactate released into the blood. This release of lactate into

AEROBIC METABOLISM

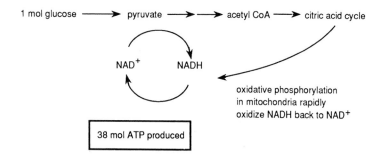

ANAEROBIC METABOLISM

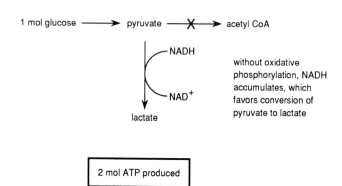

Figure 14-8. Aerobic versus anaerobic metabolism of glucose.

blood has clinical importance because the accumulation of excess lactate in blood is an early, sensitive, and quantitative indicator of the severity of oxygen deprivation (Fig. 14-9).

Regulation

Because lactate is a by-product of anaerobic metabolism, lactate is not specifically regulated, as is potassium or calcium, for example. As oxygen delivery decreases below a critical level, blood lactate concentrations rise very rapidly and indicate tissue hypoxia earlier than does pH. The liver is the major organ for removing lactate by converting lactate back to glucose by a process called *gluconeogenesis*.

Clinical Applications

Studies performed since the early 1970s have shown that measurements of blood lactate are useful for metabolic monitoring in critically ill patients, for indicating the severity of the illness, and for objectively determining patient prognosis.[26] In general, the following are indicators that a greater level of therapeutic intervention will be needed in a patient:

- High lactate level (>5–6 mmol/L) following surgery
- Lactate remains elevated despite medical treatment
- Rise in blood lactate during treatment

The preceding conditions indicate that the patient is experiencing a deficit of oxygen delivery or oxygen consumption in the tissues. Therapeutic procedures that may be used in these conditions include vasodilators, cardiac stimulants, fluid administration (volume support), ventilation, and transfusion of red cells (for anemia).

Lactate concentrations also have been used to monitor administration of nitroprusside, a fastacting, potent vasodilator that is sometimes used following surgery to increase peripheral blood flow. Because cyanide is a metabolite of nitroprusside, excess nitroprusside can block oxidative metabolism, resulting in the production of lactate. Therefore, blood lactate measurements sensitively indicate when excess nitroprusside is causing cyanide toxicity.

Determination of Lactate

Specimen Handling. Special care should be practiced when collecting and handling specimens for lactate analysis. Ideally, a tourniquet should be not be used because venous stasis will increase lactate levels. If a tourniquet is used, it should be released for 1 minute to 2 minutes before blood collection and the patient should not exercise the hand before or during collection.[27] After sample collection, glucose is converted to lactose by way of anaerobic glycolysis and should be prevented. Heparinized blood may be used but must be delivered on ice and the plasma must be quickly separated. Iodoacetate or fluoride, which inhibit glycolysis without affecting coagulation, are usually satisfactory additives, but the specific method directions must be consulted.

Methods. Although lactate is a sensitive indicator of inadequate tissue oxygenation, the use of blood lactate measurements has been hindered because older methods were slow and laborious and required relatively large volumes of blood. Thus, other means of following perfusion or oxygenation have been used, such as indwelling catheters that measure blood flow, pulse oximeters, base-excess determinations, and measurements of oxygen consumption (VO$_2$). However, more suitable methods for lactate measurement are available.

A review of methods for lactate shows a progression from those using chemical oxidation to those using specific enzyme reactions or enzyme electrodes. The early methods, based on chemical oxidation, used permanganate along with either H$_2$SO$_4$ or oxygen to produce acetaldehyde and carbon monoxide or carbon dioxide. These methods were fairly accurate but also were laborious, time consuming, and not applicable to *STAT* determinations. Current enzymatic methods make lactate determination readily available.

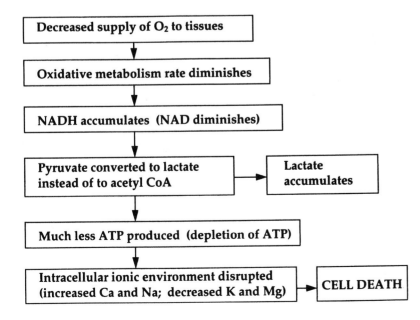

Figure 14-9. Metabolic effects of hypoxia, leading to cell death.

The most commonly used enzymatic method uses lactate oxidase to produce pyruvate and H_2O_2.

$$\text{Lactate} + O_2 \xrightarrow{\text{Lactate oxidase}} \text{pyruvate} + H_2O_2 \quad \textit{(Eq. 14–7)}$$

One of two couple reactions may then be used. Peroxidase may be used to produce a colored chromogen from H_2O_2.

$$H_2O_2 + \text{H donor} + \text{chromogen} \xrightarrow{\text{Peroxidase}}$$
$$\text{colored dye} + 2H_2O \quad \textit{(Eq. 14–8)}$$

In a second coupled reaction, the H_2O_2 is oxidized back to O_2 by a platinum anode and measured by way of pO_2 measurement.

Reference Ranges[5]
See Table 14-18.

ANION GAP

Routine measurement of electrolytes usually involves only Na^+, K^+, Cl^-, and HCO_3^- (as total CO_2). These values may be used to approximate the *anion gap (AG)*, which is the difference between unmeasured anions and unmeasured cations. There is never a "gap" between total cationic charges and anionic charges. The AG is created by the concentration difference between commonly measured cations (Na + K) and commonly measured anions (Cl + HCO_3), as shown in Figure 14-10. AG is useful in indicating an increase in one or more of the unmeasured anions in the serum. It is also helpful as a form of quality control for the analyzer used to measure these electrolytes. Consistently abnormal anion gaps in serum from healthy persons may indicate an instrument problem.[28]

There are two commonly used methods for calculating the anion gap. The first equation is

$$AG = Na^+ - (Cl^- + HCO_3^-) \quad \textit{(Eq. 14–9)}$$

It is equivalent to unmeasured anions minus the unmeasured cations in this way:

$$AG = (\text{protein} + \text{organic acids} + PO_4^- + 2SO_4^{2-})$$
$$- (K^+ + 2Ca^{+2} + Mg^{+2}) \quad \textit{(Eq. 14–10)}$$

The reference range for the AG using this calculation is 7 mmol/L to 16 mmol/L.[5] The second calculation method is

$$AG = (Na^+ + K^+) - (Cl^- + HCO_3^-) \quad \textit{(Eq. 14–11)}$$

It has a reference range of 10 mmol/L to 20 mmol/L.[5]

TABLE 14-18. Reference Ranges for Lactate
Enzymatic method, plasma
Venous: 0.5–2.2 mmol/L (4.5–19.8 mg/dL)
Arterial: 0.5–1.6 mmol/L (4.5–14.4 mg/dL)
CSF: 1.1–2.4 mmol/L (10–22 mg/dL)

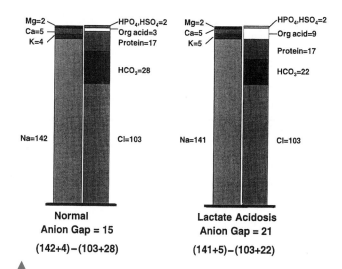

▲ *Figure 14-10.* Demonstration of anion gap from concentrations of anions and cations in normal state and in lactate acidosis.

Low anion gap values are rare but may be seen in multiple myeloma (due to abnormal proteins) or in cases of instrument error. An elevated AG may be caused by uremia, which leads to PO_4^- and SO_4^{2-} retention; ketoacidosis, as seen in cases of starvation or diabetes; poisoning due to ingestion of methanol, ethylene glycol, salicylate, or paraldehyde; lactic acidosis; severe dehydration, which causes increased plasma proteins, and instrument error.

ELECTROLYTES AND RENAL FUNCTION

The kidney is central to the regulation and conservation of electrolytes in the body. For a review of kidney structure, refer to Figure 14-11 and Chapter 21, *Renal Function*. The following is a summary of electrolyte excretion and conservation in a healthy individual:

1. Glomerulus: This portion of the nephron acts as a filter, retaining large proteins and protein-bound constituents while most of the other plasma constituents pass into the filtrate. Their concentrations in the filtered plasma should be approximately equal to the ECF without protein.
2. Renal tubules
   a. Phosphate reabsorption is inhibited by PTH and increased by 1,25-dihydroxycholecalciferol. Excretion of PO_4 is stimulated by calcitonin.
   b. Calcium is reabsorbed under the influence of PTH and 1,25-dihydroxycholecalciferol. Calcitonin stimulates excretion of calcium.
   c. Magnesium reabsorption occurs largely in the thick ascending limb of Henle's loop.
   d. Sodium reabsorption can occur through three mechanisms:
      (1) Approximately 70% of the sodium in the filtrate is reabsorbed in the proximal tubules by iso-

CASE STUDY 14-5

A 15-year-old girl in a coma was brought to the emergency room by her parents. She has been an insulin-dependent diabetic for 7 years. Her parents stated that there have been several episodes of hypoglycemia and ketoacidosis in the past, and that their daughter has often been "too busy" to take her insulin injections. The laboratory results obtained on admission are shown in Case Study Table 14-5.1.

CASE STUDY TABLE 14-5.1. Laboratory Results

	Result	Reference Range
Venous Blood		
Sodium	145 mmol/L	136–145 mmol/L
Potassium	5.8 mmol/L	3.4–5.0 mmol/L
Chloride	87 mmol/L	98–107 mmol/L
Bicarbonate	8 mmol/L	22–29 mmol/L
Glucose	1050 mg/dL	70–110 mg/dL
Urea nitrogen	35 mg/dL	7–18 mg/dL
Creatinine	1.3 mg/dL	0.5–1.3 mg/dL
Lactate	5 mmol/L	0.5–2.2 mmol/L
Osmolality	385 mOsmol/kg	275–295 mOsmol/kg

Arterial Blood		
pH	7.11	7.35–7.45
PO_2	98 mm Hg	83–100 mm Hg
PCO_2	20 mm Hg	35–45 mm Hg
Urine		Normal
Glucose	4+	Negative
Ketones	4+	Negative

Questions

1. What is the diagnosis?
2. Calculate the anion gap. What is the cause of the anion gap result in this patient?
3. Why are the chloride and bicarbonate decreased? What is the significance of the elevated potassium value?
4. What is the significance of the plasma osmolality?

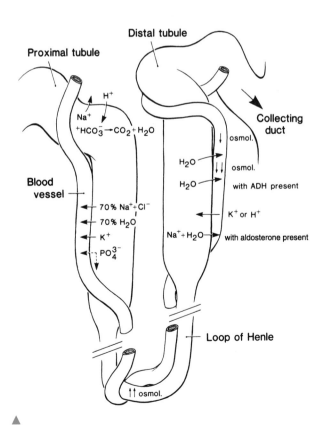

Figure 14-11. Summary of electrolyte movements in the renal tubules.

osmotic reabsorption. It is limited, however, by the availability of chloride so to maintain electrical neutrality.

(2) Sodium is reabsorbed in exchange for H^+. This reaction is linked with HCO_3^- and depends on carbonic anhydrase.

(3) Stimulated by aldosterone, Na^+ is reabsorbed in exchange for K^+ in the distal tubules. (H^+ competes with K^+ for this exchange.)

  e. Chloride is reabsorbed, in part, by passive transport in the proximal tubule along the concentration gradient created by Na^+.

  f. Potassium is reabsorbed by two mechanisms:

(1) Active reabsorption in the proximal tubule almost completely conserves K^+.

(2) Exchange with Na^+ is stimulated by aldosterone. H^+ competes with K^+ for this exchange.

  g. Bicarbonate is recovered from the glomerular filtrate and is converted to CO_2 when H^+ is excreted in the urine.

3. Henle's loop: With normal ADH function, it creates an osmotic gradient that enables water reabsorption to be increased or decreased in response to body fluid changes in osmolality.

4. Collecting ducts: Also under ADH influence, final adjustment of water excretion is made here.

SUMMARY

Electrolytes are ions capable of carrying an electric charge. They are classified as anions or cations based on the type of charge they carry. Anions have a negative charge and move toward the anode; cations have a positive charge and migrate toward the cathode. Electrolytes are an essential component in numerous processes in the body including volume and osmotic regulation, myocardial rhythm and contractility, enzyme activation, regulation of ATPase ion pumps, acid-base balance, blood coagulation, neuromuscular excitability, and the production and use of ATP from glucose. Electrolytes discussed in this chapter were sodium, potassium, chloride, bicarbonate, magnesium, calcium, phosphate, and lactate. This chapter also discussed the metabolic physiology and regulation of each electrolyte, as well as the commonly used methods of assessment. Regulation of electrolyte concentrations in the narrow physiologic ranges is primarily accomplished by the kidneys.

REVIEW QUESTIONS

1. Which is the major intracellular cation?
   a. Calcium
   b. Magnesium
   c. Sodium
   d. Potassium
2. Which is the major extracellular cation?
   a. Chloride
   b. Sodium
   c. Magnesium
   d. Calcium
3. Osmolality can be defined as a measure of the concentration of a solution based on:
   a. The number of ionic particles present
   b. The number and size of the dissolved particles
   c. The number of dissolved particles
   d. The density of the dissolved particles
4. Hyponatremia may be caused by each of the following EXCEPT:
   a. Hypomagnesemia
   b. Aldosterone deficiency
   c. Prolonged vomiting or diarrhea
   d. Acute or chronic renal failure
5. Hypokalemia may be caused by each of the following EXCEPT:
   a. Acidosis
   b. Prolonged vomiting or diarrhea
   c. Hypomagnesemia
   d. Hyperaldosteronism
6. Hyperkalemia may be caused by each of the following EXCEPT:
   a. Acute of chronic renal failure
   b. Hypoaldosteronism
   c. Alkalosis
   d. Sample hemolysis

7. The main difference between a direct ISE and an indirect ISE is:
   a. The type of membranes that are used
   b. Direct ISEs use a reference electrode, whereas indirect ISEs do not
   c. Sample is diluted in the indirect method and not diluted in the direct method
   d. Whole blood samples can be measured with the direct method and not with the indirect method
8. Which method of analysis will provide the most accurate electrolyte results if a grossly lipemic sample is used?
   a. Indirect ISE
   b. Direct ISE
   c. Flame-emission photometry
   d. Atomic absorption
9. The most frequent cause of hypermagnesemia is due to:
   a. Increased intake
   b. Hypoaldosteronism
   c. Acidosis
   d. Renal failure
10. A hemolyzed sample will cause falsely increased levels of each of the following EXCEPT:
    a. Potassium
    b. Sodium
    c. Phosphate
    d. Magnesium

REFERENCES

1. Rose BD, ed. Clinical physiology of acid-base and electrolyte disorders. 4th ed. New York: McGraw Hill, 1994.
2. Toffaletti J. Self study course: understanding the clinical uses of blood gases and electrolytes. Washington, DC: AACC Press, 1993.
3. Dufour DR. Osmometry: the rational basis for use of an underappreciated diagnostic tool. Workshop at American Association for Clinical Chemistry national meeting, 1993.
4. Tietz NW. Textbook of clinical chemistry. Philadelphia: WB Saunders, 1994.
5. Tietz NW, ed. Clinical guide to laboratory tests. Philadelphia: WB Saunders, 1995.
6. Kumar S, Tomas B. Sodium. Lancet 1998;352:352.
7. Willatts SM. Lecture notes on fluid and electrolyte balance. Oxford, England: Blackwell Scientific, 1987.
8. Vitros DT60 operator's manual. Rochester, NY: Johnson & Johnson, 1996.
9. Karselis TC. The pocket guide to clinical laboratory instrumentation. Philadelphia: FA Davis, 1994.
10. Gennari FJ. Hypokalemia. New Eng J Med 1998;339(7).
11. Fleisher M, Gladstone M, Crystal D, et al. Two whole-blood multianalyte analyzers evaluated. Clin Chem 1989;35:1532.
12. Hitachi operator's manual. Indianapolis, IN: Boehringer Mannheim, 1997.
13. Elin RJ. Magnesium: the fifth but forgotten electrolyte. Am J Clin Pathol 1994; 102(5).
14. Polancic JE. Magnesium: metabolism, clinical importance, and analysis. Clin Lab Sci 1991;4(2).
15. Whang R. Clinical disorders of magnesium metabolism. Comp Ther 1997;23(3).
16. Toffaletti J, Cooper D, Lobaugh B. The secretion of parathyroid hormone in response to changes in either ionized calcium, protein-bound calcium, or magnesium in normal adults. Metabolism 1991; 40:814.
17. American Diabetes Association. Magnesium supplementation in the treatment of diabetes. Diabetes Care 1992;15:1065.

18. Chernow B, ed. The pharmacologic approach to the critically ill patient. 2nd ed. Baltimore: Williams & Wilkins, 1988.

19. Ringer S. A further contribution regarding the influence of different constituents of blood on contractions of the heart. J Physiol 1883;4:29.

20. McLean FC, Hastings AB. A biological method for estimation of calcium ion concentration. J Biol Chem 1934;107:337.

21. Drop LJ. Ionized calcium, the heart, and hemodynamic function. Anaesth Analg 1985;64:432

22. Gray TA, Paterson CR. The clinical value of ionized calcium assays. Ann Clin Biochem 1988;25:210.

23. Bushinsky DA, Monk RD. Calcium. Lancet 1998;352:23.

24. Wandrup J. Critical analytical and clinical aspects of ionized calcium in neonates. Clin Chem 1989;35:2027.

25. King AL, Sica DA, Miller G, et al. Severe hypophosphatemia in a general hospital population. South Med J 1987;80:831.

26. Toffaletti JG. Blood lactate: biochemistry, laboratory methods, and clinical interpretation. Crit Rev Clin Lab Sci 1991;28:253.

27. Kaplan LA, Pesce AJ. Clinical chemistry: theory, analysis, and correlation. St. Louis, MO: CV Mosby, 1996.

28. Lolekha PH, Lolekha S. Value of the anion gap in clinical diagnosis and laboratory evaluation. Clin Chem 1983;29:279.

Trace Elements

Susan A. Hurgunow

Objectives

Upon completion of this chapter, the clinical laboratorian should be able to:

* *Define the terms: chelator, essential trace elements, prooxidant, trace elements, and ultra-trace elements.*
* *Discuss the homeostatic regulation of the essential trace elements.*
* *State the biologic functions of the essential trace elements.*
* *Discuss the clinical significance of the trace elements.*
* *Discuss specimen collection considerations and laboratory determination.*

KEY TERMS

Chelator	Prooxidant	Trace elements

Essential trace
  elements
	Total iron binding
	  capacity (TIBC)
		Transferrin
		Ultra-trace elements

The importance of trace elements is of far greater magnitude than their concentrations in vivo suggest. *Trace elements* are found in mg/kg concentrations, whereas *ultra-trace elements* concentrations are μg/kg or less. Some of these elements are toxic and some are essential to biologic functions (Table 15-1). Deficit of an essential trace element is associated with impairment of function and excess concentrations may be toxic. An element is considered essential only if replacement of the element corrects the biochemical or func-tional impairment. Decreased intake, impaired absorption, and genetic abnormalities are examples of conditions that could result in deficiency of trace elements. The World Health Organization has established the dietary requirement for nutrients as the smallest amount of the nutrient needed to maintain optimal function and health.

This chapter presents information on the trace elements regarding biologic function, homeostasis, clinical signifi-cance of disease states or toxicity, and laboratory determina-tion. Table 15-2 summarizes some of this information for *essential trace elements.*

IRON

Most iron in the body is incorporated into heme of hemo-globin and a smaller proportion as myoglobin, heme-containing enzymes like catalase, and iron-sulfur proteins. A significant amount of iron is stored as ferritin and hemo-siderin, primarily in the bone marrow, spleen, and liver. Al-though plasma concentration of iron is low, total body iron content is approximately 4 g.

Iron Homeostasis

In an adult male, the average loss of 1 mg of iron per day must be replaced by dietary sources.[1] Pregnant or premenopausal women and children have greater iron requirements, often obtained by dietary supplementation. To address this con-cern, in the U.S. Food and Drug Administration has raised its

TABLE 15-1. Essential Trace Elements in Clinical Chemistry

	Proven Essential	Probably Essential	Nonessential To Date
Trace	Iron		
	Zinc		
	Copper		
Ultra-trace	Manganese	Nickel	Aluminum
	Cobalt	Vanadium	Arsenic
	Selenium	Tin	Cadmium
	Molybdenum		Fluoride
	Chromium		Gold
	Iodine		Lead
			Mercury
			Silicon

standards for supplementation of foods. Because the general population includes both iron deficient individuals and those with adequate iron status, there is debate about whether dietary iron supplements may contribute to iron overload in those people with adequate iron.[2-4] Undesirable side effects associated with iron supplementation seem minimized by use of bovine hemoglobin[5] and of lactoferrin.[6]

Because iron is well conserved by the body, absorption of iron from the intestine is the primary means of regulating the amount of iron within the body. When plasma iron is decreased, more iron is absorbed from the small intestine and vice versa. Substances present in the intestinal lumen, such as anionic phosphate and phytate, may bind iron and interfere with absorption.[7]

There are different proposed modes of iron absorption; the following are three examples: (1) it is believed that, before absorption, dietary FeIII must be reduced to the more soluble FeII[8-10] by the action of HCl and reducing agents such as ascorbic acid.[1] Entry of FeII into the mucosal cell may be aided by the presence of a transmembrane enzyme[11,12] with ferric reductase activity.[13-18] (2) Recent studies have suggested that FeIII binds to mucin on the intestinal mucosal cell surface[19] and is found associated with membrane surface proteins[20,21] identified as integrin in rats. Integrin transfers iron to cytosolic mobilferrin, a protein that has been isolated from human duodenum.[2,22] Mobilferrin then complexes with paraferritin, where FeIII is reduced to FeII.[11,23-25] (3) Because no transferrin receptor has been found on the apical border of the mucosal cell, it appears that dietary iron would not enter the mucosal cell by way of the transferrin receptor route.[24,26]

In the mucosal cell of an individual with adequate plasma iron concentration, iron is stored by incorporation into ferritin, a protein comprising an apoferritin shell enclosing up to 4500 ferric atoms. FeII entering ferritin is oxidized by ferritin ferroxidase and FeIII leaving ferritin is reduced, possi-

TABLE 15-2. Trace Element Function and Abnormalities

Metals	Reference Values	Biochemical Functions	Deficiency Characteristics	Toxicity Characteristics
Iron	serum	Oxygen transport		Hepatic failure
	65–170 µg/dL, male	Respiration	Anemia	Cardiomyopathy
	50–170 µg/dL, female	Oxidative processes		Peripheral neuropathy
Zinc	serum	Hemoglobin synthesis	Impaired wound healing	Ataxia
	75–125 mg/dL	Collagen metabolism	Retarded growth	Pancreatitis
		Bone development	Skeletal abnormalities	Anemia, fever
		Growth and reproduction	Male impotence	Nausea, vomiting
Copper	serum	Pigmentation	Disorders in pigmentation	Nausea, vomiting
	70–140 µg/dL, male	Bone development	Retarded growth	Liver necrosis
	80–155 µg/dL, female	Oxygen transport	Anemia in children	Hemolytic anemia
		Nucleic acid synthesis	Wilson's disease	Renal dysfunction
		Protein synthesis	Menkes' syndrome	Neurologic dysfunction
Manganese	serum	Growth and reproduction	Disorders in spermatogenesis	Neurologic alterations
	0.5–1.0 µg/L	Oxidative phosphorylation	Bone abnormalities	Psychosis
		Cholesterol metabolism	Bleeding disorders	Speech disturbances
Cobalt	serum	Vitamin B_{12} function	Megaloblastic anemia	Gastrointestinal function
	0.11–0.45 µg/L	Methionine metabolism		Cardiomyopathy
Selenium	whole blood	Oxygen metabolism	Cardiovascular disease	Neurotoxicity
	58–234 µg/dL	Free radical protection	Muscle degeneration	Hepatotoxicity
			Carcinogenesis	
Molybdenum	serum	Xanthine metabolism	Mental disturbances	Hyperuricemia
	0.1–3.0 µg/L		Esophageal cancer	
Chromium	serum	Insulin uptake	Reduced glucose tolerance	Renal failure
	<0.5 µg/L	Glucose metabolism		Pulmonary cancer

bly by ascorbate or flavin mononucleotide.[1,27] The short life of an intestinal mucosal cell results in loss of ferritin when the cell is shed.[1,28]

When intracellular iron concentration is low, an iron regulatory protein inhibits the synthesis of apoferritin and promotes the synthesis of transferrin receptor.[26,29–32] Newly absorbed iron or iron released from ferritin is converted from FeII to FeIII by ceruloplasmin, transferred to apotransferrin at the basolateral cell membrane and released into the circulation as the holoprotein transferrin.[1] In normal states, both intracellular storage ferritin[33] and circulating transferrin[34] are only partially saturated. The very small amount of circulating ferritin is mostly apoferritin that contains little iron.[35] *Transferrin* delivers iron to the tissues for use, for example, to the bone marrow for synthesis of heme for erythrocytes. At the tissue site, transferrin receptors provide recognition sites for attachment, followed by internalization.

Biologic Function

Oxygen binds reversibly with FeII of erythrocyte hemoglobin in the lung and is transported to the tissues. Oxygen is released in the tissues and carbon dioxide is returned to the lung. Iron in hemoglobin must remain in the FeII state to carry oxygen. FeIII in hemoglobin (methemoglobin) is nonfunctional and is converted to the functional FeII by methemoglobin reductase. Decreased oxygen affinity of hemoglobin is caused by increased $[H^+]$, PCO_2, or 2,3-diphosphoglycerate, resulting in increased release of oxygen from hemoglobin in anemic conditions. Because myoglobin in tissues binds oxygen with greater affinity than does hemoglobin, myoglobin facilitates diffusion of oxygen into tissues. A decrease in myoglobin could be caused by iron deficiency, thereby decreasing oxygen diffusion into tissues.[1]

The respiratory chain comprises heme proteins, Fe-S centers, and Cu centers. Electron transport by way of the cytochromes relies on the reversible FeII/FeIII redox reaction, resulting ultimately in the production of energy as adenosine triphosphate.[1]

Peroxidases, catalase, and thyroperoxidase are iron-containing enzymes. Peroxidase and catalase convert potentially harmful hydrogen peroxide to water. Thyroperoxidase converts inorganic iodide to a reactive form that can be incorporated into hormone precursors in the thyroid gland. The cytochrome P-450 enzymes function in steroid hormone synthesis, bile acid synthesis, and drug metabolism.[1] Decreased iron concentration has been associated with changes in P-450 metabolism of some drugs. Increased iron concentration has been associated with NADPH-dependent lipid peroxidation.[36]

Iron Deficiency

In the United States, certain groups are still at risk for iron deficiency anemia: pregnant women and small numbers of young children, adolescents, and women of reproductive age. Worldwide, 15% of the population is affected; in the high-risk groups, 50% may be affected.[1,37]

Increased blood loss, decreased dietary iron intake, or decreased release from ferritin may result in iron deficiency anemia. Reduction in iron stores usually precedes a reduction in circulating iron and anemia demonstrated by decreased number of red blood cells, microcytosis, and mean corpuscular hemoglobin concentration.[1] Refer to Table 15-3.

Iron Overload

Excessive iron stores may be caused by primary hemochromatosis or a variety of secondary conditions. Primary hemochromatosis is caused by a genetic defect that causes tissue accumulation of iron and eventual damage. Secondary conditions include iron-loading anemias, repeated red cell transfusions, long-term intake of supplemental iron, some liver diseases, and porphyria cutanea tarda. Treatment may include therapeutic phlebotomy or administration of a *chelator* such as deferoxamine. Transferrin is administered in the case of atransferrinemia, and ceruloplasmin, in hereditary ceruloplasmin deficiency.[38] The laboratory parameters associated with aceruloplasminemia are discussed in the section on copper deficiency. Refer to Table 15-3.

TABLE 15-3. Laboratory Markers of Iron Status in Several Disease States

Condition	Serum Iron (65–170 µg/dL)	Transferrin (200–400 mg/dL)	% Saturation (20–55)	Ferritin (20–250 µg/L)
Iron deficiency	Decreased	Increased	Decreased	Decreased
Iron poisoning/overdose	Increased	Decreased	Increased	Increased
Hematochromatosis	Increased	Decreased	Increased	Increased
Malnutrition	Decreased	Decreased	Variable	Decreased
Malignancy	Decreased	Decreased	Decreased	Increased
Chronic infection	Decreased	Decreased	Decreased	Increased
Viral hepatitis	Increased	Increased	Normal/increased	Increased
Anemia of chronic disease	Decreased	Normal/decreased	Decreased	Normal/increased
Sideroblastic anemia	Increased	Normal/decreased	Increased	Increased

Proposed Role of Iron in Tissue Damage

Iron is a transition element capable of reduction/oxidation (redox) activity, and potential harm to the body is circumvented by binding iron to transport or storage proteins.[39] In recent years, the role of iron has been investigated as a *prooxidant,* which contributes to lipid peroxidation,[40–43] atherosclerosis,[44,45] deoxyribonucleic acid (DNA) damage,[46,47] carcinogenesis,[48–50] and neurodegenerative diseases.[51–55] FeIII, released from binding proteins, can participate in the production of free radicals by the Haber-Weiss or Fenton reaction and cause oxidative damage.[53,56–58] See Equations 15-1 to 15-3 that follow. In iron-loaded individuals with thalassemia who are treated with chelators to bind and mobilize iron, intake of ascorbic acid may lead to toxicity from oxidative processes.[53,57–60]

The Haber-Weiss or Fenton reactions are[61]

$$O_2^- + H_2O_2 \longrightarrow O_2 + OH^- + OH \quad (Eq.\ 15\text{--}1)$$

$$O_2^- + Fe \longrightarrow O_2 + Fe^{+2} \quad (Eq.\ 15\text{--}2)$$

$$Fe^{+2} + H_2O_2 \longrightarrow Fe^{+3} + OH^- + OH$$

$$(Eq.\ 15\text{--}3)$$

Laboratory Evaluation of Iron Status

Disorders of iron metabolism are evaluated with determination of packed cell volume, hemoglobin, red cell count and indices, total iron and total iron binding capacity (TIBC), percent saturation, and ferritin. Expected results are outlined in Tables 15-3 and 15-4.

A recent addition to the laboratory evaluation of iron status is the measurement of serum transferrin receptors (circulating TrfR). The circulating TrfR is produced by proteolysis of the cellular form and is primarily attached to transferrin.[62–64] Cellular transferrin receptor spans the membrane of most cells and is the means of cellular iron acquisition.[65] Tissues with highest iron requirements have the greatest concentration of transferrin receptors and the number is regulated inversely with the intracellular iron content.[66–68]

Measurement of the circulating TrfR is used primarily to differentiate types of microcytic anemia. Serum concentration of circulating TrfR increases in iron deficiency anemia once iron stores are depleted and ferritin concentration is decreased. This increase in circulating TrfR occurs before changes in mean corpuscular volume, percent saturation, and erythrocyte protoporphyrin. In inflammation accompanied by anemia of chronic disease, circulating TrfR is relatively unchanged. Serum ferritin concentration is decreased in iron deficiency anemia and increased with inflammatory processes.[66,69–72]

Total Iron Content (Serum Iron)

The specimen may be collected without anticoagulant or with heparin. Oxalate, citrate, or ethylenediaminetetraacetic acid bind Fe and are unacceptable anticoagulants. Early morning sampling is preferred due to the diurnal variation in iron concentration. Specimens with visible hemolysis should be rejected.

Spectrophotometric determinations have readily been adapted to automated analysis. These procedures have the following general steps: FeIII is released from binding proteins by acidification, reduced to FeII by ascorbate or a similar reducing agent, and complexed with a color reagent such as ferrozine, ferene, or bathophenanthroline.

Total Iron Binding Capacity

Total iron binding capacity (TIBC) refers to the amount of iron that could be bound by saturating the binding proteins present in the sample. The principle binding protein is transferrin and is typically partially (one third) saturated. TIBC is determined by the following:

- Sufficient FeIII is added to saturate the unfilled sites on binding proteins, followed by addition of $MgCO_3$ to precipitate any FeIII remaining in solution. After centrifugation to remove unbound FeIII, the supernatant solution containing the soluble saturated binding proteins is analyzed for total iron content as previously described.
- A known concentration of FeIII is added to saturate the unfilled sites on binding proteins, followed by measurement of unbound FeIII in the solution (the "excess iron"). The FeIII bound to binding proteins will not react under these conditions.

TIBC is calculated by the following formula:

$$UIBC = Fe\ added - excess\ Fe \quad (Eq.\ 15\text{--}4)$$

$$TIBC = UIBC + total\ Fe\ content \quad (Eq.\ 15\text{--}5)$$

TABLE 15-4. Reference Ranges for Parameters Used To Assess Iron Status

Patient Population	Serum Iron (µg/dL)	Transferrin (mg/dL)	Ferritin (µg/L)	% Saturation	TIBC (µg/dL)
Adult, male	65–170	200–400	20–250	20–55	250–450
Adult, female	50–170	200–400	10–120	15–50	250–450
0–5 months	40–250	130–275	50–600	12–50	100–400
6 months–6 years	50–120	150–400	7–140	12–60	100–450
7–17 years	50–150	200–400	10–150	15–55	250–450

Percent Saturation

The percent saturation is sometimes referred to as transferrin saturation, but keep in mind that transferrin is not the only protein that binds iron. Transferrin makes up approximately 70% to 90% of the binding proteins. The percent saturation is calculated as follows:

$$\% \text{ saturation} = \frac{\text{total iron content}}{\text{TIBC}} \times 100\%$$

(Eq. 15–6)

Transferrin and Ferritin

Transferrin is measured by immunochemical methods such as nephelometry. Aside from its use in evaluation of iron status, transferrin is monitored as an indicator of nutritional status. As a negative acute phase protein, its concentration decreases in inflammatory conditions.

Serum transferrin receptor measurement by enzyme-linked immunosorbent assay (ELISA) is available. Calibrators of different sources produce different reference intervals: calibrators used include purified serum transferrin receptor and transferrin-bound transferrin receptor. The transferrin-bound transferrin receptor is the predominant form of circulating transferrin receptor.[73,74]

Ferritin is measured in serum by immunochemical methods such as IRMA, ELISA, and chemiluminescent techniques. Serum ferritin concentration, a marker for iron stores, decreases before total iron content is depressed. Evaluation of storage iron status may be confounded by concurrent inflammatory or neoplastic conditions that elevate serum ferritin concentration.

COPPER

Copper Homeostasis

Significant sources of copper include shellfish, liver, nuts, and legumes.[75,76] Copper leaching from plumbing does not have much effect on copper intake, although some water supplies have been found to have appreciable copper content.[77]

In the intestine, the absorption of copper is modulated by need, resulting in a 55% to 75% rate of absorption. After diffusion across the luminal membrane, copper is bound to metallothionein in the cytosol of the mucosal cell. Intracellular metallothionein normally stores little copper, providing protection from the harmful effects of free copper. A saturable, carrier-mediated transporter through the basolateral membrane transports copper into interstitial fluid. Zinc or iron can compete with copper for this transport system; therefore, high zinc intake can interfere with the transfer of copper to the circulation.[75,76,78–80] Another mode of interference is the zinc-mediated induction of the synthesis of metallothionein. This interference increases the binding of copper to this protein, effectively trapping copper within the cell.[81] The principal route of copper excretion is by secretion into the bile.[76]

A small amount of circulating copper is bound to albumin and to the high affinity transcuprein, whereas most is incorporated into ceruloplasmin. Ceruloplasmin is synthesized in the liver and is believed to have ferroxidase activity, converting FeII to FeIII as it is incorporated into transferrin.[82] Ceruloplasmin is an acute phase reactant in which increased concentrations are thought to be a response acting as an oxygen radical scavenger.[76,83,84]

Biochemical Function

An important role for copper is as a component of enzymes or proteins involved in redox reactions. This function, as well as other functions, could be compromised by severe copper deficiency.[76] Refer to Table 15-2. Cytochrome c oxidase catalyzes the reduction of oxygen to water in the last step of the electron transport chain. The enzyme contains two heme groups and two copper atoms that mediate the electron transfer. Superoxide dismutase (SOD) is extremely important to antioxidant defense. SOD, containing copper and zinc, catalyzes the conversion of two O_2^- to O_2 and H_2O_2 in the cytosol. The peroxide produced in this reaction is destroyed by catalase containing heme or glutathione peroxidase containing selenium. Genetic mutations in SOD have been cited in the development of amyotropic lateral sclerosis.[85] (Fig. 15-1).

In connective tissue, formation of collagen cross-links is accomplished by the oxidative deamination of the lysine side chains of proteins by lysyl oxidase. Severe copper deficiency affects collagen maturation. The probability of aneurisms increases due to blood vessel defects that can occur in severe copper deficiency.[75,76] Tyrosinase, a copper-containing en-

---

CASE STUDY 15-1

A patient with thalassemia was under treatment with Desferal™ (deferoxamine mesylate). The patient failed to follow the instructions regarding concurrent use of Vitamin C, developed cardiac failure, and died 4 hours after ingesting a large amount of ascorbic acid.

Questions

1. What is the role of Desferal™ in treatment?
2. How does Vitamin C interact with iron?

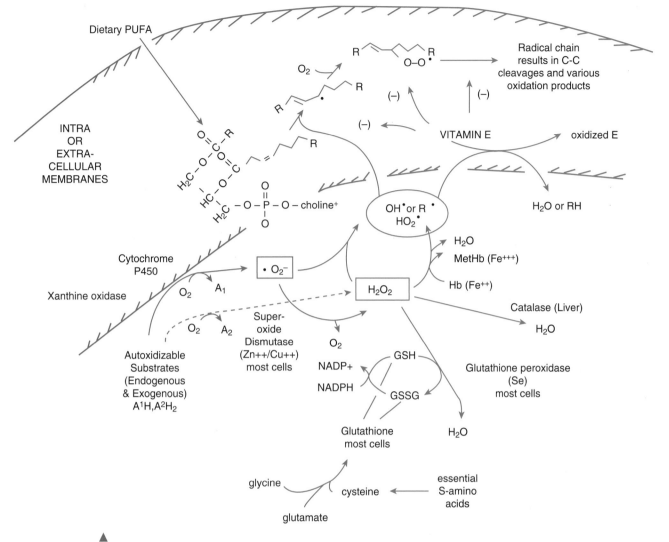

Figure 15-1. SOD catalyses the conversion of two O_2^- to O_2 and H_2O_2 in the cytosol. The peroxide produced in this reaction is destroyed by catalase containing heme or glutathione peroxidase containing selenium. From Nutritional Biochemistry and Metabolism with Clinical Applications, M. C. Linder, Ed., Elsevier Science, 1991.

zyme, is involved in the production of melanin. Dopamine β-monooxygenase catalyzes the production of the catecholamine norepinephrine, a neurotransmitter.[75]

Copper Deficiency

Copper deficiency is uncommon, but certain circumstances promote its occurrence. Examples include malnutrition, malabsorption, and conditions resulting in chronic copper loss. Zinc competes with copper for absorption from the intestine; therefore, increased intake of zinc could cause copper deficiency. Copper deficiency causes a microcytic, hypochromic anemia associated with low concentrations of ceruloplasmin. Iron accumulates in the liver and may be mobilized by administration of ceruloplasmin.[75] Logan and coworkers have reported two brothers with very low serum iron, high liver iron content, very high serum ferritin, normal TIBC, normal red blood cell count, low serum copper, and almost undetectable ceruloplasmin. Ceruloplasmin ad-

ministration caused the iron concentration to rise, confirming aceruloplasminemia.[86,87]

An early feature of copper deficiency is neutropenia, which has been observed in several conditions, including diet low in copper,[88] diarrheal loss,[89] chelation therapy in Wilson's disease,[90] peritoneal dialysis,[91] and celiac disease.[92] One study of a copper-deficient 68-year-old woman demonstrated a reduced number of mature neutrophils in marrow, erythrocyte SOD activity 20% of normal, and erythrocyte life span of 19 days versus 28 days to 40 days in healthy individuals. The lack of the copper-containing antioxidant enzyme SOD seemed related to the shortened life of erythrocytes and neutrophils.[93,94]

Menkes' syndrome is caused by a recessive X-linked genetic defect in an adenosine triphosphatase (ATPase) for copper transport. Copper is absorbed in a normal manner into the mucosal cell, accumulates there because of defective transport from the cell, and results in a copper deficiency syndrome. Mental deterioration, failure to thrive, deficient copper-containing enzyme activity, connective tissue abnormalities, kinky hair, and early death are features of this disease.[76] If started early enough, the condition may be treated with copper-histidine.[95]

Copper Excess

Accidental ingestion of copper compounds may cause acute copper toxicity and associated epigastric pain, nausea, and vomiting.[96,97] Chronic ingestion has been observed as infantile cirrhosis in which the etiology was determined to be milk boiled in brass containers. This resulted in copper intake greater than ten times the recommended intake.[96,98] Excess copper, as with excess iron, can cause free radical production and damage (see Eq. 15-1 to 15-3).[99,100]

Wilson's disease, or hepatolenticular degeneration, is a genetically determined defect in ATPase, although the transport defect is different from that of Menkes' syndrome. In Wilson's disease, copper is transported normally from the intestine into the liver, but cannot be transported out of the liver into the bile.[76] Patients develop copper overload in brain and liver,[76,101] resulting in cirrhosis of the liver and brain lesions. Copper cannot be incorporated into ceruloplasmin, thus causing a low serum ceruloplasmin concentration. This finding along with increased urinary copper excretion and copper deposits in the cornea—Kayser-Fleischer rings—are used to diagnose Wilson's disease.[76] Zinc administration may be used to reduce copper absorption.[75,79] Administration of BAL (dimercaprol), penicillamine, or ammonium tetrathiomolybdate withdraws copper from the patient by chelating copper and increasing urinary excretion. Penicillamine and BAL treatment have harmful side effects, whereas ammonium tetramolybdate blocks copper absorption and appears to preserve neurologic function.[102]

ZINC

Zinc Homeostasis

Zinc content of meat and fish[78] is high, and, as with other cationic trace elements, absorption is inhibited by increased dietary content of phytate (inositol phosphate).[78,103] Zinc induces the synthesis of metallothionine and competes for absorption with copper and iron.[79–81] Zinc is transported in the circulation by albumin and α_2-macroglobulin[78,104] and is found primarily in muscle, bone, and liver. The primary route of excretion of zinc is gastrointestinal.[105]

Biochemical Function

Zinc is a cofactor for more than 300 enzymes, and zinc-containing enzymes are found in every enzyme class. Alkaline phosphatase, alcohol dehydrogenase, carbonic anhydrase, DNA polymerase, and superoxide dismutase contain zinc.[106] Zinc enzymes are essential to growth, wound healing, reproductive function, the immune system, and protection from free radical damage.[78,106–108]

Zinc Deficiency and Toxicity

Malabsorption, chronic disease of liver and kidney, alcoholism, and diets high in phosphate or low in zinc are associated with zinc deficiency.[78,106] Tissue zinc is preserved at the expense of growth or by decreased excretion until the deficiency is severe. Plasma zinc concentration does not change significantly until zinc deficiency is pronounced.[105] Symptoms of deficiency include skin lesions, diarrhea, impotence, dwarfism, sensory alterations, and susceptibility to infection.[78,106,109] Reduced immune function of T cells has been associated with deficiency of zinc, which acts as a growth factor.[110] Zinc itself is relatively nontoxic, but its interaction with copper produces effects discussed in the copper section of this chapter.[78]

Laboratory Evaluation of Zinc Status

An assay for metallothionine has been used to evaluate body zinc status. Dietary changes in zinc concentrations were reflected in erythrocyte metallothionine, whereas plasma zinc concentration was relatively unchanged.[111]

Plasma zinc is measured by atomic absorption spectroscopy. Specimen considerations include diurnal variation[112] and hemolysis.

COBALT

As a constituent of vitamin B_{12}, cobalt is involved in folate metabolism and erythropoiesis. Inorganic cobalt salts are possibly absorbed by the same mechanism as iron and have toxic effects at high doses.[78]

CHROMIUM

Adequate chromium intake is not provided by the average U.S. diet,[113] based on the proportion of chromium to calories.[114] This, combined with urinary loss of chromium in stressful conditions such as acute exercise or trauma, may affect health.[115] In non–insulin-dependent diabetes mellitus (NIDDM), chromium supplementation has demonstrated improved glucose tolerance, reduced insulin concentrations, and decreased total cholesterol. Chromium increases the effects of insulin and decreases insulin requirements. Supplementation doses of chromium did not produce toxic effects at concentrations associated with improved glucose tolerance.[116]

FLUORINE

Prevention of dental caries by fluoride supplementation is well documented, as is the mottling of teeth by fluoride excess.[78] Fluoride is incorporated into bone crystal, increasing bone mass in the vertebrae in association with an increased incidence of vertebral and hip fractures. It is suggested that

CASE STUDY 15-2[116]

A 45-year-old female, treated for 2 years for NIDDM, relayed to her physician a continued non-compliance with dietary restrictions and suggested exercise regimen. After obtaining a laboratory report of a fasting blood glucose (FBS) of 138 mg/dL and total cholesterol of 289 mg/dL, her physician prescribed a chromium picolinate supplement of 500 µg twice daily. At the end of 4 months, her FBS decreased to 102 mg/dL and total cholesterol to 243 mg/dL.

Questions

1. What effect does chromium have on carbohydrate and lipid metabolism?
2. Is the form and dose of chromium administered related to its biologic activity?
3. Is CrIII toxic?

concurrent calcium deficiency causes redistribution of bone mineral to the growth area, and that administration of vitamin D_3 with cyclic administration of fluoride may enhance the bone formation, correct calcium deficiency, and lessen the occurrence of fracture.[117,118]

MANGANESE

The rate of absorption of manganese is low and is decreased by phosphate, phytate, calcium, and iron. Manganese is quite high in erythrocytes,[119] is transported in plasma by albumin and transferrin,[120] and is excreted in bile and pancreatic secretions. Many enzymes are associated with manganese including pyruvate carboxylase, mitochondrial superoxide dismutase, arginase, and glucokinase. The effect of manganese deficiency is minimized by substitution of other similar ions in enzymes; therefore, symptoms of manganese deficiency are extremely rare. Very high dosages, except by inhalation, are not toxic.[78] Recent reports of patients having magnetic resonance imaging using manganese agents and patients with cholestatic liver disease have demonstrated manganese deposition in brain tissue. This occurrence may be associated with neurologic symptoms.[120,121]

MOLYBDENUM

Dietary molybdenum is absorbed at a high rate and may inhibit absorption of copper and iron. It can form a thiomolybdate complex with copper and interfere with copper absorption.[78] The formation of a molybdenum–copper complex is the basis of a treatment for Wilson's disease, as discussed in the copper section of this chapter.[102]

Molybdenum is incorporated into several oxidases: xanthine dehydrogenase/xanthine oxidase, aldehyde oxidase, and sulfite oxidase. Xanthine oxidase converts hypoxanthine to uric acid, aldehyde oxidase catalyzes the conversion of acetyl CoA to acetate, and sulfite oxidase converts sulfite to sulfate.[78] Xanthine oxidase and aldehyde oxidase produce O_2^-. and H_2O_2, associated with damage in ischemia.[122,123]

SELENIUM

The availability of selenium in the soil varies widely and is reflected in the selenium content of vegetation grown there. Toxicity from selenium could result from food with high concentrations of selenium, and, conversely, selenium deficiency may occur when food sources are low in selenium.

CASE STUDY 15-3[117]

A 74-year-old female recently fractured her left femur. The patient had been treated for osteoporosis with estrogen, calcium, and fluoride for 2 years before the fracture, and serial bone density measurements have showed a continuing increase in density. At the time of hospitalization, her urinary calcium output was 0.9 mg/day.

Questions

1. What is the role of fluoride in treatment of osteoporosis?
2. Is the fluoride treatment associated with increased incidence of fractures?
3. Why was the calcium output very low in this patient?

Selenium is incorporated into proteins, notably glutathione peroxidase,[78] which may be measured to reflect selenium status. Low selenium intake has been demonstrated in the cardiomyopathy of Keshan disease and in necrosis of cartilage of Kashin-Beck disease.[124] Recent studies of effects of micronutrients on cancer risk have demonstrated that selenium supplementation is associated with a decreased risk of stomach carcinoma, lung, prostate, and total cancer. Risk of stomach cancer is reduced with concurrent intake of beta-carotene and alpha-tocopherol.[125] These findings need further study.[126]

GENERAL CONSIDERATIONS IN LABORATORY DETERMINATIONS OF TRACE ELEMENTS

Specimens for analysis must be collected with scrupulous attention to well-defined procedures aimed at minimizing contamination from the environment, collection apparatus, and body surfaces. Methodology and instrumentation must have great analytical sensitivity and specificity due to the extremely low concentrations of the trace elements found in body fluids and to the physicochemical similarity of some elements. In the clinical laboratory, the most commonly used instrument is the atomic absorption spectrometer with flameless atomization. Reference ranges for trace and ultra-trace elements are listed in Table 15-2.

SUMMARY

Trace element concentrations are tightly controlled in the healthy individual. Impaired function is often due to decreased activity of enzymes that require trace elements for optimal activity. Function can be restored to normal by replacement therapy, but care must be exercised not to induce toxicity. Laboratory evaluation of trace element status is evolving and includes determination of functional status as well as body fluid concentration. There is much research concerning trace element effects in health and disease, and many controversies that further research should resolve.

REVIEW QUESTIONS

1. Iron is physiologically active in the following substance only in the ferrous form:
   a. Cytochromes
   b. Ferritin
   c. Hemoglobin
   d. Transferrin
2. Serum iron is most likely to be decreased in:
   a. Hepatitis
   b. Polycythemia
   c. Hemochromatosis
   d. In-vitro hemolysis
3. Deficiency of the following trace metal is associated with growth retardation, dermatitis, reduced taste acuity, and impaired wound healing:
   a. Copper
   b. Iron
   c. Selenium
   d. Zinc
4. The following metals are required for optimal activity of superoxide dismutase:
   a. Iron and chromium
   b. Zinc and selenium
   c. Magnesium and manganese
   d. Copper and zinc
5. Deficiency of the following metal can cause a deficiency of iron:
   a. Copper
   b. Chromium
   c. Cobalt
   d. Zinc

CASE STUDY 15-4[128]

A 3-year-old child, after dinner meal, started vomiting and became drowsy. She was brought to the emergency room, treated with oxygen, cefotaxime, and activated charcoal. Tachycardia was observed on admission. Laboratory results included normal Na, slightly elevated K, and extremely low calcium and magnesium. The patient was then given both calcium and magnesium intravenously. During the next few hours, two episodes of ventricular fibrillation occurred. The mother reported she had found an ammonium fluoride-containing cleaner in a glass, a possible source of poisoning. A toxic concentration of urinary fluoride confirmed this source.

Questions

1. How does ammonium fluoride exert its toxic effect?
2. Why are the calcium and magnesium so low?
3. What symptoms are associated with extremely depressed calcium and magnesium?

CASE STUDY 15-5

A 36-year-old male had undergone multiple partial small bowel resections because of Crohn's disease. After recovery from surgery, he was placed on total parenteral nutrition (TPN), and after adequate rehabilitation he was discharged under the care of a home health care nurse who visited him 3 times a week to follow his overall progress and check his physical and vital conditions. Periodic routine laboratory tests were performed to assess nutritional status. His first few months postsurgery were unremarkable. Eventually, he experienced symptoms of weakness, diarrhea, general malaise, and hair loss. He visited his physician, who did a complete work-up of his problem and obtained the following laboratory values:

Na⁺	147 mmol/L	(136–146 mmol/L)
K⁺	4.9 mmol/L	(3.5–5.1 mmol/L)
Cl⁻	118 mmol/L	(98–106 mmol/L)

Glucose	102 mg/dL	(80–100 mg/dL)
Creatinine	0.6 mg/dL	(0.6–1.2 mg/dL)
Calcium	7.9 mg/dL	(8.4–10.2 mg/dL)
Hemoglobin	10.4 g/dL	(13.3–17.7 g/dL)
Hematocrit	31%	(40%–52%)
Albumin	2.5 g/dL	(3.5–5.5 g/dL)
Transferrin	112 mg/dL	(200–400 mg/dL)
Copper	38 μg/dL	(50–150 μg/dL)
Zinc	46 μg/dL	(75–125 μg/dL)

Questions

1. What do the laboratory studies suggest as a possible cause of this patient's problem?
2. What is the significance of the low copper and zinc levels? What is the underlying cause of their discrepancy?
3. What medical treatment would be instituted in this patient?

6. A metal ion required for optimal enzyme activity is best termed a (an):
   a. Accelerator
   b. Activator
   c. Coenzyme
   d. Catalyst
7. Total iron binding capacity may be calculated by:
   a. % saturation + total iron
   b. Total iron + excess iron
   c. Total iron + unsaturated binding capacity
   d. Excess iron + unsaturated iron binding capacity
8. Glucose tolerance factor contains the following trace metal:
   a. Copper
   b. Chromium
   c. Selenium
   d. Zinc
9. Menkes' syndrome is caused by accumulation of the following metal in the intestinal mucosal cell and associated low plasma concentrations:
   a. Iron
   b. Zinc
   c. Copper
   d. Manganese
10. The following metal is most associated with the "dementia of dialysis":
   a. Aluminum
   b. Cadmium
   c. Fluorine
   d. Zinc

REFERENCES

1. Beard J, Dawson B, Pinero D. Iron metabolism: a comprehensive review. Nutrev 1996;54(10):295–317.
2. Sullivan JL. Stored iron and ischemic heart disease—empirical support for a new paradigm. Circulation 1992;86:1036–1037.
3. Lauffer R. Preventive measures for the maintenance of low but adequate iron stores. In: Lauffer R, ed. Iron and human disease. Boca Raton, FL: CRC Press, 1992.
4. Natow AB, Heslin F-A. The iron counter. New York: Pocket Books, 1993.
5. Walter T, Hertrampf E, Pizarro F, et al. Effect of bovine-hemoglobin fortified cookies on iron status of schoolchildren—a nationwide program in Chile. Am J Clin Nutr 1993;57:190–194.
6. Chierici R, Sawatzki G. Tamisari L, et al. Supplementation of an adapted formula with bovine lactoferrin. 2. Effects on serum iron, ferritin, and zinc levels. Acta Paediatr 1992;81:475–479.
7. Conrad ME. Iron absorption, In: Johnson LR, ed. Physiology of the gastrointestinal tract. New York: Raven, 1987;1437–1453.
8. Wollenberg P, Rummel W. Dependence of intestinal iron absorption on the valency state of iron. Naunyn Schmiedebegs Arch Pharmacol 1987;336:578–582.
9. Raja KB, Simpson RJ, Peters TJ. Comparison of ⁵⁹Fe³⁺ uptake in vitro and in vivo by mouse duodenum. Biochim Biophys Acta 1987; 901:52–60.
10. Han O, Failla ML, Hill AD, et al. Reduction of FeIII is required for uptake of non-heme iron by CaCo₂ cells. J Nutr 1995;125: 1291–1299.
11. Goldenberg HA. Regulation of mammalian iron metabolism: current state and need for further knowledge. Crit Rev Clin Lab Sci 1997; 34(6):529–572.
12. Pountney DF, Simpson RJ, Wrigglesworth JM. Duodenal ferric reductase: purification and characterization. Biochem Soc Trans 1994; 22:284S.
13. Stremmel W, Lotz G, Niederau C, et al. Iron uptake by rat duodenal microvillous membrane vesicles: evidence for a carrier mediated transport system. Eur J Clin Invert 1987;17:136–145.
14. Raja KB, Simpson RJ, Peters TJ. Investigation of a role for reduction in ferric iron uptake by mouse duodenum. Biochim Biophys Acta 1992;1135:141–146.

15. Nunez MT, et al. Role of redox systems on Fe^{+3} uptake by transformed human intestinal epithelial (CaCo$_2$) cells. Am J Physiol 1994;267:C1582–C1588.

16. Glahn RP, et al. Bathophenanthroline disulfonic acid and sodium dithionite effectively remove surface-bound iron from CaCo$_2$ cell monolayers. J Nutr 1995;125:1833–40.

17. Riedel HD, et al. Characterization and partial purification of a ferrireductase from human duodenal microvillus membranes. Biochem J 1995;309:745–748.

18. Ekmekcioglu C, Feyertag J, Marketl W. A ferric reductase activity is found in brush border membrane vesicles isolated from CaCo$_2$ cells. J Nutr 1996;126:2209–2217.

19. Conrad ME, Umbreit JN, Moore EG. A role for mucin in the absorption of inorganic iron and other metal cations: a study in rats. Gastroenterology 1993;104:1700–1704.

20. Sturrock A, Alexander J, Lamb J, et al. Characterization of a transferrin-independent uptake system for iron in He La cells. J Biol Chem 1990;265:3139–3145.

21. Teichmann R, Stremmel W. Iron uptake by human upper intestinal microvillus membrane vesicles. Indication for a facilitated transport mechanism mediated by a membrane iron-binding protein. J Clin Invest 1990;86:2145–2153.

22. Conrad ME, Umbreit JN, Peterson RDA, et al. Function of integrin in duodenal mucosal uptake of iron. Blood 1993;81:517–521.

23. Conrad ME, et al. A newly identified iron binding protein in duodenal mucosa of rats. Purification and characterization of mobilferrin. J Biol Chem 1990;265:5273–5279.

24. Umbreit J, Conrad M, Moore E, et al. Iron absorption and cellular transport: the mobilferrin/paraferritin paradigm. Semin Hematol 1998;35(1):13–26.

25. Umbreit JN, et al. Paraferritin: a protein complex with ferrireductase activity is associated with iron absorption in rats. Biochemistry 1996;35:6460–6469.

26. Levine DS, Woods JW. Immunolocalization of transferrin-transferrin receptor in mouse small intestine absorptive cells. Histochem Cytochem 1996;38:851–858.

27. Winzerling J, Law J. Comparative nutrition of iron and copper. Annu Rev Nutr 1997;17:501–526.

28. Granick S. Ferritin. IX. Increase of the protein apoferritin in the gastrointestinal mucosa as a direct response to iron feeding. The function of ferritin in the regulation of iron absorption. J Biol Chem 1946;164:737–746.

29. Beinert H, Kennedy MC. Aconitase, a two-faced protein: enzyme and iron regulatory factor. FASEB J 193;7:1442–1449.

30. Haile DJ, Rouault TA, Harford JB, et al. Cellular regulation of the iron-responsive element binding protein: disassembly of the cubane iron-sulfur cluster results in high-affinity RNA binding. Proc Natl Acad Sci 1992;USA89:11735–11739.

31. Guo B, Yang Y, Leibold EA. Iron regulates cytoplasmic levels of a novel iron-responsive element-binding protein without aconitase activity. J Biol Chem 1994;269:24252–24222.

32. Pantopoulos K, Gray NK, Hentze MW. Differential regulation of two related RNA-binding proteins, iron regulatory protein (IRP) and IRP$_B$. RNA 1995;1:155–163.

33. Bothwell TH, Charlton RW, Cook JD, et al. Iron metabolism in man. Oxford, England: Blackwell Scientific Publications, 1979.

34. Crichton R, Ward RJ. Iron metabolism—new perspectives in view. Biochemistry 1992;31:11255–11264.

35. Linder MC, Schaeffer KJ, Hazegh-Azam M, et al. Serum ferritin: does it differ from tissue ferritin? J Gastroenterol Hepatol 1996; 11:1033–1036.

36. Guengerich F. Influence of nutrients and other dietary materials on cytochrome P-450 enzymes. Amer J Clin Nutr 1996;61S:651S–658S.

37. DeMaeyer E, Adiels-Tegman M. The prevalence of anaemia in the world. World Health Stat Q 1985;38:302–316.

38. Bottomly S. Secondary iron overload disorders. Semin Hematol 1998;35(1):77–86.

39. McCord JM. Is iron sufficiency a risk factor in ischemic heart disease? Circulation 1991;83:1112–1114.

40. Meneghini R. Iron homeostasis, oxidative stress, and DNA damage. Free Radic Biol Med 1997;23(5):783–792.

41. Smith C, Mitchinson MJ, Aruoma OI, et al. Stimulation of lipid peroxidation and hydroxyl-radical generation by the contents of human atherosclerotic lesions. Biochem J 1992;286:901–915.

42. Fuhrman B, Oiknine J, Aviram M. Iron induces lipid peroxidation in cultured macrophages, increases their ability to oxidative modify LDL, and affects their secretory properties. Atherosclerosis 1994; 286:65–78.

43. Williams RE, Zweier JL, Flaherty FT. Treatment with deferoxamine during ischemia improves functional and metabolic recovery and reduces reperfusion-induced oxygen radical generation in rabbit hearts. Circulation 1991;83:1006–1014.

44. Steinberg D, Parathasarathy S, Casew TE, et al. Beyond cholesterol: modifications of low-density lipoprotein that increases its atherogenicity. N Engl J Med 1989;320:915–924.

45. Schwartz CJ, Valente AJ, Sprague EA, et al. The pathogenesis of atherosclerosis: an overview. Clin Cardiol 1991;14:I-1–I-16.

46. Halliwell B, Aruoma OI. DNA damage by oxygen-derived species: Its mechanism and measurement in mammalian system. FEBS Lett 1991;281:9–19.

47. Meneghini R. Genotoxicity of active oxygen species in mammalian cells. Mutat Res 1988;195:215–230.

48. Siegers CP, Buman D, Trepkan HD, et al. Influence of dietary iron overload on cell proliferation and intestine tumorgenesis in mice. Cancer Lett 1992;65:245–249.

49. Weinberg ED. Cellular iron metabolism in health and disease. Drug Metab Rev 1990;22:531–579.

50. Anghileri L. Iron, intracellular calcium ion, lipid peroxidation and carcino-genesis. Anticancer Res 1995;15:1395–1400.

51. Yan SD, et al. Glycated tau protein in Alzheimer's disease: a mechanism for induction of oxidant stress. Proc Natl Acad Sci USA 1994; 91:7787–7791.

52. Smith MA, Perry G. Free radical damage, iron, and Alzheimer's disease. J Neurol Sci 1995;134S:92–94.

53. McCord J. Iron, free radicals, and oxidative injury. Sem Hematol 1998;35(1):5–12.

54. Jenner P, et al. Oxidative stress as a cause of nigral cell death in Parkinson's disease and incidental lewy body disease. Ann Neurol 1992;32:S82–S87.

55. Emard J, Thouez J, Gauvreau D. Neurodegenerative diseases and risk factors: a literature review. Soc Sci Med 1995;40(6):847–858.

56. McCord JM, Day ED Jr. Superoxide-dependent production of hydroxyl radical catalyzed by iron-EDTA complex. FEBS Lett 1978; 86:139–142.

57. Aisen P, Cohen G, Kang JO. Iron toxicosis. Int Rev Exp Pathol 1990;31:1–46.

58. Gutteridge JMC, Halliwell B. Iron and oxygen: a dangerous mixture. In: Ponka P, Schulman HM, Woodworth RC, eds. Iron transport and storage. Boca Raton, FL: CRC Press, 1990.

59. Nienhuis AW, et al. Thalassemia major: molecular and clinical aspects. Ann Int Med 1979;91:883–897.

60. Herbert V, Shaw S, Jayatilleke E. Vitamin C-driven free radical generation from iron. J Nutr 1996;126;1213S–1220S.

61. Meyers D. The iron hypothesis—does iron cause atherosclerosis. Clin Cardiol 1996;19:925–929.

62. Shih YJ, et al. Serum transferrin receptor is a truncated form of tissue receptor. J Biol Chem 1990;265:19077–19081.

63. Baynes RD, Shih YJ, Hudson BG, et al. Characterization of transferrin receptor released by K562 erythroleukemia cells. Proc Soc Exp Biol Med 1991;197:416–423.

64. Baynes RD, Shih YJ, Hudson BG, et al. Proc Soc Exp Med 1993; 204:65–69.

65. Skikane B. Circulating transferrin receptor assay—coming of age. Clin Chem 1998;44(1):7–9.

66. Baynes RD. Assessment of iron status. Clin Biochem 1996;29: 209–215.

67. Rouault T, Krishnamurthy R, Hartford J, et al. Hemin, chelatable iron, and the regulation of transferrin receptor biosynthesis. J Biol Chem 1985;269:14862–14866.

68. Kohgo Y, et al. Circulating transferrin receptor in human serum. Br J Haematol 1986;64:277–281.

69. Flowers CH, Skikne BS, Covell AM, et al. The clinical measurement of serum transferrin receptor. J Lab Clin Med 1989;114:368–377.

70. Cook JD, Baynes RD, Skikne BS. Iron deficiency and the measurement of iron status. Nutr Res Rev 1992;5:189–202.

71. Kohgo Y, et al. Quantitation and characterization of serum transferrin receptor in patients with anemias and polycythemias. Jpn J Med 1988;27:64–70.

72. Baynes RD, Skikne BS, Cook JD. Circulating transferrin receptors and assessment of iron status. J Nutr Biochem 1994;5:322–330.

73. Allen J, et al. Measurement of soluble transferrin receptor in serum of healthy adults. Clin Chem 1998;44:35–39.

74. Huebers HA, Beguin Y, Pootrakul P, et al. Intact transferrin receptors in human plasma and their relation to erythropoiesis. Blood 1990;75:102–107.

75. Linder MC. The biochemistry of copper. New York: Plenum, 1991.

76. Linder M, Hazegh-Azam M. Copper biochemistry and molecular biology. Am J Clin Nutr 1966;63:797S–811S.

77. Dunham R, Smith HE. Lead and copper—a model home approach. Proc Water Qual Technol Conf 1992;:341–352.

78. Linder MC. Nutrition and metabolism of trace elements. In: Linder MC, ed. Nutritional biochemistry and metabolism. 2nd ed. New York: Elsevier, 1991;215–276.

79. Wapnir RA, Lee SY. Dietary regulation of copper absorption and storage in rats: effects of sodium, zinc and histidine-zinc. J Am Coll Nutr 1993;12:714–719.

80. Yu S, West CE, Beymen AC. Increasing intakes of iron reduce status, absorption and biliary excretion of copper in rats. Br J Nutr 1994;71:887–895.

81. Sandstead H. Requirements and toxicity of essential trace elements, illustrated by zinc and copper. Am J Clin Nutr 1995;61S:621S–624S.

82. Walshe JM. Copper: not too little, not too much, but just right. J Royal Coll Phys 1995;29(4):280–287.

83. Miura T, Muraoka S. Ogiso T. Adriamycin-induced lipid peroxidation of erythrocyte membranes in the presence of ferritin and the inhibitory effect of ceruloplasmin. Biol Pharm Bull 1993;16:664–667.

84. Linder MC. Interactions between copper and iron in mammalian metabolism. In: Elsenhans BE, Forth W, Schumann K, eds. Metal–metal interactions. Gutersloh, Germany: Bertelsheim Foundation, 1994;11–41.

85. Yang G, Chan PH, Chen J, et al. Human copper-zinc superoxide dismutase transgenic mice are highly resistant to reperfusion injury after focal cerebral ischemia. Stroke 1994;25:165–170.

86. Harris E. The iron–copper connection: the link to ceruloplasmin grows stronger. Nutr Rev 1995;53:170–173.

87. Logan JI, Harveyson KB, Wisdom GB, et al. Hereditary caeruloplasmin deficiency, dementia and diabetes mellitus. Q J Med 1994;87:663–670.

88. Cordano A, Placko RP, Graham GG. Hypocupremia and neutropenia in copper deficiency. Blood 1966;26:289–290.

89. Phillip M, Singer A, Tal A. Copper deficiency in an infant with giardiasis fed with cow's milk. Isr J Med Sci 1990;26:289–290.

90. Brewer GJ, et al. Initial therapy of patients with Wilson's disease with tetrathiomolybdate. Arch Neurol 1991;48:42–47.

91. Becton DL, Schultz WH, Kinney TR. Severe neutropenia caused by copper deficiency in a child receiving continuous ambulatory peritoneal dialysis. J Pediatr 1986;108:735–737.

92. Goyens P, Brasseur D, Cadranel S. Copper deficiency in infants with active celiac disease. J Pediatr Gastroenterol Nutr 1985;4:677–680.

93. Percival, S. Neutropenia caused by copper deficiency: possible mechanisms of action. Nutr Rev 1995;53(3):59–66.

94. Hirase N, et al. Anemia and neutropenia in a case of copper deficiency: role of copper in normal hematopoiesis. Acta Haematol 1992;87:195–197.

95. Sakar B, Lingertat-Walsh K, Clarke JTR. Copper-histidine therapy for Menkes' disease. J Pediatr 1993;123:828–830.

96. Olivares, M. Limits of metabolic tolerance to copper and biological basis for present recommendations and regulations. Am J Clin Nutr 1996;63:846S–852S.

97. US Environmental Protection Agency, Environmental Criteria and Assessment Office. Summary review of the health effects associated with copper. EPA 600/X-84/190-1. Cincinnati, OH: Environmental Protection Agency, 1987.

98. Bhave SA, Pandit AN, Tanner MS. Comparison of feeding history of children with Indian childhood cirrhosis and paired controls. J Pediatr Gastroenterol Nutr 1987;6:562–567.

99. Cox, D. Genes of the copper pathway. Am J Hum Genet 1995;56:828–834.

100. Halliwell B. Free radicals and antioxidants: a personal view. Nutr Rev 1994;52:253–265.

101. Brewer GJ, Yuzbasiyan-Gurkan V. Wilson's disease. Medicine 1992;71:139–164.

102. Brewer G, et al. Treatment of Wilson's disease with ammonium tetramolybdate. Arch Neurol 1996;53:1017–1025.

103. Sandstrom B. Snadberg A. Inhibitory effects of isolated inositol phosphates on zinc absorption. J Trace Elem Electrolytes Health Dis 1992;6:99–103.

104. Parisi AF, Vallee BL. Isolation of a zinc alpha-2-macroglobulin from human serum. Biochemistry 1970;9:2421–2426.

105. King J. Assessment of zinc status. J Nutr 1990;120:1474–1479.

106. Vallee BL, Falchuk KH. The biochemical basis of zinc physiology. Physiol Rev 1993;73(1):79–118.

107. Prasad AS. Essentiality and toxicity of zinc. Scand J Work Environ Health 1993;19(1):134–136.

108. Cunningham-Rundles S. Zinc modulation of immune function: specificity and mechanism of interaction. J Lab Clin Med 1996;128:9–11.

109. Prasad AS, et al. Zinc metabolism in normals and patients with the syndrome of iron deficiency anemia, hepatosplenomegaly, dwarfism, and hypogonadism. J Lab Clin Med 1963;61:531–537.

110. Prasad AS, et al. Zinc deficiency affects cell cycle and deoxythymidine kinase gene expression in HUT-78 cells. J Lab Clin Med 1996;128:51–60.

111. Grider A, Bailey LB, Cousins RJ. Erythrocyte metallothionein as an index of zinc status in humans. Proc Natl Acad Sci USA 1990;87:1259–1262.

112. Pilch SM, Senti FR. Assessment of the zinc nutritional status of the US population based on data collected in the second national health and nutrition examination survey. FASEB FDA 223-223-83-2384. Bethesda, MD: Life Science Research Office, 1984.

113. Anderson RA, Kozlovsky AS. Chromium intake, absorption and excretion of subjects consuming self-selected diets. Am J Clin Nutr 1985;41:1177–1182.

114. Anderson RA, et al. Dietary chromium intake—freely chosen diets, institutional diets and individual foods. Biol Trace Elem Res 1992;117–121.

115. Anderson RA. Chromium as an essential nutrient for humans. Regul Toxicol Pharmacol 1997;26:S35–S41.

116. Anderson RA, et al. Elevated intakes of supplemental chromium improve glucose and insulin variables in individuals with type-2 diabetes. Diabetes 1997;46(11):1786–1791.

117. Dure-Smith BA, Farley SM, Linkhart SG, et al. Calcium deficiency in fluoride-treated osteoporotic patients despite calcium supplementation. J Clin Endocrinol Metab 1996;81:269–275.

118. Pak CYC, et al. Slow-release sodium fluoride in the management of postmenopausal osteoporosis. Ann Intern Med 1994;120:625–632.

119. Saric M. Manganese. In: Friberg I, Nordberg GF, Vouk VB, eds. Handbook on the toxicology of metals. Vol. II: Specific metals. Amsterdam: Elsevier, 1986;354–377.

120. Misselwitz B, Muhler A, Weinmann H-J. A toxicologic risk for using manganese complexes? A literature survey of existing data through several medical specialties. Invest Radiol 1995;30(10):611–620.

121. Krieger D, et al. Manganese and chronic hepatic encephalopathy. 1995;346(8970):270–274.

122. Repine JE. Oxidant—antioxidant balance: some observations from studies of ischemia-reperfusion in isolated perfused rat hearts. Am J Med 1991;91(3C):45S–53S.

123. Wright RM, Repine JE. The human molybdenum hydroxylase gene family: co-conspirators in metabolic free-radical generation and disease. Biochem Soc Trans 1997;25:799–804.

124. Mo DX. Pathology and selenium deficiency in Kashin-Beck disease. In: Combs GF, et al, eds. Selenium in biology and medicine. New York: Avi 1987;924–933.

125. Clark LC, et al. Effects of selenium supplementation for cancer prevention in patients with carcinoma of the skin. A randomized controlled trial. JAMA 1996;276:1957–1963.

126. Patterson RE, White E, Kristal AR, et al. Vitamin supplements and cancer risk: the epidemiologic evidence. Cancer Causes and Control 1997;8:786–802.

127. Halliwell B. Antioxidants: sense or speculation? Nutr Today 1994;29:15–19.

128. Klasner AE, et al. Marked hypocalcemia and ventricular fibrillation in two pediatric patients exposed to a fluoride-containing wheel cleaner. Ann Emerg Med 1996;28:6.

CHAPTER 16

Blood Gases, pH, and Buffer Systems

Sharon S. Ehrmeyer, Kevin D. Fallon

Objectives

Upon completion of this chapter, the clinical laboratorian should be able to:

- *Describe the principles involved in the measurement of pH, PCO₂, PO₂, and the various hemoglobin species.*

- *Outline the interrelationship of the buffering mechanisms of bicarbonate-carbonic acid and hemoglobin.*

- *Explain the clinical significance of the following pH and blood gas parameters: pH, PCO₂, PO₂, actual bicarbonate, carbonic acid, base excess, oxygen saturation, fractional oxyhemoglobin, hemoglobin oxygen (binding) capacity, oxygen content, and total CO₂.*

- *Using the Henderson-Hasselbalch equation and blood gas data, determine whether data are normal or represent metabolic or respiratory acidosis or metabolic or respiratory alkalosis. Identify whether the data represent uncompensated or compensated conditions.*

- *Identify some common causes of metabolic acidosis and alkalosis and of respiratory acidosis and alkalosis. State how the body attempts to compensate (kidney and lungs) for the various conditions.*

- *Describe the significance of the hemoglobin-oxygen dissociation curve and the impact of pH, 2,3-diphosphoglycerate (2,3-DPG), temperature, pH, and PCO₂ on its shape and how these affect the release of O₂ to the tissues.*

- *Discuss problems and precautions in collecting and handling samples for pH and blood gas analysis. Include syringes, anticoagulants, mixing, icing, and capillary*

and venous samples as well as arterial samples in the discussion.

- *Describe approaches to quality assurance including quality control (commercial liquid controls, tonometry, external [proficiency testing], and delta checks) to assess analytical quality.*

- *Given oxygen saturation data from both the blood gas analyzer and co-oximeter, discuss the reasons for possible discrepancies.*

- *Calculate partial pressures for PCO₂ and PO₂ for various percentages of carbon dioxide and oxygen. In doing the calculations, account for the barometric pressure and vapor pressure of water.*

KEY TERMS

Acid	FiO₂	acidosis and
Acidemia	Fractional	alkalosis
Acidosis	oxyhemoglobin	Oxygen content
Alkalemia	Hemoglobin oxygen	Oxygen saturation
Alkalosis	(binding) capacity	Partial pressure
Analytical error	Hemoglobin-oxygen	PCO₂
Base	dissociation	pH
Base excess	curve	pK
Bicarbonate	Henderson-	PO₂
Buffer	Hasselbalch	Preanalytical error
Calibration	Equation	Respiratory
Carbonic acid	Metabolic alkalosis	acidosis
Compensation	Metabolic	Respiratory
Electrodes	(nonrespiratory)	alkalosis

An important component of clinical biochemistry is information on acid-base/blood gas homeostasis. It forms an extensive knowledge base for handling data in life-threatening situations. Many principles and facts must be interrelated; focusing on only one result can be misleading.

The exchange of gases, specifically carbon dioxide and oxygen, is discussed. Techniques and instrumentation used in the measurement of pH and blood gases are described. Because the sample collection and handling can greatly affect the quality of the laboratory results, preanalytical considerations are discussed. Quality control (QC) approaches to blood gas analysis also are presented.

DEFINITIONS: ACID, BASE, BUFFER

A discussion of acid-base balance requires a review of several basic definitions, such as acid, base, buffer, pH, and pK. In addition, the applicability of general biochemical principles of equilibrium and the laws of mass action are relevant.

An *acid* is a substance that can yield a hydrogen ion (H^+) or hydronium ion when dissolved in water. A *base* is a substance that can yield hydroxyl ions (OH^-). The relative strengths of acids and bases as well as their ability to dissociate in water can be ranked. Dissociation constant (K value) tables can be found in most biochemistry texts. For an acid, the larger the K, the greater the tendency to dissociate into ions. The pK, defined as the negative log of the ionization constant, is that pH where the protonated and unprotonated forms are present in equal concentration. Strong acids have pK values of less than 3.0, whereas strong bases have pK values greater than 9.0. For acids, raising the pH above the pK will cause the acid to dissociate and yield a H^+. For bases, lowering the pH below the pK will cause the base to release OH^-. Many species have more than one pK, meaning they can accept or donate more than one H^+.

A *buffer*, the combination of a weak acid or weak base and its salt, is a system that resists changes in pH. The effectiveness of a buffer depends on the pK of the buffering system and the *pH* of the environment in which it is placed. In plasma, the bicarbonate-carbonic acid system, having a pK of 6.1, is one of the principal buffers.

$$H_2CO_3 \leftrightarrow HCO_3^- + H^+$$

Carbonic   Bicarbonate
acid   *(Eq. 16–1)*

When the blood plasma pH is 7.4, this buffer is more effective than the lactic acid-lactate system (pK = 3.9) or the ammonium-ammonia system, with a pK of 9.4. Weisberg cited an example to demonstrate the effectiveness of the blood buffers.[1] If the pH of 100 mL of distilled water is 7.35 and one drop of 0.05 N HCl is added, the pH will change to 7.00. To change 100 mL of "normal" blood from a pH of 7.35 to 7.00, approximately 25 mL of 0.05 N HCl is needed. With 5.5 L of blood in the average body, more than 1300 mL of HCl would be required to make this same change in pH.

ACID-BASE BALANCE

Maintenance of H⁺

The body produces 15 moles to 20 moles of H^+ per day; however, the normal concentration of H^+ in the extracellular body fluids ranges from only 36 nmol/L to 44 nmol/L. Any deviations in H^+ concentration from this range will cause alterations in the rates of chemical reactions within the cell and affect the many metabolic processes of the body. Concentrations greater than 44 nmol/L can alter consciousness and lead to coma and death, whereas H^+ concentrations less than 36 nmol/L can cause neuromuscular irritability, tetany, loss of consciousness, and death.

The logarithmic pH scale frequently expresses H^+ concentration. The two quantities are related:

$$pH = \log \frac{1}{cH^+} = -\log cH^+, \qquad \text{(Eq. 16–2)}$$

where c is concentration. The normal pH of arterial blood is 7.40 and is equivalent to an H^+ concentration of 40 nmol/L. Because of the reciprocal relationship between cH^+ and pH, an increase in H^+ decreases the pH, whereas a decrease in H^+ increases the pH. A pH below the reference range is referred to as *acidosis,* whereas a pH above the reference range is referred to as *alkalosis.* Technically, the suffix -*osis* refers to a process in the body, whereas the suffix -*emia* refers to the corresponding state in blood (-osis is the cause of the –emia).

The arterial pH is controlled by systems that regulate the production and retention of acids and bases. These include buffers, the respiratory center and lungs, and the kidneys.

Buffer Systems: Regulation of H⁺

The body's first line of defense against changes in H^+ concentration consists of the buffer systems present in all body fluids. All buffers consist of a weak acid, such as H_2CO_3, and its salt or conjugate base, HCO_3^-, for the bicarbonate-carbonic acid buffer system. H_2CO_3 is a weak acid because it does not completely dissociate into H^+ and HCO_3^-. (In contrast, a strong acid such as HCl completely dissociates into H^+ and Cl^- in solution.) When an acid is added to the bicarbonate-carbonic acid system, the HCO_3^- will combine with the H^+ from the acid to form H_2CO_3. When a base is added, H_2CO_3 will combine with the OH^- group to form H_2O and HCO_3^-. In both cases, there is a smaller change in pH than would result from adding the acid or base directly to an unbuffered solution.

Although the bicarbonate-carbonic acid system has low buffering capacity, it still is an important buffer for three reasons: (1) H_2CO_3 dissociates into CO_2 and H_2O, allowing H^+ to be eliminated as CO_2 by the lungs; (2) changes in PCO_2 modify the ventilation rate; and (3) HCO_3^- concentration can be altered by the kidneys. In addition, this buffering system immediately counters the effects of fixed nonvolatile acids (H^+A^-) by binding the dissociated hydrogen ion ($H^+A^- + HCO_3^- = H_2CO_3 + A^-$). The resultant H_2CO_3 then dissociates, and the H⁺ is neutralized by the buffering capacity of hemoglobin. Figure 16-1 shows the interrelationship of the hemoglobin and bicarbonate buffering systems.

Other buffers also are important. The phosphate buffer system ($HPO_4^{-2} - H_2PO_4^-$) plays a role in plasma and erythrocytes (red blood cells; RBCs) and is involved in the exchange of sodium ion in the urine H⁺ filtrate. Plasma proteins, especially the imidazole groups of histidine, also form an important buffer system in plasma. Most circulating proteins have a net negative charge and are capable of binding H⁺.

The lungs and kidneys play important roles in regulating blood pH. The interrelationship of the lungs and kidneys in maintaining pH is depicted with the Henderson-Hasselbalch equation (see Equation 16-4). The numerator (HCO_3^-) denotes kidney function, whereas the denominator (PCO_2 which represents H_2CO_3) denotes lung function. The lungs regulate pH through retention or elimination of CO_2 by changing the rate and volume of ventilation. The kidneys regulate pH by excreting acid, primarily in the ammonium ion, and by reclaiming HCO_3^- from the glomerular filtrate (and adding it back to the blood).

Respiratory System: Regulation of Acid-Base Balance

The end product of most aerobic metabolic processes is CO_2. Because dissolved CO_2 (dCO_2) can be more concentrated in the tissues where it is produced, it easily diffuses out of the tissue into the plasma and the RBCs in the capillaries. In plasma, small amounts of CO_2 remain as dCO_2 or combine with proteins to form carbamino compounds. Most of the CO_2 combines with H_2O to form H_2CO_3, which quickly dissociates into H⁺ and HCO_3^- (see Fig. 16-1). The reaction is accelerated by the enzyme carbonic anhydrase found in the RBC

membrane. The dissociation of H_2CO_3 causes the HCO_3^- concentration to increase in the RBCs and diffuse into the plasma. To maintain electroneutrality (the same number of positively and negatively charged ions on each side of the RBC membrane), chloride diffuses into the cell. This is known as the *chloride shift*. The freed H⁺ is buffered by plasma proteins and plasma buffers.

In the lungs, the process is reversed. Inspired O_2 diffuses from the alveoli into the blood and is bound to hemoglobin, forming oxyhemoglobin (O_2Hb). The H⁺ that was carried on the (reduced) hemoglobin in the venous blood is released to recombine with HCO_3^- to form H_2CO_3, which dissociates into H_2O and CO_2. The CO_2 diffuses into the alveoli and is eliminated through ventilation. The net effect of the interaction of these two buffering systems is a minimal change in H⁺ concentration between the venous and arterial circulation. When the lungs do not remove CO_2 at the rate of its production, it will accumulate in the blood, causing an increase in H⁺ concentration. If, on the other hand, CO_2 removal is faster than production (hyperventilation), the H⁺ concentration will be decreased. Consequently ventilation affects the pH of the blood. Even a change in the H⁺ concentration of blood due to nonrespiratory disturbances causes the respiratory center to respond by altering the rate of ventilation in an effort to restore the blood pH to normal. The lungs, by responding within seconds, provide the first of defense to changes in acid-base status.

The Kidney System: Regulation of Acid-Base Balance

The kidney is able to excrete variable amounts of acid or base making it an important player in the regulation of acid-base balance. The kidney's main role in maintaining acid-base homeostasis is to reclaim HCO_3^- from the glomerular filtrate and add it to the blood. Without this reclamation, the loss of HCO_3^- in the urine would result in an acid gain in the blood. The main site for HCO_3^- reclamation is the proximal tubules (Fig. 16-2). The glomerular filtrate contains essentially the same HCO_3^- levels as plasma. The process is not a direct transport of HCO_3^- across the tubule membrane into the blood. Instead, sodium in the glomerular filtrate is exchanged for H⁺ in the tubular cell. The H⁺ combines with HCO_3^- in the filtrate to form H_2CO_3, which is converted into H_2O and CO_2 by carbonic anhydrase. The CO_2 easily diffuses into the tubule and reacts with H_2O to reform H_2CO_3 and then HCO_3^-, which is reabsorbed into the blood along with sodium. With alkalotic conditions, the kidney excretes HCO_3^- to compensate for the elevated blood pH. (*Reabsorption* or *reclamation* refers to the process of reentering the blood. *Secretion* or *excretion* by the tubule cells concentrate or remove substances from the filtrate. These reactions determine the pH of the urine, as well as the pH of the blood.)

Under normal conditions, the body produces a net excess (50 mmol to 100 mmol) of acid each day that must be excreted by the kidney. Because the minimum urine pH is

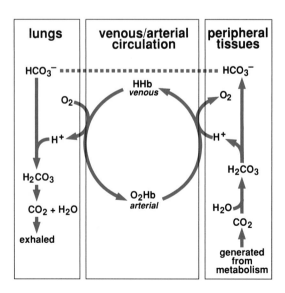

Figure 16-1. Interrelationship of the bicarbonate and hemoglobin buffering systems.

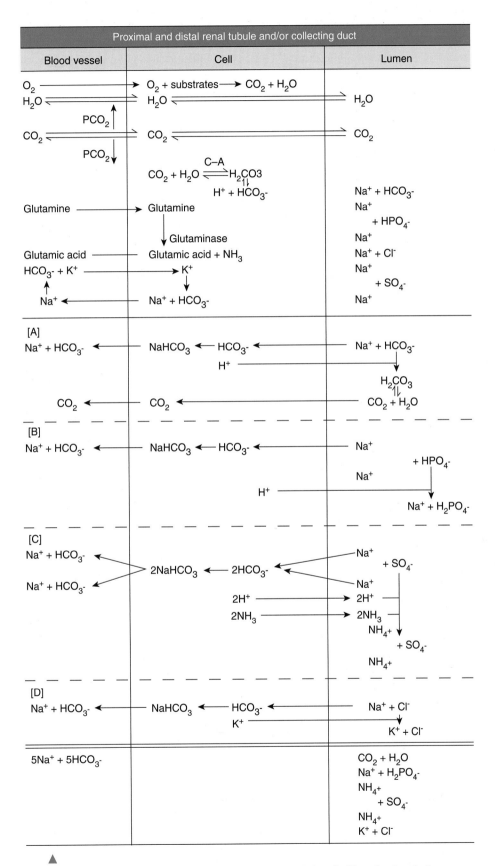

Figure 16-2. Bicarbonate reabsorption by the proximal tubule cell. CA, carbonic anhydrase.

approximately 4.5, the kidney excretes little nonbuffered H^+. The remainder of the urinary H^+ combines with dibasic phosphate ($HPO_4^=$) and ammonia (NH_3) and is excreted as dihydrogen phosphate ($H_2PO_4^-$) and ammonium (NH_4^+). The amount of $HPO_4^=$ available for combining with H^+ is fairly constant; therefore, the daily excretion of H^+ in urine largely depends on the amount of NH_4^+ formed. Because the renal tubular cells are able to generate NH_3 from glutamine and other amino acids, the concentration of NH_3 can be increased in response to a decreased blood pH.

Various factors affect the reabsorption of HCO_3^-. When the blood or plasma HCO_3^- level is higher than 26 mmol/L to 30 mmol/L, HCO_3^- will be excreted. While HCO_3^- reabsorption apparently continues, excretion also occurs. Therefore, it is unlikely that the plasma will exceed an HCO_3^- of 30 mmol/L unless these excretory capabilities fail.

The HCO_3^- level may become increased if an excessive amount of lactate, acetate, or HCO_3^- is intravenously infused. It also may increase if there is an excessive loss of chloride without replacement (as from vomiting or prolonged nasogastric suction), because the HCO_3^- will be retained by the tubule to preserve electroneutrality.

Several factors may result in decreased HCO_3^- levels. Most diuretics, regardless of mechanism of action, favor the excretion of HCO_3^-. Reduced HCO_3^- reabsorption also occurs in conditions in which there is an excessive loss of cations. In kidney dysfunction (such as chronic nephritis or infections), HCO_3^- reabsorption may be impaired.

ASSESSMENT OF ACID-BASE HOMEOSTASIS

The Bicarbonate Buffering System and the Henderson-Hasselbalch Equation

In assessing acid-base homeostasis, the bicarbonate buffering system is used. From the data, inferences can be made pertaining to the other buffers and the systems that regulate the production, retention, and excretion of acids and bases.

In the bicarbonate buffering system, the dCO_2 is in equilibrium with CO_2 gas, which can be expelled by way of the lungs. Therefore, the bicarbonate buffering system is referred to as an *open* system, and the dCO_2, which is controlled by the lungs, is the *respiratory component*. The lungs can participate rapidly in the regulation of blood pH through hypoventilation or hyperventilation. The bicarbonate concentration is controlled mainly by the kidneys, making it the *nonrespiratory*, or *metabolic, component* in evaluating acid-base homeostasis.

The *Henderson-Hasselbalch equation* expresses acid-base relationships in a mathematical formula:

$$pH = pK' + \log\frac{cA^-}{cHA}, \qquad \textit{(Eq. 16–3)}$$

where A = proton acceptor (eg, HCO_3^-), HA = proton donor, or weak acid (eg, H_2CO_3), and pK' = pH at which

there is an equal concentration of protonated and unprotonated species. Knowing any three of the variables allows for the calculation of the fourth.

In plasma and at body temperature, the pK' of the bicarbonate buffering system is 6.1. Because the equilibrium between H_2CO_3 and CO_2 in plasma is approximately 1:800, the concentration of H_2CO_3 is proportional to the partial pressure exerted by the dissolved CO_2. In plasma at 37°C, the value for the combination of the solubility constant for PCO_2 and the factor to convert mm Hg to millimoles per liter is 0.0307 mmol L^{-1} mm Hg^{-1}. Temperature and the solvent affect the constant. If either of these changes, the solubility constant also will change. Both pH and PCO_2 are measured in blood gas analysis, and the pK' is a constant; therefore, HCO_3^- can be calculated:

$$pH = pK' + \log\frac{cHCO_3^-}{0.031 \times PCO_2} \qquad \textit{(Eq. 16–4)}$$

In health, when the kidneys and lungs are functioning properly, a 20:1 ratio of HCO_3^- to H_2CO_3 will be maintained (resulting in a pH of 7.40). This is illustrated by substituting normal values (Table 16-1) for HCO_3^- and PCO_2 into the preceding equation:

$$\frac{24 \text{ mmol/L}}{(0.031 \text{ mmol/L} - \text{mm Hg}) \times 40 \text{ mm Hg}}$$

$$= \frac{24}{1.2} = \frac{20}{1} \qquad \textit{(Eq. 16–5)}$$

Adding the log of 20 (1.3) to the pK' of the bicarbonate system yields a normal pH of 7.40 (7.40 = 6.1 + 1.3).

Acid-Base Disorders

Acid-base disorders result from a variety of pathologic conditions. When the pH of blood is less than the reference range, it is termed *acidemia*. A pH greater than the reference range is termed *alkalemia*. A disorder due initially to ventilatory dysfunction (a change in the PCO_2, the respiratory component) is termed a *primary respiratory acidosis or alkalosis*. A disorder due to a change in the bicarbonate level (a renal or metabolic function) is termed a *nonrespiratory (metabolic) disorder*. Mixed respiratory and nonrespiratory disorders due to more than one pathologic process are also quite common.

TABLE 16-1. Arterial Blood Gas Reference Range at 37°C

pH	7.35–7.45
PCO_2 (mm Hg)	35–45
HCO_3^- (mmol/L)	22–26
Total CO_2 content (mmol/L)	23–27
PO_2 (mmol/L)	80–110
SO_2 (%)	>95

Because the body's cellular and metabolic activities are pH dependent, the body tries to return the pH toward normal whenever an imbalance occurs. This action by the body is termed *compensation,* and the body accomplishes this by altering the factor not primarily affected by the pathologic process. For example, if the imbalance is of nonrespiratory origin, the body will compensate by altering ventilation. For disturbances of the respiratory component, the kidneys will compensate by selectively excreting or reabsorbing anions and cations. The lungs can compensate immediately, but the response is short term and often incomplete. On the other hand, the kidneys are slower to respond (2 days to 4 days), but the response is long term and potentially complete. *Fully compensated* implies that the pH has returned to the normal range (the 20:1 ratio has been restored), and *partially compensated* implies that the pH is approaching normal. Compensation is an effort by the body to return the blood pH to normal; the primary abnormality is not corrected.

Acidosis

A primary nonrespiratory (metabolic) abnormality or a primary respiratory problem can cause acidosis. In primary nonrespiratory acidosis, there is a decrease in bicarbonate (<24 mmol/L), resulting in a decreased pH as a result of the ratio for the nonrespiratory to respiratory component in the Henderson-Hasselbalch equation being less than 20:1:

$$pH \propto \frac{\downarrow cHCO_3^-}{N(0.0307 \times PCO_2)} < \frac{20}{1}, \quad \textit{(Eq. 16–6)}$$

where N = normal value, and < indicates a decreased level.

Nonrespiratory acidosis may be caused by the direct administration of an acid-producing substance, such as ammonium chloride or calcium chloride, or by excessive formation of organic acids seen with diabetic ketoacidosis and starvation. Nonrespiratory acidosis is also seen with reduced excretion of acids, as in renal tubular acidosis, and with excessive loss of bicarbonate from diarrhea or drainage from a biliary, pancreatic, or intestinal fistula.

The body compensates for nonrespiratory acidosis through *hyperventilation,* which is an increase in the rate or depth of breathing. By "blowing off" CO_2, the base-to-acid ratio will return toward normal. Secondary compensation occurs when the "original" organ (the kidney, in this case) begins to correct the ratio by retaining bicarbonate.

Primary *respiratory acidosis* results from a decrease in alveolar ventilation (*hypoventilation),* causing a decreased elimination of CO_2 by the lungs:

$$pH \propto \frac{NcHCO_3^-}{\uparrow(0.0307 \times PCO_2)} < \frac{20}{1}, \quad \textit{(Eq. 16–7)}$$

Respiration is regulated in the medulla of the brain. Chemoreceptors present in the aortic arch and the carotid sinus respond to levels of O_2, CO_2, and pH in the blood and cerebrospinal fluid. There are several situations in which the CO_2 will not be removed in the usual manner by the lung. Many lung diseases will decrease the effective removal of CO_2. For example, in emphysema, a disease in which there is an abnormal, permanent increase in the size of the alveolar air spaces and destructive changes in the alveolar walls, the surface area available for gas exchange is reduced, and CO_2 will be retained in the blood. In bronchopneumonia, the alveoli contain secretions, white blood cells, bacteria, and fibrin, all of which impede gas exchange. Hypoventilation caused by drugs such as barbiturates, morphine, or alcohol will increase PCO_2 levels, as will mechanical obstruction or asphyxiation (strangulation or aspiration). Decreased cardiac output, such as that seen with congestive heart failure, also will result in less blood presented to the lungs for gas exchange and, therefore, an elevated PCO_2.

In primary respiratory acidosis, the compensation occurs through nonrespiratory processes. The kidneys increase the excretion of H^+ and increase the reabsorption of HCO_3^-. Although the renal compensation begins immediately, it takes days to weeks for maximal compensation to occur. With an increase in HCO_3^-, the base-to-acid ratio will be altered and the pH will return toward normal.

CASE STUDY 16-1

A 50-year-old male appeared in the ER after returning from foreign travel. His symptoms included persistent diarrhea (over the past 3 days) and rapid respiration (tachypnea). Blood gases were drawn:

>   pH = 7.21
> PCO_2 = 19 mm Hg
> PO_2 = 96 mm Hg
> HCO_3^- = 7 mmol/L
> SO_2 = 96% (calculated) (reference range >95%)

Questions

1. What is the patient's acid-base status?
2. Why is the HCO_3^- level so low?
3. Why does the patient have rapid respiration?

CASE STUDY 16-2

An 80-year-old woman fell on the ice and fractured her femur. After several hours, she arrived in the ER anxious, panting, and complaining of severe pain in her chest and not being able to breathe. Her pulse was rapid (tachycardia) as was her respiration rate (tachypnea). Blood gases were drawn and yielded the following results:

$$pH = 7.31$$
$$PCO_2 = 27 \text{ mm Hg}$$
$$PO_2 = 62 \text{ mm Hg}$$
$$HCO_3^- = 12 \text{ mmol/L}$$
$$SO_2 = 78\% \text{ (calculated) (reference range} >95\%)$$

Questions

1. What is the patient's acid-base status?
2. Why is the HCO_3^- level so low?
3. What clinically caused the acid-base imbalance?

Alkalosis

Primary *nonrespiratory alkalosis* is due to a gain in HCO_3^-, causing an increase in the nonrespiratory component and an increase in the pH:

$$pH \propto \frac{\uparrow c HCO_3^-}{N(0.0307 \times PCO_2)} > \frac{20}{1} \qquad (Eq.\ 16-8)$$

This condition may result from the excess administration of sodium bicarbonate or through ingestion of bicarbonate-producing salts such as sodium lactate, citrate, or acetate. Excessive loss of acid through vomiting, nasogastric suctioning, or prolonged use of diuretics that augment renal excretion of H^+ can produce an apparent increase in HCO_3^-. The body responds by depressing the respiratory center. The resulting hypoventilation increases the retention of CO_2.

Primary *respiratory alkalosis* from an increased rate of alveolar ventilation causes excessive elimination of CO_2 by the lungs:

$$pH \propto \frac{Nc(HCO_3^-)}{\downarrow(0.0307 \times PCO_2)} > \frac{20}{1} \qquad (Eq.\ 16-9)$$

The causes of respiratory alkalosis include chemical stimulation of the respiratory center by drugs, such as salicylates, an increase in the environmental temperature, fever, hysteria, pulmonary emboli, and pulmonary fibrosis. The kidneys compensate by excreting HCO_3^- and retaining H^+.

OXYGEN AND GAS EXCHANGE

Oxygen and Carbon Dioxide

The role of oxygen in metabolism is crucial to all life. In cell mitochondria, electron pairs from the oxidation of NADH and $FADH_2$ are transferred to molecular oxygen, causing release of the energy used to synthesize ATP from the phosphorylation of ADP. Although the measurement

---

CASE STUDY 16-3

A 24-year-old graduate student was brought to the ER comatose after being found in his room unconscious. A bottle of secobarbital was on his bedstand. He did not respond to painful stimuli, his respiration was barely perceptible, and his pulse was weak. Blood gases were drawn, and yielded the following results:

$$pH = 7.10$$
$$PCO_2 = 70 \text{ mm Hg}$$
$$PO_2 = 58 \text{ mm Hg}$$
$$HCO_3^- = 20 \text{ mmol/L}$$
$$FO_2Hb = 80\% \text{ (reference range} >95\%)$$

Questions

1. What is the patient's acid-base status?
2. What caused the profound hypoventilation?
3. Once the respiratory component returns to normal, what will be the patient's expected acid-base status?

CASE STUDY 16-4

A 24-year-old Himalayan male was accepted to graduate school in the United States. Before leaving home, he had an extensive physical exam that included a variety of blood tests. The medical staff at the U.S. University reviewed his medical records. It was noted that all of the test results were normal except the HCO_3^-, which was 15 mmol/L (reference range, 22–26 mmol/L). The HCO_3^- was done separately on a serum sample. It was not part of a blood gas panel. To rule out nonrespiratory acidosis, the university physician wanted the HCO_3^- repeated. The repeated value was 24 mmol/L.

Questions

1. Was the initial assumption of a nonrespiratory acidosis valid?
2. What would be a better description of the acid-base disturbance?
3. Why, on repeat testing, did the HCO_3^- return to normal?

---

of intracellular O_2 is not feasible with current technology, evaluation of a patient's oxygen status is possible using the partial pressure of oxygen (PO_2) measured along with pH and PCO_2 in the blood gas analysis.

For adequate tissue oxygenation, the following seven conditions are necessary: (1) available atmospheric oxygen, (2) adequate ventilation, (3) gas exchange between the lungs and arterial blood, (4) loading of O_2 onto hemoglobin, (5) adequate hemoglobin, (6) adequate transport (cardiac output), and (7) release of O_2 to the tissues. Any disturbances in these conditions can result in poor tissue oxygenation.

The amount of O_2 available in atmospheric air depends on the barometric pressure (BP). At sea level, the BP is 760 mm Hg. (In the International System of units, 1 mm Hg = 0.133 kPa, where 1 Pa = 1 N/m².) Dalton's law states that the total atmospheric pressure is the sum of the individual gas pressures. One atmosphere exerts 760 mm Hg pressure and is made up of O_2 (20.93%), CO_2 (0.03%), nitrogen (78.1%), and inert gases (approximately 1%). The percentage for each of the gases is the same at all altitudes; the *partial pressure* for each gas in the atmosphere is equal to the BP at a particular altitude times the appropriate percentage for each gas. The vapor pressure of water (47 mm Hg at 37°C) must be accounted for in calculating the partial pressure for the individual gases (Fig. 16-3). In the body, these gases are always fully saturated with water. For example,

Partial pressure of O_2 at sea level (in the body) = (760 mm Hg − 47 mm Hg) × 20.93% = 149 mm Hg (at 37°C)

Partial pressure of CO_2 at sea level (in the body) = (760 mm Hg − 47 mm Hg) × 0.03% = 0.2 mm Hg (at 37°C)

Air is moved into the lungs by the expansion of the thoracic cavity, which creates a temporary vacuum, causing air to rush into the numerous tracheal branches and the alveoli. At the beginning of inspiration, these airways still are filled with air (gas) retained from the previously expired breath. This air, termed *dead space air*, dilutes the air being inspired.

The inspired air, in addition to being somewhat diluted, is warmed to 37°C and fully saturated with water vapor. The PO_2 in the alveoli averages close to 110 mm Hg instead of the potential 150 mm Hg (no dilution and no water vapor). Three other factors can influence the PO_2 in the alveoli: (1) The percentage of O_2 in the inspired air can be increased by breathing gas mixtures up to 100% O_2. The higher the supplemental O_2 concentration inspired, the higher the fraction of inspired oxygen (FiO_2). (2) The amount of PCO_2 in the expired air dilutes the inspired air so that a patient with

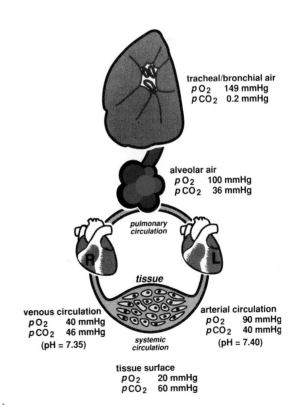

tracheal/bronchial air
pO_2 149 mmHg
pCO_2 0.2 mmHg

alveolar air
pO_2 100 mmHg
pCO_2 36 mmHg

pulmonary circulation

tissue

venous circulation
pO_2 40 mmHg
pCO_2 46 mmHg
(pH = 7.35)

arterial circulation
pO_2 90 mmHg
pCO_2 40 mmHg
(pH = 7.40)

systemic circulation

tissue surface
pO_2 20 mmHg
pCO_2 60 mmHg

Figure 16-3. Gas content in lungs, pulmonary, and systemic circulation.

increased metabolism (during exercise) may produce more CO_2 than can be eliminated, increasing both the PCO_2 in the blood and the expired gas. (3) The ratio of the volume of inspired air to the volume of the dead space air. The volume of dead space (in the airways) is usually constant because it is controlled by the person's anatomy; people breathing with very shallow breaths have less "fresh" air entering the lungs than those breathing very deeply.

There are many factors that can influence the amount of O_2 that moves through the alveoli into the blood and then to the tissues. Among the more common are

- Destruction of the alveoli. The surface area of the alveoli is normally as big as a tennis court. When the surface area is destroyed to a critically low value by diseases such as emphysema, inadequate O_2 will move into the blood.
- Pulmonary edema. Gas diffuses from the alveoli to the capillary through a very small space. With pulmonary edema, fluid "leaks" into this space, increasing the distance between the alveoli and capillary walls, which causes a barrier to diffusion.
- Airway blockage. Airways can be blocked, preventing the air from the atmosphere from reaching the alveoli. Asthma and bronchitis are more common causes of this type of problem.
- Inadequate blood supply. When the blood supply to the lung is inadequate, the amount of O_2 entering the blood is sufficient, but not enough blood is being carried away to the tissues where it is needed. This may be the consequence of a blockage in a pulmonary blood vessel (pulmonary embolism) or a failing heart.
- Movement of CO_2 and O_2. CO_2 and O_2 move together into and out of the lungs. CO_2 will diffuse 20 time faster than O_2, so it is less sensitive to problems with diffusion. Carbon dioxide production is not controlled easily, so inadequate ventilation causes CO_2 to build up. The percentage of O_2 can be increased temporarily when needed, but 100% O_2 is toxic to the lungs and its use must be limited.

Oxygen Transport

Most of the O_2 in arterial blood is transported to the tissues by hemoglobin. Each adult hemoglobin (A_1) molecule can combine reversibly with up to four molecules of O_2. The actual amount of O_2 loaded onto hemoglobin depends on the availability of O_2, the concentration and type(s) of hemoglobin present, the presence of nonoxygen substances, such as carbon monoxide (CO), the pH, the temperature of the blood, and the levels of PCO_2 and 2,3-DPG. With adequate atmospheric and alveolar O_2 available, and with normal diffusion of O_2 to the arterial blood, more than 95% of "functional" hemoglobin (hemoglobin capable of *reversibly* binding O_2) will bind O_2. Increasing the availability of O_2 to the blood further saturates the hemoglobin. However, once the hemoglobin is 100% saturated, an increase in O_2 to the alveoli serves only to increase the concentration of dissolved O_2

in the arterial blood and may cause O_2 toxicity as well as decreased ventilation and increased PCO_2.

Hemoglobin's ability to carry O_2 can be affected significantly by other molecules. Normally blood hemoglobin exists in one of four conditions:

1. Oxyhemoglobin (O_2Hb), which is O_2 reversibly bound to hemoglobin.
2. Deoxyhemoglobin (HHb; reduced hemoglobin), which is hemoglobin not bound to O_2, but is capable of forming a bond when O_2 is available.
3. Carboxyhemoglobin (COHb), which is hemoglobin bound to CO. The bond between CO and Hb is reversible, but is >200 times as strong as the bond between O_2 and Hb.
4. Methemoglobin (MetHb), which is hemoglobin unable to bind O_2 because the iron (Fe) is in the oxidized rather than the reduced state. The Fe^{3+} can be reduced by the enzyme methemoglobin reductase, which is found in RBCs.

Dedicated spectrophotometers (co-oximeters), which are discussed later in this chapter, are used to determine the relative concentrations (relative to the total hemoglobin) of each of these species of hemoglobin.

Quantities Associated With Assessing a Patient's Oxygen Status

Four parameters commonly used to assess a patient's oxygen status are oxygen saturation (SO_2), measured fractional (percent) oxyhemoglobin (FO_2Hb), trends in oxygen saturation as assessed by pulse oximetry (SpO_2), and the amount of O_2 dissolved in plasma (PO_2).

Oxygen saturation (SO_2) represents the ratio of O_2 that is bound to the carrier protein, hemoglobin, compared with the total amount of hemoglobin capable of binding O_2.[2]

$$SO_2 = \frac{cO_2Hb}{(cO_2Hb + cHHb)} \times 100$$

(Eq. 16–10)

The symbol c represents the concentration. Although the fraction of oxyhemoglobin can be determined directly on arterial blood samples, mixed-venous samples (drawn from the pulmonary artery), and venous samples through spectrophotometric measurements, software included with blood gas instruments calculate SO_2 from the PO_2, the pH, and the temperature of the sample. These calculated results can differ significantly from those determined by direct measurement due to software algorithms that assume a specific shape and location for the oxyhemoglobin dissociation curve. These algorithms do account for the presence of other hemoglobin species, such as COHb and MetHb, that are incapable of reversibly binding O_2.[2] Because of the potential for generating erroneous information, calculated SO_2 should not be used to assess patients' oxygenation status.[2,3]

Fractional (or percent) oxyhemoglobin (FO_2Hb) is the ratio of the concentration of oxyhemoglobin to the concentration of total hemoglobin ($ctHb$),

$$FO_2Hb = \frac{cO_2Hb}{ctHb} = \frac{cO_2Hb}{cO_2Hb + cHHb + cdysHb}$$

(Eq. 16–11)

where the $cdysHb$ represents the hemoglobin derivatives, such as carboxyhemoglobin (COHb), that cannot reversibly bind with O_2 but are still part of the "total" hemoglobin measurement.

These two terms, SO_2 and FO_2Hb, can be confused because, in most healthy individuals (and even those individuals with some disease states), the numerical values for SO_2 are very close to those for FO_2Hb. However, the values for FO_2Hb and SO_2 will deviate in several conditions (such as when the patient is a smoker), owing to the preferential binding of CO to hemoglobin, and the resultant loss of hemoglobin to bind O_2.

Partial pressure of oxygen dissolved in plasma (PO_2) accounts for very little of the body's O_2 stores. A healthy adult breathing room will have a PO_2 of 90 mm Hg to 95 mm Hg. Only 13.5 mL of O_2 will be available from plasma if the blood volume is 5 L.

Noninvasive measurements for following "trends" in oxygen saturation also are possible using *pulse oximetry* (SpO_2).[3,4] By passing light of two different wavelengths through the tissue of the toe, finger, or ear, the pulse oximeter differentiates between the absorption of light due to oxyhemoglobin and deoxyhemoglobin in the capillary bed and calculates oxyhemoglobin saturation. Because SpO_2 does not measure COHb, it overestimates oxygenation in its presence. The accuracy of pulse oximetry can be compromised by many factors. Dysfunctional hemoglobins (COHb, MetHb), diminished pulse due to poor perfusion, and severe anemia are but a few.

The maximum amount of O_2 that can be carried by hemoglobin in a given quantity of blood is the *hemoglobin oxygen (binding) capacity*. The molecular weight of the tetramer hemoglobin is 64,458 g/mol. One mole of a perfect gas occupies 22,414 mL. Therefore, each gram of hemoglobin carries 1.39 mL of O_2:

$$\frac{22,414 \text{ mL/mol}_4}{64,458 \text{ g/mol}} = 1.39 \text{ mL/g}$$

(Eq. 16–12)

When the total hemoglobin (tHb) is 15 g/dL and the hemoglobin is 100% saturated with O_2, the O_2 capacity is:

$$15 \text{ g/100 mL} \times 1.39 \text{ mL/g}$$
$$= 20.8 \text{ mL } O_2/100 \text{ mL of blood}$$

(Eq. 16–13)

Oxygen content is the total O_2 in blood and is the sum of the O_2 bound to hemoglobin (O_2Hb) and the amount dissolved in the plasma (PO_2). (Because PO_2 and PCO_2 are only indices of gas-exchange efficiency in the lungs, they do not reveal the *content* of either gas in the blood.) For every mm Hg PO_2, 0.00314 mL of O_2 will be dissolved in 100 mL of plasma at 37°C. For example, if the PO_2 is 100 mm Hg, 0.3 mL of O_2 will be dissolved in every 100 mL of blood plasma. The

amount of dissolved O_2 is usually not clinically significant. However, with low tHb or at hyperbaric conditions, it may become a significant source of O_2 to the tissues. Normally, 98% to 99% of the available hemoglobin is saturated with O_2. Assuming a tHb of 15 g/dL, the O_2 content for every 100 mL of blood plasma becomes

$$0.3 \text{ mL} + (20.8 \text{ mL} \times 0.97) = 20.5 \text{ mL}$$

(Eq. 16–14)

Hemoglobin-Oxygen Dissociation

In addition to adequate ventilation and gas exchange with the pulmonary circulation, O_2 must be released at the tissues. Hemoglobin transports O_2. The increased H^+ concentration and PCO_2 levels at the tissues due to cellular metabolism change the molecular configuration of O_2Hb, facilitating O_2 release.

Oxygen dissociates from adult hemoglobin (A_1) in a characteristic fashion. If this dissociation is graphed (Fig. 16-4) with the PO_2 on the *x*-axis and the percent SO_2 on the *y*-axis, the resulting curve is sigmoid, or slightly S-shaped. Hemoglobin "holds on" to O_2 until the O_2 tension is reduced to about 60 mm Hg. Below this tension, the O_2 is released rapidly. The position of the oxygen dissociation curve reflects the *affinity* that hemoglobin has for O_2 and affects the rate of this dissociation.

Several factors affect the affinity of hemoglobin for O_2. The pH affects dissociation. An increase in H^+ in the tissues leads to a decrease in affinity of hemoglobin for O_2, resulting in a shift to the right of the dissociation curve. In cases in which the hemoglobin releases the O_2 more easily or at a faster-than-normal rate, the patient benefits (provided sufficient O_2 is taken up by the lungs) and the tissues receive adequate O_2, even though the PO_2 or the hemoglobin level may not be within "normal" limits.

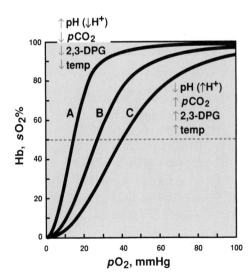

▲
Figure 16-4. Oxygen-dissociation curves. Curve *B* is the normal human curve. Curves *A* and *C* are from blood with increased affinity and decreased affinity, respectively.

As PCO_2 decreases, the blood pH becomes more alkaline in the lung, and the curve shifts slightly to the left, enhancing O_2 binding to hemoglobin. An elevated temperature results in a shift to the right and O_2 leaves the hemoglobin. Conversely, hypothermia shifts the curve to the left, and the hemoglobin binds the O_2 more tightly. An increase in CO_2 causes a shift to the right. An elevation in CO causes the curve to shift to the left. As the percentage of COHb increases, the shape of the curve loses some of its sigmoid characteristics, and the O_2 bound to hemoglobin will not be released to the tissues as easily.

2,3-DPG is a phosphate compound in RBCs that exerts an effect on the oxyhemoglobin dissociation curve. When the β chains of the hemoglobin molecule bind 2,3-DPG, a shift to the right occurs and O_2 is unloaded. DPG stimulation occurs with a decrease in intracellular pH, a decrease in O_2 level in the tissues, and adaptation to high altitude. Many patients with slow onset of anemia demonstrate elevated levels of 2,3-DPG. This may partially explain the failure of patients with extremely low hemoglobin values to lose consciousness.

The preceding discussion refers to normal adult (A_1) hemoglobin. In hemoglobinopathies and in newborns, the pattern of dissociation may differ. For example, fetal hemoglobin causes a shift to the left, but with little change in the sigmoid shape.

MEASUREMENT

Spectrophotometric (Co-Oximeter) Determination of Oxygen Saturation

The *actual percent oxyhemoglobin* (O_2Hb) can be determined spectrophotometrically using a co-oximeter designed to determine the various hemoglobin species directly. Each species of hemoglobin has a characteristic absorbance curve (Fig. 16-5). The number of hemoglobin species measured will depend on the number and specific wavelengths incorporated into the instrumentation. For example, two-wavelength instrument systems can measure only two hemo-

globin species, that is, O_2Hb and HHb, which are expressed as fraction or percentage of the total hemoglobin.

Instruments, at a minimum, should have four wavelengths for measurements of HHb, O_2Hb, and the two most common dyshemoglobins: carboxyhemoglobin (COHb) and methemoglobin (MetHb). Instruments with more than four wavelengths can recognize the presence of dyes and pigments, turbidity, sulfhemoglobin (SulfHb), and abnormal proteins. Microprocessors control the sequencing of multiple wavelengths of light through the sample and apply the necessary "matrix" equations after absorbance readings are made to calculate the percent O_2Hb, COHb, and other hemoglobin species:

$$O_2HB = a_1A_1 + a_2A_2 + \ldots + a_nA_n$$
$$HHb = b_1A_1 + b_2A_2 + \ldots + b_nA_n$$
$$COHb = c_1A_1 + c_2A_2 + \ldots + c_nA_n$$
$$MetHb = d_1A_1 + d_2A_2 + \ldots + d_nA_n,$$

(Eq. 16–15)

where a_1, a_n, b_n, and so on are coefficients that are analogues of the absorption constant a that are derived from established methods, and A_1, A_2, and so on are the absorbances of the sample. The matrix equations will change depending on the number of wavelengths of light (which is manufacturer specific) passed through the sample.[5] (The "calculation" made by these instruments should not be confused with a calculated SO_2 from a blood gas analyzer, which, in reality, *estimates* the value from a measured PO_2 and an empirical equation for the oxyhemoglobin dissociation curve. Although SO_2 and O_2Hb values are similar in the absence of COHb and MetHb, they will be vastly different in the presence of these hemoglobins. Only O_2Hb values reflect the patient's true status.)

As with any spectrophotometric measurement, there exist potential sources of error, including faulty calibration of the instrument and the presence of spectral-interfering substances. *Calibration* is usually set by the manufacturer at the factory and is modified only by a mechanical or optical failure. The user can calibrate these systems only for total hemoglobin. Co-oximeters are not subject to calibration drift as is seen with blood gas systems. The presence of any substances absorbing light at the wavelengths used in the measurement of any hemoglobin pigment has the potential of being a source of error. Product claims for specific instruments must be consulted for specific interferences.

Because the primary purpose of determining O_2Hb is to assess oxygen transport from the lungs, it is best to stabilize the patient's ventilation status before blood sample collection. Changes in supplemental O_2 or mechanical ventilation should be followed by an appropriate waiting period before the sample is redrawn. All blood samples should be collected under anaerobic conditions and mixed immediately with heparin or other appropriate anticoagulant. If the blood gas analysis is not being done on the same sample, ethylenediaminetetraacetic acid (EDTA) is an appropriate anticoagulant. Promptly analyze the sample to avoid

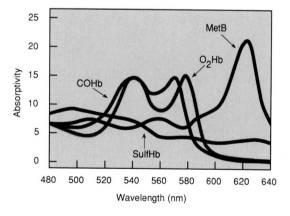

Figure 16-5. Optical absorption of hemoglobin fractions. (Reproduced with permission of Clin Chem News, January 1990.)

A 37-year-old male was admitted to the emergency room (ER). He was short of breath, dizzy, flushed (diaphoretic), sweating (hyperemic), and nauseous. Shortly after being admitted, blood gases were drawn:

$$pH = 7.48$$
$$PCO_2 = 32 \text{ mm Hg}$$
$$PO_2 = 96 \text{ mm Hg}$$
$$HCO_3^- = 24 \text{ mmol/L}$$
$$SO_2 = 98\% \text{ (calculated)}$$
$$SpO_2 = 99\% \text{ (pulse oximetry oxygen saturation)}$$

After a few hours, the patient's symptoms receded and he was released. Two weeks later, the same patient was again admitted to the ER with the same symptoms. This time, arterial blood was drawn for both blood gases and CO-oximetry measurements. The results were as follows:

$$pH = 7.49$$
$$PCO_2 = 33 \text{ mm Hg}$$
$$PO_2 = 95 \text{ mm Hg}$$
$$HCO_3^- = 23 \text{ mmol/L}$$

$$SO_2 = 98\% \text{ (calculated) (reference range} >95\%)$$
$$SpO_2 = 99\% \text{ (pulse oximetry oxygen saturation)}$$
$$\text{(reference range} >95\%)$$

Spectrophotometric (co-oximeter) measurement of hemoglobin species:

$$tHb = 13.5 \text{ g/L}$$
$$O_2Hb = 73\% \text{ (reference range} >95\%)$$
$$COHb = 22\% \text{ (reference range} <2\%; \text{ higher with smokers)}$$
$$MetHb = 1\% \text{ (reference range} <1.5\%)$$

Questions

1. Is the patient hypoxic on the first admission to the ER?
2. Considering the new laboratory data, is this patient hypoxic on the second admission to the ER?
3. Why is there a discrepancy between the calculated SO_2, SpO_2, and O_2Hb?
4. What is a possible cause of this patient's shortness of breath and low O_2Hb?

changes in saturation resulting from the use of oxygen by metabolizing cells.[6]

Blood Gas Analyzers: pH, PCO_2, and PO_2

Blood gas analyzers use *electrodes* as sensing devices to measure PO_2, PCO_2, and pH. The PO_2 measurement is amperometric, meaning that the amount of current flow is an indication of the oxygen present. The PCO_2 and pH measurements are potentiometric, in which a change in voltage indicates the activity of each analyte.

The *cathode* can be defined in at least three ways: (1) the negative electrode, (2) a site to which cations tend to travel, or (3) a site at which reduction occurs. *Reduction* is the gain of electrons by a particle (atom, molecule, or ion). The *anode* is the positive electrode, the site to which anions migrate, or the site at which oxidation occurs. *Oxidation* is the loss of electrons by a particle. An *electrochemical cell* is formed when two opposite electrodes are immersed in a liquid that will conduct the current. Several additional parameters can be calculated by the blood gas analyzer: bicarbonate, total carbon dioxide, base excess, and oxygen saturation.

Measurement of PO_2

PO_2 electrodes, called Clarke electrodes, measure the amount of current flow in a circuit that is related to the amount of O_2 being reduced at the cathode. A gas-permeable membrane

covering the tip of the electrode selectively allows the O_2 to diffuse into an electrolyte and contact the cathode. Electrons are drawn from the anode surface to the cathode surface to reduce the O_2. A small, constant polarizing potential (typically $-0.65V$) is applied between the anode and cathode. A *microammeter* placed in the circuit between the anode and cathode measures the movement of electrons (current). Four electrons are drawn for every mole of O_2 reduced, making it possible to determine the PO_2. (The semipermeable membrane also will allow other gases to pass, such as CO_2 and N_2, but these gases will not be reduced at the cathode if the polarizing voltage is tightly controlled.)

The primary source of error for PO_2 measurement is associated with the buildup of protein material on the surface of the membrane. This buildup retards diffusion and slows the electrode response. Bacterial contamination within the measuring chamber, although uncommon, will consume O_2 and cause low and drifting values. Most other errors are associated with a malfunction of the system such as incorrect calibration.

Noninstrumental concerns, including sample collection and handling, are addressed later in this chapter. However, it is particularly important not to expose the sample to room air when collecting, transporting, and making O_2 measurements. Contamination of the sample with room air ($PO_2 \cong 150$ mm Hg) can result in significant error. Even after the sample is drawn, leukocytes continue to metabolize O_2. Unless the sample is analyzed immediately after being drawn, low PO_2 values may be seen with high white blood cell counts.

Continuous measurements for PO_2 also are possible using *transcutaneous (TC) electrodes* placed directly on the skin. Measurement depends on oxygen diffusing from the capillary bed through the tissue to the electrode. Although most commonly used with neonates and infants, this noninvasive approach is not without problems. Skin thickness and tissue perfusion with arterial blood can significantly affect the results. Heating the electrode placed on the skin can enhance diffusion of O_2 to the electrode, but burns can result unless the electrodes are moved regularly. Although PO_2 measured by these electrodes may *reflect* the arterial PO_2, the two values are not equivalent because of the consumption of oxygen by the tissue where the electrode is applied and the effects of heating the tissue.

Measurement of pH and PCO₂

To understand potentiometric measurements, it is helpful to think of atoms and ions as having a chemical energy. An increased concentration or *activity* of the ions leads to an increase in force exerted by those ions.

To measure how much force—energy or potential—a given ion possesses, certain elements in the measuring device are required; namely, two electrodes (the measuring electrode responsive to the ion of interest and the reference electrode) and a voltmeter, which measures the potential difference (ΔE) between the two electrodes. The potential difference is related to the concentration of the ion of interest by the Nernst equation:

$$\Delta E = \Delta E° + \frac{0.05916}{n} \log a_i \qquad \text{at } 25°C,$$

(Eq. 16–16)

where $\Delta E°$ = standard potential of the electrochemical cell, n = charge of the analyte ion i, and a_i = activity of the analyte ion i.

To measure pH, a glass membrane sensitive to H^+ is placed around an internal Ag–AgCl electrode to form a measuring electrode. The potential that develops at the glass membrane as a result of H^+ from the unknown solution diffusing into the membrane's surface is proportional to the difference in cH^+ between the unknown sample and the buffer solution inside the electrode. For the potential developed at the glass membrane to be measured, a reference electrode must be introduced into the solution and both electrodes must be connected to a pH (volt) meter. The reference electrode [commonly either a calomel (Hg–HgCl) or an Ag–AgCl half-cell] provides a steady reference voltage against which voltage changes from the measuring electrode are compared. The pH meter reflects the potential difference between the two electrodes.

For the cell described, the Nernst equation predicts that a change of +59.16 mV, at 25°C, is the result of a ten-fold increase in H^+ activity or a decrease of an entire pH unit (*eg*, pH 7 to pH 6). Changing the temperature affects the response. At 37°C, a change of 1 pH unit elicits a 61.5mV change. The

glass membrane of the measuring electrode must be kept free from protein buildup because coating of the membrane causes sluggish or erratic responses.

PCO_2 is determined with a modified pH electrode, called a Severinghaus electrode. The glass pH electrode is covered by an outer semipermeable membrane that allows CO_2 to diffuse into a layer of electrolyte, usually a bicarbonate buffer. The CO_2 that diffuses across the membrane reacts with the buffer, forming *carbonic acid,* which then dissociates into bicarbonate plus H^+. The change in activity of the H^+ is measured by the pH electrode and related to PCO_2.

As with the other electrodes, the buildup of protein material on the membrane will affect diffusion and cause errors. PCO_2 electrodes are the slowest to respond because of the chemical reaction that must be completed. Other error sources include erroneous calibration caused by incorrect or contaminated calibration materials.

Types of Electrochemical Sensors

The *macro electrode sensors* have been used in blood gas systems since the beginnings of the clinical measurement of blood gases. These have been modified over time in an effort to simplify their use and minimize the required sample volume. *Micro electrodes* are miniaturized macro electrodes. Miniaturization became possible with better manufacturing capabilities and with the development of the sophisticated electronics that are required to handle the minute changes in signal.

Thick and thin film technology is a further modification of electrochemical sensors. The measurement principle is identical, but the sensors are reduced to tiny wires embedded in a printed circuit card. The special card has etched grooves to separate components. A special paste material containing the required components (similar in function to the electrolytes of macro electrodes) is spread over the sensors. To reduce the required sample volume, several sensors can be placed on a single small card. These sensors are less expensive to manufacture and are disposable, which reduces maintenance.

Optical Sensors

Another technology that is emerging for blood gas measurements is based on the fact that certain fluorescent dyes will react predictably with specific chemicals such as O_2, CO_2, and H^+. The dye is separated from the sample by a membrane, as with electrodes, and the analyte diffuses into the dye, causing either an increase in or a quenching of fluorescence proportional to the amount of analyte. Calibration is used to establish the relationship between concentration and fluorescence. Normally, a single calibration will suffice for long periods, because this technology is not subject to the drifts seen in electrochemical technology.

Optical technology has been applied to indwelling blood gas systems. Fiberoptic bundles carry light to sensors positioned in the tip of catheters and other bundles carry light back, allowing changes in fluorescence to be measured in a catheter within the patient's arterial system. The commercial development of indwelling systems has been limited by the increased probability of thrombogenesis and protein buildup on the membrane, separating the sample from the fluorescing dyes. This buildup impedes free sample diffusion into the measuring chamber.

Calibration

Temperature is an important factor in the measurement of pH and blood gases. The Nernst equation specifies the expected voltage output of an electrochemical cell at a given temperature. If the temperature of the measurement system changes, the output (voltage) will change. The solubility of gases in a liquid medium also depends on the temperature: as the temperature goes down, the solubility of the gas increases. Because pH and blood gas measurements are extremely sensitive to temperature, it is critical that the electrodes' sample chamber be maintained at constant temperature for all measurements. All blood gas analyzers have electrode chambers thermostatically controlled to $37 \pm 0.1°C$.

The pH electrode usually is calibrated with two buffer solutions traceable to standards prepared by the National Institute of Standards and Technology (NIST). *Traceable* usually means that the actual value of the calibrator has been determined using a NIST standard as a reference. One calibrator usually is close to 6.8 and the other is near 7.38, because most pH electrodes produce "0" voltage at this point. The calibrators must be stored at the stated temperature and not exposed to room air, because pH changes with the absorption of CO_2.

Calibration of any blood gas analyzer will vary depending on the manufacturer. Normally, two gas mixtures are used for PCO_2 and PO_2. One gas has no O_2 to set the zero point of the O_2 electrode (which is usually a very stable point). The same gas has approximately 5% CO_2 because this is the null (zero potential and very stable) point for the CO_2 electrode. The other gas sets the gain, that is, the amount of change in the electrode signal relative to the change for the analyte. The gas can have any value.

Most instruments are self-calibrating (calibrate automatically at specified time intervals) and are programmed to indicate a calibration error if the electronic signal from the electrode is inconsistent with the programmed expected value. For example, if the value(s) obtained during calibration exceed(s) a programmed tolerance limit, flagging of a *drift* error will occur at the time of calibration, and corrective action will need to be taken before patient samples can be analyzed.

Calculated Parameters

Several parameters can be calculated from the measured pH and PCO_2 values. Manufacturers of blood gas instruments include algorithms to perform the calculations. No calculated parameter is used universally and many physicians have "favorite" parameters for identifying various pathologies.

The calculation of HCO_3^- is based on the Henderson-Hasselbalch equation. This can be calculated once the pH and PCO_2 are known. One basic assumption is that the pK' of the bicarbonate buffer system in plasma at $37°C$ is 6.1.

Carbonic acid concentration can be calculated using the solubility coefficient of CO_2 in plasma at $37°C$. The solubility constant to convert PCO_2 to millimoles per liter of H_2CO_3 is 0.0307. If the temperature or the composition of plasma changes (*eg,* an increase in lipids, in which gases are more soluble), the constant will change.

Total carbon dioxide content ($ctCO_2$) is the bicarbonate plus the dissolved CO_2 (carbonic acid) plus the associated CO_2 with proteins (carbamates). A blood gas analyzer approximates total CO_2 content by adding the bicarbonate and carbonic acid values [$ctCO_2 = cHCO_3^- + (0.0307 \times PCO_2)$].

Some clinicians use *base excess* to assess the metabolic component of a patient's acid-base disorder, calculating base excess from the patient's pH, PCO_2, and hemoglobin. A positive value (base excess) indicates an excess of bicarbonate or relative deficit of noncarbonic acid and suggests *metabolic alkalosis*. A negative value (base deficit) indicates a deficit of bicarbonate or relative excess of noncarbonic acids and suggests metabolic acidosis. However, the indicated metabolic alkalosis or acidosis may be due to primary disturbances or compensatory mechanisms.

Correction for Temperature

pH, PCO_2, and PO_2 values are all temperature dependent. By convention, all of these measurements are made at $37°C$. The question becomes, When the patient's body temperature differs from $37°C$, should the blood gas values be "corrected" to the actual temperature of the patient? Although the blood gas instrument's software easily performs the correction, the data may be confusing because appropriate reference ranges *for that temperature* must be used for proper interpretation. Usually values are reported at $37°C$ as a reference when values also are reported at actual patient temperature.

QUALITY ASSURANCE

Preanalytical Considerations

Blood gas measurements, like all laboratory measurements, are subject to preanalytical, analytical, and postanalytical errors. Few other measurements, however, are as affected by preanalytical errors—those introduced during the collection and transport of samples before analysis.[6]

Figure 16-6 depicts the quality assurance cycle. The steps included in the analytical area are under the direct control of the laboratory. Much of the quality assurance cycle lies outside the laboratory. Therefore, the laboratorian must take an active role in educating *all* people involved by developing policies and procedures for controlling all the processes in the cycle to ensure quality.

Only personnel who have experience with the drawing equipment and have knowledge of the possible sources of error should draw samples for pH and blood gas analyses. Because collection may be painful and result in patient hyperventilation, which lowers the PCO_2 and increases the pH,

**BLOOD GAS ANALYSIS QUALITY
ASSURANCE CYCLE**

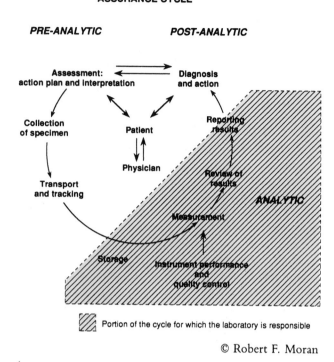

Figure 16-6. Blood gas analysis quality assurance cycle. (Reproduced with permission of Robert F. Moran.)

the ability to reassure the patient is essential. The choice of site—radial, brachial, femoral, or temporal artery—is usually customary within an institution, depending on the predominant patient population (*eg,* pediatric patients, burn patients, outpatients). The National Committee for Clinical Laboratory Standards (NCCLS) publication *Collection of Arterial Blood Specimens* is an excellent reference.[7]

The use of venous samples for pH and blood gas studies is not recommended. Peripheral venous samples can be used if pulmonary function or O_2 transport is not being assessed. The sample must be clearly identified as venous and the appropriate (venous) reference ranges included. Capillary blood may need to be used to measure pH and PCO_2. Capillary PO_2 values, even with warming of the skin before drawing the sample, do not correlate well with arterial PO_2 values, primarily due to the exposure of the sample to room air. The correlation is good for pH and PCO_2. Central venous (pulmonary artery) blood samples are obtained to assess O_2 consumption, which is calculated from the difference between the O_2 content of arterial blood and pulmonary artery blood times the cardiac output.

Sources of error in the collection and handling of blood gas specimens include the collection device, form and concentration of heparin, speed of syringe filling, maintenance of the anaerobic environment, mixing, transport, and storage time before analysis. For proper interpretation of blood gas results, the patient's status in terms of ventilation (on room air or supplemental O_2), temperature, and posture must be documented at the time the sample is collected.

Recommended collection devices for arterial blood samples include glass and plastic syringes. Glass syringes have been preferred because the plunger responds to arterial pressure and the glass is impermeable to gases that can alter PCO_2 and PO_2 values over time. Plastic syringes have become more commonly used because of the development of self-filling technology. However, plastic syringes can alter PO_2, and to some extent PCO_2, values due to room air dissolved in the syringe barrel and plunger tip.[8,9] Temperature and the PO_2 value affect the size of the error. Due to the increased solubility of O_2 with decreasing temperature, the error is magnified by icing the specimen, particularly those having PO_2 values greater than 100 mm Hg. If plastic syringes are used, it is recommended that the sample not be iced and be analyzed within 20 minutes.

Both dry and liquid heparin are acceptable anticoagulants. Because of the potential for dilution and air contamination errors, liquid heparin is not recommended.[6] The use of dry heparin eliminates these problems. However, care must be taken to ensure that the heparin is dissolved and no clot forms. Although sodium and lithium heparin are recommended for pH and blood gas analysis, other salt forms are available: ammonium, zinc, electrolyte balanced, and calcium titrated. Selection of the proper type of heparin is particularly important with instruments combining blood gas and electrolyte analyses.

Slow filling of the syringe may be caused by a mismatch of syringe and needle sizes. Although too small a needle reduces the pain and therefore the likelihood of arteriospasm and hematoma, it may produce bubbles as well as hemolysis of the sample (which is important when potassium is measured along with pH and blood gases). Maintenance of an anaerobic environment is critical to correct results.

The blood in the syringe must be mixed with the heparin anticoagulant to prevent clots from forming and being injected or aspirated into the blood gas analyzer. Adequate mixing immediately before analysis is essential.

Transport time of the sample should be minimal. Immersing the specimen in an ice water slurry prevents changes in PCO_2, pH, and PO_2 in a glass syringe. If the alveolar PO_2 is above 100 mm Hg, which could be anticipated with anyone receiving O_2 therapy, the NCCLS guidelines recommend rapid analysis of PO_2.[6]

Because sample procurement and handling are the source of many possible errors in blood gas analysis, it is necessary that procedures and policies are constructed carefully and adherence monitored to ensure quality. No QC product can monitor the preanalytical aspects of blood gas analysis.

Quality Control

QC for blood systems assesses only the analytical phase of the testing process. Traditional QC approaches for blood gases include commercial liquid controls, tonometry, and duplicate analysis of patient samples made on separate instruments. All these have limitations. The "ideal" approach would encompass some combination of the three.

Commercial liquid control materials provide the basis for most of today's QC practices. These are sold in sealed glass ampules that contain solutions equilibrated with gases. Ideally, such materials are stable, have minimal ampule-to-ampule variance, and are available in large lots that can be used for many months. Although a laboratory tries to choose a control material that closely mimics actual patient samples, this is impossible for blood gases.

At present, three general classes of commercial liquid controls are available for blood gas analysis: equilibrated aqueous solutions; blood-based solutions (solutions containing free hemoglobin); and perfluorocarbon emulsions. Controls can include additional analytes, such as sodium, potassium, chloride, and calcium. All vary in their stability and are susceptible to temperature variation in storage and handling. Each must be handled as described by the manufacturer to eliminate precision errors caused by improper handling of the ampules.

These controls are available in at least three levels corresponding to values observed with low, normal, and elevated values for each of the measured analytes. For each level, the manufacturer specifies target, or mean, values. However, a laboratory should establish its own ranges that reflect the actual testing conditions. Commercial liquid control materials have significantly different matrices than fresh whole blood. Consequently, they may not detect problems that affect patient samples, or they may detect errors induced by improper handling of the commercial controls. Aqueous base controls, the most commonly used QC material, have very low O_2 solubility, making them very sensitive to factors that affect the determination of PO_2. This is particularly obvious with interlaboratory proficiency testing data collected from different manufacturers' instruments. Aqueous controls will exaggerate some types of instrument problems and miss others, depending on the design of the blood gas system. Perfluorocarbon-based controls more closely mimic blood. They have increased O_2 solubility and thus are less sensitive to instrument design deficiencies. These materials, however, cannot be used with systems that measure electrolytes.

Tonometry, using fresh whole blood, may be the ideal approach.[6] Equilibrating, or *tonometering,* a blood sample with gases of known concentration over time and at a constant temperature is a relatively inexpensive way to check the precision and the accuracy of the PCO_2 and PO_2 measurements. Although tonometry may be the ideal QC approach for blood gases, many problems have been documented. Because many perceive tonometry to be cumbersome and time consuming, few laboratories use this approach today.

A laboratory may choose to perform *duplicate assays* using two or more instruments for simultaneous analysis of a patient sample. *Delta checks,* or the difference in values obtained on the two instruments, often pick up problems that might be missed in routine QC.[10] The allowable difference in duplicates run on split patient samples should be tighter than those observed with commercial liquid controls. Discrepancies between results provide no clue regarding which data point is wrong or which instrument is malfunctioning. Although two instruments are unlikely to have the same error simultaneously, this is not always true. Therefore, the duplicate-assay approach cannot be used as the sole method of QC. Used in conjunction with commercial liquid controls or tonometry, it can be a useful technique for detecting errors and also for troubleshooting instruments.

An effective QC scheme also includes peer review and duplicate sample analysis (on the same instrument) to help minimize the inadequacies of commercial controls that do not mimic blood.[11] Peer review is information obtained from the manufacturers of the controls. Accuracy is estimated by comparing the laboratory's mean value obtained on a lot of controls to the average value (the mean of the means) obtained by many laboratories on the same lot. The information is similar to proficiency testing but is done on a continuous basis. In addition to the standard deviation and coefficient of variation calculated from accumulated QC data, imprecision can be estimated by duplicate analysis of patient samples done throughout the workday. Changes in instrument performance that will affect patient care are quickly ascertained using this scheme.

Whatever the QC approach, the QC needs of the blood gas laboratory contrast sharply with those of the general laboratory, which analyzes many patient samples as a group and includes multiple control specimens with each run. In the blood gas laboratory, time or patient sample volume do not always allow for repeat analyses if problems exist. Consequently, the blood gas laboratory must perform *prospective* QC, because instruments must be *prequalified* to ensure proper performance before the patient sample arrives for analysis.[10,11]

External (Proficiency Testing) Quality Control

Participating in external interlaboratory surveys or proficiency testing programs assists in identifying and monitoring accuracy problems.[12] Ongoing comparisons of results through proficiency testing help to ensure that systematic (accuracy) errors do not slowly increase and go undetected by internal QC procedures. A rigorous internal QC program ensures internal consistency. Good performance in a proficiency testing program ensures the absence of significant bias relative to other laboratories and confirms the validity of a laboratory's patient results. If an individual analyzer does not produce proficiency testing results consistent with its peer laboratories (those using the same method/instrument), or if the differences between values change over time, suspicion of the instrument's performance is warranted.

Interpretation of Results

Laboratory professionals need certain knowledge, attitudes, and skills for obtaining and analyzing specimens for pH and blood gases. Although the patient's physician assimilates all results—laboratory, radiology, nuclear medicine, surgical pathology findings, and so on, along with the patient's clinical history—laboratory personnel must immediately assess patient results and make preliminary judgments about the "fit," that is, do the results make sense? Simple evaluation of the data may reveal an instrument problem (possible bubble in the sample chamber or fibrin plug) or a possible sample handling problem (PO_2 out of line with previous results and current inspired FiO_2 levels). The application of knowledge saves time. The ability to correlate data quickly reduces turnaround time and prevents mistakes.

SUMMARY

Arterial pH and blood gas measurements are ordered to facilitate the care and treatment of critically ill patients. The body maintains acid-base balance through various buffering systems. In the laboratory, the bicarbonate-carbonic acid buffer system, which works in conjunction with and reflects the status of the body's other buffering systems, is used to evaluate acid-base status. pH and PCO_2 measurements assess the patient's acid-base status. Calculated parameters, such as HCO_3^- and base excess help to differentiate metabolic (nonrespiratory) from respiratory conditions.

The PO_2 measurement also is important. In arterial blood, it directly assesses the ability of the lungs to oxygenate the blood. It is used as an indirect measurement of the body's tissue oxygenation status. However, this measurement alone can be misleading. For example, an anemic patient will have a decreased O_2 content and capacity and a normal PO_2, provided the cardiovascular and pulmonary systems are intact. Consequently, other parameters are used in conjunction with PO_2. These include FO_2Hb, identification of the presence of dyshemoglobins, and the calculated parameters of O_2 content and capacity.

REVIEW QUESTIONS

1. The presence of dyshemoglobins will cause a calculated $\%SO_2$ result to be falsely (elevated, decreased) and a pulse oximeter $\%SO_2$ value to be falsely (elevated, decreased):
   a. Elevated, elevated
   b. Decreased, decreased
   c. Elevated, decreased
   d. Decreased, elevated

2. The anticoagulant of choice for arterial blood gas measurements is:
   a. EDTA
   b. Potassium oxalate
   c. Sodium citrate
   d. Lithium heparin

3. At a pH of 7.10, the H^+ concentration is equal to:
   a. 20 nmol/L
   b. 40 nmol/L
   c. 60 nmol/L
   d. 80 nmol/L

4. The kidneys compensate for nonmetabolic alkalosis by (excretion, retention) of bicarbonate and (increased, decreased) excretion of NaH_2PO_4:
   a. Excretion, increased
   b. Retention, increased
   c. Execution, decreased
   d. Retention, decreased

5. The normal ratio of carbonic acid to bicarbonate in arterial blood is:
   a. 7.4:6.1
   b. 1:20
   c. .003:1.39
   d. 20:1

6. When arterial blood from a "normal" patient is exposed to room air, the following happens:
   a. PCO_2 decreases; PO_2 increases
   b. PCO_2 increases; PO_2 decreases
   c. PCO_2 decreases; PO_2 decreases
   d. PCO_2 increases; PO_2 increases

7. A patient's arterial blood gas results are pH 7.37; PCO_2 75 mm Hg; HCO_3^- 37 mmol/L. These values are consistent with:
   a. Partially compensated nonmetabolic acidosis
   b. Partially compensated metabolic acidosis
   c. Uncompensated nonmetabolic alkalosis
   d. Uncompensated metabolic alkalosis

8. A patient's arterial blood gas results are pH 7.48; PCO_2 54 mm Hg; HCO_3^- 38 mmol/L. These values are consistent with:
   a. Partially compensated nonmetabolic alkalosis
   b. Partially compensated metabolic alkalosis
   c. Uncompensated nonmetabolic alkalosis
   d. Uncompensated metabolic alkalosis

9. In the patient's circulatory system, bicarbonate leaves the red blood cell and enters the plasma through an exchange mechanism with:
   a. Carbonic acid
   b. Lactate
   c. Chloride
   d. Sodium

10. Hypoventilation serves to (decrease, increase) the levels of (H_2CO_3, HCO_3^-) in the blood:
    a. Decrease; H_2CO_3
    b. Decrease; HCO_3^-
    c. Increase; H_2CO_3
    d. Increase; HCO_3^-

REFERENCES

1. Weisberg HF. Water, electrolyte, and acid-base balance. 2nd ed. Baltimore: Williams & Wilkins, 1962.
2. Ehrmeyer S, et al. Fractional oxyhemoglobin, oxygen content and saturation, and related quantities in blood: terminology, measurement and reporting (C25A). Wayne, PA: National Committee for Clinical Laboratory Standards, 1997.
3. Ehrmeyer S, Ancy, J, Laessig, R. Oxygenation: measure the right thing. *Respiratory Therapy* 1998;11(3):25–28.

4. Mendelson Y. Pulse oximetry: theory and applications for noninvasive monitoring. Clin Chem 1992;38:1601–1607.

5. Moran RF, Fallon KD. Oxygen saturation, content, and the dyshemoglobins, part II. Clin Chem News 1990;16(2):8.

6. Ehrmeyer S, et al. pH and blood gas analysis (C46A). Wayne, PA: National Committee for Clinical Laboratory Standards, 1999.

7. Blonshine S, et al. Collection of arterial blood specimens (H11A3). Wayne, PA: National Committee for Clinical Laboratory Standards, 1998.

8. Mahoney JJ, et al. Changes in oxygen measurements when whole blood is stored in iced plastic or glass syringes. Clin Chem 1991;37(7):1244.

9. Müller-Plathe O, Heyduck S. Stability of blood gases, electrolytes and haemoglobin in heparinized whole blood samples influence of the type of syringe. Eur J Clin Chem Biochem 1992;30:349.

10. Ehrmeyer SS, Laessig RH. Stat labs need prospective quality control. Clin Chem News 1988;14(4):11.

11. Fallon KD. Demand for speed challenges quality control. Clin Chem News 1988;14(4):6.

12. Ehrmeyer SS, Burmeister BJ, Laessig RH. Laboratory performance in a state proficiency testing program: what can a laboratorian take home? J Clin Immunoassay 1994;17:223–230.

PART III

ASSESSMENT OF ORGAN SYSTEM FUNCTIONS

CHAPTER 17

Liver Function

Edward P. Fody

Objectives

Upon completion of this chapter, the clinical laboratorian should be able to:

- *Diagram the anatomy of the liver.*

- *Explain the physiologic functions of the liver to include bile secretion, synthetic activity, and detoxification.*

- *Discuss the basic disorders of the liver and what laboratory tests may be performed to diagnose them.*

- *Given a patient's clinical data, evaluate the information to determine any disorder.*

- *Classify the three types of jaundice and discuss their causes.*

- *Explain the principles of the tests for bilirubin.*

- *Identify the enzymes most commonly used in the assessment of hepatobiliary disease.*

- *Differentiate the various types of hepatitis to include cause (ie, bacteria or viruses), transmission, occurrence, alternate name, physiology, diagnosis, and treatment.*

KEY TERMS

Bile	Hepatoma	Prehepatic
Bilirubin	Icterus	Sinusoids
Cirrhosis	Kupffer's cells	Urobilinogen
Conjugated bilirubin	Lobule	
Hepatitis	Posthepatic	

In the past, the liver has been referred to as the center of courage, passion, temper, and love, and even as the center of the soul. It was once believed to produce "yellow bile" necessary for good health. Today, we recognize the liver to be a very complex organ responsible for many major metabolic functions in the body. More than 100 tests measuring these diverse functions have existed in the clinical laboratory at one time or another. However, many were abandoned early on in favor of those that have proven to be most clinically useful. This chapter discusses these more commonly used liver function tests with particular emphasis on current methodology.

ANATOMY

The liver is the largest and most versatile organ in the body (Fig. 17-1). It consists of two main lobes that, together, weigh from 1400 g to 1600 g in the normal adult. This organ is reddish brown and is located under the diaphragm in the right upper quadrant of the abdomen. It has an abundant blood supply, receiving approximately 15 mL per minute from two major vessels: the hepatic artery and the portal vein. The hepatic artery, a branch of the aorta, contributes 20% of the blood supply and provides most of the oxygen requirement. The portal vein, which drains the gastrointestinal tract, transports the most recently absorbed material from the intestines to the liver. Within the connective tissue of the liver, these vessels give off numerous small branches that form a vascular network around so-called lobules.[1]

Structural Unit

The *lobule,* which measures 1 mm to 2 mm in diameter, forms the structural unit of the liver (Fig. 17-2). It comprises cords of liver cells (hepatocytes) radiating from a central vein. The boundary of each lobule is formed by a portal tract made up of connective tissue that contains a branch of the hepatic artery, portal vein, and bile duct. Between the cords of liver cells are vascular spaces, called *sinusoids,* that are lined by endothelial cells and Kupffer's cells. These spaces receive blood from the small branches of the hepatic artery and portal vein, which are located in the portal tracts. The *Kupffer's cells* are phagocytic macrophages capable of ingesting bacteria or other foreign material from the blood that flows through the sinusoids. The blood from the sinusoids drains into the central veins (hepatic venule) and then to the hepatic veins and inferior vena cava. Primary bile canaliculi are conduits, 1 mm to 2 mm in diameter, located between the hepatocytes. The bile canaliculi interconnect extensively and increase in size until they connect with the larger bile ducts in the portal tracts.[2,3]

PHYSIOLOGY

The liver performs several hundred known functions each day, including numerous metabolic as well as secretory and excretory functions. The total loss of the liver usually results in death from hypoglycemia within 24 hours. Although there are many liver functions, this chapter discusses those that have prime significance in liver disease.

Excretory and Secretory Function

One of the more important liver functions, and one that is disturbed in a large number of hepatic disorders, is the excretion of bile. *Bile* comprises bile acids or salts,[4] bile pigments (primarily bilirubin esters), cholesterol, and other substances extracted from the blood. Total bile production averages about 3 L per day, although only 1 L is excreted. The primary bile acids cholic acid and chenodeoxycholic acid are formed in the liver from cholesterol. The bile acids are conjugated with the amino acids glycine or taurine, forming bile salts. Bile salts (conjugated bile acids) are excreted into the bile canaliculi by means of a carrier-mediated active transport system. During fasting and between meals, a major portion of the bile acid pool is concentrated up to ten-fold in the gallbladder. Bile acids reach the intestines when the gallbladder contracts after each meal. Approximately 500 mL to 600 mL of bile enter the duodenum each day. Here, bile is intimately involved with the digestion and absorption of lipids. When the conjugated bile acids (salts) come into contact with bacteria in the terminal ileum and colon, dehydration to secondary bile acids (deoxycholic and lithocolic) occurs, and these secondary bile acids are subsequently absorbed. The absorbed bile acids enter the portal circulation and return to the liver, where they are reconjugated and reexcreted. The enterohepatic circulation of bile occurs two to five times daily.[5–7]

Bilirubin, the principal pigment in bile, is derived from the breakdown of hemoglobin when aged red blood cells are phagocytized by the reticuloendothelial system, primarily in the spleen, liver, and bone marrow. About 80% of the bilirubin formed daily comes from the degradation of hemoglobin. The remainder comes from the destruction of heme-containing proteins (myoglobin, cytochromes, catalase) and from the catabolism of heme (Fig. 17-3).

When hemoglobin is destroyed, the protein portion—globin—is reused by the body. The iron enters the body's iron stores and is also reused. The porphyrin is broken down as a waste product and excreted. This action of splitting the porphyrin ring and releasing the iron and globin forms biliverdin, which is easily reduced to bilirubin.

Bilirubin is transported to the liver in the blood stream bound to proteins, chiefly albumin. It is then separated from

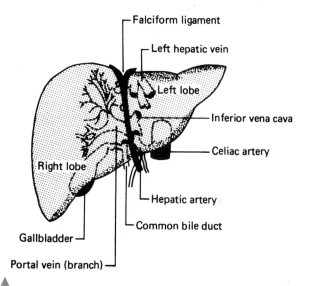

▲
Figure 17-1. Gross anatomy of the liver showing major blood vessels and bile channels. (Adapted with permission from Tietz NW. Fundamentals of clinical chemistry. Philadelphia: WB Saunders, 1976.)

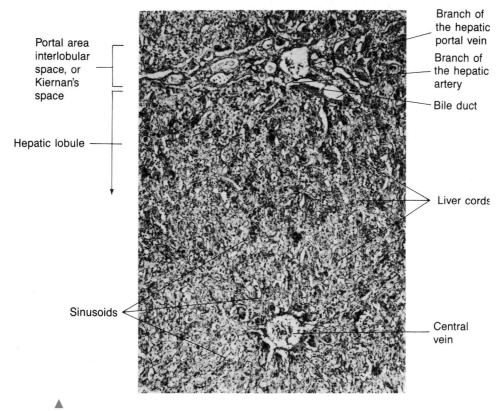

Portal area interlobular space, or Kiernan's space

Hepatic lobule

Sinusoids

Branch of the hepatic portal vein

Branch of the hepatic artery

Bile duct

Liver cords

Central vein

Figure 17-2. Liver lobule (panoramic view). (Photograph courtesy of James Furlong, MD.)

the albumin and taken up by the hepatic cells. Two non-albumin proteins, isolated from liver cell cytoplasm and designated Y and Z, account for the intracellular binding and transport of bilirubin. The conjugation (esterification) of bilirubin occurs in the endoplasmic reticulum of the hepatocyte. An enzyme, uridyldiphosphate glucuronyl transferase (UDPG-T), transfers a glucuronic acid molecule to each of the two proprionic acid side chains in bilirubin, converting bilirubin into a diglucuronide ester. This product, bilirubin diglucuronide, is referred to as conjugated bilirubin. *Conjugated bilirubin,* which is water soluble, is secreted from the hepatic cell into the bile canaliculi and then passes along with the rest of the bile into larger bile ducts and eventually into the intestines. In the lower portion of the intestinal tract, especially the colon, the bile pigments are acted on by enzymes present in the intestinal bacteria. The first product of this reaction is mesobilirubin, which is reduced to form mesobilirubinogen and then urobilinogen, which is a colorless product. The oxidation of urobilinogen produces the red-brown pigment urobilin, which is excreted in the stool. A small portion of the urobilinogen is reabsorbed into the portal circulation and returned to the liver, where it is again excreted into the bile. There is, however, a small quantity that remains in the blood. This urobilinogen is ultimately filtered by the kidney and excreted in the urine (Fig. 17-4).

A total of 200 mg to 300 mg of bilirubin is produced daily in the healthy adult. A normally functioning liver is required to eliminate this amount of bilirubin from the body. This excretory function requires that bilirubin be in the conjugated form, that is, the water-soluble diglucuronide. Almost all the bilirubin formed is eliminated in the feces, and a small amount of the colorless product urobilinogen is excreted in the urine. Under these normal circumstances, a low concentration of bilirubin (0.2–1.0 mg/dL) is found in the serum, the majority of which is in the unconjugated form. A small percentage (0.2 mg/dL) of this total bilirubin exists in normal serum as the conjugated form.[8]

When the bilirubin concentration in the blood rises, the pigment begins to be deposited in the sclera of the eyes and in the skin. This yellowish pigmentation in the skin or sclera is known as *jaundice* or *icterus*.[9,10]

Jaundice may be caused by a variety of pathophysiologic mechanisms. For example, there may be an increased bilirubin load on the liver cell or a disturbance in uptake and transport of bilirubin within the liver cell. In addition, there may be defects in conjugation or excretion of bilirubin into the bile. Further difficulties may be due to obstruction of the large bile ducts before the bilirubin reaches the intestines. Several classifications of jaundice are found in the literature. One of the more frequently used classifications is based on the presumed site of physiologic or anatomic abnormality.

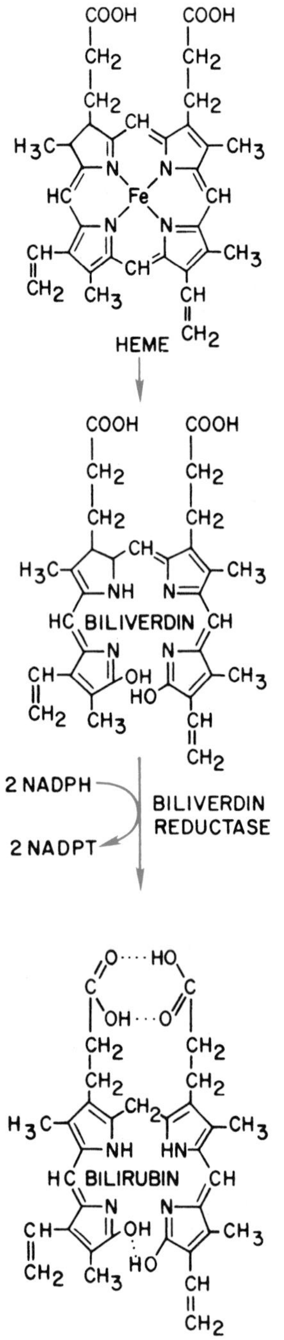

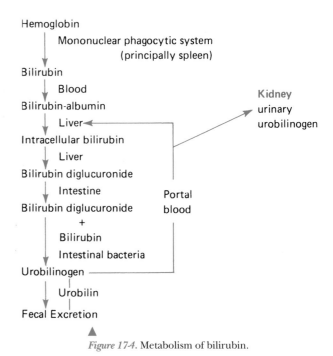

Figure 17-4. Metabolism of bilirubin.

Figure 17-3. The catabolism of heme, leading to the formation of bilirubin.

In this classification, there are predominately three types of jaundice: prehepatic, hepatic, and posthepatic.[9]

Prehepatic jaundice results when an excessive amount of bilirubin is presented to the liver for metabolism, such as in hemolytic anemia. This type of jaundice is characterized by unconjugated hyperbilirubinemia. However, the serum bilirubin levels rarely exceed 5 mg/dL because the normal liver is capable of handling most of the overload. Unconjugated bilirubin is not water soluble and is bound to albumin so that

it is not filtered out of the blood by the kidney. Therefore, bilirubin will not appear in the urine in this type of jaundice.

The largest percentage of patients with jaundice has the hepatic type. Hepatic jaundice may result from impaired cellular uptake, from defective conjugation, or from abnormal secretion of bilirubin by the liver cell.

Gilbert syndrome is a relatively common disorder characterized by impaired cellular uptake of bilirubin. Affected individuals have no symptoms but may have mild icterus. The elevated level of bilirubin is less than 3 mg/100 mL and is unconjugated. Crigler-Najjar syndrome is a more serious disorder caused by a deficiency of the enzyme UDPG-T. Two types have been described. Type I, in which there is a complete absence of the enzyme in the liver, is very rare. No conjugated bilirubin is formed, and the bile is colorless. This type is uniformly fatal. In type II Crigler-Najjar syndrome, there is a less severe deficiency of the enzyme, and some conjugated bilirubin is formed. Dubin-Johnson syndrome and Rotor's syndrome are two hereditary disorders characterized by conjugated hyperbilirubinemia from defective excretion by the liver cell. Any cause of severe hepatocellular damage also may interfere with uptake, conjugation, or secretion of bilirubin. This will lead to unconjugated as well as conjugated hyperbilirubinemia.[11–16]

Posthepatic jaundice results from the impaired excretion of bilirubin caused by mechanical obstruction of the flow of bile into the intestines. This may be due to gallstones or a tumor. When bile ceases to flow into the intestines, there is a rise in the serum level of conjugated bilirubin, and the stool loses its source of normal pigmentation and becomes clay-colored. Conjugated bilirubin appears in the urine, and urine urobilinogen levels decrease.[17] The various laboratory tests that assist in making the distinction between the different causes of jaundice are discussed later in this chapter.

Major Synthetic Activity

Among the many diverse metabolic functions carried out by the liver is the synthesis of many major biologic compounds to include proteins, carbohydrates, and lipids. The liver plays an important role in plasma protein production, synthesizing albumin and the majority of the α- and β-globulins. All the blood-clotting factors (except VIII) are also synthesized in the liver. In addition, the deamination of glutamate in the liver is the primary source of ammonia, which is then converted to urea.

The synthesis and metabolism of carbohydrates is also centered in the liver. Glucose is converted to glycogen, a portion of which is stored in the liver and later reconverted to glucose as necessary. An additional important liver function is gluconeogenesis from amino acids.[18,19]

Fat is formed from carbohydrates in the liver when nutrition is adequate and the demand for glucose is being met from dietary sources. The liver also plays a key role in the metabolism of fat. It is the major site for the removal of chylomicron remnants and for the conversion of acetyl CoA to fatty acids, triglycerides, and cholesterol. Further metabolism of cholesterol into bile acids also occurs in the liver. Very low-density lipoproteins, which are responsible for transporting triglycerides into the tissues, are synthesized primarily in the liver. High-density lipoproteins are also made in the liver, as are phospholipids.[20]

The formation of ketone bodies occurs almost exclusively in the liver. When the demand for gluconeogenesis depletes oxaloacetate and acetyl CoA cannot be converted rapidly enough to citrate, acetyl CoA accumulates, and a decyclase in the liver liberates ketone bodies into the blood.[21]

The liver is the storage site for all the fat-soluble vitamins (A, D, E, and K) and several water-soluble vitamins, such as B$_{12}$. Another vitamin-related function is the conversion of carotene into vitamin A.

The liver is the source of somatomedin (an insulin-like factor that mediates the activity of growth hormone) and angiotensinogen, and is a major site of metabolic clearance of many other hormones. As the source of transferrin, ceruloplasmin, and metallothionein, the liver plays a key role in the transport, storage, and metabolism of iron, copper, and other metals.[22]

Many enzymes are synthesized by liver cells, but not all of them have been found useful in the diagnosis of hepatobiliary disorders. Those enzymes that have been used often include aspartate aminotransferase (AST, or glutamate oxaloacetic transaminase [SGOT]) and alanine aminotransferase (ALT, or glutamate pyruvate transaminase [SGPT]), which escape into the plasma from damaged liver cells; alkaline phosphatase (ALP) and 5′-nucleotidase (5NT), which are induced or released when the canalicular membrane is damaged and biliary obstruction occurs; and γ-glutamyltransferase (GGT), which is increased in both hepatocellular and obstructive disorders.[23]

Detoxification and Drug Metabolism

Because the liver is interposed between the splanchnic circulation and the systemic blood, it serves to protect the body from potentially injurious substances absorbed from the intestinal tract and toxic by-products of metabolism. The most important mechanism in this detoxification activity is the microsomal drug-metabolizing system of the liver. This system is induced by many drugs (eg, phenobarbital) and other foreign compounds and is responsible for many of the detoxification mechanisms, including oxidation, reduction, hydrolysis, hydroxylation, carboxylation, and demethylation. These various mechanisms serve to convert many noxious or comparatively insoluble compounds into other forms that are either less toxic or more water soluble and therefore excretable by the kidney. For example, ammonia, a very toxic substance

---

CASE STUDY 17-1

The following laboratory test results were obtained in a patient with severe jaundice, right upper quadrant abdominal pain, fever, and chills (Case Study Table 17-1.1).

Question

1. What is the most likely cause of jaundice in this patient?

CASE STUDY TABLE 17-1.1. Laboratory Results

Serum alkaline phosphatase	4 times normal
Serum cholesterol	Increased
AST (SGOT)	Normal or slightly increased
5′-Nucleotidase	Increased
Total serum bilirubin	25 mg/dL
Conjugated bilirubin	19 mg/dL
Prothrombin time	Prolonged but improves with a vitamin K injection

arising in the large intestines through bacterial action on amino acids, is carried to the liver by the portal vein and converted into the innocuous compound urea by the hepatocytes.

Conjugation with moieties such as glycine, glucuronic acid, sulfuric acid, glutamine, acetate, cysteine, and glutathione occurs mainly in the cytosol or smooth endoplasmic reticulum. This mechanism is the mode of bilirubin and bile acid excretion.

DISORDERS OF THE LIVER

Jaundice

Jaundice, or *icterus,* refers to the yellowish discoloration of the skin and sclerae resulting from hyperbilirubinemia. Although the upper limit of normal for total serum bilirubin is 1 mg/dL, jaundice is not clinically apparent until the bilirubin level exceeds 2 mg/dL to 3 mg/dL. In African-American or Asian patients, yellowing of the sclerae may be the only clinical evidence of jaundice.[9] Jaundice is one of the oldest medical disorders known, having been described in ancient Greek, Roman, Chinese, and Hebrew texts. Hippocrates related jaundice to dysfunction of the liver.

Except in infants, hyperbilirubinemia is generally well tolerated and does not produce serious clinical side effects. However, in infants, hyperbilirubinemia (levels exceeding 15–20 mg/dL) may be associated with *kernicterus,* a very serious disorder of the central nervous system resulting from the increased bilirubin levels. This occurs only in infants because the immature central nervous system does not have a well-developed blood-brain barrier.[24–27]

Although all cases of jaundice result from hyperbilirubinemia, not all are caused by hepatic dysfunction. Although the majority of cases of jaundice are associated with liver disorders, hyperbilirubinemia may also result from erythrocyte destruction, or hemolysis, in patients with normal liver function. Thus, distinction between hepatic and hemolytic disease in the patient presenting with jaundice is an important task for which the attending physician will rely heavily on laboratory results. This distinction is discussed in detail later in this chapter.[28,29]

Hypercarotenemia, a disorder caused by the excessive ingestion of vitamin A, may produce skin discoloration indistinguishable from that of hyperbilirubinemia. In hypercarotenemia, however, the sclerae are usually not discolored.

Cirrhosis

Cirrhosis is derived from the Greek word that means "yellow." However, in current usage, *cirrhosis* refers to the irreversible scarring process by which normal liver architecture is transformed into abnormal nodular architecture. One way to classify cirrhosis is by the appearance of the liver, that is, by the size of the nodules. These conditions are referred to as *macronodular* and *micronodular cirrhosis,* although mixed forms occur.[30]

Another way to classify cirrhosis is by etiology. In the United States, Canada, and Western Europe, the leading cause of cirrhosis is alcohol abuse, which leads to a micronodular type of cirrhosis.[31,32] Other causes of cirrhosis include hemochromatosis, postnecrotic cirrhosis (which occurs as a late consequence of hepatitis), and primary biliary cirrhosis (which is an autoimmune disorder). Other uncommon etiologies of cirrhosis exist. About 10% to 20% of cases cannot be classified as to etiology. Cirrhosis is a serious disorder and is one of the ten leading causes of death in the United States. It causes many complications.

Portal hypertension results when the blood flow through the portal vein is obstructed by the cirrhotic liver. This may result in splenomegaly, which may not be clinically significant, and esophageal varices, which may rupture and lead to fatal hemorrhage. The synthetic ability of the liver is reduced, causing hypoalbuminemia and deficiency of the clotting factors, which may lead to hemorrhage. Ascitic fluid may accumulate in the abdomen. Although some patients with cirrhosis are capable of prolonged survival, generally this diagnosis is an ominous one.[33,34]

Tumors

On a worldwide basis, primary malignant tumors of the liver, known as *hepatocellular carcinoma, hepatocarcinoma,* or *hepatoma,* are an important cause of cancer mortality. In the United States, these tumors are relatively uncommon. Most cases of hepatocellular carcinoma can be related to previous infection with a hepatitis virus. These tumors are especially common in parts of Africa and Asia and are infrequent in North America and Western Europe. However, the liver is frequently involved secondarily by tumors arising in other organs. Metastatic tumors to the liver from primary sites such as the lung, pancreas, gastrointestinal tract, or ovary are common. Benign tumors of the liver are relatively uncommon.[35,36]

Whether primary or secondary, any malignant tumor in the liver is a serious finding with a poor prognosis. Generally, the only hope for cure relies on surgical resection, which is usually impossible. Patients with malignancies of the liver usually have a survival measured in months.[37,38]

Reye's Syndrome

Reye's syndrome is a disorder of unknown cause involving the liver and arising primarily in children, although cases in adults have been reported. It is a form of hepatic destruction that usually occurs following recovery from a viral infection, such as varicella (chickenpox) or influenza. It has been related to aspirin therapy. Shortly after the infection, the patient develops neurologic abnormalities, which may include seizures or coma. Liver functions are always abnormal, but the bilirubin level is not usually elevated. Without treatment, rapid clinical deterioration leading to death may occur.[32,39–42]

Drug- and Alcohol-Related Disorders

Many drugs and chemicals are toxic to the liver. This toxicity may take the form of overwhelming hepatic necrosis, leading to coma and death, or it may be subclinical and pass entirely unnoticed. Of all hepatic toxins, probably the most important is ethanol. In small amounts, alcohol may cause mild, inapparent injury. Heavier consumption leads to more serious damage, and prolonged heavy use may lead to cirrhosis. The exact amount of alcohol needed to cause cirrhosis is unknown, and only a minority of alcoholics develop this condition. It is, however, a leading cause of morbidity and mortality in the United States.[32]

Certain drugs, including tranquilizers such as phenothiazines, certain antibiotics, antineoplastic agents, and anti-inflammatory drugs, may cause liver injury. Usually this is mild and manifested only by elevation of liver function tests, which return to normal when the offending agent is discontinued. However, occasionally, this may lead to massive hepatic failure or cirrhosis.[43–47]

One of the most common drugs associated with serious hepatic injury is acetaminophen. This drug, when taken in massive overdose, is virtually certain to produce fatal hepatic necrosis unless rapid treatment is initiated.[48,49]

ASSESSMENT OF LIVER FUNCTION

Analysis of Bilirubin

Brief Review of Classical Methodology

Because its yellow color is detectable by the human eye, concentrations of serum bilirubin have been estimated for centuries. In 1883, Ehrlich first described a reaction in urine samples of the formation of a red or blue pigment when bilirubin was coupled with a diazotized sulfanilic acid solution. In 1913, Van den Bergh applied the Ehrlich reaction to serum bilirubins. He also used alcohol as an accelerator for the coupling of bilirubin to diazotized sulfanilic acid. Malloy and Evelyn developed the first useful quantitative technique for bilirubin in 1937 by accelerating the reaction with a 50% methanol solution, a technique that avoided the precipitation of proteins that was a source of error in the Van den Bergh method. In 1938, Jendrassik and Grof used a procedure containing caffeine-benzoate-acetate as an accelerator for the azo-coupling reaction.

Bilirubin also has been quantified by methods other than coupling to sulfanilic acid. These methods include a direct measurement of its natural color. This principle was successfully used in the development of the icterus index, which was introduced in 1919. The test involves diluting serum with saline until it visually matches the color of a .01% potassium dichromate solution. The number of times the serum must be diluted is called the *icterus index*. However, substances in the serum other than bilirubin, such as carotene, xanthophyll, and hemoglobin, also may contribute to the icterus index, limiting its clinical usefulness. This test is now obsolete. In recent years, bilirubin has been quantitated by dilution in a buffer, followed by direct measurement of the absorption, using a well-calibrated spectrophotometer. This method is used in the pediatric laboratory on newborns whose serum does not yet contain the interfering yellow lipochromes. The hemolysis that is so often a problem with pediatric specimens is "blanked out" by measuring a second wavelength.[50,51]

With several of these methods, it was determined that two types of bilirubin existed.[4] The fraction that produced a color in the Van den Bergh method in aqueous solution was described as *direct bilirubin*, whereas the fraction that produced a color only after alcohol was added was called *indirect bilirubin*. For many years, results of bilirubin determinations were reported as direct or indirect. This terminology is now outdated. Since 1956, it has been known that the direct reaction is given by the diglucuronide of bilirubin or conjugated bilirubin, which is water soluble. The indirect reaction, on the other hand, is given by unconjugated bilirubin, which is water insoluble but dissolves in alcohol to couple with the diazo reagent. Direct and indirect bilirubin should be reported as conjugated and unconjugated, respectively. Most commonly, conjugated and total bilirubin are reported.[52] Unconjugated bilirubin may be determined by subtracting conjugated bilirubin from total bilirubin.

Method Selection

Unfortunately, no single method for the determination of bilirubin will meet all the requirements of the clinical laboratory. For the evaluation of jaundice in newborns, the direct spectrophotometric method is satisfactory. The sources

---

CASE STUDY 17-2

The following laboratory test results were found in a patient with a mild weight loss and nausea and vomiting, who later developed jaundice and an enlarged liver (Case Study Table 17-2.1).

Question

1. What disease process is most likely in this patient?

CASE STUDY TABLE 17-2.1. Laboratory Results

Total serum bilirubin	20 mg/dL
Conjugated bilirubin	10 mg/dL
Alkaline phosphatase	Mildly elevated
AST (SGOT)	Markedly elevated
ALT (SGPT)	Moderately elevated
Albumin	Decreased
γ-Globulin	Increased

of error in this technique are turbidity, hemolysis, and yellow lipochrome pigments. Hemolysis and turbidity can be "blanked out" by measuring a second wavelength, but the yellow lipochromes cannot be "blanked out." This method is, therefore, only valid for newborns whose serum does not contain lipochromes. In patients older than age 1 month, a diazo-colorimetric procedure is necessary. Most investigators who have compared the different methods for the measurement of bilirubin agree that the majority of techniques available today will give accurate results for total bilirubin, provided that good standards are available. The choice of method, then, should be based on whether one prefers a manual versus automated technique and whether direct bilirubin is to be determined. The choice of a direct bilirubin method poses the greatest problem because there is no reference method or adequate standardization available.

If a manual procedure is desired, then either the Evelyn-Malloy or Jendrassik-Grof method is suitable. The Jendrassik-Grof method is slightly more complex but has the following advantages over the Evelyn-Malloy:

- It is insensitive to sample pH changes.
- It is insensitive to a 50-fold variation in protein concentration of the sample.
- It has adequate optical sensitivity even for low bilirubin concentrations.
- It has minimal turbidity and a relatively constant serum blank.
- It is not affected by hemoglobin up to 750 mg/dL.

Since the development of the original procedure by Jendrassik-Grof, a number of modifications have been made to speed up the reaction, reduce interference, and so on. There are now several commercial bilirubin procedures that use a modified Jendrassik-Grof method. It seems to be a particularly popular technique for the discrete sampler analyzers currently on the market.

The recommended methods for total bilirubin determination using all automated machines will generally give equivalent results. The techniques for direct bilirubin are unfortunately not as reliable. The best method for measuring small amounts of conjugated serum bilirubin is a research method that uses high-performance liquid chromatography, but this method is too difficult for routine laboratory use. Most clinical laboratories use either the Evelyn-Malloy or the Jendrassik-Grof method. Because this chapter does not allow for a detailed description of all the previously mentioned bilirubin test methodologies, it emphasizes only the most widely used principles for measuring adult and pediatric bilirubin.[53–56]

Jendrassik-Grof Method for Total and Conjugated Bilirubin Determination[57,58]

Principle. Serum or plasma is added to a solution of sodium acetate and caffeine–sodium benzoate, which is then added to diazotized sulfanilic acid to form a purple azobilirubin. The sodium acetate buffers the pH of the diazotization reaction,

whereas the caffeine–sodium benzoate accelerates the coupling of bilirubin with diazotized sulfanilic acid. This reaction is terminated by the addition of ascorbic acid, which destroys the excess diazo reagent. A strongly alkaline tartrate solution is then added to convert the purple azobilirubin to blue azobilirubin, and the intensity of the color is read at 600 nm.

Specimen Collection and Storage. A fasting serum specimen that is neither hemolyzed nor lipemic in nature is preferred. Before testing, serum should be stored in the dark and measured as soon as possible (within 2–3 h) after collection. Serum may be stored in the dark in a refrigerator for up to 1 week and in the freezer for 3 months without appreciable change in the bilirubin concentration.

Comments and Sources of Error. Normal blood contains no conjugated bilirubin. The reason some conjugated bilirubin is reported as normal is because current available methodology picks up some of the total bilirubin as a false-positive. The best routine methods, however, keep this technical error to a minimum and report upper limits or normal of less than 0.2 mg/dL for conjugated serum bilirubin. This method compensates for lipochrome pigments, which are present in the serum of adults and the serum of children who are older than a few months of age. However, a hemolyzed specimen will cause a decrease in serum bilirubin by this method. In addition, because lipemia causes interference, fasting blood specimens are preferable. Serious loss of bilirubin occurs after exposure to fluorescent and indirect and direct sunlight. Therefore, it is imperative that exposure of samples and standards to light be kept to a minimum and that specimens and standards be refrigerated in the dark until the tests are performed.

Reference Range. Reference ranges for infants after 1 month and adults are shown in Table 17-1.

Direct Spectrophotometric Method for Determination of Total Bilirubin in Serum[53,58]

Principle. The absorbance of bilirubin in the serum at 455 nm is proportional to its concentration. The serum of newborns does not contain lipochromes, such as carotene, that would increase the absorbance at 455 nm. The absorbance of hemoglobin at 455 nm is corrected by subtracting the absorbance at 575 nm.

Specimen. Serum is collected and stored with the same precautions as mentioned previously for adult specimens.

TABLE 17-1. Reference Ranges for Bilirubin	
Conjugated	0–0.2 mg/dL (0–3 μmol/L)
Unconjugated	0.2–0.8 mg/dL (3–14 μmol/L)
Total	0.2–1.0 mg/dL (3–17 μmol/L)

Comments and Sources of Error. Error will be introduced if the buffer is turbid. Because the method depends on the extinction coefficient of bilirubin, all volumes must be accurate, and cuvets must be flat-surfaced with a path length of exactly 1 cm. A control should be used with a level near 20 mg/dL, which is a critical decision point for the clinician, because exchange transfusion is necessary if this level is exceeded.

Precautions such as those mentioned in the previous method should also be followed for collection and storage of specimens. This method is relatively insensitive to hemolysis, which is often present in specimens obtained from infants, due to difficulty in skin puncture technique. However, it is markedly affected by the presence of lipochromes, and therefore cannot be used in infants who are older than a few months of age.[52]

Reference Range. See Table 17-2.

Urobilinogen in Urine and Feces

Urobilinogen is a colorless end product of bilirubin metabolism that is oxidized by intestinal bacteria to the brown pigment urobilin. In the normal individual, part of the urobilinogen is excreted in the feces, and the remainder is reabsorbed into the portal blood and returned to the liver. A small portion is not taken up by the hepatocytes and is excreted by the kidney as urobilinogen. Increased levels of urinary urobilinogen are found in hemolytic disease and in defective liver-cell function such as that seen in hepatitis. Absence of urobilinogen from the urine and stool is most often seen with complete biliary obstruction. Fecal urobilinogen is also decreased in biliary obstruction, as well as in hepatocellular disease.[6]

The majority of the quantitative methods for urobilinogen are based on the reaction of this substance with *p*-dimethyl aminobenzaldehyde to form a red color. This reaction was first described by Ehrlich in 1901. Many modifications of this procedure have been made over the years to improve specificity. Major improvements were made in 1925 by Terwen, who used alkaline ferrous hydroxide to reduce urobilin to urobilinogen and added sodium acetate to eliminate interference from such compounds as indole. The use of petroleum ether rather than diethyl ether for the extraction of urobilinogen was introduced in 1936 by Watson to help in the removal of other interfering substances.[59] However, because studies indicate that the quantitative methods described do not completely recover urobilinogen from the urine, most laboratories

use the less laborious, more rapid, semiquantitative method described next.[60-63]

Determination of Urine Urobilinogen (Semiquantitative)

Principle. Urobilinogen reacts with *p*-dimethyl aminobenzaldehyde (Ehrlich's reagent) to form a red color, which is then measured spectrophotometrically. Ascorbic acid is added as a reducing agent to maintain urobilinogen in the reduced state. The use of saturated sodium acetate stops the reaction and minimizes the combination of other chromogens with the Ehrlich's reagent.

Specimen. A *fresh* 2-h urine is collected. This specimen should be kept cool and protected from light.

Comments and Sources of Error.
1. The results of this test are reported in Ehrlich units rather than in milligrams of urobilinogen because substances other than urobilinogen account for some of the final color development.
2. Compounds other than urobilinogen that may be present in the urine and react with Ehrlich's reagent include porphobilinogen, sulfonamides, procaine, and 5-hydroxyindoleacetic acid. Bilirubin will form a green color and therefore must be removed, as previously described.
3. Fresh urine is necessary, and the test must be performed without delay to prevent oxidation of urobilinogen to urobilin. Similarly, the spectrophotometric readings should be made within 5 minutes after color production because the urobilinogen-aldehyde color slowly decreases in intensity.

Reference Range. Urine urobilinogen, 0.1-1.0 Ehrlich units/2 h or 0.54.0 Ehrlich units/day (0.86.8 mmol/day); 1 Ehrlich unit is equivalent to approximately 1 mg of urobilinogen.

Fecal Urobilinogen

Visual inspection of the feces usually suffices to detect decreased urobilinogen. However, the semiquantitative determination of fecal urobilinogen is available and involves the same principle described earlier for the urine.[11] It is carried out in an aqueous extract of fresh feces, and any urobilin present is reduced to urobilinogen by treatment with alkaline ferrous hydroxide before Ehrlich's reagent is added. A range of 75 Ehrlich units/100 g to 275 Ehrlich units/100 g of fresh feces or 75 to 400 Ehrlich units per 24-hour specimen is considered a normal reference range.[63]

Measurement of Serum Bile Acids

Unfortunately, complex methods are required for the analysis of bile acids in serum. These involve extraction with organic solvents, partition chromatography, gas chromatography—

Infants	Premature, Total	Full-Term, Total
24 h	1–6 mg/dL	2–6 mg/dL
48 h	6–8 mg/dL	6–7 mg/dL
3–5 days	10–12 mg/dL	4–6 mg/dL

TABLE 17-2. Reference Ranges for Infant Total Bilirubin

mass spectroscopy, spectrophotometry, ultraviolet light absorption, fluorescence, radioimmunoassay, and enzyme immunoassay methods.[64–66] Despite that serum bile acid levels are elevated in liver disease, the total concentration is extremely variable and adds no diagnostic value to other tests of liver function. The variability of the type of bile acids present in serum, together with their existence in different conjugated forms, suggests that more relevant information of liver dysfunction may be gained by examining patterns of individual bile acids and their state of conjugation. For example, it has been suggested that the ratio of the trihydroxy to dihydroxy bile acids in serum will differentiate patients with obstructive jaundice from those with hepatocellular injury and that the diagnosis of primary biliary cirrhosis and extrahepatic cholestasis can be made on the basis of the ratio of the cholic to chenodeoxycholic acids. However, the high cost of these tests, the time required to do them, and the current controversy concerning their clinical usefulness render this approach unsatisfactory for routine use.

Enzyme Tests in Liver Disease

Any injury to the liver that results in cytolysis and necrosis causes the liberation of various enzymes. The measurement of these hepatic enzymes in the serum is used to assess the extent of liver damage and to differentiate hepatocellular (functional) from obstructive (mechanical) disease. The methods used to measure these enzymes, their normal reference ranges, and other general aspects of enzymology are considered in Chapter 9 *Enzymes*. Discussion in the present chapter focuses on the characteristic changes in serum enzyme levels seen in various hepatic disorders.

The most common enzymes assayed in hepatobiliary disease include ALP and the aminotransferases. Used less often are γ-glutamyltransferase, lactate dehydrogenase (LD) and its isoenzymes, 5′-nucleotidase, ornithine carbamyltransferase, and leucine aminopeptidase.[59,67–71]

Alkaline Phosphatase

ALP is found in a number of tissues but is used most often in the clinical diagnosis of bone and liver disease. Slight to moderate increases in ALP activity occur in many patients with hepatocellular disorders, such as hepatitis and cirrhosis, and transient increases may occur in all types of liver disease. The most striking elevations occur in extrahepatic biliary obstruction, such as a stone in the common bile duct, or in intrahepatic cholestasis, such as drug cholestasis or primary biliary cirrhosis. This enzyme is almost always increased in metastatic liver disease and may be the only abnormality on routine liver function tests. Because bone is a source of the enzyme, Paget disease, bony metastases, and other diseases associated with increased osteoblastic activity may produce high levels of ALP in the absence of liver disease. The enzyme is found in placenta, and pregnant women also have elevated levels.[72–75]

Aminotransferases (Transaminases)

AST (SGOT) and ALT (SGPT) are two enzymes widely used to assess hepatocellular damage. AST (SGOT) is found in all tissues, especially the heart, liver, and skeletal muscle. ALT (SGPT) is present primarily in the liver and, to a lesser extent, in kidney and skeletal muscle, making it more "liver specific."

In the absence of acute necrosis or ischemia of other organs, elevated aminotransferase levels suggest hepatocellular damage. In severe viral hepatitis that causes extensive acute necrosis, markedly elevated serum aminotransferase levels may be found, whereas only moderate increases are found in less severe cases. Mild chronic or focal liver diseases, such as subclinical or anicteric viral hepatitis, alcoholic cirrhosis, granulomatous infiltration, and tumor invasion, may be associated with only mild abnormalities. Minimal elevations occur in biliary obstruction.

It is often helpful to conduct serial determinations of aminotransferases when following the course of a patient with acute or chronic hepatitis. However, one should exercise caution in interpreting these abnormal levels, because serum transaminases may actually decrease in some patients with severe acute hepatitis, owing to the exhaustive release of hepatocellular enzymes.[76–79]

5′-nucleotidase. 5′-nucleotidase is another phosphatase; it originates largely in the liver and is used clinically to determine whether an elevation of the ALP is caused by liver or bone disease. Levels of both 5′-nucleotidase and ALP are elevated in liver disease, whereas in primary bone disease, ALP level is elevated, but the 5′-nucleotidase level is usually normal or only slightly elevated. This enzyme is much more sensitive to metastatic liver disease than is ALP because, unlike ALP, its level is not markedly elevated in other conditions such as pregnancy or childhood. In addition, some increase in enzyme activity may be noted after abdominal surgery.[80–82]

γ-glutamyltransferase. γ-glutamyltransferase (GGT) is found in high concentrations in the kidney and the liver and is elevated in the serum of almost all patients with hepatobiliary disorders. It is not specific for any type of liver disease but is frequently the first abnormal liver function test demonstrated in the serum of heavy drinkers. The highest levels are seen in biliary obstruction. It is, therefore, a sensitive test for alcoholic liver disease. Measurement of this enzyme is also useful if jaundice is absent for the confirmation of hepatic neoplasms and is a useful test to confirm hepatic disease in patients with elevated alkaline phosphates.[83–87]

Ornithine Carbamyl Transferase. Ornithine carbamyl transferase is a urea cycle enzyme present only in liver and intestinal tissue. Elevated levels of this enzyme occur primarily in liver disease but are not specific for any particular types of liver disease. Its diagnostic usefulness is therefore limited.[88–90]

Leucyl Aminopeptidase. Leucine aminopeptidase is widely distributed in human tissues and is found in the pancreas, gastric mucosa, liver, spleen, large intestine, brain, small intestine, and kidney. Most investigators now believe that the serum activity of leucine aminopeptidase cannot be used to differentiate hepatocellular from obstructive jaundice. Furthermore, the measurement of this enzyme does not provide any useful information that cannot be obtained by other tests, such as the determination of 5'-nucleotidase or γ-glutamyltransferase.[91–93]

Lactate Dehydrogenase. Measurement of total serum LD is usually not helpful diagnostically because it is present in all organs and released into the serum from a variety of tissue injuries. However, fractionation of LD into its five tissue-specific isoenzymes may give useful information about the site of origin of the LD elevation. LD-5 is present, for the most part, in liver and skeletal muscle. An interpretation of isoenzyme patterns may be simple with an elevated LD-5 in a patient with jaundice. However, the similarity of isoenzyme patterns between different tissue-damaging states may necessitate the use of additional laboratory tests for interpretation.

Moderate elevations of total serum LD levels are common in acute viral hepatitis and in cirrhosis, whereas biliary tract disease may produce only slight elevations. High serum levels may be found in metastatic carcinoma of the liver.[94,95]

Tests Measuring Hepatic Synthetic Ability

The measurement of the end products of hepatic synthetic activity can be used to assess liver disease. Although these tests are not very sensitive to minimal liver damage, they are useful in quantitating the severity of hepatic dysfunction.

Almost all serum proteins are produced by the liver. A decreased serum albumin may be due to decreased liver protein synthesis. The albumin level correlates well with the severity of functional impairment and is found more often in chronic rather than acute liver disease. The serum α-globulins also tend to decrease with chronic liver disease. However, a very low or absent α-globulin suggests α-antitrypsin deficiency as the cause of the chronic liver disease. Serum γ-globulin levels are transiently increased in acute liver disease and remain elevated in chronic liver disease. The highest elevations are found in chronic active hepatitis and postnecrotic cirrhosis. In particular, IgG and IgM levels are more consistently elevated in chronic active hepatitis, IgM in primary biliary cirrhosis, and IgA in alcoholic cirrhosis.[96]

The prothrombin time is commonly increased in liver disease because the liver is unable to manufacture adequate amounts of clotting factors or because the disruption of bile flow results in inadequate absorption of vitamin K from the intestine. Response of the prothrombin time to the administration of vitamin K is therefore of some value in differentiating intrahepatic disease with decreased synthesizing capacity from extrahepatic obstruction with decreased absorption of fat-soluble vitamins. A marked prolongation of the prothrombin time indicates severe diffuse liver disease and a poor prognosis.[96,97]

Tests Measuring Nitrogen Metabolism

The liver plays a major role in removing ammonia from the blood stream and converting it to urea so that it can be removed by the kidneys. In liver failure, ammonia and other toxins increase in the blood stream and may ultimately cause hepatic coma.[44] In this condition, the patient becomes increasingly disoriented and gradually lapses into unconsciousness. The cause of hepatic coma is not fully defined, although ammonia is presumed to play a major role. However, the correlation between blood ammonia levels and the severity of the hepatic coma is very poor. Therefore, the ammonia level is most useful when the patient serves as his own control and multiple measurements are made over time.[98,99]

The production of glutamine by an enzymatic intracellular reaction between ammonia and glutamic acid provides a mechanism for removal of ammonia from the central nervous system. Elevations of CSF glutamine have been described in hepatic encephalopathy and in some cases of Reye's syndrome. Glutamine levels can be measured by various methods, including acid hydrolysis and laser-induced fluorescence.[100]

Hepatitis

Hepatitis means "inflammation of the liver," which may be caused by viruses, bacteria, parasites, radiation, drugs, chemicals, or toxins. Among the viruses causing hepatitis are hepatitis types A, B, C, D (or delta), and E, cytomegalovirus, Epstein-Barr virus, and probably several others (Table 17-3).

Hepatitis A

Hepatitis A, also known as *infectious hepatitis* and *short-incubation hepatitis,* is usually transmitted by contaminated food or water. Epidemiologic data have suggested that a carrier state in hepatitis A is unlikely or is of a short duration. The hepatitis A virus—officially designated as HAV—has been identified by electron microscopy as a spherical particle, 27 nm in diameter containing DNA. HAV has recently been cultured in vitro. However, this tissue-culture system is more a research tool than a diagnostic aid. The fecal shedding of the hepatitis A antigen is transient, and the antigen disappears from the stool after peak elevations of liver enzymes. The diagnosis of hepatitis A is made by the serologic detection of hepatitis A antibody. The production of hepatitis A antibody (anti-HAV) represents a specific host response to HAV and is believed to confer immunity from reinfection. Use of immune electron microscopy of the stool and radioimmunoassays for the detection of this antibody suggests that different types of antibody (both IgM and IgG) may appear at different times during the course of the illness. The IgM- specific antibody peaks in 1 week and disappears by 8 weeks, whereas the IgG antibody peaks within 1 month to 2 months and persists for years.

The documentation of anti-HAV positivity indicates exposure to HAV, absence of infectivity, and presence of immunity to recurrent infection. Anti-HAV positivity neither

TABLE 17-3. The Hepatitis Viruses

	Nucleotide	Incubation Period	Primary Mode of Transmission	Vaccine	Chronic Infection	Serologic Diagnosis Available
Hepatitis A	RNA	2–6 weeks	Fecal–oral	Yes	No	Yes
Hepatitis B	DNA	8–26 weeks	Parenteral, sexual	Yes	Yes	Yes
Hepatitis C	RNA	2–15 weeks	Parenteral, ? sexual	No	Yes	Yes
Hepatitis D	RNA	—	Parenteral, sexual	Yes	Yes	Yes
Hepatitis E	RNA	3–6 weeks	Fecal–oral	No	?	Yes*

* At the time of publication, hepatitis E tests were available on a research basis only.

implies previous clinically apparent hepatitis nor establishes an etiologic relationship between HAV and acute or chronic liver disease. The demonstration of seroconversion or fecal shedding of antigen during the acute phase of illness is the only reliable method of establishing an HAV etiology.[101–108] Many people have anti-HAV antibodies without having had a clinically apparent infection.[108–114]

Hepatitis B

Hepatitis B is also known as *serum hepatitis* or *long-incubation hepatitis.* There are three major routes of transmission: parenteral, perinatal, and sexual. An additional fecal–oral route of transmission also exists, although this is not important from an epidemiologic basis. Infected patients manifest hepatitis B in virtually all body fluids, including blood, feces, urine, saliva, semen, tears, and milk.[10]

The hepatitis B virus consists of a 42-nm double-shelved spherical particle with a central core of deoxyribonucleic acid (DNA) surrounded by a protein coat. This particle, which is present in low concentrations in the serum of patients with active viral hepatitis, was originally called the *Dane particle,* but it is now named *HBV.* Following an infection with HBV, the core of the antigen is synthesized in the nuclei of hepatocytes and then passes into the cytoplasm of the liver cell, where it is surrounded by the protein coat. An antigen present in the core of the virus (designated HBcAg) and a surface antigen present on the surface protein (designated HBsAg) have been identified by serologic studies. Another antigen, called the e antigen or HBeAG, also has been identified.[29]

The clinical course of hepatitis B is extremely variable. Approximately two thirds of the cases may be asymptomatic or may produce only a mild flulike illness. In the remaining one third of the cases, a patient may develop a hepatitis-like syndrome of malaise, irregular fevers, right upper quadrant tenderness, jaundice, and dark urine.[115–117]

In approximately 1% of infected patients, the syndrome of fulminant hepatitis may develop. This is an overwhelming clinical illness with a high mortality.

Approximately 90% of patients infected with HBV recover within 6 months. This recovery is manifested by the development of antibody to hepatitis B surface antigen. Approximately 10% of the infected patients will go on to develop chronic hepatitis, which is discussed later in the section *Chronic Hepatitis.*[115–117]

CASE STUDY 17-3

The following laboratory results are obtained from a 19-year-old college student who consulted the Student Health Service because of fatigue and lack of appetite. She adds that recently she has noted that her sclera appears somewhat yellowish and that her urine has become dark (Case Study Table 17-3.1).

Questions

1. What is the most likely diagnosis?
2. What additional factors in the patient's history should be sought?
3. What is the prognosis?

CASE STUDY TABLE 17-3.1. Laboratory Results

ALT (SGPT)	Elevated
AST (SGOT)	Elevated
Alkaline phosphatase	Minimally elevated
LD	Elevated
Serum bilirubin	5 mg/dL
Urine bilirubin	Increased
Hepatitis A antibody (IgG)	Negative
Hepatitis A antibody (IgM)	Positive
Hepatitis B surface antigen	Negative
Hepatitis B surface antibody	Negative
Hepatitis C antibody	Negative

CASE STUDY 17-4

The following results were obtained in the patient from Case Study 17-2 (Case Study Table 17-4.1).

Questions

1. What is the most likely diagnosis?
2. What is the prognosis?
3. What complications may develop?

CASE STUDY TABLE 17-4.1. Laboratory Results

Hepatitis A antibody (IgG)	Positive
Hepatitis A antibody (IgM)	Negative
Hepatitis B surface antigen	Positive
Hepatitis B surface antibody	Negative
Hepatitis Core antibody (IgM)	Positive
Hepatitis C antibody	Negative

On a worldwide basis, infection with HBV is fairly common and usually occurs at the time of birth. In some areas, such as parts of Africa, Asia, and the Pacific islands, up to 80% of the general population may show serologic evidence of past hepatitis infection. The chronic carrier rate is high. Such persons are at high risk for development of cirrhosis or hepatocellular carcinoma (hepatoma).[117]

In the United States, fewer than 10% of the population shows serologic evidence of past infection with the HBV. The chronic carrier rate is less than 1% of the general population. Persons at high risk for infection in this country include male homosexuals, intravenous drug abusers, children born to mothers who are hepatitis B surface antigen positive at the time of delivery, immigrants from endemic areas, and sexual partners and household contacts of patients who have hepatitis B. Although transmission of hepatitis B by blood products still occurs, effective screening tests now make this relatively uncommon. Health care workers, including laboratory personnel, may be at increased risk for developing hepatitis B depending on the degree of their exposure to blood and body fluids.[113]

An effective vaccine and also immunoglobulin exist for hepatitis B. The vaccine is highly effective in stimulating the production of hepatitis B surface antibody and, therefore, in rendering the recipient immune. Following an acute exposure, such as a needle-stick injury, the patient should receive both the hepatitis vaccine and immunoglobulin. A similar treatment is highly effective in preventing the development of hepatitis in infants born to infected mothers.

It is especially important that all health care workers exposed to blood and body fluids take the hepatitis vaccine. Each year thousands of cases of hepatitis B occur among health care workers, and several hundred die from its complications. This is entirely preventable.[34] Everyone who works in the clinical laboratory should take the hepatitis vaccine.[114]

Hepatitis BsAg

Hepatitis BsAg (HBsAg), previously known as the *Australia antigen* and *hepatitis-associated antigen (HAA),* is the antigen for which routine testing is performed on all donated units of blood. HBsAg is the first serologic marker to appear during the prodrome of acute hepatitis B, and it identifies infected patients before the onset of clinical illness more reliably than

any of the others. Although it has been shown that the isolated viral protein coat is not infectious, persons who chronically carry HBsAg in their serum must be considered potentially infectious, because the presence of the intact virus cannot be excluded. Patients who recover from hepatitis B develop anti-HBs following the disappearance of the HBsAg at about the time of clinical recovery (Fig. 17-5). This antibody is common in the general population and is believed to confer immunity to future reinfection with HBV.[113]

Hepatitis BcAg

The core antigen, HBcAg, has not been demonstrated in the plasma of hepatitis victims or blood donors. This antigen is seen only in the nuclei of hepatocytes during an acute infection with hepatitis B. The antibody to the core antigen, anti-HBc, usually develops earlier in the course than the antibody to the surface antigen (see Fig. 17-5). A recent serologic marker assay developed for general use is a test for IgM antibody to hepatitis B core antigen. In the proper clinical setting, this IgM assay has been shown to be specific for acute hepatitis B. Closely associated with the core antigen is a viral DN-dependent DNA polymerase. This viral enzyme is required for viral replication and is detectable in the serum early in the course of viral hepatitis, during the phase of active viral replication.[110]

Hepatitis BeAg

Another marker of HBV infection is the e antigen. This antigen appears to be more closely associated with the core than the surface of the viral particle. The presence of the e antigen appears to correlate well with both the number of infectious virus particles and the degree of infectivity of HBsAg-positive sera. The presence of HBeAg in HBsAg carriers is an unfavorable prognostic sign and predicts a severe course and chronic liver disease. Conversely, the presence of anti-HBe antibody in carriers indicates a low infectivity of the serum (Fig. 17-6). The e antigen is detected in serum only when surface antigen is present (Fig. 17-7).

Hepatitis B Viral DNA Assay

It is possible to detect actual HBV DNA in the blood using nucleic acid hybridization or polymerase chain reactions. This provides a more sensitive measurement of infectivity and

Sequence of HBV markers

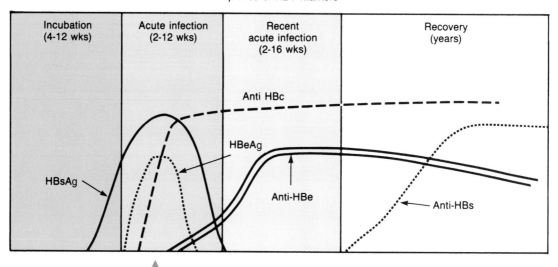

Figure 17-5. Serology of hepatitis B infection with recovery.

disease progression than serology. It may be used to monitor the effectiveness of antiviral therapy in patients with chronic HBV infection, but it supplements rather than replaces current HBV serologic assays.[117–121]

Hepatitis C

Until fairly recently, cases of viral hepatitis that could not be identified as type A or type B by serologic markers were logged in the general category of non-A or non-B hepatitis. Recently, the agent responsible for approximately 80% of these cases was identified as the ribonucleic acid (RNA) containing the *hepatitis C virus (HCV)*.

Tests for detecting those at risk for transmitting HCV are now available. They detect antibody to HCV and identify most infectious carriers.[122] Although much remains to be learned about hepatitis C, it appears most often to be transmitted parenterally. Sexual and fecal–oral routes also exist, and it may be transmitted by blood transfusion.[15,123,124]

Approximately 3% of blood donors in the United States are positive for HCV. Most of these patients are presumed to be infectious. The HCV antibody test will detect most infectious patients, although false-positive results do occur.[121,125]

Clinically, acute hepatitis C is usually mild and may be entirely inapparent. However, infection has a high rate of progression to chronic hepatitis, cirrhosis, and carcinoma. Thus, hepatitis C appears to be a major cause of chronic hepatitis in this country.[122,126]

Hepatitis C antibody is usually not detected in the first few months of the infection but will almost always be present in the later stages. It is not protective and sometimes disappears several years following the resolution of the infection.

Sequence of HBV surface markers
chronic hepatitis

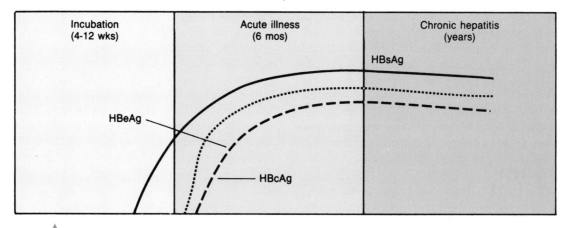

Figure 17-6. No antibody is formed against HBsAg. The persistence of HBeAg implies high infectivity and a generally poor prognosis. This patient would be likely to develop cirrhosis unless seroconversion occurs or treatment is given.

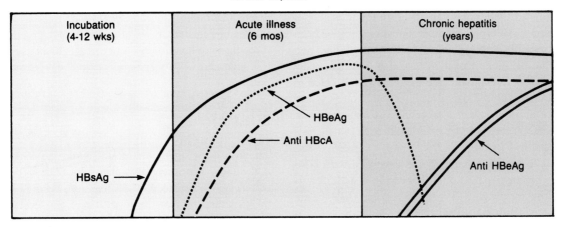

Sequence of HBV surface markers
chronic hepatitis

| Incubation (4-12 wks) | Acute illness (6 mos) | Chronic hepatitis (years) |

HBsAg

HBeAg

Anti HBcA

Anti HBeAg

Figure 17-7. Serology of chronic hepatitis with formation of antibody to HBeAg. This is a favorable sign and suggests that the chronic hepatitis may resolve. Complete recovery would be heralded by the disappearance of HBsAg and formation of its corresponding antibody.

The serologic assay for HCV antibody is a screening test, and false-positives may occur. Therefore, positive results should be confirmed by a more specific method, such as the HCV recombinant immunoblast assay.

Delta Hepatitis

Delta hepatitis, or hepatitis D (HDV), constitutes a unique example of satellite virus infection in human disease. The hepatitis delta virus causes disease only in patients who are currently infected with HBV. It is incapable of causing any illness at all in patients who do not have hepatitis B.

Hepatitis delta virus is an RNA virus with a high degree of base-pair homology with HBV. When delta virus infection occurs in a patient who already has HBV infection, the delta virus uses the HBV for replication. Then, both HBV and HDV are produced. Unlike the HBV, hepatitis delta virus appears directly toxic to human hepatocytes.

Two basic types of HDV infection exist, but both have the effect of worsening the prognosis of the hepatitis B infection. In coinfection, the patient acquires HBV and HDV simultaneously. A patient with this coinfection is more likely to develop serious complications and has a higher rate of progression to chronic hepatitis. However, the majority of the patients still recover.

The other possibility is superinfection, in which a patient who is a chronic HBV carrier is superinfected with the HDV. Fulminant hepatitis may develop, or the progression to cirrhosis may be accelerated.

Epidemiologically, HDV infection appears concentrated in countries surrounding the Mediterranean, Black, and Red seas. In the United States, between 10% and 20% of patients who are chronic HBV carriers will be serologically positive for HDV. The risk factors for HDV infection are generally the same as for HBV infection.[7,57,101–103,127–131]

CASE STUDY 17-5

A 36-year-old man consults his family physician because of liver function abnormalities, which had been noted initially during a pre-insurance physical examination 6 months ago. The following laboratory results are obtained, which are identical to those obtained 6 months previous (Case Study 17-5.1).

Questions

1. What is the most likely diagnosis?
2. What is the prognosis?
3. What complications may develop?
4. What additional tests should be done?

CASE STUDY TABLE 17-5.1. Laboratory Results

Hepatitis A antibody (IgG)	Positive
Hepatitis A antibody (IgM)	Negative
Hepatitis B surface antigen	Positive
Hepatitis B surface antibody	Negative
Hepatitis B core antibody (IgM)	Positive
Hepatitis C antibody	Negative

Hepatitis E

The RNA-containing hepatitis E virus is transmitted primarily by the fecal–oral route. This illness is found mainly in underdeveloped countries, although sporadic cases have been reported, primarily among travelers, in the United States and Western Europe. The incubation period is short, generally between 21 days and 42 days. The virus may be detected in feces and bile by about 7 days after infection.

Hepatitis E infection is generally mild, except in pregnant women, where it may be a devastating illness. Serologic tests are available for diagnosis.[106,107,132–137]

Hepatitis E is an RNA-type virus associated with both acute and chronic hepatitis on a worldwide basis. The GB group of flowlike viruses, GBV-A, GBV-B and GBV-C, are also associated with acute and chronic hepatitis. Little is known about these diseases. No diagnostic tests for them are commercially available at this time.[17,82,99,100,127,138]

Chronic Hepatitis

When evidence of hepatitis, such as elevated serum transaminase levels, is present for more than 6 months, *chronic hepatitis* is said to exist. Many different agents, including viruses, drugs, and alcohol, may cause chronic hepatitis. However, this discussion centers on the viral causes.[133]

Approximately 10% of cases of HBV will progress to chronic hepatitis. The severity of the initial illness has nothing to do with the risk of developing chronicity. Therefore, many patients may develop chronic hepatitis without even having been aware that they had hepatitis B infection. The serologic findings in patients with chronic hepatitis B are shown in Figure 17-7. These patients may be clinically quite sick, or they may be apparently entirely healthy. However, as long as HBsAg is present, they are infectious and at risk for developing complications, including cirrhosis and hepatocellular carcinoma. Patients who manifest the e antigen are highly infectious and are said to have the worst prognosis. The appearance of antibody to e antigen may herald recovery. Complete recovery occurs when hepatitis B surface antigen disappears and the corresponding antibody is then detected. These patients are then immune to further infection.[12] Although many patients with chronic hepatitis may recover spontaneously, others may require aggressive treatment.

On a worldwide basis, hepatitis B infection constitutes a major cause of morbidity and mortality. In underdeveloped countries, infection often occurs at the time of birth, and these children are at high risk for the development of cirrhosis or hepatocellular carcinoma.[21]

Another form of hepatitis with a high degree of chronicity is hepatitis C. Although patients with chronic hepatitis C infection appear to be at high risk for cirrhosis, the role of hepatitis C in the development of hepatocellular carcinoma is much less clear.[139] Interferon is currently used to treat chronic hepatitis.[15] Hepatitis A is rarely, if ever, associated with chronic disease.[21]

Other Forms of Hepatitis

Five forms of viral hepatitis (A, B, C, D, E) are well recognized. The role of G viruses is currently unclear. Hepatitis F is an enteric agent that may be transmitted to primates. Again, more needs to be learned about this agent and its role, if any, in human disease. Other forms of viral hepatitis may exist.[138]

SUMMARY

The liver is the largest and most versatile and complex organ in the body. It consists of two main lobes. The "lobule," cords of liver cells or hepatocytes radiating from a central vein, forms the structural unit of the liver. There are several hundred known functions performed by the liver each day, including metabolic, secretory, and excretory functions. One of the most important functions of the liver is the excretion of bile. Bilirubin is the principal pigment in bile and is derived from the breakdown of hemoglobin when aged red blood cells are phagocytized by the reticuloendothelial system. There are two forms of bilirubin: conjugated and unconjugated. An increase in the concentration of bilirubin is known as jaundice. Jaundice may be caused by a variety of pathophysiologic mechanisms. There are three types of jaundice: prehepatic, hepatic, and posthepatic.

Among the many diverse metabolic functions of the liver is the synthesis of proteins, carbohydrates, and lipids. Many enzymes are synthesized by the liver, but not all have been found to be useful in the diagnosis of hepatobiliary disorders. Disorders of the liver include jaundice, cirrhosis, tumors, Reye's syndrome, and drug- and alcohol-related disorders. Analysis of bilirubin concentrations has been used for centuries to assess liver function. Two commonly used methodologies to assess bilirubin are Evelyn-Malloy and the Jendrassik-Grof. Urobilinogen, a colorless end product of bilirubin metabolism, may also be measured to assess liver function. The measurement of hepatic enzymes in serum is used to assess the extent of liver damage and to differentiate hepatocellular (functional) from obstructive (mechanical) disease. The most common enzymes assayed in hepatobiliary disease include ALP and the aminotransferases (AST and ALT). Hepatitis, or inflammation of the liver may be caused by viruses, bacteria, parasites, radiation, drugs, chemicals, or toxins. Among the viruses causing hepatitis are hepatitis types A, B, C, D (or delta), and E, cytomegalovirus, Epstein-Barr virus, and probably several others.

REVIEW QUESTIONS

1. Hyperbilirubinemia in newborns usually:
   a. Results in permanent brain damage
   b. Is caused by hemolysis
   c. Is caused by biliary atresia

d. Produces no serious consequences

e. Must be treated if serum bilirubin levels exceed 5 mg/dL

2. Cirrhosis is derived from the Greek word meaning:
   a. Yellow
   b. Hard
   c. Large
   d. Tumor
   e. Liver

3. All of the following statements concerning urobilinogen are correct EXCEPT:
   a. Colorless
   b. Produced by oxidative actions of intestinal bacteria
   c. Undergoes significant enterohepatic circulation
   d. Urinary levels increased in biliary obstruction
   e. Fecal levels decreased in biliary obstruction

4. Elevated levels of CSF glutamine are found in patients with:
   a. Hepatic encephalopathy
   b. Brain tumors
   c. Strokes
   d. Schizophrenia
   e. Renal failure

5. Which form of hepatitis is caused by a DNA virus?
   a. Hepatitis A
   b. Hepatitis B
   c. Hepatitis C
   d. Hepatitis D
   e. Hepatitis E

6. In patients with hepatitis B, hepatitis e antigen is found in serum only when which of the following is also present?
   a. Surface antigen
   b. Antibody to surface antigen
   c. Antibody to hepatitis e
   d. Hepatitis c antibody (IgG)
   e. Hepatitis c antibody (IgM)

7. Which of the following enzymes is most useful in establishing the hepatic origin of an elevated serum alkaline phosphatase?
   a. Alanine aminotransferase
   b. Aspartate aminotransferase
   c. Ornithine carbamyltransferase
   d. γ-glutamyltranspeptidase
   e. Lactate dehydrogenase

8. Hepatitis E infection is likely to produce serious consequences in:
   a. Children
   b. Pregnant women
   c. Travelers in Third World countries
   d. Elderly people
   e. Patients who take aspirin

9. On a worldwide basis, most primary malignant tumors of the liver are related to:
   a. Alcoholism
   b. Gallstones
   c. Previous infection with a hepatitis virus
   d. Reye's syndrome
   e. Malaria

10. Ehrlich reagent is used in the measurement of:
   a. Bilirubin
   b. Urobilinogen
   c. Ammonia
   d. Bile acids

REFERENCES

1. Barnard SE, Becker YT, Blair KT, et al. The effect of low dose epinephrine infusion on hepatic hemodynamics. Transplant Proc 1998; 30(5):2306–2308.
2. Bismuth H. Surgical anatomy and anatomical surgery of the liver. World J Surg 1982;6(1):3–9.
3. Barwick K, Rosai, J. Ackerman's surgical pathology, 8th ed. St Louis: Mosby, 1996;857–942.
4. De Wit LT, Douma DJ, Groen AK, et al. Hepatic bile versus gallbladder bile: a comparison of protein and lipid concentration and composition in cholesterol gallstone patients. Hepatology 1998;28(1):11–16.
5. Balistreri WF. Fetal and neonatal bile acid synthesis and clinical implications. J Inherited Metab Dis 1991;14:459–477.
6. Hofmann AF. Bile acid secretion, bile flow and biliary lipid secretion in humans. Hepatology 1990;12(3, Pt 2):17S–22S;discussion 22S–25S.
7. Kools AM. Hepatitis A, B, C, D, and E. Update on testing and treatment. Postgrad Med 1992;91(3):109–114.
8. Adachi Y, Fuchi I, Katoh H, et al. Serum bilirubin fractions in healthy subjects and patients with unconjugated hyperbilirubinemia. Clin Biochem 1990;23(3):247–251.
9. Frank BB. Clinical evaluation of jaundice—a guideline of the Patient Care Committee of the American Gastroenterological Association. JAMA 1989;262(21):3031–3034.
10. Hope S, Newman TB, and Stevenson DK. Direct bilirubin measurements in jaundiced term new-born infant, a reevaluation. Am J Dis Child, 1991;145(11):1305–1309.
11. Doumas BT, Wu TW. The measurement of bilirubin fractions in serum. Crit Rev Clin Lab Sci 1991;28:415–446.
12. Mathew P. The genetic basis of Gilbert's syndrome. N Engl J Med 1996;334(12):802–803.
13. Adachi Y, Koiwai O, Sato H. The genetic basis of Gilbert's syndrome. Lancet 1996;347(9001):557–558.
14. Emi Y, Ikushiro S, Iyanagi T. Biochemical and molecular aspects of genetic disorders of bilirubin metabolism. Biochim Biophys Acta 1998;1407(3):173–184.
15. Gollan JL, Green RM. Crigler-Najjar disease type I: therapeutic approaches to genetic liver diseases into the next century. Gastroenterology 1997;112(2):649–651.
16. Berg CL. The physiology of jaundice: molecular and functional characterization of the Crigler-Najjar syndromes. Hepatology 1995;22 (4, Pt 1):1338–1340.
17. Westwood A. The analysis of bilirubin in serum. Ann Clin Biochem 1991;28(Pt 2):119–130.
18. Cherrington AD, Connolly CC, Moore MC. Autoregulation of hepatic glucose production. Eur J Endocrinol 1998;138(3):240–248.
19. Kjaer M. Hepatic glucose production during exercise. Adv Exp Med Biol 1998;441:117–127.
20. Ascher N, Carithers RL Jr, Hoofnagle JH, et al. Fulminant hepatic failure: summary of a workshop. Hepatology 1995;21(1):240–252.
21. Gorski J, Oscai LB, Palmer WK. Hepatic lipid metabolism in exercise and training. Med Sci Sports Exerc 1990;22(2):213–221.
22. Bonkovsky HL. Iron and the liver. Am J Med Sci 1991;301(1):32–43.
23. Balistreri WF, Setchell KDR. Newer liver function tests. In: Ginari G, Rozen P, Bujanover Y, et al, eds. Frontiers of gastrointestinal research: newer tests and procedures in pediatric gastroenterology. Vol 16. Basel, Switzerland: Karger, 1989;220–245.
24. Arai N, Hayashi M, Kumada S, et al. Neuropathology of the dentate nucleus in developmental disorders. Acta Neuropathol (Berl) 1997; 94(1):36–41.
25. Burns D, Perlman JM, Rogers BB. Kernicteric findings at autopsy in two sick near term infants. Pediatrics 1997;99(4):612–615.

26. Gourley, GR. Bilirubin metabolism and kernicterus. Adv Pediatr 1997;44:173–229.
27. Ostrow JD, Tiribelli C. New concepts in bilirubin and jaundice: report of the Third International Bilirubin Workshop, April 6–8, 1995, Trieste, Italy. Hepatology 1996;24(5):1296–1311.
28. Black ER. Diagnostic strategies and test algorithms in liver disease. Clin Chem 1997;43(8, Pt 2):1555–1560.
29. Neuschwander-Tetri BA. Common blood tests for liver disease. Which ones are most useful? Postgrad Med 1995;98(1):49–56,59,63.
30. Popper H. General pathology of the liver. Light microscopic aspects serving diagnosis and interpretation. Semin Liver Dis 1986;6:175–184.
31. Kosaka Y, Nakano T, Takase K, et al. Prognostic value of peritoneoscopic findings in cirrhosis of the liver. Gastrointest Endosc 1990; 36:34–38.
32. Lieber CS. Alcohol and the liver: 1994 update. Gastroenterology 1994;106(4):1085–1105.
33. Bosch J. Medical treatment of portal hypertension. Digestion 1998; 59(5):547–555.
34. Barnes DS, Geisinger MA, Henderson JM. Portal hypertension. Curr Probl Surg 1998;35(5):379–452.
35. Del Olmo JA, Escudero A, Gilabert S, et al. Incidence and risk factors for hepatocellular carcinoma in 967 patients with cirrhosis. J Cancer Res Clin Oncol 1998;124(10):560–564.
36. Carriaga MT, Henson DE. Liver, gallbladder, extrahepatic bile ducts, and pancreas. Cancer 1995;75:171–190.
37. Blair T, Blanke C, Chappam WC, et al. Hepatocellular carcinoma outcomes based on indicated treatment strategy. Am Surg 1998; 64(12):1128–1134;discussion 1134–1135.
38. Fan ST, Lau H, Ng IO, et al. Long term prognosis after hepatectomy for hepatocellular carcinoma: a survival analysis of 204 consecutive patients. Cancer 1998;83(11):2302–2311.
39. Bruce JC, Glasgow JF, Hall SM, et al. The changing clinical pattern of Reye's syndrome 1982–1990. Arch Dis Child 1996;74(5):400–405.
40. Salmona M, Tacconi MT, Visentin M. Reye's and Reye-like syndromes, drug-related diseases? (Causative agents, etiology, pathogenesis, and therapeutic approaches). Drug Metab Rev 1995;27(3): 517–539.
41. Smith TC. Reye syndrome and the use of aspirin. Scott Med J 1996;41(1):4–9.
42. Stumpf DA. Reye syndrome: an international perspective. Brain Dev 1995;17(Suppl):77–78.
43. Amacher DE. Serum transaminase elevations as indicators of hepatic injury following the administration of drugs. Regul Toxicol Pharmacol 1998;27(2):119–130.
44. Garcia Rodriguez LA, Jick H, Ruigomez A. A review of epidemiologic research on drug-induced acute liver injury using the general practice research data base in the United Kingdom. Pharmacotherapy 1997;17(4):721–728.
45. Arner P, Large V. Regulation of lipolysis in humans. Pathophysiological modulation in obesity, diabetes, and hyperlipidaemia. Diabetes Metab 1998;24(5):409–418.
46. Dossing M, Sonne J. Drug-induced hepatic disorders. Incidence, management and Avoidance. Drug Saf 1993;9(6):441–449.
47. Lee WM. Review article: Drug-induced hepatotoxicity. Aliment Pharmacol Ther 1993;7(5):477–485.
48. Anker AL, Smilkstein MJ. Acetaminophen. Concepts and controversies. Emerg Med Clin North Am 1994;12(2):335–349.
49. Lee WM. Review article: Drug-induced hepatotoxicity. Aliment Pharmacol Ther 1993;7(5):477–485.
50. Doumas BT, Wu TW. The measurement of bilirubin fractions in serum. Crit Rev Clin Lab Sci 1991;28(5–6):415–445.
51. Ou CN, Rutledge JC. Bilirubin and the laboratory. Advances in the 1980s, considerations for the 1990s. Pediatr Clin North Am 1989; 36(1):189–198.
52. Doumas BT, Lott JA. Direct and total bilirubin tests: contemporary problems. Clin Chem 1993;39(4):641–647.
53. Doumas BT, Lott JA. Direct and total bilirubin tests: contemporary problems. Clin Chem 1993 1993;39(4):641–647.
54. Slaughter, MR. Sensitivity of conventional methods for bilirubin measurements. J Lab Clin Med 1996127(2):233.
55. Ryder KW, et al. Erroneous laboratory results from hemolyzed, icteric, and lipemic specimens. Clin Chem 1993;39(1):175–176.
56. Ross JW, et al. The accuracy of laboratory measurements in clinical chemistry: a study of 11 routine chemistry analytes in the College of American Pathologists chemistry survey with fresh frozen serum, definitive methods, and reference methods. Arch Pathol Lab Med 1998; 122(7):587–608.
57. Schlebusch H, et al. Comparison of five routine methods with the candidate reference method for the determination of bilirubin in neonatal serum. J Clin Chem Clin Biochem 1990;28(4):203–210.
58. Harrison SP, et al. Three direct spectrophotometric methods for determination of total bilirubin in neonatal and adult serum, adapted to the Technicon RA-1000 Analyzer. Clin Chem 1989;35(9): 1980–1986.
59. Johnson PJ. Role of the standard "liver function tests" in current clinical practice. Ann Clin Biochem 1989;26(Pt 6):463–471.
60. Free AH, Hager CB. Urine urobilinogen as a component of routine urinalysis. Am J Med Technol 1970;36(5):227–233.
61. Binder L, Glass B, Haynes J, et al. Failure of prediction of liver function test abnormalities with the urine urobilinogen and urine bilirubin assays. Arch Pathol Lab Med 1989;113(1):73–76.
62. Fevery J, Kotal P. Quantitation of urobilinogen in feces, urine, bile and serum by direct spectrophotometry of zinc complex. Clin Chim Acta 1991;14:202(1–2):1–9.
63. Binder L, Glas B, Haynes J, et al. Abnormalities of urine urobilinogen and urine bilirubin assays and their relation to abnormal results of serum liver function tests. South Med J 1988;81(10):1229–1232.
64. Lee BL, New AL, Ong CN. Comparative analysis of conjugated bile acids in human serum using high-performance liquid chromatography and capillary electrophoresis. J Chromatogr B Biomet Sci Appl 1997;704(1–1):35–42.
65. Aldini R, Barbara L, Bazzoli F, et al. Diagnostic effectiveness of serum bile acids in liver diseases as evaluated by multivariate statistical methods. Hepatology 1983;3(5):707–713.
66. Azer SA, Klaassen CD, Stacey NH. Biochemical assay of serum bile acids: methods and applications. Br J Biomed Sci 1997;54(2)118–132.
67. A-Kader PS, Balistreri, WF, Setchell KDR, et al. New methods for assessing liver function in infants and children. Ann Clin Lab Sci 1992;22:162–174.
68. Aliberti G, Corvisieri P, de Michele LV, et al. Lactate dehydrogenase and its isoenzymes in the marrow and peripheral blood from haematologically normal subjects. Physiol Res 1997;46(6):435–438.
69. Aranda-Michel J, et al. Tests of the liver: use and misuse. Gastroenterologist 1998;6(1):34–43.
70. Sheehan M, et al. Predictive values of various liver function tests with respect to the diagnosis of liver disease. Clin Biochem 1979;12(6): 262–263.
71. Burke MD. Hepatic function testing. Postgrad Med 1978;64(3): 177–182,185.
72. Moss DW. Diagnostic aspects of alkaline phosphatase and its isoenzymer. Clin Biochem 1987;20(4):225–230.
73. Mercer DW. Serum isoenzymes in cancer diagnosis and management. Immunol Ser 1990;53:613–629.
74. Artiss JD, Epstein E, Kiechle FL, et al. The clinical use of alkaline phosphatase enzymes. Clin Lab Med 1986;6(3):491–505.
75. Moss DW. Alkaline phosphatase isoenzymes. Clin Chem 1982;28(10): 2007–2016.
76. Sherman KE,. Alanine aminotransferase in clinical practice. A review. Arch Intern Med 1991;151(2):260–265.
77. Mandell BF. Alanine aminotransferase: a nonspecific marker of liver disease. Arch Intern Med 1992;152(1):209–213.
78. Rej R. Aminotransferases in disease. Clin Lab Med 1989;9(4): 667–687.
79. Panteghini M, et al. Clinical and diagnostic significance of aspartate aminotransferase isoenzymes in sera of patients with liver diseases. J Clin Chem Clin Biochem 1984;22(2):153–158.
80. Bacq Y, et al. Liver function tests in normal pregnancy: a prospective study of 103 pregnant women and 103 matched controls. Hepatology 1996;23(5):1030–1034.
81. Panteghini M. Electrophoretic fractionation of 5'-nucleotidase. Clin Chem 1994;40(2):190–196.
82. Sunderman FW Jr. The clinical biochemistry of 5'-nucleotidase. Ann Clin Lab Sci 1990;20(2):123–139.
83. Abeloff MD, Bostick WD, Ettinger DS, et al. Biological markers as an aid in the clinical management of patients with liver metastases. J Surg Oncol 1982;20(2):83–94.
84. Fex G, Hood B, Kristensson H, et al. Comparison of gamma-glutamyltransferase and other health screening tests in average middle-aged males, heavy drinkers and alcohol non-users. Scand J Clin Lab Invest 1983;43(2):141–149.

85. Frydman MI, Lunzer MR, Ruppin DC. Value of serum gamma-glutamyltransferase activity in the diagnosis of hepatobiliary disease. Med J Aust 1982;1(10):421–424.

86. Lott JA, Nemesanszky E. Gamma-glutamyltransferase and its isoenzymes: progress and problems. Clin Chem 1985;31(6):797–803.

87. Goldberg DM. Structural, functional, and clinical aspects of gamma-glutamyltransferase. CRC Crit Rev Clin Lab Sci 1980;12(1):1–58.

88. Flendrig JG, Hermens WT, Janssen MA, et al. Measurement of ornithine carbamyl transferase (OCT) in plasma by means of enzymatic determination of ammonia. Clin Chim Acta 1991;203(2–3):395–402.

89. Seredynski L. A routine method for the determination of serum ornithine carbamyl transferase. Can J Med Technol 1970;32(3):101–111.

90. Fujiyama S, Mori M, Mori S, et al. Clinical evaluation of serum ornithine carbamoyltransferase by enzyme-linked immunosorbent assay in patients with liver diseases. Enzyme Protein 1994–1995;48(1):18–26.

91. Alhava E, Partanen K, Pasanen P, et al. Value of serum alkaline phosphatase, aminotransferases, gamma-glutamyltransferase, leucine aminopeptidase, and bilirubin in the distinction between benign and malignant diseases causing jaundice and cholestasis: results from a prospective study. Scand J Clin Lab Invest 1993;53(1):35–39.

92. Pruzanski W. The influence of steroids on serum bilirubin, leucine aminopeptidase and alkaline phosphatase in extrahepatic obstructive jaundice: the possible diagnostic importance. Am J Med Sci 1966;251(6)685–689.

93. Baier J, Borsch G, Gerhardt W, et al. Graphical analysis of laboratory data in the differential diagnosis of cholestasis: a computer-assisted prospective study. J Clin Chem Clin Biochem 1988;26(8):509–519.

94. Bates SE. The use and potential of serum tumour markers, new and old. Drugs 1989;38(1):9–18.

95. Schwartz MK. Lactic dehydrogenase. An old enzyme reborn as a cancer marker? Am J Clin Pathol 1991;96(4):441–443.

96. Chazouilleres O, Robert A. Percentage, or international normalized ratio? Paris, France: Laboratoire Central d'Hematologie et d'Immunologie, 1996.

97. Anand AC, Neuberger JM, Nightingale P. Early indicators of prognosis in fulminant hepatic failure: an assessment of the King's criteria. J Hepatol 1997;26(1):62–68.

98. Green A. When and how should we measure plasma ammonia? Ann Clin Biochem 1988;25:(PT 3):199-209.

99. Willems D, et al. Ammonia determined in plasma with a selective electrode. Clin Chem 1988;34(11):2372.

100. Tucci S, et al. Measurement of glutamine and glutamate by capillary electrophoresis and laser induced fluorescence detection in cerebrospinal fluid of meningitis sick children. Clin Biochem 1998;31(3):143–150.

101. Casey JL. Hepatitis delta virus. Genetics and pathogenesis. Clin Lab Med 1996;16(2):451–464.

102. Negro F, Rizzetto M. Diagnosis of hepatitis delta virus infection. J Hepatol 1995;22(1 Suppl):136–139.

103. Linares A, Navascues CA, Rodrigo L, et al. Epidemiology of hepatitis D virus infection: changes in the last 14 years. Am J Gastroenterol 1995;90(11):1981–1984.

104. Emerson SU, Ghabrah TM, Purcell RH, et al. Comparison of tests for antibody to hepatitis E virus. J Med Virol 1998;55(2):134–137.

105. Alter MJ, Holland PV, Mast EE. Evaluation of assays for antibody to hepatitis E virus by a serum panel. Hepatitis E Virus Antibody Serum Panel Evaluation Group. Hepatology 1998;27(3):857–861.

106. Balan V, Carpenter HA, Dawson GJ, et al. Acute hepatitis E by a new isolate acquired in the United States. Mayo Clin Proc 1997;72(12):1133–1136.

107. Irshad M. Hepatitis E virus: a global view of its seroepidemiology and transmission pattern. Trop Gastroenterol 1997;18(2):45–49.

108. Koff RS. Hepatitis A. Lancet 1998;351(9116):1643–1649.

109. Alter MJ, Bell BP, Fleenor M, et al. The diverse patterns of hepatitis A epidemiology in the United States—implications for vaccination strategies. J Infect Dis 1998;178(6):1579–1584.

110. Lemon SM. Type A viral hepatitis: epidemiology, diagnosis, and prevention. Clin Chem 1997;43(8, Pt 2):1494–1499.

111. Macedo G, Ribeiro T. Hepatitis A: Insights into new trends in epidemiology. Eur J Gastroenterol Hepatol 1998;10(2):175.

112. Dolan SA. Vaccines for hepatitis A and B. The latest recommendations on safe and extended protection. Postgrad Med 1997;102(6):74–80.

113. Kurstak C, Kurstak E, Hossain A. Progress in diagnosis of viral hepatitis A, B, C, D and E. Acta Virol 1996;40(2):107–115.

114. Kurstak C, Kurstak E, Hossain A. Progress in diagnosis of viral hepatitis A, B, C, D and E. Acta Virol 1996;40(2):107–115.

115. Gregorio GV, Mieli-Vergani G, Mowat AP. Viral hepatitis. Arch Dis Child 1994;70(4):343–348.

116. Davis GL. Hepatitis B: diagnosis and treatment. South Med J 1997;90(9):866–870; quiz 871.

117. Gitlin N. Hepatitis B: diagnosis, prevention, and treatment. Clin Chem 1997;43(8, Pt 2):1500–1506.

118. Colantoni A, De Maria N, Idilman R, et al. Pathogenesis of hepatitis B and C-induced hepatocellular carcinoma. J Viral Hepat 1998;5(5):285–299.

119. Bengtstrom M, Jalava T, Kallio A, et al. A rapid and quantitative solution hybridization method for detection of HBV DNA in serum. J Virol Methods 1992;36(2):171–180.

120. Campelo C, Cisterna R, Gorrino MT, et al. Detection of hepatitis B virus DNA in chronic carriers by the polymerase chain reaction. Eur J Clin Microbiol Infect Dis 1992;11(8):740–744.

121. Aspinall S, Mphahlele MJ, Peenze I, et al. Detection and quantitation of hepatitis B virus DNA: comparison of two commercial hybridization assays with polymerase chain reaction. J Viral Hepat 1995;2(2):107–111.

122. McHutchison JG, Person JL, et al. Improved detection of hepatitis C virus antibodies in high risk populations. Hepatology 1992;15:19–25.

123. Alter MJ, Krzysztof K, Margolis HS, et al. The natural history of community-acquired hepatitis C in the United States. N Engl J Med 1992;327:1899–1905.

124. Cardoso M, Dengler T, Kerowgan M, et al. Estimated risk of transmission of hepatitis C virus by blood transfusion. Vox Sang 1998;74(4):213–216.

125. Holland PV. Post-transfusion hepatitis: current risks and causes. Vox Sang 1998;74(Suppl 2):135–141.

126. Choo QL, De Medina M, Hasan F, et al. Hepatitis C-associated hepatocellular carcinoma. Hepatology 1990;12(3, Pt 1):589–591.

127. Vyas GN, Yang G. Immunodiagnosis of viral hepatitides A to E and non-A to -E. Clin Diagn Lab Immunol 1996;3(3):247–256.

128. Bernardez-Hermida I, Perez-Gracia MT. Serological markers for diagnosis of acute delta hepatitis infection. Eur J Clin Microbiol Infect Dis 1993;12(8):650–651.

129. Herrera JL. Serologic diagnosis of viral hepatitis. South Med J 1994;87(7):677–684.

130. McPherson RA. Laboratory diagnosis of human hepatitis viruses. J Clin Lab Anal 1994;8(6):369–377.

131. Renz M, Seelig HP, Seelir R. PCR in the diagnosis of viral hepatitis. Ann Med 1992;24(3):225–230.

132. Burkholder BT, Favorov MO, Holland PV, et al. Prevalence of and risk factors for antibody to hepatitis E virus seroreactivity among blood donors in Northern California. J Infect Dis 1997;176(1):34–40.

133. Balayan MS. Epidemiology of hepatitis E virus infection. J Viral Hepat 1997;4(3):155–15.

134. Bradley DW. Hepatitis E virus: a brief review of the biology, molecular virology, and immunology of a novel virus. J Hepatol 1995;22(1 Suppl):140–145.

135. Dawson GJ, De Guzman IJ, Holzer TJ, et al. Diagnosis of acute hepatitis E infection utilizing enzyme immunoassay. Dig Dis Sci 1994;39(8):1691–1693.

136. Castano G, Dawson GJ, Flichman D, et al. Detection of hepatitis G virus RNA in patients with acute non-A-E hepatitis. J Viral Hepat 1998;5(3):161–164.

137. Loya F. Does the hepatitis G virus cause hepatitis? Tex Med 1996;92(12):68–73.

138. Bowden DS, Locarnin SA, Moaven LD. New hepatitis viruses: are there enough letters in the alphabet? Med J Aust 1996;164(2):87–89.

139. Heubi JE. Serum bile acids as markers of liver disease in childhood. J Pediatr Gastroenterol Nutr 1982;1(4):457–458.

CHAPTER 18

Endocrinology

Shauna C. Anderson

Objectives

Upon completion of this chapter, the clinical laboratorian should be able to:

- *State the functions of the endocrine system.*

- *Define the following terms: hormone, prohormone, steroid, peptide, and amine.*

- *Relate the mechanisms involved in hormone regulation.*

- *Given a list of hormones, categorize them as steroid, protein, or amine.*

- *Explain the mechanisms of hormone action and control.*

- *Compare and contrast the various methodologies used to assay hormone levels.*

- *List two problems in assaying hormone levels.*

- *Describe the location and major hormones produced by the following endocrine glands: hypothalamus, anterior pituitary, posterior pituitary, thyroid, parathyroids, adrenal medulla, adrenal cortex, pancreas, ovaries, testes, placenta, and gastrointestinal tract.*

- *Relate the expected laboratory results associated with the following disease states: acromegaly, GH deficiency, SIADH6, diabetes insipidus, Cushing's syndrome, Addison's disease, CAH, Conn's syndrome, pheochromocytoma, hyperparathyroidism, hypoparathyroidism, infertility, hirsuitism, and amenorrhea.*

- *Discuss the difference between primary and secondary hormone disorders.*

- *State the importance of multiple measurements of some hormone levels.*

- *Discuss the importance of stimulation and suppression tests in diagnosing hormone disorders.*

- *Describe any special procedures or precautions in the collection of samples for hormone assay.*

KEY TERMS

Acromegaly
Addison's disease
Adenohypophysis
Adrenal glands
Aldosterone
Amenorrhea
Amine
Androgen
Angiotensin
Anterior pituitary
Antidiuretic
   hormone (ADH)
Autocrine
Calcitonin
Carcinoid syndrome
Conn's syndrome
Corpus albicans
Corpus luteum
Cortisol
Cushing's
   syndrome
Ectopic hormone
Endocrine gland
Enterochromaffin
   cell
Epinephrine
Erectile dysfunction
Estriol
Estrogen

Estrone
Follicular phase
Gastrin
Glucocorticoid
G protein
Gynecomastia
Hirsuitism
Hormone
Hyperaldosteronism
Hyperandrogenemia
Hypercortisolism
Hyperparathyroidism
Hyperprolactinemia
Hypoaldosteronism
Hypocortisolism
Hypogonadism
Hypoparathyroidism
Hypothalamus
Infertility
Inhibin
Islets of Langerhans
Leydig cell
Luteal phase
Metanephrine
Mineralocorticoid
Neuroblastoma
Neurohypophysis
Norepinephrine
Normetanephrine

Ovulation
Oxytocin
Panhypopituitarism
Paracrine
Peptides
Pheochromocytoma
Placenta
Posterior pituitary
Progesterone
Prohormone
Proinsulin
Receptor
Sella turcica
Serotonin
Sertoli cell
Sheehan's
   syndrome
Steroids
Testosterone
Thecal cell
Vasopressin
Virilization
Zollinger-Ellison
   syndrome
Zona fasciculata
Zona glomerulosa
Zona reticularis

The endocrine system and the manifestations of disease in endocrine glands are complex and involve the interaction of many other body systems. Because hormones produced by the *endocrine glands* circulate in the blood, they are readily accessible to laboratory evaluation. However, evaluation of endocrine function differs in two major ways from most chemistry analyses. First, the substances to be measured circulate in extremely small amounts (nanograms per milliliter in many cases). Detection of such low levels of hormones has required development of very sensitive analytical methods, which use complex techniques and sophisticated equipment. Consequently, these tests are relatively expensive to both the laboratory and the patient. In many cases, it is not economically feasible for small hospital laboratories to perform these tests and, consequently, they are usually sent to reference laboratories. Second, instead of receiving only one or two requests for a particular test on a patient in one day, the laboratory will frequently receive requests for the same test on several blood samples drawn from the same patient in a relatively short time interval. These requests will be part of a stimulation or suppression test designed to evaluate an endocrine gland's response to the usual control mechanisms. Such tests are necessary because the normal secretion of some hormones is erratic, and thus a single measurement under uncontrolled conditions is of no use; destruction of as much as 90% of the functioning mass of a gland is necessary before clinical evidence of deficiency occurs; therefore, stimulation tests designed to evaluate glandular reserve may be the only way to detect early disease; and some hormones circulate in such low levels that one unstimulated measurement cannot distinguish the normal from the hypofunctioning gland. In addition, stimulation tests allow measurement of these hormones at higher values where the technical reliability is greater.[1-3]

HORMONES

Definition

Hormones are chemical signals produced by specialized cells, secreted into the blood stream, and carried to a target tissue. The target tissue is usually, but not always, located some distance from the secretory cells and contains *receptor* sites for particular hormones. Ultimately, cellular function is regulated through hormonal interaction.

Hormones can be secreted by specialized nerve cells called neurosecretory cells. These cells are neurons that receive signals from other nerve cells and thus are capable of responding to emergency situations as well as maintaining the internal chemical environment. Interactions also occur between different portions of the endocrine system, with multiple hormones affecting one physiologic function. For example, insulin, glucagon, growth hormone (GH), cortisol, and epinephrine all affect carbohydrate metabolism. The opposite situation is also true. A single hormone (*eg,* cortisol) may affect many different organ systems, producing a variety of physiologic effects. Normal growth and development, physical and mental, and reproduction also require the correct degree of endocrine function.[2,3]

Chemical Nature

At least three chemical types of hormones have been identified: steroids, polypeptides or proteins, and substances derived from amino acids (amines). These differences in chemical structure are accompanied by differences in transport, mechanism of action, and metabolism.

Steroids

Steroid hormones are lipid molecules that have cholesterol as a common precursor. *Steroids* are produced by four organs: adrenal glands, ovaries, testes, and placenta. Regardless of the site of production, the same initial biochemical pathways are followed (Fig. 18-1). The end result depends on the enzymatic machinery that is predominant in a particular organ. Once the hormones are formed, they immediately diffuse through the cell membrane and enter the blood stream. Thus, the rate of synthesis and the amount of stored cholesterol control the rate of secretion. These hormones are water insoluble and circulate bound to a carrier protein. They have a relatively long plasma half-life, ranging from 60 minutes to 100 minutes.[2,3]

Protein

Protein hormones may be either peptides or glycoproteins. The *peptides* include insulin, glucagon, parathyroid hor-

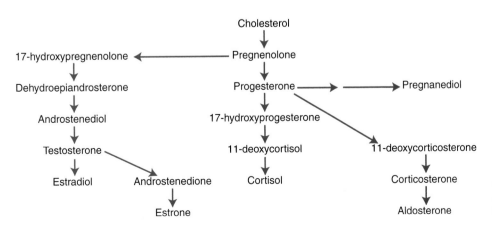

Figure 18-1. Simplified scheme of steroid synthesis from cholesterol.

mone (PTH), GH, and prolactin (PRL). These hormones are synthesized as a larger molecule called a preprohormone or *prohormone,* which is cleaved to produce the usual circulating form. As a rule, protein hormones are synthesized and stored within the cell in the form of secretory granules and are released as needed. The glycoprotein hormones, follicle-stimulating hormone (FSH), luteinizing hormone (LH), thyroid-stimulating hormone (TSH), and human chorionic gonadotropin (hCG), comprise alpha and beta subunits. The alpha subunits of all four of these hormones are immunologically identical. The functional and immunologic specificity of each are derived from the beta subunits. These hormones are water soluble and are not bound to carrier proteins in the circulation. Their half-life is short, ranging from 5 minutes to 60 minutes.[2,3]

Amines

The *amines* are derived from amino acids and include epinephrine, norepinephrine, thyroxine, and triiodothyronine. These compounds have properties intermediate between those of the steroids and protein hormones. *Epinephrine* and *norepinephrine* tend to behave like the protein hormones—that is, they circulate unbound to protein and have a short half-life and a similar mechanism of action. In contrast, thyroxine and triiodothyronine have a long half-life, circulate bound to a carrier protein, and have a mechanism of action that is similar to the steroids.[2,3]

Mechanisms of Action

The production of the appropriate response to a hormone by its target tissue depends on the interaction of the hormone and a specific receptor. The presence of high-affinity receptors accounts for the ability of the target cell to accumulate relatively high concentrations of a particular hormone. These receptors also provide for target-organ specificity, because not all cells have receptors for all hormones.

The protein hormones and catecholamines (epinephrine and norepinephrine) cannot cross the cell membrane and therefore produce their effects by interaction with a receptor on the outer surface of the target cell. The mechanism of action involves a receptor, G proteins located in the cell membrane, an enzyme or effector molecule (adenylate cyclase) attached to the intracellular side of the membrane, and a second messenger that finally causes the intracellular change.

When the hormone binds to a receptor, adenylate cyclase is activated and there is an increased production of the second messenger, cyclic adenosine monophosphate (cAMP). However, the hormone receptor and the effector enzyme do not interact directly. There are membrane proteins that act as a relay. One type of relay is called *G proteins,* the internal domains of the receptor molecule. These are guanine nucleotide-binding proteins. As the hormone binds with the specific receptor protein in the cell membrane, the subsequent complex binds with a G protein, which in turn binds with guanine triphosphate (GTP) and thus stimulates the enzyme adenylate cyclase. Adenylate cyclase converts adenosine triphosphate (ATP) to cAMP. The cAMP, or second messenger, translates the hormone binding into cellular action. In general, cAMP activates protein kinases, resulting in phosphorylation of certain enzymes. Such phosphorylation results in either activation or inhibition of the enzyme, depending on whether the G proteins are stimulatory or inhibitory. Hormones that act through cAMP produce rapid results (within minutes), which cease rapidly when the stimulus stops.

Some hormones induce responses in their target cells by way of cytoplasmic concentrations of calcium ions. This type of regulation is known as the phosphoinositide (PI) pathway. The G protein involved in this system is coupled to an enzyme known as phospholipase C. When a hormone binds to a receptor, the G protein activates phospholipase C that cleaves the membrane-bound lipid, phosphatidylinositol 4,5-bisphosphate (PIP_2) to form inositol triphosphate (IP_3) and diacylglycerol (DAG). IP_3 and DAG become second messengers and signal separate pathways. IP_3 causes the endoplasmic reticulum and the calcisome to release their stores of calcium into the interior of the cell. The calcium ions, the third messenger, may then activate a variety of enzymes, which in turn initiate a cellular response. DAG remains fixed to the cell membrane and appears to act as a cofactor, along with calcium in activating protein kinase C (PKC). This enzyme is capable of influencing many cellular functions. In contrast to the protein hormones, which cannot enter the cell, unbound steroid hormones pass through the cell membrane. These hormones bind to a receptor after they enter into the cell. This receptor is a component of a receptor-protein complex. When the hormone binds, the complex dissociates and a deoxyribonucleic acid (DNA) binding site is exposed. The hormone-receptor complex moves into the nucleus and binds to a particular segment of chromatin, inducing the formation of messenger ribonucleic acid (mRNA). The mRNA enters the cytoplasm and initiates the synthesis of proteins or peptides that carry out the action attributed to the hormone. These peptides may represent enzymes, structural proteins, secretory products, or receptors for other hormones. Because this type of hormone action depends on actual synthesis of an effector protein rather than phosphorylation of an existing enzyme system, the response of steroid hormones takes longer to occur initially and longer to stop when the stimulus for release of steroids ends. Thyroxine and triiodothyronine behave in a manner similar to the steroid hormones, except that they enter the nucleus directly without the necessity of binding to a receptor.[2,3]

Mechanisms of Control

The endocrine system is regulated primarily by means of control of hormone synthesis rather than by the rate of hormone degradation. Synthesis of a particular hormone may be stimulated by another tropic hormone or by a change in the serum level of a substance that the hormone controls. Tropic hormones have other endocrine glands as their targets. For example, insulin release is initiated by an increase in serum glucose levels. This is a simple positive control mechanism. Cessation of hormone release is accomplished either directly or indirectly by a system of negative feedback. An example of simple negative feedback mechanism would be the inhibition of PTH secretion by elevated serum calcium levels. Indirect mechanisms or complex negative feedback mechanisms involve cessation of production of tropic hormones in response to elevations of levels of the effector hormone. Lack of tropic hormone then stops production of the effector hormone. For example, increased levels of cortisol act on the hypothalamus and pituitary to decrease secretion of corticotropin releasing hormone (CRH) and adrenocorticotropic hormone (ACTH), respectively. With decreased levels of ACTH, cortisol production by the adrenal ceases and plasma cortisol levels drop. When plasma cortisol levels decrease below the desired range, low levels of this hormone stimulate the hypothalamus and pituitary to release CRH and ACTH, and adrenal production of cortisol begins again.

METHODOLOGIES

Classic Assays

Bioassays

Before the development of modern assay methods, the detection and quantitation of hormones were based on cumbersome, expensive assays that depended on a biologic response from a live animal or tissue/organ culture. Such assays required a large amount of technical time and experience. Interpretation was often based on a subjective evaluation of results that might be influenced by nonspecific factors having little or nothing to do with the desired hormonal effect.

An example of a bioassay previously used in the clinical evaluation of endocrine status is the detection of pregnancy based on corpus luteum formation in prepubertal female mice or ovulation in mature female rabbits. The assay involved injection of human urine samples into virgin animals and subsequent examination of the ovaries for corpus luteum formation. Bioassays are now used almost exclusively for research purposes and have been replaced in the hospital laboratory.

Competitive Protein Binding

Competitive protein-binding (CPB) assays are based on competition for a limited number of protein-binding sites between a known added amount of "tagged" or "labeled" hormone and the unlabeled hormone present in the patient's sample. These binding proteins may be specific antisera (immunoassay), specific receptor sites (radioreceptor assay), or serum-binding proteins (such as thyroid-binding globulin or cortisol-binding protein). Assays using serum-binding proteins, for the most part, have been replaced by immunoassay techniques. Radioreceptor assays have not been particularly useful in the clinical laboratory because the binding protein is an intact receptor in the membrane of a live cell and it is difficult to maintain the cells or receptor membrane preparations in a biologically active state.

Immunologic Assays

Radioimmunoassay

Radioimmunoassay (RIA) has been a popular method for clinical endocrine determinations. The development of an RIA for insulin by Yalow and Berson in 1960[4] was a major technologic breakthrough for hormone analysis and was rapidly followed by development of a number of endocrine assays based on these principles. Immunofluorescence or other immunoassay techniques, for the most part, have replaced this technique of hormone measurement. RIA is a CPB technique that uses radio-labeled hormone as the tagged hormone and antisera prepared against the specific hormone as a binding site. Competition between unlabeled hormone in the patient sample and the added-labeled hormone for a limited number of antibody-binding sites forms the basis of the assay.

RIA was a technologic breakthrough for endocrine testing and the technique was widely applied during the 1960s and 1970s. Although RIA is a good technique when high sensitivity is required, there are a number of drawbacks to its use. RIA is based on radioisotopes and its use in the laboratory must be licensed and regulated by radiation safety committees. In addition, radioactive materials must be disposed of or stored under special conditions and usually must be separated from normal waste. If labeled hormones are not used in a short period (about 1 month to 2 months), these substances must be discarded. The short shelf life of radioisotopes may result in waste and inefficiency in test-kit usage. Furthermore, standard curves involving complex mathematical manipulations must be run for every assay, with both patient and standards done in duplicate.

Immunoradiometric Assay

Immunoradiometric assay (IRMA) is similar to RIA in that a radiolabeled substance is used in an antibody–antigen reaction. However, the radioactive label is attached to the antibody instead of the hormone. In addition, an excess of antibody, rather than a limited quantity, is present in the assay. Because the entire unknown antigen becomes bound in IRMA rather than just a portion, as in RIA, IRMA assays are more sensitive. Both one-site and two-site IRMAs exist. In the one-site assay, the excess antibody that is not bound to the patient sample is removed by addition of a precipitating binder. In many cases, this binder is antigen (hormone) bound to some solid support. In the two-site assay ("sandwich" technique), a hormone with at least two antibody-binding sites is adsorbed onto a solid phase to which one of the antibodies is firmly attached (either the walls of the assay tube itself or beads that are added to the patient sample in assay buffer). After binding to this antibody is completed, a second antibody labeled with ^{125}I is added to the assay. This antibody reacts with the second antibody-binding site to form the "sandwich" comprising antibody-hormone-labeled antibody. In contrast to RIA and similar CPB assays, the amount of hormone present is directly proportional to the amount of radioactivity measured in the assay.

Enzyme-Linked Immunosorbent Assay

Enzyme-linked immunosorbent assay (ELISA) is similar to IRMA except that an enzyme tag is attached to the antibody instead of a radioactive label. This assay has the advantage of avoiding radioactive materials and produces an end product that can be measured in a spectrophotometer. The hormone is bound to the enzyme-labeled antibody and the excess antibody is removed as in IRMAs. Again, both one-site and two-site systems are possible. Once the excess antibody has been removed or the second antibody containing the enzyme has been added (two-site assay), the substrate and cofactors necessary for visualization of enzyme activity are added. The amount of hormone present is directly related to the amount of enzymatic activity (substrate formed) during a detection incubation. Prolonging the incubation time for producing substrate may increase the sensitivity of the assay. In some cases, substrate formed may give an optical color change so that detection of the hormone being measured can be determined by visual inspection. This type of assay has been used in pregnancy test-kits designed for home use.

Enzyme-Multiplied Immunoassay Technique

In enzyme-multiplied immunoassay technique (EMIT), enzyme tags are used instead of radiolabels but the antibody binding alters the enzyme characteristics, allowing measurement of hormone without separating the bound and free components (homogeneous assay). This technique has wide applicability in drug monitoring but has not been popular as a technique for endocrine testing because of problems with sensitivity. The enzyme is attached to the hormone or drug being tested. This enzyme-labeled antigen is incubated with the patient sample and with antibody to the hormone or drug. Binding of the antibody to the enzyme-linked hormone either physically blocks the active site of the enzyme or changes the protein conformation so that the enzyme is no longer active. After antibody binding occurs, the enzyme substrate and cofactor are added and enzyme activity is measured. If the patient sample contains the hormone of interest, it will compete with enzyme-linked hormones for antibody binding, enzyme will not be blocked by antibody, and more enzyme activity will be measurable. Thus, the greater the hormone concentration in the patient sample, the greater will be the enzyme activity.

Fluorescent Techniques

Fluorescent Immunoassay

Fluorescence as an assay technique has been used in endocrine testing for the determination of catecholamines and their metabolites.[2,5] Although capable of very high sensitivity, classical fluorescence techniques have not been very useful because many endogenous serum compounds, drugs, and dietary factors interfere by quenching or contributing to the fluorescence. Fluorescence assays for catecholamines are being replaced by high-performance liquid chromatography (HPLC)

techniques. Immunofluorescence techniques are, however, a relatively recent development. Ideally, fluorescence immunoassay represents a combination of high sensitivity (fluorescence) with high specificity (antigen–antibody reactions). In addition, microprocessors and computers have automated sample handling and data collection, permitting ease of operation and decreased technologist time. Because these assays measure relative change in fluorescence rather than absolute fluorescence, background and quenching problems can be minimized or corrected. Most applications of fluorescence immunoassay have been in the area of drug monitoring. Unlike EMIT procedures in which sensitivity appears to be limited to the microgram per milliliter range, fluorescence should permit measurements at the lower detection levels needed for hormone measurement.

Substrate-Labeled Fluorescent Immunoassay. Substrate-labeled fluorescent immunoassay was one of the first methods developed using fluorescence immunoassay. A potentially fluorescent compound is attached to a drug or hormone and is then added to the patient sample along with antibody to the hormone. If the patient sample contains the substance being tested, competition for antibody binding occurs between labeled and unlabeled hormone. After incubation, an enzyme is added that releases the fluorescent tag from the unbound labeled hormone and fluorescence is measured. When antibody binds to the tagged substance, enzymatic hydrolysis of the tagging compound is prevented and fluorescence cannot occur.

Fluorescence Polarization. Fluorescence polarization measures the change in the angle of polarized fluorescent light emitted by a fluorescent molecule. The hormone or drug, tagged with a fluorescent molecule, rotates rapidly in solution so that there is very little polarization of the fluorescence. When this tagged hormone binds to a large antibody, the rate of rotation of the tagged hormone–antibody complex becomes much slower relative to the unbound–tagged hormone and polarization increases. In the assay, patient sample is added to the tagged hormone and an initial fluorescent reading is made by the polarization fluorometer to compensate for sample quenching or background variation. Antibody to the hormone is then added. To the extent that patient material contains the substance of interest, competition occurs between unlabeled and labeled hormone for antibody-binding sites and the amount of binding of labeled hormone determines the amount of fluorescence polarization. The fluorometer can be standardized by using a series of calibrators containing different concentrations of the substance of interest. The fluorescence polarization is measured for these calibrators and patient results are calculated from this information.

Other Fluorescence Techniques. Other fluorescence techniques are being developed or are available. These include fluorescent excitation transfer immunoassay and direct quenching fluorescent immunoassay. The fluorescent excitation transfer immunoassay technique involves fluorescent tagging of both the antigen (hormone) and antibody with energy transfer from the antigen fluorescent compound to the antibody fluorescent compound when binding occurs. Direct quenching fluorescent immunoassay involves direct quenching of antigen-tagged fluorescence when unlabeled antibody binding occurs.

High-Performance Liquid Chromatography

HPLC was developed during the 1960s primarily as a research tool. It involves pumping a liquid phase at high pressure through a solid-phase column. The separation is based on differential partitioning of the compounds between the mobile liquid phase and the stationary solid phase. The solid-phase column comprises very small particles, resulting in much more efficient and faster separation than do other forms of chromatography. Partitioning is based primarily on the polarity of the solvent and the solubility in the liquid phase of the compounds to be separated. The technique offers versatility and control of the method development in that a great variety of column materials are available and the investigator can vary the solvent composition over a wide range of polarities from extremely hydrophobic to purely hydrophilic.

HPLC does not lend itself to high throughput and requires a relatively large amount of technologist time and training. The technique is currently used primarily for toxicology and drug assays and not much for endocrine testing. Catecholamine analysis is performed by HPLC. In the past, measurement of catecholamines was done by classical fluorescence methodology and was subject to a large number of background and quenching problems. The use of HPLC for separation of catecholamines and metabolites, combined with an electrochemical detection method, has resulted in more reliable measurement of these compounds.[2]

Colorimetry

Colorimetry refers to the measurement of light absorption at a specific wavelength or color. This light absorption may be a property of the substance being measured or may be possible only after the addition of reagents that form light-absorbing compounds with the hormone. As a technique for endocrine testing, colorimetry is not widely applicable because sensitivity is usually insufficient in the physiologic concentration range of hormones. However, the urinary concentrations of cortisol and other adrenal steroids and their metabolites, as well as catecholamine metabolites, are high enough to allow their colorimetric determination. Measurement of the steroid compounds by these methods is not specific for a single adrenal steroid or metabolite but measures the quantity of those steroids that possess a common functional group.

Porter-Silber Method

The Porter-Silber method is based on the yellow pigment produced by a reaction of 17,21-dihydroxy-20-ketone (dihydroxyacetone side chain) structures with phenylhydrazine in the presence of alcohol and sulfuric acid.[2] Corticosteroids with this configuration include cortisol, cortisone, 11-deoxycortisol, and the tetrahydro derivatives (metabolites). The procedure, in outline form, consists of hydrolysis of glucuronide conjugates, extraction with organic solvents, removal of interfering compounds with alkali, and color reaction with the previously mentioned reagents. Absorption occurs at 410 nm.

Zimmerman Reaction

The Zimmerman reaction procedure measures those steroids possessing a 17-keto structure or those that can be oxidized to 17-keto structures.[2] The oxidation procedure used to produce 17-ketosteroids, called the Norymberski procedure, uses sodium bismuthate as the oxidizer. Compounds that can be oxidized are called 17-ketogenic steroids. These include cortisol, cortisone, and their tetrahydro derivatives; 11-deoxycortisol and its tetrahydros; cortols; cortolones; and pregnanetriol and its 11-oxygenated derivative.

A later modification of the oxidation procedure involves reduction of C-20 ketones and C-21 methyl groups using sodium borohydride treatment before the oxidation step. This modification permits the measurement of the metabolites of the corticosteroid precursors.

The Zimmerman color-forming reaction uses m-dinitrobenzene to produce a reddish-purple color with maximal absorption at 520 nm. The procedure, like the Porter-Silber method, requires a preliminary hydrolysis, extraction, and washing before the color-forming reaction occurs.

Both the Porter-Silber and the Zimmerman reactions require a large amount of technologist time and experience in solvent extractions and processing. They are not easily adapted to high sample numbers and require a 24-hour urine sample for accurate interpretation of results. An additional drawback is that the color reaction occurs with a number of drugs and dietary by-products. These disadvantages have resulted in a shift to the use of immunoassay or chromatographic methods for measuring corticosteroids or their metabolites.

Pisano Method

Although sophisticated gas chromatography and HPLC techniques exist for measuring catecholamines and their metabolites, the classical procedure for quantitating metanephrines and *normetanephrines* is the Pisano procedure. Twenty-four–hour urine samples are subjected to acid hydrolysis and absorbed on an ion-exchange resin (amberlite CG-50). After elution with ammonium hydroxide, the compounds are converted to vanillin, reacted with periodate, and assayed spectrophotometrically at 360 nm.[6] A similar conversion to vanillin and spectrophotometric measurement are used in the analysis of another clinically important catecholamine metabolite, vanillylmandelic acid (VMA).[7]

COMPONENTS OF THE ENDOCRINE SYSTEM

Hypothalamus

The *hypothalamus* is the portion of the brain located in the walls and floor of the third ventricle. It is directly above the pituitary gland and is connected to the posterior pituitary by the pituitary stalk (infundibulum). Neurons located in the supraoptic and paraventricular nuclei of the hypothalamus send processes through the pituitary stalk, which terminate in the posterior pituitary (neurohypophysis). In addition to this direct neural connection with the posterior pituitary, other portions of the hypothalamus are in contact with the anterior pituitary (adenohypophysis) through the hypophyseal portal system. A portal system is a vascular arrangement in which venous blood flows directly from one capillary bed through a connecting vessel to another capillary bed without passing through the systemic circulation. The hypothalamus is the link between the nervous system and the endocrine system.

The supraoptic and paraventricular nuclei of the hypothalamus produce *antidiuretic hormone (ADH)* also known as *vasopressin* and oxytocin, which pass down the pituitary stalk and are stored in the posterior pituitary. The release of ADH is mediated by neural impulses from hypothalamic osmoreceptors as well as from pressure receptors in the right atrium and carotid sinus. *Oxytocin* is released in response to suckling and also may play a role in the initiation of labor.

Neurons in the anterior portion of the hypothalamus release a number of hormones that are carried by means of portal veins to the pituitary sinusoids where they either stimulate or inhibit the release of anterior pituitary hormones (Fig. 18-2). At least six such substances [thyrotropin releasing hormone (TRH); gonadotropin-releasing hormone (Gn-RH); somatostatin, also known as growth hormone inhibiting hormone (GH-IH);

growth hormone releasing hormone (GH-RH); CRH; and prolactin-inhibiting factor (PIF)] have been recognized and the chemical structures of most of these have been determined. Three of the hypophyseal hormones (TRH, Gn-RH, and somatostatin), whose structures are known, are peptides comprising 3 to 14 amino acids. CRH contains 41 amino acids and PIF has been shown to be dopamine.[2,8]

The releasing and inhibiting factors of the hypothalamus are not present in the systemic circulation in amounts that are measurable by current techniques. Therefore, it is impossible to assess hypothalamic function directly. Instead, hypothalamic function or malfunction is inferred from evaluation of the anterior or posterior pituitary and the response of the anterior pituitary to synthetic releasing factors. In some instances, it may be virtually impossible to distinguish a hypothalamic from a pituitary problem by laboratory tests.

Diseases involving the hypothalamus that may result in endocrine dysfunction include tumors, inflammatory or degenerative processes, and congenital problems. These will be manifested in excesses or, more commonly, deficiencies of anterior pituitary hormones. In addition, complex interrelationships between the hypothalamus and other portions of the brain may produce hormone irregularities in association with a variety of psychologic problems such as growth failure in emotionally deprived children and the complex endocrine abnormalities seen in patients with anorexia nervosa.[9]

Pituitary Gland

The pituitary gland is located in a small cavity in the sphenoid bone of the skull called the *sella turcica*. Fibrous tissue derived from the meninges covers the sella and is pierced by the pituitary stalk. The pituitary stalk contains nerve fibers and small blood vessels and connects the pituitary to the hypo-

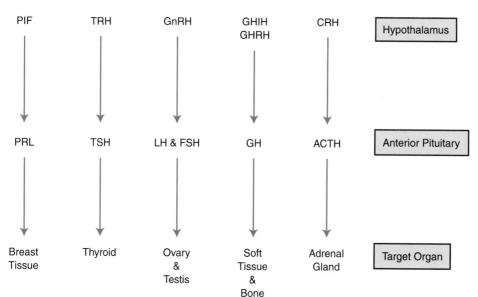

Figure 18-2. Regulation and function of the anterior pituitary.

thalamus. Anatomically and functionally, the pituitary is divided into two distinct lobes. The *posterior pituitary* comprises of nervous tissue and is also known as the *neurohypophysis*. The *anterior pituitary*, also known as the *adenohypophysis*, consists of glandular epithelial tissue.

Classically, the anterior pituitary has been described as comprising three types of cells. This classification was based on histologic staining characteristics and the cells were identified as acidophils, basophils, and chromophobes. The cells are now classified by immunochemical means, and the pituitary comprises at least five cell types. The somatotropes secrete GH, the lactotropes or mammotropes secrete PRL, the thyrotropes secrete TSH, the gonadotropes secrete LH and FSH, and the corticotropes secrete ACTH.[2,8,10] The hormones secreted are either peptides (ACTH, GH, PRL) or glycoproteins (TSH, LH, FSH).

ACTH is derived from cleavage of a large prohormone (proopiomelanocortin) into ACTH and beta-lipotropin. Beta-lipotropin, in turn, is cleaved to yield a variety of peptides, including alpha-melanocyte–stimulating hormone (MSH) and a group of peptides (endorphins) that function as endogenous opiates.[2,8,11]

Excess secretion of pituitary hormones is usually due to a pituitary tumor and almost always involves only one hormone. Isolated deficiencies of pituitary hormones occur but are rare, with GH being most common. Pituitary hormone deficiencies usually involve more than one and eventually all the anterior pituitary hormones (panhypopituitarism). These hormones are lost in a characteristic order, with GH and gonadotropins (LH and FSH) disappearing first, followed by TSH, ACTH, and PRL. Consequently, patients with this problem develop generalized endocrine dysfunction. Because these symptoms do not develop until approximately 75% of the gland has been destroyed and because the symptoms tend to develop slowly or incompletely, diagnosis is often delayed.

Panhypopituitarism may be due to a variety of processes but most commonly results from a pituitary tumor (adenoma or craniopharyngioma) or from ischemia. In the case of pituitary tumors, the expansive growth of the lesion compresses the normal gland, resulting in loss of function. In addition, surgical treatment or radiation may result in destruction of the remaining normal pituitary as well as the tumor and often leads to generalized pituitary deficiency.

The usual cause of pituitary ischemia is hemorrhage or shock in a pregnant female at the time of delivery (*Sheehan's syndrome*). The pituitary is enlarged at this time and is more susceptible than usual to ischemic change. The extent of damage and consequently, the resultant symptoms are variable. In many cases, the major symptom is loss of sexual function and secondary sex characteristics. These symptoms may be followed much later by symptoms of thyroid or adrenal insufficiency. However, in some cases, severe panhypopituitarism is present from the onset and the patients may develop Addisonian crisis.[2,8,10]

Anterior Pituitary Gland (Adenohypophysis)

Growth Hormone. GH is a peptide comprising 191 amino acids. In contrast to most of the other protein hormones, GH does not act through cAMP. Binding of GH to its membrane receptor leads to glucose uptake, amino acid transport,

---

CASE STUDY 18-1

A 27-year-old woman delivered her third child after 23 hours of labor. Delivery was complicated by severe hemorrhage requiring transfusion of 4 units of packed red cells. Because of the mother's critical condition, the infant was bottle-fed for the first few days of life. When the mother's condition improved, she wanted to breast-feed her infant as she had done with her other children. She was unable to lactate, however, and the infant continued on formula. Mother and infant were discharged 5 days after delivery.

When the infant was brought to the family practitioner for her 6-week checkup, the physician noticed that the mother appeared tired and pale. Her blood pressure was slightly low and she complained of loss of appetite, nausea, and vague abdominal pain. She was found to be mildly anemic and iron supplementation was prescribed.

Three months later, the woman returned to her physician complaining of persistent fatigue, weakness, and weight loss (20 pounds below her prepregnancy weight). She appeared thin and pale and was wearing a sweater despite the outside temperature of 80°F. Her menstrual periods had still not resumed and she had noticed thinning of her pubic hair. Physical examination revealed low blood pressure, decreased deep tendon reflexes, and thin dry skin. The patient stated that her family thought she had "postpartum depression."

Questions

1. What is the most likely diagnosis in this patient? Does it relate to her pregnancy and, if so, why?
2. Would you expect this patient to have increased skin pigmentation related to her underlying illness? Why or why not?
3. What type of additional testing should be performed on this patient and what results would you expect?

and lipolysis. Its effects are mediated through a group of second-messenger compounds called insulin-like growth factors (IGFs). These compounds are synthesized primarily by the liver under the influence of GH and have strong metabolic and mitogenic effects, especially on cartilage, adipose tissue, and striated muscle.[12]

Two hypothalamic hormones control release of GH: GH-RH and somatostatin. GH-RH appears to control the amount of GH released, whereas somatostatin governs the frequency and duration of secretory pulses.[12] GH secretion is relatively erratic and occurs primarily in short bursts. Most GH secretion occurs during sleep but secretion also increases transiently after exercise and in response to hypoglycemia. GH secretion and release are also affected by obesity, sex steroid levels, and renal and hepatic function.[13] A typical reference range for GH is given in Table 18-1. A single determination of GH provides little clinical information, however, because of the erratic nature of its release. Basal levels of GH in children are the same as for adults but children show more frequent secretory bursts. GH appears to act in conjunction with sex steroids to mediate the adolescent growth spurt. GH production decreases with age and responses to some but not all secretory stimuli are also diminished.[12,13] The overall metabolic effect of GH is to mobilize fat stores as a major energy source while conserving glucose for glucose-dependent tissues such as the brain.

Excess GH. Elevated levels of GH can be seen in patients with severe malnutrition, chronic liver or kidney disease, uncontrolled diabetes mellitus, and certain types of growth failure. However, symptomatic excess GH production is usually due to the presence of a functioning pituitary adenoma.

The consequences of excess GH secretion depend on the age of onset. If the tumor develops before the epiphyses fuse, the patient becomes a pituitary giant. Patients with this problem have grown as tall as 8 feet, 11 inches. These patients retain normal proportions but have massive generalized organomegaly. They die at a relatively early age, usually of heart failure.

In the more common situation, GH-secreting pituitary tumors develop in adults whose epiphyses are closed. Under these circumstances, continued bony proliferation results in an increase in width rather than length. These changes are particularly apparent in the small bones of the hands and feet and in the jaw and result in the clinical picture of *acromegaly*. In addition to the bony changes, the patients develop generalized organomegaly and coarsening of the facial features. Metabolic effects of GH are manifested by development of impaired glucose tolerance (due to the anti-insulin effects of GH), hypertension, and occasionally, galactorrhea. Galactorrhea may be due to lactogenic effects of GH itself or to an associated increase of PRL that occurs in 25% to 40% of patients with GH-producing pituitary adenomas.[14] As the

TABLE 18-1. Typical Reference Ranges for Selected Endocrine Tests*[2]

Hormone	Reference Range	Stimulated	Suppressed
ACTH	a.m., 8–79 pg/mL	80 ± 7 pg/mL	
	p.m., 7–30 pg/mL		
Aldosterone	Male, 6–22 ng/dL		
	Female, 5–30 ng/dL		
Catecholamines			
Urine	<100μm/m²/day		
Plasma			
Epinephrine	<140 pg/mL		
Norepinephrine	<1700 pg/mL		
Cortisol			
Plasma	a.m., 5–23 μg/dL	>20 μg/dL	<3 μg/dL
	p.m., 3–16 μg/dL		
Urinary free cortisol	10–100 μg/d		
Gastrin	<100 pg/mL		
Growth hormone	Male, <2 ng/mL	>7 ng/mL	<2 ng/mL
	Female, <10 ng/mL		
5-HIAA			
Urine	2–6 mg/day		
Metanephrines			
Urine	0.05–1.20 μg/mg creatinine		
Prolactin	Male, 0–20 ng/mL		
	Female, <23 ng/mL (follicular)		
	5–40 ng/mL (luteal)		
Parathyroid Hormone			
Midmolecule	0.29–0.85 ng/mL		

* *Results are for ambulatory adult patients. A different reference range may apply to children or elderly individuals or to pregnant women. These reference ranges are guidelines only and will differ among laboratories.*

pituitary tumor grows, these patients usually develop signs of hypopituitarism owing to compression of the normal pituitary by the expanding tumor.[2,8,14,15]

Laboratory confirmation of the diagnosis of GH excess can be made by demonstrating an elevated GH level [>7 ng/mL (>310 pmol/L)] that does not suppress after glucose administration. Measuring GH during a glucose tolerance test classically tests that an elevated GH level does not suppress after glucose administration. After administration of 100 g of glucose, GH should decrease to >2 ng/mL (220 pmol/L) by 120 minutes.[2] Serum levels of GH greater than 50 ng/mL (2210 pmol/L) may be considered diagnostic of acromegaly, whereas undetectable GH levels rule out this diagnosis.[1] Acromegaly may rarely result from ectopic secretion of GH or GH-RH by a carcinoma, usually arising in the lung or pancreas. Measurement of GH-RH, available in research and specialty laboratories, will be useful if the ectopic GH-RH production is suspected.[16–18]

Treatment of GH-secreting tumors is through surgical removal of the tumor or elimination of functioning tumor by radiation. Ectopic hormone production is also treated by removal of the neoplasm.

GH deficiency. GH deficiency may occur as an isolated problem or as part of the general picture of panhypopituitarism. Until recently, it was thought that GH deficiency in adults had no significant consequences. Recent evidence, however, indicates that some changes in body composition thought to be associated with "normal aging" may actually be reflections of GH deficiency. These changes include decreased lean body mass, increased body fat, decreased bone density, and loss of skeletal muscle mass and may be at least partially reversible with GH administration. The significance of these findings and the appropriate use of GH treatment in adults remain to be determined.[19–22]

In infants, GH deficiency may be manifested as hypoglycemia. Because there are other more common causes of hypoglycemia in infancy, however, diagnosis is rarely made at this time. More often in children, absence of GH results in pituitary dwarfism. Children with pituitary dwarfism retain normal proportions and show no abnormalities of intellectual development. These patients appear much younger than their age, not only because they are short, but also because their facial features and body proportions are immature. A similar clinical appearance occurs in Laron's dwarfism. In this situation, the patients lack IGFs, and GH levels are elevated, allowing for laboratory distinction from pituitary dwarfism.

Laboratory diagnosis of GH deficiency requires stimulation testing, because normal basal levels of GH may be essentially zero. Exercise may be a sufficient stimulus to detect normal GH production in some children.[2] The stimulation tests most often used involve infusion of arginine, L-DOPA, or insulin. GH levels should increase to >7 ng/mL in response to infusion of an appropriate dose of one of these agents. If no response is seen, repeat testing with another agent is indicated because up to 30% of normal children may show a decreased

response to a single agent. The insulin tolerance test results in elevations of plasma cortisol (by way of ACTH stimulation) as well as GH and allows for evaluation of more than one pituitary function. Consequently, in cases of suspected panhypopituitarism, it may be the test of choice. Patients must be watched very carefully during the insulin tolerance test to prevent complications of hypoglycemia.[2,12,23]

If panhypopituitarism is suspected, insulin infusion may be combined with administration of TRH and Gn-RH for a generalized test of pituitary function. In the normal patient, such a test would result in elevations of GH and cortisol in response to the insulin, TSH and PRL in response to TRH, and FSH in response to Gn-RH. If properly performed, this test may give a 1-day evaluation of anterior pituitary function. It is very expensive, however, both for the laboratory and the patient and results are frequently inconclusive.[2,8,15,24]

Isolated GH deficiency in a child is treated by administration of exogenous GH. Panhypopituitarism requires replacement of hormones stimulated by the missing pituitary tropic hormones (*ie,* thyroxine, cortisol, and estrogens or androgens).

Prolactin. Prolactin (PRL) is a protein hormone whose amino acid composition is similar to that of GH. In humans, PRL appears to function solely in the initiation and maintenance of lactation. In this regard, PRL acts in conjunction with estrogens, progesterone, corticosteroids, and insulin to promote full development of breast tissue. These additional hormones, among other functions, increase the number of PRL receptors in mammary tissue. It is thought that the high levels of estrogen present during pregnancy inhibit prolactin-induced lactation. When estrogen levels fall after delivery, PRL then combines with mammary receptors and initiates lactation.

Reference ranges for PRL are given in Table 18-1. Slightly lower levels are present in children. Prolactin secretion is controlled primarily by PIF, which is released from the hypothalamus. This factor has been shown to be dopamine. Other inhibitory factors have been identified, but their physiologic significance is unclear. Several prolactin-releasing factors have been identified including TRH, vasoactive intestinal peptide (VIP), and peptide histidine methionine. These compounds are found in the hypothalamus or anterior pituitary, but their physiologic role is uncertain.[25] Levels of PRL increase during sleep and decline during waking hours. There also appears to be a gradual increase in PRL secretion in the first half of the menstrual cycle with a decline immediately after ovulation. The level then increases again until menstruation occurs. Prolactin secretion increases during pregnancy, reaching levels as high as 200 ng/mL to 300 ng/mL (8510 pmol/L to 12,760 pmol/L) in the third trimester. In the postpartum period, stimulation of the breast produces a rapid rise in PRL concentration, resulting in lactation. Mechanical stimulation of the breast also may produce an increase in PRL concentration in some normal nonpregnant, nonlactating women.[2,8]

Hyperprolactinemia. Excessive release of PRL from the anterior pituitary may come about through a decrease in PIF, an increase in PRF (not yet demonstrated), or autonomous production of PRL by a pituitary tumor. Alterations of dopamine production by the hypothalamus can be brought about by sectioning of the pituitary stalk, irritation of the C-8 to T-5 segments of the spinal cord, or by pharmacologic manipulation. In particular, Aldomet (which interferes with dopamine synthesis) and the phenothiazines and reserpine (which interfere with dopamine release) may produce *hyperprolactinemia*. Hyperprolactinemia is also common in patients with renal failure and other endocrine diseases such as hypothyroidism and adrenal insufficiency. The pathogenesis of the hyperprolactinemia in these situations involves increased production as well as delayed metabolic breakdown in the case of renal failure.[2,8,10,25]

Pituitary tumors producing PRL are the most common type of pituitary adenoma. Many of these tumors (especially in women) are microadenomas.

In females, the presenting symptom of hyperprolactinemia is usually amenorrhea (cessation of menstrual periods). Galactorrhea (production of milk in any situation other than the postpartum period) is present in 50% to 90% of cases. Males usually have larger tumors at the time of presentation and may present with symptoms related to tumor size (headache, visual symptoms, and so forth). Impotence is also a common complaint. Galactorrhea in males may occur but is uncommon. The mechanism for amenorrhea and impotence in association with hyperprolactinemia is unclear. However, it may relate to suppression of gonadotropin release by PRL or to inhibited gonadotropin-releasing factor activity.[8,25]

The diagnosis of PRL-secreting pituitary adenoma requires the exclusion of other possible causes of hyperprolactinemia. In the presence of an enlarged sella, hyperprolactinemia is almost certainly due to a pituitary neoplasm. If radiologic evidence of a pituitary tumor is not present, the diagnosis is more difficult. Three findings suggest a pituitary tumor in these circumstances: (1) serum PRL levels >200 ng/mL (8510 pmol/L); (2) lack of response to TRH infusion (may also occur with drug-induced hyperprolactinemia); and (3) lack of response of serum PRL levels to pharmacologic manipulation of dopamine. The third category is not very helpful with methods currently available because both tumorous and nontumorous conditions may respond to dopamine manipulation.[8,15]

Treatment of hyperprolactinemia due to a pituitary adenoma is usually surgical. Pharmacologic treatment with a dopamine antagonist (bromocriptine) is also effective.

Prolactin deficiency. Prolactin deficiency is usually seen as part of the general picture of panhypopituitarism. It is particularly noticeable in acute hypopituitarism secondary to postpartum hemorrhage (Sheehan's syndrome) because lactation does not develop in these patients. Isolated PRL deficiency has been reported but is very rare.[8]

Posterior Pituitary (Neurohypophysis)

The hormones released by the posterior pituitary are actually synthesized in the supraoptic and paraventricular nuclei of the hypothalamus. These hormones are transported in membrane-bound vesicles through the neuronal axons in the pituitary stalk to the neurohypophysis, where they are stored. Release of these hormones occurs in response to stimulation of the hypothalamus by changes in serum osmolality or by suckling.[11,26]

ADH (Vasopressin). ADH is an octapeptide that is synthesized primarily in the supraoptic nucleus of the hypothalamus and is stored in the posterior pituitary until a stimulus is received for its release. It acts on the distal convoluted tubule and collecting tubule of the kidney, making them permeable to water. This action is mediated through cAMP and results in the production of concentrated urine. The hormone is metabolized primarily in the liver.[11,26,27]

Release of ADH is mediated primarily through changes in plasma osmolality, which are detected by osmoreceptors in the hypothalamus. Normally, osmolality is maintained within a very narrow range, and changes of as little as 1% to 2% will stimulate or inhibit ADH release. A decrease in blood volume or blood pressure will also stimulate ADH release. The effect of these changes is to alter the setpoints for osmoregulation to a lower level. This alteration increases blood volume at the expense of "normal" osmolality. This stimulus is mediated by baroreceptors in the thorax and requires changes in the extracellular volume of 5% to 10% before it is activated.[26,27]

In addition to tonicity and volume, ADH release may also be stimulated by nausea, pain, stress, various infectious, and vascular and neoplastic disorders. Such conditions may result in the syndrome of inappropriate ADH (SIADH). ADH release may be inhibited by cold and by numerous drugs, most commonly ethanol.[26]

Hypersecretion of ADH. Hypersecretion of ADH may be relative or absolute. Relative hypersecretion occurs in situations in which hypovolemia takes precedence over osmolality as a stimulus for ADH release. In this situation, ADH acts to retain water in the presence of low serum osmolality. This condition is associated with a low urinary excretion of sodium, which distinguishes it from SIADH.[26]

SIADH results when ADH is released despite low serum osmolality in association with a normal or increased blood volume. This situation develops in patients with tumors that produce ectopic ADH (most commonly oat-cell carcinoma of the lung) or in association with central nervous system trauma or infections; administration of certain drugs (opiates, barbiturates, clofibrate, and chlorpropamide); and some pulmonary diseases, especially tuberculosis and sarcoidosis.[25,26]

SIADH can be diagnosed by measuring serum ADH and determining if the level is too high for the serum osmolality. Diagnosis of this syndrome requires that other conditions that could result in hypo-osmolality or generalized edema (hypovolemia, hypotension, congestive heart failure, renal insuffi-

CASE STUDY 18-2

A 64-year-old man presents to his physician for a "checkup." He had been in good health and felt well until the past few weeks, when he complained of being tired and a little short of breath. He has no prior history of heart or lung problems, although he had smoked 1 pack of cigarettes per day for 40 years. He had also noticed a 15-pound weight loss in the past 3 months and had developed a nagging cough. Physical examination revealed a thin, pale man who appeared somewhat short of breath. Pulse rate was slightly increased and blood pressure was mildly elevated. Laboratory data showed hyponatremia, 125 mmol/L (135–145 mmol/L) and mild anemia.

Questions

1. What is the differential diagnosis in this patient?
2. What other diagnostic tests are needed and what do you expect the results to be?
3. What other hormones are often produced under similar conditions and how are these syndromes distinguished from primary abnormalities of endocrine glands?

ciency, and so forth) be absent.[27] ADH assays are not readily available, however, and SIADH is usually diagnosed by the presence of three factors: (1) Hypo-osmolality (hyponatremia), (2) urine osmolality greater than serum osmolality, and (3) urine Na^+ greater than 20 mEq/L to 25 mEq/L. The third factor serves to distinguish SIADH from relative excess of ADH due to volume depletion. In SIADH, expansion of extracellular volume stimulates sodium excretion even in the face of hyponatremia.[2,15,26,27]

The most appropriate therapy of SIADH is removal of the stimulus. If this is impossible, treatment by water restriction or drugs that block the renal response to ADH may be effective.

Hyposecretion of ADH. Deficiency of ADH, termed diabetes insipidus (DI), results in severe polyuria (10 L/day to 12 L/day). As long as the thirst mechanism is intact, the patient can usually compensate adequately by drinking large quantities of water. If anything occurs to disrupt the thirst mechanism or prevent the patient from drinking water, rapid development of hypernatremia will occur.

DI results from destruction of the posterior pituitary or hypothalamus secondary to neurosurgical procedures, tumor, trauma, or a degenerative or infiltrative process. These patients present with polyuria and polydipsia. To establish the diagnosis, the patient's ability to produce concentrated urine must be tested. This is accomplished by water deprivation accompanied by hourly measurements of urine osmolality. When a constant urine osmolality is obtained or the patient loses 3% to 5% of body weight, serum osmolality is measured and a subcutaneous injection of ADH is administered. Urine osmolality is again measured hourly for 2 hours to 3 hours. Normal individuals, or those with psychogenic polydipsia (excessive drinking of water because of a psychiatric disturbance), will show a urine:plasma osmolality ratio greater than 1 with water deprivation and urine osmolality will increase greater than 5% after ADH administration. Patients with DI

will show a urine:plasma osmolality less than 1 even after loss of 3% to 5% of body weight and will show a much greater response to ADH. A third category of patients who have an end-organ unresponsiveness to ADH (nephrogenic DI) will show urine and plasma osmolalities similar to DI but will not respond to ADH.[2,28,29]

Treatment of DI is replacement of ADH by injection of arginine vasopressin or administration of the vasopressin analog 1-desamino 8-D-arginine vasopressin (DDAVP, desmopressin) by nasal spray or injection. If only a partial deficiency is present, the patient may respond to therapy with drugs such as chlorpropamide or clofibrate, which increase ADH release. Nephrogenic DI is treated with diuretics or reduction of the osmotic load. A reduction of the osmotic load lowers extracellular volume or reduces the glomerular filtration rate, allowing increased reabsorption of water from the proximal tubules.[15,26,28,30]

Oxytocin

Oxytocin is a nonapeptide produced in the paraventricular nucleus of the hypothalamus. It is very similar in composition to ADH, and like ADH, it is secreted in association with a carrier protein and stored in the posterior pituitary. Oxytocin stimulates contraction of the gravid uterus at term and also results in contraction of myoepithelial cells in the breast, causing ejection of milk. It is released in response to neural stimulation of receptors in the birth canal and uterus and of touch receptors in the breast. In lactating females, emotional stimuli also may affect this pathway, resulting in stimulation or inhibition of oxytocin release and lactation. There is some evidence for a negative feedback effect of oxytocin on the posterior pituitary to inhibit further secretion.

Oxytocin exerts its physiologic effect by binding to receptors on myometrial or myoepithelial cell membranes and initiating formation of cAMP or cyclic guanosine monophosphate (cGMP). The hormone appears to be metabolized by the liver and kidneys or the lactating mammary gland. RIAs have been developed to measure oxytocin but are not in routine use.[2]

No disorders associated with excess oxytocin production have been described. Deficiency states may exist but have not been identified conclusively. Synthetic preparations of oxytocin are used to increase weak uterine contractions during labor and to aid lactation in women who have difficulties in the ejection (not production) of milk. These situations may represent deficiencies in oxytocin production.[2,15]

Thyroid

The thyroid gland and its two principal hormones, thyroxine (T$_4$) and triiodothyronine (T$_3$), are discussed in Chapter 19 *Thyroid Function*.

Calcitonin

A third thyroid hormone, *calcitonin,* is produced by the parafollicular or C cells of the thyroid. Plasma free calcium level is the primary regulator of calcitonin secretion. Increased levels of free calcium stimulate calcitonin release and decreased levels suppress its release. Other factors have been found to stimulate calcitonin release. These factors include gastrin, cholecystokinin, glucagon, estrogens, and beta-adrenergic agents. The physiologic role of calcitonin may be important in fetuses or young children but its role in adults is uncertain.[2]

Calcitonin measurement plays an important role in detection and diagnosis of thyroid cancers arising from the parafollicular (C) cells (medullary carcinoma of the thyroid). Patients with these tumors have elevated calcitonin levels and show an abnormally large release of calcitonin in response to infusion of substances known to stimulate its secretion (calcium, pentagastrin). Because many tumors of this type tend to be familial or associated with other endocrine neoplasms, stimulation of calcitonin release by pentagastrin or calcium in patients at high risk may allow for early diagnosis of medullary thyroid cancer.[31,32]

Parathyroid Glands

The parathyroid glands are usually located on or near the thyroid capsule. Most people have four parathyroid glands but some have as many as eight or as few as two. The glands may be found in other locations in the neck and also in the mediastinum.

PTH is synthesized as a pre-prohormone containing 115 amino acids. The active form of the hormone contains 84 amino acids and is secreted by the chief cells of the parathyroid glands. The intact molecule is quickly cleaved into N-terminal and C-terminal fragments. The N-terminal portion is the active portion of the molecule and has a short serum half-life. The inactive C-terminal fragment is more stable. The primary regulators of PTH secretion are free calcium levels in plasma and extracellular fluids. This is a simple negative feedback regulation. As calcium levels increase,

PTH secretion is suppressed or if calcium levels decrease, PTH is released. However, magnesium, vitamin D, and phosphate levels have been found to affect PTH secretion.[2]

PTH acts to increase serum calcium concentration by increasing resorption of calcium from bone, stimulating calcium retention by the renal tubules, and promoting 1-hydroxylation of 25-hydroxy vitamin D in the kidney. In turn, 1,25-(OH)$_2$ vitamin D increases calcium absorption from the intestine. PTH also increases renal excretion of phosphate and, to a lesser degree, that of bicarbonate. These actions are mediated by cAMP.

PTH is measured by immunoassay. Serum PTH levels were formerly determined by measurement of the N- or C-terminal fragment, but now a two-site immunoassay for intact PTH is rapidly replacing fragment-based assays and has provided a means for direct measurement of parathyroid function independent of peripheral metabolism and renal clearance of hormone fragments.[31,32]

A typical reference range is given in Table 18-1.

Hyperparathyroidism

Hyperparathyroidism may be due to three conditions: (1) autonomous release of PTH from a benign or malignant parathyroid tumor or from diffuse hyperplasia of all four glands (primary hyperparathyroidism); (2) release of increased amounts of PTH from hyperplastic glands in an attempt to maintain normal serum calcium in the face of vitamin D deficiency or renal failure (secondary hyperparathyroidism); or (3) autonomous release of PTH, developing in the setting of secondary hyperparathyroidism (tertiary hyperparathyroidism). All conditions are associated with elevated serum PTH levels.[31]

Primary hyperparathyroidism is most commonly (80%) due to the presence of a functioning parathyroid adenoma. Hyperplasia accounts for virtually all the remaining cases of primary disease with parathyroid carcinoma contributing fewer than 1% of cases. With the use of chemistry profiles as a part of a general physical examination or admission to a hospital, most cases of primary hyperparathyroidism are discovered incidentally during a routine examination or evaluation for another disease. Before chemical screening, patients commonly presented with complications due to hypercalcemia or PTH excess such as renal stones or other renal disease, peptic ulcers, pancreatitis, pathologic fractures, and so on.

The major complications of hyperparathyroidism occur in kidney and bone. High levels of serum calcium result in excess calcium excretion in the urine. Phosphaturia is also present due to the action of PTH on renal tubular cells. Consequently, calcium-containing renal stones form and renal colic is a presenting symptom in approximately 25% of patients with hyperparathyroidism. Precipitation of calcium phosphate may also occur within the renal parenchyma, resulting in decreased renal function and, occasionally, hypertension. PTH also results in activation of osteoclasts and resorption of calcium from bony trabeculae. If hyperparathyroidism goes undetected for longer periods, severe demineralization of bone may occur with replacement by fibrous tissue and formation of "cysts" (osteitis fibrosa cystica). The bone is very weak in these areas and may break with minimal stress (pathologic fracture). Other complications of hyperparathyroidism are related to increased calcium levels and include peptic ulcer disease (stimulation of hydrochloric acid and gastrin by calcium) and central nervous system symptoms.[31]

Laboratory diagnosis of primary hyperparathyroidism is based on the demonstration of increased serum calcium, decreased serum phosphorus, and increased PTH. If significant bone disease is present, alkaline phosphatase activity will be elevated because of increased osteoblastic activity. Chemical examination of urine shows increased levels of calcium and phosphorus and increased urinary hydroxyproline, reflecting increased bone turnover.[31]

Secondary hyperparathyroidism develops in response to decreased serum calcium and is reflected morphologically as diffuse hyperplasia of all four glands. The most common causes of secondary hyperparathyroidism are vitamin D deficiency and chronic renal failure. Vitamin D deficiency causes decreased intestinal absorption of calcium and a compensatory increase in PTH is required to maintain normal calcium levels. Development of increased PTH secretion in association with chronic renal failure is more complex and relates primarily to three factors: (1) phosphate retention by the failing kidney; (2) decreased functioning renal mass, resulting in decreased levels of 1,25-(OH)$_2$ vitamin D; and (3) relative resistance of bone to the effects of PTH. These factors result in a tendency toward hypocalcemia, and excess PTH is required to maintain a normal calcium concentration.[31]

Patients with secondary hyperparathyroidism frequently develop severe bone disease identical to that seen in primary hyperparathyroidism. Laboratory findings in these patients include low or normal serum calcium levels, elevated serum phosphorus levels, increased alkaline phosphatase activity (if bone disease is present), increased urinary calcium levels, decreased urinary phosphorus levels, and increased urinary hydroxyproline levels.[31]

Tertiary hyperparathyroidism occurs when patients with secondary hyperparathyroidism develop autonomous function of the hyperplastic parathyroid glands or of a parathyroid adenoma. Laboratory findings are similar to those in primary hyperparathyroidism, except that phosphate levels are normal to high. Because of elevated serum phosphorus levels, the solubility product of calcium and phosphorus is exceeded and calcium phosphate precipitates in soft tissues.[31]

Hormonal hypercalcemia of malignancy is caused by excessive amounts of PTH-related protein.[34] This protein is elaborated by solid tumors such as squamous cell carcinomas of the lung. The hypercalcemia that accompanies these types of malignancies was formerly known as ectopic hyperparathyroidism or pseudohyperparathyroidism.

Severe hypercalcemia with low or undetectable levels of PTH may also indicate malignancy. The development of osteolytic metastases or the secretions of humoral bone mobilizing factors produce the oncogenic hypercalcemia. Patients with tumor-associated malignancy generally have higher serum calcium levels and lower serum 1,25-(OH)$_2$ levels than patients with primary hyperparathyroidism. However, laboratory findings do not always provide a clear separation of these conditions.

Treatment of hyperparathyroidism begins with finding the source of PTH. If the problem is primary hyperparathyroidism, surgery may be the only curative therapy. If the problem is secondary hyperparathyroidism, the treatment is to control the underlying disease.

Hypoparathyroidism

Decreased parathyroid function is most commonly due to accidental injury to the parathyroid glands during thyroid or other neck surgery or to idiopathic atrophy. Patients with *hypoparathyroidism* are unable to maintain their serum calcium concentration without calcium supplementation and develop hypocalcemia. If serum calcium levels drop below 8 mg/dL (2.0 mmol/dL), patients develop tetany and other manifestations of altered neuromuscular activity. Latent tetany may be demonstrated in patients with borderline low calcium by tapping on the facial nerve producing jerking movements of the jaw (Chvostek's sign) or by inflating a blood pressure cuff on

CASE STUDY 18-3

A 56-year-old male has had a long-standing history of renal failure. On this admission, the patient suffered a broken wrist. The laboratory findings are as follows:

Serum calcium	7.9 mg/dL	(8.7–10.2 mg/dL)
Serum phosphorus	6.0 mg/dL	(2.7–4.5 mg/dL)
Serum creatinine	4.0 mg/dL	(0.6–1.2 mg/dL)
Urea nitrogen	48 mg/dL	(10–20 mg/dL)
Alkaline phosphatase	130 U/L	(30–90 U/L)
PTH	2.0 ng/mL	(0.29–0.85 ng/mL)

Questions

1. What is the condition called if the calcium level is decreased and the PTH level is increased?
2. What problem in the interpretation of the serum calcium might occur in this type of patient?
3. What has caused the increased secretion of PTH?
4. Explain the increased alkaline phosphatase level.
5. If the skeleton becomes unstable, what is another name for this condition?

the arm to produce ischemia and tetany in the arm (Trousseau's sign). Calcium levels below 6 mg/dL (1.5 mmol/dL) may result in laryngeal stridor and tonic–clonic seizures.[31]

Laboratory evaluation of patients with hypoparathyroidism shows hypocalcemia, hyperphosphatemia, and decreased PTH levels. Similar clinical signs and laboratory findings occur in patients with pseudohypoparathyroidism (a hereditary disorder characterized by end-organ unresponsiveness to PTH). These patients have characteristic round faces and deformities of the bones of the hand. The defect responsible for this disorder is a reduction in a regulatory protein in the adenyl cyclase enzyme complex.[31]

Adrenal Glands

The *adrenal glands* are paired organs, with one located at the upper pole of each kidney. Each gland consists of an outer cortex and an inner medulla that have different embryologic origins, different mechanisms of control, and different products.

Adrenal Cortex

The adrenal cortex comprises three layers. The outermost layer *(zona glomerulosa)* is responsible for secretion of mineralocorticoids. The *zona fasciculata* and *zona reticularis* secrete glucocorticoids and sex hormones.

All the hormones produced by the adrenal cortex are steroids derived from cholesterol. Although these hormones are under different control mechanisms, that they share a common precursor and some common enzymatic pathways results in a complex interrelationship among them. Abnormalities in production of one hormone may be accompanied by increased or decreased amounts of other adrenocortical hormones, depending on the specific lesion involved. Thus, the clinical presentation of these patients results from a combination of hormone abnormalities. The individual hormones, their mechanisms of control, and abnormalities of production are discussed as follows.

Mineralocorticoids. *Aldosterone* is the principal *mineralocorticoid* (electrolyte-regulating hormone) produced by the zona glomerulosa. The steps in the synthesis of aldosterone and the enzymes required are outlined in Figure 18-1.

Normally, aldosterone production is primarily controlled by the renin–angiotensin system (Fig. 18-3). An increased serum potassium, a decreased serum sodium, increased ACTH levels, a fall in blood volume, and a rise in estrogen levels also act as stimuli to aldosterone secretion.[35]

Renin is a protein produced by the juxtaglomerular apparatus of the kidney in response to decreased renal perfusion pressure or decreased serum sodium levels. Renin acts on *angiotensinogen,* a plasma protein produced by the liver, to produce angiotensin I. In turn, angiotensin I is converted to angiotensin II by angiotensin-converting enzyme present in the vascular endothelium. Angiotensin II is converted to angiotensin III by an aminopeptidase. Both angiotensin II and III can bind to the receptor on the target cell, which eventually stimulates the secretion of aldosterone. Aldosterone acts on renal tubular epithelium to increase retention of Na^+ (and, secondarily, chloride and water) and excretion of K^+ and H^+. Thus, the effect of the renin–angiotensin system is to increase blood pressure, both by vasoconstriction and increased plasma volume.[36,37]

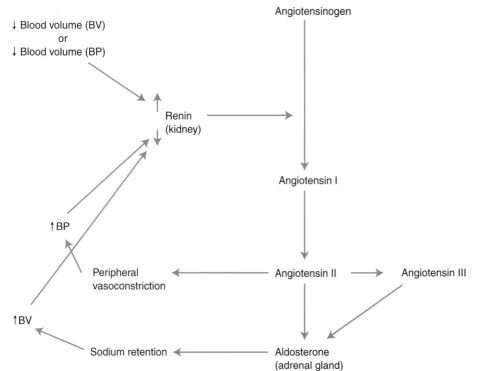

Figure 18-3. The renin–angiotensin system and its relationship to blood volume and blood pressure.

Aldosterone production also varies with dietary sodium intake. With a "normal" intake of 100 mEq to 150 mEq of Na$^+$ per day, 40 mg to 160 mg (111 nmol to 444 nmol) of aldosterone is secreted daily. Typical reference ranges for aldosterone concentration are found in Table 18-1. Serum concentration is higher in the morning than in the afternoon. Only approximately 30% are bound to plasma protein. The liver metabolizes the hormone and metabolites are excreted in the urine.[36,37]

Hyperaldosteronism. *Hyperaldosteronism* may be primary (due to an adrenal lesion) or secondary to abnormalities in the renin-angiotensin system.

Primary hyperaldosteronism is a relatively uncommon condition and is most often due to the presence of an aldosterone-secreting adrenal adenoma *(Conn's syndrome).* Cases of hyperaldosteronism due to bilateral hyperplasia of the adrenal cortex may occur and rare cases due to primary adrenal hyperplasia or aldosterone-secreting adrenal carcinomas have been reported.[38] Clinical symptoms result from hypertension or hypokalemia.

Hypertension results from volume expansion and sodium retention produced by aldosterone. Although fewer than 1% of hypertensive patients are found to have Conn's syndrome, this condition is surgically curable and should always be considered in the differential diagnosis of hypertension. From the laboratory standpoint, the first step in evaluation of the hypertensive patient is determination of serum electrolytes. Patients with primary hyperaldosteronism usually show hypokalemia associated with mild hypernatremia and metabolic alkalosis. If renal potassium wasting (>30 mEq/day) is demonstrated in a hypertensive patient who has hypokalemia and who is not on diuretic therapy, serum aldosterone levels should be determined. More than 50% of such patients are found to have primary hyperaldosteronism.[36,38]

Samples for aldosterone determination should be drawn in the morning before the patient has gotten out of bed and must be interpreted with consideration of sodium intake and posture. If serum aldosterone is above the reference range, the patient's response to manipulation of dietary sodium should be evaluated. Normally, sodium restriction results in increased plasma renin activity and aldosterone production. In the patient with an aldosterone-producing tumor, however, plasma renin activity will remain suppressed under these circumstances, and serum aldosterone levels will show no response. Sodium loading, which normally produces a decrease in plasma aldosterone levels, has no effect in patients with an aldosterone-producing tumor.[36,37]

If laboratory evidence suggests an aldosterone-secreting adrenal adenoma, the tumor can usually be localized by radiographic procedures and surgical therapy is curative. If primary hyperaldosteronism is due to adrenal hyperplasia, medical therapy with angiotensin-converting enzyme inhibitors or diuretics that block aldosterone receptors in the kidney is indicated.[38]

Secondary hyperaldosteronism results from excess production of renin. This condition may be due to renal artery stenosis (narrowing), which results in decreased renal perfusion, to malignant hypertension, or to the presence of a renin-secreting renal tumor (rare). In this situation, elevated aldosterone levels are associated with increased plasma renin activity that does not suppress with sodium loading. A vascular lesion in the main renal artery may be demonstrated by angiographic x-ray examination and is often surgically correctable. Increased plasma renin activity in patients with malignant hypertension is secondary to hypertensive damage to small vessels deep in the kidney and is not surgically correctable.[36]

Hypoaldosteronism. Aldosterone deficiency is most often due to destruction of the adrenal glands and, as such, is associated with glucocorticoid deficiency. Patients with congenital deficiency of the enzyme 21-hydroxylase may also have deficient aldosterone production that results in an inability to conserve sodium. Hyporeninemic *hypoaldosteronism* typically occurs in patients who are older than age 45 years and have chronic renal disease. The juxtaglomerular cells of the kidney are defective and the aldosterone appears to be decreased because of decreased renin release. Because of the aldosterone deficiency, hyperkalemia and metabolic acidosis may develop.

Glucocorticoids. *Cortisol,* the principal *glucocorticoid* hormone, is synthesized in the zona fasciculata and zona reticularis of the adrenal cortex as depicted in Figure 18-1. The actions of cortisol are widespread and include anti-insulin effects on carbohydrate, fat, and protein metabolism, effects on water and electrolyte balance, stabilization of lysosomal membranes, and suppression of inflammatory and allergic reactions.[2,39]

Cortisol is regulated through a feedback loop to pituitary ACTH and hypothalamic CRH secretion. ACTH is a 39-amino acid polypeptide that is synthesized as part of the prohormone, pro-opiomelanocortin. Release of ACTH from the anterior pituitary is stimulated by CRH from the hypothalamus and results in increased synthesis of cortisol by the adrenal cortex. Elevated cortisol levels then act on the pituitary and hypothalamus to shut off ACTH release, allowing serum cortisol to fall. When cortisol drops below a certain level, this inhibitory effect diminishes and ACTH is again released (negative feedback). A variety of factors other than serum cortisol levels, including pain, fever, hypoglycemia, and fear, may also stimulate release of ACTH.[39]

ACTH and cortisol normally show a diurnal variation with the highest levels present in the morning around 8 a.m. and lowest levels present in the late afternoon. Typical reference ranges for serum cortisol and ACTH are found in Table 18-1.

In the plasma, approximately 90% of serum cortisol is bound to a carrier protein (cortisol-binding globulin). Because the carrier sites on this protein are essentially saturated at normal levels of plasma cortisol, increases in free cortisol occur rapidly with increased cortisol production. Because free cortisol is readily filtered and excreted by the kidney, increased urine free cortisol levels are sensitive indicators of adrenal hyperfunction. Determination of the level of cortisol metabolites excreted in the urine may also be helpful in evaluating adrenal function.[2,39]

Hypercortisolism. The syndrome of *hypercortisolism* (Cushing's syndrome) may result from: excessive production of cortisol by an adrenal adenoma or carcinoma; excessive production of ACTH by a pituitary adenoma, leading to adrenal hyperplasia (Cushing's disease); production of "ectopic" ACTH or CRH by a tumor of nonendocrine tissue, most commonly a lung cancer; and exogenous administration of cortisol or ACTH. Because of its widespread use in the treatment of a variety of systemic inflammatory diseases, exogenous administration of cortisol is the most common cause of the group of symptoms known as *Cushing's syndrome*. In this situation, production of cortisol by the patient's adrenals is suppressed, and when exogenous cortisol is stopped abruptly, adrenal insufficiency may develop. Consequently, the usual approach to discontinuation of cortisol therapy is to decrease the dose gradually (taper the dose) over several days to allow the patient's adrenals to begin to produce cortisol again.[2,39]

The symptoms of hypercortisolism are essentially the same regardless of the etiology and include the following:[2,39] truncal obesity—weight gain predominantly in the face, neck, shoulders, and abdomen with relatively thin extremities (many patients develop a characteristic "buffalo hump," a fat pad on the upper back at the base of the neck); abnormal glucose metabolism (hyperglycemia); protein wasting, with thinning of the skin and development of striae, easy bruising, muscle wasting, and poor wound healing; hypertension; and decreased ability to limit infection with loss of delayed hypersensitivity response, decreased lymphoid tissue, and decreased circulating lymphocytes.

Laboratory evaluation of the patient with apparent Cushing's syndrome is crucial in establishing the diagnosis of hypercortisolism and in distinguishing among the adrenal, pituitary, and ectopic causes of the syndrome. The first step must be to establish a diagnosis of hypercortisolism. The best screening test is the urinary free cortisol in a 24-hour urine sample. This test should be performed on three different collection samples.[40] If the urinary free cortisol is elevated, an overnight dexamethasone suppression test is recommended. Suppression occurs in normal patients and in some with stress and depression hypercortisolism. If the cortisol level is not suppressed, a low-dose dexamethasone suppression (LDDS) test may be indicated. If, following the LDDS test, the cortisol level is not suppressed, the next step is to identify the etiology of the hypercortisolism.

Once a diagnosis of endogenous hypercortisolism is made, a variety of tests may be used in conjunction with other clinical and radiographic data to help establish where the abnormality lies. These tests include the following two:

1. Measurement of plasma ACTH. Because cortisol and ACTH interact in a feedback loop, increased plasma cortisol levels will be associated with suppression of ACTH if the syndrome is due to a primary adrenal lesion. In contrast, increased cortisol due to an ACTH-producing pituitary adenoma (Cushing's disease) or to ectopic production of ACTH would be associated with increased plasma ACTH levels. Measurement of plasma ACTH is not readily available in most hospital laboratories and must be sent to a reference lab. Consequently, turnaround time is relatively slow.
2. High-dose dexamethasone suppression (HDDS) test (administration of 2 mg dexamethasone every 6 hours for 2 days). Cushing's disease usually suppresses the urinary 17-OH steroid level to less than 35% of the baseline level and the plasma cortisol is less than 280 nmol/L. If the urinary 17-OH steroid and the plasma cortisol levels are not suppressed, adrenal tumors producing high levels of cortisol or ectopic ACTH-producing tumors are usually the etiology.[35]

Treatment of hypercortisolism depends on removal of the adrenal or pituitary tumor or of the tumor producing ectopic

---

CASE STUDY 18-4

A 20-year-old female noticed a weight gain of 40 pounds over the past 4 months. In addition, she noticed that her face was round and full and she felt extremely weak. She sought medical attention. On physical examination, it was noted that her blood pressure was 150/100 and numerous ecchymoses were present on her legs. The physician noted a "Cushinoid" appearance.

Questions

1. What are four possible etiologies of Cushing's syndrome?

2. What are the classical physical findings in Cushing's syndrome?
3. What laboratory protocol should be followed to establish the diagnosis of hypercortisolism?
4. Once the diagnosis of hypercortisolism has been made, what laboratory protocol should be followed to identify the etiology?
5. What would you expect a glucose tolerance test to show on a patient with Cushing's syndrome?

ACTH. If surgical removal cannot be accomplished, radiation therapy to the tumor or chemotherapy with an agent that inhibits cortisol formation may be helpful.

Hypocortisolism. Decreased production of serum cortisol may be associated with primary adrenal disease or may be secondary to pituitary abnormalities that result in decreased ACTH. The chronic form of *hypocortisolism* due to primary adrenal disease *(Addison's disease)* is most often due to idiopathic atrophy of the gland. This atrophy is thought to represent an autoimmune process, and Addison's disease may be associated with other autoimmune phenomena.[41] Other causes of primary adrenal hypofunction include destruction of the gland by tuberculosis or histoplasmosis, replacement of both glands by metastatic tumor, and disseminated intravascular coagulation with adrenal hemorrhage. When tuberculosis was less well controlled, tuberculous destruction of the adrenals was very common; however, this condition is now relatively rare. Similarly, although invasion of the adrenals by metastatic tumor is common, this process rarely results in sufficient destruction of both adrenals to produce hypocortisolism. When hypocortisolism is due to Addison's disease, it is usually accompanied by a deficiency in mineralocorticoids, because all three layers of the adrenal cortex are damaged.[39,42]

Secondary hypocortisolism is due to pituitary insufficiency with loss of ACTH. In this situation, mineralocorticoid deficiency is not a problem, because ACTH plays only a minor role in stimulation of the zona glomerulosa.

Patients with chronically decreased cortisol levels may present complaining of weight loss, weakness, and a variety of gastrointestinal (GI) complaints. Any stress may precipitate an acute syndrome of vascular collapse and shock. Patients with primary adrenal destruction and decreased mineralocorticoid levels have a more severe clinical presentation than those with secondary disease because of the fluid and electrolyte abnormalities associated with aldosterone deficiency.[39,42]

Laboratory findings in the patient with adrenal insufficiency include low serum sodium, bicarbonate, and glucose levels and high serum potassium and blood urea nitrogen (BUN) levels. Low plasma cortisol levels in a setting of vascular collapse and a high ACTH level is diagnostic of primary adrenal insufficiency because cortisol would normally be elevated in response to the stress associated with such circumstances.

A short ACTH stimulation test (250 mg of Synacthen [tetracosactrin] administered intramuscularly or intravenously) is performed to assess chronic adrenal insufficiency. A normal person will respond with a rise in plasma cortisol. A person with adrenal failure will not respond. The patient with adrenal failure will also demonstrate an elevated ACTH level.

A long ACTH stimulation test (1 mg of Synacthen administered intramuscularly and cortisol levels assessed at 0, 1, 2, 8, 12, and 24 hours) is required for the diagnosis of secondary adrenal insufficiency. Persons with this condition may have normal, low, or undetectable levels of ACTH.

Differentiation between hypothalamic and pituitary etiologies of secondary adrenal insufficiency requires a CRH stimulation test. If there is no ACTH response after a CRH stimulation test, the disease is pituitary in origin.

Acute adrenal insufficiency is a life-threatening medical emergency and the patient must be treated promptly with cortisol replacement and fluids.[39,42] Secondary adrenal insufficiency also requires glucocorticoid replacement therapy.

Sex Hormones. Weak *androgens* are produced by the zona fasciculata and zona reticularis. The weak androgens may serve as precursors for the production of more potent androgens and estrogens in other tissues. These sex hormones are produced in varying amounts at different stages of life even though ACTH has been described as the regulator of adrenal androgen synthesis. Another pituitary hormone called adrenal androgen-stimulating hormone has been proposed as an additional stimulator because ACTH levels do not vary in different age groups.[40] The sex hormones circulate bound to plasma proteins. Steroid hormone binding globulin (SHBG) is the main binding protein. Only a small percentage of the sex hormones are in the free state, but they are responsible for the biologic activity. The liver mainly metabolizes the sex hormones.

Congenital adrenal hyperplasia (CAH) results from the lack of an enzyme necessary for the production of cortisol. This lack causes both a deficiency in cortisol and an accumulation of metabolites before the enzymatic block. Because of the interconnecting pathways of synthesis of aldosterone, cortisol, and the sex hormones, these excess metabolites are shunted into the formation of sex hormones, especially androgens. Excessive production of androgens results in virilization, with formation of ambiguous genitalia in female infants. Male infants generally have no genital abnormalities at birth, although, if the syndrome goes undetected, they may have early development of secondary sex characteristics. Symptoms of hypocortisolism or hypertension may be present, depending on the degree of the enzyme deficiency and the mineralocorticoid activity of the accumulating metabolite.

The most common (95%) enzyme deficiency producing the syndrome of CAH is 21-hydroxylase deficiency. This enzyme is involved in the synthesis of both aldosterone and cortisol. In some variants of the disease, the enzyme is missing from both the zona glomerulosa and the zona fasciculata, resulting in severe salt wasting in association with hypocortisolism. In other variants, enzyme loss appears to be confined to the zona fasciculata, and aldosterone production is unaffected.

The presence of increased levels of androgens, decreased plasma cortisol levels, and increased ACTH levels is characteristic of CAH, regardless of the pathology. Excessive levels of circulating androgens can be confirmed by measuring plasma levels of total and free testosterone and dehydroepiandrosterone sulfate (DHEAS). Increased blood levels of 17-hydroxyprogesterone diagnose the 21-hydroxylase-enzyme deficiency.

The only other congenital enzyme deficiency that occurs in significant numbers is deficiency of 11-hydroxylase (5%

of cases of CAH). Shunting of cortisol precursors into other pathways results in increased synthesis of mineralocorticoids, as well as androgens, so that these patients often have hypertension as well as the abnormalities of sexual development described previously. In addition to the decreased cortisol, increased ACTH, and increased levels of androgens, these patients also have increased plasma 11-deoxycortisol levels.

Other enzyme deficiencies producing similar symptoms and signs have been reported but they are rare. All forms of CAH are treated by replacement of cortisol.

Adrenal Medulla

Cells composing the adrenal medulla and sympathetic nervous system are derived from the embryonic neural crest. This chromaffin tissue produces catecholamines (dopamine, norepinephrine, and epinephrine) from tyrosine. Sympathetic ganglia produce norepinephrine, which is released from nerve endings and acts as a neurotransmitter. The adrenal medulla primarily produces epinephrine, which is released into the circulation and acts as a hormone. Some adrenal norepinephrine production and release also occur. As hormones, both norepinephrine and epinephrine serve to mobilize energy stores and prepare the body for muscular activity (increase heart rate and blood pressure, increase blood sugar, and so forth). They are secreted in increased amounts with stress (pain, fear, and so forth).[2,36]

Epinephrine and norepinephrine are metabolized by the enzymes monoamine oxidase and catechol-O-methyl transferase to form metanephrines and VMA. The kidney (Fig. 18-4) excretes these metabolites.

Excess Catecholamine Production. *Pheochromocytomas* are tumors of the adrenal medulla or sympathetic ganglia that produce and release large quantities of catecholamines. The majority (90%) are located in the adrenal gland and produce both epinephrine and norepinephrine. Tumors in extra-adrenal locations produce only norepinephrine. Excessive production of these catecholamines increases cardiac output and causes peripheral vasoconstriction, leading to hypertension. Classically, the patient with a pheochromocytoma complains of "spells" of increased heart rate, headache, and a sensation of tightness in the chest, accompanied by sweating and pallor. These episodes correspond to the release of large amounts of catecholamines by the tumor. Only a small percentage of patients with these tumors actually have the classic presentation with no signs or symptoms of hypertension between attacks. These classic "spells" are seen in some form in about 45% of patients with pheochromocytomas. Of patients with this tumor, 50% have sustained hypertension and the remaining 5% are asymptomatic. Pheochromocytomas may occur in all age groups but are most common in patients in their 3rd to 5th decades.[2,36,43]

Although pheochromocytomas account for fewer than 1% of cases of hypertension, diagnosis of these tumors is important because they are surgically curable and approximately 10% are malignant. Bilateral tumors are present in about 10% of isolated cases and in a higher percentage (70%) of cases associated with a multiple endocrine neoplasia (MEN) syndrome.[2,36,43]

Diagnosis of pheochromocytoma should be considered in the following: all patients who complain of attacks similar to those described previously, especially if these attacks are associated with exercise or stress (patients with extra-adrenal tumors may present with unusual or even bizarre symptoms such as hypertensive attacks every time they urinate); children with hypertension; severe hypertension unresponsive to therapy; hypertension with diabetes mellitus or hypermetabolic states (due to the anti-insulin activity of the catecholamines); and patients with other endocrine tumors or with relatives who have pheochromocytomas.

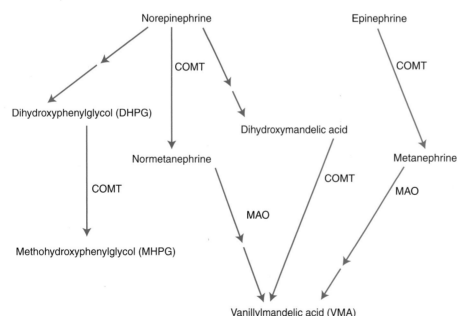

Figure 18-4. Metabolism of catecholamines. COMT, Catechol-O-methyl transferase; MAO, = Monoamine oxidase.

In the past, diagnosis depended on the precipitation of hypertensive attacks with provocative agents. Such tests were dangerous and were neither very sensitive nor very specific. These provocative tests have been replaced by measurement of urinary metanephrines and VMA and of plasma catecholamines.[2,36,43]

Urinary metanephrines and VMA are traditionally measured in a 24-hour urine collection by colorimetric or chromatographic means.[2,36,43] Determination of urinary metanephrines is considered to be the best screening test for the presence of a pheochromocytoma (see Table 18-1 for reference range). *Metanephrine* measurement is favored over VMA as a primary screen because more substances (drugs, food, and so forth) interfere with the measurement of VMA. Repetition of metanephrine analysis resulting in a second abnormal value or follow-up of abnormal metanephrine levels with the finding of elevated VMA excretion has a diagnostic accuracy of approximately 99%.[2,36,43]

Theoretically, measurement of plasma catecholamines would be preferable to analysis of urinary metabolites because such measurement would eliminate the necessity for a 24-hour urine collection and allow sampling during an attack. Measurement of plasma catecholamines is helpful in patients with sustained hypertension but is less reliable in patients who are normotensive between attacks. Even in patients with sustained hypertension, increased plasma catecholamines may be due to neurogenic or idiopathic causes. The response of these patients to administration of clonidine is necessary to separate them from those with pheochromocytoma. Patients with pheochromocytomas show no response to clonidine, whereas those with neurogenic or essential hypertension show a 50% decrease in plasma catecholamine levels. Other problems with measurement of plasma catecholamine are that the range of normal catecholamine concentration is so wide that a level that is diagnostic of pheochromocytoma is difficult to establish; blood must be obtained under carefully standardized conditions to avoid stimulation of catecholamine release; secretion of catecholamines may be episodic, and finding plasma levels within the reference range does not rule out the diagnosis; and elevated levels of catecholamines may occur in a variety of other conditions.

Some authors have suggested that epinephrine is more useful than norepinephrine in diagnosis of pheochromocytoma because epinephrine comes only from the adrenal and not from the sympathetic nervous system. Measurement of plasma catecholamines may be very useful in patients with extra-adrenal tumors because blood samples can be drawn at different levels along the inferior vena cava and help to localize the tumor.[2,15,36,43] (See Table 18-1 for typical reference ranges.) Treatment is surgical removal. The prognosis for patients with benign tumors is excellent.

Neuroblastomas. *Neuroblastomas* are malignant tumors of the adrenal medulla that occur in children. They produce catecholamines and may occasionally be associated with hypertension. Detection of elevated levels of VMA or homovanillic acid (HVA; product of dopamine metabolism) may be helpful both in diagnosis of the tumor and in detection of tumor recurrence or metastases in treated patients.[36]

Islets of Langerhans

Insulin

Insulin is a peptide hormone that is synthesized by the beta cells of the pancreatic islets. It consists of a total of 51 amino acids in two chains connected by two disulfide bridges. Insulin is synthesized as a large single chain preproinsulin that is cleaved to a more immediate precursor proinsulin in the rough endoplasmic reticulum. *Proinsulin* is then packaged into secretory granules, where it is broken down into equimolar amounts of insulin and an inactive C-peptide.[44-46]

Glucagon

Like insulin, glucagon is a protein hormone that is involved in the regulation of carbohydrate, fat, and protein metabolism. It is synthesized by the alpha cells of the pancreatic islet

---

CASE STUDY 18-5

A 38-year-old male was seen in the emergency room, complaining of a severe headache. On examination, it was noted that his blood pressure was 220/120 and that he was sweating profusely. During the previous 2 months, he had experienced one other similar episode but had not been seen by a physician. Routine laboratory tests were ordered. All tests were within normal limits, with the exception of glucose, which was 140 mg/dL.

The patient was referred to a nephrologist and hospitalized for further evaluation.

Questions

1. What classic symptoms in this patient may suggest the diagnosis of a pheochromocytoma?
2. What laboratory tests would be most valuable in the confirmation of a pheochromocytoma?
3. What are the caveats concerning VMA procedures?
4. Primary aldosteronism can also produce secondary hypertension. Describe the etiology of primary aldosteronism.
5. List classical laboratory findings associated with primary aldosteronism.

and comprises 29 amino acids. A larger proglucagon molecule is synthesized first and this precursor is cleaved within secretory granules before being released. A complete discussion of the regulation and conditions associated with insulin and glucagon is found in Chapter 10 *Carbohydrates*.

Other Hormones

Excessive production of a variety of other hormones has been described in association with islet-cell pancreatic tumors. These include gastrin, somatostatin, pancreatic polypeptide, and serotonin. The clinical syndromes produced by excessive production of somatostatin and pancreatic polypeptide are nonspecific and the diagnosis is rarely made preoperatively except in patients with multiple endocrine tumors. Measurement of somatostatin and pancreatic polypeptide in plasma can make the diagnosis of an islet-cell tumor producing one of these substances but assays for these compounds are performed primarily in reference laboratories and are generally unavailable.

Reproductive System

The embryonic gonads, both ovaries and testes, are derived from the germinal epithelium, which begins to develop on the urogenital ridge around week 4 of gestation. Primitive germ cells migrate into this area from the yolk sac and by week 6, are present in the epithelial tissue of the primitive gonad. At this point, the gonad is bipotential; that is, it may develop into either a testis or an ovary. The direction of further differentiation depends on the presence or absence of a Y chromosome. If a Y chromosome is present, the gonad develops into a testis and the presence of hCG from the placenta stimulates *Leydig cells* within the gonad to secrete testosterone. *Testosterone* then causes the male internal and external genitalia to develop and the testes to descend into the scrotum. In contrast, if the embryo has no Y chromosome, the gonad develops into an ovary. In the absence of testosterone production, the internal and external genitalia develop along female lines.[2]

Female

The female reproductive system consists of two ovaries suspended by means of ligaments from the pelvic girdle, a pear-shaped muscular organ called the uterus, and extensions of the uterus called fallopian tubes or oviducts. The hormones in the female are secreted in a cyclic pattern. The low levels of adrenal estrogens produced in the prepubertal child are sufficient to suppress pituitary gonadotropin release. At puberty, however, the pituitary becomes less sensitive to low levels of estrogen and the levels necessary for inhibition are reset.

The hypothalamus releases Gn-RH, which in turn causes release of FSH from the anterior pituitary gland. At the beginning of each menstrual cycle, an early (primordial) follicle is observed in the cortex of the ovary. This early follicle consists of a primary oocyte surrounded by a single layer of follicular cells. The follicular cells proliferate and form a layer of granulosa cells, which surround the primary oocyte. FSH causes proliferation of the granulosa cells and hence the early growth of the follicle.

The connective tissue that surrounds the primary follicle differentiates to form *thecal cells*. Estradiol is produced from testosterone that then diffuses out of the thecal cells. FSH may also enhance production of estradiol from testosterone. The release of estradiol occurs during the early to mid-follicular phase of the menstrual cycle. The increase in estradiol inhibits the production of FSH by the anterior pituitary.

The granulosa cells secrete a fluid that fills spaces between the cells and eventually displaces the oocyte to one side of the follicle. At this stage, the follicle is known as a secondary follicle. Although several follicles begin this process with each cycle, only one follicle undergoes further maturation and becomes a mature or graafian follicle. The graafian follicle produces a bulge in the outer surface of the ovary. Approximately 14 days are needed to produce this mature follicle. The 14 days are known as the *follicular phase* of the menstrual cycle.

A surge of LH produced by the anterior pituitary in response to elevated levels of estradiol causes the graafian follicle to rupture and the ovum is expelled. The process is called *ovulation*.

The graafian follicle remains in the ovary and undergoes a change and becomes known as the *corpus luteum*. LH, therefore, is necessary for ovulation and the final follicular growth. LH acts on the thecal cells to cause the synthesis of androgens and ultimately estradiol and *estrone* (estrogens). These hormones bind to albumin and SHBG to be transported in the blood. Estradiol, the most potent of the estrogens, serves as a negative feedback for FSH. *Estrogens* are responsible for growth of the uterus, fallopian tubes, and vagina, promotion of breast development, maturation of the external genitalia, deposition of body fat into the female distribution, and termination of linear growth.

In addition, LH influences a change in the granulosa cells. They are responsible for the production of progesterone. Binding to transcortin and albumin transports progesterone in the blood. *Progesterone* serves to prepare the uterus for pregnancy and the lobules of the breast for lactation. Increased levels of progesterone may serve in the negative feedback mechanism to decrease the amount of LH released from the anterior pituitary. It is thought that the granulosa cells may produce inhibin that causes a decrease in the FSH released from the anterior pituitary. Figure 18-5 summaries the hormonal regulation of the female reproductive system. Reference ranges for estrogen, progesterone, FSH, and LH at different ages and during different phases of the menstrual cycle are given in Table 18-2.[2]

The corpus luteum becomes nonfunctional and ceases to produce and secrete hormones if fertilization of the ovum does not occur after ovulation. The remaining structure becomes the *corpus albicans*. Approximately 14 days are required for the process to be completed after ovulation. This period is called the *luteal phase* of the menstrual cycle.

Simultaneous changes occur in the uterus. The changes in the levels of ovarian hormones bring about responses in the en-

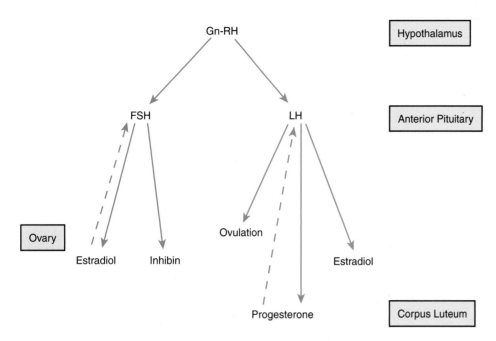

Figure 18-5. Hormonal regulation of the female reproductive system.

dometrial layer of the uterus. The endometrial lining has subsequently been prepared if the ovum is fertilized and it participates in the formation of the placenta. In the absence of hCG from an implanted ovum, the corpus luteum degenerates, and estrogen and progesterone levels drop and the endometrial lining sloughs off in a process known as menstruation.

Male

The testes, like the ovaries, are part of a hypothalamic-pituitary-gonadal axis. Gn-RH from the hypothalamus stimulates the release of LH from the anterior pituitary. LH, in turn, stimulates the production of testosterone and estradiol by the Leydig, or interstitial, cells of the testes. The testosterone that is produced is either secreted into the blood or delivered to the seminiferous tubules. The majority of testosterone in the blood is bound to SHBG. The biologic activity of testosterone is due to dihydrotestosterone, which is formed in the target tissues. It causes growth and development of the male reproductive system, prostate, and external genitalia. Increased testosterone

levels at puberty cause development of pubic, axillary, and facial hair, growth of the internal and external genitalia, rapid musculoskeletal growth and eventual epiphyseal closure, hypertrophy of the larynx with a deepening voice, initiation of spermatogenesis, and development of sex drive and potency. Testosterone or its metabolites inhibit the release of Gn-RH. Reference ranges for testosterone and gonadotropins in adult and prepubertal males are listed in Table 18-2.

The testosterone that is delivered to the seminiferous tubules stimulates the *Sertoli cells* and thus spermatogenesis. A variety of hormones are involved in spermatogenesis. FSH and testosterone stimulate these cells to produce androgen-binding protein (ABP). ABP is responsible for the transport of testosterone in the Sertoli cells where the first mitotic division of the spermatocytes occurs and to the epididymis where final maturation occurs. The action produced by FSH occurs during puberty when the initial wave of spermatogenesis occurs and does not appear to be necessary for the maintenance of the process.

TABLE 18-2. Typical Reference Ranges for Reproductive Hormones[2]

	Total Testosterone (ng/dL)	Total Estrogens (pg/mL)	Progesterone (ng/dL)	FSH (mU/mL)	LH (mU/mL)	DHEAS (μg/mL)
Children (<10 y)	<3–10	<25	7–52	<1–3	<1–5	<150
Adult females	20–75					50–540
Follicular phase		60–200	15–70	1–9	1–12	
Midcycle				6–26	16–104	
Luteal phase		160–400	200–2500	1–9	1–12	
Postmenopausal	8–35	<130		30–118	16–66	30–260
Adult males	300–1000	20–80	13–97	1–7	1–8	170–670

The Sertoli cells also produce a peptide called *inhibin*. It inhibits the release of Gn-RH from the hypothalamus and the release of FSH from the anterior pituitary. Therefore, inhibin serves as the negative feedback mechanism for FSH. Inhibin stimulates the Leydig cells to produce more testosterone. The testosterone diffuses into the Sertoli cells and is aromatized to form estradiol. The estradiol then diffuses back into the Leydig cells and may reduce the synthesis of testosterone. Figure 18-6 summaries the long-loop hormonal control of the male reproductive system.

There appears to be short-loop control system in addition to the hypothalamus-pituitary-gonadal axis in the male.[47] The Leydig cells produce adrenocorticotropic hormone/melanocyte-stimulating hormone (ACTH/MSH) peptides that stimulate the Sertoli cells. The Sertoli cells also produce a beta-endorphin, but it has an inhibitory action on the Sertoli cells.

Disorders

Infertility. Laboratory evaluation of *infertility* must involve both partners, because causes of infertility are relatively evenly split between males and females. Evaluation should include a thorough history and physical examination to rule out chronic disease or structural abnormalities that may result in failure to conceive. Schemes for laboratory evaluation of male and female infertility vary, depending on the history and physical findings. Generally, the first step for evaluation of male infertility is seminal fluid evaluation. If seminal fluid analyses are normal on several (at least four) occasions and

survival of sperm in cervical mucus is normal, no further examination of seminal fluid is indicated.

Infertility in males may classified as pretesticular, testicular, and posttesticular.[48] The causes of pretesticular infertility, or secondary hypogonadism, are commonly due to hypothalamic or pituitary lesions. In such cases, testosterone levels are decreased and FSH and LH levels are decreased or normal.

Infertility or *hypogonadism* due to testicular causes are referred to as primary hypogonadism. These conditions may be congenital (cryptorchidism, Klinefelter's syndrome, and 5-alpha-reductase deficiency) or acquired (varicocele, tumor, orchitis, and cellular damage due to chemotherapy, drugs, aging, irradiation, and infectious agents). A decreased testosterone level with increased levels of FSH and LH are present in testicular causes of infertility.

Functional impairment, mechanical obstruction of sperm transport, or disorders of sperm function are usually causes of posttesticular infertility. Testosterone, FSH, and LH levels are normal.

In women, the first step for evaluation of infertility is usually the determination of basal body temperature over several menstrual cycles to establish whether ovulation appears to be occurring. Additional tests may include endometrial biopsy and measurement of serum progesterone levels.

Infertility in the female may originate from the hypothalamus, pituitary, ovaries, or uterus. Malfunction of the hypothalamus may result in a decreased production of Gn-RH and thus decreased release of FSH and LH from the pituitary. Weight loss, stress, exercise, or chronic illness are more common causes of a disruption than tumors.[49]

Infertility may also accompany hypersecretion of PRL. This hypersecretion may be due to an adenoma of the pituitary or seen in patients with hypothyroidism or in those taking certain medications (dopamine antagonists). Amenorrhea and galactorrhea usually accompany increased PRL levels. In addition to FSH and LH levels, PRL and TSH levels may be helpful in the evaluation of infertility.

Anovulation may result from ovarian abnormalities. Infections, tumors, polycystic ovary disease (PCOD), and Turner's syndrome may prevent the production of ova. Increased levels of FSH and LH and decreased levels of estradiol and inhibin are generally associated with ovarian disorders.

Uterine and fallopian tube abnormalities may also cause infertility. Tumors in the uterus or destruction of the endometrial lining make implantation of the fetus difficult. Chronic infections and endometriosis may result in decreased patency of the fallopian tubes. In these conditions, the FSH, LH, and estradiol levels are generally normal.

Gynecomastia. The development of breast tissue in males is termed *gynecomastia*. This condition is related to the balance of estrogen to androgens. It occurs in newborn males due to exposure with maternal estrogen. It may also occur physiologically during puberty due to increased secretion of estro-

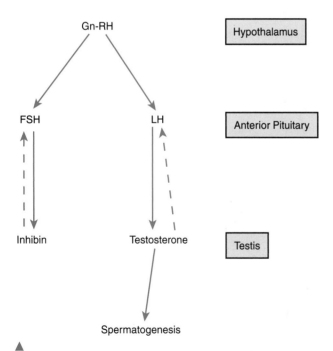

Figure 18-6. Long-loop control mechanism for the male reproductive system.

gens and in old age because of a decrease in the production of testosterone.

Pathologically, gynecomastia may be due to a decreased androgen activity. This condition may accompany primary or secondary hypogonadism. Gynecomastia may also be seen when there is an increased production of SHBG as seen in cirrhosis of the liver, hyperthyroidism, or with the use of anti-androgens. Increased estrogen activity may also cause gynecomastia. Conditions causing increased estrogen activity are estrogen-secreting tumors, gonadotropin-secreting tumors, conditions causing increased peripheral production of estrogens from androgens such as hyperthyroidism or cirrhosis of the liver, or use of drugs with estrogenic activity. Increased PRL levels and use of drugs such as methyldopa and phenothiazines may also cause gynecomastia. The etiology of gynecomastia may be assessed by the measurement of testosterone, FSH, LH, and SHBG. Additionally levels of PRL as well as liver and thyroid function tests may be helpful.

Hirsuitism. *Hirsuitism* is the excessive growth of hair with a male distribution pattern in a female. This condition may be the most common endocrine disorder in women. In most cases, it is usually benign but can indicate a serious disorder. Testosterone is primarily responsible for the initiation and growth of sexual hair. It becomes concentrated in the hair follicle and is reduced by 5-alpha-reductase to form dihydrotestosterone. Overproduction of testosterone by the ovary or the adrenal gland and increased sensitivity or activity of 5-alpha-reductase at the hair follicle should be considered in evaluating hirsuitism.

The etiology of hirsuitism can be categorized as follows: primary hyperandrogenemia, secondary hyperandrogenemia, or idiopathic hirsuitism.[50] Tumors of the adrenal gland or ovary, CAH, or PCOD are associated with primary *hyperandrogenemia*. Tumors of the adrenal gland or ovary are rare, but these tumors can produce massive amounts of androgens and a rapid onset of hirsuitism. Generally, testosterone levels of greater than 2000 ng/dL are indicative of a tumor. Virilization may also occur in these persons. *Virilization* is an abnormal development of the secondary male sex characteristics such as lowering of the voice, clitoral enlargement, acne, and changes in body mass distribution. The use of adrenal gland or ovary imaging techniques may supplement the testosterone level assessment in the diagnosis.

Without cortisol serving as a negative feedback compound in CAH, the anterior pituitary releases excessive amounts of ACTH. The adrenal gland becomes hyperplastic and excessive amounts of androgens are produced. CAH is usually manifest in childhood, but in cases in which there are only mild enzyme deficiencies, it may not be manifest until adulthood. An elevated level of DHEAS indicates that the adrenal gland is the source of the androgen. Only the adrenal gland can attach the sulfate group to steroid intermediates.

Symptoms of mild hirsuitism with normal menses to excessive hirsuitism with amenorrhea may accompany PCOD

and also is known as Stein-Leventhal syndrome. In addition, patients may be overweight. The increased levels of androgens in this condition are caused by an abnormal release of LH by the pituitary gland, which in turn causes a hyperplasia of the thecal cells in the ovary. Free testosterone levels and LH levels are increased. The LH:FSH ratio is usually greater than two in this condition.

Idiopathic or familial hirsuitism is found in approximately 10% of women with hirsuitism but not hyperandrogenemia. The cause of the hirsuitism may be either increased activity of 5-alpha-reductase or sensitivity of the receptors at the hair shaft. Laboratory values are normal in these persons.

Amenorrhea. *Amenorrhea* is the absence of vaginal bleeding.[51] The most common cause of physiologic amenorrhea is pregnancy. The pathologic causes of amenorrhea may be classified as either primary or secondary. Primary amenorrhea is menstruation having never occurred. The diagnosis of primary amenorrhea is made if a female reaches the age of 14 in the United States without showing any signs of puberty such as breast development or the appearance of pubic hair and failure to menstruate.[52] Amenorrhea with increased levels of FSH and LH are frequently found in gonadal dysgenesis or Turner's syndrome. These patients have a chromosome abnormality (45XO). Characteristic features of Turner's syndrome are short stature, webbed neck, low hairline on the neck, and broad shieldlike chest. Approximately 90% of these patients never menstruate.[52] Fibrous tissue is commonly found in the ovaries. Because there is no ovarian function, the FSH and LH are elevated.

Anatomical abnormalities may also cause amenorrhea. An example is a condition called Mullerian agenesis. In this disorder, there is a congenital malformation or absence of the fallopian tubes, uterus, or vagina.[51] FSH, LH, estradiol, and testosterone levels are generally normal.

Secondary amenorrhea is the absence of menses for 6 months or the equivalent of three previous cycle intervals, whichever is longer after previous menstrual bleeding.[52] Laboratory tests may help in categorizing the numerous causes of secondary amenorrhea. The physician may perform a progesterone challenge. If, after a dose of progesterone is administered, uterine bleeding occurs, it suggests that the pituitary-ovary-uterine pathway is intact but not functional.

Thyroid function tests may prove valuable because amenorrhea is a common finding in thyroid dysfunction. PRL levels may also be valuable. Patients with isolated hyperprolactinemia commonly present with amenorrhea and galactorrhea. Ovarian failure will be accompanied with high levels of FSH and LH and decreased levels of estradiol. On the other hand, hypothalamic or pituitary dysfunction will be accompanied with decreased levels of FSH, LH, and estradiol. Low or normal levels of FSH and increased LH levels may indicate PCOD. Serum testosterone or DHEAS levels confirm PCOD. Greatly increased levels of testosterone and low levels of FSH and LH may indicate ovarian or adrenal tumors.

Erectile Dysfunction. *Erectile dysfunction,* or impotence, is a common cause of male reproductive failure. It is the inability to achieve an erection long enough to complete satisfactory intercourse.[54] Neurogenic erectile dysfunction involves a failure to initiate erection because of the absence of autonomic pelvic nerve stimulation and corporal nerve release of endogenous neurotransmitter substances. This condition may be caused by spinal cord disorders, myelopathy, multiple sclerosis, diabetes, certain medications, use of alcohol, or anxiety. Erectile dysfunction may be classified as vascular. Atherosclerosis, microangiopathy, congestive heart failure, priapism, aging, and excessive venous outflow are associated with this type of dysfunction. Many of these patients have hypertension, diabetes, elevated serum cholesterol levels, angina, or a history of cigarette smoking.

Pregnancy

The *placenta* synthesizes and secretes a variety of protein hormones as well as the steroids estrogen and progesterone. Evaluation of maternal serum and urine concentration of these hormones may be of value not only in diagnosing pregnancy but also in monitoring placental development and fetal well-being.

hCG. hCG is a protein hormone produced by the placenta and consists of alpha and beta subunits. The alpha subunit is identical to the alpha subunits of FSH, LH, and TSH. The beta subunit, however, is sufficiently different that antibodies to this subunit show essentially no cross-reactivity with anterior pituitary hormones. Immunoassays detecting the beta subunit have served as the primary means of measuring hCG since their introduction in the early 1970s. More recently, assays of the intact molecule have become available, many investigators believe these assays are the method of choice for pregnancy testing. Measurement of the beta subunit, however, is still recommended for monitoring patients with hCG-producing neoplasms.[55]

hCG secretion serves to maintain progesterone production by the corpus luteum in the early part of pregnancy. By the time that hCG levels drop at the beginning of the second trimester, the placenta has developed sufficiently to produce enough progesterone to maintain the endometrium and to allow the pregnancy to continue. In addition, hCG stimulates development of fetal gonads and synthesis of androgen by the fetal testes.[56]

Measurement of hCG is used primarily for the diagnosis of pregnancy. With the development of monoclonal antibodies, hCG can now be detected in serum and urine as early as 1 day to 2 days after fertilization, even before the woman has missed a menstrual period. In addition, hCG determinations may be useful in the following: the diagnosis of ectopic pregnancy (implantation of the fertilized ovum in a site other than the uterus, usually the fallopian tube); pre-

diction of spontaneous abortion (lower than expected hCG values and failure of hCG to increase appropriately in the first few weeks of pregnancy indicate that the placenta is not developing correctly and have been shown to be associated with an increased incidence of spontaneous abortions); detection of multiple pregnancies (hCG levels in women carrying multiple fetuses are higher than in women carrying only one); and detection and follow-up of hCG-producing tumors (neoplasms of trophoblastic, placental, tissue produce hCG in amounts greater than those associated with normal pregnancies, and higher than expected hCG levels may be the first indication of the presence of such a tumor). After the neoplasm is diagnosed and treated, continued monitoring of hCG levels allows for the early diagnosis of recurrent or metastatic disease. hCG also may be produced by tumors arising in other sites. Although not useful in the diagnosis of a particular type of tumor, follow-up measurements of hCG in a patient who has been treated for a tumor with demonstrated hCG production is useful in the detection of recurrence or metastases.[55,57]

HPL. Human placental lactogen (HPL) is a protein hormone that is structurally, immunologically, and functionally very similar to GH and PRL. Like hCG, it is produced by the placenta and can be measured in maternal urine and serum as well as amniotic fluid. Levels rise until approximately week 37 of gestation, paralleling increasing placental weight, and decline slightly in the last few weeks of pregnancy.[58]

HPL appears to act in concert with hCG to stimulate estrogen and progesterone synthesis by the corpus luteum. In addition, it stimulates development of the mammary gland (similar to PRL) and has somatotropic actions similar to those of GH. That is, it increases maternal plasma glucose levels and mobilizes free fatty acids and promotes positive nitrogen balance.

HPL is measured by immunologic assays and may show significant cross-reactivity with GH and PRL. Measurement is useful in monitoring conditions associated with a decrease in functioning placental tissue (most commonly in maternal hypertension) and in diagnosis of intrauterine growth retardation. In these situations, HPL levels are lower than expected for the length of gestation. HPL might be expected to be increased in conditions associated with larger than normal placental mass (diabetes mellitus, erythroblastosis fetalis, and multiple pregnancies). However, most complications related to these conditions are not due to placental insufficiency, and, therefore, serial measurement of HPL in patients with these problems has not proven to be clinically useful.[59]

Steroids. The fetus and placenta function together in steroid biosynthesis. The placenta synthesizes progesterone from cholesterol. Progesterone then enters the maternal circulation, where it functions to maintain the endometrium, inhibit uterine contractions, and stimulate the lobular unit of the breast. It also enters the fetal circulation, where it is used

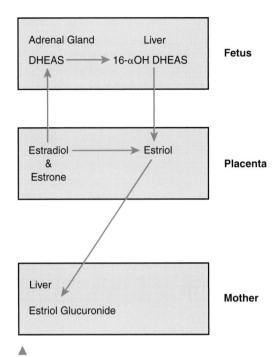

Figure 18-7. Estriol production by the fetoplacental unit.

by the fetal adrenal for synthesis of cortisol and sex hormone precursors, especially DHEA and 16-hydroxy-DHEAS. Conversion of these compounds to estrogen (Fig. 18-7) cannot proceed in the fetal adrenal because the necessary enzymatic machinery is missing. These enzymes, however, are present in the placenta, which converts DHEA to estrone and estradiol and 16-OH DHEA to *estriol*. Estriol then enters the maternal circulation, where it can be measured in plasma and urine (see Fig. 18-7). Normally, maternal plasma and urine estriol levels increase throughout pregnancy with a surge in the last 4 weeks to 6 weeks.[60]

Because measurement in plasma does not depend on maternal renal function or proper collection of a 24-hour urine specimen, plasma levels are generally preferred over urine for measurement of estriol. Serial measurements are necessary in the last 4 weeks to 8 weeks of pregnancy because the pattern of estriol levels is as important as the absolute level in the diagnosis of fetal problems. Estriol measurement may be useful in the following: evaluating the fetoplacental unit in diabetic mothers (a decrease of 40% in estriol levels compared with the mean of three preceding determinations has been shown to be an accurate predictor of fetal distress); identifying post-date gestations in which the infant is at risk for development of the postmaturity syndrome (in this case, estriol levels rise appropriately but drop off after week 40 of gestation); and diagnosis of intrauterine growth retardation (estriol levels are low in this condition and often fail to show the normal surge).

Estriol is not produced in significant quantities in the mother and is solely a reflection of fetoplacental function. Thus, estriol measurement provides valuable information about fetal well-being.[60]

Other Hormones of Clinical Interest

Gastrointestinal Hormones

A variety of hormones, including gastrin, secretin, cholecystokinin, gastric inhibitory peptide (GIP), motilin, and serotonin are produced and released by endocrine cells located along the entire mucosa. These hormones control digestion, absorption, GI motility, and release of other hormones.

Gastrin. *Gastrin* is a peptide secreted by the G cells of the antrum (lower one third) of the stomach. The major stimulus for gastrin release is the presence of amino acids but it is released in response to vagal stimulation and the presence of food in the stomach. Hypoglycemia and elevation of serum calcium may also stimulate gastrin release. Gastrin causes secretion of hydrochloric acid by parietal cells in the body (middle one third) of the stomach. As the pH of gastric contents in the antrum decreases, gastrin secretion is inhibited and acid production slows. The reference range for serum gastrin is given in Table 18-1. As with many other immunologic assays, differences in antibody specificity and laboratory techniques may make it difficult to compare values obtained in different laboratories.[61]

Excessive production of gastrin by the antrum occurs with decreasing acid production by gastric parietal cells (loss of negative feedback). Thus, elevated serum gastrin levels are seen in patients with achlorhydria for any reason (especially chronic atrophic gastritis) or in those who have had a vagotomy. Gastrin levels are also increased in chronic renal failure owing to decreased degradation and renal clearance. Measurement of serum gastrin in these situations and in the immediate postprandial period may reveal values up to five times the upper limit of the reference range.[61,62]

Gastrin also may be increased because of hyperplasia of the gastrin-producing cells in the antrum or because of ectopic production of gastrin by an islet-cell tumor—called the *Zollinger-Ellison syndrome*. The syndrome is associated with severe peptic ulceration of the stomach, often extending into the duodenum and jejunum. Serum gastrin levels in this condition may be extremely high but, more often, they overlap the values seen in the situations described previously. Therefore, an elevated serum gastrin level must be accompanied by gastric hyperacidity to make the diagnosis of Zollinger-Ellison syndrome.[61–63]

Gastrin-producing tumors are sometimes malignant. Because of the presence of metastases and also of the frequency of multiple primary tumors within the pancreas, surgical removal of the gastrin-secreting tumor may be impossible. Instead, surgical therapy of the Zollinger-Ellison syndrome is removal of the target organ (*ie,* total gastrectomy). Medical therapy with antiulcer agents is also promising and, in some patients, may make surgery unnecessary. Gastrinomas are relatively slow growing tumors.

VIP. VIP is similar in structure to secretin, GIP, and glucagon. However, it is not found in the endocrine cells of the mucosa but rather in the nerve fibers located throughout the GI tract. It acts as a neurotransmitter and causes relaxation and vasodilation of the smooth muscles. It also affects the gut by increasing water and electrolyte secretion, release of hormones, and inhibition of gastrin and gastric acid secretion.

The Verner-Morrison syndrome is associated with increased levels of VIP. The increased levels of VIP are usually caused by a pancreatic tumor (VIPoma). The syndrome consists of chronic, profuse, watery diarrhea, which leads to dehydration and hypokalemia. Attacks of flushing (vasodilation), hypotension, and achlorhydria also accompany the syndrome. Measurement of VIP plasma concentrations may be a useful test when diarrhea due to a secretory tumor is suspected.

Serotonin. *Serotonin* (5-hydroxytryptamine) is an amine derived from hydroxylation and decarboxylation of tryptophan. It is synthesized by *enterochromaffin cells,* which are located primarily in the GI tract and, to a lesser degree, in the bronchial mucosa, biliary tract, and gonads. Serotonin is secreted into the blood, where it binds to platelets and is released during coagulation. It is metabolized to 5-hydroxyindoleacetic acid (5-HIAA) which is excreted in the urine.

Tumors arising from enterochromaffin cells (carcinoid tumors) occur mainly in the appendix and the ileum, although occasionally such tumors may develop in the lung or ovary. These tumors produce large amounts of serotonin as well as other amines (especially histamine), kallikrein, and prostaglandins. If these substances are released into the circulation, they produce a constellation of symptoms known as the *carcinoid syndrome* (attacks of diarrhea, flushing, tachycardia, and hypotension). Damage of the heart valves also may occur. Detection of elevated plasma serotonin levels or urinary 5-HIAA is diagnostic of the carcinoid syndrome, provided the patient has not taken any drugs or food that produces positive or negative interference with the assay. Drugs interfering with this test include glyceryl guaiacolate, methenamine mandelate, and phenothiazines. Foods such as bananas, pineapple, walnuts, and avocados are very high in serotonin content and may produce false-positive results. The reference range for 5-HIAA in the urine is given in Table 18-1. Elevated 5-HIAA levels occur with some other GI diseases but are not as high as in the carcinoid syndrome.[64–68]

Carcinoid tumors are low-grade malignancies that grow slowly. The 5-year survival may be relatively good even if metastases are present at the time of diagnosis. Treatment is surgical removal of the primary tumor and of isolated liver metastases, if present.

Multiple Endocrine Neoplasia Syndromes

MEN syndromes are the occurrence of several tumors or hyperplasias involving two or more endocrine organs. Two main types are currently recognized. These syndromes are familial, and inheritance is in an autosomal dominant pattern.

MEN I, or Wermer's syndrome, includes tumors of pancreatic islets, often associated with peptic ulcer disease (Zollinger-Ellison syndrome); pituitary adenomas, either functional or nonfunctional; and parathyroid adenoma or hyperplasia. Other disorders in MEN I may include benign and malignant bronchial and intestinal carcinoids; tumors of nerve, adipose tissue, and thymus; and disease of the thyroid and adrenal glands.

MEN II, Sipple's syndrome, includes medullary carcinoma of the thyroid; pheochromocytomas, which are usually bilateral, rarely malignant, and often nonfunctional; and parathyroid adenoma or hyperplasia. Mucosal neuromas also may be present (MEN IIb).[66,67]

In both MEN I and MEN II, penetrance is high but expression is variable. All patients do not have the complete syndrome and many patients may present originally with one component, only to develop symptoms of other neoplasms at a later date.

Ectopic Hormone Production

Cells in sites other than the gland from which they are usually derived may occasionally produce protein hormones. This occurrence is called *ectopic hormone* production. Hormones produced in this manner appear to be immunologically identical to those produced by the usual endocrine gland. Small differences in conformation or the release of larger, inactive forms of the hormone result in varying functional activity. When these hormones are functionally active, they produce syndromes identical to those of excessive hormone production by the endocrine gland, where they are normally synthesized. The tumor most commonly associated with ectopic hormone production is oat-cell carcinoma of the lung, which not infrequently produces ACTH or ADH. Other hormones that may be produced ectopically are hCG by carcinoma of lung, GI tract, and pancreas; PTH by squamous cell carcinomas of the lung; and gastrin by islet-cell carcinoma of pancreas.[2]

Growth Factors

Growth factors (GF) are signaling peptides that act through cell-surface receptors to stimulate cell growth, proliferation, scattering, and migration.[68] GF have autocrine, paracrine, and endocrine functions. They are produced by many tissues and may exert their influence on themselves *(autocrine),* on nearby cells *(paracrine),* and on distant tissues (endocrine). Some GF affect multiple cell types. The ability to measure GF promises to revolutionize the field of clinical endocrinology. Table 18-3 provides a list of some GF and their role in the diagnosis of disease.

SUMMARY

This chapter has covered basic mechanisms of endocrine diseases as they apply to the anterior and posterior pituitary,

TABLE 18-3. Growth Factors[68]

Growth Factor	Abbreviation	Major Action
Epidermal growth factor	EGF	Epithelial proliferation
Erythropoietin	EPO	Red blood cell proliferation
Granulocyte-colony stimulating factor	G-CSF	Granulocytes
Granulocyte macrophage-colony stimulating factor	GM-CSF	Granulocytes/macrophages
Hepatocyte GF/scatter factor	HGF/SF	Cell dispersal
Interferon-α	IFN-α	Antiviral, antiproliferative
Interferon-β	IFN-β	Antiviral
Interferon-γ	IFN-γ	Antiviral, antiprotozoal
Insulin-like growth factor I	IGF-I	Mediates GH action
Insulin-like growth factor II	IGF-II	Tumorigenesis
Nerve growth factor	NGF	Neuron proliferation
Platelet-derived epidermal growth factor	PDEGF	Epidermal proliferation
Platelet-derived growth factor	PDGF	Chemotaxis/macrophages
Transforming growth factor-α	TGF-α	Tumor proliferation
Transforming growth factor-β	TGF-β	Inhibit/stimulate proliferation
Tumor necrosis factor-α	TNF-α	Inhibit tumor growth
Tumor necrosis factor-β	TNF-β	Inhibit tumor growth

parathyroid glands, adrenal glands, islets of Langerhans, reproductive system, and other hormones secreted by organs that are not technically considered part of the endocrine system. Table 18-4 summaries the endocrine glands discussed and the hormones produced. Thyroid diseases and diabetes mellitus are covered in Chapter 19 *Thyroid Function* and Chapter 10 *Carbohydrates*.

Evaluation of endocrine function differs considerably from that of other systems. Stimulation and suppression testing provide the cornerstones of endocrine evaluation and require adherence to strict protocols to ensure that results truly reflect the patient's underlying condition. As analytic methods become even more sensitive and molecular diagnostic techniques are applied to diagnosis of endocrine diseases, this field of clinical chemistry will continue to expand and tests currently restricted to reference or research laboratories will become routine.

REVIEW QUESTIONS

1. The control of adenylate cyclase and phospholipase C is through a _____ protein:
   a. P
   b. I
   c. G
   d. Y
   e. Z

2. The neurohypophysis is the:
   a. Hypothalamus
   b. Anterior pituitary
   c. Pineal gland
   d. Thyroid gland
   e. Posterior pituitary

3. The growth-promoting actions of GH are brought about by several liver-produced substances called:
   a. Insulin-like growth factors
   b. Somatotropins
   c. Somatostatins
   d. Growth inhibitory factors
   e. Endorphins

4. In primary hyperparathyroidism, one would expect to see a/an _____ serum calcium and a/an _____ serum phosphorus:
   a. Increased, increased
   b. Decreased, decreased
   c. Increased, decreased
   d. Decreased, increased
   e. Normal, normal

5. If the serum levels of estradiol do not increase after the injection of human chorionic gonadotropin, the patient has:
   a. Pituitary failure
   b. Primary ovarian failure
   c. Tertiary ovarian failure
   d. Secondary ovarian failure

6. If a patient had a luteal phase defect, which hormone would most likely be deficient?
   a. Estrogen
   b. hCG
   c. FSH
   d. Prolactin
   e. Progesterone

TABLE 18-4. Summary of Sources and Action of Hormones

Hormone	Abbreviation	Source	Action
Adrenocorticotropic hormone	ACTH	Anterior pituitary	Release of cortisol
Aldosterone	None	Adrenal cortex	Sodium regulation
Antidiuretic hormone	ADH	Posterior pituitary	Permeability of kidney tubules
Calcitonin	None	Parafollicular cells of thyroid gland	Calcium regulation
Catecholamines	None	Adrenal medulla	Mobilize energy stores
Corticotropin releasing hormone	CRH	Hypothalamus	Release of ACTH
Cortisol	None	Adrenal cortex	Carbohydrate, fat, and protein metabolism
Estradiol	E_2	Ovary	Female sexual characteristics
Follicle-stimulating hormone	FSH	Anterior pituitary	Female: Growth of the follicle Male: Initial wave of spermatogenesis
Gastrin	None	G cells of stomach	Secretion of HCl
Glucagon	None	Alpha cells of pancreatic islets	Carbohydrate metabolism
Gonadotropin releasing hormone	Gn-RH	Hypothalamus	Release of FSH and LH
Growth hormone	GH	Anterior pituitary	Growth of bone and soft tissues
Growth hormone inhibiting hormone	GH-IH	Hypothalamus	Inhibition of GH
Growth hormone releasing hormone	GH-RH	Hypothalamus	Release of GH
Human chorionic gonadotropin	hCG	Placenta	Maintain progesterone in early pregnancy
Human placental lactogen	HPL	Placenta	Stimulates estrogen and progesterone production by corpus luteum Development of mammary gland
Insulin	None	Beta cells of pancreatic islets	Carbohydrate metabolism
Luteinizing hormone	LH	Anterior pituitary	Female: Ovulation Final follicular growth Male: Production of testosterone
Oxytocin	None	Posterior pituitary	Contraction of gravid uterus Ejection of milk
Parathyroid hormone	PTH	Parathyroid	Calcium regulation
Progesterone	None	Corpus luteum	Prepare uterus for pregnancy Prepare breast for lactation
Prolactin	PRL	Anterior pituitary	Lactation
Prolactin-inhibiting factor	PIF	Hypothalamus	Inhibition of PIF
Serotonin	None	Enterochromaffin cells in GI tract	Neurotransmitter
Testosterone	None	Testis	Spermatogenesis Male sexual characteristics
Thyroid-stimulating hormone	TSH	Anterior pituitary	Release of thyroid hormones
Thyrotropin releasing hormone	TRH	Hypothalamus	Release of TSH
Vasoactive intestinal peptide	VIP	Nerve fibers in GI tract	Relaxation and vasodilation of smooth muscles Neurotransmitter

7. Which of the following substances would be found in large amounts in a patient suffering from carcinoid syndrome?
   a. Metanephrine
   b. Gastrin
   c. VIP
   d. Serotonin
   e. VMA

8. Which of the following is the precursor for estradiol formation in the placenta?
   a. Maternal testosterone
   b. Maternal progesterone
   c. Placental hCG
   d. Fetal adrenal cholesterol
   e. Fetal adrenal DHEAS

9. The best screening test to establish the diagnosis of hypercortisolism is:
   a. Urinary free cortisol
   b. Overnight dexamethasone suppression test
   c. LDDS
   d. HDDS
   e. ACTH

10. Which of the following target tissues is incapable of producing steroid hormones?
    a. Placenta
    b. Ovary
    c. Testis
    d. Adrenal cortex
    e. Adrenal medulla

REFERENCES

1. Griffin JE. Dynamic tests of endocrine function In: Wilson JD, Foster DW, eds. Williams' textbook of endocrinology. Philadelphia: WB Saunders, 1992;1663.
2. Whitley RJ, Meikle AW, Watts NB. Endocrinology. In: Burtis CA, Ashwood ER, eds. Tietz textbook of clinical chemistry. Philadelphia: WB Saunders, 1994;1645.
3. Wilson JD, Foster DW. Introduction. In: Wilson JD, Foster DW, eds. Williams' textbook of endocrinology. Philadelphia: WB Saunders, 1992;1.
4. Yalow RS, Berson SA. Immunoassay of endogenous plasma insulin in man. J Clin Invest 1960;39:1157.
5. Manger WM, Steinland OS, Nakes GG, et al. Comparison of improved fluorometric methods used to quantitate plasma catecholamines. Clin Chem 1969;15:1101.
6. Pisano JJ. A simple analysis for normetanephrine and metanephrine in urine. Clin Chim Acta 1960;5:406.
7. Pisano JJ, Crout RJ, Abraham D. Determination of 3-methoxy-4-hydroxymandelic acid in urine. Clin Chim Acta 1962;7:285.
8. Thorner MO, Vance ML, Horvath E, et al. The anterior pituitary. In: Wilson JD, Foster DW, eds. William's textbook of endocrinology. Philadelphia: WB Saunders, 1992;221.
9. Foster DW. Eating disorders: obesity, anorexia nervosa and bulimia nervosa In: Wilson JD, Foster DW, eds. Williams' textbook of endocrinology. Philadelphia: WB Saunders, 1992;1335.
10. Watts NB. Anterior pituitary. In: Noe DA, Rock RC, eds. Laboratory medicine. The selection and interpretation of clinical laboratory studies. Baltimore: Williams & Wilkins, 1994;618.
11. Reichlin S. Neuroendocrinology. In: Wilson JD, Foster DW, eds. Williams' textbook of endocrinology. Philadelphia: WB Saunders, 1985;492.
12. Cara JF. Growth hormone in adolescence. Endocrinol Metab Clin North Am 1993;22:533.
13. Cryer PE. Glucose homeostasis and hypoglycemia. In: Wilson JD, Foster DW, eds. Williams' textbook of endocrinology. Philadelphia: WB Saunders, 1992;1223.
14. Molitch ME. Clinical manifestations of acromegaly. Endocrinol Metab Clin North Am 1992;21:597.
15. Howanitz JH, Howanitz PJ, Henry JB. Evaluation of endocrine function. In: Henry JB, ed. Clinical diagnosis and management by laboratory methods. Philadelphia: WB Saunders, 1993;308.
16. Chang-Demoranville BM, Jackson IMD. Diagnosis and endocrine testing in acromegaly. Endocrinol Metab Clin North Am 1992;21:649.
17. Faglia G, Arosio M, Bazzoni N. Ectopic acromegaly. Endocrinol Metab Clin North Am 1992;21:575.
18. Losa M, Schopohl J, von Werder K. Ectopic secretion of growth hormone-releasing hormone in man. J Endocrinol Invest 1993;16:69.
19. Cuneo RC, Salomon F, McGauley GA, et al. The growth hormone deficiency syndrome. Clin Endocrinol 1992;37:387.
20. Iramanesh A, Veldhuis JD. Clinical pathophysiology of the somatotropic (GH) axis in adults. Endocrinol Metab Clin North Am 1992;21:783.
21. Jenkins D, Stewart PM. Advances in medical therapy for pituitary disease: treating patients with growth hormone excess and deficiency. J Clin Pharm Ther 1993;18:155.
22. Kaplan S. The newer uses of growth hormone in adults. Adv Intern Med 1993;38:287.
23. Corpas E, Harman SM, Blackman MR. Human growth hormone and human aging. Endocr Rev 1993;14:20.
24. Lufkin EG, Kao PC, O'Fallon WM, et al. Combined testing of anterior pituitary gland with insulin, thyrotropin-releasing hormone. Am J Med 1983;75:471.
25. Molitch ME. Pathologic hyperprolactinemia. Endocrinol Metab Clin North Am 1992;21:877.
26. Szerlip H, Palevsky P, Cox M. Sodium and water. In: Noe DA, Rock RC, eds. Laboratory medicine. The selection and interpretation of clinical laboratory studies. Baltimore: Williams & Wilkins, 1994;692.
27. Reeves WB, Andreoli TE. The posterior pituitary and water metabolism. In: Wilson JD, Foster DW, eds. Williams' textbook of endocrinology. Philadelphia: WB Saunders, 1992;311.
28. Buenocore CM, Robinson AG. The diagnosis and management of diabetes insipidus during medical emergencies. Endocrinol Metab Clin North Am 1993;22:411.
29. Saper CB, Breder CD. The neurologic basis of fever. N Engl J Med 1994;330:1880.
30. Vokes TJ, Robertson GL. Disorders of antidiuretic hormone. Endocrinol Metab Clin North Am 1988;17:281.
31. Endres DB, Rude RK. Mineral and bone metabolism. In: Burtis CA, Ashwood ER, eds. Tietz textbook of clinical chemistry. Philadelphia: WB Saunders, 1994;1887.
32. Sisson JC, Gross MD, Frietas JE, et al. Combining provocative agents of calcitonin to detect medullary carcinoma of the thyroid. Henry Ford Hosp Med J 1981;29:75.
33. Woodhead JS. The measurement of circulating parathyroid hormone. Clin Biochem 1990;23:11.
34. Seymour JF. Malignancy-associated hypercalcemia. Sci Am Science & Med 1995;Sept/Oct:48.
35. Federmann DD. The adrenal. In: Rubenstein E, Federmann DD, eds. Scientific American medicine. New York: Scientific American, 1994;6:1.
36. Kaplan N. Endocrine hypertension. In: Wilson JD, Foster DW, eds. Williams' textbook of endocrinology. Philadelphia: WB Saunders, 1992;707.
37. Weisberg L, Cox M. Potassium. In: Noe DA, Rock RC, eds. Laboratory medicine. The selection and interpretation of clinical laboratory studies. Baltimore: Williams & Wilkins, 1994;732.
38. Melby JD. Diagnosis of hyperaldosteronism. Endocrinol Metab Clin North Am 1991;20:247.
39. Orth DN, Kovacs WJ, DeBold CR. The adrenal cortex. In: Wilson JD, Foster DW, eds. Williams' textbook of endocrinology. Philadelphia: WB Saunders, 1992;489.
40. Honour JW. The investigation of adrenocortical disorders. JIFCC 1994;6:154.
41. Karlsson FA, Kampe O, Winqvist O, et al. Autoimmune endocrinopathies 5. Autoimmune disease of the adrenal cortex, pituitary, parathyroid glands, and gastric mucosa. J Intern Med 1993; 234:379.
42. Werbel SS, Ober KP. Acute adrenal insufficiency. Endocrinol Metab Clin North Am 1993;22:303.
43. Manager WM, Gifford RW. Pheochromocytoma: current diagnosis and management. Cleve Clin J Med 1993;60:365.
44. Howanitz PJ, Howanitz JH, Henry JB. Carbohydrates. In: Henry JB, ed. Clinical diagnosis and management by laboratory methods. Philadelphia: WB Saunders, 1993;172.
45. Sacks DB. Carbohydrates. In: Burtis CA, Ashwood ER, eds. Tietz textbook of clinical chemistry. Philadelphia: WB Saunders, 1994;928.
46. Saudek CD. Endocrine pancreas. In: Noe DA, Rock RC, eds. Laboratory medicine. The selection and interpretation of clinical laboratory studies. Baltimore: Williams & Wilkins, 1994;646.
47. Bardin CW. Pro-opiomelanocortin peptides in reproductive physiology. Hosp Pract 1988;23:109.
48. Wheatley JK. Evaluating male infertility. Hosp Med 1983;24:157.
49. Patten PE. Ovulatory function and dysfunction. American Clinical Laboratory 1990;10:14.
50. Comerci GD. Diagnosis: excessive growth and precocious sexual development. Hosp Med 1983;24:159.
51. Gibbons WE. Diagnosis: amenorrhea. Hosp Pract 1983;24:57.
52. Veldhuis JD. Management of amenorrhea. Hosp Pract 1988;23:40.
53. Malo JW, Bezdicek BJ. Secondary amenorrhea. Postgrad Med 1986; 79:86.
54. Guay AT. Erectile dysfunction. Postgrad Med 1995;97:127.
55. Painter PC. Discordant HCG results in pregnancy. A method in crisis. Diagn Clin Test 1989;27:20.
56. Moghissi KS, Wallach EE. Unexplained infertility. Fertil Steril 1983;39:5.
57. Norman RJ, Menabawey M, Lowings C, et al. Relationship between blood and urine concentrations of intact human chorionic gonadotropin and its free subunits in early pregnancy. Obstet Gynecol 1987;69:590.
58. Varner MW, Hauser KS. Current status of human placental lactogen. Semin Perinatol 1981;5:123.
59. Varner MW, Hauser KS. Human placental lactogen and other placental proteins as indicators of fetal well-being. Clin Obstet Gynecol 1982;25:673.
60. Rock RC. Endocrinology of the fetoplacental unit: role of estrogen assays. Lab Med 1984;15:95.

61. Henderson AR, Tietz TW, Rinker AD. Gastric, pancreatic, and intestinal function In: Burtis CA, Ashwood ER, eds. Tietz textbook of clinical chemistry. Philadelphia: WB Saunders, 1994;1576.

62. Krejs GJ. Non-insulin-secreting tumor of the gastroenteropancreatic system. In: Wilson JD, Foster DW, eds. Williams' textbook of endocrinology. Philadelphia: WB Saunders, 1992;1567.

63. McCarthy DM. Zollinger-Ellison syndrome. Annu Rev Med 1982; 33:197.

64. Roberts LJ. Carcinoid syndrome in disorders of systemic mastcell activation including systemic mastocytosis. Endocrinol Metab Clin North Am 1988;17:415.

65. Roberts LJ, Oates JA. Disorders of vasodilator hormones: the carcinoid syndrome and mastocytosis. In: Wilson JD, Foster DW, eds. Williams' textbook of endocrinology. Philadelphia: WB Saunders, 1992;1619.

66. Farhl F, Dikman SH, Lawson W, et al. Paragangliomatosis associated with multiple endocrine adenomas. Arch Pathol Lab Med 1976;100:495.

67. Sipple JH. The association of pheochromocytoma with carcinoma of the thyroid gland. Am J Med 1961;31:163.

68. Trundle, DS. Growth factors and their clinical applications. Clinical Laboratory News 1995;6:10.

CHAPTER 19

Thyroid Function

H. Jesse Guiles

Objectives

Upon completion of this chapter, the clinical laboratorian should be able to:

- *Discuss the biosynthesis, secretion, transport, and action of the thyroid hormones.*

- *On an anatomic diagram of the thyroid gland, identify the major structures.*

- *Describe the principle of the negative feedback system as it applies to thyroid hormone production.*

- *Describe the regulation of the thyroid hormones.*

- *Explain the principle of each thyroid function test discussed.*

- *Given a patient's clinical data, correlate laboratory information with regard to suspected thyroid disorders.*

- *Describe the appropriate laboratory thyroid function testing protocol to use to effectively evaluate or monitor patients with suspected thyroid disease.*

KEY TERMS

Follicular cells
Free thyroxine index (FT$_4$I)
Graves' disease
Hashimoto's disease
Hyperthyroidism
Hypothalamic-pituitary-thyroid axis (HPTA)
Hypothyroidism
Myxedema

Parathyroid glands
Perifollicular cells
Subclinical hyperthyroidism
Subclinical hypothyroidism
T$_3$ uptake
Transthyretin (TTR) or thyroxine-binding prealbumin (TBPA)
Thyroglobulin

Thyroid
Thyroxine-binding globulin (TBG)
Thyroid hormone binding ratio (THBR)
Thyroiditis
Thyroid-releasing hormone (TRH) Thyroperoxidase (TPO) antibodies

Thyrotoxicosis
Thyrotropin receptor (TSHR) antibodies
Thyrotropin (TSH)
Thyroxine-binding albumin (TBA)
Thyroxine (T$_4$)
Triiodothyronine (T$_3$)

THYROID ANATOMY AND PHYSIOLOGY

The *thyroid* gland consists of two lobes located in the lower part of the neck. The lobes are connected by a narrow band, called an isthmus, and are typically asymmetric, with the right lobe being larger than the left. The lobes weigh approximately 20 g to 30 g each and measure 4.0 cm in length and 2 cm to 2.5 cm in width and thickness (Fig. 19-1).

Two types of cells form the thyroid: the follicular (or cuboidal) and the perifollicular cells. The *follicular cells* are secretory and produce *thyroxine (T$_4$)* and *triiodothyronine (T$_3$)*. Each follicle is in the shape of a sphere and surrounds a colloid matrix called thyroglobulin. The *perifollicular cells,* or C cells, are situated in clusters along the interfollicular or interstitial spaces. The C cells produce the polypeptide calcitonin, which is involved in calcium regulation.

Adjacent to the thyroid gland are four *parathyroid glands*. Two of these glands are found in the upper portion and two are found near the lower portion of the thyroid gland. The parathyroid glands produce parathyroid hormone, which controls calcium and phosphate metabolism.

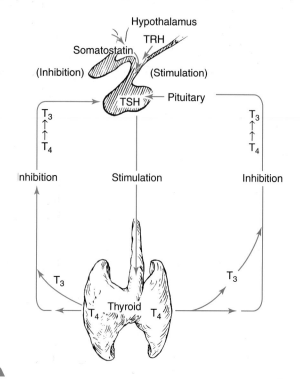

Figure 19-1. The negative feedback system regulating thyroid function.

BIOSYNTHESIS, SECRETION, TRANSPORT, AND ACTION OF THYROID HORMONES

Iodine is the most important element in the biosynthesis of thyroid hormones. Approximately 150 mg of iodide is absorbed in the intestine each day. The thyroid gland has a very high attraction for iodide and traps about 70 mg/day at the base of the cell by active transport. The iodide is then transported to the follicular lumen. The iodide molecule is oxidized by peroxidase, presumably at the interface of the cell and the lumen, to a more reactive form, I^0 or I^+. This form combines with the glycoprotein thyroglobulin. *Thyroglobulin* acts as a performed matrix containing tyrosyl groups to which the reactive iodine attaches to form the hydroxyl residues of monoiodotyrosine (MIT) and diiodotyrosine (DIT).

The next step is the enzymatic coupling of the iodinated tyrosine molecules, catalyzed by peroxidase, to form T_4 or T_3.

The coupling of two DIT molecules (Fig. 19-2) forms T_4. The coupling of one DIT molecule and one MIT molecule results in the formation of T_3 (3, 5, 3'-T_3; Fig. 19-3) or reverse T_3 (rT_3), or (3, 3', 5'-T_3; Fig. 19-4). However, most rT_3 is created by way of peripheral deiodination of T_4 rather than manufacture in the thyroid gland. T_3 contains two iodine atoms in the tyrosyl ring and one iodine atom in the phenolic ring. On the other hand, rT_3 contains two iodine atoms in the phenolic ring and one iodine atom in the tyrosyl ring. These molecules are stored in the thyroglobulin in the thyroid follicle. The stored T_4 and T_3 can be released by enzymatic cleavage of the thyroglobulin molecules. Release of the iodothyronines involves endocytosis of the thyroglobulin from the follicular lumen and hydrolysis of thyroglobulin by a protease and a peptidase in the epithelial cell to liberate free iodoamino acids (T_3, T_4, MIT, and DIT). Most iodothyronines (T_3, T_4,) are secreted into the blood stream, whereas most iodotyrosines (MIT, DIT) are deiodinated within the thyroid, and the iodide is used in producing iodoamino acids.

The serum concentration of T_4 is about 50 times greater than that of T_3. About 80% of circulating T_3 is formed following monodeiodination of T_4 in peripheral tissues, especially the liver and kidney, by 5'-deiodinase. Thus, T_4 can be considered a prohormone for T_3 production. Approximately 33% of the T_4 that is secreted by the thyroid each day undergoes monodeiodination to produce T_3. Another 40% undergoes monodeiodination in the inner ring to produce rT_3. Severe nonthyroidal illness (NTI), certain drugs, or stress can shift the monodeiodination of T_4 to favor rT_3.

Almost all circulating T_4 (~99.97%) and T_3 (~99.7%) hormones are bound to serum proteins. There are three thyroid-hormone-binding proteins: thyroxine-binding globulin (TBG), *transthyretin (TTR) or thyroxine-binding prealbumin (TBPA),* and *thyroxine-binding albumin (TBA).* T_4 binds predominantly to TBG (70%–75%), to a lesser extent to TTR (15%–20%), and to a slight extent to albumin (10%). T_3 is bound by TBG and TBA but very little by TTR. Normally, only one third of the available protein binding sites are occupied by the thyroid hormones.

The relationship between the thyroid hormones and the binding proteins can be described by the law of mass action:

$$(FT_4) \times (\text{unbound TBG}) = K \times (T_4 - TBG),$$

(Eq. 19–1)

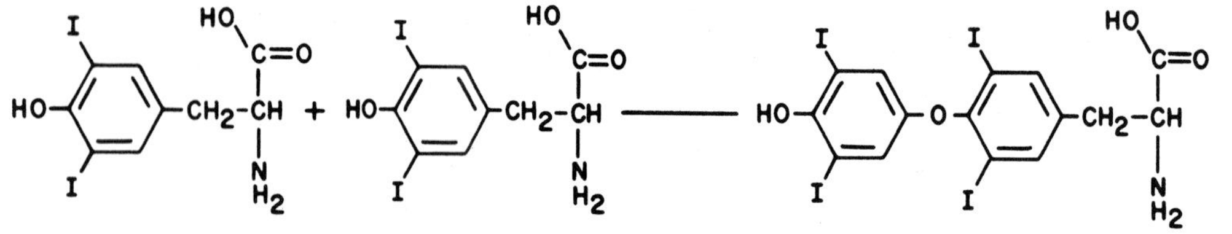

DIT + DIT ⟶ T_4

DIT · DIT · T_4

Figure 19-2. The formation of T_4.

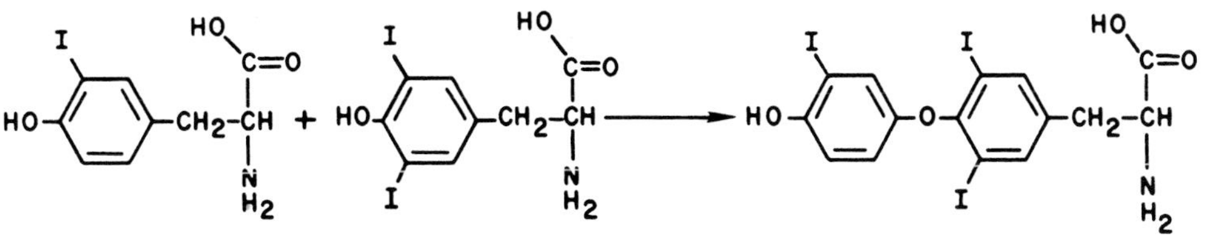

Figure 19-3. The formation of T_3.

where FT_4 is the concentration of free T_4 in serum, unbound TBG is the concentration of unbound sites on TBG, K is the reaction rate constant, and $T_4 - TBG$ is the concentration of T_4 occupying TBG binding sites.

An increase in the concentration of FT_4 or unbound TBG would drive the reaction to the right. Therefore, in situations in which there is an excess of protein, more T_4 will be bound, and in situations in which little protein is available, less T_4 will be bound. However, the concentration of free T_4 should remain stable. This relationship becomes important when hormones, drugs, or disease states affect protein balance within the body.

Only 0.03% of T_4 and 0.3% of T_3 are not bound to proteins. These fractions, called free T_4 (FT_4) and free T_3 (FT_3), are the physiologically active portions of the thyroid hormones. T_3 is the most biologically active thyroid hormone and is three to four times more potent than T_4. T_3 is more active because it is not as tightly bound to the serum proteins as is T_4, and has a greater affinity to target tissue receptors thus diffusing more easily into cells than T_4.

Thyroid hormone actions include calorigenesis and oxygen consumption by means of regulation of carbohydrate, lipid, and protein metabolism; central nervous system activity and brain development; cardiovascular stimulation; bone and tissue growth and development; gastrointestinal regulation; and sexual maturation.

REGULATION OF THYROID HORMONES

The *hypothalamic-pituitary-thyroid axis (HPTA)* is the neuroendocrine system that regulates the production and secretion of thyroid hormones (see Fig. 19-1). Regulation of the thyroid begins with the hypothalamus. *Thyrotropin-releasing hormone (TRH)* is a tripeptide released by the hypothalamus. It travels along the hypothalamic stalk to the beta cells of the anterior pituitary, where it stimulates synthesis and release of thyrotropin or thyroid-stimulating hormone. At the pituitary level, the secretion of TSH seems to be regulated by an interplay of negative feedback from circulating free T_3 and T_4, TRH, and inhibitory hypothalamus neurotransmitters, such as somatostatin and dopamine.[1,2] *Thyrotropin (TSH)* is a glycoprotein consisting of two subunits, alpha and beta, linked noncovalently. The specificity of the hormone action is conveyed by the beta subunit. The alpha subunit is comparable and interchangeable with the alpha subunits of the other glycoprotein hormones, luteinizing hormone (LH), follicle-stimulating hormone (FSH), and human chorionic gonadotropin (hCG). The alpha subunit also influences the overall shape of the glycoprotein and controls the binding to the appropriate receptor on the cell membrane of the thyroid follicle cells. TSH acts like other polypeptide hormones in that when it binds with receptor sites, it activates adenyl cyclase, which then catalyzes a reaction to produce cyclic adenosine monophosphate (AMP).[3,4] TSH is the main stimulus for the uptake of iodide by the thyroid gland and also stimulates the activation of the protease enzymes, which in turn catalyze the hydrolysis of thyroglobulin, the storage forms of T_4 and T_3. TSH also acts to increase the size and number of follicular cells.

The more active FT_3 is thought to enter the target cell, where it binds to receptor sites located in the cell nucleus, probably in the nuclear chromatin. This binding effects changes in the synthesis of particular proteins mediating or reflecting the extranuclear mode of action. There may also

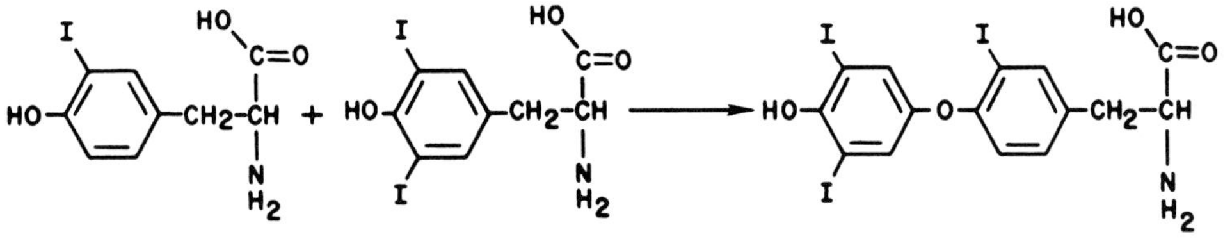

Figure 19-4. The formation of rT_3.

be thyroid-hormone binding sites in the mitochondria, membrane, and other cytoplasmic cell fractions.

The relationship between basal TSH levels and thyroid hormone concentrations is described as HPTA setpoint. A small change in thyroid hormone levels produces a logarithmic change in TSH. However, the exact extent of this relationship varies with each individual.[1,2,5]

ASSAYS FOR THYROID FUNCTION

Assays for thyroid hormones are used to determine and confirm the nature and extent of hyperthyroidism or hypothyroidism. The clinical picture of abnormal thyroid function may be nonspecific and only clarified through laboratory testing. Normal thyroid hormone levels, however, do not rule out thyroid disease, and changes in binding proteins can affect interpretation of some results. Screening of the general public for thyroid disease is not recommended, but patients in high-risk groups such as those with a family history of thyroid disease, newborns, elderly, postpartum women, and patients with autoimmune or endocrine disorders are likely candidates for thyroid testing.[6–8]

Current guidelines for initial thyroid testing call for use of the high-sensitivity thyrotropin/thyroid-stimulating hormone (s-TSH) assay supported by the free thyroxine (FT_4) direct assay or indirect measured estimate, where appropriate.[3–13] Alternately, an FT_4 index (FT_4I) can be calculated based on the total serum thyroxine (TT_4) and T_3 uptake (T_3U) test results. This FT_4I, a more traditional approach, is now seen by most experts in the field as less effective, more costly, unnecessary testing in that current indirect measured estimates of FT_4 are at least as reliable as the calculated FT_4I and require fewer procedural and data manipulations.[1,3,5,9–11] Nevertheless, the traditional assays are still routinely used in some settings. To show a clearer picture of the relationship between free and bound hormones, the effects of protein abnormalities on hormone measurements, and the evolution of laboratory procedure, the traditional thyroid function tests are reviewed.

Specimen Collection and Handling

The protocol for collection and handling for almost all thyroid function tests is generally the same. Specimens are routinely collected in a clot tube (red top/yellow top), although anticoagulants can be used. Blood from newborns can be blotted onto filter paper. The patient need not be fasting. Serum should be separated from cells as per usual laboratory protocol. If specimens need to be stored, the sample can be refrigerated for 24 hours, or frozen for 30 days. Repeated freezing and thawing should be avoided. Specimens free of lipemia or hemolysis are preferred.

Total Thyroxine (TT_4)

T_4 has routinely been measured by radioimmunoassay (RIA); however, other immunometric techniques such as immunochemiluminometric or immunoenzymometric have become

widely accepted.[3] In the *total thyroxine (TT_4)* RIA methodology, thyroid hormone is first released from the endogenous proteins by the addition of a reagent such as 8-anilino-1-naphthalene-sulfonic acid (ANS). The sample is then incubated with thyroid antibody and labeled[125] $[I]T_4$ in a barbital buffer. After incubation, the antigen–antibody bound portion and the remaining free portion are separated, and the labeled bound portion is quantified. Calibrators are run, and the unknowns and controls are extrapolated from a standard curve. The detection limit of this assay can be as low as 0.25 mg/dL.

Nonisotopic immunotechniques are widely available. The principles of these assays are similar to RIA except that the labeled analyte may be an enzyme as in the enzyme-linked immunosorbent assay (ELISA), a fluorophore, or chemiluminescence. Other immunotechniques may use a double antibody or a solid-phase system. The enzyme-multiplied immunoassay technique (EMIT) is an enzymatic method that does not require separation of the free and bound portions. Reference ranges are given in Table 19-1.

Besides rate of thyroid synthesis and release, serum TT_4 levels depend on two other major variables, the first of which is the serum concentration of T_4 binding proteins. This concentration varies with pregnancy, acute illnesses, or the effects of some drugs. Generally, when these proteins increase, more T_4 will bind, with a consequential increase of TT_4 levels. Conversely, a decrease in these proteins or a reduction in available binding sites because of competition by other substances (*eg,* drugs) results in less T_4 binding, realizing a lower TT_4 level.[2,3,5,6,9,14,19] Theoretically, the FT_4 level

TABLE 19-1. Reference Ranges for Thyroid Function Tests

Test	Reference Range
Total T_4	
Adult	4.5–13 µg/dL
Cord-blood	7.4–13.1 µg/dL
Newborn	7.0–17.0 µg/dL
THBR (T_3 uptake)	25%–30%
THBR (T_3 uptake) ratio	0.8–1.35
FT_4I (index)	4.5–12
FT_4	1.2–2.5 ng/dL
s-TSH	0.5–5.0 µU/mL
TRH stimulation (TSH)	>2 µU/mL after 20 min
Total T_3	60–220 ng/dL
rT_3 (25–62 ng/dL)	85–205 ng/dL
FT_3I (index)	15–24 µg/mL
FT_3 (0.2–0.4 ng/dL)	10–40 ng/mL
TBG	<10 ng/mL
Thyroglobulin	10%–30%
Thyroglobulin (marker)	<1:100
RAIU	<1:40
Antithyroglobulin antibodies	Varies with method
TPO antimicrosomal antibodies	
TSHR antibodies	

should remain within the reference range under these conditions, as explained by the law of mass action.

The other major influence on TT_4 levels is the peripheral conversion of T_4 to T_3 and rT_3. NTI illnesses may suppress the conversion of T_4 to T_3, or favor conversion of T_4 to rT_3. Diminished peripheral conversion of T_4 to T_3 has been reported in protein-calorie deprivation, in fasting, and after the administration of certain drugs, such as propranolol hydrochloride, glucocorticoids, propylthiouracil, and ipodate (x-ray contrast media).[1,3,6,12,14,15,19]

Thyroid Hormone Binding Ratio or T_3 Uptake (T_3U) Test

The *thyroid hormone binding ratio (THBR)* assay is used to measure the number of available binding sites of the thyroxine-binding proteins, most notably TBG. It is illustrated in Figure 19-5. Historically, the THBR assay used[125] $[I]T_3$ and consequently was commonly called the T_3 *uptake* test. To reduce the confusion between the T_3U test and measurement

of patients' endogenous T_3, the American Thyroid Association recommended the name change.

Current the THBR assay uses ^{125}I, enzyme, fluorophore, and chemiluminescent labeling techniques. In this assay, the patient's serum is mixed with labeled T_3 and a binding material such as a resin, charcoal, silicate, or antibody. Some of the labeled T_3 binds to the available binding sites of the proteins and the remainder binds to the added binding agent. The amount of labeled T_3 taken up by the binding agent is inversely proportional to the number of available binding sites on the TBG. The result is calculated by dividing the matrix-bound product/counts by the residual serum protein-bound product/counts.

The THBR is usually reported as a ratio of the patient's THBR to the average uptake of normal pooled serum. This enables the calculation of a free thyroxine index. Reference ranges are listed in Table 19-1.

The relationship of the THBR assay and the TT_4 is also illustrated in Figure 19-5. In *hyperthyroidism*, the TBG binding sites are occupied by the patient's endogenous T_4, resulting in

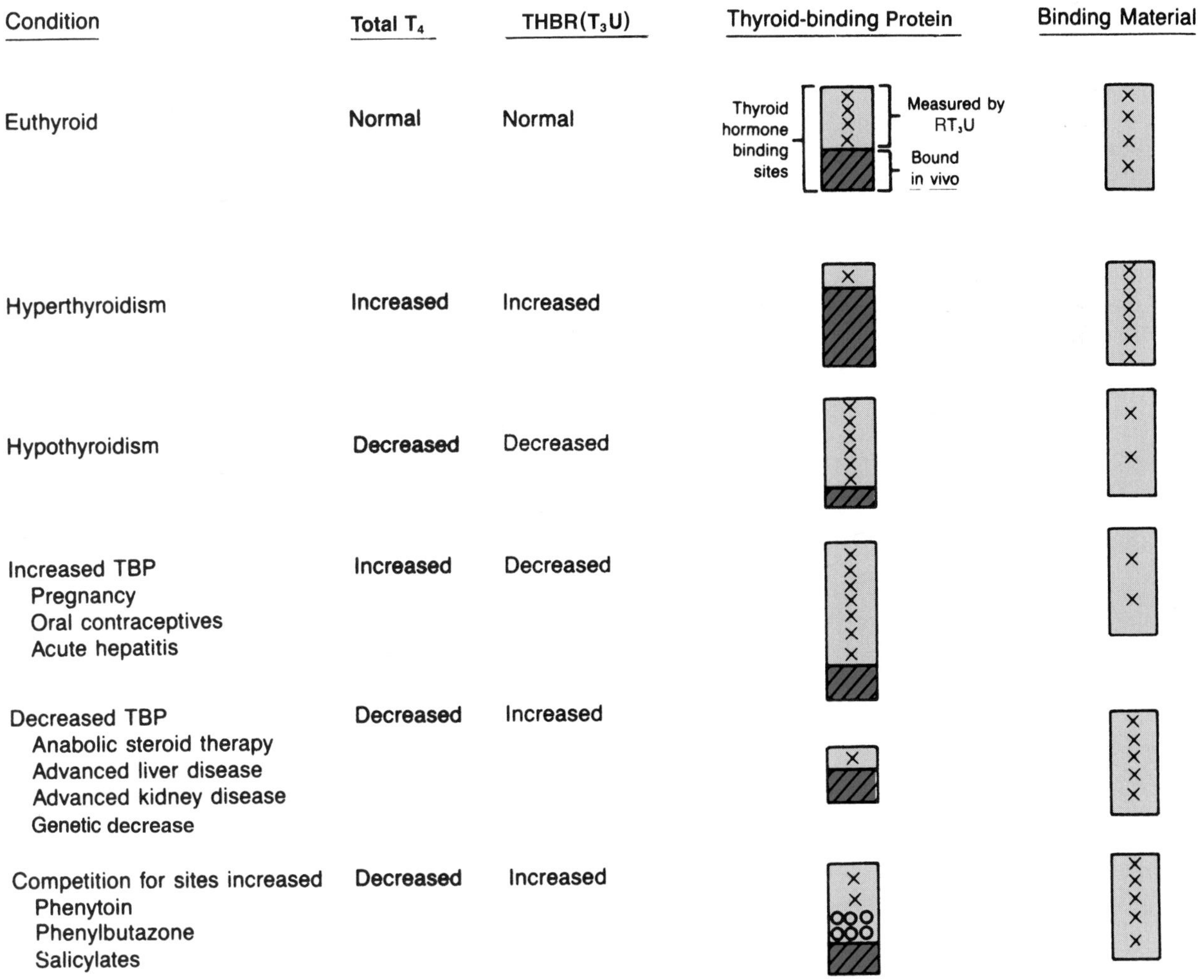

Condition	Total T_4	THBR(T_3U)	Thyroid-binding Protein	Binding Material
Euthyroid	Normal	Normal		
Hyperthyroidism	Increased	Increased		
Hypothyroidism	Decreased	Decreased		
Increased TBP Pregnancy Oral contraceptives Acute hepatitis	Increased	Decreased		
Decreased TBP Anabolic steroid therapy Advanced liver disease Advanced kidney disease Genetic decrease	Decreased	Increased		
Competition for sites increased Phenytoin Phenylbutazone Salicylates	Decreased	Increased		

Figure 19-5. The THBR (T_3U) assay and FT_4.

an increased amount of labeled T_3 being bound to the binding agent (or an increased THBR). The converse is true with hypothyroidism. Protein abnormalities also affect the THBR. Patients who take oral contraceptives, who are pregnant, or who have acute hepatitis will have a consequent increase of TBG with an increased TT_4 and a decreased THBR result. Conditions such as an anabolic steroid therapy, NTI such as advanced liver or kidney disease, or a genetic defect can result in a general decrease of TBG, causing a decreased TT_4 and increased THBR result. Competition for binding sites by such drugs as phenytoin, phenylbutazone, or salicylate reduce patient TT_4 and reagent labeled T_3 binding to TBG, resulting in an increased THBR result.[1–6,9,12,14,15]

Other assays are available that use labeled T_4 instead of T_3 to saturate the TBG and generally have a direct relationship between the amount of labeled product detected and the TBG present. These methods may be more specific than the THBR test in certain abnormal physiologic conditions (eg, dysalbuminemia), but generally convey the same type of information as the THBR.

Calculated Free Thyroxine Index (FT$_4$I)

The TT_4 and THBR test are done concurrently to make a calculated estimation of the level of free thyroxine (FT_4I). In theory, the conditions affecting thyroid hormone binding should not affect FT_4 levels. To estimate the effects of protein or interfering drugs, a calculation can be performed. The *free thyroxine index (FT$_4$I)* is an indirect measure of free hormone concentration and is based on the equilibrium relationship of bound T_4 and FT_4. The FT_4I is calculated by the following formula:

$$FT_4I = TT_4 \times THBR \qquad \textit{(Eq. 19–2)}$$

Example:

$$T_4 = 12 \text{ mg/dL}$$

$$THBR = 25\% \text{ (normal} = 30\%)$$

$$THBR \text{ ratio} = 25/30 = 0.83$$

$$FT_4I = 12 \times 0.83$$

$$= 10.0 \qquad \textit{(Eq. 19–3)}$$

The FT_4I reference range is included in Table 19-1.

The FT_4I is useful in correcting for euthyroid individuals who have altered binding-protein concentrations and who would have been misclassified on the basis of TT_4. For small or moderate variations of binding-protein concentrations, the FT_4I is an adequate indicator of thyroid status and has the advantage of being simple, rapid, and inexpensive. However, there is a curvilinear relationship between the proportion of bound T_4 and the percentage of free hormone. The direct FT_4 assay is seen as a better indicator of thyroid status than the calculated FT_4I in cases of extreme protein abnormalities, such as congenital TBG excess or deficiency, dysalbuminemia, or NTI (also known as the euthyroid sick syndrome.)[1,16,17]

Measured Free Thyroxine (FT$_4$)

Equilibrium dialysis (or ultrafiltration) is considered the direct/reference method for FT_4 determination.[16,18] It is the most reliable FT_4 test in cases in which severe NTI is concurrent with suspected thyroid disease. In this assay, the sample is separated from a buffer containing labeled T_4. An equilibrium is reached, and a ratio of detectable T_4 on both sides of the membrane is obtained. This "dialyzable" proportion is considered percent FT_4. The critical step in this procedure is separation of the free hormone from the bound hormone before analysis. Only minimum dilution is permissible, because dilution of sera can distort the equilibrium between free and bound hormone. Although reliable, the method is technically demanding, relatively expensive, and time consuming. It should be used as a confirmation test where FT_4 indirect estimation measurements are equivocal.[2,17,18]

FT_4 analogs are considered "indirect" because readings are converted into results by interpolating off a curve in which standards have been calibrated independently by equilibrium dialysis. As with the calculated $FT_4 I$, they work best when the influence of protein variability is minimal. These methods are either one or two step. In the two-step technique, FT_4 is immuno-extracted from serum onto a limited antibody. Next, a washing step eliminates proteins and any other interfering substances. A second labeled hormone is then added, which binds to the unoccupied antibody sites (eg, "back titration"). There is an inverse relationship between the patient's FT_4 and the amount of bound labeled antibody measured. The advantage of the two-step protocol is assurance that the labeled hormone cannot interact with thyroxine-binding proteins in the sample.

The one-step immunometric technique is similar to the two-step technique except the labeled analog is chemically modified to prevent it from binding to serum transport protein, while simultaneously enabling it to bind to the reagent antibody. It is maintained that FT_4 can be measured directly without significantly disturbing the natural equilibrium between the free and bound T_4 during the assay. The reliability of a one-step procedure is its ability not to upset the T_4–FT_4 equilibrium, and to minimize binding between the analog and the endogenous thyroxine-binding proteins in the sample.[18]

A fully automated competitive immunoassay using acridinium ester-labeled T_4 is currently available for measurement FT_4. The sample FT_4 competes with a chemiluminescent labeled FT_4 analog for a limited amount of antibody. Separation occurs through paramagnetic particles in the solid phase that are covalently bound to the antibody. The amount of FT_4 in the sample is inversely proportional to the amount of light detected.

FT_4, as measured by one-step analog techniques, has been shown to be affected by protein abnormalities such as dysalbuminemia, pregnancy, first-trimester with hCG elevation, severe illness, and drugs such as dopamine, amiodarone, and glucocorticoids.[1–3,6,9,12,16,19,20] With current technology, however, some FT_4 analog methods appear to be more sensitive, specific, and direct than FT_4I in differentiating euthyroid from hyperthyroid versus hypothyroid patients.

TSH

In today's health care environment of clinical pathways and reflex testing in which the focus is on cost-containment without reduction of medical effectiveness, the s-TSH assay is considered to be the first and best laboratory test for identification of thyroid abnormalities.[1,3,6,10,11,19] It has been shown to be extremely useful in the detection and confirmation of suspected primary hypothyroidism and hyperthyroidism (even in borderline cases). Subclinical hyperthyroidism— or hypothyroidism—is an abnormal TSH level and a normal TT_4 or FT_4 in the absence of overt symptoms.[4] Although the actuality of subclinical thyroid disorders is accepted, a question remains as to whether treatment of these conditions is necessary to prevent complications associated with overt hyperthyroidism or hypothyroidism.[4,7,20–23]

s-TSH has replaced the traditional monoclonal RIA technique for the hormone. The monoclonal RIA test was not sensitive enough to distinguish between the very low TSH values found in primary hyperthyroidism from those found in some healthy euthyroid patients.

Whereas the traditional RIA techniques measured an antigen–antibody complex, the high-sensitivity TSH immunometric assays (IMAs) use two or three separate monoclonal antibodies in the methodologies (Fig. 19-6). The re-

quirement of binding to more than one site gives the s-TSH its high specificity and sensitivity. The first antibody is in excess, bound to a solid phase support, and extracts the TSH molecule from the serum. This antibody is usually specific for the beta subunit of the TSH molecule. The second antibody attaches to a different portion of the TSH molecule, usually the alpha subunit, and has a label on it that may be a radioisotope, enzyme (typically peroxidase or alkaline phosphatase), fluorophor, or chemiluminescent compound. The separation of the free and the bound portion is accomplished in one or two steps. The two-step procedure has fewer interferences but is harder to achieve. The double binding results in a labeled insoluble "sandwich" with a concentration directly proportional to the amount of TSH in the specimen.[1,24,25]

A "generation" of TSH assay is characterized by an approximate tenfold sensitivity increase. Today's second-generation IMAs for TSH have sensitivities from 0.1 mU/mL to 0.01 mU/mL. Third-generation TSH tests have detection limits of 0.01 mU/mL to 0.005 mU/mL. They permit more reliable distinction between undetectable TSH in patients with hyperthyroidism and subnormal values that are occasionally found in euthyroid patients, and demonstrate superior precision in the critical subnormal range.[24,25] A key factor in improvement of sensitivity is incorporation of chemiluminescent labels instead of radioactive isotopes in the assays.[2] Currently, work is underway in the development of fourth-generation TSH tests with a sensitivity of 0.001 mU/mL.[1]

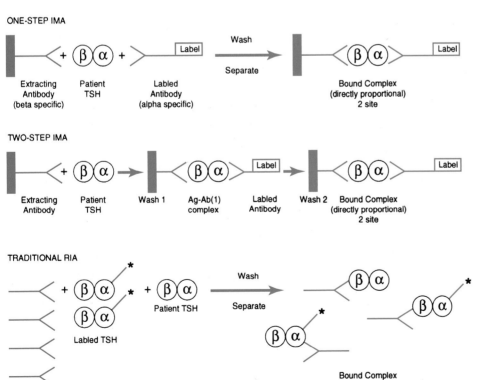

Figure 19-6. Immunometric assay (IMA) and radioimmunoassay (RIA) for TSH. (From Taylor CS, Brandt DR. Developments in thyroid-stimulating hormone testing: the pursuit of improved sensitivity. Lab Med 1994;24:337.)

It has been shown that a minor rise or fall of thyroid hormones, especially FT_4, elicits a dramatically higher inverse response in TSH by the pituitary.[1,3] Thus, the s-TSH is considered to be the most clinically sensitive assay for detection of primary thyroid disease. Furthermore, TSH concentration is not as markedly affected by physiologic alterations resulting from severe NTI or administration of therapeutic drugs as is TT_4 or FT_4.[1,3,12] However, studies have demonstrated that TSH concentration can be affected by severe NTI; pregnancy during the first trimester; treatment with dopamine, glucocorticoids, and certain other drugs; or thyroxine replacement therapy.[2,3,5,6,19] However, the s-TSH is seen as the thyroid function test least affected by these variables. Because of its exceptional clinical sensitivity (with some sacrifice of clinical specificity) along with its proven chemical sensitivity and specificity, the s-TSH assay is today considered the primary screening test for thyroid function.

Thyroid-Releasing Hormone (TRH) Stimulation Test

In the past, due to the insensitivity of the first-generation TSH tests, physicians had relied on the TRH stimulation test to distinguish between euthyroid and hyperthyroid patients who both had undetectable TSH levels. The TRH stimulation test detects residual TSH stores in the pituitary gland. In this test, TRH is injected intravenously under the direction of a physician. In euthyroid patients, the TSH will rise from 10 mU/mL to 30 mU/mL within 30 minutes. Patients with primary hyperthyroidism will not respond to the TRH and their s-TSH will remain at baseline levels. Patients with primary hypothyroidism will have an exaggerated response. However, because baseline TSH levels are typically elevated in primary hypothyroidism, the TRH test adds little information in exchange for the extra cost and pain to the patient.

With the advent of the high-sensitivity TSH assays, the TRH test has become obsolete for the confirmation of primary hyperthyroidism or detection of subclinical hyperthyroidism. The test, however, may still be useful in some special situations. In discriminating between NTI sick and hyperthyroid patients, euthyroid sick patients will generally respond to TRH, whereas hyperthyroid sick patients will not respond.[24] The TRH test may also be useful in the detection of thyroid hormone resistance syndromes.[26] Furthermore, the TRH test can be used in differentiating secondary (pituitary) from tertiary (hypothalamic) hypothyroidism.[3] In both conditions, the patients have a low FT_4 with a borderline or inappropriately normal TSH. When TRH is injected in a patient with secondary hypothyroidism, the TSH level will not increase or will be blunted, whereas a patient with tertiary hypothyroidism will respond to TRH stimulation. Some studies, however, have shown that, even though the pituitary is so damaged that it cannot secrete TSH on a sustained basis, residual thyrotrophs may have a TSH response to TRH that is within normal limits. Conversely, an abnormal response to the TRH test may occur with severe NTI and not be diagnostic.

Thyroid-Binding Proteins/TBG and Thyroglobulin

Aside from the assessment of thyroid hormone binding to thyroid-binding proteins by means of the THBR test, thyroid-binding proteins also can be measured directly by means of immunotechniques. Measurement of TBG is used to confirm results of FT_3 or FT_4 or abnormalities in the relationship of the TT_4 and THBR test. The direct measure of TBG is less technically cumbersome than the THBR and better reflects immunospecificity.

The thyroglobulin assay is normally used as a postoperative marker of thyroid cancer.[19] Low levels may indicate

---

CASE STUDY 19-1

A 34-year-old woman complained of fatigue, lethargy, constipation, nausea, amenorrhea, and anxiety. Her heart rate was increased. She had brought laboratory results with her from a community screening program that revealed several abnormal results, including increased glucose and alkaline phosphatase levels. The patient's complete blood count showed a decreased hemoglobin, hematocrit, and erythrocyte count. She stated that she had felt relatively healthy until about 3 months ago. She was not taking oral contraceptives.

Laboratory Results

$FT_4 = 8.10$
s-TSH = 5.0
(Refer to Table 19-1 for reference ranges.)

Questions

1. From the case history and the laboratory data, what is a likely diagnosis?
2. What factors in the data lead you to the conclusions in question 1?
3. What additional laboratory test(s) would be appropriate?

the presence of thyrotoxicosis factitia. The presence of circulating thyroid autoantibodies will interfere with thyroglobulin test results. Reference ranges are included in Table 19-1.

Total T_3 (TT_3), Free T_3 (FT_3), and Free T_3 Index (FT_3I)

Total serum T_3 is measured by immunometric techniques similar to that of TT_4. TT_3 is not very useful as a primary screening test for hypothyroid disorders because it generally does not give any more information than TT_4, and too many nonthyroidal factors affecting binding proteins (eg, NTI) can influence levels of TT_3. Nevertheless, T_3 is more potent than T_4, and more circulating T_3 is in the free state and available to the tissues. Thus, T_3 more closely relates to the clinical state of the patient than does T_4.[2,3]

The rate of formation from T_4 peripheral deiodination declines during almost all significant NTIs and stresses such as cirrhosis of the liver, renal failure, starvation, anorexia nervosa, cystic fibrosis, administration of certain drugs (propranolol, propylthiouracil), and some contrast media. It is estimated that TT_3 levels are decreased in approximately 70% of hospitalized patients who do not have hypothyroidism, whereas as many as 20% to 30% of patients with hypothyroidism have normal TT_3 concentrations.[8]

The value of the TT_3 or FT_3 assay is in confirming hyperthyroidism, especially in T_3 thyrotoxicosis, in which the s-TSH is decreased, but the FT_4I or FT_4 is within normal limits. Early in hyperthyroidism, TT_3 may be more elevated than TT_4.

An FT_3 level can be performed along the same manner as an FT_4; and a free T_3 index (FT_3I) is calculated similar to FT_4I by using the formula

$$FT_3I = T_3 \times THBR \text{ ratio} \qquad (Eq. 19\text{--}4)$$

FT_3 and FT_3I are usually not offered as routine tests. The tests offer no advantage over the FT_4I or FT_4 in diagnosing usual hypothyroidism and give more false-negative and false-positive results. A normal FT_3I does not exclude hypothyroidism, and a low FT_3I does not establish the diagnosis of hypothyroidism. Reference ranges for TT_3, FT_3, and FT_3I are listed in Table 19-1.

Reverse T_3 (rT_3)

Metabolically inactive reverse T_3 (rT_3) also can be measured by immunoassay. It is usually only ordered to help in the resolution of borderline or conflicting laboratory results. Almost all rT_3 comes from peripheral deiodination of T_4.

rT_3 is commonly elevated in patients with NTI in which the TT_3 level is often concurrently decreased.[20] Thus rT_3 can be used when trying to determine thyroid status of patients who are also suffering from NTI. Furthermore, rT_3 is elevated in amniotic fluid at an early stage of pregnancy. The assay for rT_3 in amniotic fluid could be useful in the diagnosis of fetal hypothyroidism and thyroiditis. However, because hypothyroidism can be diagnosed at birth and thyroid replacement therapy in utero is unnecessary, this assay has not been extensively used. The rT_3 reference range is listed in Table 19-1.

Thyroid Antibodies

Individuals with certain thyroid diseases may have circulating antibodies against several components of thyroid follicles. These antibodies can act as thyroid stimulators or blockers and are associated with hyperthyroid or hypothyroid states. They include the thyrotropin receptor (TSHR) antibodies, thyroperoxidase (TPO)—formerly known as thyroid antimicrosomal antibodies, and antithyroglobulin antibodies.

Thyrotropin receptor (TSHR) antibodies are of two functional types: those that stimulate the thyroid gland (thyroid-stimulating immunoglobulins [TSI]), and those that inhibit the action of TSH (TSH blocking antibody [TSHBAb]), or thyroxine-binding inhibitory immunoglobulins [TBII]). TBII is thought to cause certain cases of hypothyroidism.[3,27] A positive TBII indicates that antibodies are present. If the TSI is negative, the antibodies are blocking; if the TSI is positive, the antibodies are stimulating.[27] TSI can become a functional analog of TSH and can bind so strongly to the TSH receptor that TSH is stopped from binding on the thyroid membranes. Like TSH, TSHR can activate the cells to which they bind. However, unlike TSH, they are not subject to feedback control. As a result, there is overproduction of thyroid hormones. Other names for TSHR include thyroid-stimulating antibodies (TSAb), long-acting thyroid stimulator (LATS), TSH-binding immunoglobulin, and LATS-protector (LATS-P).

Thyroperoxidase (TPO) antibodies bind with the microsomal antigens associated with the follicular cells lining the microsomal membrane. They are mainly IgG and fix complement. It is believed that damage to the thyroid gland is brought about by the interaction of lymphocytic killer cells, which effect tissue destruction. TPO antibodies have been associated with patients with hyperthyroid as well as hypothyroid disorders.[2,19,27]

Thyroglobulin antibodies are mainly IgG and do not fix complement. These antibodies do not correlate well with thyroid disease and may not damage tissue. Increased levels may indicate only that there is something amiss with the immune system in relation to the gland. Occasionally, another antibody to the colloid is also found, but it seems to have no clinical significance.

Thyroid antibodies are detected by a variety of methods from hemagglutination tests to bioassays. These methods are listed in Table 19-2. Correlation of the presence of these antibodies with thyroid disease is discussed later; however, their detection does not signify the presence of overt disease, and conversely, their absence does not rule out thyroid disease.

TABLE 19-2. Antigen–Antibody Systems Involved in Humoral Responses of Thyroid Autoimmune Disease

Antigen	Antibody (Function)	Antibody Detection
Thyroglobulin	Thyroglobulin antibody (no clear function)	Precipitin technique; tanned red cell hemagglutination; immunofluorescence on fixed thyroid sections; competitive binding radioimmunoassay; co-precipitation with 125I-thyroglobulin; micro–ELISA; plaque-forming assay
Thyroperoxidase	Thyroperoxidase (microsomal) antibody (cytotoxic in conjunction with lymphocytes)	Complement fixation; immunofluorescence on unfixed thyroid sections; cytotoxicity test on cultured thyroid cells; competitive binding; radioimmunoassay; tanned red cell hemagglutination; micro–ELISA
Second colloid component	CA2 antibody (no clear function)	Immunofluorescene on fixed thyroid sections
Cell surface antigen(s)	Membrane antibodies (cytotoxic with lymphocytes)	Immunofluorescence on viable thyroid cells; hemadsorption; binding assays
Thyroxine and triiodothyronine	Thyroid hormone antibodies (bind and prevent hormone action)	Antigen-binding capacity
Antigen not defined	Growth-stimulating and growth-inhibiting antibodies (may induce or inhibit thyroid growth)	Effects on deoxyribonucleic acid content per thyroid cell nucleus or G6PD activity per cell
TSH receptor–related antigen	TSH receptor antibodies (may stimulate thyroid cells, inhibit, neither)	Stimulatory assays: current terms used include human thyroid stimulator; human thyroid stimulating immunoglobulin (TSI); thyroid stimulating antibody (TSAb) LATS bioassay; colloid droplet formation in human thyroid slices; stimulation of human thyroid adenylate cyclase in vitro cytochemical assay Binding assays: LATS protector assay; inhibition of 125I-thyrotropin binding to human thyroid membranes (thyrotropin binding inhibitor immunoglobulin [TBI]); fat cell membrane radio ligand assay Inhibitory assays: TSH receptor stimulation blocking antibody (TSBAb)

(From Volpe R. Rational use of thyroid function tests. Crit Rev Clin Lab Sci 1997;19:415–416. Adapted with permission.)

Radioactive Iodine Uptake (RAIU) and Thyroid Scan

The RAIU assay is used to measure the ability of the thyroid gland to trap iodine. Radioactive iodine is ingested by mouth, and the radioactivity is counted at various lengths of time, usually 4 hours to 6 hours and again at 24 hours. Normally, 10% to 30% of the ingested dose can be detected within 24 hours. The assay is useful in helping to determine the cause of hyperthyroidism.[20] The assay is not useful in the diagnosis of hypothyroidism because there are many other factors that can influence results and because there is overlap of euthyroid and hypothyroid responses.

Thyroid scan is used to determine the size, shape, and activity/inactivity of thyroid tissue/nodules. Thyroid scan involves in vivo use of either ^{123}I or pertechnetate (^{99m}Tc). The isotope is administered to the patient and is taken up by the thyroid gland. The distribution of the radioactive material is determined by scanning the thyroid. The thyroid scan is useful in determining the cause of hyperthyroidism, as well as detecting extrathyroid iodide-concentrating tissue, such as metastatic carcinoma of the thyroid. The most important use of thyroid scanning is to define areas of increased or decreased uptake within the gland. Nodules as small as 1 cm can be detected.

Nodules are classified according to their ability to take up iodine (*ie,* as "cold" or nonfunctioning, "warm" or normal functioning, and "hot" or hyperfunctioning). Theoretically, a malignant lesion should not be able to trap iodine and so should appear as a "cold" area.[28] However, it has been demonstrated that, although cold nodules are the ones having the best probability of being malignant, most are benign, and even a hot nodule does not exclude malignancy.

Fine-Needle Aspiration

Fine-needle aspiration provides information regarding the malignant potential of a thyroid nodule. The fine-needle aspirate is evaluated by means of histologic or cytologic examination.

THYROID DISORDERS AND CORRELATION OF LABORATORY DATA

Hypothyroidism

Hypothyroidism occurs when there are insufficient levels of thyroid hormones to provide metabolic needs at the cellular level. The incidence of hypothyroidism in the United States is about 0.5%,[6,14] and it affects females about four times as often as males. Congenital hypothyroidism exists at a rate of around 1 in 4000 births.[29] However, most patients acquire the disease between age 30 years and 60 years.

Deficiency of thyroid hormones causes many metabolic processes to slow down. Symptoms of hypothyroidism include enlargement of the thyroid gland—or goiter; impairment of cognition (including memory, speech, and attention); fatigue; slowing of mental and physical performance; change in personality; intolerance to cold; exertional dyspnea; hoarseness; constipation; decreased sweating; easy bruising; muscle cramps; paresthesias; and dry skin. Hypothyroidism has been shown to be associated with increased cholesterol, LDL, lipoprotein(a), apolipoprotein B, and increased risk of coronary heart disease.[4]

Myxedema

As the disease progresses into severe hypothyroidism, these features worsen and the condition is referred to as myxedema. *Myxedema* describes the peculiar nonpitting swelling of the skin. The skin becomes infiltrated by mucopolysaccharides so that the face is puffy, especially around the eyes. The features become coarse and the eyebrows thinned. The tongue may become enlarged and the vocal cords thickened, leading to hoarseness. The speech slows. There is usually a weight gain.

The skin tends to be dry and has a yellowish color. This discoloration may be due to carotene accumulation, because there is a reduced rate of conversion of carotene to vitamin A. Body hair is lost and not replaced. Scalp hair often becomes dry and coarse.

A history of progressive mental status dysfunction including confusion, paranoia, psychotic behavior (myxedema madness), apathy, and antisocial tendencies may occur. Myocardial contractility is reduced, and the pulse slows. Effusions may accumulate in the pericardium. Atherosclerosis is accelerated because of high levels of serum triglyceride and cholesterol.

Anemia is common and is due to a reduced rate of red-cell production resulting from hypometabolism, decreased oxygen requirement, and decreased erythropoietin. The anemia is usually normocytic and normochromic, but it can be macrocytic because of folate deficiency from malabsorption or microcytic from the menorrhagia that results in iron deficiency.

Myxedema coma is a severe presentation of hypothyroidism. The clinical symptoms include coma, cardiac dysfunction, hypothermia, hypoglycemia, hypotension, hyponatremia, and respiratory failure. Although the condition is rare, the mortality rate is high and is associated with irreversible cardiovascular collapse.[30]

Congenital Hypothyroidism/Cretinism

Most cases of *congenital hypothyroidism* result from defects in the development or function of the gland itself. Hormone dysgenesis can also cause congenital hypothyroidism. Such disorders are inherited as autosomal recessive traits that cause impaired thyroid synthesis by way of production of defective enzymes. Secondary (pituitary) or tertiary (hypothalamic) congenital hypothyroidism is rare, occurring in about 1 in 60,000 births.[29]

Numerous clinical manifestations have been associated with congenital hypothyroidism, also known as *cretinism*. Among these are puffy face; open mouth with enlarged, protruding tongue; hoarse cry; short, thick neck; narrow forehead; pug nose; short legs; distended abdomen; dry, mottled skin; yellowish skin discoloration; hirsutism; and lethargy. Skeletal and mental development is retarded (with IQ typically below 85) if treatment is not begun in early infancy.[29]

Because most cases of congenital hypothyroidism are not suspected clinically, it is necessary to screen newborns for early diagnosis. This is now a requirement in all 50 states. Either cord blood or capillary blood, usually collected as a filter-paper spot and eluted from the paper before the assay, is used. In the United States and Canada, TT_4 is used as the primary screen for neonatal hypothyroidism with TSH as the confirmatory test; in other parts of the world, TSH is the primary screening test. Using the TSH approach, subclinical cases of hypothyroidism (with less interference from NTI or drugs) are better detected; however, with the TT_4 approach, infants with TBG deficiency or with hypothalamic-pituitary hypothyroidism can be identified. Screening with both tests would probably be the best approach when methods for simultaneous measurement of both hormones on filter-paper spot blood become available.[31]

Acquired Hypothyroidism

Most cases of *acquired hypothyroidism* result from inadequate secretion of the thyroid hormones from a damaged thyroid gland. There are several causes associated with acquired hypothyroidism: chronic thyroiditis; surgical or radioactive iodine treatment of hyperthyroidism, goiter, or cancer; idiopathic atrophy; and metastatic cancer and other infiltrative disorders.

Chronic Autoimmune Thyroiditis (Hashimoto's Disease). With or without goiter, chronic autoimmune thyroiditis, or Hashimoto's disease, is the most common cause of primary hypothyroidism. The peak incidence of Hashimoto's thyroiditis occurs during the 6th decade of life. It occurs in approximately 2% of the population.[32] Patients with *Hashimoto's disease* are usually euthyroid or hypothyroid. In Hashimoto's disease, there is a massive diffuse infiltration of the thyroid by lymphocytes. In addition, plasma cells are abundant, and there is an increased amount of connective tissue. The condition is often an asymptomatic disease caused by an unidentified abnormality in the immune system. This abnormality appears to be genetically determined, because family members of persons with the condition have a high probability of developing the

---

CASE STUDY 19-2

A 32-year-old man was referred for treatment of hyperthyroidism. He had a multinodular goiter where the radioiodine uptake was 70%. His TSH was <0.01 with a FT_4 of 18 µg/dL. He was treated with radioactive ^{131}I and given a regimen of antithyroidal medication. Six months later he returned with a complaint of progressively worse fatigue. His s-TSH was >100, and was positive for antithyroid peroxidase antibodies (>1:300,000 titer). (Refer to Table 19-1 for reference ranges.)

Questions

1. Did the laboratory results indicate hyperthyroidism on the initial visit? (Explain why.)
2. What additional testing could have been done on the initial visit?
3. Do test results on the second visit indicate hyper-, hypo-, or euthyroidism? (Explain why.)
4. What additional laboratory tests could be ordered on the second visit?
5. What general disease state do these results indicate?

---

disease. Furthermore, the disease has been associated with certain histocompatibility antigens.[32] The immune system abnormality leads to the production of several thyroid autoantibodies and lymphocytes sensitized to thyroid antigens and may lead to the destruction of the thyroid tissue. TPO or antimicrosomal or antibodies can be detected in approximately 95% of patients with goitrous Hashimoto's thyroiditis.[3]

Miscellaneous Causes of Hypothyroidism

About one third of the patients receiving treatment for hypothyroidism have had thyroid surgery or radioactive iodine therapy for hyperthyroidism. Graves' disease, the major cause of thyrotoxicosis, may spontaneously terminate in hypothyroidism, presumably as the result of autoimmune thyroid damage. The incidence of hypothyroidism increases with aging, as do the levels of circulating thyroid autoantibodies. However, symptoms of hypothyroidism present more subtly in the elderly population (see Chapter 29 *Clinical Chemistry and the Geriatric Patient*).

Hypothyroidism also may result from a lack of TSH or TRH. In secondary hypothyroidism, the patient usually has a pituitary disorder. A decreased production of TSH may result from pituitary irradiation, intracellular hemorrhage, or Sheehan's syndrome (postpartum pituitary infarction). Ter-

tiary hypothyroidism or hypothalamic failure may result because of tumors, vascular insufficiency, infections, infiltrative processes, or trauma. TRH response should be normal in patients with tertiary hypothyroidism.

Subclinical Hypothyroidism

Subclinical hypothyroidism occurs when there is a mild elevation of TSH with normal levels of T_4, T_3, and FT_4 in patients showing no symptoms of thyroid disease.[4,9,20] One report stated that 20% to 50% of individuals with subclinical hypothyroidism appear to develop overt hypothyroidism within 4 years to 8 years.[4] Another concluded that 10% of patients with borderline raised TSH are at risk of becoming overt hypothyroid within 1 year.[21] However, current evidence is inconclusive as to whether subclinical hypothyroidism is linked to hyperlipidemia or cardiac dysfunction.[23] Subclinical hypothyroidism may represent an adjustment of the setpoint of the HPTA due to nonthyroidal influences (eg, NTI or pregnancy).[19] Whether patients progress to overt thyroid disease depends on age, gender, the underlying thyroid disorder, and degree of functional impairment. The presence of thyroid antibodies in cases of subclinical hypothyroidism would indicate the potential to overt thyroid disease.[1,2,4,7,9,20,23]

CASE STUDY 19-3

An apparently healthy 33-year-old woman undergoes thyroid testing with the following results: FT_4 = 1.2 ng/dL, s-TSH 15 µU/mL. No goiter or any other abnormality is noted. The physician is concerned, but decides to follow the patient for every six months. Six months later the patient's FT_4 = 1.1 ng/dL, s-TSH 25 µU/mL.

Questions

1. What clinical state is the patient apparently exhibiting? (Explain reasoning.)
2. Why is there concern over these results?
3. Why would most laboratory thyroid tests not be useful in this case?
4. What laboratory tests would be useful; why?

Laboratory Evaluation of Hypothyroidism

An algorithm for laboratory testing in the diagnosis of hypothyroidism in ambulatory patients is given in Figure 19-7. The current recommendation for testing is away from multiple thyroid tests (eg, "thyroid profile") to measurement of TSH level, complemented by appropriate FT_4 measurement.[1] The earliest laboratory abnormality noted in primary hypothyroidism is an increased TSH concentration. This may occur even before the patient is symptomatic, as in the case of subclinical hypothyroidism. Furthermore, TSH is probably the best indicator of the possibility of primary hypothyroidism coexisting with NTI.[20] An elevation of basal TSH, therefore, can be considered the single most sensitive and specific marker for primary hypothyroidism.

As hypothyroidism progresses, the indirect FT_4I and FT_4 levels will decrease. TT_4, TT_3, and rT_3 will also be decreased. Thyroid TPO and antithyroglobulin antibodies can be measured to confirm or rule out autoimmune diseases. Secondary and tertiary hypothyroidism may present with clinical hypothyroidism or decreased s-TSH TT_4 and FT_4 levels. Secondary versus tertiary hypothyroidism may be differentiated by the TRH test.[3] A summary of laboratory findings in these conditions can be found in Table 19-3.

If patients are sick, elderly, hospitalized, or on certain medications (corticosteroids, salicylate, phenytoin, dopamine, or lithium), laboratory results may be falsely decreased. Although unsuspected thyroid disease occurs in fewer than 1% of hospitalized patients, thyroid function tests may be abnormal in as many as 70% of these patients. The euthyroid sick syndrome is the most common clinical condition responsible for inaccurate test interpretation. Such abnormalities correspond with changes in thyroid hormone secretion, distribution, and metabolism. In general, the greater the severity of the nonthyroidal abnormality, the more disparate the laboratory results.

TSH levels may also rise in patients who are recovering from severe illness and in newborns. This is a transient situation, and the TSH generally returns to normal within a few days or weeks. In early treatment with amiodarone, TSH levels may increase. Hence, the FT_4 test may be appropriate to distinguish between drug induced TSH elevation and hypothyroidism.[1] In certain cases of subclinical hypothyroidism, there may be minimal change in FT_4 and TSH, thus necessitating the use of the TRH test.[3]

Administration of dopamine or high-dose glucocorticoids will suppress the TSH concentrations toward or into the normal range, causing subsequent decreased TT_4 and TT_3 levels. Furthermore, if thyroid hormone inhibitors/antibodies are in circulation, they may also complicate the findings of the profile.

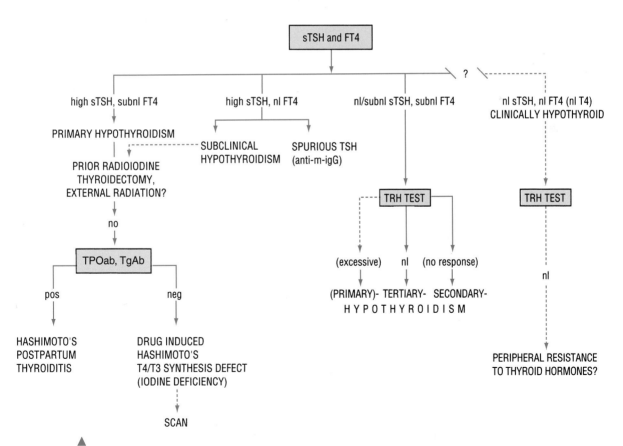

Figure 19-7. Algorithm for the diagnosis of hypothyroidism in ambulatory patients. From Bayer MF. Effective laboratory evaluation of thyroid status. Med Clin North Am 1991; 75. Philadelphia: W. B. Saunders. Reprinted with permission.

TABLE 19-3. Summary of Findings in Thyroid Function Tests in Various Conditions

Condition	TT_4	FT_4I	T_3U	T_3	TSH
Euthyroidism	N	N	N	N	N
Hyperthyroidism	↑	↑	↑	↑	↓
T_3 thyrotoxicosis	N	N	↑	↑	↓
Hypothyroidism					
Primary	↓	↓	↓	↓	↑
Secondary	↓	↓	↓	↓	↓
Tertiary	↓	↓	↓	↓	↓
Increased TBP	↑	N	↓	↑	N
Decreased TBP	↓	N	↑	↓	N
Competition for sites	↓	N	↑	↓	N

Treatment of Hypothyroidism

Treatment of hypothyroidism, except that caused by iodine deficiency, is the administration of L-thyroxine. Proper treatment of subclinical hypothyroidism has been shown to reverse lipid imbalances and improve symptom scoring and psychometric testing.[4] Treatment, however, must be individualized. Cautions regarding treatment of newborns and children must be taken to avoid overtreatment, which can lead to dangers in heart, bone, and neurologic development, and in the elderly population, in which coronary disease may already exist. Because TSH levels may adjust slowly to treatment, FT_4 should be used initially in monitoring recovery from hypothyroidism. Once the patient has reached a near euthyroid status based on clinical assessment and FT_4 levels, continued surveillance of thyroid status should be through the s-TSH assay.

Hyperthyroidism

Thyrotoxicosis

Hyperthyroidism is a disorder manifested by excessive circulating levels of thyroid hormones. The term *thyrotoxicosis* is applied to a group of syndromes caused by high levels of free thyroid hormones in the circulation. Thyrotoxicosis means only that the patient is suffering the metabolic consequences of excessive quantities of thyroid hormones. Therapeutic decisions can be made only after the etiology of the thyrotoxicosis has been determined. The symptoms found in thyrotoxicosis are related to the catabolic-hypermetabolic effect or the increased metabolic activity in various tissues and increased sensitivity to catecholamines.[33]

Symptoms vary widely from few complaints to complete incapacitation. Symptoms may include nervousness; irritability; insomnia; fine tremor; excessive sweating; heat intolerance; flushed face; pruritus; tachycardia; palpitations; frequent bowel movements (although diarrhea is unusual); hyperkinesis weight loss that occurs even though the patient eats well; gynecomastia oligomenorrhea; amenorrhea; decreased libido; loss of muscle mass; loss of fat stores, producing an increase in plasma fatty acids, a decrease in serum cholesterol, LDL, lipoprotein(a), and apolipoprotein B;[34] a tendency toward ketosis; a negative nitrogen balance; exercise intolerance; easy fatigability; and dyspnea on exertion. The skin is warm and moist with patchy depigmentation. There may be a recession of the nails from the nail beds, clubbing of the fingers and toes, and swelling of the subcutaneous tissues.[18,23]

The hematologic manifestations include a decreased white blood cell count because of a decrease in neutrophils. Lymphocytosis is also present. The patient may be anemic despite an increased red blood cell mass due to excess demand for oxygen.[4] Approximately 20% of the patients have hypercalcemia and subsequent polyuria. This probably results from the increase in bone turnover. Long-term risk of abnormal bone metabolism leading to osteoporosis, and development of cardiac atrial fibrillation associated with hyperthyroidism provides the rationale for treatment of subclinical hyperthyroidism.[4]

Thyroid Storm. Approximately 1% of patients with thyrotoxicosis experience a life-threatening complication called a *thyrotoxic crisis,* or *thyroid storm.* Laboratory testing does not

---

A 59-year-old man who has been diagnosed with acute respiratory distress syndrome showed symptoms of hypothermia, coolness in the extremities, decreased blood pressure and pulse, fatigue, and delay in reflex time. A thyroid profile was ordered with the following results:

Laboratory Results

$FT_4 = 4.0$
s-TSH = 1.0 μU/mL
(Refer to Table 19-1 for reference ranges.)

Questions

1. From the case history and the laboratory data, what is a likely diagnosis?
2. What additional laboratory test(s) would be appropriate?
3. What additional information concerning the patient would be helpful in interpreting results? Why?
4. Is treatment for thyroid disorder indicated?

differentiate the thyroid storm from hyperthyroidism. Instead, the diagnosis depends on the clinical manifestations. In addition to the classic symptoms of severe thyrotoxicosis, further signs of unregulated hypermetabolism with fever and tachycardia appear. The patient usually has a temperature greater than 100°F and exhibits prominent cardiovascular symptoms—tachycardia, arrhythmias, and congestive heart failure. Neurologic changes are the most dramatic manifestations of thyroid storm. Symptoms include agitation, delirium, confusion, insensitivity to pain, and coma.

The levels of thyroid hormones are not significantly changed during thyroid storm. Thyroid storm can occur in any patient with uncontrolled thyrotoxicosis who is subjected to stress. Infection is a common precipitating event. General anesthesia, surgery, therapeutic use of ^{131}I, or withdrawal of antithyroid drugs also can precipitate the condition. Therapy is directed at reducing thyroid hormone action and treating the precipitating cause.[30]

Causes of Thyrotoxicosis. Causes of thyrotoxicosis can be divided into two subcategories based on results of the RAIU test (Table 19-4). Uptake is increased (or normal) in Graves' disease, toxic multinodular goiter, TSH-secreting tumors, trophoblastic tumors, and toxic adenoma. RAIU is suppressed in subacute thyroiditis (SAT), silent thyroiditis, chronic thyroiditis, struma ovarii, metastatic thyroid carcinoma, or as a result of excessive circulating iodine (eg, from x-ray dyes, medication, or absorption through the skin).[15,20] Thyrotoxicosis factitia exists when excess exogenous thyroid hormone is ingested. This can occur when the hormone is used as part of a weight-reducing regimen, for control of the menstrual cycle, or in efforts to increase energy. Community outbreaks of thyrotoxicosis have been reported due to patients eating ground beef that was contaminated with bovine thyroid gland.

Graves' disease. The most common cause of thyrotoxicosis is *Graves' disease* (diffuse toxic goiter). Graves' disease occurs six times more commonly in women than in men. It occurs frequently at puberty, during pregnancy, at menopause, or following severe stress. The frequency of Graves' disease in the general population is about 0.4%.

The classic clinical features in Graves' disease consist of a triad: diffuse thyrotoxic goiter, exophthalmos (bulging eyes), and pretibial myxedema.[35] However, the full triad frequently does not develop. Exophthalmos is due to infiltration of the retro-orbital space with lymphocytes, mast cells, and mucopolysaccharides. Other characteristic clinical findings and complications are the same as those listed for thyrotoxicosis.

The pathogenesis of Graves' disease appears to be related to immunologic defects with genetic implications. It is believed to be a autoimmune disorder. TSHR autoantibodies are present in approximately 95% of patients with Graves' disease, TPO in approximately 75%.[3] Other abnormal immunoglobulins are also present in the circulation of patients with this disease. Further evidence for Graves' disease being an autoimmune disorder include that it occurs more commonly in certain families, and there is an increased frequency of HLA-Dr3 haplotypes in patients with Graves' disease than in the general population.[15] Those who are positive may never go into remission.

Graves' disease appears to be related to the production of TSHR antibody that subsequently interacts with the TSH receptor of the thyroid follicular cell membrane. Iodine uptake, synthesis of thyroid hormones, and release of the thyroid hormones into the circulation are consequently stimulated. The production of the thyroid hormones is unrelated to the body's need. TSH production is inhibited.

Neonatal Graves' disease also exists and is almost always associated with maternal disease and TSHR. The most serious complications of neonatal Graves' disease are premature closing of the fontanel, accelerated bone aging, persistent behavior problems, and mental retardation. Mortality in affected newborns can be as high as 10% to 16%.[33]

Graves' disease can be diagnosed by clinical manifestations, laboratory assays, and the measurement of antibody titers. The antibody assays may be of value in confirming that hyperthyroidism is caused by Graves' disease, in distinguishing an orbital tumor from Graves' disease with ophthalmopathy, and following treatment using antithyroid drugs.

Thyroiditis

Thyroiditis describes an inflammation of the thyroid gland. The disorder has been classified as acute (suppurative), subacute (nonsuppurative), silent (painless) or nonspecific, lymphocytic or Hashimoto's, and fibrous.[15]

Acute thyroiditis is caused by a bacteria, mycobacteria, fungi, or parasites, and is characterized by abscess and suppuration. Symptoms include tenderness and swelling of the thyroid, fever, chills, and malaise. This form of thyroiditis is rare, apparently because of the gland's inherent resistance to infection manifested by its complete encapsulation, rich blood supply, lymphatic drainage, and high iodide content. TT_4 and T_3 levels are usually normal.

SAT hyperthyroidism presumably results from release of stored hormones from the damaged follicular cells. The etiology of SAT is probably viral in nature, and the gland is very tender. The patient presents with a high fever, inflammation,

TABLE 19-4. Causes of Thyrotoxicosis	
Associated With Increased RAIU	**Associated With Decreased RAIU**
Graves' disease	Subacute thyroiditis
Toxic multinodular goiter	Chronic thyroiditis
Pituitary tumor	Silent thyroiditis
TSH secreting tumors	Struma ovarii
Trophoblastic tumors (hydatidiform mole, choriocarcinoma)	Metastatic thyroid carcinoma
	Thyrotoxicosis factitia
	Iodine excess
Toxic adenoma	

A 41-year-old woman complained of increased sweating over the previous 3 months. She also stated that she always seemed nervous and had heart palpitations and heat intolerance. She lost 12 pounds over the past 3 weeks and had noncramping diarrhea. On examination, her skin was warm and moist, she had a prominent stare, and she had tachycardia at 140 beats per minute. Her thyroid gland was diffusely enlarged with a suggestion of nodularity.

Laboratory Results

TT_4 = 21.2 µg/dL (273 nmol/L)
THBR ratio = 2.0

s-TSH = none detected
FT_4I = 42.4 ng/dL (546 pmol/L)
(Refer to Table 19-1 for reference ranges.)

Questions

1. What is the most probable cause of these symptoms and laboratory results?
2. What additional laboratory testing would be helpful in confirming the diagnosis?
3. What is the significance of the s-TSH test?
4. What causes this disorder?

pain, and an elevated erythrocyte sedimentation rate (ESR). Thyroid hormones are usually mild to moderately increased. SAT presents in three phases. Laboratory findings during the 1- to 2-month initial period indicate a mild thyrotoxicosis. During phase 2, which lasts 1 week to 2 weeks, the patient becomes euthyroid, whereas in phase 3, a transient hypothyroidism may occur that can last for 1 month to 2 months until recovery.

In silent or painless thyroiditis, the ESR is normal. The etiology of painless thyroiditis is unclear but is believed to be autoimmune, and has been associated with certain HLA types. The disease is best differentiated from Graves' disease using RAIU. The clinical course runs through three phases, similar to SAT. Recovery is the rule, but some patients may have persistent goiter, antithyroid antibodies, or chronic thyroiditis.[32,36]

About 10% of the patients with Hashimoto's disease produce excessive secretion of thyroid hormone causing hyperthyroidism, termed *chronic lymphocytic thyroiditis*. These patients have the same clinical manifestations as those with Graves' disease. The hyperthyroid mechanism involved in Hashimoto's disease may be the production of TSHR, which, in turn, stimulates the TSH receptor. The patient may exhibit true exophthalmos that is indistinguishable from that found in Graves' disease. There is some evidence to support the contention that Graves' disease, Hashimoto's thyroiditis, and hypothyroidism are different manifestations of the same underlying autoimmune condition with a shift from expression of TSHR from TSI to TBII antibodies.[27]

Riedel's thyroiditis is a condition in which sclerosing fibrosis eventually turns the thyroid into a woody or stony-hard mass, which often extends into neighboring structures such as the trachea and esophagus. Thyroid hormones are normal until extensive destruction of the gland has occurred and the patient becomes hypothyroid.[32,36]

Miscellaneous Causes of Hyperthyroidism

Other rare forms of hyperthyroidism exist. In secondary (pituitary) hyperthyroidism, the TSH will be elevated despite a high level of FT_4, due to the presence of a TSH-secreting pituitary adenoma. Excessive TSH secretion also has been found without evidence of a pituitary tumor. This condition may be caused by the hypersecretion of TRH (tertiary hyperthyroidism), loss of the inhibition of TSH secretion normally produced by excessive levels of thyroid hormones, or the presence of thyroid hormone resistance syndrome in the pituitary gland.[26]

Trophoblastic tumors may cause thyrotoxicosis. These tumors, such as hydatidiform mole, produce and secrete hCG. By virtue of the molecular structure (the alpha subunits being identical), hCG may demonstrate thyroid-stimulating ability. The hCG molecule is mistaken for TSH by the TSH receptor on the follicular cell.

In both the toxic multinodular goiter and toxic nodule, the primary defect appears to be the production of thyroid hormone without TSH stimulation. These conditions can be confirmed by thyroid scan. Toxic multinodular goiter may express itself in patients who have been euthyroid for years and then have an increase of circulating iodine concentration.[15] Thyroid nodules are normally found in up to 3% of the general population and are usually not associated with production of thyroid antibodies or alterations in thyroid function.

Thyroid cancers occur in approximately 0.004% of the population. Patients with solitary, "cold" thyroid nodules on thyroid scan have about a 20% chance of having cancer. Calcitonin levels are elevated in patients with medullary carcinoma of the thyroid. Additionally, although it also can be increased in a number of benign conditions, an elevated serum thyroglobulin level is associated with differentiated thyroid carcinoma and thus is used as a thyroid cancer "tumor marker." Metastatic thyroid carcinoma and struma ovarii are ectopic tissues that produce thyroid hormone. A struma ovarii is an ovarian teratoma that contains thyroid tissue that, on rare occasions, may cause hyperthyroidism. A radioiodine scan of the pelvis will demonstrate the ovarian tumor.

Subclinical Hyperthyroidism

Subclinical hyperthyroidism occurs in patients showing no clinical symptoms of hyperthyroidism, but whose TSH levels are less than 0.1 μU/mL to the lower limits of the normal range.[13] Like subclinical hypothyroidism, the abnormality may indicate early evidence of thyroid disease, or may be a reaction to nonthyroidal influences representing a resetting of thyroid setpoint. Subclinical hyperthyroidism can occur in 2% to 4% of euthyroid older patients. With these patients, TSH elevation is usually transient. However, Graves' disease and toxic multinodular goiter have been described as common causes of subclinical hypothyroidism.[2]

Laboratory Evaluation of Hyperthyroidism

An expanded algorithm for laboratory testing of ambulatory patients with hyperthyroidism is shown in Figure 19-8. If serum TSH is decreased, the diagnosis is usually primary hyperthyroidism, provided the patient does not have a low FT_4, is not taking thyroid hormone, and is not seriously ill.

A positive (TSHR–TSI) antibody test can help to confirm suspected hyperthyroidal states.[3,5] Other expected laboratory results include elevated TT_4, $THBR_4$, FT_4, and FT_4I (see Table 19-3). The TSH level should be clearly subnormal (<0.1 μU/L).[1,8] The presence of TSHR antibodies is diagnostic for cases of suspected Graves' disease.[15]

In cases of subclinical hyperthyroidism, look for the s-TSH to be low because high-sensitivity TSH methods can discriminate between euthyroid and hyperthyroid patients. However, the TRH stimulation test of thyrotoxic patients can still be useful as a confirmatory test in these cases.[8] The condition of hyperthyroidism in which FT_4I is normal, s-TSH is normal or suppressed, and FT_3I or FT_3 is elevated is termed *T_3 thyrotoxicosis*.[3,37]

Identification of hyperthyroidism during pregnancy and NTI is complicated because of the thyroxine-binding protein abnormalities or the pituitary-thyroid axis response. Furthermore TSH levels may be slightly depressed during pregnancy. In these cases, the FT_4 test would be useful.[1] Likewise, the diagnosis of hyperthyroidism becomes more complex with sick patients due to changes in thyroid-binding proteins and the presence of circulating inhibitors. The greater the severity of the disease, the more chance the thyroid hormone levels, including FT_4, will drop into the normal range. The key result in such cases is whether the TSH is below 0.1 μU/mL. Large doses of dopamine or corticosteroids may also result in

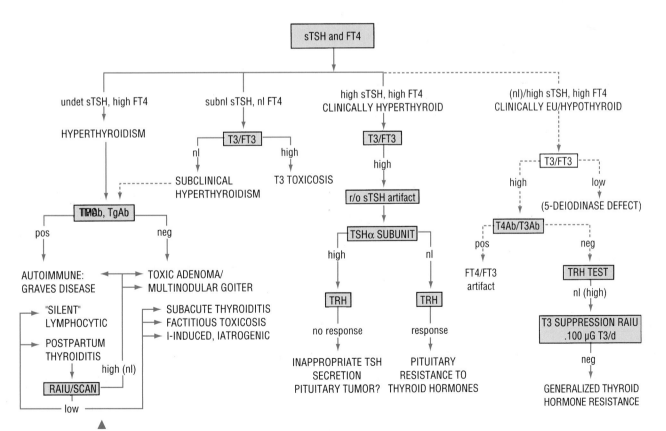

Figure 19-8. Algorithm for the diagnosis of hyperthyroidism in ambulatory patients. From Bayer MF. Effective laboratory evaluation of thyroid status. Med Clin North Am 1991;75. Philadelphia: W. B. Saunders. Reprinted with permission.

decreased TSH levels, but rarely below 0.01 mU/mL.[2,3,5,8,19] In cases of questionable results, s-TSH by equilibrium dialysis should be performed.

Normal TSH levels with elevated FT_4 is uncommon, but may occur in patients with secondary hyperthyroidism, familial dysalbuminemia, or thyroid hormone resistance syndrome. The TRH stimulation test may be useful in the diagnosis of these patients;[26,37] however, at least one major testing center has found that patients with basal TSH concentrations of less than 0.1 mU/L almost always have a subnormal s-TSH response to TRH stimulation.[5] Another source stated that almost all overt hyperthyroid patients have undetectable rather than mildly decreased s-TSH.[7]

Treatment of Hyperthyroidism

The therapy for hyperthyroidism depends on the cause, the severity, and the patient's age and general health. The objective of treatment is to lower the thyroid hormones to normal and to minimize the symptoms. Treatment can be achieved by antithyroid medication, radioiodine ablation, or thyroidectomy.

Current therapy for Graves' disease continues to be directed at the thyroid gland itself. Ablation of the gland is accomplished either by surgery or the use of radioactive iodine. The treatment with radioactive iodine (^{131}I) is currently the most popular technique. However, radioiodine treatment is contraindicated during pregnancy.[13] A major long-term side effect of ^{131}I therapy is development of hypothyroidism. Graves' disease may spontaneously revert to hypothyroidism due to thyroid damage.

Surgery is preferred in some cases of hyperthyroidism, as in adenomatous goiter, whereas in trophoblastic hyperthyroidism, the stimulus causing the hypersecretion (*eg,* placenta) is removed. Exogenous or factitious hyperthyroidism is successfully treated when the patient eliminates or reduces the intake of thyroid hormones.

Drug therapy also may be used as a treatment. Antithyroid drugs, such as propylthiouracil, methimazole, and carbimazole, have been used to inhibit the synthesis, release, or peripheral activation of thyroid hormones. However, a lasting remission with drug therapy may only be seen in 20% to 50% of patients with Graves' disease.[38] Beta-blockers, such as propranolol, may be useful as an adjunct therapy. However, beta-blockers alone do not make the patient euthyroid and may precipitate thyroid storm.[35]

TSH-suppressive treatment with L-thyroxine is the standard practice for patients with thyroid cancer and some patients with thyroid nodular disease to prevent recurrence. Doses of L-thyroxine, however, should be closely monitored by way of the s-TSH test in that L-thyroxine may produce osteopenia, other target organ damage, or lead to subclinical hyperthyroidism.[4,9,23]

Patients who have been treated with antithyroid drugs, thyroidectomy, or radioactive iodine need monitoring tests to ensure euthyroidism. The response to treatment can be best evaluated by measuring FT_4, because the s-TSH levels tend to stay undetectable for extended periods.[6,9,10]

SUMMARY

The thyroid gland consists of two lobes located in the lower part of the neck. Adjacent to the thyroid gland are four parathyroid glands. Two types of cells form the thyroid: the follicular cells and perifollicular cells. The primary hormones produced by the thyroid gland are T_3 and T_4. These hormones are regulated by the hypothalamic-pituitary-thyroid axis and, in particular, thyroid-stimulating hormone (TSH) from the pituitary gland. The function of the thyroid hormones is to control energy expenditure and regulate growth and metabolic processes. Tests for thyroid function include total thyroxine (TT_4), T_3 uptake or THBR, TSH, total T_3, and free thyroxine (FT_4). Disorders of the thyroid include both hypothyroidism and hyperthyroidism. Specific algorithms are used for the diagnosis of thyroid disorders.

REVIEW QUESTIONS

1. Which of the following statements regarding thyroid hormones is correct?
   a. Thyroxine (T_4) is a major product of the thyroid gland
   b. Thyroxine and triiodothyronine are secreted by the C cells of the thyroid
   c. The coupling of two monoiodotyrosine (MIT) molecules results in the formation of T_3
   d. The serum concentration of T_3 is about 50 times greater than that of T_4

2. Actions of the thyroid hormones include all of the following EXCEPT:
   a. Calorigenesis
   b. Central nervous system activity
   c. Bone and tissue growth and development
   d. Production of antibodies

3. Which of the following is CORRECT regarding thyroid function tests:
   a. Normal thyroid hormone levels may be used to rule out thyroid disease
   b. Screening of the general public for thyroid disease is not recommended
   c. Current guidelines for initial thyroid testing call for the utilization of the high sensitivity T_3 assay
   d. The patient must be fasting for thyroid function tests

4. Which of the following is true regarding the thyroid hormone binding ratio (THBR):
   a. It is also known as the TSH test
   b. In hyperthyroidism, the THBR will be decreased
   c. Results are reported as a ratio of the patient's THBR to the average uptake of normal pooled serum
   d. None of the above

5. A patient's T_4 is 10 mg/dL. The THBR is 24% (normal = 30%). What is the FT_4I?
   a. 8
   b. 6

   c. 4

   d. 2

6. The first and best indicator of hyper- or hypothyroidism is:

   a. THBR

   b. FT_4

   c. TSH

   d. TBG

7. Which of the following tests is used as a postoperative marker of thyroid cancer?

   a. TSH

   b. FT_4

   c. TBG

   d. Thyroglobulin (TG)

8. Hypothyroidism is associated with which of the following:

   a. Increased cholesterol

   b. Decreased LDL

   c. Increased triglycerides

   d. Increased calcium

9. Regarding Graves' disease, which of the following is CORRECT:

   a. It is more common in men than women

   b. It is an autoimmune disorder

   c. Synthesis and release of thyroid hormones is decreased

   d. The frequency of the disease is about 4%

10. The condition in which FT_4I is normal, s-TSH is normal or suppressed and the FT_3I or FT_3 is elevated is:

   a. Graves' disease

   b. Primary hypothyroidism

   c. T_3 thyrotoxicosis

   d. Subacute thyroiditis

REFERENCES

1. Masters PA, Simons RJ. Clinical use of sensitive assays for thyroid-stimulating hormone. J Gen Intern Med 1996;11:115–127.
2. LoPresti JS. Laboratory tests for thyroid disorders. Otolaryn Clin North Am 1996;29:557–574.
3. Volpe R. Rational use of thyroid function tests. Crit Rev Clin Lab Sci 1997;19:141.
4. Surks MI, Ocampo E. Subclinical thyroid disease. Am J Med 1996; 100:217–223.
5. Klee GG, Hay ID. Biochemical thyroid function testing. Mayo Clin Proc 1994;69:469–470.
6. Brody MB, Reichard RA. Thyroid screening how to interpret and apply the results. Postgrad Med 1995;98(2):54–66.
7. Helfand M, Redfern CC. Screening for thyroid disease: an update. Ann Intern Med 1998;129:144–158.
8. Surks MI, et al. American Thyroid Association guidelines for use of laboratory tests in thyroid disorders. JAMA 1990;263:1529.
9. Bartalena L, et al. Measurement of serum free thyroid hormone concentrations: an essential tool for the diagnosis of thyroid dysfunction. Horm Res 1996;45:142–147.
10. Feldcamp CS, Carey JL. An algorithmic approach to thyroid function testing in a managed care setting. Am J Clin Path 1996;20(2):11–16.
11. Rege V, et al. Comparison of Kodak Amerlite FT4 and TSH-30 with T4 and TSH as first line thyroid function tests. Clin Biochem 1996;29(1):1–4.
12. Stockigt JR. Guidelines for diagnosis and monitoring of thyroid disease: nonthyroidal illness. Clin Chem 1996;42:188–192.
13. Vanderpump MPJ, et al. Consensus statement for good practice and audit measures in the management of hypothyroidism and hyperthyroidism. Br Med J 1996;313:539–544.
14. Costa AJ. Interpreting thyroid tests. Am Fam Physician 1995;52: 2325–2330.
15. Lazaruas JH. Hyperthyroidism. Lancet 1997;349:339–343.
16. Feld RD. Clinical pathology rounds familial dysalbuminemic hyperthyroxinemia. Lab Med 1995;26:242–244.
17. Hay ID, et al. American Thyroid Association assessment of current free thyroid hormone and thyrotropin measurements and guidelines for future clinical assays. Clin Chem 1991;37:2002.
18. Chen IW, Sperling M. Estimation of free thyroxine by automated immunoassay systems. AACC In-Service Training and Continuing Education News & Views 1992;10(14):3–6.
19. Brent GA. Maternal thyroid function: interpretation of thyroid function tests in pregnancy. Clin Obstet Gynecol 1997;40(1):3–15.
20. Pittman JG. Evaluation of patients with mildly abnormal thyroid function tests. Am Fam Physician 1996;51:961–966.
21. Barron J, Rough B. Progression of patients with a borderline raised thyroid-stimulating hormone concentration. Ann Clin Biochem 1996; 33:157–158.
22. Cooper DS. Subclinical thyroid disease: a clinician's perspective [Letter]. Ann Intern Med 1998;129:135–137.
23. Hart IR. Management decisions in subclinical thyroid disease. Hosp Pract 1995;30:43–50.
24. Nicoloff JT, Spencer CA. The use and misuse of the sensitive thyrotropin assays. J Clin Endocrinol Metab 1990;71:553.
25. Taylor CS, Brandt DR. Developments in thyroid-stimulating hormone testing: the pursuit of improved sensitivity. Lab Med 1994;24:337.
26. McDermott MT. Thyroid hormone resistance syndromes. Am J Med 1993;94:424.
27. Do EL, Jones RE. Hypothyroid Graves' disease. South Med J 1997; 61:1201–1203.
28. Singer PA, et al. Treatment guidelines for patients with thyroid nodules and well-differentiated thyroid cancer. Arch Intern Med 1996; 156:2165–2172.
29. Becks GP, Burrow GN. Thyroid disease and pregnancy. Med Clin North Am 1991;75:121.
30. Gavin LA. Thyroid crises. Med Clin North Am 1991;75:179.
31. Dussault JH. Screening for congenital hypothyroidism. Clin Obstet Gynecol 1997;40:117–123.
32. Singer PA. Thyroiditis acute, subacute, and chronic. Med Clin North Am 1991;75:61.
33. Zimmerman D, Gan-Gaisano M. Hyperthyroidism in children and adolescents. Pediatr Clin North Am 1990;37:1273.
34. Engler H, Riesen WF. Effect of thyroid function on concentrations of lipoprotein(a). Clin Chem 1993;39:2466.
35. McDougall R. Graves' disease: current concepts. Med Clin North Am 1991;75:79.
36. Sakiyama R. Thyroiditis: a clinical review. Am Fam Physician 1993;48:615.
37. Bayer MF. Effective laboratory evaluation of thyroid status. Med Clin North Am 1991;75:1.
38. Francis T, Wartofsky L. Common thyroid disorders in the elderly. Postgrad Med 1992;92:225.

CHAPTER 20

Cardiac Function

Lynn Ingram

Objectives

Upon completion of this chapter, the clinical laboratorian should be able to:

- *Describe the normal cardiac system, including the heart chambers, valves, vessels, and circulation of blood through the heart and body.*
- *Explain the origin of six general symptoms of cardiac disease.*
- *Discuss the etiology and physiologic effects of the following cardiac conditions:*

  Congenital heart disease

  Hypertensive heart disease

  Infectious heart diseases

  Coronary heart disease

  Congestive heart failure
- *Identify nine risk factors for development of coronary heart disease.*
- *List six features of an ideal cardiac marker.*
- *Compare and contrast the specificity and sensitivity of the most commonly used serum cardiac markers.*
- *Assess the clinical utility of the various cardiac markers to assess the presence of myocardial infarction.*
- *Analyze the role of the clinical laboratory in the assessment of a patient with cardiac disease.*
- *Name and define the purpose of the most common types of drugs used to treat cardiac disease.*

KEY TERMS

Angina pectoris
Antiplatelet therapy
Arrhythmias
Atherosclerosis
Atrial septal defects (ASD)

Beta-adrenergic blocking drugs
Cardiac catheterization
Cardiac glycosides
Cardiac markers
Cardiomyopathy

Creatine kinase (CK)
CK-MB
CK isoforms
Coarctation of the aorta
Congestive heart failure
Diuretics
Echocardiography
Electrocardiography
Essential hypertension
Fallot's tetralogy
Infectious endocarditis
Myocardial infarction
Myocarditis

Myoglobin
Myosin heavy chains (MHC)
Myosin light chains (MLC)
Pericarditis
Persistent ductus
Rheumatic heart disease
Secondary hypertension
Thrombolytic agents
Troponin I (TnI)
Troponin T (TnT)
Vasodilators
Ventricular septal defect
   (VSD)

Heart disease is a common and debilitating condition that affects millions of patients each year, yet obtaining an accurate and timely diagnosis remains difficult. The patient's medical history and results of radiologic and laboratory tests many times do not provide enough information to assure the best medical care for each patient, especially patients presenting with chest pain. An early and accurate diagnosis would improve each patient's prognosis and quality of life, as well as reduce risks for development of further cardiac and constitutional problems.

The issues of reimbursement and managed care have stimulated interest in the development of new diagnostic markers for cardiac disease that will differentiate patients who need further procedures such as coronary artery bypass grafting, angioplasty, or thrombolytic therapy from those patients who can safely be managed medically as an outpatient. Therefore, selection of the appropriate cardiac marker or markers that provide the most cost-effective and clinically useful indicator of myocardial function is critical.

This chapter reviews the anatomy and function of the heart, commonplace cardiac diseases, diagnostic tests (both clinical laboratory and radiologic), and routine treatments for heart diseases.

THE HEART

The heart functions as a pump for the blood throughout both the pulmonary and systemic circulations. It weighs approximately 250 g to 350 g and is the size of a human fist. It is located in the mediastinum between the lungs and is enclosed in a double-walled pericardial sac with a serous membrane between the walls to provide lubricating fluid that facilitate heart movements. The middle layer of the heart is the *myocardium,* the cardiac muscle that pumps the blood throughout the body. The endocardium comprises the inner layer of the heart and also forms the heart valves that separate the four chambers of the heart. The right and left atria are the two upper chambers and the left and right ventricles form the two lower chambers. The interventricular septum separates the left and right sides of the heart. The superior and inferior vena cava and the coronary sinus provide blood to the right atrium and the pulmonary artery allows blood from the right ventricle to enter the pulmonary circulation. The pulmonary veins supply blood to the left atrium and the aorta removes blood from the left ventricle for delivery to the remainder of the body (Fig. 20-1).

There are two sets of heart valves: (1) the atrioventricular valves between the atria and ventricles and (2) the semilunar valves between the ventricles and the aorta and pulmonary artery. The atrioventricular valves comprise the tricuspid valve, with three leaflets or cusps on the right side of the heart and the mitral or bicuspid valve, with two leaflets on the left side. During diastole, these valves open into the respective ventricle, allowing blood to flow from the atria into the ventricles. The semilunar valves that sepa-rate the heart from the aorta (aortic semilunar valve) and the pulmonary artery (pulmonary semilunar valve) are both made up of three cusps. These valves open as the ventricles contract, forcing blood into the aorta or pulmonary artery, and close to prevent backflow of blood into the ventricles during diastole (Fig. 20-1).

The cardiac muscle cannot store oxygen for its own use, so it requires a constant supply of oxygen and nutrients to perform efficiently. The coronary circulation supplies these nutrients through two major arteries, the right and left coronary arteries. They receive blood from the aorta and then divide into smaller branches. The left coronary artery divides into the left anterior descending—or interventricular—artery and the left circumflex artery, both of which supply blood to the exterior of the anterior portion of the heart. The right coronary artery supplies blood to the posterior surface of the heart. It branches into the right marginal artery and the posterior interventricular artery, and then moves toward the apex of the heart. Many small branches extend inward from these major arteries to supply the myocardium and endocardium with essential nutrients (Fig. 20-2). A reduction of blood flow to the cardiac muscle through these coronary arteries will decrease heart function, but the exact consequence of this reduced perfusion depends on the specific area of the heart being supplied by that artery.

The total blood volume in a normal sized adult is approximately 5 L and is relatively constant within a particular individual. When blood volumes change, the body reestablishes this normal volume through the transfer of blood to or from the capillaries through the capillary shift, through regulation of kidney function, or through autonomic nervous

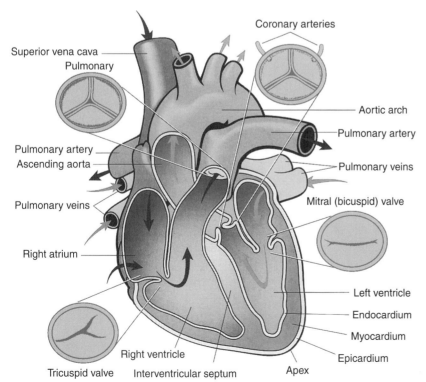

Figure 20-1. Normal heart anatomy. Chambers of the heart: The right heart pumps the venous blood into the lungs. The oxygenated blood returns from the lungs into the left atrium and is propelled by the left ventricle into the aorta. The insets show closed valves; the tricuspid valve has three leaflets, whereas the mitral valve has two leaflets. The aortic and pulmonary artery valves have three leaflets and resemble one another, except for the fact that the coronary arteries originate from behind the cusps in the aorta. (From Damjanov I. Pathology for the health related professions. Philadelphia: WB Saunders, 1996. Reprinted with permission.)

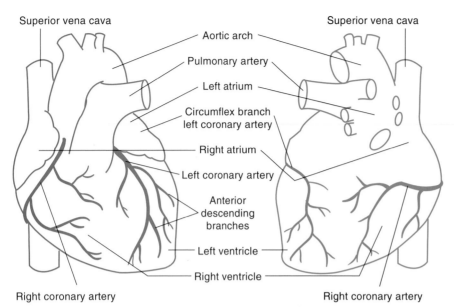

Superior vena cava

Superior vena cava

Aortic arch

Pulmonary artery

Left atrium

Circumflex branch
left coronary artery

Right atrium

Left coronary artery

Anterior
descending
branches

Left ventricle

Right ventricle

Right coronary artery

Right coronary artery

Figure 20-2. Major coronary arteries and branches. (From Miles WM, Zipes DP. Cardiovascular diseases. In: Andreoli TE, Bennett JC, Carpenter CCJ, Plum F, Smith Jr, LH, eds. Cecil essentials of medicine. 3rd ed. Philadelphia: WB Saunders, 1993. Reprinted with permission.)

control. These mechanisms ensure a consistent volume of blood to maintain sufficient circulation.

Arteries, capillaries, and veins compose a closed system for blood circulation. There are two separate circulation systems in the body: (1) the pulmonary circulation, which allows the exchange of oxygen and carbon dioxide in the lungs, and (2) the systemic circulation, which provides the exchange of nutrients and wastes between the blood and the cells throughout the body. *Arteries* transport blood away from the heart into the lungs or body tissues and *arterioles,* which are smaller branches of these arteries, control the amount of blood that flows into specific areas by contracting smooth muscles in the vessel walls. *Capillaries* are very small vessels that allow the exchange of fluids, oxygen, carbon dioxide, electrolytes, glucose, and other nutrients and wastes between the blood and interstitial fluid. Blood then returns to the heart through small venules that drain into the larger veins. *Veins* have thinner walls and less smooth muscle than arteries and contain valves that keep the blood flowing toward the heart.

The cardiac cycle describes the circulation of blood through the heart and body as the heart muscle alternates between periods of relaxation (diastole) and contraction (systole). Blood flows through this cardiac cycle in the following seven-step sequence:

1. Venous blood leaves the inferior and superior vena cava and enters the right atrium.
2. The right atrium contracts, forcing blood though the tricuspid valve into the right ventricle.
3. When the right ventricle is filled with blood, it contracts, forcing the tricuspid valve to close and the pulmonary valve to open.
4. The blood then enters the pulmonary artery and travels to the lungs.
5. In the lungs, the blood is oxygenated and is returned to the left atrium by way of four pulmonary veins.

6. The left atrium contracts, sending the oxygenated blood through the mitral valve into the left ventricle.
7. The left ventricle contracts, the mitral valve closes and the aortic valve opens, forcing blood into the aorta to continue into the general circulation.

The cardiac control center in the brain controls the heart rate and the force of each cardiac contraction. Arterial baroreceptors detect changes in blood pressure and signal the cardiac center, which responds by increasing or decreasing the rate and force of cardiac contractions as appropriate. Emotional responses, increased secretion of epinephrine, fever, and serum pH also alter heart rate.[1] Heart function is also controlled by two sets of autonomic nerves: the parasympathetic (vagus) and the sympathetic nerves. Stimulation of the parasympathetic nerves will decrease all activities of the heart, whereas stimulation of the sympathetic nerves increases cardiac functions, especially in times of stress. Most sympathetic nerves release epinephrine or norepinephrine, which increase the heart rate and increase the force of each contraction. The parasympathetic system consists of the vagus nerve, which originates in the medulla of the brain and branches to the heart and digestive organs. Parasympathetic stimulation decreases heart rate and the force of cardiac muscle contraction and releases acetylcholine.

The cardiac conduction system responds to electrochemical stimuli along specialized myocardial fibers that permit very rapid transmission of impulses, not by typical nerve cells. These specialized fibers allow all muscle fibers of the atria or ventricles to contract essentially simultaneously. This coordinated effort results in a rhythmic filling and emptying of the heart with enough force to sustain the flow of blood throughout the body. All cardiac muscle cells can initiate impulses, but the conduction pathway normally originates at the sinoatrial node, often called the pacemaker, in the right atrium. The sinoatrial node automatically generates impulses at the basic rate of approximately 70 beats per minute, called the sinus rhythm. From the sinoatrial node, the impulses then spread to both atria and are collected at the atrioventricular node, located in the floor of the right atrium. After the ventricles fill with blood, the impulses continue into the ventricles through the bundle of His (atrioventricular bundle), the right and left branches, and the Purkinje network of fibers, stimulating the simultaneous contraction of the two ventricles.[1] This orderly electrical stimulation of cardiac fibers results in normal heart function.

SYMPTOMS OF HEART DISEASE

Patients with cardiac disease are often asymptomatic until a relatively late stage in their condition. The most frequent symptoms manifested in heart disease are dyspnea, chest pain, palpitations, syncope, fatigue, and edema (Table 20-1).

Dyspnea is an awareness of or difficulty in breathing. It can be due to cardiac or respiratory disease and is a normal response during exercise in healthy individuals. Dyspnea as a result of cardiac disease may occur only on exercise or can be present at rest in advanced cardiac disease. Left ventricular heart failure causes pulmonary edema, making the lungs stiff, increasing the amount of effort necessary to breathe, and also increasing breathing rate due to stimulation of pulmonary receptors. Orthopnea, breathlessness when a patient lies flat, occurs when blood is redistributed in the supine position, which increases the pressure of abdominal contents against the diaphragm. Paroxysmal nocturnal dyspnea is an accumulation of fluid in the lungs at night, often waking the patient from sleep fighting for breath. Wheezing due to bronchial edema and a productive, blood-tinged cough is also common.[2]

Cyanosis, a bluish discoloration of the skin, is caused by an increased amount of nonoxygenated hemoglobin in the blood. Central cyanosis is due to right-to-left shunting of blood or impaired pulmonary function, whereas peripheral cyanosis is caused by shunting or to local vasoconstriction. Cyanosis becomes apparent when 4 g/dL or more of reduced hemoglobin is present.[3]

Angina pectoris is the most common symptom associated with ischemic heart disease. It is a gripping or crushing central chest pain that may be felt around or deep within the chest. The pain may radiate into the neck or jaw or, less commonly, into the back or abdomen, and is associated with heaviness, paresthesia, or pain in one (usually the left) or both arms. It is typically worsened by exercise and relieved by rest. The pain is most often caused by a lack of oxygen to the myocardium due to an inadequate coronary blood flow.[2]

A palpitation may be an increased awareness of a normal heart beat or the sensation of a slow, rapid, or irregular heart rate. The most common arrhythmias that are felt as palpitations are premature ectopic beats.[2] Premature beats are usually felt as missed beats because the early beat is followed by a pause before the next normal beat, which is rather forceful because of the longer diastolic filling period.

The most common syncopal attacks are vasovagal in nature (simple faints) and are not due to serious disease. They can be due to venous pooling or may be provoked by fright or shock and are often accompanied by dizziness, nausea, sweating, ringing in the ears, a sinking feeling, and yawning. Cardiovascular syncope is usually sudden and brief and the classic variety is due to a disturbance of cardiac rhythm, such as a slowed heart rate. Without warning, the patient falls to the ground with a very slow or absent pulse and after a few seconds, the patient recovers consciousness. There are often no sequelae.[2]

Fatigue is a common cardiac symptom but is extremely nonspecific. Lethargy is associated with heart failure, persistent cardiac arrhythmias, and cyanotic heart disease. It may be due to both poor cerebral and peripheral perfusion and poor oxygenation of blood.[2]

Retained fluid accumulates in the feet and ankles of ambulant patients and over the sacrum of bedbound patients. The edema associated with heart disease is often absent in the morning because the fluid is reabsorbed when lying down, but becomes progressively worse during the day. When severe, the calf and thigh may become edematous and ascites or a pleural effusion may develop.[2]

A cough may be the primary complaint in some patients with pulmonary congestion. It is a nonproductive cough that will help to differentiate these patients from those with infectious processes. Hemoptysis occurs in congestive heart failure and is especially common in patients with mitral stenosis.[3]

Nocturia is also common in patients with congestive heart failure. Anorexia, abdominal fullness, right upper quadrant tenderness, and weight loss are also symptoms of advanced heart failure.

CONGENITAL HEART DISEASE

Congenital heart defects are the cause of approximately 2% of all cardiac disease[4] and occur in about 1% of live births.[2] There is an overall male predominance, although some specific lesions occur more frequently in females, and it is a common cause of death in the first year of life.

TABLE 20-1. Symptoms of Heart Disease

Common Symptoms	Unusual Symptoms
Dyspnea	Cough
Syncope	Abdominal pain
Cyanosis	Hemoptysis
Pain	Headache
Palpitations	Sweating
Fatigue	Vision and speech disturbances
Edema	Weakness of extremities
	Weight loss
	Nausea/vomiting
	Fever

Congenital heart disease includes valvular defects that interfere with the normal flow of blood, septal defects that allow mixing of oxygenated blood from the pulmonary circulation with unoxygenated blood from the systemic circulation, shunts, abnormalities in position or shape of the aorta or pulmonary arteries, or a combination of these conditions.[1] Many variations and degrees of severity are possible with congenital heart disease.

The etiology of congenital cardiac disease is often unknown but most defects appear to be multifactorial and reflect a combination of both genetic and environmental influences. Because the heart develops early in embryonic life and is completely formed and functioning by week 10 of gestation, all congenital heart defects develop before week 10 of pregnancy. Factors that are closely associated with the development of congenital heart disease include maternal rubella infections, maternal alcohol abuse, drug treatment and radiation, and certain genetic and chromosomal abnormalities.

A long known cause of congenital heart defects is the rubella virus, the causative agent of German measles. Infection of the mother during the first 3 months of pregnancy is associated with a high incidence of congenital heart disease in the baby. Apparently, the virus crosses the placenta, enters the fetal circulation, and damages the developing heart.[4]

Fetal alcohol syndrome is often associated with heart defects as well. Alcohol affects the fetal heart by directly interfering with its development. It has been proposed that alcohol is toxic to fetal heart cells and destroys them, but the exact teratogenic mechanisms of alcoholism remains unknown.[4] Many therapeutic and illegal drugs ingested by the mother can cross through the placenta to harm the developing heart.

Chromosomal abnormalities are associated with several developmental syndromes, many of which include congenital heart disease. The best known example is Down's Syndrome—or trisomy 21 syndrome—which is often associated with atrial septum defects.[4]

The symptoms of congenital heart disease may be evident at birth or during early infancy, or they may not become evident until later in life. Some symptoms and signs common in many congenital heart diseases include cyanosis, pulmonary hypertension, clubbing of fingers, embolism or thrombus formation, reduced growth, or syncope.[2] The most common congenital cardiac lesions are ventricular and atrial septal defects, persistent ductus arteriosus, coarctation of the aorta, valvular defects, and Fallot's tetralogy.[2]

Ventricular septal defect (VSD), commonly known as a "hole in the heart," is the most common congenital cardiac malformation, with a prevalence of 1 in 500 live births.[2] In this condition, blood flows through the septal defect from the left ventricle to the right ventricle, causing less blood to be pumped from the left ventricle, reducing output to the systemic circulation. More blood enters the pulmonary circulation, which overloads and irreversibly damages the pulmonary vessels, causing pulmonary hypertension.[1] Some small VSDs will close spontaneously, but moderate or large VSDs should be surgically repaired before the development of severe pulmonary hypertension.[2]

Atrial septal defects (ASD) are often first diagnosed in adulthood. This abnormality causes left-to-right shunting of blood between the atria. Pulmonary hypertension and atrial arrhythmias are common if the patient is older than age 30 years, but most children with this condition are asymptomatic. A significant ASD should be surgically repaired as soon as possible after diagnosis.[2]

The ductus arteriosus connects the pulmonary artery to the descending aorta. In fetal life, the ductus turns blood away from the pulmonary circulation into the systemic circulation, where blood is reoxygenated as it passes through the placenta. At birth, the high oxygen content in the blood triggers closure of the duct. If the duct is malformed or does not contain sufficient elastic tissue, it will not close. A persistent duct produces continuous aorta-to-pulmonary artery

---

CASE STUDY 20-1

A 15-month-old female with a heart murmur since birth was evaluated for repeated pulmonary infections, failure to grow, cyanosis, and mild clubbing of fingers and toes. She had been on digitalis therapy by the referring physician. The x-ray showed a moderately enlarged heart and an enlarged pulmonary artery. Pertinent laboratory data were obtained.

Total protein (6.0–8.3 g/dL)	5.4
Albumin (3.5–5.2 g/dL)	3.0
Hemoglobin (14–18 g/dL)	19.2
Hematocrit (40%–54%)	59
Erythrocyte count (4.3–5.7 $\times$ 10^6/mm^3)	6.4

A cardiac catheterization was performed and a large ventricular septal defect was found.

Questions

1. How does this congenital defect affect the body's circulation?
2. Why are the red cell measurements increased in this patient?
3. What treatment will be suggested for this patient?
4. What is this patient's prognosis?

shunting, resulting in severe left heart failure *(persistent ductus).* Many times there are no symptoms until later in life when heart failure or infective endocarditis develops. Premature babies are often born with persistent ductus arteriosus that are anatomically normal but are immature in that they lack the mechanism to close. These babies may be treated with indomethacin, which stimulates prostaglandin production and stimulates the closure of the duct, or the duct can be corrected surgically with very little risk.[2]

Coarctation of the aorta is a narrowing of the aorta at the insertion of the ductus arteriosus. In 80% of cases, this condition is associated with a bicuspid, rather than tricuspid, aortic valve.[2] This condition may remain asymptomatic for many years and treatment is surgical excision of the coarctation.

Congenital valve problems may be classified as stenosis (narrowing of a valve that restricts the forward flow of blood) or valvular incompetence (a valve that fails to close completely, allowing blood to leak backward). Mitral valve prolapse, abnormally enlarged and floppy valve leaflets that balloon backward with pressure, is common.[1] *Fallot's tetralogy* is the most common cyanotic congenital heart abnormality in children who survive beyond the neonatal period. It is a combination of four defects: VSD, right ventricular outflow obstruction, abnormal positioning of the aorta above the VSD, and right ventricular hypertrophy (Fig. 20-3).[2] This combination of lesions leads to increased right ventricular pressure and right-to-left shunting of blood through the VSD. The pulmonary circulation receives a small amount of unoxygenated blood from the right ventricle, and the systemic circulation receives a larger amount of blood consisting of mixed oxygen and unoxygenated blood.[1] Children with this condition may present with dyspnea, fatigue, and hypoxic episodes on exertion. Complete surgical correction is possible, even in infancy.[2]

CONGESTIVE HEART FAILURE

Congestive heart failure results from an inability of the heart to pump blood effectively. It is characterized by symptoms of fluid accumulation in the lungs, throughout the body, or both. Congestive heart failure affects approximately 1 of every 100 people in the United States. It is estimated that 2 million persons are being treated for heart failure and more than 400,000 persons are diagnosed annually with congestive heart failure.[5]

When the heart is unable to pump efficiently, a lowered cardiac output occurs. If the left side of the heart fails, excess fluid accumulates in the lungs, resulting in pulmonary edema

Tetralogy of Fallot

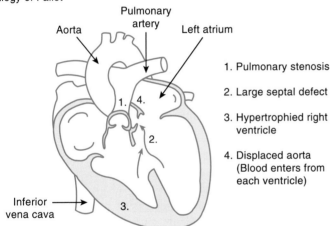

1. Pulmonary stenosis

2. Large septal defect

3. Hypertrophied right ventricle

4. Displaced aorta (Blood enters from each ventricle)

Normal anatomy of heart

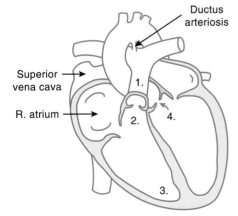

1. Pulmonary artery open

2. Complete septum

3. R. ventricle wall narrower than L.

4. Blood enters aorta from L. ventricle only.

Figure 20-3. Tetralogy of Fallot as compared to normal cardiac anatomy. (From Mulvihill ML. Human diseases: A systematic approach. 3rd ed. Norwalk, CT: Appleton & Lange, 1991. Reprinted with permission.)

and reduced output of blood to the systemic circulation. The kidneys respond to this decreased blood flow with excessive fluid retention, making the heart failure worse. If the right side of the heart fails, excess fluid accumulates in the systemic venous circulatory system and generalized edema results. There is also diminished blood flow to the lungs and to the left side of the heart, resulting in decreased cardiac output to the systemic arterial circulation.

Congestive heart failure may occur if the heart muscle itself is weak or if the heart is stressed beyond its ability to react. The most common causes of congestive heart failure are coronary artery disease, the cardiomyopathies, myocarditis, valvular disease, and cardiac arrhythmias (Table 20-2).

Coronary artery disease is the most common cause of heart failure in the United States. Management and prognosis for congestive heart failure caused by coronary artery disease is poor, with an average annual mortality of 30% to 40%.[6] Atherosclerosis of myocardial blood vessels leads to ischemia, a process causing active cardiac muscle to be replaced with fibrous material that will not function as normal cardiac muscle. Occlusion of the cardiac vessels reduces blood flow and forces the heart to use anaerobic metabolism, producing waste products that can quickly damage the tissue cells. A decrease in cardiac mass increases the load carried by the remaining viable tissue, resulting in increased cardiac stress and damage. The combination of these factors underscores the seriousness of congestive heart failure in ischemic patients.

A *cardiomyopathy* results from an abnormality of the heart muscle; cardiopathies are usually grouped as dilated cardiomyopathies, restrictive cardiomyopathies, or hypertrophic cardiomyopathies. When the heart is unable to contract efficiently, the heart dilates out of proportion, resulting in an enlarged heart with relatively thin cardiac walls.

Dilated cardiomyopathy may be the result of an autoimmune response, small blood vessel disease or direct myocardial toxicity such as seen in alcohol-induced cardiomyopathy and anthracycline cardiotoxicity. Presenting symptoms of dilated cardiomyopathy are dyspnea on exertion, orthopnea, paroxysmal nocturnal dyspnea, and chest discomfort similar to angina.

Patients with restrictive cardiomyopathy have increased diastolic pressure, resulting in inefficient filling of the ventricles and less blood being distributed to the body. This condition can be caused by an abnormal myocardium, infiltration of the myocardium with collagen or other abnormal proteins, endomyocardial disease, and space-occupying lesions (thrombus or tumor).[7] Hemochromatosis, an infiltration of iron into tissues, can also affect heart muscle and lead to loss of contractility and muscle function.

Hypertrophic cardiomyopathy is characterized by enlargement of the ventricular septum, which is out of proportion to the other ventricular walls, often called asymmetric septal hypertrophy. This condition appears to be inherited as an autosomal dominant trait in most cases.

Congestive heart failure may be a result of myocarditis caused by infections, immunologic injury to the tissues, or it may be idiopathic. It can also develop in patients with valvular heart disease. Valvular malfunctions alter the blood volume that the heart must regulate. Such patients will progress to cardiac failure as the blood volume changes alter the blood load for ventricles and atria.

Arrhythmias, or malfunctions of the cardiac conduction system, may also result in congestive heart failure. The arrhythmias may be caused by ischemia, infarction, infiltrates, electrolyte imbalances, or chemical toxins.

Clinical indications of congestive heart failure range from very mild symptoms that appear only on exertion to the most advanced conditions in which the heart is unable to function without external support. Congestive heart failure is readily detectable if it involves a patient with myocardial infarction, angina, pulmonary problems, or arrhythmias, but congestive heart failure is most commonly investigated because of dyspnea, edema, cough, or angina. Other symptoms such as exercise intolerance, fatigue, and weakness are common (Table 20-3).

TABLE 20-2. Causes of Congestive Heart Failure

Coronary Artery Disease

Cardiomyopathies
   Dilated cardiomyopathy
      Autoimmune response
      Small blood vessel disease
      Alcohol-induced
   Restrictive cardiomyopathy
      Abnormality of the myocardium
      Infiltration of myocardium with abnormal protein
      Endomyocardial disease
      Thrombus or tumor
   Hypertrophic cardiomyopathy
      Autosomal dominant inheritable condition

Inflammatory Heart Disease
   Cardiac infections
   Immunologic injury to myocardium
   Valvular incompetence

Cardiac Conduction Dysfunctions/Arrhythmias
Congenital Heart Disease

TABLE 20-3. Conditions Contributing to Congestive Heart Failure

Hypertension
Connective tissue diseases
Anemia/polycythemia
Endocrine disorders
Malnutrition
Drug/alcohol toxicity
Obesity
Pulmonary disease

CORONARY HEART DISEASE

Coronary heart disease is caused by a lack of nutrient and oxygen supplies to the heart muscle and results in myocardial ischemia. Ischemia is a reduction in the blood supply to one area of the heart often due to atherosclerosis, thrombosis, spasms, or embolisms but may also be a result of anemia, carboxyhemoglobinemia, or hypotension, which causes reduced blood flow to the heart. Increased demand for oxygen and nutrients as a result of extreme exercise or thyrotoxicosis may also cause ischemia.

Most frequently, ischemia is the result of abnormal coronary arteries, usually due to an obstruction in one or more of these arteries. *Atherosclerosis* is a thickening and hardening of the artery walls caused by deposits of cholesterol-lipid-calcium plaque in the lining of the arteries.[8] The following nine risk factors are known to predispose individuals to develop these arterial plaques:

1. Age. Atherosclerosis develops with age and becomes a more significant risk factor with increasing age. It is a more common finding after age 40 years and is found almost universally in elderly men in the Western world.[2]
2. Sex. Males tend to be more affected by atherosclerosis than premenopausal females. After menopause, women become similarly affected as men. It is thought that females are protected by higher levels of high-density lipoprotein (HDL) cholesterol until estrogen levels drop at menopause, and the use of estrogen replacement therapy in postmenopausal women may reduce the progression of atherosclerosis.[4]
3. Family history. Atherosclerosis is often found in members of the same family but a direct inheritance pattern has yet to be shown. Family lifestyles also play a role in this process and it is difficult to distinguish genetic from lifestyle factors in predicting coronary artery disease. Some conditions are directly inherited, such as familial hypercholesterolemia and familial combined hyperlipidemia.[2]
4. Hyperlipidemia. An increased serum cholesterol concentration has been shown to have a strong association with atherosclerosis. Lowering the serum cholesterol, especially the low-density lipoprotein (LDL)-cholesterol fraction, has

been shown to decrease the incidence of coronary artery disease and slow the progression of coronary atheroma.[2] The relationship between plaque formation and triglyceride levels is not as well defined.

5. Smoking. There is a direct relationship between number of cigarettes smoked and the risk of coronary artery disease in men and is related to the decrease in HDL-cholesterol levels, increased LDL-cholesterol levels, increased platelet adhesion, vasoconstriction, and increased fibrinogen and clot formation caused by smoking.[1] This relationship is not as well defined in women or in people who smoke pipes or cigars.
6. Hypertension. Both systolic and diastolic hypertension is associated with increased risk for atherosclerosis for both men and women.[2]
7. Sedentary lifestyle. Regular exercise has shown some protection against the development of heart disease and, conversely, a sedentary lifestyle is a strong factor in the development of coronary heart disease.
8. Diabetes mellitus. Because of the strong relationship between diabetes and vascular disease, there is also an increased risk for coronary artery disease in diabetic patients, especially in those whose diabetes is poorly controlled.
9. Response to stress. Aggressive, ambitious, compulsive persons have almost twice the risk for coronary disease as persons who do not express these characteristics.

Regardless of the etiology of the ischemia, there are three general results of cardiac ischemia: congestive heart failure, angina pectoris, and myocardial infarction (Fig. 20-4). Congestive heart failure will result when there is a reduced oxygen supply to the cardiac muscle, causing it to fail to pump the blood efficiently.

Angina pectoris is a symptom of inadequate perfusion of the heart muscle, resulting in chest pain. Typical angina pectoris is a frequently seen when there is pain with an increase in physical exertion or stress and usually rapidly resolves with rest. In patients with coronary artery disease, the narrowed cardiac vessels do not allow for additional blood with its nutrients and oxygen to flow into the cardiac muscle at times of additional physical or emotional stress, causing the pain. Other

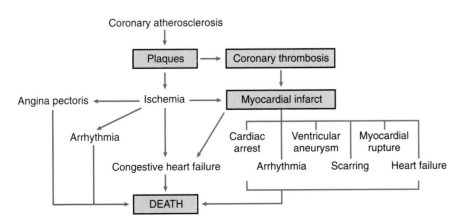

Figure 20-4. Presentation of coronary heart disease. Coronary heart disease may present as angina pectoris, congestive heart failure, or myocardial infarct. (From Damjanov I. Pathology for the health related professions. Philadelphia: WB Saunders, 1996. Reprinted with permission.)

types of angina include variant angina, nocturnal angina, and unstable angina. Variant angina is not related to exertion or any increase in stress but is caused by spasm of a coronary artery. It is often longer in duration than normal angina and the pain is more severe. Nocturnal angina generally only occurs in patients with severe coronary disease and wakes patients from sleep with severe chest pain. Unstable angina is the most severe form and is often present at rest. The pain is often provoked by very mild increases in exercise or stress and can even be initiated by ingesting a large meal. The pain is long lasting and severe and does not respond as well or as quickly to standard treatments. Many patients with unstable angina will rapidly progress to myocardial infarction. Although angina is not easily differentiated from myocardial infarction in many cases, there are no classic electrocardiogram (ECG) changes or enzyme elevations in angina and angina rarely causes cardiac damage unless it is prolonged or severe.

Myocardial infarction, or heart attack, occurs when blood flow to an area of the cardiac muscle is suddenly blocked, leading to ischemia and death of myocardial tissue. The heart tissue becomes inflamed and necrotic at the point of obstruction and is followed by the release of cellular enzymes into the blood. The damaged area of the heart quickly loses its ability to contract and conduct electrical impulses and oxygen supplies are depleted. This type of damage is irreversible and the area of necrosis is gradually replaced by fibrous scar tissue.[1]

The severity of damage from a myocardial infarction varies greatly among individuals and is primarily related to the size and location of the infarct with the location of the infarct, depending on the vessel involved. If the left anterior descending coronary artery is blocked, the anterior and lateral walls of the heart, ventricular septum, and anterolateral papillary muscle are affected. If the dominant right coronary artery is affected, damage to the inferoposterior wall, posteromedial papillary muscle, and inferior septum results. If any collateral circulation has developed because of chronic reduced blood flow in that particular area of the heart, the resulting damage will be less significant than if no collateralization has occurred.

Most infarctions involve all three layers of the heart. Patients usually present with varying degrees of severity of chest pain of several hours duration and other symptoms such as left arm pain, shortness of breath, hypotension, sweating, nausea, and vomiting. Some patients, especially diabetics, hypertensives, and elderly patients, have no chest pain and few symptoms other than gastric discomfort. Although many patients would indicate that there were no warning signs of an impending cardiac event, most will admit to a recent history of dyspnea, unstable angina, and a vague sense of ill health. Complications of a myocardial infarction are common and include sudden death due to ventricular arrhythmias and fibrillation, heart block if conduction fibers are located in the area of the infarct, and other conduction irregularities resulting in arrhythmias, congestive heart failure, and thromboembolism.[1]

HYPERTENSIVE HEART DISEASE

Hypertension is defined by the World Health Organization as a systolic pressure of greater than 160 mm Hg and diastolic pressure greater than 95 mm Hg. It is one of the most common cardiovascular diseases and it is estimated that approximately 50% of middle-aged persons have hypertension. The prevalence of hypertension increases with age from 4% in persons aged 20 to 29 years to 65% in persons older than age 80 years.[9] It is more common in the black population than in white or Hispanic groups.

Hypertension affects all organs, especially the kidneys, brain, and heart. The most important factor that determines the blood pressure is peripheral resistance and it is an increased peripheral resistance that results in heart disease. Peripheral

---

CASE STUDY 20-2

A 79-year-old male came to the emergency department of a local hospital complaining of weakness and left arm pain. He had had an episode of profound weakness, chest pain, and left arm pain the day before admission after his daily walk. He had a history of non–insulin-dependent diabetes and unstable angina.

Cardiac enzymes

	11:30 a.m., October 9	10 p.m., October 9
Total CK (54–186 U/L)	66	62
CK-MB (0–5 ng/mL)	2	1
% CK-MB (<6%)	3%	2%
Total LD (80–180 U/L)	155	164
LD-1 (14–41 U/L)	43	42
% LD-1 (16%–33%)	28%	26%
Troponin T (0–0.1 μg/L)	0.2	0.1

Questions

1. Do the symptoms and history of this patient suggest acute myocardial infarction?
2. Based on the preceding laboratory data, would this diagnosis be acute myocardial infarction? Why or why not?
3. What would the diagnosis be for this patient?

resistance increases the workload of the left ventricle, eventually resulting in hypertrophy and dilation. The increase in size of the left ventricle causes the mitral valve to allow regurgitation of blood into the left atrium, which over time results in dilatation and increased pressure in the left atrium as well. This increased pressure get transferred to the pulmonary circulation and thus affects the right side of the heart. Another complicating factor in this process is that hypertension is also associated with an increased prevalence of atherosclerosis, further increasing risk for heart disease.

There are often no symptoms associated with increased blood pressure, but dizziness, headaches, palpitations, restlessness, nervousness, and tinnitus may be present. There is a strong association between hypertension and obesity, heavy alcohol consumption, smoking, and sedentary lifestyles.

About 90% to 95% of patients with hypertension have no known cause for the condition (known as primary, or essential, hypertension).[10] Primary hypertension is multifactorial in nature, with both cardiac output factors and peripheral vasculature factors influencing the blood pressure.[10] The interplay among these many factors is poorly defined, and genetics, racial, gender, and environmental factors also complicate the picture. The remainder of persons with hypertension have an identified source of their problem and are classified as having *secondary hypertension*. Renal disease, the most common cause of secondary hypertension, is associated with sodium retention. Malignant hypertension is a serious condition seen when severe renal disease causes renal cell inflammation and destruction and is associated with high mortality unless vigorous treatment is initiated. Primary aldosteronism will result in sodium retention and, therefore, in hypertension, due to abnormal control of the renal excretion of the major electrolytes. Pheochromocytoma is an uncommon cause of striking increases in blood pressure from increased excretion of catecholamines from a chromaffin tumor.

INFECTIVE HEART DISEASE

Infectious agents continue to be implicated in a variety of heart diseases (Table 20-4). The most common infectious diseases involving the heart are rheumatic heart disease, infectious endocarditis, and pericarditis but there is increasing interest in determining if an infectious factor may be associated with cardiac conditions such as coronary heart disease.[11]

Rheumatic fever is an inflammatory disease of children and young adults that occurs as a result of an infection with Group A streptococci. A decline in the incidence of rheumatic fever from 10% of children in the 1920s to 0.01% in the 1990s reflects a reduction in all streptococcal infections and the effective use of antibiotics against these infections.[2] Although restricted to a pharyngeal involvement in most infections, in some individuals this organism progresses to affect the heart, joints, skin, and central nervous system. Rheumatic fever is not caused by a direct infection of the heart with the streptococcus organism nor is it a result of a cardiac toxin produced

TABLE 20-4. Infectious Agents Associated With Heart Disease	
Pericarditis/Myocarditis	
Mycoplasma pneumoniae	*Chlamydia trachomatis*
Mycobacterium tuberculosis	*Streptococcus pneumoniae*
Staphylococcus aureus	Enterobacteriaceae
Coxsackievirus A and B	Echovirus
Adenovirus	Influenza
Coccidioides immitis	*Aspergillus* species
Candida species	*Cryptococcus neoformans*
Histoplasma capsulatum	*Trypanosoma cruzi*
Infective Endocarditis	
Streptococcus viridans	*Streptococcus faecalis*
Staphylococcus aureus	*Staphylococcus epidermidis*
Histoplasma species	*Brucella* species
Candida species	*Aspergillus* species
Coxiella burnetii	
Rheumatic Heart Disease	
Group A β-hemolytic streptococcus	
Coronary Heart Disease	
Chlamydia pneumoniae	*Helicobacter pylori*
Cytomegalovirus	Herpes simplex virus type 2

by the microorganism, but is thought to be an autoimmune reaction stimulated by group A streptococci.[2] It is thought that the antibodies against the streptococcal antigens cross-react with similar antigens found in the heart and initiate a cell-mediated immune response involving macrophages and lymphocytes.[4]

Rheumatic heart disease affects all layers of the heart. Inflammation of the inner surface of the heart (endocarditis), especially the valves of the left heart, leads to ulceration and growth of vegetations on the heart lining and eventually to irreversible valve damage. *Myocarditis* caused by rheumatic heart disease may result in cardiac conduction problems or arrhythmias because of necrotic aggregates of lymphocytes and macrophages that are found in the myocardium.

Diagnosis of rheumatic heart disease is based on finding at least two of the following major symptoms: polyarthritis, carditis, chorea (involuntary movements caused by brain lesions), subcutaneous rheumatoid nodules, or erythema marginatum (a connective tissue and skin disease).[4] Other symptoms may include joint pain, fever, elevated erythrocyte sedimentation rate, immunologic evidence of a recent streptococcal infection, and characteristic ECG changes. Treatment options are few and surgical repair of valvular damage is often required.

Infectious endocarditis is an inflammation of the inner lining of the heart chambers and valves and may be caused by a number of microorganisms. Streptococci and staphylococci are common causes, as are gram-negative bacteria and some fungi.[4] These organisms attach to the endocardium, invade the valves and form vegetations of fibrin, platelets, blood cells, and microorganisms. These vegetations interfere with the function of the valves and may dislodge from the valves, forming emboli that can cause widespread infection or infarction of other organs. The two types of infective endocarditis are subacute and acute. Symptoms of subacute endocarditis are vague and insidious in many cases. Low-grade fevers, fatigue, anorexia, and splenomegaly are common early in the condition, and heart murmurs and congestive heart failure may develop in advanced infections. Acute endocarditis has a sudden onset of spiking fevers, chills, and drowsiness. Both types of infectious endocarditis respond to appropriate antibiotic therapy if treated early.

Pericarditis is usually secondary to another condition, often another cardiac condition. It may be caused by bacteria, viruses, or fungi and is associated with several autoimmune disorders such as systemic lupus erythematosis. Accumulation of fluid in the pericardial sac is the defining aspect of this condition and the various types of fluid within the sac will differentiate the type of pericarditis. Purulent exudates indicate bacterial infections, clear serous fluids are caused by viral infections, and a serofibrinous exudate is associated with severe damage as in rheumatic heart disease.

Symptoms of pericarditis vary with the underlying cause but often tachycardia, chest pain, shortness of breath, and cough are common. Large amounts of fluid can lead to distended neck veins, faint heart sounds, and ECG changes.[1]

Infectious agents have recently become the objects of interest in the continued search for additional causes of and effective treatments and preventive strategies for coronary artery disease. Viruses, such as herpes viruses and coxsackie B virus, and the bacteria *Chlamydia pneumoniae* and *Helicobacter pylori* have been studied widely. Although evidence for an infectious cause of atherosclerosis is only circumstantial at this point, continued efforts to determine cause and effect is ongoing.[11]

DIAGNOSIS OF HEART DISEASE

A single diagnostic laboratory test that will quickly and accurately assess cardiac function does not exist at this time, although the search continues. Such an assay would be extremely useful in evaluating many types of heart conditions but would require that the following features of an ideal myocardial marker for heart disease be met:

- The marker should be absolutely heart specific to allow reliable diagnosis of myocardial damage in the presence of skeletal muscle injury.
- The marker should be highly sensitive to detect even minor damage to the heart.
- It should be able to differentiate reversible from irreversible damage.
- In acute myocardial infarction, the marker should allow the monitoring of reperfusion therapy and the estimation of infarct size and prognosis.
- The marker should be stable and the measurement rapid, easy to perform, quantitative, and cost effective.
- The marker should not be detectable in patients who do not have any myocardial damage.[12]

Because of its dire consequences, most efforts to date have been placed on the development of such an ideal *cardiac marker* for the early and accurate diagnosis of acute myocardial infarction. Many factors must be considered in the selection of the most clinically diagnostic, effective, and cost-efficient laboratory tests for an individual patient with chest pain. The time that has elapsed after onset of chest pain, any concomitant diseases, the possibility of skeletal muscle injury, the ease of measurement and turnaround time for results, and assay specificity, sensitivity, and interferences are only a few considerations in these decisions.[13] At this point, it is impossible to find a single marker with all of these characteristics; therefore, a combination of cardiac markers is required.

Laboratory Diagnosis of Acute Myocardial Infarction

Enzymes

Creatine kinase (CK) is an enzyme that is involved in the transfer of energy in muscle metabolism. It is a dimer comprising two subunits: the B, brain form, and the M, muscle form, which results in three CK isoenzymes. The CK-BB (CK1) isoenzyme is of brain origin and is only found in the blood if the blood-brain barrier has been breached. CK-MM

CASE STUDY 20-3

A 23-year-old female presented to her physician with symptoms of fatigue and weakness, vague aches and pains, intermittent fever and chills, especially at night, and mild weight loss. The patient has a history of an atrial septal defect, which was repaired at age 7 years. Evaluation of this patient revealed a temperature of 100.6°F, pulse of 106 beats per minute, and a vague heart murmur. Blood cultures were obtained and the patient was placed on antibiotic therapy. Microbiology reports received later revealed the presence of *Staphylococcus aureus*.

Questions

1. What would this patient's diagnosis be?
2. Does the patient's history indicate a predisposition to this condition?
3. What is this patient's prognosis?

(CK3) isoenzyme accounts for almost all of the CK activity in skeletal muscle, whereas CK-MB (CK2) has the most specificity for cardiac muscle, even though it accounts for only 3% to 20% of total CK activity in the heart.[12]

CK-MB is a valuable tool for the diagnosis of acute myocardial infarction because of its relatively high specificity for cardiac injury. However, it takes at least 4 hours to 10 hours from onset of chest pain before CK-MB activities increase to significant levels in the blood. Peak levels occur with the first 24 hours and serum activities usually return to baseline levels with 2 days to 3 days (Fig. 20-5). Consider that the presence of CK-MB is not limited to the myocardium and false-positives are caused by analytical interferences as well as other clinical conditions such as muscle disease and acute or chronic muscle injuries.

During recent years, CK-MB activity assays have been increasingly replaced by CK-MB mass assays that measure the protein concentration of CK-MB rather than its catalytic activity.[12] These laboratory procedures are based on immunoassay techniques that have fewer interferences and a higher analytical sensitivity than the activity-based assays. Mass assays can detect an increased concentration of serum CK-MB about 1 hour earlier than other methods so that abnormal levels of CK-MB are found within 3 hours of onset of pain in approximately 50% of patients who develop an acute myocardial infarction and within 6 hours of pain in 80% to 100% of patients.[14] Persistently normal CK-MB mass concentrations over a period of 6 hours to 8 hours have a negative predictive value of 95%.[15] This improved immunoassay

results in better diagnostic accuracy at an earlier time than the activity-based assays.

CK isoforms are produced as a part of the normal clearance mechanism for CK isoenzymes and are present in all sera. There is only one CK-MB (CK2) and one CK-MM (CK3) isoform in myocardium. Determination of CK isoforms by high-voltage electrophoresis or isoelectric focusing is not considered a routine laboratory procedure and the results do not significantly improve the early diagnosis of acute myocardial infarction. There are inherent problems with using CK isoforms as a cardiac marker, such as the lack of heart-specificity for CK-MB, the small content of CK-MB in normal myocardium tissue, and the presence of detectable levels of CK-MB in normal individuals. However, CK isoforms may be effectively used as indicators of reperfusion after thrombolytic therapy in patients with confirmed acute myocardial infarction.[12]

Although AST was the first marker used for the laboratory diagnosis of acute myocardial infarction, it lacks cardiac specificity and it presently has no clinical significance for acute myocardial infarction diagnosis.[12] Lactate dehydrogenase (LD) is a cytoplasmic enzyme found in almost all cells of the body, including the heart, and therefore is not specific for the diagnosis of cardiac disease. Total LD activity begins to rise at 6 hours to 12 hours from the onset of chest pain, peaks at 1 day to 3 days, and slowly returns to normal within 8 days to 14 days.[12] LD is a tetramer made up of two types of subunits, the H subunit and the M subunit, resulting in five LD isoenzymes. The use of LD isoenzyme analysis improves cardiac specificity, with the LD1 and LD2 subfractions being most indicative of cardiac involvement. A patient with no cardiac damage will show a larger proportion of LD2 than LD1, whereas a patient with myocardial damage will demonstrate an increased LD1 activity over LD2 activity, often referred to as the "LD flip." This ratio is specific for myocardial damage but may be of little use in diagnosing an acute myocardial infarction unless the CK or CK-MB activity has already returned to within normal reference limits. The determination of LD1 or alpha-hydroxybutyrate dehydrogenase activity may be of clinical significance for the estimation of the size of the infarct because it has been demonstrated that there is a direct relationship between degree of elevation of LD1 and infarct size.[16]

Cardiac Proteins

Several proteins may be monitored in suspected cases of acute myocardial infarction to give significant information about damage to the heart. *Myoglobin,* an oxygen-binding heme protein that accounts for 5% to 10% of all cytoplasmic proteins, is rapidly released from striated muscles (both skeletal and cardiac muscle) when damaged. Although there is some evidence that there are several forms of myoglobin, a cardiac-specific myoglobin has not been identified. The usefulness of myoglobin as a cardiac marker was established in the mid-1970s when a radioimmunoassay technique was developed for its determination, but it was only when rapid, quantitative, and automated assays became available that myoglobin gained

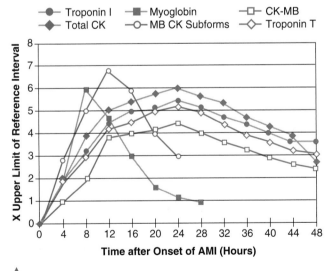

▲
Figure 20-5. Time Profiles of Characteristic Cardiac Markers after AMI. Illustrated here is the plasma temporal profiles of commonly used cardiac diagnostic markers, namely CK-MB, total CK, troponin I, troponin T, myoglobin, and CK-MB subforms. Early diagnosis of infarction (≤ 6 hours) is only potentially possible with two of the markers, namely, CK-MB subforms and myoglobin. (From Roberts R. Rapid MBCK Subform Assay and the Early Diagnosis of Myocardial Infarction. In: Keffer JH. Clinics in laboratory medicine: Recent advances in myocardial markers of injury. Vol. 17, no. 4. Philadelphia: WB Saunders, 1997. Reprinted with permission.)

acceptance as a routinely used diagnostic assay for acute myocardial infarction.[17]

Myoglobin is markedly more sensitive than CK and CK-MB activities during the first hours after chest pain onset. It starts to rise within 2 hours to 4 hours and is detectable in all acute myocardial infarction patients between 6 hours and 10 hours from chest pain onset (Fig. 20-5).[18] If the myoglobin concentration is still within the reference range 8 hours after onset of chest pain, acute myocardial infarction can essentially be ruled out. Although the early sensitivities of CK-MB mass, CK isoforms, and myoglobin are comparable, CK-MB determinations are preferable over myoglobin in patients who are admitted later than 10 hours to 12 hours after chest pain onset because the myoglobin concentration may have already returned to reference ranges within that time frame.[12] Myoglobin should not be depended on for early diagnosis of acute myocardial infarction in patients with renal disease, especially those persons in renal failure, because myoglobin is normally cleared by kidneys within 24 hours after acute myocardial infarction onset. The rapid disappearance of myoglobin from serum also allows its use as an indicator of reinfarction. A persistently normal myoglobin concentration will rule out reinfarction in patients with recurrent chest pain after acute myocardial infarction.

Muscle fibers convert the chemical energy of adenosine triphosphate (ATP) into mechanical work. As this occurs, enzymes, electrolytes, and proteins are activated or converted into materials that stimulate muscle fiber contraction. Actomyosin ATPase, calcium, actin, myosin, and a complex of three proteins known as the troponin complex are major players in this conversion. The three polypeptides of the troponin complex are troponin T, an asymmetric globular protein; troponin I, a basic globular protein; and troponin C, a dumbbell-shaped protein.[13] Troponin C is not heart specific.

Troponin T (TnT) allows for both early and late diagnosis of acute myocardial infarction. Serum concentrations of TnT begin to rise within a few hours of onset of chest pain and peak by day 2. A plateau lasting from 2 days to 5 days usually follows, and the serum TnT concentration remains elevated beyond 7 days before slowing and returning to reference values (Fig. 20-5). The early appearance of TnT gives no better diagnostic information than CK-MB or myoglobin concentrations within the first 6 hours after chest pain onset, but the sensitivity of TnT for detecting myocardial infarct is 100% from 10 hours to 5 days after chest pain onset and its diagnostic efficiency remains at 98% until 6 days after admission.[19] Also, the degree of elevation of TnT after acute myocardial infarction is significant, usually more than a 100-fold increase over the upper limit of reference intervals.[19]

TnT concentrations are particularly useful for diagnosing myocardial infarction in patients who do not seek medical attention within the usual 2- to 3-day window during which the total CK and CK-MB are elevated.[12] It may also be useful in the differential diagnosis of myocardial damage in patients with cardiac symptoms as well as skeletal muscle injury, because the TnT result will clearly and specifically indicate the extent of the cardiac damage as opposed to muscle damage.[12] Cardiac TnT also has value in monitoring patients after reperfusion of an infarct-related coronary artery. In acute myocardial infarction patients with early reperfusion, TnT appears and peaks in serum significantly earlier than in those patients with delayed or incomplete reperfusion. A greater than 6.8-fold increase in TnT 90 minutes after the start of thrombolytic therapy identifies complete reperfusion of the infarct-related coronary artery with acceptable accuracy.[20] The degree of elevation of TnT on days 3 to 4 after acute myocardial infarction can also be used as a noninvasive estimate of myocardial infarct size.[21]

Troponin I (TnI) is only found in the myocardium in adult humans, making it extremely specific for cardiac disease. It is also found in much higher concentrations than CK-MB in cardiac muscle, making it a very sensitive indicator of cardiac injury. TnI is not found in detectable amounts in the serum of patients with multiple injuries or athletes after strenuous exercise, in patients with acute or chronic skeletal muscle disease, in renal failure patients, or in patients with elevated CK-MB, unless myocardial injuries were also present.[22] TnI is a good biochemical assessment of cardiac injury in critically ill patients, those with multiple organ failure, and situations in which CK/CK-MB elevations may be difficult to interpret.

After an acute myocardial infarction, the TnI increases above the reference range between 3 hours and 8 hours after chest pain onset, peaks at 12 hours to 24 hours, and returns to within reference limits after 5 days to 10 days, depending on acute myocardial infarction size (Fig. 20-5). TnT tends to remain elevated longer and maintain higher sensitivity after day 7 after infarct than TnI.[23]

Cardiac *myosin light chains (MLC)* were at first thought to be unique myocardial proteins but recent research has determined that MLC is no more specific for cardiac injury than CK-MB or LD1 determinations. Thus, MLC will remain of limited clinical significance as a routine cardiac marker.[12]

Cardiac *myosin heavy chains (MHC)* are proteins that are structurally bound to muscle fibers and will not appear in the circulation except when muscle necrosis occurs. The significance of this protein is limited to the late diagnosis of an acute myocardial infarction or to assess the size of the infarct but gives no additional information than LD1, TnT, or TnI assays.[12]

New Research Markers

The search for the "perfect" cardiac marker continues with the development of several assays to aid in the diagnosis of myocardial injury. This "perfect" marker will be more rapidly detectable than currently available assays, be completely specific for cardiac muscle damage, and be easily performed. Some markers being investigated are as follows.

Glycogen Phosphorylase Isoenzyme BB (GPBB)

GPBB is a glycolytic enzyme that plays an essential role in the regulation of carbohydrate metabolism by mobilizing glycogen.[24] The disadvantage of using GPBB as a cardiac marker is that, although it is found primarily in the heart and brain, it has also been reported in leukocytes, platelets, spleen, kidney, bladder, testes, and digestive tract tissues.[12] However, it is significantly more sensitive than CK, CK-MB mass, myoglobin, and TnT in acute myocardial infarction patients during the first 3 hours to 4 hours after the onset of chest pain. In the majority of acute myocardial infarction patients, GPBB increases between 1 hour and 4 hours from chest pain onset and returns to within the reference level within 1 day to 2 days. Of particular significance is that GPBB is greatly affected by reperfusion with earlier and more elevated concentrations of GPBB in those patients who do not achieve adequate reperfusion.[22]

Heart Fatty Acid Binding Protein (H-FABP)

H-FABP is involved in the cellular uptake, transport, and metabolism of fatty acids.[25] This protein is found in distinctly different types in cardiomyocytes, intestine, liver, and adipose tissue, but even the so-called "heart-FABPs" are not heart specific; they are also found in significant amounts in skeletal muscle and kidneys.[26] It increases rapidly, usually within 2 hours to 4 hours, peaks within 5 hours to 10 hours, and returns to normal within 24 hours to 36 hours after onset of chest pain. It can also be used to determine recurrent infarctions, although its greatest contribution is for the early confirmation or exclusion of acute myocardial infarction.[12]

Carbonic Anhydrase (CA) Isoenzyme III

CA is a soluble protein that catalyzes the hydration of CO_2 to bicarbonate and a proton and is involved in pH regulation, transport of ions, water and electrolyte balance, and metabolism of carbohydrates, urea, and lipids.[27] There are seven CA isoenzymes with a wide range of tissue distribution but a major site of CA activity is in skeletal muscle. CAIII is not found in cardiac muscle and therefore can be used to differentiate between skeletal muscle and cardiac muscle damage when performed in conjunction with a more heart-specific analyte such as myoglobin. In the patient with actual or possible coexisting skeletal muscle injury, TnT or TnI would still be the markers of choice for confirming or excluding myocardial damage.[12]

Annexin V, enolase isoenzymes, and phosphoglyceric acid mutase isoenzyme MB (PGAM) are other analytes that are being investigated for their suitability as cardiac markers. At this point, none has demonstrated superiority to, or is likely to be used in place of, the currently accepted markers.

Patient-Focused Cardiac Tests

Another area of interest is in the development of rapid semiquantitative devices for performing whole blood cardiac assays at the patient's bedside. Currently, these devices do not provide the accuracy and precision required of such assays to obtain an early diagnosis of acute myocardial infarction.[13]

Laboratory's Role in Monitoring Heart Disease

The laboratory's role in monitoring heart function primarily involves measuring the effects of the heart on other organs such as the lungs, liver, and kidney. Arterial blood gases determine the patient's acid–base and oxygen status because respiratory acidosis with elevated PCO_2 levels is often seen in patients with heart disease. The patient with edema will develop electrolyte and osmolality changes as a result of fluid retention and ionic redistribution. An increased excretion of potassium and a decrease in urinary sodium may be early indicators of these imbalances. Decreased cardiac output results in renal sodium retention but this also causes increased fluid retention; therefore, the serum sodium generally stays within the reference ranges or is slightly decreased. Serum electrolyte determinations (including sodium, potassium, chloride, and calcium) are important to monitor diuretic and drug therapy in patients with treated heart disease.[4]

Elevations of aspartate aminotransferase (AST), alanine aminotransferase (ALT), and alkaline phosphatase (ALP) are often in patients with chronic right ventricular failure[4] and the γ-glutamyltransferase (GGT) value may be twice the value of the upper limit of the normal range in congestive heart failure, suggesting liver congestion and damage.

Lipid evaluation will assess risk for coronary artery disease. Maintenance of near-normal HDL-cholesterol, LDL-cholesterol, and triglycerides levels is highly recommended for cardiac patients.

The patient who has secondary heart failure due to thyroid dysfunction can be identified by a highly sensitive thyroid-stimulating hormone assay. The laboratory is also invaluable for monitoring therapeutic drugs following the diagnosis of heart disease.

The routine complete blood count is important for detecting anemias and infections. Hemolysis may indicate additional testing for hemoglobinuria and myoglobinuria, indicators of cardiovascular damage and myocardial disease. An increase in the number of white blood cells may indicate pericarditis, endocarditis, or valvular infections. If kidney dysfunction has occurred as a result of the heart disease, anemia due to decreased production of renal erythropoietin may develop.

An infection associated with pericarditis, endocarditis, and valvular problems should be identified by blood cultures. Cultures of material from the pericardium and endocardium might also be performed if a pericardial infection is suspected.

The laboratory's role during the treatment of heart disease may also extend to providing blood components when surgical intervention is needed. Bypass grafts, correcting valvular defects, and other surgical procedures to correct heart failure may involve the use of blood components during the surgery and the patient's recovery.

Other Diagnostic Tests

Heart size and position are important features that can be determined by radiologic techniques. A series of x-rays that include frontal, lateral, and oblique views of the chest should

An 83-year-old male with known severe coronary artery disease, diffuse small vessel disease, and significant stenosis distal to a vein graft from previous CABG surgery, was admitted when his physician referred him to the hospital after a routine office visit. His symptoms included 3+ pedal edema, jugular vein distention, and heart sound abnormalities. Significant laboratory data obtained upon admission were as follows.

Urea nitrogen (6–24 mg/dL)	53
Creatinine (0.5–1.4 mg/dL)	2.2
Total protein (6.0–8.3 g/dL)	5.8
Albumin (3.5–5.3 g/dL)	3.2
Glucose (60–110 mg/dL)	312
Calcium (4.3–5.3 mEq/L)	4.1
CK-MB (0–5 ng/L)	28
% CK-MB (<6%)	11%
Phosphorus (2.5–4.5 mg/dL)	2.4
Total CK (54–186 U/L)	252
Myoglobin (<70 µg/L)	268
Troponin T (0–0.1 µg/L)	0.9

Questions

1. Do the symptoms of this patient suggest acute myocardial infarction?
2. Based on the preceding laboratory data, would this diagnosis be acute myocardial infarction? Why or why not?
3. Based on the preceding laboratory data, are there other organ system abnormalities present?
4. What are the indicators of these organ system abnormalities?

be a part of the routine investigation of heart disease. Enlargement of the cardiac chambers, the status of the vessels—especially the aorta and pulmonary artery, and the pulmonary vasculature are also investigated with x-ray techniques. Angiography may be performed by injecting radiopaque material intravenously, intra-arterially or into one of the cardiac chambers to view the heart and great vessels.[28]

Echocardiography is a noninvasive diagnostic technique that uses high-frequency sound waves to show the heart structure on a cathode-ray tube screen. The reflection of these ultrasound waves indicates the structure and function of the heart components. Heart conditions that are routinely assessed by echocardiography are valve abnormalities, pericardial effusions, atrial thrombi, cardiac hypertrophy, and congenital cardiac malformations.[28] Doppler examination of the heart is based on the principle of shifts in transmitted sound frequencies and is generally used to evaluate the direction and velocity of blood flow and valvular abnormalities in the heart.[10]

Evaluation of the electrical stimulation of the heart is accomplished by ECG. Electrodes are placed on the body and the electric activity of the heart is transmitted through the body to the skin and measured by a galvanometer. The ECG evaluates the conduction system of the heart and also indicates heart rate and timing sequences of the heart rhythm.

Cardiac catheterization is an invasive technique in which a catheter is placed into a peripheral vessel and advanced into the heart. Right heart catheterization requires that a catheter be placed through a vein into the heart and evaluates the right atrium, ventricle, and pulmonary artery. Left heart catheterization places the catheter in an artery and assesses the left atrium, left ventricle, and aorta. Pressure measurements and oxygen contents of the vessels and chambers of the heart are measured and recorded. Cardiac catheterization

is performed to define anatomic lesions, to disclose clinically covert complicating lesions, to determine in quantitative terms the severity of these lesions, to measure the pathophysiologic alterations caused by the lesions, and to assess overall cardiac and left ventricular function.[28]

Coronary arteriography to evaluate the coronary arteries by injecting the arteries with contrast material is often used to determine if coronary artery disease is the cause of acute myocardial infarction or severe angina and can also be used to determine congenital defects of the coronary arteries or valves.

Cardiovascular nuclear medicine techniques can be performed to assess cardiovascular performance and perfusion of the myocardium and viability of the cardiac muscle.[28] These techniques require the use of radionuclides that either collect in cardiac muscle or remain in the cardiac vasculature for visualization.

TREATMENT

When a cardiac disease has been diagnosed, certain lifestyle changes to improve both quality of life and longevity are recommended. Exercise is effective in reducing the risk for further cardiac disease and for rehabilitation after myocardial infarction and other chronic cardiac disorders. Smoking cessation is an important step in reducing risks of complications and worsening of cardiac conditions. Dietary considerations are a necessary component of controlling the symptoms of cardiac disease. Weight reduction will reduce the workload of the heart and dietary restriction of sodium and fats are recommended. The use of drug therapy to reduce LDL-cholesterol concentrations may be necessary if reduced ingestion of animal and saturated fats does not lower concentrations to an acceptable level. Estrogen replacement therapy in postmeno-pausal women may reduce the risk of coronary disease by 40% to 50%.[29] Stress management is also encouraged, both to minimize anxiety associated with heart conditions and to control reactions to stressful life situations.

The strong correlation between hypertension and heart disease requires strict control of hypertension. Treatment for *essential hypertension* is aimed at reducing the risk factors such as weight reduction, salt and alcohol restriction, and increased exercise.

Drug Treatment

Multidrug treatment is routine to control the many and varied problems associated with cardiac disease. It usually consists of a combination of vasodilators, diuretics, beta-blockers, calcium channel antagonists, cardiac glycosides, and anticoagulants (Table 20-5).

Nitrates are coronary artery dilators that increase the blood supply to the heart and decrease blood pressure. Sublingual nitroglycerine is frequently used for relief of angina and is a part of the routine treatment of congestive heart failure and myocardial infarction. Topical ointments and skin patches may also be used to supply a consistent dosage of nitroglycerine to treat unstable angina and limit cardiac damage due to ischemia.

The most commonly used *vasodilators* are the ACE (angiotensin converting enzyme) inhibitors, such as hydralazine or captopril. These drugs dilate the peripheral arteries and veins, decreasing the amount of effort that the heart expends to pump blood and are often an integral part of the treatment of cardiac conditions such as congestive heart failure and myocardial infarction.

Beta-adrenergic blocking drugs reduce heart rate or the force of contractions reducing oxygen demand for the heart by blocking the beta receptors in the sinus mode and the myocardium. This group of drugs is used to treat tachycardia, ectopic beats, and arrhythmias as well as angina pain and hypertension. Propranolol is a typical beta-blocker.

Calcium channel antagonists will decrease smooth muscle contractions and result in vasodilation. These drugs bind to the calcium channel subunits to limit the ability of calcium to cross the cell membrane. They are most often used to treat hypertension, angina, and supraventricular tachycardia.[30]

Cardiac glycosides, particularly digoxin, are used to increase the contractility of the heart and slow the conduction impulses. Digoxin is now the most commonly prescribed cardiac glycoside because of its convenient pharmacokinetics, alternative routes of administration, and widespread availability of serum drug level measurements.[31] Most patients with congestive heart failure, atrial fibrillation or tachycardia, hypertension, or cardiac ischemia will receive a drug of this type to control their symptoms.

Diuretics are given to help the kidneys clear excess water and sodium to reduce blood volume and, thereby, reduce the workload placed on the heart. Diuretics provide relief of symptoms in both pulmonary edema and congestive heart failure. It is important to monitor electrolyte balance and hydration to prevent too aggressive diuresis, which can harm the patient.

Numerous trials have demonstrated that early thrombolysis-induced reperfusion reduces infarct size and mortality.[10] *Thrombolytic agents,* such as streptokinase, urokinase, or tissue plasminogen activator, may be indicated to lyse blood clots in acute myocardial infarction, acute pulmonary embolism, acute deep vein thrombosis, or arterial thrombosis or embolism.[30] The use of thrombolytic agents in patients with acute myocardial infarction, as compared with standard medical therapy,

TABLE 20-5. Drug Treatment of Cardiac Disease

Classification of Drug	Action of Drug	Example
Vasodilators	Increase blood supply to heart muscle; decrease blood pressure	Nitroglycerine; angiotensin converting enzyme inhibitors such as hydralazine or captopril
Diuretics	Reduce blood volume and edema	Furosemide; thiazides
Beta-blockers	Reduce heart rate and the force of cardiac contractions	Propranolol
Calcium channel antagonists	Decrease smooth muscle contractions	Nifedipine
Cardiac glycosides	Increase contractility of the heart and slow conduction impulses	Digoxin
Anticoagulants	Reduce formation of prothrombin and prevent thrombosis and thromboembolism	Heparin; warfarin
Antiplatelet therapy	Reduces platelet activation and aggregation	Aspirin

reduces overall mortality by 21 deaths per 1000 patients treated.[32] The greatest benefit occurs when these agents are administered within 3 hours after onset of symptoms.[32]

Antiplatelet therapy, often with aspirin, is effective in reducing the risk for progression of atherosclerosis to myocardial infarction. It reduces platelet activation and aggregation, a significant factor in thrombotic processes and, as a result, reduces risk for ischemic heart disease.[30]

The use of heparin, warfarin, or other anticoagulants reduces the formation of prothrombin or inhibits the action of thrombin. Anticoagulant therapy is recommended to prevent venous thrombosis and thromboembolism, extension and recurrence of an infarct, and to reduce mortality associated with these conditions.[28]

Surgical Treatment

Coronary artery bypass graft surgery (CABG) was introduced in the late 1960s and has become a standard treatment for ischemic heart disease since. Nearly 1 in every 1000 persons in the United States undergoes CABG every year.[33] This surgical technique relieves symptoms and decreases mortality associated with ischemic heart disease by replacing the occluded coronary arteries with unaffected artery grafts. Patients with stable and unstable angina pectoris, acute myocardial infarction, "silent" ischemia, survivors of sudden cardiac death, congenital coronary anomalies, and congestive heart failure are frequently candidates for this surgery.[33]

Percutaneous transluminal coronary angioplasty (PTCA) is a surgical procedure in which an angioplasty balloon is inserted into a coronary artery and expanded. This should open the lumen of the obstructed vessel and restore blood flow to the affected area. There is some disagreement as to the long-term benefits of PTCA as compared to those of CABG, but in single coronary vessel disease, PTCA may be the treatment of choice.

Transmyocardial laser revascularization is a promising investigational technique in which a laser creates channels in the left ventricle to provide a blood supply to the adjacent myocardium. This revascularization has been shown to decrease the size of ischemic regions in patients who were not candidates for more conventional therapies.

Orthotopic heart transplantation is an option for treatment of advanced heart disease when medical and surgical treatments have failed or when the patients have significant contraindications for these procedures. Currently the 1-year survival rate after cardiac transplantation is 85% to 90% and the 5-year survival rate is almost 70%.[33] The inavailability of sufficient donor hearts is currently the limiting factor for this treatment option.

SUMMARY

The goal of an accurate and timely diagnosis for heart conditions is currently not realized in all patients, especially those with acute chest pain. A thorough knowledge of the anatomy and function of the heart and the causes and effects of heart disease continues to be a most valuable tool in identifying and treating heart disease. The judicious use of laboratory and other diagnostic data is critical to the identification of patients who need further care and those who may safely be discharged. In patients who are exhibiting chest pain, recognizing the time-related patterns of increasing cardiac markers used today is an important factor in indicating the appropriate treatment for each individual patient. But often, the symptoms and the risk factors identified may not correlate with the laboratory and radiologic data obtained.

Advances in surgical and pharmacologic treatments for heart conditions will increase the life expectancy, as well as the quality of life, in patients with advanced heart disease. The near future will bring development of more sensitive assays that have absolute specificity for cardiac disease. Hopefully, these advancements will reduce the number of situations in which a patient with an acute myocardial infarction is sent home with antacids and the patient with indigestion is admitted into the cardiac care unit.

REVIEW QUESTIONS

1. A serum troponin T concentration is of most value to the patient with a myocardial infarction when:
   a. The onset of symptoms is within 3 hours to 6 hours of the sample being drawn
   b. The CK-MB has already peaked and returned to normal concentrations
   c. The myoglobin concentration is extremely elevated
   d. The troponin I concentration has returned to normal concentrations
2. A normal myoglobin concentration 8 hours after onset of symptoms of a suspected myocardial infarction will:
   a. Essentially rule out an acute myocardial infarction
   b. Provide a definitive diagnosis of acute myocardial infarction
   c. Must be interpreted with careful consideration of the troponin T concentration
   d. Give the same information as a total CK-MB
3. Which of the following analytes has the highest specificity for cardiac injury?
   a. Troponin I
   b. CK-MB mass assays
   c. Total CK-MB
   d. AST
4. The heart comprises four chambers and four valves between these chambers. The aortic valve allows blood to move from the:
   a. Right atrium to the right ventricle
   b. Right ventricle to the pulmonary artery
   c. Left atrium to the lungs
   d. Left ventricle to the general circulation
5. Which of the following valves is a bicuspid valve?
   a. Tricuspid valve

  b. Mitral valve

  c. Pulmonary valve

  d. Aortic valve

6. Rheumatic heart disease is a result of infection with which of the following organisms?

  a. *Staphylococcus aureus*

  b. Group A streptococci

  c. *Pseudomonas aeruginosa*

  d. *Chlamydia pneumoniae*

7. Angina, a common symptom of congestive heart failure, is often relieved by the administration of:

  a. Diuretics

  b. Beta-adrenergic blocking drugs

  c. Cardiac glycosides

  d. Nitrates

8. A factor that is closely associated with development of congenital heart defects is:

  a. Maternal age younger than 16 years

  b. Vigorous exercise during pregnancy

  c. Maternal alcohol use

  d. Maternal streptococcal infection during pregnancy

9. Lowering both systolic and diastolic hypertension has been shown to decrease the incidence of:

  a. Endocarditis

  b. Myocardial infarction

  c. Atherosclerosis

  d. Rheumatic heart disease

10. Which of the following IS NOT a feature of the non-existent, ideal cardiac marker?

  a. Absolute specificity

  b. High sensitivity

  c. Close estimation of the magnitude of cardiac damage

  d. Ability to predict future occurrence of cardiac disease

REFERENCES

1. Gould BE. Pathophysiology for the health-related professions. Philadelphia: WB Saunders, 1997.

2. Camm AJ. Cardiovascular disease. In: Kumar PJ, Clark ML, eds. Clinical medicine. 2nd ed. London: Bailliere Tindall, 1990;511.

3. Miles WM, Zipes DP. Cardiovascular diseases. In: Andreoli TE, Bennett JC, Carpenter CCJ, et al, eds. Cecil essentials of medicine. 3rd ed. Philadelphia: WB Saunders, 1993;1.

4. Damjanov I. Pathology for the health-related professions. Philadelphia: WB Saunders, 1996.

5. Braunwald E, Colucci WS, Grossman W. Clinical aspects of heart failure: high-output heart failure; pulmonary edema. In: Braunwald E, ed. Heart disease: a textbook of cardiac medicine. Vol. 1, 5th ed. Philadelphia: WB Saunders, 1997;445.

6. Smith WM. Epidemiology of congestive heart failure. Am J Cardiol 1985;55(2):3A–8A.

7. Hosenpud JD. The cardiomyopathies. In: Hosenpud JD, Greenberg BH, eds. Congestive heart failure: pathophysiology, diagnosis, and comprehensive approach to management. New York: Springer-Verlag, 1994;196.

8. Thomas CL, ed. Taber's cyclopedic medical dictionary. 18th ed. Philadelphia: FA Davis, 1997;168.

9. Wilson PWF. An epidemiologic perspective of systemic hypertension, ischemic heart disease and heart failure. Am J Cardiol 1997;80(9B):3J–7J.

10. Chesler E. Clinical cardiology. 5th ed. New York: Springer-Verlag, 1993.

11. Ellis RW. Infection and coronary heart disease. J Med Microbiol 1997;46:535–539.

12. Mair J. Progress in myocardial damage detection: new biochemical markers for clinicians. In: Hindmarsh JT, Goldberg DM, eds. Critical reviews in clinical laboratory sciences. Vol. 34. 1997;1.

13. Mercer DW. Role of cardiac markers in evaluation of suspected myocardial infarction: selecting the most clinically useful indicators. Postgrad Med 1997;102(5):113–122.

14. Mair J, et al. Equivalent early sensitivities of myoglobin, creatine kinase MB mass, creatine kinase isoform ratios, and cardiac troponins I and T for acute myocardial infarction. Clin Chem 1995;41(9):1266–1272.

15. Gibler WB, et al. A rapid diagnostic and treatment center for patients with chest pain in the emergency department. Ann Emerg Med 1995;25(1):1–7.

16. Van der Laarse A, et al. Relation between infarct size and left ventricular performance assessed in patients with first acute myocardial infarction randomized to intracoronary thrombolytic therapy or to conventional treatment. Am J Cardiol 1988;61(1):1–7.

17. Delanghe JR, Chapelle JP, Vanderschueren SC. Quantitative nephelometric assay for determining myoglobin evaluated. Clin Chem 1990;36(9):1675–1678.

18. Winter RJ, Koster RW, Sturk A, et al. Value of myoglobin, troponin T and CK-MB mass in ruling out an acute myocardial infarction in the emergency room. Circulation 1995;92(12):3401–3407.

19. Katus HA, et al. Diagnostic efficiency of troponin T measurements in acute myocardial infarction. Circulation 1991;83(3):902–911.

20. Laperche T, et al. A study of biochemical markers of reperfusion early after thrombolysis for acute myocardial infarction. Circulation 1995;92(8):2079–2085.

21. Omura T, et al. Estimation of infarct size using serum troponin T concentration in patients with acute myocardial infarction. Jpn Circ J 1993;57:1062–1069.

22. Adams III JE, et al. Cardiac troponin I: a marker with high specificity for cardiac injury. Circulation 1993;88(1):101–106.

23. Adams III JE, Schechtman KB, Landt Y, et al. Comparable detection of acute myocardial infarction by creatine kinase MB isoenzyme and cardiac troponin I. Clin Chem 1994;40(7):1291–1295.

24. Rabitzsch G, et al. Immunoenzymometric assay of human glycogen phosphorylase isoenzyme BB in diagnosis of ischemic myocardial injury. Clin Chem 1995;41(7):966–978.

25. Kleine AH, Glatz JFC, Van Nieuwenhoven FA, et al. Release of heart fatty acid-binding protein into plasma after acute myocardial infarction in man. Mol Cell Biochem 1992;16:155–162.

26. Glatz JFC, van der Vusse GJ. Cellular fatty acid-binding proteins: current concepts and future directions. Mol Cell Biochem 1990;98:237–251.

27. Sly WS, Peiyi YH. Human carbonic anhydrases and carbonic anhydrase deficiencies. Annu Rev Biochem 1995;64:375–401.

28. Goldberger E. Essentials of clinical cardiology. Philadelphia: Lippincott Williams & Wilkins, 1990.

29. Stampfer MJ, Colditz GA. Estrogen replacement therapy and coronary heart disease: a quantitative assessment of the epidemiological evidence. Prev Med 1991;20(1):47–63.

30. Opie LH. Pharmocologic options for treatment of ischemic disease. In: Smith TW, ed. Cardiovascular therapeutics: a companion to Braunwald's heart disease. Philadelphia: WB Saunders, 1996;22.

31. Kelly RA, Smith TW. The pharmacology of heart failure drugs. In: Smith TW, ed. Cardiovascular therapeutics: a companion to Braunwald's heart disease. Philadelphia: WB Saunders, 1996;176.

32. Ryan TJ. Management of acute myocardial infarction: synopsis of ACC and AHA practice guidelines. Postgrad Med 1997;102(5):84–96.

33. Solomon AJ, Gersh BJ. Ischemic heart disease: surgical options. In: Smith TW, ed. Cardiovascular therapeutics: a companion to Braunwald's heart disease. Philadelphia: WB Saunders, 1996;65.

CHAPTER 21

Renal Function

Carol J. Skarzynski, Alan H. B. Wu

Objectives

Upon completion of this chapter, the clinical laboratorian should be able to:

- *Diagram the anatomy of the nephron.*

- *Describe the physiologic role of each of the following: nephron, glomerulus, proximal tubule, Henle's loop, distal tubule, and collecting duct.*

- *In conjunction with hormones, describe the mechanisms by which the kidney maintains fluid and electrolyte balance.*

- *Discuss the concept of renal clearance and how it is measured.*

- *Distinguish between glomerular filtration rate and plasma renal flow.*

- *List the tests in a urinalysis and microscopy profile and understand the clinical significance of each.*

- *Relate the pathophysiologic role of total urine proteins, urine albumin, serum beta₂-microglobulin, the myoglobin clearance rate, and microalbumin.*

- *Describe various diseases of the glomerulus and tubules and how laboratory tests are used in these disorders.*

- *Distinguish between acute and chronic renal failure.*

- *Explain the complications of renal failure and the goals for laboratory monitoring.*

- *Discuss the therapy of chronic renal failure with regard to renal dialysis and transplantation.*

KEY TERMS

Acute renal failure	Chronic renal failure
Aldosterone	Countercurrent multiplier system
Allograft	Creatinine clearance
Antidiuretic hormone (ADH)	Cyclosporine
Beta₂-microglobulin (β₂-M)	Diabetes mellitus

Erythropoietin	Nephrotic syndrome
Glomerular filtration rate (GFR)	Plasma renal flow
Glomerulonephritis	Prostaglandins
Glomerulus	Renal threshold
Graft	Renin
Hemodialysis	Tubular reabsorption
Hemofiltration	Tubular secretion
Henle's loop	Tubule
Microalbumin	Urinalysis
Myoglobin	Vitamin D

The kidneys are vital organs that perform a variety of important functions for the body (Table 21-1). The most prominent functions are the removal from plasma of unwanted substances (both waste and surplus), homeostasis (the maintenance of equilibrium) of the body's water, electrolyte and acid-base status, and participation in endocrine regulation. Another important function is gluconeogenesis.[1] During prolonged fasting, the kidneys synthesize glucose from amino acids and other precursors. In the clinical laboratory, kidney function tests are used in the assessment of renal disease, water balance, and acid-base disorders, and in situations of trauma, head injury, surgery, and infectious diseases. This chapter focuses on renal anatomy and physiology and the analytical procedures available to diagnose, monitor, and treat renal dysfunction.

RENAL ANATOMY

The kidneys are paired, bean-shaped organs located retroperitoneally on either side of the spinal column. Macroscopically, each kidney is enclosed by a fibrous capsule of connective tissue. When dissected longitudinally, two regions can be clearly discerned: an outer region called the cortex and an

TABLE 21-1. Functions of the Kidneys[17]

- Regulation of water and inorganic ion balance
- Removal of metabolic waste products from the blood and their excretion in the urine
- Removal of foreign chemicals from the blood and their excretion in the urine
- Secretion of hormones:
  - Erythropoietin, which controls erythrocyte production
  - Renin, which controls formation of angiotensin, which influences blood pressure and sodium balance
  - 1,25-dihydroxyvitamin D_3, which influences calcium balance
- Gluconeogenesis

inner region called the medulla (Fig. 21-1*A*). The pelvis can also be seen. It is a basinlike cavity at the upper end of the ureter into which newly formed urine passes. The bilateral ureters are thick-walled canals connecting the kidneys to the urinary bladder. Urine is temporarily stored in the bladder until voided from the body by way of the urethra. Figures 21-1*B–E* show the arrangement of nephrons in the kidney, the functional units of the kidney that can only be seen microscopically. Each kidney contains approximately 1 million nephrons. Each nephron is a complex apparatus comprising the following five basic parts, which are expressed diagrammatically in Figure 21-2:

1. The *glomerulus,* which is a capillary tuft surrounded by the expanded end of a renal *tubule* known as *Bowman's capsule.* Each glomerulus is supplied by an afferent arteriole carrying the blood in and an efferent arteriole carrying the blood out. The efferent arteriole branches into peritubular capillaries that supply the tubule.
2. The proximal convoluted tubule, which is located in the cortex.
3. The long Henle's loop, which comprises the thin descending limb that spans the medulla from the corticomedullary junction to the inner medulla and the ascending limb located in both the medulla and the cortex and is composed of a thin, then thick, region.
4. The distal convoluted tubule, which is located in the cortex.
5. The collecting duct, which is formed by two or more distal convoluted tubules as they pass back down through the cortex and the medulla to collect the urine that drains from each nephron. Collecting ducts eventually merge and empty their contents into the renal pelvis.

The following sections consider how each of the five parts of the nephron normally functions.

RENAL PHYSIOLOGY

The three basic renal processes are glomerular filtration, tubular reabsorption, and tubular secretion. Figure 21-3 illustrates how three different substances are variably processed by the nephron. Substance A is filtered and secreted but not reabsorbed, substance B is filtered and a portion reabsorbed, and substance C is filtered and completely reabsorbed.[1] The following is a description of how specific substances are regulated in this manner to maintain homeostasis.

Glomerular Function

The glomerulus is the first part of the nephron to receive incoming blood and functions to filter this blood. Every substance except cells and large molecules continues into further sections of the nephron. Several factors facilitate filtration. One is the unusually high pressure in the glomerular capillaries, which is due to their position between two arterioles. This sets up a steep pressure difference across the walls. Another factor is the semipermeable glomerular basement membrane, which has a molecular size cutoff value of approximately 66,000 daltons, about the molecular size of albumin. This means that water, electrolytes, and small dissolved solutes such as glucose, amino acids, urea, and creatinine pass freely through and enter the proximal convoluted tubule; however, albumin, many plasma proteins, cellular elements, and protein-bound substances, such as lipids and bilirubin, are stopped. Another factor is that the basement membrane is negatively charged and this contributes to the filtration by repelling negatively charged molecules, such as proteins. Thus, of the 1200 mL to 1500 mL of blood that the kidneys receive each minute (approximately one quarter of the total cardiac output), the glomerulus filters out 125 mL to 130 mL of essentially protein-free, cell-free fluid. The volume of blood filtered per minute is known as the *glomerular filtration rate (GFR)* and its determination is essential in evaluating renal function, as discussed in the section on *Analytical Procedures.*

Tubular Function

Proximal Convoluted Tubule

The proximal tubule is the next part of the nephron to receive the now cell- and essentially protein-free blood. This filtrate contains both waste products, which are toxic to the body above a certain concentration, and substances that are valuable to the body. One of the functions of the proximal tubule is to return the bulk of each valuable substance back to the blood circulation. Thus, three quarters of the water, sodium, and chloride, all of the glucose (up to the renal threshold), almost all of the amino acids, vitamins, and proteins, and varying amounts of ions such as magnesium, calcium, potassium, and bicarbonate are *reabsorbed.* When the movement of the substances is from the tubular lumen to the peritubular capillary plasma, the process is called *tubular reabsorption.* With the exception of water and chloride ions, the process is active; that is, the tubular epithelial cells use energy to transport the substances across their plasma membranes to the blood. The transport processes that are involved normally have sufficient reserve for efficient reabsorption, but they are saturable, that is, there is a concentration for each substance in the filtrate

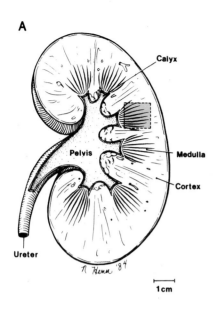

A

Calyx

Pelvis

Medulla

Cortex

Ureter

1 cm

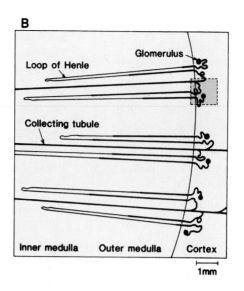

B

Loop of Henle

Glomerulus

Collecting tubule

Inner medulla   Outer medulla   Cortex

1mm

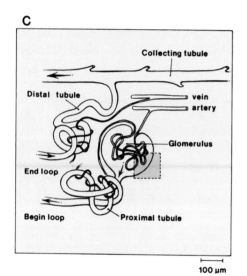

C

Collecting tubule

Distal tubule

vein

artery

Glomerulus

End loop

Begin loop   Proximal tubule

100 µm

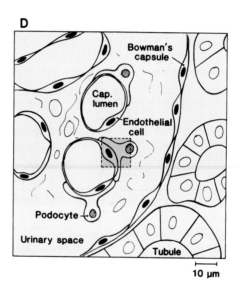

D

Bowman's capsule

Cap. lumen

Endothelial cell

Podocyte

Urinary space

Tubule

10 µm

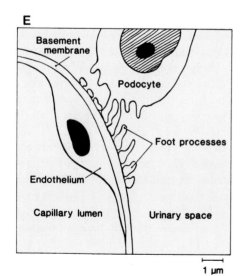

E

Basement membrane

Podocyte

Foot processes

Endothelium

Capillary lumen   Urinary space

1 µm

Figure 21-1. Successive magnifications of kidney by powers of ten.

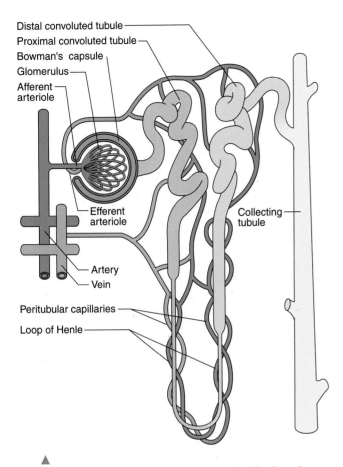

Figure 21-2. Representation of a nephron and its blood supply.

and drugs, such as penicillin. The term *tubular secretion* is used in two ways. It is used to describe the movement of substances from peritubular capillary plasma to tubular lumen. In addition, tubule cells also secrete some products of their own cellular metabolism into the filtrate in the tubular lumen, and this is also termed tubular secretion. Transport across the membrane of the cell is again either active or passive.

Almost all (98%–100%) of uric acid, a waste product, is reabsorbed actively, only to be secreted at the distal end of the proximal tubule. Urea, another waste product, is a highly diffusible molecule and passively passes out of the renal tubule and into the interstitium, where it is contributes to the osmolality gradient found in the medulla.

Henle's Loop

Countercurrent Multiplier System. In this portion of the nephron, the osmolality in the medulla, increasing steadily from the corticomedullary junction inward, facilitates the reabsorption of water, sodium, and chloride. The hyperosmolality that develops is continuously maintained by the Henle's loop. The Henle's loop forms a hairpin-like loop between the proximal tubule and the distal convoluted tubule. The opposing flows in the loop, the downward flow in the descending limb and the upward flow in the ascending limb, is termed a countercurrent flow. To understand how the hyperosmolality is maintained in the medulla, it is best to look first at what happens in the ascending limb. Along the entire length of the ascending limb, sodium and chloride are both actively and passively reabsorbed into the medullary interstitial fluid. Because the ascending limb is relatively impermeable to water, little water follows, and the medullary interstitial fluid becomes hyperosmotic compared to the fluid in the ascending limb. The descending limb, in contrast to the ascending limb, is highly permeable to water and does not reabsorb sodium and chloride. The high osmolality of the surrounding interstitial medulla fluid is the physical force that accelerates the absorption of water from the descending limb. The interstitial hyperosmolality is maintained because the ascending limb continues to pump chloride and sodium ions into it. This interaction of water leaving the descending loop and sodium and chloride leaving the ascending loop to maintain a high osmolality within the kidney medulla produces a hypoosmolal

above which the relevant transport system cannot function fast enough to bind and therefore remove the substance from the filtrate. The substance is excreted in the urine. The plasma concentration above which the substance appears in urine is known as the *renal threshold* and its determination is useful in assessing both tubular function and nonrenal disease states. A renal threshold for water does not exist because it is always transported passively through diffusion down a concentration gradient. Chloride ion in this instance diffuses in the wake of sodium.

A second function of the proximal tubule is to *secrete* products of kidney tubular cell metabolism, such as hydrogen ions,

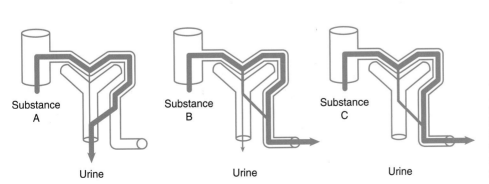

Figure 21-3. Renal processes of filtration, reabsorption, and secretion: Substance A is filtered and secreted, but not reabsorbed, whereas B is filtered and a portion reabsorbed, and C is filtered and completely reabsorbed.

urine as it leaves the loop. This process is called the countercurrent multiplier system.2

Distal Convoluted Tubule

The distal convoluted tubule is much shorter than the proximal tubule, with two or three coils that connect to a collecting duct. The filtrate entering this section of the nephron is close to its final composition. About 95% of the sodium and chloride ions and 90% of water have already been reabsorbed from the original glomerular filtrate. The function of the distal tubule is to effect small adjustments to achieve electrolyte and acid-base homeostasis. It is under the hormonal control of *aldosterone*. This hormone is produced by the adrenal cortex and its secretion is triggered by decreased blood flow in the afferent renal arteriole. It is regulated by the renin–angiotensin mechanism and to a lesser extent by adrenocorticotropic hormone (ACTH). Aldosterone stimulates sodium reabsorption by the distal tubule and potassium and hydrogen ion secretion. Hydrogen ion secretion is linked to bicarbonate regeneration and ammonia secretion, which occur here. In addition to these, small amounts of chloride are reabsorbed.

Collecting Duct

The collecting ducts are the final site for either concentrating urine or diluting it. The upper portions are under the influence of aldosterone, which acts to stimulate sodium reabsorption. Chloride and urea are also reabsorbed here. In addition, the collecting duct is under the control of *antidiuretic hormone (ADH)*. This peptide hormone is secreted by the posterior pituitary in response to neural impulses triggered mainly by increased blood osmolality or decreased intravascular volume. It has an extremely short half-life (on the order of minutes). ADH stimulates water reabsorption. The walls of the collecting duct are normally impermeable to water (like the ascending Henle's loop), but in the presence of ADH, they become permeable to water, which diffuses passively from the lumen of the collecting duct to the medulla, resulting in a more concentrated urine.

Nonprotein Nitrogenous Metabolite Elimination

Nonprotein nitrogenous (NPN) waste products are formed in the body as a result of the degradative metabolism of nucleic and amino acids and proteins. Excretion of these compounds is an important function of the kidneys. The three principal substances are urea, creatinine, and uric acid.[3,4] For a more detailed treatment of their biochemistry and disease correlations, see Chapter 12 *Nonprotein Nitrogen*.

Urea

Urea makes up the majority (>75%) of the NPN waste excreted daily as a result of the oxidative catabolism of proteins. Proteins are broken down to amino acids, which are "detoxified" by removal of the nitrogen atoms. Ammonia is formed and readily converted to urea, thereby avoiding tox-

icity. The kidney is the only significant route of excretion for urea. It has a molecular weight of 60 and, therefore, is readily filtered by the glomerulus. Instead of being excreted wholly, 40% to 60% is reabsorbed by the collecting duct. The reabsorbed urea contributes to the high osmolality in the medulla, which is one of the processes of urinary concentration mentioned earlier (see *Henle's Loop*). The amount of reabsorption depends on the GFR, the *plasma renal flow* (flow of plasma through kidneys), and the urine flow rate.

Creatinine

Muscle contains creatine phosphate, which is a reservoir of high-energy phosphoryl groups for the rapid formation of adenosine triphosphate (ATP) as catalyzed by creatine kinase. It is the first source of metabolic fuel that the muscle uses. It is formed from creatine as shown as follows:

$$creatine\ phosphate + ADP + H^+ \longrightarrow$$
$$creatine + ATP \xrightarrow{nonenzymatic} creatinine$$

(Eq. 21-1)

Every day, up to 20% of total muscle creatine (and its phosphate) spontaneously dehydrates and cycles to form the waste product creatinine. Thus, creatinine levels are a function of muscle mass and remain approximately the same from day to day unless muscle mass changes. Creatinine is inert and has a molecular weight of 113. It is thus readily filtered by the glomerulus and, unlike urea, is not reabsorbed by the tubules. However, a small amount is secreted by the kidney tubules at high serum concentrations.

Uric Acid

Uric acid is the primary waste product of purine metabolism. The purines adenine and guanine are precursors of nucleic acids ATP and guanosine triphosphate (GTP), respectively. Uric acid has a molecular weight of 168. Thus, like creatinine, it is readily filtered by the glomerulus but then it goes through a complex cycle of reabsorption and secretion as it courses through the nephron. Only 6% to 12% from the original filtrate is finally excreted. Uric acid exists in its ionized and more soluble form, usually sodium urate, at urinary pH values above 5.75 (the first pK_a of uric acid). At pH values below 5.75, it is undissociated. This fact has great clinical significance in the development of urolithiasis (formation of calculi) and gout.

Water, Electrolyte, and Acid-Base Homeostasis

Water Balance

The kidney's contribution to water balance in the body is through water loss. It is the dominant organ in this regard. The water loss is under the hormonal control of ADH. ADH responds primarily to changes in osmolality and intravascular volume. Therefore, increased osmolality stimulates secretion of ADH, which increases the permeability of the collecting ducts to water, resulting in water reabsorption and the excretion of a more concentrated urine. Decreased intravascular volume will have a similar effect. In contrast, the major system regulating water intake is thirst, which appears to be triggered by the same stimuli that trigger ADH.

In states of dehydration, the renal tubules reabsorb water at their maximal rate, resulting in production of a small amount of maximally concentrated urine (high urine osmolality,[5] 1200 mOsmol). In states of water surplus, the tubules reabsorb water at only a minimal rate, resulting in excretion of a large volume of extremely dilute urine (low urine osmolality, down to 50 mOsmol).[6–8] The continuous fine-tuning possible between these two extreme states results in the precise control of fluid balance in the body (Fig. 21-4).

Ionic Equilibria

The following description is a brief overview of the notable ions involved in maintenance of ionic equilibria within the body. For a more comprehensive treatment of this subject, refer to Chapter 14 *Electrolytes*.

Sodium. Sodium is the primary extracellular cation in the human body and is excreted principally through the kidneys. Sodium balance in the body is only controlled by way of excretion. There appears to be no control over intake. The renin-angiotensin-aldosterone hormonal system is the major mechanism for control of sodium balance.

Potassium. Potassium is the main intracellular cation in the body. The precise regulation of its concentration is of overriding importance to cellular metabolism and is controlled chiefly by renal means. Like sodium, it is freely filtered by the glomerulus and then actively reabsorbed throughout the entire nephron (except for the descending limb of the Henle's loop). Both the distal convoluted tubule and the collecting ducts can reabsorb and excrete potassium, and this excretion is controlled by aldosterone. Potassium ions can compete with hydrogen ions in their exchange with sodium (in the proximal convoluted tubule), and this process is used by the body to conserve hydrogen ions and thereby compensate in states of metabolic alkalosis (Fig. 21-5).

Chloride. Chloride is the principal extracellular anion and is involved in the maintenance of extracellular fluid balance. It is readily filtered by the glomerulus and is passively reabsorbed as a counter ion when sodium is reabsorbed in the proximal convoluted tubule. In the ascending limb of the Henle's loop, potassium is actively reabsorbed by a distinct chloride "pump," which also reabsorbs sodium. This pump can be inhibited by loop diuretics such as furosemide. As expected, the regulation of chloride is controlled by the same forces that regulate sodium.[5,8]

Phosphate, Calcium, and Magnesium. The phosphate ion occurs in approximately equal concentrations in both the intracellular and extracellular fluid environments. It exists as either a protein-bound or a non–protein-bound form; homeo-

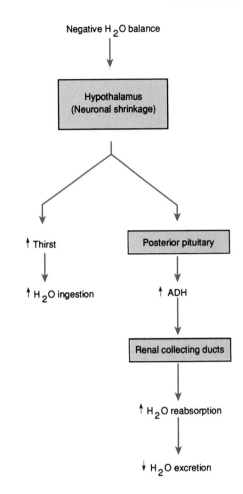

Figure 21-4. ADH control of thirst mechanism.

static balance is chiefly determined by proximal tubular reabsorption under the control of parathyroid hormone (PTH). Calcium, the second-most predominant intracellular cation, is the most important inorganic messenger in the cell. It also exists in both protein-bound and non–protein-bound states. Calcium in the non–protein-bound form is either ionized and physiologically active or nonionized and complexed to small, diffusible ligands (*eg,* phosphate and bicarbonate). The ionized form is freely filtered by the glomerulus and reabsorbed in the proximal tubule under the control of PTH. However, renal control of calcium concentration is not the major means of regulation. PTH- and calcitonin-controlled regulation of calcium absorption from the gut and bone stores is more important than renal secretion. Magnesium, a major intracellular cation, is important as an enzymatic cofactor. Like phosphate and calcium, it exists in both protein-bound and non–protein-bound states. The non–protein-bound fraction is easily filtered by the glomerulus and reabsorbed in the proximal convoluted tubule under the influence of PTH.

Acid-Base Equilibria

Many nonvolatile acidic waste products are formed by normal body metabolism each day. Carbonic acid, lactic acid, keto-

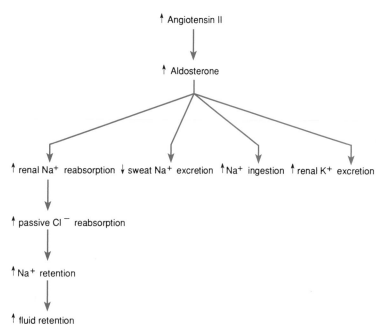

Figure 21-5. Electrolyte regulation by aldosterone.

acids, and others must be continually transported in the plasma and excreted from the body while causing only minor alterations in the physiologic pH. The renal system constitutes one of three means by which constant control of overall body pH is accomplished. The other two strategies involved in this regulation are the respiratory system and the acid-base buffering system.[9]

The kidneys manage their share of the responsibility for controlling body pH by dual means: conserving bicarbonate ions and removing metabolic acids. For a more in-depth examination of these processes, refer to Chapter 16 *Blood Gases, pH, and Buffer Systems.*

Regeneration of Bicarbonate Ions. In a complicated process, bicarbonate ions are first filtered out of the plasma by the glomerulus. Once in the lumen of the renal tubules, this bicarbonate combines with hydrogen ions to form carbonic acid, which subsequently degrades to carbon dioxide (CO_2) and water. This CO_2 then diffuses into the brush border of the proximal tubular cells, where it is reconverted by carbonic anhydrase to carbonic acid and then degrades back to hydrogen ions and regenerated bicarbonate ions. This reaction is detailed as follows:

$$H_2O + CO_2 \xleftrightarrow{CA} H_2CO_3 \longleftrightarrow H^+ HCO_3^-$$

(Eq. 21–2)

This regenerated bicarbonate is transported into the blood to replace that which was depleted by metabolism; the accompanying hydrogen ions are secreted back into the tubular lumen and from there enter the urine. Thus, filtered bicarbonate is "reabsorbed" into the circulation, thereby helping to return blood pH to its optimal level and effectively functioning as another buffering system.

Excretion of Metabolic Acids. Hydrogen ions are manufactured in the renal tubules as part of the regeneration mechanism for bicarbonate. These hydrogen ions, as well as others that are dissociated from nonvolatile organic acids, are disposed of by means of several different reactions with buffer bases.

Reaction with Ammonia (NH_3). The glomerulus does not filter NH_3. However, this substance is formed in the renal tubules when the amino acid glutamine, which serves to transport ammonia in a nontoxic form throughout the circulation to the kidneys, is deaminated by glutaminase. This NH_3 then reacts with secreted hydrogen ions to form ammonium (NH_4^+), which is unable to readily diffuse out of the tubular lumen and is therefore excreted into the urine.

$$Glutamine \xrightarrow{Glutaminase} glutamic\ acid + NH_3$$
$$NH_3 + H^+ + NaCl \longrightarrow NH_4Cl + Na^+$$

(Eq. 21-3)

This mode of acid excretion is the primary means by which the kidneys compensate for states of metabolic acidosis.

Reaction with Monohydrogen Phosphate (HPO_4^{2-}). Phosphate ions filtered by the glomerulus can exist in the tubular fluid as disodium hydrogen phosphate. This moiety can react with extant hydrogen ions to yield dihydrogen phosphate, which is then excreted and is responsible for the "titratable acidity" of the urine. The released sodium then combines with bicarbonate to yield sodium bicarbonate and is reabsorbed.

$$Na_2HPO_4 + H^+ \longleftrightarrow NaH_2PO_4 + Na^+$$

(Eq. 21–4)

These mechanisms can excrete increasing amounts of metabolic acid until a maximum urine pH of approximately 4.4 is reached. After this, the renal compensatory apparatus is unable to adjust to any further decreases in blood pH, and metabolic acidosis must ensue. Very few free hydrogen ions are excreted directly.

Endocrine Function

In addition to its numerous excretory and regulatory functions, the kidney has endocrine responsibilities as well. It is both a primary endocrine site, as the producer of its own hormones, and a secondary site, as the target locus for hormones manufactured by other endocrine organs.

Primary Endocrine Function

The kidneys synthesize renin, the prostaglandins, and erythropoietin.

Renin. *Renin* is the initial component member of the renin-angiotensin-aldosterone feedback system. It serves to catalyze the synthesis of angiotensin by means of cleavage of the circulating plasma precursor angiotensinogen. Renin is produced by the juxtaglomerular cells of the renal medulla whenever extracellular fluid volume decreases. It serves as a vasoconstrictor to increase blood pressure and is responsive to changes in both the sodium and potassium levels in the blood.[6,8] For a more rigorous look at the complexities of this feedback loop, see Chapter 18 *Endocrinology*.

Prostaglandins. The *prostaglandins* are a group of potent cyclic fatty acids formed from essential (dietary) fatty acids, primarily the unsaturated arachidonic acid. They are formed in almost all tissues and their actions are diverse. They behave like hormones but differ in that they are synthesized at the site of action. Once formed, prostaglandins exert a very short-lived effect and are rapidly catabolized. Prostaglandins can be either PG_1 or PG_2. The PG_2 type is more common. The prostaglandins produced by the kidneys such as PGA_2, PGE_2, PGI_2, and PG_{12} increase renal blood flow, sodium and water excretion, and renin release. They act to oppose renal vasoconstriction due to angiotensin and norepinephrine. Angiotensin II is believed to stimulate PGE_2. Prostaglandins have been used in antihypertensive therapy. Arachidonic acid also gives rise to *leukotrienes* and *thromboxanes* whose actions, although less well documented, have also been associated with diverse biologic activity such as inflammation and bronchoconstriction.

Erythropoietin. *Erythropoietin* is a single-chain polypeptide. It is produced by cells close to the proximal tubules and its production is regulated by blood oxygen levels. Thus, hypoxia produces increased serum concentrations within 2 hours. Erythropoietin acts on the erythroid progenitor cells in the bone marrow, causing their maturation and increasing the number of red blood cells. In chronic renal insufficiency, erythropoietin production is significantly reduced. Recently, recombinant human erythropoietin has been developed and is now in routine use in chronic renal failure patients. Before this therapy, anemia in these patients was a clinical reality.[3,6] Erythropoietin concentrations in blood can be measured by an enzyme-linked immunoassay.

Secondary Endocrine Function

The kidneys are the target locus for the action of aldosterone; for the catabolism of insulin, glucagon, and aldosterone; and as the point of activation for vitamin D metabolism.[3,5] *Vitamin D* is one of the three major hormones that determine phosphate and calcium balance and bone calcification in the human body. Chronic renal insufficiency is therefore often associated with *osteomalacia* (inadequate bone calcification, the adult form of rickets), owing to the continual distortion of normal vitamin D metabolism.

ANALYTICAL PROCEDURES

Several tests are available that assess the various aspects of nephron function including glomerular filtration, renal plasma flow, and proximal and distal tubular secretion and reabsorption.

Clearance Measurement

All laboratory methods used for evaluation of renal function rely on the measurement in blood of a waste product, usually the nitrogenous substances urea and creatinine, that accumulates once the kidneys begin to fail. However, before the concentration of either of these substances begins to rise in the blood, renal failure must be quite advanced, with only about 20% to 30% of the nephrons still functioning. Measurement of clearance affords the clinical chemist the opportunity to detect much earlier stages of the disease. *Clearance* is that volume of plasma from which a measured amount of substance can be completely eliminated into the urine per unit of time.[3]

Creatinine

Creatinine is a nearly ideal substance for the measurement of clearance for various reasons: it is an endogenous metabolic product synthesized at a constant rate for a given individual, it is cleared essentially only by glomerular filtration (it is not reabsorbed and is only slightly secreted by the proximal tubule), and it can be analyzed inexpensively by readily available colorimetric methods. As a result, creatinine clearance has become the standard laboratory assay for determination of early renal failure. This value is derived by arithmetically relating the total urine creatinine concentration to the total serum creatinine concentration excreted during a 24-hour period.

Specimen collection, therefore, must include both a 24-hour urine specimen and a serum creatinine value (ideally taken at the midpoint of the 24-hour urine collection and, realistically, no later than 24 hours *before* or *after* the urine collection). The urine container (clean, dry, and free of contaminants or preservatives) must be kept refrigerated throughout the duration of both the collection procedure and the subsequent storage period until laboratory analyses can be performed. The patient must maintain an adequate fluid intake throughout the procedure to ensure a minimum urine flow rate of at least 2 mL/min. This cancels out the possibility of negative error due to urine retention in the bladder. Additionally, coffee, tea, heavy exercise, medications, and recreational drugs should be avoided to minimize possible interferences.

Once the specimens have been collected, the concentration of creatinine in both serum and urine is measured by any of the applicable methods discussed in Chapter 12 *Nonprotein Nitrogen*. The total volume of urine is carefully measured, and the *creatinine clearance* is calculated by the following formula:

$$C_{Cr} \, (mL/min) = \frac{U_{Cr} \, (mg/dL) \times V_{Ur} \, (mL/24 \, h)}{P_{Cr} \, (mg/dL) \times 1440 \, (min/24 \, h)} \times \frac{1.73}{A},$$

(Eq. 21–5)

where as C_{Cr} = creatinine clearance
$\qquad U_{Cr}$ = urine creatinine concentration
$\qquad V_{Ur}$ = urine volume excreted in 24 hours
$\qquad P_{Cr}$ = serum creatinine concentration
$\qquad 1.73/A$ = normalization factor for body surface area

Normally, the urine output of a healthy individual averages 1500 mL/24 h; because this total volume is only about 1% of the glomerular filtrate formed each day, a large volume is reabsorbed elsewhere along the nephron.[3,6]

Most clinical chemists use the shorter, more easily remembered notation for the general clearance formula: *UV/P*. Correction for body surface area must be included in the formula, because creatinine excretion varies with regard to lean body mass. In this factor, 1.73 is the generally accepted average body surface in square meters. If the patient's body surface area varies greatly from the average (*eg,* in obese or pediatric patients), this correction prevents the accumulation of up to several hundred percent error. Nomograms for the more exact determination of body surface area from weight and height values can be found in Appendix I *Nomogram for the Determination of Body Surface Area*. Reference ranges for creatinine clearance are 97 mL/min/1.73 m² to 137 mL/min/1.73 m² and 88 mL/min/1.73 m² to 128 mL/min/1.73 m² for males and females, respectively. Creatinine clearance normally decreases with age, such that each decade of life accounts for a decrease of about 6.5 mL/min/1.73 m². Specific methodologies for the measurement of creatinine in serum and urine are given in Chapter 12 *Nonprotein Nitrogen*.

Urea

Urea clearance was one of the first clearance tests performed. Urea is freely filtered at the glomerulus and approximately 40% reabsorbed by the tubules. For this reason, it does not provide a full clearance assessment and is not as widely used today.

Older clearance tests used administration of inulin, sodium [^{125}I] iothalamate, or *p*-aminohippurate to assess glomerular filtration or tubular secretion. These tests are time-consuming, expensive, and difficult to administer and, for the most part, have been discontinued.

Urinalysis

Urinalysis (UA) is an underused tool for the overall evaluation of renal function. UA permits a detailed, in-depth assessment of renal status with an easily obtained specimen. UA also serves as a quick indicator of an individual's glucose status and hepatic-biliary function. Routine UA consists of a group of screening tests generally performed as part of a patient's admission work-up or physical examination. It includes assessment of physical characteristics, chemical analyses, and a microscopic examination of the sediment from a (random) urine specimen.

Physical Characteristics

Specimen Collection. The importance of a properly collected and stored specimen for UA cannot be overemphasized. Initial morning specimens are preferred, particularly for protein analyses, because they are more concentrated from overnight retention in the bladder. The specimen should be obtained by a clean midstream catch or catheterization. The urine should be freshly collected into a clean, dry container with a tight-fitting cover. It must be analyzed within 1 hour of collection if held at room temperature or else refrigerated at 2°C to 8°C for not more than 8 hours before analysis. If not assayed within these time limits, several changes will occur. Bacterial multiplication will cause false-positive nitrite tests, and urease-producing organisms will degrade urea to ammonia and alkalinize the pH. Loss of CO_2 by diffusion into the air adds to this pH elevation, which in turn causes cast degeneration and red-cell lysis.

The urine container must be sterile if the urine is to be cultured. For children who are not yet toilet trained, specially designed polyethylene plastic bags with adhesive backing are available to fit around the perineum so that the urine is voided directly into the bag. Specimens for routine UA are usually random, or spot, collections. Fasting and postprandial specimens are used for glucose determinations. Timed urine specimens (for clearances, and so on) should always be stored in refrigerated containers.[10]

Visual Appearance. Color intensity of urine correlates with concentration: the darker the color, the more concentrated is the specimen. The various colors observed in urine are due to different excreted pigments. Yellow and amber are generally due to urochromes (derivatives of urobilin, the end product of bilirubin degradation), whereas yellowish-brown to green are due to bile pigment oxidation. Red and brown after standing are due to porphyrins, whereas reddish-brown in fresh specimens comes from hemoglobin/red cells. Brownish-black after standing is seen in *alkaptonuria* (due to excreted homogentisic acid) and in malignant melanoma (in which the precursor melanogen oxidizes in the air to melanin). Drugs also may affect urine color; methylene blue can turn urine blue and the analgesic dye Pyridium (phenazopyridine hydrochloride) stains urine a bright orange. Some foods, such as beets, also may alter urine color.

Odor. Odor ordinarily has little diagnostic significance. The characteristic pungent odor of fresh urine is due to volatile aromatic acids, in contrast to the typical ammoniacal odor of urine that has been allowed to stand. Urinary tract infections impart a noxious, fecal smell to urine, whereas the urine of diabetics often smells fruity due to the presence of ketones.

Turbidity. The cloudiness of a urine specimen depends on both its pH and its dissolved solids composition. Turbidity in general may be due to gross bacteriuria, whereas a smoky appearance is seen in hematuria. Threadlike cloudiness is observed when the specimen is full of mucus. In alkaline urine, suspended precipitates of amorphous phosphates and carbonates may be responsible for turbidity, whereas in acidic urine, amorphous urates may be the cause.[10]

Volume. The volume of urine excreted indicates the balance between fluid ingestion and water lost from the lungs, sweat, and intestines. Most adults produce from 750 mL/24 h to 2000 mL/24 h, averaging about 1.5 L per person. (For a routine UA, a 10-mL to 12-mL aliquot from a well-mixed sample is optimal for accurate analysis of sedimentary constituents.) Polyuria is observed in cases of diabetes mellitus and insipidus (in insipidus, due to the lack of ADH), as well as in chronic renal disease, *acromegaly* (overproduction of the growth hormone somatostatin), and *myxedema* (hypothyroid edema). Anuria or *oliguria* (<200 mL/day) is found in nephritis, end-stage renal disease (ESRD), urinary tract obstruction, and acute renal failure.

Specific Gravity. The specific gravity (SG) of urine is the weight of the urine divided by the weight of the water standard (1.000). SG gives an indication of the density of a fluid, depending on the concentration of dissolved total solids, and therefore measures a property similar to that measured by osmolality. For a more detailed discussion of urine osmolality, refer to Chapter 14 *Electrolytes*. SG varies with the solute load to be excreted (consisting primarily of NaCl and urea), as well as with the urine volume. As such, it can be used as a marker for the amount of hydration/dehydration of an individual.

Laboratory methods. The analytical method most commonly encountered today consists of a refractometer, or total solids meter. This operates on the principle that the refractive index of a urine specimen will vary directly with the total amount of dissolved solids in the sample. Thus, this instrument, as typified by the Goldberg refractometer, measures the refractive index of the urine as compared with water on a scale that is calibrated directly into the ocular and

viewed while held up to a light source. Correct calibration is vital for accuracy. Most recently, an indirect colorimetric reagent strip method for assaying SG has been added to most dipstick screens. Unlike the refractometer, dipsticks measure only ionic solutes and do not take into account the presence of glucose or protein.

Disease correlation. Normal range for urinary SG is 1.005 to 1.030. Dilute specimens are classified in the range of 1.000 to 1.010, whereas concentrated samples fall between 1.025 and 1.030. SG can vary in pathologic states. Low SG can occur in diabetes insipidus, where it may never exceed the range of 1.001 to 1.003, and in pyelonephritis and glomerulonephritis, in which the renal concentrating ability has become dysfunctional. High SG can be seen in diabetes mellitus, congestive heart failure, dehydration, adrenal insufficiency, liver disease, and nephrosis. SG will increase about 0.004 unit for every 1% change in glucose concentration and about 0.003 unit for every 1% change in protein. Fixed SG (*isothenuria*) around 1.010 is observed in severe renal damage, in which the kidney excretes a urine that is iso-osmotic with the plasma. This generally occurs after an initial period of anuria, because the damaged tubules are unable to concentrate/dilute the glomerular filtrate.[7,10]

pH. Determinations of urinary pH *must* be done on fresh specimens, owing to the marked tendency of urine to alkalinize on standing. Normal urine pH falls within the range of 4.5 to 8.0. Acidity in urine (pH <7.0) is primarily caused by phosphates, which are excreted as salts conjugated to Na^+, K^+, Ca^{2+}, and N_4^+. Acidity also reflects the excretion of the nonvolatile metabolic acids pyruvate, lactate, and citrate. Owing to the Na^+/H^+ exchange pump mechanism of the renal tubules, pH (H^+ ion concentration) increases as sodium is retained. Pathologic states in which increased acidity is observed include systemic acidosis, as seen in diabetes mellitus, and renal tubular acidosis. In renal tubular acidosis, the tubules are unable to excrete excess H^+ even though the body is in metabolic acidosis, and urinary pH remains around 6. Some medications such as ammonium chloride and mandelic acid, will also acidify the urine.

Alkaline urine (pH >7.0) is observed postprandially as a normal reaction to the acidity of gastric HCl dumped into the duodenum and then into the circulation. Urinary tract infections and bacterial contamination also will alkalinize pH. Medications such as potassium citrate and sodium bicarbonate will reduce urine pH. Alkaline urine is also found in *Fanconi's syndrome,* a congenital generalized aminoaciduria due to defective proximal tubular function.

Chemical Analyses

Today, routine urine chemical analysis is rapid and easily performed with commercially available reagent strips or dipsticks. These strips are plastic coated with different reagent bands directed toward different analytes. When dipped into urine, a color change signals a deviation from normality. Colors on the dipstick bands are matched against a color chart provided with the reagents. Automated and semiautomated instruments that detect by reflectance photometry provide an alternative to the color chart and offer better precision and standardization. Abnormal results are followed up by specific quantitative or confirmatory urine assays. The analytes routinely tested are glucose, protein, ketones, nitrite, leukocyte esterase, bilirubin/ urobilinogen, and hemoglobin/blood.

Glucose and Ketones. These constituents are normally absent in urine. The testing for and pathologic significance of detecting these analytes are discussed in Chapter 10 *Carbohydrates.*

Protein. Reagent strips for UA are used as a general qualitative screen for proteinuria. They are primarily specific for albumin, and they may give false-positive results in specimens that are alkaline and highly buffered. Therefore, positive dipstick results should be confirmed either by a more specific acid precipitation test, such as the trichloroacetic acid and precipitation method described in Chapter 8 *Amino Acids and Proteins,* or more commonly by a microscopic evaluation.

Nitrite. This assay semiquantitates the amount of urinary reduction of nitrate (on the reagent strip pad) to nitrite by the enzymes of gram-negative bacteria. This scheme is shown in the following reaction:

$$\text{Nitrite} + p\text{-arsanilic acid} \longleftrightarrow \text{diazonium compound}$$
$$+ \text{ } N\text{-1-naphthylethylenediamine} \longleftrightarrow \text{pink color}$$

(Eq. 21–6)

A negative result does not mean that no bacteriuria is present, however. A gram-positive pathogen, such as the *Staphylo-, Entero-,* or *Streptococci,* may not produce nitrate-reducing enzymes; alternatively, a spot urine sample may not have been retained in the bladder long enough to pick up a sufficient number of organisms to register on the reagent strip.[10]

Leukocyte Esterase. White blood cells, especially phagocytes, contain esterases. A positive dipstick for esterases indicates possible white blood cells in a urine.

Bilirubin/Urobilinogen. Hemoglobin degradation ultimately results in the formation of the waste product bilirubin, which is then converted to urobilinogen in the gut through bacterial action. Ordinarily, the majority of this urobilinogen is excreted as urobilin in the feces. A minority, however, is excreted in urine as a colorless waste product. This amount is insufficient to be detected as a positive dipstick reaction in normal patients. If, however, hepatic or posthepatic jaundice is present (such as occurs in biliary atresia, Crigler-Najjar syndrome type II, and hepatitis), conjugated (water-soluble) bilirubin levels will rise in serum. Consequently, the kidneys will aid in excretion of this excess waste, and urinary bilirubin and urobilinogen levels will increase. A more in-depth view of this metabolic process

and these assay methods is given in Chapter 17 *Liver Function.* Reagent strip tests for bilirubin involve its diazotization to form a brown azo dye. Dipstick methods for urobilinogen differ, but most rely on a modification of the Ehrlich reaction with *p*-dimethylaminobenzaldehyde.[10]

Hemoglobin/Blood. Intact or lysed red blood cells will produce a positive dipstick result. The dipstick will be positive in cases of renal trauma/injury, infection, or obstruction due to calculi or neoplasms.

Sediment Examination

A centrifuged, decanted urine aliquot leaves behind a sediment of formed elements that is used for microscopic examination.

Cells. For cellular elements, evaluation is best accomplished by counting and then taking the average of at least ten microscopic fields.

Red blood cells. Erythrocytes greater in number than 0 to 2/high-powered field (hpf) are considered abnormal. Such hematuria may result simply from severe exercise or menstrual blood contamination. However, it also may be indicative of trauma, particularly vascular injury, renal/urinary calculi obstruction, pyelonephritis, or cystitis; in conjunction with leukocytes, it is diagnostic for infection.

White blood cells. Leukocytes greater in number than 0 to 1/hpf are considered abnormal. These cells are usually polymorphonuclear phagocytes, commonly known as *segmented neutrophils.* They are observed whenever there is acute glomerulonephritis, urinary tract infection, or inflammation of any type. In hypotonic urine (low osmotic concentration), white blood cells can become enlarged, exhibiting a sparkling effect in their cytoplasmic granules. These cells possess a noticeable Brownian motion and are called *glitter cells,* but they have no pathologic significance.

Epithelial cells. Several types of epithelial cells are frequently encountered in normal urine because they are continuously sloughed off the lining of the nephrons and urinary tract. Large, flat, squamous vaginal epithelia are often seen in urine specimens from female patients; samples heavily contaminated with vaginal discharge may show clumps or sheets of these cells. Renal epithelia are round, uninucleate cells; their presence in urine in numbers greater than 2/hpf is a clinically significant marker for active tubular injury/degeneration. Transitional bladder epithelial cells (urothelial cells) may be flat, cuboidal, or columnar and also can be observed in urine on occasion. Large numbers will be seen only in cases of urinary catheterization or bladder inflammation or neoplasia.

Miscellaneous Elements. Spermatozoa are often encountered in the urine of both males and females. They are usually not reported because they are of no pathologic significance. In males, however, their presence may indicate prostate abnormalities. Yeast cells are also frequently found in urine specimens. Because they are extremely refractile and of a similar size to red blood cells, they can easily be mistaken for them under low magnification. Higher power examination for the presence of budding or mycelial forms serves to differentiate these fungal elements from erythrocytes. Most yeasts are *Candida albicans* species, and they commonly infect women who are diabetic or on oral contraceptives. They indicate the presence of either a urinary or vaginal moniliasis. Parasites found in urine are generally contaminants from fecal or vaginal material. In fecal contaminants category, the most commonly encountered organism is *Enterobius vermicularis* (pinworm) infestation in children. In the vaginal contaminants category, the most common offender is the intensely motile flagellate *Trichomonas vaginalis*. A true urinary parasite, sometimes seen in patients from endemic areas of the world, is the ova of the trematode *Schistosoma hematobium*. This condition will usually occur in conjunction with a significant hematuria.[10]

Bacteria. Normal urine is sterile and contains no bacteria. Small numbers of organisms seen in a fresh urine specimen usually represent skin or air contamination. In fresh specimens, however, large numbers of organisms, or small numbers accompanied by white blood cells and the symptoms of urinary tract infection, are highly diagnostic for true infection. Clinically significant bacteriuria is considered to be more than 20 organisms/hpf or, alternatively, 10^5 or greater registered on a microbiologic colony count. Most pathogens seen in urine are gram-negative coliforms (microscopic "rods") such as *E. coli* and *Proteus* species. Asymptomatic bacteriuria, in which there are significant numbers of bacteria without appreciable clinical symptoms, occurs somewhat commonly in young girls, pregnant women, and diabetic patients. This condition must be taken seriously; if left untreated, it may result in pyelonephritis and, eventually, in permanent renal damage.

Casts. Casts are precipitated, cylindrical impressions of the nephrons. They comprise Tamm-Horsfall mucoprotein (*uromucoid*) from the tubular epithelia in the ascending limb of the Henle's loop. Casts form whenever there is sufficient renal stasis, increased urine salt/protein concentration, and decreased urine pH. In patients with severe renal disease, truly accurate classification of casts may require use of "cytospin" centrifugation and Papanicolaou stain test for adequate differentiation. Unlike cells, casts should be examined under low power and are most often located around the edges of the coverslip.

Hyaline. The matrix of these casts is clear and gelatinous, without cellular or particulate matter embedded in them. They may be quite difficult to visualize unless a high-intensity lamp is used. Their presence indicates glomerular leakage of protein. This leakage may be temporary (due to fever, upright posture, dehydration, or emotional stress) or may be perma-

nent; however, their occasional presence is not considered pathologic.

Granular. These casts are classified descriptively as either coarse or finely granular. The type of particulate matter embedded in them is simply a matter of the amount of degeneration that the epithelial-cell inclusions have undergone. Their occasional presence is not pathologic; however, large numbers may be found in cases of chronic lead toxicity and pyelonephritis.

Cellular. Several different types of casts are included in this category. *Red blood cell* or *erythrocytic casts* are always considered pathologic because they are diagnostic for glomerular inflammation that results in renal hematuria. They are seen in subacute bacterial endocarditis, kidney infarcts, collagen diseases, and acute glomerulonephritis. *White blood cell* or *leukocytic casts* are also always considered pathologic because they are diagnostic for inflammation of the nephrons. They are observed in pyelonephritis, nephrotic syndrome, and acute glomerulonephritis. In cases of asymptomatic pyelonephritis, these casts may be the only clue to detection. Epithelial-cell casts are sometimes formed by fusion of renal tubular epithelia after desquamation. Their occasional presence is normal. Many, however, are observed in severe desquamative processes and renal stases such as occur in heavy metal poisoning, renal toxicity, eclampsia, nephrotic syndrome, and amyloidosis. *Waxy casts* are uniformly yellowish, refractile, and brittle appearing, with sharply defined, often broken edges. They are almost always pathologic because they indicate tubular inflammation or deterioration. They are formed by renal stasis in the collecting ducts and are therefore found in the chronic renal diseases. *Fatty casts* are abnormal, coarse, granular casts with lipid inclusions that appear as refractile globules of different sizes. *Broad (renal failure) casts* may be up to two to six times wider than "regular" casts and may be cellular, waxy, or granular in composition. Like waxy casts, they are derived from the collecting ducts in cases of severe renal stasis.

Crystals

Acid environment. Crystals seen in urines with pH values of less than 7 include calcium oxalate, which are normal colorless octahedrons or "envelopes"; they may have an almost starlike appearance. Also seen are amorphous urates, which are normal yellow-red masses with a sand grainlike appearance. Uric acid crystals found in this environment are normal yellow to red-brown crystals that appear in extremely irregular shapes such as rosettes, prisms, or rhomboids. Cholesterol crystals in acid urine are clear, flat, rectangular plates with notched corners. They may be seen in nephrotic syndrome and in conditions producing chyluria. They are always considered abnormal. Cystine crystals are also sometimes observed in acid urine; they are highly pathologic and appear as colorless, refractile, nearly flat hexagons, somewhat similar to uric acid. These are observed in *cystinuria* (an inherited aminoaciduria resulting in mental retardation) and *homocystinuria* (a rare defect of cystine reabsorption resulting in renal calculi).

Alkaline environment. Crystals seen in urines with pH values greater than 7 include amorphous phosphates, which are normal crystals that appear as fine, colorless masses resembling sand. Also seen are calcium carbonate crystals, which are normal forms that appear as small, colorless dumbbells or spheres. Triple phosphate crystals are also observed in alkaline urines; they are colorless prisms of three to six sides resembling "coffin lids." Ammonium biurate crystals are normal forms occasionally found in this environment; they appear as spiny yellow-brown spheres, or "thornapples."

Other. Sulfonamide crystals are abnormal precipitates shaped like yellow-brown sheaves, clusters, or needles, formed in patients undergoing antimicrobial therapy with sulfa drugs. These drugs are seldom used today. Tyrosine/leucine crystals are abnormal types shaped like clusters of smooth, yellow needles or spheres. These are sometimes seen in patients with severe liver disease.[10]

Urine Electrophoresis

Owing to the efficiency of renal glomerular sieving and tubular reabsorption, normal urinary protein excretion is only about 50 mg/24 h to 150 mg/24 h. Proteinuria may develop when there are defects in renal reabsorption or glomerular capillary permeability or when there is a marked increase in serum immunoglobulins. As a result, urine electrophoresis in the clinical laboratory is primarily used to aid in distinguishing between acute glomerular nephropathy and tubular proteinuria and to screen for monoclonal or polyclonal dysglobulinemias. With polyclonal dysglobulinemias, urine electrophoresis can help in the differential diagnosis of multiple myeloma and Waldenström's macroglobulinemia. Positive identification and subtyping of the urinary paraproteins *(Bence Jones proteins)* detected there can be done by immunoelectrophoresis and immunofixation electrophoresis.

Beta$_2$-microglobulin

Recently, interest has arisen concerning the use of clearance of beta$_2$-*microglobulin* (β_2-M) (as an indicator of GFR. β_2-M is a small, nonglycosylated peptide of molecular weight 11,800 daltons that constitutes the constant portion of the light chain of class I major histocompatibility complex antigens on the surface of most nucleated cells. The plasma membrane sheds β_2-M as a relatively intact moiety into the surrounding extracellular fluid. This process is fairly constant in adults; thus, levels of β_2-M remain stable in normal patients. Additionally, these values do not differ between males and females. As a small endogenous peptide, β_2-M is easily filtered by the glomerulus. About 99.9% is then reabsorbed by the proximal

tubules through pinocytosis. Elevated levels in serum indicate increased cellular turn-over. This elevation can be seen in the myelo- and lympho-proliferative disorders, such as acquired immunodeficiency syndrome and multiple myeloma. Increased urine levels are then observed when such concentrations exceed the renal threshold. In the absence of increased synthesis, however, increased urinary β_2-M is due to the inability of the proximal tubules to reabsorb it. Thus, this peptide can be used to monitor the GFR as a marker for glomerular disease, particularly tubular proteinuria. Elevations are found in the intermediate and late stages of diabetic nephropathy, and thus measurements are also useful for monitoring renal transplant status, because it is not removed from the circulation by dialysis. Indeed, it has been found in some studies to be a more efficient marker of renal graft rejection than serum creatinine values, because it does not depend on lean muscle mass or daily variation in excretion. However, β_2-M is unstable in acidic urine specimens, and this poses a problem for reliable analyses. β_2-M are assayed by nephelometric techniques.[11]

Myoglobin

Recently, myoglobin clearance has been proposed as an effective early indicator of myoglobinuric acute renal failure. A high clearance or a low clearance and low serum concentration indicates low risk and a low clearance and high serum concentration indicates high risk.[12] *Myoglobin* is a low molecular weight protein (16,900 daltons) that is associated with acute skeletal and cardiac muscle injury. It can induce acute renal failure if its release from muscle is sufficient to overload the renal proximal tubules and cause myoglobinuria. For this to occur, about 200 g of muscle must be damaged. Thus, insignificant myoglobinuria is seen in an acute myocardial infarction but is significant in skeletal myoglobinuria muscle injury. Acute renal failure caused by elevated myoglobin levels is a complication whose severity may be significantly reduced or even prevented if recognized early and treated aggressively. Myoglobin functions to bind and transport oxygen from the plasma membrane to the mitochondria in muscle cells. The normal range in serum is less than 90 ng/mL. To a large extent, it is bound to plasma proteins and therefore is filtered to a much smaller degree than would be predicted from its size. Its molecular sieving coefficient (glomerular filtrate to plasma ratio) is 0.75. Serum and urine myoglobin can be measured easily and rapidly by nonisotopic immunoassays. Urine myoglobin can also be measured qualitatively by dipstick methods after removing hemoglobin, but this method suffers from a lack of sensitivity and specificity.

Microalbumin

Urine *microalbumin* measurement is important in the management of patients with diabetes mellitus. Diabetics have a serious risk of developing nephropathy over their lifetimes.

Type I have a 30% to 45% risk and type II a 30% risk. In the early stages of nephropathy, there is renal hypertrophy, hyperfunction, and increased thickness of the glomerular and tubular basement membrane. In this early stage, there are no signs of renal dysfunction. In the next 7 years to 10 years, there is progression to glomerulosclerosis with increased glomerular capillary permeability. This permeability allows small amounts (micro) of albumin to pass into the urine. If detected in this early phase, rigid glucose control, along with treatment to prevent hypertension, can be instituted and progression to ESRD may be prevented.[13] The American Diabetes Association in a recent position statement set criteria for the frequency of testing of all diabetics for urinary albumin. It recommends testing at the initial patient evaluation and yearly thereafter for all postpubertal patients who have had diabetes for at least 5 years.[14]

Urine protein analyses by dipstick methods are not sensitive enough to detect microalbumin. The sensitivity of the dipstick methods is 20 mg/dL 30 mg/dL or about 550 mg/24 h. In patients with microalbuminuria, albumin concentrations are in the range of 30 mg/24 h to 300 mg/24 h and are therefore undetectable. Quantitative albumin-specific immunoassays, usually using nephelometry, have a linear range of 0.5 mg/dL to 20.0 mg/dL. A 24-hour urine collection is preferred, but a random urine sample that uses a ratio of albumin to creatinine can be also be measured. The normal range for urinary albumin is less than 10 μg/min, less than 15 mg/24 h, or less than 0.01 mg/dL albumin to creatinine (mg/dL).[13,14]

PATHOPHYSIOLOGY

Glomerular Diseases

Disorders or diseases that directly damage the renal glomeruli may, at least initially, exhibit normal tubular function. With time, however, disease progression involves the renal tubules as well. The following syndromes have discrete symptomatologies, which are recognizable by their patterns of clinical laboratory findings.[3,6]

Acute Glomerulonephritis

Pathologic lesions in acute *glomerulonephritis* primarily involve the glomerulus. Histologic examination shows large, inflamed glomeruli with a decreased capillary lumen. Abnormal laboratory findings virtually always include rapid onset of hematuria and proteinuria (usually albumin, and generally of <3 g/day). There is usually the rapid development of a decreased GFR; anemia; elevated blood urea nitrogen (BUN) and serum creatinine; oliguria; sodium and water retention (with consequent hypertension and some localized edema); and sometimes congestive heart failure. Numerous hyaline and granular casts are generally seen on UA. The presence of actual red blood cell casts is regarded as highly suggestive of this syndrome.

Acute glomerulonephritis is often related to recent infection by group A β–hemolytic streptococci. It is theorized that circulating immune complexes trigger a strong inflammatory response in the glomerular basement membrane, resulting in a direct injury to the glomerulus itself. Other possible causes include drug-related exposures, acute kidney infections due to other bacterial (and possibly viral) agents, and other systemic immune complex diseases, such as systemic lupus erythematosus (SLE) and subacute bacterial endocarditis (SBE). For unknown reasons, the course of the syndrome becomes fulminant in some cases, quickly progressing to renal failure; this is termed *rapidly progressive glomerulonephritis.*

Chronic Glomerulonephritis

Lengthy glomerular inflammation, whether from a renal disease or from an idiopathic cause, may lead to glomerular scarring and the eventual loss of operational nephrons. This process is often undetected for lengthy periods, because only minor decreases in renal function occur at first, and only slight proteinuria and hematuria are observed. Gradual development of uremia (or *azotemia,* excess nitrogen compounds in the blood) may sometimes be the first sign of this process.

Nephrotic Syndrome

In *nephrotic syndrome,* many different etiologies can result in a specifically demonstrable type of glomerular injury: an abnormally increased permeability of the glomerular basement membrane. This defect almost always yields such abnormal findings as extremely massive proteinuria (>3.5 g/day and even up to 20 g/day), resultant hypoalbuminemia, and subsequent decreased plasma oncotic pressure, culminating in a generalized edema due to the movement of body fluids out of the vascular and into the interstitial spaces. Other hallmarks of this syndrome are hyperlipidemia (possibly secondary to lipoprotein alterations) and lipiduria. Lipiduria takes the form of oval fat bodies in the urine; these bodies are degenerated renal tubular cells containing reabsorbed lipoproteins. Owing to both the extreme proteinuria and the osmotically induced hypovolemia in these cases, disorders of several different coagulation factors are observed, and intravascular thrombus formation becomes a danger. A schematic diagram of the pathophysiologic sequence in nephrotic syndrome is shown in Figure 21-6.

Primary causes are associated directly with glomerular disease states. Secondary causes are associated with infections (SBE or syphilis); mechanical derangements of the circulatory system (constrictive pericarditis or renal vein thromboses); allergic/toxic reactions (bee venom or penicillamine); generalized disease states (carcinomas, SLE, or amyloidosis); or miscellaneous disorders (transplant rejections or severe preeclampsia).

Tubular Diseases

Tubular defects occur to a certain extent in the progression of all renal diseases as the GFR falls. In some instances, however, this aspect of the overall dysfunction becomes predominant. The result is decreased excretion/reabsorption of certain substances or reduced urinary concentrating capability. Clinically,

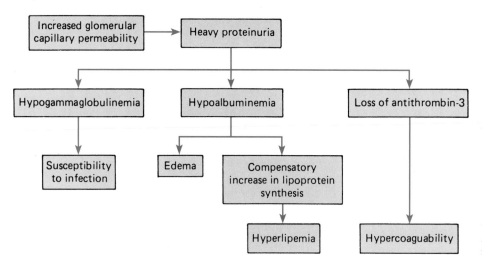

Figure 21-6. Pathophysiologic sequence in nephrotic syndrome.

the most important defect is the primary tubular disorder affecting acid-base balance: renal tubular acidosis (RTA). This disease can be classified into two types, depending on the nature of the tubular defect: *distal RTA,* in which the renal tubules are unable to keep up the vital pH gradient between the blood and tubular fluid; and *proximal RTA,* in which there is decreased bicarbonate reabsorption, resulting in hyperchloremic acidosis. In general, reduced reabsorption in the proximal tubule is manifested by findings of abnormally low serum values for phosphorus and uric acid and by the presence of glucose and amino acids in the urine. In addition, there may be some proteinuria (usually <2 g/day).

Acute inflammation of the tubules and surrounding interstitium also may occur as a result of analgesic drug or radiation toxicity, methicillin hypersensitivity reactions, renal transplant rejection, and viral-fungal-bacterial infections. Characteristic clinical findings in these cases are decreases in GFR, urinary concentrating ability, and metabolic acid excretion; the presence of leukocyte casts in the urine; and inappropriate control of sodium balance.[3,6]

Urinary Tract Infection/Obstruction

Infection

The site of infection may be either in the kidneys themselves (*pyelonephritis*) or in the urinary bladder (*cystitis*). In general, a microbiologic colony count of more than 10^5 colonies/mL is considered diagnostic for infection in either locale. *Bacteriuria* (as evidenced by positive nitrite dipstick findings for some organisms), hematuria, and *pyuria* (presence of leukocytes in the urine, as shown by positive leukocyte esterase dipstick) are all frequently encountered abnormal laboratory results in these cases. In particular, the presence of white blood cell (leukocyte) casts in the urine is considered diagnostic for pyelonephritis.[3,6,10]

Obstruction

Renal obstructions can cause disease in one of two ways. They may either gradually raise the intratubular pressure until

nephrons necrose and chronic renal failure ensues, or they may predispose the urinary tract to repeated infections.

Obstructions may be located in either the proximal or distal urinary tract. Blockages in the upper tract are characterized by histologic evidence of a constricting lesion below a dilated collecting duct. Obstructions of the lower tract are evidenced by the presence of residual urine in the bladder after cessation of *micturition* (urination); symptoms include slowness of voiding, both initially and throughout urination. Causes of obstructions are quite varied. They can include neoplasias (such as prostate/bladder carcinoma or lymph node tumors constricting ureters); acquired diseases (such as urethral strictures or renal calculi); and congenital deformities of the lower urinary tract. The clinical symptoms of advancing obstructive disease follow the course of its progression, with decreased urinary concentrating capability seen first, succeeded by diminished metabolic acid excretion, decreased GFR, and reduced renal blood flow. Laboratory tests useful in elucidating the nature of the blockage are the UA, urine culture, BUN, serum creatinine, and the complete blood count. Final diagnosis is usually made by radiologic imaging techniques.[3,6,10]

Renal Calculi

Renal calculi, commonly termed kidney stones, are formed by the combination of a variety of crystallized substances, which are listed in Table 21-2. Of these, calcium oxalate stones are by far the most commonly encountered, particularly in the tropics and subtropics.

It is currently believed that recurrence of calculi in susceptible individuals is due to a complex mixture of causes. Chief among these are a reduced urine flow rate (related to a decreased fluid intake) and saturation of the urine with large amounts of essentially insoluble substances (such as those previously mentioned). Qualitative chemical analyses of these stones permit differentiation of mixed etiologies (producing "mixed stones") in complicated diagnostic cases. Recently, specialized x-ray diffraction and infrared spec-

CASE STUDY 21-1

A 52-year-old male with a history of alcohol abuse presented to the emergency department with a 2-week history of abdominal pain, vomiting, fever, and chills. His past history was significant for manic depression. Although a toxic ingestion was suspected, the toxicology screen was negative. The clinical picture from the previous 2 weeks, however, suggested the possibility of ethylene glycol ingestion. His laboratory results are listed under Day 1 in Table 21-1.1. He was started on dialysis the next day. After a week of dialysis sessions, his urinary output increased and his electrolyte and renal function results were vastly improved. Dialysis was discontinued and the patient was discharged. Applicable laboratory results are shown in Table 21-1.1.

CASE STUDY TABLE 21-1.1 Laboratory Results

Laboratory Test	Patient Result	Reference Range
Day 1		
Na	134 mmol/L	135–145 mmol/L
K	6.7 mmol/L	3.4–5.0 mmol/L
Cl	87 mmol/L	98–107 mmol/L
CO_2	5 mmol/L	22–31 mmol/L
Glucose	126 mg/dL	70–105 mg/dL
BUN	17 mg/dL	5–18 mg/dL
Creatinine	14.2 mg/dL	0.5–1.4 mg/dL
pH	7.1	7.35–7.45
PCO_2	12 mm Hg	32–45 mm Hg
Calculated tCO_2	5 mmol/L	22–28 mmol/L
PO_2	118 mm Hg	75–95 mm Hg
Amylase	12 U/L	30–110 U/L
Lipase	8 U/L	<220 U/L
ALKP	69 U/L	20–125 U/L
AST	12 U/L	10–57 U/L
LD	229 U/L	80–160 U/L
Day 8		
Na^+	141 mmol/L	135–145 mmol/L
K^+	3.8 mmol/L	3.4–5.0 mmol/L
Cl^-	105 mmol/L	98–107 mmol/L
CO_2	30 mmol/L	22–31 mmol/L
BUN	15 mg/dL	5–18 mg/dL
Creatinine	1.9 mg/dL	0.5–1.4 mg/dL

Questions

1. What toxic substance caused this patient's renal failure?
2. How would you categorize this type of renal failure? What is the prognosis?
3. Do you think the elevated potassium on Day 1 is cause for concern? Why?
4. What is the acid-base imbalance on Day 1?
5. Why did the physician order enzyme tests? What do the results indicate?

troscopy techniques have come into more widespread use for this purpose. Clinical symptoms are, of course, similar to those encountered in other obstructive processes: hematuria, urinary tract infections, and "renal colic" (characteristic abdominal pain).[3,6,10]

TABLE 21-2. Types of Renal Calculi

Calculi Type	Etiology
Calcium oxalate	Hypercalciuria
	Hyperparathyroidism
	Vitamin D toxicity
	Sarcoidosis
	Osteoporosis
	Burnett's (milk-alkali) syndrome
	RTA
Magnesium ammonium phosphate	Infectious processes
Calcium phosphate	Alkali consumption
	RTA
	Infection by urease producers
Uric acid	Hyperuricaciduria
	Gout
	Hyperuricemia
Cystine	Primary inherited cystinuria

Renal Failure

Acute Renal Failure

Acute renal failure is a sudden, sharp decline in renal operation due to an acute toxic or hypoxic insult to the kidneys. This has been defined as occurring when the GFR is reduced to less than 10 mL/min. This syndrome is subdivided into three types, depending on the location of the precipitating defect. In prerenal failure, the defect lies in the blood supply before it reaches the kidney. Causes can include cardiovascular system failure and consequent hypovolemia. In primary renal failure, the defect involves the kidney itself. The most common cause is acute tubular necrosis; other etiologies include vascular obstructions/inflammations and glomerulonephritis. In postrenal failure, the defect lies in the urinary tract after it exits the kidney. Generally, acute renal failure occurs as the sequelae to a lower urinary tract obstruction or to rupture of the urinary bladder.

Toxic insults to the kidney severe enough to initiate acute renal failure include hemolytic transfusion reactions, heavy metal/solvent poisonings, antifreeze ingestion, and analgesic and aminoglycoside toxicities. These conditions directly damage the renal tubules. Hypoxic insults include conditions that severely compromise renal blood flow, such as septic/hemorrhagic shock, burns, and cardiac failure.

CASE STUDY 21-2

A 3½-year-old white male presented to his primary medical doctor with a 4-week history of lethargy, increased sleepiness, and decreased activity. After falling 4 weeks ago, he had been experiencing left leg pain and occasional limping. In addition, a few petechiae were noted on the right side of his neck. The physical exam revealed a temperature of 100°F, palpable lymph nodes, and splenomegaly. The boy was admitted, and intravenous fluids with bicarbonate and allopurinol were started. After further tests, he was diagnosed with acute lymphocytic leukemia. Applicable laboratory data are shown in Case Study Table 21-2.1.

Questions

1. What are two possible complications of a high serum uric acid level?
2. Why was this patient's uric acid level elevated?
3. What is allopurinol and why was it administered?

CASE STUDY TABLE 21-2.1. Laboratory Results

Laboratory Test	Patient Result	Reference Range
WBC	39,300/μL	5000–10,000/mL
Blasts	73%	0%
Lymphocytes	26%	35%–65%
Polymorphonuclear phagocytes	0%	25%–45%
Na⁺	140 mmol/L	135–145 mmol/L
K⁺	4.3 mmol/L	3.4–5.0 mmol/L
Cl⁻	107 mmol/L	98–107 mmol/L
CO_2	25 mmol/L	22–31 mmol/L
Glucose	100 mg/dL	70–105 mg/dL
BUN	15 mg/dL	5–18 mg/dL
Creatinine	0.7 mg/dL	0.3–0.7 mg/dL
Uric acid	7.4 mg/dL	2.0–5.5 mg/dL

The most commonly observed symptoms of acute renal failure are oliguria and anuria (<400 mL/day). The diminished ability to excrete electrolytes and water results in a marked increase in extracellular fluid volume, leading to peripheral edema, hypertension, and congestive heart failure. If more water is retained than sodium, hyponatremia may develop; in extreme cases, central nervous system effects are seen, progressing from profound drowsiness to seizures, coma, and death. The hyperkalemia also may become severe enough to cause dangerous cardiac arrhythmias. In addition, variable amounts of erythrocytic casts, hematuria, proteinuria, and metabolic acidosis (with resulting respiratory attempts at compensation) are seen. Generally, systemic bone disease and anemia are not as marked as they are in chronic renal failure. Most prominent, however, is the onset of the *uremic syndrome,* or *end-stage renal disease (ESRD),* in which increased BUN and serum creatinine values (azotemia) are observed in conjunction with the preceding symptoms. The outcome of this disease is either recovery or, in the case of irreversible renal damage, progression to chronic renal failure.[3,6]

Chronic Renal Failure

Chronic renal failure is a clinical syndrome that occurs when there is a gradual decline in renal operation over time. The interrelated pathophysiology of this disease process is depicted in Figure 21-7. Chronic renal failure is classified into four progressive stages. The first stage is marked by a period of silent deterioration in renal status. Kidney function decreases, but BUN and creatinine values stay within normal limits. The second stage is characterized by development of a slight renal insufficiency. A 50% reduction in normal functioning is nec-

essary before the BUN and creatinine values reflect the pathologic changes by increasing above reference ranges. The third stage is typified by impending renal failure. Anemia begins to develop (due to the constant deficit in erythropoietin production), and systemic acidosis commences (due to the faulty clearance of endogenous metabolic acids). The fourth and last stage commences with the onset of the classic symptoms of the uremic syndrome (see preceding section). The conditions that can precipitate acute renal failure also may lead to chronic renal failure.[3,11] In addition, there are several other etiologies for this syndrome, as shown in Table 21-3.

Diabetes Mellitus

Diabetes mellitus can have profound effects on the renal system. In insulin-dependent diabetes mellitus (IDDM, type 1), patients suffer from a deficit of insulin activity. Approximately 45% of these patients will develop progressive deterioration of kidney function *(diabetic nephropathy)* within 15 years to 20 years after their diagnosis. A smaller percentage of non–insulin-dependent (NIDDM, type II) diabetics also will go on to develop this condition. The lesions are primarily glomerular, but they may affect all other kidney structures as well; they are theorized to be caused by the abnormally hyperglycemic environment that constantly bathes the vascular system.[3,6]

Typically, diabetes affects the kidneys by causing them to become glucosuric, polyuric, and nocturic. These states are caused by the heavy demands *(nephromegaly)* made on the kidneys to diurese a hyperosmotic urine. In addition, a mild proteinuria *(microalbuminuria)* often develops between 10 years and 15 years after the original diagnosis is made. During the next 3 years to 5 years, this situation becomes macroprotein-

A 27-year-old man was admitted for observation after complaining of back pain radiating to the upper chest for 3 days and a rash on his trunk and upper extremities for 1 day. The patient reported being exposed to a child with chicken pox. The patient is a postrenal transplant candidate whose graft done 1 year previously, was secondary to idiopathic chronic renal failure. No problems were reported until this incident. On Day 11 postadmission, varicella was confirmed. Laboratory results are shown in Case Study Table 21-3.1.

Questions

1. Why was this patient admitted for observation?
2. Did patient suffer acute renal failure?
3. What did the β_2-M results indicate?
4. Why were a serum and urine myoglobin requested?
5. Was the patient at high or low risk of acute renal failure due to myoglobin?
6. What other organs seemed to be involved from the result?
7. Were the BUN/creatinine results helpful in indicating whether the patient was improving?

CASE STUDY TABLE 21-3.1. Laboratory Results

Day	Cr	BUN	β_2-M	Myo Ser	Myo Urine	Phos	Amy	Lip	AST	ALT	ALP	Bili	CyA
1	2.3↑	25	5.0↑	—	—	5.0↑	—	—	1218↑	697↑	105	1.7↑	<28
3	3.1↑	38↑	—	—	—	—	136↑	356↑	5856↑	2963↑	147↑	—	<28
5	3.4↑	47↑	10.4↑	299↑	178↑	—	—	—	4416↑	993↑	—	—	—
7	—	—	—	1450↑	15600↑	—	—	—	—	—	—	—	—
9	2.6↑	37↑	9.4↑	448↑	3920↑	—	563↑	5332↑	—	—	—	—	—
10	2.5↑	35↑	6.0↑	—	—	—	555↑	5928↑	—	—	—	—	—
11	2.4↑	41↑	6.5↑	—	—	—	479↑	4462↑	—	—	—	—	—

uric, and thereafter, a steady decrease in GFR is usually seen from month to month. Indeed, plotting of the reciprocal of the serum creatinine value against time often results in a straight-line approximation sufficiently sensitive to allow prediction of the deterioration rate. Hypertension often manifests itself next, further exacerbating the renal damage. Eventually, chronic renal insufficiency or nephrotic syndrome may evolve, and each may be identified by their characteristic symptoms and laboratory findings. The early appearance of this proteinuria in the progression of the disease correlates significantly with the likelihood of developing diabetic nephropathy in later life. Microalbuminuria is apparently caused by the creation of increasing numbers of large pores at the glomerulus.[15] Interventional therapy of microalbuminuric patients, such as tight control of blood glucose, may prolong the onset of chronic renal failure.

Renal Hypertension

Etiology. Renal disease-induced hypertension can be caused by either decreased perfusion to all or part of an area of the kidney or by local tissue accumulation of sodium. Chronic ischemia of any sort results in nephron dysfunction and eventual necrosis; thus, the lack of perfusion may be due to trauma damage or stenosis to a main or branching artery or to sclerosis of intrarenal arterioles. In either case, the resulting changes in blood and body fluid volumes within the kidney trigger the activation of the renin-angiotensin-aldosterone system,

setting off the vasoconstrictive responses, which are manifested as persistent hypertension.

Laboratory Tests. Renal hypertension can be evaluated by monitoring for elevations in serum aldosterone, Na^+, and plasma renin levels (particularly as renal vein differential samples). As a consequence of the increased Na^+ retention, urine K^+ levels will be increased (due to the accelerated excretion rate), and, therefore, serum K^+ levels will be decreased.

Therapy of Acute Renal Failure

In patients with acute renal failure, uremic symptoms, uncontrolled hyperkalemia, and acidosis have traditionally been indications that the kidneys are unable to excrete the body's waste products and a substitute method in the form of dialysis was necessary. Dialysis is often instituted before this stage, however. There are several forms of dialysis available but they all use a semipermeable membrane surrounded by a dialysate bath in common. In traditional *hemodialysis* (removal of waste from blood), the membrane is synthetic and outside the body. Arterial blood and dialysate are pumped at high rates (150–250 mL/min and 500 mL/min, respectively) in opposite directions. The blood is returned to the venous circulation and the dialysate discarded. The diffusion of low molecular weight solutes (<500 daltons) into the dialysate is favored by this process, but mid–molecular weight solutes (500–2000 daltons) are inadequately cleared. Creatinine clearance is roughly 150

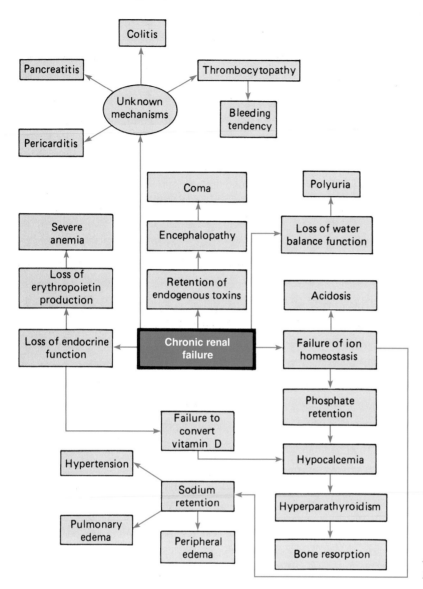

Figure 21-7. Pathophysiology of chronic renal failure.

mL/min to 160 mL/min. In peritoneal dialysis, the peritoneal wall acts as the dialysate membrane and gravity is used to introduce and remove the dialysate. Two variations of this form are available, continuous ambulatory peritoneal dialysis (CAPD) and continuous cycling peritoneal dialysis, but in both the process is continuous, being performed 24 hours a day, 7 days a week. This method is not as rigorous as the traditional method. Small solutes (*eg,* potassium) have significantly lower clearance rates compared to the traditional method, but more large solutes are cleared and steady-state levels of blood analytes are maintained. Continuous arteriovenous *hemofiltration* (ultrafiltration of blood), continuous venovenous hemofiltration, continuous arteriovenous hemodialysis, and continuous venovenous hemodialysis together make up the slow continuous renal replacement therapies developed to treat acute renal failure in critically ill patients in intensive care settings. In these methods, the semipermeable membrane is again outside the body. Solutes up to 5000 dal-

tons (the pore size of the membranes) and water are slowly (10 mL/min) and continuously filtered from the blood in the first two methods, causing minimal changes in plasma osmolality. The ultrafiltrate is lost to the body and the volume loss can be replaced in the form of parenteral nutrition and obligatory intravenous medications. The final two methods are similar to the filtration methods but a continuous trickle of dialysis fluid is pumped past the dialysis membrane, resulting in continuous diffusion. This doubles the urea clearance.

Therapy of ESRD

For patients with irreversible renal failure, dialysis and transplantation are the two therapeutic options. Initiation of either treatment occurs when the GFR falls to 5 mL/min (10–15 mL/min in patients with diabetic nephropathy).

Dialysis. Traditional hemodialysis or its more recent high-efficiency form as well as peritoneal dialysis are the available

TABLE 21-3. Etiology of Chronic Renal Failure

Etiology	Examples
Renal circulatory diseases	Renal vein thrombosis, malignant hypertension
Primary glomerular diseases	SLE, chronic glomerulonephritis
Renal sequelae to metabolic disease	Gout, diabetes mellitus, amyloidosis
Inflammatory diseases	Tuberculosis, chronic pyelonephritis
Renal obstructions	Prostatic enlargement, calculi
Congenital renal deformity	Polycystic kidneys, renal hypoplasia
Miscellaneous conditions	Radiation nephritis

methods. The clinical laboratory used in conjunction with a hemodialysis facility must be able to adequately monitor procedural efficiency in a wide variety of areas. Renal dialysis has basic goals, and specific laboratory tests should be performed to evaluate the achievement of each.

Monitoring of nitrogen balance. Rapid removal of sodium, urea, creatinine, and several other osmotically active compounds can cause *dialysis disequilibrium syndrome*. This may result in headaches, lassitude, seizures, and torpor and should be monitored by determinations of BUN, creatinine, uric acid, total protein, and albumin.

Monitoring of hydration and electrolyte balance. Swift changes in fluid and electrolyte distribution throughout the body can derange albumin concentrations and skew buffering capabilities, causing hypotension. Additionally, rapid shifts in potassium may precipitate cardiac toxicity. Important laboratory tests include Na^+, K^+, Cl^-, HCO_3^-, osmolality, and arterial pH.

Monitoring of generalized osteomalacia. Chronic alterations in bone deposition are caused by continued derangement of Ca^{2+}, phosphate, and vitamin D metabolism. A condition known as *renal osteodystrophy* can result. Laboratory analyses should assess serum phosphorus, Ca^{2+}, alkaline phosphatase, magnesium, and (at regular intervals) PTH levels. The last assay is indicated to monitor the secondary hyperparathyroidism that often ensues due to upregulation of parathyroid gland activity.

Viral hepatitis. Dialysis patients suffer from a variety of acute and chronic liver diseases including the many forms of hepatitis. The common forms of dialysis-associated hepatitis include hepatitis A, hepatitis B, hepatitis C, cytomegalovirus hepatitis, and drug-induced hepatitis. At least 50% of patients with hepatitis are asymptomatic. The type of liver disease must be established for both prognosis and therapy. Essential tests are alanine aminotransferase (ALT), aspartate aminotransferase (AST), alkaline phosphatase (ALP), bilirubin, and hepatitis B surface antigen (HBsAg). Antibody to hepatitis C virus (anti-HCV) testing has also been proposed for all patients entering dialysis programs because of the high incidence of anti-HCV positives among hemodialysis patients.[5]

Hyperlipidemia. There is a 50% incidence of hyperlipidemia in the dialysis population and the most prevalent pattern is type IV (see Chapter 11 *Lipids and Lipoproteins*). Patients with this classification have an increased incidence of ischemic heart disease. Essential tests are a lipid panel,

CASE STUDY 21-4

A 50-year-old man scheduled for surgery had preadmission blood work performed. This included a fasting glucose and urinalysis. Results are shown in Table 21-4.1.

CASE STUDY TABLE 21-4.1. Laboratory Results

Serum glucose = 220 mg/dL

Urinalysis

Color	Pale yellow
Appearance	Clear
Specific gravity	1.030
pH	5.0
Protein	1+
Glucose	100 mg/dL
Ketones	Trace
Bilirubin	Negative
Blood	Negative
Urobilinogen	Normal
Nitrite	Negative
Leukocyte	Negative

Questions

1. Is the finding of a positive urine glucose unexpected? Why?
2. What is this patient's most likely disorder?
3. Given this patient's disorder, what does the positive protein result tell you about his kidney function?
4. What measures could have been taken to prevent or delay the development of proteinuria?

calcium, and phosphorus. Calcium and phosphorus control and subsequent avoidance of hyperparathyroidism may lower the risk of ESRD-related atherosclerosis.[16]

Aluminum neurotoxicity. Aluminum neurotoxicity is well recognized in dialysis patients. Two syndromes have been defined: dialysis encephalopathy brought on by chronic oral or parenteral aluminum exposure, and acute aluminum neurotoxicity brought on by the ingestion of a combination of citrate and aluminum compounds after deferoxamine therapy or after dialysis contaminated with aluminum. Aluminum concentrations range between 100 mg/L to 150 mg/L (normal is <6 mg/L) and more than 500 mg/L in dialysis encephalopathy and acute aluminum neurotoxicity, respectively.[17]

Assessment of residual function. Evaluation of renal status after a dialysis session should be made by determinations of 24-hour urine volume and GFR. The GFR, however, usually drops temporarily on day 1 after dialysis, so measurements of creatinine clearance or other clearance tests should be made after the first 24 hours.

Therapeutic drug monitoring (TDM). Inadequate renal function, coupled with dialysis-induced changes, can dramatically derange pharmacokinetics of therapeutic agents. TDM must be undertaken to ensure the maintenance of appropriate therapeutic drug levels. In particular, the aminoglycoside antibiotics are highly nephrotoxic and must be followed using regular serum drug levels.

Transplantation. The most efficient hemodialysis techniques provide only 10% to 12% of the small solute removal of two normal kidneys and considerably less removal of larger solutes. Hence, even well-dialyzed patients have physical disabilities and a decrease in the quality of life. Kidney transplantation offers the greatest chance for a full return of a healthy, productive life. However, this option is limited by the significant shortage of donor organs. For ESRD patients, the length of time waiting for an organ can vary from several months to several years.

Renal transplantation is from a compatible donor to a recipient suffering from irreversible renal failure. The organ can be from a cadaver or a live individual (80% and 20%, respectively, of all kidney transplants in the United States). For this procedure to be successful, the body's immune response to the transplanted organ must be suppressed. Thus, the donor and recipient are carefully screened for ABO blood group, HLA (human leukocyte antigen) compatibility, and preformed HLA antibodies. The HLA system is the major inhibitor to transplantation. In addition, cytokine production is stopped by using immunosuppressive agents. Cytokines are peptide products of immune cells, which stimulate an immune response. There are four major immunosuppressive drugs or drug groups available: cyclosporine, corticosteroids, azathioprine, and the monoclonal and polyclonal antibodies. Cyclosporine has enjoyed widespread use since the 1980s, its use coinciding with improved graft (transplant tissue) survival. The

transplantation procedure itself requires considerable skill to ensure suitable positioning of the kidney(s) and vascular connections as well as attention to detail, aseptic techniques, and hemostasis, because many of these patients are anemic or malnourished at the time of surgery and the use of immunosuppressive drugs potentially compromises the healing process.

Cyclosporine is a cyclic polypeptide of fungal origin. It consists of 11 amino acids and has a molecular weight of 1203. It is neutral and insoluble in water but is soluble in organic solvents and lipids. Cyclosporine inhibits the immune response selectively by inhibiting the interleukin-2–dependent proliferation of activated T cells, which destroy the *allograft* (transplant tissue from the same species). The immune response is frozen and unresponsive. Oral cyclosporine is incompletely and variably absorbed from the gastrointestinal tract. Peak concentration is on average 3 hours and the absolute bioavailability, which increases with time, is about 30%. Thus, the amount of cyclosporine required to achieve a given blood level tends to fall with time and typically reaches a steady level within 4 weeks to 8 weeks. In the blood, one third of cyclosporine is found in plasma, bound primarily to lipoproteins with the rest bound to erythrocytes. Consequently, whole blood drug levels will typically be three-fold higher than plasma levels. Cyclosporine has a half-life of approximately 8 hours and is metabolized to at least 15 metabolites by the liver, some of which may have immunosuppressive and nephrotoxic potential. Cyclosporine is also excreted in the bile. The measurement of cyclosporine levels is an intrinsic part of the management of transplant patients because of the variation in its metabolism in individual patients and from patient to patient. There is also some correlation between drug blood levels and episodes of rejection and toxicity. Cyclosporine monitoring can be a source of much confusion because of the various assays available and the option of using different matrixes (*ie,* plasma or whole blood) for its measurement.

Laboratory support in patient management is aimed at assessing renal function and, hence, rejection status as well as toxicity due to the immunosuppressant drugs. During the early posttransplant period (usually 2 months), tests include urine volume; UA for cells, protein, and bacteria; serum creatinine and urea levels; electrolytes panel including phosphorus, calcium, and hepatic enzyme levels; complete blood count; and the blood immunosuppressive drug levels. These tests monitor for occurrence of early rejection, immunosuppressant drug toxicity, or other underlying pathologic events in the allograft. In the long-term, other tests are added to monitor for potential complications. One example is a lipid profile, because approximately 50% of patients treated with cyclosporine are hypercholesterolemic by the 3rd month after transplant. Individuals with evidence of prior hepatitis B or C infection who are clinically quiescent are eligible for transplantation, and there is some indication that immunosuppressive therapy appears to stimulate progression from benign to aggressive liver disease. Although kidney transplants have the capacity to function for decades, the mean half-life of a cadaveric transplant is approximately 7 years. The mortality rate is not significantly different from hemodialysis. Three-year graft survival figures vary from 65% to 85%, with live grafts doing better. It has been reported that there is no difference in patient survival among hemodialysis, CAPD, and cadaveric kidney transplantation. Live related-donor transplantation is associated with a better patient survival than other ESRD therapeutic options.

SUMMARY

The kidney plays a vital role in the maintenance of water and electrolyte balance, homeostasis, and removal of waste products. The kidney produces several important hormones (such as renin and the prostaglandins) needed to perform these tasks. Other renal hormones (such as erythropoietin) are important for other physiologic functions. The kidneys themselves are the target organ for catabolism of insulin, glucagon, aldosterone, and vitamin D, among others.

Renal function tests focus largely on glomerular clearances, as assessed by creatinine and urea measurements, and tubular functions, as assessed by protein measurements (eg, urine electrophoresis). The analysis of urine for analytes, such as pH, glucose, ketones, and bilirubin, continue to be important screening tests for many nonrenal diseases such as diabetes mellitus, ketoacidosis (and other acid-base imbalances), hemolysis, and liver diseases. Newer protein assays, such as for urine albumin, serum β_2-M, and serum and urine myoglobin, can provide important prognostic information useful for patient management. Microalbuminuria is useful for early detection of diabetic nephropathy, β_2-M is useful for early renal transplant rejection, and myoglobin clearance rates are helpful in predicting rhabdomyolysis-induced acute renal failure.

Common renal diseases include infectious and inflammatory processes to the glomerulus, tubules, and urinary tract, obstructions to normal kidney function, and acute and chronic renal failure. Common glomerular diseases include nephritis and nephrotic syndrome. Tubular diseases include renal tubular acidosis and Fanconi's syndrome. In situations of chronic renal failure, aggressive therapeutic approaches based on dialysis and transplantation have enabled prolonged survival of what was once a terminal condition. Variations

CASE STUDY 21-5

A 45-year-old male with ESRD secondary to hypertension was admitted for a living related-donor kidney transplant. He was on peritoneal dialysis for 1½ years before transplant surgery. Transplantation was uneventful with no complications. During surgery when the kidney was connected, urine formation occurred within a few seconds. A renal scan the next morning showed good perfusion and function with no outflow obstruction. The patient was discharged 7 days later. Applicable lab data are shown in Case Study Table 21-5.1.

Questions

1. What tests did the surgeon use to assess whether the transplanted kidney was functioning?
2. How soon after transplant was kidney function observed?.
3. Were the pretransplant BUN and creatinine values expected for a patient on peritoneal dialysis?
4. Why were cholesterol and triglyceride levels requested before surgery?
5. Is a high PTH level expected in an ESRD patient?
6. What results in the 1st week indicated the transplant was a success?

CASE STUDY TABLE 21-5.1. Laboratory Results[a]

Day	Na⁺	K⁺	Cl⁻	HCO₃⁻	Cr	BUN	β₂-M Urine	24-h	Gluc	Chol	Trig	Ca²⁺	Phos	TP	AST	ALT	ALK	Bili	PTH	CyA
1	138	5.2↑	96↓	25	27.4↑	104↑	—	—	—	132	0.6	5.0	6.8↑	6.1↓	29	21	64	0.4	281↑	—
2	141	4.6	103	17↓	22.4↑	94↑	—	—	130↑	—	—	—	—	—	—	—	—	—	—	—
3	139	5.1↑	106	22	10.3↑	59↑	6.0↑	—	141↑	—	—	—	—	—	—	—	—	—	—	—
4	138	5.3↑	107	22	4.2↑	31↑	2.9↑	2350	119↑	—	—	5.8↑	3.7	—	22	23	75	0.4	—	163
5	135	5.7↑	105	22	2.5↑	23	2.2	2100	122↑	—	—	—	—	—	—	—	—	—	—	96
6	137	4.5	105	24	2.0↑	27↑	2.2	4880	94	—	—	—	—	—	—	—	—	—	—	69
7	136	4.3	102	23	1.8↑	31↑	2.0	2500	83	—	—	—	—	—	—	—	—	—	—	108
8	139	4.8	100	26	1.7	34↑	2.2	1900	84	—	—	—	—	—	—	—	—	—	—	190
9	137	4.3	98	27	1.7	32↑	2.0	—	86	—	—	5.76↑	2.8	—	46	102↑	76	1.1	—	900

[a] Transplant surgery was performed on Day 2, before these reported results. Ca²⁺ is ionized calcium. Serum protein electrophoresis (SPEP) was ordered on Day 1, which was normal for albumin and globulins.

in dialysis techniques have made this process more available and convenient, and with the implementation of powerful immunosuppressive drugs such as cyclosporine, widespread renal transplantation is now limited only by the availability of appropriate donor organs.

REVIEW QUESTIONS

1. Calculate the creatinine clearance given the following information: serum creatinine 1.2 mg/dL, urine creatinine 120 mg/dL, urine volume 1750 mL/24 h, body surface area 1.80 m^2:
2. The proximal tubule functions to:
    a. Concentrate salts
    b. Reabsorb 75% of salt and water
    c. Form the renal threshold
    d. Reabsorb urea
3. The normal glomerular filtration rate is:
    a. 2 mL/min
    b. 125 mL/min
    c. 800 mL/min
    d. 1200 mL/min
4. Renal clearance is the:
    a. Volume of urine produced per day
    b. Volume of plasma from which a substance is removed per unit of time
    c. Amount of creatinine in urine
    d. Urine concentration of a substance divided by the urine volume per unit of time
5. Hyperuricemia can be seen in all of the following conditions EXCEPT:
    a. Leukemia
    b. Gout
    c. Renal disease
    d. Jaundice
6. Renin release by the kidney is stimulated by a decrease in:
    a. Plasma sodium concentration
    b. Sodium intake
    c. Extracellular fluid volume or pressure
    d. Renal tubular reabsorption
7. ADH acts on the:
    a. Collecting duct
    b. Henle's loop
    c. Proximal convoluted tubule
    d. Bowman's capsule
8. The set of results that most accurately reflects severe renal disease is:

	serum creatinine	creatinine clearance	BUN
a.	1.0 mg/dL	110 mL/min	17 mg/dL
b.	2.0 mg/dL	120 mL/min	14 mg/dL
c.	1.0 mg/dL	95 mL/min	43 mg/dL
d.	3.7 mg/dL	44 mL/mi.	88 mg/dL

9. The final concentration of the urine is determined within the:
    a. Proximal convoluted tubules
    b. Distal convoluted tubules
    c. Henle's loops
    d. Collecting ducts
10. Creatinine clearance results are corrected using a patient's body surface area to account for differences in:
    a. Age
    b. Dietary intake
    c. Sex
    d. Muscle mass

REFERENCES

1. Vander A, et al. Human physiology: the mechanisms of body function. 7th ed. New York: WCB McGraw Hill, 1998;503,508,519.
2. Kaplan A, et al. Clinical chemistry: interpretation and techniques. 4th ed. Baltimore: Williams & Wilkins, 1995;156–157.
3. Rock RC, Walker WG, Jennings CD. Nitrogen metabolites and renal function. In: Tietz NW, ed. Fundamentals of clinical chemistry. 3rd ed. Philadelphia: WB Saunders, 1987;669.
4. Russell PT, Sherwin JE, Obernolte R, et al. Nonprotein nitrogenous compounds. In: Kaplan LA, Pesce AJ, eds. Clinical chemistry: theory, analysis, and correlation. 2nd ed. St. Louis, MO: CV Mosby, 1989;1005.
5. Fraser D, Jones G, Kooh SW, et al. Calcium and phosphate metabolism. In: Tietz NW, ed. Fundamentals of clinical chemistry. 3rd ed. Philadelphia: WB Saunders, 1987;705.
6. First MR. Renal function. In: Kaplan LA, Pesce AJ, eds. Clinical chemistry: theory, analysis, and correlation. 3rd ed. St. Louis, MO: CV Mosby, 1996;484.
7. Kaplan LA. Measurement of colligative properties. In: Kaplan LA, Pesce AJ, eds. Clinical chemistry: theory, analysis, and correlation. 2nd ed. St. Louis, MO: CV Mosby, 1989;207.
8. Kleinman LI, Lorenz JM. Physiology and pathophysiology of body water and electrolytes. In: Kaplan LA, Pesce AJ, eds. Clinical chemistry: theory, analysis, and correlation. 2nd ed. St. Louis, MO: CV Mosby, 1989;313.
9. Sherwin JE, Bruegger BB. Acid-base control and acid-base disorders. In: Kaplan LA, Pesce AJ, eds. Clinical chemistry: theory, analysis, and correlation. 2nd ed. St. Louis, MO: CV Mosby, 1989;332.
10. Schumann GB, Schweitzer SC. Examination of urine. In: Kaplan LA, Pesce AJ, eds. Clinical chemistry: theory, analysis, and correlation. 2nd ed. St. Louis, MO: CV Mosby, 1989;820.
11. Frauenhoffer E, Demers LM. Beta$_2$-microglobulin. ASCP Check Sample Continuing Education Program, Clinical Chemistry, No. CC 865 (CC173), Chicago, IL, 1986.
12. Wu AHB, Laios I, Green S, et al. Immunoassays for serum and urine myoglobin: myoglobin clearance assessed as a risk factor for acute renal failure. Clin Chem 1994;40:796.
13. Skogen W. Urinary albumin and diabetic nephropathy. Clinical Laboratory News 1995;21(3):6–7.
14. American Diabetes Association. Position statement: standards of medical care for patients with diabetes mellitus. Diabetes Care 1994;17:616–623.
15. Van Lente F, Suit P. Assessment of renal function by serum creatinine and creatinine clearance: glomerular filtration rate estimated by four procedures. Clin Chem 1989;35:2326.
16. Golper TA. Metabolic abnormalities. In: Nissenson AR, Fine RN, eds. Dialysis therapy. 2nd ed. Philadelphia: Hanley and Belfus, 1993;261.
17. Alfrey AC. Neurological aspects of uremia. In: Nissenson AR, Fine RN, eds. Dialysis therapy. 2nd ed. Philadelphia: Hanley and Belfus, 1993;275.

CHAPTER 22

Pancreatic Function

Edward P. Fody

Objectives

Upon completion of this chapter, the clinical laboratorian should be able to:

- *Discuss the physiologic role of the pancreas in the digestive process.*

- *List the hormones excreted by the pancreas along with their physiologic roles.*

- *Describe the following pancreatic disorders and list the associated laboratory tests that would aid in their diagnosis: acute pancreatitis, chronic pancreatitis, pancreatic carcinoma, cystic fibrosis, and pancreatic malabsorption.*

KEY TERMS

Cholecystokinin (CCK)
Islets of Langerhans

Pancreatitis
Secretin
Steatorrhea

The pancreas is a large gland that is involved in the digestive process but is located outside of the gastrointestinal (GI) system. It is composed of both endocrine and exocrine tissues. Its endocrine functions include the production of insulin and glucagon; both of these hormones are involved in carbohydrate metabolism. Its exocrine function involves the production of many enzymes used in the digestive process. This chapter discusses the physiology of pancreatic function, diseases of the pancreas, and tests of pancreatic function.

PHYSIOLOGY OF PANCREATIC FUNCTION

As a digestive gland, the pancreas is only second in size to the liver and weighs about 70 to 105 g. It is located behind the peritoneal cavity across the upper abdomen at about the level of the first and second lumbar vertebrae and, thus, is 1 to 2 inches above the umbilicus. It is located in the curve made by the duodenum (Fig. 22-1). The pancreas is composed of two morphologically and functionally different tissues: endocrine tissue and exocrine tissue. The *endocrine* (hormone-releasing) component is by far the smaller of the two and consists of the *islets of Langerhans,* which are well-delineated spherical or ovoid clusters composed of at least four different cell types. The islet cells secrete at least four hormones into the blood: insulin, glucagon, gastrin, and somatostatin. The larger *exocrine* pancreatic component (enzyme secreting) secretes about 1.5 to 2 L per day of fluid that is rich in digestive enzymes into ducts that ultimately empty into the duodenum.

This digestive fluid is produced by the pancreatic acinar cells (grapelike clusters), which line the pancreas and are connected by small ducts. These small ducts empty into progressively larger ducts, eventually forming one major pancreatic duct and a smaller accessory duct. The major pancreatic duct and the common bile duct open into the duodenum at the major duodenal papilla (Fig. 22-2). Normal protein-rich pancreatic fluid is clear, colorless, and watery, with an alkaline pH that can reach up to 8.3. This alkalinity is caused by the high concentration of sodium bicarbonate present in pancreatic fluid, which is used eventually to neutralize the hydrochloric acid in gastric fluid from the stomach as it enters the duodenum. The bicarbonate and chloride concentrations vary reciprocally so that they total about 150 mmol/L.

Pancreatic fluid has about the same concentrations of potassium and sodium as does serum. The digestive enzymes or their proenzymes secreted by the pancreas are capable of digesting the three major classes of food substances (proteins, carbohydrates, and fats) and include: (1) the proteolytic enzymes trypsin, chymotrypsin, elastase, collagenase, leucine aminopeptidase, and some carboxypeptidases; (2) lipid-digesting enzymes, primarily lipase and lecithinase; (3) carbohydrate-splitting pancreatic amylase; and (4) several nucleases (ribonuclease), which separate the nitrogen-containing bases from their sugar-phosphate strands.

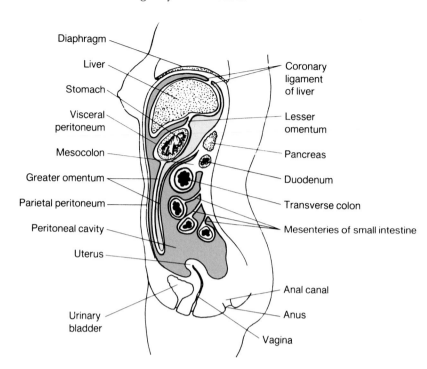

Figure 22-1. Peritoneum and mesenteries. The parietal peritoneum lines the abdominal cavity, and the visceral peritoneum covers abdominal organs. Retroperitoneal organs are covered by the parietal peritoneum. The mesenteries are membranes that connect abdominal organs to each other and to the body wall. (From Akessom. Thompson's Core Textbook of Anatomy, 2nd edition. Philadelphia: JB Lippincott, 1990;115.)

Pancreatic activity is under both nervous and endocrine control. Branches of the vagus nerve can cause a small amount of pancreatic fluid secretion when food is smelled or seen, and these secretions may increase as the bolus of food reaches the stomach. Most of the pancreatic action, however, is under the hormonal control of *secretin* and *cholecystokinin* (CCK, formerly called *pancreozymin*). Secretin is responsible for the production of bicarbonate-rich and therefore alkaline pancreatic fluid, which protects the lining of the intestine from damage. Secretin is synthesized in response to the acidic contents of the stomach reaching the duodenum. It can also affect gastrin activity in the stomach. This pancreatic fluid contains few digestive enzymes. CCK, in the presence of fats or amino acids in the duodenum, is produced by the cells of the intestinal mucosa and is responsible

for release of enzymes from the acinar cells by the pancreas into the pancreatic fluid.

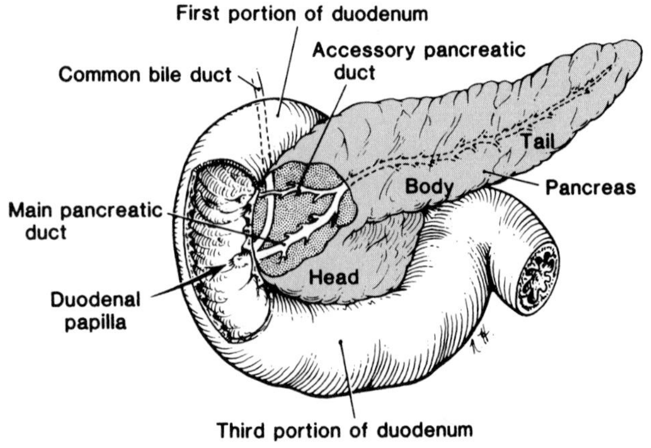

DISEASES OF THE PANCREAS

The role of the pancreas in diabetes mellitus is discussed in Chapter 10, Carbohydrates, and is not included here. Other than trauma, only three diseases cause more than 95% of the medical attention devoted to the pancreas. If they affect the endocrine function of the pancreas, these diseases can result in altered digestion and nutrient metabolism.

1. *Cystic fibrosis,* also known by various other terms, such as *fibrocystic disease of the pancreas* and *mucoviscidosis,* is inherited as an autosomal recessive disorder and is characterized by dysfunction of mucous and exocrine glands throughout the body. The disease is relatively common and occurs in about 1 in 1600 live births. It has a variety of manifestations and can initially present in such widely varying ways as intestinal obstruction of the newborn, excessive pulmonary infections in childhood, or uncommonly, as pancreatogenous malabsorption in adults. The disease causes the small and large ducts and the acini to dilate and convert into small cysts filled with mucus, eventually resulting in the prevention of pancreatic secretions reaching the duodenum or, depending on the age of the patient, a plug that blocks the lumen of the bowel, leading to obstruction. As the disease progresses, there is increased destruction and fibrous scarring of the pancreas and a corresponding decrease in function.

2. *Pancreatic carcinoma* causes about 27,000 deaths each year in the United States, which represents about 5% of all deaths from malignant neoplasms, and is the fifth most frequent form of fatal cancer. The 5-year survival rate after surgery is less than

Figure 22-2. Diagram of the pancreas and its relationship to the duodenum.

2%, and more than 90% of patients die within 1 year of diagnosis. Most pancreatic tumors arise as adenocarcinomas of the ductal epithelial. Because the pancreas has a rich supply of nerves, pain is a prominent feature of the disease. If the tumor arises in the body or tail of the pancreas, tumor detection often occurs during an advanced stage of the disease because of its central location and the associated vague symptoms. Cancer of the head of the pancreas usually is detected earlier because of its proximity to the common bile duct. These tumors make their presence known by jaundice, weight loss, anorexia, and nausea. The jaundice is associated with signs of posthepatic hyperbilirubinemia (intrahepatic cholestasis) and very low levels of fecal bilirubin resulting in clay-colored stools. However, findings are not specific for pancreatic tumors, and other causes of obstruction must be ruled out.

Islet cell tumors of the pancreas affect the endocrine capability of the pancreas. If the tumor occurs in the beta cells, hyperinsulinism is seen, resulting in very low blood glucose levels, followed by insulin shock. Pancreatic alpha-cell tumors that overproduce gastrin are called *gastrinomas* or *Zollinger-Ellison syndrome,* named after the individuals who first discovered it, and can also be duodenal in origin. These tumors are associated with watery diarrhea, recurring peptic ulcer, and marked gastric hypersecretion and hyperacidity. Pancreatic alpha-cell glucagon-secreting tumors are rare, and the hypersecretion of glucagon is associated with diabetes mellitus.

3. *Pancreatitis,* or inflammation of the pancreas, ultimately is caused by autodigestion of the pancreas as a result of reflux of bile or duodenal contents into the pancreatic duct. Pathologic changes can include acute edema with large amounts of fluid accumulating in the retroperitoneal space and an associated decrease in effective circulating blood volume; cellular infiltration, leading to necrosis of the acinar cells, with hemorrhage as a possible result of necrotic blood vessels; and intrahepatic and extrahepatic pancreatic fat necrosis. Pancreatitis is classified generally as acute (no permanent damage to the pancreas), chronic (irreversible injury), or relapsing/recurrent, which can also be acute or chronic. It commonly occurs in midlife. Painful episodes can occur intermittently, usually reaching a maximum within minutes or hours, lasting for several days or weeks, and frequently accompanied by nausea and vomiting. Pancreatitis is often associated with alcohol abuse or biliary tract disease, but patients with hyperlipoproteinemia types I, IV, and V and those with hyperparathyroidism are also at a significantly increased risk for this disease.

Other etiologic factors associated with acute pancreatitis include mumps, obstruction due to biliary tract disease, gallstones, pancreatic tumors, tissue injury, artherosclerotic disease, shock, pregnancy, hypercalcemia, hereditary pancreatitis, immunologic factors associated with postrenal transplantation, and hypersensitivity. Symptoms of acute pancreatitis include severe abdominal pain that is generalized or in the upper quadrants and often radiates toward the back or down the right or left flank. The etiology of chronic pancreatitis is similar to that of acute pancreatitis, but chronic excessive alcohol consumption appears to be the most common predisposing factor.

Laboratory findings include increased amylase, lipase, and triglycerides and hypercalcemia, which is often associated with underlying hyperparathyroidism. Hypocalcemia may be found and has been attributed to the sudden removal of large amounts of calcium from the extracellular fluid owing to impaired mobilization or as a result of calcium fixation by fatty acids that are liberated by the increased lipase action on triglycerides. Hypoproteinemia is attributable mainly to the notable loss of plasma into the retroperitoneal space. A shift of arterial blood flow from the inflamed pancreatic cells to less affected or normal cells causes oxygen deprivation and tissue hypoxia in the area of damage, including the surrounding organs and tissues.

CASE STUDY 22-1

A 38-year-old man entered the emergency room with the complaint of severe, boring, mid-abdominal pain of 6 hours' duration. A friend, who had driven him to the hospital, stated that the patient fainted three times as he was being helped into the automobile. The patient had a 15-year history of alcoholism and drank 1 to 2 pints of whiskey every day. He had last been hospitalized for acute alcoholism 3 months ago, at which time he had relatively minor abnormalities of liver function. Upon admission this time, his blood pressure was 80/40 mm Hg, pulse 110 beats/min and thready, and respirations 24 breaths/min. and shallow. Clinical laboratory test results are shown in Case Study Table 22-1.1.

CASE STUDY TABLE 22-1.1. Laboratory Results

Serum amylase	640 units (3.5–260)
Serum sodium	133 mEq/L (135–145)
Potassium	3.4 mEq/L (3.8–5.5)
Calcium	4.0 mEq/L (4.5–5.5)
Blood urea nitrogen	32 mg/dL (8–22)
White blood cell count	16,500
Hemoglobin	12 g/dL

Questions

1. What is the probable disease?
2. What is the cause for the low serum calcium?
3. What is the cause for the increased blood urea nitrogen?

All three of these conditions can result in severely diminished pancreatic exocrine function, which can significantly compromise digestion and absorption of ingested nutrients. This is the essence of the general malabsorption syndrome, which embodies abdominal bloating and discomfort, the frequent passage of bulky, malodorous feces, and weight loss. Failure to digest or absorb fats is known as *steatorrhea* and renders a greasy appearance to feces (more than 5 g of fecal fat per 24 hours). The malabsorption syndrome typically involves abnormal digestion or absorption of proteins, polysaccharides, carbohydrates, and other complex molecules as well as lipids. Severely deranged absorption and metabolism of electrolytes, water, vitamins (particularly the fat-soluble vitamins A, D, E, and K), and minerals can also occur. Malabsorption can involve a single substance, such as vitamin B_{12}, which results in a megaloblastic anemia (pernicious anemia), or lactose caused by a lactase deficiency. In addition to pancreatic exocrine deficiency, the malabsorption syndrome can be caused by biliary obstruction, which deprives the small intestines of the emulsifying effect of bile, and various diseases of the small intestine, which inhibit absorption of digested products.

TESTS OF PANCREATIC FUNCTION[1,2]

Depending on the etiology and clinical picture, pancreatic function may be suspect when there is evidence of increased amylase and lipase. The reader is referred to Chapter 9, Enzymes, for a more in-depth discussion of these enzymes. Other laboratory tests of pancreatic function include those used for detection of malabsorption (*eg,* microscopic examination of stool for excess fat, starch, and meat fibers, D-xylose test, and fecal fat analysis), tests measuring other exocrine function (*eg,* secretin, CCK, fecal fat, trypsin, and chymotrypsin), tests assessing changes associated with extrahepatic obstruction (*eg,* bilirubin), and endocrine-related tests (*eg,* gas-

trin, insulin, glucose, and cortisol) that reflect changes in the endocrine cells of the pancreas.

Direct evaluation of pancreatic fluid may include measurement of the total volume of pancreatic fluid and the amount or concentration of bicarbonate and enzymes, which requires pancreatic stimulation. Stimulation may be accomplished using a predescribed meal or administration of secretin, which allows for volume and bicarbonate evaluation, or secretin stimulation followed by CCK stimulation, which adds enzymes to the pancreatic fluid evaluation. The advantage of these tests, both of which require intubation of the patient, are that the chemical and cytologic examination are performed on actual pancreatic secretions. Cytologic examination of the fluid can often establish the presence or at least the suspicion of malignant neoplasms, although the precise localization of the primary organ of involvement (*ie,* pancreas, biliary system, ampulla of Vater, or duodenum) is not possible by duodenal aspiration.

Because of advances in imaging techniques, these stimulation tests are used less often. None of the tests has proved especially useful in diagnosis of mild or acute pancreatic disease in which the acute phase has subsided. Most of the tests have found their clinical utility in excluding the pancreas from diagnosis. The sweat test used for screening cystic fibrosis is not specific for assessing pancreatic involvement but, when used along with the clinical picture at the time of testing, can provide important diagnostic information. The following pancreatic function tests are reviewed briefly: secretin/CCK test, fecal fat analysis, sweat chloride determinations, and amylase and lipase interpretation.

Secretin/CCK Test

The secretin/CCK test is a direct determination of the exocrine secretory capacity of the pancreas. The test involves intubation of the duodenum without contamination by gas-

---

CASE STUDY 22-2

A 56-year-old alcoholic man presents with a 2-week history of mid-abdominal pain. He also describes clay-colored stools, mild icterus, nausea, vomiting, and a 10-pound weight loss. Laboratory findings are shown in Case Study Table 22-2.1.

CASE STUDY TABLE 22-2.1. Laboratory Results

Test	Result	Reference Range
Serum bilirubin	4.2 mg/dL	0.3–1.0 mg/dL
Serum lactate dehydrogenase	625 IU/IL	0–200 IU/L
Serum alanine aminotransferase	76 IU/L	0–46 IU/L
Serum alkaline phosphatase	462 IU/L	0–80 IU/L
Serum amylase	80 IU/L	0–85 IU/L
Urine bilirubin	3+	Negative

Questions

1. What organ system is primarily involved?
2. What are the major diagnostic considerations?
3. What do the laboratory results mean? What additional laboratory tests would be useful in establishing a diagnosis?
4. What other studies or procedures might be required?

tric fluid, which would neutralize any bicarbonate. The test is performed after a 6-hour or overnight fast. Pancreatic secretion is stimulated by intravenously administered secretin in a dose varying from 2 to 3 U/kg of body weight, followed by CCK administration. If a simple secretin test is desired, the higher dose of secretin is given alone.

No one protocol has been uniformly established for the test. Pancreatic secretions are collected variously for 30, 60, or 80 minutes after administration of the stimulants, either as 10-minute specimens or as a single, pooled collection. The pH, secretory rate, enzyme activities (eg, trypsin, amylase, or lipase), and amount of bicarbonate are determined. The average amount of bicarbonate excreted per hour is about 15 mM for men and 12 mM for women, with an average flow of 2 mL/kg. Assessment of enzymes must be taken in view of the total volume output. Decreased pancreatic flow is associated with pancreatic obstruction and with an increase in enzyme concentrations. Low concentrations of bicarbonate and enzymes are associated with cystic fibrosis, chronic pancreatitis, pancreatic cysts, calcification, and edema of the pancreas.[1]

Fecal Fat Analysis

Fecal lipids are derived from four sources: unabsorbed ingested lipids, lipids excreted into the intestines (predominantly in the bile), cells shed into the intestines, and metabolism of intestinal bacteria. Patients on a lipid-free diet still excrete 1 to 4 g of lipid in the feces in a 24-hour period. Even with a lipid-rich diet, the fecal fat does not normally exceed about 7 g in a 24-hour period. Normal fecal lipid is composed of about 60% fatty acids; 30% sterols, higher alcohols, and carotenoids; 10% triglycerides; and small amounts of cholesterol and phospholipids. Although significantly increased fecal fat can be caused by biliary obstruction, severe steatorrhea is almost always associated with exocrine pancreatic insufficiency or disease of the small intestines.

Qualitative Screening Test for Fecal Fat

Various screening tests have been devised for detecting steatorrhea. These tests have in common the use of fat-soluble stains (eg, Sudan III, Sudan IV, Oil Red 0, or Nile blue sulfate), which dissolve in and color lipid droplets. Of greater importance than the particular technical procedure is the level of experience and dependability of the clinical laboratorian performing the test.

Sudan Staining for Fecal Fat[3,4]

Neutral fats (triglycerides) and many other lipids stain yellow-orange to red with Sudan III because Sudan III is much more soluble in lipid than it is in water or ethanol. Free fatty acids do not stain appreciably unless the specimen is heated in the presence of the stain with 36% acetic acid. The slide may be examined warm or cool and the number of fat droplets assessed. As the slide cools, the fatty acids crystallize out in long, colorless, needlelike sheaves. Detection of meat fibers is accomplished by a third aliquot of fecal sample mixed on the slide with 10% alcohol and a solution of eosin stained for 3 minutes. The meat fibers should stain as rectangular cross-striated fibers. Splitting the sample and detecting neutral fats, fatty acids, and undigested meat fibers can provide diagnostic information. Increases in fats and undigested meat fibers are indicative of patients with steatorrhea of pancreatic origin. A representative fecal specimen is used for analysis.

Normal feces can have up to 40 or 50 small (1 to 5 mm) neutral lipid droplets per high-power microscope field. Steatorrhea is characterized by an increase in both the number and size of stainable droplets, often with some fat globules in the 50- to 100- mm range. Fatty acid assessment greater than 100 stained small droplets, along with the presence of meat fibers, is expected in patients with steatorrhea.

Quantitative Fecal Fat Analysis[3,4]

The definitive test for steatorrhea is the quantitative fecal fat determination, usually on a 72-hour stool collection, although the collection period may be increased to up to 5 days. There are two basic methods for the quantitation of fecal lipids. In the gravimetric method, fatty acid soaps (predominantly calcium and magnesium salts of fatty acids) are converted to free fatty acids, followed by extraction of most of the lipids into an organic solvent, which is then evaporated so that the lipid residue can be weighed. In titrimetric methods, lipids are saponified with hydroxide, and the fatty acid salts are converted to free fatty acids using acid. The free fatty acids, along with various unsaponified lipids, are then extracted with an organic solvent, and the fatty acids are titrated with hydroxide after evaporation of the solvent and redissolving of the residue in ethanol. The titration methods obviously measure only saponifiable fatty acids and, consequently, render results about 20% lower than those from gravimetric methods. A further objection to the titrimetric methods is that they use an assumed average molecular weight for fatty acids to convert moles of fatty acids to grams of lipid.

At one time, it was common to measure the amount of free fatty acids as a percentage of total lipids on the presumption that a high percentage of free fatty acids indicates adequate pancreatic lipase activity. This method is no longer considered reliable because of spurious results, particularly caused by lipase produced by intestinal bacteria.

It is essential that patients be placed on a lipid-rich diet for at least 2 days before instituting the fecal collection. The diet must contain at least 50 g, and preferably 100 g, of lipid each day. Fecal collections should extend for 3 or more successive days.

There are various ways to express fecal lipid excretion. Expressing lipid excretion as a percentage of wet or dry fecal weight is open to serious challenge because of the wide variations in both fecal water content and dry residue as a result of dietary intake. The most widely accepted approach is to report the grams of fecal fat excreted in a 24-hour period.

Gravimetric Method of Sobel[5] for Fecal Fat Determination (Modified)[6]

The entire fecal specimen is emulsified with water. An aliquot is acidified to convert all fatty acid soaps to free fatty acids, which are then extracted along with other soluble lipids into petroleum ether and ethanol. After evaporation of the organic solvents, the lipid residue is weighed. All feces for a 3-day period are collected in tared containers. The containers *must not* have a wax coating. The specimen must be kept refrigerated.

Total lipid does not change significantly during 5 days' storage of the specimen at refrigerator temperatures. Patients must not ingest castor oil, mineral oil, or other oily laxatives and must not use rectal suppositories containing oil or lipid for 2 days before the test and during the test.

The reference range for fecal lipids in adults is 1 to 7 g per 24 hours.

Sweat Electrolyte Determinations[6,7,8,9]

Measurement of the sodium and chloride concentration in sweat is the most useful test for the diagnosis of cystic fibrosis. Significantly elevated concentrations of both these ions occur in more than 99% of affected patients. The two- to five-fold increases of sweat sodium and chloride are diagnostic of cystic fibrosis in children. Even in adults, no other condition causes increases in sweat chloride and sodium above 80 mEq/L. Sweat potassium is also increased, but less significantly so, and is not generally relied on for diagnosis. Contrary to some assertions, sweat electrolyte determinations do not distinguish heterozygote carriers of cystic fibrosis from normal homozygotes.

Older methods for acquiring sweat specimens required skilled clinicians who performed the test frequently. Induction of sweat included applying plastic bags or wrapping the patient in blankets, which was fraught with serious risks of dehydration, electrolyte disturbances, and hyperpyrexia. In 1959, pilocarpine administration by iontophoresis was reported as an efficient method for sweat collection and stimulation.[10,11] Iontophoresis employs an electric current that causes pilocarpine to migrate into a limited skin area, usually the inside of the forearm, toward the negative electrode from a moistened pad on the positive electrode. A collection vessel is then applied to the skin. The sweat is then analyzed for chloride. For confirmation, the test should be repeated. Commercially available surface electrodes that analyze the sweat chloride are readily available. For details, the reader is referred to Chapter 24, Body Fluid Analysis.

It is widely accepted that sweat chloride concentrations in children greater than 60 mmol/L are diagnostic of cystic fibrosis.[12] It is true that sweat sodium and chloride concentrations in female patients undergo fluctuation with the menstrual cycle and reach a peak 5 to 10 days before the onset of menstruation but do not overlap with the ranges associated with cystic fibrosis.

Serum Enzymes[11]

Amylase is the serum enzyme most commonly relied on for detecting pancreatic disease. It is not, however, a function test. Amylase is particularly useful in the diagnosis of acute pancreatitis, in which significant increases in serum concentrations occur in about 75% of patients. Typically, amylase in serum increases within a few hours of the onset of the disease, reaches a peak in about 24 hours, and because of its clearance by the kidneys, returns to normal within 3 to 5 days, often making urine amylase a more sensitive indicator of acute pancreatitis. The magnitude of the enzyme elevation cannot be correlated with the severity of the disease.

Determination of the renal clearance of amylase is useful in detecting minor or intermittent increases in the serum concentration of this enzyme. To correct for diminished glomerular function, the most useful expression is the ratio of amylase clearance to creatinine clearance, as follows:

$$\frac{\% \text{ Amylase clearance}}{\text{Creatinine}} = 100 \times \frac{UA}{SA} \times \frac{SC}{UC} \qquad \textit{(Eq. 22–1)}$$

where: UA = urine amylase,
    SA = serum amylase,
    SC = serum creatinine, and
    UC = urine creatinine.

CASE STUDY 22-3

Parents brought their 7-year-old son to the pediatrician with the complaint of frequent fevers and failure to grow. The child had three bouts of pneumonia during the past 2 years and was bothered by chronic bronchitis, which caused him to cough up copious amounts of thick, yellow, mucoid sputum. Despite a big appetite, he had gained only 1 to 2 pounds in the past 2 years and was of short, frail stature. He especially liked salty foods. He usually had 3 or 4 bulky, foul-smelling bowel movements daily. A 9-year-old sister was in excellent health.

Questions

1. What is the most likely disease?
2. What clinical laboratory test would be most informative, and what results would be expected?
3. What other clinical laboratory tests would likely be abnormal?

Normal values are less than 3.1%. Significantly increased values, averaging about 8% or 9%, occur in acute pancreatitis.

The use of serum lipase in the clinical detection of pancreatic disease has been compromised in the past by technical problems inherent in the various analytic methods. Improved analytic methods appear to indicate that lipase increases in serum about as soon as amylase in acute pancreatitis and that increased levels persist somewhat longer than do those of amylase. Consequently, some physicians consider lipase more sensitive than amylase as an indicator of acute pancreatitis or other causes of pancreatic necrosis.

Both amylase and lipase may be significantly increased in serum in many other conditions (eg, opiate administration, pancreatic carcinoma, intestinal infarction, obstruction or perforation, and pancreatic trauma). Amylase levels are also frequently increased in mumps, cholecystitis, hepatitis, cirrhosis, ruptured ectopic pregnancy, and macroamylasemia. Electrophoresis of total amylase, if available, would reveal an increased p-type isoenzyme that, along with the lipase, remains elevated longer than the total amylase in acute pancreatitis. Lipase levels are often significantly increased in fractures of bones and in association with fat embolism.

Other Tests of Pancreatic Function

Case Study Table 22-2.1 summarizes several laboratory tests that might be helpful in diagnosis of pancreatic disorders. Other tests, which differentiate pancreatic from enteric malabsorption, must be performed and are discussed in detail in Chapter 23, Gastrointestinal Function. One such test is the D-xylose absorption test. D-Xylose is a pentose sugar that does not require pancreatic enzymes for absorption. In a patient with a suspected malabsorption syndrome, a normal D-xylose points toward pancreatic insufficiency.

The starch tolerance test was devised to differentiate pancreatogenous from intestinal malabsorption. In theory, a patient with inadequate delivery of pancreatic amylase into the small intestines would have a much lower increase in blood glucose levels after ingestion of a purified starch preparation than would a person with normal amounts of amylase. The results are compared with those of a standard glucose tolerance test. Unfortunately, this reference approach is confused by the fact that the glucose tolerance test is frequently abnormally flat in patients with intestinal malabsorption as well as in more than half of patients with pancreatic insufficiency. Other problems include the fact that gelation of the starch upon cooling interferes with digestion and absorption. Also, starch digestion is initiated by salivary secretion, which continues in the intestines of patients with gastric anacidity. Consequently, this test is rarely used.

Determining proteolytic enzyme activity in feces was formerly a common procedure for evaluating pancreatic exocrine function, particularly in the diagnosis of cystic fibrosis. One such procedure determines fecal proteolytic activity by its ability to digest gelatin on an x-ray film to produce clearing of the film. Unfortunately, the results are of limited reliability because many intestinal bacteria produce proteolytic enzymes, and bacteria also destroy pancreatic enzymes. More specific assays have been devised, but these have received limited acceptance.

Radiographic tests, including chest and abdominal x-rays, ultrasound, duodenography, computerized tomography, endoscopy, angiography, and pancreatic biopsy, are essential tools to proper diagnosis of pancreatic disorders.[13]

SUMMARY

The pancreas is a digestive gland and weighs about 70 to 105 g. It is composed of two morphologically and functionally different tissues: endocrine tissue and exocrine tissue. The endocrine component consists of the islets of Langerhans, which secrete at least four hormones into the blood: insulin, glucagon, gastrin, and somatostatin. The larger exocrine pancreatic component secretes digestive enzymes into ducts that ultimately empty into the duodenum. Pancreatic activity is under both nervous and endocrine control. Other than trauma, only three diseases cause more than 95% of the medical attention devoted to the pancreas: cystic fibrosis, pancreatic carcinoma, and pancreatitis. The role of the pancreas in diabetes mellitus is discussed in Chapter 10, Carbohydrates. Depending on the etiology and clinical picture, pancreatic function may be suspected when there is evidence of increased amylase and lipase. Tests of pancreatic function include those used for detection of malabsorption (eg, D-xylose, excess fat, meat fibers, and fecal fat), tests for measuring other exocrine function (eg, secretin, CCK, trypsin, and chymotrypsin), tests assessing changes associated with extrahepatic obstruction (eg, bilirubin), and endocrine-related tests that reflect changes in the endocrine cells of the pancreas (eg, gastrin, insulin, glucose, and cortisol).

REVIEW QUESTIONS

1. Laboratory findings in pancreatitis include all of the following, except:
   a. Increased amylase
   b. Increased lipase
   c. Increased triglycerides
   d. Increased cortisol
2. Which of the following tests is a direct determination of the exocrine secretory capacity of the pancreas?
   a. Amylase
   b. Quantitative fecal fat analysis
   c. Secretin/CCK test
   d. None of the above
3. Which of the following tests would be used to differentiate pancreatic insufficiency from malabsorption syndrome?
   a. Sweat electrolytes, Na and Cl
   b. Fecal fat analysis

c. D-Xylose absorption

d. None of the above

4. Cystic fibrosis is *least likely* to produce which of the following conditions?

a. Intestinal obstruction in the newborn

b. Pulmonary infections

c. Scarring of the pancreas

d. Hepatic cirrhosis

e. Intestinal malabsorption

5. The proper time period for the collection of a fecal fat specimen is:

a. 24 hours

b. 36 hours

c. 48 hours

d. 72 hours

e. 96 hours

6. The main pancreatic duct drains into the:

a. Duodenum

b. Stomach

c. Gallbladder

d. Colon

e. Liver

REFERENCES

1. Boyd EJ, Wormsley KG. Laboratory tests in the diagnosis of the chronic pancreatic diseases. Part 1. Secretagogues used in tests of pancreatic secretion. Int J Pancreatol 1987;2(3):137–148.

2. Chesner I, Lawson N. Tests of exocrine pancreatic function. Ann Clin Biochem 1994;31(Pt 4):305–314.

3. Simko V. Sudan stain and quantitative fecal fat. Gastroenterology 1990;98 (6):1722–1723.

4. Huang G, Khouri MR, Shiau YF. Sudan stain of fecal fat: New insight into an old test. Gastroenterology 1989;96(2 Pt 1):421–427.

5. Boyd EJ, Rinderknecht H, Wormsley KG. Laboratory tests in the diagnosis of the chronic pancreatic diseases. Part 3. Tests on pure pancreatic juice. Int J Pancreatol 1987;2(5–6):291–304.

6. Farrell PM, Gregg RG, Koscik R, et al. Newborn screening for cystic fibrosis in Wisconsin: Comparison of biochemical and molecular methods. Pediatrics 1997;99(6):819–824.

7. James TJ, Taylor RP. Enzymatic measurement of sweat sodium and chloride. Ann Clin Biochem 1997;34(Pt 2):211.

8. Stern RC. The diagnosis of cystic fibrosis. N Engl J Med 1997;336 (7):487–491.

9. Burnett RW, LeGrys VA. Current status of sweat testing in North America: Results of the College of American Pathologists Needs Assessment Survey. Arch Pathol Lab Med 1994;118(9):865–867.

10. Boyd EJ, Wormsley KG. Laboratory tests in the diagnosis of the chronic pancreatic diseases. Part 2. Tests of pancreatic secretion. Int J Pancreatol 1987;2(4):211–221.

11. Boyd EJ, Rinderknecht H, Wormsley KG. Laboratory tests in the diagnosis of the chronic pancreatic diseases. Part 4. Tests involving the measurement of pancreatic enzymes in body fluid. Int J Pancreatol 1988;3(1):1–16.

12. Boyd EJ, Wormsley KG. Laboratory tests in the diagnosis of the chronic pancreatic diseases. Part 5. Stool enzyme measurements. Int J Pancreatol 1988;3(2–3):101–103.

13. Boyd EJ, Rinderknecht H, Wormsley KG. Laboratory tests in the diagnosis of the chronic pancreatic diseases. Part 6. Differentiation between chronic pancreatitis and pancreatic cancer. Int J Pancreatol 1988;3(4):229–240.

SUGGESTED READINGS

Boat TF, Welsh MJ, Beaudet AL. Cystic fibrosis. In: Scriver CR, Beaudet AL, Sly WS, Valle D, eds. The metabolic basis of inherited disease. 6th ed. Vol 2. New York: McGraw-Hill, 1989;2649.

Meyer JH. Pancreatic physiology. In: Sleisinger MH, Fordtran JS, eds. Gastrointestinal disease. 4th ed. Vol 2. Philadelphia: WB Saunders, 1989;1777.

Soergel KH. Acute pancreatitis. In: Sleisinger MH, Fordtran JS, eds. Gastrointestinal disease. 4th ed. Vol 2. Philadelphia: WB Saunders, 1989;1814.

CHAPTER 23

Gastrointestinal Function

Edward P. Fody

Objectives

Upon completion of this chapter, the clinical laboratorian should be able to:

- *Describe the physiology and biochemistry of gastric secretion.*
- *List the tests used to assess gastric and intestinal function.*
- *Explain the clinical aspects of gastric analysis.*
- *Evaluate a patient's condition given clinical data.*

KEY TERMS

D-Xylose absorption test	Lactose tolerance test	Zollinger-Ellison syndrome
Intrinsic factor	Pepsin	
Gastrin	Secretagogues	

The gastrointestinal (GI) system is composed of the mouth, esophagus, stomach, small intestine, and large intestine. Digestion, which is primarily a function of the small intestines, is the process by which starches, proteins, lipids, nucleic acids, and other complex molecules are degraded to simple constituents (molecules) for absorption and use in the body. This chapter discusses the physiology and biochemistry of gastric secretion, intestinal physiology, pathologic aspects of intestinal function, and tests of both gastric and intestinal function.

PHYSIOLOGY AND BIOCHEMISTRY OF GASTRIC SECRETION[1]

Gastric secretion occurs in response to various stimuli:

- Neurogenic impulses from the brain transmitted by means of the vagal nerves (*eg,* responses to the sight, smell, or anticipation of food)
- Distention of the stomach with food or fluid

- Contact of protein breakdown products, termed *secretagogues,* with the gastric mucosa
- The hormone *gastrin,* which is the most potent stimulus to gastric secretion and is itself secreted by specialized G cells in the gastric mucosa and the duodenum in response to vagal stimulation and contact with secretagogues

Inhibitory influences include high gastric acidity, which decreases the release of gastrin by the gastric G cells. Gastric inhibitory polypeptide is secreted by K cells in the middle and distal duodenum and proximal jejunum in response to food products such as fats, glucose, and amino acids. Vasoactive intestinal polypeptide, produced by H cells in the intestinal mucosa, directly inhibits gastric secretion, gastrin release, and gastric motility.

Gastric fluid has a high content of hydrochloric acid, pepsin, and mucus. Hydrochloric acid is secreted against a hydrogen ion gradient as great as 1 million times the concentration in plasma (*ie,* gastric fluid can reach a pH of 1.2 to 1.3 under conditions of augmented or maximal stimulation). *Pepsin* refers to a group of relatively weak proteolytic enzymes with pH optima from about 1.6 to 3.6 that catalyze all native proteins except mucus. The most important component of gastric secretion in terms of body physiology is *intrinsic factor,* which greatly facilitates the absorption of vitamin B_{12} in the ileum.

CLINICAL ASPECTS OF GASTRIC ANALYSIS[2,3,4,5]

Gastric analysis is used in clinical medicine mainly for the following purposes:

- To detect anacidity, which occurs in some cases of advanced carcinoma of the stomach and in all cases of pernicious anemia in adults. In pernicious anemia, the pH of gastric fluid does not fall below 6, even with maximal stimulation.

- To detect hypersecretion characteristic of the *Zollinger-Ellison syndrome*. This syndrome involves a gastrin-secreting neoplasm, usually located in the pancreatic islets, and exceptionally high plasma gastrin concentrations. Basal 1-hour acid secretion usually exceeds 10 mEq, and the ratio of basal 1-hour to maximal secretion usually exceeds 60% (*ie*, the stomach is not really in the basal state but rather is pathologically stimulated by the high plasma gastrin level).
- To determine how much acid is secreted as an aid in determining the type of surgical procedure required for ulcer treatment

Various substances have been used to stimulate gastric secretion (*eg,* caffeine, alcohol, and test meals), but these are submaximal stimuli and thus obsolete. From 1953 until the late 1970s, histamine acid phosphate was used as a maximal stimulus to gastric secretion. Because of side effects, some of them severe, histamine has now been replaced by pentagastrin, which is a synthetic pentapeptide composed of the four C-terminal amino acids of gastrin linked to a substituted alanine derivative.

Normal gastric fluid is translucent, pale gray, and slightly viscous and often has a faintly acrid odor. Residual volume should not exceed 75 mL. Residual specimens occasionally contain flecks of blood or are green, brown, or yellow from reflux of bile during the intubation procedure. The presence of food particles is abnormal and indicates obstruction.

TESTS OF GASTRIC FUNCTION

Measurement of Gastric Acid in Basal and Maximal Secretory Tests[2,3,4,5]

Gastric analysis is usually performed, after an overnight fast, as a 1-hour basal test followed by a 1-hour stimulated test subsequent to pentagastrin administration (6 μg/kg subcutaneously). Test results reveal wide overlap among healthy subjects and diseased patients, except for anacidity (*eg,* in pernicious anemia) and the extreme hypersecretion found in Zollinger-Ellison syndrome. Gastric peptic ulcer is usually associated with normal secretory volume and acid output. Duodenal peptic ulcer is usually associated with increased secretory volume in both the basal and maximal secretory tests, but considerable overlap occurs, nevertheless, with the normal range.

Measuring Gastric Acid[2]

In stimulated-secretion specimens, the ability of the stomach to secrete against a hydrogen ion gradient is determined by measuring the pH. The total acid output in a timed interval is determined from the titratable acidities and volumes of the component specimens. After intubation, the residual secretion is aspirated and retained. Secretion for the subsequent 10 to 30 minutes is discarded to allow for adjustment of the patient to the intubation procedure. Specimens are ordinarily obtained as 15-minute collections for a period of 1 h.

The gastrin response to intravenous secretin stimulation may be used to investigate patients with mildly elevated serum gastrin levels. In this test, pure porcine secretin is injected intravenously, and gastrin levels are collected at 5-minute intervals for the next 30 minutes. In patients with Zollinger-Ellison syndrome, the gastrin level increases at least 100 pg/mL over the basal level. Patients with ordinary peptic ulceration, achlorhydria, or other conditions show a slight decrease in gastrin concentration.[8]

The volume, pH, and titratable acidity and the calculated acid output of each specimen are reported, as is the total volume and acid output for each test period (sum of the component specimens). There is considerable variation in gastric acid output among healthy subjects in both the basal and maximal secretory tests. Nevertheless, in the basal test, most healthy subjects secrete 0 to 6 mEq of acid in a total volume of 10 to 100 mL. In the maximal 1-hour test, using either histamine or pentagastrin as the stimulus, most men secrete 1 to 40 mEq of acid in a total volume of 40 to 350 mL. Women and elderly people usually secrete somewhat less acid than young men.

Plasma Gastrin[6,7,8,9]

Measurement of plasma gastrin levels is invaluable in diagnosing Zollinger-Ellison syndrome, in which fasting levels typically exceed 1000 pg/mL and can reach 400,000 pg/mL, compared with the normal range of 50 to 150 pg/mL. Gastrin is usually not increased in simple peptic ulcer disease. Increased plasma gastrin levels do occur in most pernicious anemia patients but decrease toward normal when hydrochloric acid is artificially instilled into the stomach.

INTESTINAL PHYSIOLOGY

Digestion, predominantly a function of the small intestines, is the process whereby starches, proteins, lipids, nucleic acids, and other complex molecules are degraded to monosaccharides, amino acids and oligopeptides, fatty acids, purines, pyrimidines, and other simple constituents. For most large molecules, digestion is necessary for absorption to occur. Each day, the duodenum receives about 7 to 10 L of ingested water and food and secretion from the salivary glands, stomach, pancreas, and biliary tract. The materials then enter the jejunum and ileum, where another 1 to 1.5 L of secretion is added. Ultimately, however, only about 1.5 L of fluid material reaches the *cecum,* which is the first portion of the colon or large intestines. This considerable absorptive capability is possible because the 20 feet or so of small intestines have numerous mucosal folds, minute projections from the luminal surface called *villi,* and microscopic projections on the mucosal cells called *microvilli,* all of which greatly increase the secretory and absorptive surface to an estimated 200 m². Absorption takes place by passive diffusion for some substances and by active transport for others. In addition, the small intestines actively secrete electrolytes and other metabolic products. The 5-feet-long large intestine has two major functions: water resorption, whereby the 1.5 L of fluid re-

A 34-year-old man was admitted for diagnostic evaluation with the complaint of epigastric pain, which was variously described as gnawing or burning, of 2 years' duration. He had been diagnosed as having a duodenal peptic ulcer 18 months ago, and at that time, therapy of antacids and dietary revision provided considerable alleviation of symptoms. More recently, however, the pain had become more persistent and awakened the patient four to six times each night. Radiologic studies have revealed a 2.5-cm ulcer crater in the first portion of the duodenum and a 0.5-cm ulcer in the antrum of the stomach. Serum electrolytes were normal. Hemoglobin was 8.3 g/dL with normal red blood cell indices. White blood

count was 13,100. Gastric analysis revealed 640 mL of secretion in the basal hour, with an acid output of 38 mEq, and 780 mL of secretion in the 1-hour pentagastrin stimulation test, with an acid output of 48 mEq.

Questions

1. What is the probable disease?
2. In view of the existing data, what other test would be virtually diagnostic for this disease?
3. What is the explanation for the decreased hemoglobin and the increased white blood cell count?

ceived by the cecum is reduced to about 100 to 300 mL of feces, and storage of feces before defecation. The abdominal structures that constitute the alimentary tract are shown diagrammatically in Figure 23-1.

CLINICOPATHOLOGIC ASPECTS OF INTESTINAL FUNCTION

Clinical chemistry testing of intestinal function focuses almost entirely on the evaluation of absorption and its de-

rangements in various disease states. As discussed in Chapter 22, Pancreatic Function, diseases of the exocrine pancreas and biliary tract also may cause malabsorption. Intestinal diseases that may cause the malabsorption syndrome are highly varied in terms of their etiology, pathogenesis, and severity. These intestinal diseases and disorders include tropical and nontropical or celiac sprue, Whipple's disease, Crohn's disease, primary intestinal lymphoma, small intestinal resection, intestinal lymphangiectasia, ischemia, amyloidosis, giardiasis, and many others. In addition to the malabsorption syndrome, which ordinarily causes impaired absorption of fats, proteins, carbohydrates, and other substances, specific malabsorption states also occur (*eg,* acquired deficiency of lactase, which prevents normal absorption of lactose, and the genetic disorder Hartnup syndrome, which involves deficient intestinal transport of phenylalanine and leucine).

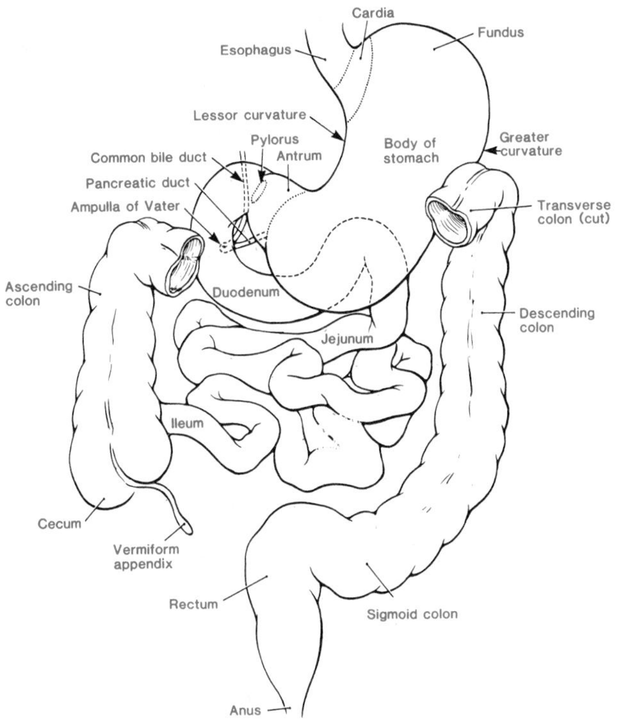

TESTS OF INTESTINAL FUNCTION

Lactose Tolerance Test[10,11]

The disaccharidases, lactase (which cleaves lactose into glucose and galactose) and sucrase (which cleaves sucrose into glucose and fructose), are produced by the mucosal cells of the small intestine. Congenital deficiencies of these enzymes are rare, but acquired deficiencies of lactase are commonly found in adults. Affected patients experience abdominal discomfort, cramps, and diarrhea after ingesting milk or milk products. About 10% to 20% of American Caucasians and 75% of African-Americans are affected. Although definitive diagnosis is established by an assay for lactase content in the intestinal mucosa, the *lactose tolerance test* is of some value in establishing the diagnosis. In one protocol, individuals drink 50 g of lactose in 200 mL of water. Blood specimens are obtained for glucose analysis before ingestion and 30, 60, and 120 minutes after ingestion. Increase in the level of blood glucose of 30 mg/dL is considered normal, a 20- to 30-mg/dL

Figure 23-1. The abdominal structures of the alimentary tract.

increase is borderline, and an increase of less than 20 mg/dL indicates lactase deficiency. People with normal lactase activity, however, may still test abnormally. This test has largely been replaced by the hydrogen breath test.

D-Xylose Absorption Test[9,12]

D-Xylose is a pentose sugar that is ordinarily not present in the blood in any significant amount. As with other monosaccharides, pentose sugars are absorbed unaltered in the proximal small intestine and thus do not require the intervention of pancreatic lytic enzymes. Therefore, the ability to absorb D-xylose is of value in differentiating malabsorption of intestinal etiology from that of exocrine pancreatic insufficiency. Because only about half the orally administered D-xylose is metabolized or lost by action of intestinal bacteria, significant amounts are excreted unchanged in the urine. Some protocols have used the measurement of only the D-xylose excreted in the urine during the 5 hours following ingestion of a 25-g dose by a fasting adult (0.5 g/kg in a child). Even with normal renal function, false-positive and false-negative results occur frequently. Blood levels measured one or more times after ingestion of the D-xylose (eg, at 30 minutes, 1 hour, 2 hours) significantly improve the diagnostic reliability of the test. Some protocols use smaller doses of D-xylose to avoid the abdominal cramps, intestinal hypermotility, and osmotic diarrhea that frequently accompany the 25-g dose.

D-Xylose Test

After ingestion of a specified solution of D-xylose, blood specimens are obtained and urine is collected for a 5-hour period to determine the extent of absorption of D-xylose. The concentration of D-xylose is determined by heating protein-free supernates of urine and plasma to convert xylose to furfural, which is then reacted with p-bromoaniline to form a pink product, the absorbance of which is measured at 520 nm. Thiourea is added as an antioxidant to prevent the formation of interfering chromogens. After an overnight fast, the patient voids and drinks a D-xylose solution: 25 g of D-xylose in 250 mL of water for adults or 0.5 g/kg for children or other dose as established. The patient drinks an equivalent amount of water during the next hour. No additional food or fluids are to be taken until the test is completed. Urine is collected for 5 hours after the D-xylose ingestion. A blood specimen is collected in potassium oxalate at 2 hours (commonly, 1 hour is chosen for children).

Normal blood concentrations of D-xylose in association with decreased urine excretion suggest impairment of renal function or incomplete urine collection. Aspirin therapy diminishes renal excretion of D-xylose, whereas indomethacin decreases intestinal absorption. After ingestion of a 25-g dose of D-xylose, healthy adults should excrete at least 4 g in the 5-hour period. For infants and children, the excretion following a dose of 0.5 g/kg for various ages expressed as per-

centages of ingested dose are shown in Table 23-1. Blood levels for healthy adults vary over a wide range, but a blood concentration of less than 25 mg/dL at 2 hours should be considered abnormal after the 25-g dose. With the 0.5 g/kg dose, infants under 6 months of age should have a blood concentration of at least 15 mg/dL at 1 hour, infants over 6 months of age and children should achieve levels of at least 30 mg/dL.[5,11]

Serum Carotenoids

Carotenoids are various yellow to orange or purple pigments that are widely distributed in animal tissues, are synthesized by many plants, and impart a yellow color to some vegetables and fruits. The major carotenoids in human serum are lycopene, xanthophyll, and beta-carotene, the chief precursor of vitamin A in humans. Being fat soluble, carotenoids are absorbed in the small intestine in association with lipids. Malabsorption of lipids typically results in a serum concentration of carotenoids lower than the reference range of 50 to 250 mg/dL. Starvation, dietary idiosyncrasies, and fever also cause diminished serum concentrations. The test does not distinguish among the various etiologies of malabsorption.

Fecal Fat Analysis

As discussed in Chapter 22, Pancreatic Function, increased fecal fat loss, or steatorrhea, is an integral manifestation of the general malabsorption syndrome. However, neither the presence of steatorrhea nor the documentation of its severity is of benefit in distinguishing among the various etiologies of malabsorption, with the exception that severe steatorrhea is rarely caused by biliary obstruction.

Other Tests of Intestinal Malabsorption

Deficiencies of numerous analytes can occur in association with intestinal malabsorption. Measurement of these analytes is usually of value, not so much in confirming the diagnosis of malabsorption as in determining the extent of nutritional deficiency and thus the need for replacement therapy. Diminished appetite and dietary intake are usually more severe in patients who have malabsorption with an intestinal etiology. Body wasting or cachexia may be severe. Frequently, loss of albumin into the intestinal lumen and diminished di-

TABLE 23-1. D-Xylose Results for Pediatric Patients[2,4]

Under 6 months	11 to 33%
6 to 12 months	20 to 32%
1 to 3 years	20 to 42%
3 to 10 years	25 to 45%
Over 10 years	25 to 50%

CASE STUDY 23-2

A 26-year-old woman appeared in the outpatient clinic with the complaint of abdominal discomfort, diarrhea, and an 18-pound unintentional weight loss during the past 2 to 3 years. She related a similar period of 5 or 6 years of abdominal distress and diarrhea in childhood, but this essentially disappeared when she was about 12 to 13 years old. She was now having three to five bowel movements daily, which were described as bulky, malodorous, and floating. She weighed 106 pounds and was 67 inches tall. She had never had any surgical procedures. Physical examination revealed poor skin turgor, general pallor, and a protuberant abdomen. Abnormal clinical laboratory values included those in Case Study Table 23-2.1.

Fecal examination revealed no ova or parasites, and bacteriologic culture revealed no pathogens.

CASE STUDY TABLE 23-2.1. Laboratory Results

Hemoglobin	8.1 g/dL
Hematocrit	30%
RBC count	$4.1 \times 10^6/\mu L$
Serum sodium	134 mEq/L
Potassium	3.4 mEq/L
Serum carotenoids	14 µg/dL
Fecal fat	22 g/24 h
D-Xylose absorption test (25-g dose)	5-h excretion of 1.3 g and blood level at 2 h of 8 mg/dL
Prothrombin time	15.8 s (12–14 s)
Activated partial thromboplastin time	56 s (30–45 s)

Questions

1. What is the disease process?
2. What is the probable etiology in this case?
3. What is the cause of the abnormal coagulation tests?
4. What is the probable major cause for the anemia, and what are other possible contributing causes?

etary intake of protein accompany the diminished absorption of oligopeptides and amino acids, so that a negative nitrogen balance occurs along with decreased serum total proteins and albumin. A serum albumin of less than 2.5 g/dL is much more characteristic of intestinal disease than of pancreatic disease. In association with severe disease of the small intestines, deficiencies of the fat-soluble vitamins A, D, E, and K occur. Vitamin K deficiency, in turn, causes deficiencies of the vitamin K–dependent coagulation factors II (prothrombin), VII (proconvertin), IX (plasma thromboplastin component), and X (Stuart-Prower factor), which are reflected in abnormal prothrombin time and partial thromboplastin time tests.

In severe small intestinal disease, such as tropical or celiac sprue, malabsorption of folate and vitamin B_{12} can occur, and megaloblastic anemia is rather common and is of some benefit in distinguishing intestinal from pancreatic disease. Absorption of iron is usually diminished, and the tendency toward low serum iron levels may be aggravated by intestinal blood loss. Intestinal absorption of calcium is often diminished as a consequence of calcium binding by unabsorbed fatty acids and accompanying vitamin D deficiency and decreased serum magnesium. Because sodium, potassium, and water absorption and metabolism may also be seriously deranged, serum sodium and potassium levels are decreased, and dehydration occurs. Impaired absorption of carbohydrates in intestinal diseases, such as sprue, results in decreased to flat blood concentration curves in glucose, lactose, and sucrose tolerance tests.

SUMMARY

Gastric secretion occurs in response to various stimuli. Gastric analysis may be performed to detect anacidity, detect the hypersecretion characteristic of Zollinger-Ellison syndrome, and assess the amount of acid being secreted for determination of the surgical procedure to be used in ulcer treatment. Measurement of plasma gastrin levels is also used in the diagnosis of Zollinger-Ellison syndrome. Digestion is predominantly a function of the small intestines. Clinical chemistry testing of intestinal function focuses almost entirely on the evaluation of absorption and its derangements in various disease states. Intestinal diseases and disorders include Whipple's disease, Crohn's disease, sprue, amyloidosis, giardiasis, and many others. Several malabsorption syndromes or states also exist (eg, acquired deficiency of lactase and Hartnup syndrome). Tests of intestinal function include the lactose tolerance test, D-xylose test, serum carotenoids, and fecal fat analysis.

REVIEW QUESTIONS

1. Which of the following is a test only of the absorptive ability of the intestine?
   a. Lactose tolerance test
   b. D-Xylose test
   c. Fecal fat (72-hour collection)
   d. Serum carotenoids
   e. Serum albumin

2. The smallest amount of D-xylose is normally absorbed by patients
   a. Under 1 year of age
   b. Between 1 and 3 years of age
   c. Between 3 and 5 years of age
   d. Between 5 and 10 years of age
   e. Who are adults

3. In a gastric acidity test, a patient produces 50 mL of gastric fluid. A 5-mL aliquot of this fluid requires 10 mL of 0.1 N NaOH to titrate to pH 7. Calculate the acid output in milliequivalents (mEq).
   a. 2
   b. 4
   c. 6
   d. 8
   e. 10

4. The results in the previous question could be found in all of the following conditions, *except:*
   a. Zollinger-Ellison syndrome
   b. Peptic ulcer
   c. Previous pentagastrin stimulation
   d. Pernicious anemia
   e. Healthy person

5. In which of the following would gastrin levels be increased:
   a. Peptic ulceration
   b. Zollinger-Ellison syndrome
   c. Achlorhydria
   d. Amyloidosis

6. Normal blood concentrations of D-xylose with decreased urine D-xylose excretion, suggests which of the following:
   a. Incomplete urine collection
   b. Impaired renal function
   c. Lactose intolerance
   d. A and B

7. A serum albumin of less than 2.5 g/dL would be most indicative of which of the following:
   a. Intestinal disease
   b. Pancreatitis
   c. Peptic ulcer
   d. Pancreatic carcinoma

8. Regarding gastric acid basal and maximal secretory tests, peptic ulcer is usually associated with:
   a. Increased secretory volume for both the basal and maximal tests
   b. Normal secretory volume and acid output
   c. Decreased secretory volume for both the basal and maximal tests
   d. None of the above

REFERENCES

1. Schubert ML, Shamburek RD. Control of acid secretion. Gastroenterol Clin North Am 1990;19(1):1–25.
2. Ausman RK, Caballero GA, Lin L, et al. Gastric secretion pH measurement: What you see is not what you get! Crit Care Med 1990;18(4):396–399.
3. Reines HD. Gastric pH measurements. Crit Care Med 1990;18(9):1047.
4. Johnston DA, Wormsley KG. Problems with the interpretation of gastric pH measurement. Clin Investig 1993;72(1):12–17.
5. Castell DO. Accurate measurement of gastric acidity. Am J Gastroenterol 1997;92(1):181.
6. McQuaid KR. Much ado about gastrin. J Clin Gastroenterol 1991;13(3):249–254.
7. Elliott DW, Stremple JF. Role of gastrin determinations in clinical surgery. Am Surg 1976;42(2):102–107.
8. Cadiot G, Mignon M. Diagnostic and therapeutic criteria in patients with Zollinger-Ellison syndrome and multiple endocrine neoplasia type 1. J Intern Med 1998;243(6):489–494.
9. Kuno C, Watanabe J, Yuasa H. Comparative assessment of D-xylose absorption between small intestine and large intestine. J Pharm Pharmacol 1997;49(1):26–29.
10. Korpela R, Peuhkuri K, Poussa T. Comparison of a portable breath hydrogen analyser (Micro H2) with a quintron microlyzer in measuring lactose maldigestion, and the evaluation of a Micro H2 for diagnosing hypolactasia. Scand J Clin Lab Invest 1998;58(3):217–224.
11. Buttery JE, Chamberlain BR, Ratnaike RN. A visual screening method for lactose maldigestion. Ann Clin Biochem 1994;31(Pt 6):566–567.
12. Hamanaka Y, Oka M, Suzuki T, et al. Oral absorption tests: Absorption site of each substrate. Nutrition 1998;14(1):7–10.

SUGGESTED READINGS

Baron JH. Clinical tests of gastric secretion: history, methodology and interpretation. New York: Oxford University Press, 1979.
Cannon DC, Freeman JA. Gastric analysis. In: Freeman JA, Beeler MF, eds. Laboratory medicine/urinalysis and medical microscopy. 2nd ed. Philadelphia: Lea & Febiger, 1983.
Kalser MH. Clinical manifestations and evaluation of malabsorption. In: Berk JF, ed. Bockus gastroenterology. 4th ed. Vol 3. Philadelphia: WB Saunders, 1985;1667.

CHAPTER 24

Body Fluid Analysis

Frank A. Sedor

Objectives

Upon completion of this chapter, the clinical laboratorian should be able to:

- *Identify the source of the following: amniotic fluid, cerebrospinal fluid, sweat, synovial fluid, pleural fluid, pericardial fluid, and peritoneal fluid.*

- *Describe the physiologic purpose of the following: amniotic fluid, cerebrospinal fluid, sweat, synovial fluid, pleural fluid, pericardial fluid, and peritoneal fluid.*

- *Discuss the clinical utility of testing each of the following: amniotic fluid, cerebrospinal fluid, sweat, synovial fluid, pleural fluid, pericardial fluid, and peritoneal fluid.*

- *Given the results of a foam stability index, L/S ratio, and sweat test, interpret the patient's status.*

- *Differentiate between a transudate and an exudate.*

KEY TERMS

Amniocentesis	Otorrhea	Serous fluid
Amniotic fluid	Pericardial fluid	Synovial fluid
Ascites	Peritoneal fluid	Thoracentesis
Effusion	Pleural fluid	Transudate
Exudate	Respiratory distress	
Hypoglycorrhachia	syndrome	
L/S ratio	Rhinorrhea	

This chapter attempts to acquaint the reader with several fluids that are often analyzed in the clinical chemistry laboratory. In general, the source, physiologic purpose, and clinical utility of laboratory measurements for each of these body fluids is emphasized.

AMNIOTIC FLUID

The amniotic sac provides an enclosed environment for fetal development. This sac is bilayered as the result of a fusion of the amnionic (inner) and chorionic (outer) membranes at an early stage of fetal development. The fetus is suspended in *amniotic fluid* (AF) within the sac. This AF provides a cushioning medium for the fetus and serves as a matrix for influx and efflux of constituents.

Obviously, the mother must be the ultimate physiologic source for AF. Depending on the interval of the gestational period, the fluid may be derived from different sources. At initiation of pregnancy, some maternal secretion across the amnion contributes to the volume. Shortly after formation of the placenta, embryo, and fusion of membranes, AF is derived in large part by transudation across the fetal skin. In the last half of pregnancy, the skin becomes substantially less permeable, and fetal micturition, or urination, becomes the major volume source. The fate of the fluid also varies with period of gestation. A bidirectional exchange is presumed to occur across the membranes and at the placenta. Similarly, during early pregnancy, the fetal skin is involved. In the last half of pregnancy, the mechanism of fetal swallowing is the major fate of AF. There is a dynamic balance established between production and clearance; fetal urination and swallowing maintain this balance. The continual swallowing maintains intimate contact of the AF with the fetal gastrointestinal tract, buccal cavity, and bronchotracheal tree. This contact is evidenced by the sloughed material from the fetus that provides us with the "window" to fetal developmental and functional stages.

Cells found in the fluid originate with the fetus, and the chemical content reflects the continual swallowing and clearance of fluid. A sample of fluid is obtained by transabdominal *amniocentesis* (amniotic sac puncture), which is performed under aseptic conditions. Before an attempt is made to obtain fluid, the positions of the placenta, fetus, and fluid pockets are visualized using ultrasonography. Aspiration of anything except fluid could lead to erroneous conclusions as well as possible harm to the fetus.

Amniocentesis and subsequent AF analysis is performed to test for (1) congenital diseases, (2) neural tube defects, (3) hemolytic disease, (4) gestational age, and (5) fetal pulmonary development. The first, diagnosis of genetic abnormality, is accomplished by cell culture. Fluid obtained between 14 and 20 weeks of pregnancy is harvested for cells of fetal origin. The cells are cultured and collected for chromosomal analysis and are lysed, then enzyme contents are determined to evaluate for metabolic defects. This procedure has been largely supplanted in recent years by the use of chorionic villus sampling and cytogenetic analysis.

Screening for neural tube defects (NTDs) is initially performed using maternal serum. The presence of elevated alpha-fetoprotein (AFP) was originally thought to indicate NTDs such as spina bifida and anencephaly. Elevated maternal serum AFP can also be closely correlated with abdominal hernias into the umbilical cord, cystic hygroma, and poor pregnancy outcome. Low maternal serum AFP is associated with an increased incidence of Down's syndrome and other aneuploidies. The protocol for usage of AFP testing is generally considered to include (1) maternal serum AFP; (2) repeat, if positive; (3) diagnostic ultrasound; and (4) amniocentesis for confirmation. Interpretation of maternal serum AFP testing is complex, being a function of age, race, weight, gestational age, and level of nutrition.

Testing of amniotic fluid AFP (AFAFP) is the confirmatory procedure. AFP is a product of first the fetal yolk sac and then the fetal liver. It is released into the fetal circulation and presumably enters the AF by transudation. Entry into the maternal circulation could be by placenta cross-over or from the AF. If there were an open defect (eg, spina bifida) that caused an increase in AFAFP, there would be a concomitant increase in maternal serum AFP. Under normal conditions, AFAFP would be cleared by fetal swallowing and metabolism. An increased presence overloads this mechanism, causing the AFAFP elevation. Assay of the protein is normally done by immunologic means. The lack of treatment of a positive presence and the lack of 100% specificity increase the need for extreme care in analysis. Societal and clinical concerns mandate that the highest levels of quality control be practiced.

This very concern generated the need for a second test to affirm NTDs and abdominal wall defects. The method used is the assay for a central nervous system (CNS)-specific acetylcholinesterase (AChE). The NTD allows direct or, at least less difficult, passage of AChE into the AF. Analysis for CNS-specific AChE in the AF then offers a degree of confirmation for AFAFP. The methods used for CNS AChE include enzymatic, immunologic, and electrophoretic with inhibition. The latter includes the use of acetylthiocholine as substrate and BW284C51, a specific CNS inhibitor, to differentiate the serum pseudocholinesterase from the CNS-specific AChE.

Analysis of AF to screen for hemolytic disease of the newborn (erythroblastosis fetalis) was the first recognized laboratory procedure performed on AF. Hemolytic disease of the newborn is a syndrome of the fetus resulting from ABO incompatibility of the maternal and fetal blood. Maternal antibodies to fetal erythrocytes cause a hemolytic reaction that can vary in severity. The resultant hemoglobin breakdown

products, predominantly bilirubin, appear in the AF and provide a measure of the severity of the incompatibility reaction.

The most commonly employed method is a direct spectrophotometric scan of undiluted AF and subsequent calculation of the relative bilirubin amount. Classically, the absorbance due to bilirubin is reported instead of a concentration of bilirubin. The method consisted of scanning AF from 700 to 350 nm against a water blank. The resultant absorbances can be used differently to derive the necessary information. A common method, the method of Liley,[1] requires the plotting of the observations at 5-nm intervals against wavelength, using semilogarithmic paper. A baseline is constructed from 550 to 350 nm; the change at 450 nm is due to bilirubin.

Care must be used in the interpretation of the spectra. A decision for treatment can be made based on the degree of hemolysis and gestational age. The rather limited treatment options are immediate delivery, intrauterine transfusion, or observation. The transfusion can be accomplished by means of the umbilical artery and titrated to desired hematocrit. Several algorithms have been proposed to aid in decision making (Fig. 24-1). An example of an uncomplicated bilirubin scan is shown in Figure 24-2.

The most commonly encountered interferences are maternal urine (from bladder interdiction), fetal or maternal blood, and meconium (fetal fecal material). Although light does not interfere per se, safeguards against exposure to light, especially sunlight, must be maintained before analysis. Light may degrade the bilirubin present, thereby causing an underestimation of hemolytic disease severity.

Examples of some interferences compared with a normal specimen are given in Figure 24-3. Each laboratory should compile its own catalog of real examples for spectrophotometric analysis. Presence of blood is identified by Soret absorbances of hemoglobin at 410 to 415 nm; urine by the broad curve and confirmed by creatinine, urea, and protein analyses; and meconium by the distinctly greenish color and flat absorbance curve.

In management of pregnancy, it is important to know the gestational age. Laboratory results are often used in age evaluation. The elucidation of an analyte to reflect gestational age has resulted in the cataloging of as many constituents as are known in serum. Four parameters have been used with some degree of frequency: creatinine, urea, uric acid, and osmolality. Creatinine is thought to reflect fetal muscle mass; urea, protein; uric acid, nucleoproteins; and osmolality, a combination of all these. The basic problem in the use of these parameters is the very wide range of concentration for given gestational ages. The ranges are so wide that significant overlap occurs, leading to inability to make accurate predictions. In practice, creatinine is the parameter used. A fluid with a level of 2 g/dL is assumed mature. Note that this is not effective in AF volume aberrations (eg, oligohydramnios or small AF volume). Ultrasonography has become the better tool to estimate fetal size and gestational age.

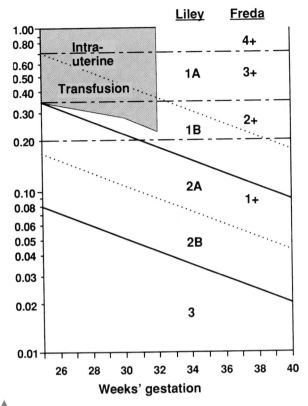

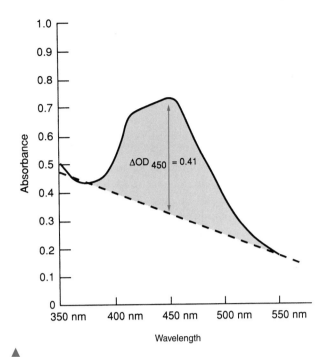

Figure 24-1. Assessment of fetal prognosis by the methods of Liley and Freda. Liley: **1A**, above broken line, condition desperate, immediate delivery or transfusion; **1B**, between broken and continuous lines, hemoglobin less than 8 g/100 mL, delivery or transfusion (stippled area) urgent; **2A**, between continuous and broken lines, hemoglobin 8–10 g/100 mL, delivery 36–37 weeks; **2B**, between broken and continuous lines, hemoglobin 11–13.9 g/100 mL, delivery 37–39 weeks; **3**, below continuous line, not anemic, delivery at term. Freda: **4+**, above upper horizontal line, fetal death imminent, immediate delivery or transfusion; **3+**, between upper and middle horizontal lines, fetus in jeopardy, death within 3 weeks, delivery or transfusion as soon as possible; **2+**, between middle and lower horizontal lines, fetal survival for at least 7–10 days, repeat test, possible indication for transfusion; **1+**, below lower horizontal line, fetus in no immediate danger of death. (Modified from Robertson JG. Am J Obstet Gynecol 1966;95:120.)

Figure 24-2. ΔOD_{450nm} from AF bilirubin scan.

Alveolar collapse in the neonatal lung may occur upon the changeover to air as an oxygen source at birth if the proper quantity and type of phospholipid (surfactant) is not present. The ensuing condition, which may vary in degree of severity, is called *respiratory distress syndrome*. It also has been referred to as *hyaline membrane disease* because of the

The primary reason for testing of AF is the need to assess fetal pulmonary maturity. All the organ systems are at jeopardy from prematurity, but the state of the fetal lungs is a priority from the clinical perspective. The availability of laboratory tests that give an indication of maturity has also fostered this emphasis. Consequently, the laboratory is asked whether sufficient specific phospholipids are reflected in the AF to prevent atelectasis (alveolar collapse) if the fetus were to be delivered. This question is important when preterm delivery is contemplated owing to other risk factors in pregnancy, such as preeclampsia or premature rupture of membranes. Risk factors to fetus or mother can be weighed against interventions, such as delay of delivery, or against at-risk postdelivery therapies, such as exogenous surfactant therapy, high-frequency ventilation, or extracorporeal membrane oxygenation.

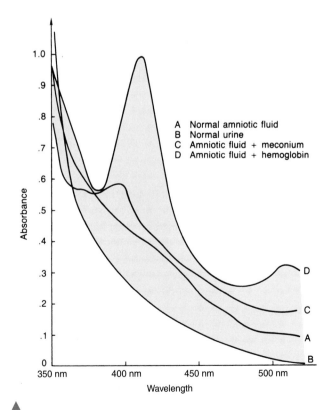

Figure 24-3. Amniotic fluid absorbance scans.

hyaline membrane found in affected lungs. Lung maturation is a function of the differentiation, beginning near the 24th week of pregnancy, of alveolar epithelial cells into type I and type II cells. The type I cells become equipped for gas exchange, and the type II cells become the producers of surfactant. As the lungs mature, increases occur in the concentration of the phospholipids, particularly the compounds phosphatidyl glycerol and lecithin, especially dipalmitoylphosphatidyl choline[2] (Fig. 24-4). These two compounds, present in 10% and 70%, respectively, of total phospholipid concentration, are most important as surfactants. Their presence in high enough levels acts in concert to allow contraction and reexpansion of the neonatal alveoli. To conceptualize their importance, remember the difficulty in blowing up a new toy balloon relative to a balloon that has been partially inflated. For the newborn, the normal amount of proper surfactant allows contraction of the alveoli without collapse. The next inspiration is the difference between a partially inflated versus flattened new balloon. Insufficient surfactant allows alveoli to collapse, thus requiring a great deal of energy to reexpand the alveoli upon inspiration. This not only creates an extreme energy demand on a newborn but also probably causes physical damage to the alveoli with each collapse. The damage may lead to "hyaline" deposition, or the newborn may not have the strength to continue inspiration at the energy cost. The end result of either can be fatal.

The approaches to assessing fetal pulmonary status may be divided into functional assays and biochemical assays. Functional assays provide a direct physical measure of AF in an attempt to assess surfactant ability to decrease surface tension. Examples of this group are surface tension, fluorescence polarization, and the foam stability index. These tests reflect the gross concentration of surfactants rather than specific phospholipid levels. Belonging to the latter group are assays that quantitate dipalmitoyl phosphatidyl choline (the major lecithin), phosphatidyl glycerols, all or most of the phospholipids, and the classical ratio of lecithins to sphingomyelins (*L/S ratio*). Each group of tests has its advocacy and extensive citations in the literature. All attempt to indicate the major changes in phospholipid concentrations that occur at about 36 weeks' gestation and indicate fetal pulmonary maturation (see Fig. 24-4). With all tests, centrifugation of the AF to remove debris is necessary.

Excessive force (anything greater than that needed to remove debris) can change the lipid profile by causing the lipids present to fractionate as a result of centrifugal force. The difference in ratios observed at 1000 versus 3000 g can radically alter clinical interpretation. Before adoption of any method for AF analysis, a protocol for centrifugal separation to include relative force (*not* revolutions per minute) and duration of centrifugation must be adopted and rigorously followed.

The foam stability index (FSI),[3] a variant of Clements' original "bubble test,"[4] appears acceptable as a rapid, inexpensive, informative assay. This qualitative, technique-dependent test requires only common equipment. The assay is based on the ability of surfactant to generate a surface tension lower than that of a 0.47-mole fraction ethanol-water solution. If sufficient surfactant is present, a stable ring of foam bubbles remains at the air–liquid interface. As surfactant increases (fetal lung maturity probability increases), a larger mole fraction of ethanol is required to overcome the surfactant-controlled surface tension. The highest mole fraction used while still maintaining a stable ring of bubbles at the air–liquid interface is reported as the FSI (Table 24-1). The test is dependent on technique and can also be skewed by contamination of any kind in the AF (*eg,* blood or meconium contamination). Interpretation of the FSI bubble patterns is difficult and technique dependent. Results can vary from one clinical laboratorian to another. Most laboratories have found an FSI of 0.47 or 0.48 to represent a borderline

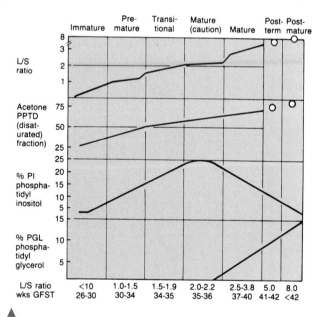

▲
Figure 24-4. The form used to report the lung profile. The four determinations are plotted on the ordinate and the weeks of gestation on the abscissa (as well as the L/S ratio as an "internal standard"). When these are plotted, they fall with a very high frequency into a given grid that then identifies the stage of development of the lung as shown in the upper part of the form. The designation "mature (caution)" refers to the patients other than those with diabetes who can be delivered *if necessary* at this time; if the patient has diabetes, she can be delivered with safety when the values fall in the "mature" grid. (Kulovich MV, Hallman MB, Gluck L. The lung profile I. Normal pregnancy. Am J Obstet Gynecol 1979;135:57. Copyright © 1977 by the Regents of the University of California.)

TABLE 24-1. FSI Determination

	TUBE 1	TUBE 2	TUBE 3
Vol AF	0.50	0.50	0.50
Vol 95% EtOH	0.51	0.53	0.55
FSI	0.47	0.48	0.49

maturity status. It is imperative that laboratory-specific reference intervals be determined. Values greater than borderline indicate increasing probability of maturity. Examples of bubble patterns are shown in Figure 24-5.

The quantitative tests were given emphasis primarily by the work of Gluck.[5] The phospholipids of importance are phosphatidyl glycerol (PG), phosphatidyl choline (lecithin, PC), and sphingomyelin (SP). The relative amounts of PG and PC increase dramatically with pulmonary maturity, whereas the SP concentration is relatively constant. Increases in PG and PC correspond to larger amounts of surfactant being produced by the alveolar type II cells as the fetal lungs mature.

The classic technique for separation and evaluation of the lipids involves thin-layer chromatography (TLC) of an extract of the AF. The extraction procedure removes most interfering substances and results in a concentrated lipid solution. Current practices use either one- or two-dimensional TLC for identification. Laboratories needs determine if a one-dimensional or a two-dimensional method is performed. An example of the phospholipids separation by one-dimensional TLC is shown in Figure 24-6.[6]

The classic breakpoint for judgment of maturity has been an L/S ratio of 2. The presence of PG is an additional marker. Initially, the presence of PG was presumed to indicate maturity, but later reports have modified this. An initial PG level of at least 3% is usually needed. This may also be method dependent. There are always exceptions and outliers, however. It is important to remember that the classic L/S ratio measures total lecithins versus sphingomyelin. Some methods include selective oxidation and removal of the nonsaturated phospholipids, so that a direct measurement of dipalmitoylphosphatidyl choline can be measured. It is difficult to compare these assays empirically.

There are still conflicting reports about decision levels for the phospholipid measurements when applied to certain pathologies, particularly diabetes. Earlier reports suggested that diabetes caused certain stresses that required an L/S ratio of greater than 2 and PG greater than 6% to be comparable with a nondiabetic situation. Whether the reports were reflecting the difficulty in management of diabetes or a real effect on fetal maturity is unknown. The most recent reports appear to agree that, in diabetes, the L/S ratio interpretation

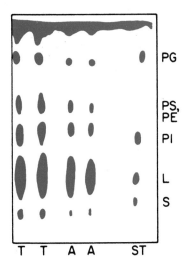

▲
Figure 24-6. Thin-layer chromatogram of amniotic fluid phospholipids. Standard phospholipids (*ST*), total extract (*T*), and acetone-precipitable compounds (*A*) in amniotic fluid are shown. The phospholipid standards contained, per liter, 2 g each of lecithin and P1, 1 g each of PG and sphingomyelin, and 0.3 g each of PS and PE. 10 mL of the standard was spotted. (Tsai MY, Marshall JG. Phosphatidylglycerol in 261 samples of amniotic fluid. Clin Chem 25[5]:683. Copyright 1979 by the American Association for Clinical Chemistry.)

is unchanged, but the lecithin fraction may have a lesser proportion of the dipalmitoyl species than is usual.

Because the analysis of AF for an L/S ratio is so terribly technique dependent, it is absolutely mandatory for the laboratory to evaluate thoroughly the methodology to be used. It is more important to relate that methodology to the particular clinical setting in developing interpretative intervals. It is not unusual to see a nominal L/S ratio maturity level of 2.0 ± 0.3 at one hospital, compared with 2.2 ± 0.2 at another center using the same method. Close cooperation with the medical staff is necessary in defining the ranges to be employed.

Application of fluorescence polarization technique to AF provides another type of measurement. This procedure, as used on a commercially available polarimeter with complete reagent system (Abbott Laboratories), offers a rapidly performed test (less than 1 hour) on a versatile analyzer that is relatively technique free. Its use has become so widespread that it has been suggested as the initial test to be used in a testing cascade.[7] The method used is based on the partition of a synthetic lecithin-like fluorescent dye between phospholipid aggregates and albumin. Dye associated with the phospholipid aggregates decreases polarization; with albumin, polarization increases. Total polarization is compared with a set of lipid:albumin standards to yield a unitless value. As the fetal lung matures, the amount of surfactant (phospholipid) increases, and therefore, the polarization is affected. The use of this system is widespread, in part, based on the accessibility to all laboratories, straightforward analytic procedure, and consistency in result interpretation.

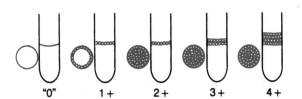

▲
Figure 24-5. Grading system used in evaluating the FS-50 test. (Statland et al. Evaluation of a modified foam stability [FS-50] test. An assay performed on amniotic fluid to predict fetal pulmonary maturity. Reproduced with permission from the Am J Clin Pathol 1978;69:51.)

A 26-year-old woman, pregnant for the first time, presented at the emergency room in possible labor. Her blood pressure was 180/110 mm Hg, and her temperature was 99.6°F. Her history revealed an anxious woman, with pregnancy of unknown duration, who had not seen a physician before. She admitted to a rapid weight gain over the past 2 weeks but said that she had been feeling well until then. Recently, she had felt faint and experienced episodes of vertigo. Urinalysis was significant for 3+ glucose and protein. Hematology revealed a low hemoglobin level and a moderately low platelet count, but leukocyte levels were normal. A chemistry panel revealed Na of 132 mmol/L, K of 3.0 mmol/L, Cl of 100 mmol/L, CO_2 of 29 mmol/L, BUN of 7 mg/dL, creatinine of 0.5 mg/dL, and glucose of 351 mg/dL. A decision was made to admit her and perform several more tests. The following chemistry tests were ordered,

with the listed results (upper limit of reference interval given):

Laboratory Results

L/S ratio = 1.8 with PG = 2%
AF creatinine = 1.5
FSI = 0.47
AST = 40(35), ALT = 40(35), ALP = 250(100)
Mg^{2+} = 1.6(2.2), Ca^{2+} = 9.0(10.5), PO_4 = 4.0(4.3),
Alb = 3.2(5.0)

After 24 hours of bed rest, the patient's BP fell to 140/90, and the contractions subsided.

Questions

1. Do the laboratory results indicate a normal pregnancy?
2. Comment on the maturity of the fetus.

The predictive value is compromised by the presence of blood or meconium (lipids) or fetal pathologies causing altered albumin levels such as urinary tract anomalies. Additionally, if the AF is centrifuged, not filtered, lipid level is altered, and therefore, polarization is decreased.

Another functional measurement recently advanced has been the quantitation of lamellar bodies. These packets of surfactant, released by the type 2 cells, are about the size of platelets and, therefore, should be quantifiable using the platelet channel of a hematology analyzer. Tentative breakpoints for an uncontaminated AF at 30,000 to 50,000 "platelet equivalents" have been suggested. Further coordinated research is required to render the assay valid, to include standardization of measurement parameters.

Reference intervals, or cut-points, are set for "normal" pregnancies. All of the tests perform reasonably well in ruling out immaturity, but all have failings with false-negative rates, specifically associated with deliveries predicted to develop respiratory distress syndrome based on an immature result that do not do so. For discussions regarding when lung maturity tests should be used and effects of conditions such as pregnancy-induced hypertension, two excellent reviews are recommended.[8,9]

CEREBROSPINAL FLUID

Cerebrospinal fluid (CSF) is the liquid that occupies the spaces of the CNS. As such, it surrounds all facets of the brain and spinal cord. These spaces are continuous and hence reflect all aspects of the CNS. CSF performs at least four major functions: (1) physical support and protection, (2) method for ex-

cretion, (3) provision of a controlled chemical environment, and (4) intracerebral and extracerebral transport (Fig. 24-7).

The major and most obvious function of CSF is as a buoyant cushion of the brain. The more dense brain floats in the less dense fluid, allowing movement within the skull. The significance of this is demonstrated by the result of a blow to the head. The initial shock is transferred to the entire brain, instead of inflicting damage to one area. It may be bruised at the side opposite the blow, depending on the force imparted.

The second major function of CSF is the maintenance of a constant gross chemical matrix for the CNS. Serum components may vary greatly, but constituent levels of CSF are maintained within narrow limits.

The excretory function is not well defined, but it is presumed to be effective, especially in pathologic states. Because there is no lymphatic system in the brain, only two paths are available for the elimination of wastes: capillary exchange and excretion by means of the CSF. The transport function is described as a neuroendocrine role. The CSF is involved in the distribution of hypophyseal hormones within the brain and the clearance of hormones from the brain to the blood.

The total CSF volume is about 150 mL, or about 8% of the total CNS cavity volume. The fluid is formed predominantly at the choroid plexus deep within the brain and by the ependymal cells lining the ventricles.

CSF is formed at an average rate of about 0.4 mL/min, or 500 mL/d. Formation is a result of selective ultrafiltration of plasma and active secretion by the epithelial membranes. Absorption of CSF occurs at outpouchings in the dura called *arachnoid villi,* and CSF drains into the venous sinuses of the dura. The villi also function to clear particulate matter, such as cellular debris. Reabsorption, like formation, is selective amination. Cloudy fluids usually require microscopic examination, whereas the presence of a yellow to brown or red color may indicate blood.

The two most common reasons for blood and hemoglobin pigments to be found in CSF are traumatic tap and subarachnoid hemorrhage. Traumatic tap is the artifactual presence of blood or derivatives due to interdiction of blood vessels during the lumbar puncture. Hemorrhage results from a breakdown of the barrier of the CNS and circulatory system from, for example, trauma. Obviously, the latter is serious. The two can be differentiated by observation and possibly testing. Bright red color and the presence of erythrocytes in decreasing number as the fluid is sampled indicate a traumatic tap. Xanthochromia or the presence of hemoglobin breakdown pigments indicates erythrocyte lysis and that metabolism has occurred previously, at least 2 hours earlier. Excluding a prior traumatic tap or hyperbilirubinemia (>20 mg/dL), xanthochromia would indicate hemorrhage.

Biochemical (chemical) analysis of CSF has led to compilations of the scope of possible constituents. In clinical practice, however, the number of useful indicators becomes very small. The tests of interest are glucose, protein (total and specific), lactate, lactate dehydrogenase, glutamine, and the acid–base parameters. Most often used are glucose and proteins. Before any analysis, the fluid should be centrifuged to avoid contamination by cellular elements. The enzyme lactate dehydrogenase has been suggested as a tumor marker, but it is relatively nonspecific. It is also elevated in bacterial infection, although other parameters are more specific. The level of glutamine should reflect the level of CNS ammonia removed by glutamine formation from glutamate. This would be elevated in the hepatic encephalopathy of Reye's syndrome. The test has largely been supplanted by the relative ease and simplicity of reliable plasma ammonia determinations. Determination of the specific acid–base parameters obtained using a blood-gas instrument is relevant because of the vital environment of the CSF and the effect on control of respiration. The lack of buffering capacity of the CSF and the stringent requirements for sample integrity in the procedures for collection and analysis render routine usage of this series of tests impractical.

The tests that have been most reliable diagnostically and accessible analytically are those for CSF glucose, total protein, and specific proteins. Glucose enters the spinal fluid predominantly by a facilitative transport as compared with a passive (diffusional) or active (energy-dependent) transport. It is carried across the epithelial membrane by a stereospecific carrier species. The carrier mechanism is responsible for transport of lipid-insoluble materials across the membrane into the CSF. Generally, this is a "downhill" process consistent with a concentration gradient. The CSF glucose concentration is roughly two thirds that of plasma.

Because an isolated CSF glucose concentration may be misleading, it is recommended that a plasma sample be obtained at the same time, so that the plasma and CSF glucose levels can be evaluated as a set. Normal CSF glucose is con-

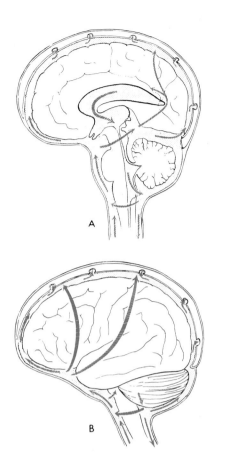

Figure 24-7. Major pathways of cerebrospinal flow. (**A**) Sagittal view; (**B**) lateral view. (Reprinted with permission from Milhorat TH. Hydrocephalus and the cerebrospinal fluid. Baltimore: Williams & Wilkins, 1972;25.)

and specific. Obviously, if a constant volume is maintained at a formation rate of 500 mL/d, resorption is constant.

Specimens of CSF are obtained by lumbar puncture, usually at the interspace of vertebrae L3 to L4 or lower, using aseptic technique. The fluid obtained is usually separated into three aliquots: (1) for chemistry and serology, (2) for bacteriology, and (3) for microscopy. It is paramount to remember that this matrix is of limited volume and should be analyzed immediately. Any remaining sample should be preserved because of its limited availability. The order of the tubes reflects the presumed order for minimalization of interference from less than optimal collection technique, with tube 3 presumably least contaminated by cells of intervening tissues.

Laboratory investigation of CSF is indicated for cases of suspected CNS infection, demyelinating disease, malignancy, and hemorrhage in the CNS. Historically, it is usually performed before radiologic procedures such as myelography, but the diagnostic utility in these cases is extremely small. As with all patient samples entering the laboratory, visual examination of the specimen is the first and often the most important observation made. The CSF, if normal, is clear, colorless, free of clots, and free of blood. Differences from these standards indicate a probable pathology and merit further ex-

A 32-year-old man was in good health until about 1 year ago, when he entered an accelerated training program in computer programming. Within the last year, he began to notice episodic blurring of vision, mild vertigo, and headache. He attributed his complaints of sensory loss in his hands and a feeling of weakness after physical exertion to being "out of shape." He decided to see his doctor upon an attack of blurred vision accompanied by a feeling of paralysis, which was followed by "pins and needles" in his left leg. An optic examination was negative. Neurologic examination led to a spinal tap being performed for laboratory findings and myelography. The latter was negative. The laboratory findings were as follows:

Laboratory Results

CSF = clear, colorless fluid, apparently free of debris; culture yields no growth

WBC = normal

Glucose = 60 mg/dL (plasma = 80 mg/dL)

IgG/Alb ratio = 1.7

IgG oligoclonal banding present

Questions

1. What is the significance of the normal CSF protein and variant IgG/Alb ratio?
2. What pathology is consistent with these results?

---

sidered to be greater than 45 mg/dL (2.5 mmol/L). Increased glucose levels are not clinically informative, usually providing only confirmation of hyperglycemia. This generality must be tempered by two factors: (1) equilibration after glucose loading usually takes 3 to 4 hours; and (2) with increasing blood glucose levels, the CSF glucose increases, but not proportionally. The first is important in cases of meals taken close to the time of sampling. The second is significant because it implies that the plasma/CSF glucose ratio decreases as gross hyperglycemia occurs.

The decreasing CSF/plasma glucose ratio as plasma glucose increases is consistent with a saturable carrier process. It would not be unusual for the ratio to be 0.4 to 0.6 with massive plasma glucose levels (more than 600 mg/dL), but a CSF glucose level of 80 mg/dL, with plasma level of 300 mg/dL, is clinically significant and would merit concern.

Decreased CSF glucose levels (*hypoglycorrhachia*) can be the result of (1) disorder in carrier-mediated transport of glucose into CSF, (2) active metabolism of glucose by cells or organisms, or (3) increased metabolism by the CNS. The mechanism of transport decrease is still under intense discussion, but it is speculated to be the cause in tuberculous meningitis and sarcoidosis. Acute purulent, amoebic, fungal, and trichinotic meningitis are examples of consumption by organisms, whereas diffuse meningeal neoplasia and brain tumor are examples of consumption by CNS tissue. Consumption of glucose is usually accompanied by an increased lactate level because of anaerobic glycolysis by organisms or cerebral tissue. An increased lactate level with a normal to decreased glucose level has been suggested as a readily accessible indicator for bacterial versus viral meningitis.[10] Analysis of glucose and lactate in CSF is easily accomplished by techniques used for plasma and serum. It is important that provision for analysis of glucose or lactate in CSF be immediate or that the specimen be preserved with an antiglycolytic, such as fluoride ion.

Protein levels in CSF reflect both the selective ultrafiltration of the CSF epithelial barrier and their secretory ability. All the protein usually found in plasma is found in CSF, except at much decreased levels. Total protein is about 0.5%, or 1/200 that of plasma. The specific protein concentrations in CSF are not proportional to the plasma levels because of the specificity of the ultrafiltration process. Correlation is best accomplished using hydrodynamic ratios of the protein species rather than molecular weight. Because of the relationship of CSF proteins to serum, serum analysis should accompany specific CSF protein analysis. A decreased level of CSF total protein can arise from (1) decreased dialysis from plasma; (2) increased protein loss (*eg,* removal of excessive volumes of CSF); or (3) leakage of CSF from a tear in the dura, otorrhea, or rhinorrhea. The last reason is most common. A dural tear can occur as a result of a previous lumbar puncture or from severe trauma. *Otorrhea* and *rhinorrhea* refer to leakage of CSF from the ear or into the nose, respectively. Identification of the source of the leak is best done by an analysis for τ-transferrin, a protein unique to the CSF.

An increased level of CSF total protein is a useful nonspecific indicator of pathologic states. Increases may be caused by (1) lysis of contaminant blood from traumatic tap, (2) increased permeability of the epithelial membrane, (3) increased production by CNS tissue, (4) obstruction, or (5) decrease in rate of removal. Contamination from blood is significant because of the 200:1 concentration ratio. The presence of any amount of blood can elevate CSF protein levels. The epithelial membrane becomes more permeable from bacterial or fungal infection or cerebral hemorrhage, whereas an increase in CNS production occurs in subacute sclerosing panencephalitis (SSPE) or multiple sclerosis. There may also be combinations of permeability and production, such as the collagen-vascular diseases. An obstructive process, such as tumor or abscess, would also cause increased protein. The last mentioned cause of increased levels of CSF total protein, decreased absorption, is theoretically possible but has not been demonstrated.

Diagnostically more sensitive information can be obtained by analysis of the protein fractions present. A comparison to the serum pattern is necessary for accurate conclusions. Under normal conditions, prealbumin and a form of transferrin (tau protein) are present in CSF in higher concentration than in serum. Although the respective proteins can be determined in both serum and CSF, the proteins of greatest interest are albumin and immunoglobulin G (IgG). Because albumin is produced solely in the liver, its presence in CSF must occur by means of membrane transport. IgG, however, can arise by local synthesis from plasma cells within the CSF. The measurement of albumin in both serum and CSF is then used to normalize the IgG values from each matrix to determine the source of the IgG. This IgG–albumin index is primarily used in the diagnosis of demyelinating diseases, such as multiple sclerosis and SSPE.

$$\frac{\text{CSF IgG/serum IgG}}{\text{CSF albumin/serum albumin}} = \text{CSF index}$$

$$\text{Normal} = 0.5 \qquad (Eq.\ 24\text{--}1)$$

Increases in serum proteins cause increases in the CSF levels because of permeability. However, increased CSF IgG without concomitant CSF albumin increase suggests local production (multiple sclerosis or SSPE). Increases in permeability and production are found with bacterial meningitis. Methods to analyze IgG and albumin CSF levels are the same as for serum but are optimized for the lower levels found.

Increased CSF protein levels or clinical suspicion usually indicates the need for electrophoretic separation of the respective proteins. At times, this separation demonstrates multiple banding of the IgG band. This observation is referred to as *oligoclonal proteins* (a small number of clones of IgG from the same cell type with nearly identical electrophoretic properties). This occurrence is usually associated with inflammatory diseases and multiple sclerosis or SSPE. These types of disorders would stimulate the immunocompetent cells. The recognition of an oligoclonal pattern supersedes the report of normal protein levels and is cause for concern if the corresponding serum separation does not demonstrate identical banding.

Another protein thought to be specific for multiple sclerosis is myelin basic protein (MBP). Initial reports suggested high specificity, but MBP also has been found in nondemyelinating disorders and does not always occur in the presence of demyelinating disorders. MBP levels are used by some to monitor therapy of multiple sclerosis.

SWEAT

The common eccrine sweat glands function in the regulation of body temperature. They are innervated by cholinergic nerve fibers and are a type of exocrine gland. Sweat has

been analyzed for its multiple inorganic and organic contents but, with one notable exception, has not proved a clinically useful model. That exception is the analysis of sweat for chloride and sodium levels in the diagnosis of cystic fibrosis (CF). The sweat test is the single most accepted diagnostic tool for the clinical identification of this disease. Normally, the coiled lower part of the sweat gland secretes a "presweat" upon cholinergic stimulation. As the pre-sweat traverses the ductal part of the gland going through the dermis, various constituents are resorbed. In CF, the electrolytes, most notably chloride and sodium ions, are improperly resorbed owing to a mutation in the cystic fibrosis transmembrane conductance regulator (CFTR) gene, which controls a cyclic AMP–regulated chloride channel.

CF (mucoviscidosis) is an autosomal recessive inherited disease that affects the exocrine glands and causes electrolyte and mucous secretion abnormalities. This exocrinopathy is present only in the homozygous state. The frequency of the carrier (heterozygous) state is estimated at 1 in 20 in the U.S. population. It is predominantly a Caucasian disease. The observed rate of expression ranks CF as the most common lethal hereditary disease in the United States, with death usually occurring by the third decade. The primary cause of death is pneumonia, secondary to the heavy, abnormally viscous secretion in the lungs. These heavy secretions cause obstruction of the microairways, predisposing the CF patient to repeated episodes of pneumonia. The third part of the diagnostic triad is pancreatic insufficiency. Again, abnormally viscous secretions obstruct pancreatic ducts. This obstruction ostensibly causes pooling and autoactivation of the pancreatic enzymes. The enzymes then cause destruction of the exocrine pancreatic tissue.

Diagnostic algorithms for CF continue to rely on abnormal sweat electrolytes, the presence of pancreatic or bronchial abnormalities, and family history. The use of blood immunoreactive trypsin, a pancreatic product, has been proposed as both a method for newborn screening and a diagnostic adjunct. The rapidly developing area of molecular genetics may eventually provide the definitive methodology. The gene defect causing CF has been localized on chromosome 7, and the most common mutation causing CF has been DNA "fingerprinted." It is anticipated that within the next decade, direct DNA analysis will catalog all mutations and provide definitive screening and diagnosis. Until that time, however, the sweat chloride test remains the signal laboratory tool for CF diagnosis even though it has been reported that a form of CF with normal sweat chloride exists.[11]

The sweat glands, although affected in their secretion, remain structurally unaffected by CF. Analysis of sweat for both sodium and chloride is valid, but historically, chloride was and is the major element, leading to use of the sweat chloride test. Because of its importance, a standard method has been suggested by the Cystic Fibrosis Foundation. It is based on the pilocarpine nitrate iontophoresis method of Gibson and Cooke.[12] Pilocarpine is a cholinergic-like drug used to stimulate the sweat glands. The sweat is absorbed on a gauze pad during the procedure. Other tests (*eg,* osmolality or conductivity) that reflect the sodium and chloride concentrations have been proposed, but the sweat chloride test remains the reference method.

After collecting the sweat by iontophoresis, chloride and sodium analysis is performed. Many methods have been suggested, and all are dependent on the requirements of the laboratory. Generally, the sweat is leached into a known volume of distilled water and analyzed for chloride (chloridometer) and sodium (flame photometry). In general, values greater than 60 mmol/L are considered positive for both ions.

Although a value of 60 mmol/L is generally recognized for the quantitative pilocarpine iontophoretic test, it is important to consider several factors in interpretation. Not only will there be analytic variation around the cutoff, there will also occur an epidemiologic borderline area. Considering this, the range of 45 to 65 mmol/L for chloride would be more appropriate in determining the need for repetition. Other variables necessarily must be considered. Age generally increases the limit—so much so that it is increasingly difficult to classify adults. Obviously, the patient's state of hydration also affects the sweat levels. Because the complete procedure is technically demanding, expertise should be developed before the test is clinically available. A complete description of sweat collection and analysis, including procedural justifications, is available for review.[13]

SYNOVIAL FLUID

Joints are classified as immovable or movable. Within the movable joint is a cavity enclosed by a capsule whose inner lining is synovial membrane. The cavity contains *synovial fluid* formed by ultrafiltration of plasma across the synovial membrane. The membrane also secretes into the dialysate a mucoprotein rich in hyaluronic acid, which causes the synovial fluid to be viscous. The membrane is composed of three different cell types: type A cells are rich in vacuoles and lysosomes and resemble phagocytes in function; type B cells are rich in rough endoplasmic reticulum and are presumed to be secretory in function; and type C cells appear to be hybrids of types A and B in appearance and function. Synovial fluid is assumed to function as a lubricant for the joints and as a transport medium for delivery of nutrients and removal of cell wastes. The volume of fluid found in a large joint, such as the knee, rarely exceeds 3 mL. Normal fluid is clear, colorless to pale yellow, and viscous, and it does not clot. Variations are indicative of pathologic conditions.

Collection of a sample is accomplished by arthrocentesis of the joint under aseptic conditions. The sample should be preserved immediately with heparin for culture, with EDTA for microscopic analysis, or with fluoride for glucose analysis. Microscopic examination is the most rewarding in terms of diagnostic significance. Although many analytes have been chemically determined, the ratio of synovial fluid to plasma glucose (normally, 0.9 to 1) remains the most useful.

Decreased ratios are found in inflammatory (*eg,* gout, rheumatoid arthritis, systemic lupus erythematosus) and purulent (bacterial, viral arthritis) conditions. Standard methods for glucose analysis are applicable.

SEROUS FLUIDS

The lungs, heart, and abdominal cavity are surrounded by single-celled membranous bilayered sacs permeable to serum constituents. The fluid formed when *serum* dialyzes across these membranes is called *serous fluid.* Specifically, they are pleural (lung), pericardial (heart), and peritoneal (abdominal) fluids.

The sacs can be pictured as a balloon containing a small amount of water. Push a football into the balloon, and the balloon expels the residual air until only water remains (Fig. 24-8). The water spreads to fill the resultant "balloon bilayer potential space." The serous membranes are analogous to the balloon; the internal organs, to the football. The pleural (lung) sac is continuous at the hilus of the bronchial tree; the pericardium, at the major vessels; and the peritoneum, around each organ. All are expandable and thus present "potential spaces" for fluid or gases to collect. The formation of serous fluid is a continuous process. Formation is driven by the hydrostatic pressure of the systemic circulation and maintenance of oncotic pressure due to protein. The potential space is usually filled; that is, no gases are present. The fluid reduces or eliminates friction due to expansion and contraction of the encased organs. A disturbance of the dynamic equilibrium that causes an increase in fluid is an abnormal state. An increase in fluid volume is called an *effusion.*

Pleural Fluid

The outer layer of the pleural sac, the parietal layer, is served by the systemic circulation; the inner, visceral layer, by the bronchial circulation. The pleural fluid is essentially interstitial fluid of the systemic circulation. Under normal condi-

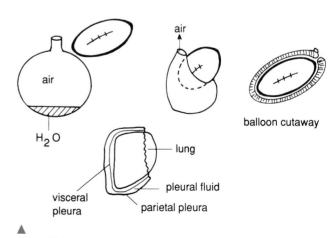

▲

Figure 24-8. Example of formation of serous fluid.

tions, there is 3 to 20 mL of pleural fluid in the pleural space. The fluid exits by drainage into the lymphatics of the visceral pleura and the visceral circulation. Any alteration in the rate of formation or removal of the pleural fluid affects the volume, causing an effusion. It is then necessary to classify the nature of the effusion by analysis of the pleural fluid. The fluid is obtained by removal from the pleural space by needle and syringe after visualization by radiology. This procedure is called *thoracentesis;* the fluid is then called *thoracentesis fluid* or *pleural fluid.* Specifically preserved aliquots of the fluid are used for future testing as follows: (1) heparinized for culture, (2) EDTA for microscopy, (3) NaF for glucose and lactate, and (4) untreated for further biochemical testing.

The classification of the fluid as transudate or exudate is crucial. *Transudates* are secondary to remote (nonpleural) pathology and indicate that treatment should begin elsewhere. An *exudate* indicates primary involvement of the pleura and lung, such as infection. Exudates demand immediate attention. For example, any mechanical disturbance in the formation of fluid (eg, hypoproteinemia causing decreased oncotic pressure) would increase pleural fluid volume. This would be a transudative process. An example of an exudative process would be obstruction of the lymphatic drainage due to malignancy, such as lymphoma (Table 24-2). Further testing, including chemical, microscopic, and culture, is then required to identify the etiology.

The assignment of fluid to either the transudate or exudate category had previously been based on the protein concentration of the fluid. This criterion has been replaced by the use of a series of fluid/plasma (F:P) ratios. Specifically, if the F:P for total protein is greater than 0.5, the ratio for lactate dehydrogenase (LD) is greater than 0.6, or if the pleural fluid LD/upper limit serum LD ratio reference interval is greater than 0.67, the fluid is an exudate. Further characterization of the exudate by the chemistry laboratory may involve analysis for glucose, lactate, amylase, triglyceride, or pH. A decrease in glucose (increase in lactate) would suggest infection or inflammation. An increase in

amylase compared with that of serum suggests pancreatitis. Grossly elevated triglyceride levels (2 to 10 × serum) could indicate chylothorax. The use of pH measurements, performed as one would a blood-gas level determination, has recently gained favor. Succinctly, pH less than 7.2 suggests infection, and pH greater than 7.4 suggests malignancy. The methodologies for these analyses are the same as those employed for the serum and blood constituents and therefore are feasible in the clinical laboratory.

Pericardial Fluid

The relationship of the pericardium, *pericardial fluid,* and the heart is quite similar to that with the lungs. Mechanisms of formation and drainage are the same, but the frequency of pericardial sampling and laboratory analysis is rare.

Peritoneal Fluid

The clinical questions to be answered are usually as follows: (1) What is causing ascites? (2) Is there infection present? (3) Is there risk for infection?[14]

The presence of excess (more than 50 mL) fluid in the peritoneal cavity indicates disease. The presence of excess fluid is called *ascites,* and the fluid is called *ascitic fluid.* The process of obtaining samples of this fluid by needle aspiration is called *paracentesis.* Usually, the fluid is visualized by ultrasound to confirm its presence and volume before paracentesis is attempted.

Theoretically, the same mechanisms that cause serous effusions in other potential spaces are operative for the peritoneal cavity. Specifically, a disturbance in the rate of dialysis secondary to a primary, remote pathology is a transudate, as compared with a primary pathology of the peritoneal membrane (exudate). The multiple factors that apply to this large space, including renal function, tend to cloud the distinction. The most common cause of ascites with a normal peritoneum is portal hypertension. Obstructions to hepatic flow, that is (1) cirrhosis (2) congestive heart failure and (3) hypoalbuminemia for any reason, demonstrate the highest incidence.

The exudative causes of ascites are predominantly metastatic ovarian cancer and infective peritonitis. The differentiation of the two is analogous to the measures of protein and LD described for pleural fluid. The ability to differentiate, however, has been challenged. The gradient of albumin concentration (serum albumin–fluid albumin) of 1.1 g/dL or more used to indicate portal hypertension is gaining adherents.

SUMMARY

In addition to serum and plasma, the clinical chemistry laboratory often analyzes other body fluids, such as AF, CSF, sweat, synovial fluid, and serous fluids. AF provides a cushioning medium for fetal development.

TABLE 24-2. Causes of Pleural Effusions	
TRANSUDATIVE	**EXUDATIVE**
Congestive heart failure*	Pneumonia,* bacterial
Nephrotic syndrome	Tuberculosis
Hypoproteinemia	Pulmonary abscess
Hepatic cirrhosis	Malignancy (lymphatic obstruction)
Chronic renal failure	Viral/fungal infection
	Pulmonary infarction
	Pleurisy
	Pulmonary malignancy
	Lymphoma
	Pleural mesothelioma

Most common cause.

Amniocentesis and subsequent AF analysis are performed to test for congenital diseases, neural tube defects, hemolytic disease, gestational age, and fetal pulmonary development. CSF is a liquid that occupies the spaces of the CNS, which includes the brain and spinal cord. Functions of the CSF include physical support and protection, method of excretion, provision of a controlled chemical environment, and intracerebral and extracerebral transport. The total volume of CSF is about 150 mL.

Specimens of CSF are obtained by lumbar puncture. Laboratory investigation of CSF is indicated for cases of suspected CNS infection, demyelinating disease, malignancy, and hemorrhage in the CNS. The tests that have been most reliable diagnostically and accessible analytically are those for CSF glucose, total protein, and specific proteins. Sweat is a product of the common eccrine sweat glands, which function in the regulation of body temperature. The sweat test is the single most accepted diagnostic tool for CF. Within the movable joint is a cavity filled with synovial fluid. This fluid is formed by ultrafiltration of plasma across the synovial membrane. Normal fluid is clear, colorless to pale yellow, and viscous; it does not clot. Any variations are indicative of pathologic conditions. Collection of this fluid is accomplished by arthrocentesis. The lungs, heart, and abdominal cavity are surrounded by single-celled membranous bilayered sac permeable to serum constituents.

The fluid formed when serum dialyzes across these membranes is called *serous fluid*. More specifically, they are pleural (lung), pericardial (heart), and peritoneal (abdominal) fluids. Formation of serous fluid is a continuous process and is driven by the hydrostatic pressure of the systemic circulation and maintenance of oncotic pressure due to protein. Under normal conditions, there is from 3 to 20 mL of pleural fluid in the pleural space. Removal of the fluid is called thoracentesis. The relationship of the pericardium and pericardial fluid with the heart is quite similar to that with the lungs. The frequency of pericardial sampling in the laboratory is rare. An excess of fluid in the peritoneal cavity indicates disease. This is called ascites. The fluid is ascitic fluid and is obtained by a procedure called paracentesis.

REVIEW QUESTIONS

1. Laboratory testing for assessment of fetal lung maturity is based on:
   a. differentiation products of type I pneumocytes
   b. production of acetylcholinesterase
   c. production of AFP
   d. products of type II pneumocytes
   e. presence of meconium
2. Amniotic fluid:
   a. provides a cushion for the fetus
   b. is a mixture of maternal and fetal fluids
   c. is assessed by umbilical catheterization
   d. a and b
   e. a, b, and c
3. Lamellar body counts reflect:
   a. platelet count of the fetus
   b. platelet count of the mother
   c. meconium count of the fetus
   d. surfactant phospholipid packets
   e. all of the above
4. Blood in CSF is commonly observed after:
   a. administration of indocyanine green
   b. traumatic tap
   c. hemolytic anemia
   d. lumbar puncture
   e. low pH
5. CSF glucose is measured to access:
   a. transport efficiency
   b. diabetes
   c. traumatic tap
   d. hemochromatosis
   e. infection
6. Increased CSF protein is pathognomonic for:
   a. multiple sclerosis
   b. collagen vascular disease
   c. fungal infection
   d. all of the above
   e. none of the above
7. CF is characterized by:
   a. elevated sweat chloride levels
   b. homozygous expression of an autosomal recessive trait
   c. pancreatic insufficiency
   d. all of the above
   e. a and c only
8. Synovial fluid
   a. is formed by plasma ultrafiltration
   b. lubricates the pneumocytes
   c. is rich in hyaluronic acid
   d. all of the above
   e. a and c only
9. Serous fluids:
   a. are derived from serum
   b. provide lubrication and protection
   c. fill the potential space
   d. all of the above
   e. a and b only
10. Pleural fluid transudate:
    a. reflects primary involvement of the pleura
    b. is characterized by an increased LD F/P ratio
    c. is characterized by an increased glucose F/P ratio
    d. is characterized by a total protein F/P ratio of 0.5
    e. all of the above
11. Analysis of paracentesis fluid is performed to:
    a. determine cause of fluid presence
    b. assess infection risk
    c. determine lung involvement
    d. all of the above
    e. a and b

12. The most common cause of ascites is:
    a. portal hypertension
    b. venous return
    c. parietal cell differentiation
    d. eccrine infection
    e. type A cell leakage

REFERENCES

1. Liley AW. Liquor amni analysis in the management of the pregnancy complicated by rhesus sensitization. Am J Obstet Gynecol 1961;82:1359.
2. Kulovich MV, Hallman MB, Gluck L. The lung profile. I. Normal pregnancy. Am J Obstet Gynecol 1979;135:57.
3. Statland BE, Freer DE. Evaluation of two assays of functional surfactant in amniotic fluid: Surface-tension lowering ability and the foam stability index test. Clin Chem 1979;25:1770.
4. Clements JA, Plataker ACG, Tierney DF, et al. Assessment of the risk of the respiratory distress syndrome by a rapid test for surfactant in amniotic fluid. N Engl J Med 1972;286:1077.
5. Gluck L, Kulovich MV, Borer RC Jr, et al. Diagnosis of the respiratory distress syndrome by amniocentesis. Am J Obstet Gynecol 1971;109:440.
6. Tsai MY, Marshall JG. Phosphatidylglycerol in 261 samples of amniotic fluid from normal and diabetic pregnancies, as measured by one-dimensional thin layer chromatography. Clin Chem 1979;25:682.
7. Herbert WNP, Chapman JF, Schnoor MM. Role of the TDX FLM assay in fetal lung maturity. Am J Obstet Gynecol 1993;168:808.
8. ACOG Educational Bulletin, Number 230, November 1996, Assessment of Fetal Lung Maturity.
9. Field NT, Gilbert WM. Current status of amniotic fluid tests of fetal lung maturity. Clin Obstet Gynecol 1977;40(2):366–386.
10. Bailey EM, Domenico P, Cunha BA. Bacterial or viral meningitis. Postgrad Med 1990;88:217.
11. Highsmith WE, Burch L, Zhou Z, et al. A novel mutation in the cystic fibrosis gene in patients with pulmonary disease but normal sweat chloride concentrations. N Engl J Med 1994;331:974.
12. Gibson LE, Cooke RE. A test for concentration of electrolytes in cystic fibrosis of the pancreas utilizing pilocarpine by iontophoresis. Pediatrics 1959;23:545.
13. NCCLS. Sweat testing: Sample collection and quantitative analysis approved guidelines. NCCLS Document C34-A. Villanova, PA: National Committee for Clinical Laboratory Standards, 1994.
14. Habeeb, KS, Herrera, JL. Management of ascites: Paracentesis as a guide. Postgrad Med 1997;101(1):191–2, 195–200.

SUGGESTED READINGS

American Society of Human Genetics. Policy statement for maternal serum alpha-fetoprotein screening programs and quality control for laboratories performing maternal serum and amniotic fluid alpha-fetoprotein assays. Am J Hum Genet 1987;40:75.

Brown LM, DuckChong CG. Methods of evaluating fetal lung maturity. Crit Rev Clin Lab Sci 1982;17(1):85.

Committee for a Study of Evaluation of Testing for Cystic Fibrosis: Report. J Pediatr 1976;88(4):711.

Daveson H. Physiology of the cerebrospinal fluid. London: Churchill, 1967.

Fairweather DVI, Eskes TKAB, eds. Amniotic fluid research and clinical application. 2nd ed. New York: Excerpta Medica, 1978.

Freeman JA, Beeler MF, eds. Laboratory medicine/urinalysis and medical microscopy. 2nd ed. Philadelphia: Lea & Febiger, 1983.

Freer DE, Statland BE. Measurement of amniotic fluid surfactant. Clin Chem 1981;27:1629.

Guyton AC. Textbook of medical physiology. 5th ed. Philadelphia: WB Saunders, 1976.

Halsted JA, Halsted CH, eds. The laboratory in clinical medicine. 2nd ed. Philadelphia: WB Saunders, 1981.

Health and Public Policy Committee, American College of Physicians. Diagnostic thoracentesis and pleural biopsy in pleural effusions. Ann Intern Med 1985;103:799.

Platt PN. Examination of synovial fluid in the role of the laboratory in rheumatology. Clin Rheum Dis 1983;9(1):51.

Queenan JT, ed. Modern management of the Rh problem. 2nd ed. Hagerstown, MD: Harper & Row, 1977.

Rocco VK, Ware AJ. Cirrhotic ascites. Ann Intern Med 1986;105:573.

Rodriguez EM, van Wimersma Greidanus TB. Frontiers of hormone research. Vol 9. Cerebrospinal fluid and peptide hormones. New York: S Karger, 1982.

Russell JC, Cooper CC, Ketchum CH, et al. Multicenter evaluation of TDX test for assessing fetal lung maturity. Clin Chem 1989;35:1005.

Teloh HA. Clinical pathology of synovial fluid. Ann Clin Lab Sci 1975;5(4):282.

Webster HL. Laboratory diagnosis of cystic fibrosis. Crit Rev Clin Lab Sci 1983;18(4):313.

Wiswell TE, Mediola J Jr. Respiratory distress syndrome in the newborn: Innovative therapies. Am Fam Physician 1993;47:407.

Wood JH, ed. Neurobiology of the cerebrospinal fluid. Vol 2. New York: Plenum Press, 1983.

PART IV

SPECIALTY AREAS OF CLINICAL CHEMISTRY

CHAPTER 25

Therapeutic Drug Monitoring

David P. Thorne

Objectives

Upon completion of this chapter, the clinical laboratorian should be able to:

- *Discuss the characteristics of a drug that make therapeutic drug monitoring essential.*
- *Identify the factors that influence the absorption of an orally administered drug.*
- *Relate the factors that influence the rate of elimination of a drug.*
- *Define drug distribution and the factors that influence it.*
- *Calculate volume of distribution, elimination constant, and half-life of a drug.*
- *Relate the concentration of a circulating drug to pharmacokinetic parameters.*
- *Name the therapeutic category of each of the drugs presented in this chapter.*

- *Describe the major toxicities of each of the drugs presented in this chapter.*
- *Identify features of each of the drugs presented in this chapter that may influence their serum drug concentration.*

KEY TERMS

Drug absorption	Pharmacokinetics
Drug distribution	Therapeutic drug monitoring
Drug elimination	Therapeutic range
Peak drug level	Trough drug levels

*T*herapeutic drug monitoring (TDM) involves the analysis, assessment, and evaluation of circulating concentrations of drugs in serum, plasma, or whole blood. The purpose of these actions is to ensure that a given dosage of a drug produces maximal therapeutic benefit and minimal toxic side effects.[1,2] For most drugs, dosage regimens have been established that are safe and effective in most of the population. Thus, TDM is unneeded with most drug therapies. With some drugs, however, the correlation between dosage and therapeutic effects or toxic outcomes is weak. Thus, it is hard to predict what dose should be used. In these situations, trial and error in conjunction with direct observation may work. For example, if the standard dose of a drug does not display therapeutic benefit and increasing the dose does, without toxic side effects, an appropriate dosage adjustment has been made. Unfortunately, this system may be inappropriate in all situations. If overdosing or underdosing results in severe consequences to the patient, trial and error is not justified. If this is the case, TDM based on strong correlations between circulating concentrations of drug and therapeutic benefit or toxic side effects assists in the determination of an appropriate dosage regimen. The standard dosage is statistically derived from observations in a healthy population.[2] Disease states may produce altered physiologic conditions in which the standard dose does not produce the predicted response.[3] In these cases, individualizing a dosage regimen is warranted.[4] Also, TDM provides a basis for establishing a rational dosage regimen to fit individual situations.[5]

The following are the common reasons that TDM is indicated:

- The consequences of overdosing and underdosing are serious.
- There is a small difference between a therapeutic and toxic dose.
- There is a poor relationship between the dose of drug and circulating concentrations but a good correlation between circulating concentrations and therapeutic or toxic effects.
- There is a change in the patient's physiologic state that may unpredictably affect circulating drug concentrations.
- A drug interaction is or may be occurring.
- TDM helps in monitoring patient compliance.

A common feature of all aspects of TDM is the quantitative evaluation of circulating concentrations of drugs. When taken together with clinical context, this provides a basis for rational problem solving and optimizing patient outcomes.[6] This process requires that several key factors be taken into consideration, such as the route of administration, rate of absorption, distribution of drug within the body, and rate of elimination (Fig. 25-1). This chapter begins by focusing on these factors and how they influence the circulating concentration of a drug. The remainder of the chapter surveys selected drugs that are commonly subject to TDM.

ROUTES OF ADMINISTRATION

For a drug to cause a therapeutic benefit, it must be at the appropriate concentration at its site of action. In many situations, that site of action is not easily accessed. The circulatory system offers a convenient route that can effectively de-

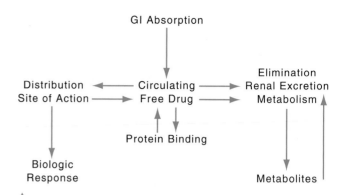

Figure 25-1. Overview of factors that influence the circulating concentration of an orally administered drug

liver most drugs to its site of action. The goal of most therapeutic regimens is to acquire a blood, plasma, or serum concentration that has been correlated with an effective concentration at the site of action. Drugs can be administered by several routes. They can be injected directly into the circulation (intravenous, IV), into muscles (intramuscular, IM), or just under the skin (subcutaneous, SC). They can also be inhaled or absorbed through the skin (transcutaneous). Rectal delivery (suppository) is commonly used in infants and in situations in which oral delivery is unavailable. Oral administration is the most common route of delivery. Each route presents with different kinetics that determine circulating concentrations. The focus of the current discussion is on oral and IV administration.

ABSORPTION

The efficiency of *drug absorption* from the gastrointestinal tract is dependent on many factors. The formulation of the drug is a key issue. Tablets and capsules require dissolution before being absorbed. Liquid solutions have a tendency to be more rapidly absorbed. Some drugs are subject to uptake by transport mechanisms intended for dietary constituents; however, most are absorbed by passive diffusion. This process requires that the drug be in a hydrophobic (nonionized) state. Because of gastric acidity, weak acids are efficiently absorbed in the stomach. Weak bases are preferentially absorbed in the intestine, where the pH is more neutral. For most drugs, absorption from the gastrointestinal tract occurs in a consistent manner in healthy people. Changes in intestinal motility and stomach pH, as well as the presence of food or other drugs, however, may dramatically change absorption characteristics.[7]

All substances absorbed from the intestine, including drugs, enter the hepatic portal system. In this system, all the blood from the gastrointestinal tract is routed through the liver before it enters the general circulation. Certain drugs are subject to significant hepatic uptake and metabolism during this passage through the liver. This process is known as *first-pass metabolism*.[8]

For some drugs, there may be a wide degree of variance in these processes, even within a normal population. In addition, many of the absorption characteristics of a drug may change with age, pregnancy, or pathologic conditions. In these instances, predicting the final circulating concentration from a standard oral dose can be difficult. With the use of TDM to determine what dosage results in a measured circulating concentration, however, effective oral dosage regimens can be determined.

FREE VERSUS BOUND DRUGS

Most drugs in circulation are subject to binding with serum constituents. Even though many potential species may be formed, most are drug–protein complexes. An important aspect needed to understand drug dynamics is that only the free fraction can interact with its site of action and result in a biologic response. Thus, it is the free fraction that best correlates with both the therapeutic and toxic effects of a drug. For many drugs, the percentage free is dependent on physiologic and biochemical parameters. At a standard dose, total plasma content may be within the *therapeutic range,* but with the patient experiencing toxic side effects (high free fraction) or not realizing a therapeutic benefit (low free fraction). The percentage free can change due to changes in serum proteins. Changes in serum-binding proteins may occur during inflammation, malignancies, pregnancy, hepatic disease, nephrotic syndrome, and malnutrition. The percentage free may also be influenced by the concentration of substances that compete for binding sites, which may be other drugs or endogenous substances such as urea, bilirubin, or hormones. Free drug measurements should be considered for drugs that are highly protein bound and for which clinical signs are inconsistent with total drug concentrations.

DRUG DISTRIBUTION

The free fraction of circulating drugs is subject to diffusion out of the vasculature into interstitial and intracellular spaces. The ability to leave circulation is largely dependent on the lipid solubility of the drug. Drugs that are highly hydrophobic can easily traverse cellular membranes and partition into lipid compartments, such as adipose and nerve cells. Drugs that are polar but not ionized also cross cell membranes but do not sequester into lipid compartments. Ionized species diffuse out of the vasculature but at a slow rate. The volume of distribution (V_d) is an index used to describe the distribution characteristics of a drug. It is expressed mathematically as follows:

$$V_d = D/C_t \qquad \textit{(Eq. 25–1)}$$

where: V_d = volume of distribution (in liters),

D = an injected dose (milligrams [mg] or grams [g]),

and

C = concentration in plasma (mg/L or g/L)

Drugs that are very hydrophobic can have a very large V_d. Substances that are ionized or are primarily bound in plasma have very small V_d values.

DRUG ELIMINATION

Drugs can be cleared from the body by a variety of mechanisms. Independent of the clearance mechanism, decreases in the serum concentration of drugs most often occur as a first-order process (exponential rate of loss). This implies that the rate of change of drug concentration over time varies continuously in relation to the concentration of the drug itself. First-order elimination follows the following general equation:

$$\mu C/\mu t = -kC \qquad \textit{(Eq. 25–2)}$$

This equation defines how the change in concentration per unit time ($\mu C/\mu t$) is directly related to the concentration of drug (C) and the constant (k). The k value is a simple proportionality factor that describes the percentage change (negative because it is decreasing) per unit time; it is commonly referred to as the *elimination constant* or the *rate of elimination*. The graphic solution to this equation is an exponential function that declines in the predicted curvilinear manner, asymptotically approaching zero (Fig. 25–2). The graph shown in Figure 25–2 illustrates a rapid rate of change at high drug concentrations and slow rates of change at low drug concentrations. Plotting it in semilogarithmic dimensions (Fig. 25–3) can linearize this function.

Either hepatic metabolism, renal filtration, or a combination of the two eliminates most drugs. For some drugs, elimination by these routes is highly variable. In addition, functional changes in these organs may result in changes in the rate of elimination. In these situations, information regarding elimination rate and estimating the circulating concentration of a drug after a given time period are important factors in establishing a dosage regimen that is effective and safe. Equation 25–2 and Figures 25–2 and 25–3 are useful in determining the rate of elimination and the concentration of a drug after some period of time. The following equation illustrates how the equation is used.

Integration of Equation 25–2 gives the following:

$$C_T = C_0 e^{-kT} \qquad \textit{(Eq. 25–3)}$$

where: C_0 = the initial concentration of drug,

C_T = the concentration of drug after the time period (T),

k = the elimination constant, and

T = the time period evaluated.

This is the most useful form of the elimination equation. From it, we can calculate the elimination constant, or if k is known, we can determine the amount of drug that will be present after a certain time period.

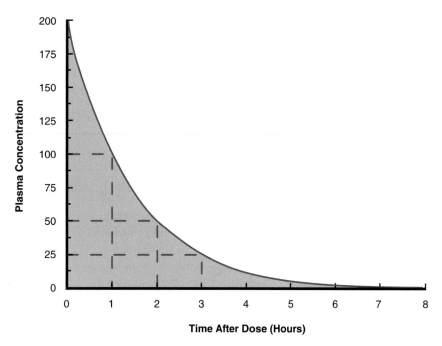

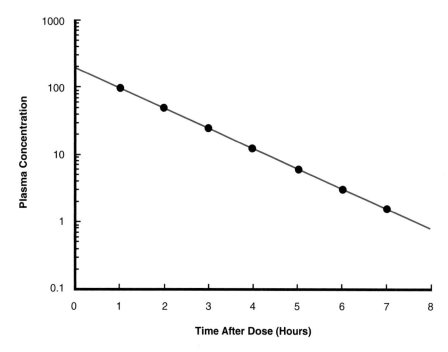

Figure 25-2. First order drug elimination. This graph demonstrates exponential rate of loss on a linear scale. Hash marked lines are representative of half-life.

Example:

The concentration of gentamicin is 10 μg/mL at 12:00. At 16:00, the gentamicin concentration is 6 μg/mL. What is the elimination constant (k) for gentamicin in this patient?

Using Equation 25-3:

$C_0 = 10$

$C_T = 6$

$T = 4$ hours

Substituting these values into the equation produces:

$6 = 10 \ e^{-k \ (4 \ hours)}$

Dividing both sides by 10 produces:

$0.6 = e^{-k \ (4 \ hours)}$

To eliminate the exponential sign, we would take the natural logarithm of both sides:

$\ln 0.6 = -k \ (4 \ hours)$

Solving the natural log:

$-0.51 = -k \ (4 \ hours)$

Multiplying through by −1:

$0.51 = k \ (4 \ hours)$

Dividing both sides by 4 hours:

$0.13/h = k$

This calculated value for k indicates the patient is eliminating 13% of serum gentamicin per hour.

In this same patient on the same day, what would be the predicted serum concentration of gentamicin at midnight (24:00)?

Figure 25-3. Semi logarithmic plot of exponential rate of drug elimination. The slope of this plot is equal to the rate of elimination (k).

For C_0, we can use either the 12:00 or 16:00 value as long as the correct corresponding time value is used. In this example, we will use the 16:00 value of 6 μg/mL.

$C_0 = 6$ μg/mL

T = 8 hours

k = 0.13/h

Substituting into Equation 25-3:

$C_T = 6\ e^{-0.13/h\ (8\ hours)}$

Solving for the exponent:

$C_t = 6\ e^{-1.04}$

Note that the time unit (hours) has canceled. Solving for the exponent:

$C_T = 6\ (0.35)$

$C_t = 2.1$ μg/mL

Note that the concentration units have carried through.

Even though the elimination constant (k) is a useful value, it is not common nomenclature in the clinical setting. Instead, the term *half-life* is used. This represents the time needed for the serum concentration to decrease by one half. It can be determined graphically (see Fig. 25-2) or by conversion of the elimination constant (k) to half-life ($T_{1/2}$) using the formula given in Equation 25-4. Of these two methods, the calculation provides an easy and accurate way to determine half-life.

$$T_{1/2} = 0.693/k \qquad \textit{(Eq. 25–4)}$$

Metabolic Clearance

Xenobiotics are substances that are not normally found within human systems and are capable of entering biochemical pathways intended for endogenous substances. Most drugs are xenobiotics. There are many potential biochemical pathways in which drugs can be acted on. The biochemical pathway responsible for the greatest portion of drug metabolism is the hepatic mixed function oxidase (MFO) system. The basic function of this system involves taking hydrophobic substances and, through a series of enzymatic reactions, converting them into water-soluble substances. These products are then either pumped into the bile or released into the general circulation, where they are eliminated by renal filtration.

There are many enzymes involved in the MFO system. They are commonly divided into two functional groups or phases. Phase I reactions produce reactive intermediates. Phase II reactions conjugate functional groups to these reactive sites, the products of which are water soluble. The groups that the reactive intermediates can be conjugated with are variable; glutathione, glycine, phosphate, and sulfate are common. The MFO system is a nonspecific system that allows many different endogenous and exogenous substances to go through this general scheme of reactions. Even though there are many potential substrates for this pathway, the products formed from an individual substance are specific. For example, acetaminophen is a substrate for MFO and always forms a glutathione conjugate. In addition, if the conjugating group for a given drug becomes depleted, the phase I products continue to be produced. In this situation, accumulation of phase I products may result in toxic side effects.

It is also noteworthy that the MFO system is inducible. This is seen as an increase in the synthesis and activity of the rate-limiting enzyme within this pathway. The most common inducers are xenobiotics which are substrates for this pathway. Because many potential substrates can enter this pathway, many drug–drug interactions can occur within this pathway.[9] Many of these interactions are competitive; however, noncompetitive inhibition can also occur. In both cases, this results in altered rates of elimination of the involved drugs.[10]

Implied in the discussion is that changes in hepatic status can result in changes in circulating drugs eliminated by this pathway.[11] Induction of the MFO system typically results in accelerated clearance and a corresponding shorter half-life. In the opposite manner, hepatic disease states characterized by a loss of functional tissue may result in slow rates of clearance and corresponding longer half-lives.[8] In these situations, TDM aides in dosage adjustment.

For some drugs, there is considerable variance in the rate of hepatic and nonhepatic drug metabolism within a normal population. This results in atypical clearance, even in the absence of disease. Establishing dosage regimens for these drugs is in many instances aided by the use of TDM. With the use of molecular genetics, it is now possible to identify common genetic variants of some drug-metabolizing pathways.[12] Identification of these individuals may assist in establishing an individualized dosage regimen.

Renal Clearance

The plasma free fraction of parent drugs or their metabolites is subject to glomerular filtration, renal secretion, or both.[13] For those drugs not secreted or subject to reabsorption, the elimination rate of free drug directly relates to the creatinine clearance rate. Thus, decreases in glomerular filtration rate directly results in increased serum half-life and concentration. The aminoglycoside antibiotics vancomycin and cyclosporine are examples of drugs with this behavior.

PHARMACOKINETICS

Pharmacokinetics is the mathematic modeling of changes in circulating drug concentration that assists in establishing or modifying a dosage regimen. This process takes into consideration all factors that determine the concentration of a serum drug and its rate of change. Many of the factors previously discussed in this chapter would be included in this field of study. Figure 25-3 is an idealized plot of elimination after an IV bolus. It assumes there is no distribution of this drug. A drug that does distribute outside of vascular space would produce an elimination graph such as seen in Figure 25-4. The rapid rate of change seen immediately after the

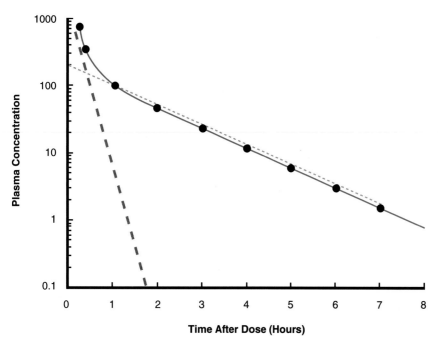

Figure 25-4. Semi logarithmic elimination plot of a drug subject to distribution. Initial rate of elimination is influenced by distribution (dashed line) and terminal elimination rate (dotted line). After distribution is complete (1.5 hr), elimination is first order.

initial IV bolus is due to distribution and elimination. The rate of elimination (k) can only be determined after distribution is complete. Figure 25-5 is a plot of serum concentration as it would appear after oral administration of a drug. As absorbed drug enters the circulation, it is subject to simultaneous distribution and elimination. Serum concentrations rise when the rate of absorption exceeds distribution and elimination. The concentration declines as the rate of elimination and distribution exceeds absorption. The rate of elimination can only be determined after absorption and distribution are complete.

Most drugs are not administered as a single bolus but instead are delivered on a scheduled basis (*eg*, once every 8 hours). With this type of administration, serum drug concentration oscillates between a maximum (*peak drug level*) and a minimum (*trough drug level*). The goal of a multiple dosage regimen is to achieve a trough that is in the therapeutic range and a peak that is not in the toxic range. Evaluation of this oscillating function cannot be done immediately after initiation of a scheduled dosage regimen. About seven doses are required before a steady-state oscillation is acquired. The basis of this number (seven doses) is demonstrated in Figure 25-6.

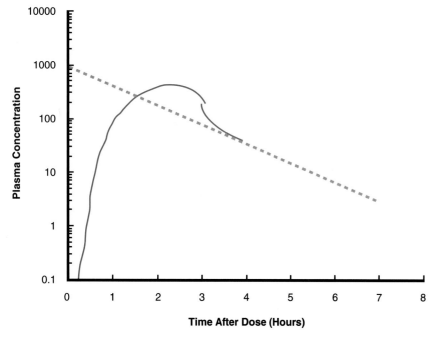

Figure 25-5. Plasma concentration of a drug after oral administration. After oral administration at time zero, serum concentration increases after a brief lag period (solid line). Plasma concentrations peak when rate of elimination and distribution exceed rate of absorption. First order elimination (dotted line) occurs when absorption and distribution are complete.

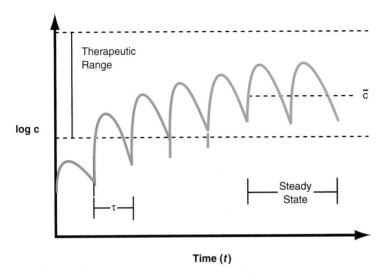

Figure 25-6. Steady state kinetics in a multiple dosage regimen. The dosage interval is indicated by the character τ. Equal doses at this interval reach steady state after 6 or 7 dosage intervals.

After the first oral dose, absorption and distribution occur, followed by elimination only. Before the concentration of drug drops significantly, the second dose is given. The peak of the second dose is additive to what remained from the first dose. Because elimination is first order, the higher concentration produces an increased rate of elimination. The third through seventh scheduled doses all have the same effect; increasing serum concentration and increasing the rate of elimination. By the end of the seventh dose, the amount of drug administered in a single dose is equal to the amount eliminated during the dosage period. At this point, steady state is established, and peak and trough concentrations can be evaluated.

SAMPLE COLLECTION

Timing of specimen collection is the single most important factor in TDM. In general, trough concentrations for most drugs are drawn right before the next dose; peak concentrations are drawn 1 hour after an orally administered dose. This rule of thumb must always be used within the clinical context of the situation. Drugs that are absorbed slowly may require several hours before peak drug levels can be evaluated. In all situations, determination of serum concentrations should be done after only steady state has been achieved.

Serum or plasma is the specimen of choice for the determination of circulating concentrations of most drugs. Care must be taken that the appropriate container is used when collecting these specimens. Some drugs have a tendency to be absorbed into the gel of some serum-separator–type collection tubes. Thus, it is necessary to follow vendor recommendations when this effect is possible. Failure to do so may result in falsely low values. Heparinized plasma is suitable for most drug analysis. The calcium-binding anticoagulants add a variety of anions and cations that may interfere with analysis or cause a drug to distribute differently between cells and plasma. As a result, EDTA and citrated and oxalated plasma are not usually acceptable specimens.

CARDIOACTIVE DRUGS

Many cardiac conditions are treated with drugs. Of these drugs, only a few require TDM.[14] The cardiac glycosides and the antiarrhythmics are two classes of drugs for which assessment of serum concentration aids in decisions regarding their dosage regime.[15]

Digoxin

Digoxin is a cardiac glycoside used in the treatment of congestive heart failure.[16] It functions by inhibiting membrane Na^+-K^+-ATPase. This causes a decrease in intracellular potassium that results in an increased intracellular calcium in cardiac myocytes. The increased presence of calcium improves cardiac contractility (inotropic effect). This effect is seen in the serum concentration range of 0.8 to 2 ng/mL. Higher serum concentrations (3 ng/mL) decrease the rate of ventricular depolarization. This level can be used to control ventricular tachycardia, but this is done infrequently because of toxic side effects that become apparent at serum concentrations of greater than 2 ng/mL. Digoxin toxicity affects many organ and cell types. Nausea, vomiting, and visual disturbances are common. Cardiac effects, such as premature ventricular contractions (PVCs) and atrioventricular node blockage, are also common.

Absorption of orally administered digoxin is variable. It is influenced by dietary factors, gastrointestinal motility, and the formulation of the drug. In circulation, about 25% is protein bound. The nonbound (free) form of serum digoxin is sequestered into muscle cells. At equilibrium, the tissue concentration is 15 to 30 times greater then plasma. Elimination of digoxin occurs primarily by renal filtration of the plasma free form. The remainder is metabolized to several products by the liver. The half-life of plasma digoxin is 38 hours in an average adult. The major contributing factor to the half-life is the slow release of tissue digoxin back into circulation.

Because of variable gastrointestinal absorption of digoxin, establishing a dosage regimen usually requires assessment of serum concentrations after initial dosing to ensure that effective and nontoxic serum concentrations are achieved.[17] Because changes in glomerular filtration rate can have a dramatic effect on serum concentration, frequent dosage adjustments, in conjunction with serum levels, need to be done in patients with renal disease. In addition, the therapeutic actions and toxicities of digoxin can be influenced by the concentration of serum electrolytes. Low serum potassium and magnesium potentiate digoxin actions. In these conditions, adjustment of serum concentrations below the therapeutic range may be necessary to avoid toxicity. Thyroid status may also influence digoxin actions. Hyperthyroid patients display a resistance to digoxin actions; hypothyroid patients are more sensitive.

The timing for evaluation of peak digoxin levels is crucial. In an average adult, serum levels peak between 2 to 3 hours after an oral dose. However, uptake into tissue is a relatively slow process. As a result, peak serum concentrations do not correlate with tissue concentrations. It has been established that the serum concentration 8 hours after an orally administered dose correlate with tissue concentration. Thus, peak levels are usually evaluated at this time. Levels collected before 6 to 8 hours after the last dose give misleading values and are not valid.

Immunoassay is used to measure total digoxin concentration in serum. With most commercial assays, cross-reactivity with hepatic metabolites is minimal; however, newborns, pregnant women, and patients with uremia or late-stage liver disease produce an endogenous substance that cross-reacts with the antibodies used to measure serum digoxin.[18] Thus, in patients with these digoxin-like immunoreactive substances, falsely elevated concentrations are common.

Lidocaine

Lidocaine is used to correct ventricular arrhythmia and to prevent ventricular fibrillation. This is particularly important in patients with acute myocardial infarction. Lidocaine administration cannot be oral owing to almost complete hepatic removal of the absorbed drug by hepatic metabolism (first-pass metabolism). Lidocaine is commonly delivered by continuous intravenous infusion after a loading dose. Thus, plasma levels remain relatively constant during administration. The primary reasons for monitoring plasma lidocaine concentration are to ensure it is in the therapeutic range of 1.5 to 4 µg/mL and to avoid toxicity, which exists just above the therapeutic range. Plasma lidocaine in the range of 4 to 8 µg/mL is associated with central nervous system depression. Plasma concentrations greater then 8 µg/mL are associated with seizures and severe decreases in blood pressure and cardiac output. Lidocaine is primary eliminated by hepatic metabolism. Thus, the concentration of plasma lidocaine is dependent on the rate of administration and the rate of hepatic clearance. Changes in renal function have little effect on plasma lidocaine. One of the primary products formed is monoethylglycinexylidide (MEGX). This metabolite is nearly as toxic as the parent drug. In many instances, the sum of lidocaine and MEGX must be taken into consideration when evaluating possible toxic reactions. Lidocaine can be assayed by chromatography or immunoassay.

Quinidine

Quinidine is a naturally occurring drug that can be used to treat a variety of cardiac arrhythmic situations.[19] The two most common clinical formulations are quinidine sulfate and

---

CASE STUDY 25-1

A patient with congestive heart failure has been successfully treated with digoxin for several years. Laboratory records indicate semi-annual peak digoxin levels have all been in the therapeutic range. This patient recently developed renal failure, and admission testing was performed. Selected serum or blood laboratory results from this specimen were as follows:

CASE STUDY TABLE 25-1.1. Laboratory Results

Test	Result	Reference Range
Sodium	129	135–145 mEq/L
Potassium	5.5	3.5–5 mEq/L
Chloride	113	97–107 mEq/L
Blood pH	7.25	7.35–7.45
TCO₂	16	21–31 mmol/L
Urea Nitrogen	180	5–20 mg/dL
Creatinine	4.5	0.6–1 mg/dL
Osmolality	275	282–300 mOsm/kg
Digoxin	2.5	0.9–2 ng/mL

Even though the digoxin is high, the physician indicates the patient is not exhibiting signs or symptoms of toxicity.

Questions

1. If these results were derived from a random specimen, how may the time since the last dose effect the interpretation of the digoxin results?
2. Other than time, what additional factors should be taken into consideration when interpreting the digoxin results?
3. What additional laboratory test would aid in the interpretation of this case?

quinidine gluconate. Oral intake is the most common route of administration. Gastrointestinal absorption is complete and rapid for the sulfate. Peak serum concentrations are reached about 2 hours after an oral dose of the sulfate. The gluconate is a slow-release formulation. Peak serum concentration is reached 4 to 5 hours after an oral dose. The most predominant toxic side effects of quinidine are nausea, vomiting, and abdominal discomfort. Cardiovascular toxicity, such as PVCs, may be seen at twice the upper limit of the therapeutic range. In most instances, monitoring of quinidine involves only determination of the trough level to ensure it is within the therapeutic range. Peak assessment is performed only when symptoms of toxicity are present. Because of its slow rate of absorption, trough levels of the gluconate are usually drawn 1 hour after the last dose.

Absorbed quinidine is about 70% bound to serum proteins. Most is eliminated by the hepatic MFO system. Induction of this system, such as by barbiturates, increases the clearance rate. Impairment of this system, as seen in late-stage liver disease, may extend the half-life of this drug. Plasma quinidine concentration can be determined by chromatography or immunoassay.

Procainamide

Procainamide is used to treat a variety of cardiac arrhythmias. Oral intake is the most common route of administration. Gastrointestinal absorption is rapid and complete. Peak plasma concentrations occur at about 1 hour. Absorbed procainamide is about 20% bound to plasma proteins. It is eliminated by a combination of renal filtration and hepatic metabolism. Alteration in the function of either of these organs may lead to increased serum concentration of the parent drug and its metabolites. Increased concentration results in myocardial depression and arrhythmia.[20] N-acetyl procainamide is one of the hepatic metabolites of the parent drug. It has antiarrhythmic activity similar to that of procainamide. Thus, the total antiarrhythmic potential of this drug must take into consideration the parent drug and this metabolite. Both can be measured by immunoassay.

Disopyramide

Disopyramide is a drug used to treat a variety of cardiac arrhythmias. It is commonly used as a quinidine substitute when quinidine side effects are excessive. It is most commonly administered as an oral preparation. Gastrointestinal absorption is complete and rapid. It binds to several plasma proteins. Binding is highly variable within individuals and is also concentration dependent: as serum concentration increases, so does the percentage free. As a result, it is difficult to correlate total serum concentration with therapeutic benefit and toxicity. In most patients, total serum concentrations in the range of 3 to 5 μg/mL have been determined to be effective and nontoxic; however, interpretation of disopyramide results should take the clinical perspective into consideration. The primary toxicities of disopyramide are dose dependent. Anticholinergic effects, such as dry mouth and constipation, may be seen at serum concentrations greater than 4.5 μg/mL. Cardiac effects, such as bradycardia and atrioventricular node blockage, are usually seen at serum concentrations greater than 10 μg/mL. Disopyramide is eliminated primarily by renal filtration and to a lesser extent by hepatic metabolism. With renal disease, the half-life is prolonged, and serum concentrations rise. Plasma disopyramide concentration can be determined by chromatography or immunoassay.

ANTIBIOTICS[21]

Aminoglycosides

The aminoglycosides are a group of chemically related antibiotics used for the treatment of infections with gram-negative bacteria that are resistant to less toxic antibiotics. There are many individual agents within this classification. The most commonly encountered in a clinical setting are gentamicin, tobramycin, amikacin, and kanamycin.[22] All share a common mechanism of action but vary in their effectiveness against different strains of bacteria. All have a common toxicity. Nephrotoxicity and ototoxicity are the most frequent. The ototoxic effect involves disruption of inner ear cochlear

---

CASE STUDY 25-2

A patient is receiving procainamide for treatment of cardiac arrhythmia. An intravenous loading dose resulted in a serum concentration of 6.0 μg/mL. The therapeutic range for procainamide is 4 to 8 μg/mL, and its half-life is 4 hours. Four hours after the initial loading dose, another equivalent dose was given as an intravenous bolus. This resulted in a serum concentration of 7.5 μg/mL.

Questions

1. Does the serum concentration after the second dose seem appropriate? If not, what would be the predicted serum concentration at this time?
2. What factors would influence the rate of elimination of this drug?

and vestibular membranes, which results in hearing and balance impairment.[23] These effects are irreversible. Cumulative effects may be seen with repeated high-level exposure. Nephrotoxicity is also of major concern. Aminoglycosides impair the function of proximal tubules of the kidney, which may result in electrolyte imbalance and possibly proteinuria. These effects are usually reversible; however, extended high-level exposure may result in necrosis of these cells and subsequent renal failure. Toxic concentrations are usually considered any concentration above the therapeutic range.

Because aminoglycosides are not well absorbed from the gastrointestinal tract, administration is limited to the IV or IM route. Thus, these drugs are not used in an outpatient setting. Aminoglycosides are eliminated by renal filtration. In patients with compromised renal function, appropriate adjustments based on serum concentrations must be made. Chromatography and immunoassay are the primary methods used for aminoglycoside determinations.

Vancomycin

Vancomycin is a glycopeptide antibiotic that is effective against gram-positive cocci and bacilli. Because of its poor oral absorption, vancomycin is administer by IV infusion. Unlike other drugs, a clear relationship between serum concentration and toxic side effects has not been firmly established. Indeed, many of the toxic side effects occur in the therapeutic range (5 to 10 µg/mL). The major toxicities of vancomycin are "red-man syndrome," nephrotoxicity, and ototoxicity. Red-man syndrome is characterized by an erythemic flushing of the extremities. The renal and hearing effects are similar to those of the aminoglycosides. It appears that the nephrotoxic effects occur more frequently at trough concentrations that are greater than 10 µg/mL. The ototoxic effect occurs more frequently when peak serum concentrations exceed 40 µg/mL. Because of these circumstances and because vancomycin has a long distribution phase, in most instances, only trough levels are monitored to ensure the serum drug concentration is within the therapeutic range.[24] Vancomycin is primary eliminated by renal filtration and excretion. It is assayed by immunoassay and chromatographic methods.

ANTIEPILEPTIC DRUGS

Epilepsy, convulsions, and seizures are prevalent neurologic disorders. Therapeutic ranges for this class of drugs are considered guidelines. Effective concentrations are determined as the concentration that works with no or acceptable side effects.[25] Most antiepileptic drugs can be analyzed by immunoassay or chromatography.

Phenobarbital

Phenobarbital is a slow-acting barbiturate that effectively controls several types of seizures. Absorption of oral phenobarbital is slow but complete. For most patients, peak serum concentration is reached about 10 hours after an oral dose. Circulating phenobarbital is 50% bound. It is eliminated primarily by hepatic metabolism. However, renal filtration is also significant. With compromised renal or hepatic function, the rate of elimination is decreased. The half-life of serum phenobarbital is 70 to 100 hours. Because of the slow absorption and long half-life, serum concentrations do not change dramatically within a dosing interval. Therefore, only trough levels are usually evaluated unless toxicity is suspected. Toxic side effects of phenobarbital include drowsiness, fatigue, depression, and reduced mental capacity.

Phenobarbital is a potent inducer of the hepatic MFO system, which is responsible for the metabolism of many drugs and many endogenous substances. Thus, after initiation of therapy, a dosage adjustment period is required to account for the degree of induction that expedites its clearance.

Primidone is an inactive proform of phenobarbital. After absorption of an oral dose, this drug is rapidly converted to its active form, phenobarbital. Primidone is used in preference of phenobarbital when steady-state kinetics need to be established quickly. Primidone is rapidly absorbed and converted to the active drug. Both primidone and phenobarbital need to be measured to assess the total potential amount of phenobarbital in circulation.

Phenytoin

Phenytoin (Dilantin) is used to treat a variety of seizure disorders. It is also used as a short-term prophylactic agent in brain injury to prevent loss of functional tissue. Phenytoin is primarily administered as an oral preparation. Gastrointestinal absorption is variable and sometimes incomplete. Circulating phenytoin has a high but variable degree of protein binding (87% to 97%). Like most drugs, the unbound (free) fraction is the biologically active portion of total serum concentration. Reduced protein binding may occur with anemia, with hypoalbuminemia, and in the presence of other drugs. Thus, toxicity may be observed when the total serum drug concentration is within the therapeutic range. The major toxicity of phenytoin is initiation of seizures. Thus, seizures in a patient being treated with phenytoin may be due to subtherapeutic or toxic levels. Additional adverse effects of phenytoin include hirsutism, gingival hyperplasia, vitamin D, and folate deficiency. Phenytoin is eliminated by hepatic metabolism in a somewhat unique manner. At therapeutic concentrations, this elimination pathway may become saturated (zero-order kinetics). Thus, relatively small changes in dosage or elimination may have dramatic effects on plasma concentration.

For most patients, total serum concentrations of 10 to 20 g/mL are effective. In many situations, however, the effective range of total serum concentration must be individualized to suit the clinical situation. The therapeutic range for free serum phenytoin is 1 to 2 µg/mL. This has been well correlated with the pharmacologic actions of this drug. In patients with altered serum protein binding, determination of the free fraction aids in dosage adjustment.

Fosphenytoin is an injectable proform of phenytoin that is rapidly metabolized in serum, releasing the parent drug.[26] It takes about 75 minutes for this conversion to take place. Most immunoassays for phenytoin do not detect this proform. Thus, peak levels should be evaluated only after the conversion to the active drug is complete.

Valproic Acid

Valproic acid is used for the treatment of petite mal and absence seizure.[27] It is administered as an oral preparation. Gastrointestinal absorption is rapid and complete. Circulating valproic acid is highly protein bound (93%). The percentage bound decreases in renal failure, in late liver disease, and in the presence of other drugs that may compete for its binding site. It is eliminated by hepatic metabolism. The therapeutic range for valproic acid is relatively wide (50 to 120 µg/mL). Determination of serum concentration is primarily done to ensure that toxic levels (more than 120 µg/mL) are not present. Nausea, lethargy, and weight gain are the most common adverse effects. Pancreatitis, hyperammonemia, and hallucinations have been associated with high serum levels (more than 200 µg/mL). Hepatic dysfunction occasionally occurs in some patients even at therapeutic serum concentrations. Thus, hepatic indicators should be checked frequently for the first 6 months after initiation of therapy. Many factors may influence the nonbound (free) fraction of total serum valproic acid. Therefore, determination of the free fraction provides a more reliable index of therapeutic and toxic concentrations.

Carbamazepine

Carbamazepine is an effective treatment in a wide variety of seizure disorders. Because of its serious toxic side effects, it is less frequently used, except when patients do not respond to other drugs. Orally administered carbamazepine is absorbed with a high degree of variability. Circulating carbamazepine is 70% to 80% protein bound. It is eliminated by hepatic metabolism. Many forms of liver dysfunction may result in serum accumulation. Carbamazepine is an inducer of its own metabolism. Thus, frequent plasma levels must be analyzed upon initiation of therapy until the induction period has come to completion.

Carbamazepine toxicity is diverse. Some effects occur in a dose-dependent manner; others do not. There are several non–dose-dependent, idiosyncratic effects of carbamazepine that may require discontinuation of the drug. These include rashes, leukopenia, nausea, vertigo, and febrile reactions. Of these, leukopenia is the most serious. Leukocyte counts are commonly done during the first 2 weeks of therapy to detect this possible toxic effect. Other toxic effects occur in a dose-dependent manner. The therapeutic range for carbamazepine is 4 to 12 µg/mL.[28] Plasma concentrations greater than 15 µg/mL are associated with hematologic dyscrasias and possible aplastic anemia. At therapeutic concentrations, carbamazepine may also cause mild liver dysfunction, which is usually self-limiting. Increases of liver indices that are three times the upper limit of the normal reference range are an indication for discontinuation.

Ethosuximide

Ethosuximide is used for control of petite mal seizure. It is administered as an oral preparation. The therapeutic range is 40 to 100 µg/mL. The toxicities associated with high plasma concentrations are rare, tolerable, and self-limiting. TDM of ethosuximide is done to ensure that serum concentrations are in the therapeutic range.

PSYCHOACTIVE DRUGS

Lithium

Lithium is an orally administered drug used to treat manic-depressive illness (bipolar disorders). Absorption is complete and rapid. Lithium is a cationic metal that does not bind to proteins.

Distribution is uniform throughout the total body water. It is eliminated predominately by renal filtration and is sub-

---

CASE STUDY 25-3

A child, who has been successfully treated for seizure disorders with oral phenytoin for several years has had severe diarrhea for the past 2 weeks. Subsequent to this, the patient had a seizure. Evaluation of serum phenytoin at the time of the seizure revealed a low value. The dose was increased until serum concentrations were within the therapeutic range. The diarrhea was resolved. Several days after this the patient had another seizure.

Questions

1. What is the most probable cause of the initial low serum phenytoin?
2. Would determination of free serum phenytoin aid in resolving the cause of the initial seizure?
3. What assays other than the determination of serum phenytoin would aid in this situation?
4. What is the most probable cause of the seizure after the diarrhea has been resolved?

ject to reabsorption. Compromises in renal function usually result in accumulation. Correlations between serum concentration and therapeutic response have not been well established. However, serum concentrations in the range of 0.8 to 1.2 mmol/L are effective in a large portion of the patient population. The purpose of TDM for lithium is to avoid serum concentrations that have been associated with toxic side effects.[29] Serum concentrations in the range of 1.2 to 2 mmol/L may cause apathy, lethargy, speech difficulties, and muscle weakness. Serum concentrations greater then 2 mmol/L are associated with muscle rigidity, seizures, and possible coma. Determination of serum lithium is most commonly done by ion-selective electrode. Flame emission photometry and atomic absorption are also viable methods.

Tricyclic Antidepressants

Tricyclic antidepressants (TCAs) are a class of drugs used to treat depression, insomnia, extreme apathy, and loss of libido. From a clinical laboratory perspective, imipramine, amitriptyline, and doxepin are the most relevant.[24] Desipramine and nortriptyline are active metabolic products of imipramine and amitriptyline, respectively, and thus must also be included. The TCAs are orally administered drugs with a variable degree of absorption. In many patients, they slow gastric emptying and intestinal motility, which significantly slows their rate of absorption. As a result, peak serum concentrations are reached in the range of 2 to 12 hours.

The TCAs are highly protein bound (85% to 95%). For most TCAs, the therapeutic effects are not seen for the first 2 to 4 weeks after initiation of therapy. The correlations between serum concentration and therapeutic effects of most TCAs is moderate to weak. They are eliminated by hepatic metabolism. Many of the products formed have therapeutic actions. The rate of metabolism of these agents is variable and influenced by a wide variety of factors. As a result, the half-life of TCAs vary considerably among patients. The rate of elimination can also be influenced by the coadministration of other drugs that are eliminated by hepatic metabolism. The toxicity of TCAs is dose dependent. At serum concentrations about twice the upper limit of the therapeutic range, drowsiness, constipation, blurred vision, and memory loss are common adverse effects. Higher levels may cause seizure, cardiac arrhythmia, and unconsciousness.

Because of the high variability in half-life and absorption, plasma concentrations of the TCAs should not be evaluated until a steady state has been achieved. At this point, therapeutic efficacy is determined from clinical evaluation of the patient, and potential toxicity is determined by serum concentration. Many of the immunoassays for TCAs use polyclonal antibodies, which cross-react among the different TCAs and their metabolites. In this analytic system, the results are reported out as "total tricyclics." Other immunoassays use an extraction step to separate parent drugs from the metabolites. Interpretation of these results after extraction requires an in-depth understanding of the assay. Chromato-

graphic methods provide simultaneous evaluation of both the parent drugs and metabolites, which provides unambiguous interpretation of results.[30]

BRONCHODILATORS

Theophylline

Theophylline is used in the treatment of asthma and other chronic obstructive pulmonary diseases. It is effective in both acute situations and prophylactically. Theophylline therapy is usually initiated intravenously, then changed to oral administration. Oral theophylline is absorbed completely but at a variable rate, which is dependent on the formulation of the drug and dietary factors. Absorbed theophylline is 50% protein bound in plasma. It is eliminated by a combination of renal filtration and hepatic metabolism. Thus, small changes in serum protein content and glomerular filtration rate have little influence on plasma concentrations. The primary reason for monitoring serum theophylline is to ensure that concentrations are not in the toxic range. The therapeutic range for theophylline is 10 to 20 μg/mL. Toxic effects are noted at serum concentrations greater the 20 μg/mL. Symptoms of toxicity include nausea, vomiting, and diarrhea. Serum concentration greater than 30 μg/mL are associated with cardiac arrhythmia, seizures, and a poor prognosis.

IMMUNOSUPPRESSIVE DRUGS

Transplantation medicine is a rapidly emerging discipline within clinical medicine. The clinical laboratory plays many important roles that determine the success of any transplantation program.[31] Among these responsibilities, monitoring of the immunosuppressive drugs used to prevent rejection is of key concern. Most of these drugs require establishment of individual dosage regimens to optimize therapeutic outcomes and minimize toxicity.[32]

Cyclosporine

Cyclosporine is a cyclic polypeptide that has potent immunosuppressive activity. Its primary clinical use is suppression of host-versus-graft rejection of heterotopic transplanted organs. It is administered as an oral preparation. Absorption of cyclosporine is in the range of 5% to 50%. Because of this high variability, the relationship between oral dose and blood concentration is poor. Thus, TDM is an important part of establishing an initial dosage regime. Circulating cyclosporine sequesters in cells, including erythrocytes. Erythrocyte content is highly temperature dependent; thus, evaluation of plasma concentration requires rigorous control of specimen temperature. To avoid this preanalytic variable, whole blood is the specimen of choice. Correlations have been established between whole blood concentration and therapeutic and toxic effects. Cyclosporine is eliminated by hepatic metabolism to inactive products.

Immunosuppression requirements differ depending on the organ transplanted. Cardiac, liver, and pancreas transplants have the highest requirement (300 ng/mL). Whole blood concentrations in the range of 350 to 400 ng/mL have been associated with toxic effects. The toxic effects of cyclosporine are primarily renal tubular and glomerular dysfunction, which may result in hypertension. Several immunoassays are available for determination of whole blood cyclosporine concentration. Many cross-react with inactive metabolites. Chromatographic methods are available; they provide separation and quantitation of the parent drug from metabolites.

Tacrolimus

Tacrolimus (FK-506) is an orally administered immunosuppressive drug that is 100 times more potent than cyclosporine; thus, the dosage is far less than that of cyclosporine.[33] Early use of tacrolimus suggested a low degree of toxicity compared with cyclosporine at therapeutic concentrations. After extensive use in clinical practice, however, it has been demonstrated that both have comparable degrees of nephrotoxicity at therapeutic concentrations. The renal effects of tacrolimus are similar to those of cyclosporine. Tacrolimus has also been associated with thrombus formation.

Many aspects of tacrolimus pharmacokinetics are similar to cyclosporine. Gastrointestinal uptake is highly variable. Whole blood concentrations correlate well with therapeutic and toxic effects. Tacrolimus is eliminated almost exclusively by hepatic metabolism. Metabolic products are primarily secreted into the bile. Increases in immunoreactive tacrolimus may be seen in cholestasis owing to cross-reactivity with several of these products. Because of the high potency of tacrolimus, circulating therapeutic concentrations are very low. This limits the methodologies capable of measuring whole blood concentrations. The most original method is HPLC/MS; however, several immunoassays are also available.[34]

ANTINEOPLASTICS

Assessment of therapeutic benefit and toxicity of most antineoplastic drugs is not aided by TDM.[35] This is because correlations between plasma concentration and therapeutic benefit are hard to establish with many of these agents. Many of these agents are rapidly metabolized or incorporated into cellular macromolecular structures within seconds to minutes of their administration. In addition, the therapeutic range for many of these drugs also includes the concentrations associated with toxic side effects. Considering that most antineoplastic agents are administered intravenously as a single bolus, the actual delivered dose is more important than circulating concentrations.

Methotrexate

Methotrexate is one of the few antineoplastic drugs in which TDM offers benefits to a therapeutic regimen.[36] High-dose methotrexate followed by leucovorin rescue has been shown to be an effective therapy for a variety of neoplastic conditions. The basis of this therapy involves the relative rate of mitosis of normal versus neoplastic cells. In general, neoplastic cells divide more rapidly than normal cells. Methotrexate inhibits DNA synthesis in all cells. Neoplastic cells, owing to their rapid rate of division, have a higher requirement for DNA and are thus susceptible to depravation of this essential constituent sooner than normal cells. The efficacy of methotrexate therapy is dependent on a controlled period of inhibition, one that is selectively detrimental to neoplastic cells. This is accomplished by administration of leucovorin, which reverses the actions of methotrexate at a specific time after methotrexate infusion. This is referred to as *leucovorin rescue*. Failure to stop methotrexate actions results in cytotoxic effects to most cells. Evaluation of serum methotrexate concentration, after the appropriate time period has passed, is used to determine how much leucovorin is needed to counteract many of the toxic effects of methotrexate.

SUMMARY

TDM is a process used to generate indices that are used as the basis for establishing a rational, individualized drug regimen to ensure optimal patient outcomes.[37] For most drugs, this process is unneeded. Some drugs, however, can produce severe side effects at dosages very close to those which result in therapeutic benefit. In these instances, trial and error may be an inappropriate method to establish a safe and effective dosage regimen. This is especially true if the drug is administered orally and if the rate of drug absorption is highly variable.

Monitoring serum concentrations of these drugs when establishing a dosage regimen is an important component of therapy with these agents. TDM allows for the safe use of drugs that would otherwise be potentially toxic. This expands the number of drugs available to treat disease and in many instances improves patient care.[38] In a similar manner, standard drug dosages are established for healthy, average people. For individuals who vary from these characteristics, however, adsorption and distribution of a drug may be unpredictable.[39] Again, if the therapeutic range of the drug is narrow and potentially toxic side effects exist, TDM is important in establishing a dosage regimen.

Many drugs are subject to biotransformation (metabolism). For most drugs, the products of these reactions are neither toxic nor pharmacologically active. Some drugs, however, present with active or toxic products that need to be taken into consideration when evaluating their therapeutic or toxic effects.

Specimen timing, collection, and transport are important factors that determine the usefulness and validity of data derived from TDM. Lack of control of these preanalytic variables may result in inappropriate patient outcomes.

Even though only a small number of drugs prescribed are commonly subject to TDM, the scope of this field is expanding as toxic side effects become better defined and therapeutic ranges are further refined. The basic principles of TDM, which address absorption, distribution, and elimination, can also be applied to nontherapeutic substances that have entered the body. Indeed, the use of these concepts is central to the study of poisons and toxicology.

REVIEW QUESTIONS

1. Drug X has a half-life ($T_{1/2}$) of 2 days. The concentration at noon today is 10 μg/mL. What would be the expected concentration of drug X tomorrow at noon?
   a. 7.5 μg/mL
   b. 7 μg/mL
   c. 5 μg/mL
   d. 3.5 μg/mL
   e. Cannot be determined with the information provided
2. Salicylic acid is a common component of many over-the-counter drugs. In a patient suffering from gastric achlorhydria, what would be the predicted serum concentration of this drug after a standard dose?
   a. Greater then expected
   b. Less then expected
   c. No change
   d. Not enough information provided
3. Which of the following drugs would be correctly classified as an antiepileptic?
   a. Digoxin
   b. Disopyramide
   c. Chloramphenicol
   d. Phenytoin
   e. Tacrolimus
4. Of the following, which would be the most appropriate time for evaluation of a peak digoxin level after oral administration?
   a. Immediately before the next dose
   b. Immediately after a dose
   c. 8 hours after a dose
   d. 3 days after a dose
   e. None of the above; digoxin is not administered orally.
5. Of the following statements concerning lidocaine, which is/are true?
   a. Lidocaine is only administered as an oral preparation.
   b. Phenobarbital is one of the products of lidocaine metabolism.
   c. The toxicity of lidocaine is related to the concentration of the parent drug and one of its metabolites—MEGX.
   d. All of the above are true.
   e. Only a and c are true.

6. Of the following statements concerning procainamide, which is/are true?
   a. Procainamide is an antibiotic.
   b. *N*-acetylprocainamide is an active product of procainamide metabolism.
   c. The primary toxicity of procainamide is bone marrow suppression.
   d. All of the above are true.
   e. Only a and c are true.
7. Of the following statements concerning lithium, which is/are true?
   a. Lithium is an element.
   b. Lithium is used as a drug to treat depression and mania.
   c. Lithium concentration in serum is most commonly evaluated by ion-specific electrode.
   d. All of the above are true.
   e. Only a and c are true.
8. What is the purpose for the determination of serum concentrations of the antineoplastic drug methotrexate?
   a. To ensure that serum concentrations are in the therapeutic range
   b. To ensure that serum concentrations are not in the toxic range
   c. To determine the amount of leucovorin needed to halt methotrexate action
   d. All of the above are true.
   e. Only a and c are true.
9. I am an immunosuppressive drug used to control host-versus-graft rejection of transplanted organs. Renal toxicity is my primary problem. I am commonly assayed for by chromatographic techniques using whole blood as the specimen. What am I?
   a. Cyclosporine
   b. Carbamazepine
   c. Tacrolimus
   d. I could be any of the above.
   e. I could be either a or c.
10. A patient, who has been successfully receiving gentamicin for the past 2 weeks, has suddenly developed renal failure. What would be the expected adjustment in dosage in response to this?
    a. The dosage should be increased.
    b. The time interval between dosages should be increased.
    c. Phenobarbital should be coadministered to stimulate hepatic metabolism.
    d. No dosage adjustment is required.
    e. The drug should be discontinued.
11. Salicylate and bilirubin compete for the same binding site on serum albumin. What effect would prehepatic jaundice have on salicylate?
    a. An increase in the rate of clearance of salicylate
    b. A decrease in the pharmacologic response to salicylate
    c. An increase in the free concentration of salicylate
    d. All of the above are true.
    e. Only a and c are true.

12. A drug with a very small volume of distribution:
    a. is confined to the vasculature.
    b. diffuses out of the vasculature into interstitial space.
    c. diffuses out of the vasculature into intracellular space.
    d. selectively partitions into the fatty compartment.
    e. is rapidly eliminated in exhaled breath

13. Twenty milligrams of a drug is injected intravenously. One hour after injection, blood is drawn and assayed for the drug. The concentration in this specimen was 0.4 mg/L. What is the volume of distribution for this drug?
    a. 0.8 L
    b. 8 L
    c. 20 L
    d. 50 L
    e. Unable to determine with the data provided

14. A new orally administered drug has been introduced in your institution. It is unclear whether TDM is needed for this drug. What factors should be taken into consideration when addressing this question?
    a. Consequences of a subtherapeutic concentration in circulation
    b. Severity of toxic side effects
    c. Predictability of serum concentrations after a standard oral dose
    d. Proximity of toxic range to therapeutic range
    e. All of the above should be taken into consideration.

REFERENCES

1. Gentry CA, Rodvold KA. How important is TDM in the prediction and avoidance of adverse reactions? Drug Saf 1995;12:359–363.
2. Samara E, Granneman R. Role of population pharmacokinetics in drug development. Clin Pharmacokinet 1997;32:294–312.
3. Walson PD. Therapeutic drug monitoring in special populations. Clin Chem 1998;44:415–419.
4. Marshall A. Laying the foundations for personalized medicines. Nat Biotechnol 1997;15:954–957.
5. Schumacher GE, Barr JT. Total testing process applied to therapeutic drug monitoring: impact on patient outcomes and economics. Clin Chem 1998;44:370–374.
6. Tonkin AL, Bocher F. Therapeutic drug and patient outcomes: A review of the issues. Clin Pharmacokinet 1994;27:169–174.
7. Feldman EB. How grapefruit juice potentiates drug bioavailability. Nutr Rev 1997;55:398–400.
8. Kwan KC. Oral bioavailability and the first-pass effect. Drug Metab Dispos 1997;25:1329–1336.
9. Fraser AG. Pharmacokinetic interactions between alcohol and other drugs. Clin Pharmacokinet 1997;33:79–90.
10. Paintaud G, Bechtel Y, Brientini MP, et al. Effects of liver disease on drug metabolism. Therapie 1996;51:384–389.
11. Huet PM, Villeneuve JP, Fenyves D. Drug elimination in chronic liver disease. J Hepatol 1997;26(Suppl 2):63–72.
12. Kalow W. Pharmacogenetics. Pharmacol Rev 1997;49:369–379.
13. Lam YW, Banjerji S, Hatfield C, et al. Principals of drug administration in renal insufficiency. Clin Pharmacokinet 1997;32:30–57.
14. Valdes R, Jortani SA, Gheorhiade M. Standards of laboratory practice: cardiac drug monitoring. National Academy of Clinical Biochemistry. Clin Chem 1998;44:1096–1109.
15. Keyler DE, VanDeVoort JT, Howard JE, et al. Monitoring blood levels of selected drugs: remember to factor in many confounding variables. Postgrad Med 1998;103:209–212, 215–219.
16. Cohn JN. Overview of the treatment of heart failure. Am J Cardiol 1997;80:2L–6L.
17. Caufield JS, Gums JG, Grauer K. The serum digoxin concentration: ten questions to ask. Am Fam Physicians 1997;56:495–503.
18. Jortani SA, Valdes R. Digoxin and its related endogenous factors. Crit Rev Clin Lab Sci 1997;34:225–274.
19. Grace AA, Camm AJ. Quinidine. N Engl J Med 1998;338:35–45.
20. Kolecki PF, Curry SC. Poisoning by sodium channel blocking agents. Crit Care Clin 1997;13:829–848.
21. Hammett CA, Johns T. Laboratory guidelines for monitoring antimicrobial drugs. National Academy of Clinical Biochemistry. Clin Chem 1998;44:1129–1140.
22. Beggs EJ, Barclay ML. Aminoglycosides: 50 years on. Br J Clin Pharmacol 1995;39:597–603.
23. Priuska EM, Schacht J. Mechanism and prevention of aminoglycocide ototoxicity. Ear Nose Throat J 1997;76:164–171.
24. Andres I, Lopex R, Pou L, et al. Vancomycin monitoring: one or two serum levels? Ther Drug Monit 1997;19:614–619.
25. Brodie MJ, Ditcher MA. Established antiepileptic drugs. Seizure 1997;6:159–174.
26. Luer MS. Fosphenytoin. Neurol Res 1998;20:178–182.
27. Sundqvist A, Tomson T, Lundkvist B. Valproate as monotherapy for juvenile myoclonic epilepsy: dose effect study. Ther Drug Monit 1998;20:149–157.
28. Bialer M, Levy RH, Perucca E. Does carbamazepine have a narrow therapeutic plasma concentration range? Ther Drug Monit 1998;20:56–59.
29. Linder MW, Keck PE. Standards of laboratory practice: antidepressant drug monitoring. National Academy of Clinical Biochemistry. Clin Chem 1998;44:1073–1084.
30. de la Torre, Ortuno J, Pascual JA, et al. Quantitative determination of tricyclic antidepressants and their metabolites in plasma by solid-phase extraction and separation by capillary gas chromatography with nitrogen-phosphorus detector. Ther Drug Monit 1998;20:340–346.
31. Hickman PE, Poter JM, Pesce Aj. Clinical chemistry and post-liver-transplant monitoring. Clin Chem 1997;43:1546–1554.
32. Shaw LM, Kaplan B, Kaufman D. Toxic effects of immunosuppressive drugs: mechanisms and strategies for controlling them. Clin Chem 1996; 42(8 Pt 2);1316–1321.
33. Jusko WJ, Thomson AW, Fung J, et al. Consensus document: therapeutic monitoring of tacrolimus (FK-506). Ther Drug Monit 1995;17:606–614.
34. Cogill JL, Taylor PJ, Westly IS, et al. Evaluation of the tacrolimus II micro particle enzyme assay in liver and kidney transplant recipients. Clin Chem 1998;44:1942–1946.
35. McLeod HL. Therapeutic drug monitoring opportunities in cancer therapy. Pharmacol Ther 1997;74:39–54.
36. Lovell DJ. Ten years experience with methotrexate. Rev Rheum Engl Ed 1997;64 Suppl 10:186S–188S.
37. Schumacher GE, Barr JT. Bayesian and threshold probabilities in therapeutic drug monitoring: when can serum drug concentration alter clinical decisions. Am J Hosp Pharm 1994;51:321–327.
38. Moyer TP, Oliver LK. Supporting pharmaceutical studies from FDA submissions: diversifying the drug monitoring laboratory. Clin Chem 1998;44:433–436.
39. Koren G. Therapeutic drug monitoring principals in the neonate. Clin Chem 1997;43:222–227.

SUGGESTED READINGS

Goodman LS, Milinoff PB, Limbird LE, et al. Goodman & Gilman's the pharmacological basis of therapeutics. 9th ed. New York: Pergamon Press.
Greenblatt DJ, Shader RI. Pharmacokinetics in clinical practice. Philadelphia: WB Saunders, 1985.
Wong SH, Sunshine I. Handbook of analytic therapeutic drug monitoring and toxicology. Boca Raton: CRC Press, 1997.

CHAPTER 26

Toxicology

David P. Thorne

Objectives

Upon completion of this chapter, the clinical laboratorian should be able to:

- *Define the term* toxicology.
- *List the major toxicants.*
- *Define the pathologic mechanisms of the toxicants discussed in the chapter.*
- *Discuss the laboratory methods used to evaluate toxicity.*
- *Explain the difference between quantitative and qualitative tests in toxicology.*
- *Critically evaluate clinical laboratory data in poisoning cases and provide recommendations for further testing.*
- *Define the role of the clinical laboratory in the evaluation of exposure to poisons.*

KEY TERMS

Dose response relationship	Drugs of abuse Poison	TD_{50} Toxicology

*T*oxicology is the study of poisons. The scope of this field is very broad. There are four major disciplines within toxicology: mechanistic, descriptive, forensic, and clinical toxicology. Mechanistic toxicology elucidates the cellular and biochemical effects of toxins. These studies provide a basis for rationale therapy design and the development of tests to assess the degree of exposure of poisoned individuals. Descriptive toxicology uses the results from animal experiments to predict what level of exposure will cause harm in humans. This process is known as *risk assessment*. Regulatory toxicologists are responsible for interpreting the data from mechanistic and descriptive studies to establish standards that define the level of exposure that will not pose a risk to public health or safety. Typically, these toxicologists work for, or in conjunction with, government agencies. Forensic toxicology is primarily concerned with the medicolegal consequences of exposure to a toxin. A major focus of this area is establishing and validating the analytic performance of the methods used to generate evidence in legal situations, including the cause of death. Clinical toxicology is the study of interrelationships between toxin exposure and disease states. This area emphasizes not only diagnostic testing but also therapeutic intervention.

Within the organizational scheme of a typical medical laboratory, toxicology is usually considered to be part of chemistry, mainly because the methods used to evaluate toxins qualitatively and quantitatively are best suited to this area. However, appropriate diagnosis and management of poisoning victims, in many instances, requires an integrated approach from all sections of the clinical laboratory.[1]

EXPOSURE TO TOXINS

Exposure to toxic agents can occur for a variety of reasons. From a clinical standpoint, about 50% of poisoning cases are intentional suicide attempts. Accidental exposure accounts for about 30% of cases. The remainder of cases are due to homicide or occupational exposure. Of these, suicide has the highest mortality rate. Accidental exposure occurs most frequently in children; however, an accidental drug overdose of either therapeutic or illicit drugs is relatively common in adults. Occupational exposure primarily occurs in industrial and agricultural settings.

ROUTES OF EXPOSURE

Toxins can enter the body by several routes. Ingestion, inhalation, and transdermal absorption are the most common. Of these, ingestion is the most often seen in a clinical setting. For most toxins to exert a systemic effect, they must be absorbed into circulation. Absorption of toxins from the gastrointestinal tract occurs by several mechanisms. Some are taken up by processes intended for dietary nutrients. However, most are absorbed by passive diffusion. This process requires that the substance cross cellular barriers. Hydrophobic substances have the ability to diffuse across cell membranes and thus can be absorbed anywhere along the gastrointestinal tract. Ionized substances cannot passively diffuse across membranes. Weak acids can become protonated in gastric acid. The result is a non-ionized species, which can be absorbed in the stomach. In a similar manner, weak bases favor absorption in the intestine, where the pH is largely neutral or slightly alkaline. Other factors can influence absorbance of toxins from the gastrointestinal tract, including rate of dissolution, gastrointestinal motility, resistance to degradation in the gastrointestinal tract, and interactions with other substances. Toxins that are not absorbed from the gastrointestinal tract do not produce systemic effects but may produce local effects, such as diarrhea, bleeding, or malabsorption of nutrients, which may cause systemic effects secondary to toxin exposure.

DOSE RESPONSE RELATIONSHIP

A *poison* can be defined as any substance that causes a harmful effect upon exposure. Even though this basic definition is useful, other factors must be taken into consideration. Among these, dose is a key issue. The concept that any substance has the potential to cause harm if given at the correct dosage (even water) is a central theme in toxicology. Thus, there is a need to establish an index of the relative toxicity of substances to allow assessment of their potential to cause pathologic effects. Several systems are available. Most correlate the dose of a toxin that will result in a harmful response. One such system correlates a single acute oral dose range with the probability of a lethal outcome in an average 70-kg man (Table 26-1). This is a useful system to compare the relative toxicities of substances. The predicted response in this system is death, which is valid. However, many toxins can express pathologic effects at different relative concentrations compared to death. Thus, other indices have been developed.

A more in-depth characterization can be acquired by evaluating data from a cumulative frequency histogram of toxic responses over a range of doses. This experimental approach is typically used to evaluate several responses over a wide range

TABLE 26-1. Toxicity Rating System

Toxicity Rating	Lethal Oral Dose in Average Adult
Super toxic	<5 mg/kg
Extremely toxic	5–50 mg/kg
Very toxic	50–500 mg/kg
Moderately toxic	0.5–5 g/kg
Slightly toxic	5–15 g/kg
Practically nontoxic	>15 g/kg

Adapted from Klaassen CD. Principles of toxicology. In: Klaassen CD. Amdur MO, Doull J (eds). Toxicology: the basic science of poisons. 3rd ed. New York: Macmillan, 1986; 13.

of concentrations. One response monitored is the toxic response. This is the response that causes an early pathologic effect at lower than lethal doses. This response has been determined to be a indicator of the toxic effects specific for that toxin. Thus, for a substance that exerts early toxic effects by damaging liver cells, the response monitored may be increases in serum alanine aminotransferase (ALT) or gamma-glutamyl-transferase (GGT) activity. The dose response relationship implies that there will be an increase in the toxic response as the dose is increased. However, not all individuals display a toxic response at the same dose. The population variance can be seen in a cumulative frequency histogram of the percentage of people producing a toxic response over a range of concentrations (Fig. 26-1). The TD_{50} is the dose that would be predicted to produce a toxic response in 50% of the population. If the monitored response is death, the LD_{50} is the dose that would predict death in 50% of the population. Similar experiments can be used to evaluate the doses of therapeutic drugs. The ED_{50} is the dose that would be predicted to be effective or have a therapeutic benefit in 50% of the population. The therapeutic index is the ratio of the TD_{50} to the ED_{50}. Drugs

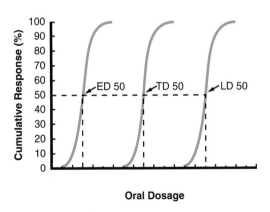

▲

Figure 26-1. **Dose-response relationship.** Comparison of responses of a therapeutic drug over a range of doses. The ED_{50} is the dose of drug in which 50% of treated individuals will experience benefit. The TD_{50} is the dose of drug in which 50% of individuals will experience toxic side-effects. The LD_{50} is the dose of drug in which 50% of individuals will result in morbidity.

with a large therapeutic index have few toxic side effects when the dose of drug is in the effective range.

Acute and Chronic Toxicity

Acute toxicity and *chronic toxicity* are terms used to relate the duration and frequency of exposure to the toxic effects seen. Acute toxicity is associated with a single, short-term exposure to a substance, the dose of which is sufficient to cause a toxic effect. Chronic toxicity is associated with repeated exposure for extended periods of time, usually at doses that are insufficient to cause an acute response. In many instances, chronic exposure is due to an accumulation of the toxicant or the toxic effects. Dose response relationships have been establish for many toxic substances in both acute and chronic situations.

ANALYSIS OF TOXIC AGENTS

In most instances, analysis of toxic agents in a clinical setting is a two-step procedure.[2] The first step is a screening test. Screening tests are rapid, simple, qualitative procedures intended to detect specific drugs or classes of drugs or toxicants. In general, these procedures are sufficiently sensitive but lack specificity. Thus, a negative result can rule out the presence of a drug or toxicant. However, a positive result should be considered a presumptive positive until confirmed by a second method that is more specific.

A variety of analytic methods can be used for screening and confirmatory testing. Immunoassays are commonly used to screen specimens for drugs. In some instances, these assays are specific for a single drug (*eg*, tetrahydrocannabinol [THC]). In most cases, however, drugs within general classes are detected (*eg*, barbiturates, opiates). Thin-layer chromatography is a relatively simple, inexpensive method of detecting a wide variety of drugs and other organic compounds. Gas chromatography is a widely used, well-established technique for the qualitative and quantitative determination of a wide number of volatile substances. The reference method for the qualitative identification of most organic compounds is gas chromatography using a mass spectrometer as the detector.

TOXICOLOGY OF SPECIFIC AGENTS

Many chemical agents encountered on a regular basis have potential adverse effects. The focus of this section is to survey the commonly encountered nondrug toxins seen in a clinical setting as well as those that present as medical emergencies with acute exposure.

Alcohols

The toxic effects of alcohols are both general and specific. Exposure to alcohols, like exposure to most volatile organic solvents, initially causes disorientation, confusion, and euphoria,

which can progress to unconsciousness, paralysis, and, with high-level exposure, even death. Most alcohols display these effects at about equivalent molar concentrations. This similarity suggests a common depressant effect on the central nervous system (CNS) that appears to be mediated by changes in membrane properties. In most cases, recovery from CNS effects are rapid and complete after cessation of exposure.

Distinct from the general CNS effects are the specific toxicities of each type of alcohol, which are usually mediated by biotransformation of alcohols to toxic products. There are several pathways by which short-chain aliphatic alcohols can be metabolized. Of these, hepatic conversion to an aldehyde, by alcohol dehydrogenase (ADH), and further conversion to an acid, by hepatic aldehyde dehydrogenase (ALDH), is the most significant.

$$\text{Alcohol} \xrightarrow{\text{ADH}} \text{Aldehyde} \xrightarrow{\text{ALDH}} \text{Acid}$$

(Eq. 26–1)

Ethanol exposure is very common.[3] Excessive ethanol consumption, with its associated consequences, is a leading cause of economic, social, and medical problems in industrialized countries. The economic impact is estimated to exceed $100 billion per year in terms of lost wages and productivity. Many social and family problems are associated with excessive ethanol consumption. The burden to the health care system is significant. Ethanol-related disorders are consistently one of the top-10 leading causes of hospital admissions. About 20% of all hospital admissions have some degree of alcohol-related problems. It is estimated that 80,000 Americans die each year either directly or indirectly as a result of abusive alcohol consumption. This correlates to about a five-fold increase in premature mortality. In addition, consumption of ethanol during pregnancy may lead to fetal alcohol syndrome or fetal alcohol effects, both of which are associated with delayed motor and mental development in children.

Correlations have been made between blood-alcohol concentration and the clinical signs and symptoms of acute intoxication. A blood-alcohol level in the range of 80 to 100 mg/dL has been established as the statutory limit for operation of a motor vehicle in most states. This is associated with a diminution of judgment and motor performance in most people. The determination of blood ethanol concentration by the laboratory in cases of drunk driving requires an appropriate chain of custody, documentation of quality control, and proficiency records.[4]

In addition to the short-term effects of ethanol, most pathophysiologic consequences of ethanol abuse are associated with chronic consumption over a long period of time. In an average adult, this correlates to the consumption of about 50 grams of ethanol per day for about 10 years. Consumption to this degree has been associated with compromised function in a wide variety of organ, tissue, and cell types. In most people, however, the liver is the most sensitive organ system to the toxic effects of excessive chronic ethanol consumption. The pathologic sequence starts with the accumulation of lipids in hepatocytes. With continued consumption, this may progress to alcoholic hepatitis. About 20% of individuals with long-term, high-level intake develop cirrhosis. Cirrhosis can be characterized as an irreversible loss of functional hepatic mass. Progress through this sequence is associated with changes in many laboratory tests related to hepatic function.

Several laboratory indicators of excessive ethanol consumption have sufficient sensitivity and specificity to identify excessive ethanol consumption as the cause of a disease state.[5] Most are related to the progression of ethanol-induced liver disease. Table 26-2 list some of the common laboratory indicators of prolonged hazardous consumption.

Several mechanisms have been proposed to mediate the pathologic effects of long-term ethanol consumption. Of these, adduct formation with acetaldehyde appears to play a key role. Hepatic metabolism of ethanol is a two-step enzymatic reaction. The final product is acetic acid. Acetaldehyde is a reactive intermediate in this pathway. Most ethanol is converted to acetic acid in this pathway; however, a significant portion of the intermediate is released in the free state.

$$\text{Ethanol} \longrightarrow \text{Acetaldehyde} \longrightarrow \text{Acetate}$$
$$\longrightarrow \text{Acetaldehyde adducts}$$

(Eq. 26–2)

Extracellular acetaldehyde is a very transient species owing to rapid adduct formation with amine groups of proteins. Formation of acetaldehyde adducts has been shown to change the structure and function of a wide variety of proteins. Many of the pathologic effects of ethanol have been correlated with the formation of these adducts.

Methanol is a common solvent. It may be ingested accidentally as a component of many commercial products or as a contaminant of homemade liquors. Methanol is initially metabolized by hepatic ADH to the intermediate formaldehyde. Formaldehyde is rapidly converted to formic acid by hepatic ALDH. The formation of formic acid causes a severe acidosis, which may lead to death. Formic acid is also responsible for an optic neuropathy that may lead to blindness.

Isopropanol, also known as rubbing alcohol, is commonly available. It is metabolized by hepatic ADH to acetone, which is its primary metabolic end product. Both isopropanol and acetone have CNS depressant effects similar to ethanol. However, acetone has a very long half-life. Thus, intoxication with isopropanol may result in severe acute-phase ethanol-like symptoms that may persist for an extended period of time.

Ethylene glycol (1, 2-ethanediol) is a common component of hydraulic fluid and antifreeze. Ingestion by children is relatively common owing to its sweet taste. The immediate effects of ethylene glycol ingestion are similar to those of ethanol. However, metabolism by hepatic ADH and ALDH results in the formation of several toxic species, including oxalic acid and glycolic acid, which causes a severe metabolic acidosis. This is complicated by the rapid formation and deposition of calcium oxalate crystals in renal tubules. With high levels of consumption, calcium oxalate crystal formation may result in renal tubular damage.

Determination of Alcohols

From a medicolegal perspective, determination of blood ethanol concentration must be accurate and precise.[6] Serum, plasma, or whole blood are acceptable specimens. Correlations have been established between ethanol concentration in these specimens and impairment of psychomotor function. Because ethanol uniformly distributes itself in total body water, serum, which has a greater water content than whole blood, has a higher concentration per unit volume. Most states have standardized the acceptable specimen types admissible as evidence.

When acquiring a specimen for ethanol determination, the venipuncture site should be cleaned with an alcohol-free disinfectant. Because of the volatile nature of short-chain aliphatic alcohols, specimens must be capped at all times to avoid evaporation. Sealed specimens can be stored refrigerated or at room temperature for up to 14 days without loss of ethanol. Nonsterile specimens or those intended to be stored for long periods of time should be preserved with sodium fluoride to avoid increases in ethanol content as a result of fermentation.

Several analytic methods can be used for the determination of ethanol in serum. Among these, the enzymatic, gas chromatography, and osmometry methods are the most commonly used. When osmolarity is measured by freezing point depression, increases in serum osmolarity correlates well with increases in serum ethanol concentration. The degree of increase in osmolality due to ethanol is expressed as the difference between the measured and the calculated osmolality; the difference is called the *osmolal gap*. Serum osmolality increases by about 10 mOsm/kg for each 60 mg/dL increase in serum ethanol.

TABLE 26-2. Common Indicators of Ethanol Abuse

Test	Comments
GGT	Increases can be seen before the onset of pathologic consequences. Increases in serum activity can occur in many non–ethanol-related conditions.
AST	Increases occur in conjunction with ethanol-related liver disease. Increases in serum activity can occur in many non–ethanol-related conditions.
AST/ALT ratio	A ratio of greater than 2.0 is highly specific for ethanol-related liver disease.
HDL	High serum HDL is very specific for ethanol consumption.
MCV	Increased erythrocyte MCV is commonly seen with excessive ethanol consumption. Increases are not related to folate or vitamin B_{12} deficiency.

Osmolar gap = measured osmolarity

$$- \text{ calculated osmolarity} \quad (Eq.\ 26\text{--}3)$$

This relationship is not specific for ethanol. Increases in the osmolar gap can also occur with some metabolic imbalances; therefore, use of the osmolar gap for the determination of serum or blood ethanol concentration lacks analytic sensitivity and specificity. However, it is a useful screening test.

Gas chromatography is the reference method for the determination of ethanol. This method can simultaneously quantitate other alcohols, such as methanol and isopropanol. This analysis starts with the dilution of the serum or blood sample with a saturated solution of sodium chloride in a closed container. Volatiles within the liquid specimen partition into the air space (head space) of the closed container. Sampling of this head space provides clean specimens with little or no matrix effect. Quantitation of peaks can be done by constructing a standard curve or by ratio to an internal standard as shown in Figure 26-2 (n-propanol).

Enzymatic methods for the determination of ethanol are common. The enzyme used in this assay is a nonhuman form of ADH. This enzyme oxidizes ethanol to acetaldehyde with reduction of NAD^+ to NADH.

$$\text{Ethanol} + NAD^+ \xrightarrow{\text{ADH}} \text{acetaldehyde} + \text{NADH} \quad (Eq.\ 26\text{--}4)$$

The NADH produced can be monitored directly by absorbance at 340 nm or can be coupled to an indicator reaction. This form of ADH is relatively specific for ethanol (see Table 26-1). Intoxication with methanol or isopropanol produces a negative or very low result. Thus, a negative result by this method does not rule out ingestion of other common alcohols. There is good agreement between the enzymatic reactions of ethanol and gas chromatography. The enzymatic reactions can be fully automated and do not require specialized instrumentation.

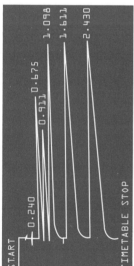

Retention Time (min)	Analyte
0.675	Methanol
0.911	Acetone
1.098	Ethanol
1.611	Isopropanol
2.430	n-Propanol

▲
Figure 26-2. **Headspace gas chromatography of alcohols.** The concentration of each alcohol can be determined by comparison to the response from the internal standard n-propanol.

Carbon Monoxide

Carbon monoxide is produced by incomplete combustion of carbon-containing substances. The primary environmental sources of carbon monoxide include gasoline engines, improperly ventilated furnaces, and wood or plastic fires. Carbon monoxide is a colorless, odorless, and tasteless gas that is rapidly absorbed into blood through inspired air. Carbon monoxide expresses its toxic effects by high-affinity binding to divalent iron within heme proteins, such as cytochromes, myoglobin, and hemoglobin.[7] Of these, binding to hemoglobin produces the most significant toxic outcome.

When carbon monoxide binds to hemoglobin, it is called carboxyhemoglobin (CO-Hb). The affinity of carbon monoxide for hemoglobin is 245 times greater than that for oxygen. Air is about 20% oxygen by volume. Thus, if inspired air contained 0.1% carbon monoxide by volume, this would result in a 50% carboxyhemoglobinemia at equilibrium. For this reason, carbon monoxide is considered a very toxic substance. Because both carbon monoxide and oxygen compete for the same binding site, exposure to carbon monoxide results in a decrease in the concentration of oxyhemoglobin. Furthermore, binding of carbon monoxide to hemoglobin increases the affinity of oxygen to hemoglobin, a shift to the left on the hemoglobin–oxygen dissociation curve. The net effect of carbon monoxide exposure is a decrease in the amount of oxygen delivered to tissues, thereby producing hypoxia. Thus, the major toxic effects of carbon monoxide exposure are seen in organs with high oxygen demand, such as the brain and heart.[8] The concentration of carboxyhemoglobin (expressed as the percentage of CO-Hb present to the capacity of the specimen to form CO-Hb) and corresponding symptoms are detailed in Table 26-3.

Several methods are available to determine carbon monoxide poisoning. Carboxyhemoglobin has a cherry-red appearance. This is the basis of a spot test for excessive carbon monoxide exposure. To 5 mL of a 1/20 aqueous dilution of whole blood is added 5 mL of 40% NaOH. Persistence of a pink solution is consistent with a carboxyhemoglobin level of 20% or greater. There are two primary quantitative assays for carboxyhemoglobin: differential spectrophotometry and gas chromatography.

Gas chromatography is accurate, precise, and the reference method for the determination of carboxyhemoglobin. Carbon monoxide is released from hemoglobin after treatment with potassium ferricyanide. After analytic separation, carbon monoxide is detected by changes in thermal conductivity. Spectrophotometric methods work on the principle that different forms of hemoglobin present with different spectral absorbency curves. By measuring absorbance at four to six different wavelengths, the concentration of the different species of hemoglobin (including carboxyhemoglobin) can be determined by calculation. This is the most common method used and is the basis for several automated systems.

CASE STUDY 26-1

A patient with a provisional diagnosis of depression is sent to the laboratory for a routine work-up. The complete blood cell count was unremarkable except for an elevated erythrocyte mean cell volume (MCV). Results of urinalysis were unremarkable. The serum chemistry testing revealed slightly increased aspartate amino transferase (AST), total bilirubin, and high-density lipoprotein (HDL) levels. All other chemistry results, including glucose, urea, creatinine, cholesterol, pH, pCO$_2$, alanine aminotransferase (ALT), cholecystokinin (CK), sodium, and potassium, were within the normal reference range. The physician suspects ethanol abuse; however, the patient claims to be a nonconsumer. Subsequent testing revealed a serum GGT three times the upper limit of normal. No ethanol was detected in serum. Screening tests for infectious forms of hepatitis were negative.

Questions

1. Are the above results consistent with a patient who is consuming hazardous quantities of ethanol?
2. Is further testing needed to rule in or out ethanol abuse? If so, what tests would you recommend?

Caustic Agents

Caustic agents are found in many household products and occupational settings. Even though any exposure to a strong acid or alkaline substance is associated with injury, aspiration and ingestion present the greatest hazard. Aspiration is usually associated with pulmonary edema and shock, which can rapidly progress to death. Ingestion produces lesions in the esophagus and gastrointestinal tract, which may produce perforations. This results in hematemesis, abdominal pain, and possibly shock. Onset of metabolic acidosis or alkalosis occurs rapidly after ingestion. Corrective therapy for ingestion is usually by dilution.

Cyanide

Cyanide is classified as a supertoxic substance.[9] It can exist as a gas, as a solid, or in solution. Exposure can occur by inhalation, ingestion, or transdermal absorption. Cyanide is used in many industrial processes.[10] It is also a component of some insecticides and rodenticides. Cyanide is also produced as a py-

rolysis product from the burning of some plastics, such as urea foams, which have been used as insulation in homes. Thus, carbon monoxide and cyanide exposure may account for a significant portion of the toxicities associated with smoke inhalation. Ingestion of cyanide is a common suicide agent.

Cyanide expresses its toxicity by binding to heme iron. Binding to mitochondrial cytochrome oxidase causes an uncoupling of oxidative phosphorylation. This results in rapid depletion of cellular adenosine triphosphate owing to the inability of oxygen to accept electrons. Increases in cellular oxygen tension and venous pO$_2$ occur as a result of lack of oxygen utilization. At low levels of exposure, patients experience headaches, dizziness, and respiratory depression, which can rapidly progress to seizure, coma, and death at slightly greater doses. Cyanide clearance is primary mediated by rapid enzymatic conversion to thiocyanate, a nontoxic product rapidly cleared by renal filtration. Cyanide toxicity is associated with acute exposure at concentrations sufficient to exceed the removal by this enzymatic process.

Evaluation of cyanide exposure requires a rapid turnaround time. There are several methods available. Ion-specific electrode methods and photometric analysis following two-well microdiffusion separation are the most commonly used. Chronic low-level exposure can be evaluated by determination of urinary thiocyanate concentration.

Metals

Arsenic

Arsenic may exist bound to or as a primary constituent of many different organic and inorganic compounds. It exists in both naturally occurring and manmade substances. Thus, exposure to arsenic may occur in a variety of settings. Environmental exposure through air and water is prevalent in many industrialized areas.[11] Occupational exposure occurs in agriculture and the smelting industries. It is also a common homicide and suicide agent.

TABLE 26-3. Symptoms of Carboxyhemoglobinemia

CO-Hb (%)	Symptoms and Comments
0.5	Typical in nonsmokers
5–15	Range of values seen in smokers
10	Shortness of breath with vigorous exercise
20	Shortness of breath with moderate exercise
30	Severe headaches, fatigue, impairment of judgment
40–50	Confusion, fainting on exertion
60–70	Unconsciousness, respiratory failure, death with continuous exposure
80	Immediately fatal

Absorption of arsenic depends on the form of the compound. Organic arsenic containing compounds are rapidly absorbed by passive diffusion. Other forms are absorbed at a slower rate. Clearance of arsenic is primarily by renal filtration of the free, ionized state. Arsenic expresses its toxic effects by high-affinity binding to the thiol groups in proteins. Thus, the portion available for filtration in serum is very low. This results in a very long half-life in the body; thus, the whole body content of arsenic may be cumulative with chronic exposure.

Arsenic binding to proteins in many cases results in a change in structure and function. Because many proteins are capable of binding arsenic, the toxic symptoms of arsenic poisoning are nonspecific. Many cellular and organ systems are effected. Fever, anorexia, and gastrointestinal distress are seen with chronic or acute ingestion at low levels. Peripheral and central damage to the nervous system, renal effects, hemopoietic effects, and vascular disease leading to death are associated with high levels of exposure.

Analysis of arsenic is done by atomic absorption spectrophotometry. Blood and urine are acceptable specimens to evaluate short-term exposure. Hair and fingernail content have been found useful in the assessment of long-term exposure.

Cadmium

Cadmium is a metal found in many industrial processes. Its main use is in electroplating and galvanizing. It is commonly encountered during the mining and processing of many metals. Cadmium is a pigment found in paints and plastics and is the cathodal material of nickel-cadmium batteries. It is a significant environmental pollutant.[12]

Exposure occurs most frequently by inhalation of cadmium particulates in industry and by ingestion of contaminated food. Cadmium expresses its toxicity primarily by binding to proteins; however it can also bind to other cellular constituents. Cadmium distributes itself throughout the body but has a tendency to accumulate in the kidney, where it expresses most of its toxic effects.[13] An early finding of cadmium toxicity is manifested by renal tubular dysfunction. Tubular proteinuria, glucosuria, and amino aciduria are typically seen. Evaluation of excessive cadmium is accomplished by determination of whole blood or urinary cadmium using atomic absorption spectrophotometry.

Lead

Lead is a common environmental contaminant.[14] It was a common constituent of household paints before 1972 and is still found in commercial and art paints. Gasoline was leaded until 1978. Residuals from automobile exhaust can still be found in high concentrations on highways. Plumbing constructed of lead pipes or joined with leaded connectors can significantly contribute to the lead concentration of water. Lead is a by-product or component of many industrial processes. Exposure and absorption of lead can occur by any route; however, ingestion of contaminated dietary constituents accounts for the majority of lead toxicity. The lead content of foods is highly variable. In the United States, average daily intake for an adult is between 75 and 120 μg/d. This level of intake is not associated with overt toxicity.[15] Because lead is present in all biologic systems, and because no physiologic or biochemical function has been found for it, the key issue is what dose causes a toxic effect. Susceptibility to lead toxicity is dependent primarily on age. Adults are largely tolerant to the effects of lead compared with children.[16]

Gastrointestinal absorption of lead is influenced by a variety of factors. Adults absorb 5% to 15% of ingested lead. Children have a greater degree of absorption. Infants absorb 30% to 40%. Factors controlling rate of absorption are unclear.[17] Absorbed lead binds with high affinity to a wide variety of macromolecular structures. It distributes itself throughout the body. Elimination of lead occurs primarily by renal filtration. There are two theoretical compartments that lead distributes into. One is the skeleton, which is the largest pool. Lead combines with the matrix of bone and can persist in this compartment for a long period of time. The half-life of lead in bone is longer than 20 years. The other theoretical compartment is soft tissue. Half-lives of lead in this compartment are somewhat variable. The average half-life in soft tissue is 120 days. Considering the relatively constant rate of exposure and the slow elimination rate, total body lead accumulates over a lifetime. The largest accumulation occurs in bone. Significant accumulation also occurs in kidney, bone marrow, circulating erythrocytes, and peripheral and central nerves.

Lead toxicity is multifaceted and occurs in a dose-dependent manner (Fig. 26-3). Most of the toxic effects of lead are due to binding to proteins, which results in a change in structure and function. The neurologic effects of lead are of particular importance. Lead exposure causes an encephalopathy characterized by a cerebral edema and hypoxia. Severe lead poisoning can result in stupor, convulsions, and coma. Lower levels of exposure may not present with observable symptoms. However, low-level exposure may result in subclinical effects typified by behavioral changes, hyperactivity, attentional deficit disorder, and a decrease in intelligence quotient scores. Children appear particularly sensitive to these effects[9] and are now evaluated for lead poisoning before entry into school. Lead also causes a demyelinization of peripheral nerves, which results in a decrease in nerve conduction velocity.

Lead is a potent inhibitor of many enzymes; this inhibition mediates many of its toxic effects. Noteworthy are the effects on vitamin D metabolism and the heme synthetic pathway. This results in changes in bone and calcium metabolism and in anemia. Decreased serum concentrations of both 25-hydroxy and 1-25-dihydroxy vitamin D are seen in excessive lead exposure. The anemia is primarily caused by an inhibition of the heme synthetic pathway. This results in increases in the concentration of several intermediates in this pathway including aminolevulinic acid and protoporphyrin. Increases in protoporphyrin result in high concentrations of zinc protoporphyrin in circulating erythrocytes. Zinc protoporphyrin is a highly fluorescent compound. Measurement of this fluorescence can be used in the detection of lead toxicity. Increased urinary aminolevulinic acid is a highly sensi-

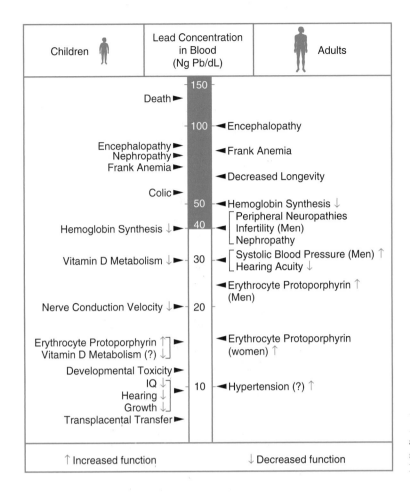

Figure 26-3. **Comparison of effects of lead on children and adults** (From Royce SE, Needleman HL. Eds. Case studies in environmental medicine: Lead toxicity. U.S. Public Health Service, ATSDR, 1990.)

tive and specific indicator of lead toxicity that correlates well with blood levels. Another hematologic finding is the presence of basophilic stippling in erythrocytes. This is the result of inhibition of erythrocytic pyrimidine nucleotidase. This enzyme is responsible for removal of residual DNA after extrusion of the nucleus. The presence of basophilic stippling is a sensitive indicator of lead exposure.

Excessive lead exposure has also been associated with hypertension, carcinogenesis, birth defects, and compromised immunity. Lead causes several toxic renal effects.[18] Early stages are associated with tubular dysfunction. This results in glycosuria, aminoaciduria, and hyperphosphaturia. Late stages are associated with tubular atrophy and glomerular fibrosis. The fibrosis may result in a decreased glomerular filtration rate.

Treatment of lead poisoning involves removal from exposure and treatment with therapeutic chelators, such as ethylenediaminetetraacetic acid (EDTA) and dimercaptosuccinic acid (DMSA). These substances are capable of removing lead from soft tissue and bone by forming low-molecular-weight, high-affinity complexes that can be cleared by renal filtration. The efficacy of this therapy is determined by monitoring the urinary concentration of lead.

The assessment of total body burden of lead poisoning is best evaluated by the quantitative determination of lead concentration in whole blood. The use of urine is also valid but correlates closer to the level of recent exposure. Care must be taken during specimen collection to ensure that the specimen does not become contaminated from exogenous sources. Lead-free containers are recommended for this purpose.

Several methods can be used to measure lead concentration. Chromogenic reactions and anodic stripping voltammetry are methods that have been used but lack clinical utility owing to their lack of analytic sensitivity. At present, graphite furnace atomic absorption spectrophotometry (AAS) is the most common method used. Inductively coupled plasma emission spectrophotometry (ICPES) is a method with performance characteristics comparable to AAS.

Mercury

Mercury is a metal that exists in three forms: elemental (which is a liquid), inorganic salts, or a component of organic compounds. Exposure occurs primarily by inhalation and ingestion. Consumption of contaminated foods is the major source of exposure in the general population. Inhalation and accidental ingestion of inorganic and organic forms in industrial settings is the most common reason for toxic levels.[19] Each of the forms of mercury has different toxicologic characteristics. Elemental mercury (Hg^0) can be ingested without significant effects. Inhalation of elemental mercury is insignificant owing to its low vapor pressure. Cationic mercury (Hg^{2+}) is moderately toxic. Organic mercury, such as methyl mercury (CH_3Hg^+), is very toxic.[20] Considering that the most common route of exposure to mercury is by ingestion, the primary factor that determines these differences in toxicity is gastrointestinal absorbance.

Elemental mercury is largely not absorbed owing to its viscous liquid nature. Inorganic mercury is only partially absorbed. Even though not significantly absorbed, inorganic mercury still has significant local toxicity in the gastrointestinal tract. The organic forms of mercury are rapidly and efficiently absorbed by passive diffusion. Systemic organic mercury partitions into hydrophobic compartments. This results in high concentrations in brain and peripheral nerves. In these lipophilic compartments, organic mercury is biotransformed to the divalent state, allowing it to bind to neuronal proteins.[21] Absorbed inorganic mercury binds to many proteins and is distributed throughout the body. Elimination of systemic mercury occurs primarily by renal filtration of bound low-molecular-weight species or the free (ionized) state. Considering that most mercury is bound to proteins, the elimination rate is slow. Therefore, chronic exposure exerts a cumulative effect.

The toxicity of mercury is due to protein binding that results in a change of structure and function. The most significant result of this interaction is the inhibition of a wide number of enzymes. Binding to intestinal proteins after ingestion of inorganic mercury results in acute gastrointestinal disturbances. Ingestion of moderate amounts may result in severe bloody diarrhea owing to ulceration and necrosis of the gastrointestinal tract. In severe cases, this may lead to shock and death. The absorbed portion of ingested inorganic mercury affects many organs. Clinical findings include tachycardia, tremors, thyroiditis, and most significantly, a disruption of renal function. The renal effect is associated with glomerular proteinuria and a loss of tubular function. Organic mercury may also have a renal effect at high levels of exposure. However, neurologic symptoms are the primary toxic effects of this hydrophobic form. Low levels of exposure cause tremors, behavioral changes, mumbling speech, and loss of balance. Higher levels of exposure result in hyporeflexia, hypotension, bradycardia, renal dysfunction, and death. Analysis of mercury is done by atomic absorption using whole blood or an aliquot of a 24-hour urine specimen.

Pesticides

Pesticides are substances that have been intentionally added to the environment to kill or harm an undesirable life form. Pesticides can be classified into several subcategories, including insecticides and herbicides. These agents have been applied to the control of vector-borne disease, improvement of agricultural productivity, and control of urban pests. Pesticides can be found in both occupational settings and in the home. Thus, there are frequent opportunities for exposure. Contamination of food is the major route of exposure for the general population. Inhalation, transdermal absorption, and ingestion as a result of hand-to-mouth contact are common occupational and accidental routes of exposure.[22]

Ideally, the actions of pesticides would be target specific. Unfortunately, most are nonselective and result in toxic effects to many nontarget species, including humans. Pesticides come in many different forms with a wide range of potential toxic effects.[23] The health effects of short-term, low-level exposure to most of these agents have yet to be well elucidated. Chronic exposure to low levels may result in disease conditions or altered laboratory test values. Of primary concern is acute high-level exposure, which may result in frank disease states or death. The most common victims of acute poisoning are people who are applying pesticides and do not take appropriate precautions to avoid exposure. Ingestion by children at home is also common. Pesticide ingestion is also a common suicide vehicle.

There is a wide degree of variation in the chemical configuration of pesticides, ranging from simple salts of heavy metals to complex high-molecular-weight organic compounds. Insecticides are the most prevalent of the pesticides. Based on chemical configuration, the organophosphates, carbamates, and halogenated hydrocarbons are the most common insecticides. Organophosphates are the most abundant pesticides and are responsible for about one third of all pesticide poisonings.

Organophosphates and carbamates both function by inhibition of acetylcholinesterase, an enzyme present in both insects and mammals. In mammals, acetylcholine is a neurotransmitter found in both central and peripheral nerve. It is also responsible for stimulation of muscle cells and several endocrine and exocrine glands. The actions of acetylcholine are terminated by the actions of membrane-bound postsynaptic acetylcholinesterase. Inhibition of this enzyme by these agents results the prolonged presence of acetylcholine on its receptor. This produces a wide variety of systemic effects. Low levels of exposure are associated with salivation, lacrimation, and involuntary urination and defecation. Higher levels of exposure result in bradycardia, muscular twitching, cramps, apathy, slurred speech, and behavioral changes. Death due to respiratory failure may also occur.

Absorbed organophosphates bind with very high affinity to several proteins, including acetylcholinesterase. Protein-binding prevents direct analysis of organophosphates. Thus, exposure is evaluated indirectly by measurement of acetylcholinesterase inhibition. Inhibition of this enzyme has been found to be a sensitive and specific indicator of organophosphate exposure. Because acetylcholinesterase is a membrane-bound enzyme, serum activity is very low. To increase the analytic sensitivity of this assay, erythrocytes that have high surface activity are commonly used. Evaluation of erythrocytic acetylcholinesterase activity for detection of organophosphate exposure, however, is not commonly available because of low demand and the lack of an automated method.

An alternative test that has become commonly available is measurement of serum pseudocholinesterase (SChE) activity. This enzyme is inhibited by organophosphates in a similar manner to the erythrocytic enzyme. Unlike the erythrocytic enzyme, however, changes in the serum activity of SChE lack sensitivity and specificity for organophosphate exposure. Pseudocholinesterase is found in liver, pancreas, brain, and serum. The biologic function of this enzyme is unknown. Decreased levels of SchE can occur in acute infection, pulmonary embolism, hepatitis, and cirrhosis. There are also several variants of this enzyme that demonstrate diminished activity. Thus, decreases in SChE are not specific for organophosphate poisoning. The normal reference range for SChE is between 4,000 and 12,000 U/L. The intraindividual variation (the degree of variance within an average individual) is about 700 U/L. Symptoms associated with organophosphate toxicity occur at about a 40% reduction in activity. Thus, an individual whose normal SChE exists on the high side of the normal reference range and who has been exposed to toxic levels of organophosphates may still have an SChE activity in the normal reference range. Because of these factors, determination of SChE activity lacks sensitivity in the diagnosis of organophosphate poisoning. Therefore, SChE is considered a screening test, and clinical context must be taken into consideration when interpreting the results. Immediate antidotal therapy can be initiated in cases of suspected organophosphate poisoning with decreased activity of SChE. However, continuation of therapy and documentation of such poisoning should be confirmed by testing of the erythrocytic enzyme.

TOXICOLOGY OF THERAPEUTIC DRUGS

Many overdose situations are the result of either accidental or intentional excessive dosage of pharmaceutical drugs. All drugs are capable of toxic effects at the right dosage. The focus of this discussion is on those therapeutic drugs most commonly seen in clinical overdose situations.

Salicylates

Aspirin (acetylsalicylic acid) is a commonly used analgesic, antipyretic, and anti-inflammatory drug. It functions by decreasing thromboxane and prostaglandin formation through inhibition of cyclooxygenase. At recommended doses, there are several noteworthy side effects, including interference with platelet aggregation and gastrointestinal disturbances.

There is also an epidemiologic relationship between aspirin, childhood viral infections (eg, varicella and influenza), and the onset of Reye's syndrome.

Acute ingestion of high doses of aspirin is associated with a variety of toxic effects through several different mechanisms.[24] Because it is an acid, excessive salicylate ingestion is associated with a metabolic acidosis. Salicylate is also a direct stimulator of the respiratory center. The hyperventilation produces a respiratory alkalosis. In many instances, the net result of both is a mixed acid–base disturbance. Salicylates also inhibit the Krebs cycle. This results in excess conversion of pyruvate to lactate. In addition, at high levels of exposure, salicylates stimulate mobilization and use of free fatty acid, resulting in excess ketone body formation. All these factors contribute to a metabolic acidosis that may lead to death. Treatment for overdose involves neutralizing and eliminating this excess of acid and maintaining electrolyte balance.

Correlations have been established between serum concentrations of salicylates and toxic outcomes. Several methods are available for the quantitative determination of salicylate in serum. Gas or liquid chromatography methods provide the highest analytic sensitivity and specificity but have not found clinical utility owing to equipment expense and technical difficulty. Several immunoassay methods are available. The most common method is a chromogenic assay known as the *Trinder reaction,* which reacts salicylate with ferric nitrate to form a colored complex that is then evaluated spectrophotometrically.

Acetaminophen

Acetaminophen, either solely or in combination with other compounds, is a commonly used analgesic drug. In healthy subjects, therapeutic dosages have very few side effects. Overdose of acetaminophen, however, is associated with a severe hepatotoxicity (Fig. 26-4).

Absorbed acetaminophen is bound with high affinity to a variety of proteins. This results in a very low free fraction. Thus, renal filtration of the parent drug is minimal. Most is eliminated by hepatic uptake, biotransformation, conjugation, and excretion. Acetaminophen can follow several different

---

CASE STUDY 26-2

An emergency room case with a provisional diagnosis of overdose with an over-the-counter cold medicine undergoes a drug screen. Test results from immunoassay screening were negative for opiates, barbiturates, benzodiazapines, THC, and cocaine but positive for amphetamines. The salicylate level was 15 times the upper limit of the therapeutic range. Results for acetaminophen and ethanol were negative.

Questions

1. What would be the expected results of arterial blood gas analysis?
2. What would be the expected results of a routine urinalysis?
3. What are some of the possible reasons the amphetamine screen is positive?

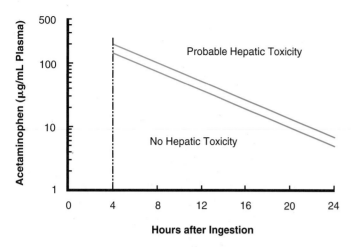

Figure 26-4. **Rumack-Matthew Nomogram. Prediction of aceta-minophen induced hepatic damage based on serum concentration.** (Rumack BH, Matthew H. Acetaminophen poisoning and toxicity. Pediatrics 1975;55:871.)

pathways through this process. Each forms a different product. The pathway of major concern is the hepatic mixed-function oxidase system. In this system, acetaminophen is first transformed to reactive intermediates, which are then conjugated with reduced glutathione. In overdose situations, glutathione can become depleted, yet reactive intermediates continue to be produced. This results in reactive intermediates accumulating inside the cell. Because some of these intermediates are free radicals, this results in a toxic effect to the cell that leads to necrosis of the liver, the organ in which these reactions are occurring.[25]

The time frame for the onset of hepatocyte damage is relatively long. In an average adult, serum indicators of hepatic damage do not become abnormal until 3 to 5 days after ingestion of a toxic dose. The initial symptoms of acetaminophen toxicity are vague, nonspecific, and not predictive of hepatic necrosis. The serum concentration of acetaminophen that results in depletion of glutathione has been determined for an average adult. Acetaminophen is rapidly cleared from serum, however, and many times determination of serum acetaminophen is made many hours after ingestion. In these situations, it is unknown whether toxic concentrations of acetaminophen were present at some previous time. Nomograms are available that predict hepatotoxicity based on the serum concentrations of acetaminophen at a known time after ingestion. Chronic heavy consumers of ethanol metabolize acetaminophen at a more rapid rate than average, resulting in a more rapid formation of reactive intermediates and an increased possibility of depleting glutathione at a lower dose than normal. Therefore, alcoholic patients are more susceptible to acetaminophen toxicity, and using the nomogram for interpretation in these patients is inappropriate.[26]

The reference method for the quantitation of acetaminophen in serum is high-performance liquid chromatography. This method, however, is not widely used in clinical settings because of its expense and technical difficulty. Immunoassay is currently the most common analytic method used for serum acetaminophen determination. Competitive enzyme or fluorescence polarization immunoassay systems are the most frequently used.

TOXICOLOGY OF DRUGS OF ABUSE

Assessment of drug abuse is of medical interest for many reasons. In cases of drug overdose, it is essential to identify the responsible agent to provide appropriate treatment. In a similar manner, identification of drug abuse in nonoverdose situations provides a rationale for treatment for addiction. For these reasons, testing for *drugs of abuse* is commonly done. This typically involves screening of a single urine specimen for many substances by qualitative screening procedures. In most instances, this protocol is only of detecting only recent drug use. Thus, with abstinence of relatively short duration, many abusing patients may not be identified. In addition, a positive drug screen cannot discriminate between a single casual use and chronic abuse. Identification of chronic abuse usually involves several positive test results in conjunction with clinical evaluation. In a similar manner, a positive drug screen does not determine the time frame or dose of the drug taken. In most cases, a positive drug screen is usually sufficient to justify administration of an antidote or initiation of therapy.

Drug abuse or overdose can occur with prescription, over-the counter, or illicit drugs. The focus of this discussion is on substances with an addictive potential.

The use of drugs for recreational or performance enhancement purposes is relatively common. The National Institute on Drug Abuse reports that about 30% of the population beyond high school age have used an illicit drug. Testing for drug abuse has become commonplace in professional, industrial, and athletic settings. The potential punitive measures associated with this testing may involve or result in civil or criminal litigation. Therefore, the laboratory must ensure that data are legally admissible and defendable. This requires the use of analytic methods that have been validated as accurate and precise. It also requires documentation of specimen security. Protocols and procedures must be established that prevent and detect specimen adulteration and that may prevent drug detection. Measurement of urinary temperature, pH, specific gravity, and creatinine is commonly done to ensure that these specimens have not been diluted or treated with substances that may interfere with testing. Specimen collection should be monitored and a chain-of-custody established to guard against specimen exchange.

Testing for drugs of abuse can be done by several methods. A two-tiered approach is usually employed and involves screening and confirmation.[27] Screening procedures should be simple, rapid, inexpensive, and capable of being automated. They are often referred to as *spot tests*. In general, screening procedures have good analytic sensitivity with marginal specificity. Thus, a negative result can rule out the presence of an analyte with a reasonable degree of certainty. These methods usually detect classes of drugs based on similarities in chemical configuration. This allows detection of parent compounds and congeners, which have similar effects. Considering that many designer drugs are modified forms of established drugs of abuse, these methods increase the scope of the screening process. A drawback to this type of analysis is that it may also detect chemically related substances that have no or low abuse potential. Thus, interpretation of positive test results requires integration of clinical context and further testing. Confirmation testing uses methods that have high sensitivity and specificity; many of these tests provide quantitative as well as qualitative information. Confirmatory testing requires the use of a method different from that used in the screening procedure.

There are several general analytic procedures commonly used for analysis of drugs of abuse. Chromogenic reactions, the generation of a colored product usually by a chemical reaction, are occasionally used as screening procedures. Immunoassay-based procedures are widely used both as screening and confirmatory assays. In general, immunoassays offer a high degree of sensitivity and are easily automated. A wide variety of chromatography techniques are used for the qualitative identification and quantitation of drugs. Thin-layer chromatography is an inexpensive method for the screening of a wide number of drugs and has the advantage that no instrumentation is required. Liquid and gas chromatography allows complex mixtures of drugs to be separated and quantitated. These methods are generally labor intensive and not well suited to screening.

Many drugs have the potential for abuse.[28] Trends in drug abuse vary geographically and between different socioeconomic groups. For a clinical laboratory to provide an effective toxicology service requires knowledge of the drug or drug groups likely to be found within the patient population it serves. Fortunately, the process of selecting which drugs to test for has been aided by national studies that have identified the drugs of abuse most commonly seen in the population (Table 26-4). This provides the basis for test selection in most situations. The following discussion focuses on select drugs with a high frequency of abuse.

Amphetamines

Amphetamine and methamphetamine are therapeutic drugs used for narcolepsy and attentional deficit disorder. These drugs are stimulants with a very high abuse potential.[29] They produce an initial sense of increased mental and physical capacity along with a perception of well-being. These initial effects are followed by restlessness, irritability, and possibly

TABLE 26-4. Prevalence of Common Drugs of Abuse

Substance	Prevalence (%)*
Alcohol	75–80
Marijuana	20–26
Cocaine	5–13
Benzodiazapines	1–5
Barbiturates	0.5–5
Opiates	0.1–2
Phencyclidine	0.1–2
Amphetamines	0.1–1
Other stimulants	0.8–2
Other sedative hypnotics	0.6–2

* This table provides approximate frequencies of relevant drugs of abuse encountered in the clinical laboratory. The percentage values estimate the prevalence of use in college-aged individuals, the most common users, who by survey claim to have used drugs within the last 30 days, and individuals who tested positive in the same age group.

(Adapted from NIDA Capsules and Infofax Web sites: URL http://www.nida.nih.gov; accessed 6/8/98.)

psychosis. Abatement of these late effects is often countered with repeated use. Tolerance and psychological dependence develop with chronic use. Overdose, even though rare in experienced users, results in hypertension, cardiac arrhythmias, convulsions, and possibly death. A variety of compounds chemically related to amphetamines are components of over-the-counter medications. Examples of these are ephedrine, pseudoephedrine, and phenylpropanolamine. These amphetamine-like compounds are common in allergy and cold medications.

Identification of amphetamine abuse involves analysis of urine for the presence of the parent drugs. Immunoassay systems are commonly used as the screening procedure. Because of variable cross-reactivity with over-the-counter medications that contain amphetamine-like compounds, a positive result by immunoassay is considered presumptive. Confirmation of immunoassay-positive tests are most commonly done by liquid or gas chromatography.

Anabolic Steroids

Anabolic steroids are a group of compounds that are related chemically to the male sex hormone testosterone. These artificial substances were developed in the 1930s as a therapy for male hypogonadism. It was soon discovered that use of these compounds in healthy subjects increases muscle mass. In many instances, this results in an improvement in athletic performance. Recent studies have shown that 6.5% of adolescent boys and 1.9% of girls reported the use of steroids without a prescription.[30]

Most illicit steroids are obtained through the black market from underground laboratories and foreign sources. The quality and purity of these drugs is highly variable. The toxic effects of these drugs in many instances is related to inconsistent formulation, which may result in very high dosages and the presence of impurities. A wide variety of both physical and psychological effects have been associated with steroid abuse. Chronic use of steroids has been associated with a toxic hepatitis. Chronic use has also been associated with accelerated atherosclerosis and abnormal aggregation of platelets, both of which predispose to stroke and myocardial infarction. In addition, steroid abuse causes an enlargement of the heart. In this condition, heart muscle cells develop faster than the associated vasculature. This may lead to a hypoxia of heart muscle cells, which predisposes cardiac arrhythmias and possible sudden death. In males, chronic steroid use is associated with testicular atrophy, sterility, and impotence. In females, it causes development of masculine traits, breast reduction, and sterility.

Identification of steroid abusers by laboratory testing is limited to chromatographic methods. Both gas and liquid systems have been used.[31] Gas chromatography with mass spectrometry is the most commonly used method.

Cannabinoids

Cannabinoids are a group of psychoactive compounds found in marijuana.[32] Of these, THC is the most potent and abundant. Marijuana, or its processed product hashish, can be smoked or ingested. THC induces a sense of well-being and euphoria. It is also associated with impairment of short-term memory and intellectual functions. Effects of chronic use have not been well established. Overdose has not been associated with specific physiologic toxic outcomes. Tolerance and a mild dependence may develop with chronic use. THC is a lipophilic substance. It is rapidly removed from circulation by passive distribution into hydrophobic compartments, such as brain and fat. This results in slow elimination owing to redistribution back into circulation and subsequent hepatic metabolism.

The half-life of THC in circulation is 1 day after a single use and 3 to 5 days in chronic heavy consumers. Hepatic metabolism of THC produces several products that are primarily eliminated in urine. The major urinary metabolite is 11-nor-delta-tetrahydrocannabinol (THC-COOH). This metabolite can be detected in urine for 3 to 5 days after a single use or for up to 4 weeks in a chronic heavy consumer after abstinence. Immunoassay for THC-COOH is the basis of the screening test for marijuana consumption. Gas chromatography with mass spectrometry is used for confirmation. Both of these methods are very sensitive and specific. Because of the low limit of detection of these methods, it is possible to find THC-COOH in urine as a result of passive inhalation. Urinary concentration standards have been established that can discriminate between passive and direct inhalation.

Cocaine

Cocaine is an effective local anesthetic with few side effects at therapeutic concentrations. At higher circulating concentrations, it is a potent CNS stimulator that elicits a sense of excitement and euphoria.[33] Cocaine is an alkaloid salt that can be administered directly (eg, by insufflation or intravenous injection) or inhaled as a vapor when smoked in the free-base form (crack). It has high abuse potential. The half-life of the circulating cocaine is brief: 0.5 to 1 hour. Acute cocaine toxicity is associated with hypertension, arrhythmias, seizures, and myocardial infarction. Both the subjective and toxic effects are expressed when circulating concentrations are rising. Because of its short half-life, maintaining the subjective effects over a single extended period of time requires repeated dosages of increasing quantity. Thus, correlations between serum concentration and the subjective or toxic effects cannot be established. Because rate of change is more important than serum concentration, a primary factor that determines the toxicity of cocaine is the dose and route of administration. Intravenous administration presents with the greatest hazard, closely followed by smoking.

Cocaine's short half-life is due to rapid hepatic hydrolysis to inactive metabolites. This is the major route of elimination. Only a small portion of an administered dose can be found in urine. The primary product of hepatic metabolism is benzoylecgonine, which is primarily eliminated in urine. The half-life of benzoylecgonine is 4 to 7 hours. The presence of this metabolite in urine is a sensitive and specific indicator of cocaine use. It can be detected in urine for up to 3 days after a single use. In chronic heavy abusers, it can be detected in urine for up to 20 days after the last dose. The primary screening procedure for identification of cocaine use is detection of benzoylecgonine in urine by immunoassay. Confirmation testing is done by gas chromatography with mass spectrometry.

Opiates

Opiates are a class of substances capable of analgesia, sedation, and anesthesia.[34] All are derived from or chemically related to substances derived from the opium poppy. The naturally occurring substances include opium, morphine, and codeine. Heroin, hydromorphone (Dilaudid), and oxycodone (Percodan) are chemically modified forms of the naturally occurring opiates. Meperidine (Demerol), methadone (Dolophine), propoxyphene (Darvon), pentazocine (Talwin), and fentanyl (Sublimaze) are the common synthetic opiates. Opiates have a high abuse potential. Chronic use leads to tolerance with physical and psychological dependence. Acute over-dose presents with respiratory acidosis due to depression of respiratory centers, myoglobinuria, and possibly an increase in serum indicators of cardiac damage (CKMB, troponin). High-level opiate overdose may lead to death due to cardiopulmonary failure. Treatment of overdose includes the use of the opiate antagonist naloxone.

Laboratory testing for opiates usually involves initial detection (screening) by immunoassay. Most of these immunoassays are primarily designed to detect morphine and codeine. However, cross-reactivity due to similarities in chemical structure allows detection of many of the opiates: naturally occurring, chemically modified, and synthetic. Gas chromatography with mass spectrometry is the confirmatory method of choice.

Phencyclidine

Phencyclidine (PCP) is an illicit drug with stimulant, depressant, anesthetic, and hallucinogenic properties. It has high abuse potential. Adverse effects are commonly noted at doses that produce the desired subjective effects, such as agitation, hostility, and paranoia. Overdose is associated with stupor and coma. PCP can be ingested or inhaled by smoking PCP-laced tobacco or marijuana. It is a lipophilic drug that rapidly distributes into fat and brain. Elimination is slow owing to redistribution into circulation and hepatic metabolism. About 10% to 15% of an administered dose is eliminated unchanged in urine. Hepatic metabolism forms a variety of products. Identification of PCP abuse is by detection of the parent drug in urine. In chronic heavy users, PCP can be detected 7 to 30 days after abstinence. Immunoassay is used as the screening procedure. Gas chromatography with mass spectrometry is the confirmatory method.

Sedative Hypnotics

Many therapeutic drugs can be classified as sedative hypnotics or tranquilizers. All members of this class are CNS depressants. They have a wide range of therapeutic roles and are commonly used. Most of these drugs have abuse potential, ranging from high to low. These drugs become available for illegal use through diversion from approved sources. Barbiturates and benzodiazepines are the most common type of sedative hypnotics abused. Even though barbiturates have a higher abuse potential, benzodiazepines are more commonly found in abuse and overdose situations. This appears to be due to availability. There are many individual drugs within the barbiturate and benzodiazepine classification. Secobarbital, pentobarbital, and phenobarbital are the more commonly abused barbiturates. Diazepam (Valium), chlordiazepoxide (Librium), and lorazepam (Ativan) are commonly abused benzodiazepines. Overdose with sedative hypnotics initially presents with lethargy and slurred speech, which can rapidly progress to coma. Respiratory depression is the most serious toxic effect of most of these agents. Hypotension can occur with barbiturates. The toxicity of many of these agents is potentiated by ethanol.

Immunoassay is the most common screening procedure for both barbiturates and benzodiazepines. Broad cross-reactivity within members of each group allows for detection of many individual drugs. Liquid or gas chromatography can be used for confirmatory testing.

SUMMARY

Poisoning cases account for a significant number of hospital admissions and visits to doctors' offices.[35] The clinical laboratorian serves multiple roles that affect patient outcome in these cases. Identification and quantitation of the toxin is a primary responsibility. In addition, the laboratorian should be ready to suggest testing regimens that may further define the diagnosis and provide a basis for monitoring the efficacy of therapy. Providing an effective clinical toxicology service requires an understanding of the patient population served and a basic understanding of the toxic mechanisms of the poisons commonly encountered.[36]

REVIEW QUESTIONS

1. Compound A is reported to have an oral LD_{50} of 5 mg/kg body weight. Compound B is reported to have an LD_{50} of 50 mg/kg body weight. Of the following statements regarding the relative toxicity of these two compounds, which is true?
   a. Ingestion of low amounts of compound A would be predicted to cause more deaths then an equal dose of compound B.
   b. Ingestion of compound B would be expected to produce no toxic effects at a dose of greater than 100 mg/kg body weight.
   c. Neither compound A nor compound B is toxic at any level of oral exposure.
   d. Compound A is more rapidly adsorbed from the gastrointestinal tract than compound B.
   e. Compound B would be predicted to be more toxic than compound A if the exposure route were transdermal.

2. Which of the following statements best describes the TD_{50} of a compound?
   a. The dosage of a substance that is lethal to 50% of the population
   b. The dosage of a substance that would produce therapeutic benefit in 50% of the population
   c. The dosage of a substance that would be predicted to cause a toxic effect in 50% of the population
   d. The percentage of individuals who would experience a toxic response at 50% of the lethal dose
   e. The percentage of the population who would experience a toxic response after an oral dosage of 50 mg

3. Of the following analytic methods, which is most commonly used as the confirmatory method for identification of drugs of abuse?
   a. Scanning differential colorimetry
   b. Ion-specific electrode
   c. Gas chromatography with mass spectrometry
   d. Immunoassay
   e. Nephelometry

4. A weakly acidic toxin (pK = 4.0) that is ingested will:
   a. not be absorbed because it is ionized.
   b. not be absorbed unless a specific transporter is present.
   c. be passively absorbed in the colon (pH = 7.5).
   d. be passively absorbed in the stomach (pH = 3.0).
   e. be absorbed only if a weak base is ingested at the same time.

5. What is the primary product of methanol metabolism by the alcohol-aldehyde dehydrogenase system?
   a. Acetone
   b. Acetaldehyde
   c. Oxalic acid
   d. Formaldehyde
   e. Formic acid

6. Which of the following statements concerning cyanide toxicity is/are true?
   a. Inhalation of smoke from burning plastic is a common cause of cyanide exposure.
   b. Cyanide is a relatively nontoxic compound that requires chronic exposure to produce a toxic effect.
   c. Cyanide expresses its toxicity by inhibition of oxidative phosphorylation.
   d. All of the above are true.
   e. Only a and c are true.

7. Which of the following laboratory results would be consistent with acute high-level oral exposure to an inorganic form of mercury (Hg^{2+})?
   a. High concentrations of mercury in whole blood and urine
   b. Proteinuria
   c. Positive occult blood in stool
   d. All of the above are true.
   e. All of the above are false.

8. A child presents with a microcytic, hypochromia anemia. The physician suspects iron-deficiency anemia. Further laboratory testing reveals a normal total serum iron and iron-binding capacity; however, the zinc protoporphyrin level was very high. A urinary screen for porphyrins was positive. Erythrocytic basophilic stippling was noted on the peripheral smear. Which of the following laboratory tests would be best applied to this case?
   a. Urinary thiocyanate
   b. Carboxyhemoglobin
   c. Whole blood lead
   d. Urinary anabolic steroids
   e. Urinary benzoylecgonine

9. A patient with suspected organophosphate poisoning presents with a low SChE level. However, the confirmatory test, erythrocyte acetylcholinesterase, presents with a normal result. Excluding analytic error, which of the following may explain these conflicting results?
   a. The patient has late-stage hepatic cirrhosis.
   b. The patient was exposed to very low levels of organophosphates.
   c. The patient has a variant of SChE that displays low activity.
   d. All of the above are correct.
   e. Only a and c are correct.

10. A patient enters the emergency room in a coma. The physician suspects a drug overdose. Immunoassay screening tests for opiates, barbiturates, benzodiazepines, THC, amphetamines, and PCP were all negative. No ethanol was detected in serum. Can the physician rule out drug overdose as the cause of this coma with these results?
    a. Yes
    b. No
    c. Maybe

REFERENCES

1. Kirk M, Pace S. Pearls, pitfalls and updates in toxicology. Emerg Med Clin North Am 1997;15:427.
2. Brettell TA, Saferstein R. Forensic science. Anal Chem 1997;69:123R.
3. McKenna M, Chick J, Buxton M, et al. The SECCAT survey: the cost and consequences of alcoholism. Alcohol 1996;31:565.
4. Urry FM, Wong Y. Current issues in alcohol testing. Lab Med 1995;26:194.
5. Anastasio AM, Mehdi T. Laboratory markers of ethanol intake and abuse: a critical appraisal. Am J Med Sci 1992;303:415.
6. Church AS, Witting MD. Laboratory testing in ethanol, methanol, ethylene glycol and isopropanol toxicities. J Emerg Med 1997;15:687.
7. Jaffe FA. Pathogenicity of carbon monoxide. Am J Forensic Med Pathol 1997;18:406.
8. Balzan MV, Agius G, Galea A. Carbon monoxide poisoning: easy to treat but difficult to recognize. Postgrad Med J 1996;72:470.
9. Borron SW, Baud FJ. Acute cyanide poisoning: clinical spectrum, diagnosis and treatment. Arh Hig Rada Toksikol 1996;47:307.
10. Herber RF, Christensen JM, Sabbioni E. Critical evaluation of cadmium concentration in blood for use in occupational health. Int Arch Occup Environ Health 1997;69:372.
11. Peters GR, McCurdy RF, Hindmarsh JT. Environmental aspects of arsenic toxicity. Crit Rev Clin Lab Sci 1996;33:457.
12. Cabrera C, Ortega E, Lorenzo ML, et al. Cadmium contamination of vegetable crops, farmlands and irrigation waters. Rev Environ Contam Toxicol 1998;154:55.
13. Lauwerys RR, Bernard AM, Roels HA, et al. Cadmium: exposure markers as predictive indicators of nephrotoxic effects. Clin Chem 1994;40:1391.
14. Silbergeld EK. Preventing lead poisoning in children. Annu Rev Public Health 1997;18:187.
15. Silbergeld EK. Lead poisoning: implications of current biomedical knowledge for public policy. Md Med J 1996;45:209.
16. Berlin CM. Lead poisoning in children. Curr Opin Pediatr 1997;9:173.
17. Diamond GL, Goodrum PE, Felter SP, et al. Gastrointestinal absorption of metals. Drug Chem Toxicol 1997;20:345.
18. Loghman M. Renal effects of environmental and occupational lead exposure. Environ Health Perspect 1997;105:928.
19. Ratcliffe HE, Swanson GM, Fischer LJ. Human exposure to mercury: a critical assessment of the evidence of adverse health effects. J Toxicol Environ Health 1996;49:221.
20. Watanabe C, Satoh H. Evolution of our understanding of methyl mercury as a health threat. Environ Health Perspect 1996;104(Suppl 2):367.
21. Mottet NK, Vahter ME, Charleston JS, et al. Metabolism of methyl mercury in the brain and its toxic significance. Met Ions Biol Syst 1997;34:371.
22. Blondell J. Epidemiology of pesticide poisoning in the United States, with special reference to occupational cases. Occup Med 1997;12:209.
23. Lusk SL, Connon C. Monitoring for pesticide exposure. AAOHN J 1996;44:599.
24. Temple AR. Acute and chronic effects of aspirin toxicity and their treatment. Arch Intern Med 1981;141:P364.
25. Larsen LC, Fuller SH. Management of acetaminophen toxicity. Am Fam Physicans 1996;53:185.
26. Johnson SC, Pelletier LL. Enhanced hepatotoxicity of acetaminophen in the alcoholic patient. Medicine 1997;76:185.
27. Eshridge KD, Gutherie SK. Clinical issues associated with urine testing of substances of abuse. Pharmacotherapy 1997;17:497.
28. Repetto MR, Repetto M. Habitual, toxic and lethal concentrations of 103 drugs of abuse in humans. J Toxicol Clin Toxicol 1997;37:1.
29. Cho AK, Segal DS, eds. Amphetamines and its analogs: psychopharmacology, toxicology and abuse. New York: Academic Press, 1994.

30. Yesalis CE, Barsukiewicz CK, Kopstein AN, et al. Trends in anabolic steroid use among adolescents. Arch Pediatr Adolesc Med 1997; 151:1197.

31. Bowers LD. Analytic advances in detection of performance-enhancing compounds. Clin Chem 1997;43:1299.

32. Adams IB, Martin BR. Cannabis: pharmacology and toxicology in animals and humans. Addiction 1996;91:1585.

33. Boghdadi MS, Henning RJ. Cocaine: pathophysiology and clinical toxicology. Heart Lung 1997;26:484.

34. Kalant H. Opium revisited: a brief review of its nature, composition nonmedical use and the relative risks. Addiction 1997;92:267.

35. Vernon DD, Gleich MC. Poisoning and drug overdose. Crit Care Clin 1997;13:647.

36. Liang HK. Clinical evaluation of the poisoned patient and toxic syndromes. Clin Chem 1996;42:1350.

SUGGESTED READINGS

Casarett LJ, Amdur MO, Klaassen CD, eds. Casarett and Doull's toxicology: the basic science of poisons. 5th Ed. New York: Mcgraw-Hill, 1995.

Chang LW, Magos L, Suzuki T. Toxicology of metals. Boca Raton, FL: Lewis Publishers, CRC Press, 1996.

Sipes IG, McQueen CA, Gandolfi AJ. Comprehensive toxicology. New York: Pergamon, 1997.

CHAPTER 27

Tumor Markers

Anthony W. Butch, Nicole A. Massoll, Alex A. Pappas

Objectives

Upon completion of this chapter, the clinical laboratorian should be able to:

- *Discuss the incidence of cancer in the United States.*
- *Explain the role of tumor markers in cancer management.*
- *Identify the characteristics or properties of an ideal tumor marker.*
- *State the major clinical value of tumor markers.*
- *Name the major tumors and their associated markers.*
- *Describe the major properties, methods of analysis, and clinical use of CEA, AFP, CA 125, CA 19-9, PSA, β-hCG, and PALP.*
- *Explain the use of enzymes and hormones as tumor markers.*

KEY TERMS

Cancer	Oncogenes	Tumor-associated
DNA index (DI)	Protooncogenes	antigens
Neoplasm	Staging	Tumor-specific
Oncofetal antigens	Tumor markers	antigens

Cancer is second only to heart disease as a major health issue in the United States, accounting for about 23% of all deaths annually. Although the cancer mortality rate has remained relatively constant for several decades, it has been projected that cancer will soon be the leading cause of death in the United States, supplanting cardiovascular disease, which has had a steadily declining mortality rate.[1] The American Cancer Society has estimated that in 1998 there were 1,228,600 new cancer cases and that 564,000 people will have died from cancer.[2] These figures do not take into account basal and squamous cell skin cancers or carcinoma in situ, which are expected to account for an additional 1 million or more newly diagnosed cancer cases.[2]

There are more than 50 different types of human malignancies, with about 54% of new cancer cases occurring in the lung-bronchus, colon-rectum, breast, prostate, and uterus. Cancers of the prostate, lung-bronchus, colon-rectum, and urinary bladder are most commonly seen in men, whereas breast, lung-bronchus, colon-rectum, and uterine cancers are the most frequently reported cancers in women.[2] Lung-bronchus cancer is the most common cause of death from cancer among both sexes, followed by prostate cancer in men and breast cancer in women.[2]

Three major approaches have been taken in the United States to deal with cancer: (1) prevention by identification and elimination of the cancer-causing agents or factors, (2) early disease detection when it is localized in primary sites and is more responsive to therapy, and (3) development of more effective therapeutic modalities. Of course, the best and most ideal way is to prevent the occurrence of cancer, which could be carried out by identifying and eliminating exposure to all carcinogens and by enhancing host resistance to cancer. Despite the large amount of research and health care dollars spent to implement this strategy, it is unrealistic and difficult (if not impossible) to avoid exposure to all environmental and biologic cancer-associated risk factors. For instance, deaths due to lung cancer could be reduced at least 20-fold if people would abstain from smoking.[3,4] Therefore, early detection is the most promising approach to controlling cancer through early intervention.

A number of methods are currently available to detect cancer in its early stages. Biophysical methods, such as conventional radiographs, nuclear medicine, rectilinear scanners, ultrasonography, computed tomography (CT), and magnetic resonance imaging (MRI), all play an important role in early detection and treatment of cancer. Clinical laboratory testing for tumor markers can also be used as an aid in early cancer detection. *Tumor marker tests* measure either tumor-associated antigens or other substances present in cancer patients that aid in diagnosis, *staging,* disease progression, monitoring response to therapy, and detection of recurrent disease.[5,6]

CLASSIFICATION OF TUMOR MARKERS

Tumor markers can be defined as biologic substances synthesized and released by cancer cells or substances produced by the host in response to cancerous tissue. Tumor markers can be present in the circulation, body cavity fluids, cell membranes, or the cytoplasm or nucleus of the cell. Tumor markers differ from substances produced by normal cells in quantity and quality. Tumor markers that circulate in the blood can be measured biochemically by employing chemical reactions, radioimmunoassays, or enzyme immunoassays. The markers present on cell membranes or in the cytoplasm are measured routinely by immunohistologic or immunocytologic techniques (*eg,* immunofluorescence, immunoperoxidase, and flow cytometry). Nuclear DNA content can be measured routinely by propidium iodine staining and flow cytometry for numerous solid and hematologic malignancies. The detection of chromosomal alterations can now be achieved using molecular diagnostic techniques.[7-11] Clinically useful tumor markers that aid in the diagnosis and management of cancer patients are classified into groups, as shown in Table 27-1.

TUMOR MARKERS IN CANCER MANAGEMENT

Theoretically, tumor markers should be extremely useful in the detection, management, and monitoring of cancer patients, as illustrated in Table 27-2. Unfortunately, almost all available tumor marker tests do not possess sufficient specificity to be useful screening tools in a cost-effective manner. Even highly specific tumor marker tests suffer from poor predictive value because the prevalence of a particular cancer is relatively low in the general population. Most available tumor marker tests are not useful in diagnosing cancer in patients who have symptoms because elevated levels are also seen in a variety of benign diseases. The major clinical value of tumor markers is in tumor staging, monitoring therapeutic responses, predicting patient outcomes, and detecting cancer recurrence. The usefulness of tumor markers in these clinical settings varies greatly depending on disease-related variables, such as therapeutic efficacy. For example, if treatment of a particular cancer is effective, a tumor marker may be useful for monitoring tumor burden and prognosis but would not be helpful

TABLE 27-1. Classification of Several Clinically Useful Tumor Markers

Oncofetal Antigens
  Carcinoembryonic antigen (CEA)
  Alpha-fetoprotein (AFP)
  CA 125
  CA 19-9
  CA 15-3
  CA 549
  Tissue polypeptide antigen (TPA)
  Prostate specific antigen (PSA)

Placental Proteins
  Human chorionic gonadotropin (hCG, intact and β-hCG)
  Human placental lactogen (HPL)
  Placental alkaline phosphatase (Regan isoenzyme)

Enzymes and Isoenzymes as Tumor Markers
  Prostatic acid phosphatase (PAP)
  Creatine kinase (CK)-BB isoenzyme
  Alkaline phosphatase
  Neuron-specific enolase (NSE)
  Lactate dehydrogenase (LD1) isoenzyme
  Lysozyme (Muramidase)

Hormones
  Eutopic hormones (normally secreted by the tissue)
    Human chorionic gonadotropin (hCG)-trophoblastic
      tumors, nonseminomatous testicular tumors
    Epinephrine and norepinephrine—pheochromocytoma
      and related malignancies
    Gastrin-Zollinger-Ellison syndrome (gastrinoma)
    Calcitonin—medullary carcinoma of the thyroid
  Ectopic hormones (not normally synthesized or secreted by
    the tissue).
    Adrenocorticotropic hormone (ACTH)—small cell
      carcinoma of the lung
    Antidiuretic hormone (ADH)
    Parathyroid-related peptide
    Erythropoietin

Monoclonal Immunoglobulins
  IgG, IgA, IgM, IgD, IgE
  Kappa and lambda light chains

Steroid Receptors
  Estrogen and progesterone receptors
  Androgen receptors
  Corticosteroid receptors

Immunophenotyping
  Lymphoid cells
  Myeloid cells

DNA Analysis

Molecular Diagnosis

for monitoring response to therapy. On the other hand, if therapy is highly effective, then a sensitive tumor marker would be of value in monitoring treatment response and disease recurrence but of little value as a prognostic indicator. In the case of gestational trophoblastic tumors, human chorionic gonadotropin (hCG) is a sensitive and specific marker useful in staging disease and selecting appropriate therapy but is not

TABLE 27-2. Potential Applications of Tumor Markers in Cancer

Screen for disease in the general population
Diagnosis in patients with symptoms
Adjunct in clinical staging
Indicator of tumor volume
Aid for selecting appropriate therapy
Monitor response to therapy
Prognostic indicator
Early detection of disease recurrence

helpful in patients undergoing intensive therapy because hCG levels do not correlate with prognosis. Thus, the application of tumor markers requires a thorough understanding of their capabilities and limitations if they are to be used wisely in the management of cancer patients.

PROPERTIES OF AN IDEAL TUMOR MARKER

An ideal tumor marker test should possess characteristics that meet both analytical and clinical requirements, as outlined in Table 27-3. Analytically, the method used to measure the tumor marker should be capable of detecting very small quantities (ie, analytically sensitive) and should not measure unrelated or closely related substances (ie, analytically specific). The true quantity of the substance should be measured by the test method, and the correct value should be obtained every time the method is performed (the assay should be precise). Operationally, the test should be rapid, easy to perform, and of relatively low cost to the laboratory.

TABLE 27-3. Characteristics of an Ideal Tumor Marker

Analytical Requirements
High analytical sensitivity
High analytical specificity
Accuracy
Precision
Rapid turn-around time
Easy to measure at a low cost

Clinical Requirements
High sensitivity for disease (no false-negative results, ability to detect micrometastasis)
High specificity for disease (no false-positive results, negative in disease-free individuals)
Levels should reflect tumor burden
Levels should remain relatively constant and not fluctuate in patients with stable disease
Should be undetectable or low in patients in complete remission
Should predict outcome in patients with stable disease

Clinically, an ideal tumor marker should be highly sensitive for disease. In other words, the tumor marker should be positive in all patients with that specific cancer. Elevations in tumor marker levels above normal should also be evident in the presence of micrometastasis and should predict and precede disease recurrence before it is clinically evident by other parameters used to detect disease. The tumor marker should also have high disease specificity and should be undetectable or at low levels in healthy people, patients with benign diseases, and patients with tumors other than the one in question. If the marker is normally present in the circulation, then it should exist at significantly lower concentrations than those detected in patients with all stages of cancer. The levels should not fluctuate in patients with stable disease and should parallel clinical and pathophysiologic changes in disease activity. For instance, tumor marker levels should correlate with tumor burden, should increase over time as the tumor load increases, and should diminish in patients undergoing successful chemotherapy. Patients in complete remission should have low to undetectable tumor marker levels as long as they are disease free, whereas increasing levels should reflect relapse. Finally, initial and serial tumor marker levels should have prognostic value when medical treatment for the malignancy is ineffective (see Table 27-3).

TUMOR ANTIGENS

During the past two decades, tests based on tumor antigens have greatly contributed to the clinical management of cancer patients. There are basically two types of tumor antigens: *tumor-specific antigens* and *tumor-associated antigens*. Tumor-specific antigens are thought to be a direct product of oncogenesis induced by viral oncogenes, radiation, chemical carcinogens, or unknown risk factors. The oncogenic process causes disorganization of the genetic information, leading to synthesis of antigens that are unique to the specific tumor cell. Mechanisms of tumor-specific antigen production have been demonstrated in animal models. Tumor-specific antigens play an important role in clinical oncology.[6,12]

Oncofetal Antigens

Most tumor antigens are not specific for individual tumors but rather are usually associated with different tumors derived from the same or closely related tissue. Oncofetal proteins are an example of tumor-associated antigens. They are present in both embryonic-fetal tissue and in cancer cells. *Oncofetal antigens* are produced in large amounts during fetal life and are released into the fetal circulation. After birth, the production of oncofetal antigens is repressed, and only minute quantities are present in the circulation of adults. It has been postulated that oncofetal protein production is reactivated with the onset of malignant cell transformation. This would explain the reappearance of oncofetal proteins in the circulation of cancer patients. Examples of embryonic-fetal antigens that are expressed by tumor cells and can be found in the circulation of patients with cancer are listed in Table 27-1. These tumor markers are routinely used in clinical oncology.[13]

Carcinoembryonic Antigen

Carcinoembryonic antigen (CEA) was identified in 1965 by Gold and Freedman as a tumor-specific antigen present in extracts of tumor tissue from patients with adenocarcinoma of the digestive tract. It is also found in normal fetal gastrointestinal tract epithelial cells. Its name is derived from the fact that it is present in both carcinoma and embryonic tissue. CEA was initially considered to be highly specific and sensitive for diagnosing colorectal carcinoma. As clinical data and experience accumulated, however, this marker was also found to be elevated in patients with other malignancies as well as several noncancerous disorders. Cigarette smokers also have elevations in serum CEA compared with nonsmokers. CEA is the most widely investigated and most frequently used tumor marker in cancer management.[14–18]

Properties

CEA comprises a large family of cell-surface glycoproteins with common antigenic determinants residing in the protein portion of the molecule. CEA is composed of a single polypeptide chain consisting of 30 amino acids. More than 35 different glycoproteins belong to the CEA family. CEA has a molecular mass between 160 and 300 kd, with 45% to 57% of the molecule consisting of carbohydrate. The ratio of protein to carbohydrate in CEA can vary as much as five-fold among different tumors.

CEA is present in both endodermally derived tissues (intestinal mucosa, lung, and pancreas) and nonendodermally derived tissues. It is found in the digestive tract of fetuses after 12 weeks' gestation and is present in embryonic liver and pancreas. CEA is a component of mucin and other mucopolysaccharides found on the brush border of normal mucosal cells and in the cytoplasm of colonic carcinoma cells. The concentration of CEA in the circulation depends on the number of cells producing CEA, the rate of synthesis, and

hepatic clearance from the circulation. CEA normally disappears from the circulation about 3 weeks after successful removal of a CEA-producing tumor.

Methods of Analysis

The two most commonly used methods for CEA analysis are enzyme immunoassay (EIA) and radioimmunoassay (RIA), both of which are commercially available. Because of the antigenic heterogeneity in CEA molecules from various tumors, different methods do not always produce equivalent results. Therefore, when switching analytical methods, it is important to establish a new baseline value when monitoring cancer patients by serial CEA measurements. In addition, reference ranges provided by the manufacturer for CEA should be used only as a guideline. Each laboratory should carefully evaluate the performance of the method employed and establish its own reference ranges.

Clinical Uses

Although CEA has achieved widespread clinical application, accumulated clinical data have shown that CEA levels are of limited use in detecting colorectal, pancreas, and liver cancer. For instance, about 35% of patients with colorectal cancer do not have an elevated CEA level.[19] This lack of sensitivity (high false-negative rate) makes CEA testing impractical for ruling out the presence of cancer.

Furthermore, abnormally elevated CEA values have been observed in patients with pulmonary emphysema, acute ulcerative colitis, alcoholic liver cirrhosis, hepatitis, cholecystitis, benign breast disease, and rectal polyps as well as in heavy smokers.[17,20–23] Elevated levels of CEA do not normally exceed four to five times the upper limit of the reference range in benign disorders and return to normal after the acute reaction has subsided. In contrast, continuously elevated CEA levels in serial samples is highly suggestive of a malignant disease process. Elevated CEA values are also seen more often in patients with advanced and disseminated diseases than in patients with localized disease. This severely limits the use of CEA for establishing an early diagnosis of cancer. Because CEA lacks disease sensitivity and specificity, it cannot be used for screening the general symptom-free

---

CASE STUDY 27-1

A 72-year-old woman complained of intermittent diarrhea and constipation for the last 2 weeks and a weight loss of 22 lb over the last 6 months. Physical examination was unremarkable except for a guaiac-positive stool. A colonoscopy was performed, which identified a circumferential mass in the sigmoid colon. A biopsy was performed, which identified the mass as an adenocarcinoma. A carcinoembryonic antigen (CEA) level was obtained as part of her initial surgical work-up.

Questions

1. Is the CEA test useful as a screening test for colon carcinoma?
2. What are some other conditions that can result in an elevated CEA level?
3. How is CEA used to monitor patients after surgery for colon cancer?
4. What other laboratory tests should be considered in the initial surgical work-up of this patient?

population or a subpopulation at high risk for malignancies or for independently diagnosing cancer. CEA can be used, however, as an aid in diagnosis, as an adjunct in clinical staging, to detect recurrence in patients who have undergone surgery, as a prognostic indicator, and to monitor the therapeutic response in patients undergoing chemotherapy or radiotherapy. In patients with cancer symptoms, an elevated CEA value, along with other clinical data and diagnostic methods, strongly suggests the presence of cancer.

The CEA level and its degree of elevation have been shown to correlate with the pathologic stage of the cancer lesion. For instance, Wanebo and colleagues found that only 27.5% of patients with Dukes' stage A colon cancer had elevated CEA levels, whereas 83.8% of stage D patients had an elevation.[24] The degree of CEA elevation also correlated with staging: only 9% of patients with stage B lesions had CEA elevations greater than four times the upper limit of the reference range, and 69% of patients with stage D had similar CEA elevations. Colorectal cancer patients with initially high CEA values are generally monitored by serial CEA measurements because values normalize after complete and successful tumor resection. CEA values generally remain normal in the absence of recurrent disease; however, a small number of patients (6%) develop recurrent disease without a rise in CEA.[25] In postoperative patients, a rise in CEA is indicative of recurrent disease in about 66% of cases. An elevation of CEA often precedes other clinical signs in 75% of patients developing recurrent disease by several months.[26,27] It has been suggested that postoperative follow-up should include CEA measurements every 6 weeks for a period of 2 years to increase resectability of recurring tumors, which will extend patient survival.[28]

CEA is also useful as a prognostic indicator. In colorectal carcinoma, patients with normal pretreatment CEA levels had a lower incidence of developing metastasis, whereas most patients with initially high CEA levels developed metastasis. Also, CEA-producing tumors have the tendency to metastasize earlier, exhibit aggressive biologic disease activity, and show shorter disease-free intervals after therapy. Serial CEA measurements in patients undergoing chemotherapy or radiotherapy also show a good correlation between cancer activity and CEA values.[26] The National Cancer Institute Panel on CEA has recommended that CEA *not* be used independently to establish a diagnosis of cancer and concluded that serial CEA measurements are the best available noninvasive tool for the postoperative monitoring of patients undergoing surgery for colorectal cancer. In addition to colon cancer, CEA has also been reported to be a useful tumor marker for breast cancer, bronchogenic carcinoma, pancreatic carcinoma, gastric carcinoma, urinary bladder cancer, prostatic cancer, ovarian cancer, neuroblastoma, and carcinoma of the thyroid.[29]

Alpha-Fetoprotein

Alpha-fetoprotein (AFP) was the first tumor-associated antigen to be used as a tumor marker in the clinical management of human malignancies. AFP is an oncofetal antigen that was detected in human fetal sera in 1956.[30] It was later shown to be present in mice with spontaneous or transplanted hepatomas.[31] Interest in AFP became widespread (in 1964) when AFP was found in the sera of human patients with hepatocellular carcinoma. Further investigations revealed that serum AFP levels are also elevated in patients with ovarian, testicular, and presacral teratocarcinomas.[32]

Properties

AFP is a single-chain glycoprotein with a molecular mass of 67 to 74 kd comprising 95% protein and 5% carbohydrate.[33] Variations in molecular mass may reflect variations in glycosylation that have been observed in AFP derived from different tumors. AFP is closely related to albumin, both chemically and physically. AFP preparations have considerable charge heterogeneity, and numerous forms have been detected on isoelectric focusing in the pH range of 4.5 to 5.2.[33,34]

AFP is synthesized by the liver, yolk sac, and gastrointestinal tract of the developing fetus. This glycoprotein is released into the circulation and is normally found in sera from human fetuses older than 6 weeks of age. The highest concentration of AFP occurs during embryonic fetal life. The fetal serum AFP concentration may reach a peak of 2 to 3 mg/mL by the 12th to 15th week of gestation, constituting about one third of the total fetal serum protein. At birth, the serum AFP declines to about 50 μg/mL, and by 1 year of age, the AFP level is at very low concentrations. AFP is found in trace amounts by 2 years of age and is similar to concentrations observed during adult life (usually less than 20 ng/mL).

Methods of Analysis

Initially, large-scale studies were performed to screen for liver cancer in high-risk areas, such as Asia and Africa, using the insensitive technique of immunodiffusion to measure AFP. In these studies, elevated serum AFP values were observed in 50% to 90% of patients with hepatocellular carcinoma, whereas apparently healthy individuals and patients with nonmalignant liver disease or other cancers did not have elevated AFP levels.[35,36] Similar studies in the United States revealed that this method is highly specific but lacks adequate sensitivity. More sensitive methods, such as EIA and RIA, are currently used in the United States and are commercially available. Depending on the assay method selected, each laboratory should establish its own reference ranges for serum AFP.[37]

Clinical Uses

Elevations in AFP (using sensitive and specific RIA or EIA assays) are observed in a wide variety of malignant and benign disorders. For instance, elevated AFP levels are seen in patients with hepatocellular carcinoma, testicular and ovarian germ cell tumors, pancreatic carcinoma, gastric carcinoma, and colonic carcinoma. AFP is the best tumor marker for managing patients with hepatocellular carcinoma. Nevertheless, elevations of serum AFP, similar to levels observed in malignant disease, are seen in nonmalignant hepatic disorders such as viral hepatitis and chronic active hepatitis. Serum AFP values can also be elevated in ataxia telangiectasia and tyrosinemia.[37–40]

Serum AFP levels are the most specific biochemical test for primary carcinoma of the liver. A National Institutes of Health consensus panel has recommended that serum AFP be measured as a screening test (along with ultrasound) in all patients in whom liver cancer, cirrhosis, or chronic active hepatitis is suspected.[41] AFP levels can also be useful in patients who have a change in clinical status. Measurement of AFP levels is helpful in monitoring the therapeutic response in cancer patients undergoing treatment. Postsurgical AFP levels that fail to return to normal may indicate incomplete surgical resection or metastatic disease. AFP levels also serve as a prognostic indicator, with higher values being associated with a larger tumor mass and a shorter survival time.[42]

AFP is also normally elevated in patients with testicular tumors of germ cell origin (the most common is testicular cancer).[43] Levels of hCG are also elevated in testicular cancer; therefore, AFP and hCG are commonly used in combination for managing patients with this type of cancer. AFP (as well as hCG) is frequently not elevated in small testicular tumors, making this test unreliable as a screening tool for cancer.[44]

Germ cell testicular tumors are classified as either seminomatous or nonseminomatous. AFP is not elevated in seminomatous testicular tumors, whereas most patients with nonseminomatous tumors have elevated levels of AFP. The nonseminomatous tumors may be further divided into four major types: choriocarcinoma, yolk sac tumor (endodermal sinus tumor), embryonal carcinoma, and teratoma. Testicular germ cell tumors often contain two or more different tumor types. AFP is specifically elevated in embryonal carcinoma and yolk sac tumors, whereas hCG is elevated in choriocarcinoma and in a few seminoma cases that contain syncytiotrophoblastic giant cells. The combined use of AFP and hCG as tumor markers for testicular cancer has proved essential in monitoring the response to therapy and in detecting disease recurrence. In one study, AFP was positive in 58% of nonseminomatous testicular cancer cases, and hCG was elevated in 77%. In 91% of the cases, either AFP or hCG was elevated.[45,46]

CA 125

Properties

CA 125 is an oncofetal antigen with a high molecular mass (more than 200 kd) that is produced by ovarian epithelial cells.[47] CA 125 is a glycoprotein, defined in 1981 by reactivity with a monoclonal antibody named OC125. This antibody was developed by immunizing mice with ascitic fluid from a patient with papillary serous cystadenocarcinoma of the ovary.[48] Early studies found that less than 1% of apparently healthy individuals and about 6% of patients with benign disease had elevations of CA 125. More importantly, greater than 80% of patients with ovarian carcinoma had elevated CA 125 values that appeared to correlate with tumor burden and extent of disease.

Methods of Analysis

Numerous commercial kits are available to measure CA 125 by EIA and RIA,[49] and automated immunoassay systems have recently been introduced for measuring CA 125. However, results obtained by the automated CA 125 method from Abbott Laboratories were shown to be 20% lower in apparently healthy women and at least 50% higher in patients with benign or malignant ovarian disorders when compared with RIA.[50] This illustrates the importance of establishing a new baseline value when changing the method used to measure a specific tumor marker level.

Clinical Uses

CA 125 levels are elevated in more than 80% of patients with nonmucinous epithelial ovarian cancer. More than 50% of the newly diagnosed cases of ovarian carcinoma are epithelial in origin. Elevated CA 125 levels are also observed in patients with advanced nonovarian gynecologic malignancies, including endometrial cancer, and in those with metastatic malignancies of the breast, colon, pancreas, and lung. Mild CA 125 elevations are observed in several benign conditions, including menstruation, pregnancy, benign cysts, endometriosis, and cirrhosis.[51] In patients with a palpable pelvic mass, CA 125 levels higher than 35 U/mL (10 times normal) could be detected in 78% of patients with malignant pelvic tumors and in 22% of patients with benign tumors.[52] Other studies have found that CA 125 is not highly sensitive for cancer in the early stages of disease, whereas still others reported that CA 125 cannot differentiate between benign and malignant disease.[53]

The major clinical use for CA 125 is in monitoring the treatment response of ovarian carcinoma. CA 125 levels that do not normalize after initial therapy are indicative of residual disease. Rising CA 125 levels after surgery or chemotherapy almost always represent cancer recurrence. In most cases, CA 125 levels above 35 U/mL are indicative of tumor recurrence.[54,55] A normal CA 125 level, however, does not always indicate the absence of tumor recurrence because CA 125 elevations are not reliability observed with tumors less than 2 cm in size.[56] CA 125 is also useful as a prognostic indicator because the degree of elevation correlates with tumor stage and size.[57] Also, increased levels after surgery are associated with shortened survival.

CA 19-9

Properties

CA 19-9 is an asialylated Lewis blood group antigen that was initially identified in a human colorectal carcinoma cell line in 1981.[58] This tumor marker is elevated in numerous malignancies, including colon, gastric, hepatobiliary, and pancreatic cancer. CA 19-9 can also be elevated in benign disorders, such as pancreatitis, extrahepatic cholestasis, and cirrhotic liver disease. CA 19-9 is measured by standard RIA and EIA methods.[59]

Clinical Uses

Much like the other oncofetal tumor markers, CA 19-9 is not useful as a screening test for cancer. CA 19-9 has been found to be more sensitive and specific than CEA for pancreatic cancer, but it cannot identify patients with early resectable pancreatic cancer.[60] Using a cutoff of 40 U/mL, CA 19-9 was elevated in 81% of patients with pancreatic cancer, whereas using a cutoff of 100 U/mL improved the sensitivity to 96%.[61,62] CA 19-9 is also elevated in 50% to 60% of patients with gastric carcinoma, 60% of patients with hepatobiliary cancer, and 30% of patients with colorectal carcinoma. The specificity of CA 19-9 for pancreatic cancer is further lowered by increased CA 19-9 levels typically found in patients with pancreatitis and liver disease. Because patients with benign pancreatitis usually have serum CA 19-9 elevations that do not normally exceed 120 U/mL, the specificity of CA 19-9 can be improved by increasing the cutoff value for pancreatic cancer.[63] A cutoff of 120 U/mL would severely limit the use of CA 19-9 for pancreatic cancer because patients in the early stages of disease (when surgical resection would be most beneficial) have CA 19-9 levels of less than 120 U/mL.

CA 19-9 is often used in conjunction with CEA to monitor recurrence of disease, and there is good correlation between increasing CA 19-9 levels and recurrence of disease.[64,65] CA 19-9 levels return to normal after successful surgical resection, and increasing levels can detect tumor recurrence 1 to 7 months before a clinical diagnosis of relapse is made.[66] Because treatment is ineffective after early detection of relapse, however, the clinical usefulness of CA 19-9 in these settings is somewhat limited.

CA 15-3

CA 15-3 is a high-molecular-weight glycoprotein mucin present on mammary epithelium known as *episialin*. An RIA method has been developed to detect CA 15-3 using antibodies against a glycoprotein found in membrane-enriched extracts of human breast cancer and human milk-fat globule membrane.[67] CA 15-3 is elevated in patients with metastatic breast, pancreatic, lung, colorectal, and liver cancer. It is also elevated in benign diseases of the breast and liver. Several studies have found that CA 15-3 is elevated in less than 33% of patients with stage I or II breast cancer.[68] CA 15-3 is more useful in identifying patients with advanced stages of breast cancer because 79% to 92% of patients with progressive disease have elevations in CA 15-3. CA 15-3 is not recommended for monitoring therapeutic responses because declining levels do not always occur in patients undergoing successful therapy.[69] Serum CA 15-3 levels correlate with tumor burden and are more sensitive than CEA for detecting disease recurrence.[70]

CA 549

CA 549 is an acidic glycoprotein located in the cell membranes and luminal surfaces of normal breast and other epithelial tissues. This tumor marker is found in serum as part of a cancer-associated mucin. EIA and RIA methods to measure CA 549 have been developed using antibodies against a human breast tumor cell line T 417 and human milk-fat globule membrane.[71] CA 549 is detectable in sera of healthy men and women. Although levels can be high in benign disease, increased values are most often seen in a variety of malignant diseases, such as breast, lung, prostate, and colon cancer.[72] Elevations in CA 549 are not normally observed in the early stages of breast cancer but rather occur during disease progression to a more advanced stage.[73] CA 549 levels also correlate with tumor burden and may prove useful for monitoring treatment and disease progression.

Tissue Polypeptide Antigen

Tissue polypeptide antigen (TPA) is a single-chain polypeptide related to the cytokeratin family. It consists of four protein subunits (A1, B1, B2, and C2) with individual molecular weights between 20 to 45 kd. TPA is synthesized during the S to M phase of the cell cycle and is an indicator of cell generation time and tumor aggressiveness.[74] An RIA method has been developed against the M3-specific epitope of TPA.[75] Serum levels of TPA may be increased in pregnancy and in several nonmalignant diseases, such as bacterial and viral infections, acute hepatitis, and autoimmune disorders. TPA levels are also elevated in sera of patients with several malignancies, including those of the lung, breast, pancreas, colon-rectum, stomach, prostate, bladder, and uterus. For lung and breast cancer, serum levels of TPA appear to correlate with tumor burden and may be clinically useful in staging and prognosis.[76,77]

Prostate-Specific Antigen

Between 1966 and 1978, three independent laboratories identified proteins in seminal fluid that were named gamma-seminoprotein, E1 antigen, and p30.[78] In 1979, Wang and colleagues purified a protein from benign and malignant prostatic epithelial cells and named it *prostate-specific antigen* (PSA).[79] Later studies revealed that all four substances were either the same or closely related molecules. PSA is a 30- to 34-kd serine protease that is secreted by prostatic epithelial cells. Although high levels of PSA are found in seminal fluid, very little PSA is found in the circulation of healthy men. In addition to the prostate, PSA is also produced by the peri-urethral and perirectal glands.[80] Most PSA in the blood circulates complexed to the serine protease inhibitors, alpha-1-antichymotrypsin (ACT) and alpha-2-macroglobulin, and functions by digesting gel-forming proteins into smaller peptides.[81] A minor fraction of PSA exists in the unbound form and lacks significant proteolytic activity. Trace amounts of PSA can also be found in the circulation complexed to alpha-1-antitrypsin inter-alpha-trypsin.[82] The complexing of PSA with serum proteins probably contributes to the relatively long PSA serum and plasma half-life of 2 to 3 days. PSA has replaced prostatic acid phosphatase in the management of known or suspected prostate cancer patients.[79]

Methods of Analysis

The first commercial immunoassay for PSA was approved by the Food and Drug Administration in 1986 for monitoring patients with prostate cancer. Since then, numerous companies have developed automated immunoassays to measure PSA using enzyme and chemiluminescent detection techniques. There are five antibody-binding sites on the PSA molecule, and two molecular forms of PSA are routinely detected by immunoassays currently in clinical use.[83] One form of PSA is free or uncomplexed and the other is PSA that has formed a complex with ACT. PSA complexed to ACT (PSA-ACT) accounts for about 70% to 80% of the total immunoreactive PSA.[82] Although a fraction of PSA normally exists in serum and plasma complexed to alpha-2-macroglobulin, this form is not detected by conventional immunoassays because the PSA antibody binding sites are located on the inside of the complex.[84]

Using a combination of antibodies that recognize different antigen-binding sites on the PSA molecule, it is possible to detect either free PSA, complexed PSA, or both immunoreactive forms (total PSA). Ultrasensitive assays have also been developed that are capable of measuring total PSA levels below 0.1 ng/mL. These newer PSA assays have a potential advantage over standard, less sensitive assays, for monitoring cancer recurrence because posttherapeutic PSA levels are usually less than 0.1 ng/mL.[85] Not all antibodies, however, recognize both forms of PSA with comparable affinity, and this can lead to discordant results when using different assays. Differences in assay performance can also result from variables such as assay method (monoclonal versus polyclonal antibodies), type of solid phase, incubation time, pH, and composition of PSA calibrators and their assigned values.

A comparison of the Ciba Corning ACS 180 with the DPC Immunlite assay for total PSA revealed that values obtained with the Ciba Corning ACS 180 were, on average, 1.97-fold higher.[86] Despite these differences, there was good correlation between the two methods, allowing the results from one assay to be normalized to the other, provided the user is aware of this problem. Not all PSA assays show good correlation, however, making it difficult to follow trends when analyzing samples over time using different assays. Moreover, discrepancies among assays are further magnified when the free-to-total PSA ratio (free PSA%) is calculated based on two separate measurements. For these reasons, it is imperative that clinical laboratories carefully evaluate PSA immunoassays and establish assay-specific reference ranges for their own patient population.

Clinical Uses

PSA lacks sufficient sensitivity and specificity to be used alone as a screening test for prostate cancer.[87] Because PSA is only organ specific and not tumor specific, there is considerable overlap in values obtained in patients with prostate cancer and those with benign conditions, such as benign prostatic hyperplasia and prostatitis. Blood samples for PSA testing can be obtained any time of day because PSA does not exhibit diurnal variation.[88] Although clinically significant elevations in serum PSA do not occur after routine digital examination,[89,90] vigorous prostate massage and needle biopsies can lead to significant increases in circulating PSA levels.[88,91] It is generally accepted that PSA can significantly improve prostate cancer detection when used in conjunction with digital rectal examination and transrectal ultrasound.[92] However, 38% to 48% of patients with organ-confined prostate cancer can have a normal PSA value using a cutoff of 4 ng/mL.[93,94]

PSA can be used to monitor patients with known prostatic carcinoma. The sensitivity of PSA can also be correlated with disease staging: 10% sensitivity for stage A; 24% for stage B; 53% for stage C; and 92% for stage D.[95,96] After radical prostatectomy, radiation therapy, or androgen therapy, PSA values normally decline to undetectable levels. Failure of PSA levels to become undetectable after therapy has been correlated with advanced disease stage and a poor prognosis. A rising PSA level after therapy precedes other indicators of tumor recurrence by 2 to 6 months. The 50% survival rate in postoperative patients (stages A2 to D2) was 40 weeks for those with PSA concentrations that were one to two times normal and 18 weeks for those with PSA concentrations that were greater than two times normal.[97]

To improve the performance of PSA as a diagnostic test, several approaches have been developed to enhance both sensitivity and specificity. PSA density was introduced in 1992 to aid in distinguishing increases in PSA due to prostate cancer from increases associated with benign conditions.[98] PSA density was defined as the ratio of serum PSA concentration to prostate volume, as determined by transrectal ultrasonography. The cost and invasiveness of transrectal ultrasonography has limited the use of this approach. Furthermore, a large screening study found that PSA density was no better than PSA alone, using a PSA cutoff of 4 ng/mL.[99]

A change in PSA concentration with time, designated *PSA velocity,* has also been used to improve prostate cancer detection. An increase in PSA of 0.75 ng/mL or greater per year was used as a threshold for suspicion of prostate cancer.[100] However, this approach is useful only when the same assay is used to measure PSA levels. In addition, each specimen must be obtained under similar conditions to limit preanalytical variability. It has been suggested that increases between two consecutive PSA levels of less than 20% to 46% may be due to biologic or analytical variation.[101] These practical limitations minimize the utility of PSA velocity.

In view of the observation that serum PSA levels increase with advancing age, Ruckle and associates advocate the use of age-specific reference ranges.[90] Age-specific ranges are designed to increase PSA sensitivity in younger men by reducing the false-negative rate and to improve specificity in older men by decreasing the number of false-positive results. Although this concept has been confirmed by other groups,[102] it is complicated by the observation that age-specific reference ranges are different for white, African American, and Asian men.[103] In addition, the use of age-specific ranges in men older than 60 years of age has been disputed because a PSA cutoff of 4 ng/mL (compared with 4.5 and 6.5 for white men 60 to 69 and 70 to 79 years old, respectively) was found to be superior for detecting cancer in this group of men.[104]

The discovery that PSA exists in two immunoreactive forms in blood prompted numerous clinical studies investigating the clinical utility of measuring free PSA, complexed PSA, or the ratio of these forms to total PSA for the early detection of prostate cancer. A preliminary study in 1991 demonstrated that complexed PSA was higher in patients with prostate cancer than in patients with benign prostatic hyperplasia and in healthy subjects.[105] The ratio of complexed to total PSA (also called *complexed PSA%*) was reported to discriminate between benign and malignant disease and to reduce the false-positive rate by 50%, thereby eliminating many unnecessary biopsies.[105] Although these findings were subsequently confirmed by another group, these investigators suggested that the ratio of free PSA to total PSA (also called *free PSA%*), was more sensitive than total PSA for identifying patients with prostate cancer.[106] They indicated that a cutoff of 0.18 for free PSA% resulted in a specificity of 73%, as compared with 55% for total PSA.[106]

The improvement in clinical performance of free PSA% over total PSA for detecting prostate cancer has been amply documented by several recent studies.[107] Furthermore, not only is free PSA reflex testing recommended when the total PSA is more than 4 ng/mL, it also has been advocated when the total PSA is in the upper range of normal, above 2.5 or 3 ng/mL, to improve cancer detection.[103] Unlike free PSA and total PSA, which both differ with age, free PSA% is not significantly different between the 3rd and 8th decades of life in healthy men, eliminating the need for age-adjusted reference ranges.[108,109] A consensus cutoff for free PSA% does not currently exist, however, and the cutoff can range from 14% to 26%, depending on study design and PSA assay, emphasizing the need for standardization.[107] Free PSA% only poorly correlates with stage of disease, limiting its usefulness in iden-

tifying prostate cancer patients who would benefit from radical prostatectomy.[110]

Because most PSA in prostate cancer is complexed to ACT, assays that recognize only the complexed form of PSA may be of more benefit when the total PSA is low because only a minor fraction is in the free form. Another advantage is that complexed PSA appears to be more stable than free PSA in blood samples.[111] A recent study compared free PSA to complexed PSA and found that complexed PSA% provides better specificity than free PSA% at cutoffs yielding 90% to 100% sensitivity, and could potentially eliminate 20% to 30% of the unnecessary prostate biopsies.[112] Complexed PSA% in patients with benign prostatic hyperplasia does not vary with age, eliminating the need for age-adjusted reference ranges.[113] When complexed PSA alone was compared with free PSA%, specificity was better using complexed PSA at several sensitivity thresholds.[113] Although additional studies are needed to confirm the superiority of complexed PSA over free PSA and PSA ratios, this approach eliminates the need to perform two measurements (a second measurement for total PSA when expressed as a ratio) and, therefore, would result in cost savings.

In summary, measurement of either free or complexed PSA results in a significant increase in specificity over total PSA assays for detecting prostate cancer. Further studies are needed to resolve the issue of universally accepted cutoff values and to establish whether complexed PSA is the optimal test for detection of prostate cancer.

PLACENTAL PROTEINS

Human Chorionic Gonadotropin

Properties

The heterodimer hCG has a molecular mass of 40 kd, consisting of a noncovalently linked alpha and beta subunit. The alpha subunit is common to other glycoprotein hormones, such as luteinizing hormone, follicle-stimulating hormone,

---

CASE STUDY 27-2

A 69-year-old Caucasian man presented to the medical clinic with complaints of increased frequency of urination and difficulty urinating. Further questioning revealed the need to urinate several times a night. Although the problem has been present for some time, it has recently become worse. A serum prostate-specific antigen (PSA) was obtained and a digital rectal examination was performed. The PSA level was elevated at 15 ng/mL (reference range <4 ng/mL) and the digital exam revealed an enlarged prostate without palpable nodules.

Questions

1. Can serum PSA be used alone as a screening test for prostate cancer?
2. What are some causes for an elevated serum PSA level?
3. If this patient has cancer limited to the prostate, should PSA levels be monitored after radical prostatectomy and/or radiation therapy?
4. What other laboratory tests should be performed in the initial work-up of this patient?

and thyroid-stimulating hormone. The beta subunit of hCG is homologous to the beta subunit of luteinizing hormone except for an additional 24 amino acids at the carboxyl-terminal end that confers immunogenic specificity.[114,115] During pregnancy, hCG is synthesized by the syncytiotrophoblastic cells of the placenta. It is found in trace amounts in healthy men and nonpregnant women and becomes markedly elevated in the circulation during pregnancy, reaching a peak 8 to 10 weeks after conception. Serum or urine hCG is routinely measured to confirm pregnancy.

Methods of Analysis

Early RIA techniques using antisera against intact hCG and the radioreceptor assay lack specificity and are no longer in use. There are several qualitative and quantitative commercially available EIA and RIA methods for measuring intact hCG and β-hCG. The specificity of the two antibodies (polyclonal or monoclonal) used in the EIA generally determine which forms of hCG are measured. For instance, if one antibody is against the alpha subunit and the other is against the beta subunit, then only intact hCG is measured. When both antibodies recognize the beta subunit, both intact hCG and β-hCG are detected; however, binding to both forms of hCG may not be equal. When intact hCG and β-hCG are recognized with equal affinity, the assay is usually designated as a total β-hCG assay. These distinctions are extremely important because hCG values for a specific patient can vary tremendously depending on assay design and relative proportions of intact hCG versus free β-hCG present. This issue is further complicated because reference standards used to calibrate hCG assays differ in relative proportions of intact and β-hCG, which can lead to different results, depending on which forms are measured and how the assay is calibrated. Nevertheless, even with all of these variables, both intact hCG and β-hCG are used equally as tumor markers and in other clinical situations.[116]

Clinical Uses

As a tumor marker, β-hCG is used for the diagnosis and management of gestational trophoblastic disease, which consists of complete and partial hydatidiform mole, gestational choriocarcinoma, and placental-site trophoblastic tumor. It is also useful for accurate staging and for monitoring the response to treatment in germ cell tumors of the testis and ovary. The frequency of patients with elevated β-hCG has been estimated to be 15% for seminoma, 50% for embryonal carcinoma, 42% for teratocarcinoma, and nearly 100% for choriocarcinoma. Elevated β-hCG levels have also been observed in ovarian cancer, breast cancer, melanoma, gastrointestinal tract *neoplasms,* sarcoma, lung cancer, renal cancer, and hematopoietic malignancies. Serial measurements of β-hCG are best for monitoring the clinical course and response to therapy. Persistent elevations of β-hCG after therapy indicates residual disease. HCG (intact and the beta subunit) is normally undetectable 4 to 6 weeks postpartum or after abortion. Persistent postpartum elevations may indicate trophoblastic disease.[117]

Human Placental Lactogen

Human placental lactogen (HPL) is a trophoblastic hormone that is normally produced by syncytiotrophoblastic cells of the placenta. HPL has lactogenic properties, is present in the serum of pregnant women, and has been used as an index of placental function and fetal well-being. Levels of HPL correlate with fetal and placental weight, reaching a plateau in the last month of pregnancy and becoming extremely low or undetectable 24 to 48 hours postpartum.

Elevated levels of HPL are associated with trophoblastic neoplasms (gestational, gonadal, or extragonadal), although levels are not as high as those observed in pregnancy. In trophoblastic malignancies, a rapid decline is normally observed after chemotherapy. HPL can be useful for differentiating trophoblastic disease from pregnancy, with low values being associated with disease and high values indicating pregnancy.[118]

Placental Alkaline Phosphatase (Regan Isoenzyme)

Properties

Placental alkaline phosphatase (PALP) is an isoenzyme of alkaline phosphatase synthesized by placental trophoblasts. PALP is normally found in the serum of pregnant women. In 1968, the Regan isoenzyme was identified in serum and tumor tissue from a patient with lung cancer.[119] In adults, the PALP gene is normally repressed, but Regan isoenzymes can be produced in various diseases as a slightly modified form of PALP. PALP (Regan isoenzyme) can be differentiated from other isoenzymes of alkaline phosphatase (liver, bone, intestine) by its remarkable heat stability or by using chemical inhibitors such as urea and *l*-phenylalanine. PALP (Regan isoenzyme) is resistant to urea treatment but sensitive to *l*-phenylalanine (80% inhibition in activity).[120]

Clinical Uses

PALP is elevated in about 10% to 20% of patients with a wide variety of malignancies, such as lung, ovarian, gastrointestinal, and trophoblastic cancers. PALP is not specific for malignancy and is also observed in nonmalignant disorders, such as cirrhosis, ulcerative colitis, and diverticulitis. Its low incidence in cancer and poor disease specificity limits its usefulness as a tumor marker, although levels may correlate with therapeutic response in individual patients.

ENZYMES AND ISOENZYMES AS TUMOR MARKERS

Various enzymes can be increased in malignancy; however, enzyme elevations are not specific for the type of cancer, its location, or the presence or absence of metastasis. Mechanisms accounting for elevated serum enzyme levels are as follows: altered membrane permeability or necrosis of malignant cells resulting in enzyme leakage into the blood stream, blockage of the biliary or pancreatic tract resulting in choleostasis, and altered renal clearance. Enzymes that can be used as tumor markers are shown in Table 27-1.

Prostatic Acid Phosphatase

Properties

Discovered more than 50 years ago, acid phosphatase was one of the first laboratory tests used for the diagnosis of cancer.[121] Although acid phosphatase is produced primarily by the prostate gland, it is also found in bone, liver, spleen, kidney, bone marrow, and several cell types. Prostatic acid phosphatase (PAP) can be distinguished from other sources in enzymatic assays by using substrates, such as thymolphthalein monophosphate, that are more specific for PAP. EIA and RIA are also commercially available to measure PAP. This marker is one of the few enzymes showing organ and cell-type specificity.[122]

Clinical Uses

PAP has had limited clinical utility since PSA emerged as a tumor marker for detecting localized prostate disease and for monitoring therapeutic response. PAP can be elevated in prostate cancer, benign prostatic hyperplasia, or acute prostatitis and after prostatic massage. PAP is not sensitive enough to be used as a screening test and is not suitable for detecting prostate cancer in the early stages of disease. For instance, only 12% of patients with stage A disease and 20% of patients with stage B disease who had localized and potentially curable prostate cancer had elevated PAP levels.[123] The only area in which PAP may be superior to PSA is in selecting those patients with metastatic disease who are not candidates for surgery. Elevated PAP values were found in 84% to 100% of patients with prostate cancer exhibiting extracapsular extension and regional lymph node metastases.[124,125] In view of the limited role of PAP in the management of prostate cancer and the proven advantages of PSA, it is not surprising that PAP has been almost completely replaced by PSA for detection and monitoring of prostate cancer.

Creatine Kinase-BB Isoenzyme

Creatine kinase (CK) levels are helpful in establishing a diagnosis of myocardial infarction. CK is a dimer composed of two subunits designated M and B. CK-MM (CK3) is present in skeletal muscle; CK-MB (CK2) is found primarily in cardiac muscle; and CK-BB (CK1) is found in the brain, gastrointestinal tract, uterus, and prostate. Elevations of CK-BB (and total CK) can be found in prostatic carcinoma and metastatic cancer of the stomach.[126]

Alkaline Phosphatase

Alkaline phosphatase (ALP) has been used to detect primary malignancies in bone and liver and to detect metastases to these organs. Osteoblastic lesions in the bone produced by prostate cancer metastases and osteogenic sarcoma give rise to enormous elevations in ALP, whereas osteolytic lesions produced by metastatic breast cancer cause only mild or no ele-

vations in ALP.[127] During successful treatment, there may be a paradoxical rise in ALP owing to repair of damaged bone.

Other causes of elevated ALP include extrahepatic obstruction of the biliary tract, which usually results in a twofold increase in ALP levels. Diseases such as leukemia that infiltrate the liver ("space-occupying lesions") can cause marked elevations in ALP levels that correlate with the extent of liver involvement. The source of the elevated ALP (bone or liver) can be identified by measuring other liver enzymes, such as gamma-glutamyl-transferase and 5′ nucleotidase, or by measuring ALP activity after heat inactivation of the bone isoenzyme.[128] A much more specific test to measure bone-specific alkaline phosphatase has recently been developed that should be extremely helpful in managing patients with various bone disorders.[129]

Neuron-Specific Enolase

Neuron-specific enolase (NSE) is a glycolytic enzyme that is found in neurons, neuroendocrine cells, and neurogenic tumors.[130] NSE is an enzyme marker for tumors associated with the neuroendocrine system, such as small cell carcinoma of the lung, neuroblastoma, carcinoid tumors, and pancreatic islet tumors. Elevations of NSE are found in 70% of patients with small cell lung cancer. NSE levels decline after successful therapy and rise during recurrence of disease.[131] Elevations of NSE are also found in 90% of children with neuroblastoma.[132] An NSE RIA method is commercially available to monitor patients with these types of cancers.

Lactate Dehydrogenase-1 Isoenzyme

Lactate dehydrogenase (LD) is composed of four peptide chains of either the M or H type that combine to produce five isoenzymes. Elevations in LD occur in 70% of patients with liver metastasis and in 20% to 60% of cancer patients without liver involvement. The LD-5 isoenzyme is normally elevated in breast, stomach, lung, and colon cancers. In leukemias, an increase in total LD occurs in about 50% of patients and is usually associated with LD-3 isoenzyme elevations. LD can also be used to monitor response to therapy.[133]

The LD-1 isoenzyme has been shown to be markedly elevated in germ cell tumors such as teratoma and seminoma of the testis. LD-1 isoenzyme elevations are also observed in benign conditions, such as acute myocardial infarction, hemolytic disorders, and renal infarction, and should not be used independently to diagnose cancer.

HORMONES

Eutopic hormones are secreted by endocrine tissue under normal physiologic conditions and in excess in certain neoplastic disorders (Table 27-4). Ectopic hormones are produced *de novo* by endocrine or nonendocrine malignant tissues that normally do not produce the hormone. Inappropriate synthesis and release of hormones by neoplasms produce severe systemic effects that usually occur distant from the primary can-

CASE STUDY 27-3

A 25-year-old man was found to have an elevated lactate dehydrogenase (LD) of 350 IU/L (reference range 93–139 IU/L) that was confirmed with a second sample. LD electrophoresis was performed and revealed an elevation in the LD-1 isoenzyme above the normal range. Electrocardiogram (ECG) studies were negative for an acute myocardial infarct. Creatine kinase (CK) electrophoresis showed only a CK-MM band. On physical examination, the man was found to have a tumor mass in the left scrotum. Serum alpha-fetoprotein (AFP) and beta-human chorionic gonadotropin (β-hCG) tests were ordered.

Questions

1. What is the most likely primary diagnosis for this patient?
2. What kind of benign diseases should be considered in the differential diagnosis? Why?
3. Can the type of testicular tumor be determined based on the serum AFP and β-hCG results?
4. Can a final diagnosis be made based only on the tumor marker findings? If not, why not?

---

cer or metastasis. These specific effects are described as *paraneoplastic endocrine syndromes*. Clinical manifestations of excess hormonal secretion and the associated malignancies of endocrine paraneoplastic syndromes are detailed in Table 27-5.

The symptoms resulting from paraneoplastic syndromes may be a patient's only presenting indication of malignancy and can be life-threatening. About 20% of cancer patients have a paraneoplastic syndrome. Paraneoplastic syndromes are most commonly associated with ectopic hormone production but may manifest as connective tissue or dermatologic disorders, neurologic syndromes, or a consequence of vascular, hematologic, or immunologic complications.[134] Hormones, such as estrogen, progesterone, and testosterone, are now being measured routinely in the clinical laboratory by EIA and RIA techniques that have excellent analytical sensitivity and specificity.

STEROID RECEPTORS

Hormonally dependent organs, such as breast, prostate, and uterus, give rise to malignancies that may in themselves be hormonally dependent. Eliminating hormonal stimulation by the use of drugs or removal of the hormone-producing organ or tissue, often leads to a regression of the respective malignancy. However, not all cancers arising from hormone-dependent tissues respond to such therapy, and removing the hormonal stimulation carries considerable risk to the patient.

Predicting which tumors are likely to respond to hormonal manipulation can be achieved by measuring steroid hormone receptors on tumors.[135]

Properties

Steroid hormones (estrogen, progesterone, testosterone) are transported in plasma by a number of proteins, including testosterone-estradiol–binding globulin and corticosteroid-binding globulin. Unbound steroid enters the cell, apparently by passive diffusion, and complexes with its respective receptor protein. Receptor proteins are found in a variety of concentrations in target organs but are virtually absent in nontarget tissues. Receptor proteins have a high affinity and specificity for their respective steroid hormones. Complexing of the steroid receptor in the cytoplasm induces translocation into the nucleus of the cell, where it associates with chromatin accepter sites. This initiates RNA synthesis for proteins important in cellular growth and differentiation.

Methods of Analysis

Tissue for cytosolic steroid receptor analysis must be representative of the malignancy. The tissue is first homogenized to release the hormone receptors that can then be quantitated by analyzing the supernatant. Sensitive EIAs are presently available to measure estrogen and progesterone receptors and can be performed with considerably less tissue than the older radioreceptor assay.[136]

TABLE 27-4. Eutopic Hormone Production

Hormone	Clinical Symptom	Tumor Site
β-hCG	Menstrual irregularities	Trophoblastic tumors
Calcitonin	Hypocalcemia	Thyroid-medullary carcinoma
Gastrin	Zollinger-Ellison syndrome	Non–β-cell pancreatic tumor
Catecholamines	Hypertension	Pheochromocytoma

TABLE 27-5. Ectopic Hormone Production

Hormone	Clinical Symptom	Tumor Site
ACTH	Cushing's syndrome	Lung, thyroid, pancreas
ADH	Hyponatremia	Oat cell carcinoma
Serotonin	Carcinoid syndrome	Enterochromaffin cell
Erythropoietin	Erythrocytosis	Kidney
TSH	Hyperthyroidism	Trophoblastic tumors

Clinical Uses

Estrogen receptor (E^R) and progesterone receptor (Pg^R) assays are used to predict the therapeutic response of breast cancer patients to hormonal treatment. Levels greater than 10 fmol/mg for both E^R and Pg^R are considered positive. About 80% of breast cancer patients who test positive for both E^R and Pg^R respond to hormonal therapy. The higher the concentration of E^R and Pg^R, the more predictable the therapeutic response. About 30% of patients who test positive for E^R and negative for Pg^R respond to hormonal therapy, whereas 40% to 50% of patients with E^R-negative and Pg^R-positive cancers respond. As expected, only a small fraction of patients (2%) with breast cancer lacking both receptors respond to hormonal therapy.[137,138]

Receptors for other types of cancers are being studied and appear to correlate with clinical response to hormonal manipulation. These include androgen and progesterone receptors for prostate cancer, estrogen and progesterone receptors for endometrial cancer, and binding of acute lymphoblastic leukemia cells to dexamethasone.[139–141]

IMMUNOPHENOTYPING

With the development of highly specific fluorescent monoclonal antibodies and a new generation of flow cytometers, it is now possible to detect routinely a variety of antigens and receptors on the cell surface. This can aid in classifying hematopoietic cells along their respective lineages and in identifying their degree of maturation.[8,142]

Immunophenotyping, using comprehensive antibody panels, can serve as confirmatory evidence of the morphologic diagnosis in leukemias and lymphomas. In general, human leukemic (lymphoid) cells are phenotypically similar to their normal hematopoietic counterpart. In lymphoid leukemias and lymphomas, identification of tumor cells as either T or B lymphocytes can be an important predictor of clinical outcome. The use of monoclonal antibodies in acute nonlymphoid (myelogenous) leukemia is particularly helpful in differentiating acute myelogenous leukemia from acute lymphoblastic leukemia.

Methods of Analysis

Various panels are used to differentiate among the various acute leukemias. Multiple monoclonal antibodies, recognizing different antigens, should be used to identify a particular cell lineage and the maturational stage. A listing of commonly used commercially available monoclonal antibodies is shown in Table 27-6. Single-cell suspensions are required for analysis, with blood and tissue being the most common source of cells. Mononuclear cells can be isolated from whole blood by density gradient separation (centrifugation through Ficoll-Hypaque technique) or by lysis of red blood cells. Solid or semisolid tissues, such as lymph nodes and bone marrow aspirates, require gentle deaggregation.

Single-cell suspensions are stained either directly or indirectly with fluorescent monoclonal antibody. Direct staining uses an antibody conjugated to a fluorescent label, whereas indirect staining uses a cell-surface–specific antibody as the first step followed by a fluorescently labeled antibody against immunoglobulin G (IgG) or IgM. Commonly used fluorochromes include fluorescein isothiocyanate, phycoerythrin, and rhodamine.

Fluorescently labeled cells are analyzed by flow cytometry. This method quantitates the number of cells that are positive for the antigen of interest (based on the amount of light emitted after excitation of the fluorochrome) and also provides information about cellular size and nuclear complexity. Two or more monoclonal antibodies that recognize different antigens (labeled with different fluorochromes) can be used simultaneously to determine the total number of T lymphocytes (CD3) and the proportion of T lymphocytes expressing the CD4 (T helper cell phenotype) and CD8 (T suppressor cell phenotype) surface antigens.[143]

Clinical Uses

Surface antigen phenotyping is used to classify acute leukemias, to provide prognostic data, and to aid in selecting the appropriate treatment regimen. Other techniques involving monoclonal antibodies and flow cytometry that are currently being used in the clinical laboratory include measuring CD4/CD8 T-lymphocyte subset ratios, reticulocyte counting, and detection of platelet-associated IgG in immune thrombocytopenia and antineutrophil antibodies.[144]

DNA ANALYSES

Measuring nuclear DNA content (ploidy) and proliferative capacity (S-phase fraction) of malignant cells is useful for predicting the overall survival and disease-free interval in can-

TABLE 27-6. Commonly Used Monoclonal Antibodies

Cluster	Antibody/Common Name	Specificity
Lymphoid Antibodies		
CD1	Leu-6, OKT-6, T6	Thymocytes
CD2	Leu-5, OKT-11, T11	Pan T cell (sheep RBC receptor)
CD3	Leu-4, OKT-3, T3	Pan T cell (T-cell receptor complex)
CD4	Leu-3, OKT-4, T4	T helper/inducer
CD5	Leu-1, OKT-1, T1	Thymocytes, B-cell subset
CD8	Leu-2, OKT-8, T8	T suppressor cells
CD10	CALLA	Common acute lymphoblastic leukemia antigen
CD19	Leu-12, B-4	Immature B cells, all malignant B-cell neoplasms except multiple myeloma
CD20	Leu-16, B-1	B lymphocytes, all malignant B cells except multiple myeloma
CD24	BA-1	Peripheral B cells
CD25	Tac	Interleukin-2 receptor
CD45	HLe-1	Pan leukocyte antigen
—	PCA1	Plasma cells
—	HLA-DR, DQ	MHC class II antigen
Myeloid Antibodies		
CD11	Leu-15, My-7	Granulocytes, monocytes
CD14	Leu-M3, My-4	Monocytes
CD15	Leu-1, My-1	Granulocytes, monocytes
CD33	Leu-6, My-9	Early myeloid
—	Glycophorin	Erythroblasts
CD34	HPCA, My-10	Early myeloid
CD45	HLe-1	Pan leukocyte

cer patients. DNA analysis can help discriminate between benign and malignant disease, monitor disease progression, and predict response to treatment. In humans, nonreplicating normal cells in G_0 or G_1 phase of the cell cycle are referred to as *diploid*. During the synthetic phase (S phase), DNA content increases (aneuploid), which is followed by mitosis (M phase) and cell division. Malignant cells can be nonreplicating (diploid) or contain increased amounts of DNA (aneuploid). Ploidy status, or *DNA index* (DI), of malignancies is defined by the amount of measured DNA in cancer cells relative to that in normal cells and is calculated by the following equation:

$$DI = \frac{\text{peak channel number of aneuploid } G_0/G_1 \text{ peak}}{\text{peak channel number of diploid } G_0/G_1 \text{ peak}}$$

(Eq. 27–1)

A diploid population has a DI equal to 1 by definition. Any DI not equal to 1 is termed *aneuploid*. A DI of less than 1 is *hypodiploid,* whereas a DI of more than 1 is considered *hyperdiploid*. The percentage of cells in S phase of the cell cycle can also be determined by flow cytometry; more than 5% is considered elevated.[9,145]

Methods of Analysis

The determination of DNA content is one of the most common clinical applications of flow cytometry. The tissue may be fresh, fixed with alcohol, or removed from paraffin tissue blocks and rehydrated before analysis. Cells are permeabilized, and DNA within nuclei are stained with DNA inter-calating agents, such as propidium iodide and ethidium bromide. The DNA content is determined by fluorescence intensity (channel number).

Clinical Uses

DNA flow cytometry measurements of ploidy (DI) and S-phase percentages, or fractions, are prognostic indicators in cancer patients. In breast cancer, both aneuploidy and high S-phase fraction correlate with the lack of estrogen and progesterone receptors. These cases are associated with a worse prognosis when compared with diploid, low-S-phase, receptor-positive malignancies. In general, aneuploidy is seen in advanced disease and shortened disease-free intervals. Aneuploidy in ovarian cancer is associated with a shorter survival and in prostate cancer predicts the likelihood of regional spread or metastasis. DNA analysis has also been used to aid in the diagnosis and monitoring of disease and the response to therapy. Flow cytometric results of bladder washings appear to correlate well with bladder biopsies. In colorectal cancer, most malignancies are aneuploid, and those with a high S-phase fraction have a poorer prognosis.

Molecular Diagnosis

Molecular diagnosis employs techniques that can detect alterations at the level of the gene (DNA). It is firmly established that when mutations occur within DNA from chemicals, radiation, or viruses, cancer may arise. DNA analysis in the past decade has made a quantum leap from the research laboratory to the clinical setting and has been used in microbiology, genetics, and forensic pathology. It is rapidly emerging in the area of cancer diagnosis.[146]

TABLE 27-7. Major Tumors and Associated Tumor Markers

Tumor Type	Associated Tumor Marker
Hepatocellular carcinoma	AFP, ferritin, ALP liver isoenzyme, LD-5
Germ cell tumors of the testis and ovary	AFP, β-hCG, PTH, calcitonin, CK-BB, PALP (Regan isoenzyme), prolactin
Gastrointestinal cancer	CEA, AFP, fetal sulfoglycoprotein
Breast cancer	CEA, β-hCG, ferritin, casein, mammary tumor–associated glycoprotein (MTGP), calcitonin, TPA
Pancreatic cancer	CEA, pancreatic oncofetal antigen (POA), ACTH
Prostatic cancer	Prostatic acid phosphatase, CEA, LD-5 isoenzyme
Trophoblastic diseases (choriocarcinoma or hydatidiform mole)	β-hCG, LD, LD1, SP-1
Renal and bladder cancer	CEA, LD
Multiple myeloma, Waldenström's macroglobulinemia	IgG, IgA, IgM, IgD, and IgE immunoglobulins, free lambda and kappa light chains (Bence-Jones protein)

Viruses have been associated with a number of cancers in animals and humans. Viral DNA segments that can transform normal cells into malignant cells are known as *oncogenes*. Normal cellular genes that play essential roles in cell differentiation and proliferation and potentially can become oncogenic are known as *protooncogenes*. The transformation of protooncogenes into oncogenes can occur by single-point mutations, translocation, and amplification. Only 20% to 25% of human malignancies are associated with activated oncogenes. In addition, genes that normally repress cellular proliferation can become inactivated, resulting in uncontrolled cellular growth. Cancer suppressor genes have been identified in retinoblastoma, Wilms' tumor, and hepatic carcinoma.

Methods of Analysis. Early molecular diagnostic techniques were time consuming, difficult to perform, and involved the use of ^{32}P labeling. With the advent of nonisotopic tags, automation, the polymerase chain reaction, and the growing market of vendors, molecular diagnostic techniques will rapidly become a routine test in the clinical laboratory.

SUMMARY AND FUTURE DEVELOPMENTS

During the last two decades, numerous clinically useful tumor markers have been identified and used in the management of cancer patients. Currently used tumor markers and their associated tumors are shown in Table 27-7. Unfortunately, the majority of available tumor markers are also elevated in nonneoplastic disease processes. In addition, a normal tumor marker value does not necessarily exclude a diagnosis of cancer, since most tumor markers have poor disease sensitivity in the early stages of cancer. Consequently, with few exceptions, none of the tumor markers can be used as a screening tool or can make an independent diagnosis of cancer. Introduction of molecular techniques into the clinical arena, and the rapidly growing number of methods currently available, may make it

possible to identify patients at risk for the development of cancer. Perhaps molecular probes will fulfill the criteria of the ideal tumor marker in the not-too-distant future.

REVIEW QUESTIONS

1. Cancer accounts for what percentage of deaths in the United States annually?
   a. 5%
   b. 13%
   c. 20%
   d. 23%

2. Tumor marker tests are used to:
   a. Aid in staging of cancer
   b. Monitor response to therapy
   c. Detect recurrent disease
   d. All of the above

3. Tumor markers may be defined as:
   a. Analytical tests (*eg,* flow cytometry) used to mark cancer cells
   b. Radioactive substances and chemicals used to help the physician identify cancer cells
   c. Biological substances synthesized and released by cancer cells or substances produced by the host in response to cancer cells
   d. None of the above

4. Which of the following is an oncofetal antigen?
   a. Alpha-fetoprotein
   b. LD
   c. PAP
   d. Neuron-specific enolase

5. Which of the following tumor markers is found in both carcinoma and embryonic tissue and is useful as a prognostic indicator in colorectal carcinoma?
   a. AFP
   b. CEA
   c. PAP
   d. CA-120

6. The major clinical use for CA-125 is monitoring treatment response of:
   a. Ovarian carcinoma
   b. Colorectal cancer
   c. Prostatic cancer
   d. Breast cancer
7. Which of the following tumor markers is used in the management of patients with prostatic cancer?
   a. Prostatic acid phosphatase
   b. Prostate-specific antigen
   c. CA 549
   d. Tissue polypeptide antigen
8. Which of the following enzymes is commonly used as a tumor marker?
   a. Lipase
   b. LD
   c. Aldolase
   d. Catalase
9. A tumor marker used in the assessment of choriocarcinoma or hydatidiform mole is:
   a. β-hCG
   b. CEA
   c. AFP
   d. IgG
10. DNA analysis (measuring nuclear DNA content) may be used to:
   a. Diagnose cancer
   b. Mark cancer cells for removal by surgery
   c. Discriminate between benign and malignant disease
   d. Detect alterations in genes

REFERENCES

1. Henderson BE, Ross RK, Pike MC. Toward the primary prevention of cancer. Science 1991;254:1131.
2. Landis SH, Murray T, Bolden S, et al. Cancer statistics, 1998. CA Cancer J Clin 1998;48:6.
3. American Cancer Society. 1989 survey of physicians attitudes and practices in early cancer detection. CA Cancer J Clin 1990;2:77.
4. Statland BE. Tumor markers. Diagn Med 1981;4:21.
5. Fritsche HA. Tumor marker test in patient monitoring. Lab Med 1982;13:528.
6. Reynoso G. CEA, basic concepts, clinical applications. Diagn Med 1981;4:41.
7. Bates SE, Longo DL. Tumor markers: value and limitations in the management of cancer patients. Cancer Treat Rev 1985;12:163.
8. Bray RA, Landay AL. Identification and functional characterization of mononuclear cells by flow cytometry. Arch Pathol Lab Med 1989;113:579.
9. Dressler LG, Bartow SA. DNA flow cytometry in solid tumors: practical aspects and clinical applications. Semin Diagn Pathol 1989;6:55.
10. Foon KA, Todd RF. Immunologic classification of leukemia and lymphoma. Blood 1986;68:1.
11. Garret PE, Kurtz SR. Clinical utility of oncofetal proteins and hormones as tumor markers. Med Clin North Am 1986;70:1295.
12. Virji MA, Mercer DW, Herberman RB. Tumor markers in cancer diagnosis and prognosis. CA Cancer J Clin 1988;38:104.
13. Statland BE. The challenge of cancer testing. Diagn Med 1981;4:13.
14. Bacchus H. Serum glycoproteins and malignant neoplastic disorder. Crit Rev Clin Lab Sci 1977;8:333.
15. Dieksheide WC. CEA assay precautions. [Letter]. Diagn Med 1981;4:114.
16. FDA drug bulletin. 1981;11:617.
17. Gardner RC, Feinerman AE, Kantrowitz PA, et al. Serial carcinoembryonic antigen (CEA) blood levels in patients with ulcerative colitis. Am J Dig Dis 1978;23:129.
18. Klee GG, Go VLW. Serum tumor markers. Mayo Clin Proc 1982;57:129.
19. Go VL. Carcinoembryonic antigen: clinical application. Cancer 1976;37:562.
20. Hansen HJ, Snyder LJ, Miller E, et al. Carcinoembryonic antigen (CEA) assay: a laboratory adjunct in the diagnosis and management of cancer. J Hum Pathol 1974;5:139.
21. Khoo SK, MacKay IR. Carcinoembryonic antigen in serum in diseases of the liver and pancreas. J Clin Pathol 1973;26:470.
22. Khoo SK, Warner NL, Lie JT, et al. Carcinoembryonic antigen activity of tissue extracts: A qualitative study of malignant and benign neoplasms, cirrhotic liver, normal adult and fetal organs. Int J Cancer 1973;11:681.
23. Lurie BB, Lowenstein MS, Zamcheck N. Elevated carcinoembryonic antigen levels and biliary tract obstruction. J Am Med Assoc 1975;233:326.
24. Wanebo HJ, Rao B, Pinsky CM, et al. Preoperative carcinoembryonic antigen level as a prognostic indicator in colorectal cancer. N Engl J Med 1978;299:448.
25. Wang JY, Tang R, Chiang JM. Value of carcinoembryonic antigen in the management of colorectal cancer. Dis Colon Rectum 1994;37:272.
26. Lunde OCH, Havig O. Clinical significance of carcinoembryonic antigen (CEA) in patients with adenocarcinoma in colon and rectum. Acta Chir Scand 1982;148:189.
27. Teramoto YA, Mariani R, Wunderlich D, et al. The immunochemical reactivity of a human monoclonal antibody with tissue sections of human mammary tumors. Cancer 1982;50:241.
28. Martin EW Jr, Cooperman M, Carey LC, et al. Sixty second-look procedures indicated primarily by rise in serial carcinoembryonic antigen. J Surg Res 1980;28:389.
29. Seleznick MJ. Tumor markers. Cancer Diagn Treat 1992;19:715.
30. Bergstrand CG, Czar B. Paper electrophoretic study of human fetal serum protein with demonstration of a new protein fraction. Scand J Clin Lab Invest 1957;9:277.
31. Abelev GI, Perova SD, Kharmkova NI, et al. Production of embryonic α-globulin by transplantable mouse hepatomas. Transplantation 1963;1:174.
32. Masopust J, Keithier K, Ridl I, et al. Occurrence of fetoprotein in patients with neoplasms and non neoplastic diseases. Int J Cancer 1968;3:364.
33. Brummund W, Arvan DA, Mennuti MT, et al. Alpha-fetoprotein in the routine clinical laboratory: evaluation of a simple radioimmunoassay and review of current concepts in its clinical application. Clin Chem Acta 1980;105:25.
34. Deutsch HF. Chemistry and biology of α-fetoprotein. Adv Cancer Res 1991;56:253.
35. Kew MC. Clinical, pathologic, and etiologic heterogeneity in hepatocellular carcinoma: evidence from South Africa. Hepatology 1981;1:366.
36. McMahon BJ, London T. Workshop on screening for hepatocellular carcinoma. J Natl Cancer Inst 1991;83:916.
37. Waldman TA, McIntire KR. The use of a radioimmunoassay for alpha-fetoprotein in the diagnosis of malignancy. Cancer 1974;34:1510.
38. Bosl GJ, Lange PH, Fraley EE. Human chorionic gonadotropin and alpha-fetoprotein in the staging of nonseminomatous, testicular cancer. Cancer 1981;47:328.
39. Liu FJ, Fritsche HA, Trujillo JM, et al. Serum alpha-fetoprotein (AFP); beta subunits of human chorionic gonadotropin (HCG) and lactate dehydrogenase 1 isoenzyme (LD1) in patients with germ cell tumors of the testis. Am J Clin Pathol 1983;80:120.
40. Talerman A, Haije WG, Baggerman L. Alpha1 antitrypsin (AAT) and alpha-fetoprotein (AFP) in sera of patients with germ cell neoplasms: value as tumor markers in patients with endodermal sinus tumor (yolk sac tumor). Int J Cancer 1977;19:741.
41. Di Bisceglie AM, Rustgi VK, Hoofnagle JH, et al. Hepatocellular carcinoma. Ann Intern Med 1988;108:390.
42. Masseyeff R. Human alpha-fetoprotein. Pathol Biol 1972;20:703.
43. Altug MU, Akdas A, Ruacan S, et al. Tissue alpha-fetoprotein and human chorionic gonadotrophin in non-seminomatous testicular tumors. Br J Urol 1987;59:458.

44. Bates SE, Longo DL. Clinical applications of serum tumor markers in cancer diagnosis and management. Semin Oncol 1987;14:102.

45. Scardino PT, Cox PD, Waldman TA, et al. The value of serum tumor markers in the staging and prognosis of germ cell tumors of the testis. J Urol 1977;118:994.

46. Skinner DG, Scardino PT, Daniels JR. Testicular cancer. Annu Rev Med 1981;32:543.

47. Bast RC Jr, Feeney M, Lazarus H, et al. Reactivity of a monoclonal antibody with human ovarian carcinoma. J Clin Invest 1981;68:133.

48. Bast RC, Feeney M, Lazarus H, et al. Reactivity of a monoclonal antibody with human ovarian cancer. J Clin Invest 1981;68:1331.

49. Bast RC, Klug TL, St John E, et al. Radioimmunoassay using a monoclonal antibody to monitor the course of epithelial ovarian cancer. N Engl J Med 1983;309:883.

50. Thomas CMG, Massuger LFAG, Segers MFG, et al. Analytical and clinical performance of improved Abbott Imx CA 125 Assay: comparison with Abbott CA 125 RIA. Clin Chem 1995;41:211.

51. Jacobs I, Bast RC Jr. The CA 125 tumor-associated antigen: a review of the literature. Hum Reprod 1989;4:1.

52. Vasilev SA, Schlaerth JB, Campeau J, et al. Serum CA 125 levels in preoperative evaluation of pelvic masses. Obstet Gynecol 1988;71:751.

53. Buahmah PK, Skillen AW. Serum CA 125 concentrations in patients with benign ovarian tumors. J Surg Oncol 1994;56:71.

54. Malkasian GD, Knapp RC, Lavin PT, et al. Preoperative evaluation of serum CA-125 levels in premenopausal and postmenopausal patients with pelvic masses: discrimination of benign from malignant disease. Am J Obstet Gynecol 1988;159:341.

55. Niloff JM, Knapp RC, Schaetzl E, et al. CA-125 antigen levels in obstetric and gynecologic patients. Obstet Gynecol 1984;64:703.

56. Atack DB, Nisker JA, Allen HH, et al. CA 125 surveillance and second-look laparotomy in ovarian carcinoma. Am J Obstet Gynecol 1986;154:287.

57. Cruickshank DJ, Fullerton WT, Klopper A. The clinical significance of pre-operative serum CA 125 in ovarian cancer. Br J Obstet Gynecol 1987;94:196.

58. Koprowski H, Herlyn M, Steplewise Z, et al. Specific antigen in serum of patients with colon carcinoma. Science 1981;212:53.

59. Del Villano BC, Brennan S, Brook P, et al. Radioimmunometric assay for a monoclonal antibody defined tumor marker. Clin Chem 1983;29:549.

60. Bakemeier RF, Krachov SF, Adams JT, et al. Cancer of the major digestive glands: pancreas, liver, bile ducts, and gallbladder. In: Rubin P. Clinical oncology. Chicago: American Cancer Society, 1983;178.

61. Ritts RE, DelVillano BC, Go VLW, et al. Initial clinical evaluation of an immunoradiometric assay for CA 199 using the NCI serum bank. Int J Cancer 1984;33:339.

62. Welfare M, Tanner A. CA 199 estimation in the assessment of pancreatic or biliary tract malignancy: what is the most efficient threshold value? Eur J Gastroenterol Hepatol 1991;3:539.

63. Haglund C, Roberts PJ, Kuusela P, et al. Tumor markers in pancreatic cancer. Scand J Gastroenterol Suppl 1986;126:75.

64. Safi F, Roscher R, Beger HG. The clinical relevance of the tumor marker CA 199 in the diagnosis and monitoring of pancreatic carcinoma. Bull Cancer 1990;77:83.

65. Satake K, Kanazawa G, Kho I, et al. Evaluation of serum pancreatic enzymes, carbohydrate antigen 19-9, and carcinoembryonic antigen in various pancreatic diseases. Am J Gastroenterol 1985;80:630.

66. Berretta E, Malesci A, Zerbi A, et al. Serum CA 19-9 in the postsurgical follow-up of patients with pancreatic cancer. Cancer 1987;60:2428.

67. Kufe DW, Inghirami G, Abe M, et al. Differential reactivity of a novel monoclonal antibody (DF3) with human malignant versus benign breast tumors. Hybridoma 1984;3:223.

68. Werner M, Faser C, Silverberg M. Clinical utility and validation of emerging biochemical markers for mammary adenocarcinoma. Clin Chem 1993;39:2386.

69. Tondini C, Hayes DF, Gelman R, et al. Use of various epithelial tumor markers and a stromal marker in assessment of cervical carcinoma. Obstet Gynecol 1991;77:566.

70. Safi F, Kohler I, Rottinger E, et al. The value of the tumor marker CA 15–3 in diagnosis and monitoring breast cancer: a comparative study with carcinoembryonic antigen. Cancer 1991;68:574.

71. Bray KR, Koda JE, Gaur C, et al. Serum levels and biochemical characteristics of cancer-associated antigen CA 549, a circulating breast cancer marker. Cancer Res 1987;47:5853.

72. Dnistrian AM, Schwartz MK, Greenberg EJ, et al. CA 549 as a marker of breast cancer. Int J Biol Markers 1991;6:139.

73. Chan DW, Beveridge RA, Bhargava A, et al. Breast cancer marker CA 549: a multicenter study. Am J Clin Pathol 1994;101:465.

74. Björklund B, Björklund V. Specificity and basis of the tissue of polypeptide antigen. Cancer Detect Prev 1983;6:41.

75. Björklund B, Björklund V, Brunkener M; et al. The enigma of human tumor marker: TPA revisited. In: Cimino F, Birkmayer GD, Klavins JV, Pimentel E, Salvatore F, eds. Human tumor markers. Berlin: Walter de Gruyter, 1987;169.

76. Buccheri G, Ferrigno D. Prognostic value of the tissue polypeptide antigen in lung cancer. Chest 1992;101:1287.

77. Eskelinen M, Hippeläinen M, Kettunen J, et al. Clinical value of serum tumor markers TPA, TPS, TAG 12, CA 15-3 and MCA in breast cancer diagnosis: results from a prospective study. Anticancer Res 1994;14:699.

78. Hara M, Koyanagi Y, Inoue T, et al. Some physico-chemical characterestics of gamma-seminoprotein: an antigenic component specific for human seminal plasma. Jpn J Leg Med 1971;25:322.

79. Wang MC, Valenzuala LA, Murphy GP, et al. Purification of a human prostate-specific antigen. Invest Urol 1979;17:159.

80. Kamashida S, Tsutsumi Y. Extraprostatic localization of prostatic acid phosphatase and prostate-specific antigen: distribution in cloacogenic glandular epithelium and sex-dependent expression in human anal gland. Hum Pathol 1990;21:1108.

81. Christensson A, Laurell CB, Lilja H. Enzymatic activity of prostate-specific antigen and its reactions with extracellular serine proteinase inhibitors. Eur J Biochem 1990;194:755.

82. McCormack RT, Rittenhouse HG, Finlay JA, et al. Molecular forms of prostate-specific antigen and the human kallikrein gene family: a new era. Urology 1995;45:729.

83. Pettersson K, Piironen T, Seppala M, et al. Free and complexed prostate-specific antigen (PSA): in vitro stability, epitope map, and development of immunofluorometric assays for specific and sensitive detection of free PSA and PSA-alpha1-antichymotrypsin complex. Clin Chem 1995;41:1480.

84. Schreiber WE. Free prostate-specific antigen: does it measure up? Am J Clin Pathol 1998;109:511.

85. Vessella RL. Trends in immunoassays of prostate-specific antigen: serum complexes and ultrasensitivity. Clin Chem 1993;39:2035.

86. Trinkler FB, Schmid DM, Hauri D, et al. Free/total prostate-specific antigen ratio can prevent unnecessary prostate biopsies. Urology 1998;52:479.

87. Crawford ED, DeAntoni EP. PSA as a screening test for prostate cancer. Urol Clin North Am 1993;20:637.

88. Klein LT, Franklin CL. The effects of prostatic manipulation on prostate-specific antigen levels. Urol Clin North Am 1997;24:293.

89. Brawer MK, Schifman RB, Ahmann FR, et al. The effect of digital rectal examination on serum levels of prostate-specific antigen. Arch Pathol Lab Med 1988;112:1110.

90. Ruckle HC, Klee GG, Oesterling JE. Prostate-specific antigen: critical issues for the practicing physician. Mayo Clin Proc 1994;69:59.

91. Bossens MMF, Van Straalen JP, De Reijke TM, et al. Kinetics of prostate-specific antigen after manipulation of the prostate. Eur J Cancer 1995;31A:682.

92. Catalona WJ, Smith DS, Ratliff TL, et al. Measurement of prostate-specific antigen in serum as a screening test for prostate cancer. N Engl J Med 1991;321:1156.

93. Hudson MA, Bahnson RR, Catalona WJ. Clinical use of prostate-specific antigen in patients with prostate cancer. J Urol 1989;142:1011.

94. Lange PH, Ercole CJ, Lightner DJ, et al. The value of serum prostate-specific antigen determinations before and after radical prostatectomy. J Urol 1989;141:873.

95. Gillatt DA, Ferro MA, Gingell JC, et al. Prostate-specific antigen in adenocarcinoma of the prostate. 1988;318:992.

96. Guinan P, Bhatti R, Ray P. An evaluation of prostate-specific antigen in prostatic cancer. J Urol 1987;137:686.

97. Chu TM, Murphy GP. What's new in tumor markers for prostate cancer. Urology 1986;27:487.

98. Benson MC, Whang IS, Olsson CA, et al. The use of prostate specific antigen density to enhance the predictive value of intermediate levels of serum prostate specific antigen. J Urol 1992;147:817.

99. Catalona WJ, Richie JP, deKernion JB, et al. Comparison of prostate specific antigen concentration versus prostate specific antigen density

in the early detection of prostate cancer: receiver operating characteristic curves. J Urol 1994;152:2031.

100. Carter HB, Pearson JD, Metter EJ, et al. Longitudinal evaluation of prostate-specific antigen levels in men with and without prostate disease. JAMA 1992;267:2215.

101. Nixon RG, Wener MH, Smith KM, et al. Biological variation of prostate specific antigen levels in serum: an evaluation of day-to-day physiological fluctuations in a well-defined cohort of 24 patients. J Urol 1997;157:2183.

102. Collins GN, Lee RJ, McKelvie GB, et al. Relationship between prostate specific antigen, prostate volume and age in the benign prostate. Br J Urol 1993;71:445.

103. Vashi AR, Oesterling JE. Percent free prostate-specific antigen: entering a new era in the detection of prostate cancer. Mayo Clin Proc 1997;72:337.

104. Catalona WJ, Hudson MA, Scardino PT, et al. Selection of optimal prostate-specific antigen cutoffs for early detection of prostate cancer: receiver operating characteristic curves. J Urol 1994;152:2037.

105. Stenman UH, Leinonen J, Alfthan H, et al. A complex between prostate-specific antigen and α_1-antichymotrypsin is the major form of prostate-specific antigen in serum of patients with prostatic cancer: assay of the complex improves clinical sensitivity for cancer. Cancer Res 1991;51:222.

106. Christensson A, Bjork T, Nilsson O, et al. Serum prostate specific antigen complexed to alpha1-antichymotrypsin as an indicator of prostate cancer. J Urol 1993;150:100.

107. Woodrum DL, Brawer MK, Partin AW, et al. Interpretation of free prostate specific antigen clinical research studies for the detection of prostate cancer. J Urol 1998;159:5.

108. Catalona WJ, Colberg JW, Smith DS, et al. Measurement of percent-free PSA improves specificity for lower PSA cutoffs in prostate cancer screening. J Urol 1996;155(Suppl):422A.

109. Lein M, Koenig F, Jung K, et al. The percentage of free prostate specific antigen is an age-independent tumour marker for prostate cancer: establishment of reference ranges in a large population of healthy men. Br J Urol 1998;82:231.

110. Henricks WH, England BG, Giacherio DA, et al. Serum percent-free PSA does not predict extraprostatic spread of prostate cancer. Am J Clin Pathol 1998;109:533.

111. Piironen T, Pettersson K, Suonpaa M, et al. In vitro stability of free prostate-specific antigen (PSA) and prostate-specific antigen (PSA) complexed to alpha 1-antichymotrypsin in blood samples. Urol 1996;48(Suppl):81.

112. Espana F, Royo M, Martinez M, et al. Free and complexed prostate-specific antigen in the differentiation of benign prostatic hyperplasia and prostate cancer: studies in serum and plasma samples. J Urol 1998;160:2081.

113. Brawer MK, Meyer GE, Letran JL, et al. Measurement of complexed PSA improves specificity for early detection of prostate cancer. Urol 1998;52:372.

114. Bohn H, Dati F. Placental-related proteins. In: Ritzman SE, Killingsworth LM, eds. Proteins in body fluids, amino acids, and tumor markers: diagnostic and clinical aspects. New York: Alan R Liss, 1983;333.

115. Stigbrand T, Engall E. Placental proteins as tumor markers. In: Sell S, Wahren B, eds. Human cancer markers. Clifton, NJ: Humana Press, 1982;275.

116. Pardue MA, Taylor EH, London S, et al. Clinical comparison of immunoradiometric assays for intact versus beta-specific human chorionic gonadotropin. Am J Clin Pathol 1990;93:347.

117. Zarate A, MacGregor C. Beta subunit HCG and the control of trophoblastic disease. Semin Oncol 1982;9:187.

118. Rosen SW, Weintraub BD, Vaitukatis JL, et al. Placental proteins and their subunits as tumor markers. Ann Intern Med 1975;82:71.

119. Fishman WH, Inglis NR, Stolbach LL, et al. A serum alkaline phosphatase isoenzyme of human neoplastic cell origin. Cancer Res 1968;28:150.

120. Fishman WH. Perspectives on alkaline phosphatase isoenzymes. Am J Med 1974;56:617.

121. Gutman AB, Gutman EB. "Acid" phosphatase occurring in serum of patients with metastasizing carcinoma of prostate gland. J Clin Invest 1938;17:473.

122. Pappas AA, Gadsden RH. Acid phosphatase, clinical utility in detection, assessment and monitoring of prostatic carcinoma. Ann Clin Lab Sci 1984;14:285.

123. Heller JE. Prostatic acid phosphatase: its current clinical status. J Urol 1987;137:1091.

124. Bahnson RR, Catalona WJ. Adverse implications of acid phosphatase levels in the upper range of normal. J Urol 1987;137:427.

125. Oesterling JE. Clinically useful serum markers for adenocarcinoma of the prostate II: prostate-specific antigen. Am Urol Assoc Update Series 1991;10:137.

126. Silverman LM, Dermer GB, Zweig MH, et al. Creatinine kinase BB: a new tumor associated marker. Clin Chem 1979;25:1432.

127. Tatarinov Y. New data on the embryo-specific antigen components of human blood serum. Vopr Med Khim 1964;10:584.

128. Narayanan S. Alkaline phosphatase as a tumor marker. Ann Clin Lab Sci 1983;12:133.

129. Panigraphi K, Delmas PD, Singer F, et al. Characteristics of a two-site immunoradiometric assay for human skeletal alkaline phosphatase in serum. Clin Chem 1994;40:822.

130. Schmechel D, Marangos PJ, Brightman M. Neuron-specific enolase is a molecular marker for peripheral and central neurendocrine cells. Nature 1978;276:834.

131. Carney DN, Ihde DC, Cohen MM, et al. Serum neuron specific enolase. Lancet 1982;1:583.

132. Zeltzer PM, Marangos PJ, Evans AE, et al. Serum neuron-specific enolase in children with neuroblastoma. Cancer 1986;57:1230.

133. Zondag HA, Klein F. Clinical applications of lactate dehydrogenase isoenzymes: alterations in malignancy. Ann N Y Acad Sci 1968; 151:578.

134. DeWys WD, Killen JY. The paraneoplastic syndromes. In: Rubin P, Bakemeier RF, Krachov SK, eds. Clinical oncology: a multidisciplinary approach. 6th ed. Chicago: American Cancer Society, 1983;112.

135. Witliff JL. Steroid-hormone receptors in breast cancer. Cancer 1984;53:630.

136. DiFronzo G, Miodini P, Brivio M, et al. Comparison of immunochemical and radioligand binding assays for estrogen receptors in human breast tumors. Cancer Res 1986;46:4278.

137. Jaing NS. Breast cancer: estrogen and progesterone receptor assays as a guide to therapy. Mayo Clin Proc 1983;58:64.

138. Lesser ML, Rosen PP, Senie RT, et al. Estrogen and progesterone receptors in breast carcinoma: correlation with epidemiology and pathology. Cancer 1981;48:299.

139. Geller J. Hormone dependency of prostate cancer. In: Thompson BE, Lippman ME, eds. Steroid receptors and the management of cancer. Boca Raton, FL: CRC Press, 1979;113.

140. Lippman ME, Yarbo GK, Leventhal BG, et al. Clinical correlations of glucocorticoid receptors in human acute lymphoblastic leukemia. In: Thompson BE, Lippman ME, eds. Steroid receptors and the management of cancer. Boca Raton, FL: CRC Press, 1979;67.

141. Young PCM, Ehrlich CE. Progesterone receptors in human endometrial cancer. In: Thompson BE, Lippman ME, eds. Steroid receptors and the management of cancer. Boca Raton, FL: CRC Press, 1979;135.

142. Drexler HG, Gignac SM, Minowada J. Routine immunophenotyping of acute leukemias Blut: J Clin Exp Hematology 1988;57:327.

143. McCoy JP, Lovett EJ. Basic principles in clinical flow cytometry. In: Keren DF, ed. Flow cytometry in clinical diagnosis. Chicago: ASCP Press, 1989;12.

144. Hansen CA. Nontraditional applications of flow cytometry. In: Keren DF, ed. Flow cytometry in clinical diagnosis. Chicago: ASCP Press, 1989;280.

145. Shapiro HM. Flow cytometry of DNA content and other indicators of proliferative activity. Arch Pathol Lab Med 1989;113:591.

146. Grody WA, Gatti WW, Faramarz N. Diagnostic molecular pathology. Mod Pathol 1989;6:553.

CHAPTER 28

Vitamins and Nutritional Assessment

A. Michael Spiekerman

Objectives

Upon completion of this chapter, the clinical laboratorian should be able to:

- *Describe the biochemical roles of vitamins.*

- *For each vitamin, list the signs and symptoms that characterize deficient and toxic states.*

- *Correlate alterations in vitamin status with circumstances of increased metabolic requirements, age-related physiologic changes, or pathologic conditions.*

- *Describe drug–nutrient interactions that influence vitamin status.*

- *State the clinical evidence and biochemical bases for therapeutic and prophylactic use of vitamin supplements.*

- *Recognize how vitamins, in excess, may affect biochemical tests for various analytes.*

- *Delineate and appraise the clinical laboratory procedures used in the assessment of vitamin status.*

- *Recognize and be able to correct for variables that interfere with vitamin measurements by the clinical laboratory.*

- *Discuss the present and projected role of the laboratory in nutritional assessment and monitoring as a function of population demographics and health care economics.*

- *Define the terms anthropometric, kwashiorkor, marasmus, nutritional assessment, and total parenteral nutrition (TPN).*

- *List the populations at risk for malnutrition.*

- *Identify the anthropomorphic measures of nutritional status.*

- *Relate the plasma proteins of interest in nutritional assessment.*

- *Discuss the guidelines used to monitor patients on TPN.*

- *Describe some of the electrolyte and mineral abnormalities associated with TPN.*

- *Given clinical data, evaluate a patient's vitamin and nutritional status.*

KEY TERMS

Anthropometric methods	Kwashiorkor	Nyctalopia
Antioxidant	Malnutrition	Osteomalacia
Beriberi	Marasmus	Pellagra
Hypervitaminosis	Megavitamins	RDA
Hypovitaminosis	Nutritional assessment	Rickets
		Vitamin

The current emphasis on nutrition and health has resulted in the increasing importance of vitamin and *nutritional assessment* in the clinical laboratory. Clinicians and the public alike are now more aware of how nutrition can affect both health and healing. Nutritional components of the diet include both macronutrients and micronutrients. Macronutrients are generally used for energy and include carbohydrates, proteins, and fats. Micronutrients are required by the body in only minute concentrations of milligrams or less. Vitamins, minerals, and the trace elements are classified as micronutrients. The first part of this chapter discusses the micronutrients, including their biochemistry, deficiency, toxicity, and methods of analysis. Various minerals (*ie,* inorganic elements or compounds) and trace elements (*eg,* iron, zinc, copper) are discussed in Chapter 15. The second part of this chapter discusses assessment of the macronutrients and other general means of assessing nutritional status.

VITAMINS

Vitamins are low-molecular-weight compounds with a wide range of functions in biologic tissues. They serve as cofactors in many enzymatic reactions. A variety of enzymes become inactive and cannot aid in cellular reactions if vitamins are not present. These compounds and their biologically inactive precursors must be partially obtained from food sources and in some instances from bacterial synthesis. Inadequate diet or inadequate intestinal absorption is termed *vitamin deficiency*. Abnormal increases of metabolism requiring high supplies of one of these cofactors may be termed *vitamin insufficiency* or *vitamin dependency,* depending on the level of supply demanded for physiologic function.[1,2]

Variabilities in clinical expression of vitamin abnormalities result from differences in specific cause, degree, and duration of vitamin inadequacy. Additional factors contributing to this variability include the genetic constitution of the individual, the simultaneous presence of multiple nutritional insufficiencies, or the increased metabolic demands imposed by conditions such as pregnancy, infection, and cancer. There is still limited knowledge on the mechanisms of absorption, storage, transport, and metabolism of vitamins in humans. Also, there needs to be a continued investigation into the effects of vitamins on disease processes, therapies, and exogenous drugs and agents.

It is now understood that the clinical symptoms of vitamin deficiencies are usually nonspecific and vague, often delaying definitive diagnosis. This is especially true in the early stages of vitamin deficiency and in mild chronic deficiency states. A combination of dietary history, physical examination, and laboratory measurements is often required to diagnose a vitamin deficiency. Vitamin metabolism is complex and interrelated; hence, vitamin supplementation without proper studies may lead to other nutrient deficiencies or toxicities or to an irreversible state of untreated deficiency with symptoms masked by inappropriate therapy.

For simplicity, vitamins of diverse chemical structure are classified as either water soluble or fat soluble. The fat-soluble vitamins include A, D, E, and K. Those soluble in water include the B complex of vitamins (thiamine, riboflavin, niacin, vitamins B_6 and B_{12}, biotin, folate, and vitamin C). Water-soluble vitamins are readily excreted in the urine and are less likely to accumulate to toxic levels in the body than are fat-soluble vitamins. Vitamins, classified as fat or water soluble, their functions, and the symptoms usually seen in deficient and toxic states are shown in the Table 28-1.[3]

Investigating the dietary deficiency of vitamins (*hypovitaminosis*) is sustained primarily from knowledge of dietary sources and dietary practices that produce inadequate intake or absorption. Recommended dietary allowances are defined by the Food and Nutrition Board as levels of intake of essential nutrients based on available scientific knowledge to be sufficient to meet the nutritional needs of healthy individuals.[1,2]

Information on the dietary intake requirements needed to avoid deficiency symptoms and to maintain specific functions define the dietary allowances of these vitamins. The recommended dietary allowances are established to meet the needs of 97.5% of the population; in some instances, they are higher if the nutrient is poorly absorbed or insufficiently used.[3]

Chemical determination of human vitamin states has been approached in the following ways:

- Measurement of the active cofactors or precursors in biologic fluids or blood cells

- Measurement of urinary metabolites of the vitamin
- Measurement of a biochemical function requiring the vitamin (such as enzymatic activity) with and without in vitro addition of the co-factor form
- Measurement of urinary excretion of vitamin or metabolites after a test load of the vitamin
- Measurement of urinary metabolites of a substance, the metabolism of which requires the vitamin after administration of a test load of the substance

Reduced serum concentrations of a vitamin do not always indicate a deficiency that interrupts cellular function. Conversely, values within the reference interval do not always reflect adequate function. Interpretation of laboratory values must be done with knowledge of the biochemistry of vitamins and methodologic factors used in identifying possible physiologic abnormalities.[3–6]

The Fat-Soluble Vitamins

Vitamin A

Retinol and retinoic acid are derived directly from dietary sources, primarily as retinyl esters, or from metabolism of dietary carotenoids (provitamin A), primarily beta-carotene. Major dietary sources of these compounds include animal products and pigmented fruits and vegetables (carotenoids). A schematic of vitamin A transport and metabolism is shown in Figure 28-1.[6,7]

A clearly defined physiologic role for retinol is in vision. Retinol is oxidized in the rods of the eye to retinal, which, when complexed with opsin, forms rhodopsin, allowing dim-light vision. In vitamin A deficiency states, epithelial cells (cells in the outer skin layers and cells in the lining of the gastrointestinal, respiratory, and urogenital tracts) become dry and keratinized. An isomer of vitamin A, retinoid-13-*cis*-retinoic acid (Accutane) has become the therapy of choice for several forms of acne. Accutane, however, must be used with great caution because large doses of retinoid taken during pregnancy can result in congenital malfunctions. The anticarcinogenic role of high-dosage retinoids has been well established in animals. An inverse correlation has been noted between serum vitamin A and the incidence of lung cancer, indicating that vitamin A may have a role in prevention of this cancer.[7]

Vitamin A and related retinoic acids are a group of compounds essential for vision, cellular differentiation, growth, reproduction, and immune system function. Fruits and vegetables contain carotene, which is a precursor of retinol. Carotenes provide more than half of the retinol requirement in the American diet. Vitamin A deficiency is most common among children living in nonindustrialized countries and is usually due to insufficient dietary intake. Deficiency may also occur as a result of chronic fat malabsorption. Vitamin A deficiency leads to night blindness (*nyctalopia*) and when prolonged may cause total blindness. Vitamin A deficiency is often associated with disorders of fat absorption. Premature infants are

TABLE 28-1. Vitamin Functions and Symptoms Seen in Deficient and Toxic States

Vitamin Name	Chemical Name	Source	Function	Clinical Deficiency	Assay Methodology
Fat-Soluble Vitamins					
Vitamin A₁	Retinol	Animal—fish-liver oils, liver, butter, cream, whole milk, whole milk cheeses, egg yolk Plant—dark-green leafy vegetables, yellow vegetables, yellow fruit Fortified margarines	Vision, growth, immune responses, reproduction, cancer prevention	Night blindness, growth retardation, abnormal taste response, dermatitis, recurrent infections	Photometric, fluorometric, RIA, HPLC
Vitamin E	Tocopherols, α, β, γ, δ	Plant tissues—vegetable oils, wheat germ, rice germ, green leafy vegetables, nuts, legumes Animal foods are poor sources	Antioxidant (membrane stability), cardiovascular, disease prevention	Mild hemolytic anemia (newborn), red blood cell fragility, ataxia	Photometric, HPLC
Vitamin D₂	Ergocalciferol, cholecarciferol	Fish-liver oils, fortified milk	Calcification of bone and teeth	Rickets (young), osteomalacia (adults)	Competitive protein binding, HPLC
Vitamin D₃		Activated sterols; exposure to light; very small amounts in butter, liver, egg yolk, salmon, sardines			
Vitamin K	Phylloquinones, menaquinones	Green leaves such as alfalfa, spinach, cabbage; liver; synthesis in intestine	Blood clotting, osteocalcins	Hemorrhage (ranging from easy bruising to massive bruising), especially post-traumatic bleeding	Photometric, HPLC, prothrombin time, RIA for abnormal prothrombin
Water-Soluble Vitamins					
Vitamin B₁	Thiamine	Whole grain and enriched breads, cereals, flours; organ meats, pork, other meats, poultry, fish, legumes, nuts, milk, green vegetables	Carbohydrate metabolism (decarboxylations), nervous function	Infants: dyspnea, cyanosis, diarrhea, vomiting Adults: beriberi (fatigue, peripheral neuritis), Wernicke-Korsakoff syndrome (apathy, ataxia, visual problems)	Fluorometric, microbiologic, enzymatic
Vitamin B₂	Riboflavin	Milk, organ meats, eggs, green leafy vegetables	Oxidation, reduction enzymatic reactions	Angular stomatitis (mouth lesions), dermatitis, photophobia, neurologic changes	Fluorometric, HPLC, microbiologic, glutathione reductase
Vitamin B₆	Pyridoxine, pyridoxal, pyridoxamine	Meat, poultry, fish, potatoes, sweet potatoes, vegetables	Enzyme systems involving amino acid transaminases, phosphorylases, decarboxylases	Infants: irritability, seizures, anemia, vomiting, weakness Adults: facial seborrhea	Microbiologic, HPLC, tyrosine, decarboxylase
Niacin Niacinamide	Nicotinic acid Nicotinamide	Meat, poultry, fish; whole grain and enriched breads, flours, cereals, nuts, legumes; tryptophan as a precursor	Oxidation-reduction reactions	Pellagra (dermatitis, mucous membrane inflammation, weight loss, disorientation)	Microbiologic, fluorometric
Folic acid	Pteroylglutamic acid	Organ meats, deep-green vegetables; muscle meats, poultry, fish, eggs; whole grain cereals	Amino acid and nucleic acid biosynthesis, one-carbon transfers	Megaloblastic anemia	Competitive protein-binding enzyme immunoassay

(continued)

TABLE 28-1. Vitamin Functions and Symptoms Seen in Deficient and Toxic States (*Continued*)

Vitamin Name	Chemical Name	Source	Function	Clinical Deficiency	Assay Methodology
Vitamin B_{12}	Cyanocobalamin	In animal foods only; organ meats, muscle meats, poultry, fish, eggs, milk	Myelin formation, branched-chain keto acid metabolism, folate interconversions, DNA synthesis	Megaloblastic anemia, neurologic abnormalities	Competitive protein binding, enzyme immunoassay
Biotin	—	Organ meats, egg yolks, nuts, legumes	Coenzyme for CO_2 carboxylation reactions and for carboxyl group exchange	Dermatitis can progress to mental and neurologic changes, nausea, anorexia, peripheral vasoconstriction	Colorimetric, microbiological, CO_2 measurements
Pantothenic acid	—	Meat, poultry, fish, whole grain cereals, legumes, smaller amount in fruits and vegetables	Acyl-group transfer reactions (as part of coenzyme A and acyl carrier protein)	Normally not seen alone but with chemical agonist: depression, depressed immune system, muscle weakness	Microbiologic
Vitamin C	Ascorbic acid	Citrus fruits, tomatoes, melons, cabbage, broccoli, strawberries, fresh potatoes, green leafy vegetables	Connective tissue formation, catecholamine synthesis, cholesterol metabolism, antioxidant	First, vague aches and pains; if long-term, scurvy (hemorrhages into skin, alimentary and urinary tract, anemia, wound healing delayed)	Photometric
Carnitine	—	Meat	Energy, metabolism and acyl group transport	Muscle weakness, fatigue	RIA

born with lower sodium retinol and retinol-binding protein (RBP) levels as well as low hepatic stores of retinol. Sick newborns are thus treated with vitamin A as a preventive measure.[8]

Low levels of vitamin A are found in diseases that impair small intestinal function or liver function. When ingested in high doses either chronically or acutely, vitamin A causes many toxic manifestations and may ultimately lead to liver damage due to *hypervitaminosis*. High doses of vitamin A may be obtained from ingestion of excess vitamin supplements or large amounts of liver or fish oils, which are rich in vitamin A. Carotenoids, on the other hand, are not known to be toxic. This is due to a reduced efficiency of carotene absorption at high doses and limited conversion to vitamin A. The *recommended daily allowance (RDA)* of vitamin A for an adult male is 1000 µg/d and that for an adult female is 800 µg/d. Measurement of retinol is the most common means of assessing vitamin A status in the clinical setting. Retinol is most commonly measured by high performance liquid chromatography. The toxicity is usually assessed by measuring retinyl ester levels in serum rather than retinol. This is accomplished by high-performance liquid chromatography (HPLC).[9]

Vitamin E

Vitamin E functions in the body as a powerful *antioxidant* and is the primary defense against potentially harmful oxidations that cause disease and aging. It has been shown to prevent the oxidation of unsaturated fatty acids by trapping free radicals. Vitamin E is given the generic name *tocopherol* and shown to include several biologically active isomers. The word tocopherol is a Greek derivation, meaning an "oil that brings forth in childbirth." Alpha-tocopherol is the predominant isomer in plasma and is the most potent isomer by current biologic assays. Whether the tocopherol isomer has separate physiologic effects is unknown. Dietary sources of tocopherols include vegetable oils, fresh leafy vegetables, egg yolk, legumes, peanuts, and margarine. Diets suspect for vitamin E deficiency are those low in vegetable oils or fresh green vegetables or those low in unsaturated fats.[10]

The absorption, transport, storage, and metabolism of tocopherols are only partially understood. Absorption is believed to be associated with intestinal fat absorption. About 40% of ingested tocopherol is absorbed; the percentage is affected by the amount and by the degree of unsaturation of

Small Intestine

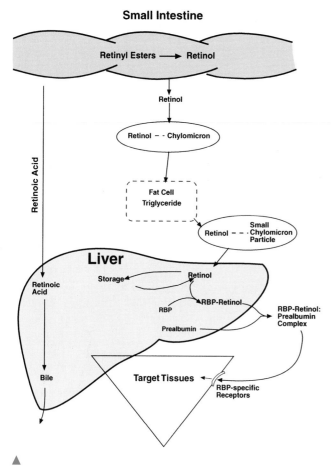

Figure 28-1. Retinol metabolism. RBP, Retinol-binding protein.

ognized between vitamin E deficiency and progressive loss of neurologic function in infants and children with chronic cholestasis. At the present time, assessment of vitamin E status is primarily indicated in newborns, in patients with fat-malabsorption states, and in patients receiving synthetic diets.

Toxicity may result from chronic voluntary overdoses (*megavitamins*). Premature infants receiving vitamin E sufficient to sustain serum levels above 30 mg/L have an increased incidence of sepsis and necrotizing enterocolitis.[11] Patients receiving synthetic diets should be monitored to avoid vitamin E toxicity. This process is aided by two other essential nutrients, selenium and ascorbic acid. Vitamin E has been shown to strengthen cell membranes and augment such functions as drug metabolism, heme biosynthesis, and neuromuscular function. Absorption of dietary vitamin E is most efficient in the jejunum, where it is combined with lipoproteins and transported through the lymphatics. The storage sites for vitamin E are the liver and other tissues with high lipid content. Vitamin E is excreted principally in the feces. Deficiency is associated with increased platelet aggregation, which leads to hemolytic anemia and neurologic degeneration. Vitamin E deficiency commonly occurs in two groups: premature, very-low-birthweight infants and patients who do not absorb fat normally. Megadoses of vitamin E do not produce toxic effects but may significantly impair the absorption of vitamins D and K. Megadoses are not recommended because health benefits from high doses have not been shown. The RDA of vitamin E for adult males is 10 mg/d and for adult females 8 mg/d.[12] The most widely distributed form of vitamin E is alpha-tocopherol. It is also the most biologically active form and the form commonly measured in the laboratory. Vitamin E is measured using HPLC methods.

Vitamin D

Vitamin D refers to a group of related metabolites. Vitamin D is used for proper formation of the skeleton and for mineral homeostasis. Exposure of the skin to sunlight (ultraviolet light) catalyzes the formation of cholecalciferol from 7-dehydrocholesterol. The other major form of the vitamin is ergocalciferol (vitamin D_2). Vitamin D in foods occurs as cholecalciferol or ergocalciferol. The most active metabolite of vitamin D is $1,25(OH)_2D_3$. It stimulates intestinal absorption of calcium and phosphate for bone growth and metabolism and, together with parathyroid hormone, stimulates bone to increase the mobilization of calcium and phosphate. The transport and metabolism of vitamin D are shown in Figure 28-2.

Major dietary sources of vitamin D include irradiated foods and commercially prepared milk. Small amounts occur in butter, egg yolk, liver, tuna fish, and salmon. Vitamin D is absorbed in the small intestine and requires bile salts for absorption. It is stored in the liver and excreted in the bile. In children, severe deficiency causes a deformation of the skeleton called *rickets*. In adults, the deficiency leads to undermineralization of bone matrix resulting in excessive bone loss and *osteomalacia*. Low levels of vitamin D are reported in small bowel disease, chronic renal failure, hepatobiliary disease, pancreatic insufficiency, hypoparathyroidism, and the

dietary fat as well as by the isomer type. The physiologic requirement of vitamin E increases with increasing polyunsaturated fatty acids in the diet. Absorbed vitamin E is first associated with circulating chylomicrons and very-low-density lipoprotein, and some is transferred to adipose tissue during triglyceride hydrolysis. The remaining vitamin E in chylomicron remnants is transported to the liver.

Vitamin E functions as an antioxidant, protecting unsaturated lipids from peroxidation (cleavage of fatty acids at unsaturated sites by oxygen addition across the double bond and formation of free radicals). The role of vitamin E in protecting the erythrocyte membrane from oxidant stress is presently the major documented role of vitamin E in human physiology. There is evidence for preventive roles of vitamin E in retrolental fibroplasia, intraventricular hemorrhage, and mortality of small premature infants.

The major symptom of vitamin E deficiency is hemolytic anemia. Although such vitamin E use is still controversial, premature newborns are commonly supplemented with vitamin E to stabilize their red blood cells and prevent hemolytic anemia. Patients with conditions resulting in fat malabsorption, especially cystic fibrosis and abetalipoproteinemia, are susceptible to vitamin E deficiency. Also, a relationship has been

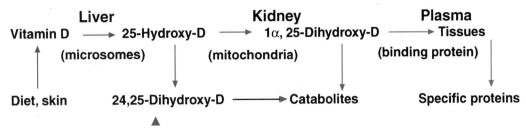

Figure 28-2. Transport and metabolism of vitamin D.

use of anticonvulsant drugs. Vitamin D can be toxic, especially in children. When vitamin D is taken in excess, it produces hypercalcemia and hypercalciuria, which can lead to calcium deposits in soft tissue and irreversible renal and cardiac damage.[6]

One of the more active metabolic forms of vitamin D, $1,25(OH)_2D_3$, is a reflection of calcium synthesis by the body. It is important to measure parathyroid hormone and calcium levels in conjunction with $1,25(OH)_2D_3$ levels in diagnosing primary hyperparathyroidism, in diagnosing different types of rickets, in monitoring patients with chronic renal failure, and

in assessing patients on $1,25(OH)_2D_3$ therapy. The RDA of vitamin D for adults is 5 μg/d. Elevated levels of vitamin D are present in hyperparathyroidism and hypophosphatemia and during pregnancy.

Two forms of vitamin D most commonly measured in the clinical laboratory are 25 $(OH)D_3$ and $1,25(OH)_2D_3$. The $25(OH)D_3$ is the major circulating form of vitamin D, and its measurement is a good indicator of vitamin D nutritional status as well as vitamin D intoxication. The reference range for $25(OH)D_3$ is 22 to 42 ng/mL and that for $1,25(OH)_2D_3$ is 30 to 53 pg/mL. Quantitation of the metabolites of vitamin D

CASE STUDY 28-1

A 4-year-old boy comes to an Alaskan children's clinic by the Social Services System of Alaska. The child has been living in a foster home. He has been outdoors very infrequently and has had a diet consisting of very few green vegetables, dairy products, or meats. He began walking at 14 months, and his legs showed some bowing. There is tetany or convulsions. He is a small child for his age, 27 pounds and 90.6-cm tall, third percentile. The following are his laboratory test results:

CASE STUDY TABLE 28-1.1. Laboratory Results

Test	Result	Reference Range
Hgb	14.1 g/dL	12.0–18.0
Hct	46%	34.0–52.0
WBC	8.4 × 10³/μL	4.0–11.0
	Normal differential	
Urinalysis		
Specific gravity	1.010	1.003–1.030
pH	6.8	5.0–9.0
Glucose	negative	negative
Protein	negative	negative
Microscopic	negative	negative
Amino acid screen	negative	negative
Serum		
Na	140 mEq/L	132–143
K	4.0 mEq/L	3.2–5.7
Cl	101 mEq/L	98–116
CO₂	14.3 mg/dL	13–29
Ca	7.0 mg/dL	8.9–10.3
Phos	1.1 mg/dL	3.4–5.9
Alkaline phosphatase	23 units	15–20
Total protein	6.6 g/dL	5.6–7.7
Albumin	3.4 g/dL	3.1–4.8

Questions

1. This disorder is probably related to:
   a. Wilson's disease
   b. Maple syrup urine disease
   c. Galactosemia and galactosuria
   d. Some form of vitamin or mineral metabolism disorder, probably related to some vitamin D deficiency

2. The causative agent in this disease is:
   a. Inadequate intake of lipids and fatty acids
   b. A viral disease related to a rhinovirus
   c. An abnormality related to either vitamin D or inadequate intake or vitamin D resistance
   d. Inadequate intake of sugar

3. Therapy that might help in this situation is:
   a. Vitamin D, 10,000 IU/d
   b. 5% Glucose and water
   c. Normal Ringer's lactate solution
   d. A diet with a high level of unsaturated fatty acids

should be performed using radioimmunoassay (RIA) or HPLC in conjunction with competitive protein binding.[13]

Vitamin K

Vitamin K (German: Koagulation) is the group of substances that are essential for the formation of prothrombin and at least five other coagulation proteins, including factors VII, IX, and X and proteins C and S. The quinone-containing compounds are a generic description for menadione and derivatives exhibiting this activity. Vitamin K helps in the process of converting precursor forms to functional forms of these coagulation proteins, and this transformation occurs in the liver. Dietary vitamin K is absorbed primarily in the terminal ileum and possibly in the colon. Vitamin K is synthesized by intestinal bacteria, and the intestinal synthesis provides 50% of the vitamin K requirement. Major dietary sources are cabbage, cauliflower, spinach and other leafy vegetables, pork, liver, soybeans, and vegetable oils. Uncomplicated dietary vitamin K deficiency is considered rare in healthy children and adults.

Vitamin K deficiency may be caused by antibiotic therapy. This is a result of decreased synthesis of the vitamin by intestinal bacteria. When vitamin K antagonists, such as warfarin sodium (Coumadin), are used for anticoagulant therapy, anticoagulant factors II, VII, IX, and X are synthesized but are nonfunctional. An apparent vitamin K deficiency may lead to a hemorrhagic episode or may result when anticoagulants such as warfarin sodium are used.[14,15]

Determination of prothrombin time (velocity of clotting after addition of thromboplastin and calcium to citrated plasma) is an excellent index of prothrombin adequacy. This time is prolonged in deficiency of vitamin K and also in liver diseases characterized by decreased synthesis of prothrombin. Deficiency of vitamin K also results in prolongation of the partial thromboplastin time, but the thrombin time is within the reference interval.

Toxicity from vitamin K is not commonly seen in adults. Large doses in infants may result in hyperbilirubinemia. The adult RDA of vitamin K is 80 μg/d for males and 65 μg/d for females. For most laboratories, the vitamin K is not assayed, but prothrombin time is used as a functional indicator of vitamin K status. The normal prothrombin time is 11 to 15 seconds, and this varies with method. With a vitamin K deficiency, the prothrombin time is prolonged.[16]

The Water-Soluble Vitamins

Thiamine

Thiamine (also referred to as vitamin B_1) functions as a coenzyme in energy metabolism, particularly carbohydrate metabolism. The coenzyme thiamine pyrophosphate (TPP) is required for the activity of transketolase in the pentose-phosphate pathway. In its thiamine pyrophosphate co-factor form, thiamine catalyzes the decarboxylation of alpha-ketoacids (pyruvate and alpha-ketoglutarate), the oxidative decarboxylation by alpha-ketoacid dehydrogenases, and the for-

mation of ketols. TPP functions in major carbohydrate pathways and in the metabolism of branched-chain amino acids.

Thiamine is rapidly absorbed from food in the small intestine and is also excreted in the urine. The clinical condition associated with the chronic deficiency of thiamine is called *beriberi*. It is found in underdeveloped countries of the world and is characterized by symptoms involving the nervous and cardiovascular systems. Thiamine deficiency is most frequently found among chronic alcoholics in the United States. Decreased intake, impaired absorption, and increased requirements all appear to play a role in the development of thiamine deficiency in alcoholics. Excess thiamine is easily cleared by the kidneys, and there is no evidence of thiamine toxicity by oral administration. The RDA of thiamine for an adult male is 1.5 mg/d and for an adult female 1.1 mg/d.[17]

The concentration of thiamine in serum or urine can be measured manually by the fluorometric thiochrome method or by HPLC. It is now thought that neither serum nor urine concentration is as prognostic as the measurement of erythrocyte transketolase (ETK) activity. The enzymatic activity assay ETK provides a functional measurement of thiamine. The sensitivity of the ETK activity assay can be further improved by measuring the enzyme activity before and after the addition of TPP. The presence of TPP ensures optimal enzyme activity. If the increase in activity after addition of TPP is greater than 25%, thiamine deficiency is present.

The reference ranges for thiamine are as follows[6]:

Serum concentration: 0.21 to 0.43 μg/dL
Urine concentration: more than 100 μg/24 hours
ETK activity increase with TPP: 0% to 14%
Marginal deficiency: 15% to 24%
Deficiency: more than 25%

Riboflavin

Riboflavin (vitamin B_2) functions primarily as a component of two coenzymes, flavin mononucleotide and flavin adenine dinucleotide (FAD). These two coenzymes catalyze a variety of oxidation-reduction reactions. Dietary riboflavin is absorbed in the small intestine. The body stores of a well-nourished person are adequate to prevent riboflavin deficiency for 5 months. Excess riboflavin is excreted in the urine. Foods high in riboflavin include milk, liver, eggs, meat, and leafy vegetables.[18]

Riboflavin deficiency commonly occurs along with other nutritional deficiencies. Conditions such as alcoholism and starvation, which lead to a deficiency of the B vitamins in general, also usually lead to riboflavin deficiency. Chronic diarrhea and malabsorption syndromes can produce riboflavin deficiency. Certain drugs antagonize the action or metabolism of riboflavin, including phenothiazine, oral contraceptives, and tricyclic antidepressants.[19] Because riboflavin is water soluble, its toxicity is not known. The RDA of riboflavin for an adult male is 1.7 mg/d and for an adult female is 1.3 mg/d.

Most flavoproteins such as riboflavin catalyze removal of either hydride ion or a pair of hydrogen atoms from substrate. Riboflavin has key roles in respiratory enzymes, such as d-amino acid oxidase, pyruvate dehydrogenase, xanthine oxidase, glutathione reductase, and NADH dehydrogenase. The concentration of riboflavin in serum or urine can be measured by HPLC; however, neither blood nor urine levels of riboflavin are as sensitive an indicator of riboflavin status as erythrocyte glutathione reductase activity stimulated by the coenzyme FAD. Glutathione reductase activity is a reflection of the functional status of riboflavin. Reference ranges for riboflavin are as follows[6]:

Serum levels: 4 to 24 µg/dL
Urine levels: greater than 120 µg/dL
Glutathione reductase increases with FAD: 0% to 20%
Marginal deficiency: 20% to 40% of glutathione reductase
Deficiency of glutathione reductase: greater than 40%

Pyridoxine

Pyridoxine, also called vitamin B$_6$, is the collective name given to three related compounds: pyridoxine, pyridoxal, and pyridoxamine. Pyridoxine occurs mainly in plants, whereas pyridoxal and pyridoxamine are present mainly in animal products. Major dietary sources of vitamin B$_6$ are meat, poultry, fish, potatoes, and vegetables; dairy products and grains contribute lesser amounts. The pyridoxine co-factor forms act in more than 50 different enzyme systems catalyzing a variety of reaction types, including the transaminases aspartate transaminase and alanine transaminase. Some of these enzymes are important in lipoprotein metabolism.[20] The best-known functions of the pyridoxine cofactors are their roles in the conversion of tryptophan to 5-hydroxytryptamine (serotonin) and in the separate pathway of tryptophan to nicotinic acid ribonucleotide. Vitamin B$_6$ plays an essential role in maintaining the operational integrity of the brain. It is required for the formation of delta-aminolevulinic acid, an intermediate in porphyrin synthesis, as well as for the maintenance of the immune response and endocrine metabolism. Vitamin B$_6$ is readily absorbed from the intestinal tract and excreted in the urine in the form of metabolites.[21]

Vitamin B$_6$ deficiency rarely occurs alone and is more commonly seen in patients deficient in several B vitamins. Those particularly at risk for deficiency are patients with uremia, liver disease, absorption syndromes or malignancies, or chronic alcoholism. High intake of proteins increases the requirements for vitamin B$_6$. Therefore, a high-protein diet may cause a rapid onset of deficiency if the increased requirements are not met. Pregnancy, some calcium-channel blockers, and oral contraceptives also cause decreased vitamin B$_6$. Pyridoxine deficiency has been reported to cause convulsions, dermatitis, and a form of sideroblastic anemia. Drugs known to antagonize pyridoxine include isonicotinic acid hydrazide (isoniazid, INH), steroids, and penicillamine. Vitamin B$_6$ has very low toxicity because of its water-soluble nature, but extremely high doses may cause peripheral neuropathy. The RDA of vitamin B$_6$ for an adult male is 2.0 mg/d and for an adult female is 1.6 mg/d.[22]

Niacin

Niacin is a water-soluble vitamin whose requirement in humans is met to some extent by the conversion of dietary tryptophan to niacin. Niacin is the generic term used for both nicotinic acid and nicotinamide. Niacin functions as a component of the two coenzymes NAD and NADP, which are necessary for many metabolic processes, including tissue respiration, lipid metabolism, fatty acid metabolism, and glycolysis. Niacin is absorbed in the small intestine, and excess is excreted in the form of metabolites in the urine.[23]

NAD and NADH are involved in a large number of oxidation-reduction reactions catalyzed by dehydrogenases, including alcohol, glutamate, glucose-6-phosphate, and glycerol-3-phosphate dehydrogenase. Reduction yields dihydronicotinamide (NADH or NADPH), which has a strong absorption at 340 nm, a feature widely used in assays of pyridine nucleotide-dependent enzymes.

The clinical syndrome resulting from niacin deficiency is called *pellagra*. Pellagra is associated with diarrhea, dementia, dermatitis, and death. This deficiency syndrome is associated with diets providing very low levels of niacin and tryptophan. Niacin deficiency may also be seen in alcoholism. To decrease lipid levels, pharmacologic doses of nicotinic acid are given therapeutically. The toxicity of niacin is low. When large doses are ingested, however, as often occurs during lipid-lowering therapies, flushing of the skin and vasodilation may occur. The RDA of niacin for an adult male is 19 mg/d and for an adult female is 15 mg/d.[6] Blood or urinary niacin levels are of value in assessing niacin nutritional status. HPLC methods are available to measure the metabolites of niacin in urine.

Folate

Folate is the generic term used for components that are nutritionally and chemically similar to folic acid. Folates function metabolically as coenzymes that are involved in a variety of one-carbon transfer reactions. Folate and vitamin B$_{12}$ are closely related metabolically. The hematologic changes that result from deficiency of either vitamin are indistinguishable. Folate in the diet is absorbed in the jejunum, and the excess is excreted in the urine and feces. Large quantities of folate are also synthesized by bacteria in the colon. Structural relatives of pteroylglutamic acid (folic acid) are the metabolically active compounds usually referred to as folates. Food folates are primarily found in green and leafy vegetables, fruits, organ meats, and yeast. Boiling of foods and use of large quantities of water result in folate destruction. The average American diet may be inadequate in folate for adolescents and for pregnant or lactating women.[24]

The major clinical symptom of folate deficiency is megaloblastic anemia. Chemical indices of deficiency are, in order of occurrence, low serum folate, hypersegmentation of neutrophils, high urinary FIGLU (a histidine metabolite accumulating in the absence of folate), low erythrocyte folate, macro-ovalocytosis, megaloblastic marrow, and finally, anemia. Serum folate levels, although an early index of deficiency, can frequently be low despite normal tissue stores. Because most folate storage occurs after the vitamin B_{12}–dependent step, erythrocyte folate can also be reduced in deficiency of either vitamin B_{12} or folate. Despite this overlap, erythrocyte folate concentration is accepted as the best laboratory index of folate deficiency. Most physicians order both serum and erythrocyte folate levels because serum levels indicate circulating folate and erythrocyte levels better approximate stores. Homocysteine elevation in serum or urine occurs in folate deficiency. Total homocysteine is generally measured, which is the sum of all homocysteine species, both free and protein-bound forms.[25–27]

Folate requirement is increased during pregnancy and especially during lactation. The increase during lactation results in part from the presence in milk of high-affinity folate binders. Folate supplementation of the diet of pregnant women reduces the incidence of fetal neural tube defects. Other instances of increased folate requirement include hemolytic anemia, iron deficiency, prematurity, and multiple myeloma. Patients receiving dialysis treatment lose folate rapidly. Folate deficiency as a result of malabsorption can occur in sprue, celiac disease, inflammatory bowel diseases, cardiac failure, and systemic bacterial infections. Folate deficiency of dietary origin commonly occurs in the elderly. Phenytoin (Dilantin) therapy accelerates folate excretion and interferes with folate absorption and metabolism. Alcohol interferes with folate's enterohepatic circulation, whereas the chemotherapeutic agent methotrexate inhibits the enzyme dihydrofolate reductase. Low levels of serum folate can occur with use of oral contraceptives.

Deficiency of folate leads to impaired cell division and alternates in protein synthesis. Folate deficiency is a common vitamin deficiency and is found in a variety of clinical conditions, including megaloblastic anemia, alcoholism, malabsorption syndrome, carcinomas, liver disease, chronic hemodialysis, and hemolytic and sideroblastic anemia. Some anticonvulsant drugs used to treat epilepsy lead to development of folate deficiency. Several other drugs have been shown to interfere with folate metabolism, including sulfasalazine, isoniazid, and cycloserine, which is used for tuberculosis therapy. There are no known cases of toxicity. The requirement of folate for an adult male is 200 µg/d and for an adult female is 180 µg/d.

The reference ranges are as follows:

Serum: 3 to 16 ng/mL
Erythrocyte: 130 to 630 ng/mL
Deficient stores: less than 140 ng/mL

Folate levels may be measured in serum using a microbiologic assay employing *Lactobacillus caseii* or a competitive protein-binding assay for levels in serum and erythrocytes. In development of folate deficiency, serum levels fall first, followed by a decrease in erythrocyte folate levels and ultimate hematologic manifestation. It is helpful to measure both serum and erythrocyte levels because serum levels indicate circulating folate and erythrocyte levels better approximate stores.[25]

The more common two-stage competitive assay for folate is more sensitive and involves a noncompetitive sequential incubation. In the first step of the procedure, the patient sample is incubated with the protein binder. After an incubation step, the temperature is decreased to 4°C to minimize dissociation of folate from the protein binder, and then radiolabeled folate is added to occupy any remaining binding sites. Separation of bound and unbound radiolabeled folate is performed with dextran-coated charcoal or another binder. Nonradioactive competitive protein-binding methods have been developed to permit automated analysis of folate in serum. An assay based on a nonradio-isotope sandwich-type methodology for folate has been developed by Chiron. This procedure is on both the ACS-180 and the Chiron Centaur. It employs a chemiluminescent acridinium ester–labeled folate.[28]

Most of the nonisotopic folate assays are heterogeneous assays similar to the one described. However, the Abbott fluorescence polarization procedure is a homogeneous assay. Serum contains endogenous binding proteins that can bind folate and result in falsely low serum folate concentrations being measured. These endogenous binding proteins may be inactivated by chemicals or by boiling at a high pH with 2-mercaptoethanol. Care must be taken, however, that the endogenous folate is not destroyed in these extraction steps. Studies have shown that measurement of red blood cell folate concentrations are of greater clinical utility for the diagnosis of megaloblastic anemia. Analytical problems may result, however, from the various forms of folate in erythrocytes and the problem of adequately converting all the forms to the reduced monoglutamate methyltetrahydrofolate during analysis. Folate in serum is almost exclusively present in the monoglutamate form. In red blood cells, however, it is present as the polyglutamate form and as high-molecular-weight complexes.[25]

Vitamin B_{12}

Vitamin B_{12}, or cobalamin, refers to a large group of cobalt-containing compounds. The intestinal absorption of vitamin B_{12} takes place in the ileum and is mediated by a unique binding protein called *intrinsic factor,* which is secreted by the stomach. Vitamin B_{12} participates as a coenzyme in enzymatic reactions necessary for hematopoiesis and fatty acid metabolism. Excess vitamin B_{12} is excreted in the urine. Vitamin B_{12} bears a corrin ring (containing pyrroles similar to porphyrin) linked to a central cobalt atom. Different corrinoid compounds, or cobalamins, are distinguished by the substituent linked to the cobalt. The active co-factor forms of vitamin B_{12} are methylcobalamin and deoxyadenosylcobalamin. Dietary sources of vitamin B_{12} are of animal origin (meat, eggs, milk) and few plant products. Total vegetarian diets are therefore likely to have deficiencies of vitamin B_{12}. Animals derive their vitamin B_{12} from intestinal microbial synthesis. The average daily diet contains 3 to 30 µg of vitamin B_{12}, of which 1 to 5 µg is absorbed. The frequency of dietary deficiency increases with age, occurring in more than 0.5% of people over 60 years of age,[29,30] although the symptoms resulting from dietary deficiency are rare.

Most vitamin B_{12} absorption occurs through a complex with intrinsic factor, a protein secreted by gastric parietal cells. This intrinsic factor–B_{12} complex binds with specific ileal receptors. "Blocking" intrinsic factor antibodies prevent binding of vitamin B_{12} to intrinsic factor, and "binding" antibodies can combine with either free intrinsic factor or the intrinsic factor–B_{12} complex, thus preventing attachment of the complex to ileal receptors and intestinal uptake of the vitamin.

Parietal cell antibodies have also been identified as a cause of pernicious anemia. After being released from the intrinsic factor complex within the mucosal cell, vitamin B_{12} circulates in plasma bound to specific transport proteins and is deposited in liver, bone marrow, and other tissues. There is a significant enterohepatic circulation of vitamin B_{12}. A person with normal B_{12} stores but lacking intrinsic factor requires less time (1 to 4 years) for deficiency to become evident. Transcobalamin II appears to be the major serum protein transporting exogenous vitamin B_{12} to tissues. Cobalophilin (previously R, or rapidly migrating, binding protein, or transcobalamin I) transports endogenous B_{12} and is the binder of food cobalamins. Saliva, breast milk, and granulocytes contain large amounts of this binding protein compared with relatively small amounts of cobalamin. Plasma contains both types of transport proteins and the three forms of vitamin B_{12} (hydroxycobalamin, methylcobalamin and deoxyadenosylcobalamin). The absorption of dietary vitamin B_{12} is shown in Figure 28-3 along with its transport to storage sites.[31,32]

In the Schilling test, the patient receives an oral dose of a small amount of radiolabeled vitamin B_{12}. Parental B_{12} is given simultaneously to saturate binding sites. Serum and urine are collected at intervals, and labeled B_{12} is measured in the specimens. Patients who cannot absorb vitamin B_{12} (most often due to a deficiency of intrinsic factor, as in pernicious anemia) cannot absorb the labeled B_{12} and therefore have low levels in the blood and urine.

Inadequate secretion of intrinsic factor may accompany lesions of the gastric mucosa, gastric atrophy, gastrectomy, iron deficiency, and some endocrine disorders. The intrinsic factor–B_{12} complex may be inadequately formed in pancreatic insufficiency because there is insufficient pancreatic protease activity to split the dietary vitamin B_{12} from cobalophilin in the duodenum. The intrinsic factor–B_{12} complex may be inadequately absorbed in ileal malfunction (sprue, enteritis, ileal resection, neoplasias, and granulomas). The term *pernicious anemia* is now most commonly applied to vitamin B_{12} deficiency resulting from lack of intrinsic factor. Antibodies to intrinsic factor and to parietal cells are common in pernicious anemia patients, in their healthy relatives, and in patients with other autoimmune disorders.

One of the most common causes of vitamin B_{12} deficiency is due to a defect in the secretion of intrinsic factor.

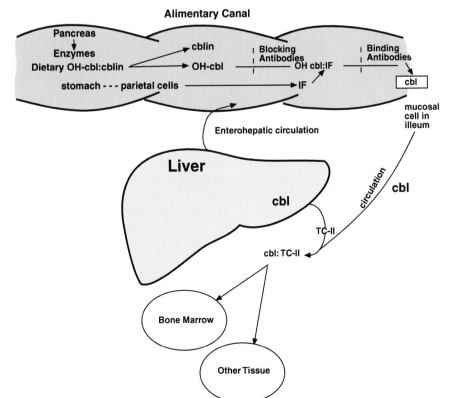

Figure 28-3. Absorption of dietary vitamin B_{12} and transport to storage sites.

Pernicious (megaloblastic) anemia is the disease characterized by B_{12} deficiency and results from a lack of intrinsic factor or antibodies against the factor. Deficiency of B_{12} can occur as a result of dietary deficiency, and this can sometimes occur in strict vegetarians. Also, a loss of B_{12} can occur when individuals are infected with fish tapeworm or as a result of malabsorption diseases, such as sprue or celiac disease. Low vitamin B_{12} levels can occur in folate deficiency, and B_{12} deficiency can be masked when a patient starts taking large doses of folate. Toxicity of vitamin B_{12} has not been reported. The RDA of vitamin B_{12} for adults is 2 μg/d. Assay methods for B_{12} are either a microbiologic assay using *Lactobacillus leichmanii* competitive protein-binding radioimmunoassay or an enzyme immunoassay.

Deficiency of vitamin B_{12} causes two major disorders—megaloblastic anemia (pernicious anemia) and a neurologic disorder called *combined systems disorder*. The neurologic manifestations are variable and may be subtle. For this reason, vitamin B_{12} deficiency should be considered a cause of any unexplained macrocytic anemia or neurologic disorder, especially in an elderly person. Serum vitamin B_{12} may be used in the initial assessment, but because the lower limit of the reference interval is not well defined, homocysteine and methylmalonic acid levels or a Schilling test may be needed. Patients with pernicious anemia usually have atrophic gastritis and have an increased incidence of gastric carcinoma.

The reference range for vitamin B_{12} is 110 to 800 pg/mL.[31] The most common methods currently in use for determination of vitamin B_{12} are the competitive protein-binding radioimmunoassays. These assays are based on the principle that vitamin B_{12}, which has been released from endogenous binding proteins, can be measured by its competition with co-labeled B_{12} for a limited amount of specific binding protein. The binding proteins typically used are animal intrinsic factors. Special measures must be taken to eliminate the interference caused by other, nonspecific protein binders of vitamin B_{12}. These nonspecific protein binders are commonly referred to as *R-protein* or *R-binders*. "R" stands for rapid mobility on electrophoresis. These proteins are found in human gastric juice and have a high binding affinity for vitamin B_{12}. The commercially available assays typically use one of two techniques to address the issue of B_{12}-binding proteins. Some assays employ vitamin B_{12} analogs as R-protein blockers, whereas others use heat denaturation in boiling water to remove interfering compounds. Some assays use chemical denaturation of interfering compounds and thus eliminate the boiling step. It has been found that the latter no-boil procedures do not completely denature anti–intrinsic factor antibodies. Some assay procedures may exhibit a high degree of nonspecific protein binding as a result of incomplete inactivation of endogenous binding proteins in serum or use of nonspecific binders of vitamin B_{12}.

Several nonradioisotopic assays for vitamin B_{12} have been developed for routine laboratory use. One procedure uses a competitive magnetic-separation immunoassay and is based on competitive protein binding. In the first phase of the assay, vitamin B_{12} is released from endogenous binding proteins using a pretreatment reagent. After this, the released vitamin B_{12} is reacted with reagent containing intrinsic factor. After an incubation step, vitamin B_{12} that is conjugated to the enzyme alkaline phosphatase is introduced into the system. This B_{12}–enzyme conjugate competes with vitamin B_{12} from the sample for protein-binding sites on the intrinsic factor. Separation of bound and free labeled vitamin B_{12} is accomplished with use of monoclonal antibodies bound to a magnetic particle. The monoclonal magnetic particles bind to the intrinsic factor–B_{12} complexes. In the final step of the assay, substrate for the enzyme label (*para-* or *p*-nitrophenol phosphate) is added. The substrate is cleaved, producing *p*-nitrophenoxide, which can be monitored spectrophotometrically at 405 nm. Samples containing no vitamin B_{12} have maximum binding of enzyme-labeled vitamin B_{12} to intrinsic factor and thus exhibit maximum absorbance readings, whereas samples with increased B_{12} concentrations have minimum bound label.[33,34]

Another assay based on similar principles is the chemiluminescence assay available from Ciba Corning.[33] This procedure, automated on the ACS-180, employs a chemiluminescence label, acridinium ester–labeled B_{12}.

Biotin

Biotin, sometimes called vitamin H, is a coenzyme for several enzymes that transport carboxyl units in tissue and plays an integral role in gluconeogenesis, lipogenesis, and fatty acid synthesis. Dietary biotin is absorbed in the small intestine. Biotin is also synthesized in the gut by bacteria. Numerous foods contain biotin, although no food is especially rich (up to 20 μg/100 g). Dietary intake is low in the neonatal period despite the fact that concentrations increase as newborns switch from colostrum to mature breast milk. Biotin is excreted in the urine and feces.

In humans, biotin deficiency can be produced by ingestion of large amounts of avidin, which is found in raw egg whites and binds biotin, rendering it metabolically inactive. Biotin deficiency has been noted in patients receiving long-term parenteral nutrition without biotin supplementation. Biotin deficiency also occurs in infants with genetic defects of carboxylase and biotinidase enzymes. No toxicity to biotin has been described. The RDA of biotin has not been established, but a provisional range of 30 to 100 is μg/d is recommended. Levels of biotin are rarely measured in the clinical setting. Assays for determining biotin in blood are used in research settings and include a microbiologic method using *Ochromonas dancia,* a cellulose-binding assay, and most recently, a method employing chemiluminescence.[6,35]

Pantothenic Acid

A growth factor occurring in all types of animal and plant tissue was first designated *vitamin B3* and later named *pantothenic acid* (Greek meaning "from everywhere"). Dietary sources include liver and other organ meats, milk, eggs, peanuts, legumes, mushrooms, salmon, and whole grains. Approxi-

A 65-year-old woman in mild congestive heart failure was admitted to the hospital. She had been seen in the family practice clinic with complaints of numbness, tingling in the calves and feet, and weight loss. The physical examination revealed a slightly confused and depressed, pale woman with a blood pressure of 110/70 mm Hg. Faint scleral icterus was present. There was 1+ pitting and ankle edema. The neurologic examination revealed loss of vibratory sensation in both legs with exaggerated ankle and knee reflexes. Initial laboratory results are shown in Case Study Table 28-2.1.

CASE STUDY TABLE 28-2.1. Laboratory Results

Test	Result	Reference Range
Hemoglobin	9.3 g/dL	12–16 g/dL
Hematocrit	28%	38–47%
MCH	35 pg	27–31 pg
MCV	108 fL	80–96 fL
MCHC	32.4 g/dL	32–36 g/dL
Chem. Panel		
Na	141 mEq/L	136–145 mEq/L
K	4.2 mEq/L	3.5–5.3 mEq/L
Cl	102 mEq/L	96–106 mEq/L
CO_2	26 mg/dL	22–33 mg/dL
Ca	9.7 mg/dL	8.4–10.3 mg/dL
Glucose	100 mg/dL	70–110 mg/dL
BUN	14 mg/dL	10–20 mg/dL
Creat	1.0 mg/dL	0.4–1.4 mg/dL
Serum B_{12}	130 pg/mL	180–900 pg/mL
Serum folate	6 ng/mL	5–12 ng/mL
RBC folate	105 ng/mL	200–700 ng/mL

Questions

1. What is a biologically active form of cobalamin in plasma?
   a. Cyanocobalamin
   b. Hydrocobalamin
   c. Deoxyadenosylcobalamin
   d. Aquocobalamin
   e. Transcobalamin
2. Where is intrinsic factor made in the body?
   a. Stomach
   b. Esophagus
   c. Small intestine
   d. Large intestine
3. What is the binding protein for vitamin B_{12}?
   a. Retinol-binding protein
   b. Extrinsic factor
   c. Corticotropin
   d. Transcobalamin II
4. List some food substances that contain vitamin B_{12}.
5. List some food substances that contain folate.

mately 50% of pantothenate in food is available for absorption. Pantothenate is metabolically converted to 4′-phosphopantothenine, which becomes covalently bound to either serum acyl carrier protein or to coenzyme A. Little is known about pantothenate metabolism. Free pantothenate is the major form in both urine and serum, whereas coenzyme A is the major erythrocytic form. Coenzyme A is a highly important acyl-group transfer coenzyme that is involved in a large number of reactions of a great variety of reaction types. Whole blood pantothenate of less than 1000 mg/L and urinary excretion of less than 10 mg/d are regarded as indicative of deficiency. Most past measurements have used biologic assays for free pantothenate, releasing pantothenate from coenzyme A by multiple enzymatic treatments. Low urinary excretion and reduced blood levels of pantothenate have been reported in patients with chronic *malnutrition,* acute alcoholism, and acute rheumatism. Patients with circulatory and cardiovascular diseases and those with peptic ulcers have reduced circulating pantothenic acid; chronic alcoholics have increased excretion, an indication of impaired use. Pantothenate has been given postoperatively to stimulate the gastrointestinal tract and also to treat streptomycin-induced neuropathy. No toxicity is known.[36]

Ascorbic Acid

The most commonly discussed vitamin, ascorbic acid (vitamin C), is thought to function in hydrogen transfer and regulation of intracellular oxidation reduction potentials. Ascorbic acid is required for normal amino acid metabolism of certain drugs. By virtue of its enediol group, ascorbate is a strong reducing compound. Although plants and most animals can synthesize this vitamin, humans cannot, and dietary ingestion is essential. Major dietary sources include fruits (especially citrus) and vegetables (tomatoes, green peppers, cabbage, leafy greens, potatoes).

Ascorbic acid is important in the formation and stabilization of collagen by hydroxylation of proline and lysine for cross-linking and the conversion of tyrosine to catecholamines (by dopamine β-hydrolase). It increases the absorption of certain minerals, such as iron. Ascorbic acid is absorbed in the upper small intestine and distributed throughout the water-soluble compartments of the body. Ascorbic acid deficiency is known as scurvy. This condition is characterized by hemorrhagic disorders, including swollen bleeding gums, as well as impaired wound healing and anemia.[37]

Urine is the primary route of excretion. Measurement of urinary ascorbate is not recommended for status assessment

CASE STUDY 28-3

A 27-year-old man was diagnosed with a carcinoid tumor in the lower portion of his small intestine. The tumor was debulked, with removal of a portion of the lower section of the small intestine. His recovery course was somewhat complicated by weight loss. Seven months after the surgery, he underwent a laparotomy, which showed that all of the carcinoid tumor had not been removed; thus, more of the small bowel was removed because of the obstructing adhesions. He recovered from the second surgery, and tube feeding was discontinued. He came back to the clinic 18 months after the initial surgery slightly pale and stating that he was having trouble maintaining weight. His laboratory evaluation showed a slight hypochromic, macrocytic anemia, normal renal function, and normal liver function. A stool specimen was negative for ova, parasites, and enteric pathogens. He was readmitted for intravenous fluid and electrolyte replacement.

CASE STUDY TABLE 28-3.1. Laboratory Results

Test	Result	Reference Range
Albumin	3.4 g/dL	3.5–5 g/dL
Prealbumin	15 mg/dL	18–40 mg/dL

Sodium	139 mmol/L	136–145 mmol/L
Potassium	3.7 mmol/L	3.5–5 mmol/L
Chloride	101 mmol/L	99–109 mmol/L
Bicarbonate	23 mmol/L	22–28 mmol/L
Calcium	8.6 mg/dL	8.5–10.5 mg/dL
Magnesium	1.7 mg/dL	1.5–2.5 mg/dL
Phosphate	3 mg/dL	2.8–4 mg/dL
Prothrombin time	17 seconds	
Ferritin	22 ng/mL	20–250 ng/mL
Vitamin A	207 µg/L	300–800 mg/L
Vitamin D	6 ng/mL	30–53 ng/mL
Vitamin E	2 mg/L	5–18 mg/L
Vitamin C	0.4 mg/dL	0.4–0.6 mg/dL
Vitamin B_{12}	100 pg/mL	110–800 pg/mL
Fecal fat (72 hours)	30 g/day	<6 g/d

Questions

1. What biochemical evidence exists for fat malabsorption?
2. What nutritional parameters would be affected by the fat malabsorption?
3. Identify the fat-soluble vitamins.
4. What condition results from vitamin B_{12} deficiency?

because it reflects recent intake and has numerous analytical difficulties. Drugs known to increase the urinary excretion of ascorbate include aspirin, aminopyrine, barbiturates, hydantoin, and paraldehyde. Ascorbic acid requirements are increased during pregnancy and during oral contraceptive use.

Deficient dietary intake can lead to low levels of ascorbic acid; however, chronic inflammatory diseases and acute and chronic infections can also lead to decreased blood levels. The reference range for ascorbic acid is 0.4 to 0.6 mg/dL. The usefulness of ascorbate in preventing colds has lead to the common practice of megadose intakes of vitamin C. Although there is little toxicity of this ascorbate, excess intake may interfere with metabolism of vitamin B_{12} and with drug actions (aminosalicylic acid, tricyclic antidepressants, and anticoagulants).[38]

Ascorbic acid is a physiologic reducing agent, and many methods are based on the oxidation of ascorbic acid to dehydroascorbic acid or on its reducing properties. The most widely used assay for ascorbic acid is the 2,4-dinitrophenylhydrazine method. In this procedure, ascorbic acid is first oxidized to dehydroascorbic acid and 2,3-diketogulonic acid. Copper is the most common agent chosen for oxidation. After the initial step of oxidizing ascorbic acid, 2,4-dinitrophenylhydrazine is added, forming bis-2,4-dinitrophenylhydrazine. Treatment with sulfuric acid results in the formation of a colored product that absorbs at 520 nm. This method measures the total vitamin C content of the sample because ascorbic acid, dehydroascorbic acid, and diketogulonic acid are also measured. This assay is subject to interference from amino acids and thiosulfates. HPLC has been developed to give increased sensitivity and specificity. HPLC methods can measure ascorbic acid, dehydroascorbic acid, and the stereoisomer of ascorbic acid, isoascorbic acid. These HPLC methods use either ultraviolet detection or electrochemical detection.[39]

Carnitine

Carnitine, which includes L-carnitine and its fatty acid esters (acylcarnitine), is described as a conditionally essential nutrient. Meat, poultry, fish, and dairy products are the major dietary sources. Foods of plant origin generally contain little carnitine, except for peanut butter and asparagus. Normal diets provide more than half the human requirement, but strict vegetarian diets provide only 10% of the total carnitine needed by humans. Synthesis occurs in liver, brain, and kidney. L-Carnitine facilitates entry of long-chain fatty acids into mitochondria for oxidation and energy production.[40] The major signs of carnitine deficiency are muscle weakness and fatigue. Chemical indices of deficiency include decreased total or free carnitine in serum, urine, or tissues; measurements are generally radioenzymatic. Total carnitine is measured after hydrolysis of ester forms to free carnitine.

Human deficiency can be either hereditary or acquired. Acquired deficiency can be caused by inadequate intake, increased requirement (pregnancy and breast feeding) or increased urinary loss (valproic acid therapy). Infants, patients following a course of long-term parenteral nutrition, and perhaps children are groups most vulnerable to deficiency as judged by decreased circulating levels and altered indicators of energy metabolism. Patients receiving hemodialysis may lose carnitine in the dialysis fluid.[41]

Recommended Dietary Allowances

The RDA, for use in the United States, is the level of intake of essential nutrients, on the basis of available scientific knowledge, to be sufficient to meet the nutritional needs of practically all healthy individuals in the general population. They are not considered to be adequate to cover the special therapeutic needs of people who have infections, metabolic disorders, or chronic diseases. Neither are they considered to meet the needs of premature infants, nor of people who use certain pharmaceutical preparations, such as oral contraceptives.[2]

The RDA should not be interpreted as nutrient requirements for individuals; however, they may be used in assessing individual dietary intakes. When they are, it must be understood that because they are intended to meet the nutrient needs of practically everybody, they may be in excess of the needs for some people. Thus, although an individual's diet does not meet the allowances, it must not be assumed that this person is malnourished nor that the diet is deficient unless there is evidence of some clinical or biochemical abnormalities.

Factors Affecting Vitamin Metabolism

Although it is true that all people are somewhat equal with respect to their nutrient requirements, there are numerous factors affecting vitamin and nutritional requirements, including genetic differences, acquired factors (eg, pregnancy, disease, hypermetabolic states) and others (eg, surgery, drugs, alcohol).

The primary cause of latent malnutrition is inadequate food intake. Secondary causes must also be considered. These include problems related to nutrient needs, such as poor absorption, inefficient utilization, impaired transportation, increased requirement, and excessive excretion of nutrients.

Drugs and oral contraceptives also have a definite impact on vitamin metabolism. Table 28-2 summarizes the actions of oral contraceptives and drugs on vitamin metabolism.

Diet and Cancer

A neoplasm is a multifactorial process that can be broadly categorized into five etiologies: genetic, viral (eg, oncogenes), chemical (eg, zenobiotic carcinogens), physical (radiation injury), and inflammatory (chronic inflammation). The latter three are linked to free radicals that can readily induce genomic damage.[42]

TABLE 28-2. Actions of Oral Contraceptives and Drugs on Vitamin Metabolism

Drug or Nutrient

Pyridoxine—antagonized by isoniazid, steroids, penicillamine

Riboflavin—antagonized by phenothiazines, some antibiotics

Folate—antagonized by phenytoin, alcohol, methotrexate, trimethoprim

Ascorbate—antagonized by (increased excretion) aspirin, barbiturates, hydantoins

Ascorbate excess—interferes with actions of aminosalicylic acid, tricyclic antidepressants, anticoagulants; may cause "rebound scurvy" on withdrawal

Oral Contraceptive Agents Cause:

Increased serum vitamin A, RBP

Decreased requirement for vitamins K and C

Decreased vitamin B_6 status indices

Decreased riboflavin use

Increased niacin pathway of tryptophan

Decreased induction of thiamin deficiency

Decreased serum folate (cycle-day dependent)

Decreased induction of cervical folate deficiency

A wide variety of basic investigations using free radical scavengers and other antioxidants, as well as epidemiologic studies relating diet to cancer, complement the free radical data obtained from genomic damage. It has been repeatedly shown that obesity, as well as a high-fat diet, is associated with an increased risk for developing cancer (and coronary artery disease). On the other hand, a wide variety of studies have linked cancer with a low intake of vitamins C, E, and beta-carotene. As noted previously, low serum selenium levels also have been linked to increased cancer rates.[43]

Special Diets

The current trend in the eating habits of some young adults away from the familiar Western food patterns toward vegetarianism has caused concern about the nutritional implications of such changes. This is a legitimate concern, shared by the parents of teenagers and many other Americans. Most nutritionists agree that vegetarian diets can be adequate if sufficient care is taken in planning them. Both the quantity and the quality of protein are of central concern in all diets. The quality of proteins in plant foods, notably cereal grains, is generally lower than that of animal proteins. If mixing of plant proteins is done judiciously, combinations of lower-quality protein foods can give mixtures of about the same nutritional value as high-quality animal protein foods.[3,4]

Vitamin B_{12} intake is usually low in most vegetarian diets. Therefore, when anemia is found, chances are it is due to inadequate intake of B_{12}. More drastic diets, such as the Zen macrobiotic diet, may cause more severe nutritional deficiencies. The Zen macrobiotic diet is perhaps the most dangerous of the diets for growing children. Ten stages of dietary restriction progress from −3 to +7, with gradual elimination of animal products, fruits, and vegetables. The lower-level diets can meet nutritional needs, but the highest-level diet is composed only of cereals and restricts nutritional balance that is inherent in more diverse diets. In addition, caloric intake is usually low. Strict adherence to the more rigid diets can result in scurvy, anemia, hypoproteinemia, hypocalcemia, emaciation, or even death. Poor growth is the main clinical finding in infancy. Other nutrients likely to be of marginal content in all-plant diets are calcium, iron, riboflavin, and vitamin B_{12}, and for children not exposed to sunlight, vitamin D.[3-6]

NUTRITIONAL ASSESSMENT

Nutritional assessment, which is the evaluation of a patient's metabolic and nutritional needs, is performed by diurnal, empirical, laboratory, and other objective measurements of nutritional status. Malnutrition is a state of decreased intake of calories or micronutrients (vitamins and trace elements) resulting in a risk of impaired physiologic function associated with increased morbidity and mortality.[44-46] Patients who are chronically calorie malnourished suffer from a loss of both adipose and muscle tissue as a result of increased lipolytic and gluconeogenic activity, respectively. These patients, however, do not demonstrate protein deficiency as reflected by normal levels of serum transport proteins. This type of malnourished state is referred to as *marasmus*. Acute protein-calorie malnutrition, known as *kwashiorkor*, is associated with physiologic stress and results in impaired protein synthesis and lower serum transport proteins.[47] Decreases in vitamin and trace metal intake are usually associated with calorie and protein deficiencies. They may occur independently, however, as a result of dietary inadequacy, as in alcoholism, or as a result of malabsorption syndromes. Malnourished states can be assessed by the measurements of anthropomorphic methods. The *anthropomorphic methods* (height, weight, etc) are the common measurements that dietitians use in evaluating patients. These methods are shown in Table 28-3.[48]

Those at Risk for Malnutrition

Malnutrition has been estimated to affect more than 1.3 billion people worldwide.[49] It is a major co-morbid cause of death. The estimates of malnutrition in the United States range from 10 to 20 million individuals. Poverty is certainly a major predisposing factor for the risk of malnutrition in the nonhospitalized individual, especially worldwide. However, malnutrition is erroneously believed to be associated only with low socioeconomic status. On the contrary, the prevalence of significant malnutrition in the United States among nursing home patients, elderly people, and medical and surgical hospitalized patients is no longer challenged. Those at risk for malnutrition are listed in Table 28-4.

It was recently documented that 30% to 50% of hospitalized patients may be malnourished.[48,50] Malnutrition in the hospital setting may occur when oral nutritional intake is inadequate or impossible because of treatment procedures or complications that follow surgery. Chronic malnutrition may occur before hospitalization as a result of a chronic disease process or a medical condition that leads to poor food choices, protein depletion, or low protein intake. Chronic diseases leading to chronic malnutrition include malignancy, gastrointestinal disease including malabsorption syndrome, infectious diseases such as acquired immunodeficiency syndrome (AIDS) and tuberculosis, alcoholism, and liver disease.[44,47,48]

The hospitalized patient may undergo acute malnutrition as a result of decreased food intake secondary to the disease, chemotherapy, or depression, or reduced intake due to nausea.[51] The hypermetabolic state may cause acute malnutrition secondary to an increased energy expenditure. The hypermetabolic condition is associated with major trauma, sepsis, surgery, burns, multiple organ failure, and acute renal failure.[52]

A severely ill patient with hypermetabolism (kwashiorkor-like) and multiorgan failure is metabolically different from the starved hypometabolic patient (marasmus-like), although a clear distinction in clinical presentation is not seen frequently. The long-term malnourished hypometabolic patient with primary loss of body fat uses ketogenic fatty acid, not carbohydrates, as a primary fuel. These individuals tend to be thin, with a wasted appearance, and have little adipose tissue reserves.

In the hypometabolic starved patient, aggressive feeding, especially with carbohydrates, may not be desirable. In these patients, with the body's ability to adapt to chronic starvation by reducing the basal metabolic rate, cardiac output, temperature, and physical activity, overzealous nutritional support may induce a refeeding syndrome.[53] In the refeeding syndrome, life-threatening fluid and electrolyte shifts occur after the initiation of aggressive nutritional support therapies. Refeeding syndrome can reduce cardiac mass, leading to cardiac insufficiency conditions such as hypophosphatemia and hypokalemia. It can also reduce respiratory muscle mass and decrease adenosine triphosphate content, leading to respiratory failure.[54] The hypermetabolic patient, on the other hand, has an increased energy need and may require aggressive nutritional support to minimize catabolism and protein losses and to support immune response. When malnutrition is not detected and treated in patients during their hospital stay, there is increased morbidity and mortality.[55]

Patients with protein-energy malnutrition have impaired wound healing and increased rates of infection[50] with extended lengths of hospital stay and higher mortality rates. Of the patients who are identified as malnourished, too few receive nutritional support early enough. The early identification of malnourished patients and the establishment of nutritional intervention have become a quality of care issue.[56,57]

CASE STUDY 28-4

Ms. W. is a single, 57-year-old executive admitted to the hospital for collapsing at the pool while on vacation. She has had a recent loss of weight and loss of appetite. She had a physical examination 9 months ago, and her doctor gave her a "clean bill of health." Since her physical, Ms. W. was demoted in her job from Regional Sales Director to Sales Representative for five states. Her new job requires her to travel 5 days per week, and her income is now based on sales commission rather than salary. She tells you that she is very nervous about being on sick leave because she is not making money. She has three children in college and has considerable debt. Several years ago, she became a vegetarian. She eats mainly vegetables, grains, and fruits. She does not drink milk or eat eggs, meat, chicken, or fish. She takes some herbal supplements.

Questions

1. How has Ms. W's nutritional status changed in the past 9 months? Specify lab results and suggest possible nutrition-related problems.
2. What factors do you think have affected her current medical profile?
3. What additional lab work would you recommend and why?

CASE STUDY TABLE 28-4.1. Medical Profile

	9 Months Ago	Today
Height	5' 6"	5' 6"
Weight	145 lb	123 lb
Albumin	4.0 g/dL	2.8 g/dL
Hemoglobin/hematocrit	12 g/dL 36%	11 g/dL 33%
Mean cell volume	90 μ³	99 μ³
Mean cell hemoglobin	30 pg	38 pg
Serum Fe	82 mg/dL	56 mg/dL
Transferrin	NA	290 μg/dL
Schilling test	NA	290 μg/dL
Stage I		18%
Folate	NA	42 μg/mL
Stage 2		34%

Reference Range

Albumin	3.5–5.5 g/dL
Hemoglobin/hematocrit	12–16 g/dL 37%–47%
Mean cell volume	81–99 μ³
Mean cell hemoglobin	27–34 μg
Serum iron	39–150 μg/dL
Transferrin	204–306 mg/dL
Folate serum	3–14 ng/dL
RBC	160–700 ng/dL
Schilling test	>7% excretion in 24 hours

CASE STUDY TABLE 28-4.2. Selected Laboratory Results in Iron Deficiency, Macrocytic Anemia, and Anemia of Chronic Disease

Test	Reference Range	Fe Deficiency	Macrocytic Anemia (Folate or B₁₂)	Anemia of Chronic Disease
Hemoglobin	F—12–16 g/dL	<12 g	<12 g	<12 g
	M—14–18 g/dL	<14 g	<14 g	<14 g
Hematocrit	F—33%–43%	<33%	<33%	<33%
	M—39%–49%	<39%	<39%	<39%
Mean cell volume	80–95 μ³	<80 μ³	>95 μ³	Normal
Mean cell hemoglobin	27–31 pg	<27 pg	>31 pg	Normal
Mean cell hemoglobin concentration	32–36 g/dL	<32 g/dL	Normal	Normal
Serum Fe	60–190 μg/dL	<60 μg/dL	>190 μg/dL	<60 μg/dL
Total iron-binding capacity	25–420 μg/dL	>420 μg/dL	—	<25 μg/dL
Transferrin	200–400 μg/dL	>400 μg/dL	<200 μg/dL	<200 μg/dL
Ferritin	F—10–150 ng/mL	<10 ng/mL	>150 ng/mL	<10 ng/mL
	M—12–300 ng/mL	<12 ng/mL	>300 ng/mL	<12 ng/mL
% Transferrin saturation	30%–40%	<30%	—	Normal
Vitamin B₁₂ (Schilling test)	8%–40%	—	<8% Pernicious anemia	Normal or decreased
Folate	5–20 μg/mL	—	>5 μg/dL Pernicious anemia <5 μg/mL Folate deficiency	Normal or decreased
Prealbumin	18–40 mg/mL	—	Moderately malnourished: 12–18	Severely malnourished: 11 or less

TABLE 28-3. Anthropomorphic Measures of Nutritional Status

Height
Weight
Skinfold thickness
Mid-arm muscular circumference
Wrist circumference

Assessment of Nutritional Status

Many of the methods used for nutritional assessment are designed to provide information regarding body composition abnormalities. In addition to clinical evaluation, they include straightforward physical measurement (body weight and anthropometric measurements), biochemical or immunologic determinations, and sophisticated laboratory measurements.

Some authors suggest that clinical judgment is a desirable method to evaluate nutritionally associated complications. This type of nutritional assessment is based on past nutritional intake, disease process, the extent of catabolic disease, functional status, edema, skin rash, and neuropathology.[58,59] Other groups have shown that clinical judgment alone cannot reliably identify the malnourished patient.[60,61]

One method used for nutritional assessment is anthropometric indices. In this procedure, the thickness of the biceps skinfold of the nondominant arm is taken to represent body fat mass. Despite its limitations, some usefulness has been established for anthropometrics and skinfold thickness.[62]

Measurements from arm muscle circumference are thought to provide an indication of the body's muscle mass and hence its main protein reserve. Arm muscle circumference can be derived from total arm muscle circumference and triceps skinfold thickness, but a 15% to 25% overestimation of lean body mass by anthropometry in arm muscle can be incorrectly obtained.[63] Other studies found no correlation between the circumference of arm muscle and the initial total

TABLE 28-4. Groups at Risk for Malnutrition

Depressed or mentally ill people who fail to eat
Elderly people, especially those in nursing homes
People of low socioeconomic status
Population with loss of fluids and nutrients from long-term diarrhea, draining fistula, or wounds
Stroke victims
Patients in the postoperative state and those with ileus with long periods without oral intake
Patients with hypermetabolic conditions: head trauma, multiple trauma, major burns, organic failure, and sepsis
Patients with cancer of the gastrointestinal tract or cachexia
Patients with unintentional weight loss exceeding 10% in 6 months
Patients with chronic disease, long-term dialysis
Patients with pancreatitis

body nitrogen content in patients receiving parenteral nutrition. These studies further showed that no changes in total body nitrogen were found after completion of total parenteral nutrition.[64,65]

Because it was shown that a preoperative weight loss of more than 20% was associated with 33% mortality, whereas a 4% mortality rate occurred in those with less than a 20% weight loss, the estimation of weight loss is a key indicator in the nutritional assessment of surgical patients.[66] Recent weight change is a commonly used index of malnutrition. More than 10% loss over any time period has been taken as evidence of malnutrition, and a useful correlation has been proposed between the extent of weight loss and the time over which it develops,[67] as illustrated in Table 28-5. A weight loss 10 days prior to surgery of more than 4.5 kg is highly predictive of surgical mortality.[68,69]

Creatinine/Height Index

Because creatine is present almost entirely within muscle (as creatine phosphate) and is converted to creatinine at a relatively constant rate, the excretion rate of urinary creatinine may be indicative of total muscle mass.[70] This is reasonably accurate in patients without renal failure, if it is based on multiple measurements.[71] Urinary creatinine excretion correlates with the total body nitrogen content except in cancer patients. In these patients, creatinine excretion remains stable when total body nitrogen decreases, indicating increased creatinine loss as the disease progresses.

Immunologic Testing

Immunologic testing has been useful in predicting morbidity and identifying the effects of malnutrition in disease states. For example, the peripheral blood absolute lymphocyte count and the skin response to injected antigen have been used as indicators of nutritional status. A lymphocyte count below $300/mm^3$ reflects immune deficiency[68] and is associated with a negative response to an injected antigen.[72]

A strong association between morbidity and decreased skin test reactivity has also been shown.[73,74] The proportion of patients with negative skin test reactivity (no response) who died (50%) was significantly higher than the proportion of those with normal skin test reactivity who died (23%).[75] In surgical patients studied, those who were anergic and

TABLE 28-5. Weight Change Evaluation*

Time	Significant Weight Loss (%)	Severe Weight Loss (%)
1 week	1–2	>2
1 month	5	>5
3 months	7.5	>7.5
6 months	10	>10

* Values used are percent weight change (%W): %W = (Usual weight − Actual Weight)/(Usual Weight) × 100.

showed no skin response to injected antigens preoperatively had a 29% incidence of sepsis and a 29.9% mortality rate, compared with 7.5% and 4.6%, respectively, for those classified as immunologically normal.[76] Correcting anergy and providing a positive immunologic response with parenteral nutrition have been shown to improve the outcome of nutritionally depleted patients.[77]

Multivariable Index

Multiple regression equations from a number of anthropomorphic and laboratory tests have been used to predict surgical outcomes as affected by nutritional variables. The prognostic nutritional index is based on serum albumin, triceps skinfold, serum transferrin, and cutaneous reactivity to three recall antigens.[45] The best use of the prognostic nutritional index is in relating complications and assessing hazards of individual patients.[78] A discriminate function test for the treatment of cancer patients has been developed based on serum albumin, skin test response, and the presence of sepsis at the time of assessment.[75]

Body Composition

Measuring body composition has proved useful in monitoring nutritional states because of marked changes in body composition that occur with malnutrition. Body composition can be described in either biochemical or cellular terms. The composition of an adequately nourished 75-kg adult man is illustrated in Figure 28-4 in both biochemical and cellular terms.[79]

The values may vary from one person to another. Body mass described in cellular terms is divided into two mass components: lean body mass (tissue devoid of all fat) and body fat (sum of all ether-extractable substances). Lean body mass comprises the metabolically active tissues, which are known as *body cell mass,* and the metabolically inactive components, which are known as the *extracellular mass.* The body cell mass is the portion of the body responsible for all of the oxygen consumption and carbon dioxide production. This is directly linked to energy requirements.[79,80] The primary function of the extracellular mass is intracellular transport and structural support.

Malnourished hospitalized patients have a body cell mass that is 40.5% less than healthy nonhospitalized patients of the same age group. Obesity is associated with little increase in body cell mass and extracellular mass and greatly expanded body fat. Body cell mass declines with age and is greater in men than women. Body cell mass is affected by race, heredity, and stature.[81]

Body Component Status

The body is composed of three major compartments: fat, protein, and water. Although much body fat is subcutaneous, the abdomen contains a considerable quantity that is not easily measured and is not comparably used during starvation. Endocrine studies show that abdominal fat is used more slowly during long-time starvation. The relationship between components within each compartment may be defined during health, but these relationships change during disease and lean muscle depletion.

Body composition determinations use a variety of techniques, including multiple isotope detection methods, proton-gamma analysis, and neuron activation, to measure total body nitrogen.[79] These methods are generally complicated and time-consuming and thus are not suitable for routine use in patient care or for measurement of relatively short-term changes. However, they are the standard against which simpler measurements must be validated.[82,83]

Bioelectrical Impedance Analysis

Standard body composition methods are complex and are not widely available. For this reason, bedside techniques to assess body composition are constantly sought. The body impedance analysis procedure is based on a measurement of changes in the conduction of an applied electrical current

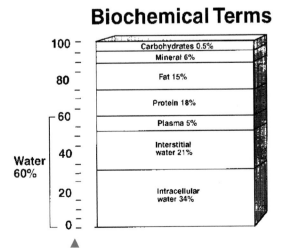
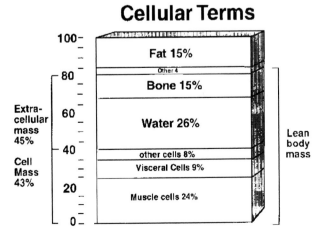

Figure 28-4. Biochemical and cellular terms used to describe body mass and the percentage contribution to total mass in a well-nourished 75-kg male.

through the body. Intracellular and extracellular fluids are thought to act as electroconductors, and all membranes are considered to act as electrocapacitors. Both, however, are regarded as imperfect reactive elements. Because lean body tissue has far greater electrolyte content than fat, this marked difference in ionic content permits the estimation of lean body mass by the assessment of body electrical impedance.[84,85] In one study of healthy subjects, bioelectric impedance measurements were found to be reliable in the estimation of body density, total body water, and total body fat.[86] Bioelectric impedance is a new technique that has been found to be reliable in determining nutritional state and hydration status in the sick surgical patient.[87,88]

Functional Tests

Muscle function has been evaluated as an index of nutritional status because it is susceptible to the effects of withdrawing nutrients and refeeding. One method of evaluating muscle function is a hand-grip dynamometry.[89] In this new approach, grip strength had a sensitivity of 90% in predicting postoperative complications. The hand-grip dynamometry involved some motivation, which may or may not be constant from patient to patient or with the same patient during the course of an illness. To obviate this problem in very ill patients, an electrical stimulation technique of the ulnar nerve of the wrist has been adapted and is shown in Figure 28-5. The result is the contraction of the abductor pollicis longus muscle, which negates the cooperation of the patient and does not appear to be affected by sepsis, drugs, trauma, surgical intervention, or anesthesia.[90,91] Figure 28-6 shows the anatomic location of the abductor pollicis longus muscle in the wrist. The rate of relaxation of the electrically stimulated abductor pollicis longus muscle is a sensitive and specific measurement of nutrient status in both chronic states and anorexia and during nutrient withdrawal and refeeding in the postoperative period.[92–95]

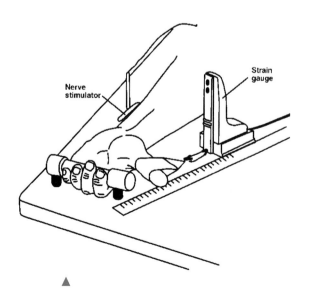

▲
Figure 28-5. Muscle function testing device.

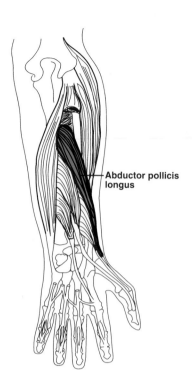

▲
Figure 28-6. Anatomical site of abductive pollicis longus muscle.

Protein Markers in Nutritional Assessment

The primary objective of nutritional assessment is to identify the patient who is malnourished and then, through nutritional therapy, to preserve or replenish the protein component of the body. Laboratory nutritional assessment is best accomplished by monitoring selected serum proteins. These proteins should have a short biologic half-life and should reflect protein status by measurable concentration changes in the serum. The proteins should have a concentration that is relatively small, a rapid rate of synthesis, and a constant catabolic rate and should be responsive only to protein and energy restrictions.

Protein deficiency states in humans are prolonged and severe if insensitive markers are used because it takes a long period before significant change occurs in concentrations of these proteins. The concentration of insensitive proteins may be affected by protein deficiency as well as by other factors, such as hepatic density, renal disease, and severe infection. Measurements of the concentration of selected individual proteins not only may provide a sensitive index on protein states but also may offer a valuable indication of morbidity. Visceral protein measurements also remain a reliable and relatively easy method of assessing the energy state of the patient.[96] Table 28-6 offers detailed information about these protein markers.

Albumin

Albumin has long been used in the assessment of hospitalized patients. Low levels of serum albumin may reflect low hepatic production or protein loss from the vascular com-

TABLE 28-6. Characteristics of Plasma Proteins of Nutritional Interest

Protein	Molecular Weight (kd)	Half-Life	Reference Range
Albumin	65,000	20 days	33–48 g/L
Fibronectin	250,000	15 hours	220–400 mg/L
Prealbumin (transthyretin)	54,980	48 hours	160–350 mg/L
Retinol binding protein	21,000	12 hours	30–60 mg/L
Somatomedin C (insulin growth factor-1)	7,650	2 hours	0.10–0.40 mg/L
Transferrin	76,000	9 days	1.6–3.6 g/L

ponent. Low albumin levels have been identified as a common abnormality in patients in long-term care facilities. Hospitalized patients with low serum albumin levels experienced a four-fold increase in morbidity and a six-fold increase in mortality.[97,98]

The albumin concentration in the body is influenced by albumin synthesis, degradation, and distribution. Albumin from the extravascular pool can be mobilized during periods of protein depletion resulting from the stress of major surgery or infection, so that serum albumin concentrations may not decline for a long period.[99–101] The degeneration rate of the albumin is proportional to the size of the extravascular pool, which allows the concentration in the serum to be relatively constant. The long biologic half-life of albumin (20 days) allows changes in the serum concentration only after long periods of malnutrition. Because there are many mechanisms that can potentially produce a depressed albumin concentration, an isolated serum albumin level may be of limited value in evaluating liver synthesis rates in patients who are critically ill.[102]

Serum albumin is not a good indicator of short-term protein and energy deprivation; however, albumin levels are good indicators of chronic deficiency. Traditionally, albumin has been used to help in determining two important nutritional states. First, it helps identify chronic protein deficiency under conditions of adequate non–protein-calorie intake, which leads to marked hypoalbuminemia. This may result from the net loss of albumin from both the intravascular and extravascular pools, causing kwashiorkor. Second, albumin concentrations may help define marasmus. This is caused by caloric insufficiency without protein insufficiency, so that the serum albumin level remains normal but there is considerable loss of body weight.

Studies have classified various levels of malnutrition by using albumin levels. Serum albumin levels of 35 g/L or greater are considered normal.[103] Albumin levels of 28 to 35 g/L indicate mild malnutrition, levels of 21 to 27 g/L indicate moderate malnutrition, and levels less than 21 g/L indicate severely depleted levels of albumin.[104] Serum albumin is an accurate marker of the catabolic stress of infection.[105,106] This finding in children is in agreement with studies in adults. Levels of albumin have been interpreted in various ways; for

example, a level of 32 g/L or less indicates that a patient has a 75% chance of developing decubitus ulcers if he or she is in the hospital for up to 10 days. Serum albumin levels of less than 25 g/L can be used as an accurate measure of predicting survival prognosis in 90% of critically ill patients.[107]

Transferrin

Transferrin is a glycoprotein with a biologic half-life of 9 days (shorter than that of albumin). It is synthesized in the liver and binds and transports ferric iron. Transferrin synthesis is regulated by iron stores. When hepatocyte iron is absent or low, transferrin levels rise in proportion to the deficiency. It is an early indicator of iron deficiency, and the elevated transferrin is the last analyte to return to normal when iron deficiency is corrected.[108,109]

The half-life of transferrin is one half that of albumin, and the body pool is smaller than that of albumin; hence, transferrin is more likely to indicate protein depletion before serum albumin concentration changes. The usefulness of transferrin in diagnosis of subclinical, marginal, or moderate malnutrition is questionable, however, because a wide range of values have been reported in various studies. Conditional status can alter transferrin levels.[110]

Birth control pills, pregnancy, iron deficiency, and acute hepatitis are associated with increased levels of transferrin. Protein-losing states and end-stage liver disease are to known to decrease transferrin. Large fluid shifts and postoperative severe stress in patients with complications have effects on the level of transferrin that have not been clearly defined. Transferrin levels can be lowered by factors other than protein or energy deficiency, such as nephrotic syndrome, liver disorders, anemia, and neoplastic disease.[111]

In hospital and nursing home settings, transferrin levels have been used as indices of morbidity and mortality.[45,112] Important levels of transferrin have been established for assessing nutritional status, as shown in Table 28-7. Information shows, however, that serum transferrin concentrations are not sufficiently sensitive to detect a change in nutritional status that occurs after 2 weeks of total parenteral nutrition.[113] In addition to being responsive to serum iron concentrations, transferrin is uniquely sensitive to some antibiotics and fungicides. Transferrin levels are altered when

TABLE 28-7. Transferrin Levels Used in Assessing Nutritional Status

Transferrin Concentrations (mg/dL)	Indication
<100	Severe visceral protein deficiency
100–150	Moderate protein depletion
150–200	Mild nutritional deficiency
>200	Adequate nutritional state

patients are taking high doses of aminoglycosides, tetracycline, and some cephalosporins.[114]

Transthyretin

Transthyretin is sometimes called *prealbumin* because it migrates ahead of albumin in the customary electrophoresis of serum or plasma proteins. In normal situations, each transthyretin subunit contains one binding site for RBP. Transthyretin and RBP are considered the major transport proteins for thyroxine and vitamin A, respectively.

Transthyretin has a high concentration of the amino acid tryptophan, and tryptophan plays a major role in the initiation of protein synthesis.[115] Transthyretin has one of the highest proportions of essential to nonessential amino acids of any protein in the body.[116] Because of its high tryptophan content, short half-life, high proportion of essential to nonessential amino acids, and small body pool, transthyretin is a better indicator of visceral protein status and positive nitrogen balance than are albumin and transferrin.[117] Transthyretin and other carrier proteins are better indicators for monitoring short-term effects of nutritional therapy. The concentration of transthyretin and RBP complex is greatly decreased in protein-energy malnutrition and returns toward normal values after nutritional replenishment. Transthyretin has a low pool concentration in the serum, a half-life of 2 days, and a rapid response to low energy intake, even when protein intake is inadequate for as few as 4 days.[118] Serum transthyretin concentrations are decreased postoperatively by 50 to 90 mg/L in the first week and have the ability to double in a week or at least increase 40 to 50 mg/L in response to adequate nutritional support. If the transthyretin response increases less than 20 mg/L in 1 week as an outcome measure, this indicates either an inadequate nutritional support or an inadequate response.[101,119]

When transthyretin decreases to levels of less than 80 mg/L, there is development of severe protein-calorie malnutrition; however, nutritional support can cause a daily increase in transthyretin of up to 10 mg/L.[48] These concentrations do not appear to be significantly influenced by fluctuations in the hydration state. Although end-stage liver disease appears to affect all protein levels in the body, liver disease does not affect transthyretin as early or to the same extent as it affects other serum protein markers, particularly RBP. Although transthyretin levels may be elevated in patients with renal disease, if a trend in the direction of change

is noted, the changes are likely to reflect alteration in nutritional status and nitrogen balance. Steroids can cause a slight elevation in transthyretin, but the nutritional trend can still be followed because transthyretin responds to both overfeeding and underfeeding.[120]

Serum transthyretin concentration above 110 mg/L is an entropy value that has to be obtained for any patient being transitioned from parenteral to enteral or oral feeding.[121] A new use of serum transthyretin is its ability to be a predictor or indicator of the adequacy of a nutritional feeding plan.[48]

Changes in plasma protein have also been correlated with nitrogen balance. Transthyretin concentrations have been shown to increase significantly during a week of intensive tube feeding when the patient was in positive nitrogen balance.[121] Serum albumin concentrations, because of the long half-life of albumin and its regulation by osmotic conditions in the body, do not change during this same period. Transthyretin concentrations increase in patients with positive nitrogen balance and decrease in patients with negative nitrogen balance. When the transthyretin level is at 180 mg/L or more, this correlates with a positive nitrogen balance and indicates a return to adequate nutritional status. Transthyretin has been shown in both the pediatric and neonate population to be a highly accurate and relatively inexpensive marker for nutritional status (Fig. 28-7).[122,123] It is also accurate in high-risk, sick infants. Transthyretin has been found to be the most sensitive and helpful indicator when looking at the nutritional status of very ill patients.[124,125]

In summary, transthyretin effectively demonstrates an anabolic response to feeding and is a good marker for visceral protein synthesis in patients receiving metabolic or nutritional support.[126–130]

Retinol-Binding Protein

RBP has been used in monitoring short-term changes in nutritional status.[131] Its usefulness as a metabolic marker is based on its biologic half-life of 12 hours and its small body pool size. RBP reportedly responds quickly to both energy and protein deprivation and has been correlated with nitrogen balance in severe burn patients.[132,133]

As a single polypeptide chain, RBP interacts strongly with plasma transthyretin and circulates in the plasma as a 1:1 mol/L transthyretin–RBP complex.[134] A potential problem exists in using RBP as a nutritional marker, however. Although RBP has a shorter half-life than transthyretin (12 hours compared with 2 days), it is excreted in urine, and its concentration increases more significantly than transthyretin in patients with renal failure. In chronic renal failure, the plasma concentrations of RBP are elevated because of its decreased catabolism by the kidney. In contrast to RBP, the catabolism of transthyretin occurs only to a limited extent in the kidney, which accounts for only moderately elevated concentrations being measured in advanced chronic renal insufficiency. In a patient with chronic renal failure, the marked increase in concentration of RBP may make it unreliable in monitoring nutritional status.[135,136] The initial high levels of RBP and the moderately elevated transthyretin levels in patients with acute

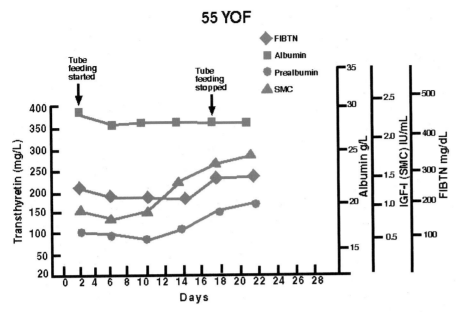

Figure 28-7. Sequential measurements of transthyretin, albumin, insulin-like growth factor-I (somatomedin C, SMC), and fibronectin (FIBTN) in a 55-year-old female with poor nutritional status. Short half-life proteins: transthyretin and SMC.

renal failure may be difficult to interpret with respect to nutritional status. In patients with relatively stable renal failure, the rate of change of the concentration of RBP and transthyretin, rather than an isolated value, can be better used to monitor the adequacy of nutritional therapy.[137]

Somatomedin C

Somatomedin C (SMC) is also known as *insulin growth factor-I*. It plays an important role in the stimulation of biologic growth. The molecular size and structure of SMC is similar to that of proinsulin.[138] SMC serum concentrations are regulated by both growth hormone and nutritional intake. Growth hormone stimulates liver and other tissues to produce SMC. SMC is bound in the serum to specific proteins. These binding proteins are responsible for maintaining a relatively constant concentration of SMC in the serum. The proteins also help to minimize any rapid decreases in this short half-life protein. The binding proteins modulate the biologic effect of SMC under certain circumstances, causing both decreases and increases in its biologic activity.[139]

SMC is thought to mediate the many effects of growth hormone and is not subject to diurnal variation or acute influences of stress, sleep, exercise, or changes in circulating levels of nutrients. For these reasons, SMC may be a more sensitive index of rapid changes in nitrogen balance than albumin, transferrin, and possibly transthyretin, all of which have a longer half-life. The nutritional intake and plasma concentration of SMC appear tightly controlled by dietary uptake. An important property of SMC that makes it a good nutritional marker is the 2- to 4-hour half-life of the bound form.[140,141]

SMC is produced by many types of cells, but most circulating SMC is of hepatic origin. Although the concentration of SMC in circulation can be estimated by radioimmunoassay on nonextracted plasma or serum samples using high-affinity antibodies, the binding proteins interfere with measurements in these direct assays. In healthy subjects, the binding proteins have less effect on the calculated values if these values are calculated against a serum pool standard that has been treated in a manner similar to that of the unknown sample.

Assays to measure insulin-like growth factor-I must have the capability of performing some method to remove binding proteins that might interfere with the assay. These extraction procedures have consisted of either acid gel chromatography, acid ethanol extraction, or chromatography of samples on octadecasalicylic acid columns.[142] In some patients, binding proteins can interfere to a variable degree with the quantitation of SMC in serum. For this reason, the measurement of SMC after extraction is recommended.[143,144]

Circulating SMC levels are regulated by nutritional status. These levels are particularly altered by the protein content of the diet. SMC levels fall during protein starvation and rise rapidly upon refeeding.[145] This relationship of SMC level to malnutrition has been thoroughly investigated in undernourished children who fail to grow.[146] Low circulating levels of SMC in children rise rapidly within 1 day of refeeding. SMC levels in healthy humans respond rapidly to dietary changes in protein and calorie intake, and SMC levels are much more reliable than albumin, transferrin, and lymphocyte count in documenting a malnourished patient's replenishment of nutritional status.[146,147] SMC levels can be variable in patients with liver disease, kidney failure, and other autoimmune diseases unless the carrier proteins are completely removed during the chromatography or extraction procedure in the assay. SMC remains the most reliable nutritional marker in children (Fig. 28-8).[148]

Fibronectin

Fibronectin is an opsonic glycoprotein that has a half-life of about 15 hours in humans. This protein has been classified as a paradigm of a modular protein. It has been shown to have repeating blocks of a homogeneous sequence of amino acids. It is an alpha-2-glycoprotein that serves important roles in

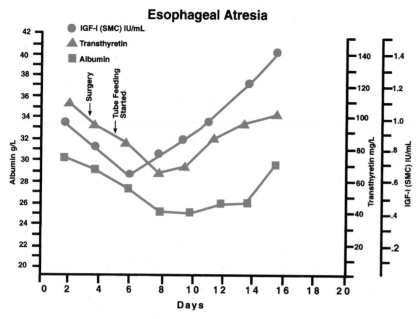

Esophageal Atresia

● IGF-I (SMC) IU/mL
▲ Transthyretin
■ Albumin

Figure 28-8. Sequential protein markers on a 4-week-old that had a surgical correction for a congenital birth defect, partial esophageal atresia. The protein markers utilized were transthyretin or prealbumin, albumin, and SMC. Note that upon feeding on day 8, SMC increased its concentration and showed nutritional repletion much sooner than did albumin or transthyretin. SMC is a very sensitive protein nutritional marker in the pediatric population.

cell-to-cell adherence and tissue differentiation, wound healing, microvascular integrity, and opsonization of particulate matter. It is considered the major protein regulating phagocytosis. Sites of synthesis include endothelial cells, peritoneal macrophages, fibroblasts and the liver. Fibronectin concentrations may decrease after physiologic damage caused by severe shock, burns, or infection. Levels return to normal on recovery. Fibronectin is useful in nutritional panels because it is one of the few nutritional markers that is not exclusively synthesized in the liver.[149] Fibronectin showed response to both caloric deprivation and repletion in healthy obese subjects who fasted 3 weeks and then were refed.[150–152]

Fibronectin levels have been shown to decrease during infection or severe stress due, in part, to its opsonic property. The infection or trauma, however, does not significantly decrease fibronectin concentration.[153] Fibronectin does not decrease during mild fever and only increases in severe infection, in which it remains flat and does not increase as soon as SMC does. There are two factors that potentially can diminish fibronectin's usefulness as a nutritional marker. First, heavy glucose composition in a feeding formula has been shown to decrease fibronectin concentration.[154] Also, bacterial infections, such as *Staphylococcus aureus,* which may cause tissue degradation at the site of the injury, can inhibit the concentration of fibronectin by reducing its ability to bind to actin, fibrin, collagen, and gelatin.[155] Fibronectin is a reliable index in monitoring short-term nutritional support in infants with protein malnutrition.[156,157]

Nitrogen Balance

Another nutritional evaluation tool, *nitrogen balance,* is the difference between nitrogen intake and nitrogen excretion. It is one of the most widely used indicators of protein change. In the healthy population, anabolic and catabolic rates are in equilibrium, and the nitrogen balance approaches zero. During stress, trauma, or burns, the nutritional intake decreases, and the nitrogen loss may exceed intake, leading to a negative nitrogen balance. During recovery from illness, the nitrogen balance should become positive with nutritional support. In humans, 90% to 95% of the daily nitrogen loss is accounted for by elimination through the kidneys. About 90% of this loss is in the form of urea. Therefore, the determination of 24-hour urinary urea nitrogen is a method for estimating the amount of nitrogen excretion. The nitrogen balance is calculated as follows:[158]

$$\text{Nitrogen balance} = \frac{\text{24-h protein intake (g)}}{6.25}$$
$$- \frac{\text{24-h UUN} + 4}{\text{Total volume (L)}}$$

(Eq. 28–1)

Protein intake includes grams of protein that are provided either by intravenous amino acids or by enteral feeding. Protein intake is converted into grams of nitrogen by dividing by 6.25. The factor of 4 in the equation represents an estimation of nonurinary nitrogen loss (*eg,* from skin, feces, hair, nails).[158] Nitrogen balance, as calculated by this equation, is not valid in patients with severe stress or sepsis, as can be seen in critical care areas or in patients with renal disease. Determining the validity of this equation in other clinical conditions involving normally high nitrogen losses may be difficult or even incorrect.

C-Reactive Protein

C-reactive protein is an acute-phase protein that increases dramatically under conditions of sepsis, inflammation, and infection. C-reactive protein can increase dramatically up to 1000 times after tissue injury, which is more than two or three orders of magnitude greater than any other acute-phase reactant. C-reactive protein rises in concentration 4 to 6 hours before other acute-phase reactants begin to rise.[159,160]

A generally accepted concept now is that major trauma or critical illness can be a direct cause of problematic malnutrition. From this knowledge has come the current medical practice of selective nutritional support. Immediately after the onset of trauma, sepsis, or critical illness, there is a breakdown of cells with the loss of intracellular ions and nitrogen.[161] Critical care physicians now characterize the injured patient with early shock as in the *ebb phase* followed by the *catabolic flow phase*. The flow phase is a period of marked catabolism that presents itself clinically with tachycardia, fever, increased respiratory rate, and increased cardiac output.[162,163] During this time, synthesis rates of C-reactive protein and other acute-phase proteins increase, and albumin and prealbumin decrease. Even with this increase in acute-phase proteins, a significant negative nitrogen balance usually occurs secondary to the greater protein catabolism. Whether this catabolic state produces a clinically defined malnutrition or a separate entity is unknown, but it certainly produces weight loss and decreases albumin and prealbumin levels. An example of a condition known as *acute-phase reaction* is shown in Figure 28-9.[164]

Interleukins

Nutrition research has recently focused on the interleukins, which are a complex group of proteins and glycoproteins that can exert pleiotropic effects on a number of different target cells. Most of the interleukins are produced by macrophages and T lymphocytes, in response to antigenic or mitogenic stimulation, and affect primary T-lymphocyte function. Most of the nutritional investigations have been performed on interleukin-1 (IL-1),[165,166] which is involved in the induction of fever, slow-wave sleep, bone resorption, and muscle pro-

teolysis. Recent research has shown that interleukins need adequate amounts of fat-soluble vitamins and pyridoxine to exert adequate biologic effects.[167,168] It has been postulated that if IL-1 and IL-2 are present in adequate amounts, the fat-soluble and B complex vitamins have an adequate concentration in the body.[169]

Total Parenteral Nutrition

Total parenteral nutrition (TPN) is a widely used means of intense nutritional support for patients who are malnourished, or in danger of becoming malnourished, because they are unable to consume required nutrients. Those requiring TPN include: (1) presurgical patients in a poor nutritional state who require repletion of nutrients to prevent or limit postsurgical complications (eg, impaired wound healing and infection); (2) patients with injuries that result in impaired mental and physical capacities (trauma or sepsis) who may require TPN to maintain adequate nutrition when basal metabolic rates are greatly increased due to a hypercatabolic state[170]; and (3) patients with chronic gastrointestinal disease (eg, cancer, short bowel syndrome, or surgical procedures, including bypass or resection) who are often candidates for long-term therapy.[171]

Parenteral nutrition therapy involves administering appropriate amounts of carbohydrate, amino acid, and lipid solutions as well as electrolytes, vitamins, minerals, and trace elements to meet the caloric, protein, and nutrient requirements while maintaining water and electrolyte balance.[172] Parenteral nutritional preparations usually are administered by enteral or parenteral routes. Enterally, a nasogastric tube is placed to deliver nutrients directly into the stomach or duodenum, or patients having gastrointestinal surgery may have a feeding gastrostomy or jejunostomy catheter put in place during the

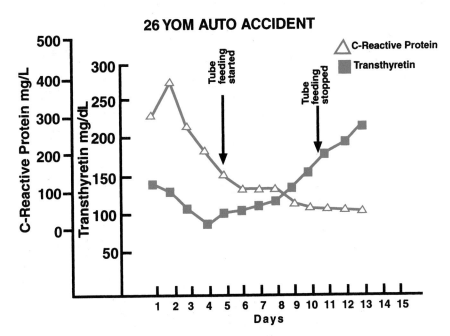

Figure 28-9. Sequential levels of C-reactive protein and transthyretin were obtained on a 26-year-old man who was involved in an automobile accident. This pattern shows an initial elevation of C-reactive protein, which is increasing because of the inflammatory response due to the automobile accident. As it begins to decrease, transthyretin, an inverse acute phase reactant, decreases. As day 5, acute phase response is over and transthyretin becomes a nutritional marker. At this time, tube feeding is started because the patient shows both biochemical and physical signs of malnutrition. Tube feeding is stopped once the transthyretin reaches a level of 180 mg/L. This illustrates the use of transthyretin in an acute phase response situation.

surgical procedure. Feeding solutions are then administered through the catheter. The parenteral route uses an intravenous catheter (most often a Hickman catheter) that is threaded through the right subclavian vein to the superior vena cava and positioned above the right atrium.[173,174] This is shown in Figure 28-10.

Because TPN administration bypasses the normal absorption and circulation routes, careful laboratory monitoring of these patients is critical. By whatever routes nutrients are administered, the goal is to provide optimal nutritional status. Often, this requires administering high levels of nutrients during short periods of time. When a patient is considered for TPN, accurate dietary and medical histories are needed to ensure adequate nutritional support. Anthropometric information is used to assess the general nutritional status of the patient. These determinations include arm circumference and triceps skinfold measurement to evaluate muscle mass and fat stores. The patient's weight is used as a general indicator of possible disease or malnutrition. An unintended weight loss of more than 10% to 12% leads to suspicion of either disease or malnutrition. The patient's height, age, and activity level are also considered.[175]

An important part of the initial TPN work-up involves acquiring laboratory measurements that provide information relating to the patient's initial protein and lipid status, electrolyte balance, and renal and liver function. From this information, a TPN solution is prepared to meet specific caloric (lipid or glucose) and nutritional (protein) needs, which may change weekly or even daily. It is important to monitor the TPN patient to avoid possible complications. Such laboratory monitoring provides necessary information to administer TPN therapy properly. Suggested guidelines for frequency of monitoring of metabolic variables are given in Table 28-8.

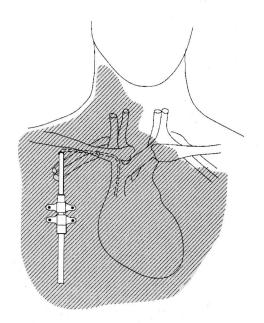

▲
Figure 28-10. Insertion path of catheter for total parenteral nutrition.

TABLE 28-8. Guidelines for Test Monitoring Frequency in Patients on TPN

Baseline Studies
CBC, glucose
Sodium, potassium, chloride, CO_2
Bilirubin, aspartate aminotransferase, alkaline phosphatase
BUN, creatinine, ammonia
Calcium (or ionized calcium), phosphorus, magnesium
Thyroid-binding prealbumin (transthyretin)
Cholesterol, triglycerides
Urinalysis, serum acid–base status
UUN
Zinc, copper, iron (optional)

Measurements Obtained Twice Weekly
Glucose, BUN, creatinine
Sodium, potassium, chloride, CO_2
Thyroid-binding prealbumin (transthyretin)
Calcium (or ionized calcium), phosphorus, magnesium
Bilirubin, aspartate aminotransferase, alkaline phosphatase

Results Obtained Every Week
CBC, urinalysis, serum acid–base status
Zinc, copper, iron, magnesium
Cholesterol, triglycerides (if infusing lipids)
UUN
Ammonia, creatinine

Measurements Obtained Every Two Weeks
Zinc, copper, iron, folate, vitamin B_{12}
In unstable clinical conditions, laboratory studies may be required several times a day

Urine Testing

Many patients receiving TPN have normal kidney function, and laboratory monitoring of urine composition provides valuable information. Routine orders normally include volume, specific gravity or osmolarity, and an estimate of glycosuria every 8 hours. Glycosuria should be minimal. In very small premature infants, glycosuria during the early phase of parenteral nutrition is a signal that glucose infusion is too rapid. If glucose appears in the urine of small infants after glucose tolerance has been established, however, the clinician should question the presence of respiratory disease, sepsis, or cardiovascular changes.[176,177]

Tests to Monitor Electrolyte Disturbances

Sodium regulation during TPN is a problem in children.[178] Daily sodium requirements may vary depending on renal maturity and the ability of the child's body to regulate sodium.[179] Factors that increase the amount of sodium necessary to maintain normal serum sodium concentrations in both children and adults are glycosuria, diuretic use, diarrhea or other excessive gastrointestinal losses, and increased postoperative fluid losses.[180]

Hyperkalemia is a common problem in children when blood is obtained by heel stick. The squeezing of the heel may cause red cell hemolysis, resulting in a falsely high serum potassium. Although adequate nutrition may be supplied to promote anabolism, hypokalemia may develop as the extra-cellular supply is used for cell synthesis.

The primary function of chloride is osmotic regulation. Hyperchloremia metabolic acidosis is a problem when crystal amino acid solutions are used, but such acidosis can be prevented or treated by altering the amount of chloride salt in the parenteral nutrition solution. Supplying some of the sodium and potassium requirements as acetate or phosphate salts can reduce the required amount of chloride.[61] The reformation of synthetic amino acid solutions by commercial manufacturers has helped to avoid this serious complication. If hyperchloremia metabolic acidosis does occur, treatment with sodium and potassium acetate solutions is used because acetate is metabolized rapidly to bicarbonate. Acetate salts are compatible with all other common parenteral nutrition components and thus are ideal to use when acidosis is present. Sodium bicarbonate cannot be used in parenteral nutrition solutions containing calcium because calcium carbonate readily precipitates. The use of acetate not only increases serum bicarbonate but also decreases the amount of chloride delivered to the patient.[181]

Mineral Tests to Monitor

One of the most important aspects of TPN monitoring is determining deficiencies and excesses of calcium, phosphorus, and magnesium. When regulated inadequately, these minerals not only affect bone mass but also can precipitate life-threatening situations. Calcium and phosphorus are related closely in the important role of bone mineralization. Calcium is present in serum in two forms—protein-bound, or non-diffusible, calcium and ionized diffusible calcium. Ionized calcium is the physiologically active form and constitutes only 25% of total serum calcium. Regardless of the total serum calcium, a decrease in ionized calcium may result in tetany.[182] A decrease in ionized calcium often is caused by an increase in blood pH (alkalosis). It is important to monitor ionized serum

calcium and blood pH, especially in a patient on TPN who is receiving calcium supplementation along with ingredients in the TPN solution that may alter blood pH.[183] Although calcium imbalance is frequent in newborns undergoing TPN, it is much less common in adolescents and adults.[180] Hypercalciuria with nephrolithiasis has been reported, however, as a complication in patients on long-term TPN.[184–186]

A reciprocal relationship exists between calcium and phosphorus. Intracellular phosphate is necessary to promote protein synthesis and other cellular functions. Calcium and phosphorus must be monitored carefully to maintain the correct balance between these two minerals. Severe hypophosphatemia has been reported in patients undergoing prolonged TPN.[187] Magnesium, as a TPN solution additive, is related closely to calcium and phosphorus. A reciprocal relationship exists between magnesium and calcium and, in certain situations, between magnesium and phosphorus. Low levels of magnesium can cause tetany, whereas high levels can increase cardiac atrioventricular conduction time.[61] Some electrolyte and mineral abnormalities associated with parenteral nutrition are shown in Table 28-9.[188,189]

Trace Elements to Monitor

The diets of most patients on TPN must be supplemented to maintain optimal levels of several trace elements; these elements also must be monitored to prevent deficiency or toxicity. Copper and zinc are the most common trace elements added to TPN solutions.[190] Pallor, decreased pigmentation, vein enlargement, and rashes resembling seborrheic dermatitis are the major clinical signs of copper deficiency, which at times go unnoticed.[191] Some other abnormalities include recurrent leukopenia (white blood cell count less than $5 \times 10^9/L$) and neutropenia (neutrophils less than $1.5 \times 10^9/L$).[192–194]

The diagnosis of copper deficiency is confirmed when both serum copper and ceruloplasmin (the copper-binding glycoprotein) are low. It is difficult to make this diagnosis in

TABLE 28-9. Electrolyte and Mineral Abnormalities Associated with TPN

Abnormality	Manifestations	Usual Causes
Hypernatremia	Edema, hypertension, thirst, intracranial hemorrhage	Inappropriate sodium intake in relation to water intake, especially with abnormal losses (diarrhea, diuretic use)
Hyponatremia	Weakness, hypotension, oliguria, tachycardia	Inadequate sodium intake relative to water intake
Hyperkalemia	Weakness, paresthesia, cardiac arrhythmias	Acidosis, renal failure, excessive potassium intake
Hypokalemia	Weakness, alkalosis, cardiac abnormalities	Insufficient potassium intake associated with protein anabolism
Hyperchloremia	Metabolic acidosis	Excessive chloride intake, amino acid solutions with high chloride content
Hypercalcemia	Renal failure, aberrant ossification	Inadequate phosphorus intake, excessive vitamin D intake
Hypocalcemia	Tetany, seizures, rickets, bone demineralization	Inadequate calcium, phosphorus, and/or vitamin D intake
Hypophosphatemia	Weakness, bone pain, bone demineralization	Insufficient phosphorus intake in relation to calcium intake
Hypomagnesemia	Seizure, neuritis	Inadequate intake of magnesium

premature infants, however, because their serum copper levels remain depressed until about 9 weeks of age. Low copper levels have also been reported in malabsorption syndrome, protein-wasting intestinal diseases, nephrotic syndrome, severe trauma, and burns.[191]

Patients on TPN may develop acute zinc deficiency.[195,196] They initially suffer from a massive urinary loss of zinc during phases of catabolism. When weight gain begins, the zinc-deficient patient may experience diarrhea, perioral dermatitis, and alopecia.[197] Premature newborns are particularly predisposed to zinc deficiency because zinc normally is acquired at the rate of about 500 mg/d during the final month of gestation. To compensate for this deficiency, zinc supplements for the premature newborn should be 50% higher than for the full-term newborn. This concentration is then decreased gradually until it is the same as that for a full-term infant.[198]

Serum zinc and copper levels should be monitored weekly, or at least bimonthly. Zinc should be monitored even more frequently in patients with ongoing gastrointestinal losses, even if they are receiving zinc supplements in their parenteral solutions.[199]

Chromium deficiency has been described in patients on long-term parenteral nutrition.[200] Initial signs and symptoms include weight loss, increased carbohydrate intolerance, and neuropathy.[201] The diagnosis is supported by low levels of serum chromium and by the clinical response to chromium administration. Evidence indicates that plasma chromium levels are reduced not only in deficiency but also in acute illnesses.[202] They are elevated by increased insulin and glucose loads.[203]

Selenium deficiency has been described in long-term TPN.[204] It has been associated with both cardiomyopathy and malabsorption. The cardiomyopathy can be severe, and death has been reported.[205,206] More work is needed to establish its significance in TPN, however.

Use of Panel Testing

To regulate the metabolic state of critically ill patients on TPN, it is necessary to monitor their nutritional therapy consistently. The constant demand for monitoring these patients produces an ideal situation for laboratory panels. By requesting these panels, a group of tests can be performed in the most cost-efficient manner. This allows the laboratory the option of test batching and automated analysis, which results in less specimen processing, more efficient chemical analysis, and therefore cost containment. Panels should consist only of routine tests that are ordered repeatedly.

Two such panels designed for intensive care in newborns and adults on TPN are shown in Table 28-10. These panels should be ordered on a weekly basis, or more frequently if the need arises.

SUMMARY

The current emphasis on nutrition and health has resulted in the increasing importance of vitamin and nutritional assessment in the clinical laboratory. Vitamins are low-molecular-

TABLE 28-10. Panel Testing for Patients Receiving TPN

Adults	Newborns
CBC	CBC
Prothrombin time	Prothrombin time
Glucose	Glucose
Electrolytes	BUN
Calcium	Electrolytes
Magnesium	Calcium
Phosphate	Magnesium
Copper	Phosphate
Zinc	Aspirate aminotransferase
Transthyretin (prealbumin)	Bilirubin (total and direct)
	Alkaline phosphatase
	Triglycerides
	Cholesterol
	Transthyretin (prealbumin)

weight compounds with a wide range of functions in biologic tissues. They must be partially obtained from food sources and in some instances from bacterial synthesis. Inadequate diet or inadequate intestinal absorption is termed a vitamin deficiency. Nutritional assessment is the evaluation of a patient's metabolic and nutritional needs. It is performed by diurnal, empirical, laboratory, and other objective measurements of overall nutritional status. Nutritional assessment has become increasingly important in medical care, especially in the treatment of critically ill patients and of chronically emaciated patients. Unfortunately, identifying patients who are at risk of developing nutritionally related complications in the course of an illness is uncommon, and there are no standards for monitoring the effectiveness of feeding plans to replace nutritional losses.

Malnutrition is a state of decreased intake of calories or micronutrients (vitamins and trace elements) resulting in a risk for impaired physiologic function associated with increased morbidity and mortality.[44–46] It has been estimated that 10 to 20 million people in the United States are malnourished. Several methods may be used to assess nutritional status. These include clinical evaluation, physical measurement (body weight and anthropometric measurements), biochemical or immunologic determinations, and sophisticated laboratory measurements. TPN is a widely used means of intense nutritional support for patients who are malnourished, or in danger of becoming malnourished, because they are unable to consume required nutrients. Parenteral nutrition therapy involves the administration of the appropriate amounts of carbohydrate, amino acid, and lipid solutions as well as electrolytes, vitamins, minerals, and trace elements. Parenteral nutrition therapy is usually administered by enteral or parenteral routes. However, because TPN administration bypasses the normal absorption and circulation routes, careful laboratory monitoring of these patients is critical.

CASE STUDY 28-5

A 66-year-old postmenopausal woman complained of severe weakness and dyspnea on exertion during the past 6 months. She also noted that her appetite decreased, with resultant weight loss. Her past medical history revealed a history of ulcers. On admission to the hospital, she was found to have normal blood pressure and pulse. Skin, conjunctiva, and mucous membranes were pale. Lungs were clear at auscultation. A grade ⅗ systolic murmur heard best at the lower left sternal border and radiating to the carotids and axilla was present. Stool guaiac examination was 2+. Nail beds were pale with no pedal edema. Laboratory data on admission are as follows:

Test	Reference Range
CBC	**CBC**
Hct: 15%	36–48%
Hb: 3.8 g/dL	12–16 g/dL
RBC: $2.79 \times 10^6/\mu L$	3.6–$5.0 \times 10^6/\mu L$
MCV: 53.8 FL	82–98 FL
MCHC: 25.3 g/dL	31–37 g/dL
WBC: $8.2 \times 10^3/\mu L$	WBC: 4.0-$11.0 \times 10^3/\mu L$
Seg: 80%	Neutrophils: 40–80%
Lymphs: 20%	Lymphs: 15–40%
Reticulocyte count: 5.5%	Reticulocyte count: 0.5–1.5%
Reticulocyte index: 0.8%	Reticulocyte index: >3

Chemistry	**Chemistry**
BUN: 15 mg/dL	BUN: 7–18 mg/dL
Glucose: 150 mg/dL	Glucose:
Electrolytes: normal	fasting 70–100 mg/dL
	non fasting 70–150 mg/dL
Bilirubin: 1.0 mg/dL total	Bilirubin: total 0.2–1 mg/dL
0.4 mg/dL direct	0–0.2 mg/dL direct

T_3 and T_4: normal

Serum Fe: 15 μg%   Serum Fe: 30–150 μg/dL

TIBC: 439 μg%   TIBC: 241–421 μg/dL

Questions

1. This patient's anemia is probably best explained by:
   a. Dietary habits
   b. Chronic blood loss
   c. Chronic intravascular hemolysis
   d. Chronic inflammatory disease
2. Clinical manifestations of her anemia may include all of the following except:
   a. Glossitis
   b. Pica
   c. Spoon nails (koilonychia)
   d. Peripheral neuropathy
3. The bone marrow iron stores are:
   a. Reduced
   b. Normal
   c. Absent
   d. Increased but present only in reticuloendothelial cells
4. If the correct treatment for this anemia is with oral ferrous sulfate, the treatment should:
   a. Be stopped as soon as the hematocrit returns to normal
   b. Be continued indefinitely, even if the cause of the deficiency has been corrected
   c. Be continued for 3–6 months after the hematocrit returns to normal to replace the body iron stores
   d. Be continued only until there is a brisk response seen in the reticulocyte index and hematocrit

REVIEW QUESTIONS

1. Match the vitamin in the left column with its appropriate category (ie, water soluble or fat soluble).
   Vitamin A ___   A. Fat soluble
   Vitamin E ___   B. Water soluble
   Folate ___
   Biotin ___
   Vitamin K ___
2. Which of the following describes the correct source, function, and deficient state of the vitamin listed?
   a. Vitamin E—plant tissues, antioxidant, osteomalacia
   b. Thiamine (B_1)—whole grains, carbohydrate metabolism, beriberi
   c. Niacin—meat, oxidation-reduction reactions, scurvy
   d. Folic acid—dairy products, myelin formation

3. Which vitamin would be affected if a patient was diagnosed with a disorder involving fat absorption?
   a. Vitamin B_{12}
   b. Ascorbic acid
   c. Thiamine
   d. Vitamin K
4. Which vitamin is a powerful antioxidant, protects the erythrocyte membrane from oxidative stress, and is found primarily in vegetable oils?
   a. Vitamin K
   b. Vitamin C
   c. Vitamin E
   d. Folic acid
5. A 70-year-old man presented to his physician with a broken arm. Laboratory work indicated an elevated prothrombin time, with all other laboratory results being

normal. The man was also taking an antibiotic for an earlier respiratory infection. Which, if any, of the following vitamins might be involved?

   a. Vitamin K
   b. Vitamin D
   c. Biotin
   d. None of the above

6. The most commonly used method for determination of vitamin B_{12} is:

   a. Chemiluminescence assay
   b. Magnetic-separation immunoassay
   c. Competitive protein-binding radioimmunoassay
   d. HPLC

7. The term describing patients who are chronically calorie malnourished and lose both adipose and muscle tissue, but who do not demonstrate a protein deficiency, is:

   a. Kwashiorkor
   b. Marasmus
   c. Debilitated
   d. None of the above

8. The percentage of hospitalized patients who may be malnourished is:

   a. 5% to 8%
   b. 10% to 15%
   c. 20% to 25%
   d. 30% to 50%

9. Which of the following nutritional markers has been found to be the most sensitive and helpful indicator of nutritional status in very ill patients?

   a. Transthyretin
   b. Transferrin
   c. Albumin
   d. Somatomedin C

10. Laboratory monitoring of the patient on TPN therapy is important to avoid possible complications. Which of the following trace elements should be monitored on a weekly basis?

   a. Copper
   b. Selenium
   c. Molybdenum
   d. Chromium

REFERENCES

1. Briggs MH, ed. Vitamins in human biology and medicine. Boca Raton, FL: CRC Press, 1981.
2. Food and Nutrition Board. Recommended dietary allowances. 10th ed. Washington, DC: National Academy of Science, 1989.
3. Calabrese EJ. Nutrition and environmental health. Vol 1. The vitamins. New York, John Wiley & Sons, 1980.
4. Pereira GR, Zucker A. Nutritional deficiencies in the neonate. Clin Perinatal 1986;13:175–189.
5. Rosenthal MJ, Goodwin JS. Cognitive effects of nutritional deficiency. Adv Nutr Res 1985;7:71–100.
6. Ensminger AH, Ensminger ME, Konlande JE, et al. Foods and nutrition encyclopedia. 2nd ed. Boca Raton, FL: CRC Press, 1994; 2257–2260.
7. Garry PJ. Vitamin A. In Labbe RF, ed. Clinics in laboratory medicine. Vol 1. Laboratory assessment of nutritional status. Philadelphia, WB Saunders, 1981.
8. Sklan D. Vitamin A in human nutrition. Prog Food Nutr Sci 1987; 11:39–55.
9. Underwood BA. Methods for assessment of vitamin A status. J Nutr 1990;120(Suppl 11):1459–1463.
10. Sokol RJ. Vitamin E and neurological deficits. Adv Pediatr 1990; 37:119–148.
11. Sokol RJ, Guggenheim MA, Henbi JE, et al. Frequency and clinical progression of the vitamin E deficiency neurologic disorder in children with prolonged neonatal cholestasis. Am J Dis Child 1985;139: 1211–1215.
12. Bieri JG, Evarts RP, Thorp S. Factors affecting the exchange of tocopherol between red cells and plasma. Am J Clin Nutr 1977;30:686.
13. Holick MF. The use and interpretation of assays for vitamin D and its metabolites. J Nutr 1990;120(Suppl 11):1464–1469.
14. Suttie JW. Role of vitamin K in the synthesis of clotting factors. In Draper HH, ed. Advances in nutritional research. Vol 1. New York: Plenum, 1977.
15. Hazell K, Baloch KH. Vitamin K deficiency in the elderly. Gerontol Clin 1970;12:10–17.
16. Wallach J. Interpretation of diagnostic tests. 6th ed. Boston: Little, Brown, 1996;465–475.
17. Flint DM, Prinsley DM. Vitamin status of the elderly. In Briggs MH, ed. Vitamins in human biology and medicine. Boca Raton, FL: CRC Press, 1981.
18. Bates CJ. Human riboflavin requirements and metabolic consequences of deficiency in men and animals. World Rev Nutr Diet 1987;50:215–266.
19. Pinto J, Raiczyk GB, Huang YP, et al. New approaches to the possible prevention of side effects of chemotherapy by nutrition. Cancer 1986;58:1911–1924.
20. Vermaak WJ, Bernard HC, Potgieter GM, et al. Vitamin B_6 and coronary artery disease: epidemiological observations and case studies. Atherosclerosis 1987;63:235–238.
21. Shuster K, Bailey LB, Madan CS: Vitamin B_6 status of low-income adolescent and adult pregnant women and the condition of their infants at birth. Am J Clin Nutr 1981;34:1731.
22. Leklem JE. Vitamin B_6: a status report. J Nutr 1990; 120(Suppl 11):1503–1507.
23. Wahlqvist ML. Effects on plasma cholesterol of nicotinic acid and its analogues. In Briggs MH, ed. Vitamins in human biology and medicine. Boca Raton, FL: CRC Press, 1981.
24. Scott JM, Weir DG. Folate composition, synthesis and function in materials. Clin Haematol 1976;5:547–568.
25. Bailey LB. Folate status assessment. J Nutr 1990;120(Suppl 11): 1508–1511.
26. Ueland PM, Refsum H, Stabler SP, et al: Total homocysteine in plasma or serum: methods and clinical applications. Clin Chem 1993;39:1764–1779.
27. Brown RD, Jun R, Hughes W, et al: Red cell folate assays: some answers to current problems with radioassay variability. Pathology 1990;22:82–87.
28. Chiron ACS. 180: Test description, folate. East Walpole, MA: Chiron Diagnostic Corporation, 1995;1–4.
29. Lee DSC, Griffiths BW: Human serum vitamin B_{12} assay method: a review. Clin Biochem 1985;18:261–264.
30. Steinkamp RC. Vitamin B_{12} and folic acid: clinical and pathophysiological considerations. In Brewster MA, Naito IIK, eds. Nutritional elements and clinical biochemistry. New York: Plenum, 1980.
31. Allen RH. Current status of serum cobalamin (vitamin B_{12}) assays, syllabus, seventh annual meeting. Clinical Radioassay Society, 1981;85–109.
32. Bailey LB. Folate status assessment. J Nutr 1990;120(Suppl 11): 1508–1511.
33. van der Weide J, Homan HC, Cozijnsen-van Rheenen E, et al. Nonisotopic binding assay for measuring vitamin B_{12} and folate in serum. Clin Chem 1992;38(5):766–768.
34. Reynoso G, Fontelo P, Konopka S, et al. Ligand assay method for serum cobalamin. Ligand Q 1980;3:34–40.
35. Roth KS. Biotin in clinical medicine: a review. Am J Clin Nutr 1981;34:1967.
36. Schwabedal PE, Pietrzik K, Wittkowski W: Pantothenic acid deficiency as a factor contributing to the development of hypertension. Cardiology 1985;71(Suppl 1):187–189.
37. Englard S, Seifter S. The biochemical functions of ascorbic acid. Annu Rev Nutr 1986;6:265–304.

38. Jacob RA. Assessment of human vitamin C status. J Nutr 1990; 120(Suppl 11):1480–1485.
39. Tietz NW. Clinical guide to laboratory tests: Ascorbic acid testing. Philadelphia: WB Saunders, 1995;643–644.
40. Rebouche CJ. Carnitine function and requirements during the life cycle. FASEB J 1992;6:3379–3386.
41. Tanphaichitr V, Leelahagul P. Carnitine metabolism and human carnitine deficiency. Nutrition 1993;9:246–254.
42. Cheeseman KH, Slater TF. An introduction to free radical biochemistry. Br Med Bull 1993;49:481–493.
43. Manson JE, Gaziano JM, Jonas MA, et al. Antioxidants and cardiovascular disease: a review. J Am Coll Nutr 1993;12:426–432.
44. Meguid MM, Mughal MM, Meguid V, et al. Risk-benefit analysis of malnutrition and preoperative nutrition support: a review. Nutr Int 1987;3:25.
45. Mullen JL, Buzby GP, Waldman MT, et al. Prediction of operative morbidity and mortality by preoperative nutritional assessment. Surg Forum 1979;30:80.
46. Mullen JL, Gertner MH, Buzby GP, et al. Implications of malnutrition in the surgical patient. Arch Surg 1979;114:121.
47. Bistrian BR, Blackburn GL, Vitale J, et al. Protein status in general medical patients. JAMA 1976;235:1567.
48. Spiekerman AM, Rudolph RA, Bernstein LH. Determination of malnutrition in hospitalized patients with the use of a group-based reference. Arch Pathol Lab Med 1993;117:184.
49. 1991 World Health Statistics Annual. Geneva, Switzerland: WHO, 1992.
50. Reilly JJ Jr, Hull SF, et al. Economic impact of malnutrition: a model system for hospitalized patients. J Parenter Enteral Nutr 1988;12:371.
51. Weinsier RL, Hunker EM, Krumdieck GL, et al. A prospective evaluation of general medical patients during the course of hospitalization. Am J Clin Nutr 1979;32:419.
52. Bistrian BR, Blackburn GL, Hollwell H, et al. Protein status of general surgical patients. JAMA 1974;230:858.
53. Reinhardt GF, Myskofski JW, Wilkins DB, et al. Incidence and mortality of hypoalbuminemic patients in hospitalized veterans. J Parenter Enteral Nutr 1980;4:357.
54. Cannon PR, Wissler RW, Woolridge RL, et al. The relationship of protein deficiency to surgical infection. Ann Surg 1944;120:514.
55. Klidjian AM, Archer TJ, Foster KG, et al. Detection of dangerous malnutrition. J Parenter Enteral Nutr 1982;6:119.
56. Bernstein LH, Shaw-Stiffel TA, Schorow M, et al. Financial implications of malnutrition. In: Labbe R, ed. Clinics in laboratory medicine. Philadelphia: WB Saunders, 1993;1.
57. Buzby GP, et al. Perioperative total parenteral nutrition in surgical patients: VA TPN cooperative study group. N Engl J Med 1991;325:525.
58. Baker JP, Detsky AS, Wesson DE, et al. Nutritional assessment: a comparison of clinical judgment and objective measurements. N Engl J Med 1982;306:969.
59. Detsky AS, Baker JP, Mendelson RA, et al. Evaluation of accuracy of nutritional assessment techniques applied to hospitalized patients: methodology and comparisons. J Parenter Enteral Nutr 1984;8:153.
60. Pettigrew RA, Charlesworth PM, Farmilo RW, et al. Assessment of nutritional depletion and immune competence: a comparison of clinical examination and objective measurements. J Parenter Enteral Nutr 1984;8:21.
61. Quinby GE, Nowak MM, Andrew BF. Parenteral nutrition in the neonate. Clin Perinatol 1975;2:59.
62. Durnin JUGA, Womersley J. Body fat assessed from total body density and its estimation from skinfold thickness: measurements on 481 men and women aged from 16 to 72 years. Br J Nutr 1974;32:77.
63. Heymsfield SB, Olafson RP, Utner MH, et al. A radiographic method of quantifying protein-calorie malnutrition. Am J Clin Nutr 1979;32:693.
64. Jeejeebhoy KN, Baker JP, Wolman SL, et al. Critical evaluation of the role of clinical assessment and body composition studies in patients with malnutrition and after total parenteral nutrition. Am J Clin Nutr 1982;35(Suppl):1117.
65. Collins JP, McCarthy ID, Hill GL. Assessment of protein nutrition in surgical patients: the value of anthropometrics. Am J Clin Nutr 1979;32:1527.
66. Studley HO. Percentage of weight loss: a basic indicator of surgical risk in patients with chronic peptic ulcer. JAMA 1936;106:458.
67. Morgan DB, Hill GL, Burkinshaw L. The assessment of weight loss from a single measurement of body weight: the problems and limitations. Am J Clin Nutr 1980;33:210.
68. Blackburn GL, Bistrian BR, Maini BS, et al. Nutritional and metabolic assessment of the hospitalized patient. J Parenter Enteral Nutr 1977;1:11.
69. Shamberger RJ. Vitamin A and retinol binding protein alterations in disease. In: Brewster MA, ed. Nutritional elements and clinical biochemistry. New York: Plenum, 1980;117.
70. Greenblatt DJ, Ransil BJ, Harmatz JS, et al. Variability of 24-hour urinary creatinine excretion by normal subjects. J Clin Pharmacol 1976;16:321.
71. Forbes GB, Bruining GJ. Urinary creatinine excretion and lean body mass. Am J Clin Nutr 1976;20:1359.
72. Johnson WC, Ulrich F, Meguid MM, et al. Role of delayed hypersensitivity in predicting postoperative morbidity and mortality. Am J Surg 1979;137:536.
73. Daly JM, Dudrick SJ, Copeland EM. Effects of protein depletion and repletion on cell mediated immunity in experimental animals. Ann Surg 1978;188:791.
74. Haffejee AA, Angorn IB. Nutritional status and the nonspecific cellular and humoral immune response in esophageal carcinoma. Ann Surg 1979;189:475.
75. Harvey KB, Moldawer LL, Bistrian BR, et al. Biological measures for the formulation of a hospital prognostic index. Am J Clin Nutr 1981;34:2013.
76. Christou NV, Meakins JL. Neutrophil function in anergic surgical patients: neutrophil adherence and chemotaxis. Ann Surg 1979; 190:557.
77. Copeland EM, MacFadyen BV, Dudrick SJ. Effect of intravenous hyperalimentation on established delayed hypersensitivity in the cancer patient. Ann Surg 1976;184:60.
78. Buzby GP, Mullen JP, Matthews DC, et al. Prognostic nutritional index in gastrointestinal surgery. Am J Surg 1980;139:160.
79. Moore FD, Oleson KH, McMurphy JD, et al. The body cell mass and its supporting environment: body composition in health and disease. Philadelphia: WB Saunders, 1963.
80. Friis-Hansen B. Body composition in growth. Pediatrics 1971; 47:264.
81. Shizgal HM. Nutritional assessment with body composition measurements. J Parenter Enteral Nutr 1987;11:42S.
82. Forse RA, Shizgal HM. The assessment of malnutrition. Surgery 1980;88:17.
83. Almond DJ, Burkinshaw L, Laughland A, et al. Potassium depletion in surgical patients: intracellular cation deficiency is independent of loss of body protein. Clin Nutr 1987;6:45.
84. Lukaski HC, Bolonchuk W, Hall CB, et al. Estimation of fat free mass in humans using the bioelectrical impedance method: a validation study. J Appl Physiol 1986;60:1327.
85. Meguid MM, Campos AC, Lukaski HC, et al. A new single cell in vitro model to determine volume and sodium concentration changes by bioelectrical impedance analysis. Nutrition 1988;4:363.
86. Segal KR, Butin B, Presta E, et al. Estimations of human body composition by bioelectrical impedance methods: a comparative study. J Appl Physiol 1985;58:1565.
87. Campos ACL, Chen M, Meguid MM. Comparisons of body composition derived from anthropomorphic and bioelectrical impedance methods. J Am Coll Nutr 1989;8:484.
88. Kushner RF, Scholler DA. Estimation of total body water by bioelectrical impedance analysis. Am J Clin Nutr 1986;44:417.
89. Klidjian AM, Foster KJ, Kammerling RM, et al. Anthropometric and dynamometric variables to serious postoperative complications. Br Med J 1980;281:899.
90. Russell DMcR, Leiter LA, Whitwell J, et al. Skeletal muscle function during hypocaloric diets and fasting: a comparison with standard nutritional assessment parameters. Am J Clin Nutr 1983;37:133.
91. Russell DMcR, Walker PM, Leiter LA, et al. Metabolic and structural changes in skeletal muscle during hypocaloric dieting. Am J Clin Nutr 1984;39:503.
92. Antonas KN, Curtas S, Meguid MM. Use of serum CPK-MM to monitor response to nutritional intervention in catabolic surgical patients. J Surg Res 1987;42:219.
93. Fraser IM, Russel DMcR, Whittaker JS, et al. Skeletal and diaphragmatic muscle function in malnourished patients with chronic obstructive lung disease [Abstract]. Am Rev Respir Dis 1984;129:A269.

94. DeLone JB, Curtas S, Jeejeebhoy JN, et al. Effect of operation and nutrient intake on muscle function and enzymes. Surg Forum 1987;37:36.

95. Meguid MM, Curtas S, Chen M, et al. Adductor pollicis muscle tests to detect and correct subclinical malnutrition in preoperative cancer patients [Abstract]. Am J Clin Nutr 1987;45:843.

96. Fischer JE. Plasma proteins as indicators of nutritional status. In Levenson SM, ed. Nutritional assessment, present status, future direction and prospects: report of the second cross conference on medical research. Columbus, OH: Ross Laboratories, 1982;25.

97. Seltzer MH, Bastidas JA, Cooper DM, et al. Instant nutritional assessment. J Parenter Enteral Nutr 1979;3:157.

98. Forse RA, Shizgal HM. Serum albumin and nutritional status. J Parenteral Enteral Nutr 1980;4:450.

99. Dahn MS, Jacobs LA, Smith S, et al. The significance of hypoalbuminemia following injury and infection. Am Surg 1985;51:340.

100. Hay RW, Whitehead RG, Spicer CC. Serum-albumin as a prognostic indicator in edematous malnutrition. Lancet 1975;ii:427.

101. Ingenbleek Y, De Visscher M, De Nayer P. Measurement of prealbumin as an index of protein-calorie malnutrition. Lancet 1972; 2:106.

102. Reeds PJ, Laditan AAO. Serum albumin and transferrin in protein-energy malnutrition. Br J Nutr 1976;36:255.

103. Anderson CF, Wochos DN. The utility of serum albumin values in the nutritional assessment of hospitalized patients. Mayo Clin Proc 1982;57:181.

104. Starker PM, Gump FE, Askanazi J, et al. Serum albumin levels as an index of nutritional support. Surgery 1982;91:194.

105. Grant JP, Custer PB, Thurlow J. Current techniques of nutritional assessment. Surg Clin North Am 1981;61:437.

106. Royle GT, Kettlewell MGW. Liver function tests in surgical infection and malnutrition. Ann Surg 1980;192:192.

107. Apelgren KN, Rombeau JL, Twomey PL, et al. Comparison of nutritional indices and outcome in critically ill patients. Crit Care Med 1982;10:305.

108. Irie S, Tavassoli M. Transferrin-mediated cellular iron uptake [Review]. Am J Med Sci 1987;293:103.

109. Klausner RD. From receptor to genes: insights from molecular iron metabolism. Clin Res 1988;36:494.

110. Ingenbleek Y, Van Den Schrieck H-G, De Nayer P, et al. Albumin, transferrin and thyroxine-binding prealbumin/retinol-binding protein (TBPA-RBP) complex in assessment of malnutrition. Clin Chem Acta 1975;63:61.

111. Roza AM, Tuitt D, Shizgal HM. Transferrin: a poor measure of nutritional status. J Parenter Enteral Nutr 1984;8:523.

112. Rainey-Macdonald CG, Holliday RL, Wells GA, et al. Validity of a two-variable nutritional index for use in selecting candidates for nutritional support. J Parenter Enteral Nutr 1983;7:15.

113. Georgieff MK, Amarnath UM, Murphy EL, et al. Serum transferrin levels in the longitudinal assessment of protein-energy status in preterm infants. J Pediatr Gastroenterol Nutr 1989;8:234.

114. Rubin J, Deraps GD, Walsh D, et al. Protein losses and tobramycin absorption in peritonitis: treated by hourly peritoneal dialysis. Am J Kidney Dis 1986;8:124.

115. Smith FR, Goodman DS, Zaklama MS, et al. Serum vitamin A, retinol-binding protein, and prealbumin concentrations in protein-calorie malnutrition. I. A functional defect in hepatic retinol release. Am J Clin Nutr 1973;26:973.

116. Gofferje H. Prealbumin and retinol-binding protein [hm3u] = m[hm3u]: highly sensitive parameters for the nutritional state in respect to protein. Med Lab 1978;5:38.

117. Smith JE, Goodman DS. Retinol-binding protein and the regulation of vitamin A transport. Federation Proc 1979;38:2504.

118. Bernstein LH, Leukhardt-Fairfield CJ, Pleban W, et al. Usefulness of data on albumin and prealbumin concentrations in determining effectiveness of nutritional support. Clin Chem 1989;35:271.

119. Large S, Neal G, Glover J, et al. The early changes in retinol-binding protein and prealbumin concentrations in plasma of protein-energy malnourished children after treatment with retinol and an improved diet. Br J Nutr 1980;43:393.

120. Spiekerman AM. Some recognized nutritional markers in liver and kidney disease: poster session at the 1989 American Association of Clinical Chemistry Meeting. San Francisco, CA, August 1989 (Beckman Brochure).

121. Vanlandingham S, Spiekerman AM, Newmark SR. Prealbumin: a parameter of visceral protein levels during albumin infusion. J Parenter Enteral Nutr 1982;6:230.

122. Georgieff MK, Sasanow SR, Pereira GR. Serum transthyretin levels and protein intake as predictors of weight gain velocity in premature infants. J Pediatr Gastroenterol Nutr 1987;6:775.

123. Giacoia GP, Watson S, West K. Rapid turnover transport proteins, plasma albumin, and growth in low birth weight infants. J Parenter Enteral Nutr 1984;8:367.

124. Moskowitz SR, Pereira G, Spitzer A, et al. Prealbumin as a biochemical marker of nutritional adequacy in premature infants. J Pediatr 1983;102:749.

125. Bourry J, Milano G, Caldani C, et al. Assessment of nutritional proteins during the parenteral nutrition of cancer patients. Ann Clin Lab Sci 1982;12:158.

126. Katz MD, Lor E, Norris K, et al. Comparison of serum prealbumin and transferrin for nutritional assessment of TPN patients: a preliminary study. Nutr Support Serv 1986;6:22.

127. Winkler MF, Pomp A, Caldwell MD, et al. Transitional feeding: the relationship between nutritional intake and plasma protein concentrations. J Am Diet Assoc 1989;89:969.

128. Leider Z. Nutritional assessment present and future. Open Forum Nutritional Support Services 1988;8:8.

129. Church JM, Hill GL. Assessing the efficacy of intravenous nutrition in general surgical patients: dynamic nutritional assessment with plasma proteins. J Parenteral Enteral Nutr 1987;11:135.

130. Tuten MB, Wogt S, Dasse F, et al. Utilization of prealbumin as a nutritional parameter. J Parenter Enteral Nutr 1985;9:709.

131. Cavarocchi NC, Au FC, Dalal FR, et al. Rapid turnover proteins as nutritional indicators. World J Surg 1986;10:468.

132. Carlson DE, Cioffi WG Jr, Mason AD Jr, et al. Evaluation of serum visceral protein levels as indicators of nitrogen balance in thermally injured patients. J Parenteral Enteral Nutr 1991;15:440.

133. Ingenbleek Y, Van Den Schrieck HG, De Nayer P, et al. The role of retinol-binding protein in protein-calorie malnutrition. Metabolism 1975;24:633.

134. Smith FR, Suskind R, Thanangkul O, et al. Plasma vitamin A, retinol-binding protein and prealbumin concentrations in protein-calorie malnutrition. III. Response to varying dietary treatments. Am J Clin Nutr 1975;28:732.

135. Spiekerman AM. Retinol binding protein and prealbumin levels in renal failure, abstracted. American Society of Parenteral and Enteral Nutrition, Sixth Clinical Congress, San Francisco, 1982;6:341.

136. Gerlach TH, Zile MH. Metabolism and secretion of retinol transport complex in acute renal failure. J Lipid Res 1991;32:515.

137. Smith FR, Goodman DS. The effects of disease of the liver, thyroid, and kidney on the transport of vitamin A in human plasma. J Clin Invest 1971;50:2426.

138. Baxter RC. The somatomedins: Insulin-like growth factors. Adv Clin Chem 1986;25:49.

139. Hall K, Tally M. The somatomedin-insulin-like growth factors. J Intern Med 1989;225:47.

140. Ballard J, Baxter R, Binoux M, et al. On the nomenclature of IgF binding proteins. Acta Endocrinol 1989;121:751.

141. Phillips LS. Nutrition, metabolism and growth. In: Daughaday WH, ed. Endocrine control of growth. New York: Elsevier, 1981;121.

142. Powell DR, Rosenfeld RG, Baker BK, et al. Serum somatomedin levels in adults with chronic renal failure: the importance of measuring insulin-like growth factor I (IGF-I) and IGF-II in acid-chromatographed uremic serum. J Clin Endocrinol Metab 1986;63:1186.

143. Chatelain PG, Van Wyk JJ, Copeland KC, et al. Effect of in vitro action of serum proteases or exposure to acid on measurable immunoreactive somatomedin-C in serum. J Clin Endocrinol Metab 1982;56:376.

144. Isley WL, Lyman B, Pemberton B. Somatomedin-C as a nutritional marker in traumatized patients. Crit Care Med 1990;18:795.

145. Clemmons DR, Underwood LE, Dickerson RN, et al. Use of plasma somatomedin-C/insulin-like growth factor I measurements to monitor the response to nutritional repletion in malnourished patients. Am J Clin Nutr 1985;41:191.

146. Unterman TG, Vazquez RM, Slas AJ, et al. Nutrition and somatomedin. XIII. Usefulness of somatomedin-C in nutritional assessment. Am J Med 1985;78:228.

147. Hawker FH, Stewart PM, Baxter RC, et al. Relationship of somatomedin-C/insulin-like growth factor I levels to conventional nutritional indices in critically ill patients. Crit Care Med 1987;15:732.

148. Kirschner BS, Sutton MM. Somatomedin-C levels in growth-impaired children and adolescents with chronic inflammatory bowel disease. Gastroenterology 1986;91:830.

149. Mosher DF. Physiology of fibronectin. Annu Rev Med 1984;35:561.

150. McKone TK, Davis AT, Dean RE. Fibronectin: a new nutritional parameter. Am Surg 1985;51:336.

151. Chadwick SJD, Sim AJW, Dudley HAF. Changes in plasma fibronectin during acute nutritional deprivation in healthy human subjects. Br J Nutr 1986;55:7.

152. Scott RL, Sohmer PR, MacDonald MG. The effect of starvation and repletion on plasma fibronectin in man. JAMA 1982;248:2025.

153. Blackburn GL, Bistrian BR, Harvey K. Indices of protein-calorie malnutrition as predictors of survival. In: Levenson SM, ed. Nutritional assessment: present status, future directions and prospects. Columbus, OH: Ross Laboratories, 1981;131.

154. Saba TM, Blumenstock FA, Shah DM, et al. Reversal of opsonic deficiency in surgical, trauma, and burn patients by infusion of purified human plasma fibronectin. Am J Med 1986;80:229.

155. Horowitz GD, Groeger JS, Legaspi A, et al. The response of fibronectin to differing parenteral caloric sources in normal man. J Parenter Enteral Nutr 1985;9:435.

156. Yoder MC, Anderson DC, Gopalakrishna GS, et al. Comparison of serum fibronectin, prealbumin, and albumin concentrations during nutritional repletion in protein-calorie malnourished infant. J Pediatr Gastroenterol Nutr 1987;6:84.

157. Sandberg LB, Owens AJ, VanReken DE, et al. Improvement in plasma protein concentrations with fibronectin treatment in severe malnutrition. Am J Clin Nutr 1990;52:651.

158. Spiekerman AM. Laboratory tests for monitoring total parenteral nutrition (TPN). Clin Chem 1987;27:1.

159. Deodhar SD. C-Reactive protein: the best laboratory indicator available for monitoring disease activity. Cleve Clin J Med 1989;56:126.

160. Hokama Y, Nakamura RM. C-Reactive protein: current status and future perspectives. J Clin Lab Anal 1987;1:15.

161. Wilmore DW, Black PR, Muhlbacher F. Injured man: trauma and sepsis. In: Winters RW, Greene M, eds. Nutritional support of the seriously ill patient. New York: Academic Press, 1983;33.

162. Cerra FB. Hypermetabolism, organ failure, and metabolic support. Surgery 1987;101:1.

163. Watters JM, Bessey PQ, Dinarello CA, et al. Both inflammatory and endocrine mediators stimulate host responses to sepsis. Arch Surg 1986;121:179.

164. Fleck A. Acute phase response: implications for nutrition and recovery. Nutrition 1988;4:109.

165. Whicher JT, Evans SW. Cytokines in disease. Clin Chem 1990; 36:1269.

166. Dinarello CA. Biology of interleukin 1. FASEB J 1988;2:108.

167. Meydani SN, Ribaya-Mercado JD, Russell RM, et al. Vitamin B-6 deficiency impairs interleukin 2 production and lymphocyte proliferation in elderly adults. Am J Clin Nutr 1991;53:1275.

168. Payette H, Rola-Pleszynski M, Ghadirian P. Nutrition factors in relation to cellular and regulatory immune variables in a free-living elderly population. Am J Clin Nutr 1990;52:927.

169. Cannon JG, Meydani SN, Fielding RA, et al. Acute phase response in exercise. II. Associations between vitamin E, cytokines, and muscle proteolysis. Am Physiol Soc 1991;260:R1235.

170. Abumrad NN, Schneider AJ, Steel DR, et al. Acquired molybdenum deficiency. Clin Res 1979;27:774A.

171. Canizaro PC. Methods of nutritional support in surgical patients. In Yarborough MF, ed. Surgical nutrition. New York: Churchill Livingstone, 1981;13.

172. Dudrick SJ. A clinical review of nutritional support of the patient. J Parenter Enteral Nutr 1979;3:444.

173. Phillips GD, Odgers CL. Parenteral nutrition: Current status and concepts. Drugs 1982;23:276.

174. Mitchel L, Serrano A, Mart RA. Nutritional support of hospitalized patients. N Engl J Med 1981;304:1147.

175. Howard L, Meguid MM. Nutritional assessment in total parenteral nutrition. Clin Lab Med 1981;1:611.

176. Zlotkin SH, Stallings VA, Pencharz PB. Total parenteral nutrition in children. Pediatr Clin North Am 1985;32:381.

177. Borresen HC. Urine electrolytes and body weight changes in the routine monitoring of total intravenous feeding. Scand J Clin Lab Invest 1979;39:591.

178. Perkins RM, Levine DL. Common fluid and electrolyte problems in pediatric intensive care unit. Pediatr Clin North Am 1980;27:567.

179. Committee on Nutrition of the American Academy of Pediatrician. Sodium intake of infants in the United States. Pediatrics 1981;68:444.

180. Kerner JA, Sunshine P. Parenteral alimentation. Semin Perinatol 1979;3:417.

181. Forlaw L. Parenteral nutrition in the critically ill child. Crit Care Quart 1981;3:1.

182. Eggert LD, Rusho WJ, MacKay MW, et al. Calcium and phosphorus compatibility in parenteral nutrition solutions for neonates. Am J Hosp Pharm 1982;39:49.

183. Crottogini AJ, Siggaard-Anderson O. Plasma ionized calcium in the critically ill on total parenteral nutrition. Scand J Clin Lab Invest 1981;41:49.

184. Adelman RD, Abern SB, Merten D. Hypercalciuria with nephrolithiasis: a complication of total parenteral nutrition. Pediatrics 1977; 59:473.

185. Miller RR, Menke JA, Mentser MI. Hypercalcemia associated with phosphate depletion in the neonate. J Pediatr 1984;105:814.

186. Tsang RC, Steichen JJ, Brown DR. Perinatal calcium homeostasis: neonatal hypocalcemia and bone mineralization. Clin Perinatol 1977;4:385.

187. Takala J, Neuvonen P, Klossner J. Hypophosphatemia in hypercatabolic patients. Acta Anaesthesiol Scand 1985;29:65.

188. Knochel JP. Hypophosphatemia. West J Med 1981;134:15.

189. Tovey SJ, Benton KGF, Lee HA. Hypophosphatemia and phosphorus requirements during intravenous nutrition. Postgrad Med 1977; 53:289.

190. Shils ME, Burke AW, Greene HL, et al. Guidelines for trace element preparation for parenteral use. JAMA 1979;241:2051.

191. Shike M. Copper in parenteral nutrition. Bull N Y Acad Med 1984;60:132.

192. Joffe G, Etzioni A, Levy J, et al. A patient with copper deficiency anemia while on prolonged intravenous feeding. Clin Pediatr 1981; 20:226.

193. Zidar BL, Shadduck RK, Zeigler Z, et al. Observations on the anemia and neutropenia of human copper deficiency. Am J Hematol 1977;3:177.

194. Heller RM, Kirchner SG, O'Neill JA, et al. Skeletal changes of copper deficiency in infants receiving prolonged total parenteral nutrition. J Pediatr 1978;92:947.

195. Gordon EF, Gordon RC, Passal DB. Zinc metabolism: basic clinical and behavioral aspects. J Pediatr 1981;99:341.

196. Jeejeebhoy KN: Zinc and chromium in parenteral nutrition. Bull N Y Acad Med 1984;60:118.

197. Van Vloten WA, Bos LP. Skin lesions in acquired zinc deficiency due to parenteral nutrition. Dermatologica 1978;156:175.

198. Sann L, Rigal D, Galy G, et al. Serum copper and zinc concentration in premature and small-for-date infants. Pediatr Res 1980;14:1040.

199. Lockitch G, Godophin W, Pendray MR, et al. Serum zinc, copper, retinol-binding protein, prealbumin and ceruloplasmin concentrations in infants receiving intravenous zinc and copper supplement. J Pediatr 1983;102:304.

200. Freund H, Atamian S, Fischer JE. Chromium deficiency during total parenteral nutrition. JAMA 1979;241:496.

201. Jeejeebhoy KN, Chu RC, Marliss EB, et al. Chromium deficiency, glucose intolerance, and neuropathy reversed by chromium supplementation in a patient receiving long-term total parenteral nutrition. Am J Clin Nutr 1977;30:531.

202. Pekarek RS, Haver EC, Rayfield EJ, et al. Relationship between serum chromium concentrations and glucose utilization in normal and infected subjects. Diabetes 1975;24:350.

203. Glinsmann WH, Feldman FJ, Metz W. Plasma chromium after glucose administration. Science 1966;152:1243.

204. Van Rij AM, Thompson CD, McKenzie JM, et al. Selenium deficiency in total parenteral nutrition. Am J Clin Nutr 1979;32:2076.

205. Fleming CR, Lie JT, McCall JT, et al. Selenium deficiency and fatal cardiomyopathy in a patient on home parenteral nutrition. Gastroenterology 1982;83:689.

206. Sjils ME, Levander OA, Alcock NW. Selenium levels in long-term TPN patients. Am J Clin Nutr 1982;35:838.

CHAPTER 29

Clinical Chemistry and the Geriatric Patient

Janet L. Duben-Engelkirk, Sharon M. Miller

Objectives

Upon completion of this chapter, the clinical laboratorian should be able to:

- *Define aging, apoptosis, atherosclerosis, free radical, geriatrics, gerontology, homeostasis, menopause, osteoporosis.*

- *Discuss the impact of geriatric patients on the clinical laboratory.*

- *Describe the current theories of aging.*

- *Appraise the physiologic changes that occur with the aging process.*

- *Identify the age-related changes in clinical chemistry analytes.*

- *Explain the problems associated with establishing reference intervals for the elderly.*

- *Describe the effects of medications on clinical chemistry results in the elderly.*

- *Discuss the effects of exercise and nutrition on chemistry results in the elderly.*

- *Correlate age-related physiologic changes and laboratory results with pathologic conditions.*

KEY TERMS

Aging	Free radical	Homeostasis
Apoptosis	Geriatrics	Menopause
Atherosclerosis	Gerontology	Osteoporosis

Aging is a complex process that is not very well understood. There is no universally accepted definition of aging. One acceptable definition of *aging* is, "a progressive unfavorable loss of adaptation, leading to increased vulnerability, decreased viability, and decreased life expectancy."[1] How is this "progressive unfavorable loss of adaptation" reflected in clinical laboratory results, specifically clinical chemistry results? This chapter focuses on the clinical chemistry laboratory and the biochemical and physiologic processes of aging.

THE IMPACT OF GERIATRIC PATIENTS ON THE CLINICAL LABORATORY

In addition to the definition of aging, the reader should be familiar with the terms gerontology and geriatrics. *Gerontology* is the study of the aging process, whereas *geriatrics* is the branch of general medicine dealing with physiologic, psychological, economic, and sociologic problems of the elderly. Although there is no specific physiologic basis (*ie*, puberty or menopause) for the distinction in our society, the terms *elderly people, seniors,* and *geriatric patients* are generally considered to include anyone over the age of 65 years. For the purpose of this chapter, information primarily relates to individuals 65 years of age and older, except where indicated.

Because of improved medical care, better nutrition, and an emphasis on exercise, an ever-growing number of people are living to 65 years of age and older. Thus, the overall percentage of elderly people in the U.S. population continues to increase. Among the elderly, it is the "oldest-old," (those 85 years of age and older) that are the fastest growing age group. According to a U.S. Census Bureau report, the number of people 65 years of age and over increased by a factor of 11 between 1900 and 1994, from 3.1 million to 33.2 million. In 1994, 1 in 8 Americans was elderly, but the projection for the year 2030 is that 1 in 5 Americans will be elderly.[2] Figure 29-1 shows the increase in the percentage of the U.S. population aged 65 years and older from 1900 to 2040.

The increase in the percentage of elderly people presents a major challenge to the nation's health care and social systems as well as to the clinical laboratory. Clinical laboratorians must familiarize themselves with problems unique to or especially common in the geriatric population. They must become aware of special considerations regarding the collection of blood samples, development of reference intervals, effect of medications on chemistry results, and diagnosis of diseases in the elderly. Most important, however, they must thoroughly understand the effects of aging on laboratory values.

THEORIES OF AGING

To understand the biochemical and physiologic changes that occur during the aging process, it is important to first be familiar with the theories of aging. Although several theories have been proposed, many experts now believe that aging is not the result of a single process but rather of many phenomena occurring in concert. Current theories (Table 29-1)

include (1) random genetic damage, (2) glycation, (3) developmental processes involving the immune and neuroendocrine systems, (4) genetic programming, and (5) free radical damage.[3,4]

The first theory, involving damage or alteration of genetic materials, relates to the idea that mutagens, background radiation (ultraviolet), or both can cause chromosome or DNA damage.[3] This damage can affect all cells, and because the damage is cumulative, critical cell functions fail, and the cells die. Supporting this theory is the fact that experiments with laboratory animals exposed to radiation have shown a shorter life span.[3,5]

A second theory involving genetic materials, the "error catastrophe" theory, deals with random errors in chromosomal translation or transcription that lead to genetic abnormalities and ultimate death of the cell.[3] This theory is one of the least supported theories of aging.[3]

The glycation theory contends that the nonenzymatic interaction of glucose with numerous proteins forms glycated end products and cross-linked protein molecules. These modified proteins may eventually accumulate and interfere with both cell structure and function. This process may then result in a variety of problems characteristic of the elderly, (eg, stiffening or loss of flexibility).[3,6]

The developmental theories of aging involve the immune and neuroendocrine systems. It has been observed that immune system capability declines with age.[3] In particular, the thymus atrophies during the aging process, resulting in a reduction of the T-cell population. Because B-cell function is dependent on T cells, there is also a loss of B cells and a re-

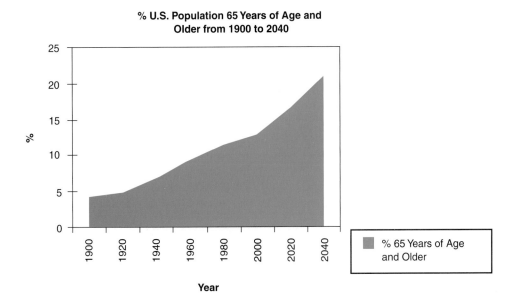

▲
Figure 29-1. Percent of U.S. population 65 years of age and older from 1900 to 2040. Numbers for 2000, 2020 and 2040 are projections. (Source: U.S. Bureau of the Census. Current population reports, special studies, P23-190, 65+ in the United States. Washington, DC: U.S. Government Printing Office; 1996.)

TABLE 29-1. Some Current Theories of Aging

- Random genetic damage
  Mutagen or background radiation damage
  Errors in chromosomal translation or transcription
- Glycation of proteins
- Developmental
  Immune system decline
  Neuroendocrine
- Genetically programmed
  Pre-programmed cell death (apoptosis)
- Free radical (eg, OH·, O$_2^-$) damage

Adapted from Knight JA. Laboratory medicine and the aging process. Chicago: American Society of Clinical Pathologists, 1996.

sultant decreased response to new antigens. This may be one reason that older people are more prone to various infectious diseases and disorders (*eg,* autoimmune disorders, lymphocytic leukemia, and cancer).[3,6,7]

The aging theory involving the neuroendocrine system is also well documented and has been around for a while. This theory focuses on the hypothalamic-pituitary system and its target glands. The most notable changes involving decline in endocrine function occur in postmenopausal women and include a loss of estrogen and bone calcium. In men, plasma testosterone levels decrease with age.[1] Other endocrine-related changes are discussed later in this chapter. Whether neuroendocrine decline is a primary reason for aging or a secondary outcome is not clear at this time.[3]

The genetically programmed theory of aging suggests that genes play a role in the aging process and that everyone is "programmed" by their genes to live a certain number of years.[3] Genes carry instructions not only for growth and development but also for cell destruction, causing decline of the body and its ultimate death. Preprogrammed cell death is called *apoptosis* from the Greek term for "dropping out." Support for this theory comes from general observations of life span in families and various aging syndromes (*eg,* progeria, Werner's syndrome, and Down's syndrome).[3]

The basis of the free radical theory is that oxygen free radicals cause progressive, random damage to cellular components. A *free radical* is an atom or molecule with one or more unpaired electrons. Thus, free radicals have an odd number of electrons, resulting in an open bond, or a half bond, making them highly reactive. A free radical may be represented by a superscript dot which signifies the unpaired electron, for example, $H_2O = HO· + H·$. The hydroxyl ion ($HO·$) is a highly reactive free radical and perhaps one of the most damaging to cells. Free radicals are also electrophilic and attack sites of increased electron density (*eg,* DNA, RNA, proteins, membranes). Eventually, free radical damage to cellular components causes death of the cell.[3,5,8]

Superoxide (O_2^-) is another free radical that is generated in the body by several reactions, (*eg,* oxidative phosphorylation and cytoplasmic reactions). Fortunately, the body has a

way of handling most of these free radicals. For example, superoxide dismutase, an enzyme present in all body cells, converts superoxide to hydrogen peroxide. Other enzymes (*eg,* glutathione peroxidase and catalase) then inactivate the hydrogen peroxide. Hydroxyl radicals are neutralized by such nonenzymatic compounds as vitamins C and E and the provitamin A, beta-carotene (the antioxidants).[3,6,8] Recently, there has been much support for and research into this particular theory of aging.[8]

In summary, although many theories of aging have been proposed, no one mechanism has been fully borne out by research. In fact, in studies of multiple species, the only intervention known to delay aging is caloric restriction. In rodents, for example, caloric restriction increased average life expectancy and maximum life span and delayed the onset of some typical age-associated diseases as well as the deterioration of physiologic processes (*eg,* immune system responsiveness and glucose metabolism).[4] The reasons for these effects appears to be related solely to caloric restriction and *not* to the reduction of any one dietary factor (*eg,* fat intake) or dietary supplement such as vitamins or antioxidants. Unfortunately, the impact of caloric restriction on aging in humans is still not known.[4]

BIOCHEMICAL AND PHYSIOLOGIC CHANGES OF AGING

As previously stated, although many theories of aging have been proposed, the aging process is still not well understood. In fact, the theories overlap and are not mutually exclusive. What is understood is that as an individual ages, many biochemical and physiologic changes occur. In general, aging is associated with a decreasing efficiency in maintenance of *homeostasis* (ie, a state of equilibrium). As long as the aging systems are not subjected to excessive physiologic stress, they may continue to function adequately. Generally, however, the body's ability to cope successfully with stress decreases with advancing age. Individuals differ greatly in how well and for how long they can continue to cope with internal and external changes. The extent and rate at which the ability to restore homeostasis declines depends on many factors (*eg,* heredity, life-style, nutrition); thus, it becomes difficult to generalize about the complex process of aging. There is a decrease in total body water, a reduction in muscle mass, an increase in lipids (*eg,* cholesterol, high-density lipoprotein [HDL] cholesterol, and triglycerides), and a gradual decline in respiratory, cardiovascular, kidney, liver, gastrointestinal, immune, neurologic, and endocrine system functions.[1,4,6,9,10]

Aging is also typically associated with the development of several diseases and disorders.[4,11] Table 29-2 shows some of the more common diseases and disorders associated with the aging process. The leading causes of death in people 65 years of age and over are shown in Table 29-3.

The specific biochemical and physiologic changes of the aging process, as they relate to clinical chemistry tests, are discussed in the following sections. Table 29-4 summarizes

TABLE 29-2. Diseases and Disorders Commonly Associated with Aging[3,11]

- Atherosclerosis (*eg*, myocardial infarct, renal disease, stroke)
- Cancer
- Diabetes mellitus
- Hyperparathyroidism
- Hyperthyroidism
- Hypothyroidism
- Monoclonal gammopathies (*eg*, multiple myeloma)
- Osteoporosis

TABLE 29-4. Changes in Selected Clinical Chemistry Analytes With Age[1,3,4,6,18,19,20,25,28]

Increases
  GGT
  Alkaline phosphatase, women only
  Alpha-1-antitrypsin
  Amylase
  AST
  BUN
  Creatine kinase, slight
  Gamma-globulin, slight
  Glucose, fasting
  (HDL)
  Inorganic phosphate
  Lactate dehydrogenase (LD)
  pCO_2
  Potassium, slight
  Total cholesterol
  Triglycerides
  TSH, slight
  Uric acid

Decreases
  Albumin
  Aldosterone
  Bilirubin
  Creatinine clearance
  DHEA
  Growth hormone
  pO_2
  T_3
  Total protein
  Transferrin

Unchanged
  Chloride
  Cortisol
  Free T_4
  Haptoglobin
  Insulin, fasting
  pCO_2, or slight increase
  pH, or slight decrease
  Sodium
  T_4, or slight decrease
  Thyroid-binding globulin (TBG)

Note: Changes in analytes are those generally cited in the literature. Some variability in the results of studies on aging and laboratory results may exist (ie, one author may report no significant change for an analyte, whereas another may report a slight decrease, or increase, for the same analyte).

changes in chemistry analytes that are associated with the aging process. Changes in analytes are those generally cited in the literature.[4,10] Some variability in the results of studies on aging and laboratory values may exist (*ie*, one author may report no significant change for an analyte, whereas another may report a slight decrease for the same analyte). Continued research is needed to determine which laboratory values are a result of "true aging" and which are associated with an underlying disease or disorder. Whenever possible, laboratorians should examine the possibility of establishing age-adjusted reference intervals based on analyte values determined for healthy older adults. Until there is sufficient research to differentiate between age-related laboratory values and those associated with underlying disease, the most medically useful data are likely to be provided by analysis of a person's previous test values collected over their adult life.

Endocrine Function Changes

It has long been known that endocrine-related abnormalities are common in the elderly and tend to increase in frequency during the aging process. Not only are there the obvious changes in the production of hormones by the sex organs, but there are also changes in thyroid, pituitary, and adrenal func-

TABLE 29-3. Top 10 Leading Causes of Death in People 65 Years of Age and Older

1. Diseases of the heart
2. Malignant neoplasms
3. Cerebrovascular diseases
4. Chronic obstructive pulmonary diseases and related conditions
5. Pneumonia and influenza
6. Diabetes mellitus
7. Accidents and adverse effects
8. Nephritis, nephrotic syndrome, and nephrosis
9. Alzheimer's disease
10. Septicemia

From: Monthly Vital Statistics Report, NCHS, Vol. 46, No. 1 (S2), Birth and Deaths: United States, 1996.

tion. The most notable changes relate to the gonadal and thyroid hormones.

A variety of significant and complex hormonal changes relating to gonadal function occur in both men and women. In men, the rate of testosterone production declines gradually with age. Between the ages of 60 and 80 years, the ratio of testosterone to estradiol falls from 12:1 to 2:1.[1,4] The decrease in testosterone is linked primarily to diminished testicular function.[1,12]

In women, endocrine system changes are primarily related to menopause, when there is a cessation of ovarian estrogen production. *Menopause* is the permanent cessation of menstruation caused by a decline in ovarian follicular activity; it generally occurs between the ages of 35 and 58 years. The major consequences of estrogen deficiency are osteoporosis and coronary heart disease (CHD).[13,14]

Osteoporosis is a problem in both older men and women, but particularly in women after menopause (at about 50 years of age). It has been estimated that about 30% of elderly women and 20% of elderly men have osteoporosis.[14,15] Osteoporosis involves a gradual loss of bone mass. The skeleton becomes weak and less dense, most likely due to increased bone resorption and decreased bone formation.[15,16] This gradual loss of skeletal mass may then lead to pain, skeletal deformity, and fractures. Common clinical chemistry values in osteoporosis are generally normal, including serum calcium, phosphorus, magnesium, alkaline phosphatase, and the parathyroid and thyroid hormones.[15,16] After a diagnosis of osteoporosis has been made, laboratory tests to assess bone metabolism (turnover) are helpful in following its progress in response to therapy. Postmenopausal Caucasian women with a life history of a calcium deficiency are most likely to be affected by osteoporosis. Major risk factors include diet, inactive life-style, genetic predisposition, smoking, endocrine disturbances, and medications.[17] Osteoporosis, then, is a significant problem among the elderly, resulting in increased morbidity and health care costs.

The thyroid gland is an important endocrine organ, participating in body development and the regulation of metabolic processes (*ie,* regulation of metabolic rate). Although there is little evidence of changes in thyroid function in the elderly, the incidence of both hypothyroidism and hyperthyroidism increases.[1,3,4] It has been reported that the prevalence of hypothyroidism in the elderly is between 0.5% and 4.4%, whereas the prevalence of hyperthyroidism is between 0.5% and 3%.[18]

Hypothyroidism, although more common in the elderly, is often more difficult to diagnose.[19] Typically, the signs and symptoms of hypothyroidism may be easily misinterpreted as just "old age."[3] In subclinical hypothyroidism, for exam-

ple, patients have few or no clinical symptoms, but have a normal thyroxine (T_4) level with an elevated thyroid-stimulating hormone (TSH) level.[19] Whether these patients should be treated with thyroid hormone or followed with periodic thyroid tests is controversial.

Interpretation of thyroid function in hospitalized and very ill patients is also difficult because of the effect of non–thyroid-related illness on common thyroid function tests.[19] Illness can depress the serum concentrations of both triiodothyronine (T_3) and T_4, whereas TSH remains normal or is even decreased. Alterations in protein binding, thyroid hormone metabolism, and suppression of the pituitary release of TSH may account for these findings in non–thyroid-related illness. Generally, these patients are considered to be euthyroid (*ie,* have normal thyroid function), and no hormone supplementation is prescribed.[19].

Many early studies attributed changes in thyroid function to the natural aging process. More recent studies, however, indicate that abnormal thyroid function is probably secondary to some underlying or associated disorder and not just old age.[3]

Because thyroid disorders may present subtly and are often difficult to diagnose, laboratory evaluation becomes very important. In healthy elderly people, there is essentially no change in T_4, free T_4, thyroid-binding globulin, and reverse T_3 levels.[6,18,19] However, some studies have shown a significant decrease in T_3 levels after 50 years of age, with a slight increase in TSH levels (perhaps as a normal response to the low T_3 level). It is still unclear whether these changes are age related or are due to some underlying disease.[6,18,19]

There are few morphologic changes of the pancreas in the elderly, besides some degree of atrophy and an increased incidence of tumors.[20] The major change associated with the pancreas is related to glucose metabolism. Glucose tolerance declines with age, with elderly people having a slightly higher fasting serum glucose level than younger adults. Fasting glucose increases about 1 to 2 mg/dL (0.11 mmol/L) per decade throughout life.[6,20] The renal threshold (the point at which glucose spills in the urine) likewise increases with age. There

---

CASE STUDY 29-1

A 65-year-old woman hospitalized for pneumonia and uncontrolled diabetes had the following thyroid test results:

CASE STUDY TABLE 29-1.1. Laboratory Results

Test	Result	Reference Range
Serum TSH	1.5 µU/mL	0.5–5 µU/mL
Total T_4	3.8 µg/dL	4.5–12 µg/dL
Total T_3	55 ng/dL	60–220 ng/dL

Questions

1. Based on the patient's status and the laboratory test results, how might the thyroid data be explained?
2. Should any additional testing be done?

is also an altered insulin response to glucose.[1,3] Diabetes mellitus is a common problem in the elderly, with a prevalence of about 18.4% in those 65 years of age and older.[20,21]

Other aspects of endocrine function, such as the hypothalamus–anterior pituitary system and adrenal glands, exhibit less dramatic change. There is essentially no change in cortisol production.[22,23] However, aldosterone and dehydroepiandrosterone (DHEA) both decline with age.[22,23] The prevalence of hypertension also increases with age, with about 60% of people older than 60 years having the condition.[24] Causes include increased peripheral resistance due to atherosclerosis, chronic renal and endocrine disorders, and multiple medications. In general, there is also a decline in the efficiency of homeostatic regulation.[4,23]

Renal Function Changes

All aspects of renal function are affected by the aging process; it is age of onset, specific changes, and consequences that vary in elderly people.[25] Renal function begins to decline after the age of 30 years and by 60 years of age is further reduced to half.[25] This decline has been attributed to the gradual loss of nephrons, decreased enzymatic and metabolic activity of tubular cells, and increased incidence of pathologic processes (eg, atherosclerosis).[25]

At any age, renal function may be assessed by a number of clinical tests, including urine volume, analysis of constituents and concentration, blood urea nitrogen (BUN), uric acid, and several clearance tests.[25] Creatinine clearance, glomerular filtration rate (GFR), and renal plasma flow all decrease with age.[6,25] The analytes, BUN, uric acid, and inorganic phosphate that reflect GFR are found to increase.[6]

In general, elderly people have a decreased capacity to conserve water through the kidneys and a significantly lower sensation of thirst.[26] This, of course, can lead to dehydration, a common and underappreciated problem in the elderly. Not only is dehydration a common finding in the elderly, it also can be a very serious one, leading to increased mortality rates.[3] Clinical laboratory tests indicative of dehydration include hypernatremia, increased BUN/creatinine ratio, increased serum osmolality, and increased urine specific gravity.[3]

In addition to the previously mentioned renal changes of aging, there is an increased incidence of renal disease and a reduced ability to handle the excretion of drugs. Problems affecting renal function are related to damage from infections or drugs (medications), hypertension, or disorders such as diabetes mellitus, tuberculosis, and nephritis.[25]

Hepatic Function Changes

In general, atrophy and decreased liver weight, as well as a decline in liver function, are common in the elderly.[3,27] Although the liver performs many functions (see Chap. 17, *Liver Function*), three of importance (synthetic, excretory and secretory, and detoxification and drug metabolism) are addressed in this section in relation to the aging process.

The synthetic function of the liver can be monitored by means of concentrations of plasma proteins. Tietz and colleagues reported a slight decrease of total protein in "fit" aging people.[22] Albumin and transferrin show a decline as well. Gamma-globulin and alpha-1-antitrypsin, however, increase slightly with age, whereas haptoglobin remains essentially the same as in young adults.[22]

Certain enzymes that are part of the liver's excretory and detoxification function also change. Alkaline phosphatase (ALP) and lactate dehydrogenase (LD), for example, increase in both men and women. In addition, urea synthesis, a process that occurs in the hepatocytes, and bilirubin metabolism both decline with age.[3,22]

The liver's detoxification or drug metabolism function is important when considering the increased numbers of medications that are usually prescribed to elderly patients. Even though people 65 years of age and older compose about 12% of the U.S. population, they receive about one third of all prescribed medications.[3] In addition to possible drug interactions, drug toxicity is a potential problem. As mentioned previously, there is some atrophy of the liver as well as reduced hepatic blood flow with the aging process. This in

---

CASE STUDY 29-2

An otherwise healthy 70-year-old man entered the hospital for abdominal surgery. Preoperative chemistry test results were as follows:

CASE STUDY TABLE 29-2.1. Laboratory Results

Test	Result	Reference Range
Albumin	53 g/L	35–50 g/L
BUN	40 mg/dL	8–26 mg/dL
Creatinine	1.6 mg/dL	0.9–1.5 mg/dL
Serum osmolality	330 mOsm/kg	275–295 mOsm/kg
Sodium	150 mmol/L	135–145 mmol/L

Questions

1. What is the BUN/creatinine ratio for this patient?
2. What do these data suggest?
3. Which test results support this conclusion?
4. How common is this condition in the elderly?

turn may lead to the accumulation of some drugs (eg, lidocaine and morphine) and the possibility of toxicity.[3] Medications and the elderly are discussed in more detail later.

Pulmonary Function and Electrolyte Changes

A number of known anatomic and physiologic changes occur in cardiopulmonary function during the aging process. Pulmonary function actually begins to decline after the age of 25 years owing to changes in the lung, thoracic cage, respiratory muscles, and respiratory centers in the central nervous system.[28] Of course, pulmonary function is perhaps more affected by personal habits, such as smoking, and by environmental pollution.[28]

The more frequently described changes relating to pulmonary function in the elderly include the pO_2 and pCO_2 values. These changes reflect the decreased vital (lung) capacity found in most elderly people.[28] Arterial pO_2, for example, decreases during the aging process, whereas pCO_2 is reported to increase slightly or remain the same.[4,28] The blood pH value is reported to remain fairly constant or to decrease only slightly.[6,22,28]

The electrolytes sodium, potassium, and chloride all show little change in healthy elderly people from values seen in younger adults.[22] Sodium remains fairly constant from young adulthood to older age. Chloride values are also fairly constant but have been found to be slightly higher in people older than 90 years of age. Potassium, however, increases slightly from 60 to 90 years of age.[22]

Respiratory-related diseases are prevalent in elderly people and account for 25% of all deaths in those over 85 years of age.[28] Respiratory diseases of the elderly include chronic bronchitis, emphysema, neoplasia, and lung infections, particularly tuberculosis and pneumonia.[28]

Cardiovascular and Lipid Changes

Cardiovascular disease continues to be an important cause of death in old age in both men and women. Atherosclerosis, a type of arteriosclerosis, is the major cause of death from cardiovascular disease in the United States.[29] *Atherosclerosis,* a progressive disease process that begins early in life, involves vascular alterations characterized by fatty accumulations in the vascular walls.[3,29] Atherosclerosis develops slowly over the years. It gradually destroys the arteries, thereby preventing the exchange of gases and nutrients necessary for the body's normal functioning. Some of the results of atherosclerosis include hypertension, hemorrhage, thrombosis, stroke, and CHD.[29]

CHD is a consequence of atherosclerosis, involving the coronary arteries. Also called ischemic heart disease, it continues to be a major cause of disability and death in the United States, its prevalence increasing with advancing age.[29] Some of the risk factors for CHD include age, gender, genetic predisposition, obesity, hypertension, poor fitness, diabetes mellitus, cigarette smoking, and hyperlipidemia.[3,29]

Lipids shown to play a major role in the atherosclerotic process and risk for CHD are HDL and low-density lipoprotein (LDL) cholesterol, total cholesterol, and triglycerides. In a study by Tietz and colleagues of "fit" elderly, total cholesterol, HDL cholesterol and triglycerides were found to increase as a part of the aging process.[22] HDL cholesterol, or "good" cholesterol, however, is considered as an important *inverse* risk factor for CHD, with values less than about 35 mg/dL indicating a high risk and values more than 35 mg/dL indicating a low risk for CHD.[3]

Enzyme Changes

Enzymes are specific proteins that catalyze various chemical reactions in the body. Changes in enzyme levels during the aging process have been studied extensively; they are varied and complex. Enzyme concentration and synthesis are under genetic control and are affected by hormones, substrates, and other factors. In general, enzymes do not appear to follow any particular age-related pattern; they may increase, decrease, or remain the same during the aging process. However, the ability to initiate adaptive changes in the activity of enzymes has been found to be impaired with increasing age.[30]

Enzymes reported to change in healthy elderly people include aspartate aminotransferase (AST), alanine aminotransferase (ALT), ALP, gamma-glutamyltransferase (GGT), creatine kinase (CK), lactate dehydrogenase, and amylase.[22] AST, GGT, LD, and amylase show increases in both men and women. ALT levels show only a marginal increase in men, whereas values in women show no change. ALP, on the other hand, shows a marked increase in women, whereas men show no increase until 90 years of age. CK values in men increase slightly between 60 and 69 years of age. Between ages 70 and 90 years, however, the values decrease. In women, CK values also increase slightly from 60 to 70 years of age but decrease in those older than 70 years of age. Lipase increases only slightly, if at all, in those 60 to 90 years old, but definitely increases in those over 90 years of age.[22]

CLINICAL CHEMISTRY RESULTS AND AGING

In addition to a knowledge of the basic biochemical and physiologic changes of aging, clinical laboratorians must be aware of other factors that may affect clinical chemistry results in the elderly. For example, does the aging process affect laboratory test results sufficiently to warrant separate reference intervals for the elderly? What preanalytical variables (eg, diet, posture medications) relating to the elderly affect clinical chemistry results? How does the aging process affect interpretation of drug levels in the elderly? What are the effects of exercise and nutrition on the elderly and chemistry results? Table 29-5 shows

TABLE 29-5. Some Factors to Consider in the Interpretation of Clinical Laboratory Results for the Elderly[3,22,33]

- Exercise
    - Duration
    - Type
- Medications
    - Polypharmacy
- Mobility
    - Immobility
    - Posture
- Nutritional status
- Personal habits
    - Alcohol use
    - Smoking
- Presence of multiple chronic or subclinical disorders
- Reference interval validity
- Specimen collection variables
    - Site
    - Trauma
    - Volume

some of the factors laboratorians should consider when interpreting laboratory values for the elderly.

Establishing Reference Intervals for the Elderly

Interpretation of test results and reference intervals for the elderly can be confusing and complex. In addition, there appears to be some confusion on the topic of reference intervals for the elderly among experts in the field of clinical chemistry and aging.

As discussed in previous sections, many reference intervals for analytes vary from the reference intervals of younger adults. Some of these, like the levels of male and female hormones, are the undisputed result of aging organs. With other analytes, however, the case may not be so clear. Some variance in analyte levels may be due to various secondary conditions (eg, subclinical diseases, drugs, inactivity, nutrition) and not the aging process alone; for example, the relationship between non–insulin-dependent diabetes mellitus and a rise in serum glucose levels.

Frequently, abnormal test results in the elderly are interpreted as "normal" for the individual's age and not a sign of a disorder or disease.[11] Research is now showing, however, that in many instances, abnormal laboratory results in so-called healthy individuals, may, in fact, be associated with unrecognized subclinical disorders or with other secondary conditions.[3,11] Adding to the problem is the fact that reference intervals are not easily determined, even for younger healthy adult populations. Frequently, reference intervals are poorly defined and not always determined by a uniform process.[31] The establishment of reference intervals involves a well-defined protocol including careful selection of reference individuals, control of any preanalytical factors, a comprehensive list of analytical interferences, careful and consistent collection of specimens for a given analyte, analysis of specimens under well-defined conditions, and identification of any data errors.[31] It is apparent, then, that determination of reference intervals for the elderly can be problematic because a large percentage of elderly have some subclinical or obvious pathologic abnormality. The National Committee for Clinical Laboratory Standards (NCCLS) publishes an important guideline, C28-A, dealing with the determination of valid reference intervals for quantitative clinical laboratory tests.[32]

Although the aging process does affect some selected analytes, the relationship between aging and changes in other analytes is less well understood. Generally, clinicians and laboratorians agree that separate reference intervals for the elderly are needed to prevent false abnormal results. Knight recommends that until we have a better understanding of the relationships among true aging, age-associated disorders, and various laboratory results, clinicians and laboratorians should maintain the "normal" reference intervals established for healthy younger adults.[11] As mentioned previously, however, until there is a better understanding of what analyte changes are the consequence of age-related phenomenon and *not* disease, it may be more useful to compare an elderly person's current values with values obtained throughout their adult life. In any case, clinicians and clinical laboratory scientists should keep in mind all factors affecting the interpretation of laboratory test results for the elderly.

Preanalytical Variables, the Elderly, and Chemistry Results

Many preanalytical variables may affect the interpretation of chemistry results and the establishment of reference intervals in all age groups. However, preanalytical variables may have a much greater affect on the elderly. Preanalytical variables relating to the patient include such factors as diet, gender, posture (ie, sitting or lying), personal habits (eg, smoking and alcohol consumption), body composition, physical activity, and prescribed medications.[33] Any of these may affect the concentration of various analytes. For example, body composition changes with age. In general, in healthy people, there is an increase in body fat and a decrease in lean muscle mass with age. Body mass and height also decrease after about 60 years of age.[33] These changes in turn may affect the levels of various analytes (eg, creatinine). The effects of medications, diet, and exercise are discussed in the next sections.

Other preanalytical variables affecting laboratory results for the elderly involve the collection of specimens for analysis. A number of physical and physiologic changes of aging can affect the collection and quality of a specimen and make phlebotomy a challenge to the phlebotomist. In many instances, elderly patients suffer from diseases such as arthritis, malnutrition, or dehydration. These conditions, in addition to the aging process itself, can cause a decrease in muscle tone and skin elasticity (ie, "flabby" skin). Veins may be difficult to find or inappropriate to use. This, in turn, may make phlebotomy difficult, causing hemolysis and an increased likelihood of an elevated plasma hemoglobin and lactate dehydrogenase level.[33] In addition, veins of the elderly may provide poor blood flow, resulting in a less-than-adequate sample for some laboratory tests.[34]

Therapeutic Drug Monitoring in the Elderly

Therapeutic drug monitoring is much more critical in the elderly than in the younger adult population. Elderly people are far more likely to be taking medications (and often, several medications together) than any other age group. And, unfortunately, overdosing and adverse drug reactions (from multiple medications) are problematic in the elderly.

With the normal process of aging, there are many changes in how the body handles drugs. The absorption, distribution, metabolism, and excretion of a drug are all affected by the aging process in some way. (See Chapter 25, *Therapeutic Drug Monitoring*, for more information on the pharmacokinetics of drugs.) For example, gastric emptying time may be prolonged in the elderly, thereby causing a delay in the absorption of drugs.[35,36] Absorption by intramuscular injections may also be impaired because of decreased blood flow in the elderly.[35] Metabolism or biotransformation of drugs, which is handled primarily by the liver, may be impaired as well due to an age-related decrease in hepatic mass and blood flow.[35]

The most significant pharmacokinetic change in the elderly is related to the elimination of drugs through the kidneys. Renal mass and blood flow both decrease with advancing age.[35] Glomerular filtration rate (as measured by creatinine clearance) also declines linearly with age.[4,35] This means that for drugs principally excreted in the urine, caution should be exercised in treatment to prevent any overdosing or toxic side effects.[36]

Psychosocial factors, such as depression, dementia, poverty, and loneliness, common among the elderly, are also factors contributing to medication problems. Elderly patients, for example, may be less compliant (eg, because of poverty) or incapable of following medication instructions, thereby adding to the problem.[35] Because the effects of drugs are more likely to be exaggerated in the elderly, laboratorians should be knowledgeable of the principles of therapeutic drug monitoring and the affects of aging on therapeutic drug monitoring. To provide the best quality care to the elderly, frequent drug monitoring and the development of therapeutic ranges for the elderly may be necessary.[36]

The Effects of Exercise and Nutrition on the Elderly and Chemistry Results

Many studies indicate that exercise is an excellent way for the elderly to maintain health and increase longevity. As an example, Rowe reported that in a study of more than 40,000 postmenopausal women over a 7-year period, those who regularly exercised were 20% less likely to die than those who were sedentary.[26] In another study, Shephard concludes that the sedentary elderly population, aged 75 to 80 years, may very well benefit from exercise by improving their overall health, increasing their social contacts, and improving their cerebral function.[37] Obesity, as a result of metabolic changes, reduced activity, and age-related alterations in muscle, is a common problem in elderly people that is also helped by exercise.[38] Other studies have indicated that (1) the more frequent the exercise, the greater the benefit, (2) moderate exercise (eg, walking) is nearly as beneficial as vigorous exercise, and (3) exercise can negate the adverse effects of other risk factors, such as high blood pressure and hyperglycemia.[26] The health benefits from exercise include reduced cardiovascular risk, weight control, increased functional capacity, improved nutrient intake, and better sleep.[3]

As increasing numbers of elderly people are encouraged to participate in exercise programs, laboratorians will need a better understanding of how exercise affects the test results of these individuals. For example, exercise affects lipids by lowering triglycerides and increasing HDL cholesterol. In addition, insulin is lowered, and growth hormone is increased.[38]

Consideration should also be given to the type of exercise, the timing of specimen collection, and prescribed medications.[38] Glucose and insulin, for example, do not respond in the same manner to different types of exercise. Both essentially remain stable during isometric exercise with large muscle groups, whereas insulin decreases during dynamic exercise of moderate to high intensity.[38] Timing of specimen collection is important because glucagon secretion is stimulated after intense exercise in patients with type 2 diabetes. This results in a transient elevation of glucose levels for about 1 hour after exercise.[38] The effect of medications on test results was discussed previously.

Nutritional problems are a common occurrence in elderly patients. Although relatively few studies have been conducted in the past, there is a growing body of information on the nutritional needs of the elderly population. Elderly people are at increased risk for poor nutritional status (eg, protein-calorie malnutrition) compared with younger adults, owing to both physiologic and psychological factors, including age-related changes in taste and smell, malabsorption due to medications or changes in stomach acidity, problems with mobility, disabilities, depression, and poverty.[39] The incidence of malnutrition among elderly residents in long-term health care facilities is reported to be 50% to 85%; in acute care hospitals, the percentage may range from 17% to 65%.[40]

The relationships among nutrition, general health, aging, and disease are important and need to be investigated further. A protein-calorie deficit can lead to decreased resistance to infections and lymphopenia. Excessive calories result in obesity and the probability of type 2 diabetes. Deficiencies in vitamins A, C, and E (the antioxidants) may lead to atherosclerosis and an increased risk for cancer.[8] Low-fiber diets may lead to diverticulitis and colon cancer.[3]

Through nutritional assessment, it may be possible to prevent and identify nutritional deficiencies in elderly people, thereby preventing many of the chronic diseases of aging and also promoting better health.[40] In the clinical laboratory, nutritional assessment includes the measurement of various proteins (*eg,* albumin, prealbumin, transferrin, and retinol-binding protein) and vitamin levels. A detailed discussion of nutritional assessment and vitamins can be found in Chapter 28, *Vitamins and Nutritional Assessment.*

SUMMARY

The proportion of elderly in the U.S. population is increasing, presenting a major challenge for the health care system. With this increase, it will become increasingly important for clinical laboratorians to be knowledgeable of the aging process and its effects on clinical chemistry results. The aging process has been described as a loss of adaptation with a decrease in viability and life expectancy. A knowledge of the various theories of aging is the basis for understanding changes in laboratory values. Although several theories of aging have been described, no one theory is adequate.

Aging is associated with several physiologic changes. There is a decreasing efficiency in maintenance of homeostasis; a reduction in muscle mass; and a decline in respiratory, cardiovascular, kidney, liver, immune, neurologic, and endocrine system functions. Carbohydrate, protein, lipid, and calcium metabolism all change with age. The specific changes in selected chemistry analytes are shown in Table 29-4. In addition to a knowledge of the basic biochemical and physiologic changes of aging, clinical laboratorians must be aware of other factors that may affect clinical chemistry test results. These factors include the establishment and interpretation of reference intervals, some preanalytical variables, prescribed medications, and the effects of exercise and nutrition.

The clinical laboratorian should remember that aging is not fixed by an individual's biology or by a specific time frame, however. There is a great deal of variability among individuals with regard to how they age and the rate at which aging occurs. Environmental and social factors are also important and play a role in the aging process. With an increasing proportion of elderly in the population, it will become even more important to research the physiology *and* psychosocial aspects of aging.

REVIEW QUESTIONS

1. Which of the following serum analytes essentially remains unchanged in healthy elderly adults?
   a. Triglycerides
   b. Glucose
   c. Chloride
   d. Bilirubin
2. Which of the following is a common and underappreciated condition in the elderly?
   a. Hyponatremia
   b. Dehydration
   c. Amyloidosis
   d. Multiple myeloma
3. Of the following, which is *not* one of the top 10 causes of death in people older than 65 years of age?
   a. Alzheimer's disease
   b. Pneumonia
   c. Diabetes mellitus
   d. Diverticulitis
4. When interpreting laboratory test results from the elderly, the clinical laboratorian must consider all of the following, *except:*
   a. Multiple medications (polypharmacy)
   b. Poor nutrition or malnutrition
   c. Subclinical disorders
   d. Intelligence quotient (IQ)
5. Which of the following theories of aging involves the accumulation of "waste" products that prevent the cell from carrying out its normal functions?
   a. Genetic programming theory
   b. Glycation of protein theory
   c. Random genetic damage theory
   d. Free radical theory

REFERENCES

1. Liew CC. Biochemical aspects of aging. In Gornall AG, ed. Applied biochemistry of clinical disorders. 2nd ed. Philadelphia: Lippincott-Raven, 1986;558–565.
2. U.S. Bureau of the Census. Current population reports, special studies, P23-190, 65+ in the United States. Washington, DC: U.S. Government Printing Office, 1996.
3. Knight JA. Laboratory medicine and the aging process. Chicago: American Society of Clinical Pathologists, 1996.
4. Resnick NM. Part 1: Introduction to clinical medicine. Chapter 9, Geriatric medicine. Harrison's online: www.harrisonsonline.com. New York: McGraw-Hill, 1998.
5. Strehler BL. A critique of theories of biological aging. In Dietz AA, ed. Aging: its chemistry. Washington, DC: American Association of Clinical Chemistry, 1980;25–45.
6. Gorman LS. Aging: laboratory testing and theories. Clin Lab Sci 1995;8:24–30.
7. Cearlock DM, Laude-Flaws M. Stress, immune function and the older adult. Med Lab Observer 1997;29(10):36–46.
8. Miller SM. Antioxidants and aging. Med Lab Observer 1997;29: 42–51.
9. Etnyre-Zacher P, Isabel JM. The impact of an aging population on the clinical laboratory. Med Lab Observer 1997;29(4):48–54.
10. Young, DS. Effects of preanalytical variables on clinical laboratory tests. 1st ed. Washington, DC: AACC Press, 1993.
11. Knight JA. Laboratory issues regarding geriatric patients. Lab Med 1997;28(7):458–461.
12. Merry BJ, Holehan AM. Aging of the male reproductive system. In Timiras PS, ed. Physiological basis of aging and geriatrics. 2nd ed. Boca Raton, FL: CRC Press, 1994;171–178.
13. Merry BJ, Holehan AM. Aging of the female reproductive system: the menopause. In Timiras PS, ed. Physiological basis of aging and geriatrics. 2nd ed. Boca Raton, FL: CRC Press, 1994;147–170.
14. Byyny RL, Speroff L. Menopause and postmenopause hormone therapy. In Byyny RL, Speroff L, eds. A clinical guide for the care of older women. 2nd ed. Baltimore: Williams & Wilkins, 1996;161–226.
15. Kane RL, Ouslander JG, Abrass IB. Essentials of clinical geriatrics. 3rd ed. New York: McGraw-Hill, 1994.

16. Meier DE. Osteoporosis and other disorders of skeletal aging. In Cassel CK, et al, eds. Geriatric medicine. 3rd ed. New York: Springer-Verlag, 1997;411–432.

17. Davies MD, Bliziotes M. Osteoporosis: risk factors, diagnosis and therapy. Lab Med 1998;29:418–421.

18. Timiras PS. Aging of the thyroid gland and basal metabolism. In Timiras PS, ed. Physiological basis of aging and geriatrics. 2nd ed. Boca Raton, FL: CRC Press, 1994;179–189.

19. Simons RJ, Demers LM. Thyroid disorders in the elderly. In Faulkner WR, Meites S, eds. Geriatric clinical chemistry: reference values. Washington, DC: AACC Press, 1994;103–116.

20. Timiras PS. The endocrine, pancreas and carbohydrate metabolism. In Timiras PS, eds. Physiological basis of aging and geriatrics. 2nd ed. Boca Raton, FL: CRC Press, 1994;191–197.

21. Miller SM. Diabetes: Targeting an old adversary. Med Lab Observer 1998;30:30–41.

22. Tietz NW, Shuey DF, Wekstein DR. Laboratory values in fit aging individuals: sexagenarians through centenarians. In Faulkner WR, Meites S, eds. Geriatric clinical chemistry: reference values. Washington, DC: AACC Press, 1994;145–184.

23. Timiras PS. Aging of the adrenals and pituitary. In Timiras PS, ed. Physiological basis of aging and geriatrics. 2nd ed. Boca Raton, FL: CRC Press, 1994;133–146.

24. Timiras PS. Cardiovascular alterations with age: atherosclerosis, coronary heart disease, hypertension. In Timiras PS, ed. Physiological basis of aging and geriatrics. 2nd ed. Boca Raton, FL: CRC Press, 1994; 199–214.

25. Timiras ML. The kidney, the lower urinary tract, the prostate, and body fluids. In Timiras PS, ed. Physiological basis of aging and geriatrics. 2nd ed. Boca Raton, FL: CRC Press, 1994;235–246.

26. Rowe JW, Kahn RL. Successful aging. New York: Pantheon Books, 1998.

27. Timiras PS. Aging of the gastrointestinal tract and liver. In Timiras PS, ed. Physiological basis of aging and geriatrics. 2nd ed. Boca Raton, FL: CRC Press, 1994;247–257.

28. Timiras PS. Aging of respiration, erythrocytes, and the hematopoietic system. In Timiras PS, ed. Physiological basis of aging and geriatrics. 2nd ed. Boca Raton, FL: CRC Press, 1994;225–233.

29. Timiras PS. Cardiovascular alterations with age: atherosclerosis, coronary heart disease, hypertension. In Timiras PS. Physiological basis of aging and geriatrics. 2nd ed. Boca Raton, FL: CRC Press, 1994; 199–214.

30. Timiras PS. Degenerative changes in cells and cell death. In Timiras PS, ed. Physiological basis of aging and geriatrics. 2nd ed. Boca Raton, FL: CRC Press, 1994;47–59.

31. Sasse, EA. The determination of reference intervals. In: Faulkner, WR and Meites, S. Geriatric clinical chemistry: Reference values. Washington, D.C. AACC Press 1994;6–17.

32. NCCLS Approved Guideline. How to define and determine reference intervals in the clinical laboratory. Villanova, PA: National Committee for Clinical Laboratory Standards, 1995.

33. Young DS. Pre-analytical variability in the elderly. In Faulkner WR, Meites S. Geriatric clinical chemistry: reference values. Washington, DC: AACC Press, 1994;19–39.

34. Klosinski DD. Collecting specimens from the elderly patient. Lab Med 1997;28(8):518–522.

35. Beizer JL, Timiras ML. Pharmacology and drug management in the elderly. In Timiras PS, ed. Physiological basis of aging and geriatrics. 2nd ed. Boca Raton, FL: CRC Press, 1994;279–284.

36. Warner A. Therapeutic drug monitoring in the elderly. In: Faulkner WR, Meites S. Geriatric clinical chemistry: reference values. Washington DC: AACC Press, 1994;134–144.

37. Shephard RJ. The scientific basis of exercise prescribing for the very old. J Am Geriatr Soc 1990;38:62–70.

38. Cearlock DM, Nuzzo NA. (1997). Evaluating the benefits and hazards of exercise in the older adult. Med Lab Observer 1997;29(6):40–49.

39. Garry PJ. Nutrition and aging. In Faulkner WR, Meites S, eds Geriatric clinical chemistry: reference values. Washington, DC: AACC Press, 1994;48–72.

40. Miller SM. Nutrition, the elderly, and the laboratory. Med Lab Observer 1997;29(3):23–30.

CHAPTER 30

Pediatric Clinical Chemistry

John E. Sherwin, Juan R. Sobenes

Objectives

Upon completion of this chapter, the clinical laboratorian should be able to:

- *Define the following terms: extrauterine, infancy, neonate, and pediatric.*

- *Describe some of the adaptive changes that occur upon the birth of an infant.*

- *Discuss some of the problems with blood collection from children.*

- *Summarize some of the changes that occur in children with regard to electrolyte and water balance, energy metabolism, hormone balance, and humoral and cellular immunity.*

- *Explain how drug treatment and pharmacokinetics are different in children and adults.*

- *Relate the procedures used to identify inherited metabolic disorders in children.*

KEY TERMS

Adaptive immune system	Hypoxemia
Common variable	Infancy
agammaglobulinemia	Innate immune system
Congenital	Microchemistry
agammaglobulinemia	Neonate
Extrauterine	Pediatric
Hyperoxemia	

AGE-RELATED DEVELOPMENT CHANGES

Much of the value of laboratory medicine in diagnosis and therapy is based on the fact that specific organ system function can be assessed by serum concentrations of substances produced by the organ system. This premise relies on the fact that people normally have consistent reference values for body fluid analyte concentrations ("normal values").[1,2] Because humans grow and develop throughout life, those reference values may vary with age. This is particularly true during *infancy* and adolescence because significant growth and developmental changes occur during this period.

The birth of an infant is attended by rapid adaptation to *extrauterine* life.[3] The infant adapts by initiating active respiration, increasing kidney function, closure of the patent ductus arteriosus of the heart, and eating. Generally, infants lose weight following delivery because of insensible water loss. This loss is quickly offset by weight gain due to caloric intake. Weight gain is about 15 g/d for the first year, and the infant's birthweight of about 3200 g doubles in 4 to 6 months. Kidney function increases during the first year until it approximates adult values. Liver function matures, and portal circulation becomes patent within 30 to 60 days. Motor development increases linearly during at least the first 12 months. This development is accompanied by changes in the electroencephalogram, growth and differentiation of the brain, and increasing visual acuity. In addition, several specific changes are of diagnostic importance.

Thyroid function changes abruptly following birth.[3] Serum thyrotropin-stimulating hormone concentration decreases rapidly. Hemoglobin shifts over a period of 90 days from fetal to adult hemoglobin.[3,4] This, in turn, frequently causes increased bilirubin concentrations.

During the adolescent growth spurt, bone growth results in a large increase in serum alkaline phosphatase.[5] Sexual maturation in girls is associated with chronic blood loss due to menstruation. The cyclic production of female sexual hormones also requires special attention. Sexual maturation in boys is associated with fewer large changes in serum analytes, but nonetheless, the changes result in differentiation of male and female values for such analytes as creatine kinase and uric acid as well as the steroid hormones. Therefore, it is important that evaluation of illness in infants include the use of sex- and age-related reference values.[1,2]

PHLEBOTOMY AND MICROCHEMISTRY FOR THE PEDIATRIC PATIENT

Blood collection from the *pediatric* patient is complicated not only by the patient's size but also often by his or her ability to communicate. This can complicate identification of the patient. The frequent presence of parents during blood collection from infants also can make blood collection more difficult because of behavior problems. Venipuncture in pediatric patients varies little, if any, from venipuncture in adults.[6]

It is important to remember that the availability of appropriate veins for therapy is often minimal in infants because of their size. Therefore, these veins must be reserved for therapy rather than phlebotomy. This is a particular problem in the newborn. For this reason, blood collection by skin puncture is often preferable to venipuncture.

Skin puncture for blood collection is done on a finger in older children or on the heel in infants. The National Committee for Clinical Laboratory Standards (NCCLS) has developed a standard protocol for skin puncture.[7] Before performing skin puncture, the skin should be "arterialized" using either a warm towel or a histamine cream. Care should be taken not to use dry heat or a water temperature greater than 42°C. Some reports have questioned the efficacy of histamine creams in inducing local hyperemia, but they are still commonly used.

Fingers are commonly used in children for skin puncture. The palmar surface of the distal phalanx of the middle finger should be used. In infants, the heel is preferable to the fingers because of size. The specimen should be collected on the outer portions of the heel to avoid the calcaneus. In larger infants, skin puncture can be done using the appropriate infant lancet, whereas the appropriate premature infant lancet should be used for small or premature infants because of the shorter puncture point. Punctures with these devices generally yield 300- to 400-mL specimens after the first drop of blood has been discarded owing to tissue contamination. Several studies have demonstrated that with proper technique, hemolysis and excessive tissue fluid contamination can be avoided.

After completion of the skin puncture, it is important that the phlebotomist not leave the patient until the bleeding has stopped. In small children and infants, the usual adhesive bandages should be avoided because of the possibility of ingestion or aspiration by the patient. Collection of microsamples (less than 500 mL) presents unique problems for the laboratory with regard to labeling, centrifugation, and separation. The production of microcollection systems, such as those of Sarstedt or Becton Dickinson, has simplified the task of specimen collection; however, no completely satisfactory system for labeling of these small containers has been developed. If adhesive labels are used, care must be taken that the size permits the container to fit into the centrifuge. Because of the smaller tube size, mixing of anticoagulated specimens is particularly important to prevent formation of small clots. This can be a particularly vexing problem with blood-gas specimens collected in capillary tubes.

After the specimens have been separated, they can be analyzed using any of a variety of automated instruments. In the past, *microchemistry* was a separate discipline because most analyzers required several milliliters of sample for a complete analysis. Today, however, virtually all analyzers require only a few microliters of sample for each test. Therefore, the analysis of microsamples is no longer a large problem. Nonetheless, some instruments do require additional specimen to account for "dead volume."

ACID–BASE BALANCE IN THE INFANT

Acid–Base Balance Disorders

The trauma of labor during delivery often induces an acidosis in the newborn as a result of significant asphyxia.[3,4,8] In such cases, blood pH may be an indication for surgical intervention if the clinical presentation warrants it. For this reason, some practitioners have advocated the collection and analysis of fetal scalp blood specimens as a means of monitoring fetal distress. In the hour after birth, the pH rapidly returns to the normal value (Table 30-1). Newborn infants with acidosis require additional evaluation. Many are normal because the acidosis was the result of labor; however, a small number of infants are acidotic and clinically depressed. These infants may be suffering from inborn errors of metabolism. On the other hand, clinically depressed infants with a normal pH also present a diagnostic challenge.

Infants who are born premature and have respiratory distress syndrome rapidly become acidotic and exhibit an elevated PCO_2.[3,9] In most cases, they respond to oxygenation; however, therapy generally cannot be effectively designed to return the blood-gas parameters to normal, and the patients remain slightly acidotic with a pH of 7.20 to 7.25 and an elevated PCO_2. The PO_2 is kept above 40 mm Hg. In most nurseries, the oxygenation status is measured with the transcutaneous O_2 (tcO_2) monitor. During the years since the introduction of a successful transcutaneous oxygen electrode, the close correlation between the PaO_2 and the $tcPO_2$ has

TABLE 30-1. Reference Ranges for pH and Blood-Gas Parameters in the Newborn

	Normal Values	Panic Values
pH		
Arterial	7.27–7.47	Less than 7.10, greater than 7.55
Venous	7.33–7.43	—
PCO$_2$ (mm Hg)		
Arterial	27–40	Greater than 80
Venous	35–40	—
PO$_2$ (mmHg)		
Arterial	65–80	Less than 40, greater than 250
Venous	30–50	—
Bicarbonate (mmol/L)		
Arterial	16–23	—
Serum, venous	18–25	—

been repeatedly verified. The PO$_2$ measured by skin puncture is about 5 mm Hg lower than the tcPO$_2$.

Continuous transcutaneous oxygen monitoring has also provided surprises. Transient hyperoxemia (PO$_2$ > 100 mm Hg) and *hypoxemia* (PO$_2$ < 50 mm Hg) are induced easily by infant handling, which is routine in many neonatal nurseries. Hypoxemia can be induced by crying, bowel movements, feeding, diaper changes, blood sampling, and physical examination. The relatively banal treatment of rubbing of the heel with a gel is sufficient in some infants to cause a prolonged decrease in the tcPO$_2$. In one infant, however, this same treatment actually improved the oxygen status. Hyperoxemia can also result from abrupt changes in shunt blood flow patterns and improvement in lung function, owing to resolving respiratory distress, and from briefly breathing concentrated O$_2$ to alleviate a spell of apnea.

In some instances, the transcutaneous monitor does not accurately reflect arterial PO$_2$. Fortunately, these instances are relatively rare, being restricted to infants in whom the blood pressure falls 2 standard deviations below normal and to infants given tolazoline, a pulmonary vasodilator, or anesthetics, such as nitrous oxide and halothane. These changes are transient, of only 5 to 10 minutes' duration, and sometimes shorter. These changes, which can be as large as 20 mm Hg, would be missed by traditional phlebotomy and laboratory analysis practices. For this reason, transcutaneous monitors are in widespread use.

Transcutaneous CO$_2$ (tcCO$_2$) monitors have also been developed and are in routine use. The tcCO$_2$ exhibits smaller changes than the tcO$_2$ but can provide valuable trend information. The use of these monitors has reduced the need for frequent blood-gas analysis. In many intensive care units, it has resulted in a 50% decrease in the number of requests for blood-gas analysis.

The calculation of bicarbonate from the pH and PCO$_2$ using the Henderson-Hasselbalch equation is generally adequate for therapeutic intervention. Most blood-gas instruments do this automatically using a pK of 6.1 and a solubility coefficient of 0.0302. Although these parameters have been reported to vary with acute illness, there is no compelling evidence that for medical practice, the constants are inadequate.

If the acidosis is not respiratory in nature, as described previously, the cause needs to be defined and treated, rather than merely treating the acidosis. Metabolic acidosis in infants can have any of several etiologies (Table 30-2).

Analysis of Acid–Base Deficits

The analyses of pH, PCO$_2$, and PO$_2$ are independent of the patient's age and size.[3,4,9] Few of the blood-gas analyzers, however, easily accommodate capillary blood samples. This can be a particular problem in a busy neonatal intensive care unit, which may generate as many as 200 samples daily. Capillary samples can easily clot and coat electrode membranes with protein, thereby distorting values and increasing the required maintenance. Experience suggests that manufacturers' recommended maintenance periods must be reduced by at least half if performance is to be optimized.

ELECTROLYTE AND WATER BALANCE

The immediate postnatal period is attended by growth and developmental changes that adapt the infant to extrauterine life. The ability of the infant to adapt is a function of the intrauterine environment and the infant's developmental maturity. The normal infant adapts with little difficulty. On the other hand, a 26-week gestation infant weighing 600 g requires intensive care. Fluid balance in these infants is precarious and requires careful monitoring.

The *neonate* (infant up to 1 month of age) has a relatively high surface-to-volume ratio.[3] This leads to increased insensible water loss at birth. Renal function is immature, and glomerular filtration rates are low. Additionally, water balance is complicated by the fact that the body water content is normally being reduced by contracting the extracellular fluid volume at this time. In fact, this accounts in large part for the observed normal weight loss following birth. Electrolyte balance is also precarious in these infants not only

TABLE 30-2. The Etiologies of Metabolic Acidosis in the Infant

Birth trauma	Excess intralipid
Liver dysfunction	Fanconi's syndrome
Lactic acidosis	Renal tubular acidosis
Shock	Sepsis
Diarrhea	Glycogen storage diseases
Diabetes	Hyperthyroidism

because of fluid volume changes but also because of immaturity in renal filtration and reabsorption of electrolytes.

Developmental Changes in Body Fluid and Electrolyte Composition During the Neonatal Period[3,5]

The total body water content of the normal newborn is about 80%: 55% intracellular fluid and 45% extracellular fluid. The extracellular fluid is about 20% plasma water and 80% interstitial fluid. During the first weeks of postnatal life, the total body water content decreases to about 60%, primarily at the expense of interstitial fluid.

Renal function is immature even in the normal neonate. The glomerular filtration rate is only about 25% of the adult and matures slowly during the first 6 months of life (Table 30-3). The maximal concentrating ability of the kidney is only about 78% of the adult because the loop of Henle is shorter and does not penetrate the medulla as deeply as in the adult kidney. Response to antidiuretic hormone in the neonatal period appears to be normal but is limited by the renal concentrating ability.

Insensible water loss is inversely proportional to birthweight or gestational age. This is a result of increased permeability of the skin epidermis to water, larger body surface per unit of weight, and greater skin blood flow relative to metabolic rate. The higher respiratory rate of infants may also be a contributing factor. Elevation of ambient temperature by 1°C results in a three- to four-fold increase in insensible water loss. Use of a radiant warmer or phototherapy may also increase insensible water loss by as much as 50%. In sick newborn infants, these considerations are very important because the infant's fluid intake is derived solely from parenteral sources. Fluid overload may lead to congestive heart failure secondary to patent ductus arteriosus with a left to right shunt. Factors that may reduce insensible water loss include the use of head shields, high relative humidity in the ambient environment, maintaining an infant in a thermoneutral environment, and keeping an infant in an isolette.

In infants receiving enteral fluids, water and electrolytes are absorbed in the proximal jejunum. Intestinal disorders resulting in diarrhea can rapidly cause dehydration in the neonate. Administration of hyperosmotic enteral fluids may also result in an increased rate of water secretion by the mucosa of the small intestine. This, in turn, may result in necrotizing enterocolitis. Administration of parenteral fluids does not ordinarily alter the secretion of water by the gastrointestinal tract in the infant.

The electrolytes of concern here are the four major ones: sodium, potassium, chloride, and bicarbonate. The reference ranges for these electrolytes are nearly the same as for the adult, although the concentrations are less rigidly controlled in the neonate (Table 30-4). Serum sodium concentrations change most commonly in response to alterations in water homeostasis. Hypernatremia is usually associated with water loss, but it may be accompanied by an increased extracellular fluid. It is particularly common in the preterm infant because of high insensible water loss, a reduced ability to respond to shifts in sodium balance, and a physical inability to replace water loss ad libitum. Iatrogenic hypernatremia can result from osmotic diuresis as a result of hyperosmolar fluid or glucose solution administration. Hypernatremia can also result from excessive sodium intake from formula, parenteral fluids, or sodium bicarbonate administration. These sodium overload states are often accompanied by increased extracellular fluid volume, and the infant becomes edematous.

Hyponatremia with hypovolemia results in large part from the fact that the kidneys of premature infants have a low capacity to retain sodium when they are in negative sodium balance. In this instance, the urinary sodium is high. It is important to distinguish this renal immaturity from congenital adrenal hyperplasia, which may present with similar electrolyte values. Enteric fluid loss by vomiting or diarrhea due to systemic infection or other cause can also result in hypo-

TABLE 30-3. Glomerular Filtration Rates of Infants and Adults

Age	Glomerular Filtration Rate (mL/min/1.73 m²)	
	Mean	Range
1 day	24	3–38
2–8 days	38	17–60
10–22 days	50	32–69
37–95 days	58	30–86
1.5–2 years	115	95–135
Adult	100	85–115

Note: Adult values are achieved by 12 to 18 months of age.

TABLE 30-4. Age-Related Reference Ranges for Serum Electrolytes

Analyte	Age	Concentration Range (mmol/L)
Sodium	2–5 days	135–148
	1–12 months	130–145
	Greater than 12 months	135–147
Potassium	Newborn to 3 years	3.5–5.1
	3–8 years	3.6–5.0
	8–15 years	3.5–4.9
	Greater than 12 years	3.5–4.7
Chloride	Newborn to 3 years	100–110
	3–10 years	99–109
	10–19 years	99–107
	Greater than 19 years	100–106
Bicarbonate	Infant	19–24
	Child	21–27
	Adult	23–32

natremia with a decreased extracellular volume.[10] Hyponatremia accompanied by edema due to increased extracellular volume is generally a result of inappropriate antidiuretic hormone secretion or administration of hypoosmolar or dextrose solutions. Hyperkalemia can result from cell destruction, impaired renal function, or excessive intake. Symptoms of hyperkalemia include muscle weakness and cardiac conduction changes manifested in the electrocardiogram. Clinical manifestations of hyperkalemia are apparent at 6.5 mEq/L. Hypokalemia may be caused by decreased intake, increased urinary losses due to renal tubular acidosis or hyperaldosteronism, or increased gastrointestinal losses due to vomiting, diarrhea, or diuretics. Hypokalemia results in muscle weakness, loss of abdominal sounds, and abdominal distention and can be accompanied by polyuria.

Under normal circumstances, sodium and chloride are excreted in equimolar amounts. In most instances, sodium is absorbed as neutral sodium chloride across the intestinal brush border. Hypochloremia can accompany states of hyponatremia, (eg, inappropriate antidiuretic hormone syndrome), but it also may selectively occur in infants who vomit excessively. This hypochloremia is due to loss of chloride in the stomach acid. Hyperchloremia occurs primarily as a result of increased intake and is associated with hypernatremia.

Newborn infants are susceptible to acidosis because of a limited capacity to excrete hydrogen ion. Renal reabsorption of bicarbonate appears to be only slightly reduced in the neonate, which is reflected in the somewhat lower plasma bicarbonate concentration of 22 mmol/L. This approaches the normal adult value of 24 to 27 mmol/L by 1 to 2 months of age. The proximal tubule reabsorbs most bicarbonate, and any remaining bicarbonate is filtered by the distal tubule. Therefore, two types of renal tubular acidosis can occur, one involving the proximal tubule and the other, more common, involving the distal tubule. Neonatal acidosis is usually treated with sodium bicarbonate.

After the first 12 to 18 months of life, the electrolyte balance approximates that of the adult.[11,12] Nonetheless, in preadolescent children, it is important to consider fluid and electrolyte balance in any serious illness because their condition tends to be more fragile and communication of fluid loss may be incomplete. The approach to managing fluids and electrolytes is similar in all age groups.

Fluid and Electrolyte Management

Maintenance of fluid and electrolyte balance requires assessment of three parameters: (1) estimation of the quantity of the fluid and electrolyte deficit, (2) calculation of the amounts required to replace losses and correct the abnormal deficit, and (3) implementation of therapy and monitoring and modifying, if necessary, this therapy.[13] If an accurate measure of weight loss is available, this can be used as a direct reflection of fluid loss. If not, clinical assessment of dehydration can be used. The physical signs of skin turgor, blood pressure, activity, and mucous membrane hydration are used to differentiate hypotonic (serum sodium less than 130 mmol/L), isotonic, and hypertonic (serum sodium greater than 150 mmol/L) dehydration. Laboratory measurements of hematocrit, blood urea nitrogen, creatinine, and total protein are useful in conjunction with serum electrolytes for assessment of dehydration. Replacement of fluids is ordinarily done in the same manner as the fluid was lost (eg, if dehydration is of rapid onset, fluid replacement should also be rapid). The one exception is in hypertonic dehydration, where rapid rehydration may cause convulsions. The assessment of electrolyte deficit is based on physical examination and laboratory measurements of serum electrolytes.

Replacement of fluids is ordinarily done over at least 48 hours, with 50% of the fluids administered in the first 8 hours, except in the case of hypertonic dehydration, in which replacement should be done evenly over 3 to 5 days. Calculations are done to return the plasma osmolality to normal, 280 mOsm/kg. It must be remembered that 50% of the osmolality deficit is due to cations and 50% is due to anions.

After the fluid and electrolyte deficit has been estimated and maintenance requirements have been considered, therapy can be initiated using appropriate combinations of the readily available fluids.[13] Maintenance therapy in the neonate requires an additional 75 to 80 mL/kg per 24 hours containing 2 to

CASE STUDY 30-1

At 36 weeks of uneventful pregnancy, a 22-year-old primigravida woman goes into spontaneous labor. Delivery is unremarkable, and the apparently normal male infant is taken to the special care nursery because of mild respiratory insufficiency. Oxygen, fluids, and electrolytes are administered.

At 3 days of age, the infant is doing well except for mild volume depletion. However, mild hyperkalemia and hyponatremia are noted on a routine chemistry profile.

Questions

1. What are the possible causes of these findings?
2. What items in the infant's history should be investigated?
3. What additional testing should be performed?

3 mmol of potassium and sodium. Obviously, this fluid therapy will also require the administration of calories as glucose. Abnormal losses are added to the maintenance requirements for the total amount required.

Monitoring therapy in the neonate is crucial to an appropriate outcome because the neonate has higher insensible water loss owing to increased permeability of the epidermis, greater skin blood flow, and a higher respiratory rate, as well as a larger surface-to-volume ratio, than the adult. Fluid overload in these patients may lead to congestive heart failure secondary to a patent ductus arteriosus with a left-to-right shunt.

Therapy should be monitored regularly.[14] Laboratory tests include serum sodium, potassium, chloride, osmolality, and urine specific gravity. Other laboratory tests that may be required in specific disorders include blood urea nitrogen, serum creatinine, and urinary electrolytes. These parameters should be monitored periodically until they return to normal.

ENERGY METABOLISM

Carbohydrates[3,4]

The carbohydrates hold an important place in the production of metabolic calories. They are usually ingested as complex polymers (eg, starch, saccharose) and are broken down into monosaccharides, which are absorbed by the intestine and enter the portal vein to be transformed into glucose-phosphate in the hepatocyte. This compound can follow any of five different metabolic pathways, depending on the physiologic needs of the moment.

1. It can be hydrolyzed to glucose and phosphate by the enzyme glucose phosphatase, which is present exclusively in the hepatocyte. The glucose leaves the liver cells and circulates in blood, nourishing the cells of all the organs.
2. A small portion is metabolized in the hepatic cell to produce energy.
3. It may follow the pentose phosphate pathway to supply the nucleotides of pyrimidine reduced pyridine (NADPH), which will be used in the synthesis of certain compounds, such as cholesterol and fatty acids.
4. It may be stored in the hepatocytes as glycogen, which is composed of thousands of glucose molecules, but exerts the same osmotic pressure as a single mole of glucose. Thus, the storage of glucose in the liver is done without altering the body's intracellular osmotic pressure.
5. When all the preceding avenues are saturated by an excessive availability of glucose, it is transformed into lipids after going through the two carbon acetyl coenzyme A (CoA) stage.

The glucose that gains intracellular access to any organ is immediately transformed into glucose-phosphate, which is unable to traverse the cellular membrane. Only hepatocytes contain glucose-phosphatase and are capable of freeing glucose.

The blood glucose is captured by all cells, which use it either to derive energy by metabolizing it to CO_2 and water, transform it into polymers and store it (glycogen of muscle cells), or incorporate it into macromolecules such as glycoproteins. Thus, the importance of maintaining a normal glucose level and its narrow hormonal regulation is obvious.

During periods of muscular contraction, the myocytes use chemical energy derived from adenosine triphosphate (ATP) and convert glycogen into glucose, which is catabolized to lactic acid. Lactic acid leaves the muscle and travels by means of the blood to the liver, where it is resynthesized to glucose (Cori's cycle).

Between meals, when the glycemia tends to decrease, the hepatic glycogen frees units of glucose (glycogenolysis). If the degradation of glycogen is insufficient to maintain a normal glycemia to satisfy the organs that are glucose dependent (brain, red blood cells), the process of gluconeogenesis goes into effect. Gluconeogenesis is the synthesis of glucose in the liver using noncarbohydrate substances such as alanine and other glycogenic amino acids or lactic acid that comes from muscles or from glycerol (lipids).

Hormonal Control of Carbohydrate Metabolism

Many hormones are involved in the control of carbohydrate metabolism. Their effects are aimed at ensuring that glucose leaves the liver in exactly the same amount as glucose being used by other organs. This balance is based on maintaining the level of blood glucose within a relatively narrow range. The main hormones that control plasma glucose and that have apparent antagonistic effects are glucagon and insulin. Glucagon produces hyperglycemia and insulin produces hypoglycemia.

Glucagon produces glycogenolysis, activating the hepatic phosphorylase under the influence of cyclic AMP. A second mechanism is the hyperglycemic effect of the hypophysis through growth hormone and adrenocorticotropic hormone, which activates the gluconeogenesis mechanism and the production of glucose from amino acids, glycerol, and triglycerides. At the same time, the fatty acids provide the necessary energy. Note that these are the "stress" hormones and as such can influence test results done under nonbasal conditions.

Insulin plays an essential and continuous role in the use of glucose derived from the reactions just described. It is secreted by the beta cells of the islets of Langerhans of the pancreas, where it is primarily synthesized as a larger inactive precursor composed of a single polypeptide chain called *proinsulin*. This polypeptide is cleaved at several points by peptidases, which break up the polypeptide in an N-terminal fraction, and a central protein called *peptide C*.

The amount of peptide C freed into blood is equimolecular with the amount of insulin produced. The C peptide is not metabolized by the liver. Its assessment allows us to evaluate the rate of endogenous insulin secretion. The molecule of active insulin is composed of two polypeptide chains A and B joined by two disulfide bonds. It attaches to specific receptors in the membrane of peripheral cells and activates the process of membrane penetration of glucose. It also inhibits the migration of hepatic glucose by the action of glucose phosphatase and activates the synthesis of glycogen.

Diabetes[15,16]

Diabetes is classified into the following three types:

Type 1: Insulin-dependent diabetes mellitus (IDDM)
Type 2: Non–insulin-dependent diabetes mellitus (NIDDM)
Type 3: Gestational diabetes mellitus

Diabetes affects more than 6 million Americans and generates direct and indirect costs of more than $12 billion a year. Screening for disease is only justified when early treatment in symptom-free patients is more effective than treatment begun after symptoms have appeared. Currently, early treatment is available only for gestational diabetes.

Gestational diabetes is glucose intolerance that appears for the first time during pregnancy. It occurs in about 3% of all pregnancies. Untreated gestational diabetes mellitus can cause macrosomia of the fetus, increased risk for birth trauma, hyperbilirubinemia, hypercalcemia, hypoglycemia, or respiratory distress syndrome. Control of plasma glucose in the mother usually can be accomplished by diet alone, although insulin treatment is required in 10% to 15% of patients with gestational diabetes (oral hypoglycemic agents are contraindicated in gestational diabetes because of possible teratogenesis). Desirable fasting blood glucose levels are less than 115 mg/dL.

The etiologic considerations of diabetes suggest an underlying hereditary role. Hyperglycemic syndromes can be the result of pancreatic insults that reduce its capacity to secrete insulin (*eg,* pancreatitis, carcinoma of the pancreas, hemochromatosis, pancreatectomy) or can be the result of hyperfunction of antagonistic hormones (*eg,* acromegaly, hypercortisolism, pheochromocytoma, glucagonoma). Hyperglycemias can also be iatrogenic as a result of corticotherapy or the use of progestational estrogens or certain drugs that sometimes can elicit a latent anomaly to glucose tolerance.

Usually, IDDM is seen in young people (infantile or juvenile diabetes); occasionally, it is seen in older people. It appears to be related to diverse insults, particularly viral infections of the beta cells of the islets of Langerhans that exhibit an exaggerated immunologic reaction and a genetic background characterized by the predominance of certain HLA groups (Dw3, Dw4, B8, and Bw15). NIDDM, type II, or maturity-onset diabetes mellitus, is by far the more common (80% of cases) and is accompanied in most cases by obesity. It is manifested as a peripheral resistance to insulin associated with an anomaly of insulin secretion that is responsible for glucose intolerance.

The end result of diabetes, regardless of its type, is related to diffuse tissue damage involving chiefly the vascular basal membranes that is responsible for the microangiopathy in the retina and glomeruli and involvement of the nervous system. Additionally, diabetes and glucose intolerance are fac-

tors in the risk for atheroma formation. It has been postulated that glycosylated hemoglobin binds oxygen with great affinity and probably deprives the endothelium of adequate oxygenation. The concept of reversible functional angiopathy with normalization of glycemia reinforces the need to control diabetes mellitus as effectively as possible.

One of the consequences of the altered glucose metabolism is ketosis, with the possibility of ketoacidosis. It is related to the acute deficiency of insulin, the absence of glycolysis, and the resulting absence of oxaloacetic acid.

Nitrogen Metabolism[3,4]

The liver plays a central role in nitrogen metabolism. It is involved in the metabolic interconversions of amino acids and the synthesis of nonessential amino acids. The liver synthesizes most body proteins, such as albumin, beta-macroglobulin, and orosomucoid, but does not synthesize the immunoglobulins. Hemoglobin is synthesized in the liver of neonates before birth but not in the liver of adults. The liver is also largely responsible for the production of metabolic end products, such as creatinine, urea, ammonia, and uric acid, which can be more easily excreted. The liver is also involved in the detoxification of bilirubin produced from hemoglobin. The metabolic reactions are primarily concerned with adding glucuronic acid to bilirubin to increase its aqueous solubility. These metabolites are usually excreted more easily in the urine or intestinal lumen.

Table 30-5. lists the reference ranges for creatinine, ammonia, urea, and uric acid and the effect of age on their concentration. Blood ammonia decreases after the neonatal period because of the completion of the development of the hepatic portal circulation. Urea and uric acid increase because of dietary changes during development. Creatinine increases slightly owing to growth, which increases muscle mass. This analyte is relatively independent of diet and thus can be used as a measure of renal function.

Bilirubin[3,4,8]

The breakdown of heme-containing proteins, primarily hemoglobin, results in the production of about 250 mg/d of bilirubin. This occurs in the spleen as a result of hydrolysis of heme to release the iron and form the intermediate biliverdin,

TABLE 30-5. Reference Ranges for Selected Nitrogenous Metabolites in Plasma Analyte

Age (y)	Ammonia (μmol/L)	Urea (mg/L)	Creatinine (mg/L)	Uric Acid (mg/L)
0–1	—	60–450	2–10	10–76
1–5	10–40	50–170	2–10	18–50
5–19	11–35	80–220	4–13	30–60
Adult male	11–35	100–210	5–12	40–90
Adult female	11–35	100–210	4–10	30–60

which is reduced to bilirubin. The iron is bound to transferrin and is reused for heme synthesis. The bilirubin is bound to albumin and is transported to the liver. This serves to detoxify bilirubin during transport. In some instances, bilirubin can become covalently bound to albumin, which is referred to as *delta-bilirubin*. This complex is apparently nontoxic and has been suggested as a measure of the recovery phase in obstructive liver disease. Bilirubin binding occurs at two to three sites on albumin, one with a high affinity for bilirubin and one or two weak sites. This bilirubin can be displaced by drugs such as salicylate, which can cause nerve damage.

After the bilirubin is transported to the liver, it is released from albumin and reacts with one or two molecules of glucuronic acid. This increases the water solubility of bilirubin so that it can be excreted in the bile. Most of the bilirubin (about 85%) is excreted as diglucuronide and the remainder as monoglucuronide. After excretion in the bile, the conjugated bilirubin is cleaved to bilirubin, which is converted by intestinal bacteria to a series of urobilinogens. These metabolic steps result in steady-state concentrations in the blood and urine of bilirubin and its metabolites, as indicated in Table 30-6.

Increases in bilirubin result from either overproduction or impaired excretion. Overproduction is the result of hemolysis, whereas impaired excretion signals liver dysfunction. Increased production is the result of accelerated red blood cell hemolysis. Total serum bilirubin generally does not exceed 50 mg/L in adults but may be higher in infants. It is virtually all unconjugated bilirubin. The laboratory analysis should include a complete blood count as a means of identifying the source of the hemolysis. The distinction between these alternatives is accomplished by clinical assessment and laboratory evaluation.

If the clinical features of the disease suggest liver dysfunction, bilirubin analysis can be helpful in differentiating the cause of jaundice. Prehepatic jaundice results in a large increase in unconjugated bilirubin because of the increased release and metabolism of hemoglobin after hemolysis. No increase or only a slight increase in conjugated bilirubin is observed because the transport of bilirubin into the liver and the formation of the glucuronide conjugated become rate limiting. Additionally, because of the increased levels of unconjugated bilirubin excreted by the liver, urinary urobilinogen and fecal urobilin concentrations are elevated, but urinary bilirubin (which is only the freely soluble, conju-

gated form) is absent. In contrast, posthepatic obstructive jaundice is characterized by large increases in serum conjugated bilirubin. The accumulation of bilirubin in the serum is the result of decreased biliary excretion after the conjugation of bilirubin in the liver rather than of an increased bilirubin load caused by hemolysis. Excretion of bilirubin metabolites is low, and urinary bilirubin can usually be demonstrated. Hepatic jaundice presents an intermediate pattern wherein both conjugated and unconjugated serum bilirubin levels are increased to the same degree and conjugated bilirubin is present in the urine. However, the fecal concentration of urobilin is generally decreased.

Urea and Ammonia Metabolism in the Liver[4]

Metabolic pools of amino acids are present in the hepatocytes of the liver. From these pools, amino acids are drawn for the synthesis of proteins. When a protein is degraded, the bulk of the constituent amino acids is returned to these intracellular pools. The released amino acids can also be used in gluconeogenesis, transamination, or deamination reactions or can be reincorporated into new proteins. Important transamination reactions are catalyzed by the enzymes alanine aminotransferase (ALT, or SGPT) and aspartate aminotransferase (AST, or SGOT). Normally, the amino groups of excess amino acids in serum are converted to ammonia or urea for excretion, and the carbon skeletons are used for glycogen formation or gluconeogenesis. Some amino acids are also excreted unchanged in the urine.

Urea, creatinine, and ammonia account for 70% to 75% of the serum nonprotein nitrogen; urea accounts for 60% of the total. Most of the metabolism of nonprotein nitrogen occurs in the liver. Urea is produced in the liver from ammonia by means of the Krebs-Henseleit urea cycle. Production of urea is restricted to the liver because arginase, the enzyme that converts arginine to urea and ornithine, is present only in the liver. Most excess nitrogen is converted to urea. Because urea is the principal nitrogenous compound excreted by the body, most of this formed urea is eliminated in the urine.

Urea Metabolism in Liver Disease

Because urea is synthesized in the liver, liver disease without renal impairment results in a low serum urea nitrogen, although the urea-to-creatinine ratio may remain normal. Other causes of a decreased urea include malnutrition, normal pregnancy, inappropriate ADH secretion, and overhydration. An elevated serum urea nitrogen level does not necessarily imply renal damage because dehydration may result in urea nitrogen levels as high as 600 mg/L. Infants receiving high-protein formulas may exhibit urea nitrogen levels of 250 to 300 mg/L. Other causes of an elevated urea level include congestive heart failure, hypotension, and renal diseases, such as acute glomerulonephritis, chronic nephritis, polycystic kidney, and renal necrosis.

The urea-to-creatinine ratio is normally between 15:1 and 24:1. Measurement of this ratio can be helpful in assessing the source of azotemia. In cases of retention of urea due to pre-

TABLE 30-6. Reference Values for Bilirubin

Age	Total Bilirubin (mg/L)	Conjugated Bilirubin (mg/L)	Delta-Bilirubin (mg/L)
Up to 1 week	1–126	0–12	Not detected
1 week to 1 year	2–12	0–5	3–6
Greater than 1 year	2–14	0–2	3–6

renal causes, such as intestinal bleeding, the ratio is increased, sometimes as high as 40:1, because creatinine clearance is normal but the urea load is elevated. In cases of severe renal tubular damage, the ratio decreases to as low as 10:1.

Ammonia[3]

Although the blood-ammonia concentration is normally low (less than 35 mmol/L), ammonia is an important intermediate in amino acid synthesis. Sources of ammonia include hepatic oxidation of glutamate to ketoglutarate, the transamination and oxidative deamination of amino acids and catecholamines, and bacterial breakdown of urea in the gut. The primary mechanism for the metabolic disposal of ammonia is the synthesis of glutamate, glutamine, and carbamyl phosphate. Glutamate and glutamine are both excreted into the urine when in excess. If required, however, they can be reabsorbed and returned to the amino acid pool. Carbamyl phosphate may be used to synthesize orotic acid and ultimately pyrimidines for nucleic acids or to synthesize urea.

In addition to being converted to urea, ammonia is also excreted in the urine as the ammonium ion (NH_4^+). This ammonium salt formation from ammonia serves as a significant mechanism for excretion of excess protons produced during metabolism. This acid–base balance function of ammonia is important, particularly in diseases associated with acidosis.

The blood-ammonia concentration is higher in infants than adults because the development of the hepatic circulation is not completed until after birth. Hyperammonemia results infrequently from congenital defects of the urea cycle. The most common of these inborn errors of metabolism is a deficiency of the enzyme ornithine transcarbamylase. A much more frequent cause of hyperammonemia in infants is hyperalimentation. Most hyperalimentation fluids contain amino acid concentrations above the infant's nutritional nitrogen requirements. The amino acids are used for energy, and the excess ammonia is excreted in the urine. Ammonia concentrations can be as high as 60 mmol/L. Reye's syndrome is frequently diagnosed by an elevated blood-ammonia level in the absence of any other demonstrable cause. Reye's syndrome is usually seen in children rather than infants. This may be the result of maternally acquired immunity in the infant. Elevation of plasma ammonia above five times normal is associated with significant mortality in Reye's syndrome.

Some patients exhibit elevated blood-ammonia levels in the terminal stages of liver cirrhosis, hepatic failure, and acute and subacute liver necrosis. This is due to the failure of the liver to produce urea. Ammonia is a toxin that acts as a central nervous system depressant and can cause coma. Urinary ammonia excretion is increased in acidosis and decreased in alkalosis because ammonia salt formation is a significant mechanism for excretion of excess protons. Damage to the renal distal tubules, as occurs in renal failure, glomerulonephritis, hypercorticoidism, and Addison's disease, results in decreased ammonia excretion in the urine.

Creatinine is formed by the dehydration of creatine. Creatine is produced by the reversible transamidination of the guanidino group of arginine to glycine. Creatine is produced in the liver and in muscle tissue. It is phosphorylated with ATP to produce creatine phosphate, a reserve energy source. The irreversible dehydration of creatine to produce creatinine results in a metabolic end product that is filtered by the kidney and excreted in the urine.

Creatinine concentration is relatively constant in the serum at 5 to 12 mg/L. The amount excreted daily is a function of individual body muscle mass. Therefore, women and children tend to have lower total urinary creatinine values than men. Serum and urine creatinine measurements are most useful in evaluating renal function. Calculation of a creatinine clearance by determining the ratio of serum to urine creatinine and correcting for body surface area is a good measure of kidney function because, as noted earlier, creatinine does not vary appreciably with diet. The urea clearance is no longer used for this purpose because of dietary-induced fluctuations in clearance results.

Uric Acid and Cell Turnover

Uric acid is the end product of purine metabolism. As cells die and are replaced, a portion of the DNA and RNA is degraded into the individual purines and pyrimidines. The purines are converted to uric acid as shown in Figure 30-1.

The production of uric acid in humans is relatively constant; however, dietary intake of foods high in purines, such as liver and shellfish, can increase serum uric acid concentrations. Men, because of their increased muscle mass, exhibit a higher concentration of uric acid than do women and children. Continuous urinary excretion of uric acid is im-

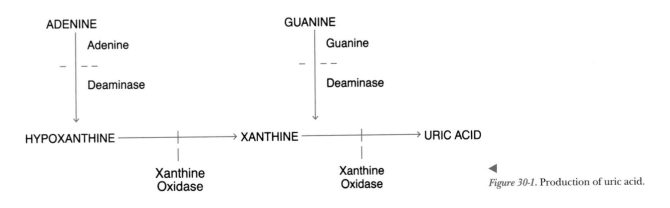

Figure 30-1. Production of uric acid.

portant because uric acid is not highly water soluble and the normal concentration of 55 to 65 mg/L is near the solubility limit for uric acid. Therefore, increased dietary load or renal dysfunction can quickly result in crystallization of uric acid in the bladder and kidney, resulting in stones, or in crystallization in the joints of the extremities, resulting in gout.

Uric acid is useful in assessing cellular turnover and destruction. Therefore, uric acid is increased in idiopathic hyperuricemia, gout, chronic renal disease, tissue neurosis, metabolic acidosis, and hematologic conditions such as leukemia, lymphoma, and anemia.

Lipid Metabolism[8,17,18]

The common American diet contains about 40% fat. The fat is predominately animal fat (ie, saturated fat), whereas the remainder is vegetable or polyunsaturated fat. These fats are composed primarily of triglycerides. The remainder is cholesterol, steroids, phospholipids, and monoglycerides and diglycerides. The triglycerides contain primarily the fatty acids myristic, palmitic, stearic, oleic, and linoleic acid. Three of the polyunsaturated fatty acids are considered to be essential because they cannot be synthesized by the human body.

Ingested fat absorption occurs in three stages: the intraluminal phase, in which fats are hydrolyzed; the cellular or absorptive phase, in which the hydrolyzed materials are absorbed by mucosal cells and then the triglycerides are reformed; and the transport phase, during which the lipids are transported throughout the body. Bile acids produced from cholesterol in the liver and stored in the gallbladder play an important role in the intraluminal phase by emulsifying the lipids and allowing attack by lipases. The lipids are converted to glycerides and fatty acids. During the absorptive phase, these compounds are taken up by mucosal cells of the small intestine and are reformed into triglycerides. After the triglycerides are reformed, they are packaged with apolipoproteins (Table 30-7). These lipoproteins, primarily chylomicrons,

TABLE 30-7. The Plasma Apolipoproteins

Apolipoprotein	Plasma Concentration (mg/mL)	Lipoprotein Class
AI	1.2–1.4	Chylomicrons, HDL
AII	0.25–0.40	HDL
AIV	0.14	Chylomicrons, HDL
BI	0.9	VLDL, LDL
BII	?	Chylomicrons, VLDL
CI	0.07	Chylomicrons, VLDL, HDL
CII	0.04	Chylomicrons, VLDL, HDL
CIII	0.100	Chylomicrons, VLDL, HDL
D	0.100	HDL
E	0.045	VLDL, HDL

are transported in the lymphatic vessels and then in the systemic circulation.

Metabolism of chylomicrons is complex and involves sequential delipidation of the chylomicrons. This, in turn, results in the formation of the various classes of lipoproteins. Chylomicrons and very-low-density lipoproteins (VLDLs) are depleted of triglycerides by a lipoprotein lipase. VLDLs are sequentially depleted of apoprotein C and triglycerides and ultimately form low-density lipoproteins (LDLs). LDLs and other lipoprotein remnants are then converted to high-density lipoproteins (HDLs) by action of lipoprotein lipase (CII). This conversion occurs by two pathways: the receptor pathway and the scavenger cell pathway. The receptor pathway ultimately transfers LDLs to liposomes of adipose tissue. Both apoproteins B and E are important in this pathway. LDLs are probably the major vehicle for the catabolism of cholesterol esters resulting from lecithin cholesterol acyl transferase (AI) action. Esters are transferred from either VLDLs or HDLs, and this is mediated by apoprotein D.

Fatty acids are metabolized by the liver in the beta-oxidation cascade. This, in turn, results in the production of acetyl CoA and, subsequently, ATP and NADH. In cases of starvation or diabetes, the formation of acetyl CoA exceeds the ability to form citrate in the tricarboxylic acid cycle. This excess acetyl CoA is shunted into forming the ketone bodies acetone acetoacetate and beta-hydroxybutyrate, and metabolic acidosis develops. Subsequent replenishment of oxaloacetate by carbohydrate metabolism allows normal metabolism of acetyl CoA and relief of the acidosis.

The metabolism of lipids in infants is essentially the same as in adults; however, the expected values are different (Table 30-8). In the case of atherosclerosis, it is now established that lower values of cholesterol (less than 170 mg/dL) are required in childhood if atherosclerosis is to be minimized in adulthood. Additionally, in early infancy, hydroxybutyrate can be used by the brain as an energy source. This is not so in adults.

Disorders of lipid metabolism in childhood are either genetic or acquired. Acquired disorders include hepatitis, which causes a rise in cholesterol in the early stages of the disease. Other causes of obstructive liver disease cause the same increase. Malabsorption results in decreased triglycerides and cholesterol. The syndrome is characterized by abnormal fecal fat loss (more than 6 g/d). The clinical diagnosis is often confirmed by a 72-hour fecal fat measurement. Malabsorption, in turn, can result in impaired absorption of the fat-soluble vitamins A, D, E, and K. Three general defects cause malabsorption in infants: (1) a defect in intraluminal absorption, which can result from pancreatic insufficiency, bacterial overgrowth, or hepatobiliary disease; (2) a defect in mucosal cell transport resulting from disaccharidase deficiency, beta-lipoproteinemia, vitamin B_{12} malabsorption, or endocrine disorders; and (3) intestinal lymphatic obstruction due to congestive heart failure, constrictive pancarditis, or intestinal lymphangiectasia.

Genetic disorders of lipid metabolism include the sphingolipid storage diseases (Table 30-9), mitochondrial and elec-

TABLE 30-8. Age-Related Reference Ranges for Serum Lipids

Age (y)	Cholesterol (mg/dL)	HDL Cholesterol (mg/dL)	LDL Cholesterol (mg/dL)	Triglycerides (mg/dL)
<1	90–260	—	—	20–120
1–5	95–215	35–82	55–160	20–120
5–10	95–240	35–65	75–165	20–190
10–20	110–230	35–65	50–160	25–270

tron transport defects, defects of beta-oxidation, and the dyslipoproteinemias. The storage diseases are all autosomal recessive mutations, except Fabry's disease, which is X-linked. They are almost all degenerative diseases with mental retardation and early demise. Table 30-9 summarizes the storage diseases. Defects of mitochondrial transport of lipids and electron transport, although rare, have been reported. In some instances, treatment with carnitine may have a palliative effect. Defects of beta-oxidation are known to occur and are the result of mutation in the enzymes of the beta-oxidation cascade.

The dyslipoproteinemias include both the hypolipidemias and the hyperlipidemias. Hypolipidemias include abetalipoproteinemia, which results in degeneration of the cerebellar tract. Tangier disease is due to the absence of HDL and results in the accumulation of cholesterol in the tissues. Malabsorption often results in hypolipidosis.

Hypercholesterolemia has a strong genetic component, and evidence is accumulating that early intervention and therapy can lower the incidence of coronary heart disease. Mass screening of children for hypercholesterolemia remains controversial. The American Academy of Pediatrics recommends that only children older than 2 years and with a family history of myocardial infarction at an early age be screened for hypercholesterolemia. Children of this group with serum cholesterol levels greater than 170 mg/dL require follow-up care. Before initiating either dietary or drug therapy, secondary causes of hypercholesterolemia should be ruled out. These include liver disease, hypothyroidism, nephritic syndrome, anorexia nervosa, and systemic lupus erythematosus. If dietary therapy is to be used, other coronary heart disease risk factors, such as obesity, smoking, high blood pressure, and sedentary lifestyle, should be assessed and modified if

TABLE 30-9. The Sphingolipidoses

Disease	Signs and Symptoms	Enzyme Defect	Mode of Inheritance	Variants Identified
Farber's disease	Hoarseness, dermatitis, skeletal deformation, mental retardation	Ceramidase	Autosomal recessive	—
Gaucher's disease	Spleen and liver enlargement, erosion of long bones and pelvis, mental retardation only in infantile form	Glucocerebroside, β-glucosidase	Autosomal recessive	3
Niemann-Pick disease	Liver and spleen enlargement, mental retardation, about 30% with red spot in retina	Sphingomyelinase	Autosomal recessive	5
Krabbe's disease (globoid)	Mental retardation, almost total absence of myelin	Galactocerebroside, β-galactosidase	Autosomal recessive	—
Metachromatic leukodystrophy	Mental retardation, psychological disturbances in adult form, nerves stain yellow-brown with cresyl violet dye	Sulfatidase	Autosomal recessive	4
Fabry's disease	Reddish purple skin rash, kidney failure, pain in lower extremities	Ceramidetrihexoside, α-galactosidase	X-linked	1
Tay-Sachs disease	Mental retardation, red spot in retina, blindness, muscular weakness	Hexosaminidase A	Autosomal recessive	1
Sandhoff's disease	Same as Tay-Sachs disease, but progressing more rapidly	Hexosaminidase A and B	Autosomal recessive	1
Generalized gangliosidosis	Mental retardation, liver enlargement, skeletal deformities, about 50% with red spot in retina	β-Galactosidase	Autosomal recessive	2
Fucosidosis	Cerebral degeneration, muscle spasticity, thick skin	α-Fucosidase	Autosomal recessive	—

possible. Drug therapy should be reserved only for the most severe cases in which dietary intervention has not been effective.

DEVELOPMENTAL CHANGES IN HORMONE BALANCE[4,5,8]

Hormones are substances that are formed by one group of cells, secreted into blood, and carried to other cells, where they manifest their effects. Some of them are proteins, and others are steroids or amino acid derivatives. The endocrine system is involved in the control of many aspects of growth and development.

Regulation of Secretion

The endocrine output can be stimulated or suppressed by other hormones or regulatory factors. The production of cortisol and thyroid and sex hormones is under the influence of trophic hormones from the anterior pituitary. These, in turn, are regulated by hypothalamic factors; for example, thyrotropin-releasing hormone (TRH) from the hypothalamus stimulates the pituitary to produce and secrete thyroid-stimulating hormone (TSH), which, in turn, stimulates the production of thyroid hormones that serve as a signal to the pituitary to increase or decrease the production of TSH. Thyroid hormone evidently also mediates the inhibition of TRH. This kind of control is called *negative feedback,* and many endocrine glands are regulated in this fashion. For example, parathyroid hormone secretion is directly stimulated by the level of Ca^{2+}.

Hormones act on target cells by interacting with specific receptors located in the cell membrane in the protein hormone systems, whereas receptors for most of the nonprotein hormones are intracellular and interact directly with chromosomal acceptors. Depending on the system, specific DNA is transcribed, and mRNA for that message is increased in the cell, resulting in the synthesis of a specific protein.

Peptide and Polypeptide Hormones

The peptide and polypeptide hormones include TRH, gonadotropin-releasing hormone (GnRH), corticotropin-releasing hormone (CRH), and somatostatin, as well as the protein hormones from the anterior pituitary: ACTH, growth hormone (GH), TSH, follicle-stimulating hormone (FSH), luteinizing hormone (LH), and melanocyte-stimulating hormone (MSH). Also included are the pancreatic hormones insulin and glucagon; parathyroid hormone (PTH); the medullary cells of the thyroid, calcitonin; and the gonads, inhibin. Many of these hormones exert their effects by increasing intracellular cyclic adenosine monophosphate (cAMP) in their respective target cells. The binding of the hormone to the receptor site leads to activation of membrane-bound adenylate cyclase, which catalyzes the conversion of ATP to cAMP.

Derivatives of Amino Acids

Both thyroid hormone and catecholamines are derivatives of the amino acid tyrosine. Epinephrine interacts with adrenergic receptors on the cell membrane and alters adenylate cyclase in a manner similar to peptide hormones. Thyroid hormones bind to nuclear receptors, increasing the synthesis of specific mRNAs and the production of specific proteins. Triiodothyronine (T_3) and thyroxine (T_4) are carried in the blood bound reversibly to thyroid-binding globulin (TBG) and thyroxine-binding prealbumin (TBPA). Only the small free hormone fractions interact with their specific receptors.

Steroid Hormones

Steroids are lipid-soluble hormones that are carried in the blood stream bound to protein carriers and readily cross the outer cell membrane to enter the target cell, where specific receptors are intracellular. There are six general categories of steroid hormones: vitamin D and its derivatives, glucocorticoids, mineralocorticoids, estrogens, progestins, and androgens. They are all derived from cholesterol, and all, except for vitamin D, are synthesized by a common metabolic pathway. Transcortin (cortisol-binding globulin) has high affinity for both progesterone and cortisol. Sex hormone–binding globulin binds dihydroxytestosterone, testosterone, and estradiol with high affinity.

Role of Hormones in Normal Growth

Different tissues mature at different rates; thus, the growth of a child is a complex series of changes. One of the problems pediatricians face is the child who is very short and who may have a treatable disorder of growth.[19] Standardized growth curves are available for heights and weights. The 50th percentile at a given age is that height at which 50% of the children fall above the curve and 50% fall below the curve.

The major factor contributing to growth is lengthening of the skeleton. Several tables exist to predict adult height from a comparison of height and bone age, but such predictions are only rough estimates. Most of the hormones involved in growth (GH, TSH, glucocorticoids, sex hormones, and somatomedins) are partly or completely regulated by the hypothalamic neurons and are secreted and carried to the anterior pituitary. The factors involved in controlling pituitary synthesis and secretion of growth hormone have been isolated and purified and their structures elucidated. Somatostatin, TRH, and GnRH are available for clinical studies. CRH and GnRH are currently available only for research.

The inhibiting factors somatostatin and dopamine inhibit release of the appropriate pituitary hormones (GH, prolactin). Those pituitary trophic hormones acting on their respective target tissues activate the synthesis and release of the final hormone: somatomedin C, GH, TSH, glucocorticoids (ACTH), and sex hormones (LH, FSH). The final hormone products produce a negative-feedback effect on the hypothalamus, pituitary gland, or both, thus controlling formation of the releasing factors and the pituitary trophic hormones.

Growth Hormone

GH promotes normal growth from birth to adulthood. GH is present and produced in the fetus but is not indispensable, as seen in normal-weight anencephalic infants born without pituitary glands. Children who are deficient in GH grow at a much lower rate and reach an adult height of about 130 cm with normal body proportions. Growth hormone appears to act indirectly on target tissues by means of somatomedins. The terms *somatomedin* and *insulin-like growth factor* (IGF) are now used as synonyms. Human serum contains IGF-I, which is immunologically identical to somatomedin C and IGF-II. Their biologic responses are mediated by interaction with the insulin receptor or with specific IGF receptors, depending on the tissue. The IGF is tightly bound to carrier proteins. IGF-I–related peptides reach a peak during puberty and fall during old age. It is known that the production of IGF-I–related peptides is regulated by GH; the trophic hormone for IGF-II peptides is unknown.

In addition to IGF peptides, many other growth factors are in the tissues and blood. Factors such as epidermal growth factor, erythropoietin, nerve growth factor, platelet-derived growth factor, and fibroblast pneumocyte factor are all involved with growth.

Thyroid Hormone

Although it does not appear to play a direct role in growth of the fetus, TSH is essential for protein synthesis in the brain, for normal development of the central nervous system and skeleton, and for dental maturation. Severe hypothyroidism results in nearly complete lack of growth. T_3 increases the concentration of mRNA for GH in GH-secreting cells. In general, thyroid hormone stimulates metabolism, which results in growth and development of the organism.

Sex Hormones

Testosterone and its metabolite, dihydrotestosterone (DHT), are powerful anabolic hormones that promote growth, weight gain, and muscle mass. The presence of growth hormone is necessary for these effects. Estrogens in physiologic amounts stimulate growth and epiphyseal maturation; in excessive amounts, their effect may be inhibitory. Administration of pharmacologic doses of estrogen to excessively tall girls leads to cessation of their growth due to epiphyseal maturation and fusion.

Glucocorticoids

Glucocorticoids, as opposed to the growth-promoting effects of other hormones, have an inhibitory effect of growth. It takes only a two- or three-fold amount above the average of cortisol to arrest growth. GH administered under this set of circumstances does not reverse the growth retardation. Glucocorticoids impede growth by inhibiting thymidine incorporation into DNA of a variety of tissues.

Insulin

There is evidence that suggests that insulin may function as a stimulator of growth in addition to its role in carbohydrate metabolism. Insulin deficiency or insulin resistance is associated with intrauterine growth retardation. In tissue cultures, high concentrations of insulin are required to stimulate DNA synthesis and mitosis. Insulin promotes the uptake of glucose and amino acids, and its role in growth may be to provide the energy and building blocks for growth rather than directly influencing it.

Primary Disorders of Growth: Short Stature[20,21]

Skeletal Disorders[22]

Achondroplastic dwarfism is due to mutation of a single gene and is inherited as an autosomal dominant condition. These patients have normal intelligence and health.

Chromosomal Abnormalities

In children with chromosomal abnormalities, growth is retarded, but bodily proportions are normal or near normal. The most common chromosomal disorder is trisomy of chromosome 21 (Down's syndrome). Shorter stature, mental retardation, and characteristic physical features are seen. The lack of the second X chromosome produces Turner's syndrome. Any girl with growth retardation should be tested for this disorder.

Intrauterine Growth Retardation

Abnormal implantation of the placenta, vascular malformation or disease, malnutrition, toxemia, severe diabetes, or drug abuse can be the cause of intrauterine growth retardation in small children.

Genetic Short Stature

A genetic cause of short stature should be considered only when all other causes of growth retardation have been excluded. When the diagnosis is made, the question of GH therapy is usually raised. Currently, the policy is to reserve human GH for those children demonstrated to lack biologically active GH.

Secondary Disorders of Growth: Endocrine

Growth Hormone Deficiency

A deficiency of GH secretion or diminished action on the target organ may be the result of impaired pituitary function, hypothalamic dysfunction, or target cell unresponsiveness. Secretion of GH occurs in a series of irregular bursts throughout the day and night. Sleep, exercise, a rapid decrease in glucose concentration, administration of certain amino acids, and stress can bring about bursts of GH release. Sleep-associated GH release is closely associated with the development of slow-wave rhythm in sleep. Grandma was probably right when she said, "If you don't take your nap, you will not grow."

Neurotransmitters play an important role in the regulation of GH secretion. The administration of 500 mg of L-dopa by mouth leads to the secretion of GH between 45 minutes and 2 hours later. Administration of 5 mg of hydroxytryptophan, a serotonin precursor, increases GH release.

Pituitary hypofunction secondary to hypothalamic damage may be caused by infections, granulomas, hydrocephalus, and hypothalamic tumors. The prototype of the GH-resistant syndrome is Lorain's dwarfism, believed to have an autosomal recessive mode of transmission. These children have elevated basal GH and exaggerated GH response to provocative stimuli. The somatomedin concentrations are low and do not increase in response to GH. Resting levels of GH are low. Provocative tests are laborious and expensive. In the exercise stimulation test, the child is asked to exercise vigorously (running up and down stairs) for 20 minutes after an overnight or 4-hour fast. An increase in GH of 7 ng/mL or more is considered normal. Probably one of the most reliable provocative tests is the induction of hypoglycemia with intravenous insulin. The risk for severe hypoglycemia necessitates the presence of a physician while performing the test. An intravenous line with concentrated glucose should be available during this test.

The peak of blood glucose occurs 20 to 30 minutes after insulin administration, but GH response is observed later. Blood is collected every 15 minutes for 2.5 hours, and blood glucose is monitored by the use of Chemstrips during the procedure. The samples obtained are analyzed for glucose, GH, and cortisol. LH, FSH, and thyroid hormone should be checked in the initial blood sample.

Arginine is also an excellent stimulus for GH secretions. Recently, GHRH has become available, but it is not as readily available as TRH. There is a poor correlation between levels of serum somatomedin C and growth rate. Very low values (less than 0.25 U/mL) should be of some concern. Malnutrition also lowers somatomedin C. Hypercortisolism (excess amounts of glucocorticoids, either endogenous or exogenous) suppresses linear growth. Sex hormone deficiency and pseudohypoparathyroidism type I are two other conditions in which short stature can be manifested.

Excessive Growth

Genetic tall stature, Klinefelter's syndrome, Marfan's syndrome, and homocystinuria patients are usually well above average in height.

Growth Hormone Excess

Elevation of the mean GH level above 10 ng/mL is usually accompanied by signs and symptoms. Elevations are usually caused by somatotropic adenomas of the pituitary gland. When manifest in childhood, this produces gigantism. Excess GH after fusion of the epiphyses results in acromegaly. Impaired glucose tolerance is present in about half of the cases of acromegaly. Insulin resistance is present in varying degrees in all patients. Radiographs of the skulls, hands, and feet are helpful in making the diagnosis of acromegaly. Computed tomographic scans of the area of the sella turcica may show the tumor.

Fetal Gigantism

Fetal gigantism is characterized by excessive growth in utero with no evidence of endocrine abnormalities, although some patients have hypoglycemia, probably due to hyperinsulinemia. The infant of a diabetic mother is another example of fetal overgrowth.

Hyperthyroidism

Linear growth may be accelerated somewhat in hyperthyroidism, and the bone age may be advanced.

Precocious Secretion of Sex Hormones[23]

In precocious secretion of sex hormone, the rapid growth is accompanied by an inappropriate bone maturation for age. The final adult stature is diminished. Regardless of the origin of excess sex hormones when secondary to tumor (gonad,

---

CASE STUDY 30-2

A 9-year-old boy is brought to his pediatrician by his parents because of growth failure. The past history is unremarkable except for an episode of head trauma suffered 2 years previously, when he was struck by an automobile. Neurosurgical intervention was required because of a basalar skull fracture and intracranial bleeding. However, the child seemed to recover fully. He continued to do well at school, but his parents state that he has grown very little since the accident. Measurement of his height and weight reveal that he is indeed about 2 years below the median for his age.

Questions

1. What conditions may be associated with growth retardation?
2. What condition should be primarily considered because of this child's history?
3. What laboratory tests should be performed to confirm this?
4. What additional testing should be done?

adrenal, hypothalamus), a surgical approach should be employed for removal of the tumor.

Posterior Pituitary Disorders[24]

Posterior pituitary disorders are characterized by either excessive or deficient secretion of vasopressin or a deficient response of target cells. Clinical disorders of oxytocin secretion, however, have not been identified. The hormones of the posterior pituitary (oxytocin and vasopressin [ADH]) are synthesized in the hypothalamic nuclei and then transported along the neurohypophyseal tract to the posterior pituitary, where they are stored and released.

Vasopressin

Osmoreceptors located near the supraoptic nuclei and the thirst center influence both drinking behavior and vasopressin secretion; increased osmolality results in vasopressin release and increased osmolality of the urine. Vasopressin release is also influenced by intravascular volume, which inhibits release of vasopressin during hypervolemia, whereas baroreceptors stimulate release during hypotension. The half-life of vasopressin is about 15 to 20 minutes. Vasopressin receptors are present in cells of the collecting duct of the kidney. Deficiency of vasopressin causes diabetes insipidus, which is diagnosed when urine osmolality is inappropriately low, or when serum osmolality is raised.

Oxytocin

Oxytocin is primarily involved in milk ejection and binds to myoepithelial cells of the mammary gland. Naturally, its secretion is stimulated by suckling. Oxytocin has been used in obstetrics to induce labor because of its uterus-contracting properties, but oxytocin-deficient women present normal labor.

Thyroid Gland[4,5]

The synthesis and release of T_4 and T_3 are controlled by TSH from the anterior pituitary, which, like other glycoprotein hormones, is composed of an alpha and a beta subunit. The beta subunit confers specificity on the hormone. The normal range is 0.5 to 4.5 mU/mL. The synthesis and release of TSH is governed by TRH, which is stimulatory, and by T_4, which is inhibitory. TRH in the blood stream has a half-life of 5½ minutes, and appropriate doses (200 to 500 mg) cause a rise in serum TSH, with a peak in 30 minutes; when T_4 concentration in the blood is low, the response of thyrotrophs to TRH is exaggerated. T_4 and T_3 circulate bound to TBG (70%), TBPA (20%), thyroxine-binding prealbumin, and albumin. Only about 0.03% of T_4 and 0.3% of T_3 are not bound, constituting the most "bioavailable" hormones for entry into target cells. Hormone carried by albumin is weakly bound and readily provides additional bioavailable hormone.

Testing Thyroid Function

Serum total T_4 includes both free and protein-bound protein. An increase in TBG increases the total T_4, and in these cases, estimation of the free thyroid hormone is necessary. The advent of high-sensitivity TSH assays have made this test the single laboratory test of choice in initial thyroid gland function assessment.

$$FTI = T_4 \times \% \text{ uptake}$$

or

$$\text{"Normalized" FTI} = T_4 \times \frac{\text{patient uptake}}{\text{pooled normal serum uptake}}$$

(Eq. 30–1)

TBG can be measured directly by radioimmunoassay if hereditary defects are suspected. Free T_3 and free T_4 determinations can also be performed by equilibrium dialysis or immunoassay. As noted previously, high-sensitivity TSH assays are increasingly used to evaluate thyroid function.

Thyroid Function Disorders

Hypothyroidism

Hypothyroidism is a clinical state of thyroid hormone insufficiency. Complete or partial lack of thyroid hormone may occur, and it can be congenital or acquired. Congenital hypothyroidism can be due to dysgenesis, ectopic positioning, inborn errors of thyroid hormone synthesis (dyshormonogenesis), and defects of the hypothalamic-pituitary system.[24,25] Also, iodine deficiency, as seen in high-altitude areas such as the Andes and Himalayas, goitrogens, or peripheral resistance to thyroid hormone can cause hypothyroidism. Congenital hypothyroidism occurs in about 1 in 3500 to 4000 infants. Twice as many girls as boys are affected.

Screening programs for hypothyroidism were begun in 1973.[26] Since then, many countries have made this screening mandatory. Usually, it is done using dry blood specimens in filter paper obtained within the first days of life. Infants with low concentrations of T_4 undergo further measurements to assess TSH levels. If the testing of T_4 and TSH both suggest hypothyroidism, the physician is informed of this result. The infant should be checked for signs of hypothyroidism, and blood should be drawn for additional thyroid hormone analysis, generally T_4, T_3, free thyroxine index (FTI), and TSH. In congenital TBG deficiency, the T_4 is low, but the FTI is usually normal. The deficiency can be confirmed by TBG analysis.

Infants with hypothyroidism should receive L-thyroxine therapy. Therapy should not be delayed awaiting further testing (scan). If the scan is normal, the thyroid levels are borderline low normal, and the child does not have symptoms, a period of close observation can be instituted, with frequent periodic assessment of thyroid function and reinstitution of therapy as soon as indicated.

Recent review of results of early therapy in those hypothyroid infants picked up in screening programs are encouraging. Early therapy appears to be important in preventing mental retardation, and the expense-to-benefit ratio of screening programs is heavily in favor of patient benefit and state and federal savings by avoiding potential underemployment and need for institutional care.

DRUG THERAPY AND PHARMACOKINETICS

Drug therapy in children is markedly different than in adults,[27] although the goal is the same: a therapeutic event that improves patient outcome. Adjustment of adult drug dosages solely on the basis of the child's weight is inadequate to account for the observed differences in clinical response between pediatric and adult patients. Children and infants exhibit different pharmacodynamics and pharmacogenetics for most drugs than do adults.[28]

These therapeutic changes stem from developmental differences between adults and children. Drug absorption is often different because gastric pH is higher in infants than in adults and gastric emptying time is prolonged in infants relative to adults. At birth, the gastric pH is near neutral, but it rapidly falls to below 3. Gastric acidity is not stable in infants until nearly 3 years of age owing to development of the gastric mucosa. This difference in pH affects absorption of acid labile drugs, such as the penicillins. Delayed gastric emptying can increase the concentration of circulating drug because of an increase in drug absorption. This effect is generally observed only in the first 3 to 6 months of life, at which time gastric emptying approaches that of adults.

Distribution of drugs can also be different in infants and children when compared with adults. This is due primarily to differences in protein binding of drugs and in body water content and compartmentalization. Protein binding is generally reduced in infants and neonates compared with adults. In the neonate, binding of drugs to albumin may be decreased because of bilirubin binding to albumin. An alternative concern relates to drug displacement of bilirubin and the associated increased risk for neurologic damage due to hyperbilirubinemia. Other proteins that bind drugs may actually be present in smaller concentrations in the neonate and in-

fant than in the adult. This decreased binding implies that smaller drug dosages will result in a therapeutic outcome and that the standard dosage may have toxic results.

Drug distribution is also altered by body water and fat content. Following birth, the total body water decreases, and fat tissue is only about 15% of body mass, whereas in the adult it is nearer 20%. Skeletal muscle mass is about 25% of body weight. These differences generally result in larger apparent volumes of distribution in infants and children.

Elimination of drugs is also altered in children. For those drugs excreted in the urine without transformation, the elimination rate is determined primarily by renal function. Renal function develops during the first 12 to 18 months of life, when it approaches the adult value. This altered renal function can result in a significantly prolonged half-life of some drugs. At the same time, drugs, such as the barbiturates, that require hepatic biotransformation by the cytochrome P 450 enzyme system also exhibit a prolonged half-life. This is due to the immaturity of the liver in infants. In addition to decreased drug hydroxylation, infants exhibit a decreased ability to inactivate drugs by reaction with carbohydrates; however, sulfation reactions appear to be similar to those of adults.

All of these differences are compounded by the issue of drug compliance in children.[29] Therefore, it is obvious from these observed differences that a comprehensive program of therapeutic drug monitoring is an essential aspect of pediatric biochemistry. Table 30-10 indicates the suggested specimen collection time following drug administration for several commonly administered drugs. Table 30-11 summarizes the therapeutic ranges from several drug classes.

INBORN ERRORS OF METABOLISM

Metabolic Screening Procedures[25]

Procedures for identifying inherited metabolic disorders span the spectrum from broad screening procedures to specific tests for individual defects. Screening tests are generally oriented toward biochemical systems, and the first-line specimen of choice is urine.[30] Timed urine specimens are the most useful because they permit estimation of the amount of

---

CASE STUDY 30-3

A normal-appearing term infant is found to have a low thyroxine level on routine neonatal screening.

Questions

1. What additional testing should be performed?
2. What therapy is indicated?
3. What are the consequences of untreated hypothyroidism?

TABLE 30-10. Suggested Drug Specimen Collection Times for Commonly Prescribed Drugs

Drug Formation	Administration Route	Collection Time
Amikacin	If administered IV, infuse over 30 min	Peak—60 min after beginning of dose (60 min after IM dose)
		Trough—just prior to next dose
Carbamazepine	Oral dosage	Trough—just prior to next dose
Digoxin	N/A	6–24 hours after dose
Disopyramide	PO for 2 days	Trough—just prior to next dose
Ethosuximide	PO for 1 week	Peak—2–4 h after dose
		Trough—just prior to next dose
Gentamicin/tobramycin	If administered IV, infuse over 30 min	Peak—60 min after beginning of dose (60 min after IM dose)
		Trough—just prior to next dose
Lidocaine	IV	12 h after initiating therapy and every 12 h thereafter
Phenobarbital	N/A	Patient should be at steady state (10–25 d)
Phenytoin	N/A	Patients should be at steady state (2 d)
Primidone	N/A	Trough—just prior to next dose
Procainamide	PO	Peak—60–75 min after dose
		Trough—just prior to next dose
Propranolol	N/A	Patient should be at steady state (minimum 24 h)
		Trough—just prior to next dose
Quinidine	N/A	6–24 h after dose
Theophylline	IV therapy	Prior to intravenous infusion if patient has received theophylline
		30 min after completion of loading dose
		12–18 hours after beginning infusion
		Every 24 h thereafter
	Oral therapy	Peak—2 h after liquid or rapid dissolution dose regardless of name of drug
		4 h after Slo-Phylline dose
		4–6 h after Theodur dose
		Trough—just prior to next dose
Valproic acid	PO (since syrup is absorbed more rapidly than tablets, use the shorter peak draw time if syrup is used)	Peak—60–180 min after dose
		Trough—just prior to next dose
Chloramphenicol	IV	Peak—IV push: 60 min after dose IV infusion: 60 min after infusion was begun (30-min drug infusion)
	PO	2 h after dose
	All	Trough—just prior to next dose (toxic trough concentrations have not been established)
Vancomycin	IV	Peak—2 h after start of dose
		Trough—just prior to next dose

metabolite being produced per unit time. Blood specimens are generally used for follow-up testing. Many metabolic disorders become apparent only during periods of significant stress. Urine specimens collected during these periods can be particularly enlightening.

The clinician is presented with a difficult problem in differentiating metabolic causes of disease from other organic causes because of the nonspecificity of the clinical signs and symptoms. Symptoms may include lethargy, convulsions, fever, and hepatosplenomegaly, among others. Clearly, these symptoms also can be present in serious infections.

When a clinical suspicion has developed that an inherited metabolic disorder is the source of the illness, a routine urinalysis can be helpful. The odor of the urine can be distinctive (Table 30-12), such as the sweet odor associated with maple syrup urine disease. Urine color may change upon standing, such as occurs in alcaptonuria. Crystals in urine also may be instructive. Cystinuria can easily be recognized by

the presence of characteristic crystals of cystine in the urine. Other useful information can be the anion gap: the organic acids will increase the anion gap in the presence of normal electrolyte balance. Blood ammonia that is persistently elevated in the absence of identifiable causes should lead to consideration of inherited urea cycle defects, such as argininosuccinic acidemia. There are often physical signs of inherited metabolic diseases. In the case of argininosuccinic acidemia, patients have peculiarly kinky, brittle hair.

Hypoglycemia, particularly with vomiting, should lead the clinician to consider galactosemia, fructose intolerance, or branched-chain amino acid disorders. The presence of hypercalcemia or hypophosphatemia may herald the presence of renal tubular disorders. Abnormal liver function may also indicate disorders of the urea cycle, galactosemia, glycogenosis, or aminoacidopathy. Most urinalyses include a test for ketones using nitroprusside. Positive results commonly indicate diabetic ketoacidosis but also may indicate maple

TABLE 30-11. Therapeutic Ranges (in µg/mL) for Commonly Used Drugs in Pediatric Biochemistry

Anticonvulsants
Phenobarbital	15–40
Valproic acid	50–100
Carbamazepine	8–12
Phenytoin	10–20 (neonates, 5–15)

Antimicrobials
Tobramycin	Peak—6–10; trough—up to 2
Gentamicin	Peak—6–10; trough—up to 2
Chloramphenicol	Peak—10–20; trough—up to 10

Antidepressants
Diazepam	0.10–1.50
Propranolol	0.05–0.10
Procainamide	5–10

Antiasthmatics
Theophylline	Neonatal apnea: 5–15
Caffeine	Asthma: 10–20
	Neonatal apnea: 5–20

syrup urine disease or glucose-phosphate dehydrogenase deficiency.

Screening tests for inborn errors of metabolism are now mandated by many states. Screening for phenylketonuria (PKU) has been common for many years. Many mandated programs include testing for hypothyroidism and galactosemia as well. These tests are generally performed on all newborns about 3 days after birth and after the beginning of protein feedings. These tests have identified and successfully treated these three inborn errors of metabolism. There are, however, a multitude of other inherited metabolic diseases that have been recognized and require diagnosis.[23]

These diseases require broad-spectrum screening programs. The ferric chloride test is usually included because, in addition to identifying phenylketonuria, it can identify tyrosinemia, maple syrup urine disease, ketosis, and alcaptonuria (homogentisic acid). The interpretation of results of this test is complicated by the fact that both salicylates and phenothiazines react with ferric chloride. Certainly, a test for reducing substances is required. The Clinitest™ tablet is the most commonly used and can be helpful in identifying those diseases listed in Table 30-13. Nitroprusside is included to identify cystinemia or homocystinemia. A test for glycosaminoglycans to identify the mucopolysaccharidoses is an essential aspect of a good screening program. These tests involve color changes due to dye binding with cetylpyridinium chloride or azure A. Dinitrophenylhydrazine is used to identify the organic acidemias. This test has poor sensitivity, and contrast dyes cause significant interference.

Positive results from the screen require confirmation and follow-up testing to identify the specific defect. Organic acids are most commonly identified by gas chromatography and increasingly includes mass spectrometry for specific identification. On the other hand, amino acid disorders are confirmed by any of several techniques, including high-voltage electrophoresis, thin-layer chromatography (either one or two dimensional), and high performance liquid chromatography using either precolumn or postcolumn derivatization with a variety of compounds, such a orthophthaldehyde or ninhydrin. Normal ranges for amino acids are shown in Table

TABLE 30-12. Odors Associated With Inherited Metabolic Disorders

Disorder	Description
Diabetic ketoacidosis	Decomposing fruit, acetone like
Hawkinsinuria	Swimming pool (chlorine like)
Isovaleric acidemia	Sweaty feet
Maple syrup urine disease	Pancake syrup or burnt sugar
Methionine malabsorption (Oasthouse syndrome)	Malt or hops
β-Methylcrotonic aciduria	Tomcat urine
Methylmalonic acidemia	Ammoniacal
Phenylketonuria	Musty; mouse urine
Propionic acidemia	Ammoniacal
Trimethylaminuria	Fishy
Tyrosinemia	Methionine (cabbage like)
Urea-cycle defects	Ammoniacal

TABLE 30-13. Metabolic Diseases With Positive Reducing Substances

Glucosuria
- Diabetes mellitus
- Renal glycosuria
- Cystinosis
- Fanconi's syndrome
- Thyrotoxicosis
- Hyperadrenal states
- Phenochromocytoma

Galactosuria
- Galactosemia
- Severe liver damage

Fructosuria
- Hereditary fructose intolerance
- Essential fructosuria
- Severe liver damage

Lactosuria
- Lactose intolerance
- Severe liver damage
- Severe intestinal disease

Positive Benedict's Test Only
- Alcaptonuria (oxidation product of homogentisic acid)
- Essential pentosuria
- Pentosuria ingestion of fruit
- Sialic acid
- Cephalothin
- Ampicillin

30-14. Recently, the use of tandem mass spectroscopy has been advocated for identification of aminoacidopathies. This technique has the ability to identify a variety of aminoacidopathies as well as acyl dehydrogenase deficiencies. This broader applicability has made it attractive for use in some screening programs.[31]

Many inherited metabolic disorders can be identified only by measuring the specific activity of the enzyme. This often requires cultured cells, either white blood cells or skin fibroblasts. The complexity of many of these tests has restricted their performance to a few centralized laboratories. These laboratories not only are the repository of most experience with these assays but also often include the most extensive clinical experience with these rare disorders. Therefore, most laboratories refer these specimens and often patients for work-up by these specialized centers.

Cystic fibrosis is one of most common inborn errors of metabolism, with an incidence of 1 in 2400 live births.[23] Therapy for this disease is palliative and includes both pulmonary lavage and antibiotic therapy as well as physical therapy. Diagnosis of cystic fibrosis is made by the presence of clinical symptoms of repeated respiratory infections, meconium ileus

at birth, and pancreatic insufficiency. If some of these are present, a sweat chloride test should be performed according to the Cystic Fibrosis Foundation guidelines. The results of this test should be positive on two separate occasions.[3-5]

The gene deletion responsible for cystic fibrosis has been identified as delta F508 in about 70% of cases. This has made the diagnosis of cystic fibrosis more comprehensive, and polymerase chain reaction assays for delta F508 and other gene deletions have already been developed. This identification of the primary cystic fibrosis gene location has permitted the identification of carriers and the prenatal diagnosis of cystic fibrosis. This will permit effective genetic counseling. This assay may be as significant a tool as the sweat test in the diagnosis of cystic fibrosis because of its ease of performance.

HUMORAL AND CELLULAR IMMUNITY

Basic Concepts

Humans are continually exposed to infectious agents. It is evident that most infections normally are of limited duration and leave little permanent damage. This is due to the individual

TABLE 30-14. Reference Values (µmol/L; Mean ± SD) for Amino Acids in Serum

Amino Acid, Plasma Serum	Age			
	1–12 Months (N = 31)	1–5 Years (N = 30)	5–12 Years (N = 32)	12–18 Years (N = 22)
Phosphoserine	34 ± 17	32 ± 16	36 ± 14	26 ± 10
Taurine	100 ± 64	104 ± 61	104 ± 56	114 ± 70
Phosphoethanolamine	11 ± 21	8 ± 14	7 ± 9	20 ± 29
Aspartic acid	31 ± 25	23 ± 14	25 ± 11	30 ± 16
Threonine	155 ± 82	127 ± 43	122 ± 34	146 ± 83
Serine	270 ± 208	188 ± 52	181 ± 54	164 ± 67
Asparagine	40 ± 17	42 ± 22	20 ± 14	48 ± 33
Glutamic acid	206 ± 148	184 ± 106	227 ± 61	177 ± 114
Glutamine	379 ± 162	331 ± 116	255 ± 76	311 ± 127
Proline	206 ± 129	197 ± 74	223 ± 105	209 ± 82
Glycine	292 ± 153	260 ± 46	243 ± 56	249 ± 103
Alanine	514 ± 249	492 ± 128	485 ± 134	468 ± 184
Citruline	24 ± 15	30 ± 15	27 ± 16	29 ± 18
α-Aminobutyric acid	10 ± 5	20 ± 24	16 ± 21	9 ± 5
Valine	186 ± 77	210 ± 82	191 ± 68	186 ± 51
Cystine	10 ± 10	9 ± 10	10 ± 12	11 ± 12
Methionine	31 ± 16	21 ± 13	19 ± 10	20 ± 11
Isoleucine	65 ± 27	65 ± 29	66 ± 27	66 ± 26
Leucine	135 ± 53	142 ± 55	140 ± 47	137 ± 48
Tyrosine	94 ± 35	78 ± 27	75 ± 29	74 ± 25
Phenylalanine	73 ± 37	63 ± 22	63 ± 22	70 ± 26
β-Alanine	11 ± 10	7 ± 10	4 ± 5	6 ± 10
Tryptophan	95 ± 29	85 ± 40	75 ± 33	65 ± 54
Ornithine	187 ± 128	154 ± 53	179 ± 59	157 ± 60
Lysine	167 ± 76	174 ± 54	184 ± 55	178 ± 51
Histidine	107 ± 52	103 ± 29	98 ± 24	99 ± 26
Arginine	54 ± 36	46 ± 31	29 ± 20	39 ± 25

From Meites S, ed. Pediatric clinical chemistry. 3rd ed. Washington, DC: American Association for Clinical Chemistry, 1988.

patient's immune system. The immune system is divided into two functional divisions, namely, the *innate immune system* and the *adaptive immune system*. The innate immune system is the first line of defense. The adaptive immune system produces a specific reaction to each infectious agent that normally eradicates that agent and remembers that particular infectious agent, preventing it from causing disease later. For example, measles and diphtheria produce a lifelong immunity after an infection. The functions of innate and adaptive immunity are depicted in Table 30-15.

The Innate Immune System

Skin. Skin is an effective barrier to most microorganisms. Most infections enter the body by means of the nasopharynx, gut, lungs, and genitourinary tract. A variety of physical and biochemical defenses protect these areas; for example, lysozyme is an enzyme distributed widely in different secretions that is capable of splitting a bond found in the cell walls of many bacteria.

Phagocytes. When an organism penetrates an epithelial surface, it encounters phagocytic cells or the reticuloendothelial system. Phagocytes are of different types, but they are all derived from bone marrow. They engulf particles, including infectious agents. The blood phagocytes include the polymorphonuclear cells (PMNs) and monocytes. Both can migrate out of the blood into tissues in response to suitable stimuli. The PMNs are short-lived, whereas the monocytes develop into tissue macrophages.

Natural Killer Cells and Soluble Factors. Natural killer (NK) cells are leukocytes capable of recognizing cell-surface changes on virally infected cells. The NK cells bind to these target cells and can kill them. These cells may be activated by minor stimuli, such as minimal trauma or surgical invasion.[32] The natural killer cells are activated by interferons, which are themselves components of the innate immune system. Interferons are produced by virally infected cells and sometimes also by lymphocytes. Apart from their action on NK cells, interferons induce a state of viral resistance in uninfected tissue cells. Interferons are produced early in infection and are the first line of resistance against many viruses.

In bacterial infections, some of the serum proteins increase by many fold (2- to 100-fold). These are called *acute-phase proteins*. An example of these is C-reactive protein (CRP),

so called because of its ability to bind the C protein of pneumococci. C-reactive protein bound to bacteria promotes the binding of complement, which facilitates their uptake by phagocytes. This process is called *opsonization*. Another group of proteins (about 20) that interact with each other and with other components of the innate and adaptive immune systems is called *complement*. The complement system, which is composed of a cascade similar to the blood-clotting system, is activated spontaneously by the surface of a number of microorganisms by the alternative complement pathway. After activation, some complement components can cause opsonization, whereas others attract phagocytes to the site of infection. An additional group of complement components causes direct lysis of the cell membranes of bacteria by the lytic pathway. All these mechanisms, although described separately, act in concert in vivo.

For the purposes of simplification, the immune system can be divided into four groups or compartments: the antibody or B-lymphocyte system, the T-lymphocyte system, the complement system, and the phagocytic system. Predisposition to recurrent infections may occur when any one of the immune systems is deficient. Table 30-16 is a list of immunodeficiencies.

The Antibody or B-Lymphocyte System

B-Cell Compartment. B cells are lymphocytes characterized by the presence of surface immunoglobulins and by the ability to differentiate into plasma cells that produce immunoglobulins. The activation, proliferation, and differentiation of B cells is aided by signals provided by T cells. Upon binding antigen, antibodies can activate the complement cascade, producing lysis of the organisms (by fixing C1 to C9) or facilitating their phagocytosis by neutrophils and monocytes.

Immunoglobulins are of five major isotypes based on their structure and function: IgG, IgM, IgA, IgD, and IgE. Their main properties are summarized in Table 30-17. The IgG group is the major antibody in blood and tissues and represents the main antibody response to antigenic stimulation. IgG is the only immunoglobulin isotype that can cross the placenta and represents an important temporary maternal contribution to the immunologic defense of the newborn, as discussed later.

IgM antibody consists of five immunoglobulin units linked by disulfide bonds. These antibodies are usually secreted as a primary response to an immunologic challenge.

TABLE 30-15. Functions of Innate and Adaptive Immunity

	Innate Immune System	Adaptive Immune System
	Resistance not improved by repeated infection	Resistance improved by repeated infection
Soluble factors	Lysozyme, complement, acute-phase proteins, *eg*, CRP interferon	Antibody
Cells	Phagocytes, natural killer cells	T lymphocytes

TABLE 30-16. Immunodeficiency Diseases

Predominant Antibody Defect
  X-linked agammaglobulinemia
  X-linked hypogammaglobulinemia with growth hormone deficiency
  Autosomal recessive agammaglobulinemia
  Immunoglobulin deficiency with increased IgM (and IgG)
  IgA deficiency
  Selective deficiency of other immunoglobulin isotypes
  Kappa-chain deficiency
  Antibody deficiency with normal gammaglobulin levels or hypergammaglobulinemia
  Immunodeficiency with thymoma
  Transient hypogammaglobulinemia of infancy
  Common variable immunodeficiency with predominant B cell defect:
    Nearly normal B-cell number with $\mu^+\delta^+$, without $\mu^+\delta^-$, $\mu^+\gamma^+$, γ, or α^+ cells
    Very low B-cell numbers
    $\mu^+\gamma^+$ or γ^+ "nonsecretory" B-cells with plasma cells
    Normal or increased B-cell number with $\mu^+\delta^+\gamma^+$, $\mu^+\delta^+\alpha^+$, γ^+, and α^+ B cells
  Common variable immunodeficiency with predominant immunoregulatory T-cell disorder:
    Deficiency of helper T cells
    Presence of activated suppressor T cells
  Common variable immunodeficiency with autoantibodies to B or T cells

Predominant Defects of Cell-Mediated Immunity
  Combined immunodeficiency with predominant T-cell defect
  Purine-nucleoside phosphorylase deficiency
  Severe combined immunodeficiency with adenosine deaminase deficiency
  Severe combined immunodeficiency:
    Reticular dysgenesis
    Low T- and B-cell numbers
    Low T-cell and normal B-cell numbers (Swiss)
    "Bare lymphocyte" syndrome
  Immunodeficiency with unusual response to Epstein-Barr virus

Immunodeficiency Associated With Other Defects
  Transcobalamin 2 deficiency
  Wiskott-Aldrich syndrome
  Ataxia telangiectasia
  Third- and fourth-pouch/arch (DiGeorge's) syndrome

From Rosen FS, Cooper WD, Wedgewood RJP. Medical progress: TH primary immunodeficiencies. N Engl J Med 1984;311:235.
©1984 Massachusetts Medical Society. All rights reserved.

The IgM–producing B cells later switch to the production of IgG, IgA, and IgE isotypes. IgA is the predominant antibody in secretions. Two such classes of IgA, IgA1 and IgA2, are recognized. In serum, the predominant IgA is IgA1 (90%), and IgA2 represents only 10%. In secretions, both IgA1 and IgA2 are almost equal. IgD is present in very small quantities in serum. Its role is not well understood. IgE antibodies have the capacity to bind to high–affinity receptors on mast cells and basophils and promote degranulation of these cells. Degranulation releases preformed mediators, such as histamine, eosinophil chemotactic factors of anaphylaxis, and platelet-activating factor, and also triggers the synthesis and

TABLE 30-17. Properties of Immunoglobulins

Property	IgG				IgA	IgM	IgD	IgE
	IgG$_1$	IgG$_2$	IgG$_3$	IgG$_4$				
Molecular weight ($\times 10^{-3}$)	150	150	150	150	150	150–600	900	190
Present in secretions					++	–		
Crosses placenta	+	±	+	+	–	–	–	–
Fixes complement	++	+	++	–	–	++		–
Binds to mast cells	–	–	–	+	–	–	–	++
Binds to macrophages	++	+	++	±		–		

release of leukotrienes. These substances cause allergic reactions by increasing vascular permeability and smooth muscle contraction.

Neonatal and Infant Antibody Production. Antibodies synthesized by the human fetus are mainly IgM and, to a much lesser degree, IgA. In the normal fetus, the bulk of circulating antibodies represent transfer of maternal IgG by means of the placenta. Virtually no IgA and IgM cross the placenta.

Maternal IgG is slowly catabolized, with a half-life of about 30 days, and by the third month, the infant has lost about 87% of maternal IgG. Because of a delayed onset of IgG synthesis in the human infant, a physiologic hypogammaglobulinemia stage occurs between 4 and 9 months. Adult levels are attained by the end of the first year of life. Severe hypogammaglobulinemia may be seen in premature infants owing to birth before completion of transplacental transfer of gamma-globulin.

Humoral Immunity Disorders

Transient hypogammaglobulinemia of infancy is characterized by an abnormal prolongation and degree of the physiologic hypogammaglobulinemia. This condition affects both male and female infants, and sometimes there is a familial occurrence. Affected infants present with recurrent infections of the upper and lower respiratory tracts and rarely of skin or meninges. In some cases, this disorder is associated with food allergies.

Lymph nodes from these patients display very small or no germinal centers with marked reduction in the number of plasma cells. Because these patients have a normal number of circulating B cells, the defect presumably involves terminal differentiation of B cells into antibody-producing plasma cells. In patients with recurrent infection, gamma-globulin replacement therapy is indicated, usually for 12 to 36 months. It is discontinued when IgG synthesis is adequate, which occurs spontaneously at about 4 years of age.

X-linked agammaglobulinemia is also called Bruton's disease or *congenital agammaglobulinemia*. This disorder presents with early onset of recurrent pyogenic infections. These patients have no circulating B cells, low concentrations of all circulating immunoglobulin classes, and absence of plasma cells in all lymphoid tissue. T cells are not involved. This disease is transmitted in an X-linked recessive pattern. Patients are usually symptom free during the first 6 months of life, probably because of protection conferred by maternal gamma-globulin. The most common infections are of the lower and upper respiratory tracts, causing pneumonia, otitis, purulent sinusitis and bronchiectasis, meningitis, sepsis, pyoderma, and osteomyelitis. Without early replacement of gamma-globulin, many of these children develop bronchiectasis and die from pulmonary complications. Cellular immunity in these patients is intact, with normal delayed hypersensitivity, normal T-cell number, and normal in vitro response to mitogens, antigens, and allogenic cells.

Acquired agammaglobulinemia is also known as *common variable agammaglobulinemia* and represents a group of disorders characterized by hypogammaglobulinemia. Patients have immunodeficiency that varies in time of onset as well as clinical and immunologic pattern. The etiology of common variable agammaglobulinemia is unknown. There is no clear genetic influence. Both sexes are affected equally, and symptoms rarely occur before the age of 6 years. The disorder should be suspected in any patient with chronic progressive bronchiectasis. Noncaseating granulomas of the lung are also a frequent occurrence. IgG levels are usually less than 500 mg/100 mL, and IgA and IgM levels are less than 50 mg/100 mL. Although most patients have normal numbers of circulating B cells, plasma cells are rarely found. In addition to B-cell deficiency, many of these patients have T-cell abnormalities. More than half eventually exhibit cutaneous anergy.

In addition to these immunodeficiencies, there are also selective IgE subclass deficiencies, selective IgA deficiency, and immunodeficiency with increased IgM. In the latter, IgA and IgG are deficient, but IgM is normal or elevated.[33]

The T-Lymphocyte System

T-Cell Compartment. T cells (thymus-derived cells) are involved in several immune mechanisms.[4,34] One subset of T cells, called helper or inducer cells, helps B cells in the process of differentiation to antibody secreting cells. In addition, helper T cells are involved in expanding another subset of cytotoxic and suppressor T cells. The function of cytotoxic cells is to bind and destroy target cells, such as tumor cells or virus-infected cells. T cells are responsible for contact sensitivity (*eg,* poison ivy dermatitis), delayed hypersensitivity reactions (*eg,* tuberculosis test), immunity to intracellular organisms (viruses, tuberculosis, brucellosis),[35] immunity to fungal organisms, and graft-versus-host disease.

The T-cell population can be subdivided into two subsets, one of which has helper T function and the other of which has suppressor T activity. Release of soluble factors by activated T cells is important in the immune response. These lymphokines include macrophage chemotactic and activation factors, B-cell growth and differentiation factors, and gamma interferons.[8] Activated suppressor T cells, in turn, release soluble factors that suppress helper T cells, dampening the immune response.

T-Cell Immunodeficiency Diseases

T-cell immunodeficiency can result in severe fungal and viral infections. Because T cells pass through multiple stages in thymic development and T-cell activation is a multistage process, we can find defects in both maturation and activation that ultimately would result in functional T-cell deficiencies. The investigation of T-cell function includes a complete blood count because lymphopenia is often seen. Chest radiographs to assess thymic shadow are also useful. Delayed hypersensitivity skin tests are helpful in ensuring T-cell function. (Tetanus toxoid, *Monilia* species, and mumps are the usual skin tests performed.) Total T-cell and T-cell subset distribution can be readily seen by use of monoclonal antibodies: T_3 for total T cells, T_4 for helper cells, and T_8 for suppressor T cells.

Primary T-Cell Deficiency. When T-cell deficiency is severe, both T- and B-cell compartments are deficient, resulting in severe combined T- and B-cell deficiency. This immunodeficiency can be inherited as an autosomal recessive or an X-linked recessive trait; 75% of patients with these disorders are male. In some cases, the T cells do not express HLA-A and HLA-B antigens. These "bare" lymphocytes are functionally incompetent. These patients usually reveal absence of tonsils and lymph nodes, evidence of wasting and oral thrush, absence of thymic shadow on radiographs, lymphopenia, and markedly decreased immunoglobulins. Delayed skin tests are always negative. Proliferative responses to mitogens and antigens are absent. Acquired immunodeficiency syndrome (AIDS) must be ruled out in the differential diagnosis. Stem cell transplantation can cure many primary immunodeficiencies. In many instances, unless bone marrow transplantation is performed, death usually occurs within the first few weeks of life.

DiGeorge's Syndrome (Thymic Hypoplasia, Third and Fourth Pouch Syndrome). DiGeorge's syndrome is probably a developmental rather than an inherited disease. It is manifested by dysmorphogenesis of the third and fourth pharyngeal pouches, which results in aplasia or severe hypoplasia of the thymus and parathyroid glands. It is associated with abnormalities of the great vessels, atrial and ventricular septal defects, esophageal atresia, bifid vulva, short philtrum, mandibular hypoplasia, hypertelorism, and low-set, notched ears. These patients usually develop hypocalcemic seizures. Cardiac defects for which these children are usually evaluated initially are the most common cause of death.

Graft-Versus-Host Disease. When a T-cell–deficient host receives a transfusion of HLA-incompatible immunocompetent T cells, necrosis of skin, liver, and gastrointestinal tract can result. This can be precipitated by intrauterine transfusions, blood transfusions, or more commonly, bone marrow transplantation.

The acute disease begins 7 to 14 days after grafting and usually starts with a maculopapular skin eruption or with "scalded skin" syndrome and alopecia. The epidermis shows coagulation necrosis. Liver changes can result in liver failure. Chronic diarrhea and malabsorption are poor prognosticators. Latent viral infections and opportunistic infections can manifest and may be lethal.

Recent approaches to preventing acute graft-versus-host disease associated with bone marrow transplantation consist of depleting the donor bone marrow of T cells by use of monoclonal antibodies such as antithymocyte globulins and complement by absorption on lectin columns.[34] Patients who recover from this acute episode may present dermatofibrosis, progressive hepatitis, or Sjögren's syndrome weeks or years later.

Chronic Mucocutaneous Candidiasis. Chronic mucocutaneous candidiasis is not a single clinical entity but a collection of related syndromes. Cell-mediated immunity is most frequently impaired. Other deficiencies include low secretory IgA, abnormal complement system, and impaired macrophage-monocyte function. Some children show concomitant polyendocrine failure.

Wiskott-Aldrich Syndrome.[36] This syndrome is characterized by recurrent infections, thrombocytopenia, and eczema. It is X-linked and is present in infancy. The basic defect has not been identified. Initially, patients present with otitis, upper respiratory tract infections, sepsis, and meningitis. When T-cell function diminishes, viral, fungal, and *Pneumocystis carinii* infections may occur. In vitro mitogen responses are often normal. Responses to antigens are usually absent. The diagnosis is confirmed by the presence of platelets that are half the normal size. IgG levels are normal or slightly low, IgM is slightly low, and IgA and IgE are elevated.

Ataxia Telangiectasia. This is an autosomal recessive disorder characterized by progressive cerebellar ataxia associated with increasing telangiectasia beginning in the bulbar conjunctiva and later spreading to the skin; infections are common. Tumors, usually of the lymphoreticular system, are the most common cause of death. Numerous chromosomal breaks and translocations involving chromosomes 7 and 14 have been reported.

Hyper-IgE Syndrome.[34,37,38] Hyper-IgE syndrome is a rare primary immunologic deficiency characterized by recurrent staphylococcal abscesses and markedly elevated IgE. Patients have a lifelong history of severe staphyloccal infections involving the skin, lungs, joints, and other sites. This syndrome is a multisystem disorder that is inherited as an autosomal dominant trait. There is an abnormally low anamnestic antibody response to booster immunizations and poor antibody and cell-mediated responses to neoantigens. Serum IgE is usually greater than 10,000 IU/mL. Lymphocyte proliferative responses to the antigens *Candida albicans* and tetanus toxoid are absent or very low in those tested.

AIDS in Children. Transplacental acquisition of AIDS causing virus results in a syndrome that usually involves B-cell dysfunction. These children grow poorly and present with lymphadenopathy. Often, parotid hypertrophy and otitis, sepsis pneumonitis, pneumonia due to Epstein-Barr virus and cytomegalovirus, hypergammaglobulinemia, and HIV antibody are present.

The Complement System

The primary deficiencies of the complement system include C1q, C1r, C2, C3, C4, C5, C6, C7, C8, and C9 as well as C1 inhibitor deficiency (hereditary) and angioneurotic edema. The main clinical features of these inherited diseases are summarized in Table 30-18.

TABLE 30-18. Inherited Deficiencies in Complement and Complement-Related Protein

Deficient Protein	Observed Pattern of Inheritance at Clinical Level	Reported Major Clinical Correlates*
C1q	Autosomal recessive	Glomerulonephritis, systemic lupus erythematosus
C1r	Probably autosomal recessive	Syndrome resembling systemic lupus erythematosus
C1s	Found in combination with C1r deficiency	Systemic lupus erythematosus
C4	Probably autosomal recessive (two separate loci, C4A and C4B)†	Syndromes resembling systemic lupus erythematosus
C2	Autosomal recessive, HLA-linked	Systemic lupus erythematosus, discoid lupus erythematosus, juvenile rheumatoid arthritis, glomerulonephritis
C3	Autosomal recessive	Recurrent pyogenic infections, glomerulonephritis
C5	Autosomal recessive	Recurrent disseminated neisserial infections, systemic lupus erythematosus
C6	Autosomal recessive	Recurrent disseminated neisserial infections
C7	Autosomal recessive	Recurrent disseminated neisserial infections, Raynaud's phenomenon
C8 beta-chain or C8 alpha-γ chains	Autosomal recessive	Recurrent disseminated neisserial infections
C9	Autosomal recessive	None identified
Properdin	X-linked recessive	Recurrent pyogenic infections, fulminant meningococcemia
Factor $\overline{D}$	?	Recurrent pyogenic infections
C1 inhibitor	Autosomal dominant	Hereditary angioedema, increased incidence of several autoimmune diseases‡
Factor H	Autosomal recessive	Glomerulonephritis
Factor I	Autosomal recessive	Recurrent pyogenic infections
CR1	Autosomal recessive§	Association between low numbers of erythrocyte CR1 and systemic lupus erythematosus
CR3	Autosomal recessive¶	Leukocytosis, recurrent pyogenic infections, delayed umbilical-cord separation

*Note that many people with complement deficiencies, especially of C2 and the terminal components, are clinically well. A substantial number of patients with defects in C5 through C9 have had autoimmune disease.

†The deficiency in people lacking C4A or C4rB is referred to as q = 0, for quantity zero. Thus, such a deficiency can be designated C4Aq0 or C4bq0. People with such deficiencies are reported to have a higher than normal incidence of autoimmune diseases. Similarly, heterozygous C2-deficient people are reported to have an increased incidence of autoimmune disease.

‡About 85% of cases involve silent alleles, and 15% involve alleles encoding for dysfunctional variant C1-inhibitor protein.

§Homozygosity for a low (not absent) numerical expression of CP1 on erythrocytes is detectable in vitro and appears to be associated with systemic lupus erythematosus. An acquired defect in the number of CP1 receptors may also be operative.

¶Low but not absent levels of leukocyte CR3 are detectable in both parents of most CRF3-deficient children.

From Henry JB. Clinical diagnosis and management by laboratory methods. 18th ed. Philadelphia, Saunders, 1991.

The Phagocytic System

Phagocytes include neutrophils and monocyte-macrophages. Their primary role is to engulf and kill microbial organisms. Phagocytosis is enhanced by opsonization. Phagocytic disorders can be classified as quantitative disorders and functional disorders.

Quantitative Disorders. When the neutrophil count is below 500/mm³, there is increased risk for infection. This decrease can be secondary to antimetabolite therapy or to autoimmune disease. Neutropenia may be congenital.

Qualitative Disorders. Defects in chemotaxis result in superficial infections of skin and mucous membranes. Disorders of chemotaxis can be secondary to circulating inhibitors, as seen in some patients with Hodgkin's disease and hepatic cirrhosis. They are also seen in immotile cilia syndrome and Schichman's syndrome.

Defects in phagocytosis. Three cell-surface molecules have been described: MO1 (on monocytes), LFA1 (lymphocyte functional on T and B cells), and P150/90 (on monocytes). They are all involved in cell adhesion. MO1 has been shown to be the receptor for the fragment of the third component of complement (C3bi), and its expression is essential for bacterial phagocytosis, chemotaxis, and adherence. LFA1 is important for T cell–monocyte interaction.

Patients with defective phagocytosis have severe recurrent pyogenic infections. The diagnosis is made by demonstrating absence of MO1 antigen by fluorescence technique and by the failure of neutrophils to ingest opsonized bacteria adequately. The severity of the defect is variable. Patients with the lowest expression of LEV CAM suffer the most severe, potentially fatal, recurrent infections.

Defects in intracellular killing. The prototype is chronic granulomatous disease. Opsonized bacteria are phagocytized, but the mechanism of producing superoxide (which is needed for intracellular killing) is defective. This condition is usually inherited in an X-linked recessive manner and is extremely rare.

The organisms most prone to be involved in this type of defect are the ones that do not produce hydrogen peroxide or that produce catalase, such as *Klebsiella, Serratia,* and *Salmonella* species and certain fungi such as *Aspergillus* and *Candida* species. When the organism involved is not killed by the neutrophils, macrophages are summoned to the scene and form a granuloma. When these reach large sizes, they can obstruct lumina. Pulmonary disorders occur in nearly all children with chronic granulomatous disease and include hilar lymphadenopathy, bronchopneumonia empyema, and lung abscess; gastrointestinal disorders are common. Perianal fistula, steatorrhea, and vitamin B_{12} malabsorption are seen. Osteomyelitis, particularly of small bones of the hands and feet, is also seen. Chronic granulomatous disease is distinguished by an inability of leukocytes to change nitroblue tetrazolium (NBT) from colorless to deep blue during phagocytosis. Bone marrow transplantation has been successful in correcting the defect.

Chédiak-Higashi Syndrome.[39] The clinical features of Chédiak-Higashi syndrome consist of partial albinism, hypopigmentation, recurrent fever and infection, mostly with *Staphylococcus aureus,* and abnormal leukocyte morphology. There is also hepatosplenomegaly, lymphadenopathy, neutropenia, anemia, and perivascular infiltration. Death usually occurs as a result of infection (65%), bleeding (15%), or respiratory failure due to lymphohistiocytic and reticular cell infiltrate. Ascorbic acid and agents that increase cAMP have both been found to help some patients.

SUMMARY

Pediatrics is a branch of medicine that deals with the treatment of children. The birth of an infant is attended by rapid adaptation to extrauterine life. The infant adapts by initiating active respiration, increasing kidney function, closure of the patent ductus arteriosus of the heart, and eating. Generally, there is a loss of weight after delivery followed by a weight gain. Kidney function increases during the first year, and liver function matures within 30 to 60 days. Thyroid function changes abruptly following birth, and hemoglobin shifts over a period of 90 days from fetal to adult hemoglobin. Many other hormonal related changes occur during adolescence. Because of all of these changes, it is important that evaluation of illness in infants include the use of sex- and age-related reference values. Treatment of the pediatric patient often requires the collection of blood samples, which may be complicated by the patient's size and inability to communicate.

Clinical laboratorians need to make themselves aware of the special requirements and precautions regarding blood collection in pediatric patients. There are several aspects of infant development or physiology that also need to be considered in the treatment of pediatric patients. These include acid–base balance, electrolyte and water balance, energy metabolism, developmental changes in hormone balance, drug therapy and pharmacokinetics, inborn errors of metabolism, and humoral and cellular immunity. Details of each of these physiologic aspects are discussed in the chapter.

REVIEW QUESTIONS

1. Which of the following occurs in an infant immediately after birth:
   a. Kidney function decreases to adult values
   b. TSH increases
   c. Hemoglobin shifts from fetal to adult hemoglobin
   d. Bilirubin decreases
2. Which of the following should be used to collect blood from a small or premature infant?
   a. Needle and butterfly
   b. Appropriate lancet
   c. Needle and syringe
   d. Blood should not be drawn from a premature infant.
3. Which of the following blood-gas results would be expected in a premature infant with respiratory distress syndrome?
   a. Elevated pCO_2, pH of 7.20 to 7.25
   b. Normal pCO_2, normal pH
   c. Decreased pCO_2, pH of 7.20 to 7.25
   d. Normal pCO_2, pH of 7.45 to 7.55
4. A frequent cause of hyperammonemia in infants is:
   a. Deficiency of the enzyme ornithine transcarbamylase
   b. Hepatic failure
   c. Hyperalimentation
   d. None of the above
5. The reference range for triglycerides in children 1 to 5 years of age is:
   a. 25 to 270
   b. 90 to 260
   c. 95 to 240
   d. 20 to 120

6. *Distribution* of drugs in infants and children is different from adults because of:
   a. Protein binding of drugs
   b. Decreased renal function
   c. Immaturity of the liver
   d. Higher gastric acidity
7. A fairly common inborn error of metabolism that has an incidence of 1 in 2400 live births is:
   a. PKU
   b. Galactosemia
   c. Hypothyroidism
   d. Cystic fibrosis
8. The main antibody synthesized by the human fetus is:
   a. IgA
   b. IgM
   c. IgG
   d. IgD

REFERENCES

1. Soldin SJ, Hicks JMB. Pediatric clinical chemistry reference values. Washington DC: American Association of Clinical Chemistry, 1995.
2. National Committee for Clinical Laboratory Standards (NCCLS). C28-A, How to define, determine and utilize reference intervals in the clinical laboratory. Villanova, PA: NCCLS, 1995.
3. Soldin SJ, Rifai N, Hicks SMB, eds. The biochemical bases of pediatric disease. Washington DC: American Association of Clinical Chemistry, 1992.
4. Tilton RC, Balows, Hohnadel DC, et al, eds. Clinical laboratory medicine. St. Louis: Mosby–Year Book, 1992.
5. Behrman RE, Behrman RE, Kliegman RM, et al, eds. Nelson's textbook of pediatrics. 15th ed. Philadelphia: WB Saunders, 1995.
6. National Committee for Clinical Laboratory Standards (NCCLS). H3-A3, Procedures for the collection of diagnostic blood specimens by venipuncture. Villanova, PA: NCCLS, 1991.
7. National Committee for Clinical Laboratory Standards (NCCLS). H4-A3, Procedures for the collection of diagnostic blood specimens by skin punctures. 3rd ed. Villanova, PA: NCCLS, 1991.
8. Kaplan LA, Pesce AJ, eds. Clinical chemistry: theory, analysis, correlation. 3rd ed. St. Louis: Mosby–Year Book, 1996.
9. National Committee for Clinical Laboratory Standards (NCCLS). C32-P, Considerations in the simultaneous measurement of blood gases, electrolytes and related analytes in whole blood: proposed guideline. Villanova, PA: NCCLS, 1993.
10. Pflederer TA. Emergency fluid management for hypovolemia. Postgrad Med 1996;100:243–251.
11. Fann BD. Fluid and electrolyte balance in the pediatric patient. J Intravenous Nurs 1998;21:153–159.
12. Cornell S. Maintaining a fluid balance: appropriate management of dehydration in children. Adv Nurse Pract 1997;5:43–44.
13. Hugger J, Harkless G, Rentschler D. Oral rehydration therapy for children with acute diarrhea. Nurse Pract 1998;23:57–62.
14. Liebelt EL. Clinical and laboratory evaluation and management of children with vomiting diarrhea and dehydration. Curr Opin Pediatr 1998;10:461–469.
15. Glaser NS. Noninsulin dependent diabetes mellitus in childhood and adolescence. Pediatr Clin North Am 1997;44:307–337.
16. Sperling MA. Aspects of the etiology, prediction, and prevention of insulin-dependent diabetes mellitus in childhood. Pediatr Clin North Am 1997;44:269–284.
17. Joukendrup AE, Saris WH, Wagenmakers AJ. Fat metabolism during exercise: a review. Part I: fatty acid metabolism and muscle metabolism. Int J Sports Med 1998;19:231–234.
18. Hamosh M. Lipid metabolism in pediatric nutrition. Pediatr Clin North Am 1995;42:839–849.
19. Ogilvy-Stuart AL. Endocrinology of the neonate. Br J Hosp Med 1995;54:207–211.
20. Audi L, Granada ML, Carrascosa A. Growth hormone secretion assessment in the diagnosis of short stature. J Pediatr Endocrinol Metab 1996;3:313–324.
21. Bell JJ, Dana K. Lack of correlation between growth hormone provocative test results and subsequent growth rates during growth hormone therapy. Pediatrics 1998;102:518–520.
22. Styne DM. New aspects in the diagnosis and treatment of pubertal disorders. Ped Clinics of North Amer 1997; 44:505–529.
23. Styne DM. New aspects in the diagnosis and treatment of pubertal disorders. Pediatr Clin North Am 1997;44:505–529.
24. Schriver CR, Beaudet AL, Sly WS, eds. The metabolic and molecular basis of inherited disease. 7th ed. New York: McGraw-Hill, 1995.
25. Swerloff RS, Wang C. Influence of pituitary disease on sexual development and functioning. Psychother Psychosomat 1998;67:173–180.
26. DeLange F. Neonatal screening for congenital hypothyroidism: results and perspectives. Horm Res 1997;48:51–61.
27. Pagliaro LA, Pagliaro AM, eds. Problems in pediatric drug therapy. 3rd ed. Hamilton, IL: Drug Intelligence Publications, 1995.
28. Leeder SJ, Kearns GL. Pharmacogenetics in pediatrics, Pediatr Clin North Am 1997;44:55–77.
29. Matsui DM. Drug compliance in pediatrics: clinical and research issues. Pediatr Clin North Am 1997;44:1–14.
30. National Committee for Clinical Laboratory Standards (NCCLS). GP16-A, Routine urinalysis and collection transportation and preservation of urine specimens. Villanova, PA: NCCLS, 1995.
31. Chace D, Sherwin JE, Lorey F, et al. Use of phenylalanine-to-tyrosine ratio by tandem mass spectrometry to screen for phenylketonuria in specimens collected in first 24 Hours. Clin Chem 1998;44:2405–2409.
32. Walker CBJ, Bruce DM, Heys SD, et al. Minimal modulation of lymphocyte and natural killer cell subsets following minimal access surgery. Am J Surg 1999;117:48–54.
33. Grimbacher B, et al. Hyper-IgE syndrome with recurrent infections: an autosomal dominant multisystem disorder. N Engl J Med 1999;340:696–702.
34. Kapoor N, Crooks G, Kohn DB, et al. Hematopoietic stem cell transplantation for primary lymphoid immunodeficiencies. Semin Hematol 1998;35:346–353.
35. Selin LK, Varga SM, Wong IC, et al. Protective heterologous antiviral immunity and enhanced immunopathogenesis mediated by memory T cell populations. J Exp Med 1998;188:1705–1715.
36. Abo A. Understanding the molecular basis of Wiskott-Aldrich syndrome. Cell Mol Life Sci 1998;Oct 54 (10) 1145–1153.
37. Salaria M, Singh S, Kumar L. Hyperimmunoglobulin E syndrome. Indian Pediatr 1997;34:827–829.
38. Pasic S, Lilic D, Pejnovic N, et al. Disseminated bacillus Calmette-Guérin infection in a girl with hyperimmunoglobulin E syndrome. Acta Paediatr 1998;87:702–704.
39. Faigle W, et al. Deficient peptide loading and MHC class II endosomal sorting in a human genetic immunodeficiency disease: the Chediak-Higashi syndrome. J Cell Biol 1998;141:1121–1134.

Appendices

For additional information, refer to the latest edition of one of the following excellent references:
Bold, AM, and Wilding, P. Clinical chemistry companion. Oxford: Blackwell Scientific Publications.
Weast, RC. CRC handbook of chemistry and physics, Boca Raton, FL: CRC Press. Werner, M.
CRC handbook of clinical chemistry. Vol. 1. Boca Raton, FL: CRC Press.

APPENDIX A. Selected Laboratory Abbreviations

A	Ampere, current		M.	*misce, mistura* (Latin), mix, mixture
ABC	Arterial blood gases		min.	minute(s)
AC	Alternating current		mo	months
a.c.	*ante cibum* (Latin), before meals		NA	Not applicable
ad lib	*ad libitum* (Latin), at will		n/m	Not measured
A/G	Albumin-globulin ratio		NPO	*non per orum* (Latin), nothing by mouth
AMAP	As much as possible		o.d.	*omni die* (Latin), everyday
AOAP	As often as possible		OR	Operating room
BBT	Basal body temperature		p.c.	*post cibum* (Latin), after meals
BFP	Biologic false positive		PO	*per os* (Latin), by mouth
b.i.d.	*bis in die* (Latin), twice daily		ppm	Parts per million
BMR	Basal metabolic rate		Ppt.	Precipitate
BP	Blood pressure		p.r.n.	*pro re nata* (Latin), as required
bp	Boiling point		psi	Pounds per square inch
BS	Blood sugar		Px	Physical examination
BSA	Body surface area		QA	Quality Assurance
C	Coulomb, quantity of electricity		q.a.m.	*quaque ante meridiem* (Latin), every morning
CCU	Coronary care unit		QC	Quality Control
CLIA	Clinical Laboratory Improvement Act		q.d.	*quaque die* (Latin), every day
DC	Direct current		q.h.	*quaque hora* (Latin), every hour
DOA	Dead on arrival		q.i.d.	*quater in die* (Latin), 4 times a day
DRG	Diagnostic-related group		q.n.s.	*quantum non satis* (Latin), quantity not sufficient
Dx	Diagnosis		q.s.	*quantum satis* (Latin), quantity sufficient
ECG,EKG	Electrocardiogram		rad	Dose of radiation absorbed by tissue
EEG	Electroencephalogram		RDA	Recommended dietary allowance
EM	Electron microscopy		rem	Roentgen equivalent man
ER	Emergency Room		R/O	Rule out
FBS	Fasting blood sugar		RR	Recovery room
Fp	Freezing point		Rx	*recipe* (Latin), prescription
FVU	First voided urine		SBP	Systolic blood pressure
GTT	Glucose tolerance test		SE	Standard error
gtt.	*guttae* (Latin), drops		SI	Système International d'Unités
h, hr	hour		SQ	Subcutaneous
h.s.	hour of sleep, at bedtime		stat.	*statim* (Latin), immediately
Hx	history		TBW	Total body weight
IB	Inclusion body		T/D	Treatment discontinued
ICU	Intensive care unit		t.i.d.	*ter in die* (Latin), 3 times a day
ICCU	Intensive coronary care unit		TLV	Threshold limit value
ID	Initial dose		TO	Telephone order
I & O	Intake and output		Tx	Treatment
IR	Infrared		UV	Ultraviolet
IV	Intravenous		V	Volt or volume
J	Joule, energy		W	Watt, power
KUB	Kidney, ureter, bladder		wk	week(s)
LD50	Lethal dose (median)		WNL	Within normal limits
LED	Light-emitting diode		W/V	Weight by volume
LFT	Liver function test			

APPENDIX B. Basic SI Units

Measurement	Name	Symbol
Length	Meter	m
Mass	Kilogram	kg
Quantity of substance	Mole	mol
Time	Second	s
Electric current	Ampere	A
Thermodynamic temperature	Kelvin	K
Luminous intensity	Candela	cd

Note: SI (Systèms International d'Unitès) are units having a definition recognized by international agreement. Note that some SI units have capitalized symbols. This is to avoid confusion with SI prefixes using the same latter symbol.

APPENDIX C. Prefixes to be Used with SI Units

Factor	Prefix	Symbol
10^{-18}	atto	a
10^{-15}	femto	f
10^{-12}	pico	p
10^{-9}	nano	n
10^{-6}	micro	μ
10^{-3}	milli	m
10^{-2}	centi	c
10^{-1}	deci	d
10^{1}	deka	da
10^{2}	hecto	h
10^{3}	kilo	k
10^{6}	mega	M
10^{9}	giga	G
10^{12}	tera	T
10^{15}	peta	P
10^{18}	exa	E

Note: Prefixes are used to indicate a subunit or multiple of a basic SI unit.

APPENDIX D. Basic Clinical Laboratory Conversions

Length, Volume, Weight Conversions

To Convert	Into	Multiply By
Inches	Centimeters	2.54
Centimeters	Inches	.39
Yards	Meters	.91
Meters	Yards	1.09
Gallons (US)	Liters	3.78
Liters	Gallons (US)	.26
Fluid ounces (US)	Milliliters	29.6
Milliliters	Fluid ounces (US)	.034
Ounces	Grams	28.4
Grams	Ounces	.035
Pounds	Kilograms	.45
Kilograms	Pounds	2.2

Temperature Conversions

To Convert	Into	Use
Centigrade (°C)	Kelvin (°K)	$°K = °C + 273$
Centigrade (°C)	Fahrenheit (°F)	$°F = (°C \times 1.8) + 32$
Fahrenheit (°F)	Centigrade (°C)	$°C = (°F - 32) \times 0.556$

Concentration Conversions

To Convert	Into	Use
% w/v	Molarity (M)	$M = \dfrac{\% \text{ w/v} \times 10}{\text{GMW}}$
% w/v	Normality (N)	$N = \dfrac{\% \text{ w/v} \times 10}{\text{eq wt}}$
mg/dL	mEq/L	$\text{mEq/L} = \dfrac{\text{mg/dL} \times 10}{\text{eq wt}}$
Molarity	Normality	$N = M \times \text{valence}$

APPENDIX E. Conversion of Traditional Units to SI Units for Common Chemistry Analytes*

	Conventional/ Current	SI Unit	Conversion Factor
Albumin	g/100 mL	g/L	10
Aspartate aminotransferase (AST)	U/L (mU/mL)	μkat/L	0.0167
Ammonia	μg/dL	μmol/L	0.587
Bicarbonate (HCO$_3$)	mEq/L	mmol/L	1.0
Bilirubin	mg/dL	μmol/L	17.1
BUN	mg/dL	mmol/L	0.357
Calcium	mg/dL	mmol/L	0.25
Chloride	mEq/L	mmol/L	1.0
Cholesterol	mg/dL	mmol/L	0.026
Cortisol	μg/dL	μmol/L	0.0276
Creatinine	mg/dL	μmol/L	88.4
Creatinine clearance	mL/min	mL/s	0.0167
Folic acid	ng/mL	nmol/L	2.27
Glucose	mg/dL	mmol/L	0.0555
Hemoglobin	g/dL	g/L	10
Iron	mg/dL	μmol/L	0.179
Lithium	mEq/L	μmol/L	1.0
Magnesium	mEq/L	mmol/L	0.5
Osmolality	mOsm/kg	mmol/kg	1.0
Phosphorus	mg/dL	mmol/L	0.323
Potassium	mEq/L	mmol/L	1.0
Sodium	mEq/L	mmol/L	1.0
Thyroxine (T$_4$)	μg/dL	nmol/L	12.9
Total protein	g/dL	g/L	10
Triglyceride	mg/dL	mmol/L	0.0113
Uric acid	mg/dL	mmol/L	0.0595
Vitamin B$_{12}$	ng/mL	pmol/L	0.0738
PCO$_2$	mm/Hg	kPa	0.133
PO$_2$	mm/Hg	kPa	0.133

* To obtain SI unit, multiply current unit by conversion factor. To obtain the conventional or current unit, divide the SI unit by the conversion factor.

APPENDIX F. Greek Alphabet Table

Greek Letter Upper, Lower Case	Greek Name	English Equivalent
A α	Alpha	a
B β	Beta	b
Γ γ	Gamma	g
Δ δ	Delta	d
E ε	Epsilon	e
Z ζ	Zeta	z
H η	Eta	ē
Θ θ	Theta	th
I ι	Iota	i
K κ	Kappa	k
Λ λ	Lambda	l
M μ	Mu	m
N ν	Nu	n
Ξ ξ	Xi	x
O o	Omicron	o
Π π	Pi	p
P ρ	Rho	r
Σ σ	Sigma	s
T τ	Tau	t
Y υ	Upsilon	ũ
Φ φ	Phi	ph
X χ	Chi	ch
Ψ ψ	Psi	ps
Ω ω	Omega	ō

Representative examples:

α_1 antitrypsin ϵ (molar absorptivity) Σ (sum of)

β globulin λ (wavelength) π (= 3.14...)

ΔA (change in) $X \mu L$ (volume) Ω (Ohm, resistance)

APPENDIX G. Concentrations of Commonly Used Acids and Bases with Related Formulas

Chemical Substance	Chemical Formula	Formula Weight/GMW	Equivalent Weight	Average Specific Gravity Conc Reagent*	Average % Purity CONC Reagent*	Normality CONC Reagent (Approximate)
Ammonium hydroxide	NH_4OH	35.05	35.05	0.90	28.0	15
Acetic acid (glacial)	CH_5COOH	60.05	60.05	1.06	99.5	18
Formic acid	$HCOOH$	46.03	46.03	1.20	88.0	23
Hydrochloric acid	HCL	36.46	36.46	1.19	37.0	12
Nitric acid	HNO_3	63.02	63.02	1.42	70.0	16
Perchloric acid	$HCLO_4$	100.46	100.46	1.67	71.0	12
Phosphoric acid	H_3PO_4	98.00	32.67	1.69	85.0	44
Sulfuric acid	H_2SO_4	98.08	49.04	1.84	96.0	36

* Varies according to lot and/or manufacturer.

Related formulas:

$$Molarity\ (M) = \frac{g/L}{GMW}$$

$$Normality\ (N) = \frac{g/L}{ef\ wt}$$

$$Equivalent\ weight\ (eq\ wt) = \frac{GMW}{valence}$$

Specific gravity (sp gr):

$$sp\ gr \times \%\ purity\ (in\ decimal\ form) = g\ of\ solute/mL$$

APPENDIX H. Examples of Incompatible Chemicals

Chemical	Is Incompatible With
Acetic acid	Chromic acid, nitric acid, hydroxyl compounds, ethylene glycol, perchloric acid, peroxides, permanganates
Acetylene	Chlorine, bromine, copper, fluorine, silver, mercury
Acetone	Concentrated nitric and sulfuric acid mixtures
Alkali and alkaline earth metals (such as powdered aluminum or magnesium, calcium, lithium, sodium, potassium)	Water, carbon tetrachloride or other chlorinated hydrocarbons, carbon dioxide, halogens
Ammonia (anhydrous)	Mercury (in manometers, for example), chlorine, calcium hypochlorite, iodine, bromine, hydrofluoric acid (anhydrous)
Ammonium nitrate	Acids, powdered metals, flammable liquids, chlorates, nitrites, sulfur, finely divided organic or combustible materials
Aniline	Nitric acid, hydrogen peroxide
Arsenical materials	Any reducing agent
Azides	Acids
Bromine	See Chlorine
Calcium oxide	Water
Carbon (activated)	Calcium hypochlorite, all oxidizing agents
Carbon tetrachloride	Sodium
Chlorates	Ammonium salts, acids, powdered metals, sulfur, finely divided organic or combustible materials
Chromic acid and chromium trioxide	Acetic acid, naphthalene, camphor, glycerol, alcohol, flammable liquids in general
Chlorine	Ammonia, acetylene, butadiene, butane, methane, propane (or other petroleum gases), hydrogen, sodium carbide, benzene, finely divided metals, turpentine
Chlorine dioxide	Ammonia, methane, phosphine, hydrogen sulfide
Copper	Acetylene, hydrogen peroxide
Cumene hydroperoxide	Acids (organic or inorganic)
Cyanides	Acids
Flammable liquids	Ammonium nitrate, chromic acid, hydrogen peroxide, nitric acid, sodium peroxide, halogens
Fluorine	Everything
Hydrocarbons (such as butane, propane, benzene)	Fluorine, chlorine, bromine, chromic acid, sodium peroxide
Hydrocyanic acid	Nitric acid, alkali
Hydrofluoric acid (anhydrous)	Ammonia (aqueous or anhydrous)
Hydrogen peroxide	Copper, chromium, iron, most metals or their salts, alcohols, acetone, organic materials, aniline, nitromethane, combustible materials
Hydrogen sulfide	Fuming nitric acid, oxidizing gases
Hypochlorites	Acids, activated carbon
Iodine	Acetylene, ammonia (aqueous or anhydrous), hydrogen
Mercury	Acetylene, fulminic acid, ammonia
Nitrates	Sulfuric acid
Nitric acid (concentrated)	Acetic acid, aniline, chromic acid, hydrocyanic acid, hydrogen sulfide, flammable liquids, flammable gases, copper, brass, any heavy metals
Nitrites	Acids
Nitroparaffins	Inorganic bases, amines
Oxalic acid	Silver, mercury
Oxygen	Oils, grease, hydrogen, flammable liquids, solids, or gases
Perchloric acid	Acetic anhydride, bismuth and its alloys, alcohol, paper, wood, grease, oils
Peroxides, organic	Acids (organic or mineral), avoid friction, store cold
Phosphorus (white)	Air, oxygen, alkalis, reducing agents
Potassium	Carbon tetrachloride, carbon dioxide, water
Potassium chlorate	Sulfuric and other acids
Potassium perchlorate (see also chlorates)	Sulfuric and other acids
Potassium permanganate	Glycerol, ethylene glycol, benzaldehyde, sulfuric acid
Selenides	Reducing agents
Silver	Acetylene, oxalic acid, tartaric acid, ammonium compounds, fulminic acid
Sodium	Carbon tetrachloride, carbon dioxide, water
Sodium nitrite	Ammonium nitrate and other ammonium salts
Sodium peroxide	Ethyl or methyl alcohol, glacial acetic acid, acetic anhydride, benzaldehyde, carbon disulfide, glycerin, ethylene glycol, ethyl acetate, methyl acetate, furfural
Sulfides	Acids
Sulfuric acid	Potassium chlorate, potassium perchlorate, potassium permanganate (similar compounds of light metals, such as sodium, lithium)
Tellurides	Reducing agents

(Reprinted with permission from National Research Council, Committee on Hazardous Substances in the Laboratory. Prudent practices for handling hazardous chemicals in laboratories. Washington: National Academy Press, 1981. For additional information see Pipitone, DA. Safe storage of laboratory chemicals. 2nd ed. New York: John Wiley & Sons, 1991.

APPENDIX I. Nomogram for the Determination of Body Surface Area

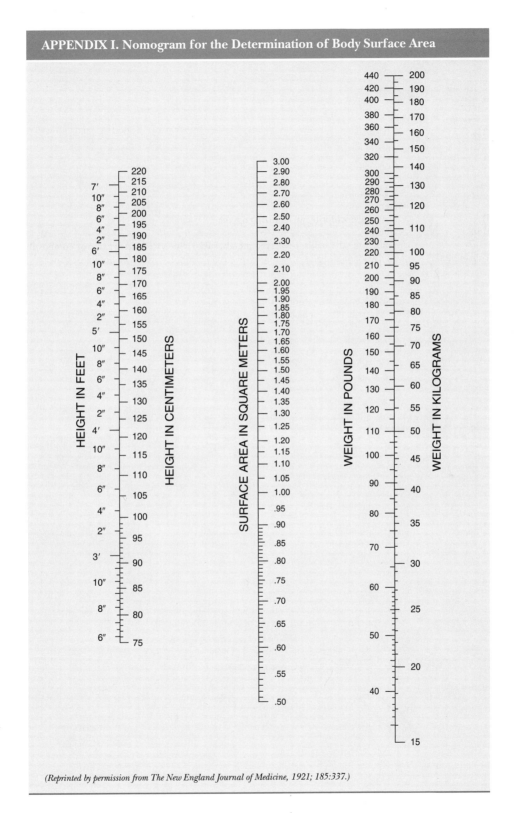

(Reprinted by permission from The New England Journal of Medicine, 1921; 185:337.)

APPENDIX J. Relative Centrifugal Force Nomograph

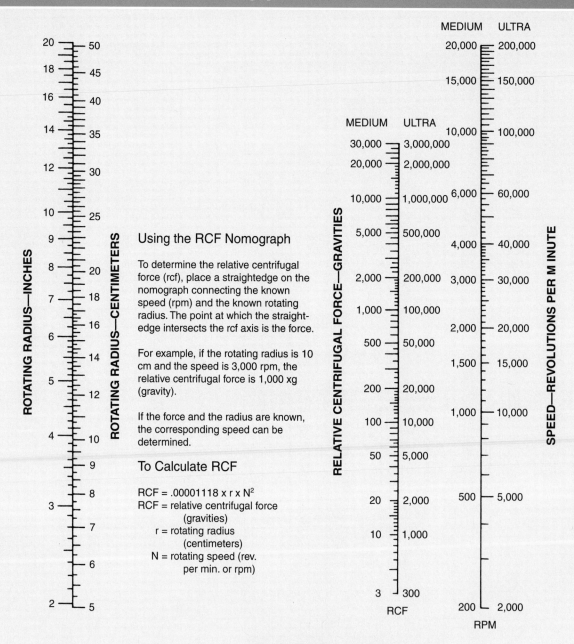

Using the RCF Nomograph

To determine the relative centrifugal force (rcf), place a straightedge on the nomograph connecting the known speed (rpm) and the known rotating radius. The point at which the straight-edge intersects the rcf axis is the force.

For example, if the rotating radius is 10 cm and the speed is 3,000 rpm, the relative centrifugal force is 1,000 xg (gravity).

If the force and the radius are known, the corresponding speed can be determined.

To Calculate RCF

RCF = .00001118 x r x N²
RCF = relative centrifugal force (gravities)
 r = rotating radius (centimeters)
 N = rotating speed (rev. per min. or rpm)

ROTATING TIP RADIUS

The distance measured from the rotor axis to the tip of the liquid inside the tubes at the greatest horizontal distance from the rotor axis is the rotating tip radius.

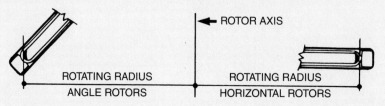

(Reprinted by permission from International Equipment Co., Damon Corporation)

APPENDIX K. Centrifugation—Speed and Time Adjustment

Use the following formula to calculate the new speed necessary to achieve the relative centrifugal force specified in the original procedure, given a different rotor with a different radius.

$$RPM = 1000 \sqrt{\frac{RCF}{1.12\ r}}$$

where RPM = revolutions per minute (speed)
RCF = relative centrifugal force
r = rotating tip radius in millimeters (check manufacturer's rotor specifications)

Use this formula to calculate the new centrifugation time necessary when a different rotor is used.

$$t_N = \frac{t_0 \times RCF_0}{RCF_N}$$

where t_N = centrifugation time necessary with different rotor
t_0 = time (minutes) specified in original procedure
RCF_N = relative centrifugal force of different rotor
RCF_0 = relative centrifugal force specified in original procedure

APPENDIX L. Percent Transmittance–Absorbance Conversion Table

%T	Absorbance* .00	.25	.50	.75	%T	Absorbance* .00	.25	.50	.75
1	2.000	1.903	1.824	1.757	52	.284	.282	.280	.278
2	1.690	1.648	1.602	1.561	53	.276	.274	.272	.270
3	1.523	1.488	1.456	1.426	54	.268	.266	.264	.262
4	1.398	1.372	1.347	1.323	55	.260	.258	.256	.254
5	1.301	1.280	1.260	1.240	56	.252	.250	.248	.246
6	1.222	1.204	1.187	1.171	57	.244	.242	.240	.238
7	1.155	1.140	1.126	1.112	58	.237	.235	.233	.231
8	1.097	1.083	1.071	1.059	59	.229	.227	.226	.224
9	1.046	1.034	1.022	1.011	60	.222	.220	.218	.216
10	1.000	.989	.979	.969	61	.215	.213	.211	.209
11	.959	.949	.939	.930	62	.208	.206	.204	.202
12	.921	.912	.903	.894	63	.201	.199	.197	.196
13	.886	.878	.870	.862	64	.194	.192	.191	.189
14	.854	.846	.838	.831	65	.187	.186	.184	.182
15	.824	.817	.810	.803	66	.181	.179	.177	.176
16	.796	.789	.782	.776	67	.174	.172	.171	.169
17	.770	.763	.757	.751	68	.168	.166	.164	.163
18	.745	.739	.733	.727	69	.161	.160	.158	.157
19	.721	.716	.710	.704	70	.155	.153	.152	.150
20	.699	.694	.688	.683	71	.149	.147	.146	.144
21	.678	.673	.668	.663	72	.143	.141	.140	.138
22	.658	.653	.648	.643	73	.137	.135	.134	.132
23	.638	.634	.629	.624	74	.131	.129	.128	.126
24	.620	.615	.611	.606	75	.125	.124	.122	.121
25	.602	.598	.594	.589	76	.119	.118	.116	.115
26	.585	.581	.577	.573	77	.114	.112	.111	.100
27	.569	.565	.561	.557	78	.108	.107	.105	.104
28	.553	.549	.545	.542	79	.102	.101	.100	.098
29	.538	.534	.530	.527	80	.097	.096	.094	.093
30	.523	.520	.516	.512	81	.092	.090	.089	.088
31	.509	.505	.502	.498	82	.086	.085	.084	.082
32	.495	.491	.488	.485	83	.081	.080	.078	.077
33	.482	.478	.475	.472	84	.076	.074	.073	.072
34	.469	.465	.462	.459	85	.071	.069	.068	.067
35	.456	.453	.450	.447	86	.066	.064	.063	.062
36	.444	.441	.438	.435	87	.061	.059	.058	.057
37	.432	.429	.426	.423	88	.056	.054	.053	.052
38	.420	.417	.414	.412	89	.051	.049	.048	.047
39	.409	.406	.403	.401	90	.046	.045	.043	.042
40	.398	.395	.392	.390	91	.041	.040	.039	.037
41	.387	.385	.382	.380	92	.036	.035	.034	.033
42	.377	.374	.372	.369	93	.032	.030	.029	.028
43	.367	.364	.362	.359	94	.027	.026	.025	.024
44	.357	.354	.352	.349	95	.022	.021	.020	.019
45	.347	.344	.342	.340	96	.018	.017	.016	.014
46	.337	.335	.332	.330	97	.013	.012	.011	.010
47	.328	.325	.323	.321	98	.009	.008	.007	.006
48	.319	.317	.314	.312	99	.004	.003	.002	.001
49	.310	.308	.305	.303	100	.0000	.0000	.0000	.0000
50	.301	.299	.297	.295					
51	.292	.290	.288	.286					

*Absorbance = 2 − log % T.

APPENDIX M. Selected Atomic Weights

Element	Symbol	Atomic Number	Atomic Weight*	Valence
Aluminum	Al	13	26.98	3
Antimony	Sb	51	121.75	3,5
Argon	Ar	18	39.95	0
Arsenic	As	33	74.92	3,5
Barium	Ba	56	137.34	2
Beryllium	Be	4	9.01	2
Bismuth	Bi	83	208.98	3,5
Boron	B	5	10.81	3
Bromine	Br	35	79.90	1,3,5,7
Cadmium	Cd	48	112.40	2
Calcium	Ca	20	40.08	2
Carbon	C	6	12.01	2,4
Cerium	Ce	58	140.12	3,4
Cesium	Cs	55	132.91	1
Chlorine	Cl	17	35.45	1,3,5,7
Chromium	Cr	24	51.99	2,3,6
Cobalt	Co	27	58.93	2,3
Copper	Cu	29	63.55	1,2
Fluorine	F	9	18.99	1
Gold	Au	79	196.97	1,3
Helium	He	2	4.00	0
Hydrogen	H	1	1.01	1
Iodine	I	53	126.90	1,3,5,7
Iron	Fe	26	55.85	2,3
Lead	Pb	82	207.19	2,4
Lithium	Li	3	6.94	1
Magnesium	Mg	12	24.31	2
Manganese	Mn	25	54.94	2,3,4,6,7
Mercury	Hg	80	200.59	1,2
Molybdenum	Mo	42	95.94	3,4,6
Nickel	Ni	28	58.71	2,3
Nitrogen	N	7	14.01	3,5
Oxygen	O	8	16.00	2
Phosphorus	P	15	30.97	3,5
Platinum	Pt	78	195.09	2,4
Potassium	K	19	39.10	1
Rubidium	Rb	37	85.47	1
Selenium	Se	34	78.96	2,4,6
Silicon	Si	14	28.09	4
Silver	Ag	47	107.87	1
Sodium	Na	11	22.99	1
Strontium	Sr	38	87.62	2
Sulfur	S	16	32.06	2,4,6
Tellurium	Te	52	127.60	2,4,6
Tin	Sn	50	118.69	2,4
Tungsten	W	74	183.85	6
Uranium	U	92	238.03	4,6
Xenon	Xe	54	131.30	0
Zinc	Zn	30	65.37	2
Zirconium	Zr	40	91.22	4

* Atomic weights are based on C^{12} and have been rounded to two decimal places.

APPENDIX N. Selected Radioactive Elements

The following list of radioactive elements represents those of diagnostic or clinical interest.

Element	Symbol	Atomic No.	Atomic Weight	Half-Life, $t_{1/2}$	
Bismuth	Bi	83	210	5	days
Calcium	Ca	20	45	165	days
Carbon	C	6	14	5730	years
Chromium	Cr	24	51	27.8	days
Cobalt	Co	27	57	270	days
			58	72	days
			60	5.3	years
Hydrogen	H	1	3	12.3	years
Iodine	I	53	125	60	days
			131	8.0	days
Iron	Fe	26	59	45.6	days
Mercury	Hg	80	203	46.9	days
Nickel	Ni	28	63	125	years
Phosphorus	P	15	32	14.3	days
Potassium	K	19	42	12.4	hours
Rubidium	Rb	37	86	18.7	days
Selenium	Se	34	75	127	days
Sodium	Na	11	22	2.6	years
			24	15	hours
Strontium	Sr	38	85	64	days
			90	27.7	years
Sulfur	S	16	35	87.9	days
Technetium	Te	43	99	6.0	hours
Tin	Sn	50	113	115	days
Xenon	Xe	54	133	5.27	days
Zinc	Zn	30	65	245	days

APPENDIX O. Characteristics of Types of Glass

Category	Type of Material	Common or Brand Names	Routine Uses	Limitations
High thermal resistance	Borosilicate with low alkaline content	Pyrex Kimax	All purpose, all types of beakers, flasks, etc. Can tolerate heating and sterilization for lengthy periods of time to 510°C	Should not be cooled too quickly after heating. May cloud after use with a strong alkali. Subject to scratching
	Aluminosilicate	Corex	Centrifuge tubes a thermometers. Extremely strong and hard. Temperature stability to 672°C; short-term use to 850°C	Resists scratching. Subject to some acid or alkali attack at temperature of 100°C
		Vycor	Ashing and ignition techniques. Can withstand very high temperature (900–1200°C), as well as drastic changes in temperature. Most are alkali resistant in this category.	
High silica	96% silica		Cuvets and thermometers. Can be used at high temperatures (900–1200°C) and withstand a sharp change in temperature. Can be considered optically pure (cuvets, thermometers)	
High resistance to alkali	Aluminosilicate		Can be used with strong alkali and suffer minimal attack (0.09 mg/cm² vs. 1.4 mg/cm² for borosilicate or 0.35 mg/cm² for regular aluminosilicate	Must be heated and cooled with care. Highest temperature for safe use is 578°C

APPENDIX P. Characteristics of Types of Plastic

Plastic	Temperature Limit (°C)	Transparency	Autoclavable	Flexibility	Usage Examples
Polystyrene (PS)	70	Clear	No	Rigid	Disposables
Polyethylene Conventional (CPE)	80	Translucent	No	Excellent	All-purpose Reagent bottles Test-tube rack Carboys Droppers
Linear (LPE)	120	Opaque	With caution	Rigid	Specimen transport containers Reagent bottles
Polypropylene (PP)	135	Translucent	Yes	Rigid	Screw-cap closures Bottles
Tygon	95	Translucent	Yes	Excellent	Tubing
Teflon FEP	205	Clear Translucent	Yes	Excellent	Stopcocks Wash bottles Beakers
Polycarbonate (PC)	135	Very clear	Yes	Rigid	All-purpose Large reagent containers Carboys Test-tube rack
Polyvinylchloride* (PVC)	70	Clear	No	Rigid	Bottles/tubing

PVC tubing can be heated to 120°C, can be autoclaved, and is very flexible.

APPENDIX Q. Chemical Resistance of Types of Plastic

Plastic	Chemical Resistance*
Polystyrene	Useful with water and aqueous salt solutions. It is not recommended for use with acids, aldehydes, ketones, ethers, hydrocarbons, or essential oils. Alcohols and bases can be used, but storage beyond 24 h is discouraged.
Polyethylene	Both classifications of polyethylene (*i.e.*, conventional and linear) have similar chemical resistances. They have excellent chemical resistance to most substances, with the exception of aldehydes, amines, ethers, hydrocarbons, and essential oils. For conventional polyethylene, the exceptions should also include lubricating oil and silicones. The usage of any of the above-named chemical groups should be limited to 24 h at room at room temperature.
Polypropylene	Has the same chemical resistance as linear polyethylene.
Teflon	This resin possesses excellent chemical resistance to almost all chemicals used in the clinical laboratory.
Polycarbonate	Very susceptible to damage by most chemicals. It is resistant to water, aqueous salts, food, and inorganic acids for a long period of time.

It should be noted that this information is based on room temperature (22°C) and normal atmospheric pressure. Resistance to chemicals decreases as the temperature of the resin nears its maximum. Chemical resistance will also vary as the concentration of the chemical increases.

APPENDIX R. Cleaning Labware

Glassware "Problem"	Cleaning Technique
General usage (procedure 1 is recommended for routine washing needs)	1. *Dirty* glassware should be immediately placed in a soapy or dilute bleach solution and allowed to soak. Wash using any detergent designed for labware. Rinse with tap water 3 times, followed by 1 rinse with distilled water. Dry in an oven at temperature less than 140°C. 2. *Acid dichromate.* Dissolve 50 g technical-grade sodium dichromate in 50 mL of distilled water. Add this mixture to 500 mL of technical-grade concentrated sulfuric acid. This solution is useful until a green color develops. Store in a covered glass jar. Soak glassware overnight and then rinse with dilute ammonia. Rewash glassware according to procedure 1. 3. *Nitric acid* (20%). Soak for 12–24 h. Wash according to procedure 1.
Blood clots	4. *Sodium hydroxide* (10%). Soak for 12–24 h; then follow routine procedure. Dry micropipets using an acetone rinse.
New pipets (S1 alkaline)	5. *Rinse* with 5% hydrochloric acid or 5% nitric acid. Wash following routine procedure.
Metal ion determinations	6. *Acid soak* (20% nitric acid), for 12–24 h. Rinse with distilled water 3 to 4 times. Water should be fresh for each rinsing step. Dry.
Grease	7. *Soak* in any organic solvent. 8. *Dissolve* 100 g potassium hydroxide in 100 mL of distilled water. Allow to cool. Add 900 mL commercial-grade 10% ethanol. Not to be used for delicate glassware. 9. *Contrad 70* (manufactured by Harleco).
Permanganate stains	10. *50% Hydrochloric acid.* Rinse with tap water. Wash. 11. *Dissolve* 1% ferrous sulfate in 25% sulfuric acid.

APPENDIX S. Summary Table of Pharmacokinetic Parameters

	Therapeutic Range, per mL Plasma	Toxic Concentrations, per mL Plasma	Time to Peak Conc., Hours	Half-Life Hours	% Protein Bound	Volume of Distribution, L/kg	Oral Bioavailability, %	% Excreted in Urine Unchanged
Cardioactive drugs								
Amiodarone	1.0–3 µg		2–6	15–100 d	95–98	70–150	22–88	0
Digitoxin	15–30 ng	>35		2.4–16.4 d	90	0.6	95	30–50
Digoxin	0.8–2 ng	>2.4	1–5	36–51	20–40	5–10	Tablets, 60–75 Elixir, 80 Capsules, 95 IM, 80	60–80
Disopyramide	2–5 µg	>7	0.5–3.0	5–6	10–80	0.8–2	80	
Lidocaine	1.5–5 µg 0.5–1.5 free	>5	15–30[a]	1–2	70	1.3	25–50 [b]	5–10
Procainamide NAPA Total	4–10 µg 15–25 µg 5–30 µg	>12	1–2	2.5–4.7 4.3–15	15	1.7–2.2	70–95	50
Propranolol	50–100 ng	Variable	1–2	2–6	90–95	4–6	20–40[b]	1–4
Quinidine	2–6 µg	>6	1–2 4–8[c]	6–8	70–90	2–3	70–80[b]	10–30
Antiepileptic drugs								
Carbamazepine	6–12 µg	>15	6–12	18–54[d] 10–25[e]	72–75	0.8–1.4	75–85	2
Ethosuximide	40–100 µg	>150	1–4	40–60	<10	0.6–0.9	100	10–20
Phenobarbital	15–40 µg	>40	6–18	50–120	49–58	0.6	80–100	10–30
Phenytoin	10–20 µg	>20	4–8	7–42	87–93	0.5–0.8	85–95	5
Primidone	5–12 µg	>15	2–4	3.3–19	0–20	0.6–1	80–90	45–50
Valproic Acid	50–100 µg	>100	1–2	8–20	85–95	0.1–0.5	85–100	3
Bronchodilator								
Theophylline	10–20 µg	>20	2–3	6–12	55–65	0.3–0.7	95–100	9–11
Antibiotics								
Aminoglycosides			0.5–IM[a]					
Amikacin								
Peak	20–25 µg	>32						
Trough	1–4 µg	>5						
Gentamicin								
Peak	5–10 µg	>12						
Trough	0.5–1.5 µg	>2						
Kanamycin								
Peak	20–25/µg	>30						
Trough	1–4 µg	>10						
Netilmicin								
Peak	5–12 µg	>12						
Trough	0.5–1.5 µg	>2						
Streptomycin								
Peak	20–25 µg	>30						
Trough	1–3 µg	>10						
Tobramycin								
Peak	5–12 µg	>12						
Trough	0.5–1.5 µg	>2						
Vancomycin								
Peak	30–40 µg	>80		3–9	50	0.5–0.8	<2	80–90
Trough	5–10 µg	>20						
Chloramphenicol	10–20 µg	>25	2	1.5–3	50	0.5–1	90	10

(*continued*)

APPENDIX S. (*Continued*)

	Therapeutic Range, per mL. Plasma	Toxic Concentrations, per mL Plasma	Time to Peak Conc., Hours	Half-Life Hours	% Protein Bound	Volume of Distribution, L/kg	Oral Bioavailability, %	% Excreted in Urine Unchanged
Psychoactive drugs								
Amitriptylline	125–250 ng	>500	1–5	17–40	82–96	6.4–36	56–70	
Nortriptylline	50–150 ng	>500	3–12	16–88	87–95	14–38	46–70	2–5
Imipramine	150–250 ng	>500[f]	1.5–3	6–34	63–96	9–23	29–77	1–4
Desipramine	150–300 ng	>500	3–6	11–46	73–92	15–60	31–51	1–4
Doxepin	110–250 ng		1–4	8–36	68–82	9–52	13–45	
Protriptylline	70–260 ng		6–12	54–198	90–94	15–31	75–90	
Lithium	0.8–1.4 µEq	>2	1–3	8–35[g]	0	0.5–1.0	85–95	100
Immunosuppressants								
Cyclosporine (HPLC)	100–300 ng	>400	1–8	4–60	98	3.5–4.5	4–90	<6
Antineoplastics								
Methotrexate	After 24 h > 10^{-5} M After 48 h > 10^{-6} M After 72 h > 10^{-7} M		1–2	Variable	50–70	0.75–0.8	30	90

[a] *Varies dependent on dosage regimen, immediately after IV infusion.*

[b] *Much of the drug metabolized on first pass through the liver.*

[c] *Slow-release preparation.*

[d] *After single dose.*

[e] *After multiple doses.*

[f] *Imipramine + desipramine.*

[g] *Variable with renal function.*

APPENDIX T. Selected Information on Commonly Abused Drugs

Drug	Street Name (Trade Name)	Route of Ingestion	Duration of Effect (h)	Half-Life (h)	Excreted Unchanged in Urine	Principal Urinary Metabolites	Symptomatology
Stimulants							
Cocaine	Coke, crack, snow, flake	Nasal, oral, IV, smoked	1–2	2–5	<10%	Benzoyl ecogonine; ecogonine; ecogonine methyl ester	Anesthesia, euphoria, confusion, depression, convulsions, cardiotoxicity
Amphetamine	Bennies, dexies, uppers	Oral, IV	2–4	4–24	~30%	Benzoic acid; *p*-hydroxyamphetamine; *p*-hydroxynorephedrine; phenylacetone	Insomnia, anorexia, euphoria, tolerance and dependence, paranoid psychosis
Methamphetamine	Meth, speed, crystal	Oral, IV	2–4	9–24	10–20%	4-Hydroxymethamphetamine; amphetamine; 4-thyroxyamphetamine; norephedrine	Euphoria, agitation, psychosis, depression, exhaustion
Narcotics							
Heroin	Horse, smack, white lady, scag	IV, nasal, smoked	3–6	1–1.5	<1%	6-Acetylmorphine; morphine; morphine glucuronide	Euphoria, drowsiness, respiratory depression, convulsions, coma
Codeine	Oral, IV, IM	Oral, IV, IM	3–6	2–4	5–20%	Morphine; norcodeine; conjugates	Sedation, convulsions, respiratory failure
Morphine	Junk, white stuff, morpho, M	IV, IM, oral, smoked	3–6	2–4	<10%	Morphine-3-glucuronide; morphine-6-glucuronide; morphine sulfate; normorphine; codeine	Analgesia, euphoria, nausea, respiratory coma
Methadone	Methadose	Oral, IV, IM	12–24	15–60	5–50%	2-Ethylidene-1, 5-dimethyl-3,3-diphenylpyrroline; 2-ethyl-5-methyl-3,3-diphenyl-pyrroline methadol; normethadol; conjugates	Analgesia, sedation, respiratory depression, coma
Meperidine	(Demerol)	IV, oral	3–6	2–5	5%	Normeperidine; meperidinic acid; normeperidinic acid	Analgesia, stupor, respiratory depression, hypotension, coma
Propoxyphene	Yellow footballs (Darvon)	Oral	1–6	8–24	<1%	Norpropoxyphene; dinorpropoxyphene	Analgesia, stupor, respiratory depression, coma

Drug	Street names	Route	Duration (hr)	Half-life (hr)	Excreted unchanged	Metabolites	Effects of overdose
Hallucinogens							
Phencyclidine (PCP)	PCP, angel dust, hog, killer weed	IV, oral, nasal, smoked	2–4; psychoses may last weeks	7–16	30–50%	4-Phenyl-4-piperidinocyclohexanol; 1-(1-phenylcyclohexyl)-4-hydroxy-piperidine; glucuronide conjugates	Dissociative anesthesia, depression, psychosis, stupor, coma, seizures
LSD	Acid, LSD-25, white lightning, microdots	Oral	8–12	3–4	1%	N-Desmethyllysergide; 13-hydroxylysergide	Hallucinations, flashbacks, psychosis, vomiting, paralysis, respiratory depression
Marijuana, hashish	Pot, THC, mary jane, grass, hash	Oral, smoked, IV	2–4	14–38	<1%	11-Nor-9-carboxy-Δ^9-THC; 11-hydroxytetrahydrocannabinol	Altered perception, memory loss, disorientation, psychosis
Benzodiazepines							
Chlordiazepoxide	(Librium)	Oral, IM	4–8	6–27	<1%	Norchlordiazepoxide; demoxepam; nordiazepam; oxazepam; glucuronide conjugates	Drowsiness, muscle relaxation, coma
Diazepam	(Valium)	Oral, IV, IM	4–8	20–50	<1%	Nordiazepam; oxazepam; 3-hydroxydiazepam; glucoronide conjugates	Drowsiness, dizziness, muscle relaxation
Sedatives/ Depressants							
Pentobarbital	Yellow, nembies, yellow jackets	Oral, IV, IM	3–6	15–48	1%	3-Hydroxypentobarbital; N-hydroxypentobarbital; 3-carboxypentobarbital	Sedation, respiratory collapse
Amobarbital	Rainbows, blues, bluebirds	Oral, IV, IM	3–24	12–60	<1%	3-Hydroxyamobarbital; N-glucosyl amobarbital	Exhilaration, sedation, disorientation, respiratory depression, coma
Secobarbital	Reds, seccies, red devils, M & M's	Oral, IV, IM	3–6	15–40	5%	3-Hydroxysecobarbital secodiol; 5-(1-methylbutyl) barbituric acid	Sedation, lethargy, coma, respiratory collapse
Ethanol		Oral	2–6	2–14	2–10%	Acetaldehyde; acetic acid	Slurred speech, loss of equilibrium, drowsiness, coma, respiratory collapse
Methaqualone	Ludes, soapers	Oral	4–8	20–60	<1%	3',4', and 6-hydroxymethaqualone; respective glucuronide	Sedation, dizziness, paresthesias, convulsions, respiratory and circulatory depression
Chloral hydrate	Joy juice	Oral, rectal	5–8	<1	<1%	Trichloroethanol; trichloracetic acid; conjugates	Sedation, GI distress, hypotension, respiratory depression

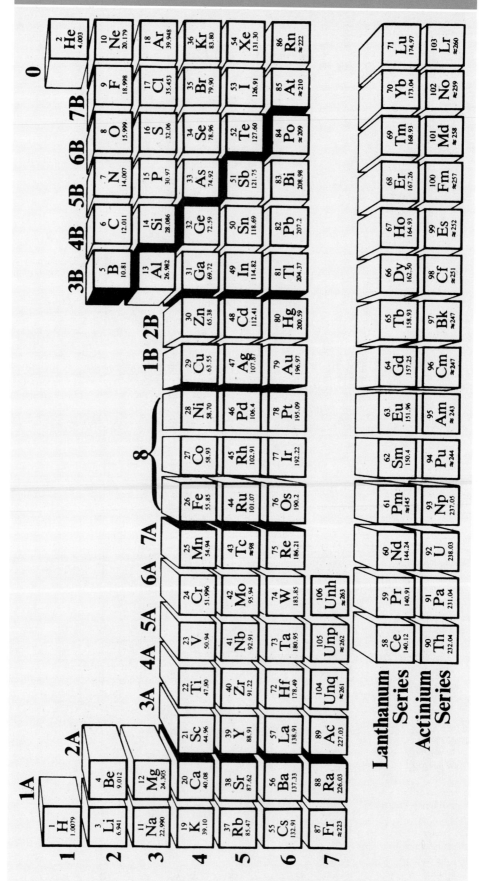

* Source: Monroe M, Abrams K. Experimental chemistry: A laboratory course. Belmont, CA: Star Publishing Company, 1991. Reprinted with permission.

Glossary

A

Accuracy   without error; closeness to the true value.

Acid   a substance that can yield a hydrogen ion or hydronium ion when dissolved in water.

Acidemia   a pH of blood less than the reference range.

Acidosis   a pH below the reference range.

Acromegaly   chronic disease of middle-aged individuals typified by elongation and enlargement of the extremities and certain head bones.

ACTH   see adrenocorticotropic hormone

Activation energy   energy required to raise all molecules in one mole of a compound at a certain temperature to the transition state at the peak of the energy barrier.

Activator   a substance that converts an inactive substance to an active one; induces activity.

Active transport   a mechanism that requires energy in order to move ions across cellular membranes.

Acute renal failure   a sudden, sharp decline in renal operation due to an acute toxic or hypoxic insult to the kidneys. This has been defined as occurring when the glomerular filtration rate (GFR) is reduced to <10 mL/min.

Adaptive immune system   one of two functional parts of the immune system; it produces a specific reaction to each infectious agent, then normally eradicates that agent and remembers that particular infectious agent, preventing it from causing disease later. For example, measles and diphtheria produce a lifelong immunity following an infection.

Addison's disease   disease stemming from a deficiency in the secretion of adrenocortical hormones.

ADH   see antidiuretic hormone

Adrenal glands   paired organs located one at the upper pole of each kidney. Each gland consists of an outer cortex and an inner medulla, which have different embryological origins, different mechanisms of control, and different products.

Adrenocorticotropic hormone (ACTH)   a peptide hormone secreted by the anterior pituitary. It stimulates the cortex of the adrenal glands to produce adrenal cortical hormones.

Affinity   attraction or force causing two substances to unite.

Aging   maturing; a progressive loss of adaptation leading to decreased viability and life expectancy and increased vulnerability.

Airborne pathogen   any infectious agent transmissible by air (eg, tuberculosis).

Albumin   the main protein in plasma.

Aldosterone   the principal mineralocorticoid (electrolyte-regulating hormone) produced by the zona glomerulosa of the adrenal cortex.

Alkalemia   a blood pH greater than the reference range (~7.35–7.45).

Alkalosis   a pH above the reference range.

Allantoin   produced by the oxidation of uric acid by the enzyme uricase. It is the end product of purine metabolism.

Allograft   transplant tissue from the same species, (eg, kidney).

Amenorrhea   cessation of menstruation.

Amine   any one of a group of nitrogen-containing organic compounds. The amine hormones include epinephrine, norepinephrine, thyroxine, and tri-iodothyronine.

Amino acid   small biomolecules with a tetrahedral carbon covalently bound to an amino group, a carboxyl group, a variable (R) group and a hydrogen atom.

Aminoacidopathies   inherited disorders of amino acid metabolism.

Ammonia NH_3; formed in vivo from breakdown of amino acids.

Amniocentesis   puncture of the amniotic sac to obtain fluid for analysis.

Amniotic fluid   a fluid in which the fetus is suspended; it provides a cushioning medium for the fetus and serves as a matrix for influx and efflux of constituents.

Amphoteric   having two or more ionizable sites that can result in either a net positive or a net negative charge depending on the pH of the environment.

Amplicon   amplified target DNA sequences.

Analgesic   a drug that relieves pain.

Analyte   a biologic solute or constituent, (eg, calcium, glucose, sodium)

Analytical error   errors due to the instrument's, procedure's or laboratory scientist's handling of a specimen during testing.

Analytical variations   non-identical measurements due to diverse causes, including instrument, reagent, and operator variations.

Androgen   any substance (hormone) stimulating the development of male characteristics (eg, testosterone).

Angina pectoris   pain and a feeling of constriction around the heart; may radiate down arm and into jaw. Caused by deficiency of oxygen to heart muscle.

Angiotensin   a polypeptide produced when renin is released from the kidney; a vasopressor substance.

Anhydrous   without water.

Anion   a negatively charged ion; anions move toward the anode (positive pole) because of its positive charge.

Anion gap   the difference between unmeasured anions and unmeasured cations.

Antecubital fossa   area of forearm at the bend of the elbow; most commonly used for venipuncture.

Anterior pituitary   adenohypophysis. The pituitary is located in a small cavity in the sphenoid bone of the skull called the sella turcica. The tropic hormones of the anterior pituitary are mediated by negative feedback, which involves interaction of the effector hormones with the hypothalamus, as well as with cells of the anterior pituitary.

Anthropometric methods   used to assess the general nutritional status of a patient. Includes skin fold test, arm circumference and height and weight measurements.

Antibody   glycoproteins (immunoglobulins) secreted by plasma cells, which in turn are under the control of many lymphocytes and their cytokines. Antibodies are produced in response to antigens.

Anticoagulant   inhibits the blood's clotting action; yields specimens containing intact clotting factors.

Antidiuretic hormone (ADH)   vasopressin; produced by the hypothalamus.

Antigen   agents that are recognized as foreign by the immune system. The immune system produces antibodies in response.

Antigenic determinant   a part of an antigens structure that is recognized as foreign by the immune system. This structural domain is also referred to as an epitope.

Antioxidant   substance (*eg*, vitamin) that inhibits or prevents oxidation.

Antiplatelet therapy   therapy to destroy platelets.

Apoenzyme   protein portion of an enzyme.

Apoptosis   programmed death of cells; disintegration of body cells into membrane-bound particles that can then be phagocytized by other cells.

Arrhythmia   irregular heartbeat or action.

Arterial blood   blood from arteries.

Arterial septal defect   heart abnormality which causes left-to-right shunting of blood between the atria.

Arteriosclerosis   includes a number of pathological conditions in which there is a thickening or hardening of the walls of the arteries.

Ascites   excess fluid in the peritoneal cavity; the fluid is called ascitic fluid.

Atherosclerosis   a disease in which there is an accumulation of lipid material in the veins and arteries.

Autocrine system   cell secretions that act to influence only their own development.

Atomic absorption   analytical technique that measures concentration of analyte by detecting absorption of electromagnetic radiation by atoms rather than by molecules. Instrument is atomic absorption spectrophotometer.

Autoimmune disorder   disease or disorder in which the body produces antibodies (immunological response) against itself.

Automation   mechanization of the steps in a procedure. Manufacturers of clinical chemistry analyzers design their instruments to mimic the manual techniques in an analytical procedure.

Avidity   strength of bond of antigen-antibody complex; attraction.

Azotemia   elevated level of urea in blood.

B

Bar code   a set of vertical bars of varying width used to encode information. Used most frequently in the clinical laboratory for patient and specimen information.

Basal state   early morning before the patient has eaten or become physically active. This is a good time to draw blood specimens because the body is at rest and food has not been ingested during the night.

Base   a substance that can yield hydroxyl ions (OH⁻).

Base excess   (BE) the theoretical amount of titratable acid or base required to return the plasma pH to 7.40 at a PCO_2 of 40 mm Hg at 37°C.

Beer's law   mathematically establishes the relationship between concentration and absorbance in photometric determinations; expressed as: A = *abc*.

Beriberi   chronic deficiency of the vitamin thiamin produces the disease beriberi.

Beta-adrenergic blocking drug   drug which reduces heart rate and/or the force of contractions reducing the oxygen demand for the heart by blocking the beta receptors in the sinus mode and the myocardium (*eg*, propranolol).

Beta₂-microglobulin   a small, nonglycosylated peptide; used as an indicator of GFR.

Bicarbonate   the HCO₃ anion.

Bile   a fluid produced by the liver and composed of bile acids or salts, bile pigments (primarily bilirubin esters), cholesterol, and other substances extracted from the blood. Total bile production averages about 3 L per day, although only 1 L is excreted.

Bilirubin   the principal pigment in bile; derived from the breakdown of hemoglobin when aged red blood cells are phagocytized by the reticuloendothelial system, primarily in the spleen, liver, and bone marrow.

Biohazard   anything harmful or potentially harmful to man, other organisms, or the environment. Examples include blood or blood products, and contaminated laboratory waste.

Bloodborne pathogen   any infectious agent or pathogen transmissable by means of blood or blood products.

Buffer   a substance which minimizes any change in hydrogen ion concentration; a weak acid or base and its conjugate salt.

BUN/creatinine ratio   mg/dL plasma or serum urea nitrogen/mg/dL plasma or serum creatinine.

Buret   a wide, long, graduated pipette with a stopcock at one end.

C

CAH   see congenital adrenal hyperplasia

Calcitonin   hormone produced by the thyroid gland; important in bone and calcium metabolism.

Calibration   standardization or the determination of the accuracy of an instrument.

Cancer   the uncontrolled growth of cells from normal tissue. Cancer cells can grow and spread thereby killing the host.

Capillary blood   blood from minute blood vessels.

Carbohydrate   polyhydroxy aldehydes or polyhydroxy ketones, or multimeric units of such compounds. The general formula of a carbohydrate is $(CH_2O)_n$.

Carbonic acid H_2CO_3

Carcinogen   cancer causing agent.

Carcinoid syndrome   syndrome produced by metastatic carcinoid tumors which secrete excessive amounts of serotonin.

Cardiac catheterization   invasive technique where a catheter is placed into a peripheral vessel and advanced into the heart.

Cardiac glycosides   drugs used to increase the contractility of the heart and slow the conduction impulses (*eg*, digoxin).

Cardiac markers   diagnostic test or analyte used to assess cardiac function.

Cardiac radiology   use of x-rays to assess heart size and position, etc.

Cardiomyopathy   disease of the myocardium.

Cardiovascular nuclear imaging   technique using radionuclides to assess cardiovascular performance and perfusion of the myocardium and viability of the cardiac muscle.

Cation   a positively charged ion; cations migrate in the direction of the cathode because of their positive charge.

CCK   see cholecystokinin

Centrifugal analysis   analytical technique that uses the force generated by centrifugation to transfer and then contain liquids in separate cuvets for measurement at the perimeter of a spinning rotor.

Centrifugation   a process whereby centrifugal force is used to separate solid matter from a liquid suspension.

Cerebral spinal fluid (CSF)   a selective ultrafiltrate of the plasma that surrounds the brain and spinal cord.

Ceruloplasmin   an alpha₂-glycoprotein to which copper is attached; more than 90%–95% of the circulating copper in the plasma is bound to ceruloplasmin.

Chain-of-custody   records each step and each person who handled a sample (for toxicological analysis) from time of collection to time of analysis.

Channel   on an automated analyzer, a path or passage for reagents, specimens, or electrical impulses. Automated analyzers may be single or multiple channel analyzers.

Character   the number to the left of the decimal point in a logarithmic expression.

Chelator   causing the joining of an ion (*eg*, metal) and a ring structured chemical.

Chemical hygiene plan   procedures and work practices for regulating exposure of laboratory personnel to hazardous chemicals.

Chemiluminescence   light produced as a result of a chemical reaction. Most important chemiluminescence reactions are oxidation reactions of luminol, acridinium esters, and dioxetanes and are characterized by a rapid increase in intensity of emitted light followed by a gradual decay.

Cholecystokinin (CCK)   formerly called pancreozymin; a hormone produced by the pancreas. CCK, in the presence of fats and/or amino acids in the duodenum, is produced by the cells of the intestinal mucosa and responsible for release of enzymes from the acinar cells by the pancreas into the pancreatic juice.

Cholesterol an unsaturated steroid alcohol of high molecular weight, consisting of a perhydrocyclopentanthroline ring and a side chain of eight carbon atoms. In its esterified form, it contains one fatty acid molecule.

Chronic renal failure a clinical syndrome that occurs when there is a gradual decline in renal operation over time.

Chylomicrons large triglyceride-rich particles.

Cirrhosis derived from the Greek word that means "yellow." However, in current usage, cirrhosis refers to the irreversible scarring process by which normal liver architecture is transformed into abnormal nodular architecture.

Clearance volume of plasma filtered by glomeruli per unit time.

Clinical Laboratory Improvement Amendment (CLIA) regulations signed into federal law in 1988; mandates standards in clinical laboratory operations and testing.

Closed collection system a type of collection system in which blood is taken directly from a patient's vein into a stoppered tube; the sample is completely contained, thereby reducing the risk of outside contaminants to the sample and reducing the hazard of the collector's exposure to the blood; also known as the evacuated-tube system.

Coarctation of aorta narrowing of the aorta at the insertion of the ductus arteriosus.

Coenzyme an enzyme activator (*eg*, coenzyme A)

Cofactor a nonprotein molecule which may be necessary for enzyme activity.

Colligative property the properties of osmotic pressure, freezing point, boiling point, and vapor pressure.

Common variable agammaglobulinemia also known as acquired agammaglobulinemia. This disorder represents a group of disorders characterized by hypogammaglobulinemia.

Compensation the body's attempt to return the pH toward normal whenever an imbalance occurs.

Competitive protein binding also competitive immunoassay. Labeled and unlabeled antigen compete for limited antibody sites. Labeled antigen bound to antibody will be inversely proportional to the concentration of antigen.

Conductivity relates to the ease in which electricity passes through a solution.

Congenital adrenal hyperplasia (CAH) results from the lack of an enzyme necessary for the production of cortisol.

Congenital agammaglobulinemia X-linked agammaglobulinemia, also called Bruton disease. This disorder presents with early onset of recurrent pyogenic infections. These patients have no circulating B cells, low concentrations of all circulating immunoglobulin classes, and absence of plasma cells in all lymphoid tissue. T cells are not involved.

Congestive heart disease (also congestive heart failure) results from an inability of the heart to pump blood effectively.

Conjugated bilirubin bilirubin diglucuronide; it is water-soluble, and is secreted from the hepatic cell into the bile canaliculi and then passes along with the rest of the bile into larger bile ducts and eventually into the intestines.

Conjugated protein composed of a protein (amino acids) and a non-protein moiety.

Conn's syndrome aldosterone-secreting adrenal adenoma.

Continuous flow an approach to automated analysis in which liquids (reagents, diluents, and samples) are pumped through a system of continuous tubing. Samples are introduced in a sequential manner, following each other through the same network. A series of air bubbles at regular intervals serve as separating and cleaning media.

Control a substance or material of determined value, used to monitor the accuracy and precision of a test. Controls are run with the patients' specimens.

Control rule criterion for judging whether an analytical process is out of control; error detection criteria.

Corpus albicans fibrous tissue which replaces a degenerating corpus luteum.

Corpus luteum small body that develops within a ruptured ovarian follicle; secretes progesterone.

Corrosive chemical chemicals injurious to the skin or eyes by direct contact or to the tissues of the respiratory and gastrointestinal tracts if inhaled or ingested. Examples include acids (acetic, sulfuric, nitric, and hydrochloric) and bases (ammonium hydroxide, potassium hydroxide, and sodium hydroxide).

Corticotropin-releasing hormone (CRH) A hormone released from the hypothalamus that acts on the anterior pituitary to increase ACTH secretion.

Cortisol a steroid hormone produced by the adrenal glands.

Countercurrent multiplier system process occurring in the Loop of Henle whereby a high osmolality is maintained within the kidney and a hypoosmolal urine is produced.

Counterimmunoelectrophoresis an immune precipitation method that uses an electrical field to cause an antigen and antibody to migrate toward each other

Coupled enzymatic method use of several sequential enzymatic reactions which produce a product which can be detected spectrophotometrically and whose concentration will be related to the analyte in question.

Creatine compound found in muscle synthesized from several amino acids. It combines with high energy phosphate to form creatine phosphate which functions as an energy compound in muscle.

Creatinine compound formed when creatine or creatine phosphate spontaneously loses water or phosphoric acid. It is excreted into the plasma at a relatively constant rate in a given individual and excreted in the urine.

Creatinine clearance rate of removal of creatinine from plasma. Calculated as urine creatinine concentration times 24 h urine volume divided by serum creatinine concentration or UV/P, expressed in mL/min. Usually corrected to normal body surface area.

CRH see corticotropin-releasing hormone

Cross reactivity capability of an antibody to react with an antigen that is structurally similar to the homologous antigen.

Cryogenic material material brought to low temperatures, such as liquefied gases.

CSF see cerebral spinal fluid

Cushing's syndrome syndrome resulting from excessive production of glucocorticoids by adrenal cortex.

Cyclosporine a cyclic polypeptide of fungal origin. It consists of 11 amino acids and has a molecular weight of 1203. It inhibits the immune response selectively by inhibiting the interleukin-2 dependent proliferation of activated T cells which destroy the allograft. The immune response is frozen and unresponsive.

Cytochrome a pigment which plays a role in respiration, such as hemoglobin or myoglobin.

Cytokines extracellular factors produced by a variety of cells including monocytes, lymphocytes and other non-lymphoid cells. They are important in controlling local and systemic inflammatory responses.

D

D-Xylose absorption test an analytical technique which assesses the ability to absorb D-xylose; it is of value in differentiating malabsorption of intestinal etiology from that of exocrine pancreatic insufficiency.

Dehydroepiandrosterone (DHEA) An androgen primarily derived from the adrenal gland.

Deionized water water purified by ion exchange.

Deliquescent substances compounds that absorb enough water from the atmosphere to cause dissolution.

Delta absorbance difference in absorbance, known as delta absorbance or ΔA.

Delta check an algorithm in which the most recent result of a patient is compared with the previously determined value.

Denaturation alteration of a substance (*eg*, proteins) to alter physical and chemical properties.

Density weight of a substance compared with a standard; expressed in terms of mass per unit volume

Descriptive statistics statistics or values (*eg*, mean, median, mode) used to summarize the important features of a group of data.

Desiccant causing dryness; materials that can remove moisture from the air as well as from other materials.

Desiccator a closed chamber for drying substances.

DHEA see dehydroepiandrosterone

Diabetes mellitus a diverse group of hyperglycemic disorders with different etiologies and clinical pictures.

Dialysis a method for separating macromolecules from a solvent

Diffusion the movement of the molecules of a substance from a location of higher concentration to one of lesser concentration, as in gel diffusion precipitation.

Dilution a dilution represents the ratio of concentrated or stock material to the total final volume of a solution and consists of the volume or weight of the concentrate plus the volume of the diluent (the concentration units remaining the same).

Direct immunofluorescence detection of antigens with fluorescent labeled antibody.

Disaccharide separate carbohydrates, or monosaccharides joined together.

Discrete analysis an approach to automated analysis in which each sample and accompanying reagents is in a separate container. Discrete analyzers have the capability of running multiple tests one sample at a time or multiple samples one test at a time.

Dispersion the spread of data; most simply estimated by the range, the difference between the largest and smallest observations. The most commonly used statistic for describing the dispersion of groups of single observations is the standard deviation, which is usually represented by the symbol *s*.

Distal tubule portion of renal tubule extending from ascending limb of loop of Henle to the collecting duct.

Distilled water water solely purified by distillation results

Diuretics agents that increase the secretion of urine.

DNA index (DI) ploidy status of malignancies; the amount of measured DNA in cancer cells relative to that in normal cells.

DNA probe a known fragment of DNA molecule used to join with or locate an unknown or comparable DNA strand. Human papilloma virus DNA has been detected using DNA probes.

Dose Nomogram a chart that approximates drug toxicity given the time of ingestion and blood drug level.

Dose-response relationship comparison of the dose of a substance (*ie*, drug or chemical) with its potential pathologic effects. Dose-response relationship implies that there will be an increase in toxic response with an increased dose.

Drug absorption uptake of a drug in the gastrointestinal tract into the body.

Drug disposition the way in which the body handles a foreign compound, the drug. The mechanisms used by the body to handle a drug can be explained in terms of four general processes: absorption, distribution, metabolism, and excretion.

Drug distribution the circulation and diffusion of a drug into the interstitial and intracellular spaces.

Drug elimination clearance of a drug from the body by a variety of mechanisms.

Drugs of abuse drugs used illegally or inappropriately; many drugs have the potential for abuse.

Dry chemistry slide a multi-layered film technology (dry chemicals) used by the Vitros series of automated analyzers. All reagents necessary for a particular test are contained on the "slide."

Dyslipidemias diseases associated with abnormal lipid concentrations.

E

ECG see electrocardiography

Echocardiography noninvasive diagnostic technique that uses high frequency sound waves to show the heart structure on a CRT screen.

Ectopic in an abnormal position or location.

Ectopic hormone production hormones produced by cells in sites other than the gland from which they are usually derived.

EDTA ethylenediaminetetra-acetic acid; anticoagulant used in lavender and royal blue stopper blood collection tubes. Commonly used for whole blood hematology studies.

Effusion abnormal accumulations of pleural or pericardial fluid.

Electrocardiography (ECG) test used to evaluate the electrical stimulation of the heart.

Electrochemistry the use of galvanic and electrolytic cells (electrochemical cells) for chemical analysis. Examples include potentiometry, amperometry, coulometry, and polarography.

Electrodes electronic sensing devices to measure PO_2, PCO_2, and pH.

Electrolyte ions capable of carrying an electric charge.

Electrophoresis migration of charged solutes or particles in an electrical field.

Embden–Meyerhof pathway series of steps involving the anaerobic metabolism of glucose, glycogen, or starch to lactic acid; principal means of producing energy in man.

Encephalopathy disorder or dysfunction of brain.

Endocrine gland a ductless gland that produces a secretion (hormone) into the blood or lymph to be carried by the circulation to other parts of the body.

Endogenous triglycerides synthesized in liver and other tissues.

Enzyme specific biologically synthesized proteins that catalyze biochemical reactions without altering the equilibrium point of the reaction or being consumed or undergoing changes in composition.

Enzyme-substrate complex a physical binding of a substrate to the active site of an enzyme.

Epinephrine an amine hormone. The adrenal medulla primarily produces epinephrine.

Epitope component of an antigen that functions as an antigenic determinant, allowing the attachment of certain antibodies.

Equivalent weight equal to the molecular weight of a substance divided by its valence.

Erectile dysfunction inability to have or maintain an erection.

Erythropoietin a hormone that stimulates red blood cell production.

Essential element an element that is absolutely necessary for life. Deficiency or absence of the element will cause a severe alteration of function and/or eventually lead to death.

Essential hypertension hypertension with no known cause.

Estriol an estrogenic hormone; estriol is not produced in significant quantities in the mother and is solely a reflection of feto-placental function. Thus estriol measurement provides valuable information about fetal wellbeing.

Estrogen any of the group of substances (hormones) that induces estrogenic activity; specifically the estrogenic hormones, estradiol and estrone, produced by the ovary.

Estrone an estrogenic hormone found in the urine of pregnant women.

Evacuated tube sample collection tube with a vacuum.

Exocrine gland any gland that excretes externally through a duct. Secretions of exocrine glands do not directly enter the circulation.

Exogenous triglycerides obtained from dietary sources.

Extracellular fluid water (fluid) outside the cell; can be subdivided in the intravascular extracellular fluid (plasma) and the interstitial cell fluid (ISCF) that surrounds the cells in the tissues.

Extrauterine outside the uterus or womb.

Exudate accumulation of fluid in a cavity; also the production of pus or serum. In comparison with a transudate, an exudate contains more cells and protein. Exudates demand immediate attention.

F

F-Test statistical test used to compare the features of two or more groups of data.

Fallot's tetralogy a congenital condition of the heart which includes septal defects, stenosis of the pulmonary artery, dextroposition of the aorta, and hypertrophy of the right ventricle.

Fasting specimens a blood specimen taken after the patient has not eaten for at least 12 hours.

Fatty acids major constituents of triglycerides and phospholipids. There are short-chain (4–6 carbon atoms), medium-chain (8–12 carbon atoms), and long-chain fatty acids (>12 carbon atoms).

Ferritin a spherical protein shell composed of 24 subunits with a molecular mass of 500 kDa. The protein can bind up to 4,000 iron molecules, making it a large potential source of iron.

Filtrate liquid that passes through filter paper is called the filtrate.

Filtration separation of solids from liquids.

FiO$_2$ the fraction of inspired oxygen; can be as much as 100% when oxygen is being supplied.

Fire tetrahedron a three-dimensional pyramid representing the element of fire; previously the fire triangle.

First order kinetics point in enzyme reaction where the rate of the reaction is dependent on enzyme concentration only.

Fisher projection diagrammatic model that can be used to represent carbohydrates. The Fisher projection of a carbohydrate has the aldehyde or ketone at the top of the drawing. The carbons are numbered starting at the aldehyde or ketone end and the compound can be represented either as a straight chain or cyclic, hemiacetal form.

Flag on automated analyzers, a computerized printed warning of an error or instrument problem.

Flame photometry analytical technique that measures the wavelength and intensity of light emitted from a burning solution (patient specimen).

Flow cytometry the use of immunofluorescent labels to identify specific antigens on live cells in suspension. Stained cell suspensions are transported under pressure, past a laser beam and emitted fluorescence (at 90° relative to the beam) is measured and computer analyzed with this technique. Using multiple labels, cells can be identified and either electronically or physically sorted. The technique has been used to analyze a subpopulation of lymphocyte cells in various clinical diagnoses.

Fluorometry analytical technique used to measure fluorescence (light emitted as a result of energy absorbed).

Follicular cells (or cuboidal) one of two types of cells of the thyroid; they are secretory cells and produce thyroxine (T$_4$) and triiodothyronine (T$_3$)

Follicular phase the first half of the menstrual cycle, when estrogen effect is unopposed by progesterone; also called the proliferative phase.

Follicle-stimulating hormone (FSH) a protein hormone secreted by the anterior pituitary.

Forensic pertaining to the law or legal matters (eg, toxicology and the law).

Forensic medicine medical knowledge or results that apply to questions of law affecting life or property. A specimen result might be used as evidence in a court of law to prove cause of death, innocence or guilt of an accused individual, or possibly to prove alcohol or drug abuse.

Fractional oxyhemoglobin (FO$_2$Hb) is the ratio of the concentration of oxyhemoglobin to the concentration of total hemoglobin (ctHb).

Free radical a highly reactive molecule containing an open bond or half a bond; free radicals are harmful to the body (eg, OH·).

Free thyroxime index (FT$_4$I) an indirect measure of free hormone concentration and is based on the equilibrium relationship of bound T$_4$ and FT$_4$. The FT$_4$I is calculated by the following formula: FT$_4$I = T$_4$ × T$_3$U ratio.

Friedewald calculation calculation used to estimate LDL cholesterol in routine clinical practice.

FSH see follicle-stimulating hormone

G

Gas chromatography analytical technique used to separate mixtures of compounds that are volatile or can be made volatile. Gas chromatography may be gas-solid chromatography (GSC), with a solid stationary phase, or gas-liquid chromatography (GLC), with a nonvolatile liquid stationary phase.

Gastrin a gastrointestinal hormone. It is a peptide secreted by the G cells of the antrum (lower one third) of the stomach. It is released in response to contact of food.

Geriatrics branch of general medicine dealing with remedial and preventable clinical problems in the elderly, as well as the social consequences of such illness.

Gerontology study of the aging process in the human body.

GFR see glomerular filtration rate

GH see growth hormone

Globulins heterogeneous group of proteins that can be separated by electrophoresis into alpha$_1$, alpha$_2$, and gamma fractions.

Glomerular filtrate filtrate of plasma containing water and small molecules, but lacking cells and large molecules such as most proteins.

Glomerular filtration rate (GFR) rate at which plasma is filtered by glomerulus expressed in mL/minute.

Glomerulonephritis inflammation of the glomeruli of the kidney. May be acute, subacute, or chronic.

Glomerulus small blood vessels in the nephron which project into the capsular end of the proximal tubule and serves as filtering mechanism.

Glucagon a protein hormone that is involved in the regulation of carbohydrate, fat, and protein metabolism. It is synthesized by the α cells of the pancreatic islet and is composed of 29 amino acids.

Glucocorticoid a general classification of hormones synthesized in the zona fasciculata of the adrenal cortex. Cortisol is the principal glucocorticoid hormone.

Gluconeogenesis the conversion of amino acids by the liver, and other specialized tissues such as the kidney, to substrates that can be converted to glucose. Gluconeogenesis also encompasses the conversion of glycerol, lactate, and pyruvate to glucose.

Glycogen a polysaccharide, similar to starch; form in which carbohydrates are stored.

Glycogenolysis process by which glycogen is converted back to glucose-6-phosphate for entry into the glycolytic pathway.

Glycolipids sugar-containing lipids that consist of a sphingosine molecule with a fatty acid attached to its amino group, and a sugar linked to the primary alcohol group.

Glycolysis hydrolysis of glucose by an enzyme into pyruvate or lactate; the process is anaerobic.

Glycolytic inhibitor a substance that prevents the hydrolysis of sugar. Sodium fluoride is a glycolytic inhibitor.

Gout arthritis associated with increased levels of uric acid in the blood which then become deposited in the joints or tissues causing painful swelling. It is more common in men than in women.

Graft tissue that is transplanted.

Graves' disease diffuse toxic goiter. Graves' disease occurs six times more commonly in women than in men. It occurs frequently at puberty, during pregnancy, at menopause, or following severe stress.

Growth hormone (GH) a peptide composed of 191 amino acids. In contrast to most of the other protein hormones, GH does not act through cAMP. Binding of GH to its membrane receptor leads to glucose uptake, amino acid transport, and lipolysis. Secreted by the anterior pituitary.

Gynecomastia abnormally large mammary gland development in the male.

H

Hapten a substance which can bind with an antibody, but cannot initiate an immune response unless bound to a carrier.

Hashimoto's disease chronic autoimmune thyroiditis; it is the most common cause of primary hypothyroidism.

Haworth projection represents glucose in a cyclic form which is more representative of the actual structure. When glucose is drawn in a Haworth projection, the form of D-glucopyranose is represented by the hydroxy group of carbon one oriented downwards or below the plane of the paper.

Hazard communication standard based on the fact that all employees must be informed of any health risks involving the use of chemicals; from the Hazardous Communication Standard of 1987 (Right to Know Law).

Hazardous material a material which may potentially cause personal injury or damage if handled.

Hazardous waste any potentially dangerous waste material.

HBA$_{1C}$ see hemoglobin A$_{1C}$.

HCG see human chorionic gonadotropin

HDL see high density lipoproteins

Hemodialysis a technique or procedure which provides the function of the kidneys when one or both are damaged. The patient's blood is circulated through membranes to remove wastes.

Hemofiltration   an ultrafiltration procedure or technique (similar to hemodialysis) used to remove an excess accumulation of normal metabolic products from the blood.

Hemoglobin A$_{1C}$ (HbA$_{1C}$)   largest sub-fraction of normal HbA in both diabetic and non-diabetic subjects. It is formed by the reaction of the b chain of HbA with glucose. It reflects the concentration of glucose present in the body over a prolonged time period related to the 60-day half-life of erythrocytes.

Hemoglobin-oxygen (binding) capacity   the maximum amount of oxygen that can be carried by hemoglobin in a given quantity of blood.

Hemoglobin-oxygen dissociation curve   a graphic representation ("S"-shaped) of oxygen content as percent oxygen saturation against PO$_2$ based on the principle that oxygen dissociates from adult hemoglobin in a characteristic fashion.

Hemoglobinopathy   a disorder associated with the presence of an abnormal hemoglobin.

Hemolysis   damage to erythrocyte membranes, causing release of cellular constituents (*eg*, hemoglobin) into the blood plasma or serum.

Henderson–Hasselbalch equation   equation that mathematically describes the dissociation characteristics of weak acids and bases and the effect on pH; pH= pK$_a$ (6.1) + log of the ratio of bicarbonate to carbon dioxide (HCO$_3$/H$_2$CO$_3$)

HEPA filter   High Efficiency Particulate Air filter; a respirator.

Heparin   an anticoagulant used in blood collection.

Hepatitis   "inflammation of the liver;" may be caused by a virus, bacteria, parasites, radiation, drugs, chemicals, or toxins. Among the viruses causing hepatitis are hepatitis types A, B, C, D (or delta), and E, cytomegalovirus, Epstein-Barr virus, and probably several others.

Hepatoma   primary malignant tumors of the liver; also known as hepatocellular carcinoma or hepatocarcinoma.

Heterogeneous assay   a technique (*eg*, radioimmunoassay) where it is necessary to physically separate labeled antigen or hapten bound to antibody from a labeled antigen or hapten that remains free in solution.

High density lipoproteins (HDL)   a "clean-up crew" which gathers up extra cholesterol for transport back to the liver.

Histogram   graphical representation of data where the number or frequency of each result is placed on the y axis and the value of the result is plotted on the x axis.

Holoenzyme   enzyme consisting of a protein portion and a non-amino acid portion or prosthetic group.

Homeostasis   state of equilibrium in the body maintained by dynamic processes.

Homogeneous assay   a technique (*eg*, EMIT) that does not require the physical separation of the bound and free labeled antigen.

Hormone   a chemical substance that is produced and secreted into the blood by an organ or tissue and has a specific effect on a target tissue.

HPL   see human placental lactogen

HPTA   see hypothalamic-pituitary-thyroid axis

Human chorionic gonadotropin (HCG)   A protein hormone produced by the placenta and consisting of α and β subunits.

Human placental lactogen (HPL)   a protein hormone that is structurally, immunologically, and functionally very similar to growth hormone and prolactin. Like HCG, it is produced by the placenta and can be measured in maternal urine and serum as well as amniotic fluid.

Hybridization   the production of hybrids.

Hydrate   compound and its associated water.

Hydrolase   an enzyme that catalyzes hydrolysis of ether, ester, acid–anhydride, glycosyl, C—C, C—halide, or P—N bonds.

Hygroscopic   substances that take up water on exposure to atmospheric conditions.

Hyperaldosteronism   excessive production of aldosterone by the adrenal gland. It may be primary (due to an adrenal lesion) or secondary to abnormalities in the renin-angiotensin system.

Hypercalcemia   elevated levels of calcium in the blood.

Hyperchloremia   elevated levels of chloride in the blood.

Hypercoagulability   increased ability (*ie*, of blood) to coagulate.

Hypercortisolism   increased levels of cortisol.

Hyperglucagonemia   excessive production of glucagon. Hyperglucagonemia due to tumors must be differentiated from other causes of increased glucagon, which include diabetes, pancreatitis, and trauma.

Hyperglycemic   increased level of blood sugar.

Hyperinsulinemia   excessive and/or inappropriate release of insulin.

Hyperkalemia   elevated levels of potassium in the blood.

Hypermagnesemia   elevated levels of magnesium in the blood.

Hypernatremia   elevated levels of sodium in the blood.

Hyperoxemia   increased oxygen in the blood.

Hyperparathyroidism   condition resulting from the increased activity of the parathyroid glands.

Hyperphosphatemia   elevated levels of phosphorus in the blood.

Hyperprolactinemia   excess secretion of prolactin due to hypothalamic-pituitary dysfunction. Generally due to a pituitary neoplasm.

Hyperproteinemia   total protein level in the blood that is higher than the reference interval.

Hyperthyroidism   excessive secretion of the thyroid glands.

Hyperuricemia   plasma levels of uric acid greater than 7.0 mg/dL in males and 6.0 mg/dL in females.

Hypervitaminosis   a condition that results from excessive intake of a vitamin(s).

Hypoaldosteronism   decreased level of aldosterone.

Hypocalcemia   decreased levels of calcium.

Hypochloremia   decreased levels of chloride in the blood.

Hypocortisolism   low or decreased levels of cortisol.

Hypogonadism   aberrant internal secretion of the gonads.

Hypoglycemic   decreased level of blood sugar.

Hypoglycorrhachia   decreased CSF glucose levels.

Hypoinsulinemia   decreased level of insulin.

Hypokalemia   decreased levels of potassium.

Hypomagnesemia   decreased levels of magnesium in the blood.

Hyponatremia   decreased levels of sodium in the blood

Hypoparathyroidism   most often due to destruction of the adrenal glands and, as such, is associated with glucocorticoid deficiency.

Hypophosphatemia   decreased levels of phosphorus in the blood.

Hypoproteinemia   total protein level in the blood that is below the reference interval.

Hypothalamic-pituitary-thyroid axis (HPTA)   the neuroendocrine system that regulates the production and secretion of thyroid hormones.

Hypothalamus   the portion of the brain located in the walls and floor of the third ventricle. It is directly above the pituitary gland and is connected to the posterior pituitary by the pituitary stalk.

Hypothyroidism   decreased thyroid secretion.

Hypouricemia   plasma levels of uric acid less than 2.0 mg/dL.

Hypovitaminosis   a condition that results from a lack or deficiency of vitamins in the diet.

Hypovolemia   decreased blood volume.

Hypoxemia   insufficient or decreased oxygenation of the blood.

I

ICF   see intracellular fluid

Icterus   yellow pigmentation in the skin or sclera (including tissues, membranes and secretions); also known as jaundice. A result of excess bilirubin concentration in the blood.

Icterus index   a test which involves diluting serum with saline until it visually matches the color of a 0.01% potassium dichromate solution. The number of times the serum must be diluted is called the icterus index.

Immune system   a complex series of events in the body that protect the individual from external, harmful agents. Individual survival depends on a properly functioning system.

Immunoassay   a technique that measure the rate of immune complex formation. Immunoassays can be labeled or non-labeled.

Immunoblot   a technique for analysis and identification of antigens; western blotting.

Immunocytochemistry   use of antibodies to detect antigens in cells.

Immunoelectrophoresis   a technique combining electrophoresis of proteins and immunodiffusion.

Immunofixation or immunofixation electrophoresis (IFE); a technique in which an immune precipitate is trapped (fixed) on the an electrophoretic support medium. This method has superseded immunoelectrophoresis because of ease and speed.

Immunohistochemistry use of antibodies to detect antigens in tissues.

Immunophenotyping use of flow cytometer to detect intracellular and cell surface antigens.

Indirect immunofluorescence a technique in which serum antibody reacts with antigen fixed on a slide and the antibody in turn reacts with conjugated antihuman globulins. If the patient serum contains antibody of interest this will be observed under a fluorescent microscope.

Infancy infant; period in life where child is unable to walk or feed itself.

Infectious endocarditis inflammation of the inner lining of the heart chambers and valves; caused by a number of microorganisms.

Inferential statistics values or statistics used to compare the features of two or more groups of data.

Infertility inability or diminished ability to produce offspring.

Inhibin a testicular hormone that inhibits luteinizing hormone.

Innate immune system one of two functional divisions of the immune system; it is the first line of defense.

Insulin a peptide hormone that is synthesized in the β cells of the islets of Langerhans in the pancreas.

International unit IU, the amount of enzyme that will catalyze the reaction of one micromole of substrate per minute under specified conditions of temperature, pH, substrates, and activators.

Intracellular fluid (ICF) fluid inside the cells.

Intrinsic factor a substance normally present in gastric juice that allows absorption of vitamin B$_{12}$.

Inulin an exogenous plant polysaccharide derived from artichokes and dahlias. It is completely filtered by the glomeruli and neither secreted nor reabsorbed by the tubules. It is the most accurate of all GFR assays.

Ion selective electrodes the half-cell or electrode (indicator) which responds to a specific ion in a solution.

Ionic strength concentration or activity of ions in a solution or buffer.

Islets of Langerhans clusters of cells (alpha, beta and delta) in the pancreas; insulin is a peptide hormone that is synthesized by the beta cells of the pancreatic islets.

Isoelectric point (pI) the pH at which the molecule has no net change.

Isoenzyme different forms of an enzyme which may originate from genetic or nongenetic causes and may be differentiated from each other based on certain physical properties, such as electrophoretic mobility, solubility, or resistance to inactivation.

K

Ketone a compound containing a carbonyl group (C=O) attached to two carbon atoms.

Kinetic assay a type of test or procedure in which there is a reaction proceeding at a particular rate.

Kinetic method of measurement quantitation by determining the rate at which a reaction occurs or a product is formed.

Km see Michaelis-Menten constant

Kupffer's cells phagocytic macrophages capable of ingesting bacteria or other foreign material from the blood that flows through the sinusoids.

Kwashiorkor acute protein calories malnutrition.

L

Laboratory standard a rule or criterion related to the laboratory.

Lactescence resembling milk; milky appearance.

Lactose tolerance test an assay to determine the lactase content in the intestinal mucosa. The enzyme lactase is essential to the absorption of lactose from the intestinal tract.

Lancet a surgical device used to puncture the skin for blood collection.

Lateral to the side.

Lactate dehydrogenase flipped pattern situation in which serum levels of LD-1 increase to a point where they are present in greater concentration than LD-2.

LDL see low density lipoproteins.

Lecithins/sphingomyelins ratio (L/S Ratio) a classic test that assesses the ratio of lecithins to sphingomyelins to determine fetal lung maturity.

Leutinizing hormone (LH) A glycoprotein hormone secreted by the anterior pituitary.

Leydig Cells cells of the testicles that produce testosterone.

LH see leutinizing hormone.

Ligase chain reaction probe amplification technique which uses 2 pairs of labeled probes that are complementary for 2 short target DNA sequences in close proximity.

Lipoprotein protein bound with lipid components (ie, cholesterol, phospholipid, and triglyceride). May be classified as very low density (VLDL), low density (LDL), and high density (HDL).

Lipoprotein A [Lp(a)] LDL-like lipoprotein particles.

Liquid chromatography separation technique in which the mobile phase is a liquid.

Lobule forms the structural unit of the liver, which measures 1 to 2 mm in diameter. It is composed of cords of liver cells (hepatocytes) radiating from a central vein.

Loop of Henle descending and ascending loops of the renal tubule.

Low density lipoproteins (LDL) rich in cholesterol, are the "empty tankers" which remain after the triglycerides have been deposited.

Lp(a) see lipoprotein A

L/S ratio see lecithins/sphingomyelins ratio

Luteal phase phase in menstrual cycle; progesterone is synthesized by the corpus luteum during this phase. Also secretory phase

M

Malnutrition a state of decreased intake of calories or micronutrients (vitamins and trace elements) resulting in a risk of impaired physiologic function; associated with increased morbidity and mortality.

Mantissa that portion of the logarithm to the right of the decimal point, derived from the number itself.

Marasmus a condition caused by caloric insufficiency without protein insufficiency so that the serum albumin level remains normal; there is considerable loss of body weight.

Material safety data sheets (MSDS) a major source of safety information for employees who may use hazardous materials in their occupations.

Mechanical hazard any potential danger from equipment such as centrifuges, autoclaves, and homogenizers.

Medial in the middle.

Medical waste material which is infectious or physically dangerous; includes discarded blood, tissues, fluids, body parts, sharps, etc.

Megavitamin intake of a vitamin or vitamins in extreme excess of daily requirements.

MEN see multiple endocrine neoplasia.

Menopause the permanent cessation of menstrual activity.

Metabolic (nonrespiratory) acidosis and alkalosis a disorder due to a change in the bicarbonate level (a renal or metabolic function).

Metabolite any product of metabolism as in the derivative of a drug.

Metanephrine a metabolic product of epinephrine and norepinephrine.

Michaelis-Menten constant (Km) Constant for a specific enzyme and substrate under defined reaction conditions and is an expression of the relationship between the velocity of an enzymatic reaction and substrate concentration.

Microalbuminuria small quantities of albumin in the urine; microalbumin concentrations are between 20–300 mg/d.

Microchemistry clinical chemistry analyses involving only several microliters of sample.

Mineralocorticoid a group of substances produced by the adrenal cortex. Aldosterone is the principal mineralocorticoid (electrolyte-regulating hormone) produced by the zona glomerulosa.

Molality represents the amount of solute per one kilogram of solvent.

Molarity number of moles per liter of solution.

Monoclonal arising from one line of cells.

Monosaccharide a simple carbohydrate; it cannot be decomposed by hydrolysis. Examples include glucose, galactose, and fructose.

MSDS   see material safety data sheets

Multiple endocrine neoplasia (MEN)   the occurrence of several tumors or hyperplasias involving diverse endocrine organs.

Myocardial infarction   also heart attack; occurs when blood flow to an area of the cardiac muscle is suddenly blocked, leading to ischemia and death of myocardial tissue.

Myocarditis   inflammation of the myocardium.

Myoglobin   myoglobin is a heme protein found only in skeletal and cardiac muscle in humans. It can reversibly bind oxygen in a manner similar to the hemoglobin molecule, but myoglobin is unable to release oxygen, except under very low oxygen tension.

Myosin heavy/light chains   myocardial proteins.

Myxedema   condition resulting from hypofunction of the thyroid. The term is used to describe the peculiar nonpitting swelling of the skin.

N

Narcotic   any group of substances (drug) that produces the opioid group of substances; encompass not only heroin, morphine, and codeine, but also several synthetic compounds such as meperidine, methadone, propoxyphene, pentazocine, and others. All have some potential for addiction.

NCCLS   National Committee for Clinical Laboratory Standards; an agency that establishes laboratory standards.

Neonate   a newborn up to six weeks of age.

Neoplasm   a new and abnormal growth of cells or tissue; also tumor.

Nephelometry   analytical technique that measures the amount of light scattered by particles (immune complexes) in a solution. Measurements are made at 5–90° incident to the beam.

Nephrotic syndrome   glomerular injury; an abnormally increased permeability of the glomerular basement membrane. Can be a result of many different etiologies.

Neuroblastoma   malignant tumors of the adrenal medulla that occur in children. They produce catecholamines and may occasionally be associated with hypertension.

Neurohypophysis   posterior portion of the pituitary gland.

NFPA   National Fire Protection Association.

Nitrogen balance   equilibrium between protein anabolism and catabolism.

Noncompetitive immunoassay   use of labeled reagent antibody to detect an antigen; also immunometric immunoassay.

Non-protein nitrogen (NPN)   Nitrogen containing compounds remaining in a blood sample after the removal of protein constituents.

Norepinephrine   a hormone (and catecholamine) produced by the adrenal medulla. As hormones, both norepinephrine and epinephrine serve to mobilize energy stores and prepare the body for muscular activity (increase heart rate and blood pressure, increase blood sugar, etc.). They are secreted in increased amounts with stress (pain, fear, etc.).

Normality   number of gram equivalent weights per liter of solution.

Normetanephrine   a metabolite of epinephrine.

Northern blot   technique for detection of RNA molecules or species with defined sequences.

NPN   see non-protein nitrogen

Nucleic acid probes   use of nucleic acids to investigate cellular changes; nucleic acids store all genetic information and direct the synthesis of specific proteins.

Nutritional assessment   evaluation of a patient's metabolic and dietary (nutritional) needs.

Nyctalopia   poor vision in dim light due to vitamin A deficiency. This condition is also known as night blindness.

O

Oncofetal antigen   a protein produced in large amounts during fetal life and released into the fetal circulation. After birth, the production of oncofetal antigens is repressed, and only minute quantities are present in the circulation of adults.

Oncogenes   viral DNA segments that can transform normal cells into malignant cells.

One point calibration   (or calculation) a term which refers to the calculation of the comparison of a known standard/calibrator concentration and its corresponding absorbance to the absorbance of an unknown value.

Opsonization   action of opsonins to facilitate phagocytosis.

OSHA   Occupational Safety and Health Act; enacted by Congress in 1970. The goal of this federal regulation was to provide all employees (clinical laboratory personnel included) with a safe work environment.

Osmolal gap   difference between the measured osmolality and the calculated osmolality. The osmolal gap indirectly indicates the presence of osmotically active substances other than sodium, urea, or glucose, such as ethanol, methanol, ethylene glycol, lactate, or β-hydroxybutyrate.

Osmolality   physical property of a solution, based on the concentration of solutes (expressed as millimoles) per kilogram of solvent.

Osmolarity   concentration of osmotically active particles in solution reported in milliosmoles per liter; not routinely used.

Osmometer   laboratory instrument used to measure osmolality, or concentration of solute per kilogram of solvent.

Osmotic pressure   pressure that allows solvent flow between a semipermeable membrane to establish an equilibrium between compartments of different osmolality.

Osteomalacia   condition resulting from a deficiency of vitamin D. Deficiency of vitamin D causes bones to be soft and brittle. It is the adult form of rickets.

Osteoporosis   a disease involving the gradual loss of bone mass resulting in a less dense and weak skeleton.

Otorrhea   discharge from the ear; also leakage of CSF from the ear.

Ovulation   the periodic discharge of an ovum from the ovary.

Oxidized   loss of electrons; combined with oxygen.

Oxidizing agent   substance that accepts electrons.

Oxidoreductase   an enzyme that catalyzes an oxidation–reduction reaction between two substrates.

Oxygen content   sum of the oxygen bound to hemoglobin as O_2Hb and the amount dissolved in the blood.

Oxygen saturation   (SO_2) represents the ratio of oxygen that is bound to the carrier protein—hemoglobin—compared with the total amount that the hemoglobin could bind.

Oxytocin   a hormone produced in the hypothalamus. It stimulates contraction of the gravid uterus at term and also results in contraction of myoepithelial cells in the breast, causing ejection of milk.

P

P_{50}   represents the partial pressure of oxygen at which the hemoglobin oxygen saturation (SO_2) is 50%. The P_{50} is a measure of the O_2Hb binding characteristics and identifies the position of the oxygen–hemoglobin dissociation curve at half saturation.

Pancreatitis   inflammation of the pancreas ultimately caused by autodigestion of the pancreas as a result of reflux of bile or duodenal contents into the pancreatic duct.

Panhypopituitarism   a condition resulting from pituitary hormone deficiencies; it usually involves more than one and eventually all the anterior pituitary hormones. These hormones are lost in a characteristic order, with growth hormone and gonadotropins disappearing first, followed by TSH, ACTH, and prolactin.

Paracentesis   aspiration of fluid through the skin; as in removal or aspiration of pericardial, pleural, and peritoneal fluids.

Paracrine   secretion of a hormone from other than a endocrine gland.

Parathyroid glands   four glands adjacent to the thyroid gland. Two of these glands are found in the upper portion and two are found near the lower portion of the thyroid gland. The parathyroid glands produce parathyroid hormone, which controls calcium and phosphate metabolism.

Parathyroid hormone (PTH)   hormone synthesized as a prohormone containing 115 amino acids. The active form of the hormone contains 84 amino acids and is secreted by the chief cells of the parathyroid glands.

Partial pressure   the pressure exerted by an individual gas in the atmosphere; equal to the BP at a particular altitude times the appropriate percentage for each gas.

PCO₂ the partial pressure of carbon dioxide.

PCR see polymerase chain reaction

Peak drug level the time after administration until a drug reaches peak concentration in the body. A rule of thumb suggests that, for peak drug levels, the specimen should be collected 1 hour after the dose is administered.

Pediatric regarding the treatment of children.

Pellagra a condition resulting from a deficiency of niacin. Initial signs of pellagra include anorexia, headaches, weakness, irritability, indigestion, and sleeplessness. This progresses to the classic "four Ds" of advanced pellagra: dermatitis, diarrhea, dementia, and death.

Pepsin a group of relatively weak proteolytic enzymes with pH optima from about 1.6 to 3.6 that catalyze all native proteins except mucus.

Peptide bond linkage combining the carboxyl group of one amino acid to the amino group of another amino acid.

Peptides compounds formed by the cleavage of peptones, and which contain two or more amino acids. The peptide hormones include insulin, glucagon, parathyroid hormone, growth hormone, and prolactin.

Percent solution the amount of solute per 100 total units of solution.

Pericardial fluid fluid surrounding and protecting the heart. The frequency of pericardial sampling and laboratory analysis is rare.

Pericarditis inflammation of the pericardium.

Perifollicular cells one of two types of cells forming the thyroid gland. Also C cells; they are situated in clusters along the interfollicular or interstitial spaces. The C cells produce the polypeptide calcitonin, which is involved in calcium regulation.

Peritoneal fluid a clear to straw-colored fluid secreted by the cells of the peritoneum (abdominal cavity). It serves to moisten the surfaces of the viscera.

PG see prostaglandins

pH represents the negative or inverse log of the hydrogen ion concentration; -log [H⁺].

Phalanx any of the bones of the fingers or toes.

Pharmacokinetics characterization, mathematically, of the disposition of a drug over time in order to better understand and interpret blood levels and to effectively adjust dosage amount and interval for best therapeutic results with minimal toxic effects.

Phenylketonuria (PKU) phenylpyruvic acid in the urine; a recessive hereditary disease.

Pheochromocytoma tumors of the adrenal medulla or sympathetic ganglia that produce and release large quantities of catecholamines.

Phlebotomist an individual who obtains or draws blood samples.

Phlebotomy procedure for withdrawing blood from the body.

Phospholipids formed by the conjugation of two fatty acids and a phosphorylated glycerol. Phospholipids are amphipathic, which means they contain polar hydrophilic (water-loving) head groups and non-polar hydrophobic (water-hating) fatty acid side chains.

Pipet utensils made of glass or plastic that are used to transfer liquids; they may be reusable or disposable.

pK the negative log of the ionization constant.

PKU see phenylketonuria

Placenta a structure in the uterus through which the fetus derives nourishment. The placenta synthesizes and secretes a variety of protein hormones, as well as the steroids estrogen and progesterone. Evaluation of maternal serum and urine concentration of these hormones may be of value not only in diagnosing pregnancy but also in monitoring placental development and fetal well-being.

Plasma liquid portion of blood which contains clotting factors.

Plasma renal flow renal secretory capability.

Pleural fluid essentially interstitial fluid of the systemic circulation; it is contained in a membrane that surrounds the lungs.

PO₂ the partial pressure of oxygen.

Point-of-care testing (POCT) analytical testing of patient specimens performed outside the physical laboratory and at the site of patient care.

Poison any substance which causes a harmful effect upon exposure.

Polyclonal arising from different cell lines.

Polydipsia excess H_2O intake due to chronic thirst.

Polymerase chain reaction (PCR) an in vitro process used to replicate unlimited specific short regions of DNA

Polysaccharide complex carbohydrates; polysaccharides are the most abundant organic molecules in nature.

Porphyria disorders which result from disturbances in heme synthesis.

Porphyrin chemical intermediates in the synthesis of hemoglobin, myoglobin, and other respiratory pigments called cytochromes.

Porphyrinogens the reduced form of porphyrins.

Porphyrinuria increased amount of porphyrins in the urine.

Postrenal obstruction in the flow of urine from kidney to bladder and its excretion.

Posterior pituitary a portion of the pituitary; also the neurohypophysis.

Posthepatic extrahepatic disturbance in the excretion of bilirubin. Posthepatic jaundice results from the impaired excretion of bilirubin caused by mechanical obstruction of the flow of bile into the intestines. This may be due to gallstones or a tumor.

Postzone in immunoprecipitation reactions, antigen concentration is in excess and crosslinking is decreased.

Prerenal prior to plasma reaching the kidney.

Preanalytical error mistakes introduced during the collection and transport of samples prior to analysis.

Precision the closeness of repeated results; quantitatively expressed as standard deviation or coefficient of variation.

Precocious puberty onset of puberty (normal sexual development) earlier than the expected time (generally 10 to 14 years old).

Predictive value theory referring to diagnostic sensitivity, specificity and predictive value. The predictive value of a test can be expressed as a function of sensitivity, specificity, and disease prevalence.

Prehepatic before the liver. Prehepatic jaundice results when an excessive amount of bilirubin is presented to the liver for metabolism, such as in hemolytic anemia. This type of jaundice is characterized by unconjugated hyperbilirubinemia.

Primary standard a highly purified chemical that can be measured directly to produce a substance of exact known concentration.

Probe on an automated analyzer a mechanized device which automatically dips into a sample cup and aspirates a portion of the liquid.

Proficiency testing confirmation of the quality of laboratory testing by means of "unknown" samples.

Progesterone a steroid hormone produced by the corpus luteum and placenta. Progesterone serves to prepare the uterus for pregnancy and the lobules of the breast for lactation.

Prohormone a precursor to the active hormone.

Proinsulin precursor to insulin; it is packaged into secretory granules, where it is broken down into equimolar amounts of insulin and an inactive C-peptide.

Prolactin a protein hormone whose amino acid composition is similar to that of GH. It is produced by the pituitary gland. In humans, it appears to function solely in the initiation and maintenance of lactation.

Prostaglandins (PG) a group of biologically active unsaturated fatty acids; metabolites of arachidonic acid.

Protein-free filtrate clear filtrate of whole blood or plasma prepared by adding acid or other ions to remove proteins.

Proteinuria protein in the urine.

Proto-oncogenes normal cellular genes that play essential roles in cell differentiation and proliferation and potentially can become oncogenic are known as proto-oncogenes. Transformation of proto-oncogenes into oncogenes can occur by single-point mutations, translocation, and amplification.

Proximal tubule portion of renal tubule beginning at Bowman's capsule and extending to loop of Henle.

Prozone in immunoprecipitation reactions, antibody concentration is in excess and cross-linking is decreased.

PTH see parathyroid hormone.

Pyrrole a heterocyclic structure or compound that is the basis for substances such as hemoglobin.

Q

Quality assurance system or process that encompasses (in the laboratory) preanalytical, analytical, and postanalytical factors. Quality control is part of a quality-assurance system.

Quality control   system for recognizing and minimizing (analytical) errors. The purpose of the quality-control system is to monitor analytical processes, detect analytical errors during analysis, and prevent the reporting of incorrect patient values. Quality control is one component of the quality-assurance system.

Quality improvement   activities and systems designed to evaluate and improve patient care.

Quaternary structure   the arrangement of two or more polypeptide chains to form a functional protein molecule.

R

Radial immunodiffusion (RID)   immune precipitation technique used to quantitate a protein (the antigen).

Radioactive material   any material capable of emitting radiant energy (rays or particles).

Random access   the capability of an automated analyzer to process samples independently of other samples on the analyzer. Random access analyzers may be programmed to run individual tests or a panel of tests without operator intervention.

Random error   a type of analytical error; random error affects precision and is the basis for disagreement between repeated measurements. Increases in random error may be caused by factors such as technique and temperature fluctuations.

Random urine specimen   a urine specimen in which collection time and volume are not important. The first morning void is often requested because it is the most concentrated; generally used for routine urinalysis (pH, glucose, protein, specific gravity, and osmolality).

RDA   see recommended dietary allowance

Reabsorption   process of absorbing again.

Reactive chemical   substances that, under certain conditions, can spontaneously explode or ignite or that evolve heat and/or flammable or explosive gases.

Receptor   site of hormone action on a cell; receptors provide for target-organ specificity, since not all cells have receptors for all hormones.

Recommended dietary allowance (RDA).   The amount of a vitamin that should be ingested by a healthy individual to meet routine metabolic needs and allow for biologic variation, maintain normal serum concentrations, prevent depletion of body stores, and thus preserve normal function and health.

Redox potential   a measure of a solution's ability to accept or donate electrons.

Reduced   substance which has gained electrons.

Reference interval   the usual values for a healthy population; also normal range.

Reference method   an analytical method used for comparison. It is a method with negligible inaccuracy in comparison with its imprecision.

Renal threshold   plasma concentration above which a substance appears in the urine.

Renal tubular secretion   process which transports substances from plasma into tubular filtrate for excretion into urine.

Renin   enzyme produced by the kidney which acts on angiotensin to form angiotensin I.

Respiratory acidosis and alkalosis   a disorder due to ventilatory dysfunction (a change in the PCO_2, the respiratory component).

Respiratory distress syndrome   a condition that may occur upon the changeover to air as an oxygen source at birth if the proper quantity and type of phospholipid (surfactant) is not present. Also referred to as *hyaline membrane disease* because of the hyaline membrane found in affected lungs.

Restriction fragment length polymorphisms (RFLP)   a technique to evaluate differences in genomic DNA sequences.

Reverse transcriptase-polymerase chain reaction   conversion of RNA to DNA by reverse transcriptase; the complementary DNA (cDNA) can then be analyzed by PCR.

Rhinorrhea   discharge from the nose; also leakage of CSF into the nose.

RID   see radial immunodiffusion

Rickets   the classic vitamin D deficiency disease of children. It may be nutritional or metabolic in origin.

Robotics   front end automation to "handle" a specimen through the processing steps and load the specimen onto the analyzer.

Rocket immunoelectrophoresis   combination of immunodiffusion and electrophoresis used to quantify antigens on the basis of rocket-shaped precipitin bands.

Rotor   on some automated analyzers, a round device which holds sample cups and is capable of spinning.

S

Secondary hypertension   hypertension with an identified source.

Secondary standard   a substance of lower purity whose concentration is determined by comparison with a primary standard.

Secretagogues   an agent which stimulates or causes secretion.

Secretin   secretin is synthesized by cells in the small intestine in response to the acidic contents of the stomach reaching the duodenum. It can control gastrin activity in the stomach and cause the production of alkaline bicarbonate rich pancreatic juice thereby protecting the lining of the intestine from damage.

Secretion   process whereby glandular organ cells produce substances from blood.

Self-Sustained sequence replication   target amplification method which detects target RNA and involves continuous isothermic cycles of reverse transcription.

Sella turcica   a small cavity in the sphenoid bone of the skull; the pituitary is located in this cavity.

Serial dilution   multiple progressive dilutions ranging from more concentrated solutions to less concentrated solutions.

Serotonin   (5-OH tryptamine) an amine derived from hydroxylation and decarboxylation of tryptophan. It is synthesized by enterochromaffin cells, which are located primarily in the gastrointestinal tract and, to a lesser degree, in the bronchial mucosa, biliary tract, and gonads.

Serous fluid   liquid of the body similar to blood serum; in part secreted by serous membranes.

Sertoli cell   cell of seminiferous tubules that nourish spermatids.

Serum   liquid portion of the blood without clotting factors.

Sheehan's syndrome   hypopituitarism arising from an infarct (necrosis) of the pituitary.

Shift   a sudden change in data and the mean.

SI   see Système International d'Unités

SIADH   syndrome of inappropriate ADH; it results when ADH is released despite low serum osmolality in association with a normal or increased blood volume.

Significant figures   the minimum number of digits needed to express a particular value in scientific notation without loss of accuracy.

Simple protein   composed only of amino acids.

Sinusoids   spaces between the cords of liver cells; they are lined by endothelial cells and Kupffer's cells.

Skin puncture   an open collection system. Blood is brought to the surface of the skin by applying pressure to the site. The sample is dripped into a capillary blood collector (instead of vacuum pressure pulling the sample into the tube). Specimen contains both venous and arterial blood.

Solid phase   in radioimmunoassay (RIA), solid particles or tubes onto which antibody is adsorbed.

Solute   a substance that is dissolved in a liquid or solvent.

Solution   a liquid containing a dissolved substance; the combination of solute and solvent.

Solvent   liquid in which the solute is dissolved.

Southern blot   a technique for detecting specific DNA sequences using a mixture of DNA molecules.

Specific gravity   term used to express density.

Specificity   in regard to quality control, the ability of an analytical method to quantitate one analyte in the presence of others in a mixture such as serum.

Spectrophotometry   analytical technique to measure the light absorbed by a solution. A spectrophotometer is used to measure the light transmitted by a solution in order to determine the concentration of the light-absorbing substance in the solution.

SRM   see standard reference materials

Staging   determination of the period in the course of a disease. The major clinical value of tumor markers is in tumor staging, monitoring therapeutic responses, predicting patient outcomes, and detecting cancer recurrence.

Standard   a substance or solution in which the concentration is determined. Standards are used in the calibration of an instrument or method.

Standard precautions   guidelines which consider blood and other body fluids from all patients as infective; include hand washing, gloves, eye protection, and so on.

Standard reference materials (SRM)   established by authority; used for comparison of measurement.

Steatorrhea   failure to digest and/or absorb fats.

Steroids   classification of hormones; they are all synthesized from cholesterol, using the same initial biochemical pathways. The end result depends on the enzymatic machinery that is predominant in a particular organ.

Synovial fluid   fluid formed by ultrafiltration of plasma across the synovial membrane of a joint. The membrane also secretes into the dialysate a mucoprotein rich in hyaluronic acid, which causes the synovial fluid to be viscous.

Systematic error   a type of analytical error which arises from factors that contribute a constant difference, either positive or negative, and directly affects the estimate of the mean. Increases in systematic error can be caused by poorly made standards or reagents, failing instrumentation, poorly written procedures, etc.

Systeme international d'Unités (SI)   internationally adopted system of measurement. Established in 1960 and is the only system used in many countries. The units of the system are referred to as SI units.

T

TBA   see thyroxin binding albumin

TBG   see thyroid-binding globulin.

TBPA   see thyroxin binding prealbumin

TD$_{50}$ dose of a drug which would be predicted to produce a toxic response in 50% of the population.

TDM   see therapeutic drug monitoring

T-Test   used to determine whether there is a statistically significant difference between the means of two groups of data.

T$_3$ uptake   assay used to measure the number of available binding sites of the thyroxine-binding proteins, most notably TBG. It should not be confused with the T$_3$ assay.

Thyroxine (T$_4$)   a hormone produced by the thyroid gland.

Teratogen   anything that may cause abnormal development of an embryo.

Testosterone   a steroid hormone synthesized from cholesterol. It causes growth and development of the male reproductive system, prostate, and external genitalia.

Tetany   irregular muscle spasms.

Thalassemia   disorder involving a defect in rate and quantity of production of hemoglobin.

THBR   see thyroid-hormone-binding ratio

Therapeutic drug monitoring (TDM)   the determination of serum drug levels in order to produce a desirable effect.

Therapeutic range   concentration range of a drug which is beneficial to the patient without being toxic.

Thermistor   electronic thermometer.

Thoracentesis   removal of fluid from the pleural space by needle and syringe after visualization by radiology.

Thrombolytic agents   substances which break up a thrombus (blood clot) (eg, streptokinase).

Thyroglobulin   an iodine containing protein secreted by the thyroid gland.

Thyroid   a gland consisting of two lobes located in the lower part of the neck. The lobes are connected by a narrow band called an isthmus and are typically asymmetrical, with the right lobe being larger than the left.

Thyroid-binding globulin (TBG)   a protein that binds thyroid hormones. The TBG assay is used to confirm results of FT$_3$ or FT$_4$ or abnormalities in the relationship of the TT$_4$ and T$_3$U test; or as a postoperative marker of thyroid cancer.

Thyroid-hormone-binding ratio (THBR)   test used to measure available binding sites of the thyroxine-binding proteins; also T$_3$-uptake test.

Thyrotropin-releasing hormone (TRH)   a tripeptide released by the hypothalamus. It travels along the hypothalamic stalk to the beta cells of the anterior pituitary, where it stimulates the synthesis and release of thyrotropin or thyroid-stimulating hormone (TSH).

Thyroiditis   inflammation of the thyroid gland.

Thyroperoxidase (TPO) antibodies   thyroid antibodies; formerly known as Thyroid Antimicrosomal antibodies.

Thyrotoxicosis   a group of syndromes caused by high levels of free thyroid hormones in the circulation. Thyrotoxicosis means only that the patient is suffering the metabolic consequences of excessive quantities of thyroid hormones.

Thyrotropin (TSH)   thyrotropin, or thyroid stimulating hormone (TSH), is a glycoprotein consisting of two subunits, alpha and beta, linked noncovalently. It is released by the anterior pituitary.

Thyrotropin receptor (TSHR) antibodies   thyroid antibodies associated with hyper- or hypothyroid states.

Thyroxin binding albumin (TBA)   a protein which binds thyroxine.

Thyroxin binding prealbumin (TBPA)   transport protein of thyroxine; also transthyretin (TTR).

Thyroxine (T$_4$)   hormone produced by the thyroid gland.

TIBC   see total iron binding capacity

Timed urine specimen   specimens collected at specific intervals, such as before and after meals, or specimens to be collected over specific periods of time. For example, discrete samples collected over a period of time are used for tolerance tests (eg, glucose).

Titer   the highest dilution of serum that shows a positive reaction in the presence of antigen (eg, precipitation of antigen-antibody complex).

Total iron binding capacity (TIBC)   an estimate of serum transferrin levels; obtained by measuring the total iron binding capability of a patient's serum. Since transferrin represents most of the iron-binding capacity of serum, TIBC is generally a good estimate of serum transferrin levels.

Total laboratory automation   automated devices and robots integrated with existing analyzers to perform all phases of laboratory testing.

Total parenteral nutrition (TPN)   a widely used means of intense nutritional support for patients who are malnourished, or in danger of becoming malnourished, because they are unable to consume required nutrients.

Toxicology   the study of poisons, their actions, their detection, and the treatment of the conditions produced by them.

Trace element   an element that occurs in biological systems at concentrations of mg/kg amounts or less (parts per million). Typically the daily requirement of such an element is a few milligrams per day.

Tracer   radioactive isotope used to tag or mark a molecule; also called label.

Transferase   an enzyme that catalyzes the transfer of a group other than hydrogen from one substrate to another.

Transferrin saturation   percent of transferrin molecules that have iron bound. A ratio of serum iron (actual iron in the serum) and serum transferrin or TIBC (potential quantity of iron that can be bound). Also percent saturation.

Transudate   fluid that passes through a membrane; in comparison to an exudate it has fewer cells and is of lower specific gravity. Transudates are secondary to remote (nonpleural) pathology and indicate that treatment should begin elsewhere.

Trend   a gradual change in data and the mean.

TRH   see thyrotropin-releasing hormone

Triglycerides   consists of one molecule of glycerol with three fatty acid molecules attached (usually three different fatty acids including both saturated and unsaturated molecules).

Triiodothyronine (T$_3$)   a hormone produced by the thyroid gland.

Triose   a monosaccharide having three carbons.

Troponin I   globular protein; specific marker for cardiac disease.

Troponin T   asymmetrical globular protein; cardiac marker which allows for both early and late diagnosis of AMI.

Trough drug level   the lowest concentration of drug obtained in the blood. Trough levels should be drawn immediately before the next dose.

TSH   see thyrotropin

Tubular reabsorption   process in which movement of a substance (*eg,* calcium) is from the tubular lumen to the peritubular capillary plasma.

Tubular secretion   movement of substances from peritubular capillary plasma to tubular lumen; also secretion of some products of cellular metabolism into the filtrate in the tubular lumen.

Tubules   tubes or canals that make up a part of the kidney; as in convoluted tubules.

Tumor-associated antigen   an antigen associated with a tumor; it is derived from the same or closely related tissue. Oncofetal proteins are an example of tumor-associated antigens. They are present in both embryonic/fetal tissue and cancer cells.

Tumor markers   biological substances synthesized and released by cancer cells or substances produced by the host in response to cancerous tissue. Tumor markers can be present in the circulation, body cavity fluids, cell membranes, or the cytoplasm/nucleus of the cell.

Tumor-specific antigen   a tumor antigen; thought to be a direct product of oncogenesis induced by viral oncogenes, radiation, chemical carcinogens, or unknown risk factors.

Turbidimetry   an analytical technique that measures the decreased amount of light transmitted through a solution as a result of light scatter by particles. Measurements are made at 180° to the incident beam (unscattered light).

U

Ultratrace element   present in tissues at concentrations of mg/kg amounts or less (parts per billion) and has extremely low daily requirements (usually less than 1 mg).

Urea   compound synthesized in the liver from ammonia and carbon dioxide and excreted in the urine.

Uremia or uremic syndrome   very high levels of urea in blood accompanied by renal failure.

Uric Acid   end product of the breakdown of purines from nucleic acids in humans.

Urinalysis   a group of screening tests generally performed as part of a patient's admission work-up or physical examination. It includes assessment of physical characteristics, chemical analyses, and a microscopic examination of the sediment from a (random) urine specimen.

Urobilinogen   colorless product or derivative of bilirubin formed by the action of bacteria.

V

Valence   mass of material that can combine with or replace one mole of hydrogen ions.

Vanillylmandelic acid (VMA)   epinephrine and norepinephrine are metabolized by the enzymes monoamine oxidase and catechol-O-methyl transferase to form metanephrines and vanillylmandelic acid.

Vasodilator drugs   drugs which dilate the peripheral arteries and veins, decreasing the amount of effort that the heart expends to pump blood.

Vasopressin   a hormone secreted by the hypothalamus. Involved in blood pressure regulation.

Venipuncture   puncture of a vein. For example, to obtain blood for analysis.

Venous blood   blood obtained from a vein.

Ventricular septal defect   defect in the septum between the left and right ventricles of the heart.

Very-low density lipoproteins (VLDL)   a group of lipoproteins that carry triglycerides assembled in the liver out to cells for energy needs or storage as fat.

Virilization   development of masculine sex characteristics in the female.

Vitamin   organic molecules required by the body in amounts ranging from micrograms to milligrams per day for maintenance of structural integrity and normal metabolism. They perform a variety of functions in the body.

VLDL   see very-low density lipoproteins

VMA   see vanillylmandelic acid

W

Waived tests   the simplest complexity listing in CLIA. It involves primarily test systems approved by the Food and Drug Administration for home use. The requirements are that there be no reasonable risk of harm to the patient if the test is performed incorrectly; that the likelihood of erroneous results be negligible; that the test method be simple and uncomplicated; and it be available for home use.

Western blot   transfer technique used for analyzing protein antigens. Antigens are separated by electrophoresis and transferred to a new medium by absorption or covalent bonding and then detected by a wide range of antibody probes. The probe could be labeled with a radioactive tag, an enzyme that can produce a visual product, or a fluorescent or chemiluminescent label. Detection would be accomplished with specialized instrumentation, spectrophotometer, fluorometer and luminometer, respectively. Used to detect the presence of human immunodeficiency virus (HIV).

Whole blood   complete blood, containing the liquid portion (plasma) and the cellular elements.

Z

Zero-order kinetics   point in enzyme reaction when product is formed and the resultant free enzyme immediately combines with excess free substrate; the reaction rate is dependent on enzyme concentration only.

Zollinger-Ellison syndrome   a gastrin-secreting neoplasm, usually located in the pancreatic islets, associated with exceptionally high plasma gastrin concentrations. The fasting plasma gastrin levels typically exceed 1000 pg/mL and can reach 400,000 pg/mL, compared with the normal range of 50 to 150 pg/mL.

Zona fasciculata   the adrenal cortex.

Zona glomerulosa   outer portion of the adrenal cortex.

Zona reticularis   inner layer of the cortex of the adrenal gland.

Zymogens   inactivated forms of enzymes. must be converted into active forms for biological function.

Answers to Review Questions, Exercises, Practice Problems, and Case Studies

CHAPTER 1

Exercises

1. a. 2.19 M (NaCl = 58.5 g)
   b. Normality = molarity
   c. 12.8%
   d. 1/7.8
2. a. 5 mg/dL
   b. 4.50×10^{-4} (CaCl$_2$ = 111 g)
   c. 9.01×10^{-4}
3. 5.83 mL/L
4. 5.75×10^{-5}
5. a. ~15/1
   b. 4.46
6. 4.03
7.

Std (g/dL)	(A) Dilution Factor	(B) Volume of Stock (mL)	(B) Volume of Diluent (mL)
1.0	1/30	0.5	14.5
2.0	2/30(1/15)	1	14
4.0	4/30(2/15)	2	13
6.0	5/30(1/5)	3	12
8.0	8/30(4/15)	4	11
10.0	10/30(1/3)	5	10

Total volume desired is 15 mL.

8. 8×10^6; 4×10^{-2}; 0.13; 0.2; 5×10^{-3}; 5×10^{-2}; 40
9. 0.114 mL
10. a. 160 mg/dL
    b. 1.6 mg/mL
    c. 1.92 g/24 h
11. 312.5 g
12. 363 U
13. Hemolysis will adversely affect the potassium, AST, and LD. The CK isn't affected by very slight hemolysis; however, release of substances from the RBC's may interfere in testing when gross hemolysis is present.
14. The answer to this question lies in the laboratory's protocols and procedures. Since this is a situation that could easily be predicted, a policy addressing this situation must be in place. In some laboratories, the phlebotomist would be asked to positively identify and certify (in writing) that the sample belonged to the patient in question, in other laboratories, the sample would be rejected.
15. The request would most likely be denied since fluoride inhibits the enzyme, urease, that is most frequently used in urea (BUN) determinations.

Review Questions

1. b
2. c
3. d
4. b
5. b

CHAPTER 3

Practice Problems

Problem 3-1

$$\text{Sensitivity} = \frac{100\text{TP}}{\text{TP} + \text{FN}} = \frac{100 \times 5}{5 + 3} = 62.5\%$$

$$\text{Specificity} = \frac{100\text{TN}}{\text{TN} + \text{FP}} = \frac{100(843)}{843 + 4} = 99.5\%$$

$$\text{PV}^+ (\%) = \frac{100\text{TP}}{\text{TP} + \text{FP}} = \frac{100(5)}{5 + 4} = 55.5\%$$

$$\text{Efficiency} = \frac{100(\text{TP} + \text{TN})}{\text{TP} + \text{TN} + \text{FP} + \text{FN}}$$

$$= \frac{100(5 + 843)}{5 + 843 + 4 + 3}$$

$$= 99.2\%$$

Problem 3-2

1. See Figure A3–1.

HIGH   JANUARY

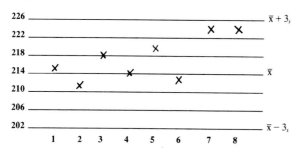

LOW

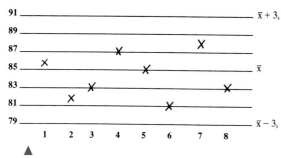

Figure A3-1. Plot of daily control values for January.

2. The values of the high concentration control material appear to be shifted. You are not getting an even distribution about the mean.
3. Systematic error is occurring. This is indicated by 2_{2_s} rule violation.
4. Because the low concentration control is within the control limits of the test and all the patient values are in that range, you could make the decision to report the patient data. Meanwhile, troubleshoot the systematic error.

Problem 3-3

1. The problem appears to be in the preparation of the syringes for drawing a heparinized sample or in the mixing of the samples after collection.
2. If you are the laboratory supervisor or section head, you need to discuss the situation with the supervisor of the MICU. The problem needs to be explained, and the laboratory supervisor and MICU supervisor should solve the problem. If you are the blood gas technologist, you need to discuss the situation with the supervisor in your section and explain the problem. Then the problem should be addressed supervisor-to-supervisor. This is a quality-management problem.
3. This is an extra-analytical problem, and the problem is occurring in the preanalytical phase of sample collection. Quality-control systems that use reference samples detect only analytical errors in the testing procedure.

Problem 3-4

$\bar{Y}_1 = 120.3$ mg/dL
$s_{TM1} = 2.43$ mg/dL
$CV_1 = 2.0\%$

$\bar{Y}_2 = 300.5$ mg/dL
$s_{TM2} = 5.45$ mg/dL
$CV_2 = 1.8\%$

Problem 3-5

Concentration added = 50 mg/dL	100 mg/dL
Recovery	Recovery
102%	97%
98%	97%
100%	99%
96%	96%
90%	95%
Percent recovery: 97.2%	96.8%

Average recovery: $R = 97\%$

Problem 3-6

Ascorbic acid concentration tested = 15 mg/dL
Interference for each patient:

A	-9 mg/dL
B	-8 mg/dL
C	-10 mg/dL
D	-10 mg/dL
E	-11 mg/dL

Average: -9.4 mg/dL

Problem 3-7

1. See Figure A3–2.
2. a.

x_i	y_i	x_i^2	y_i^2	$x_i y_i$
191	192	36,481	36,864	36,672
97	96	9,409	9,216	9,312
83	85	6,889	7,225	7,005
71	72	5,041	5,184	5,112
295	299	87,025	89,401	88,205
63	61	3,969	3,721	3,843
127	131	16,129	17,161	16,637
110	114	12,100	12,996	12,540
320	316	102,400	99,856	101,120
146	141	21,316	19,881	20,586

b.

Σx_i	Σy_i	Σy_i^2	Σy_i^2	$\Sigma x_i y_i$
1,503	1,507	300,759	301,505	301,032

c. $\bar{x} = 150.3$ mg/dL
   $\bar{y} = 150.7$ mg/dL

$$m = \frac{10(301,032) - 1,503(1,507)}{10(300,759) - (1,503)^2}$$

$$= \frac{3,010,320 - 2,265,021}{3,007,590 - 2,259,009} = \frac{745,299}{748,581}$$

$$= 0.996$$

$y_0 = 150.7 - (0.996)(150.3)$
$= 1.001$ mg/dL

d. $Y_i = 1.001 + 0.996x_i$

Y_i	$y_i - Y_i$	$(y_i - Y_i)^2$
191	1	1
98	-2	4
84	1	1
72	0	0
295	4	16
64	-3	9
127	4	16
111	3	9
320	-4	16
146	-5	25

$\Sigma(y_i - Y_i)^2 = 97$

e. $S_{y/x} = \sqrt{\dfrac{\Sigma(y_i - Y_i)^2}{n-2}} = \sqrt{\dfrac{97}{8}} = 3.5$

f. $r = \dfrac{10(301,032) - (1,503)(1,507)}{\sqrt{[10(300,759) - (1,503)^2][10(301,505) - (1,507)^2]}}$

$= \dfrac{3,010,320 - 2,265,021}{\sqrt{(3,007,590 - 2,259,009)(3,015,050 - 2,271,049)}}$

$= \dfrac{745,299}{\sqrt{(748,581)(744,001)}}$

$\sqrt{5.5694501 \times 10^{11}} = 746,287$

$= \dfrac{745,799}{746,287}$

$= 0.999$

3. $y_0 = 1.001$ (mg/dL)
$m = 0.999$
$S_{y/x} = 3.5$ (mg/dL)
$r = 0.996$

Problem 3-8

1. RE = $1.96s_{TM}$
   X_{c1}:   RE = 4.76 mg/dL
   X_{c2}:   RE = 10.68 mg/dL
2. CE = y−intercept = 1.001 mg/dL
   PE = $1.000 - bX_c$
   X_{c1}:   PE = 0.48 mg/dL
   X_{c2}:   PE = 1.20 mg/dL
3. SE = $(y_0 + mX_c) - X_c = (1.001 + 0.996X_c) - X_c$
   X_{c1}:   SE = 0.52 mg/dL
   X_{c2}:   SE = 0.20 mg/dL
   X_{c1}, CE (1.001 > 0.48)
   X_{c2}, PE (1.2 > 1.001)
4. X_{c1}:   TE = 5.28 mg/dL
   X_{c2}:   TE = 10.88 mg/dL
5. a. $s_{y/x}$
   b. $s_{y/x} = 3.5$ mg/dL
6. a. The following are the E_a for glucose:
   $X_{c1} = 120$ mg/dL $E_{a1} = 10$ mg/dL
   $X_{c2} = 300$ mg/dL $E_{a2} = 25$ mg/dL

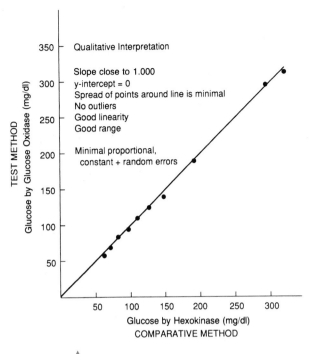

Figure A3-2. Graph of Problem 3-7 data.

At X_{c1} all errors are <10 mg/dL.
At X_{c2} all errors are <25 mg/dL.

b. The test method is acceptable.

Problem 3-9

1. Serving on this team might be: laboratory medical director, laboratory manager/director, and quality assurance representatives from the laboratory, medical-surgical unit, and nursing, as well as representation from individuals who will be performing the test. Instrument manufacturers should have representatives who can interface with this group.

2. The following must be in place before implementing any point of care testing: (Modified with permission from Jones BA, Testing at the patient's bedside. Clin Lab Med 1994;14:473.)

   a. Clinical laboratory oversight for the organization, administration and quality assurance of the testing program

   b. Characterization of the analytical limitations of the POC device including the bias

   c. Procedure manual for all aspects of the program

   d. Training program and continuing education for instrument operators

   e. Quality control testing on each instrument

   f. System to monitor and document operator competence

   g. Internal proficiency testing to compare the POC result with main laboratory's

   h. External proficiency testing

   i. Scheduled instrument maintenance and cleaning

3. The following are desirable features of POC analyzers (Reprinted with permission from Cembrowski GS, Kiechle

FL. Point of care testing: Critical analysis and practical application. Adv Pathol Lab Med 1994;7:3).

Review Questions

1. b
2. d, f
3. b
4. c
5. a
6. d
7. a
8. b
9. d
10. b

CHAPTER 4

Review Questions

1. d
2. b
3. a
4. d
5. a
6. b
7. d
8. a
9. d
10. a
11. c

CHAPTER 5

Review Questions

1. c
2. b
3. b
4. a
5. c
6 a. true
   b. true
   c. false
   d. true
7. a
8. a. 1
   b. 4
   c. 3
   d. 2
   e. 5
9. b
10. b

CHAPTER 6

Review Questions

1. d
2. a
3. c
4. d
5. a
6. c
7. c
8. c
9. a
10. a

CHAPTER 7

Review Questions

1. d
2. d
3. c
4. b
5. d
6. a
7. b
8. a

CHAPTER 8

Review Questions

1. b
2. c
3. d
4. b
5. c
6. a
7. c
8. d
9. b
10. a

Case Studies

Case Study 8-1

1. Increases in serum aminotransferases and blood ammonia levels are the most prominent laboratory indicators of Reye's syndrome. Leukocytosis, hypoglycemia, ketonuria, and an elevated BUN may also be present. In this case, the patient's serum aminotransferases were elevated, but the blood ammonia level remained in normal range.
2. With the serum and urine showing the presence of ketoacids, the patient is acidotic. Investigation of the

blood gas results shows an acidic pH and a decreased PCO_2. Multiplying the PCO_2 by the solubility coefficient for CO_2 (0.0306), the amount of dissolved CO_2 (dCO_2) is found to be 0.70 mmol/L. The HCO_3^- content can be calculated by subtracting the dCO_2 from the TCO_2. The HCO_3^- content is 9.3 mmol/L (normal values: 23-28 mmol/L). Therefore, the calculated ratio of HCO_3^-/dCO_2 is decreased ($\approx 13/1$) due to a HCO_3^- deficit, indicating that the patient is in a state of metabolic acidosis. Since ketosis is more commonly seen in impaired carbohydrate metabolism, the serum glucose does not agree with these findings. A defect in protein metabolism is then suspected.

3. A metabolic screen of the urine followed by a quantitation of plasma amino acids would be needed. High levels of the amino acids valine, leucine, and isoleucine were found (Case Study Table 8-1.1). The results are seen in maple syrup urine disease.

Maple syrup urine disease (MSUD) is a rare (1 in 216,000 live births) autosomal recessive trait in which there is a branched-chain amino acid disorder that causes the build-up of ketoacids in plasma and urine. In the classic form of the disease, this buildup results in mental retardation and eventually leads to death in the early stages of infancy. However, several variants of MSUD have been found that do not present this clear picture. These variants have been classified as "intermittent" MSUD, "intermediate" MSUD, "thiamine-responsive" MSUD, and "E3 deficiency" MSUD.

In intermittent MSUD, the disease is manifested as a result of a variety of stresses on relatively normal children. However, each episode is potentially fatal. In intermediate MSUD, there is a continuous accumulation of ketoacids and branched-chain amino acids, but the neurologic problems are not as severe and there is an increase in longevity among untreated cases. When the level of thiamine within the body drops below a critical point, thiamine-responsive MSUD will appear. The diet must then be supplemented with overdoses of thiamine for treatment and prevention. The multi-enzyme complex responsible for the conversion of branched-chain amino acids contains three known enzymes. It is the lack of the third enzyme (E3) in the complex that results in the accumulations of ketoacids in E3 deficiency MSUD. The prognosis for such patients is poor after seven months of age.

CASE STUDY TABLE 8-1.1. Amino Acid Values (Reference Range)

Urine metabolic screen: Ketoacids 2+ (negative)
   (Presence of valine and leucine or isoleucine)
Plasma quantitation

Isoleucine	70 µmol/dL	(3.7–14)
Leucine	163 µmol/dL	(7.0–17)
Valine	134 µmol/dL	(16–35)
Alloisoleucine	35 µmol/dL	(0-trace)

The patient in this case report has intermediate MSUD. The presence of an upper respiratory tract infection coupled with ketonuria and hypoglycemia produced from the undetected MSUD gave initial symptoms which resembled Reye's syndrome. There have been other documented cases in which patients had initial symptoms that did not suggest MSUD. Only after further testing was the disease discovered.

One important point to keep in mind is that the urine in MSUD will have a distinct maple syrup odor—whether the form of the disease is classic or variant. Noting this odor may provide a clue to early detection of these variants.

Case Study 8-2

1. The mother has an increased α_2-globulin. This may be due to a residual increase of ceruloplasmin and α_2-macroglobulin that occurred in pregnancy. The cord-blood serum, representing the baby, has an abnormally low α_1-globulin fraction. This is associated with α_1-antitrypsin deficiency.

This case illustrates α_1-antitrypsin deficiency. The mother was known to be heterozygous prior to admission to the hospital. Her serum α_1-antitrypsin concentration is within the expected range. The reason for this concentration even in the MZ phenotype was her pregnancy. Pregnant women are reported to have concentrations that may reach twice the reference values.

2. The baby had α_1-antitrypsin deficiency of a severe phenotype, ZZ. Although the electrophoretic pattern with no discernible α_1-globulin fraction may be found in premature infants, term newborns have concentrations of α_1-antitrypsin in the range of adult levels. This deficiency is associated with juvenile pulmonary emphysema and infantile hepatitis. Most babies with α_1-antitrypsin deficiency, however, do not develop symptoms until later in life, when the predisposition to pulmonary emphysema materializes.

3. Other tests that were performed to confirm the α_1-antitrypsin deficiency were the quantitation by radial immunodiffusion and phenotyping of α_1-antitrypsin by isoelectric focusing. The α_1-antitrypsin concentration obtained by radial immunodiffusion of the mother's blood was 225 mg/dL, while that of the baby's blood was 42 mg/dL.

Subsequently, the mother was phenotyped by isoelectric focusing and found to be of the MZ phenotype, while the baby girl was found to be ZZ phenotype. The baby girl is now 10-years-old and is thriving, although she has been slightly slower than average in the development of motor skills.

Case Study 8-3

1. Troponin I is a late marker for myocardial infarctions. Troponin I elevates within 3 to 6 hours following myocardial damage and remains elevated for up to 10 days whereas CK-MB returns to normal in 2-3 days and total CK in 3-4 days. While Troponin I is elevated in other

heart abnormalities such as in chronic heart disease, the degree of the elevation in this patient (8 times the upper reference limit) points to a myocardial infarction that had occurred greater than 3 days prior to hospital admission.

2. Myoglobin elevates when muscle tissue is damaged. In myocardial infarctions the myoglobin returns to normal levels within 18 to 30 hours. With the age of the infarction in this patient, the kidney would have already cleared the myoglobin that originated from the heart muscle. The source of the myoglobin in this instance is likely from muscle destruction arising from the gangrenous toe.

3. The low pre-albumin is an indication of protein-caloric malnutrition. Chronically inadequate food intake, as indicated in this patient, or malabsorption leads to depletion of skeletal muscle and catabolism of visceral and plasma proteins. The low total serum protein is due almost entirely to a low level of albumin and is sometimes partially obscured by a relatively high level of gamma globulins. Albumin, however, is a poor indicator of recent malnutrition because of its long half-life. Albumin is also lowered by several other conditions independent of nutritional status. In contrast, pre-albumin, with a short half-life, responds rapidly to changes in protein-caloric intake. It is, therefore, also very useful in monitoring the effectiveness of nutritional support therapy.

Case Study 8-4

1. The clinical diagnosis that is consistent with the patient's presentation is nephrotic syndrome. The syndrome can be caused by diabetes mellitus, connective tissue disease, glomerular disease and circulatory disease. It is characterized by hypoproteinemia, hypoalbuminemia, edema, hyperlipidemia, and proteinuria.

2. Increased permeability of the glomerular membrane permits passage of albumin and other low molecular weight proteins (i.e. alpha-1 antitrypsin) into the glomerular filtrate. In contrast, the α_2 and beta globulin fractions consist of high molecular weight proteins such as α_2-macroglobulin, haptoglobin polymers and low density lipoprotein. These proteins are retained due to their size.

3. Extensive loss of serum proteins in this condition results in edema due to reduced intravascular oncotic pressure. The edema is aggravated by sodium and water retention that is secondary to aldosterone secretion stimulated by intravascular hypovolemia. The edema is characteristically soft and pitting, and is most prominent in the legs and around the eyes, but may be quite massive.

Case Study 8-5

1. The presence of a monoclonal band in serum does not necessarily indicate that the patient's tumor was still viable because it may take weeks to months for an immunoglobulin protein to clear from the circulation even after it is no longer being synthesized and released from the plasmacytoma.

2. Presence of free light chains in the urine is stronger evidence for current synthesis by tumor cells as the light chains are smaller in molecular size and normally clear from the circulation in a much shorter time than required for the complete immunoglobulin molecule.

3. The presence of urinary light chains (Bence-Jones Protein) can be confirmed by immunofixation electrophoresis.

CHAPTER 9

Review Questions

1. c
2. d
3. a
4. b
5. c
6. b
7. c
8. d
9. a
10. c

Case Studies

Case Study 9-1

1. Blood was drawn on this patient 5 days after onset of symptoms. The CK if elevated has returned to normal. The most important result here is the positive LD flip with LD1 greater than LD 2. This gives a definitive diagnosis in this patient for a myocardial infarct. Not all patients present with crushing chest pain that radiates down the left arm. This patient presents with a common complaint of "indigestion" which is really chest pain. The patient did have a myocardial infarct.

2. The patient could be followed for several days with LD and AST assays. Perhaps a troponin should be run.

3. This patient should be followed either as an inpatient to follow up on the extent of damage to his myocardium and the extent of arteriosclerosis.

Case Study 9-2

1. The most likely diagnosis is a growth spurt, perhaps related to an early puberty.

2. The alkaline phosphatase is usually elevated in children during the growth years, especially during growth spurts.

3. No treatment is indicated. The muscle pain is related to fast growth. Many children report "growing pains."

Case Study 9-3

1. To rule out appendicitis, a CBC looking for an elevated WBC, a rectal exam and possibly an ultrasound of the abdomen is indicated.

2. Amylase and lipase on serum and perhaps a 2 hour urine amylase.

3. Supportive therapy is all that is indicated for this patient.

CHAPTER 10

Review Questions

1. b
2. a
3. c
4. b
5. b
6. b
7. a
8. c
9. b
10. a

Case Studies

Case Study 10-1

1. Diabetic in ketoacidosis
2. Hb A_1c
3. Ketosis occurs as a result of diabetes mellitus (which results from a lack of insulin) the lack of intracellular glucose needed for energy leads to an increased breakdown of fatty acids and triglycerides for energy; the breakdown products of fatty acids and triglycerides are ketoacids (β-hydroxybutyric acid and acetoacetic acid).
4) Acetest (sodium nitroprusside) detects acetoacetic acid. If contains glycine, also detects acetone.

Case Study 10-2

1. Type 2 diabetes mellitus
2. 287 mOsm/L
3. Indicates the presence of other anions in large concentrations-ketones, lactic acid, etc
4. It is also increased confirming the presence of other anions-besides Cl and CO_2 in large concentrations.
5. Test for the other anions. Do ketones on blood and urine or lactic acid level
6. Metabolic acidosis

Case Study 10-3

1. No, according to the ADA, 2 of 3 criteria must be met-the patient has 1 of those criteria, a fasting plasma glucose greater than 126
2. A random plasma glucose, fasting plasma glucose, or 2 hour OGTT
3. Symptoms of diabetes plus a random plasma glucose level of ≥200 mg/dl, a fasting plasma glucose of ≥126 mg/dL, or an oral glucose tolerance test (OGTT) with a 2 hour postload (75-g glucose load) level ≥200 mg/dL. Must be confirmed on a subsequent day by any one of the three methods

4. Type 1 based on the moderate ketones and age of patient. Type 2 diabetics do not usually form ketones and are usually older.

Case Study 10-4

1. The mother is diabetic.
2. In the neonate of a diabetic mother, fetal insulin secretion is stimulated. However, when the infant is born and the umbilical cord is severed, the infant's oversupply of glucose is abruptly terminated, causing severe hypoglycemia
3. The glucose concentration in whole blood is approximately 15% lower than the glucose concentration in serum or plasma.
4. The screening tests include the measurement of plasma glucose at 1-hour postload (50-g glucose load). If the value is ≥140 mg/dL (7.8 mmol/L), then the need to perform a 3 hour OGTT using a 100-g glucose load is indicated. GDM is diagnosed when any two of the following four values are met or exceeded: fasting >105 mg/dL, 1 hr >190 mg/dL, 2 hr ≥165 mg/dL or 3 hr ≥145 mg/dL.

Case Study 10-5

1. An FBG of 90 mg/dL is entirely normal
2. A lean Caucasian with no family history of type 2 diabetes has no additional risks for diabetes and should be screened for diabetes with an FPG in another 3 years according to the 1997 ADA guidelines.
3. FBG
4. Obesity, family history of diabetes in a first-degree relative, a Member of a high-risk minority population (*eg*, African American, Hispanic American, Native American, or Asian American), history of GDM or delivering a baby >9 lb. (>4.1 kg), Hypertension (140/90), Low high-density lipoprotein (HDL) cholesterol concentrations (*eg*, <35 mg/dL), Elevated triglyceride concentrations (*eg*, 250 mg/dL), a history of IFG or IGT.

CHAPTER 11

Case Study 11-1

1. The triglyceride and HDL-cholesterol values.
2. The estimated LDL cholesterol, 167 mg/dL (4.3 mmol/L), is considered high.
3. One could not estimate the LDL cholesterol concentration, since the triglycerides were >400 mg/dL (4.5 mmol/L). Lipoproteins would need to be measured in a specialty laboratory following ultracentrifugation.

Case Study 11-2

1. Cardiac enzyme CK and CK isoenzyme, LD and LD isoenzyme would be useful. Fasting total cholesterol, triglycerides, HDL cholesterol, and LDL cholesterol

could also be performed, but the values should be interpreted with caution, since the patient is under stress and may have suffered an MI. The fasting lipids should be repeated after he has been released from the hospital. If he has suffered an MI, it will be approximately 6 weeks before the values represent his true levels.

2. With a total cholesterol level of approximately 280 mg/dL (7.2 mmol/L) and normal triglycerides and HDL cholesterol, his LDL cholesterol is estimated to be >190 mg/dL (4.9 mmol/L), which is extremely elevated. His physician presumably would have started him on a Step II diet prior to leaving the hospital, based on the in-hospital values. If the 6 to 8 week values do not show significant improvement, the physician would probably start him on medication. The patient has a family history of CHD, and presumably the patient has now been diagnosed as having CHD himself. Therefore the goal would be to reduce his LDL cholesterol value to <100 mg/dL (2.6 mmol/L).

Case Study 11-3
1. Diagnosis-FH heterozygote
2. Yes, he needs an exercise stress test and carotid ultrasound to evaluate his current risk
3. Lp(a) and homocysteine
4. Yes; a more potent statin, or combination drug therapy or a retrial of niacin to further lower his LDL cholesterol, but his liver function tests would need to be monitored carefully; other options, low dose aspirin and vitamin E

Case Study 11-4
1. Every assay except potassium, cholesterol, and glucose.
2. Her kidney function is grossly abnormal, and the associated hypertriglyceridemia is classically seen when the kidney is malfunctioning. Because the patient is a 60-year-old woman, however, it should also be determined if she is taking replacement estrogen, since that can also elevate triglycerides in some women.
3. Nephrotic syndrome.

Case Study 11-5
1. Triglycerides, Eruptive xanthomas
2. Yes
   Yes, she has diabetes mellitus
3. Cessation of estrogen replacement therapy, with consideration of transdermal estrogen therapy, triglyceride-lowering diet and medication, hypoglycemic medication
   Most acute risk—pancreatitis

Case Study 11-6
For each of the three patients seen in clinic:
1. a. 135 mg/dL; b. 123 mg/dL; c. can't be determined
2. Patient c, who has triglycerides ≥250 mg/dL
3. a. 3; b. 0; c. 5

4. a. yes; more than two risk factors and an LDL cholesterol ≥130 mg/dL; b. no; c yes; more than two risk factors, an HDL cholesterol <35 mg/dL, and triglycerides ≥200 mg/dL.

CHAPTER 12

Review Questions
1. d
2. a
3. a
4. d
5. c
6. d
7. a
8. d
9. b
10. b

Case Studies

Case Study 12-1
1. The most likely cause of the patient's elevated BUN is *prerenal* because the BUN-creatinine ratio is greater than 10, and the creatinine is only minimally elevated. This is substantiated by the return to normal with improved cardiac output.

Case Study 12-2
1. The most likely cause is chronic renal disease. Supporting data are the essentially normal BUN-creatinine ratio and the marked elevation of all NPN values. There was no marked improvement when cardiac function improved, thus further eliminating congestive heart failure as a cause of the elevated BUN.
2. If the patient had an elevated level of acetone and other alpha-ketoacids, as might be found in a diabetic, her elevated creatinine levels could have been an erroneous result, since alpha-ketoacids are known to cause a positive bias when creatinine is measured by a kinetic Jaffee reaction, the most commonly used assay method. However, the normal glucose level and abnormal values for other NPN substances make this unlikely.

Case Study 12-3
1. The increased uric acid is due to the marked increase in nuclear breakdown in the presence of a very high WBC. It is not from renal disease because the BUN and creatinine are normal.
2. Chemotherapy has reduced the WBC to below normal levels, and the patient is taking allopurinol.
3. It is probably due to decreased intake (patient is unable to eat); a determination of total serum protein and albumin would be helpful.

CHAPTER 13

Review Questions

1. d
2. a
3. b
4. c
5. a
6. d
7. c
8. b
9. a
10. d

Case Studies

Case Study 13-1

1. Porphyria cutanea tarda (PCT)
2. High performance liquid chromatography would be helpful to confirm the diagnosis.
3. Liver damage due to alcoholism
4. Estrogen therapy, oral contraceptives, insecticide poisoning, hemodialysis
5. The case presented in adulthood; was most probably acquired due to alcoholism
6. VP and HCP also produce neuropsychiatric symptoms; CEP is very rare and symptoms appear shortly after birth; EP demonstrates abnormal porphyrins in the erythrocytes.

Case Study 13-2

1. She had variegate porphyria and had an acute attack because of her ingestion of the barbiturates for a week prior to the surgery.
2. Yes; since this disease is inherited as an autosomal dominant trait, a number of her family and relatives could have this disease.
3. The enzyme that causes this disease is protoporphyrinogen oxidase.
4. The feces could be tested for the presence of coproporphyrin and protoporphyrin. Family history also revealed that she had skin that was photosensitive.

Case Study 13-3

1. This woman has sickle cell trait.
2. No; other than a persistent anemia, this patient should live a normal life.
3. If the father of the child is hemoglobin AA, then there is a 1:2 chance that the child will also acquire sickle cell trait. If the father has hemoglobin AS, there is a 1:4 chance that the child will have sickle cell anemia and a 1:2 chance that the child will have the trait.
4. Yes, these values are normal. A slightly elevated hemoglobin A_2 is not uncommon in sickle cell trait.

Case Study 13-4

1. The cellulose-acetate electrophoresis shows two bands of about equal magnitude migrating in the areas of HbS and HbC, indicating hemoglobin SC disease. The citrate-agar electrophoresis confirmed these findings.
2. In sickle cell anemia (homozygous HbSS), infarction in the spleen is so common that after childhood the spleen usually becomes very small and nonfunctional because of scarring (autosplenectomy). The spleen was enlarged in this patient due to sequestration of sickled cells.
3. Hemoglobins E and A_2 migrate in the same position as hemoglobin C on cellulose acetate at an alkaline pH. These hemoglobins may be differentiated by hemoglobin electrophoresis on citrate agar at an acid pH.
4. Hemoglobin SC disease has a wide range of severity but generally is usually less severe than hemoglobin SS disease, and life expectancy is only slightly shortened. HbSC patients usually are less anemic and have fewer symptoms than HbSS patients, and some cases go undiagnosed throughout life. The mixture of hemoglobin S and hemoglobin C within the erythrocyte does not allow it to sickle as readily. However, of all the sickling syndromes, HbSC disease is the most common cause of sudden death, generally ascribed to infarction of vital tissue. Because HbSC patients tend to be less anemic, the resulting increased blood viscosity can cause more massive infarcts in larger vessels if sickling does occur. Examination of lung postmortem in this patient showed multiple, small emboli of necrotic bone marrow. The presence of large amounts of dead tissue initiated DIC. The hip pain in this patient was due to necrosis of the femoral head, which is common in HbSC disease.

Case Study 13-5

1. A combination of thalassemia from the father and Hb C from the mother.
2. The mother is heterozygous for Hb C and the siblings were heterozygous for either thalassemia or Hb C, thus exhibiting no clinical symptoms. The patient inherited an abnormal gene from both parents which produced a clinically significant disorder.

    Hemoglobin electrophoresis of the parents and other siblings revealed that the father and two siblings had beta thalassemia minor, the mother and one sibling had Hb C trait, and one sibling inherited a normal gene from both parents resulting in normal Hemoglobin A.
3. The type of β-thalassemia minor must have been the β^0 variant in which there is a complete absence of beta chain production.
4. Different test methods; the alkali denaturation test is more accurate than quantitation from hemoglobin electrophoresis by densitometry.
5. Hemoglobin C is found primarily in individuals of African heritage. It is found in 17–28% of West Africans and in 2–4% of African Americans. Isolated cases of Hb C have also been seen in Italians and South African natives

of European descent. The patient's mother was of Sicilian descent.

Case Study 13-6

1. The disorder is beta thalassemia minor. It is often confused with iron deficiency because it produces a microcytic anemia.
2. This patient had a decreased MCV indicating a microcytic anemia, but the RBC count was elevated and the RDW was within normal range. The RBC is usually decreased in iron deficiency and the RDW is increased. Also beta thalassemia minor characteristically has normal to very slightly decreased MCHC as demonstrated in this patient, in contrast to a more definite hypochromia in iron deficiency
3. The iron treatment will not help the anemia. Also, it is important for the patient to know for genetic counseling purposes. If she marries an individual who also has thalassemia trait there is the potential to produce a child with beta thalassemia major which is a serious life threatening disease.

CHAPTER 14

Review Questions

1. d
2. b
3. c
4. a
5. a
6. c
7. c
8. b
9. d
10. b

Case Studies

Case Study 14-1

1. Loss of sodium, chloride and bicarbonate is due to the prolonged vomiting.
2. The urine sodium result is <20 mmol/L, which indicates that the loss of sodium is nonrenal. This concurs with the prolonged vomiting, or GI loss. The physical symptoms (skin turgor and dry mucous membranes) also indicate hypovolemia.

Case Study 14-2

1. Hypokalemia, hypocalcemia and hypomagnesemia are all possible causes for a cardiac arrhythmia. In addition, low magnesium and potassium levels can cause symptoms of digitalis toxicity.
2. The prolonged diuretic use can lead to magnesium loss.

3. Hypomagnesemia can cause decreased levels of potassium and calcium. The exact mechanism for hypokalemia is not completely understood, but it is known that magnesium is required for normal $Na^+ - K^+$ pump activity, which is responsible for active transport of K^+. Magnesium deficiency can impair PTH release and target tissue response, thus leading to hypocalcemia.
4. Providing magnesium therapy alone may correct the hypokalemia and hypocalcemia. Replenishment of either potassium or calcium alone often does not remedy the disorder unless magnesium therapy is provided.

Case Study 14-3

1. The patient's symptoms are relatively classic signs of hypermagnesemia—GI symptoms, decreased respiration, hypotension, bradycardia, and the warm and flushed skin.
2. The patient population that is at greatest risk for hypermagnesemia are those in nursing homes who have combined renal problems and are given magnesium-containing medications, such as antacids, enemas or cathartics. This patient had been treated for constipation within the last 24 hours. It would have been wise for a creatinine clearance test to have been performed prior to administration of the laxative to insure adequate renal function to clear the increased intake of magnesium. This patient was dialyzed to decrease the magnesium level.
3. Elevated magnesium can inhibit PTH release and target tissue response, thus causing hypocalcemia. A second cause may be the hypoalbuminemia. Since total calcium measurement assesses both the free and bound calcium, the bound fraction may be decreased due to the decrease in albumin.

Case Study 14-4

1. A - primary hyperparathyroidism
   B - hypomagnesemic hypocalcemia
   C - malignancy

Case Study 14-5

1. Diabetic ketoacidosis
2. $(Na^+ + K^+) - (Cl^- + HCO_3^-) = (145 + 5.8) - (87 + 8) = 55.8$ mmol/L
   The calculated anion gap exceeds the normal of 10–20 mmol/L. The cause is due to the acidosis.
3. The bicarbonate is decreased due to its use to buffer the increased acid production (H^+). The chloride is decreased due to the increase in unmeasured anions, lactate and ketones (β-hydroxybutyrate).
4. Plasma osmolality is increased due to increased amounts of glucose and urea.

CHAPTER 15

Review Questions

1. c
2. b
3. d
4. d
5. a
6. b
7. c
8. b
9. c
10. a

Case Studies

Case Study 15-1

1. Desferal chelates iron, forming a complex which can be excreted from the body.
2. Vitamin C causes the mobilizations of iron from iron stores. The increase in plasma iron can saturate the binding proteins and result in free−iron damage to the heart. If Desferal is given with Vitamin C in proper dosage, this does not occur.

Case Study 15-2

1. Chromium potentiates the action of insulin and increases the number of insulin receptors.
2. In the form of chromium picolinate or combined with nicotinic acid and glutathione, a better response is obtained versus that obtained with chromium chloride. A dose of 200 μg/day was not effective, while a dose of 1000 μg/day was effective.
3. At 1000 μg/day, no evidence of toxicity has been demonstrated.

Case Study 15-3

1. Fluoride increases spinal bone density and is incorporated into the crystalline matrix of bone.
2. There is controversy: there are reports of decreased, no change, and increased incidence of fracture.
3. Some patients demonstrate a large response to fluoride treatment, increasing osteoblastic activity and requiring increased intake of calcium to mineralize the bone.

Case Study 15-4

1. Ammonium fluoride reacts to form hydrofluoric acid and fluoride ion. Hydrofluoric acid is corrosive and may cause tissue damage. Fluoride ion inhibits Na/K ATPase and acetylcholinesterase.
2. Fluoride complexes with calcium and with magnesium, causing a drop in their concentration.
3. When Ca and/or Mg are extremely low, membrane excitability is increased. This may affect the heart, muscles, and nerves.

Case Study 15-5

1. Low albumin and transferrin values are hallmarks for malnutrition. Transferrin is a very sensitive indicator of inadequate nutritional status. With cases of severe malnutrition, levels are often lower than 50 mg/dL. Low serum and/or copper are commonly observed in patients on long-term parenteral nutrition (due to inadequate supplementation of administered fluids).
2. Zinc and/or copper are commonly observed in patients receiving long-term parenteral nutrition. It can take several months for these deficiencies to become evident. They can produce impairment of several biological functions. Low zinc is associated with impaired growth, particularly in children, diarrhea, dermatitis, and other abnormalities including hair loss (as observed in this case). Copper deficiency commonly produces defective pigmentation of hair and skin and mental cognitive impairment.
3. Treatment would be aimed at ensuring adequate nutritional intake of all essential components. Patients on total parenteral nutrition are susceptible to deficiency of essential fatty acids, protein, trace metals, and vitamins.

CHAPTER 16

Review Questions

1. a
2. d
3. d
4. c
5. b
6. b
7. a
8. b
9. c
10. c

Case Studies

Case Study 16-1

1. Nonrespiratory acidosis.
2. Persistent diarrhea causes significant loss of HCO_3^-.
3. The patient's rapid respirations reflect a compensatory mechanism to decrease the PCO_2 level and restore the 20:1 ratio (HCO_3^-/PCO_2) and the pH to 7.4.

Case Study 16-2

1. Partially compensated nonrespiratory acidosis.
2. The HCO_3^- level is the primary contributor to the nonrespiratory acidosis. The PCO_2 level is low due to the patient hyperventilating and blowing of CO_2 in an attempt to restore the 20:1 ratio and return the pH to 7.4.

3. A major and serious complication of long-bone fractures in the elderly is pulmonary fat emboli. Tachycardia, tachypnea, and low PO_2 values along with chest pain are the classic signs. Globules of fatty marrow from the fracture enter small veins in the area of the fracture and travel to the lung, obstructing the pulmonary circulation.

Case Study 16-3

1. The patient's blood gas data indicate a respiratory and nonrespiratory acidosis. Respiratory acidosis is indicated by the elevated PCO_2; nonrespiratory acidosis is indicated by the decreased HCO_3^- level.
2. Secobarbital depresses the breathing center, consequently CO_2 is not eliminated effectively nor is sufficient O_2 taken into the lungs.
3. Nonrespiratory acidosis. Appropriate ventilation will correct the respiratory component, returning the PCO_2 and PO_2 to normal. The decreased HCO_3^- level will take longer to return to normal.

Case Study 16-4

1. No. Diagnosis of acid–base imbalances cannot be based on just one parameter. A low HCO_3^- level does not necessarily mean that the pH is decreased. Since this measurement was not a part of the blood gas panel, information on the other blood gas analytes was not available.
2. Respiratory alkalosis. People living at extreme elevations breathe atmospheric air containing low O_2 levels. To compensate, they increase their ventilation, which lowers their PCO_2. Renal mechanisms kick in to restore the blood pH to normal. This is achieved by decreasing the loss of H^+ in the urine and decreasing the reabsorption of HCO_3^- in the blood.
3. When the student was retested in the U.S., the HCO_3^- returned to normal. Since the student was breathing normal atmospheric O_2 levels and there was no underlying pathologic condition, the student no longer hyperventilated and the kidneys no longer needed to employ compensatory mechanisms.

Case Study 16-5

1. Based on the blood gas data drawn during the first admission to the ER, the patient is *not* hypoxic. The PO_2, SO_2 and SpO_2 are all normal. No further tests were run and the patient recovered after staying several hours in the ER.
2. Considering the new data from the co-oximeter obtained on the second admission, the patient is hypoxic. The O_2Hb determined spectrophotometrically reflects the amount of O_2 actually bound to hemoglobin.
3. The SO_2 is from a calculation based on the assumption that only normal hemoglobin is present. SpO_2 assessment of oxygen saturation from a pulse oximeter also assumes

that only normal hemoglobin is present. O_2Hb is an actual measurement that considers all hemoglobin species and reflects the amount of oxygen actually bound to hemoglobin.
4. The presence of CO as reflected in the elevated $COHb$ value. This patient was seen in the Fall shortly after turning the furnace on in his home. He was suffering from CO poisoning.

CHAPTER 17

Review Questions

1. d
2. a
3. d
4. a
5. d
6. c
7. d
8. b
9. c
10. b

Case Studies

Case Study 17-1

1. These laboratory test results suggest jaundice due to extrahepatic obstruction either from a stone in the common bile duct, a carcinoma of the head of the pancreas, or, possibly, a postoperative stricture.

Case Study 17-2

1. This laboratory profile suggests hepatocellular damage resulting from either viral hepatitis, alcoholic hepatitis, or a toxin such as carbon tetrachloride or phosphorous. A specific diagnosis of viral hepatitis can be made with additional serologic tests such as HepBSAg determination, etc. Other causes of suspected hepatocellular jaundice may require a needle biopsy of the liver for confirmation.

Case Study 17-3

1. The chemical and serologic findings are typical of hepatitis A infection. The enzymes associated with liver function, serum bilirubin, and the urine bilirubin are all elevated. The patient's history is typical for hepatitis A infection.
2. The patient should be questioned for recent contact with other infected individuals, foreign travel, or exposure to unsafe drinking water. In this case, the patient gave a history of a recent camping trip to Mexico.
3. The prognosis is good. The patient can be reassured that she will gradually begin to feel better. No further therapy is indicated.

Case Study 17-4

1. The presence of hepatitis B surface antigen and IgM hepatitis core antibody and the absence of hepatitis B surface antibody all point to the diagnosis of acute hepatitis B. The hepatitis A studies indicate remote infection and current immunity to that disorder. The test for hepatitis C is negative.

2. The prognosis is good. Approximately 90% of patients with hepatitis B recover uneventfully.

3. Fewer than 1% of patients with acute hepatitis B develop the fulminant form of the disease. Many of these patients die. Therefore, the patient should be observed carefully until improvement occurs. Of greater concern is the fact that about 10% of patients with acute hepatitis B will develop chronic hepatitis. While many of these patients will eventually recover, those who do not are at risk to develop cirrhosis, liver failure, and hepatocellular carcinoma.

Case Study 17-5

1. These findings are diagnostic of hepatitis B infection. It is not possible on the serologic basis to distinguish acute from chronic hepatitis B. However, by definition, since the hepatitis B surface antigen has been present for 6 months, this patient has chronic hepatitis B. It should also be noted that the patient has a previous infection with hepatitis A and is now immune. This plays no part in his current illness.

2. The prognosis for hepatitis B is variable. A liver biopsy may be necessary to establish the degree of damage and to determine whether interferon therapy is needed.

3. Patients with chronic hepatitis B infection are at high risk for the development of cirrhosis and also hepatocellular carcinoma.

4. A test for hepatitis Be antigen and antibody should be performed. The presence of hepatitis Be antigen indicates that the patient is highly infectious and has a poor prognosis. More favorable prognosis is indicated by the presence of antibody to hepatitis Be antigen. The patient should also be tested for Delta hepatitis infection. Co−infection with Delta hepatitis worsens the prognosis of hepatitis B.

CHAPTER 18

Review Questions

1. c
2. e
3. a
4. c
5. b
6. e
7. d
8. e
9. a
10. e

Case Studies

Case Study 18-1

1. This patient most likely has panhypopituitarism secondary to severe hemorrhage during delivery also known as Sheehan's syndrome. The pituitary is enlarged during pregnancy and is especially susceptible to ischemia at the time of delivery. Failure to lactate after delivery is very suggestive of this condition when taken in conjunction with the history of hemorrhage and successful breast feeding after previous deliveries.

2. No. Increased skin pigmentation is seen in patients with primary adrenal insufficiency because of the melanocyte stimulating activity of ACTH. In primary adrenal insufficiency, ACTH levels in plasma are increased in an effort to stimulate the unresponsive adrenal cortex. This patient has secondary adrenal insufficiency characterized by low ACTH. Therefore, increased pigmentation would not be present.

3. This patient should receive a complete pituitary endocrine workup beginning with determination of thyroid and adrenal hormone levels. In a patient with panhypopituitarism, thyroxine, free thyroxine, and cortisol would all be decreased as would TSH and ACTH. Stimulation testing should then be performed to confirm the diagnosis. Although several different agents could be used, this patient could be a good candidate for an insulin tolerance test combined with administration of TRH and GnRH. This combination of agents should normally result in increased levels of growth hormone, cortisol and ACTH, thyroxine and TSH, prolactin, and FSH. Since the suspicion of panhypopituitarism is very strong in this patient, if the insulin tolerance test is used, the physician would have to watch this patient very closely to prevent complications of hypoglycemia. Other combinations of tests would take longer to perform but have less risk of complication. These would include arginine infusion (GH), TRH infusion (thyroxine and prolactin), GnRH infusion (FSH), and CRH infusion or an ACTH stimulation test (ACTH or cortisol).

Case Study 18-2

1. This 64-year-old smoker presents with symptoms of fatigue, weight loss, and shortness of breath. The differential diagnosis in this setting would include congestive heart failure (due to coronary artery disease, hypertension, etc.), lung disease (emphysema or bronchitis), or malignancy (most likely lung).

2. Evaluation of the patient's cardiac status would require a stress test and possibly additional radiologic studies, up to and including cardiac catheterization. Other less invasive and costly procedures can be done first to evaluate the possibility of pulmonary disease or a lung tumor. Chest x-ray would be a good place to start and in this case showed a widening of the mediastinum with a probable mass in the hilum of the left lung. Additional chemistry

tests that should be performed include evaluation of other electrolytes, serum osmolality, urine osmolality and urine sodium excretion. In this case, the serum osmolality was decreased while urine osmolality and urine sodium were increased. These data are strongly suggestive of the syndrome of inappropriate ADH. Cytological examination of the patient's sputum revealed a diagnosis of small cell undifferentiated carcinoma of the lung.

3. Numerous examples of ectopic hormone production have been described. The most common are associated with lung neoplasms, especially small cell carcinoma. These tumors frequently produce ACTH or ADH resulting in Cushing's syndrome or SIADH, respectively. Other hormones that may be produced ectopically include HCG, PTH, gastrin, GH, etc. Differentiating the ectopic production of a hormone from a primary endocrine abnormality usually involves a combination of suppression or stimulation testing and radiologic procedures.

Case Study 18-3

1. Secondary hyperparathyroidism.
2. One would expect that a patient with a long-standing history of renal disease to have serum protein abnormalities. To evaluate total serum calcium results, correction must be applied for the serum protein status. For each change in serum albumin of 1.0 g/dL, total serum calcium changes in parallel by approximately 0.8 mg/dL. The serum calcium should be adjusted to correct for a decreased albumin before interpretation.
3. The decreased serum calcium causes secretion of PTH.
4. PTH acts on bones causing increased stimulation of the osteoclasts and excessive resorption of calcium and phosphorus. The skeleton becomes unstable. In an attempt to repair the destroyed bone, there is an increase in osteoblastic activity and therefore alkaline phosphatase levels rise.
5. Renal osteodystrophy.

Case Study 18-4

1. Excessive production of cortisol by the adrenal gland, excessive production of ACTH by the pituitary gland, production of "ectopic" ACTH or CRH, and exogenous administration of cortisol or ACTH.
2. Truncal obesity, hyperglycemia, protein wasting, easy bruisability, muscle wasting, and poor wound healing.
3. The best screening test is urinary free cortisol in a 24-h urine sample performed on three different collection samples. If the UFC is elevated, an overnight dexamethasone suppression test is recommended. If the cortisol level is not suppressed a low dose dexamethasone suppression test may be indicated. If the cortisol level is not suppressed, the patient has hypercortisolism.
4. Measurement of ACTH will help classify the hypercortisolism as either non-ACTH dependent or ACTH dependent. In addition, a high-dose dexamethasone suppression test is performed. If the ACTH level is decreased and the cortisol level is not suppressed by the HDDS test, the etiology is an adrenal tumor. If the ACTH is increased and the cortisol level is suppressed after the HDDS test, the etiology is most likely a tumor of the pituitary gland. If the ACTH is increased and the cortisol is not suppressed after the HDDS test, the etiology is probably ectopic ACTH production.
5. Since there is an increased gluconeogenesis and glycogenolysis, a glucose tolerance test may reveal glucose intolerance.

Case Study 18-5

1. New onset hypertension, severe headache, and excessive sweating.
2. The test for urinary metanephrines is considered to be the best screening test for the presence of a pheochromocytoma. Follow-up testing may include a second metanephrines measurement or VMA.
3. An accurate 24-h urine must be collected (the same is true for the metanephrines procedure) and there are substances that interfere with the colorimetric measurement of VMA such as chocolate or walnuts.
4. Primary aldosteronism results from oversecretion of aldosterone from the adrenal cortex.
5. Increased aldosterone, decreased renin, and decreased serum potassium.

CHAPTER 19

Review Questions

1. a
2. d
3. b
4. c
5. a
6. c
7. d
8. a
9. b
10. c

Case Study 19-1

1. Pregnancy.
2. Symptoms and increased TT_4, decreased T_3U, and normal FT_4I and TSH. The abnormal tests are due to pregnancy. Also, the abnormal glucose result may be significant.
3. Pregnancy test, beta HCG. If negative, then a more detailed history is in order to exclude other artifactual causes.

Case Study 19-2

1. They indicate hyperthyroidism: sTSH very low, FT_4 very high with an increased uptake of 131I.
2. Antithyroid Peroxidase Antibodies, Thyroid Receptor Antibodies (TSHR) could have been ordered at the first visit.
3. Second visit results indicate hypothyroidism. sTSH very high.

4. Additional testing could include FT_4, preferably by equilibrium dialysis, TSHR antibodies
5. Results indicate Hashimoto's thyroiditis.

Case Study 19-3

1. Subclinical hypothyroidism. sTSH is elevated, FT_4 is within normal limits and patient shows no overt signs of thyroid disorder.
2. Subclinical hypothyroidism can progress to overt hypothyroidism with possible long-term effects of hyperlipidemia and/or cardiac dysfunction.
3. Routine Thyroid function tests are typically within the normal reference range in cases of subclinical hypothyroidism.
4. Thyroperoxidase Antibodies (TPO) is a valuable marker in determining whether subclinical hypothyroidism will progress to overt hypothyroidism, or is merely a "resetting" of the Hypothalamic-Pituitary-Thyroid Axis.

Case Study 19-4

1. Euthyroid sick syndrome (or NTI).
2. FT_4 by equilibrium dialysis; perhaps FT_3 and rT_3.
3. Whether the patient is on drug therapy, especially dopamine or corticosteroids. These drugs decrease thyroid hormone levels.
4. Treatment with L-thyroxine is not indicated.

Case Study 19-5

1. Graves' disease causing hyperthyroidism.
2. Thyroid antibody tests, FT_4.
3. The high-sensitivity TSH is undetected. This is a good indication of primary hyperthyroidism and not another disease or drug effect.
4. Graves' disease is believed to be an autoimmune disorder whereby antibodies attach to TSH receptor sites causing increased production of thyroid hormones.

CHAPTER 20

Review Questions

1. b
2. a
3. a
4. d.
5. b
6. b
7. d
8. c
9. d
10. d

Case Studies

Case Study 20-1

1. A ventricular septal defect allows blood to flow from the left ventricle to the right ventricle. This reduces the amount of blood which is pumped from the left ventricle to the systemic circulation and increases the amount of blood entering the pulmonary circulation, usually causing pulmonary hypertension. This explains the presence of the enlarged pulmonary artery in this patient.
2. A decreased flow of oxygenated blood to the body results in a compensatory response by the kidneys to increase the cellular capacity to carry oxygen by increasing the release of erythropoeitin. This will increase red blood cell counts, hematocrit and hemoglobin.
3. Surgery to repair the septal defect.
4. Very good after surgery.

Case Study 20-2

1. Yes
2. No; no significant elevation of total CK, CK-MB, troponin T, or LD.
3. Unstable angina.

Case Study 20-3

1. Bacterial endocarditis, caused by Staphylococcus aureus.
2. Yes, prior heart disease is a risk factor for the development of endocarditis.
3. Good; endocarditis will usually respond well to appropriate antibiotic therapy.

Case Study 20-4

1. No; no chest pain, left arm pain, weakness, or dyspnea at this time.
2. Yes; total CK, CK-MB, troponin T and myoglobin concentrations are all elevated which would indicate myocardial infarction.
3. Yes; renal dysfunction is probably a result of congestive heart failure.
4. Edema, increased urea nitrogen and creatinine, decreased total protein, albumin, calcium and phosphorous all indicate reduced perfusion of organs such as seen with congestive heart failure.

CHAPTER 21

Review Questions

1. 117 mL/min.
2. b
3. b
4. b
5. d
6. c
7. a
8. d
9. d
10. d

Case Studies

Case Study 21-1

1. The ingestion of ethylene glycol caused damage to the kidney tubules.
2. The patient experienced acute renal failure due to ethylene glycol toxicity. The outcome is either recovery or progression to chronic renal failure. Fortunately, the damage to the kidney was reversible, and the kidneys began to regain normal function.
3. The elevated potassium is serious cause for concern. Potassium is directly involved in the excitability of nerve and muscle tissue. Either an increase or decrease in extracellular potassium concentration can cause abnormal rhythms of the heart and abnormalities of skeletal muscle contraction.
4. The patient is in a state of metabolic acidosis due to renal toxicity with loss of bicarbonate.
5. The physician had to determine the cause of the abdominal pain which might be due to pancreatitis or obstructive or inflammatory liver disease. The normal amylase and lipase indicate that the pancreas is normal. The enzymes point to normal liver function. The elevated LD suggests tissue damage, nonspecific for liver.

Case Study 21-2

1. Two possible complications are kidney stones and gout.
2. Uric acid is formed from the breakdown of the purine nucleosides adenosine and guanosine. In any condition where there is increased turnover of nucleated cells, as in leukemia, elevated uric acid levels are possible.
3. Allopurinol is a drug which inhibits the synthesis of uric acid by replacing it with more soluble metabolites. Since this patient has an elevated uric acid level, it was administered to prevent the complications of gout and kidney stones.

Case Study 21-3

1. Because he was immunocompromised from cyclosporine therapy.
2. A mild renal failure is present in this case.
3. Renal tubular cell injury.
4. To determine the functional capability of the kidneys to clear myoglobin.
5. As urine myoglobin greatly exceeds serum values, adequate clearance of myoglobin is observed. In this case, acute renal failure is not due to myoglobin precipitation within the tubules.
6. Liver and pancreas, however moderate increases in amylase are seen in renal insufficiency since this is a major route of amylase excretion.
7. The ratio indicated that the patient did not have pre-renal azotemia.

Case Study 21-4

1. No. The patient is hyperglycemic. Glucose is freely filtered by the glomeruli and actively reabsorbed by the tubules. Glucose is normally present in the urine in only trace amounts. In this patient the glucose level in the blood exceeds the renal threshold (160–180 mg/dL) and glucose spills over into the urine.
2. This patient's disorder is diabetes mellitus. The presence of ketones indicates he has been in poor control of his diabetes.
3. The presence of protein in the urine indicates the patient is suffering from nephropathy, a common complication of diabetes. He is in danger of progressing to renal failure and eventual hemodialysis.
4. Yearly testing for microalbumin may have prevented or delayed the onset. If detected early, the patient is counseled to maintain tight control over his blood glucose levels through proper diet and self monitoring of his glucose levels.

Case Study 21-5

1. Urine formation which implied a good blood flow in the kidney.
2. Immediately as reflected by the formation of urine.
3. Yes, peritoneal dialysis is much less efficient than the traditional hemodialysis, which in turn is much less efficient than a normal kidney.
4. Cholesterol and triglyceride levels are often elevated in dialysis. It was also ordered for a baseline to compare post-transplant values.
5. Yes, renal failure is associated with increased PTH due to increased secretion caused by a decreased Ca^{++} and also reduced renal clearance, particularly of the C-terminal fragment, which is normally degraded by the renal tubules after filtration by the glomeruli.
6. The decreasing BUN, CR, K^+, and P, β_2-microglobulin and increasing urine volumes indicated graft success.

CHAPTER 22

Review Questions

1. d
2. c
3. c
4. d
5. d
6. a

Case Studies

Case Study 22-1

1. Acute pancreatitis.
2. Enzymatic fat necrosis and digestion to result in free fatty acids in the adipose tissue of the abdomen. The fatty acids then bind calcium as they form fatty acid salts.
3. Shock resulting in prerenal azotemia.

Case Study 22-2

1. Mid-abdominal pain in the presence of a history of alcoholism is highly suggestive of pancreatic disease.

2. Alcoholics are prone to acute and chronic pancreatitis. The two week history indicates that this is a chronic rather than acute problem. However, pancreatic carcinoma, with or without pancreatitis, is also a consideration.

3. The elevated bilirubin corresponds to the observed icterus. An elevated LD is nonspecific. The alkaline phosphatase is indicative of biliary obstruction. The mildly elevated alanine aminotransferase is a nonspecific indicator of hepatic injury. Bilirubin appears in the urine because its normal biliary excretion is blocked. The normal serum amylase is not surprising; this test is often normal in chronic pancreatitis and pancreatic carcinoma. A 24-hour urine amylase may be abnormal, but laboratory tests are not particularly effective in distinguishing between pancreatic carcinoma and pancreatitis. Since chronic pancreatitis predisposes to pancreatic carcinoma, the two conditions may exist simultaneously.

4. Imaging studies of the pancreas may be performed to search for an abnormal pancreatic mass. If this is detected, biopsy may be necessary to determine whether this constitutes scarring or a carcinoma

Case Study 22-3

1. Cystic fibrosis.
2. Sweat electrolyte determination—greatly increased chloride and sodium concentrations.
3. Increased levels of fecal lipids, decreased levels of serum total proteins and albumin, diabetic glucose tolerance test, prolonged prothrombin time, and greatly diminished response to secretin.

CHAPTER 23

Review Questions

1. b
2. a
3. e
4. d
5. b
6. d
7. a
8. b

Case Studies

Case Study 23-1

1. Zollinger-Ellison syndrome (high acid output and volume of secretion in both the basal and stimulated test and the high ratio of basal acid output to stimulated acid output).
2. Plasma gastrin, which should be very high.
3. Hemoglobin is decreased as a result of blood loss. White blood count is slightly high as a reaction to stress and blood loss.

Case Study 23-2

1. Malabsorption syndrome.
2. Intestinal malabsorption characteristic of nontropical or celiac sprue.
3. Malabsorption of vitamin K and resulting deficiencies of coagulation factors II, VII, IX, and X.
4. Iron deficiency (hypochromic, microcytic and red blood cell indices) is the probable cause for the anemia. This is caused by malabsorption of iron. Poor dietary intake of iron may be a contributing factor. Intestinal blood loss is unlikely to be an important factor. Other possible contributing causes are deficiencies of folate and vitamin B_{12} as a result of malabsorption.

CHAPTER 24

Review Questions

1. d
2. d
3. d
4. b
5. e
6. e
7. d
8. e
9. d
10. d
11. e
12. a

Case Studies

Case Study 24-1

1. The patient presented with a combination of two common complications of pregnancy, hypertension and diabetes. The two are often linked. The rapid weight gain, proteinuria, and electrolyte values suggest edema as well, consistent with hypertension. A first pregnancy appears especially prone to these two complications.

2. Delivery of the fetus at this time would be with some risk. The L/S ratio is not at a safe decision level for the complications. The PG%, FSI, and creatinine all suggest borderline situations. Since the contractions subsided, no further action was taken. The enzymes are relatively normal for the situation. Magnesium levels will be carefully monitored for hypertension, and dietary intervention will be used to treat the diabetes, which was probably secondary to the pregnancy.

Case Study 24-2

1. The CSF total protein is at the upper limit of the reference interval. It may or may not be elevated for this specific patient. The ratio elevation, however, definitely suggests increased permeability.

2. The results are consistent with demyelinating processes such as MS. The elevated ratio and oligoclonal banding present in the CSF but not serum electrophoresis support this hypothesis.

CHAPTER 25

Review Questions

1. b
2. b
3. d
4. c
5. c
6. c
7. b
8. d
9. c
10. e
11. b
12. e
13. a
14. d
15. e

Case Studies

Case Study 25-1

1. The peak level of digoxin is evaluated about 8 hours after an oral dose. Levels measured prior to this may present with a concentration greater than the upper limit of the therapeutic range and yet not reflect the tissue concentration associated with toxic effects.
2. The therapeutic actions of digoxin may be influenced by serum electrolyte concentration. In particular, high serum potassium may attenuate digoxin actions. Thus, the patient may not be displaying digoxin toxicity due to this effect.
3. Digoxin like immunoreactive factors may be observed in uremic patients. The measured digoxin in this case may not truly reflect the circulating concentration of drug present.

Case Study 25-2

1. After one half-life has passed, the predicted serum concentration should have fallen to one half of its initial concentration. Thus, after one half-life the predicted serum concentration of procainamide would be 3 µg/mL. The serum concentration after the second dose should be the sum of what remains from the first dose plus what is being added. The first dose resulted in a serum concentration of 6 µg/mL. Thus, the second dose should have resulted in a serum concentration of 9 µg/mL. A disparity exists between the predicted and observed concentration after the second dose. Procainamide is eliminated by hepatic metabolism. A plausible explanation for these results is that this patient is metabolizing procainamide at about twice the normal rate (half-life = 2 hours). If this were the case,

4 hours after the initial dose the predicted serum concentration would be 1.5 µg/mL. The predicted serum concentration after the second dose would then be 7.5 µg/mL.
2. The rate of hepatic metabolism may be influenced by a wide variety of factors. Procainamide is eliminated by acetylation. There is a range of activity of the enzyme responsible for this reaction in the population. Individuals that are fast acetylators are relatively common.

Case Study 25-3

1. Gastrointestinal absorption of phenytoin, like most drugs, occurs by passive diffusion. In these systems many factors may influence the degree of absorption. In this case a key element may be time. In diarrhea associated with hyper motility may not allow sufficient time for absorption to occur. Due to the fact that phenytoin is eliminated according to zero order kinetics small changes in absorption can have large changes in serum concentration.
2. Determination of free phenytoin would probably not provide information beneficial to patient care in this situation. The low total phenytoin in conjunction with the onset of seizure is probably sufficient to assume the free fraction is also low. Further supporting the notion that the determination of free drug in not needed is that upon increasing the dose to the therapeutic range, no further seizures were seen.
3. Severe diarrhea is associated with changes in a wide variety of serum analytes, many of which may affect the binding characteristics of phenytoin. Possibilities include serum total protein, serum albumin, blood pH or the presence of other substances which may compete for the phenytoin binding site.
4. In the presence of excessive loss due to diarrhea, the dosage was increased. After the diarrhea has stopped the dosage of phenytoin needs to be adjusted again to reflect the changes in the degree of absorption.

CHAPTER 26

Review Questions

1. a
2. c
3. c
4. d
5. e
6. e
7. d
8. c
9. e
10. b

Case Studies

Case Study 26-1

The results on this patient are consistent with excessive ethanol consumption. The lack of detectable ethanol in serum does not rule-out the possibility of alcoholism. The combined results of increased GGT, MCV, AST and HDL with no other apparent disease states provide sufficient diagnostic sensitivity and specificity to identify ethanol abuse as the etiology of this condition. Further diagnostic testing would probably not be warranted or cost effective.

Case Study 26-2

The primary toxic effect of acute salicylate overdose is severe acid/base disturbance. This effect is mediated by the acidic nature of the drug, a stimulation of lactate and ketone production, as well as a direct stimulation of ventilation which contributes a respiratory alkalosis. In high level overdose, the net effect is an apparent metabolic acidosis. Arterial pH and TCO_2 will be very low. The anion gap will be high. Urine will be acidic with ketones present but glucose absent. The amphetamine screen may be positive due to the drug taken.

CHAPTER 27

Review Questions

1. d
2. d
3. c
4. a
5. b
6. a
7. b
8. b
9. a
10. c

Case Studies

Case Study 27-1

1. No. Like the majority of tumor markers, CEA is not sensitive or specific enough to be used as a screening test for malignancy. In colorectal cancer, less than 33% of patients present with CEA elevations that are two times the upper limit of normal, and 50% of patients present with normal levels of CEA.
2. CEA is a glycoprotein that is found in normal fetal gastrointestinal tract epithelial cells. Elevations of CEA also occur in colorectal, pancreas, and liver cancer. CEA is also elevated in pulmonary emphysema, acute ulcerative colitis, alcoholic liver cirrhosis, hepatitis, cholecystitis, benign breast disease, rectal polyps, and in heavy smokers.
3. CEA levels should return to normal after complete and successful resection of colorectal tumors in patients with initially elevated CEA levels. CEA values generally remain normal in the absence of recurrent disease, and a rise in serial CEA levels is indicative of recurrent disease in approximately 66% of the cases. An elevated CEA often precedes other clinical signs in 75% of patients developing recurrent disease by at least several months.

4. Any patient who has long-term changes in bowel habits, especially diarrhea, should have their electrolyte levels monitored. Potassium levels can rapidly decrease in patients with diarrhea. Blood urea nitrogen and creatinine can also be used to help determine if dehydration is present. A complete blood count (CBC) should be performed and may show a low hemoglobin and hematocrit from blood loss (if the patient was guaiac positive) due to anemia or chronic disease. Liver function tests (ALT, AST, alkaline phosphatase) may be elevated if the cancer has metastasized to the liver.

Case Study 27-2

1. No. PSA lacks sufficient sensitivity and specificity to be used alone as a screening test for prostate cancer. However, when used in conjunction with digital rectal exam and transrectal ultrasound, PSA can significantly improve prostate cancer detection.
2. Besides prostate cancer, elevations in PSA occur in benign prostatic hypertrophy, prostatitis, and intraepithelial neoplasia. Significant elevations in PSA do not normally occur following routine digital examination. A diagnosis of cancer can only be made by histological examination of prostate tissue.
3. Yes. PSA is extremely useful for monitoring cancer recurrence in patients following therapeutic intervention for prostate cancer. After therapy, PSA levels should be extremely low or undetectable, with rising levels indicating recurrent disease. There is a direct relationship between the serum PSA concentration and the volume of the prostate gland.
4. Additional laboratory tests, such as creatinine and blood urea nitrogen measurements, should be performed to assess renal function. Prolonged obstruction of the urinary tract can damage the renal parenchyma. Electrolytes should also be measured. Urine analysis may also be performed to rule out a possible urinary tract infection, which is often seen in patients with obstructive symptoms.

Case Study 27-3

1. The most likely primary diagnosis for this patient is a testicular germ-cell tumor. Given the ECG findings and the age of the patient, an acute myocardial infarct is highly unlikely. Before surgery, this patient's serum AFP level was normal, whereas his β-hCG was elevated. After surgical removal of the tumor, the elevated serum β-hCG as well as the elevated LD-1 isoenzyme returned to normal. Histologic examination of the tumor mass showed a seminoma containing syncytiotrophoblastic giant cells.
2. Any benign condition associated with an elevated LD-1 isoenzyme value should be considered in the differential diagnosis. A tumor mass of the testicle could be a benign nodule or cyst (although rare). Benign conditions associated with an abnormal LD-1 isoenzyme may be observed in patients with a myocardial infarct, *in vivo* or *in vitro* hemolytic disorders, ineffective erythropoiesis, renal infarct, or muscular dystrophy.

3. Yes. If serum AFP and β-hCG levels are normal and all other possible benign conditions associated with an elevated LD-1 are ruled out, the germ-cell tumor is most likely a pure seminoma. An elevated serum AFP level would suggest an embryonal carcinoma and/or yolk-sac tumor. Elevations of serum β-hCG would suggest a pure seminoma containing syncytiotrophoblastic giant cells, a choriocarcinoma, or a mixture of these two types. Combined elevations of both serum AFP and β-hCG usually indicate a mixed germ-cell tumor.

4. No. A final tumor diagnosis can be made only by histopathologic examination of the tumor tissue removed during surgery. None of the currently available tumor marker tests can be used independently to establish a diagnosis because they lack high specificity for disease.

CHAPTER 28

Review Questions

1. a
   a
   b
   b
   a
2. b
3. d
4. c
5. a
6. c
7. b
8. d
9. a
10. a

Case Studies

Case Study 28-1
1. d
2. c
3. a

Case Study 28-2
1. c
2. a
3. b
4. Animal foods only: organ meats, muscle meats, poultry, fish, eggs, milk
5. Organ meats, deep-green vegetables, muscle meats, poultry, fish, eggs, whole grain cereals

Case Study 28-3
1. Biochemical evidence for fat malabsorption is decreased, 72-hours fecal fat levels and low levels of fat soluble vitamins

2. Nutritional parameters that are affected by fat absorption are low serum levels of fat soluble vitamins and low serum bile salt levels
3. Vitamins A, D, E, K
4. Pernicious (megaloblastic) anemia is the disease characterized by B_{12} deficiency and results from a lack of intrinsic factor or antibodies against the factor

Case Study 28-4
1. She has lost 22 pounds from 145 pounds to 123 pounds. Her albumin level has gone from 4 g/dL to 2.8 g/dL. Her hematocrit has decreased and her serum iron has also decreased. Other laboratory tests which may help her nutritionally-related problem are monitoring twice weekly serum transthyretin (prealbumin) or another short half-life nutritional marker such as insulin-like growth factor 1 (somatomedin C or fibronectin). Another test which might help define her nutritional status would be a thyroid profile consisting of a free-T_4 and a TSH.
2. The fact that she now has to travel in her job continuously and with her status as a vegetarian, she may not be able to consume the vegetables, grains and fruits that she normally might eat because to the demands of her new job. She is predisposed to malnutrition in that she does not drink milk or eat eggs or any meat products such as red meat, fish or chicken.
3. A thyroid profile, including a free-T_4 and TSH, a serum ferritin, sequential protein nutrition markers which show that she is in positive nitrogen balance, a serum B_{12} and a folate level would be helpful.

Case Study 28-5
1. b
2. d
3. a
4. c

CHAPTER 29

Review Questions

1. c
2. b
3. d
4. d
5. b

Case Studies

Case Study 29-1
1. The patient is very ill. Illness can affect thyroid function tests. In such cases, the T_4 and T_3 may be depressed with a normal or decreased TSH. This patient's T_4 and T_3 are both low (suggesting hypothyroidism) but with a normal TSH. In hypothyroidism, the TSH would be elevated.

2. The thyroid tests should be repeated after the patient has recovered from her illness to rule out hypothyroidism.

Case Study 29-2

1. The BUN/Creatinine ratio for this patient is 25:1 indicating decreased GFR or prerenal causes such as dehydration. The normal BUN/Creatinine ratio is 10:1 to 15:1.
2. These data suggest the patient is in a state of dehydration (ie, increased osmolality, BUN/Creatinine ration and sodium.
3. The elevated sodium, BUN/Creatinine ratio and serum osmolality are all suggestive of dehydration.
4. Dehydration is a common and serious problem in the elderly leading to increased hospitalization and mortality.

CHAPTER 30

Review Questions

1. c
2. b
3. a
4. c
5. d
6. a
7. d
8. b

Case Study 30-1

1. Hyponatremia, often with hypovolemia, is fairly common in premature infants. The immature kidney has a poor capacity to retain sodium. Vomiting, diarrhea, and infection can cause these findings. Hyponatremia with increased volume may result from excessive fluid therapy with hypoosmotic or glucose solutions or from the syndrome of inappropriate antidiuretic hormone secretion.

 Hyperkalemia is less common and may result from renal failure, cellular destruction, or excessive therapy.

 An important consideration when these findings are observed is the adrenogenital syndrome.

2. The fluid and electrolyte therapy were carefully reviewed and found to be in order. Renal function was normal for a neonate. There was no clinical or laboratory evidence of infection. The infant had not experienced vomiting or diarrhea.

3. At this point, it was decided to begin a workup for the adrenogenital syndrome. Several forms exist, but the most common is due to deficiency of the enzyme 21-hydroxylase, which is involved in cortisol synthesis. Cortisol deficiency results in lack of negative feedback on pituitary ACTH secretion, resulting in androgen excess. Cortisol and aldosterone deficiency occur, resulting in the electrolyte abnormalities observed in this case. The diagnosis may be confirmed by finding elevated levels of progesterone (a cortisol precursor) and adrenal androgen (DHEA) in the infant's serum.

The adrenogenital syndrome (also known as congenital adrenal hyperplasia) occurs in about 1 in 10,000 live births. It is due to an abnormality of the p-450 c21 gene. Affected male infants are usually normal in appearance, while females undergo pronounced virilization due to the excessive androgens and may have ambiguous genitalia. Therapy is based on cortisol replacement. Because of the genetic basis of this disorder, parental counseling and investigation of siblings is indicated.

Case Study 30-2

1. Many different conditions may be associated with growth retardation. Various skeletal dysplasias such as anchondroplasia, chromosomal abnormalities such as gonadal dysgenesis, and dysmorphic syndromes such as the Prader-Willi syndrome may be associated with short stature. Many children with these disorders have a characteristic physical appearance or other findings which may serve as diagnostic clues. Additionally, nutritional deficiencies, malabsorption, renal failure, hemoglobinopathy, diabetes mellitus and a variety of inborn errors of metabolism may cause growth retardation.
2. This child's history points to an intracranial cause, specifically, growth hormone deficiency.
3. A random growth hormone level is of limited value; indeed, it was normal in this case. However, after insulin and arginine stimulation, the serum growth hormone did not increase. This is diagnostic of growth hormone deficiency.
4. A deficiency of growth hormone may be secondary to pituitary or hypothalamic failure, in this case, as a result of previous head trauma. Infusion of GHRH resulted in no increase in serum growth hormone levels, indicating pituitary failure. Synthetic growth hormone was administered to this patient, resulting in a growth spurt. Since the pituitary produces other hormones, they may also be deficient. Thyroid and adrenal function should be assessed due to possible deficiencies of TSH and ACTH. If puberty does not occur in a few years, gonadotropin levels should be determined.

Case Study 30-3

1. Serum thyroxine T_3, TSH, and the FTI should all be measured. Some screening T_4 results are not confirmed by low serum levels. In hypothyroidism, the T_3 level is decreased and the TSH level is elevated. Rarely, hypothyroidism may be caused by pituitary insufficiency, in which case the TSH level will be low.

 TBG deficiency, a rare condition, may be excluded by direct measurement of TBS. A thyroid scan may be useful to detect any anatomic defects.

2. Hypothyroidism is treated with thyroxine.
3. Neonatal screening is necessary because even severe hypothyroidism is not evident at birth, since the infant receives maternal hormone. Untreated, neonatal hypothyroidism leads to severe growth and mental retardation, a syndrome known at cretinism. Once these changes have developed, they are usually irreversible, but they may be entirely prevented by thyroxin therapy.

Index

Lowercase *f* following a page number indicates a figure; *t* following a page number indicates tabular material.